国家出版基金项目
NATIONAL PUBLICATION FOUNDATION

Flora *of* Pan-Himalaya

Volume 46

Lentibulariaceae, Acanthaceae, Bignoniaceae, Verbenaceae, Martyniaceae, Stemonuraceae, Cardiopteridaceae

Yun-Fei DENG

Key Laboratory of Plant Resources Conservation and Sustainable Utilization, South China Botanical Garden, Chinese Academy of Sciences.
Email: yfdeng@scbg.ac.cn

Zhen-Yu LI

State Key Laboratory of Systematic and Evolutionary Botany, Institute of Botany, Chinese Academy of Sciences. Email: lizy@ibcas.ac.cn

Qiang WANG

State Key Laboratory of Systematic and Evolutionary Botany, Institute of Botany, Chinese Academy of Sciences. College of Life Sciences, University of Chinese Academy of Sciences. Email: wangqiang@ibcas.ac.cn

Hua PENG

Key Laboratory for Plant Diversity and Biogeography of East Asia Kunming Institute of Botany, Chinese Academy of Sciences. Email: hpeng@mail.kib.ac.cn

SCIENCE PRESS
Beijing

CAMBRIDGE
UNIVERSITY PRESS

Flora *of* Pan-Himalaya

Volume 46

Lentibulariaceae, Acanthaceae, Bignoniaceae, Verbenaceae, Martyniaceae, Stemonuraceae, Cardiopteridaceae

Authors:

Yun-Fei DENG
Key Laboratory of Plant Resources Conservation and Sustainable Utilization, South China Botanical Garden, Chinese Academy of Sciences
Email: yfdeng@scbg.ac.cn

Qiang WANG
State Key Laboratory of Systematic and Evolutionary Botany, Institute of Botany, Chinese Academy of Sciences. College of Life Sciences, University of Chinese Academy of Sciences
Email: wangqiang@ibcas.ac.cn

Zhen-Yu LI
State Key Laboratory of Systematic and Evolutionary Botany, Institute of Botany, Chinese Academy of Sciences
Email: lizy@ibcas.ac.cn

Hua PENG
Key Laboratory for Plant Diversity and Biogeography of East Asia Kunming Institute of Botany, Chinese Academy of Sciences
Email: hpeng@mail.kib.ac.cn

Editor-in-Chief:

De-Yuan HONG

Map Approval No.: GS (2019) 1135

Published by Science Press (Beijing)

ISBN 978-7-03-062355-3

Flora *of* Pan-Himalaya

Office: Dan PENG
Tel: 86-10-62836084
Fax: 86-10-62590843

Email: flph@ibcas.ac.cn
Website: www.flph.org

It is with great pleasure that we acknowledge the generous financial support from the following:

Ministry of Science and Technology of the People's Republic of China
National Natural Science Foundation of China
Chinese Academy of Sciences

Shenzhen Donghua Landscape Co., Ltd

Foreword

We are delighted to welcome the appearance of this highly significant regional flora, and congratulate De-Yuan HONG and all of the others that worked together to make it possible. The Himalaya is one of the most beautiful, extensive, and revered ranges on earth, containing nine of our ten highest peaks, and its plants deserve the kind of inclusive and collaborative treatment that we have been presented in these fine volumes.

The mighty Himalaya was formed as the basaltic Indian plate plowed into the underbelly of Asia. Carrying the lighter rocks that would become the mountain ranges we see now, it broke out of the gigantic southern continent of Gondwanaland approximately 90 million years ago and started its journey northward to its current position. In the south, any part of India that rose above the ocean would have been covered with archaic southern vegetation similar to that which survives today in southern Australia and South America, New Zealand, and New Caledonia. As the fragment that includes India journeyed northward, it passed from the south across the equator and started crowding into the Asian plate perhaps 55–50 million years ago, continuing to thrust up the Himalaya from that time onward. The same movement folded what is now China into a series of accordion-like pleats that became the current mountain ranges of southern and southwestern China and its neighboring countries.

Crossing so many latitudes seems to have led to the extinction of virtually all of the plants and animals that occupied India when its position lay south of the equator. However, when it came within reach of Asia and under the influence of a tropical and subtropical climate from the early Eocene Period onward, the plants and animals of southern Asia colonized the new lands, and more migrated from the north as the mountains of the region were thrust ever upward. Importantly, there was a great opportunity for the evolution of endemic plants and animals in the new mountains, and the mixture of all of these rich biological strands have combined to form the incredible biological richness of the area as we see it today.

Both because of the political subdivisions of the region and its complex floristic composition, it took very courageous and forward-looking leadership for De-Yuan HONG and his colleagues to undertake this monumental flora. Except for China, there are relatively few endemics in most of the countries of the Himalayan ranges and their bordering lowlands, but many in the region. This lack of national endemics, however, stems mainly from the artificial presence of countries in the area, which taken as a whole is an extremely rich one for endemics in virtually all groups of plants, animals, and fungi that occur there. The plants of some of the countries are reasonably well known, those of others poorly known, and our historical efforts to bring the available information together and improve its quality by transnational comparisons have been inadequate.

That situation is all the more unfortunate because of the severe threats that the ecosystems of the regions facing from rapid human population growth and the even more rapidly rising expectations for increased consumption and especially from the climate change that threatens all of the ecosystems of the area, including the possible destruction of its glaciers and the disruptions of the monsoons on which so many depend for the food and livelihood. A sound knowledge of the plants and ecosystems of the region is the only way to win sustainability, even if population, consumption levels, and the runaway use of malign technologies like burning coal are brought under control.

For all of these reasons, we declare once more our delight in the appearance of this flora and congratulate those who have worked together to make it a reality. It is a critically important first step in improving our knowledge of the plants of the area on a regional basis, and one of great and general importance.

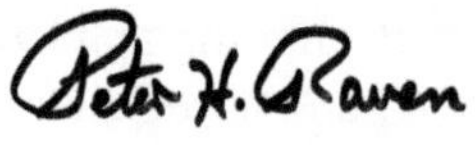

Peter H. Raven,
President Emeritus,
Missouri Botanical Garden

Introduction

— General Guideline of the *Flora of Pan-Himalaya*

The Himalaya and adjacent regions (Pan-Himalaya) form a natural phytogeographical unit, from the Vākhān Corridor eastwards to the Hengduan Mountains via the Karakorum and the Himalaya. This region covers the northeastern corner of Afghanistan, northern Pakistan, northern India, Nepal, Bhutan, northern Myanmar, and southwestern China (S Xizang, SE Qinghai, SE Gansu, W Sichuan and NW Yunnan). This region is divided into 17 subregions: Vākhān, N Pakistan, Kashmir, U Ganga & Indus, U Yarlung Zangbo, W Nepal, C Nepal, E Nepal, Sikkim & Darjiling, Bhutan, M Yarlung Zangbo, L Yarlung Zangbo, Yarlung Zangbo-Brahmaputra, Tangut, N Hengduan, S Hengduan, and U Irrawaddy (see the map 'Subdivisions of the Pan-Himalaya').

The *Flora of Pan-Himalaya* (FLPH) will be published in English in two editions: printed and online. It treats all the native and naturalized vascular plants of this region.

The Flora will be published in 50 volumes with ca. 80 books, with the larger families divided into two to seven parts. The classification systems used in this Flora will reflect our current understanding of phylogenetic or phylogenomic relationships of the relevant groups. The third edition of the Angiosperm Phylogeny Group system (APG III) will be adopted for the angiosperms, and up-to-date phylogenetic or phylogenomic systems for the gymnosperms, ferns, and lycophytes will be followed for the treatments of those groups.

The natural and rational delimitation of species is the critical task of taxonomy, and is the most important criterion for judging the scientific value and standard of a flora. Population concepts should be employed in determining the taxonomy; character analysis should be considered as the basis for a rational taxonomic treatment. Therefore, field observations at the population level, and statistic analysis based on broad population sampling and collections are greatly encouraged to investigate the variation of characters and evaluate their taxonomic significance for the delimitation of taxa. Herbarium specimens kept at the following herbaria must be examined and identified: PE, KUN, K, BM, E, CDBI, and SZ. The authors are also encouraged to visit the herbaria of A, B, CAL, CAS, G, KATH, LE, P, TI, W, and HNWP.

Delimitation of closely related species should be based on correlated discontinuous (including statistically discontinuous) variation of at least two characters. As to apomicts (such as *Taraxacum* spp., *Sorbus* spp.) and vegetatively reproducing plants, macrospecies should be adopted instead of microspecies.

The use of subspecies is encouraged in infraspecific classifications, but varieties only in special cases. Subspecies are geographical races and horizontally or vertically vicarious, whilst varieties are prominent ecotypes. Genotype variants within populations should not be recognized as different taxa. The description of a species should cover all its elements. Variants which do not merit formal taxonomic recognition may be noted below the description of the species.

Larger families and genera may be concisely subdivided. These systems should reflect the current understanding of phylogenetic or phylogenomic relationships, especially those generated from solid molecular and genomic data.

The concept of the genus should reflect the current understanding of phylogenies. Generic concepts strongly supported by both molecular/genomic and morphological evidence should be adopted, but those with weak support in molecular/genomic analysis and have no sound morphological evidence should not be accepted. The treatment of the more poorly known groups should be conservative and any perceived problems indicated.

Identification keys are one of the most important products of classification, and the best reflection of its usefulness. Authors should construct a key based on the examination of all relevant specimens. Where possible, keys should not be based on a single character, and must avoid vague words, such as relatively large vs. relatively small, longer vs. shorter. It is important to use easily visible characters in keys. Dichotomous indented keys such as those in the *Flora of China* are to be used in this flora.

Description of species should be concise with emphasis on diagnostic characters, and should include habitat information and the geographical distribution, plus chromosome numbers, phenology, and other biological features, as well as reliable economic uses and conservation information where available.

Literature citation for accepted names and their basionyms must be provided. Important references, such as monographs and floras, should be cited, but not more than three titles (at most five if absolutely necessary). All synonyms based on types from within the flora area should be listed along with other synonyms that have been widely used in relevant publications.

Detailed type information should be indicated after the nomenclatural citation of taxa. Designation of types is highly encouraged in this flora.

A distribution map is given for each species. This will be a dot map based on the specimens seen for the flora. The vouchers are to be cited to record geographical distribution for each species, subspecies (or variety). At least one representative specimen is needed for each distribution unit (county in China, district in the other countries). For the more widely distributed taxa (across nine or more subdivisions), a larger distribution unit can be used, no citation of vouchers is needed, and the distribution outlines are also acceptable.

The nomenclatural novelties (new taxa, new combinations, and *nomina nova*) may be published in the flora or elsewhere.

Sterile hybrids, and those hybrids not forming populations, are to be briefly discussed after the descriptions of their parents. Plants only known from cultivation may be indicated at the end of the treatment of the genus, but should not be included in the identification keys.

All genera, and at least one third of the species, should be illustrated with good quality line drawings with scale bars to indicate magnifications. The illustrations can be drawn from herbarium materials, living materials or images available, selected by the authors, or taken from publications with permission. Wherever possible the material used should be indicated. Diagnostic characters of the species should be clearly illustrated. One plate may hold illustrations of one, two, or three species. For larger genera composite plates of diagnostic characters are encouraged.

The web edition of the flora will follow the corresponding printed edition. Color photographs, detailed citation of vouchers, and other sources of information are to be included only in the web edition.

The physical geography, history of studies of vascular plants from this region, biogeography, biodiversity conservation, general references, and key to families are to be included in the first (introductory) volume. A comprehensive index to taxa and other subjects are to be published in the last volume.

More detailed guidelines and a model treatment for preparation of flora accounts will be provided by the Editorial Committee.

The Editorial Committee of *Flora of Pan-Himalaya*

Preface

Volume 46 of the *Flora of Pan-Himalaya* is the ninth book in the series of 50 volumes with about 80 parts. It treats seven families (Lentibulariaceae, Acanthaceae, Bignoniaceae, Verbenaceae, Martyniaceae, Stemonuraceae, Cardiopteridaceae), 52 genera and 245 species, of which 62 species (25%) are endemic to the Pan-Himalaya.

During the work on this volume, the authors and their teams made several expeditions to the Pan-Himalaya for field observations and collection of materials. We visited more than 40 herbaria in the world. A new genus, *Wuacanthus* Y. F. Deng, N. H. Xia & H. Peng, and five new species were described, 11 new combinations were proposed, and 18 names are reduced to the synonymy. The nomenclatural novelties in this volume include one name at new rank and three names of new sections. The new classifications of the genera *Peristrophe* and *Phlogacanthus* are proposed. In the present volume, 624 scientific names of new taxa are involved, among which 473 had been already typified, while 151 need to be typified. Of these names, the present authors designated 136 lectotypes and one neotype for 137 names of species or infraspecific taxa and three generitypes for three names of genus in all the seven families. Unfortunately, 11 names, i.e. 7.3% for the total, are still necessary to be typified because we did not trace the original material.

The figures of this volume were drawn by several artists. Thirty-three figures were redrawn or modified from *Flora Reipublicae Popularis Sinicae*, 38 from *Icomographia Cormophytorum Sinicorum*, 18 from *Flora of China*, one from *Flora Yunnanica*, one from *Flora Xizangica* and one from *Chinese Endemic Genera of Seed Plants*. Ai-Li LI edited or modified the figures. John R. I. WOOD of University of Oxford kindly permits to use his published illustrations. The following journal granted permission to republish in this book several illustrations that were previously published in their journals: *Botanical Journal of the Linnean Society*, *Journal of Tropical and Subtropical Botany*, *Novon*, *Phytotaxa* and *Plant Diversity*. Some new figures were drawn by Ding-Han CUI, Shun-Qing HE, Ai-Li LI, Yun-Xiao LIU, Ling WANG, Xi-Lin WU, Han-Ping YU and Yun-Xi ZHU. The distribution maps were automatically produced by Mr. Jing XUAN based on the data from the cited voucher specimens.

Acknowledgements

Partial funding of the *Flora of Pan-Himalaya* has been provided by the International Partnership Program of Chinese Academy of Sciences (Grant No. 151111KYSB20170021), the National Natural Science Foundation of China (Grant Nos. 31620103902, 31110103911), China Ministry of Science and Technology (Grant No. 2013FY112100), the Chinese Academy of Sciences (Grant No. KSCX2-EW-Z-1), the State Key Laboratory of Systematic and Evolutionary Botany Institute of Botany, Chinese Academy of Sciences, and Shenzhen Donghua Landscape Co., Ltd.

The directors, curators and librarians of the following institutions are warmly thanked for their assistance in sending loans and/or providing images of specimens, and/or providing the authors with herbarium and library facilities: A, AAU, ABD, B, BKF, BM, C, CAL, CAS, CDBI, CGE, E, HITBC, HNWP, IBSC, IFP, K, KATH, KUN, KYO, L, LE, LIV, MBK, MO, NAS, NY, OXF, P, PE, SING, SYS, SZ, TI, TNS, TUCH, UC, US and W. We are grateful to Prof. De-Yuan HONG for his valuable suggestions on improving the manuscript.

Yun-Fei DENG would like to thank the following persons for their kind assistance during the field work: Dr. Ya-Ping CHEN, Prof. Xue-Jun GE, Dr. Zhe-Li LIN, Prof. Hua PENG and Dr. Yi TONG.

The present volume was financially supported also by the National Natural Science Foundation of China (Grant Nos. 30870152, 31070175, 31270247, 31470302, 31670191, 31870181, 31970208), the Knowledge Innovation Program of the Chinese Academy of Sciences (Grant No. KSCX2-YW-Z-021), the Biological Resources Programme, Chinese Academy of Sciences (KFJ-BRP-017-11), Youth Innovation Promotion Association, Chinese Academy of Sciences (grant 2018105), Southeast Asia Biodiversity Research Institute, Chinese Academy of Sciences (Grant No. Y4ZK111B01) and "One-Three-Five" Strategic Planning of SCBG, CAS to Yun-Fei Deng.

Contents

LENTIBULARIACEAE

Zhen-Yu LI

Herbs perennial or annual, carnivorous, terrestrial, epiphytic or aquatic; stems often modified into rhizoids, stolons, or rhizomes, rarely developed. Traps (bladders) present or absent. Stipules absent. Leaves alternate or in a basal rosette, simple or variously divided. Inflorescences usually racemose or reduced to a solitary flower. Calyx 2–5-lobed or -parted; lobes imbricate, persistent. Corolla gamopetalous; tube short, limb 2-lipped or equally 5-lobed; lower lip entire or 2- or 3(–6)-lobed; lobes imbricate; upper lip entire or 2- or 3-lobed; spur subulate, cylindric, conic, or saccate. Stamens 2, inserted at base of corolla tube, included. Ovary superior, 1-locular, placenta free-central or basal. Fruit a capsule, loculicidal, or circumscissile. Seeds few to numerous, or rarely 1 per capsule, without endosperm.

Recent references: Li & Cheek, 2011. Lentibulariaceae. In: Wu, Raven & Hong (eds.), Fl. China 19: 480–491. Science Press, Beijing & Missouri Botanic Gardens Press, St. Louis.

Three genera and ca. 290 species: cosmopolitan but mostly in tropical regions; two genera and 20 species (5 endemic) in the Pan-Himalaya.

1a. Leaves with margin usually involute, adaxially with viscid glands; traps absent; calyx 5-parted; corolla open at throat ·· 1. *Pinguicula*
1b. Leaves with margin not involute, adaxially without viscid glands; traps present; calyx 2(–4)-parted; corolla with variously raised palate at throat ·· 2. *Utricularia*

1. PINGUICULA L.

Linnaeus, Sp. Pl. 1: 17. 1753; Taylor in Hara et al., Enum. Fl. Pl. Nepal 3: 131. 1982. Type: *Pinguicula vulgaris* L.

Herbs perennial or annual, terrestrial, epilithic or epiphytic, rhizomatous, without rhizoids, stolons, or traps. Roots fibrous. Leaves in a basal rosette, simple, petiolate or sessile, soft and fleshy; leaf blade adaxially usually with numerous viscid glands, veins pinnate, margin entire and usually involute. Inflorescences with a solitary flower, rarely 2–3-flowered raceme, without scales, bracts, or bracteoles. Calyx 2-lipped; lower lip 2-lobed; upper lip 3-lobed. Corolla open at throat, lower lip larger than upper lip, spurred, 3-lobed, upper lip 2-lobed. Capsule loculicidal, opening by 2 valves. Seeds many per capsule.

About 55 species; N temperate regions, most diverse in Central America, extending Southwards to Patagonia; one species in the Pan-Himalaya.

1. Pinguicula alpina L., Sp. Pl. 1: 17. 1753; Handel-Mazzetti, Symb. Sin. 7: 872. 1936; Taylor in Hara et al., Enum. Fl. Pl. Nepal 3: 131. 1982; Li & Cheek in Wu, Raven & Hong (eds.), Fl. China 19: 480. 2011. Type: [*icon*] Linnaeus, Fl. Lapponica: 11, t. 12, f. 3. 1737 (lectotype designated by Blanca & Jarvis, 1999).

Herbs perennial. Roots numerous, 0.4–1 mm thick. Leaves shortly petiolate; leaf blade pale green, elliptic-oblong to lanceolate-oblong, 1–4.5 cm long, 0.5–1.7 cm broad, abaxially glabrous, lateral veins 5–7 pairs, base broadly cuneate, margin involute, apex obtuse to rounded. Flowers 1–5 from rosette; pedicel 2.5–13 cm long, glabrous. Calyx 2-lipped, glabrous; lower lip 1–2.5 mm long, 2-lobed; lobes obovate; upper lip 2–4 mm long, 3-lobed; lobes triangular-ovate. Corolla mostly white with yellow spots on palate, 0.9–2 cm long; tube shortly funnelform-tubular, 3–7 mm long, outside glabrous, inside puberulent; spur yellowish, cylindric, straight, about as long as corolla tube, apex rounded; lobes circular to broadly ovate; filaments curved, 1.4–1.6 mm long, anther-thecae confluent. Stigma lower lip circular and ciliate, upper lip deltoid. Capsule ovoid-oblong, 5–7 mm long, glabrous. Seeds narrowly ellipsoid, 0.6–0.8 mm long. **Chromosomes** 2n = 32. Figure 46-1.

Phenology: flowering from May to July; fruiting from July to September.

Habitats: bogs and wet places on mountains; 3000–4900 m.

Distribution: U Ganga & Indus, U Yarlung Zangbo, W Nepal, C Nepal, Sikkim & Darjiling, Bhutan, Yarlung Zangbo-Brahmaputra, N Hengduan and S Hengduan (Map 46-1); also in Mongolia, Russia, Europe, and other parts of N Myanmar and China (Guizhou, W Hubei, E Qinghai, S Shaanxi, Sichuan).

U GANGA & INDUS: Kumaon, *R. Strachey & J. E. Winterbottom s. n.* (K). **U YARLUNG ZANGBO: Gyirong**, Gyirong Town, 3700 m, 2013, *PE-Xizang Exped. 3838* (PE). **W NEPAL:** s. loc., *J. F. Duthie 5871* (K). **C NEPAL: Dhawalagiri**, Muktinath Himal, 1954, *J. D. A. Stainton, W. R. Sykes & L. H. J. Williams 1336* (BM); Mustang, Chhairogaon, 1954, *J. D. A. Stainton et al. 773* (BM); Mustang, Syangboche-Ghemi, 2001, *S. Noshiro et al. 20104107* (E, TI). **E NEPAL: Mechi**, Topke Gola, 1956, *J. D. A. Stainton 660* (BM). **SIKKIM & DARJILING: Sikkim**, 1849, *J. D. Hooker s. n.* (K); Lasha Chhu Valley, 1996, *Edinburgh Expediton to Northern Sikkim (1996) 329* (E). **BHUTAN: Thimphu**, 4519 m, 2010, *JML15-33* (MO). **YARLUNG ZANGBO-BRAHMAPUTRA: Cona**, Lampug, 2013, *Y. S. Chen et al. 13-0907* (PE); **Zayü**, Zhuwagengzhaku, 4200 m, Jul. 2010, *X. H. Jin*

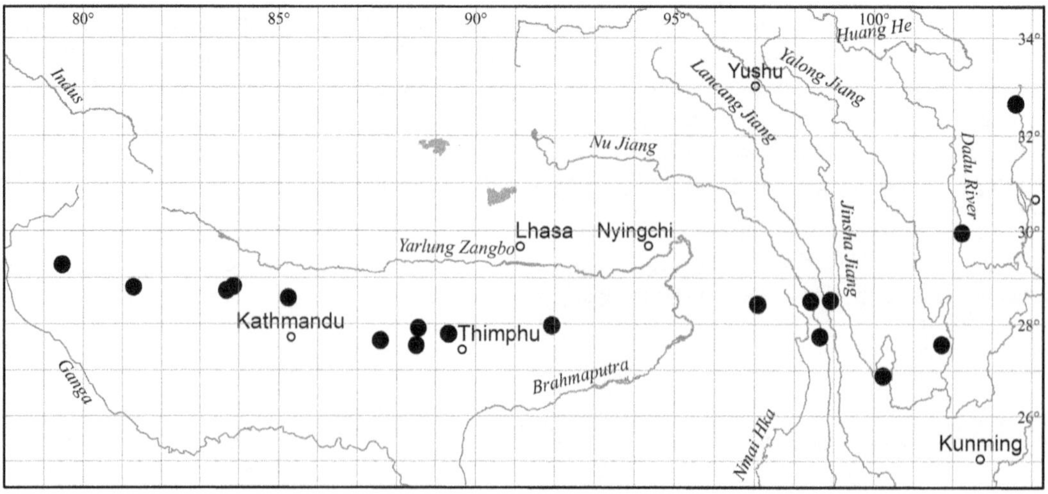

Map 46-1: **Pinguicula alpina** L.

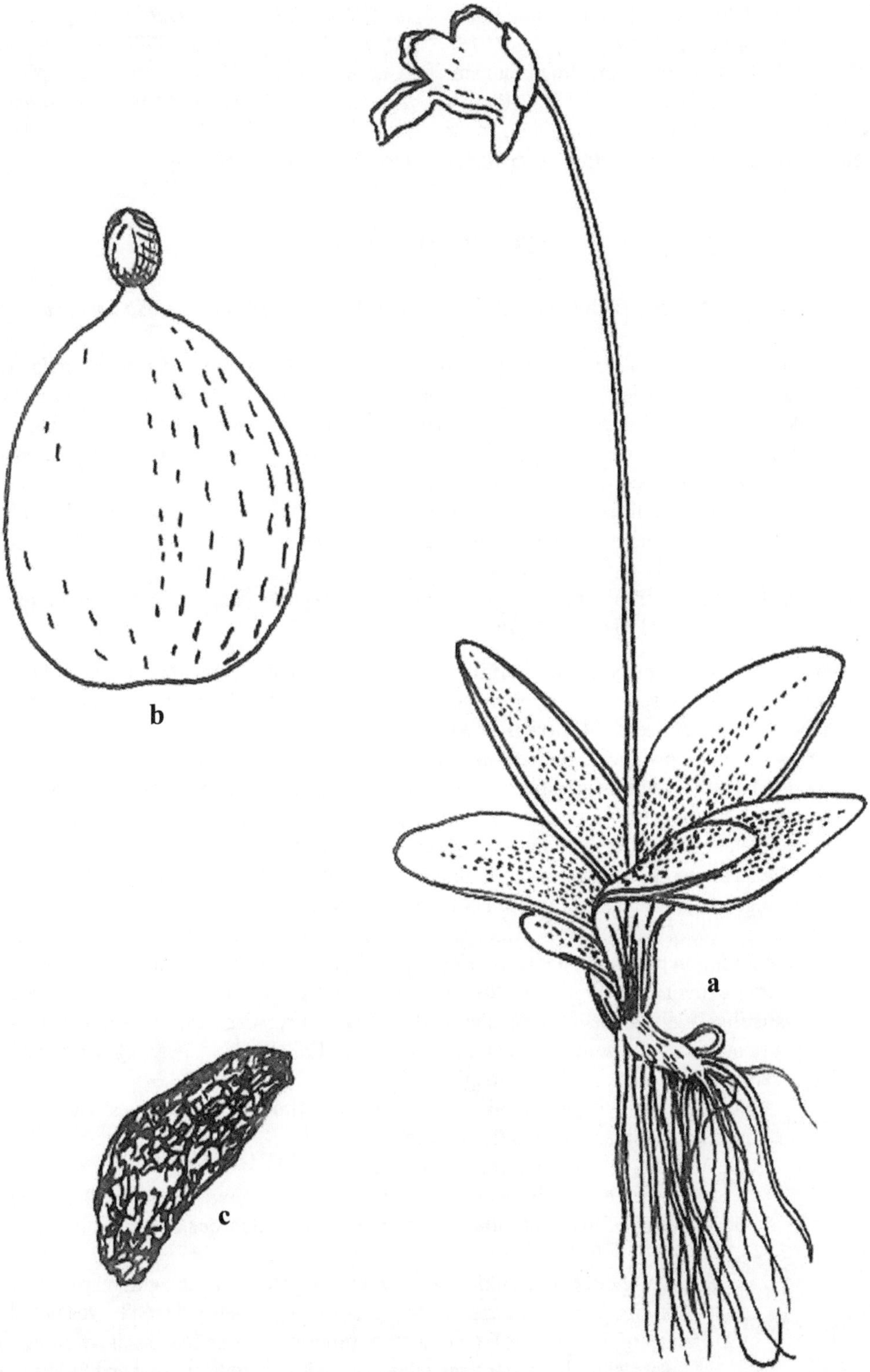

Figure 46-1: **Pinguicula alpina** L.: **a,** habit; **b,** capsule; **c,** seed. Drawn by Chun-Rong LIU; reproduced from *Fl. Reipubl. Popularis Sin.* 69: 587, pl. 162: 1–3. 1990.

et al. STET 0967 (PE); Zayü, near Ridong, 3800–4000 m, 1982, *Qinghai-Xizang Exped. 10587* (PE). **N HENGDUAN: Songpan**, Huanglongshi, 3200 m, 1983, *K. Y. Lang et al. 1751* (PE). **S HENGDUAN: Gongshan**, Dulongjiang, near the Yako of Kongdang, 3400 m, 2008, *X. H. Jin et al. DLT-ET 2200* (PE); **Luding**, Eastern slope of Gonggashan, Yanzhigou, 3500–3700 m, 1982, *K. Y. Lang et al. 290* (PE); **Shangri-la** (Zhongdian), Tianbaoshan, 3700 m, 1981, *Qinghai-Xizang Exped. 1536* (PE); **Yulong** (Lijiang), 1915, *H. Handel-Mazzetti 6807* (W).

2. UTRICULARIA L.

Linnaeus, Sp. Pl. 1: 18. 1753; Taylor, Kew Bull. Add. Ser. 14: 75. 1989. Type: *Utricularia vulgaris* L.

Herbs, perennial or annual, terrestrial, epiphytic, or aquatic, without true roots; stems modified into rhizoids and stolons, rarely developed. Traps on rhizoids, stolons, or leaves. Leaves alternate or in a basal rosette, simple to many times divided. Inflorescences racemose or flowers solitary, erect to twining. Bracts and bracteole often present, scale-like, sometimes basisolute. Calyx 2-parted from base into 2 equal or unequal lobes. Corolla 2-lipped, lower lip larger than upper one, entire or 2- or 3(–6)-lobed, spurred, palate variously raised; upper lip entire or 2- or 3-lobed. Capsule adaxially loculicidal, both abaxially and adaxially loculicidal, or circumscissile, rarely indehiscent. Seeds few, many, or rarely 1 per capsule.

About 220 species, cosmopolitan but mostly in tropical regions, a few in N temperate regions; 19 species (5 endemic) in the Pan-Himalaya.

1a. Leaves divided into narrowly linear to capillary segments, ultimate segments apically and laterally setulose; bracteoles absent; capsule globose or ellipsoid, not dorsiventrally compressed; seeds lenticular or prismatic (sect. IV. **Utricularia**).
  2a. Scales absent on peduncle; pedicels dorsiventrally compressed, deflexed in fruit.
    3a. Floats usually absent, fusiform if present, whorled at or near base of peduncle; pedicels 0.4–2 cm long, not winged but thickened distally in fruit; persistent style greatly enlarged and elongated, often equaling to or exceeding capsule in length ·················· 18. *U. aurea*
    3b. Floats ellipsoid, usually whorled above the middle of peduncle; pedicels 1–5 mm long, winged in fruit; persistent style about or less than 1/2 as long as capsule ....19. *U. stellaris*
  2b. Scales present on peduncle; pedicels terete, spreading to decurved in fruit.
    4a. Turions (overwintering buds) absent; corolla with a few short stipitate glands; capsule laterally 2-valvate; seeds lenticular, with a broadly irregular wing ············· 13. *U. gibba*
    4b. Turions usually present; corolla without external stipitate glands; capsule circumscissile; seeds prismatic, (4- or)5- or 6-angled.
      5a. Vegetative shoots similar, all with green, many times divided leaves with traps, leaf segment capillary or slightly flattened.
        6a. Margins of lower corolla lip deflexed; palate hairy; bracts ovate ····················· ··· 14. *U. vulgaris*
        6b. Margins of lower corolla lip spreading; palate glabrous.; bracts suborbicular ·· ··· 15. *U. australis*
      5b. Vegetative shoots dimorphic, bearing either green leaves with narrowly linear segments with few or no traps or bearing colorless, much reduced leaves with traps.
        7a. Palate of lower corolla lip with a prominent rounded basal swelling; spur subulate, slightly shorter than lower corolla lip; rhizoids present and filiform ········ ··· 16. *U. intermedia*

7b. Palate of lower corolla lip without a prominent rounded swelling but much elongated with a raised marginal rim; spur saccate to obtusely broadly conic, ca. 1/2 as long as lower corolla lip; rhizoids absent ························ 17. *U. minor*

1b. Leaves simple, orbicular to linear, glabrous, margin entire or rarely divided into broader lobes; bracteoles present; capsule globose, ovoid, or ellipsoid, often dorsiventrally compressed; seeds ovate in outline, compressed, obovoid, or ellipsoid.

8a. Bracts and bracteoles basisolute, attached just above to below the middle; pedicels not winged; corolla violet, pink, or white.

9a. Leaves linear to narrowly obovate, 1-nerved; traps mouth with a beaklike dorsal appendage; calyx lobes about equal in size, upper lobe shorter than upper corolla lip (sect. I. **Nigrescentes**)··· 1. *U. caerulea*

9b. Leaves reniform, orbicular, or obovate, veins dichotomously branched; traps mouth with a flabellate, or 2-cleft dorsal appendage; calyx lobes very unequal in size, upper lobe much longer than upper corolla lip (sect. III. **Phyllaria)**.

10a. Plants with a basal tuber.

11a. Spur of corolla about as long as or shorter than lower corolla lip.

12a. Leaf blade reniform, 2.5–6 mm wide; spur of corolla 3–5 times as long as lower calyx lobe··· 5. *U. brachiata*

12b. Leaf blade orbicular to obovate, 1–1.5 mm wide; spur of corolla less than twice as long as lower calyx lobe ································ 6. *U. christopheri*

11b. Spur of corolla much longer than lower corolla lip ··················· 7. *U. forrestii*

10b. Plants without basal tuber.

13a. Trap mouth with a dorsal deeply 2-cleft appendage; corolla spur subulate, about as long as lower corolla lip, apex acute; seeds covered with glochidiate processes.

14a. Lower corolla lip 4-lobed, considerably shorter than spur .... 11. *U. furcellata*

14b. Lower corolla lip 5-lobed, about as long as spur................. 12. *U. striatula*

13b. Trap mouth with a dorsal flabellate appendage; corolla spur cylindric, much shorter than lower corolla lip, apex obtuse; seeds processes not glochidiate.

15a. Middle lobe of lower corolla lip rounded; spur ca. 2 times as long as upper calyx lobe .... 10. *U. salwinensis*

15b. Middle lobe of lower corolla lip emarginate; spur as long as or slightly longer than upper calyx lobe.

16a. Lower corolla lip 5-lobed; seeds with tufts of sometimes long bristles at both ends..8. *U. kumaonensis*

16b. Lower corolla lip 3-lobed; seeds with a tuft of short bristles at distal end only ....9. *U. multicaulis*

8b. Bracts and bracteoles basifixed; pedicels winged in fruit; corolla yellow, mauve or violet; traps with 2 subulate dorsal appendages (sect. II. **Oligocista**).

17a. Leaves 1-nerved; corolla yellow.

18a. Peduncle erect; pedicel strongly recurved in fruit; upper calyx lobe shorter than upper corolla lip, apex obtuse.. 2. *U. bifida*

18b. Peduncle twining to erect; pedicel erect in fruit; upper calyx lobe longer than upper corolla lip, apex shortly acuminate·········· 4. *U. scandens* subsp. *firmula*

17b. Leaves 3-nerved; corolla mauve or violet····································3. *U. graminifolia*

I. Utricularia sect. **Nigrescentes** (Oliv.) Komiya, J. Jap. Bot. 48: 151. 1973. ≡ *Utricularia* (unranked) *Nigrescentes* Oliv. , J. Linn. Soc. Bot. 3: 184. 1859. Type: *Utricularia caerulea* L.

Small terrestrials. Leaves narrowly obovate, 1-nerved. Traps on stolons and leaves, with a beaklike dorsal appendage. Scales, bracts and bracteoles basisolute. Calyx lobes subequal. Corolla violet, white or yellow. Capsule dorsiventrally compressed, dehiscing by a ventral slit. Seeds obovoid, with testa cells slightly elongate, verrucose or rugulose.

Three species; one in the Pan-Himalaya.

1. Utricularia caerulea L., Sp. Pl. 1: 18. 1753; Taylor in Hara et al., Enum. Fl. Pl. Nepal 3: 132. 1982; Taylor, Kew Bull. Add. Ser. 14: 187. 1989; Li & Cheek in Wu, Raven & Hong (eds.), Fl. China 19: 483. 2011. Type: Ceylon (Sri Lanka), *P. Hermann s. n.* [lectotype designated by the Committee for Spermatophyta, 1989: BM (BM000621547)].

Utricularia capillaris D. Don, Prodr. Fl. Nepal.: 85. 1825. Type: [C] Nepal, *N Wallich s. n.*, probably *Wallich Cat. 1497* (cited by Taylor, 1989).

Utricularia racemosa Wall. ex Walp. in Meyon, Obs. Bot. Nov. Acta 19: 401. 1845. Type: Bangladesh, Sylhet, *Bruce s. n. in Wallich Cat. 1496* (holotype: B destroyed; isotypes: BR, C, E, G, K, NY, P, PH, W).

Utricularia filicaulis Wall. ex A. DC. in de Candolle (ed.), Prodr. 8: 21. 1844. ≡ *Utricularia racemosa* var. *filicaulis* (Wall. ex A. DC.) C. B. Clarke in Hooker (ed.), Fl. Brit. India 4: 333. 1884. ≡ *Utricularia caerulea* L. var. *filicaulis* (Wall. ex A. DC.) Haines, Bot. Bihar & Orissa: 645. 1922. Type: Burma (Myanmar), Tavoy, *Gomez s. n. in Wallich Cat. 1501* (lectotype designated by Taylor, 1989: G; isotypes: BR, C, K, P).

Annuals, terrestrial. Rhizoids and stolons capillary, branched. Traps on stolons and leaves, ovoid, 0.2–1 mm long, with a breaklike dorsal appendage. Leaves from peduncle base and stolon nodes, glabrous; leaf blade linear to linear-obovate, 0.3–2 cm long, 0.5–2 mm broad, veins 3, margin entire. Inflorescences erect, 5–44 cm long, 1–15-flowered; peduncle 0.3–1.2 mm thick; scales and bracts basisolute, oblanceolate, bracteoles basisolute, ovate-lanceolate; pedicel erect at anthesis but spreading to deflexed in fruit, 0.2–1 mm long, minutely papillose. Calyx lower lobe suborbicular, smaller than upper lobe, apex retuse; upper lobe ovate-oblong, apex rounded. Corolla violet, pink, or white, marked with yellow at throat, 4–10 mm long; lower lip broadly ovate, with a basal 4-ridged swelling, apex rounded to emarginate; spur narrowly conic, usually longer than lower corolla lip, upper lip oblong to ovate-oblong, apex retuse. Capsule globose to ellipsoid, 1.5–2 mm long, dehiscing by a short longitudinal ventral slit. Seeds obovoid to oblong-ellipsoid, 0.2–0.3 mm long; with prominent elongate reticulations and papillae. **Chromosomes** 2n = 36, 40. Figure 46-2.

Phenology: flowering from June to January; fruiting from July to February.

Habitats: wet rocks, beside streams, wet grassland, swamps, damp places; 1100–2000 m.

Distribution: C Nepal, E Nepal, S Hengduan and U Irrawaddy (Map 46-2); also in Bangladesh, Cambodia, Indonesia, Japan, Korea, Laos, Malaysia, Papua New Guinea, the Philippines, Sri Lanka, Thailand, Vietnam, Australia, Madagascar, Pacific islands (Guam, Palau), and other parts of India, Myanmar and China (S Fujian, Guangdong, Guangxi, Guizhou, Hainan, Hunan, Shandong, Taiwan).

C NEPAL: Janakpur, Dolakha, 1954. *A. Zimmermann 1170* (G). **E NEPAL:** s. loc., *J. Poelt s. n.* (GZU). **S HENGDUAN: Huaping**, Xingjie, 1700 m, 1934, *H. Handel-Mazzetti 13026* (W). **U IRRAWADDY: Tengchong**, Houqiao, 1100 m, 2012, *X. H. Jin et al. 13511* (PE).

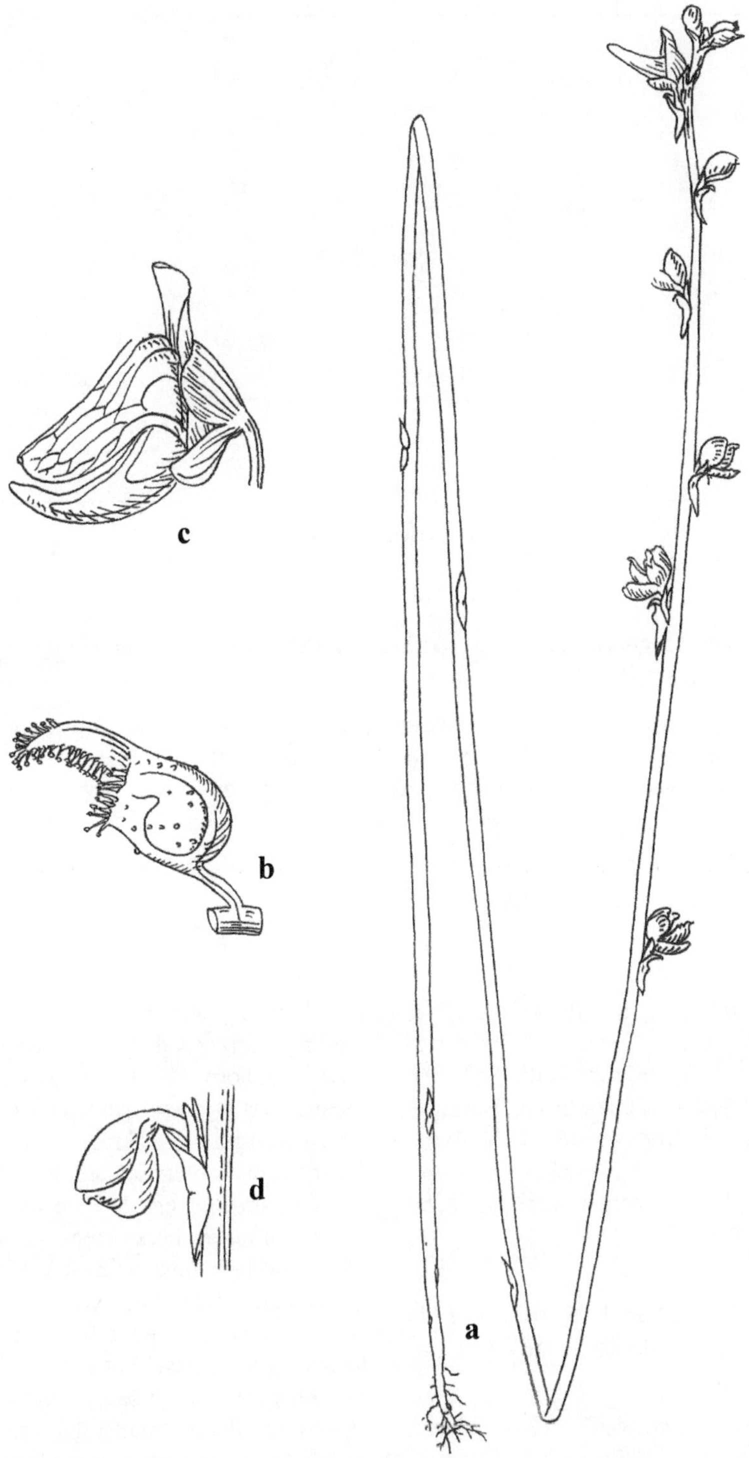

Figure 46-2: **Utricularia caerulea** L.: **a,** habit; **b,** trap; **c,** corolla; **d,** capsule, persistent bract, bracteoles and calyx. Drawn by Jin-Yong FENG; reproduced from *Fl. Reipubl. Popularis Sin.* 69: 596, pl. 163: 8–11. 1990.

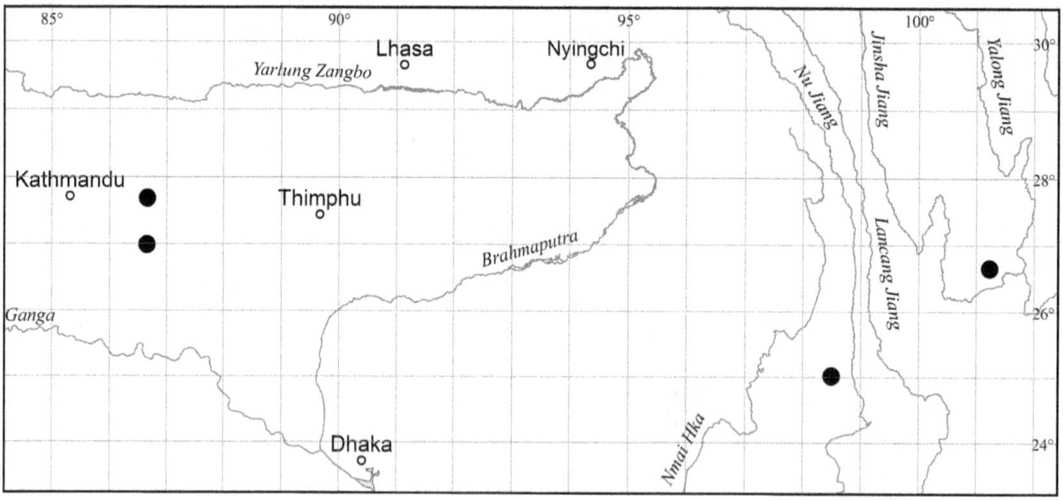

Map 46-2: **Utricularia caerulea** L.

II. Utricularia sect. **Oligocista** A. DC., Prodr. 8: 12. 1844. Type: *Utricularia bifida* L. (lectotype designated by Taylor, 1989).

Small to medium-sized terrestrials. Leaves narrowly linear or obovate, 1- or 3- or many-nerved. Traps on stolons and leaves, with 2 subulate dorsal appendages. Scales, bracts and bracteoles basifixed. Calyx lobes mostly subsimilar. Corolla yellow, mauve or violet, rarely blue. Capsule ovoid or broadly ellipsoid, ± dorsiventrally compressed, dehiscing by a dorsal, or both dorsal and ventral slits. Seeds globose, broadly ovoid or obovoid.

Thirty-six species; three in the Pan-Himalaya.

2. Utricularia bifida L., Sp. Pl. 1: 18. 1753; Taylor in Hara et al., Enum. Fl. Pl. Nepal 3: 173. 1982; Taylor, Kew Bull. Add. Ser. 14: 307. 1989; Li & Cheek in Wu, Raven & Hong (eds.), Fl. China 19: 484. 2011. Type: China, Guangdong, *P. Osbeck s. n. in Herb. Linn. No. 34.7* (lectotype designated by Taylor, 1989: LINN).

Utricularia wallichiana Benj., Bot. Zeit. 3: 213. 1845. Type: Nepal, *N. Wallich 1498* (holotype: "Herb. Lehm", HBG? n. v.; isotypes: G, K).

Utricularia antirrhinoides Wall. ex Oliv., J. Linn. Soc. Bot. 3: 182. 1859, in syn. Type: C Nepal, *N. Wallich 1498 B* (holotype: K).

Utricularia biflora Wall., Numer. List no. 1498. 1829, nom. invalid., non Lam. (1791).

Annuals, terrestrial. Rhizoids and stolons capillary, branched. Traps on stolons and leaves, globose, 0.5–1 mm long, mouth basal; appendages 2, dorsal, subulate, curved. Leaves from stolon nodes, glabrous; leaf blade narrowly linear to linear-oblanceolate, 0.7–2(–3) cm long, 1–4 mm broad, vein 1, base attenuate, margin entire. Inflorescences erect, 2–40 cm long, 1–16-flowered; peduncle 0.3–1.2 mm thick; scales and bracts basifixed, ovate; bracteoles subulate, much shorter than bracts; pedicel spreading at anthesis but decurved in fruit, 2–5 mm long, capillary, broadly winged. Calyx lower lobe ovate, slightly smaller than upper lobes, apex rounded to emarginate; upper lobe broadly ovate, apex obtuse. Corolla yellow, 6–10 mm long; lower lip suborbicular, base with a prominent rounded swelling, apex rounded; spur subulate, about as long as lower corolla lip; palate margin ciliate; upper lip oblong to oblong-obovate, apex rounded. Capsule broadly ellipsoid, 2.5–3 mm long, dorsiventrally compressed, dehiscing by longitudinal dorsal and ventral slits. Seeds obliquely obovoid, 0.4–0.6 mm long, with prominent elongate reticulations. Figure 46-3.

Phenology: flowering from June to December; fruiting from July to January.

Habitats: wet soil and rocks, rice fields; 350–2250 m.

Distribution: C Nepal, E Nepal, Sikkim & Darjiling, S Hengduan and U Irrawaddy (Map 46-3); also in Bangladesh, Cambodia, Indonesia, Japan, Korea, Laos, Malaysia, Papua New Guinea, the Philippines, Sri Lanka, Thailand, Vietnam; N Australia, Pacific islands (Guam, Palau), and other parts of India, Myanmar and China (Anhui, Chongqing, Fujian, Guangdong, Guangxi, Guizhou, Hainan, Henan, Hubei, Hunan, Jiangsu, Jiangxi, Shandong, Taiwan, Yunnan and Zhejiang).

C NEPAL: Bagmati, Langtan, Tanguni-Chipling, 2000 m, 1969, *H. Kanai & S. B. Malla 673565* (KATH); Dupche Ghyang, 2250 m, 1969, *H. Kanai & S. B. Malla 674607* (KATH); Jumla Khola, 600 m, 1954, *J. D. A. Stainton, W. R. Sykes & L. H. J. Williams 7124* (E). **E NEPAL:** Risingo to Pheda, 1220 m, *M. L. Banerji 969* (KATH). **SIKKIM & DARJILING: Darjiling**, Katambari, 150 m [from Noltie in Grierson & Long, Flora of Bhutan 2(3): 1340. 2001]. **S HENGDUAN: Huaping**, Xingjie, 1700 m, 1934, *H. Handel-Mazzetti 13030* (W). **U IRRAWADDY: Putao**, Makohwa, ca. 450 m, 2014, *X. H. Jin et al. PT-ET 1394* (PE).

3. Utricularia graminifolia Vahl, Enum. 1: 195. 1804; Taylor, Kew Bull. Add. Ser. 14: 370. 1989; Li & Cheek in Wu, Raven & Hong (eds.), Fl. China 19: 484. 2011. ≡ *Utricularia caerulea* L. var. *graminifolia* (Vahl) Bhattacharyya, Bull. Bot. Soc. Bengal 30: 76. 1976. Type: "India orientalis", 1776, *J. G. Konig s. n. in Herb. Vahl* (holotype: C).

Utricularia subrecta Lace, Bull. Misc. Inf. Kew 1915: 405. 1915. Type: Burma (Myanmar), Shan Hills Plateau, Koni, *A. G. Collett 469* (syntype: K); Maymyo Plateau, Apr. 1908, *J. H. Lace 3151* (syntypes: E, K); Sep. 1909, *J. H. Lace 4845* (syntypes: E, K); Oct. 1911, *J. H. Lace 5488* (syntypes: E, K); Oct. 1911, *J. H. Lace 5505* (syntypes: E, K).

Utricularia reticulata auct. non Sm.: Wu, Index Fl. Yunnan. 2: 1635. 1984.

Perennials or perhaps sometimes annuals, terrestrial. Rhizoids and stolons capillary, branched. Traps on rhizoids, stolons and leaves, globose, 0.5–1.3 mm long, mouth basal; appendages 2, dorsal, simple, subulate. Leaves from stolon nodes, glabrous; leaf blade linear to narrowly obovate, 0.4–2 cm long, 0.8–3 mm broad, veins 3, base attenuate, margin entire. Inflorescences erect, 2.5–30 cm long, 1–6-flowered; peduncle 0.4–1 mm thick; scales and bracts basifixed, ovate; bracteoles subulate, slightly shorter than bracts. Pedicel ascending, 3–13 mm long, filiform, narrowly winged. Calyx lower lobe slightly smaller than upper lobe, apex shortly 2-dentate; upper lobe apex acute to acuminate. Corolla mauve or violet, 0.7–1.3 cm long; lower lip suborbicular, base with a conspicuous swelling, apex rounded; spur subulate, about as long as lower corolla lip; palate with a raised ciliate margin; upper lip oblong, apex rounded. Capsule ellipsoid, 2–3 mm long, slightly dorsiventrally compressed, dehiscing by a longitudinal ventral slit. Seeds ovoid to ellipsoid, 0.3–0.4 mm long, with prominent elongate reticulations. Figure 46-4.

Phenology: flowering from May to December; fruiting from June to January.

Habitats: marshes, swamps, stream banks, wet soil among rocks; 1600–2300 m.

Distribution: S Hengduan and U Irrawaddy (Map 46-4); also in Sri Lanka, Thailand, and other parts of India, Myanmar and China (S Fujian, Guangdong and Yunnan).

S HENGDUAN: Eryuan, Sanyang, 2250 m, 1995, *W. M. Chu & Z. R. He* (PE); **Heqing**, *s. coll. s. n.* (KUN); **Yulong**, Eastern flank of Lijiang range, 1906, *G. Forrest 2969* (E). **U IRRAWADDY: Tengchong**, Houqiao, 1680 m, 1964, *S. K. Wu 6767* (PE).

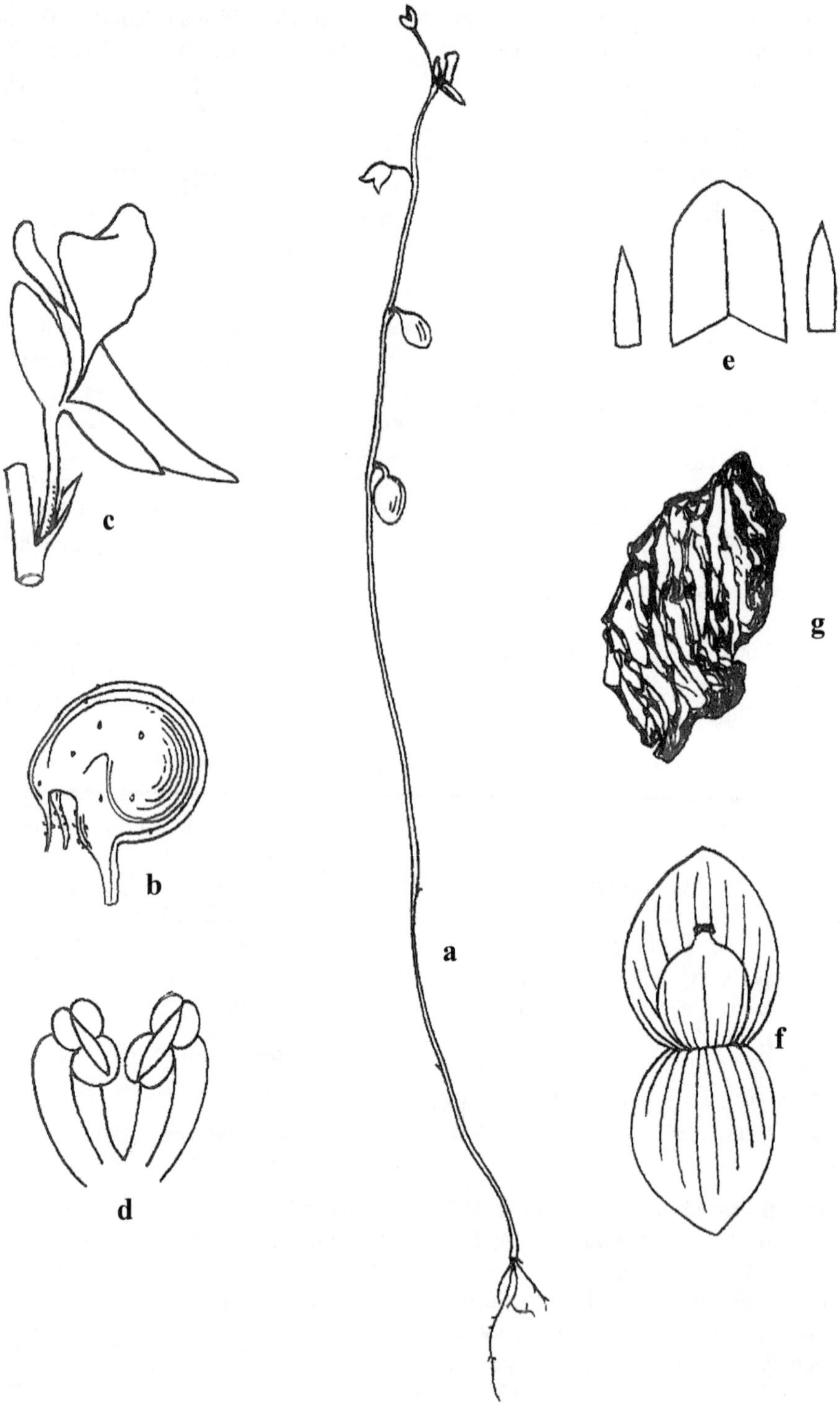

Figure 46-3: **Utricularia bifida** L.: **a,** habit; **b,** trap; **c,** flower; **d,** stamens; **e,** bract and bracteoles; **f,** capsule and persistent calyx; **g,** seed. Drawn by Chun-Rong LIU; reproduced from *Fl. Reipubl. Popularis Sin.* 69: 585, pl. 162: 4–10. 1990.

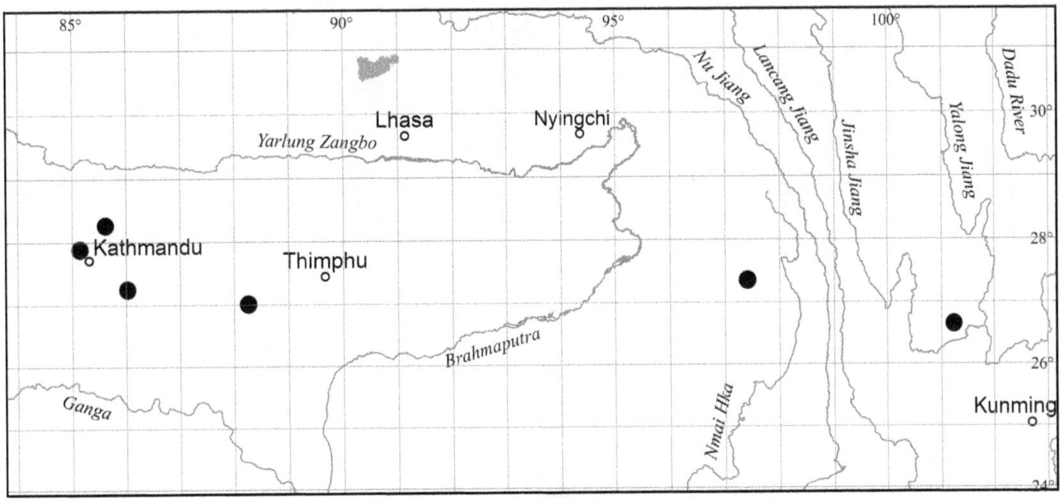

Map 46-3: **Utricularia bifida** L.

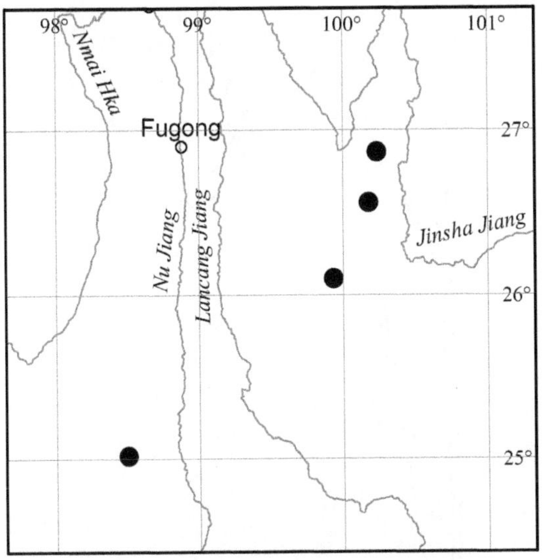

Map 46-4: **Utricularia graminifolia** Vahl

4. Utricularia scandens Benj., Linnaea 20: 309. 1847; Subramanyam, Bull. Bot. Surv. India 23: 161. 1981; Taylor, Kew Bull. Add. Ser. 14: 379, 1989; Li & Cheek in Wu, Raven & Raven & Hong (eds.), Fl. China 19: 485. 2011. Type: India, Tamil Nadu, Arcot, s. d., *Chuter s. n.* (holotype: K).

Annuals, terrestrial. Rhizoids and stolons capillary, branched. Traps on rhizoids, stolons and leaves, globose, 0.5–1 mm long, mouth basal; appendages 2, dorsal, subulate, glandular. Leaves from stolon nodes, glabrous; leaf blade narrowly linear, 5–10 mm long, 0.4–1 mm broad, vein 1, base attenuate, margin entire. Inflorescences erect or twining, 1.5–35 cm long, (1–)3–8-flowered, glabrous; peduncle 0.2–0.6 mm thick; scales and bracts basifixed, broadly ovate-deltoid, apex acute to acuminate; bracteoles narrowly linear to subulate, as long as or slightly shorter than bracts, pedicel suberect, 2.5–3 mm long, filiform, winged. Calyx lobes ovate to elliptic, lower lobe subequal to or slightly smaller than upper lobe, apex shortly 2-cleft; upper lobe apex acute to acuminate. Corolla yellow, 5–11 mm long; lower lip suborbicular, base distinctly swollen, apex rounded to obscurely crenate; spur subulate, longer than lower corolla lip, apex acute; palate papillose; upper lip oblong, apex rounded to emarginate. Capsule ovoid, 2–2.5 mm long, dorsiventrally compressed, dehiscing by a longitudinal ventral slit. Seeds obliquely obovoid, 0.2–0.3 mm long, with prominent elongate reticulations.

Distribution: W Nepal, C Nepal, E Nepal, Sikkim & Darjiling, Bhutan, S Hengduan and U Irrawaddy (Map 46-5); also in Bangladesh, Indonesia, Laos, Malaysia, Papua New Guinea, Sri Lanka, Thailand, Vietnam; Africa, N Australia, Madagascar, and other parts of India, Myanmar and China (S Yunnan and SW Guizhou).

Figure 46-4: **Utricularia graminifolia** Vahl: **a,** habit; **b,** flower. Drawn by Chun-Rong LIU; reproduced from *Fl. Reipubl. Popularis Sin.* 69: 585, pl. 162: 16–17. 1990.

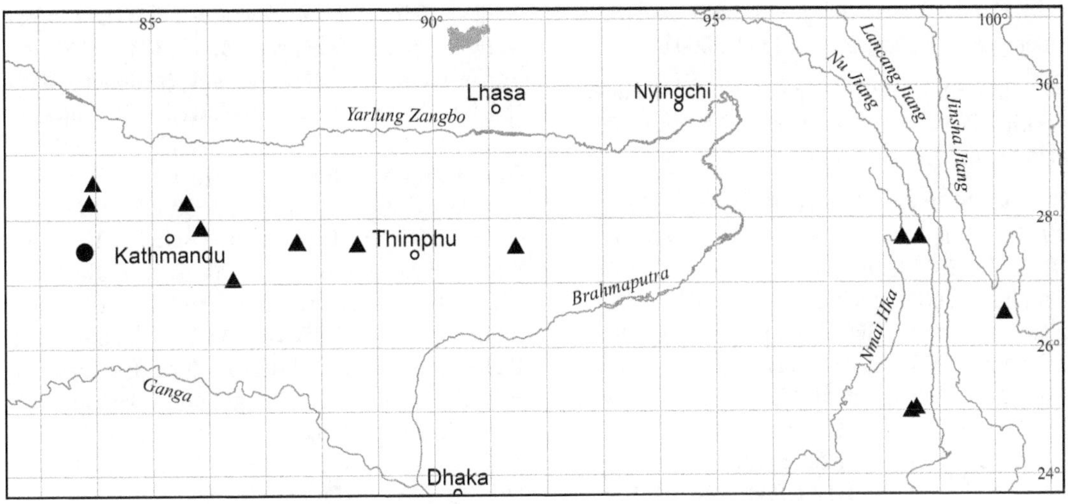

Map 46-5: **Utricularia scandens** Benj. subsp. **scandens** (●); subsp. **firmula** (▲)

1a. Inflorescences twining to straight; calyx lobes subequal, ovate at anthesis and in fruit .... 4a. subsp. *scandens*

1b. Inflorescences straight; calyx lobes unequal, upper lobe broadly ovate at anthesis, becoming rhombic-elliptic in fruit, lower lobe ovate-elliptic at anthesis, becoming ovate-oblong in fruit .... 4b. subsp. *firmula*

4a. Utricularia scandens subsp. **scandens**

Utricularia wallichiana Wight, Icon. Pl. Ind. Or. 4(4): 9. 1850. ≡ *Utricularia bifida* var. *wallichiana* (Wight) Thwaites, Enum. Pl. Zeyl.: 171. 1860. Type: India, Tamil Nadu, *Herb. Wight* (syntypes: K).

Inflorescences twining to straight; calyx lobes subequal, ovate at anthesis and in fruit. **Chromosomes** 2n = 12, 14.

Phenology: flowering from August to September; fruiting from September to October.

Habitats: bogs, meadows; at low altitude.

Distribution: C Nepal; also in Bangladesh, Indonesia, Laos, Malaysia, Papua New Guinea, Sri Lanka, Thailand, Vietnam, Africa, N Australia, Madagascar, and other parts of India, Myanmar and China (S Yunnan, SW Guizhou).

C NEPAL: s. loc., *Pradhan & Thapa 6489 p. p.* (n. v., from Taylor in Hara et al., Enum. Fl. Pl. Nepal 3: 132. 1982).

4b. Utricularia scandens subsp. **firmula** (Oliv.) Z. Y. Li, Bull. Bot. Res. Harbin 8(2): 29. 1988; Li & Cheek in Wu, Raven & Hong (eds.), Fl. China 19: 485. 2011. ≡ *Utricularia wallichiana* var. *firmula* Oliv., J. Linn. Soc. Bot. 3: 182. 1859. ≡ *Utricularia scandens* var. *firmula* (Oliv.) Subramanyam & Banerjee, Bull. Bot. Surv. India 10: 106. 1968; Taylor in Hara et al., Enum. Fl. Pl. Nepal 3: 173. 1982.Type: India, Khasia, s. d., *J. D. Hooker & Thomson s. n.* (lectotype **here designated**: K); India, Sikkim, Lachung, 1849, *J. D. Hooker s. n.* (remaining syntype: K).

Utricularia recta P. Taylor, Kew Bull. 41: 10. 1986. Type: Bhutan, Trashiyangse Dzong, Aug. 1949, *F. Ludlow, G. Sherriff & J. H. Hicks 21023* (holotype: K; isotype: BM).

Inflorescences straight. Calyx lobes unequal; lower lobe ovate-elliptic at anthesis, becoming ovate-oblong in fruit; upper lobe broadly ovate at anthesis, becoming rhombic-elliptic in fruit.

Phenology: flowering from June to September; fruiting from July to October.

Habitats: bogs, marshes, swamps, wet rocks, among grasses along streams; 1300–3400 m.

Distribution: W Nepal, C Nepal, E Nepal, Sikkim, Bhutan, S Hengduan and U Irrawaddy; also in other parts of India.

W NEPAL: Jangla Banyang, 3800 m, 1973, *Einarsson, Skarby & Wetterall 3792* (BM). **C NEPAL: Bagmati**, Langtang, between Dhaibung and Ramche, 1700 m, 1966, *D. H. Nicolson 2341* (BM); Sindhupalchok, N of Thagaon, 2011, *M. F. Watson et al. 21* (E); Kaski, Annapurna Himal, Mardi Khola, 1954, *J. D. A. Stainton, W. R. Sykes & L. H. J. Williams 8533* (E); Rasuwa, Khola kharka, 1949, *O. Polunin 1485* (E). **E NEPAL: Kosi**, Tamur Valley, 1956, *J. D. A. Stainton 1300* (E); Tamur Valley, 1963, *H. Hara et al. 6303032* (TI); **Mechi**, Topke Gola, 1977, *H. Ohashi & H. Kanai 773407* (KUN). **SIKKIM & DARJILING: Sikkim**, Lachung, 2950–3600 m, 1849, *D. Hooker s. n.* (K). **BHUTAN: Tashi Yangtsi**, 1949, *F. Ludlow, G. Sherriff & J. H. Hicks 21023* (K, BM, E). **S HENGDUAN: Gongshan**, Salwin Kiukiang divide, East of Tehahtu, 2900 m, *T. T. Yu 20276* (PE); **Heqing**, Xiangshuihe, 1929, *R. C. Ching 24356* (KUN). **U IRRAWADDY: Drungjiang**, Dulongjiang region, 3400–3600 m, Jul. 2013, *X. H. Jin ST1362* (PE); **Tengchong**, Base of hills to northwest of Tengyueh (Tengchong), 2166 m, 1912, *G. Forrest 8051* (PE); Houqiao, 1100 m, Oct. 2012, *X. H. Jin 13512* (PE).

III. Utricularia sect. **Phyllaria** (Kurz) Kamieński in Engler & Diels, Nat. Pflanzenfam. 4(3b): 120. 1891. ≡ *Utricularia* subg. *Phyllaria* Kurz, J. Bot 12: 54. 1874. Type: *Utricularia striatula* Sm. (lectotype designated by Taylor, 1989).

Small lithophytes or epiphytes. Tubers sometimes present. Leaves petiolate, obovate to reniform, with dichotomous nerves. Traps on rhizoids and stolons but never on leaves, the mouth with a single dorsal, dorsiventrally flattened, flabellate, entire of bifid appendage. Scales absent or very few, similar to bracts. Bracts and bracteoles basisolute or rarely bracteoles basifixed, subsimilar. Calyx lobes very unequal, upper lobe larger. Corolla white or violet. Capsule dorsiventrally compressed dehiscing by a ventral longitudinal slit. Seeds ovate in outline, compressed, with rounded, conical, clavate, hair-like or glochidiate processes.

Thirteen species; eight (five endemic) in the Pan-Himalaya.

5. Utricularia brachiata Oliv., J. Linn. Soc. Bot. 3: 187. 1859; Polunin & Stainton, Fl. Himal.: 306. 1984; Taylor, Kew Bull. Add. Ser. 14: 462. 1989; Li & Cheek in Wu, Raven & Hong (eds.), Fl. China 19: 484. 2011. Type: [Sikkim & Darjiling] India, Sikkim, Lachen, 2600–3600 m, 1849, *J. D. Hooker s. n.* (holotype: K; isotypes: G, L, W).

Perennials, lithophytic. Tuber solitary at peduncle base, globose to ellipsoid, 1–2 mm long. Rhizoids and stolons capillary, simple. Traps on rhizoids and stolons, ovoid, ca. 1 mm long, mouth lateral; appendage 1, dorsal, transversely elliptic to flabellate, fringed with multicellular stipitate glands. Leaves from tuber, glabrous; leaf blade reniform, 1–3 cm long, 2.5–6 mm broad, veins dichotomously branched, with long petiole, margin entire. Inflorescences erect, 3–8 cm long, 1–2-flowered, glabrous; peduncle 0.2–0.4 mm thick; scales absent. Bracts and bracteoles basisolute, oblong, base truncate to emarginate, margin erosely denticulate; pedicel spreading, 2–8 mm long, filiform, dorsiventrally compressed. Calyx glabrous, lower lobe oblong, much smaller than upper lobe, apex emarginate; upper lobe orbicular, apex emarginate. Corolla white, with a yellow spot at base of lower lip and violet streaks on upper lip; lower lip suborbicular, distinctly 5-lobed; spur narrowly cylindric, 3–5 times as long as upper calyx lobe, apex obtuse; palate with a slightly raised rim, fringed with moniliform hairs; upper lip much shorter than upper calyx lobe, apex 2-lobed. Capsule globose with a ventral keel, ca. 2.5 mm long, dehiscing by a longitudinal ventral slit. Seeds ovate in outline, compressed, ca. 0.5 mm long, with a tuft of hairs of variable length at each end usually about as long as body of seeds and a few shorter hairs on distal surface. Figure 46-5.

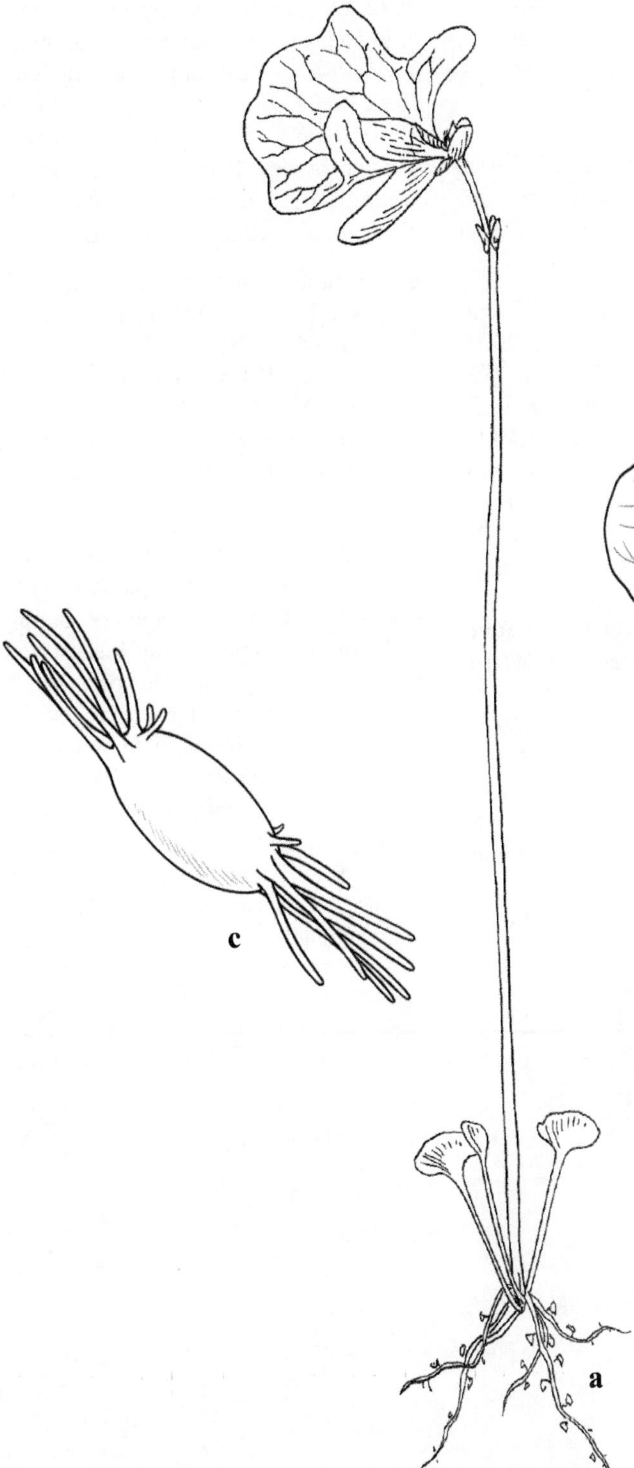

Figure 46-5: **Utricularia brachiata** Oliv.: **a,** habit; **b,** leaf; **c,** seed. a, reproduced from *Fl. Reipubl. Popularis Sin.* 69. 1990; b, c, drawn by Ai-li LI.

Phenology: flowering from July to September; fruiting from August to October.

Habitats: forests among bryophytes on rocks; 2600–4200 m.

Distribution: C Nepal, E Nepal, Sikkim & Darjiling, Yarlung Zangbo-Brahmaputra, S Hengduan and U Irrawaddy (Map 46-6); also in other parts of China (W Sichuan).

C NEPAL: Bagmati, Langtang, 3770 m, 1949, *O. Polunin 1958* (E); Rasuwa, 2001, *Edinburgh Nepal Exped. (2001) 336* (E); **Gandaki**, Lamjung, 1954, *J. D. A. Stainton, W. R. Sykes & L. H. J. Williams 6380* (E). **E NEPAL: Mechi**, Topke Gola, 1956, *J. D. A. Stainton 830* (BM); Sagarmatha, Jorsale, 2006, *Third Darwin Nepal Fieldwork Training Exped. BX 25* (E). **SIKKIM & DARJILING: Darjiling**, Sandakphu (from Noltie in Grierson & Long, Flora of Bhutan 2(3): 1338. 2001); **Bhutan**: Thimphu and Upper Mo Chu (from Noltie in l. c.). **YARLUNG ZANGBO-BRAHMAPUTRA: Mêdog**, Bomi-Mêdog Road 80 to 52 km, 2009, *X. H. Jin et al. SET-ET 1389* (PE). **S HENGDUAN: Dêqên**, Mekong Salwin divide, Sila, 3500 m, 1938, *T. T. Yu 21216* (PE); l. c., 4000 m, 1938, *T. T. Yu 22432* (PE); **Gongshan**, Changputung, 3150 m, 1916, *H. Handel-Mazzetti 9925* (W); Judulong, Jibu to Donshaofang, 2600–2700 m, 1982, *Qinghai-Xizang Exped. 8347* (PE); Salwin Kiukiang divide, Lunguailaka, 3300 m, 1938, *T. T. Yu 20255* (PE); Taron Taru divide, valley of Bucahwang, 2800 m, 1938, *T. T. Yu 20112* (PE). **U IRRAWADDY: Chimili Valley**, near the Yunnan border, 1919, *R. J. Farrer 1146* (E, K); **Drungjiang**, Dulongjiang, 1200–2000 m, 2011, *X. H. Jin & L. Zhang 11552* (PE).

6. Utricularia christopheri P. Taylor, Kew Bull. 41: 12. 1986; Taylor, Kew Bull. Add. Ser. 14: 464. 1989; Noltie in Grierson & Long (eds.), Fl. Bhutan 2(3): 1339. 2001. Type: [E] Nepal, Koshi Zone, Sankhuwasabba District, upper Kashwa Khola Dhap, May 1981, *C. Grey-Wilson et al. 4365d* (holotype: K).

Perennial lithophyte. Tuber solitary just below the peduncle base, globose, 1–2 mm long. Rhizoids and stolons capillary, simple. Leaves 2–5 from the upper part of a short vertical stem between the tuber and the peduncle base, long petiolate; leaf blade orbicular to obovate, 4–10 mm long, 1–1.5 mm broad, with a few dichotomous nerves. Traps numerous on the rhizoids, stolons ovoid, stalked, mouth lateral with a small dorsal broadly flabellate appendage, distal margin rounded and fringed with multicellular stipitate glands. Inflorescences erect, solitary, 2.5–4 cm long; peduncle terete, 0.3–0.4 mm thick, glabrous;

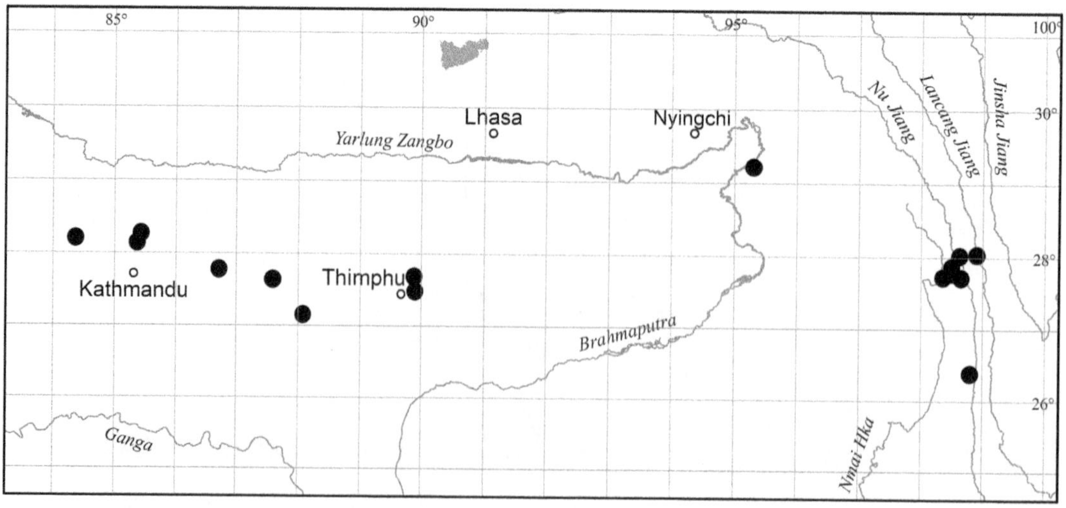

Map 46-6: **Utricularia brachiata** Oliv.

scales absent. Bracts and bracteoles narrowly ovate-oblong, basisolute. Flowers solitary; pedicel erect, 3–5 mm long. Calyx lobes unequal, upper lobe orbicular, apex emarginate; lower lobe much smaller, oblong, apex emarginate. Corolla white with a yellow spot at base of lower lip 4.5–9 mm long; upper lip approximately quadrate, apex obscurely and very shortly 2-lobed, much shorter than upper calyx lobe; lower lip approximately circular in outline, unequally 5-lobed; palate with a slightly raised rim fringed with hairs; spur cylindric, apex shortly bifid, less than twice as long as lower calyx lobe. Capsule and seeds unknown.

Phenology: flowering from May to July.

Habitats: on mossy rocks, usually growing with *U. brachiata* Oliv. and *U. multicaulis* Oliv.; 3600–3960 m.

Distribution: **endemic**; C Nepal, E Nepal and Sikkim & Darjiling (Map 46-7).

C NEPAL: Bagmati, (Langtang Valley). **SIKKIM & DARJILING: Sikkim**, just above Changu, 3960 m [from Taylor, 1989 and Noltie in Grierson & Long (eds). Fl. Bhutan 2(3): 1339. 2001].

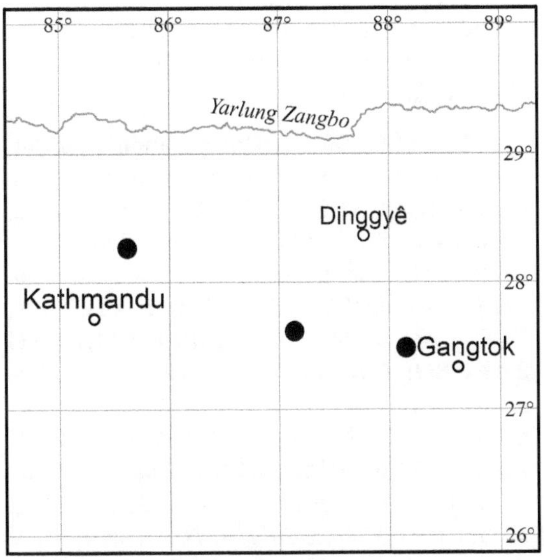

Map 46-7: **Utricularia christopheri** P. Taylor

7. Utricularia forrestii P. Taylor, Kew Bull. 41: 13. 1986; Taylor, Kew Bull. Add. Ser. 14: 466. 1989; Li & Cheek in Wu, Raven & Hong (eds.), Fl. China 19: 486. 2011. Type: [U Irrawaddy] China, Yunnan, hills northwest of Tengchong (Tengyue), 1 Sep. 1924, *G. Forrest 25232* (holotype: E; isotype: K).

Perennials, lithophytic. Tuber solitary at peduncle base, globose, 1–2 mm long. Rhizoids and stolons capillary, simple. Traps on rhizoids and stolons, ovoid, ca. 1 mm long, mouth lateral; appendage 1, dorsal, broadly flabellate, fringed with multicellular stipitate glands. Leaves from a short vertical stem between tuber and peduncle base, glabrous; leaf blade broadly obovate, orbicular or reniform, 4–8 mm long, 1–3 mm broad, veins dichotomously branched, base broadly cuneate to subcordate, margin entire. Inflorescences erect, 2–4 cm long, 1- or 2-flowered, glabrous; peduncle terete, 0.2–0.3 mm thick; scales absent. Bracts and bracteoles basisolute, narrowly ovate, base truncate, apex subacute; pedicel spreading, 2–8 mm long, filiform, dorsiventrally compressed. Calyx lobes glabrous; lower lobe oblong to obovate, much smaller than upper lobe, apex emarginate; upper lobe orbicular to broadly transversely elliptic, apex emarginate. Corolla mauve to violet, 1.2–1.5 cm long; lower lip suborbicular, apex 3-lobed about to the middle; lobes retuse to shortly 2-cleft; spur filiform, 0.9–1.5 cm long, curved; palate with a slightly raised rim fringed with hairs; upper lip suborbicular, shorter than upper calyx lobe, apex shortly 2-lobed. Capsule globose, ca. 2 mm, with a ventral keel, dehiscing by a single longitudinal ventral slit. Seeds narrowly elliptic, ca. 0.5 mm, with a hollow obscurely apically dentate multicellular appendage at each end.

Phenology: flowering from July to August; fruiting from August to September.

Habitats: among mosses on rocks; 2100–3000 m.

Distribution: endemic; S Hengduan and U Irrawaddy (Map 46-8).

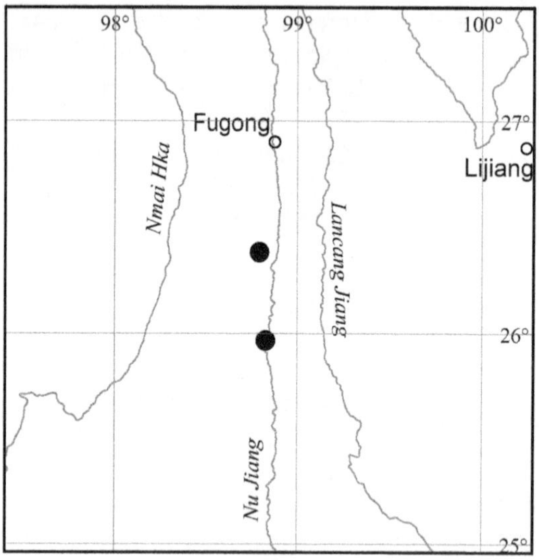

Map 46-8: **Utricularia forrestii** P. Taylor

S Hengduan: Lushui, Shweli-Salween divide, 1912, *G. Forrest 8909* (E); 1917, *G. Forrest 18451* (E, K). **U IRRAWADDY: Chimili Valley**, near the Yunnan border, 1919, *R. J. Farrer 1210* (E).

8. Utricularia kumaonensis Oliv., J. Linn. Soc. Bot. 3: 189. 1859; Strachey, Cat. Pl. Kumaon 130. 1906; Taylor, Kew Bull. Add. Ser. 14: 468. 1989; Li & Cheek in Wu, Raven & Hong (eds.), Fl. China 19: 487. 2011. Type: [U Ganga & Indus] India, Kumaon, near Pilti River, 2460 m, *R. Strachey & J. E. Winterbottom s. n.* (holotype: K; isotype: BM).

Diurospermum album Edgew., Proc. Linn. Soc. 1: 351. 1847, non *Utricularia alba* Hoffmg. ex Link (1820). Type: "in Vishnugangetis Valle, Himala; at 8000 ped", *M. P. Edgeworth s. n.* (not located).

Annuals, epiphytic, lithophytic, or terrestrial. Rhizoids and stolons capillary, simple. Traps on rhizoids and stolons, ovoid, ca. 1 mm long, mouth lateral; appendage 1, dorsal, flabellate, fringed with multicellular stipitate glands. Leaves rosulate at peduncle base, petiolate, glabrous; leaf blade broadly obovate, orbicular, or reniform, 2–6 mm long, 1.5–3 mm broad, veins dichotomously branched, base broadly cuneate, margin entire. Inflorescences erect, 2–7 cm long, 1–3-flowered, glabrous; peduncle terete, 0.2–0.3 mm thick; scales absent. Bracts and bracteoles basisolute, ovate-oblong, base truncate, apex obtuse. Pedicel erect at anthesis but decurved in fruit, terete, 2–6 mm long, filiform. Calyx lower lobe oblong, much smaller than upper lobe, apex truncate and emarginate to erose; upper lobe obcordate, apex deeply 2-parted. Corolla white, with a basal yellow spot and mauve lobes on lower lip, 3–5(–7) mm long; lower lip suborbicular, 5-lobed, middle lobe much larger than lateral lobes; spur broadly cylindric, about as long as upper calyx lobe, slightly curved, apex obtuse; palate with a slightly raised rim, fringed with short hairs; upper lip subquadrate, ca. 1/2 as long as upper calyx lobe, apex truncate to obscurely 2-lobed. Capsule obliquely ovoid, 2–2.5 mm long, dehiscing by a single longitudinal ventral slit. Seeds few per capsule, ellipsoid, ca. 0.5 mm long, with a tuft of hairs at each end, hairs usually longer than body of seed; seed coat with short and obtuse papillae.

Phenology: flowering from July to August; fruiting from August to September.

Habitats: among mosses on rocks, cliffs, fallen trees; 2600–2700 m.

Distribution: **endemic**; confined to U Ganga & Indus, U Yarlung Zangbo, E Nepal, Bhutan, Yarlung Zangbo-Brahmaputra, S Hengduan and U Irrawaddy (Map 46-9).

U GANGA & INDUS: Kumaon, near Pilti River, 2460 m, *R. Strachey & J. E. Winterbottom s. n.* (BM, K). **E NEPAL: Koshi**, Sankhuwasabha, 2070–2450 m, 1998, *Flora of East Nepal 9820057* (E); Barun Khola, 1981, *C. Grey-Wilson 4128 & 4396* (K). **BHUTAN: Punakha**, 1999, *G. & S. Miehe 99-234-31* (BM). **U YARLUNG ZANGBO: Nyalam**, Zham, near 707 Bridge, 3205 m, *PE-Xizang Exped. 4374* (PE); l. c., 2641 m, 2011, *Y. S. Chen et al. 701* (PE). **YARLUNG ZANGBO-BRAHMAPUTRA: Cona**, Lampug, 2013, *Y. S. Chen et al., 13-0909* (PE); **Mêdog**, 52 km, 3600–3700 m, 2012, *X. H. Jin, W. T. Jin & S. Z. Xu 13184* (PE). **S HENGDUAN: Dali**, Cangshan, 3300–3500 m, 2012, *X. H. Jin, W. T. Jin & S. Z. Xu 13334* (PE); **Eryuan**, Dengchuan, Huangjiapin, 1940, *R. C. Ching 24872* (PE). **U**

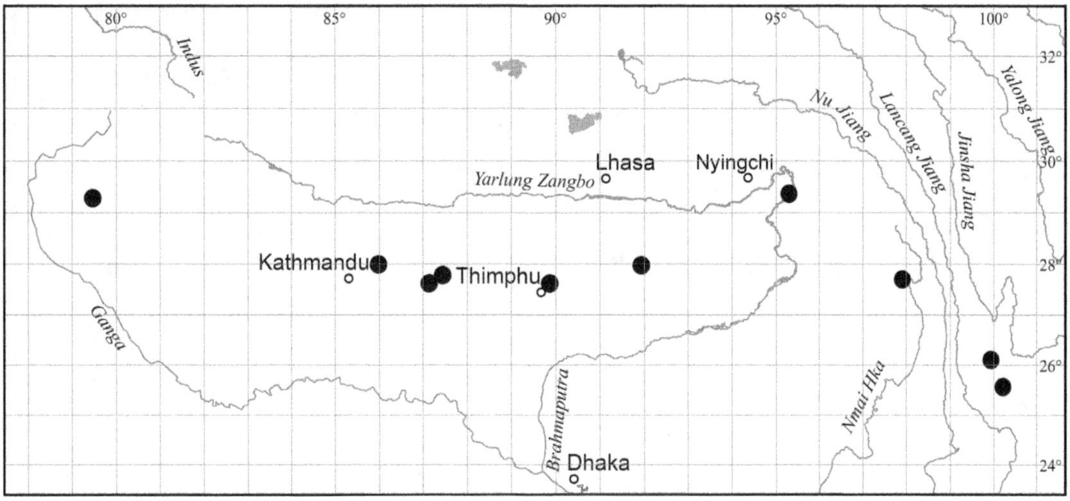

Map 46-9: **Utricularia kumaonensis** Oliv.

IRRAWADDY: Mungku Hkyet, 2440–2740 m, 1937, *F. Kingdon Ward 13037* (BM).

9. Utricularia multicaulis Oliv., J. Linn. Soc. Bot. 3: 188. 1859; Taylor, Kew Bull. Add. Ser. 14: 470. 1989; Li & Cheek in Wu, Raven & Hong (eds.), Fl. China 19: 487. 2011. Type: [Sikkim & Darjiling] India, Sikkim, 1849, *J. D. Hooker s. n.* (holotype: K; isotypes: G, L, P, W).

Annuals, terrestrial or lithophytic. Rhizoids and stolons capillary, simple. Traps on rhizoids and stolons, shortly stalked, ovoid, ca. 1 mm long, mouth lateral; appendage 1, dorsal, flabellate, fringed with multicellular stipitate glands. Leaves rosulate at peduncle base, long petiolate, glabrous; leaf blade obovate, 2–8 mm long, 1–1.5 mm broad, rather fleshy, veins dichotomously branched, base attenuate, margin entire. Inflorescences erect, 1–5 cm long, 1–3-flowered, glabrous; peduncle terete, 0.3–0.4 mm thick; scales absent. Bracts and bracteoles basisolute, oblong, 0.6–1.2 mm, base truncate, apex obtuse to acute; pedicel spreading at anthesis but decurved and distally thickened in fruit, terete, 1–5 mm long, filiform. Calyx lower lobe oblong, much smaller than upper lobe, apex entire, emarginate or erose; upper lobe orbicular to broadly transversely elliptic, apex deeply emarginate. Corolla white or pale mauve, with a yellow spot at base of lower lip, 2–5 mm long; lower lip 3-lobed, middle lobe much longer than lateral lobes, apex rounded to emarginate; spur cylindric, slightly longer than upper calyx lobe, straight or almost so, apex obtuse; palate with a slightly raised rim fringed with short hairs; upper lip subquadrate, much shorter than upper calyx lobe, apex shortly 2-lobed. Capsule globose to obliquely ovoid, 1.5–2 mm long, dehiscing by a single longitudinal ventral slit. Seeds ovate in outline, compressed, 0.7–0.9 mm long, distal end with a tuft of short hairs ca. 1/2 as long as body of seed; seeds coat with regular longitudinal rows of acute papillae.

Phenology: flowering from July to September; fruiting from August to October.

Habitats: moist rocks, open swampy meadows, among bryophytes and species of the Cyperaceae; 1200–4200 m.

Distribution: endemic; confined to C Nepal, E Nepal, Sikkim & Darjiling, Bhutan, L Yarlung Zangbo and S Hengduan (Map 46-10).

C NEPAL: Bagmati, Langtang, Latsu, 2440 m, 1969, *H. Kanai & S. B. Malla 673588* (KATH); Langur-Tsedang Pokhari, 2240–3000 m, 1969, *H. Kanai & S. B. Malla 674646* (KATH); Sindhupalchok, 2011, *M. F. Watson et al. 85* (E); **Gandaki**, Gorkha, Ganesh Himal, 1994, *Central Nepal Exped. 9400169* (E). **E NEPAL: Kosi**,

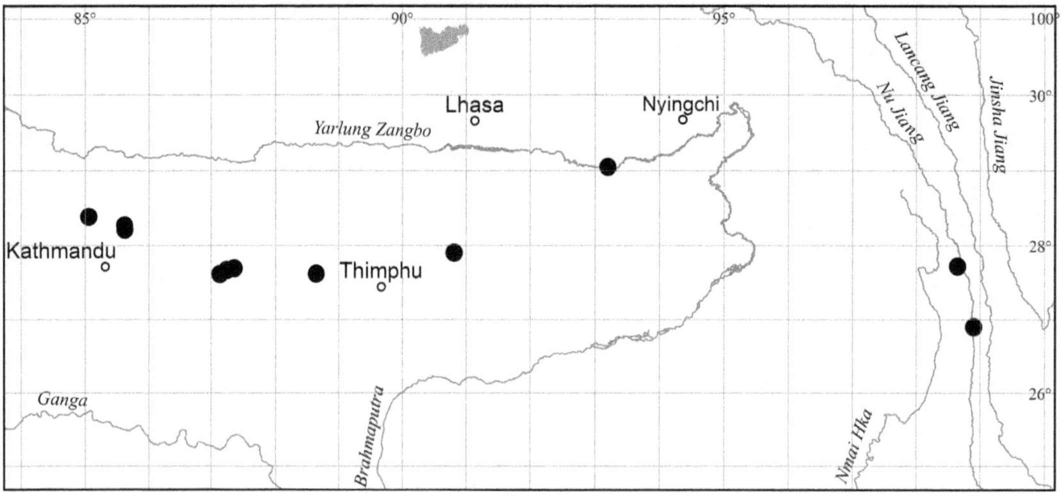

Map 46-10: **Utricularia multicaulis** Oliv.

Sankhuwasabha, 2070–2450 m, 1998, *Flora of East Nepal 9820055* (E); Barun Khola, 1981, *C. Grey-Wilson 4136* (K); Upper Kashwa Khola Dhap, 1981, *C. Grey-Wilson 4365b* (K). **SIKKIM & DARJILING:** Sikkim, Lachung, 1970–2300 m, 1849, *J. D. Hooker s. n.* (G, K, L, P, W). **BHUTAN: Bumthang** (Shabjethang), 1949, *F. Ludlow, G. Sherriff & J. H. Hicks 19508* (BM, E, K). **L YARLUNG ZANGBO: Nangxian** (Kongbo), 1938, *F. Ludlow, G. Sherriff & J. H. Hicks 5802* (K). **S HENGDUAN: Fugong**, Yaping, Lumadeng, 3400 m, 2011, *X. H. Jin & L. Zhang 11091* (PE); **Gongshan**, Gomba-La, 3900 m, 1916, *H. Handel-Mazzetti 9887* (W).

10. Utricularia salwinensis Hand.-Mazz., Symb. Sin. 7: 873. 1936; Taylor, Kew Bull. Add. Ser. 14: 472. 1989; Li & Cheek in Wu, Raven & Hong (eds.), Fl. China 19: 487. 2011. Type: [S Hengduan] China, Yunnan, Gongshan, "Tschiangschel" Pass between Salwin and Taron (Irrawadi) divide near the China and Burma (Myanmar) borders at 27°52′E, 3275 m, Sep. 1916, *H. Handel-Mazzetti 9223* (holotype: W; isotype: K).

Probably perennials, terrestrial or lithophytic. Rhizoids capillary, simple. Stolons capillary, branched. Traps on rhizoids and stolons, ovoid, ca. 1 mm long, mouth lateral; appendage 1, dorsal, flabellate, fringed with multicellular stipitate glands. Leaves from peduncle base and stolons, long petiolate, glabrous; leaf blade obovate to orbicular, 2–12 mm long, 1–2 mm broad, veins dichotomously branched, base attenuate, margin entire. Inflorescences erect, 3–7 mm long, 1- flowered, glabrous; peduncle terete, 0.2–0.4 mm thick; scales absent. Bracts and bracteoles basisolute, oblong, base truncate, margin 2-cleft or erose, apex acute to truncate; pedicel erect, spreading, or about recurved, 2–10 mm long. Calyx lower lobe oblong, much smaller than upper lobe, apex emarginate; upper lobe suborbicular, apex emarginate. Corolla white or pale pink, with a yellow spot at base of lower lip; lower lip 3-lobed, middle lobe much larger than lateral lobes, apex rounded; spur narrowly cylindric, ca. 2 times as long as upper calyx lobe, apex obtuse; palate with a slightly raised rim, fringed with moniliform hairs; upper lip subquadrate, much shorter than upper calyx lobe, apex emarginate. Capsule and seeds not seen.

Phenology: flowering from July to August.

Habitats: among mosses on wet cliffs and in bogs; 3200–4000 m.

Distribution: **endemic**; confined to Yarlung Zangbo-Brahmaputra, S Hengduan and U Irrawaddy (Map 46-11).

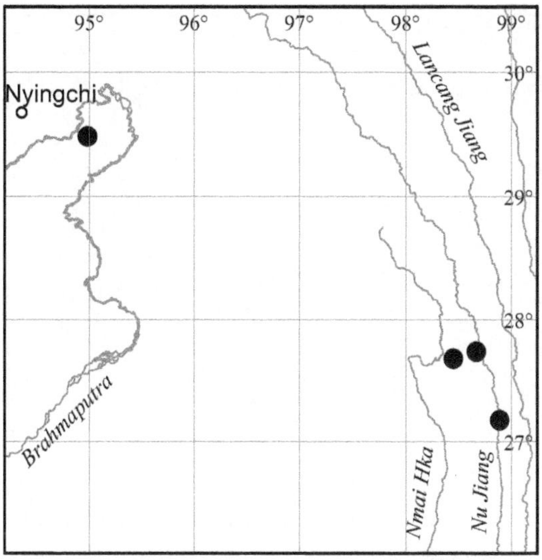

Map 46-11: **Utricularia salwinensis** Hand.-Mazz.

YARLUNG ZANGBO-BRAHMAPUTRA: Lhünzê, Tsangpo Valley, 1938, *F. Ludlow, G. Sherriff & J. H. Hicks 5264* (K); l. c., 1947, *F. Ludlow, G. Sherriff & J. H. Hicks 14304* (BM). **S HENGDUAN: Gongshan**, Mowangshan, Dongshaofang, 1997, *Gaoligongshan Exped. 9438* (KUN); Pihe Village to Gongshan, 1997, *Gaoligongshan Exped. 18629* (KUN). **U IRRAWADDY: Drungjiang**, Dulongjiang, 3500 m, 1982, *Qinhai-Xizang Exped. 8523* (PE).

11. Utricularia furcellata Oliv., J. Linn. Soc. Bot. 3: 189. 1859; Taylor, Kew Bull. Add. Ser. 14: 476. 1989; Hu, Chen & Long, Acta Bot. Yunnan. 30(2): 146. 2008. Type: India, Khasia Mts, Boga Pani, Oct. 1850, *J. D. Hooker 2534* (lectotype designated by Taylor, 1989: K; isotypes: G, M).

Utricularia furcellata var. *minor* C. B. Clarke in Hooker (ed.), Fl. Brit. Ind. 4: 334. 1884. Type: [Sikkim & Darjiling] India, Sikkim, Lachen, on wet rocks, 1849, *J. D. Hooker s. n.* (holotype: K).

Annual, lithophyte. Rhizoids and capillary, simple. Leaves rosulate at the peduncle base, long petiolate, leaf blade circular to transversely elliptic, 3–7 mm long, up to 4 mm broad, with dichotomous nerves. Traps on stolons, ovoid, stalked, mostly ca. 1 mm long, mouth lateral with a dorsal, bifid appendage, distal margin of which is fringed with multicellular stipitate glands. Inflorescences erect, 1.5–4(–6) cm long, 1–3(–5)-flowered; peduncle terete, 0.1–0.2 mm thick, glabrous; scales absent. Bracts and bracteoles basisolute, attached near base; pedicels spreading, 1–5 mm long. Calyx lobes unequal, obscurely nerved; upper lobe circular, 2–2.5 mm long, base rounded, apex emarginate; lower lobe much smaller, oblong, apex emarginate. Corolla ca. 6mm long, white, violet or pink, with yellow palate; upper lip approximately quadrate with apex shortly 2-lobed, about as long as upper calyx lobe; lower lip oblong in outline, 4-lobed, lateral lobes much smaller than apical pair; palate with a slightly raised rim, fringed with hairs; spur subulate, with apex acute, straight or slightly curved, 2–4 mm long. Capsule globose with a ventral keel, 1–1.5 mm long, dehiscing by a single, longitudinal, ventral slit. Seeds obovoid, ca. 0.4 mm long, minor end subacute, major end densely covered with glochidiate processes.

Phenology: flowering from August to October.

Habitats: moist rocks; 1100–3000 m.

Distribution: Sikkim & Darjiling, S Hengduan and U Irrawaddy (Map 46-12); also in other parts of India (Meghalaya).

SIKKIM & DARJILING: Darjiling, Rungbool, 2300–2950 m, *W. Griffith 4084* (K); **Sikkim**, Lachen, 1849, *J. D. Hooker s. n.* (K). **S HENGDUAN:** On the way from **Fugong** to **Gongshan**, 2011, *S. X. Yu & W. Wang 6470* (PE). **U IRRAWADDY: Tengchong**, Gudong, Yunfengshan, 1760 m, Aug. 2007, *G. W. Hu 24480* (KUN); l. c., 2350 m, 2007, *G. W. Hu 24481* (KUN); Houqiao, 1100 m, 2012, *X. H. Jin 13509 & 13510* (PE).

12. Utricularia striatula Sm. in Rees, Cyclop. 37: No. 17. 1819; Taylor, Kew Bull. Add. Ser. 14: 479. 1989; Li & Cheek in Wu, Raven & Hong (eds.), Fl. China 19: 488. 2011. Type: Sierra Leone, *A. Afzelius s. n.* (holotype: LINN; isotypes: BM, BR, LD, M,S).

Utricularia orbiculata Wall. ex A. DC. in de Candolle (ed.), Prodr. 8: 18. 1844; Clarke in Hooker (ed.), Fl. Brit. India 4: 334. 1884. Type: Nepal, 1821, *N. Wallich 1500* (holotype: G; isotype: K).

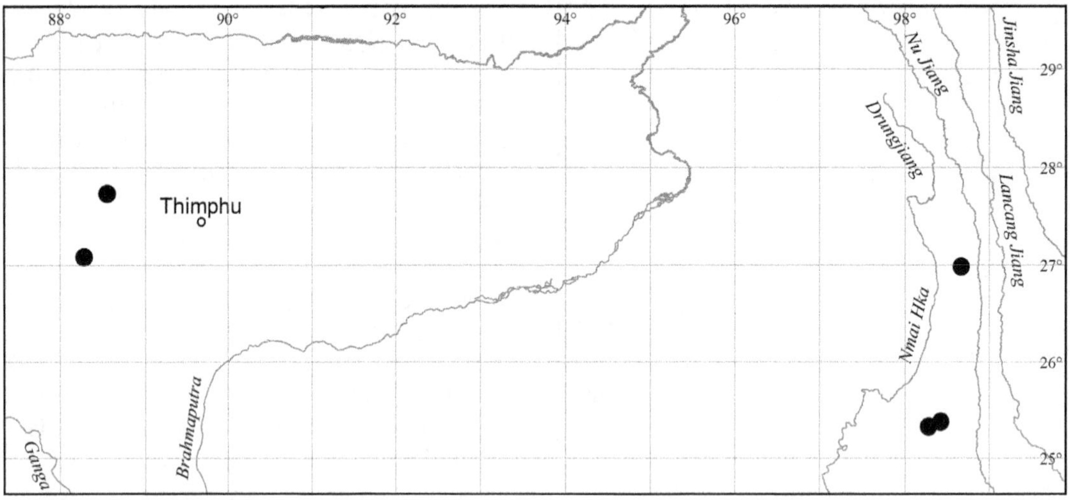

Map 46-12: **Utricularia furcellata** Oliv.

Perennials, epiphytic or lithophytic. Rhizoids and stolons capillary, simple. Traps on rhizoids and stolons, ovoid, 0.6–0.8 mm long, mouth lateral; appendage 1, dorsal, deeply 2-cleft, fringed with multicellular stipitate glands. Leaves from peduncle base and stolons, subsessile to petiolate; leaf blade obovate, orbicular, or transversely elliptic, 3–10 mm long, 2–6 mm broad, veins dichotomously branched. Inflorescences erect, 1–15 cm long, 1–10-flowered; peduncle terete, 0.2–0.4 mm thick, glabrous; scales, bracts and bracteoles basisolute, narrowly ovate to oblong; pedicel spreading at anthesis but often pendulous and withered in fruit, 2–6 mm long. Calyx lower lobe oblong-elliptic, much smaller than upper lobe, apex rounded to emarginate; upper lobe suborbicular to broadly obcordate, 1.5–2.5 mm long, apex rounded to emarginate. Corolla white or violet, with a yellow spot at base of lower lip, 3–10 mm long or much smaller when cleistogamous; lower lip transversely elliptic to suborbicular, apex 5-lobed; spur subulate, about as long as lower lip, straight or curved, apex acute; palate fringed with hairs; upper lip subovate, much shorter than calyx, apex shortly 2-cleft. Capsule globose, ca. 2.5 mm in diameter, dorsiventrally compressed, dehiscing by a single longitudinal ventral slit. Seeds obovoid, 0.2–0.4 mm long, densely covered with glochidiate processes except proximal end. Figure 46-6.

Phenology: flowering from June to October; fruiting from July to November.

Habitats: damp rocks, tree trunks; 400–3600 m.

Distribution: U Yarlung Zangbo, W Nepal, C Nepal, E Nepal, Sikkim & Darjiling, Bhutan, Yarlung Zangbo-Brahmaputra, S Hengduan and U Irrawaddy (Map 46-13); also in Indonesia, Malaysia, Papua New Guinea, the Philippines, Sri Lanka, Thailand, Vietnam; tropical Africa and Indian Ocean islands (N Andaman Islands), and other parts of India, Myanmar and China (S Anhui, Chongqing, Fujian, Guangdong, Guangxi, Guizhou, Hainan, Hubei, Hunan, Jiangxi, Sichuan, Taiwan, Yunnan and Zhejiang).

U YARLUNG ZANGBO: Gyirong, Rigelamu, 3200 m, *Tibet Chinese Mater. Medic. & Herbs. Exped. 359* (HNWP). **W NEPAL: Karnali**, Jumla, Giri Daha, 1952, *C. Polunin, W. R. Sykes & L. H. J. Williams 580* (E). **C NEPAL: Bagmali**, Langtang, between Dhaibung and Ramche, 1700 m, 1966, *D. H. Nicolson 2342* (BM); Ghoptc-Thade, 3180 m, Jul., 1967, *S. B. Malla 9553* (KATH); Sindhupalchok, N of Thagaon, 2011, *M. F. Watson et al. 22* (E). **E NEPAL: Mechi**, Topke Gola, 1100–1500 m, 1977, *H. Ohashi et al. 770280* (KUN); **Shemma**, 1400 m, 1977, *H. Ohashi et al. 770280* (KUN); **Taplejung**, 1956, *J. D. A. Stainton 1225* (BM). **SIKKIM & DARJILING: Sikkim**, Phodong

Figure 46-6: **Utricularia striatula** Sm.: **a,** habit; **b,** trap; **c,** flowers; **d,** capsule and persistent calyx; e, seed. Drawn by Jin-Yong FENG; reproduced from *Fl. Reipubl. Popularis Sin.* 69: 596, pl. 163: 1–5. 1990.

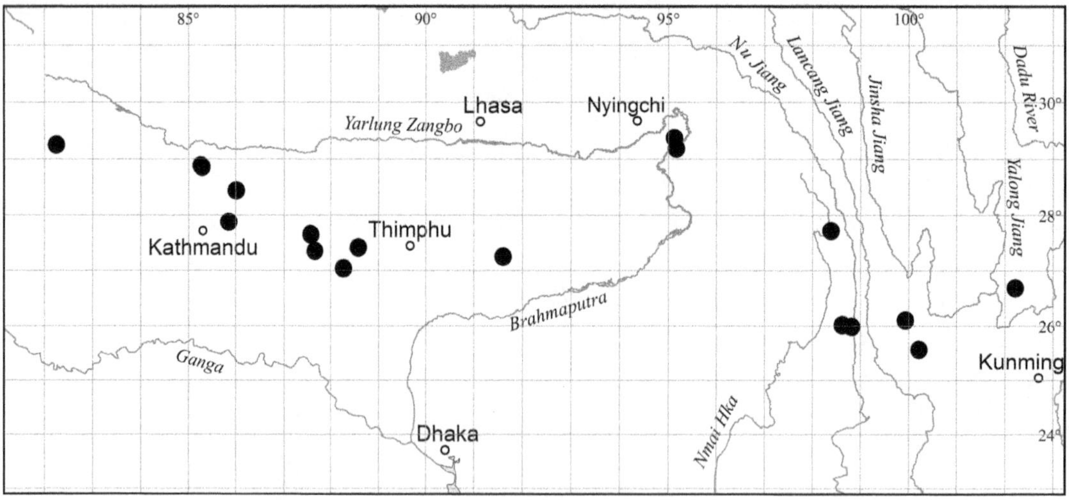

Map 46-13: **Utricularia striatula** Sm.

Gompa; **Darjiling**, Rungbool [from Noltie in Grierson & Long (eds.), Fl. Bhutan 2(3): 1337. 2001]. **BHUTAN: Mongar**, Khoma, Lhuntse, 2000, *S. Bowes Lyon 13067* (E). **YARLUNG ZANGBO-BRAHMAPUTRA: Mêdog**, Dexing, 800 m, 2010, *X. H. Jin et al. STET 2039* (PE); Hanmi to Aliqian, 2000 m, 1974, *K. Y. Lang s. n.* (PE-02237029); near Lage, 3200 m, 1974, *K. Y. Lang 747-31* (PE). **S HENGDUAN: Dali**, *Dept. Ecology and Geobotany 02695* (PYU); **Eryuan**, Dengchuan, Huangjiaping, 1940, *R. C. Ching 24872* (PE); **Gongshan**, Kiukiang, 1600 m, 1938, *T. T. Yu 19926* (KUN, PE); **Huili**, Hsiaodun, 1800 m, 1914, *H. Handel-Mazzetti 5098* (W); **Lushui**, Pianma, 3600 m, 1964, *S. K. Wu 8404* (KUN). **U IRRAWADDY: Drujiang**, Dulongjiang, 1200–2000 m, 2011, *X. H. Jin & L. Zhang 115511* (PE).

IV. Utricularia sect. **Utricularia**. Type: *Utricularia vulgaris* L.

Suspended or affixed aquatics. Stolons usually long. Leaves divided into narrowly linear to capillary segments, ultimate segments usually apically and laterally setulose, sometimes leaves reduced. Traps lateral on leave segments, with two setiform appendages. Scales present or absent on peduncle. Bracts basifixed, rarely basisolute, bracteoles absent. Calyx lobes subsimilar. Corolla yellow, sometimes white or pink. Capsule globose, or ellipsoid, laterally bivalvate, circumscissile or indehiscent. Seeds globose, lenticular, or prismatic.

Thirty-two species; seven species in the Pan-Himalaya.

13. Utricularia gibba L., Sp. Pl. 1: 18. 1753; Taylor, Kew Bull. Add. Ser. 14: 572. 1989; Li & Cheek in Wu, Raven & Hong (eds.), Fl. China 19: 489. 2011. Type: USA, Virginia, s. d., *J. Clayton 515 & 517* [lectotype designated by Taylor, 1964: BM (BM 000098329)].

Utricularia exoleta R. Br., Prodr.: 430. 1810; Taylor in Hara et al., Enum. Fl. Nepal 3: 132. 1982. ≡ *Utricularia gibba* L. subsp. *exoleta* (R. Br.) P. Taylor, Mitt. Bot. Staatss. Munchen 4: 101. 1961. Type: Australia, New South Wales, Hawksburg, s. n., *R. Brown s. n.* (holotype: BM).

Utricularia khasiana Joseph & Mani, Bull. Bot. Surv. India 25: 192. 1983; Joseph & Joseph, Insectiv. Pl. Khasi & Jaintia Hills: 16. 1986. Type: India, Meghalaya, Khasi Hills, s. d., *J. Joseph 76944* (holotype: CAL).

Utricularia elegans Wall., Numer. List no.: 1502. 1828, nom. invalid.

Annuals or perennials, affixed or suspended aquatic. Rhizoids absent or present, filiform, branched. Stolons filiform, much branched, often mat-forming. Traps lateral on leaf segments, ovoid, 1–2.5 mm long, mouth lateral; appendages 2, dorsal, branched, setiform, with shorter setae. Leaves on stolons, 0.5–1.5 cm long; primary segments (1 or) 2, unbranched or sparsely dichotomously branched into 3–8 ultimate segments; ultimate segments capillary, slightly flattened, margin entire or sparsely denticulate, apex and teeth setulose. Inflorescences erect, 2–15 cm long, 1–3(–6)-flowered; peduncle terete, 0.3–0.5 mm thick, glabrous; scale 1, similar to bracts. Bracts basifixed, semiorbicular, ca. 1 mm, minutely glandular, apex truncate and obscurely dentate; pedicel erect to spreading, 2–12 mm long, filiform; bracteoles absent. Calyx lobes subequal, broadly ovate to orbicular, apex rounded. Corolla yellow, 4–8 mm long; lower lip slightly smaller than upper lip, base with a prominent 2-lobed swelling, apex rounded; spur narrowly conic to cylindric from a conic base, shorter or longer than corolla lower lip, distal part sparsely stipitate glandular, apex obtuse; palate densely pubescent; upper lip broadly ovate to suborbicular, ca. 2 times as long as upper calyx lobe, apex obscurely 3- lobed. Capsule globose, 2–3 mm in diameter, 2-valvate. Seeds lenticular, 0.8–1 mm in diameter, margin broadly winged, wing shallowly and irregularly dentate, with small prominent reticulations. **Chromosomes** 2n = 28. Figure 46-7.

Phenology: flowering from April to November; fruiting from May to December.

Habitats: shallow still or slowly flowing water, pools, lakes, bogs, swamps, marshes, rice fields; 200–1600 m.

Distribution: C Nepal, E Nepal (Map 46-14); widely distributed in Africa, America, SW to E Asia, Europe, Indian Ocean islands and Pacific islands (New Caledonia, Palau).

C NEPAL: *N. Wallich 1502* (K); **Dhauwalagiri**, Parbat, 1952, *J. D. A. Stainton, W. R. Sykes & L. H. J. Williams 64* (BM). **E NEPAL: Jhapa**, Mai Pokhari, 1600 m, 1963, *H. Hara et al. 6303034* (TI).

14. Utricularia vulgaris L., Sp. Pl. 1: 18. 1753; Taylor, Kew Bull. Add. Ser. 14: 591. 1989; Li & Cheek in Wu, Raven & Hong (eds.), Fl. China 19: 489. 2011. Type: Europe, 'Lapland', *Herb. Linn. No. 34.2* (lectotype designated by Taylor in Kew Bull. 18: 171. 1964: LINN).

Perennials, suspended aquatic. Stolons ca. 0.5 mm thick, branched. Traps on leaf segments, ovoid, 1.5–5 mm long, mouth lateral; dorsal appendages 2, simple or branched, setiform, with shorter simple lateral setae. Leaves on stolons, 1.5–6 cm long, divided from base into 2 unequal primary segments; primary segments ovate, elliptic, or ovate-oblong in outline, pinnately divided into secondary segments; secondary segments dichotomously divided into many further segments; ultimate segments capillary, slightly flattened, apically and laterally minutely setulose, without or sometimes with short marginal teeth. Inflorescences erect, 10–30 cm long, 3–12-flowered, glabrous; peduncle terete, 1–2.4 mm thick; scales and bracts basifixed, broadly ovate; base cordate to auriculate; bracteoles absent; pedicel erect at anthesis but decurved in fruit, terete, 0.6–1.5 cm long. Calyx lobes ovate; lower lobe slightly shorter than upper lobe, apex obtuse to emarginate; upper lobe apex acute to subacute. Corolla yellow, 1.2–2 cm long; lower lip transversely elliptic to broadly ovate, base with prominent swelling, margin strongly deflexed, apex retuse; spur narrowly

Figure 46-7: **Utricularia gibba** L.: **a,** habit; **b,** flowers; **c,** seed. Drawn by Jin-Yong FENG; reproduced from *Fl. Reipubl. Popularis Sin.* 69: 600, pl. 164: 8–11. 1990.

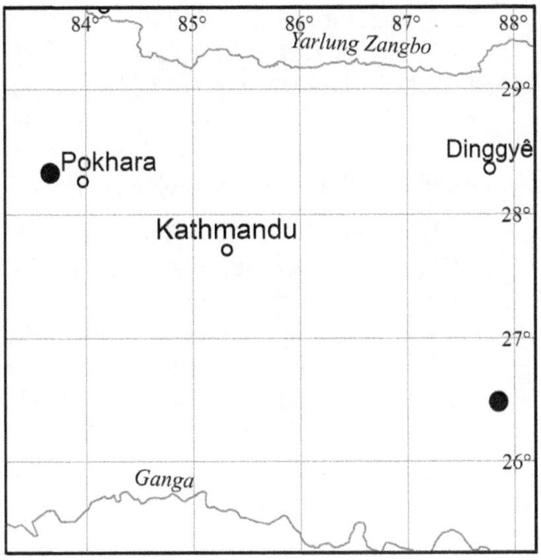

Map 46-14: **Utricularia gibba** L.

conic to narrowly cylindric from a conic base, shorter than or about as long as corolla lower lip, straight or curved upwards, apex slightly obtuse; palate distal 1/2 covered with short hairs and stipitate glands; upper lip broadly ovate, ca. 2 times as long as upper calyx lobe, apex retuse to truncate. Capsule globose, ca. 5 mm in diameter, circumscissile. Seeds prismatic, 0.3–0.4 mm in diameter, 4–6-angled.

Distribution: N Pakistan, U Ganga & Indus, N Hengduan (Map 46-15); also in Kazakhstan, Mongolia, Russia, Uzbekistan; N Africa, SW Asia, Europe, North America, and other parts of Afghanistan, Pakistan and China (Gansu, Hebei, Heilongjiang, Henan, Jilin, Liaoning, Nei Mongol, Ningxia, Qinghai, Shaanxi, Shandong, Shanxi, Sichuan, Xinjiang).

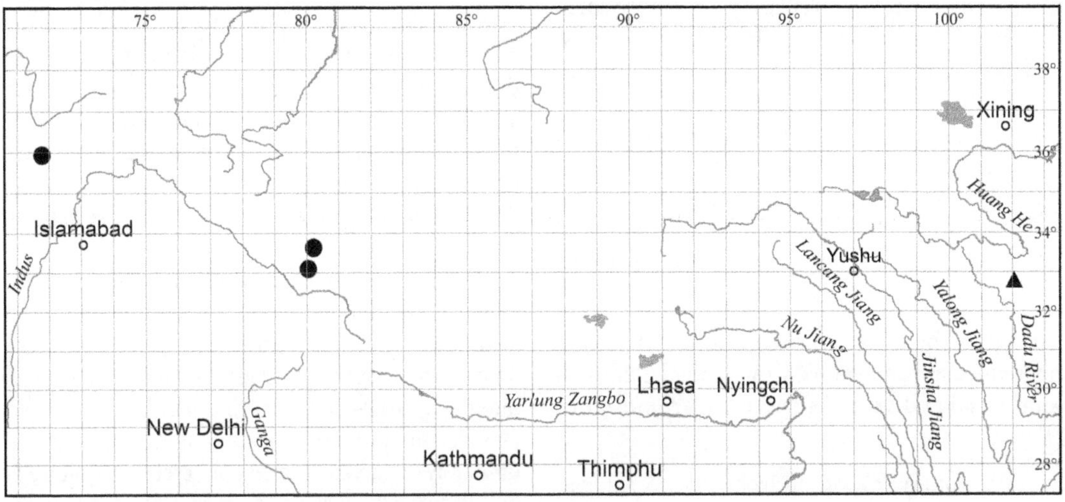

Map 46-15: **Utricularia vulgaris** L. subsp. **vulgaris** (●), subsp. **macrorhiza** (▲)

1a. Spur conspicuously shorter than corolla lower lip, more or less straight, with internal glands on dorsal surface only .... 14a. subsp. ***vulgaris***

1b. Spur about as long as corolla low lip, markedly curved upwards, with internal glands on both dorsal and ventral surfaces .... 14b. subsp. ***macrorhiza***

14a. Utricularia vulgaris subsp. **vulgaris**

Spur conspicuously shorter than corolla lower lip, more or less straight, with internal glands on dorsal surface only. **Chromosomes** 2n = 40, 44.

Phenology: flowering from July to August; fruiting from August to September.

Habitats: lakes, marsh, pods; 2900–4400 m.

Distribution: N Pakistan and U Ganga & Indus; also in Kazakhstan, Russia, Uzbekistan, N Africa, SW Asia, Europe, and other parts of Afghanistan.

N PAKISTAN: Chitral, 3600 m, 1958, *J. D. A. Stainton 3023* (BM). **UGANGA & INDUS: Rutog**, Kuba, 4350 m, 1987, *Qinghai-Xizang Exped. 01603* (PE); Meimarnan marshland, 1987, *B. S. Li, D. Zheng & B. P. Zhang 11098* (PE).

14b. Utricularia vulgaris L. subsp. **macrorhiza** (Le Conte) Clausen, Mem. Cornell Univ. Agr. Exp. Sta. 291: 9. 1949; Li & Cheek in Wu, Raven & Hong (eds.), Fl. China 19: 490. 2011. ≡ *Utricularia macrorhiza* Le Conte, Ann. Lyc. New York 1: 73, t. 6, fig. 2. 1824; Taylor, Kew Bull. Add. Ser. 14: 594. 1989. Type: [*icon*] USA, *Le Conte drawing*, t. 35 of Nieuwland lectotype designated by Taylor, 1989: ND.

Spur about as long as corolla lower lip, markedly curved upwards, with internal glands on both dorsal and ventral surfaces. **Chromosomes** 2n = 44.

Phenology: flowering from July to August; fruiting from August to September.

Habitats: lakes, pools, ditches, rivers, rice fields; 3000–4500 m.

Distribution: N Hengduan; also in Mongolia, Russia (Altai to Kamchatka), North America, and other parts of China (Gansu, Hebei, Heilongjiang, Henan, Jilin, Liaoning, Nei Mongol, Ningxia, Qinghai, Shaanxi, Shandong, Shanxi, Sichuan and Xinjiang).

N HENGDUAN: Hongyuan, 4500 m, 1985, *Z. S. Dao, Z. P. Pu & K. Q. Li 2135* (PE).

15. Utricularia australis R. Br., Prodr: 430. 1810; Taylor in Hara et al., Enum. Pl. Nepal 3: 173. 1982; Taylor, Kew Bull. Add. Ser. 14: 598. 1989; Li & Cheek in Wu, Raven & Hong (eds.), Fl. China 19: 490. 2011. Type: Australia, New South Wales, between Hawksbury & Paramatta, *R. Brown s. n.* (holotype: BM).

Utricularia neglecta Lehm., Nov. Stirp. Pug. 1: 38. 1828; Handel-Mazzetti, Symb. Sin. 7: 874. 1936. ≡ *Utricularia vulgaris* var. *neglecta* (Lehm.) Cosson & Germain, Fl. Env. Paris, ed. 2: 375. 1861. Type: Germany, "in agro Hamburgensl, praesertim prope Eppendof", *J. G. C. Lehmann s. n.* (holotype: HBG).

Utricularia flexuosa auct. non Vahl: Oliver, J. Linn. Soc. Bot. 3: 175. 1859, p. p.; Clarke in Hooker (ed.), Fl. Brit. India 4: 329. 1884, p. p.

Perennials, suspended aquatic. Rhizoids usually present, filiform, bearing numerous short dichotomously divided branches. Stolons filiform, branched. Traps on leaf segments, ovoid, 0.5–2.5 mm long, mouth lateral or basal; appendages 2, dorsal, simple or branched, setiform, with shorter simple lateral setae. Leaves numerous, 1.5–4 cm long, divided from base into 2 ovate to ovate-oblong primary segments; primary segments, pinnately divided into secondary segments; secondary segments dichotomously divided into further segments; ultimate segments capillary, slightly flattened, margin sparsely denticulate, apex and teeth setulose. Inflorescences erect, (5–)10–30 cm long, 3–8-flowered, glabrous; peduncle terete, 1–2 mm thick; scales and bracts basifixed, suborbicular, 2–5 mm long, base auriculate, apex rounded and obscurely tridentate; pedicel erect at anthesis but decurved in fruit, terete, 1–2.5 cm long; bracteoles absent. Calyx lobes ovate, subequal; lower lobe apex emarginate; upper lobe apex rounded. Corolla yellow, 1.2–1.5 cm long; lower lip transversely elliptic, base with a prominent swelling, apex rounded to retuse; spur broadly conic, considerably shorter than corolla lower lip, slightly curved, apex obtuse; palate glabrous; upper lip broadly ovate, apex retuse. Capsule globose, 3–4 mm in diameter, circumscissile. Seeds prismatic, 0.5–0.6 mm in diameter, 4–6 angled, narrowly winged on all angles, with small prominent slightly elongate reticulations. **Chromosomes** 2n = 40, 44. Figure 46-8.

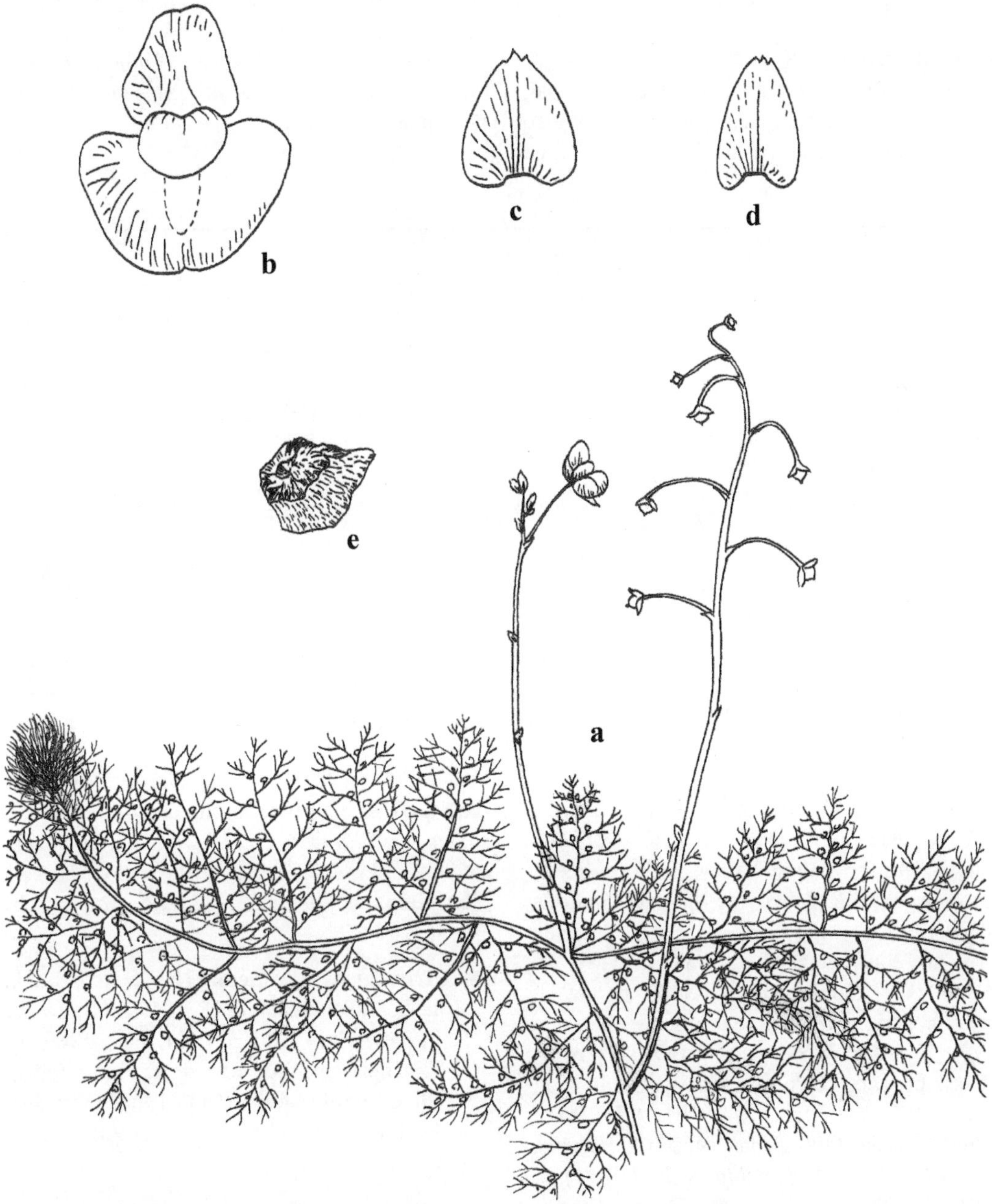

Figure 46-8: **Utricularia australis** R. Br.: **a,** habit; **b,** corolla; **c,** scale; **d,** bract; **e,** seed. Drawn by Zhang-Hua WU; reproduced from *Fl. Reipubl. Popularis Sin.* 69: pl. 165: 603, 8–12. 1990.

Phenology: flowering from July to September; fruiting from August to October.

Habitats: lakes, pools, river backwaters, ditches, rice fields; 1300–3900 m.

Distribution: N Pakistan, Jammu & Kashmir, U Yarlung Zangbo, W Nepal, C Nepal, N Hengduan and S Hengduan (Map 46-16); also in Indonesia, Japan, Korea, Mongolia, Papua New Guinea, the Philippines, Russia (Sakhalin), Sri Lanka, Africa, SW Asia, Australia, Europe, Pacific islands (New Zealand), and other parts of Afghanistan, Pakistan, India, Myanmar and China (Anhui, Chongqing, Fujian, Guangdong, Guangxi, Guizhou, Hainan, Hubei, Hunan, Jiangsu, Jiangxi, S Shaanxi, Sichuan, Taiwan, Yunnan and Zhejiang).

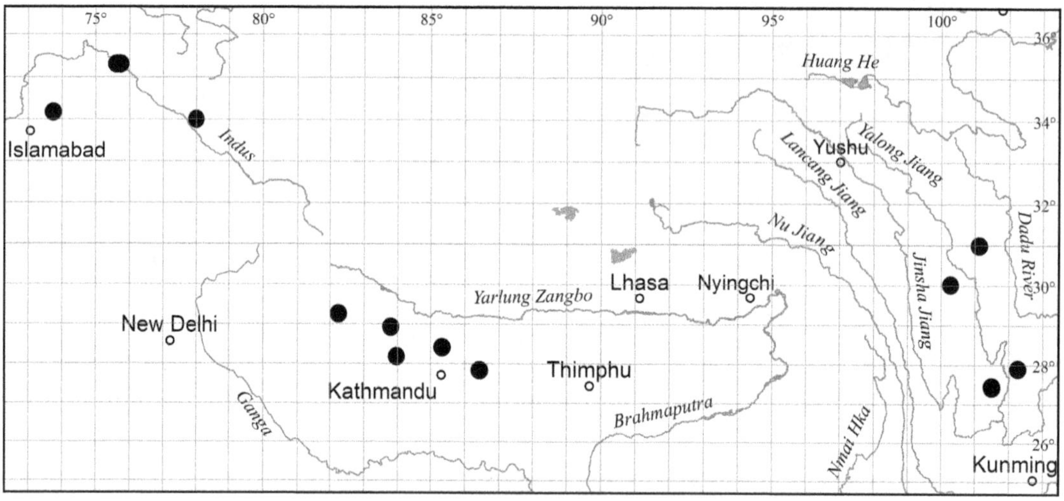

Map 46-16: **Utricularia australis** R. Br.

N PAKISTAN: Baltistan, east of Skardu, 2460 m, 1955, *G. L. Webster 6581* (K); **Dal Lake**, **Rambur**, Jhelum Valley, 1680 m, 1917, *R. R. Stewart 3218* (RAW, fide Malik & Ghafoor, 1988); 1310 m, 1917, *R. R. Stewart 1213a* (RAW, fide Malik & Ghafoor, 1988). **JAMMU & KASHMIR: Ladak**, Spitug, 3400 m, 1931, *W. Koelz 2658* (RAW, fide Malik & Ghafoor, 1988). **U YARLUNG ZANGBO: Gyirong**, between Gyirong town and Naixia, 2800–3351 m, 2010, *PE-Tibet Exped. 818* (PE). **W NEPAL: Karnali**, Jumla, Giri Daha, 1952, *O. Polunin, W. R. Sykes & L. H. J. Williams 5011* (BM). **C NEPAL: Bagmati**, Rasuwa, Khola Kharka, 1949, *O. Polunin 1487* (BM); **Dhawalagiri**, Mustang, Muktinath-Khingar, 2000, *Iokawa et al. 20020270* (TI); Chhusang-Gyu La, 2000, *Iokawa et al. 20020175* (TI). **N HENGDUAN: Dawu**, 1960, *Sichuan Medic Exped. 15854* (NAS); **Litang**, *Geobotany Section 315-3* (YUN). **S HENGDUAN: Butuo**, Huolie, 2800 m, 1985, *Z. S. Dao, Z. P. Pu & K. Q. Li 2179* (PE); **Yanyuan**, 2500 m, 1985, *Z. S. Dao, Z. P. Pu & K. Q. Li 2165* (PE); **Xichang**, Qiongxi, 1982, *Z. S. Dao & Y. W. Zhang 1292* (PE).

16. Utricularia intermedia Hayne in Schrad., J. Bot. (Schrader) 3: 18, t. 5. 1800; Taylor in Kew Bull. Add. Ser. 14: 605. 1989; Li & Cheek in Wu, Raven & Hong (eds.), Fl. China 19: 490. 2011. Type: Germany, Berlin, "*Ehrh. Herb. no. 91*" (holotype: B, not seen).

Perennials, usually affixed aquatic. Rhizoids usually present, filiform, with short dichotomously divided branches. Stolons filiform, sparsely branched, some green, others without chlorophyll. Traps lateral on reduced leaves on colorless stolons, ovoid, 1.5–4 mm long, mouth lateral; appendages 2, dorsal, branched, setiform, with a few lateral simple setae. Leaves numerous on green stolons, suborbicular to elliptic in outline, 0.3–1.5(–2) cm long, 0.4–1.8 cm broad, palmately divided from base into 2 or 3 primary segments; primary segments dichotomously divided from base into 2–5 further segments; ultimate segments flattened, narrowly linear, 0.2–0.7 mm broad, margin entire or sparsely denticulate, laterally and apically setulose. Inflorescences erect, 8–20 cm long, 2–5-flowered, glabrous; peduncle terete, 0.4–0.8 mm thick; scales and bracts basifixed, broadly ovate to ovate-deltoid, base conspicuously auriculate, apex acute; pedicel erect, terete, 0.5–1.5 cm long; bracteoles absent. Calyx lobes ovate, lower lobe slightly shorter than upper lobe, apex obtuse, shortly 2-cleft, or truncate; upper lobe apex acute. Corolla yellow, 0.9–1.5 cm long; lower lip transversely elliptic, base with a prominent rounded swelling, apex rounded; spur subulate, slightly shorter than lower corolla lip, apex acute; palate glabrous; upper lip broadly ovate. Capsule globose, 2.5–3 mm in diameter, circumscissile. Seeds not seen. **Chromosomes** 2n = 44. Figure 46-9.

Phenology: flowering from June to September.

Habitats: bogs, lakes, ponds; 3200–4000 m.

Distribution: U Yarlung Zangbo, Yarlung Zangbo-Brahmaputra, N Hengduan and S Hengduan (Map 46-17); also in Japan, Korea, Kyrgyzstan, Mongolia, Russia, Europe and North America, and other parts of China (Heilongjiang, Jilin, Nei Mongol and W Sichuan).

U YARLUNG ZANGBO: **Gyirong**, Rugalam, 3200 m, 1972, *Xizang Chinese Mater. Medic. & Herbs Exped. 359* (PE). **N HENGDUAN: Dawu,** *Sichuan Medic. Resources Exped. 15807* (NAS); **Hongyuan**, 3400 m, 1964, *B. C. Ni et al. 398* (CDBI); **Litang**, Labo, *Sichuan Vegetation Exped. 3682* (CDBI, PE). **S HENGDUAN: Cawarong (Tsawarong),** 3600 m, 1935, *C. W. Wang 65862* (PE); l. c., 3400 m, 1935, *C. W. Wang 66012* (PE); **Kangding,** Tachienlu, Yulingkong, 2900 m, 1937, *H. Smith 10766* (PE).

17. Utricularia minor L., Sp. Pl. 1: 18. 1753; Clarke in Hooker (ed.), Fl. Brit. India 4: 330. 1884; Casper in Rechinger, Fl. Iranica 58: 2. 1969; Taylor, Kew Bull. Add. Ser. 14: 616. 1989; Li & Cheek in Wu, Raven & Hong (eds.), Fl. China 19: 490. 2011. Type: Sweden, *Herb. Linn. No. 34.3* (lectotype designated by Thor, 1979: LINN).

Utricularia nepalensis Kitam., Acta Phytotax. Geobot. 15: 133. 1954. Type: [C] Nepal, Gandaki Zone, Manang District, Annapuma Himal, near Pisang, 3150 m, 26, May 1953, *S. Nakao s. n.* (holotype: KYO, n. v.).

Utricularia ?exoleta auct. non R. Br.: Handel-Mazzetti, Symb. Sin. 7: 874. 1936; Wu, Index Fl. Yunnan. 2: 1635. 1984.

Perennials, usually affixed aquatic. Rhizoids absent. Stolons filiform, terete, sparsely branched, some green, others without chlorophyll. Traps lateral on leaf segments, ovoid, 0.8–2.5 mm long, mouth lateral; appendages 2, dorsal, long, branched, setiform, with a few lateral simple setae. Leaves numerous on green stolons, semicircular to suborbicular in outline, 2–15 mm long, 3–15 mm broad, palmately divided from base into 2 or 3 primary segments; primary segment dichotomously divided from base into 2–11 further segments; ultimate segments flattened, narrowly linear, 0.1–0.5 mm broad, margin entire or sparsely denticulate, apex acute, apex and teeth with or without microscopic setulae. Inflorescences erect, 5–20(–25) cm long, 2–8-flowered, glabrous; peduncle terete, 0.3–1 mm thick; scales and bracts basifixed, broadly ovate to ovate-deltoid, base conspicuously auriculate, apex acute to obtuse; pedicel erect at anthesis but spreading and decurved in fruit, terete, 4–8 mm long; bracteoles absent. Calyx lobes broadly ovate, subequal; lower lobe smaller, apex narrowly truncate; upper lobe apex subacute. Corolla lemon-yellow, 6–8 mm; lower lip broadly obovate, margin deflexed, apex rounded to retuse; spur saccate to obtusely broadly conic; palate elongate, with a raised marginal rim, distally narrowed and glandular; upper lip ovate to ovate-oblong, apex retuse. Capsule globose, 2–3 mm in diameter, circumscissile. Seeds lenticular-prismatic, 0.3–0.4 mm in diameter, scarcely winged on angles, with small prominent about isodiametric reticulations. **Chromosomes** 2n = 40, 44. Figure 46-10.

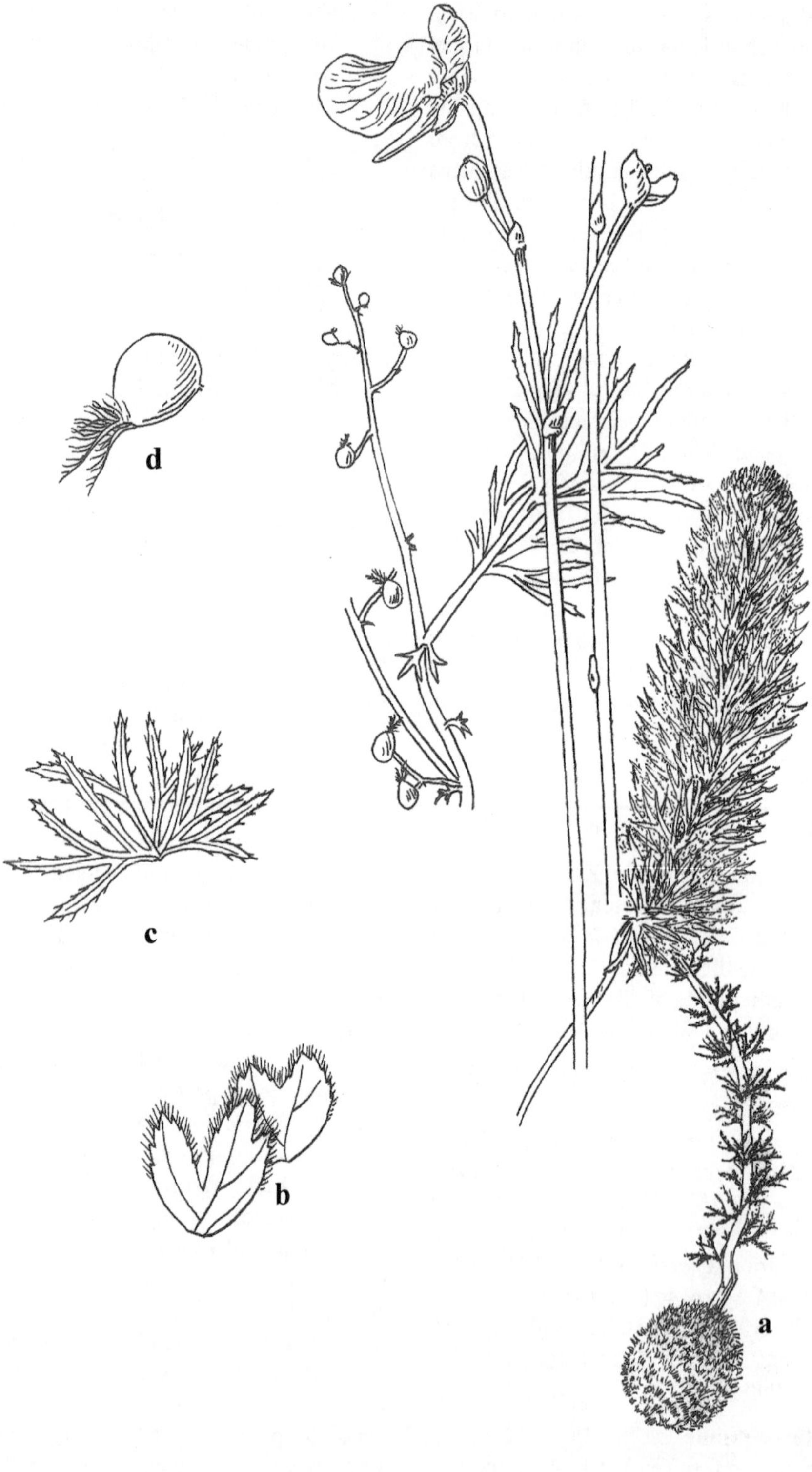

Figure 46-9: **Utricularia intermedia** Hayne: **a,** habit; **b,** leaves of winter bud; **c,** leaf; **d,** trap. Drawn by Jin-Yong FENG; reproduced from *Fl. Reipubl. Popularis Sin.* 69: 600, pl. 164: 1–4. 1990.

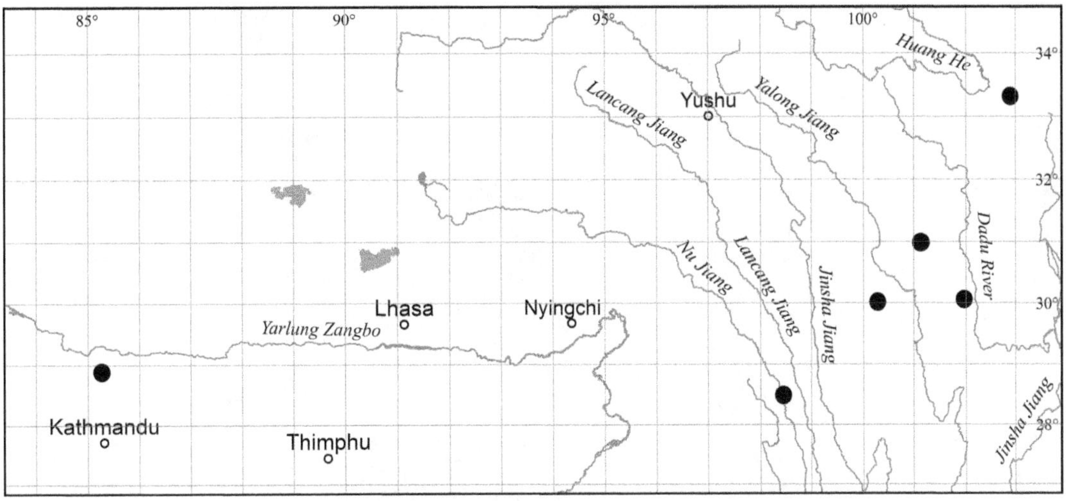

Map 46-17: **Utricularia intermedia** Hayne

Phenology: flowering from June to September; fruiting from July to October.

Habitats: bogs, shallow ponds, marshes; 2700–4300 m.

Distribution: Jammu & Kashmir, C Nepal, Bhutan, L Yarlung Zangbo, Yarlung Zangbo-Brahmaputra, Tangut and N & S Hengduan (Map 46-18); also in Japan, Kyrgyzstan, Mongolia, Papua New Guinea, Russia, Uzbekistan; SW Asia, Europe, North America, and other parts of Afghanistan, Pakistan, Myanmar and China (Heilongjiang, Jilin, Nei Mongol, Shanxi, Sichuan and N Xinjiang).

JAMMU & KASHMIR: Nubra Valley, 3600 m, 1849. *T. Thomson s. n.* (K). **C NEPAL: Dhawalagiri**, Mustang, Jargeng Khola, 1950, *D. G. Lowndes 1382* (BM); **Gandaki**, Gorkha, Ganesh Himal, 1994, *Central Nepal Exped. 9470735* (E). **BHUTAN: Bumthang** and Upper Kulong Chu, 2700–3960 m [from Noltie in Grierson & Long, Flora of Bhutan 2(3): 1339. 2001]. **L YARLUNG ZANGBO: Nyingchi**, Lulang, 3100 m, 1986, *T. Naito et al. 1156* (PE). **YARLUNG ZANGBO-BRAHMAPUTRA: Gutang**, 3650 m, 1982, *B. S. Li & S. Z. Cheng 761* (PE). **TANGUT: Qamdu**, near Yanduo, 1951, *Y. W. Tsui 5743 & 5746* (PE). **N HENGDUAN: Baiyü**, Acha-Garze, 4197 m, 2010, *Kham Zang Exped. 10-0803* (PE). **S HENGDUAN: Weixi**, Lidiping (Litiping), 3450 m, 1916, *H. Handel-Mazzetti 8491* (W).

18. Utricularia aurea Lour., Fl. Cochinch. 1: 26. 1790; Handel-Mazzetti, Symb. Sin. 7: 874. 1936; Taylor in Hara et al., Enum. Fl. Pl. Nepal 3: 132. 1982; Taylor, Kew Bull. Add. Ser. 14: 623. 1989; Li & Cheek in Wu, Raven & Hong (eds.), Fl. China 19: 491. 2011. Type: Vietnam, "Cochin-China", 1774?, *J. de Loureiro s. n.* (not located, cited by Taylor, 1989).

Utricularia flexuosa Vahl, Enum. 1: 198. 1804. Type: India, *J. G. König s. n.* (holotype: C; isotype: W).

Utricularia confervifolia Jacks. ex D. Don, Prodr. Fl. Nepal.: 84. 1825. Type: Nepal, Narainhetty, *Hamilton s. n.* (holotype: BM).

Perennials or annuals, suspended aquatic. Rhizoids usually present, verticillate at base or shortly above base of peduncle, fusiform, inflated, with filiform branches. Stolons filiform to relatively thick, branched. Traps on leaf segments, obliquely ovoid, 1–4 mm long, mouth

Figure 46-10: **Utricularia minor** L.: **a,** branches; **b,** flower; **c,** trap and leaf segments. Drawn by Jin-Yong FENG; reproduced from *Fl. Reipubl. Popularis Sin.* 69: 600, pl. 164: 5–7. 1990.

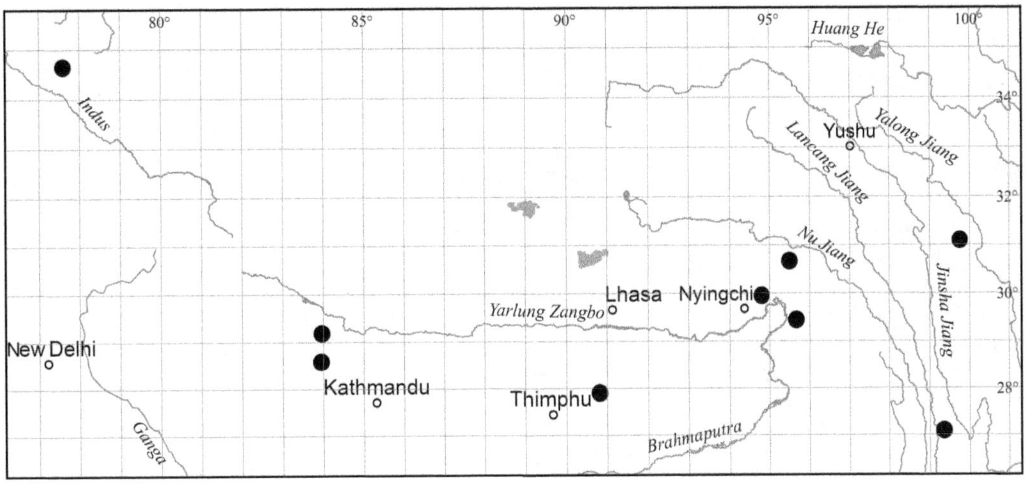

Map 46-18: **Utricularia minor** L.

lateral or basal; appendages 2 or sometimes absent, dorsal, sparsely branched, setiform. Leaves on stolons, 2–8 cm long, divided from base into 3 or 4 (or 5) semiverticillate primary segments; primary segments ovate to oblong-lanceolate in outline, 1.5–3 cm broad, pinnately divided into secondary segments; secondary segments dichotomously divided from base into numerous further segments; ultimate segments capillary, slightly flattened, laterally and apically setulose; stipule-like auricles usually present at base of primary segments, semicircular in outline, divided dichotomously into filiform sparsely setulose segments. Inflorescences erect, 5–25 cm long, 3–10-flowered, glabrous; peduncle terete, 0.5–1.5 mm thick; scales absent. Bracts basifixed, broadly ovate to orbicular, apex rounded; pedicel erect at anthesis but sharply deflexed and thickened in fruit, 0.4–2 cm long, dorsiventrally compressed; bracteoles absent. Calyx lobes ovate, subequal, apex rounded to subacute; lower lobe often slightly broader than upper lobe, 2–3 mm long at anthesis, becoming fleshy, 7–9 mm long and spreading to reflexed in fruit. Corolla pale yellow, 1–1.5 cm long, pubescent or glabrous; lower lip transversely elliptic, base with a prominent 2-lobed swelling, apex rounded to emarginate; spur cylindric from a narrowly conic base, about as long as and parallel with lower lip, straight or slightly curved, apex obtuse to subacute; palate pubescent; upper lip broadly ovate, apex rounded. Capsule globose, 4–5 mm in diameter, circumscissile; persistent style greatly enlarged and elongated, often equaling or longer than capsule. Seeds prismatic, 1–2 mm in diameter, 5- or 6-angled, narrowly winged on all angles, with about isodiametric obscure reticulations. **Chromosomes** 2n = 80. Figure 46-11.

Phenology: flowering from June to November; fruiting from July to December.

Habitats: pools, lakes, swamps, river backwaters, ditches, rice fields; 100–4000 m.

Distribution: C Nepal, E Nepal, Sikkim & Darjiling, L Yarlung Zangbo and S Hengduan (Map 46-19); also in Cambodia, Indonesia, Japan, Korea, Laos, Papua New Guinea, the Philippines, Sri Lanka, Thailand, Vietnam, Australia, and other parts of India, S Pakistan and China (Anhui, Fujian, Guangdong, Guangxi, Guizhou, Hainan, Hubei, Hunan, Jiangsu, Jiangxi, Shandong, Taiwan, Yunnan and Zhejiang).

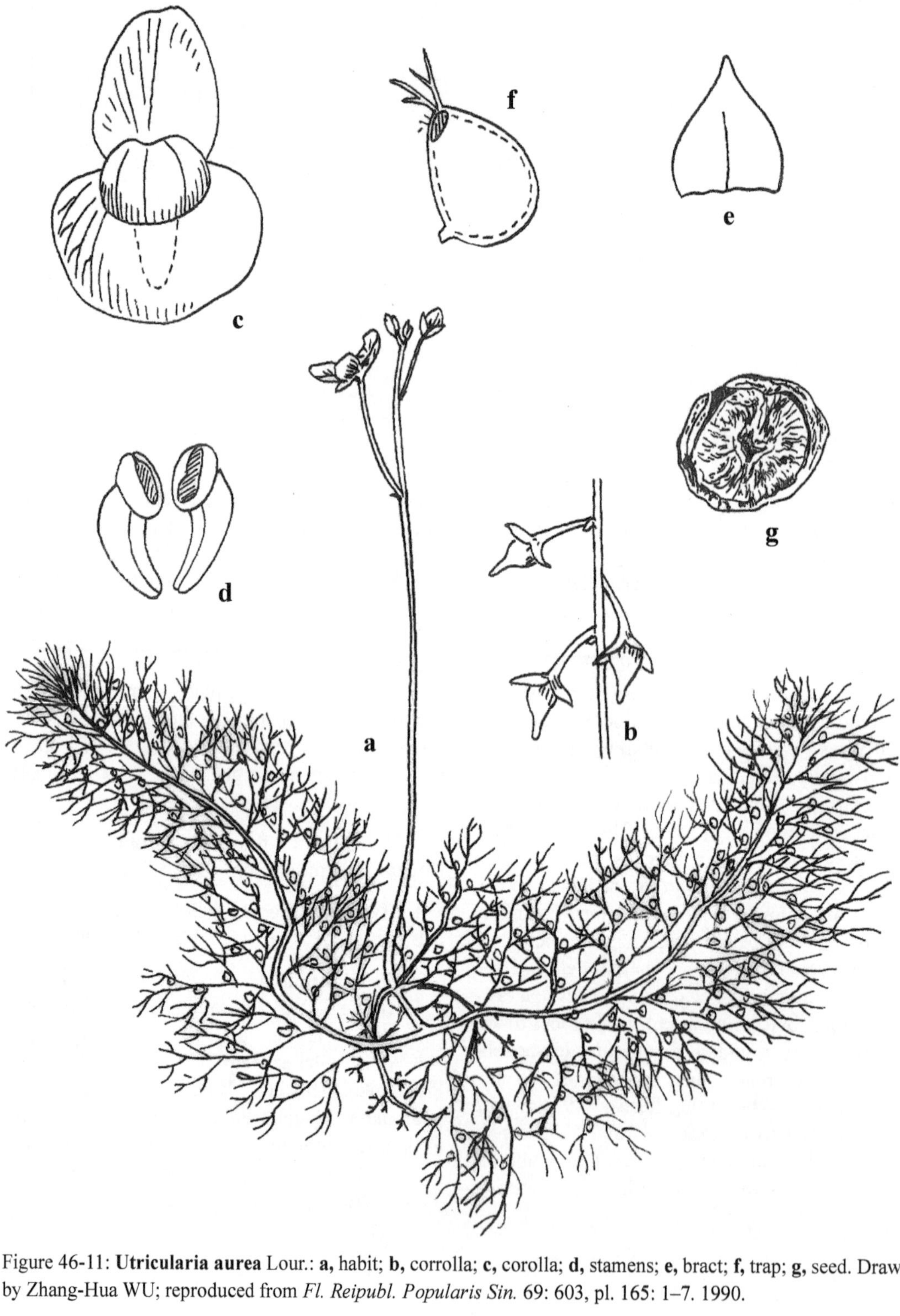

Figure 46-11: **Utricularia aurea** Lour.: **a,** habit; **b,** corrolla; **c,** corolla; **d,** stamens; **e,** bract; **f,** trap; **g,** seed. Drawn by Zhang-Hua WU; reproduced from *Fl. Reipubl. Popularis Sin.* 69: 603, pl. 165: 1–7. 1990.

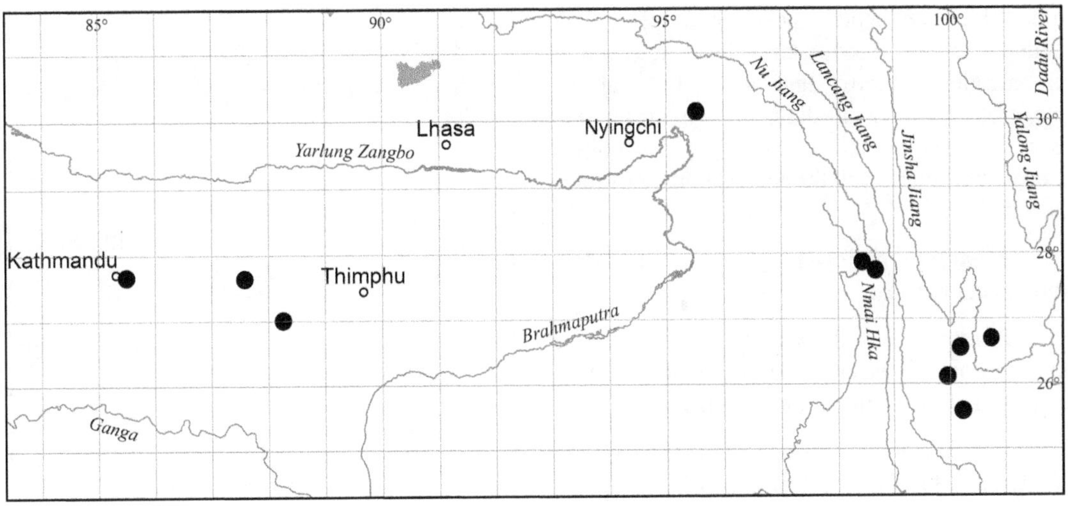

Map 46-19: **Utricularia aurea** Lour.

C NEPAL: 1821, *N. Wallich 1494, p. p.* (K); **Bagmati**, 1966, *D. H. Nicolson 2693* (BM). **E NEPAL: Mechi**, Topke Gola, 1972, *H. Kanai et al. 722025* (TI, KUN); l. c. 1972, *H. Kanai 721973* (KUN). **SIKKIM & DARJILING:** Darjiling, Siliguri, 150 m [from Noltie in Grierson & Long, Flora of Bhutan 2(3): 1339. 2001]. **L YARLUNG ZANGBO: Bomi**, Yi'ong, 3900 m, 1976, *C. Y. Wu 5728* (HITBC). **S HENGDUAN: Dali**, Changshan, 3100 m, 1962, *Zhondian Exped. s. n.* (KUN); Xiaguan, Erhai Lake, Sep. 1975, *B. Y. Qiu 60828* (KUN); **Eryuan**, Cibihu, 1975, *B. Y. Qiu 60884* (IBK); **Gongshan**, Mekong-Salwin divide, 3500 m, 1938, *T. T. Yu 22216* (KUN); **Heqing**, Mekong-Salwin divide, Sila, 4000 m, 1938, *T. T. Yu 22432* (KUN); Taron-Kiukiang divide, Lunguailaka, 2600 m, 1938, *T. T. Yu 20255* (KUN); **Yongsheng**, Chenhai, 1976, *Water and Soil Conservation Section 087* (KUN).

19. Utricularia stellaris L. f., Suppl. Pl.: 86. 1781 (Apr. 1782); Subramanyam, Bull. Bot. Surv. India 23: 155. 1981; Taylor in Hara et al., Enum. Fl. Pl. Nepal 3: 133. 1982; Taylor in Kew Bull. Add. Ser. 14: 631. 1989. ≡ *Utricularia inflexa* var. *stellaris* (L. f.) P. Taylor, Mitt. Bot. Staatssamml. München 4: 96. 1961. Type: "India orientalis", *J. G. Konig s. n.* (holotype: BM; isotypes: LD, S).
Utricularia macrocarpa Wall., Numer. List no.: 1494. 1828, nom. invalid., p. p.

Possibly perennial, suspended aquatic. Stolons 0.5–1.5 mm thick, branched, glabrous, with apical, elliptic scales. Leaves semicircular in outline, 1–6 cm long, divided from base into 3–6 semiverticillate primary segments, secondary segments pinnate, alternate, each pinna dichotomously divided into numerous further segments, ultimate segments capillary, somewhat flattened, laterally and apically setulose. Traps lateral on secondary segments, obliquely ovoid, stalked, 1–3 mm long, mouth lateral, naked or with 2 dorsal, simple or sparsely branched, setiform appendages. Inflorescences erect, 3–30 cm long, 2–12-flowered; peduncle 0.5–1.5 mm thick, with a whorl of 3–8 narrowly ovoid floats usually above the middle; scales absent. Bracts basifixed, broadly ovate to circular; bracteoles absent; pedicels 1–5 mm long, dorsiventrally compressed and winged, erect or spreading at anthesis, deflexing and elongating and widening in fruit. Calyx lobes subequal, broadly ovate to circular, ca. 3 mm long at anthesis, upper lobe apex rounded, lower lobe apex retuse. Corolla yellow, 0.7–1 cm long, densely covered externally with fine glandular-hairs; upper lip broadly ovate apex retuse or emarginate; lower lip limb transversely elliptic, base with a bigibbous swelling, apex rounded or emarginate; palate pubescent; spur cylindric from a broad, conical base, apex retuse or emarginate, shorter than lower lip. Capsule globose, 3–4 mm in diameter, circumscissile. Seeds disk-shaped, 0.8–1.2 mm wide, 4–6 angled, narrowly winged on all the angles. **Chromosomes** 2n = 42.

Phenology: flowering more or less throughout the year.

Habitats: in shallow to quite deep, still or slowly flowing water in lakes, pools, ditches, swamps and rice field; at low altitude.

Distribution: C Nepal (Map 46-20); also in Africa, Bangladesh, Sri Lanka, Vietnam to Australia, and other parts of India and Myanmar.

C NEPAL: 1821, *N. Wallich 1494, p. p.* (K.).

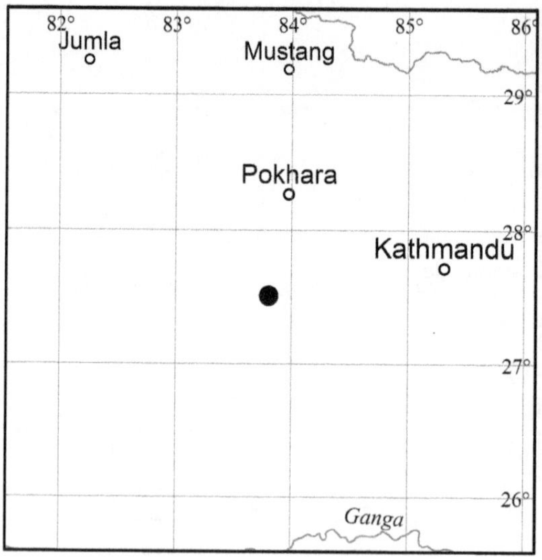

Map 46-20: **Utricularia stellaris** L. f.

ACANTHACEAE

Yun-Fei DENG

Annual or perennial herbs, subshrubs, shrubs; usually with or rarely without cystoliths. Stipules absent. Leaves opposite, rarely alternate or whorled. Inflorescences terminal or axillary spikes, racemes, panicles or dense clusters, rarely of solitary flowers. Bracts foliose to minute, rarely absent; bracteoles usually 2 per flower, or rarely absent. Flowers bisexual, zygomorphic to subactinomorphic. Calyx usually 4- or 5-lobed, rarely reduced to an entire cupular ring or 10–20-dentate. Corolla sympetalous, sometimes resupinate 180° by twisting of corolla tube; tube cylindric or funnelform; limb subequally 5-lobed, or either 2-lipped with upper lip subentire to 2-lobed and lower lip 3-lobed, or rarely 1-lipped with 3 lobes; lobes ascending or descending cochlear, quincuncial, contorted or open in bud. Stamens 2 or 4; filaments distinct, connate in pairs or monadelphous basally via a sheath; anthers monothecous or bithecous; thecae parallel to perpendicular, equally inserted to superposed, spherical to linear, base muticous or spurred, usually longitudinally dehiscent; staminodes 0–2. Ovary superior, 2-locular, placentation axile, with 2 to many ovules per locule; style filiform, simple; stigma funnel-shaped or 2-lobed. Fruit a loculicidal, 2-valved, explosively dehiscent capsule, 2-many-seeded, apex with or without a beak; septa remaining attached to or separating from inner wall of mature capsule. Seeds usually ovate in outline, laterally compressed; retinacula hook-like or lacking.

Recent references: Bremekamp, 1953. The delimitation of the Acanthaceae. *Proc. Kon. Ned. Akad. Wetens. Ser. C, Biol. Med. Sci.* 56: 533–546; Grant, 1955. A cytogenetic study in the Acanthaceae. *Brittonia* 8: 121–149; Raj, 1961. Pollen morphological studies in the Acanthaceae. *Grana Palynol.* 3: 3–108; Bremekamp, 1965. Delimitation and subdivision of the Acanthaceae. *Bull. Bot. Surv. India* 7: 21–30; Barker, 1986. A taxonomic revision of Australian Acanthaceae. *J. Adelaide Bot. Gard.* 9: 1–286; Malik & Ghafoor, 1988. Acanthaceae. In: Basir & Ali (eds.), *Fl. Pakistan* 188: 1–79. Islamabad & Karachi: National Herbarium, (Stewart Coll.) Pakistan Agriculture Research Council, & Dept. Bot. Karachi University; Scotland, 1992. Systematics, simililarity and Acanthaceae pollen morphology. *Bot. J. Linn. Soc.* 109: 529–544; Scotland, 1993. Pollen morphology of Contortae (Acanthaceae). *Bot. J. Linn. Soc.* 111: 471–504; Scotland et al., 1994. Corolla ontogeny and aestivation in the Acanthaceae. *Bot. J. Linn. Soc.* 114: 49–65; Scotland & Vollesen, 2000. Classification of Acanthaceae. *Kew Bull.* 55: 513–589; Wood, 2001. Acanthaceae. In: Grierson & Long (eds.), *Fl. Bhutan* 2(3): 1243–1292. Edinburgh & Thimphu: Royal Botanical Garden Edinburgh & Royal Government of Bhutan; McDade et al., 2008. Toward a comprehensive understanding of phylogenetic relationships among lineages of Acanthaceae s. loc. (Lamiales). *Amer. J. Bot.* 95(9): 1136–1152; Begum, 2008. Acanthaceae. In: Sarma (ed.), *Encyl. Fl. Fauna. Bangladesh.* 6: 1–75. Dhaka: Asiatic Society of Bangladesh; Hu, Deng, Daniel & Wood, 2011. Acanthaceae. In: Wu, Raven & Hong (eds.), *Fl. China* 19: 369–477. Beijing & St. Louis: Science Press & Missouri Botanic Garden Press.

About 220 genera and ca. 4000 species: mostly pantropical and subtropical; 33 genera and 188 (43 endemic) species in the Pan-Himalaya.

1a. Plants without cystoliths; retinacula lacking.
  2a. Prostrate or erect herbs, shrubs or rarely small trees; calyx 4-5-lobed; capsule lacking a prominent apical sterile beak (subfam. 1. **Nelsonioideae**).
    3a. Prostrate herbs; bracteoles absent; anterior two calyx lobes connate to 2/3; stamens 2 .... 1. *Nelsonia*

3b. Erect herbs; bracteoles present; calyx 5-lobed almost to base; stamens 4 .... 2. *Staurogyne*

2b. Vines; calyx cupular with 10–20 lobes or reduced to an entire ring; capsule with a prominent apical sterile beak (subfam. 2. **Thunbergioideae**) .... 3. *Thunbergia*

1b. Plants with or without cystoliths; retinacula hook-like (subfam. 3 **Acanthoideae**).

4a. Cystoliths absent; corolla 1-lipped, upper lip absent, lower lip 3-lobed; ovules 2 per locule (tribe 1. **Acantheae**).

5a. Leaves opposite; lower filaments without an apical appendage; seeds lacking hygroscopic trichomes .... 4. *Acanthus*

5b. Leaves usually in pseudo-whorl of 4; lower filaments with an apical appendage; seeds with hygroscopic trichomes.. 5. *Blepharis*

4b. Cystoliths present; corolla 2-lipped or subequally 5-lobed; ovules 2 or many per locule.

6a. Calyx lobes heteromorphic, posterior lobe usually largest, anterior lobes connate to a greater degree than other lobes, lateral lobes smallest and innermost; corolla lobes aestivation quincuncial (tribe 2. **Barlerieae**).

7a. Inflorescences usually not dorsiventral or of dense heads or racemoide cymes; corolla longer than 3 cm, upper lip usually 4-lobed and lower lip 1-lobed .... 6. *Barleria*

7b. Inflorescences distinctly dorsiventral with flowers and bracts on one side and sterile bracts in other side or of dense globose heads or 4-sided cymes; corolla shorter than 2 cm, upper lip 2-lobed and lower lip 3-lobed .... 7. *Lepidagathis*

6b. Calyx lobes homomorphic or nearly so, sometimes partially fused to form a bipartite or tripartite calyx, not as described above; corolla lobe aestivation not quincuncial.

8a. Corolla lobes contorted in bud; stamens 4, or 2 with staminodes 2 but corolla subequally 5-lobed (tribe 3. **Ruellieae**).

9a. Corolla salverform, basal cylindric tube longer than limb; fertile stamens 2, staminodes 2 .... 8. *Eranthemum*

9b. Corolla campanulate, basal cylindric tube shorter than limb; fertile stamens 4, rarely 2, staminodes 0–2.

10a. Ovules 2 per locule; seeds 4 per capsule; septa with attached retinacula separating from inner wall of mature capsule.

11a. Corolla 3–3.5 cm long; filaments monadelphous basally via a sheath .... 9. *Petalidium*

11b. Corolla ca. 6 mm long; filaments each two coherent at base........ 10. *Phaulopsis*

10b. Ovules 4 or more per locule; seeds more than 4 per capsule; septa with attached retinacula separating from inner wall of mature capsule.

12a. Style held in place by trichomes on one side of corolla tube; stamens monadelphous by a sheath at base.................................... 11. *Strobilanthes*

12b. Style not held in place by trichomes on one side of corolla tube; stamens not monadelphous at base.

13a. Corolla limb distinctly 2-lipped, lobes usually unequal; calyx tube often rupturing in weak hyaline regions between lobes........... 12. *Hygrophila*

13b. Corolla limb nearly regular; lobes equal to subequal; mature calyx tube not rupturing in weak hyaline regions between lobes.

14a. Leaves entire; anther-thecae spurred at base ...... 13. *Echinacanthus*

14b. Leaves toothed; anther-thecae muticous at base.

15a. Stems shorter than 5 cm; leaves in a basal rosette; anther-thecae usually widely separated by a broad connective .... 14. *Pararuellia*

15b. Stems longer than 20 cm; leaves cauline; anther-thecae not separated by a broad connective .... 15. *Ruellia*

8b. Corolla lobes ascending cochlear in bud; stamens 2 or 4 but corolla 2-lipped, staminodes absent or 2.

16a. Ovules 3 to many, rarely 2, per locule; seeds 6 to many, rarely 4; pollen with apertural margins conspicuously thickened (tribe 4. **Andrographideae**).

17a. Capsules linear and quadrangular, not compressed perpendicular to septum; seeds strongly compressed, glabrous or pubescent.............. 16. *Phlogacanthus*

17b. Capsules compressed perpendicular to septum; seeds subglobose, glabrous.

18a. Corolla tube straight; stamens exserted ....17. *Andrographis*

18b. Corolla tube curved; stamens included................................ 18. *Haplanthus*

16b. Ovules 2 per locule; seeds 4 per capsule; pollen with apertural margins not thickened (tribe 5. **Justicieae**).

19a. Stamens 4; staminodes absent.

20a. Inflorescences usually racemose; calyx 5-lobed almost to base ....
....19. *Asystasia*

20b. Inflorescences usually paniculate; calyx united to at least 1/3 above base..
....20. *Phialacanthus*

19b. Stamens 2; staminodes 2 or absent.

21a. Staminodes 2; anthers bithecous, thecae at different height.

22a. Corolla salverform; tube slender and cylindric, conspicuously longer than limb, not or only slightly widened distally ....
....21. *Pseuderanthemum*

22b. Corolla campanulate to strongly 2-lipped; tube not longer than limb, gradually widened distally.

23a. Corolla 2.5–4 cm long; anther-thecae 4–5 mm long .. 22. *Mackaya*

23b. Corolla less than 2 cm long; anther-thecae less than 3 mm long.

24a. Corolla slightly 2-lipped; tube very short; anther-thecae muticous at base... 23. *Codonacanthus*

24b. Corolla strongly 2-lipped; tube subequaling to limb; anther-thecae both spurred at base .... 24. *Wuacanthus*

21b. Staminodes absent; anthers bithecous, thecae at different height, or monothecous.

25a. Flowers subtended by involucre of 2 or more pairs of floral bracteoles; corolla resupinate 180°, lip in upper position 3-lobed and lip in lower position entire to 2-lobed.

26a. Involcrate bracts 4; anthers monothecous .... 25. *Hypoestes*

26b. Involcrate bracts 2; anthers bithecous.

27a. Anther-thecae spherical; septa with attached retinacula separating from inner wall of mature capsule......................... 26. *Dicliptera*

27b. Anther-thecae linear; septa with attached retinacula remaining attached to inner wall of mature capsule .... 27. *Peristrophe*

25b. Flowers subtended by a single pair of bracteoles; corolla not resupinate 180°, upper lip entire to 2-lobed, lower lip 3-lobed.

28a. Upper lip of corolla rugulate, with a furrow on inner surface of lip in which style resides; anther-thecae one or at least lower one with a basal appendage.

29a. Septa with attached retinacula separating from inner wall of mature capsule; inflorescences often with imbricate bracts 2-ranked, or 4-ranked but only 2 ranks fertile .... 28. *Rungia*

29b. Septa with attached retinacula remaining attached to inner wall of mature capsule; inflorescences not as above .... 29. *Justicia*

28b. Upper lip of corolla without a stylar furrow on internal surface; anther-thecae muticous at base.

30a. Corolla white, greenish or purple; tube narrowly cylindric, 3–4 cm long, upper lip lanceolate, much shorter than lower lip, entire, acuminate.. 30. *Rhinacanthus*

30b. Corolla variously coloured and shaped; tube less than 3 cm, upper lip oblong to broadly ovate, equal to or slightly shorter than lower lip, 2-lobed; lobes rounded to acute at apex.

31a. Corolla tube slightly inflated, constricted at throat .... 31. *Sphinctacanthus*

31b. Corolla tube cylindric, not constricted at throat.

32a. Corolla 1.7–4.5 cm; anther-thecae subequally to unequally inserted .... 32. *Isoglossa*

32b. Corolla less than 1.0 cm; anther-thecae equally to subequally inserted............................ 33. *Leptostachya*

Subfam. 1. Nelsonioideae Lindl. ex Pfeiff., Nomencl. Bot. 1(1): 10. 1871; Tsui in Hu (ed.), Fl. Reipubl. Popularis Sin. 70: 31. 2002. Type: *Nelsonia* R. Br.

Plants without cystoliths. Leaves opposite, rarely whorled. Flowers in spikes or racemes. Bracts spirally arranged; calyx more or less equally 4–5-lobed; corolla 2-lipped or equally 5-lobed; lobes imbricate in buds; stamens 2 or 4, anthers bithecous; staminodes absent or 2; anthers bithecous; ovules numerous per locule in 2 or 4 rows. Capsules beaked or pointed; retincula absent.

Recent references: Bremekamp, 1955. A revision of the Malaysian Nelsonieae (Scrophulariaceae). *Reinwardtia* 3: 157–261; Ahmad, 1974. Cuticular studies in some Nelsonioideae (Acanthaceae). *Bot. J. Linn. Soc.* 68: 73–80; Sreemadhavan, 1977. Diagnosis of some new taxa and some new combinations in Bignoniales. *Phytologia* 37: 412–416; Reddy et al., 1980. Floral anatomy of Nelsonieae (Acanthaceae) with a note on its taxonomic status. *Bot. Not.* 133: 311–317; Wenk & Daniel, 2009. Molecular phylogeny of Nelsonioideae (Acanthaceae) and phylogeography of *Elytraria*. *Proc. Calif. Acad. Sci.* 60: 53–68; McDade et al., 2012. Phylogenetic placement, delimitation and relationships among genera of the enigmatic Nelsonioideae (Lamiales: Acanthaceae). *Taxon* 61: 637–651; Daniel & McDade, 2014. Nelsonioideae (Lamiales: Acanthaceae): Revision of genera and catalog of species. *Aliso* 32: 1–45.

Five genera and about 170 species: pantropical regions; two genera and four species in the Pan-Himalaya.

1. NELSONIA R. Br.

Brown, Prodr.: 480. 1810; Clarke in Hooker (ed.), Fl. Brit. India 4: 394. 1884; Lindau in Engler & Prantl (eds.), Nat. Pflanzenfam. 4(3b): 289. 1895; Wood in Grierson & Long (eds.), Fl. Bhutan 2(3): 1250. 2001; Tsui in Hu (ed.), Fl. Reipubl. Popularis Sin. 70: 43. 2002; Hu, Deng & Daniel in Wu, Raven & Hong (eds.), Fl. China 19: 371. 2011. Type (designated by Baker, *J. Adelaide Bot. Gard.* 9: 55. 1986): *Nelsonia campestris* R. Br. [= *Nelsonia canescens* (Lam.) Spreng.].

Herbs, pubescent, without cystoliths. Leaves opposite, shortly petiolate; leaf blade pinnately veined. Inflorescences axillary or terminal spikes. Bracts spirally imbricate, persistent; bracteoles absent. Calyx 5-lobed; lobes unequal, anterior two connate to 2/3. Corolla tube slender, sometimes widening into a throat; limb 2-lipped, upper lip 2-lobed, lower lip 3-lobed; lobes subequal, descending cochlear in bud. Stamens 2, included or slightly exserted; filaments basally pubescent; anthers bithecous; thecae subspherical, muticous; staminodes absent. Ovary with 8–28 ovules in 2–4 rows per locule; stigma usually unequally 2-lobed. Capsules conical; retinacula absent. Seeds subspherical, small, flattened on one face, lacking trichomes.

Recent references: Bremekamp, 1964. On the systematic position of the Australian nelsonias and thunbergias and the *Ruellia* species which by Domin were referred to *Aporuellia* Clarke. *Proc. Kon. Ned. Akad. Wetens. Ser. C, Biol. Med. Sci.* 67: 301–306; Morton, 1979. An overlooked species of *Nelsonia* (Acanthaceae) from Africa. *Kew Bull.* 33: 399–402; Hossain & Emumwen, 1981. Apropos of *Nelsonia canescens* and *N. smithii* (Acanthaceae). *Kew Bull.* 36: 565–568; Hossain, 1984. Taxonomic notes on the *Nelsonia canescens* complex (Acanthaceae). *Willdenowia* 14: 397–403; Vollesen, 1994. Taxonomy, ecology and distribution of *Nelsonia* (Acanthaceae) in Africa. pp. 315–325. In: Seyani & Chikuni (eds.), *Proceedings of the XIIIth Plenary Meeting of AETFAT, Zomba, Malawi, 2–11 Apr. 1991.* Vol. 1. Zomba: National Herbarium and Botanic Gardens of Malawi.

Two to five species: widely occurring in tropical Africa, Asia, Australia and South America (naturalized elsewhere); one species in the Pan-Himalaya.

1. Nelsonia canescens (Lam.) Spreng., Syst. Veg. (ed. 16) 1: 42. 1824; Nees in de Candolle (ed.), Prodr. 11: 67. 1847; Bremekamp, Reinwardtia 3: 248. 1955; Wood in Grierson & Long (eds.), Fl. Bhutan 2(3): 1250. 2001; Tsui in Hu (ed.), Fl. Reipubl. Popularis Sin. 70: 43. 2002; Hu, Deng & Daniel in Wu, Raven & Hong (eds.), Fl. China 19: 372. 2011; Daniel & McDade, Aliso 27. 2014. ≡ *Justicia canescens* Lam., Tab. Encycl. 1: 41. 1791. Type: Senegal, 1789, D. Roussillon 53 [holotype: P-LAM (P00435346); isotype: P-JU (P00435347)].

Nelsonia campestris R. Br., Prodr.: 481. 1810; Clarke in Hooker (ed.), Fl. Brit. India 4: 394. 1884. Type: Australia, Sweer's Island and Bentinck's Island, 18 Nov. 1802, *R. Brown s. n.* [lectotype designated by Barker, J. Adelaide Bot. Gard. 9: 55. 1986: BM (BM001041157); isotypes: BM, CANB, DAO, K (K000357704), MEL, NSW, P (P00650308, P00650309, P00650310), TUM].

Justicia tomentosa Roxb., Fl. Ind. (ed. 1820) 1: 132. 1820, nom. illeg., non *Justicia tomentosa* Vahl, Enum. Pl. 1: 160. 1804. ≡ *Nelsonia vestita* Roem. & Schult., Syst. Veg. (ed. 15 bis) 1: 145. 1822. ≡ *Justicia bengalensis* Spreng., Syst. Veg. 1: 82. 1825, nom. illeg. (Art. 52.1. & 52.2). ≡ *Nelsonia tomentosa* A. Dietr., Sp. Pl. 1: 419. 1831, nom. illeg. (Art. 52.1. & 52.2). ≡ *Nelsonia campestris* var. *vestita* (Roem. & Schult.) C. B. Clarke in Hooker (ed.), Fl. Brit. India 4: 395. 1884. ≡ *Nelsonia canescens* var. *vestita* (Roem. & Schult.) E. Hossain, Willdenowia 14: 403. 1984. Type: [*icon*] *Roxburgh Icon. 519* (lectotype designated by Vollesen in Seyani & Chikuni (eds.), Proc. XIII Plenary Meet. AETFAT, Zomba, Malawi, 1991 1: 323. 1994: K).

Annual herbs, 10–20 cm high; stems creeping, sprawling, prostrate or decumbent, villous, often rooting at nodes. Petiole 0.2–3(–4) cm long, villous; leaf blade elliptic to ovate, (1–)6–12 cm long, (0.5–)3.5–5 cm broad, both surfaces villous, base cuneate, margin entire, apex acute. Spikes 1.5–4 cm long. Bracts elliptic, 6–7.5 mm long, 3–4 mm broad, 5–7-veined. Calyx 2–3 mm long, anterior two lobes 1.6–2 mm long, 0.5–0.6 mm broad, connate to 2/3 above base, posterior lobe 2.5–3 mm long, 0.8–1 mm broad, lateral lobes 1.8–2 mm long, 0.4–0.5 mm broad. Corolla bluish-purple or white, outside glabrous; tube cylindric, 1.2–1.5 mm long; lower lip 2–2.3 mm long; upper lip 1.8–2 mm. Filaments 0.5–0.8 mm long, glabrous. Ovary glabrous; ovules 4–8 per locule. Capsules 4–5 mm long, 1.5–2 mm broad, 8–16-seeded. Seeds broadly ellipsoid, granulate. **Chromosomes** 2n = 34+B, 36. Figure 46-12.

Figure 46-12: **Nelsonia canescens** (Lam.) Spreng.: **a**, portion of plant; **b**, bract; **c**, calyx; **d**, corolla; **e**, stamen; **f**, capsule; **g**, opened capsule showing seeds. Drawn by Quan LIN; reproduced from *Icon. Cormophyt. Sin.* 4: 152, fig. 5717. 1975.

Phenology: flowering from October to March; fruiting from March to May.

Habitats: wet places, open forests, roadside, stream banks; 100–2000 m.

Distribution: U Ganges & Indus, W Nepal, C Nepal, E Nepal, Sikkim & Darjiling, Bhutan, Yarlung Zangbo-Brahmaputra (Map 46-21); widespread in tropical region of the world.

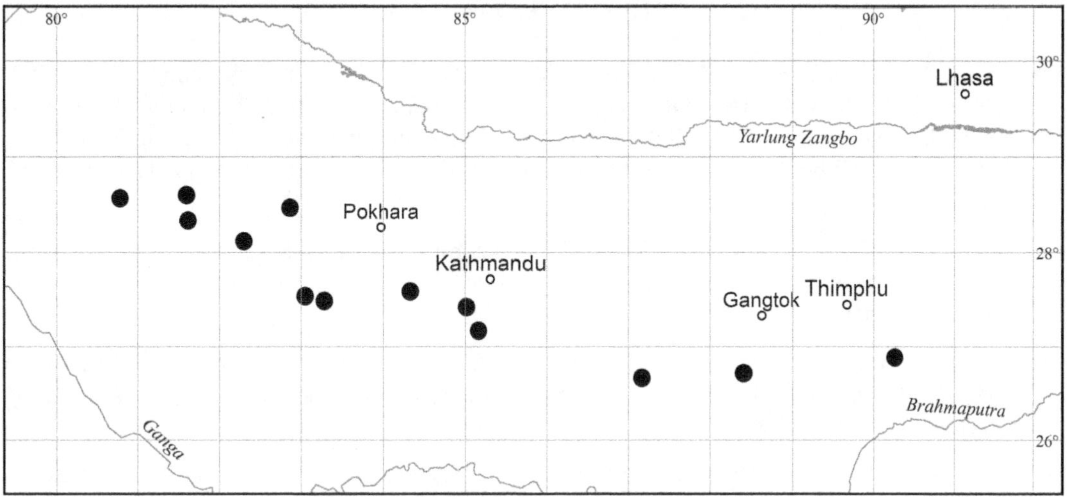

Map 46-21: **Nelsonia canescens** (Lam.) Spreng.

Notes: The name *Nelsonia brunelloides* (Lam.) Kuntze (basionym: *Justicia brunelloides* Lam.) was misapplied to this species by several authors (e.g., Kuntze, Revis. Gen. Pl. 2: 493. 1891; Lindau in Engler & Prantl (eds.), Nat. Pflanzenfam. 4(3b): 289. 1895). However, it is a species of *Strobilanthes*, *S. brunelloides* (Lam.) J. R. I. Wood.

U GANGA & INDUS: Hamirpur, Belatal, 2 Apr. 1962, *U. C. Bhattacharyya 21029* (MO). **W NEPAL: Bheri**, Banke, Babai, 180 m, 23 Mar. 1976, *K. J. Malla et al. 896* (KATH); Surkhet, Kalyan, 870 m, 23 Feb. 1991, *N. P. Manandhar 185-91* (KATH); **Rapti**, Dang, Amrai, South east of Bijami, 610 m, 25 Mar. 1952, *O. V. Polunin, W. R. Sykes & L. H. J. Williams 3645* (BM); Rapti Nadi, Hatauda, 460 m, 9 Mar. 1966, *J. D. A. Stainton 5217* (BM); **Seti**, Kailali, Dhangarhi, 1 Apr. 1974, *N. P. Manandhar et al. 24* (KATH). **C NEPAL: Lumbini**, Kapilvastu, Tilaurakot, 85 m, 6 Apr. 2007, *P. P. Kurmi 10122* (KATH); Rupandehi, Butwal, 240 m, 25 Feb. 1974, *D. P. Joshi & M. M. Amatya 741230* (KATH); **Narayani**, Bara, Patraia-Sekti Mohar Khola, 200 m, 18 Apr. 1969, *H. Kanai & M. S. Bista 11462* (KATH); Chitwan, Ram Nagar, 500 m, 1994, *K. Wesche 1238* (BM); Chitwan, Ramnagar, 300 m, 1995, *K. Weschi 7186* (E); Chitwan, Sauraha to Kasara, 13 Jan. 1996, *M. Suzuki et al. 9614006* (E, KATH); Chitwan, Rapti Valley, Tikoli, 350 m, 12 Mar. 1966, *J. D. A. Stainton 5239* (BM); Makwanpur, Hetauda, 520 m, 10 Mar. 1966, *T. B. Shrestha 3912* (BM, KATH); Makwanpur, Rapti Valley, Hetauda, 500 m, 9 Mar. 1966, *J. D. A. Stainton 5217* (BM, KATH). **E NEPAL: Koshi**, Sunsari, Barampur, near Dhoram, 230 m, 19 May 1969, *L. H. J. Williams 80* (BM). **SIKKIM & DARJILING: Darjiling**, Siliguri, 4 May 1913, *G. H. Cave s. n.* (E); Panchkilla, 320 m, 1 Jan. 1875, *C. B. Clarke 26531* (BM); Panchkilla (Sukua), 14 Mar. 1938, *K. Biswas 3679* (A); Sukna, 14 Mar. 1939, *K. Biswas 3698* (CAL); Terin, 160 m, 18 Dec. 1876, *C. B. Clarke 31800* (K); **Sikkim**, 1300 m, *J. D. Hooker s.*

n. (K). **BHUTAN: Samchi**, Buduni, 380 m, 1 Mar. 1982, *A. J. C. Grierson & D. G. Long 3372* (E); Balu Khola, 7 km W of Phipsoo, 200 m, 19 Mar. 1982, *A. J. C. Grierson & D. G. Long 3865* (E); **Sarbhang**, SW of Sarbhang, 390 m, 9 Mar. 1982, *A. J. C. Grierson & D. G. Long 3626* (E). **YARLUNG ZANGBO-BRAHMAPUTRA:** Aka Hills, 1934, *N. L. Bor 17295* (K).

2. STAUROGYNE Wall.

Wallich, Pl. Asiat. Rar. 2: 80. 1831; Lindau in Engler & Prantl (eds.), Nat. Pflanzenfam. 4(3b): 288. 1895; J. R. I. Wood in Grierson & Long (eds.), Fl. Bhutan 2(3): 1250. 2001; Tsui in Hu (ed.), Fl. Reipubl. Popularis Sin. 70: 31. 2002; Hu, Deng & Daniel in Wu, Raven & Hong (eds.), Fl. China 19: 372. 2011. Type (designated by Leonard, Contr. U. S. Natl. Herb. 31: 5. 1951): *Staurogyne argentea* Wall.

Ebermaiera Nees, Pl. Asiat. Rar. 3: 75, 80. 1832; Clarke in Hooker (ed.), Fl. Brit. India 4: 395. 1884. Type (designated by Leonard, Contr. U. S. Natl. Herb. 31: 5. 1951): *Ebermaiera humilis* Nees [≡ *Staurogyne humilis* (Nees) Kuntze].

Erythracanthus Nees, Pl. Asiat. Rar. 3: 75, 80. 1832; Endlicher, Gen. Pl. 1: 698. 1839; Nees in de Candolle (ed.), Prodr. 11: 78. 1847. Type: *Erythracanthus racemosus* Nees [≡ *Staurogyne racemosa* (Nees) Kuntze].

Herbs, subshrubs or shrubs; without cystoliths. Leaves rosette or cauline, opposite or sometimes alternate on upper portion of stems, petiolate or sessile. Inflorescences terminal or axillary, racemes or spikes. Bracts opposite or alternate; bracteoles present. Calyx 5-lobed almost to base; lobes equal or unequal. Corolla tube cylindric or basally cylindric; limb 5-lobed; lobes subequal, descending cochlear in bud. Stamens 4, didynamous, included or anterior pair slightly exserted; filaments usually hairy; anthers bithecous, thecae parallel, usually equal; staminode 1, sometimes absent. Ovary with usually 12–60 ovules in 2 or 4 rows per locule; style glabrous; stigma 2-lobed, sometimes repeatedly 2-lobed. Capsules oblong, many-seeded; retinacula absent. Seeds minute, subspherical or ovoid. **Chromosomes** 2n = 30.

About 140 species: tropical regions of the world; three species in the Pan-Himalaya.

1a. Stems viscous; leaves on upper portion of stems alternate; blade spatulate .... 1. *S. spatulata*
1b. Stems not viscous; all leaves opposite; blade elliptic to ovate-oblong.
  2a. Leaves sessile; blade ovate to elliptic; bracts shorter than flowers .... 2. *S. humilis*
  2b. Leaves petiolate; blade lanceolate-oblong to narrowly elliptic; bracts longer than flowers......... .... 3. *S. argentea*

1. **Staurogyne spatulata** (Blume) Koord., Exkurs.-Fl. Java 3: 211. 1912; Bremekamp, Reinwardtia 3: 170. 1955; Wood in Grierson & Long (eds.), Fl. Bhutan 2(3): 1251. 2001. ≡ *Adenosma spatulata* Blume, Bijdr. Fl. Ned. Ind. 14: 757. 1826. ≡ *Ebermaiera glauca* var. *spatulata* (Blume) C. B. Clarke in Hooker (ed.), Fl. Brit. India 4: 396. 1884. ≡ *Staurogyne glauca* var. *spatulata* (Blume) C. B. Clarke, Bull. Herb. Boiss. 5: 716. 1905. Type: Indonesia, Java, s. d., *C. Blume s. n.* [holotype: L (L0002943)].

Ebermaiera glauca Nees in Hooker, Comp. Bot. Mag. 2: 310. 1836; Clarke in Hooker (ed.), Fl. Brit. India 4: 395. 1884. ≡ *Staurogyne glauca* (Nees) Kuntze, Revis. Gen. Pl. 2: 497. 1891. Type: India, Travancore, *Herb. Wight 1932* [lectotype designated by Wood, 1994: E (E00273450)].

Ebermaiera ligulata Bedd., Icon. Pl. Ind. Or.: 59. 1864. Type: India, "Coorg, about Kiggatuad", s. d., *Herb. Beddome 20* (holotype: BM).

Staurogyne rivularis Merr., Phillipp. J. Sci., C, 7: 248. 1912; Tsui in Hu (ed.), Fl. Reipubl. Popularis Sin. 70: 37. 2002; Hu, Deng & Daniel in Wu, Raven & Hong (eds.), Fl. China 19: 374. 2011. Type: the Philippines, Luzon Island, Quezon, Manila & Vicinity, Nov. 1910, *E. D. Merrill 7396* [holotype: PNH; isotypes: K (K000882396), P (P00650254), US (US00136397)].

Herbs, 30–40 cm high; stems patently viscous, pubescent upwards. Leaves alternate on upper portion of stem; petiole 3–5 mm long; leaf blade spatulate-oblong, 2.5–8 cm long, 1–2 cm broad, slightly viscous-pubescent or glabrous, base attenuate, margin entire, apex obtuse. Spikes 5–15 cm long, lower flowers on main shoot subtended by ordinary leaves; bracteoles linear to oblong, apex obtuse. Calyx lobes unequal, sparsely glandular-hairy, longest one much broader, linear-oblong, 7–8 mm long, apex obtuse, others smaller, linear-subulate, 5–7 mm long. Corolla 5–8 mm long, purplish with rose or yellow stripes, straight; tube cylindric; lobes subequal, suborbicular, 1.3–1.6 mm long, 1.2–1.4 mm broad, apex rounded. Filament pubescent, longer pair 2.6–3.2 mm long, shorter pair 1.8–2.2 mm long; anther-thecae 0.8–1 mm long. Ovary ellipsoid, 1.8–2 mm long, glabrous; ovules 12–16 in four rows; style 5–6 mm long, glabrous; stigma bifid. Capsules oblong, 4–6 mm long, glabrous, 15–30-seeded. Seeds globose, foveolate.

Phenology: flowering and fruiting from August to January of next year.

Habitats: wet places; below 600 m.

Distribution: U Ganga & Indus, Nepal, Sikkim & Darjiling, Yarlung Zangbo-Brahmaputra (Map 46-22); also in tropical regions of Asia, and Australia.

U GANGA & INDUS: Chamba, 1889, *J. F. Duthie 9695* (BM, K). **NEPAL:** s. loc., *s. coll. 9083B* (E). **SIKKIM & DARJILING: Darjiling**, Phansiowa, 150 m, 18 Dec. 1876, *C. B. Clarke 31800A* (K); *J. M. Cowan s. n.* (E). **YARLUNG ZANGBO-BRAHMAPUTRA:** Aka Hills, Aug. 1934, *N. L. Bor 15263* (K).

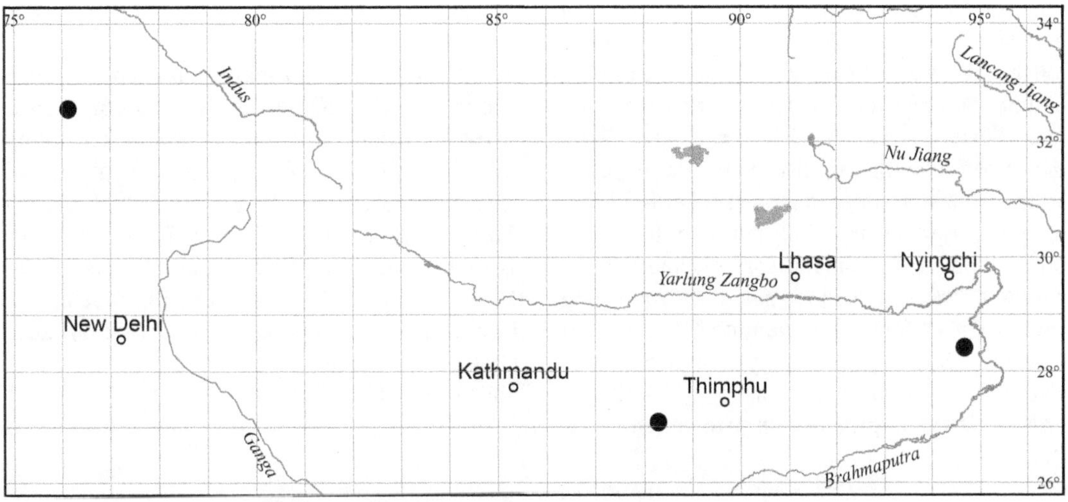

Map 46-22: **Staurogyne spatulata** (Blume) Koord.

2. Staurogyne humilis (Wall. ex Nees) Kuntze, Revis. Gen. Pl. 497. 1891. ≡ *Ebermaiera humilis* Wall. [Numer. list. N. 7149. 1832] ex Nees, Pl. Asiat. Rar. 3: 80. 1832; Nees in de Candolle (ed.), Prodr. 11: 74. 1847. ≡ *Staurogyne polybotrya* var. *humilis* (Wall. ex Nees) E. Hossain, Notes Roy. Bot. Gard. Edinburgh 31: 380. 1972. Type: Myanmar, Mt. Prome, 1826, *Wallich Cat. 7148* [lectotype **here designated**: K (K000838884); isolectotype: K].

Ebermaiera polybotrya Nees in de Candolle (ed.), Prodr. 11: 74. 1847; Clarke in Hooker (ed.), Fl. Brit. India 4: 396. 1884. ≡ *Staurogyne polybotrya* (Nees) Kuntze, Revis. Gen. Pl.2: 497. 1891; Wood in Grierson & Long (eds.), Fl. Bhutan 2(3): 1250. 2001. Type: Myanmar, Mergu, s. d., *W. Griffith 6084* [holotype: K (K000838886)].

Slender herbs, 30–70 cm high; stems, glabrescent below, adpressed grey pubescent upwards. Leaves all opposite, sessile; blade ovate to elliptic, 2–4.5 cm long, 1–2.5 cm broad, abaxially glabrous except for pubescent along veins, base cuneate, margin subentire, apex obtuse. Spikes terminal, 1–4 cm long, densely flowered. Bracts oblong-linear, 6–9 mm long, glandular-pubescent; bracteoles linear, 4–5 mm long, glandular-pubescent. Calyx 6–9 mm long, glandular-pubescent; lobes unequal, one much longer and broader than others, larger one oblong-elliptic, 5–7 mm long, 1–1.5 mm broad, smaller ones linear-oblong, 3.5–5.5 mm long, 0.4–0.5 mm broad. Corolla purplish with rose or yellow stripes, 6–8 mm long, glabrous; tube 5–5.5 mm long; lobes orbicular, 1.5–2 mm long, 0.5–1 mm broad. Stamens 4; filaments sparsely hairy, longer pair 3–4 mm long, shorter pair 2–2.5 mm long; anthers 0.8–1 mm long; staminodes 1, 0.5–0.6 mm long. Ovary glabrous; styles 4–5 mm long. Capsules ellipsoid, 5–6 mm long, 1–1.2 mm broad, glabrous, grooved on two sides. Seeds subspherical to ellipsoid, 0.2–0.3 mm in diameter, brown, falsi-foveate.

Phenology: flowering and fruiting from November to April of next year.

Habitats: moist swamps; 100–1600 m.

Distribution: C Nepal, Sikkim & Darjiling (Map 46-23); also in Thailand, and other parts of India and Myanmar.

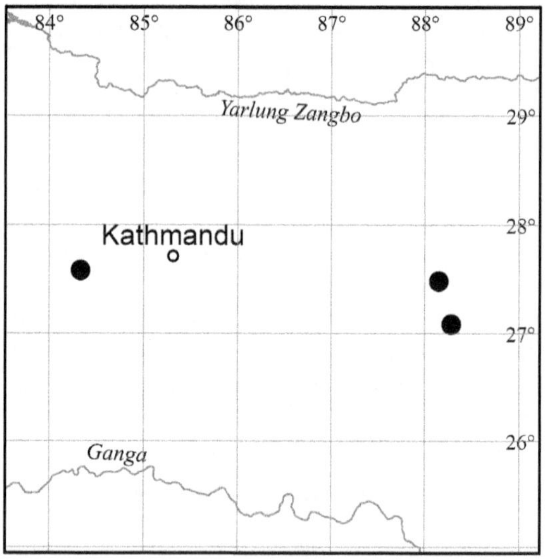

Map 46-23: **Staurogyne humilis** (Wall. ex Nees) Kuntze

C NEPAL: Narayani, Chitwan, Chitwan National Park, 150 m, 23 Nov. 2004, *Second Darwin Nepal Fieldwork Training Expedition DNEP2 A93* (E, KATH); Makwanpur, Jaspal to Gothdanda, 290–620 m, 20 Apr. 1969, *H. Kanai & M. S. Bista 12220* (TI). **SIKKIM & DARJILING: Darjiling**, Silgori, 19 Dec. 1876, *C. B. Clarke 31785* (K); Darjiling Terai, 18 Dec. 1876, *C. B. Clarke 31800B* (K); **Sikkim**, Dulkajhar, 1600 m, 14 Oct. 1884, *C. B. Clarke 36954* (BM, CAL, K).

3. Staurogyne argentea Wall., Pl. Asiat. Rar. 2: 80. t. 186. 1831. ≡ *Ebermaiera argentea* (Wall.) Nees, Prodr. 11: 76. 1847. Type: Bangladesh, Syllet, *F. de Silva in Wallich Cat. 4905* [lectotype **here designated**: K (K000838872)].

Staurogyne angustifolia Wall., Pl. Asiat. Rar. 2: 81. 1831. ≡ *Ebermaiera angustifolia* (Nees) T. Anderson, J. Linn. Soc., Bot. 9: 451. 1867; Clarke in Hooker (ed.), Fl. Brit. India 4: 398. 1884. ≡ *Ebermaiera staurogyne* var. *angustifolia* (Wall.) Nees, Prodr. 11: 75. 1847. Type: Myanmar, Tavoy, 1827, *W. Gomez in Wallich Cat. 4906* [holotype: K-W (K001104184)].

Ebermaiera staurogyne Nees, Prodr. 11: 75. 1847; Anderson, J. Linn. Soc., Bot. 9: 451. 1867; Clarke in Hooker (ed.), Fl. Brit. India 4: 398. 1884. Type: [U Ganga & Indus] India, Simla, s. d., *Fielding s. n.* (annotated "herb. Hook.") (lectotype **here designated**: K).

Herbs, 15–30 cm high; stems creeping and rooting at base, pubescent upwards. Petiole 2–3 cm long, pubescent; leaf blade lanceolate-oblong to narrowly elliptic, 3–10 cm long, 2–4.5 cm broad, adaxially glabrous, abaxially puberulent along veins, base rounded to attenuate, margin entire, apex acute. Spikes terminal or axillary, 1–3 cm long, cylindric or sub-capitate, dense-flowered. Bracts leaf-like, lanceolate, 1.5–2 cm long, 3-veined, reddish, pubescent; bracteoles subulate. Calyx 1.2–1.5 cm long, pubescent; lobes subequal, linear-lanceolate. Corolla 2.0–2.2 cm long, purplish, glabrous; lobes spreading, patent. Stamens 4, included; filaments puberulent, longer pair 1.7–2 cm long, shorter pair 1.2–1.5 cm long; anthers ovoid, 1.6–2 mm long. Ovary oblong, glabrous; ovules numerous per locule in two rows; style glabrous; stigma bifid. Capsules ellipsoid, 5–6 mm long, 1.5–2 mm broad, 15–30-seeded. Seeds orbicular in outline, compressed, 0.1–0.2 mm in diameter, foveolate, puncticulate. Figure 46-13.

Phenology: flowering and fruiting from October to April of next year.

Habitats: wet places; below 800 m.

Distribution: U Ganga & Indus, Yarlung Zangbo-Brahmaputra (Map 46-24); also in Thailand, Malaysia, and other parts of India and Myanmar.

U GANGA & INDUS: Simla, *Fielding s. n.* (K). **YARLUNG ZANGBO-BRAHMAPUTRA: Walong**, 750 m, 21 Oct. 1871, *C. B. Clarke 16222* (CAL, K).

Subfam. 2. Thunbergioideae Kostel., Allg. Med.-Pharm. Fl. 3: 923. 1834; Anderson in Thwaites, Enum. Pl. Zeyl.: 223. 1860; Tsui in Hu (ed.), Fl. Reipubl. Popularis Sin. 70: 22. 2002. Type: *Thunbergia* Retz.

Vine or shrubs; plants without cystoliths. Leaves opposite. Flowers in terminal panicle or solitary or fasciculate in leaf axil. Bracts absent, bracteoles large; calyx reduced to a subentire to undulate rim or with 10–20 linear irregular lobes; corolla not contorted; stamens 4, anthers bithecous, thecae slightly displaced. Ovules 2 per locule; retinacula absent. Seeds 2–4, large, subglobose, hollow or with large scar on one side. **Chromosomes** 2n = 18, 28.

Recent references: Borg et al., 2008. Molecular phylogenetics and morphological evolution of Thunbergioideae (Acanthaceae). *Taxon* 57: 811–822.

Five to seven genera and about 200 species: pantropical; one genus and six (one endemic) species in the Pan-Himalaya.

Figure 46-13: **Staurogyne argentea** Wall.: **a**, habit; **b**, flower; **c**, bract, bracteoles, calyx and style; **d**, opened corolla showing stamens; **e-f**, anthers. Drawn by Yun-Xiao LIU.

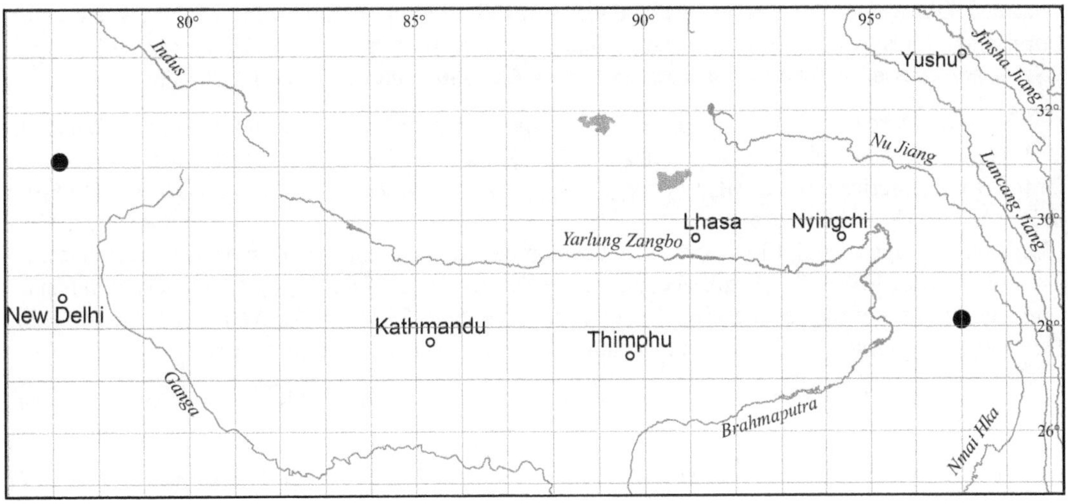

Map 46-24: **Staurogyne argentea** Wall.

3. THUNBERGIA Retz.

Retzius, Physiogr. Sälsk. Handl. 1(3): 163. 1780, nom. cons., non *Thunbergia* Montin, Kongl. Vetensk. Acad. Handl. 34: 288, t. 11. 1773, nom. rej.; Nees in de Candolle (ed.), Prodr. 11: 54. 1847; Bentham & Hooker, Gen. Pl. 2: 1072. 1876; Clarke in Hooker (ed.), Fl. Brit. India 4: 390. 1884; Lindau in Engler & Prantl (eds.), Nat. Pflanzenfam. 4(3b): 291. 1895; Nasir & Ali in Malik & Ghafoor (eds.), Fl. Pakistan 188: 71. 1988; Wood in Grierson & Long (eds.), Fl. Bhutan 2(3): 1246. 2001; Tsui in Hu (ed.), Fl. Reipubl. Popularis Sin. 70: 23. 2002; Hu, Deng & Daniel in Wu, Raven & Hong (eds.), Fl. China 19: 377. 2011. Type: *Thunbergia capensis* Retz.

Flemingia Roxb. ex Rottl., Ges. Naturf. Freunde Berlin Neue Schriften 4: 202. 1803, *nom. rej.* Type: *Flemingia grandiflora* Roxb. ex Rottl., nom. illeg. (Art. 52.1 & 52.2). (≡ *Thunbergia fragrans* Roxb.).

Hexacentris Nees, Pl. Asiat. Rar. 3: 74, 78. 1832. ≡ *Thunbergia* subg. *Hexacentris* (Nees) Benth. & Hook. f., Gen. Pl. 2: 1073. 1876. ≡ *Thunbergia* sect. *Hexacentris* (Nees) Lindau in Engler & Prantl (eds.), Pflanzenfam. 4(3b): 291. 1895. Type [designated by Bremekamp, Verh. Kon. Ned. Akad. Wetensch., Afd. Natuurk., Sect. 2 50(4): 16. 1955]: *Hexacentris coccinea* (Wall.) Nees (≡ *Thunbergia coccinea* Wall.).

Vines or shrubs; without cystoliths. Leaves petiolate; blade entire to lobed to dentate. Flowers solitary or paired in leaf axils, or in terminal racemes, pedunculated. Bracts absent; bracteoles leaf-like, paired, enclosing calyx and most of corolla tube, distinct or fused along one side, persistent. Calyx cupular with 10–20 subulate lobes or reduced to an entire ring. Corolla funnel-shaped; tube base shortly cylindric; limb subequally 5-lobed. Stamens 4, included, often didynamous, inserted near tube base; anthers bithecous; thecae oblong or ovoid, parallel, spurred or not at base, sometimes bearded. Ovary with 2 ovules per locule, collateral; style glabrous or pubescent; stigma 2-cleft, entire or fringed. Capsules basally subglobose, apically prominently beaked, 2–4-seeded; retinacula absent. Seeds compressed to spherical, lacking trichomes. **Chromosomes** 2n = 16, 18, 20, 28.

Recent references: Bremekamp, 1955. The *Thunbergia* species of the Malesian area. *Verh. Kon. Ned. Akad. Wetensch., Afd. Natuurk.*, Sect. 2 50(4): 1–90; Kartesz & Gandhi, 1991. Nomenclatural notes for the North American Flora. VII. *Phytologia* 71(2): 87–100; Wood, 1994. Notes relating to the Flora of Bhutan: XXIX Acanthaceae, with special reference to *Strobilanthes*. *Edinburgh J. Bot.* 51(2): 175–273; Schonenberger, 1998. Floral structure, development and diversity in *Thunbergia* (Acanthaceae). *Bot. J. Linn. Soc.* 130: 1–36; Adhikari et al., 2013. An account of *Thunbergia* (Acanthaceae) in Nepal, with a description of the new species *T. nepalensis*. *Kew Bull.* 68: 651–661; Adhikari & Wood, 2020. *Thunbergia kasajuana*, a new species of Acanthaceae from Nepal. *Kew Bull.* 75: 26.

More than 100 species: Old World tropical regions; six (two endemic) species in the Pan-Himalaya.

1a. Flowers several in terminal or axillary racemes, rarely deduced to single flower; calyx reduced to an entire ring (subg. I. **Hexacentris**).
  2a. Racemes pendulous; corolla red or orange-red; ovary and capsules glabrous........1. *T. coccinea*
  2b. Racemes not pendulous; corolla bluish or bluish white; ovary and capsules pubescent.
    3a. Woody climber; corolla lobes 2.5–3 cm long, 2.2–2.5 cm broad..................2. *T. grandiflora*
    3b. Herbaceous climber; corolla lobes 0.5–0.8 long, 0.5–1 cm broad.................. 3. *T. kasajuana*
1b. Flowers usually solitary, axillary; calyx 5–17-toothed.
  4a. Leaf blade pinnately veined, base cuneate and decurrent onto petiole; basal part of anther-thecae bordered with conical bristles (subg. II. **Coniostephanus**) ....4. *T. lutea*
  4b. Leaf blade palmately veined, base truncate to cordate; basal part of anther-thecae not bordered with conical bristles.
    5a. Calyx 10–17-toothed; corolla tube cylindric; anthers muticous at base (subg. III. **Adelphia**) ....5. *T. fragrans*
    5b. Calyx 5–7-toothed; corolla tube gradually widened from base; anthers spurred at base (subg. IV. **Parahexacentris**)..6. *T. nepalensis*

I. Thunbergia subg. **Hexacentris** (Nees) Benth. & Hook. f., Gen. Pl. 2: 1073. 1876. ≡ *Hexacentris* Nees, Pl. Asiat. Rar. 3: 74, 78. 1832. ≡ *Thunbergia* sect. *Hexacentris* (Nees) Lindau in Engler & Prantl (eds.), Pflanzenfam. 4(3b): 291. 1895. Type [designated by Bremekamp, Verh. Kon. Nederl. Akad. Wetensch., Afd. Natuurk. Sect. 2 50(4): 16. 1955]: *Thunbergia coccinea* Wall.

Woody or herbaceous vines. Leaves palmately nerved. Flowers several in terminal or axillary racemes, rarely deduced to single flower. Calyx reduced to an entire ring. Stamens didynamous; anthers oblong, with long gently curved spurs. Stigmas with two subequal flattened erect lobes.

About ten species in tropical Asia; three species in the Pan-Himalaya.

1. Thunbergia coccinea Wall., Tent. Fl. Nepal. 1: 48. 1826; Anderson, J. Linn. Soc. 9: 448. 1867; Clarke in Hooker (ed.), Fl. Brit. India 4: 393. 1884; Nasir & Ali in Malik & Ghafoor (eds.), Fl. Pakistan 188: 73. 1988; Wood in Grierson & Long (eds.), Fl. Bhutan 2(3): 1248. 2001; Tsui in Hu (ed.), Fl. Reipubl. Popularis Sin. 70: 25. 2002; Hu, Deng & Daniel in Wu, Raven & Hong (eds.), Fl. China 19: 378. 2011. ≡ *Hexacentris coccinea* (Wall.) Nees, Pl. Asiat. Rar. 3: 78. 1832; Nees in de Candolle (ed.), Prodr. 11: 61. 1847. Type: Nepal, "Napalia", *E. Gardner 1818* [lectotype designated by Adhikari et al., Kew Bull. 68: 659. 2013: BM (BM000834068); isolectotype: E (E00273460)].

Vines, woody; stems 9-angled, subglabrous, pubescent at nodes. Petiole 2–7 cm long, grooved; leaf blade broadly ovate, ovate or lanceolate, 8–15 cm long, 3.5–11 cm broad, both surfaces pubescent, palmately 3–5-veined, base rounded to cordate, margin undulate or remotely toothed, apex acuminate. Racemes axillary or terminal, 20–35 cm long, pendulous; peduncle 3–4 cm long, pubescent. Bracteoles oblong, 2.2–2.6 cm long, 1.1–1.5 cm broad, apex acute. Calyx 1.5–2 mm long, reduced to a minute rim. Corolla red or orange-red, 6–8 cm long; tube base cylindric, 5–6 mm long, throat 1.5–1.6 cm long; lobes suborbicular, 6–7 mm in diameter. Stamens 4; filaments 1.2–1.5 cm long, glabrous except a tuft of trichomes at base; anther-thecae unequal, 4–5.5 mm long, base spurred. Ovary glabrous; style glabrous, exserted; stigma 2-cleft. Capsules glabrous, basal fertile portion 1–1.2 cm long, 1.5–2 cm broad, beak 1.5–2.3 cm long. Seeds ovate in outline, compressed, verrucose. **Chromosomes** 2n = 20. Figure 46-14.

Phenology: flowering from August to March; fruiting from November to May.

Habitats: climber in tropical and subtropical forests; 100–2900 m.

Distribution: N Pakistan, W Nepal, C Nepal, E Nepal, Sikkim & Darjiling, Bhutan, Yarlung Zangbo-Brahmaputra, S Hengduan, U Irrawaddy (Map 46-25); also in Laos, Thailand, and other part of Myanmar and China (S Yunnan).

N PAKISTAN: Hazara, 900 m, s. d., *R. Strachey & J. E. Winterbottom s. n.* (BM). **W NEPAL: Bheri**, Bardiya, Harya, 1650 m, 1977, *K. J. Malla & P. R. Sharma 1480* (KATH); Jajarkot, Gad Rangchi, 1650 m, 16 Oct. 1952, *O. V. Polunin, W. R. Sykes & L. H. J. Williams 5750* (BM); Surkhet, 1978, *K. J. Malla 1552* (KATH); Surkhet, Harre to Chinchu, 1800 m, 1980, *T. K. Bhathacharya 3423* (KATH); Surkhet, Khola Pani, 1800 m, 2 Oct. 1982, *P. P. Kurmi & D. K. Shrestha 4965* (KATH); Surkhet, Kalyan, 870 m, 24 Feb. 1991, *N. P. Manadhar 298-91* (KATH); Surkhet, Kalyan, 870 m, 24 Feb. 1991, *N. P. Manandhar 316-91B* (KATH); **Mahakali**, Dadeldhura, Jogbudha, 350 m, 18 Jan. 1983, *I. Sharma & R. Joshi 7071* (KATH); **Rapti**, Pyuthan, Thapa Dada, 850 m, 12 Feb. 1983, *U. R. Poudyal et al. 5324* (KATH); **Rukum**, Sisneri, 900 m, 1973, *D. P. Joshi & M. M. Amatya 73/950* (KATH); **Seti**, S Gunarpunk, 1300–2000 m, s. d., *D. Pround 79179* (BM); Kailali, Chaumala, 23 Mar. 1929, *Biskam 56* (A). **C NEPAL: Bagmati**, Bhaktapur, Nagarkot, 2150 m, 21 Dec. 1966, *Prabha & Ramola 6374* (KATH); Dhading, Bhimbang Lekh, 2000 m, 1988, *N. P. Manandhar 12771* (KATH); Dhading, Tharlang, 1550 m, 1989, *s. coll. 13109* (KATH); Kathmandu, Jawal Tole, 1500 m, 21 Dec. 1966, *D. H. Nicolson 2878* (BM, KATH); Kathmandu, Shiopuri, 2500 m, 28 Sep. 1978, *H. Idzumi & M. Togashi s. n.* (TI); Kathmandu, Naubise to Mugling, 5 Nov. 2000, *T. Watanabe et al. NPL-SP001105(033)* (MBK); Lalitpur, Godawari, 1600 m, 20 Dec. 1965, *A. D. Schilling 711* (K, TI); Lilatipur, Godamachaurl, 480 m, 13 Dec. 2001, *T. Watanabe et al. NPL-SP011213(121)* (MBK); Rasuwa, Between Ramche and Ghrang, *H. Tabata et al. 8588* (KATH, KUN); **Dhawalagiri**, Myagdi, Takam, 1820 m, 27 Oct. 1954, *J. D. A. Stainton, W. R. Sykes & L. H. J. Williams 9164* (A, BM, E); Myagdi, Tatopani to Beni, 29 Sep. 1995, *M. Mikage et al. 9550529* (TI); Rusuwa, Pati-Bangjang to Gul-Bangjang, 16 Oct. 1984, *T. Watanabe NPL-SP841016(004)* (MBK); **Gandaki**, Kaski, Ghorepani, 2834 m, 1 Jul. 2002, *T. Watanabe et al. NPL-SP020701(067)* (MBK); Kaski, Pothana to Tolka, 1750 m, 20 Aug. 1988, *M. Suzuki et al. 8840431* (BM, TI); Kaski,

Figure 46-14: **Thunbergia coccinea** Wall. **a**, flowering branch; **b**, corolla with bracteoles; **c**, stamens; **d**, capsule. Drawn by Quan LIN; reproduced from *Icon. Cormophyt. Sin.* 4: 152, fig. 5718. 1975.

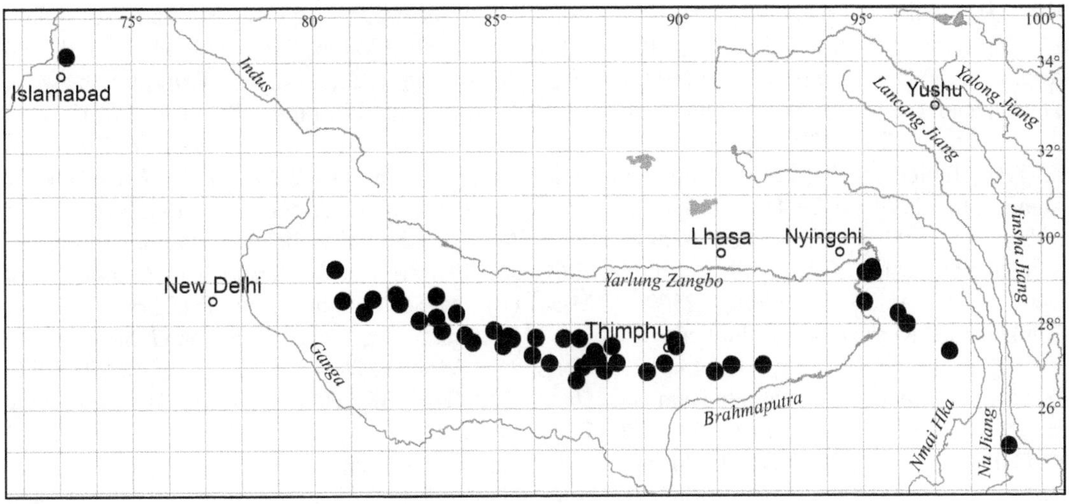

Map 46-25: **Thunbergia coccinea** Nees.

Yangjakot, 1500 m, *J. D. A. Stainton W. R. Sykes & L. H. J. Williams 8966* (BM); Kaski, between Lamduli and Ulleri, 1580 m, 1976, *H. Tabata et al. 6031* (A); Kaski, Lumle, 1675 m, 23 Jan. 1976, *G. Dawson 506* (BM); Lamjung, Jagat, 1350 m, 1983, *N. P. Manandhar 9589* (KATH); Janakpur, Dolakha, Torikhet, 950 m, 1988, *N. K. Bhattarai & L. P. Kattel 88/193* (KATH); Sindhuli, Jaeago, 570 m, 1975, *P. R. Shakya & K. R. Rajbhandari 3604* (KATH); Sindhuli, Hariharpur Gadhi, 300 m, 14 Feb. 1967, *J. D. A. Stainton 5635* (BM, KATH); **Lumbini,** Gulmi, Badigad Khola, near Rupakot, 1520 m, 11 Oct. 1954, *J. D. A. Stainton, W. R. Sykes & L. H. J. Williams 8883* (BM, E); Nawalparasi, Tribeni, s. d., *Bailey* s. n. (BM); Palpa, Dowmma Danda, 400 m, 1975, *D. P. Joshi & I. Bojracharya 75/3590* (KATH); **Narayani**, Chitwan, Rapti Valley, Lothar, 380 m, 17 Mar. 1966, *J. D. A. Stainton 5281* (BM, KATH); Chitwan, Sauraha, 250 m, 26 Dec. 1980, *M. Allison* 75 (E); Chitwan, Lothar, 400 m, 1966, *M. L. Banerji 1756* (A); Makwanpur, Bara, near Badahikim Bridge, Hetauda to Birgunj, 670 m, 27 Nov. 2004, *C. A. Pendry et al. A166* (E, KATH); Makwanpur, Torke, 1590 m, 1985, *N. P. Manandhar 12681* (KATH). **E NEPAL: Koshi**, Dhankuta, Basantapur, 2160 m, 28 Aug. 2014, *B. Adhikari et al. 15* (E); Dhankuta, Sindua to Teku Nala, 800–2100 m, 3 Sep. 1977, *H. Ohashi et al. 773111* (KATH, MO, TI); Dhankuta, Nudhe, 2500 m, 1978, *P. Pradhan et al. 809* (KATH); Dhankuta, Sindhuwa, 2100 m, 1977, *N. P. Manandhar 97* (KATH); Dhankuta, Murhay, 2000 m, 23 Oct. 1963, *H. Hara et al. 6303593* (TI); Sankhuwasabha, Arun Valley, SW Bhotebas, 2010 m, 19 Sep. 1991, *D. G. Long et al. EMAK60* (E, KATH); Sankhuwasabha, Sabha Khola, 1950 m, 7 Sep. 1956, *J. D. A. Stainton 1603* (BM, E); Sankhuwasabha, Pahakhola to Baidep, 12 Aug. 1991, *H. Ohba et al. 9120408* (TI); Sankhuwasabha, Nurbu Gaon to Dungameti, 1900 m, 21 Aug. 1997, *S. Noshiro et al. 9760230* (A, E, TI); Sunsari, Koshi Tappu Wild Life, 100 m, 25 Nov. 2012, *B. Adhikari & D. R. Kandel 87* (E); Terhathum, Mulpani, 1800 m, 1976, *N. P. Manandhar 206* (KATH); Terathum, ridge between Oyakjung and Sankranti, 2000 m, 9 Oct. 1981, *C. Grey-Wilson et al. 4696* (K); **Mechi,** Ilam, Mahabharat Lekh, 1950 m, 13 Sep. 1967, *L. H. J. Williams & J. D. A. Stainton 8478* (BM, K, TI); Ilam, Ilam to Ranga, 8 Dec. 1963, *H. Hara et al. 6303594* (KATH, TI); Ilam, Hanspokhari, 1760 m, 1977, *P. Pradhan et al. 179* (KATH); Panchthar, Phalaicha, 1900 m, *R. M. Kunwar et al. C208* (E); Panchthar, Falaincha-6, Tinte, 1900 m, 27 Sep. 2007, *K. K. Shrestha et al. D208* (KATH); Panchthar, Hangthwa, Phidim, 700 m, 1979, *P. Pradhan et al. 1009* (KATH); Taplejung, Sanghu, T. Gulley, 2000 m, 4 Oct. 1961, *A. H. Norkett 5614* (BM, E); Taplejung, Barundin, 25 Nov.

1963, *H. Hara et al. 6303590* (BM, K, KATH, TI); Taplejung, Nesum to Surke Pati, 2100 m, 29 Aug. 1977, *H. Ohashi et al. 772923* (BM, KATH, TI); Taplejung, Environs Yamphudin, 1650 m, 25 Sep. 1989, *C. Grey-Wilson et al. 957* (E, K, KATH); Taplejung, between Yamphudin and Mamamkhe, 1700 m, 29 Dec. 1989, *C. Grey-Wilson et al. 1097* (E, K, KATH); **Sagarmatha**, Solukhumbu, Dudh Koshi Valley, 1900 m, 26 Oct. 1977, *A. D. Schilling 2307* (BM); Solukhumbu, Kharikhola to Nunthala, 1790 m, 30 Aug. 1997, *M. Wakabayashi et al. 9730404* (KATH); Udyapur, Mul Pokhari to Dumhan, 1500 m, 30 Oct. 1963, *H. Hara et al. 6303579* (TI). **SIKKIM & DARJILING: Darjiling**, East Burch Hill Road, 2200 m, 24 Jul. 1969, *H. Hara et al. 69106* (TI); Rungbee, 1160 m, 1869, *C. B. Clarke 8788* (K); East Birch Hill Road, 2100 m, 22 Jul. 1972, *H. Ohashi et al. 72722* (A, KUN, TI); Lebong, 1660 m, 1881, *J. S. Gamble 9781* (K); **Sikkim**, Jowbari, 2330 m, 21 Sep. 1968, *N. C. Mazumder 618* (CAL); between Karponang and Gangtok, 2350 m, 3 Nov. 1949, *F. Ludlow, G. Sherriff & J. H. Hicks 17567* (BM, K, TI); Zamha, 1881, *H. Collet 1* (K). **BHUTAN:** Diwangiri, 1000 m, 26 Nov. 1938, *F. Ludlow, G. Sherriff & J. H. Hicks 6800* (BM, E, TI); **Chukka**, Chasilakha, 2020 m, 21 Feb. 1982, *A. J. C. Grierson & D. G. Long 3107* (E); **Lhuentse**, Khinay Lhakang, 2000 m, 24 Jul. 1949, *F. Ludlow, G. Sherriff & J. H. Hicks 20944* (A, BM, E, TI); Khinay Lhakang, Ungar Chu, 1830 m, 5 Oct. 1949, *F. Ludlow, G. Sherriff & J. H. Hicks 21252* (BM, TI); **Pemagatsel**, Satsalor to Diwangiri, 800 m, 14 Nov. 1938, *F. Ludlow, G. Sherriff & G. Taylor 7278* (BM, E, TI); **Phuntsholing**, 300 m, 3 Apr. 1967, *H. Hara et al. 1894* (TI); **Punakha**, Tilagong, 2660 m, 17 Aug. 1914, *R. E. Cooper 2667* (BM, E); Punakha, 2330 m, 1917, *R. E. Cooper 3218* (BM, E); Lomitsawa, 2000 m, 13 Aug. 1989, *J. R. I. Wood 7082* (K); Punakha to Sinchula, 2160 m, 13 Sep. 1967, *S. Boweslyon 5050* (BM); **Samchi**, between Samchi and Dorokha, 1000 m, 20 Nov. 1991, *J. R. I. Wood 7515* (E); **Tashi Yangtse**, Shali, Trashiyangsi Valley, 2160 m, 1934, *F. Ludlow & G. Sherriff 1052* (BM, E); **Thimphu,** between Lobeysa and Thilegang, 1900 m, 12 Jul. 1979, *A. J. C. Grierson & D. G. Long 2679* (E, K); **Tsongsa**, 2160 m, 30 Aug. 1949, *F. Ludlow, G. Sherriff & J. H. Hicks 19664* (A, BM, E, TI); Mangde Chu valley below Tsongsa, 2100 m, 10 Jul. 1979, *A. J. C. Grierson & D. G. Long 2642* (E); betweem Tsongsa & Trashilong, Tsauha, 2500 m, 5 Aug. 1949, *F. Ludlow, G. Sherriff & J. H. Hicks 17044* (BM, TI). **YARLUNG ZANGBO-BRAHMAPUTRA: Baibung**, way to Gelin, 1110 m, 12 Nov. 2019, *Y. F. Deng 28902* (IBSC); to Mêdog, 850 m, 27 Sep. 2003, *X. F. Gao et al. 8026* (CDBI); Baibung to Hanmi, 840 m, 2010, *South Tibet Expedition Team 2276* (PE); Jiefangqiao, 18 Jul. 1973, *Exped. Xizang 1482* (HNWP); way to Mêdog, 960 m, 4 Aug. 1974, *Exped. Qinghai-Xizang 1560* (KUN); **Dexing**, along Yarlung Zangbo, 800 m, 2010, *South Tibet Expedition Team 2011* (PE); near bridge by Yarlung Zangbo, 730 m, 13 Nov. 2019, *Y. F. Deng 28916* (IBSC); **Digdong**, Baibung to Didong, 800 m, 1992, *H. Sun et al. ETM-460* (KUN); near Didong Valley, 650 m, 1992, *H. Sun et al. ETM-2237* (KUN); **Dirang**, 1964, *J. Joseph 40055* (CAL); **Lohit Valley**, 670–1000 m, 3 Dec. 1926, *F. Kingdon-Ward 7692* (K); Theronliang, 670 m, 7 Mar. 1928, *F. Kingdon-Ward 7931* (K); 23 Sep. 1969, *A. S. Rao 48024* (CAL); **Mêdog**, way to Rattan Bridge, 23 Aug. 1990, *G. Yao et al. 3532* (NAS); near Mêdog, 1200 m, 1992, *H. Sun et al. 2934* (KUN); Mêdog to Dexing, 900 m, 26 Aug. 1974, *Exped. Qinghai-Xizang 74-4562* (KUN); **Mishimi Hill**, *Herb. Griffith 6093* (K); **Riga**, 1958, *R. S. Rao 17594* (CAL). **S HENGDUAN: Longyang**, Baihualing, 1620 m, *X. C. Shi & S. X. Yang 527* (KUN); Mangkuan, Yutang, Baihualing, 1460 m, *H. Li 14099* (CAS, KUN). **U IRRAWADDY: Putao**, Upper Shankaung to Wasadam, 2014, *Putao Expedition Team PT-ET 188* (PE).

2. Thunbergia grandiflora Roxb., Bot. Reg. 6: t. 495. 1820; Nees in Wallich (ed.), Pl. Asiat. Rar. 3: 77. 1832; Nees in de Candolle (ed.), Prodr. 11: 54. 1847; Anderson, J. Linn. Soc., Bot. 9: 447. 1867; Clarke in Hooker (ed.), Fl. Brit. India 4: 392. 1884; Nasir & Ali in Malik & Ghafoor (eds.), Fl. Pakistan 188: 73. 1988; Wood in Grierson & Long (eds.), Fl. Bhutan 2(3): 1248. 2001; Tsui in Hu (ed.), Fl. Reipubl. Popularis Sin. 70: 26. 2002; Hu, Deng & Daniel in Wu, Raven & Hong (eds.), Fl. China 19: 377. 2011. Type: [*icon*] Bot. Reg. 6: t. 495. 1820 (lectotype designated by Kartesz & Gandhi, Phytologia 71: 89. 1991).

Thunbergia adenophora W. W. Sm., Notes Roy. Bot. Gard. Edinburgh 10: 74. 1917. Type: Myanmar, Kachin State, valley of the Taping, 2000 ft., Feb. 1914, *G. Forrest 12193* [holotype: E; isotype: K (K000898161)].

Thunbergia lacei Gamble, Bull. Misc. Inform., Kew 1913: 116. 1913. Type: Myanmar, Upper Burma, Maymyo Plateau, *J. H. Lace 5419* [lectotype **here designated**: K (K000898163); isolectotypes: E (E00095250, E00095251, E00095252, E00095253)].

Thunbergia clarkei Yamazaki, J. Jap. Bot. 46(2): 52. 1971. Type: Bhutan, Phuntsholing, 300 m, 3 Jun. 1968, *H. Kanai et al. 1069* [holotype: TI (TI00010897); isotypes: BM, TI (TI00010898)].

Thunbergia grandiflora var. *racemosa* Nees, Prodr. 11: 54. 1847. Type: India, Gualpara, 15 Jun. 1808, *Buchman-Hamilton in Wallich Cat. 768g* [lectotype **here designated**: K-W (K001111851); isolectotypes: K-W (K001111849, K001111850)].

Vines to 10 m or longer, woody; stems quadrangular, sulcate, pubescent. Petiole 1–7 cm long, grooved, pubescent; leaf blade ovate to triangular-ovate, 5–10 cm long, 4–8 cm broad, papery, both surfaces pubescent, palmately 3–7-veined, base subcordate to truncate, margin undulate, irregularly angular on basal half or rarely entire, apex acuminate to acute. Flowers in terminal racemes with 2–4 flowers per node, sometimes few flowers solitary or paired in leaf axils; peduncle 4–7 cm long, sulcate, pubescent; rachis pubescent with large cyathiform glands. Bracteoles oblong to ovate, 2.5–4 cm long, 1.5–2.2 cm broad, pubescent, 5–7-veined, base truncate, margin entire or ciliate, apex acute with a short mucro. Calyx 1.8–2 mm long, annular, unlobed, densely pubescent. Corolla bluish with yellowish on throat, 4–6 cm long, outside glabrous; tube base cylindric, 5–7 mm long, 2–3 mm broad, upwards gradually widened to 4.5–5 cm broad at throat; lobes ovate, 2.5–3 cm long, 2.2–2.5 cm broad. Stamens 4; filaments 7–9 mm long; anther-thecae pubescent, base spurred. Style glabrous; stigma subequal 2-lobed. Capsules pubescent, basal fertile portion 1.2–1.5 cm long, 1.3–1.8 cm in diameter, beak 2.2–2.5 cm long. Seeds ovate in outline, compressed, verrucose. **Chromosomes** 2n = 56. Figure 46-15.

Phenology: flowering from July to November; fruiting from July to December.

Habitats: climber on forest margins and road sides; 0–2000 m.

Distribution: W Nepal, C Nepal, E Nepal, Sikkim & Darjiling, Bhutan, Yarlung Zangbo-Brahmaputra, U Irrawaddy (Map 46-26); also in Laos, Thailand, Vietnam, and other parts of India, Myanmar and China (Fujian, Guangxi, Hainan, S Yunnan), and also widely cultivated and naturalized in tropical regions of the world.

W NEPAL: Seti, Bajang, Pujarigaon, 2000 m, 1972, *M. S. Bista & D. P. Joshi 303* (KATH). **C NEPAL: Bagmati**, Kathmandu, 1300 m, 21 Aug. 1954, *J. D. A. Stainton, W. R. Sykes & L. H. J. Williams 6966* (A, BM); **Janakpur**, Sindhuli, Gijeni, 1250 m, 1988, *N. P. Manandhar 12528* (KATH); Sindhuli, Kalapani, 320 m, 1987, *N. P. Manandhar 11911* (KATH); Sindhuli, Kalapani, 320 m, 1987, *N. P. Manandhar & L. P. Katel 11945* (KATH); **Narayani**, Bara, Nijgadh (Nijgar), 200 m, 1969, *H. Kanai & M. S. Bista 11468* (KATH); Chitwan, Royal Chitwan National Park, S of Rapti River, 250 m, 5 Jul. 1976, *R. G. Troth 897* (BM, KATH, TI). **E NEPAL: Koshi**, Sunsari, Prakashpur, 27 May 1993, *P. R. Shakya et al. 69* (KATH); Sunsari, Koshi Tappu Wildlife Reserve, 60–65 m, 22 Oct. 1995, *M. Mikage et al. 9552964* (E, TI); Sunsari, S Dharan, 200 m, 1 Jun. 1972,

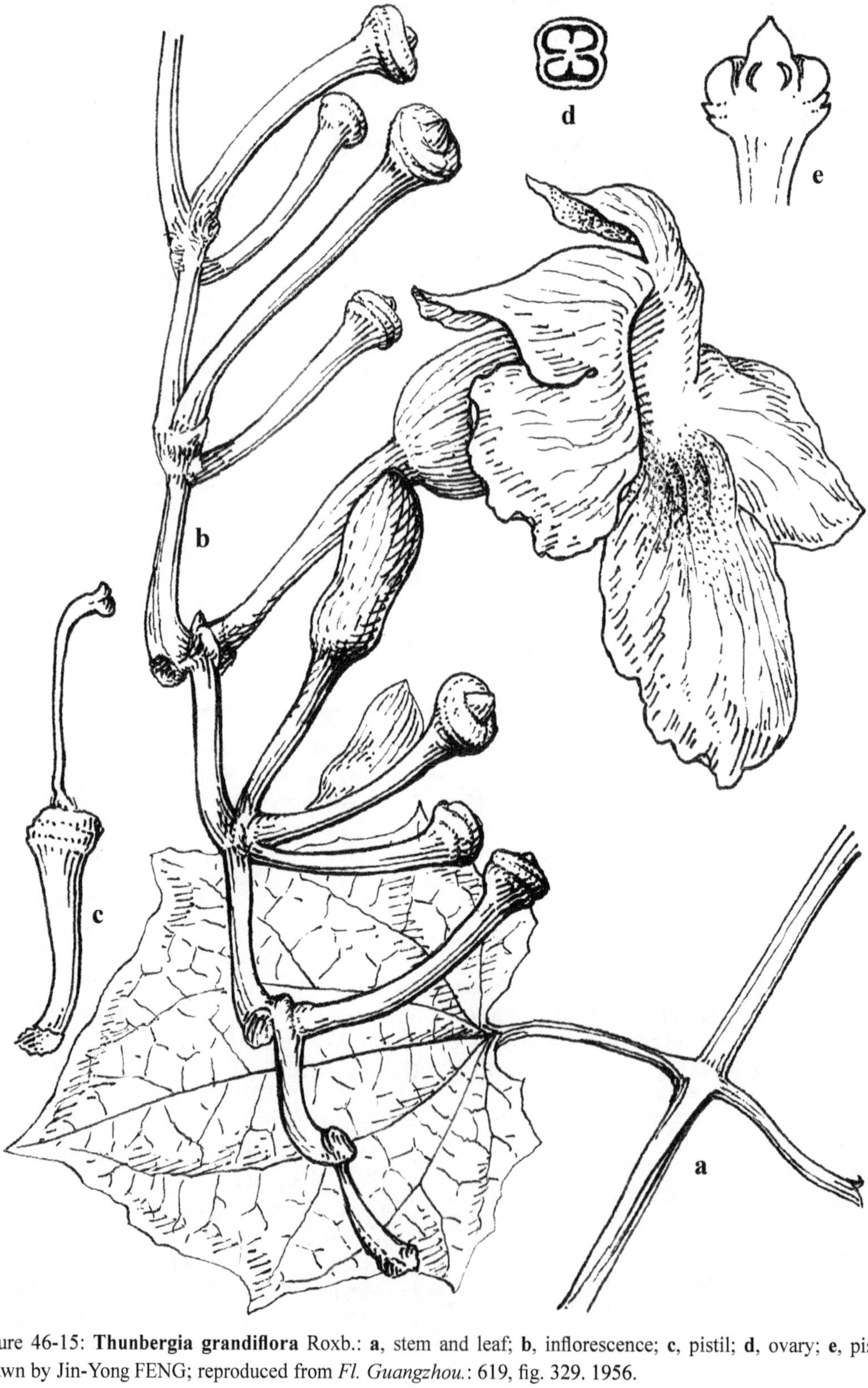

Figure 46-15: **Thunbergia grandiflora** Roxb.: **a**, stem and leaf; **b**, inflorescence; **c**, pistil; **d**, ovary; **e**, pistil. Drawn by Jin-Yong FENG; reproduced from *Fl. Guangzhou.*: 619, fig. 329. 1956.

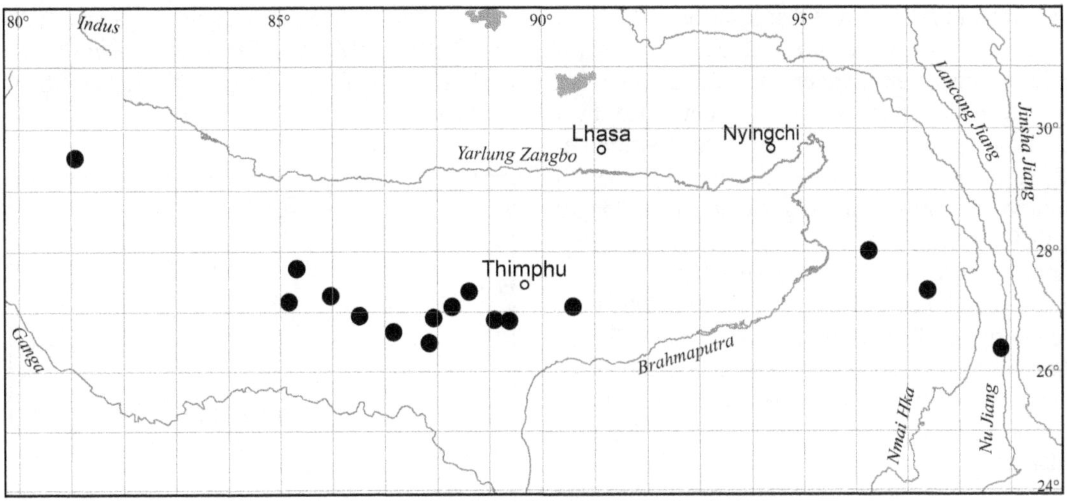

Map 46-26: **Thunbergia grandiflora** Roxb.

H. Kanai et al. s. n. (TI); Sunsari, S Dharan, 200 m, 1 Jun. 1972, *H. Kanai et al. 10011554* (TI); Sunsari, Tarahara Forests, 230 m, 15 Jun. 1974, *P. Pradhan et al. 324/74* (KATH); Tarhara, 270 m, 1972, *T. B. Shrestha & T. K. Bhattacharya 72-427* (KATH); **Mechi**, Ilam, Ragapani, 9 Dec. 1963, *H. Hara et al. 6303578* (BM, TI); Ilam, Ratuwa Khola, 300 m, 31 May 1969, *L. H. J. Williams 312* (BM, KATH); Ilam, Around Soktim, 500 m, 11 Jun. 1974, *P. Pradhan 257/74* (KATH); Jhapa, Dharan, Chulachuli, 300 m, 14 May 1967, *J. D. A. Stainton 5712* (BM); **Sagarmatha**, Udyapur, Trijuga Khola, 200 m, 16 Feb. 1982, *J. D. A. Stainton* 8506 (E). **SIKKIM & DARJILING: Darjiling**, Tista, 350 m, 20 Jun. 1960, *H. Hara et al. 5908* (TI); Manjitar, 450 m, 6 Aug. 1992, *D. G. Long et al. ESIK1226* (E); **Sikkim**, Bamonpokri, 29 Aug. 1857, *T. Thomson s. n.* (CAL); Bamonpokri, 8 Jul. 1862, *T. Anderson s. n.* (CAL); Bamonpokri, 600–1000 m, 1 Aug. 1872, *G. King 257* (CAL); Bamonpokri, 25 Nov. 1913, *E. A. Modder 103* (CAL); Rungeet Valley, 19 Nov. 1874, *D. Treutler 360* (K). **BHUTAN: Gaylegphug**, Mao valley, 300 m, 28 May 1979, *A. J. C. Grieson & D. G. Long 1401* (E, K); **Phuntsholing**, ca. 4 km above Kharbandi, 900 m, 25 Aug. 1988, *J. R. I. Wood 6579* (E); between Phuntsholing and Kharbandi, 600 m, 2 Jun. 1991, *J. R. I. Wood 7381* (E); Phuntsholing, 300 m, 3 Jun. 1968, *H. Kanai et al. 1069* (BM, TI); **Samchi**, west side of Charmurchi River, 300 m, 22 Jun. 1990, *J. R. I. Wood 7266* (E); **Samdrup Jongkhar** to **Deothang** road, 7 km above Samdrup Jongkhar, 550 m, 21 Jun. 1979, *A. J. C. Grieson & D. G. Long 2121*(A, E, K, TI). **YARLUNG ZANGBO-BRAHMAPUTRA: Lohit Valley**, 9 Sep. 1969, *A. S. Rao47729* (CAL); 150–350 m, 1 Mar. 1928, *F. Kingdon-Ward 7926* (K). **U IRRAWADDY: Chipwi**, Naungkhio, *F. G. Dickson 9546* (A); W Tang Hpre, 350 m, s. d., *J. Keenan et al. 3993* (A); **Putao**, way to Myitkyina, 1914, *F. Kingdon-Ward 1998* (E).

3. Thunbergia kasajuana Bh. Adhikari & J. R. I. Wood, Kew Bull. 75(26): 3. 2020. Type: Nepal, Bhaktapur, Nagarkot, 2020 m, 6 Sep. 2018, *S. Kasaju 133* (holotype: KATH; isotype: E).

Herbaceous vine, 2–4 m long. Stem angular, sulcate, hirsute, more densely so towards nodes. Leaves petiolate; petiole 2–5 cm long; blade ovate or ovate-triangular, 5–9 cm long, 3.5–6.5 cm broad, palmately 5–7-veined, both surfaces sparsely strigose, especially along veins, base cordate to hastate, apex acuminate, margin irregularly and shallowly dentate. Racemes terminal, 3–12 cm long; pedicels 1–1.5 cm. Bracts small, ovate, 5–10 mm long, 2–4 mm broad, sparsely pubescent; bracteoles green, ovate, 1.5–2 cm long, 1–1.5 cm broad, sparsely pubescent, apex mucronat. Calyx reduced to annular ring, 1.5–2 mm long, accrescent to ca. 3 mm in fruit. Corolla white, throat yellowish with pinkish veins, glabrous; tube 2.5–3.5 cm long, constricted at base for 4–5 mm broad, then gradually widened to 0.8–1 cm broad at mouth; lobes obovate to rounded, 5–8 mm long, 5–10 mm broad. Filaments slightly unequal, longer pair 9–12 mm long, shorter pair 8–10 mm long; anther-thecae bearded, base spurred, spur 2–3 mm long, acuminate. Ovary glabrous; style 2.5–3 cm long; stigma bilobed, lobes subequal, upper folded inwards, lower spreading. Capsule glabrous or sparsely pubescent, basal fertile portion 1–1.5 cm broad, beak 2–2.5 cm long. Seeds 5–7 mm in diameter.

Phenology: flowering from August to September; fruiting from September to December.

Habitats: slopes; 2000–2100 m.

Distribution: endemic; C Nepal (Map 46-27).

C NEPAL: Bagmati: Bhaktapur, Nagarkot, 2020 m, 12 Dec. 2017, *S. Kasaju 1010* (KATH); Bhaktapur, Nagarkot, 2020 m, 21 Aug. 2018, *S. Kasaju 120* (KATH); Bhaktapur, Nagarkot, 2020 m, 23 Sep. 2018, *B. Adhikari & S. Kasaju 111* (E, KATH); Bhaktapur, Nagarkot, 2020 m, 6 Sep. 2018, *S. Kasaju 133* (E, KATH).

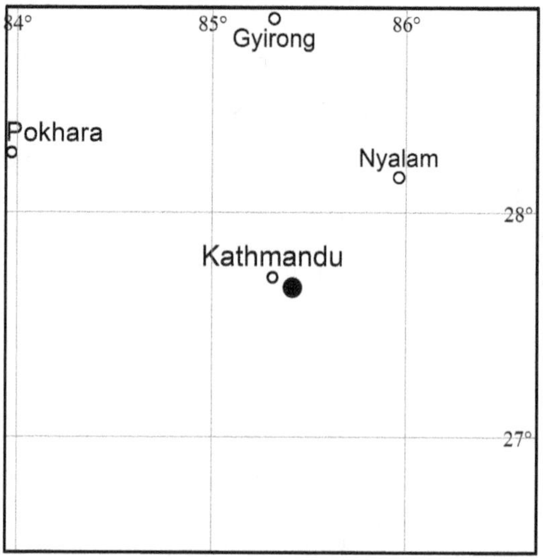

Map 46-27: **Thunbergia kasajuana** Bh. Adhikari & J. R. I. Wood

II. **Thunbergia** subg. **Coniostephanus** Bremek., Verh. Kon. Nederl. Akad. Wetensch., Afd. Natuurk. Sect. 2 50(4): 15. 1955. Type: *Thunbergia erecta* (Benth.) T. Anderson.

Erect or scandent shrubs, or vines. Leaves pinnately nerved. Flowers solitary, axillary. Calyx with linear or narrowly triangular teeth. Corolla with short tube and long campanulate throat. Anther-thecae unequal, one longer than another, bordered with conical bristles at base.

About 30 species: tropical regions of Asia and Africa; one species in the Pan-Himalaya.

4. Thunbergia lutea T. Anderson, J. Linn. Soc., Bot. 9: 448. 1867; Clarke in Hooker (ed.), Fl. Brit. India 4: 392. 1884; Wood in Grierson & Long (eds.), Fl. Bhutan 2(3): 1247. 2001; Tsui in Hu (ed.), Fl. Reipubl. Popularis Sin. 70: 29. 2002; Hu, Deng & Daniel in Wu, Raven & Hong (eds.), Fl. China 19: 378. 2011. Type: [Sikkim & Darjiling] India, Sikkim, *J. D. Hooker s. n.* [lectotype designated by Adhikari et al., Kew Bull. 68: 657. 2013: K (K000357692); isolectotypes: A, K (K000357693)].

Thunbergia salwenensis W. W. Sm., Notes Roy. Bot. Gard. Edinburgh 12: 224. 1920. Type: [S Hengduan] China, Yunnan, Shweli-Salwin Divided, 25°30'N, 7000-8000 ft., Jul.-Aug. 1918, *G. Forrest 17645* [lectotype **here designated**: E (E00265801); isolectotype: K].

Thunbergia maculata Lace, Bull. Misc. Inform. Kew 1914: 154. 1914, **syn. nov.** Type: Myanmar, Ruby Mines, Kyatpin to Mogok, 4000 ft., 23 Oct. 1912, *J. H. Lace 6000* [holotype: K (K000898165); isotypes: E (E00273457, E00273458, E00273459)].

Vines to 5 m or longer, climbing; stems glabrous except for pubescent on nodes. Petiole 5–7.5 cm long, glabrous; leaf blade ovate to lanceolate-ovate, 4–13 cm long, 2–6 cm broad, abaxially subglabrous, adaxially pubescent or setose, base cuneate and decurrent onto petiole, margin obscurely toothed to sometimes irregularly denticulate, apex acuminate. Flowers axillary, solitary. Bracteoles elliptic, 2.1–2.4 cm long, 1.5–1.7 cm broad, abaxially glabrous, adaxially glandular-hairy, palmately 7-veined, apex mucronate. Calyx 10-toothed. Corolla pinkish or white; tube base cylindric, 6–7 mm long, throat 3.2–3.5 cm long; lobes orbicular, 1.0–1.2 cm in diameter. Filaments glabrous, 1.8–2 cm long; anther-thecae 5–7 mm long, barbate, spurred at base, spurs bordered with conical bristles. Ovary glabrous; style 2.2–2.5 cm long, glabrous; stigma funnel-shaped, 2-cleft. Capsules glabrous, basal fertile portion 0.8–1 cm long, 1.6–1.8 cm broad, beak 1.6–1.8 cm long. Seeds reniform, 6–8 mm long, 4–5 mm broad. Figure 46-16.

Phenology: flowering from August to October; fruiting from September to November.

Habitats: forests; 1000–2700 m.

Distribution: E Nepal, Sikkim & Darjiling, S Hengduan, U Irrawaddy (Map 46-28); also in Thailand, and other parts of India, Myanmar and China (S Yunnan).

E NEPAL: Koshi, Dhankuta, Sindhuwa, 1100 m, 24 Oct. 1963, *H. Hara et al. 6303596* (BM, KATH, TI); Dhankuta, Citre to Teku Nala, 8 Jul. 1972, *H. Kanai et al. 72708* (KATH, TI); Dhankuta, Tute to Sindua, 2 Sep. 1977, *H. Ohashi et al. 773093* (TI); Dhankuta, Tute to Sindua, 2100–2300 m, 2 Sep. 1977, *H. Ohashi et al. 771287* (TI); Terathum, Tinjure Danda, 2270 m, 6 Sep. 1967, *L. H. J. Williams& J. D. A. Stainton 8380* (BM, K). **SIKKIM & DARJILING: Darjiling**, Senchal to Ghum, 2300–2400 m, 6 Jul. 1969, *H. Hara et al. 69109* (TI); Palmajua to Batasi, 8 Aug. 1972, *H. Kanai et al. 881271* (KUN); Three Miles (Simkuna) to Mungpoo, 2000–2200 m, 24 Sep. 1977, *H. Ohashi et al. 775762* (A, BM, KATH, KUN, MO, TI); Rungbo, 2000 m, 17 Aug. 1869, *C. B. Clarke 8653* (K); Dumrong, 2000 m, 22 Sep. 1869, *C. B. Clarke 9217* (LE); Goom Pahar, 2350 m, 12 Sep. 1875, *C. B. Clarke 27567* (BM, CAL, K); **Sikkim**, Lepehajigat, 2300 m, 27 Jul. 1889, *H. H. Haines 120* (K); Khubrong, 1830 m, 26 Sep. 1884, *C. B. Clarke 35907* (K, LE). **S HENGDUAN: Longyang**, Lujiang, Nankang, 2200 m, 21 Oct. 1998, *H. Li 10677* (KUN); Nankang, Lihuipo, 2150 m, 19 Nov. 2000, *H. Li 13232* (CAS, KUN); Lihuipo, 2200 m, 21 Sep. 2013, *Y. F. Deng 24700* (IBSC); **Gongshan**, Shweili-Salwin Divide, s. d., *G. Forrest 18286* (BM, E); Shweili-Salwin Divide, Aug. 1925, *G. Forrest 27154* (IBSC, PE). **U IRRAWADDY: Putao**, Upper Shankaung to Wasadam, 1000–1300 m, 13 Oct. 2014, *Putao Expedition Team PT-ET 205* (PE); Upper Shankaung to Wasadam, 1000–1300 m, 13 Oct. 2014, *Putao Expedition Team PT-ET 209* (PE); Wasadam to Upper Shankaung, 1400–1600 m,

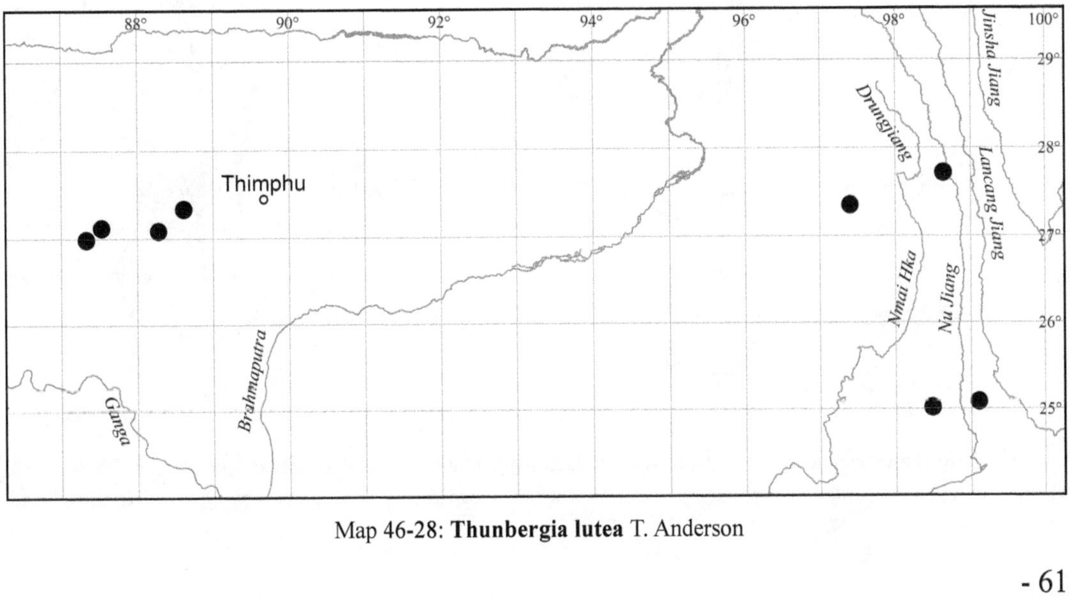

Map 46-28: **Thunbergia lutea** T. Anderson

Figure 46-16: **Thunbergia lutea** T. Anderson: **a**, flowering branch; **b**, leaf blade; **c**, opened corolla showing stamens and pistil; **d**, anther; **e**, appendage of anther-thecae; **f**, capsule; **g**, seed. Drawn by Quan LIN; reproduced from *Icon. Cormophyt. Sin.* 4: 154, fig. 5721. 1975.

23 Oct. 2014, *Putao Expedition Team PT-ET 1211* (PE); **Tengchong**, Mangbang, Dahaoping, 2340 m, 1963, *Z. Z. Ding & G. J. Zhu 138* (NAS); Puchuan, 1880–2020 m, 4 Oct. 1960, *W. Q. Yin 60-1325* (KUN); Gaoligongshan, 2150 m, 29 Dec. 1958, *X. Zhou 221* (KUN); Sanyun, Tanchang, 2210 m, 3 Sep. 2003, *Gaoligong Shan Biodiversity Survey 18665* (CAS, KUN).

III. Thunbergia subg. **Adelphia** Bremek., Verh. Kon. Nederl. Akad. Wetensch., Afd. Natuurk. Sect. 2 50(4): 18. 1955. Type: *Thunbergia fragrans* Roxb.

Herbaceous vine. Leaves palmately nerved. Flowers solitary, axillary. Calyx 5–23-toothed. Corolla cylindric. Anthers muticous at base.

About 30 species in tropical and subtropical regions of Asian and Australia; one species in the Pan-Himalaya.

5. Thunbergia fragrans Roxb., Pl. Coromandel 1: 47. 1795; Nees in Wallilch, Pl. Asiat. Rar. 3: 77, 78. 1832; Nees in de Candolle (ed.), Prodr. 11: 57. 1847; Anderson, J. Linn. Soc., Bot. 9: 448. 1867; Clarke in Hooker (ed.), Fl. Brit. India 4: 390. 1884, p. p. ; Nasir & Ali in Malik & Ghafoor (eds.), Fl. Pakistan 188: 72. 1988; Wood in Grierson & Long (eds.), Fl. Bhutan 2(3): 1247. 2001; Tsui in Hu (ed.), Fl. Reipubl. Popularis Sin. 70: 30. 2002; Hu et al., Fl. China 19: 378. 2011. Type: [*icon*] in Roxburgh, Pl. Coromandel 1: 47, tab. 67. 1796. (lectotype designated by Wasshausen, Fl. Guianas 23: 137. 2006).

Flemingia grandiflora Roxb. ex Rottl., Neue Schriften Ges. Naturf. Freunde Berlin 4: 202. 1803. Type: [*icon*] *Roxburgh Icon. 959* (lectotype **here designated**: K).

Vines, herbaceous; stems hirsute. Petiole 0.5–4.5 cm long, hirsute; leaf blade oblong-ovate to broadly ovate, 3–14 cm long, 1.8–7 cm broad, both surfaces puberulent to subglabrous, palmately 3–5-veined, base rounded to sometimescuneeate or cordate, margin irregularly sinuate or shallowly coarsely dentate, apex acute to acuminate. Flowers axillary, solitary; peduncle 1.5–5.5 cm long. Bracteoles ovate, 1.5–2.5 cm long, 0.8–1.5 cm broad, puberulent, apex acute. Calyx 3–5 mm long, unequally 10–17-dentate, glabrous. Corolla white, 3–5 cm long, sparsely puberulent; tube base cylindric, 4–7 mm long, throat 1.8–2.3 cm long; lobes obovate, 1.3–2.5 cm long, 1.5–2.3 cm broad. Filaments 6–10 mm long, glabrous; anther-thecae 2–3 mm long, glabrous, base muticous. Ovary glabrous; style 1.5–2 cm long; stigma funnel-shaped, 1.5–2 mm long. Capsules glabrous, basal fertile portion 6–7 mm long, 1–1.3 cm broad, beak 1.5–1.9 cm long. Seeds subglobose, 4–5 mm in diameter, smooth or with scales. Figure 46-17.

Phenology: flowering from July to November; fruiting from July to December.

Habitats: forest margins and roadsides; 200–2300(–2900) m.

Distribution: W Nepal, C Nepal, E Nepal, Sikkim & Darjiling, Bhutan, S Hengduan, U Irrawaddy (Map 46-29); also in Cambodia, Indonesia, Laos, Malaysia, the Philippines, Thailand, Vietnam, and other parts of India, Myanmar and China (S Yunnan, SE Sichuan, Guizhou, Guangxi, Hainan, Guangdong), and also widely cultivated throughout the tropics or sometimes naturalized.

Figure 46-17: **Thunbergia fragrans** Roxb.: **a**, portion of plant; **b**, leaf blade; **c**, capsule; **d**, opened capsule and bracteole; **e**, seed. Drawn by Quan LIN; reproduced from *Icon. Cormophyt. Sin.* 4: 153, fig. 5719. 1975.

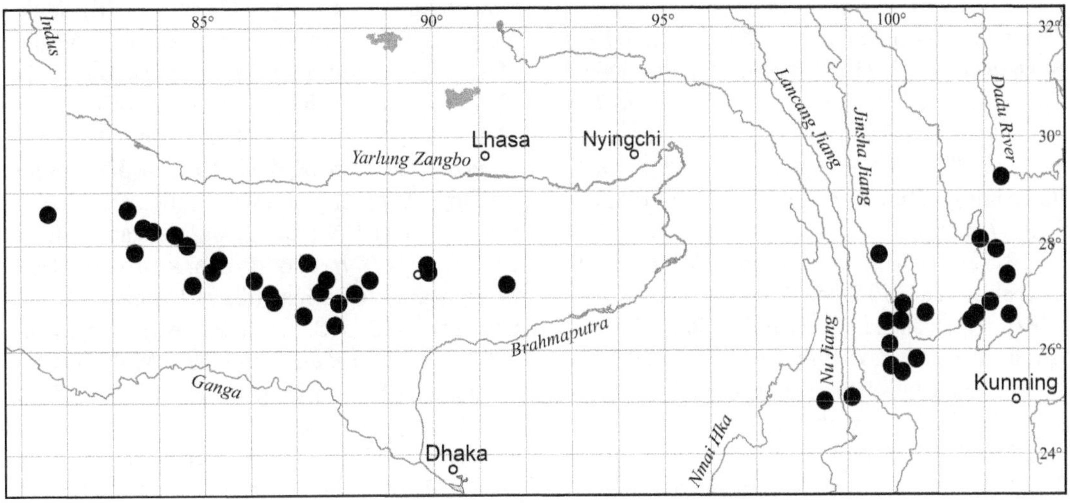

Map 46-29: **Thunbergia fragrans** Roxb.

W NEPAL: Bheri, Dailekh, Gairi, 1400 m, 1977, *S. K. Dhungana 49. 12/34* (KATH); Surkhet, Ranimatta, 1981, *N. P. Manandhar & D. P. Joshi 8410* (KATH); **Seti**, Doti, Baklekh, 2000 m, 1982, *L. P. Kattel 1122* (KATH). **C NEPAL: Bagmati**, Kathmandu, Jawal Tole, 1300 m, 22 Dec. 1966, *D. H. Nicolson 2885* (BM, KATH); Kathmandu, 1360 m, 28 Aug. 1935, *F.M. Bailey s. n.* (E); Kathmandu Pouwa, 9 Oct. 1963, *H. Hara et al. 6303576* (TI); Rasuwa, Dhunche, 1520 m, 1965, *J. D. A. Stainton 5180* (BM); Rasuwa, Trisuli, Syabro, 2000 m, 29 Sep. 1974, *J. D. A. Stainton 7276* (E, K); Rasuwa, Sim Chautara to Gram, 900–1900 m, 20 Aug. 1972, *H. Kanai et al. 721661* (KATH, KUN, TI); Rasuwa, Manigaun to Ramche, 1830 m, 1977, *H. K. Saiju & P. M. Amatya 7/77* (KATH); Sindupalchok, Gulu Bhanjyang, 2000 m, 7 Aug. 1974, *J. H. de Haas 2337* (BM, L); **Daulagiri**, Myagdi, Tatopani, 1200 m, 30 Aug. 1954, *J. D. A. Stainton, W. R. Sykes & L. H. J. Williams 9582* (BM, E); Myagdi, Dandagaun, 1760 m, 1992, *N. P. Manandhar & S. K. Acharya 430/92* (KATH); Parbat, Birethati, 1250 m, 1974, *D. P. Joshi & T. K. Bhattacharya 74/1710* (KATH); Parbat, Khya Argali Danda, 1050 m, 1973, *D. P. Joshi 0256* (KATH); Parbat, Shikha to Tatopani, 15 Jul. 1983, *H. Ohba et al. 8330451* (TI); **Gandaki**, Gorka, Negua Dovan, 750 m, 1981, *P. R. Shakya 6988* (KATH); Gorkha, Ganesh Himal, 1975, *B. Yorl 445* (E); Gorkha, Arkhet Bazar, 560 m, 20 Jul. 1994, *M. Suzuki et al. 9485076* (TI); Kaski, Rotepani, 1000 m, 1986, *N. P. Manandhar & L. P. Katel 10854* (KATH); Kaski, Birethanti, 1100 m, 25 Jul. 1996, *T. Hoshino et al. 9662008* (A, E, TI); Kaski, vicinity of Tamage, 1680 m, 8 Aug. 1999, *M. Mikage et al. 9964021* (KATH, TI); Kaski, Puranchaur, 1050 m, 1954, *J. D. A. Stainton, W. R. Sykes & L. H. J. Williams 6494* (BM); Kaski, Besi, 1230–1630 m, 1985, *D. P. Joshi et al. 35/85* (KATH); Kaski, Dhampus, 1650 m, 1992, *N. P. Manandhar & S. K. Acharya 117-92* (KATH); Lamjung, Seilung, 1500 m, 18 Nov. 1969, *T. Wraber 36635* (BM); **Janakpur**, Ramechhap, 1430 m, 1977, *K. R. Rajbhandarii 2241* (KATH); **Lumbini**, Palpa, Ridikot, 1870 m, 1973, *D. P. Joshi 0209* (KATH); **Narayani**, Makwanpur, Kuli Khani, 1500 m, 2 Sep. 1970, *H. Kanai & T. B. Shresta 672706* (KATH, TI); Makwanpur, Kulekhini to Deorali, 2000, 1970, *T. B. Shrestha 17343* (KATH); Parsa, Parsa Wildlife Reserve, along Mahadev Khola, 190–200 m, 9 Oct. 1995, *M. Mikage et al. 9552777* (A, E, TI). **E NEPAL: Koshi**, Sankhuwasabha, Num, Arun Valley, 1800 m, 1956, *J. D. A. Stainton 1609* (BM, E, K); Sankhuwasabha, Mani Bhanjyang, 1100 m, 28 Jul. 1990, *M. Minaki et al. 9080055* (E, TI); Sankhuwasabha, Num, 1525 m, 13 Aug. 1998, *S. Noshiro et al. 9830002* (A, E, TI); Sankhuwasabha, Num, Patria, 850 m, 31 Jul.

1977, *H. Ohashi et al. 773367* (E, TI); Sunsari, Dhara Pani to Dharan Bazar, 1000 m, 5 Sep. 1977, *H. Ohashi et al. 771383* (TI); Sunsari, Panchain to Dharan, 270–300 m, 1972, *T. B. Shrestha & T. K. Battacharya 72-281* (KATH); Terhathum, 1650 m, 1976, *N. P. Manandhar 346* (KATH); **Mechi**, Ilam, Ranga Pani to Ghorwa, 9 Dec. 1963, *H. Hara et al. 6303577* (TI); Ilam, Dhara Pani, 1000 m, 1977, *Ohashi et al. 771383* (BM, TI); Ilam, Maibeli to Ilam, 1050 m, 1977, *P. Pradhan et al. 242* (KATH); Jhapa, Chulachuli, 300 m, 1972, *T. B. Shrestha & T. K. Bhattacharya 72-27* (KATH); Jhapa, Chulachuli, Bisnetar to Kawal Dhar, 390 m, 1972, *T. B. Shrestha & T. K. Bhattachaya 72-236* (KATH); Taplejung, Phembu, 1350 m, 3 Aug. 1956, *J. D. A. Stainton1240* (BM, E); **Sagarmatha**, Khotang, Regmitar to Khani Danda, 200 m, 24 Oct. 1995, *M. Mikage et al. 9552999* (E, TI); Udyapur, Mai Pokhari, 2000 m, 28 Sep. 1971, *J. F. Dobremez 1232* (E). **SIKKIM & DARJILING: Darjiling**, below Chumbhati, 500 m, 10 Sep. 1964, *H. Hara 1561* (TI); Selim, 450 m, 7 Sep. 1869, *C. B. Clarke 9069* (BM); Selim, 1050 m, 7 Sep. 1869, *C. B. Clarke 9108* (BM); **Sikkim**, Kinsing, 1857, *T. Thomson s. n.* (CAL); Banumpohue, Aug. 1873, *J. S. Gamble 3533a* (K); Banumpohia, 1879, *J. S. Gamble 7365* (K); Selim, Terai, 450 m, 18 Oct. 1884, *C. B. Clarke 36667* (CAL, K). **BHUTAN:** Saly, 1200 m, 28 Aug. 1915, *R. E. Cooper 4725* (BM, E); **Mongar**, 1990, *R. Reynolds s. n.* (E); **Punakha**, ca. 20 km S of Wangdi Phodrang, 1000 m, 8 Oct. 1989, *J. R. I. Wood 7207* (E); **Thimphu**, 2900 m, 1989, *J. R. I. Wood 7209* (E). **S HENGDUAN: Binchuan**, Jizushan, 2100 m, 26 Aug. 1982, *Y. H. Li & Y. Chen 33804* (HITBC); Jizushan, 2250 m, 7 Sep. 1966, *Y. Y. Hu & Y. F. Qi 662214* (YIM); **Dali**, near Gantongsi, 2000 m, 20 Jul. 2006, *Y. F. Deng 19099* (IBSC); Tali valley, 2010 m, Jun.–Jul. 1906, *G. Forrest 4477* (E, K, PE); Malongfeng, Jun. 1941, *T. N. Liou 1146* (PE); in the city, 27 Nov. 1946, *T. N. Liou 17410* (PE); Liujufeng, 12 Sep. 1932, *H. C. Wang 4089* (IBSC, PE); **Eryuan**, Zhujiaying, Dawan, 20 Jul. 1939, *R. C. Ching 23319* (KUN, PE); **Heqing**, Baiya, 7 Sep. 1929, *R. C. Ching 24424* (KUN, PE); Junle, Yingwulan to Shangdapigzi, 2100 m, 16 Aug. 1963, *Exped. Dianxibei Jinshajiang 6553* (KUN, PE); **Huidong**, in the city, 1750 m, 4 Jul. 1978, *Q. S. Zhao 5829* (SZ); 1750 m, 5 Jul. 1978, *Q. S. Zhao & K. H. Mou 5896* (CDBI, SZ); **Jianchuan**, Cuifeng, Mingxing, Dahu, 7 Aug. 1975, *P. Y. Chiu 60360* (KUN); **Longyang**, Wayao, 1140 m, 23 Sep. 1977, *Y. H. Li & P. H. Yu 20401* (HITBC); **Miyi**, near Puwei, Guanmiao, 21 Jun. 1958, *S. Y. Chen et al. 10285* (SZ); near Puwei, 1720 m, 1 Jul. 1958, *S. Y. Chen et al. 10556* (SZ); **Muli**, Huying, 1750 m, 16 Aug. 1978, *Q. S. Zhao 7996* (CDBI, SZ); Dsengo, 2350 m, 9 Aug. 1915, *H. Handel-Mazzetti 7566* (E, WU); **Panzhihua**, Renhe, Datian, Baoxingshan, 1700 m, 27 Sep. 2002, *X. F. Gao et Y. L. Peng 4093* (CDBI); **Puge**, Li'an, 18 Aug. 1959, *CHUAN JINNG LIANG 5416* (KUN); **Shangri-la** (Zhongdian), Washua to Zhisi, 2 Sep. 1939, *K. M. Feng 2254* (KUN); between Bodo to Haba, 11 Sep. 1939, *K. M. Feng 2345* (KUN); Zhongdian to Langda, near Luojihe, 2300 m, 20 Aug. 1939, *K. M. Feng 2711* (A, KUN, PE); Chiaotou, 1850 m, 28 Oct. 1939, *K. M. Feng 3164* (A, KUN); **Shimian**, Guowan Bridge, 1100 m, 1980, *Q. Q. Wang 22615* (IBSC); **Xichang**, 2000 m, Aug. 1932, *T. T. Yü 1226* (A, PE); **Yanbian**, Yongxing, 1340 m, 28 Jul. 2006, *Y. F. Deng 18882* (IBSC); Yongxing, Tianshengqiao, 1280 m, 29 Jul. 2003, *X. F. Gao et al. 6097* (CDBI); **Yangbi**, way to Xiaguan, 1800 m, 11 Aug. 1963, *Exped. Zhongdian 63-4295* (KUN); Yangpi Mountains, 2700 m, Sep. 1913, *G. Forrest 11170* (BM, PE); **Yongsheng**, Songping, 2200 m, 5 Aug. 1976, *C. Y. Wu 4094* (KUN); **Yulong** (Lijiang), Jiazi, bank of Baishuihe, 2000 m, 16 Sep. 1955, *K. M. Feng 21563* (KUN, SZ); Taku, 13 Aug. 1914, *C. Schneider 2157* (A, K); W slope of Likiang Snow Range, *J. F. Rock 5413* (E); Longpan, Lunan, 1800–2000 m, 24 Aug. 2007, *Y. F. Deng 19634* (IBSC). **U Irrawaddy: Tengchong**, E Tengyueh, 1800–2100 m, Jun. 1912, *G. Forrest 8263* (BM, E, IBSC, K); bank of Dieshuihe, 1820 m, 8 Aug. 1980, *S.T. Li 80-570* (KUN).

IV. Thunbergia subg. **Parahexacentris** Bremek., Verh. Kon. Nederl. Akad. Wetensch., Afd. Natuurk. Sect. 2 50(4): 15. 1955. Type: *Thunbergia alata* Boj. ex Sims.

Herbaceous vines. Leaves palmatly nerved. Flowers solitary, axillary. Calyx 5–10-toothed. Corolla tube gradually widened from base. Anther-thecae spurred at base.

About 40 species in tropical regions of Asia and Africa; one species in the Pan-Himalaya.

6. Thunbergia nepalensis B. Adhikari & J. R. I. Wood, Kew Bull. 68: 654. 2013. Type: [W] Nepal, Seti, Doti, Trail from Silgadi to Khaptad National Park, 81°00′46″E, 29°17′55″N, 2105 m, 30 Jun. 2009, *H. Ikeda et al. 20919006* [holotype: E (E00625010); isotypes: KATH, TI].

Twining herbs, to 3 m long; stems terete, pubescent, especially at nodes. Petiole 1.5–3.5 cm long; leaf blade ovate, 5–8 cm long, 2–5.5 cm broad, palmately 5-veined, strigillose-pilose, base truncate, shallowly cordate or hastate, margin minutely dentate, apex acuminate. Flowers solitary, axillary; pedicels 2–3.5 cm long, glabrous. Bracteoles ovate or oblong-ovate, 2–3 cm long, 1–2 cm long, slightly pubescent to glabrous, apex acute. Calyx irregularly 5–7-toothed, teeth 1 mm long. Corolla white with yellow throat, glabrous; tube gradually widened from base, 3–4 cm long; lobes orbicular, oblong or broadly obovate, 1–1.5 cm long, 0.8–1 cm broad. Filaments subequal, 10–12 mm long, with tuft of hairs at base, anthers bearded, 5–6 mm long, base spurred. Ovary glabrous, style 2–2.5 cm long, stigma bilobed. Capsules unknown.

Phenology: flowering from June to August; fruiting from July to September.

Habitats: degraded forests; 1200–2100 m.

Distribution: endemic; W Nepal, C Nepal (Map 46-30).

W NEPAL: Seti, Doti, 2100 m, 30 Jun. 2009, *H. Ikeda et al. 20919006* (E, KATH, TI); Bajura, 29 Aug. 1991, *M. Suzuki et al. 9161132* (TI). **C NEPAL: Gandaki**, Kaski, Tikhedhunga to Kande, 1240 m, 7 Sep. 1988, *M. Suzuki et al. 8800826* (TI).

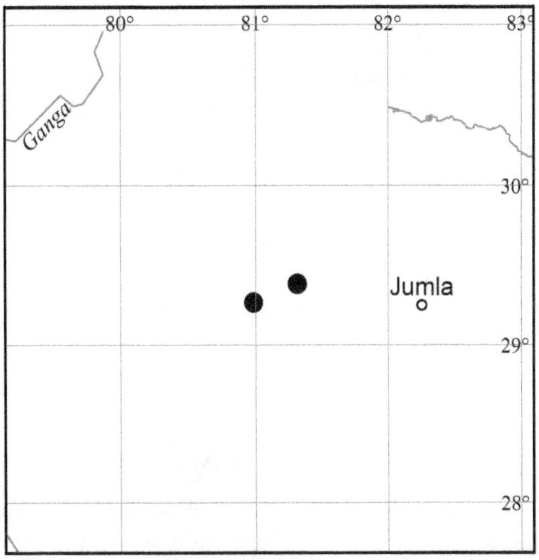

Map 46-30: **Thunbergia nepalensis** B. Adhikari & J. R. I. Wood

Subfam. 3. Acanthoideae Link, Handbuch. 1: 500. 1829, "Acanthaceae"; Lindau, Bot. Jahrb. Engler 18: 43. 1893; Hu, Fl. Reipubl. Popularis 70: 44. 2002. Type: *Acanthus* L.

Plants with or rarely without cystoliths. Leaves opposite or whorled. Bracts opposite; bracteoles 2 or absent. Calyx 4–5-lobed. Stamens 2 or 4, anthers monothecous or bithecous. Ovules 2–several per locule. Retinacula hook-like.

About 220 genera and 4000 species; 30 genera and 178 species in the Pan-Himalaya.

Tribe 1. Acantheae Dumort., Anal. Fam. Pl.: 23. 1829. Type: *Acanthus* L.

Plant without cystoliths. Leaves opposite or whorled. Bracts opposite. Calyx 4- or 5-lobed, chaffy or bony. Corolla unilabiate, upper lip absent, lower lip 3–5-lobed. Stamens 4, anther monothecous.

Ovules 2 per locule. Capsules with retinacula. Seeds 2–4.

Recent references: McDade et al., 2005. Phylogenetic relationships among Acantheae (Acanthaceae): major lineages present contrasting patterns of molecular evolution and morphological differentiation. *Syst. Bot.* 30: 834–862.

Twenty genera and about 500 species in the tropical regions of the world; two genera and three species (one endemic) in the Pan-Himalaya.

4. ACANTHUS L.

Linnaeus, Sp. Pl. 2: 639. 1753; Nees in de Candolle (ed.), Prodr. 11: 269. 1847; Clarke in Hooker (ed.), Fl. Brit. India 4: 480. 1884; Lindau in Engler & Prantl (eds.), Pflanzenfam. 4(3b): 318. 1895; Wood in Grierson & Long (eds.), Fl. Bhutan 2(3): 1279. 2001; Tsui in Hu (ed.), Fl. Reipubl. Popularus Sin. 70: 44. 2002; Hu, Deng & Daniel in Wu, Raven & Hong (eds.), Fl. China 19: 379. 2011. ≡ *Acanthodus* Raf., Princ. Fond. Somiol. 25. 1814 ("1813"), nom. illeg. (Art. 52.1 & 52.2). Type (designated by Green, Prop. Brit. Bot.: 170. 1929): *Acanthus mollis* L.

Shrubs or perennial herbs; without cystoliths. Leaves opposite, rarely whorled or rosette, petiolate; blade margin entire, pinnatifid or spinescent toothed. Inflorescences terminal or axillary, uninterrupted spikes. Bracts imbricate, margin spiny; bracteoles 2 or sometimes absent. Calyx 4-lobed, posterior and anterior lobes larger, lateral pair smaller. Corolla tube very short, horny; limb 1-lipped, upper lip reduced or sometimes rudimentary, lower lip elongate, obtusely 3-lobed. Stamens 4, paired, inserted at top of corolla tube, exserted; filaments stout, thickened, bony; anthers monothecous; theca linear-oblong, usually densely bearded, base muticous; staminodes absent. Ovary with 2 ovules per locule; style slender; stigma 2-cleft. Capsules ellipsoid, shiny, 2–4-seeded; retinacula present. Seeds discoid, glabrous or sericeous-puberulous. **Chromosomes** 2n = 44, 56, 80.

Recent references: Furness, 1996. Pollen morphology of *Acanthopsis* Harvey, *Acanthus* L. and *Blepharis* Jussieu (Acanthaceae: Acantheae). *Rev. Palaeobot. Palynol.* 92(3-4): 253–268; Chhetri, 2002. A haptophyte in the Himalaya. *Curr. Sci.* 83: 1199–1200; Baruah et al., 2013. Notes on two interesting angiosperms from Assam, India. *Pleione* 7(2): 401–405.

About 25 species from S Europe to the Pacific and south throughout tropical regions of Asia and Africa; two species (one endemic) in the Pan-Himalaya. Notes: Acanthus ilicifolius was reported by Chhetri (2002) and Ghosh & Mallick (Flora of Darjiling Himalaya and foothills (Angiosperms): 15. 2014) from Darjiling. However, the image provided by the authors indicates that it might be misidentified for Acanthus montanus L. which is a widely cultivated species as ornamental. It was also reported from Sikkim by Yamazaki (in Bull. Univ. Museum, Univ. Tokyo 2: 1–393. 1971), but no representative collection was cited. I haven't seen any collections from Sikkim, and the distribution in Sikkim need to be confirmed in the future.

1a. Herbs; stems rufous-villous; leaves spiny-serrate, sometimes sinuate; bracts less than 1 cm broad.. .... 1. *A. leucostachyus*
1b. Shrubs; stems glabrous; leaves deeply pinnatifid; bracts usually 1.5–3 cm broad .... 2. *A. carduaceus*

1. **Acanthus leucostachyus** Wall. ex Nees in Wallich (ed.), Pl. Asiat. Rar. 3: 98. 1832; Need in de Candolle (ed.), Prodr. 11. 270. 1847; Anderson, J. Linn. Soc., Bot. 9: 501. 1867; Clarke in Hooker (ed.), Fl. Brit. India 4: 48. 1884; Wood in Grierson & Long (eds.), Fl. Bhutan 2(3): 1280. 2001; Tsui in Hu (ed.), Fl. Reipubl. Popularis Sin. 70: 45. 2002; Hu, Deng & Daniel in Wu, Raven & Hong (eds.), Fl. China 19: 380. 2011. Type: Bangladesh, Sillet, 1830, *F. de Silva in Wallich Cat. 2512* [lectotype **here designated**: K-W (K000639334); isolectotypes: BM (BM000948485), C, CGE, E, GZU (GZU000251367), LE, M, NY, S].

Herbs, 15–50 cm high; stem brown, hairy. Petiole 0.8–2.5 cm; leaf blade elliptic, ovate-lanceolate, oblanceolate or lanceolate, 9–23 cm long, 2–4.5 cm broad, abaxially glabrescent or brown pubescent along veins, adaxially glabrous, lateral veins 5–8 pairs, base rounded to subcordate, margin slightly spinose-dentate, sometimes inconspicuously dentate or occasionally entire, apex acute. Spikes terminal, 5–16 cm long, 4-ranked; rachis greyish pubescent. Bracts ovate to obovate, 1–1.5 cm long, 0.5–0.8 cm broad, abaxially pubescent, base rounded, margin apically spiny, apex acute to acuminate; bracteoles lanceolate to oblanceolate, 1.3–2 cm long, 0.2–0.4 cm broad, abaxially pubescent, base rounded, margin entire and ciliate, apex rounded with a short mucro. Calyx 1.8–2 cm long, pubescent; lobes unequal, oblanceolate to lanceolate, 1.2–2 cm long, 0.5–2.5 mm broad, margin ciliate. Corolla white, 1.8–2.1 cm long, villous; lobes orbicular, 4–5 mm long, 4–5 mm broad, apex emarginate. Filaments 6–8 mm long, glabrous; anther-thecae oblong, 4–4.5 mm long, pilose. Ovary 2.5–3 mm long; style 1–1.2 cm long. Capsules ellipsoid, 9–11 mm long, 2.5–3 mm broad, glabrous. Seeds reniform, 2–2.5 mm long, 2.5–3 mm broad, glabrous. Figure 46-18.

Phenology: flowering from February to May; fruiting from August to September.

Habitats: shady places in dense forests, wet places; 200–1000 m.

Distribution: U Yarlung Zangbo, Bhutan, Yarlung Zangbo-Brahmaputra, U Irrawaddy (Map 46-31); also in Laos, Thailand, Vietnam, and other parts of India, Myanmar and China (S & SW Yunnan).

U YARLUNG ZANGBO: Yadong, Chumbi Valley, Apr. 1905, *G. L. Searight s. n.* (BM). **BHUTAN: Deothang**, *Griffith 2415* (BM). **YARLUNG ZANGBO-BRAHMAPUTRA: Abor Hills**, Pasi Ghat, Dihang Valley, 450 m, 20 Feb. 1928, *F. Kingdon-Ward 7888* (K); **Dalpha Hills**, Aka Hills, 1934, *N. L. Bor 15260* (K); **Mishimi**, *Herbs Griffith 6147* (C, K). **U IRRAWADDY: Putao**, Tamu Hka, Nmai Hka Valley, 600 m, 7 Apr. 1939, *F. Kingdon-Ward 510* (GH); Makohwa, 2014, *Putao Expedtion Team PT-ET1415* (PE).

2. **Acanthus carduaceus** Griff., Itin. Pl. Khasyah Mts. 144, No. 688. 1848; Anderson, J. Linn. Soc. 9: 500. 1867; Clarke in Hooker (ed.), Fl. Brit. India 4 : 480. 1884; Wood in Grierson & Long (eds.), Fl. Bhutan 2(3): 1279. 2001. Type: Bhutan, *Herb. Griffith 6416* (lectotype **here designated**: K; isolectotypes: GH, LE, M, P).

Shrubs, 2–3 m high; stems glabrous. Leaves sessile; blade elliptic, 30–60 cm long, 10–20 cm broad, glabrous, base broadly cuneate or truncate, margin irregularly pinnatifid with narrowly triangular lobes for 3–12 cm long, apex mucronate. Spikes terminal and axillary, 10–30 cm long, terminal ones often sessile, axillary ones pedunculated; peduncles (0–)8–20 cm long. Bracts ovate, 2–4 cm long, 1.5–3 cm broad, pilose to lanate, margin spiny, apex acuminate; bracteoles linear-lanceolate, 1.5–3 cm long, 0.5–0.8 cm broad, villous, apex acuminate, margin

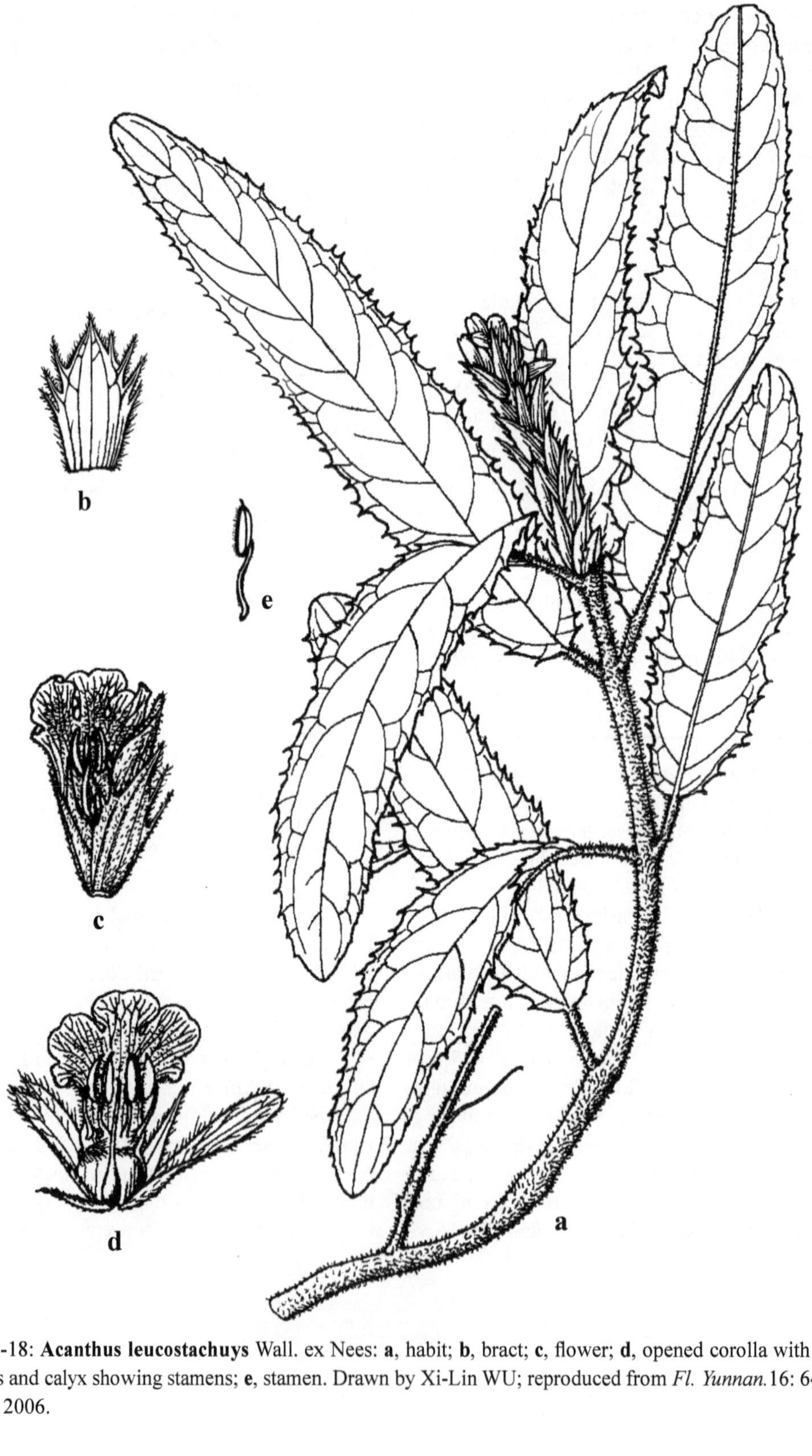

Figure 46-18: **Acanthus leucostachuys** Wall. ex Nees: **a**, habit; **b**, bract; **c**, flower; **d**, opened corolla with bract, bracteoles and calyx showing stamens; **e**, stamen. Drawn by Xi-Lin WU; reproduced from *Fl. Yunnan.*16: 649, pl. 184: 1–5. 2006.

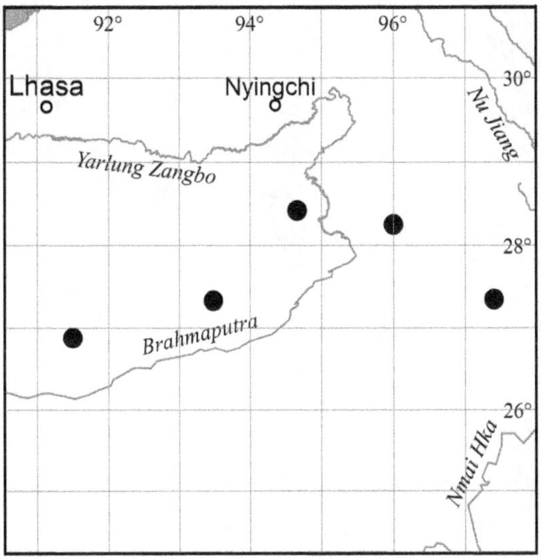

Map 46-31: **Acanthus leucostachyus** Wall. ex Nees

ciliate. Calyx heteromorphic, outer pair ovate-elliptic, 2.2–3.0 cm long, 0.7–1.0 cm broad, inner pair linear-lanceolate, 1.5–1.8 cm long, 0.4–0.5 cm broad, margin ciliate, anterior one shortly 2-toothed at apex. Corolla white, 2.5–3 cm long, outside glabrous, inside pilose; tube 0.8–1 cm long, lower lip 3-lobbed, suborbicular, margin slightly crenate. Filaments 0.8–1 cm long, swollen towards base; anther-thecae oblong, 2–3 mm long. Ovary ellipsoid, glabrous; style 1.5–1.7 cm long, glabrous; stigma bifid. Capsules oblong, 2.5–3 cm long, 3–4 mm broad, brown, glabrous. Seeds 4, triangular, 6–8 mm long and broad, laterally compressed, purple-brown, glabrous. Figure 46-19.

Phenology: flowering and fruiting from November to March.

Habitats: scrubby cliff slopes and rocky gullies; 900–2200 m.

Distribution: endemic; Sikkim & Darjiling, Bhutan (Map 46-32).

Notes: Flora of Darjiling Himalaya Foothills (Angiosperms).

SIKKIM & DARJILING: Darjiling, 1980 m, 20 Nov. 1962, *J. K. Maheswari 4956* (CAL); Jalpaiguri, 1350 m, 29 Nov. 1975, *J. K. Sikdar 4034* (CAL); Birch Hill, 2200 m, *G. Mutata & H. Ohashi 2756* (KUN); below Birch Hill, 2100 m, 4 Aug. 1992, *D. G. Long et al. 1104* (E). **BHUTAN: Mongar**, between Lingmethang and Namning, 1800 m, 10 Nov. 1991, *J. R. I. Wood 7510* (E); **Sankosh**, halfway between Goshi and Daga Dzong, 1700 m, 19 May 1988, *J. R. I. Wood 6322* (E); **Tongsa**, near Pertimi, 1400 m, 3 Apr. 1982, *A. J. C. Grierson & D. G. Long 4266* (E, GH, K).

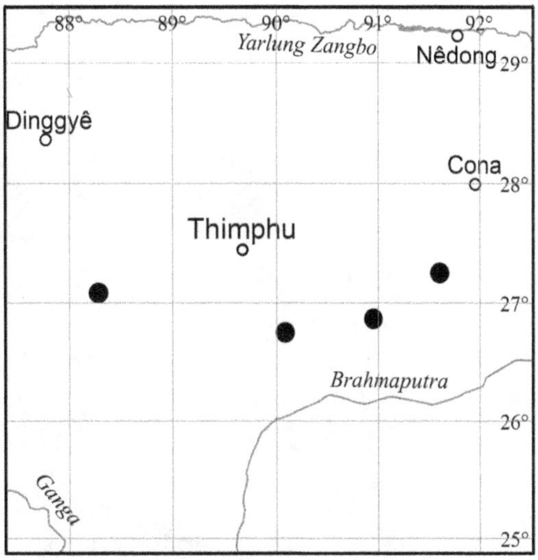

Map 46-32: **Acanthus carduaceus** Griff.

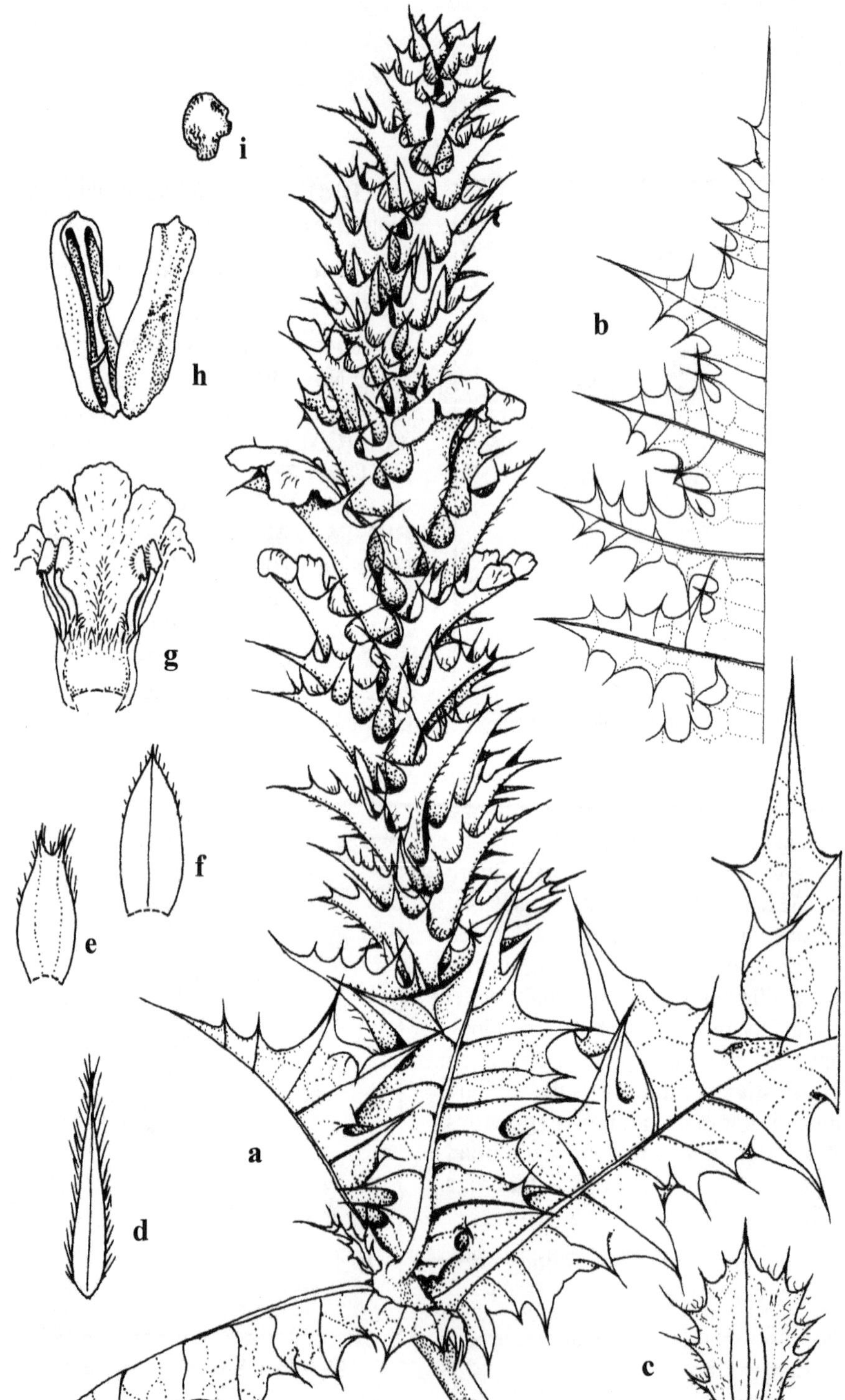

Figure 46-19: **Acanthus carduaceus** Griff.: **a**, habit; **b**, leaf; **c**, bract; **d**, bracteole; **e**, upper calyx segment; **f**, lower calyx segment; **g**, opened corolla; **h**, capsule; **i**, seed. Drawn by Louise OLLEY; reproduced from *Fl. Bhutan* 2(3): 1276, fig. 105: m-u. 2001.

5. BLEPHARIS Juss.

Jussieu, Gen. Pl.: 103. 1789; Nees in de Candolle (ed.), Prodr. 11: 265. 1847; Bentham & Hooker, Gen. Pl. 2: 1089. 1873; Clarke in Hooker (ed.), Fl. Brit. India 4: 478. 1884; Lindau in Engler & Prantl (eds.), Pflanzenfam. 4(3b): 316. 1895; Tsui in Hu (ed.), Fl. Reipubl. Popularis Sin. 70: 48. 2002; Hu, Deng & Daniel in Wu, Raven & Hong (eds.), Fl. China 19: 380. 2011. Type: *Blepharis boerhaviifolia* Pers. nom. illeg. (Art. 52.1 & 52.2.) [≡ *Blepharis maderaspatensis* (L.) Heyne ex Roth].

Herbs or shrubs; without cystoliths, anisophyllous. Leaves usually in pseudo-whorls of 4 or rarely opposite, sessile to subsessile. Flowers axillary and solitary or several in terminal dense head-like spikes. Bracts often clustered in up to 5 pairs, imbricate, margin often bristly dentate with bristles barbed; bracteoles absent or 2. Calyx 4-lobed; lobes in 2 unequal opposite pairs, posterior and anterior lobes larger, 2–7-veined, lateral pair smaller, 1-veined. Corolla tube shorter than limb, expanded from base and constricted near apex; limb 1-lipped; upper lip absent; lower lip elongate, 3-lobed. Stamens 4, didynamous, inserted at or near apex of corolla tube; filaments bony, usually bifurcate apically into a branch; anthers monothecous; theca bearded. Ovary glabrous, with 2 ovules per locule; style linear; stigma 2-cleft. Capsules oblong to ellipsoid, 2(–4)-seeded; retinacula present. Seeds discoid, covered with long branched hygroscopic trichomes. **Chromosomes** 2n = 24, 26, 34, 46.

Recent references: Kumar & Gupta, 1990. *Blepharis maderaspatensis* (L.) Heyne ex Roth—a new record for Himachal Pradesh. *J. Econ. Taxon. Bot.* 14(1): 221–222; Xing & Li, 1991. *Blepharis* Juss. (Acanthaceae), a new record genus from China. *Bull. Bot. Res. Harbin* 11(1): 49–51; Furness, 1996. Pollen morphology of *Acanthopsis* Harvey, *Acanthus* L. and *Blepharis* Jussieu (Acanthaceae: Acantheae). *Rev. Palaeobot. Palynol.* 92(3–4): 253–268; Vollesen, 2000. *Blepharis (Acanthaceae): a taxonomic revision.* Kew: Royal Botanical Gardens, Kew. 342 pp; Fisher, 2015. Evolutionary history of *Blepharis* (Acanthaceae) and the origin of C4 photosynthesis in section *Acanthodium. Int. J. Plant Sci.* 176: 770–790.

About 130 species: Old World tropical and subtropical regions; one species in the Pan-Himalaya.

1. Blepharis maderaspatensis (L.) B. Heyne ex Roth, Nov. Pl. Sp.: 320. 1821; Malik & Ghafoor in Nasir & Ali (eds.), Fl. Pakistan 188: 7. 1988; Tsui in Hu (ed.), Fl. Reipubl. Popularis Sin. 70: 48. 2002; Hu, Deng & Daniel in Wu, Raven & Hong (eds.), Fl. China 19: 380. 2011. ≡ *Acanthus maderaspatensis* L., Sp. Pl. 2: 639. 1753. ≡ *Blepharis boerhaviifolia* Pers., Syn. Pl. 2: 180. 1806, nom. superfl. illeg. (Art. 52); Nees in de Candolle (ed.), Prodr. 11: 266. 1847; Clarke in Hooker (ed.), Fl. Brit. India 4: 478. 1884. ≡ *Blepharis boerhaviifolia* var. *maderaspatensis* (L.) Nees, Pl. Asiat. Rar. 3: 97. 1832; Nees in de Candolle (ed.), Prodr. 11: 267. 1847. Type: India, *Herb. Linn. 816.10* (neotype designated by Nasir & Ali in Nasir & Ali (eds.), Fl. Pakistan 188: 7. 1988: LINN).

Herbs perennial, 20–30 cm high; stems trailing or scrambling to decumbent, scabrous or pubescent. Leaves usually 4 per pseudo-whorl of 2 unequal pairs; petiole 1–2 mm long, puberulent; blade elliptic to obovate-lanceolate, 2.5–5 cm long, 1–2 cm broad, puberulent, lateral veins 3–5 pairs, base cuneate, margin undulate or dentate, apex acute. Flowers axillary, solitary or in spikes of few-flowers. Bracts broadly spatulate to obovate-oblong, 4–13 mm long, 2–6 mm broad, unequal, inner pair larger than outer pair, margin bristly ciliate. Calyx puberulent, lobes unequal, posterior and anterior lobes oblong, 1.2–1.8 cm long, 0.4–0.6 cm broad, 3-veined, anterior one often 2-cleft at apex, lateral ones linear-lanceolate, 9–12 mm long, 2–3 mm broad, margin ciliate. Corolla whitish, often with pink or purple, lower lip obovate, 1.6–1.7 cm long, 6–8 mm broad, apex 3-lobed; lobes ovate-oblong. Filaments 4–5 mm long, glabrous; anthers oblong, 2.5–3 mm long. Style ca. 1.2 cm long. Capsules ellipsoid, 6–8 mm long, glabrous. Seeds orbicular, compressed, 2.5–3 mm long, 2–2.5 mm broad, echinate. **Chromosomes** 2n = 26. Figure 46-20.

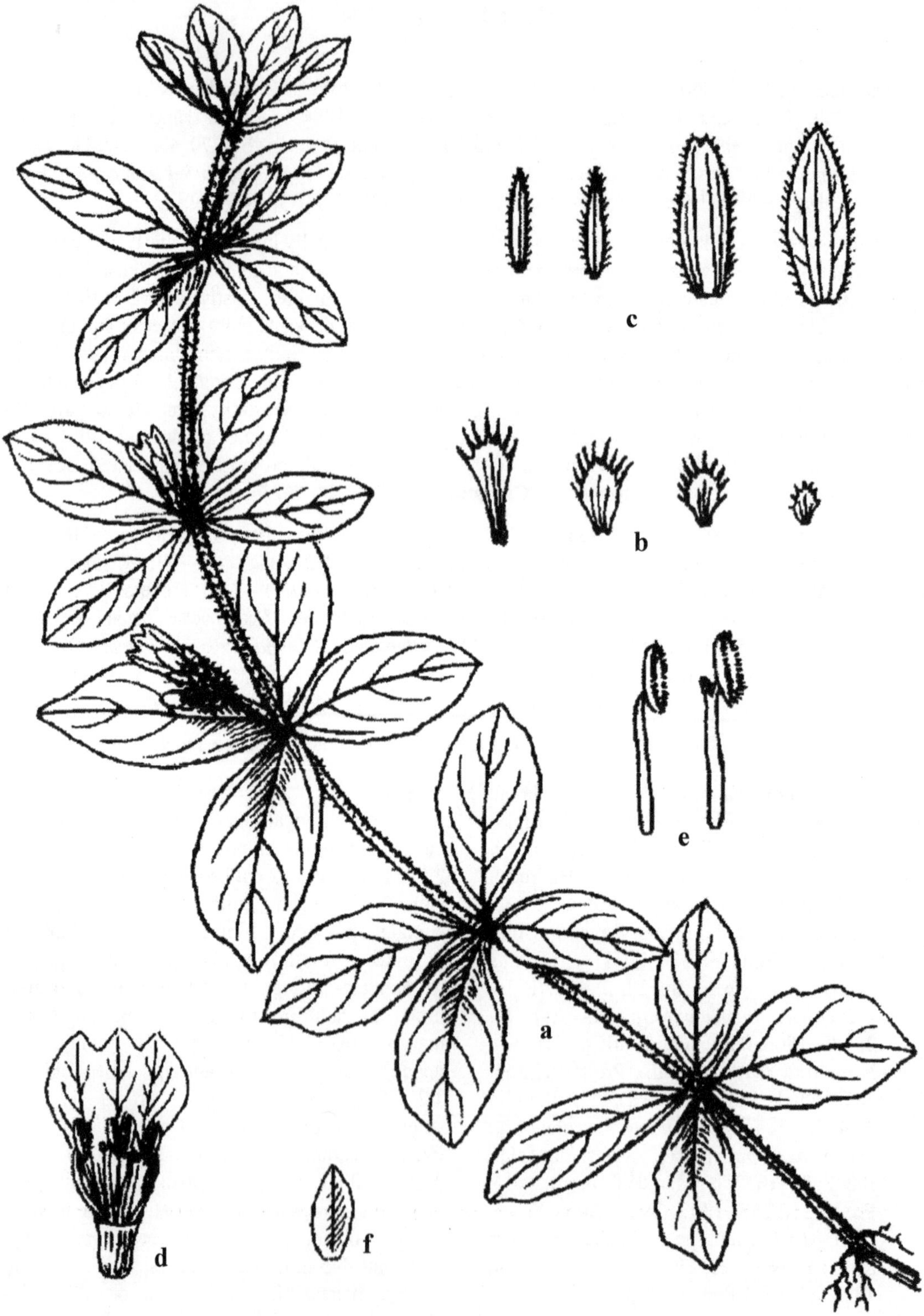

Figure 46-20: **Blepharis maderaspatensis** (L.) B. Heyne ex Roth: **a**, habit; **b**, bracts; **c**, calyx-lobes; **d**, corolla; **e**, stamens; **f**, capsule. Drawn by Han-Ping YU.

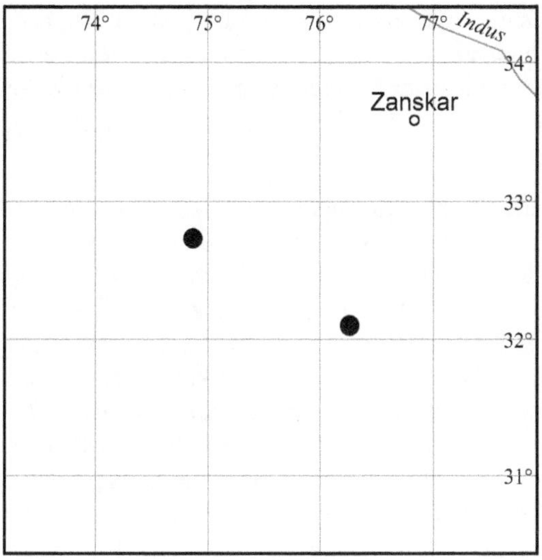

Map 46-33: **Blepharis maderaspatensis** (L.) B. Heyne ex Roth

Phenology: flowering from September to January.

Habitats: limestone cliffs or rocks.

Distribution: Jammu & Kashmir, U Ganga & Indus (Map 46-33); also in Bangladesh, Cambodia, Sri Lanka, Thailand, Vietnam, tropical and subtropical Africa, SW Asia, Madagascar, and other parts of Pakistan, India, Myanmar and China (Hainan).

JAMMU & KASHMIR: Jammu, Ram Nagar, 1969, *B. M. Sharma 152* (RRLH). **U GANGA & INDUS: Kangra**, Una to Dhundala, 1989, *V. Kumar 1254* (BSD, DD, fide Kumar & Gupta, 1990).

Tribe 2. Barlerieae Nees in Martius, Fl. Bras. 9: 7, 65. 1847. Type: *Barleria* L.

Plant with cystoliths. Leaves opposite. Calyx unequal, 4-lobed, outer 2 lobes larger, anterior lobe sometimes 2-cleft, inner 2 lobes smaller. Corolla quincuncial, funnel-shaped or bilabiate. Stamens 4, inserted at base of throat, didynamous, shorter pair usually sterile; anther bithecous or one pair monothecous. Capsules 2- or 4-seeded.

Nine genera and about 150 species in the tropical regions of the world; two genera and nine (one endemic) species in the Pan-Himalaya.

6. BARLERIA L.

Linnaeus, Sp. Pl. 2: 636. 1753; Nees in de Candolle (ed.), Prodr. 11: 222. 1847; Bentham & Hooker, Gen. Pl. 2: 1091. 1876; Clarke in Hooker (ed.), Fl. Brit. India 4: 482. 1884; Lindau in Engler & Prantl (eds.), Pflanzenfam. 4(3b): 313. 1895; Nasir & Ali in Malik & Ghafoor (eds.), Fl. Pakistan 188: 44. 1988; Wood in Grierson & Long (eds.), Fl. Bhutan 2(3): 1280. 2001; Tsui in Hu (ed.), Fl. Reipubl. Popularis Sin. 70: 62. 2002; Hu, Deng & Daniel in Wu, Raven & Hong (eds.), Fl. China 19: 468. 2011. Type (designated by Green, Prop. Brit. Bot.: 169. 1929): *Barleria cristata* L.

Barleriacanthus Oerst., Vidensk. Meddel. Dansk Naturhist. Foren. Kjobenhavn: 24. 1854. Type (designated by Balkwill & Balkwill, Kew Bull. 52: 555. 1997): *Barleriacanthus acanthoides* (Vahl) Oerst. ≡ *Barleria acanthoides* Vahl.

Prionitis Oerst., Vidensk. Meddel. Dansk Naturhist. Foren. Kjobenhavn 24: 137. 1854, nom. illeg., non *Prionitis* Adans., Fam. 2: 499, 594. 1763 (nom. rej.), nec *Prionitis* J. G. Agardh, Sp. Alg. 2: 185. 1851 (nom. cons.). ≡ *Barleria* subg. *Prionites* C. B. Clarke in Hooker (ed.), Fl. Brit. India 4: 482. 1884. Type: *Prionitis hystrix* Miq. (≡ *Barleria prionitis* L.).

Shrubs, subshrubs or perennial herbs; with cystoliths. Leaves opposite, petiolate. Inflorescences axillary cymes, sometimes in dense clusters or terminal spikes, or rarely flowers solitary. Bracts present or absent; bracteoles 2, sometimes spinose. Calyx deeply 4-lobed, lobes heteromorphic, outer 2 lobes larger, anterior one sometimes 2-cleft, inner 2 lobes smaller. Corolla funnel-shaped; limb 2-lipped, upper lip 4-lobed, lower lip 1-lobed; lobes subequal, quincuncial in bud. Stamens 4, didynamous, or 2; anthers bithecous, thecae subequal, base muticous; staminodes 1 or 3. Ovary with 2 ovules per locule; stigma 2-cleft or entire. Capsules subellipsoid to ovoid, 2–4-seeded; retinacula present. Seeds discoid, pubescent with appressed hygroscopic trichomes. **Chromosomes** 2n = 32, 38, 40.

Recent references: Cui, 1990. Materiae ad floram Acanthaceae sinicarum I. *Acta Bot. Yunnan.* 12: 269–278; Balkwill & Balkwill, 1996. Problems with generic delimitation and subdivision in a large genus, *Barleria* (Acanthaceae). pp. 393–408. In: Maesen et al. (eds.), *The Biodiversity of African Plants: Proceedings XIVth AETFAT Congress, 22–27 Aug. 1994, Wageningen, The Netherlands.* Dordrecht: Kluwer Academic; Balkwill & Balkwill, 1997. Delimitation and infrageneric classification of *Barleria* (Acanthaceae). *Kew Bulletin* 52: 535–573; Balkwill & Balkwill, 1998. A preliminary analysis of distribution pattern in a large, pantropical genus, *Barleria* L. (Acanthaceae). *J. Biogeogr.* 25: 95–110; Shendage & Yadav, 2010. Revision of the genus *Barleria* (Acanthaceae) in India. *Rheedea* 20: 81–130; Darbyshire et al., 2019. Phylogenetic relationships among species of Barleria (Acanthaceae, Lamiales): Molecular data reveal complex patterns of morphological evolution and support a revised classification. *Taxon* 68: 92–111.

80–120 species: tropical regions of Africa and Asia; five species (one endemic) in the Pan-Himalaya.

1a. Corolla golden yellow; capsules 2-seeded, with a hollow apical beak (subg. I. **Prionitis**) .... 1. *B. prionitis*
1b. Corolla white or blue; capsules 4-seeded, without an apical beak (subg. II. **Barleria**).
  2a. Stems and branches with sharp spines in or near leaf axils; corolla cream or white .... 2. *B. acanthoides*
  2b. Stems and branches without spines; corolla purplish-blue.
    3a. Bracteoles elliptic, 1–1.2 cm long; calyx purple .... 3. *B. strigosa*
    3b. Bracteoles linear to linear-lanceolate, 2.4–6.5 cm long; calyx green.
      4a. Leaf blade elliptic to ovate, 5.5–15 cm long, 2.2–5.5 cm broad; margin of calyx lobes serrate or spiny .... 4. *B. cristata*
      4b. Leaf blade lanceolate, 3–4.5 cm long, 0.7–1.8 cm broad; margin of calyx lobes entire .... 5. *B. integrisepala*

I. **Barleria** subg. **Prionitis** (Nees) C. B. Clarke in Hooker (ed.), Fl. Brit. India 4: 482. 1884. ≡ *Prionitis* Oerst., Vidensk. Meddel. Dansk Naturhist. Foren. Kjobenhavn 24: 137. 1854, non *Prionitis* Adans. 1763 (nom. rej.), nec *Prionitis* J. G. Agardh 1851 (nom. cons.). ≡ *Barleria* [unranked] *Prionitis* Nees in de Candolle (ed.), Prodr. 11: 237. 1847. Type: *Prionitis hystrix* Miq. (≡ *Barleria prionitis* L.).

Corolla golden yellow; capsules 2-seeded, with a hollow apical beak.

About 45 species: most diverse in NE and E Africa and Madagascar, few species in Asia; one species in the Pan-Himalaya.

1. Barleria prionitis L., Sp. Pl. 2: 636. 1753; Clarke in Hooker (ed.), Fl. Brit. India 4: 482. 1884; Nasir & Ali in Malik & Ghafoor (eds.), Fl. Pakistan 188: 46. 1988; Wood in Grierson & Long (eds.), Fl. Bhutan 2(3): 1280. 2001; Tsui in Hu (ed.), Fl. Reipubl. Popularis Sin. 70: 66. 2002; Hu, Deng & Daniel in Wu, Raven & Hong (eds.), Fl. China 19: 469. 2011. Type: India, *Herb. Linn. 262.13* (lectotype designated by Wood et al., Kew Bull. 38: 436. 1983: S).

Barleria hystrix L., Mant. 1: 89. 1767. ≡ *Prionitis hystrix* (L.) Miq., Fl. Ind. Bat. 2: 809. 1858.

≡ *Barleria prionitis* L. var. *hystrix* (L.) Kuntze, Revis. Gen. Pl. 2: 483. 1891. Type: India, "Habitats in India", *D. D. Royen s. n.* [holotype: LINN (LINN805.5)].

Shrubs to 2 m tall, with branched spines in lower leaf axils; stems and branches glabrous. Petiole 1–2.5 cm long; leaf blade elliptic to ovate, 3.5–13.5 cm long, 1–5.5 cm broad, both surfaces pubescent or glabrescent, base attenuate and decurrent onto petiole, margin entire, apex acute. Flowers clustered in upper leaf axil. Bracts linear-oblong, 1.2–2.2 cm long, 0.2–0.8 cm broad, margin ciliate, apex abruptly acuminate; bracteoles linear-lanceolate, 1–1.4 cm long, 2–3 mm broad, cuspidate. Outer calyx lobes ovate-oblong, 1.2–1.5 cm long, 0.3–0.4 cm broad, apex mucronate; inner calyx lobes linear-lanceolate, 1.1–1.3 cm long, 0.2–0.25 cm broad, apex mucronate. Corolla yellow to orange, 2.4–4 cm long, outside pilose; tube with narrow basal portion slightly shorter than throat; lobes ovate-oblong to orbicular, 8–10 mm long, 6–7 mm broad, recurved. Stamens 4; filaments of longer pair 0.8–1.1 cm long, shorter pair 1.5–2 mm long; anthers of longer pair 3–3.2 mm long, anthers of shorter pair 1–1.2 mm long. Ovary ovoid; stigma slightly enlarged, 2-cleft. Capsules ovoid, 1.6–1.8 cm long, contracted gradually towards apex forming a beak, 2-seeded. Seeds ovate in outline, compressed, 5–7 mm long, 4–5 mm broad. **Chromosomes** $2n = 40$.

Phenology: flowering from October to December; fruiting from December to February.

Habitats: roadsides, thickets, dry places in evergreen broad-leaved forests; below 600 m.

Distribution: N Pakistan, Jammu & Kashmir, U Ganga & Indus, W Nepal (Map 46-34); also in Sri Lanka, Thailand, Laos, Vietnam, tropical Africa, Madagascar, and other parts of India, Myanmar and China (S Yunnan, Guangxi, Guangdong, Hainan).

Uses: medicine.

N PAKISTAN: Mirpur, Bhimbar, 7 Jan. 1958, *J. Mohammad 14* (RAW, fide Malik & Ghafoor, 1988). **JAMMU & KASHMIR: Jammu**, Gummat, 2001, *B. K. Kapahi & S. N. Sharma 17564* (RRLH). **U GANGA & INDUS: Dehradun**, Nov. 1932, *Raizada 87447* (MH). **W NEPAL: Bheri**, Banke, Napelganj, 180 m, 18 Nov.1972, *N. P. Manandhar 9472* (KATH); Napelganj, 180 m, 1979, *T. K. Bhattacharya T23* (KATH); Bardiya, Bandarbharia, 180 m, 19 Feb. 1993, *P. P. Kurmi KB617* (KATH).

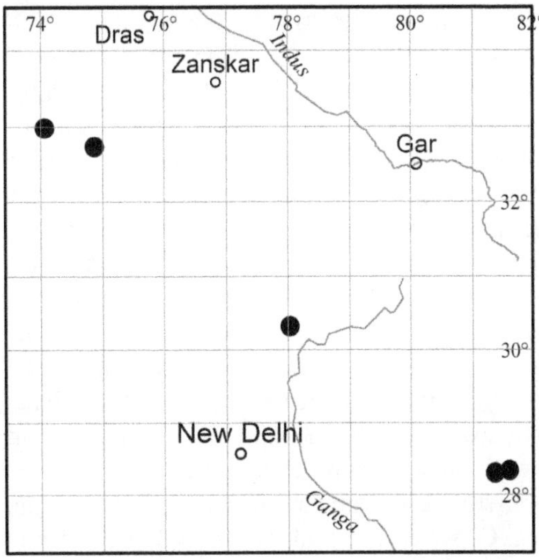

Map 46-34: **Barleria prionitis** L.

II. **Barleria** subg. **Barleria**

Corolla white or blue. Capsules 4-seeded, without an apical beak.

About 100 species in tropical regions of Asia and Africa; four species in the Pan-Himalaya.

2. **Barleria acanthoides** Vahl, Symb. Bot. 1: 47. 1790; Clarke in Hooker (ed.), Fl. Brit. India 4: 484. 1884; Nasir & Ali in Malik & Ghafoor (eds.), Fl. Pakistan 188: 45. 1988. ≡ *Barleriacanthus acanthoides* (Vahl) Oerst., Vidensk. Meddel. Dansk Naturhist. Foren. Kjøbenhavn: 136. 1855. Type: Yemen, *P. Forsskal s. n.* [lectotype designated by Darbyshire et al., Fl. Trop. E. Afr. Acanth. 2: 349. 2010: C (C10001836)].

Spiny shrubs, 20–60 cm high, sparsely to much branched; stems densely greyish-pubescent, armed with simple spins in leaf axils, spines paired, 5–6 mm long, margin pinnately spinose. Petiole 4–7 mm long, hairy; leaf blade oblong-ovate to obovate, 1.2–5(–7.5) cm long, 0.6–2 cm broad, pubescent on both surfaces, base attenuate orcuneate, margin entire, apex obtusely mucronate with spine for 1–1.5 mm long. Flowers solitary, or often in Cymes axillary, scorpioid, 1–4.5 cm long, 3–11-flowered. Bracts elliptic, 1.2–1.5(–2) cm long, membranous, pinnatifid-spinescent; bracteoles linear-lanceolate, 0.7–2.7 cm long, 0.5–2.0 cm broad, margin denticulate to pinnately spinose, 1–4 teeth or spines each side. Outer calyx lobes broadly ovate, 1.2–1.7 cm long, 0.8–1.0 cm broad, yellowish-brown, subequal, apex sharply mucronate; inner lobes lanceolate, 7–8 mm long. Corolla white or cream, 6.5–10 cm long; tube cylindric 5–6(–8) cm long; lobes ovate, 0.6–1.0 cm long, apex acute. Stamens 2; filaments 3–4.5 cm long; anthers bithecous, thecae 3–4 mm long; staminodes 3, 0.8–2.3 cm long. Ovary villous towards apex, style minutely crisped hairy toward base; stigma subcapitate to clavate, 0.5–0.8 mm long. Capsules ellipsoid, 1–1.4 cm long, 4–5 mm broad, attenuate at both ends, villous towards apex, 4-seeded. Seeds ovate in outline, compressed, 3–4 mm in diameter, silky hairy. **Chromosomes** 2n = 40.

Phenology: flowering and fruiting from October to March.

Habitats: rocky places and dry hillslopes.

Distribution: N Pakistan (Map 46-35); also in Egypt, Sudan, Ethiopia, Somalia, Djibouti, and other parts of Pakistan and India.

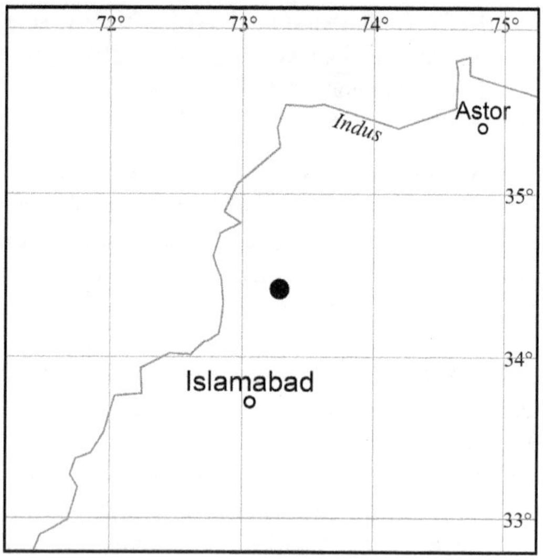

Map 46-35: ***Barleria acanthoides*** Vahl.

N PAKISTAN: Hazara, Naran Kaghan valley, *S. I. Ali 951* (KUH, fide Malik & Ghafoor, 1988).

3. **Barleria strigosa** Willd., Sp. Pl. 3: 379. 1800; Clarke in Hooker (ed.), Fl. Brit. India 4: 489. 1884; Wood in Grierson & Long (eds.), Fl. Bhutan 2(3): 1281. 2001; Hu, Deng & Daniel in Wu, Raven & Hong (eds.), Fl. China 19: 469. 2011. Type: India, 13 Dec. 1798, *Hort. Heyniana 707/ Hot. Miss. 514* [holotype: B (B-W 11664-101)].

Barleria caerulea Roxb., Fl. Ind. 3: 39. 1832; Nees in Wallich (ed.), Pl. Asiat. Rar. 3: 91. 1832; Nees in de Candolle (ed.), Prodr. 11: 226. 1847; Anderson, J. Linn. Soc., Bot. 9: 491. 1867. ≡ *Pseudobarleria coerulea* (Roxb.) Oerst., Vidensk. Meddel. Naturhist. Foren. Kjøbenhavn 135. 1854. Type: India, s. d., *Wallich Cat. 2509a* [lectotype **here designated**: K-W (K001116276)].

Barleria strigosa Willd. var. *subglabra* Kuntze, Revis. Gen. Pl. 1(2): 483. 1891. Type: India, s. d., *Herb. Kuntze s. n.* (holotype: NY).

Subshrubs, 30–80 cm high; stems coarsely fulvous strigose. Petiole 0.7–2.5 cm long, brownish yellowish strigose; leaf blade elliptic to ovate, 5.5–15 cm long, 2.2–5.5 cm broad, both surfaces fulvous strigose especially along veins, lateral veins 6–8 pairs, base cuneate and decurrent onto petiole, apex acute. Spikes axillary and termina, 3–5 cm long. Bracts oblong to elliptic-oblong, abaxially shortly strigose along veins, apex sharply acute; bracteoles elliptic, 1–1.2 cm long, 1-veined, margin ciliate, apex acute. Outer calyx lobes purple, ovate, 3–3.2 cm long, 1.6–1.8 cm broad, base rounded, margin subentire to dentate, apex acute; inner calyx lobes yellowish brown, lanceolate, 1.2–1.5 cm long, 2–2.5 mm broad, strigose, apex acute. Corolla purplish red, 5–6 cm long; tube base cylindric, 3.2–3.7 cm long, throat 1–1.2 cm long; lobes obovate-oblong, 1.6–2 cm long, 1–1.5 cm broad, subequal. Stamens 4; filaments basally pilose, longer pair 1–1.3 cm long, shorter pair 3–4 mm long; anthers of longer pair 5–6 mm long, anthers of shorter pair 1.5–2 mm; staminode 1. Ovary ovoid, 2–3 mm long; style 2.2–2.5 cm long. Capsules ellipsoid, 1.4–1.8 cm long, 0.4–0.5 cm broad, glabrous, shining, 4-seeded. Seed ovate in outline, 5–7 mm long and broad. **Chromosomes** 2n = 40.

Phenology: flowering from October to November; fruiting from December to February.

Habitats: forests; below 1400 m.

Distribution: U Ganga & Indus, W, C and E Nepal, Bhutan, Sikkim & Darjiling (Map 46-36); also in Cambodia, Indonesia, Laos, Malaysia, Sri Lanka, Thailand, Vietnam, and other parts of India, Myanmar, China (S Yunnan).

U GANGA & INDUS: Chamba, 12 Jan. 1890, *J. F. Duthie 9650* (LE); **Kulu,** Dec. 1852, *Charles Ritchie 582* (E). **W NEPAL: Rapti**, Dang, Rapi Nadi, Rapti Valley, 300 m, 27 Nov. 1952, *O. V. Polunin, W. R. Sykes&L. H. J. Williams 5887* (BM, E); **Seti**, Kailali, Godavari Kailali, above Godavari, 820 m, 11 Nov. 2012, *B. Adhikari & G. D. Bhatt 48* (E). **C NEPAL: Lumbini**, Nawalparasi, Daunne, 420 m, 8 Nov. 2012, *B. Adhikari & G. D. Bhatt 23* (E); Rupandehi, Alarch, near Butwal, 610 m, 6 Oct. 1954, *J. D. A. Stainton, W. R. Sykes & L. H. J. Williams 8768* (BM); **Narayani**, Bara, Lower Nijgad, near Juwa, 140 m, 1993, *L. P. Kattel 125/93* (KATH); Bara, Tamagadhi Forests, 1993, *L. P. Kattel 267/93* (KATH); Chitwan, Ram Naga, 500 m, June 1994, *K. Wesche 5141* (BM); Chitwan, Ram Nagar, 300 m, 1995, *K. Wesche 7186* (BM); Chitwan, Tungati, 380 m, 15 Nov. 1989, *N. P. Manandhar 13855* (KATH). **E NEPAL: Koshi**, Dhankuta, Dhara Pani to Tambur Bridge to Dhankuta, 17 Oct. 1963, *H. Hara et al. 6303715* (KYO, TI); Sankhuwasabha, Chirkhua Khola to Tumlingtar, 270 m, 12 Nov. 1981, *P. R. Shakya 7414* (KATH); Sunsari, Phusri-Sanguri Bhanjang-Dhara Pani, 16 Oct. 1963, *H. Hara et al. 6303643* (TI); **Mechi**, Ilam, Garuwa to Sanichare, 10 Dec. 1963, *H. Hara et al. 6303644* (BM, KATH, TI); Ilam, Raga

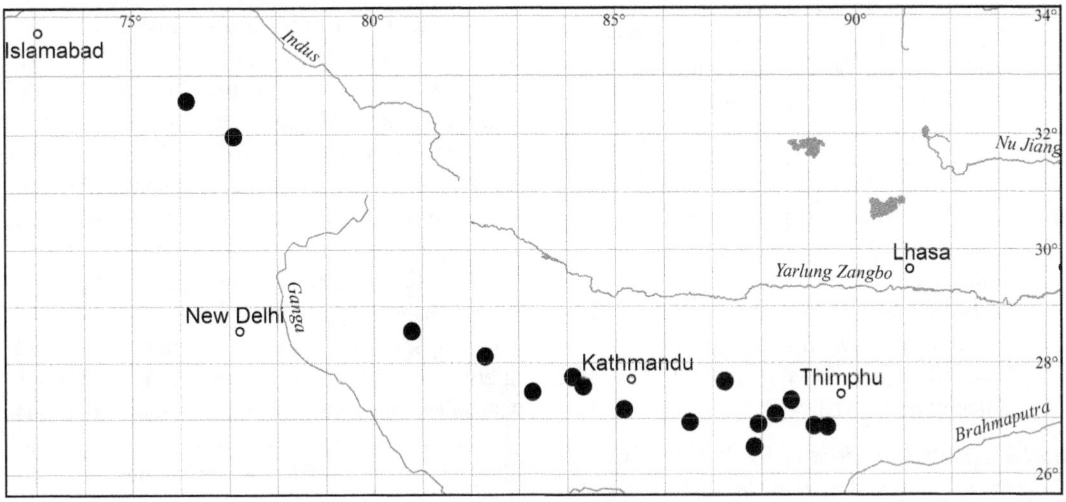

Map 46-36: **Barleria strigosa** Willd.

Pani to Ghorwa, 9 Dec. 1963, *H. Hara et al. 6303645* (A, E, K, KATH, KYO, TI); Jhapa, Jal-Thal Forests, near Prithwinagar, 100 m, 26 Jan. 2003, *N. Thapa et al. 2073* (KATH); Jhapa, Lenti Khola, 500 m, 1977, *P. R. Shakya 4651* (KATH); **Sagarmatha**, Khotang, Regmitar to Khani Danda, 360 m, 28 Oct. 1995, *M. Mikage et al. 9554185* (TI); Udayapur, 310 m, 24 Oct. 1995, *M. Mikage et al. 9554100* (E, TI); Udayapur, Udaipur Garhi, Panchawati, 610 m, 3 Nov. 1969, *J. D. A. Stainton 6640* (BM, KATH, TI); Udayapur, Balaute, 400 m, 6 Nov. 1954, *A. Zimmermann 2075A* (E). **BHUTAN:** Onduphodung, 3 May 1906, *J. C. White 107* (E); **Phuntsholing**, Torsa River, 200 m, 2 May 1979, *A. J. C. Grierson & D. G. Long 793* (E); **Samchi**, Khagra valley near Gokti, 420 m, 2 Mar. 1982, *A. J. C. Grierson & D. G. Long 3382* (A, K). **SIKKIM & DARJILING: Darjiling**, Punkabari, 300 m, 4 Nov. 1869, *C. B. Clarke 10356* (BM); Chonbultee, 600 m, 1 Dec. 1875, *C. B. Clarke 26488* (K); **Sikkim**, Popina, 1200 m, *Herbier de Ch. d'Alleizette s. n.* (L); 600–1200 m, *J. D. Hooker s. n.* (A, BM, CGE, K); Punkabari, 300 m, *C. B. Clarke 13836* (K, LE); Teesta, 150 m, *C. B. Clarke 13933* (LE); Teesta valley, 12 Oct. 1908, *Smith 430* (CAL).

4. Barleria cristata L., Sp. Pl. 2: 636. 1753; Nees in de Candolle (ed.), Prodr. 11: 229. 1847; Clarke in Hooker (ed.), Fl. Brit. India 4: 489. 1884; Nasir & Ali in Malik & Ghafoor (eds.), Fl. Pakistan 188: 50. 1988; Wood in Grierson & Long (eds.), Fl. Bhutan 2(3): 1281. 2001; Tsui in Hu (ed.), Fl. Reipubl. Popularis Sin. 70: 63. 2002; Hu, Deng & Daniel in Wu, Raven & Hong (eds.), Fl. China 19: 468. 2011. Type: India, *Herb. Linn. 805.12* (designated by Brummitt & Vollesen, Taxon 41: 558. 1992: LINN).

Barleria ciliata Roxb., [Hort. Bengal.: 45. 1814, nom. invalid.]; Fl. Ind. 3: 38. 1832; Nees in Wallich (ed.), Pl. Asiat. Rar. 3:92. 1832; Nees in de Candolle (ed.), Prodr. 11: 228. 1847. Type: India, *Wallich Cat. 2507a* [lectotype **here designated**: K-W (K001116259); isolectotype: CAL].

Barleria dichotoma Roxb., Fl. Ind. (ed. Carey) 3: 39. 1832; Nees in Wallich (ed.), Pl. Asiat. Rar. 3: 91. 1832. ≡ *Barleria cristata* var. *dichotoma* (Roxb.) Prain, Bengal Pl.: 812. 1903. Type: India, *Herb. Roxb. s. n.* [lectotype **here designated**: BR (BR000694312); isotypes: BR, LIV, OXF].

Barleria napalensis Nees, Pl. Asiat. Rar. 3: 91. 1832; Nees in de Candolle (ed.), Prodr. 11: 228. 1847. Type: [C] Nepal, 1821, *Wallich Cat. 2508E* [lectotype **here designated**: K-W (K001116269); isotypes: CAL, CGE].

Barleria laciniata Nees, Pl. Asiat. Rar. 3: 91. 1832. Type: Myanmar, Tong Dong, 1826, *Wallich Cat. 7144* [lectotype **here designated**: K-W (K001126780); isolectotype: CAL].

Barleria mairei H. Lév., Repert. Spec. Nov. Regni Veg. 12: 285. 1913. ≡ *Barleria cristata* var. *mairei* (H. Lév.) H. P. Tsui, Fl. Reipubl. Popularis Sin. 70: 65. 2002. Type: China, Yunnan, "Rives du fleuve Bleu", 400 m, May 1912, *E. E. Maire s. n.* [holotype: E (E00265795)].

Subshrubs to 2 m high; stems softly pubescent. Petiole (0–)0.3–1 cm long; leaf blade elliptic to oblong to ovate, 5–5.5 cm long, 2.2–5.5 cm broad, both surfaces villous especially along veins, lateral veins 4–7 pairs, base cuneate and decurrent onto petiole, apex acute to sometimes acuminate. Cymes axillary, dense-flowered, shortly pedunculate. Bracts absent; bracteoles linear to linear-lanceolate, 2.4–6.5 cm long, 0.5–1.5 cm broad, 3–7-veined, base cuneate, margin usually spiny, sometimes bristly pilose, apex acuminate. Outer calyx lobes ovate to narrowly elliptic to lanceolate, 1.2–2.5 cm long, 0.5–1.3 cm broad, pilose, margin spiny, apex mucronate; inner calyx lobes linear to lanceolate, 6–12 mm long, 1-veined, margin scarious. Corolla purplish blue, 4.5–6.5 cm long, outside pilose; tube base narrowly cylindric upwards gradually widened; lobes oblong-elliptic, 1.2–1.5 cm long, 1–1.5 cm broad. Stamens 4, didynamous; filaments pilose especially toward base; staminode 1, filament sparsely pilose. Ovary oblong-ellipsoid, glabrous; style 2.2–2.5 cm long, glabrous. Capsules ellipsoid, 1.2–1.8 cm long, glabrous, 4-seeded. Seeds subspherical to ovoid, 4–5 mm long, 3–4 mm broad. **Chromosomes** 2n = 40. Figure 46-21.

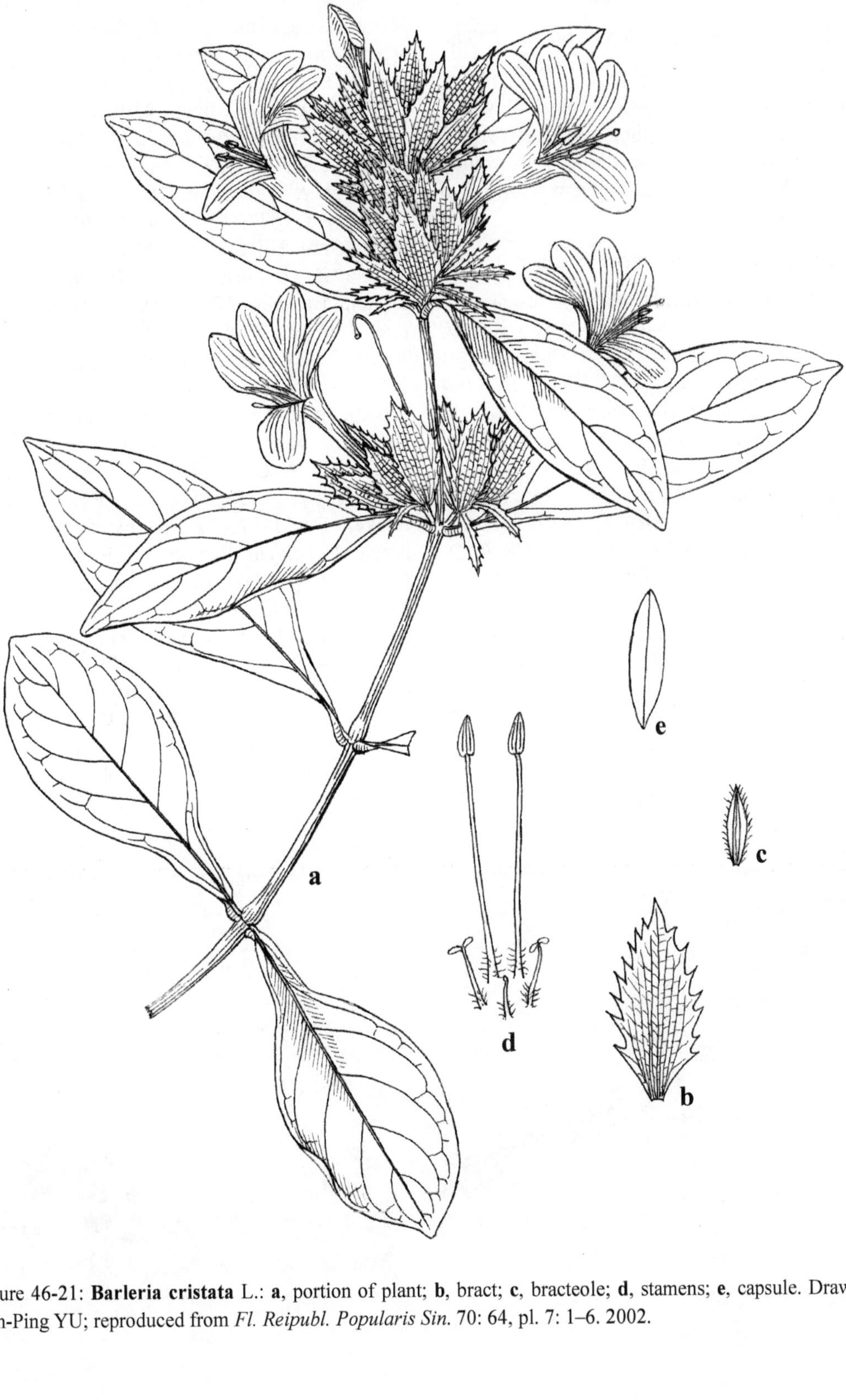

Figure 46-21: **Barleria cristata** L.: **a**, portion of plant; **b**, bract; **c**, bracteole; **d**, stamens; **e**, capsule. Drawn by Han-Ping YU; reproduced from *Fl. Reipubl. Popularis Sin.* 70: 64, pl. 7: 1–6. 2002.

Phenology: flowering from August to December; fruiting from October to January.

Habitats: roadsides, slopes; below 2700 m.

Distribution: N Pakistan, Jammu & Kashmir, U Ganga & Indus, W Nepal, C Nepal, E Nepal, Sikkim & Darjiling, Bhutan, Yarlung Zangbo-Brahmaputra, S Hengduan, U Irrawaddy (Map 46-37); also in tropical and subtropical regions of Asia.

Uses: medicinal and ornamental.

N PAKISTAN: Azad Kashmir, Muzaffarabad, Neelem, Punch River, Neelem valley road, *M. Qaiser & A. Ghafoor 4902* (KUH, fide Nasir & Ali, 1988); **Chitral**, Aug. 1921, *Hasan Din 199* (RAW, fide Malik & Ghafoor, 1988); **Hazara**, Abbotabad, Nathia Gali to Abbotabad, *S. A. Faruqi & M. Qaiser 3426* (KUH, fide Nasir & Ali, 1988); Murree Hills, above Tret, Murree Road, 1998, *s. coll. 4144* (RAW, fide Nasir & Ali, 1988); **Swat**, Swat Valley, 1800 m, 1938, *T. D. Weatherhead 54* (BM); Swat Valley, 1800 m, 28 Jul. 1938, *T. D. Weatherhead 55* (BM); Manglore, 1350 m, 16 Sep. 1976, *Dilawar & Naeem Shah s. n.* (K). **JAMMU & KASHMIR: Jammu**, Vaishno Devi, 400 m, 1977, *T. N. Srivastava & B. K. Kapahi 11171* (RRKH). **U GANGA & INDUS: Chamba**, Rakh to Chamba, *J. H. Lace 1688* (E); **Dehradun**, Nalapani, Mar. 1884, *J. S. Gamble 24598* (CAL); **Kumaon**, Gangoli, 1350 m, *R. Strachey & J. E. Winterbottom 1* (K); **Mussorie**, Sep. 1955, *K. Shamber s. n.* (KYO); **Nainital**, 1800 m, s. d., *R. Strachey & J. E. Winterbottom 4* (K); **Pauri Garhwal**, Garhwal, Above Ghats, 1650 m, 21 Sep. 1885, *J. R. Reid s. n.* (E); Góbeser to Ókimath, 1500–2040 m, 14–16 Sep. 1855, *Tohlagintweil 8872* (A); **Pithoragarh**, Kalapani, Sep. 1891, *J. S. Gamble s. n.* (BSI, fide Shendage & Yadav, 2010); **Simla**, Saiuj Giri Valley, Sep. 1877, *J. S. Gamble 5406* (K); Sutle Valley, Simla Hill States, 1500 m, 20 Jun. 1939, *F. Ludlow & G. Sherriff 7321* (BM); Parada to Rampur, Bashebr, 900–1500 m, 6 Oct. 1881, *J. H. Lace 1091* (E); Sani, Mai, 900 m, Oct. 1887, *G. Watt 8857* (E); **Solan**, Kalka Road, below Kasanli, 1350 m, 5 Nov. 1877, *J. S. Gamble 5746* (K); Kasanli to Kolka, *J. R. Drummond 26433* (E); Kasanli to Kolka, *J. R. Drummond 22684* (K); below Siki, *H. Collett 167* (K); **Tehri Garhwal**, *P. P. Huggins A5* (BM). **W NEPAL: Bheri**, Banke, 7 Dec. 1977, *C. P. Rauniyar 2104* (KATH); Banke, Napelganj, Samsergunj, 9 Oct. 1977, *R. Rana 1897* (KATH); Banke, Napelganj, Kamdi, 1973, *N. P. Manandher & K. J. Malla 17162* (KATH); Bardiya, Guleriya, 180 m, 14 Oct. 1981, *T. K. Bhattachaya 3840* (KATH); Bardiya, near Babai, 19 Dec. 360 m, 1979, *T. K. Bhattachaya 2582* (KATH); Bardiya, Royal Bardia National Park, Tiger top, 3 Oct. 1996, *M. Mikage et al. 9689187* (TI); Dailekh, Gairi, 900 m, 25 Feb. 1991, *N. P. Manandhar 470-91* (KATH); Dailekh,

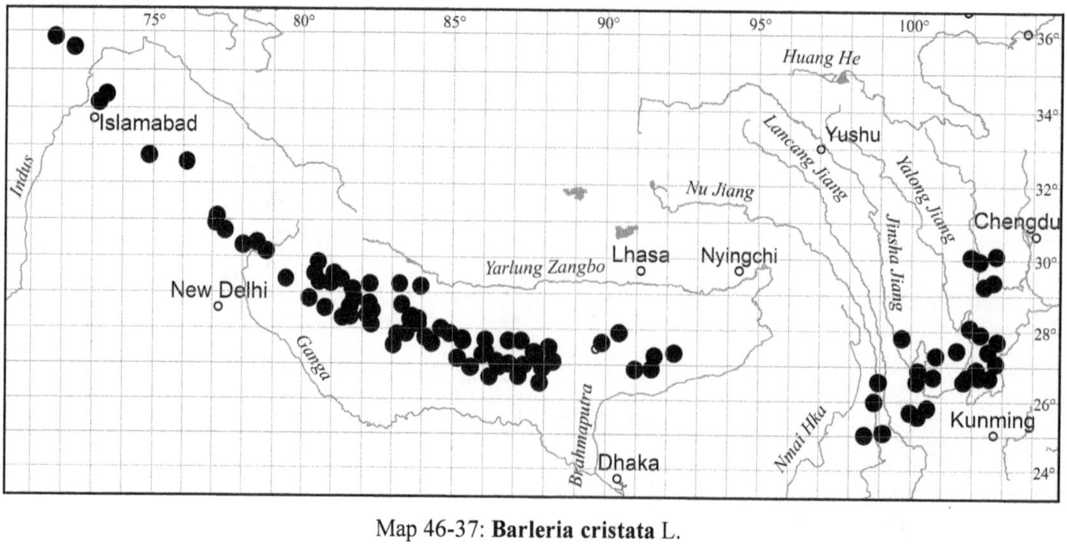

Map 46-37: **Barleria cristata** L.

Solay, 1260 m, 1982, *Shukla & Paudyal 4107* (KATH); Dailekh, Dandamdi to Panipokjari, 1300 m, 3 Aug. 1991, *M. Suzuki et al. 9192139* (TI); Jajarkot, 1060 m, 17 Oct. 1952, *O. V. Polunin, W. R. Sykes & L. H. J. Williams 5803* (BM); Jajarkot, Pali, 800 m, 19 Apr. 1974, *J. F. Dobremez 2617* (*N. P. Manandhar 74-246*) (E, KATH); Surkhet, Surkhet to Ratanangla, 1380 m, 6 Sep. 2014, *B. Adhikari & R. Basnet 44* (E); **Karnali**, Dolpa, Namdu, 1300 m, 25 Apr. 1963, *K. Yoda s. n.* (KYO); Dolpa, below Lunh, Bheri River, 1670 m, 8 Sep. 1952, *O. V. Polunin, W. R. Sykes & L. H. J. Williams 3284* (BM); Kalikot, near Ramni, Tila Valley, 1820 m, 19 Apr. 1952, *O. V. Polunin, W. R. Sykes & L. H. J. Williams 1912* (BM); Kalikot, Lapha, Karnali Valley, 1200 m, 27 Apr. 1952, *O. V. Polunin, W. R. Sykes & L. H. J. Williams 3994* (BM); Kalikot, Mugraha, Padma Ghat, 1520 m, 1985, *P. R. Shakya et al. 8463* (KATH); Kalikot, Padma, 1430 m, 1982, *N. P. Manandher & N. K. Bhattarai 9113* (KATH); Jumla, Above Doti-Silgarhi, 1350 m, *M. Ecker-Racz 26* (LE); Jumla, Ratopani, 22 Aug. 1976, *H. Tabata et al. 2786* (KYO); **Mahakali**, Baitadi, 1000 m, 21 Sep. 1981, *I. Sharma et al. 641* (KATH); Baitadi, 1500 m, 22 Sep. 1981, *I. Sharma et al. 655* (KATH); Dadeldhura, 1730 m, 1981, *I Sharma et al. 537* (KATH); Darchuala, Paribagad to Makarigad, 1050 m, 2 Sep. 1982, *M. M. Amatya & P. M. Regmi 651/82* (KATH); Darchuala, Darchula to Dalilekh, 840–1810 m, 1982, *M. M. Amatya & D.M. Regmi W536/82* (KATH); Darchuala, Darchula to Huti, 1000 m, 24 Aug. 1980, *K. R. Rajbhandari & K. J. Malla 5528* (KATH); Kanchanpur, Royal Suklaphanta Wildlife Reserve, Majhgaon, 1996, *M. Mikage 9688249* (KATH); Kanchanpur, Royal Suklaphanta Wildlife Reserve, Majhgaon, 1996, *M. Mikage 9688254* (KATH); **Rapti**, Dang, Kwera Pani, 10 Mar. 1976, *N. P. Manandhar & P. M. Regmi 288* (KATH); Dang, Ghorahi, 640 m, 16 Apr. 1985, *P. M. Regmi et al. 4336* (KATH); Dang, Koilabas, Darbaja, 800 m, 1974, *K. J. Malla 9281* (KATH); Rolpa, Bainja, 1430 m, 1 Sep. 1982, *N. P. Manandhar & M. K. Bhattarai 8660* (KATH); Salyan, near Phulgaon, 1650 m, 25 Aug. 1952, *O. V. Polunin, W. R. Sykes & L. H. J. Williams 589* (BM); **Seti**, Bajhang, between Chainpur and Rishini, 20 Jul. 1976, *H. Tabata et al. 1406* (KYO); Bajhang, Suni Gad, 1400 m, 1973, *J. F. Dobremez 2165* (BM, E); Bajhang, Kinara, 24 Aug. 1991, *M. Suzuki et al. 9160982* (A, E, TI); Bajura, Basala, 15 Aug. 1976, *H. Tabata et al. 2487* (KYO); Bajura, Tolebhir to Berma, 9 Aug. 1991, *M. Suzuki et al. 9160425* (TI); Bajura, Baudi Khola, S of Kunu, 1770 m, 21 Aug. 2017, *s. coll. B175* (E); Doti, Silgari, 21 Apr. 1929, *Bis Ram 292* (A, BM); Doti, near Palebas, 1820 m, 21 Sep. 1982, *L. R. Sharma & P. P. Kurmi 4811* (KATH). **C NEPAL: Bagmati**, Dhading, Chyangli, 1100 m, 10 Nov. 1989, *N. P. Manandhar 13290* (KATH); Dhading, 2270 m, 1928, *L. Dhwoj 238* (E); Kathmandu, Sunderjal, 20 Sep. 1963, *H. Hara et al. 6303639* (TI); Kathmandu, Kathmandu Airport, 13 Oct. 1963, *H. Hara et al. 6303640* (TI); Kathmandu, Gokarna, 1330 m, 10 Sep. 1966, *P. Pradhan & R. Thapa 6464* (KATH); Kathmandu, Sundarijal Waterfall, 1600 m, 20 Sep. 1963, *H. Hara et al. 6303651* (TI); Kavre Palanchok, Bokse Community Forests, Panchkhal, 2001, *S. R. Baral 1055* (KATH); Kavre Palanchok, Dumbreghari Leasehold Forests, Baluwa, 2001, *S. R. Baral & P. P. Kurmi 1101* (KATH); Kavre Palanchok, en Montant a Chyauban, 1070 m, 1954, *A. Zimmermann 1070* (K); Lalitpur, Godavari, 1590 m, *A. D. Schiling 1056* (K, LE); Lalitpur, Godavari, 1500 m, 1 Sep. 1966, *D. H. Nicolson 2194* (BM, KATH); Lalitpur, Godavari, 1500 m, 8 Sep. 1970, *C. Chuma 674765* (KATH, KYO, TI); Nuwakot, Labshi, 180 m, 16 Sep. 1927, *C. Wigram 134* (E); Nuwakot, Trisuli, 1971, *J. F. Dobremez 1140* (E); Rasuwa, Chalis, 1060 m, 1935, *F. M. Baikey s. n.* (BM); Rasuwa, Langtang Valley, Gangja La to Palchock Danda, 18 Aug. 1670 m, 10-17 Sep. 1949, *O. V. O. V. Polunin 1940* (A, BM); Sindhupalchok, Khosori Khabre–Barabise, 1000 m, *H. Kanai et al. 674764* (TI); **Dhawalagiri**, Baglung, Daplang, 850 m, 1991, *N. P. Manandher 883-31* (KATH); Baglung, Haliya, 1100 m, 1992, *N. P. Manandhar & S. K. Achaya 610-92* (KATH); Mustang, Kali Gandaki Valley, S of Gharsa, 2000 m, 1973, *Geyhilson & Phillips 865B* (K); Mustang, Kali Gandaki, 760 m, 1 Nov. 1954, *J. D. A. Stainton, W. R. Sykes & L. H. J. Williams 9223* (BM);

Myagdi, Beni, 2 Sep. 1996, *M. Mikage et al. 9684022* (A, BM, TI); Myagdi, Dharapani, 1450 m, *M. Mikage et al. 9684306* (A, BM, TI); Myagdi, Babiyachaur, 960–1550 m, 3 Sep. 1996, *M. Mikage et al. 9685092* (A, BM, TI); Myagdi, Takam, near Takumsibang, 1520 m, 30 Aug. 1954, *J. D. A. Stainton, W. R. Sykes & L. H. J. Williams 4078* (BM); Parbat, Kusma, Kali Gandaki River, 910 m, *J. D. A. Stainton, W. R. Sykes & L. H. J. Williams 6981* (A, BM); **Gandaki**, Gorkha, Darondi Khola, S of Himal Chola, 910 m, 14 Oct. 1967, *J. D. A. Stainton 6052* (BM); Gorkha, Jogat to Hosdobar, 1400 m, 30 Jul. 1953, *S. Nakao s. n.* (KYO); Gorkha, Ganesh Himal, Apr.–May 1975, *B. Yori 443* (E); Kaski, Nayapur, 960–1460 m, 1995, *M. Mikage et al. 9552102* (A, TI); Kaski, Lamjung Himal, between Simalshapur and Midam Khola river, 1969, *T. Wraber 36654* (BM); Kaski, Kuli Khani, 1450 m, 2 Sep. 1970, *H. Kanai & T. B. Shresta 672702* (KATH, TI); **Janakpur**, Dolakha, between Chedi and Kirantichap, 1280 m, 1975, *Schiling 2116* (K); Dolakha, Tamba Kosi, Pikhuti to Totlabari, 28 Aug. 1983, *H. Ohba et al. 8331571* (BM, KYO, TI); Dolakha, Khare Khola, Khanigaon to Suri Dhoban, 15 Sep. 1983, *H. Ohba 8332136* (BM, E, TI); Dolakha, Bush to Yarea, 860 m, 20 Apr. 1978, *I. Sharma et al. 78/195* (KATH); Dolakha, Bhorle, 1000 m, 1979, *N. P. Manandher & M. R. Adhikera 1427* (KATH); Ramechhap, 1985, *H. Ohba et al. 8571745* (A, E, KYO, TI); Sarlahi, Sahejpur, 20 Jun. 1976, *H. Tabata et al. 607* (KATH); **Sindhuli**, Bhandar to Sete, 18 Aug. 1985, *H. Ohba et al. 8530903* (BM, KATH, TI); **Lumbini**, Arghakhachi, Dakakhola, Sandhikharka, 1330 m, 15 Jul. 2016, *B. Adhikari & G. Parmar BG57* (E); Kapilvastu, between Chandrauta and Bhalubang, 270 m, 9 Nov. 2012, *B. Adhikari & G. D. Bhatt 26* (E); Nawalparasi, Daunne, 270 m, 8 Nov. 2012, *B. Adhikari & G. D. Bhatt 20* (E); Nawalparasi, Daunne, 420 m, 8 Nov. 2012, *B. Adhikari & G. D. Bhatt 23* (E); Palpa, Marek, near Butral, 600 m, 6 Oct. 1954, *J. D. A. Stainton, W. R. Sykes & L. H. J. Williams 8812* (BM); Palpa, Shreepus Danda, 1500 m, 1975, *D. P. Joshi et al. 75/3662* (KATH); **Narayani**, Bara, Badahikim Bridge, Hetauda to Birgunj, 480 m, 2 Nov. 2004, *C. A. Pendry et al. A143* (KATH); Bara, Nijgadh, 150 m, 26 Nov. 2012, *B. Adhikari & D. R. Kandel 96* (E); Chitwan, Chitwan National Park Buffer Zone, Bis Hajar Tal, 180 m, 24 Nov. 2004, *C. A. Pendry et al. A105* (KATH); Chitwan, Sauraha, 160 m, 13 Jan. 1996, *M. Suzuki et al. 9611010* (A, BM, TI); Chitwan, Kasara, 2000 m, 1975, *D. P. Joshi et al. 75/3402* (KATH); Chitawan, Sauraha to Tal, 18 Jan. 1996, *M. Mikage et al. 9614164* (TI). **E NEPAL: Kosi**, Dhankuta, 910 m, *J. D. A. Stainton 1802* (A, BM, E); Dhankuta, 1200 m, 28 Oct. 1963, *H. Hara et al. 6303649* (KYO, TI); Dhankuta, 1200 m, 20 Sep. 1963, *H. Hara et al. 6303650* (A, BM, E, KATH, KYO, TI); Sunsari, Phusri to Sanguri, 1963, 16 Oct. *H. Hara et al. 6303638* (BM, K, KYO, TI); Sunsari, near Sanguri, 1200 m, 1981, *P. R. Shakya 6465* (KATH); Sunsari, Dharan to Dharapani, 1972, *T. B. Shrestha & T. K. Bhattacharya 72-314* (KATH); Sankhuwasabha, Arun Valley, between the airport and the Arun Bridge, 470 m, 8 Nov. 1972, *T. Wraber 34788* (BM); Sankhuwasabha, Tumlingtar, Sabbaya, 550 m, 26 Dec. 1961, *A. H. Norkett 8132* (BM); **Mechi**, Ilam, Dhara Pani, 5 Dec. 1963, *H. Hara et al. 6303635* (KATH, TI); Ilam, Dhara Pani to Tambur Bridge, 17 Oct. 1963, *H. Hara et al. 6303648* (A, BM, E, K, KATH, KYO, TI); Ilam, Maibeli to Ilam, 1977, *P. Pradhan et al. 246* (KATH); Jhapa, Happen Chok, Shantinagar, 300 m, 20 Nov. 2012, *B. Adhikari & D. R. Kandel 61* (E); Panchthar, Yektin, 1300 m, 2 Oct. 1971, *J. F. Bremez 1292* (E); Panchtchar, Yektin, 1100 m, 29 Nov. 1963, *H. Hara et al. 6303634* (TI); Taplejung, Amur River between Sinwa and Tawa, 1000 m, 2 Sep. 1989, *C. Grey-Wilson et al. 160* (E, K, KATH); Taplejung, Khebang to Bharomdin, 24 Nov. 1963, *H. Hara et al. 6303632* (KYO, TI); Taplejung, Dumhan to Taplejung, 1 Nov. 1963, *H. Kanai et al. 6303636* (BM, KATH, KYO, TI); **Sagarmatha**, Khotang, Dihigaon to Regmotar, 220 m, 27 Oct. 1995, *M. Mikage et al. 9554167* (E, TI); Siraha, Sogarat, 130 m, 9 Nov. 1954, *A. Zimmermann 21477* (K); Solukhumbu,

Namokhil, 1985, *H. Ohba et al. 8581497* (BM, KATH); Solukhumbu, Kensa to Deorali, 1500 m, 1995, *F. Miyamoto et al. 9584348* (CBM, E, TI); Solukhumbu, Bhote Kosi, Khari Khola, 1200 m, 22 Jun. 1964, *J. D. A. Stainton 4657* (BM, TI); Udayapur, 220 m, 26 Oct. 1995, *M.Mikage et al. 9554147* (TI); Udyapur, Trijuea Khola, 230 m, 7 Nov. 1969, *J. D. A. Stainton 6654* (BM, TI). **SIKKIM & DARJILING: Darjiling**, *J. M. Cowan s. n.* (SING); **Sikkim**, Bamanpokie, Nov. 1879, *J. S. Gamble 7371* (E); 600–1200 m, *J. D. Hooker s. n.* (BM, C, CE, E, K); Punkabam, 300 m, 15 Nov. 1870, *C. B. Clarke 13836* (K). **BHUTAN: Mangde Chu**, Banjormani, lower Mangde Chu Valley, 1330 m, 25 May 1979, *A. J. C. Grierson & D. G. Long 1335* (E, K); **Deothang**, between the Polytechnic and Deothang Village, 1000 m, 18 Oct. 1987, *J. R. I. Wood 5960* (E); **Monggar**, Lingmethang, Shongar Chu, 950 m, 2 Jul. 1979, *A. J. C. Grierson & D. G. Long 2427* (E); **Punakha**, Samtigong, Augduphorang, 1350 m, 1911, *R. E. Cooper 2337* (BM, E); Damoitsi, 1950 m, 1915, *R. E. Cooper 4592* (BM, E); around Lobesa, 1300 m, 1987, *J. R. I. Wood 5931* (E); between Wangdu Phodrang and Chusom, Tang Chu, 1350 m, *I. W. J. Sinclair & D. G. Long 5597* (E, K); **Tongsa**, Tintibi Bridge, Mangde Chu, 540 m, 31 Mar. 1982, *A. J. C. Grierson & D. G. Long 4173* (E, K); **Trashiyangtse**, Shali, 1950 m, 11 Oct. 1934, *F. Ludlow & G. Sherriff 1045* (BM, E). **YARLUNG ZANGBO-BRAHMAPUTRA: Dirang**, 1500–1800 m, 14 Oct. 1935, *F. Kingdon-Ward 12426* (BM). **S HENGDUAN: Bijiang**, Pihe to Liuku, 1000 m, 7 Jun. 1978, *Nujiang Exped. 0455* (KUN); **Binchuan**, Longyi to Lushang, 13 Dec. 1940, *T. N. Liou 17663* (PE); Xiayang to Bangying, 14 Oct. 1946, *T. N. Liou 21581* (PE); **Butuo**, Jiaojihe, Dadiba, 1250 m, 20 Aug. 1959, *Chuan Jiang Liang 5625* (CDBI, KUN, PE); Niujiaowan, Xixihe, 680 m, 11 Oct. 2014, *Y. F. Deng 25861* (IBSC); **Dali**, western flank of the Tali Range, 3000 m, 1913, *G. Forrest 11648* (AAU, IBSC, PE); S flank of the Tali Range, *G. Forrest 9203* (IBSC); Hongyanpo, 1800 m, 13 Oct. 1985, *Z. S. Yue 85-144* (KUN); **Hanyuan**, bank of Daduhe, 1000 m, 3 Aug. 2006, *Y. F. Deng 19078* (IBSC); Xigou, 780 m, 15 Sep. 1993, *Z. C. Zhao 336* (PE); Dadu, 900 m, 17 Sep. 1993, *Z. C. Zhao 442* (PE); Fulin to Wusihe, 750 m, 18 Jul. 1980, *Sichuan Vegitat. Exped. 22696* (CDBI); **Heqing**, Baiyan, 11 Sep. 1929, *R. C. Ching 24517* (KUN, PE); Ta pin tze, 3 Oct. 1887, *M. Delavay 2982* (NAS); **Huidong**, Dacong, 11 Oct. 2014, *Y. F. Deng 25876* (IBSC); **Huili**, Luchang, Hongchuanhe, 1750 m, 10 Oct. 1958, *C. Ho et al. 11581* (NAS, SZ); s. loc., 1932, *T. T. Yü 1626* (A, IBSC, PE); between Homöndschang and Bödschagwan, 900–1700m, 18 Mar. 1918, *H. Handel-Mazzetti 695* (W, WU); **Kangding**, Wasigou, 1300 m, 5 Oct. 1976, *C. Y. Wu et al. 6269* (KUN); between Luding and Kangding along Dadu River, 24 Aug. 1997, *D. E. Boufford et al. 27493* (A, CAS, E, MO); Kangting (Tachienlu), Vaszeko, ca. 1600 m, 12 Nov. 1934, *H. Smith 13357* (MO, PE); Xiayutong, 1640 m, 28 Jul. 1933, *T. H. Tu 4541* (IBK, IBSC, PE); **Longyang**, Lijiang, Bawan, 1000 m, 31 Aug. 2003, *Gaoligong Shan Biodiversity Survey 18241* (CAS, KUN); Mangkuan, ca. 700 m, 1 Jun. 2005, *Gaoligong Shan Biodiversity Survey 26179* (KUN); Huitongqiao, 700 m, 5 Oct. 1977, *Y. H. Li 20145* (HITBC); **Luding**, Dewei, Kuiwu, Wayaoping, 1500 m, 7 Jun. 1982, *G. H. Xu 41085* (CDBI); Mozigou, 1300 m, 9 Sep. 1953, *W. K. Hu 35504* (SZ); Huangjinping, 1500 m, 11 Aug. 1959, *S. Jiang & C. L. Jin 2772* (KUN, SZ); Mozigou, 9 Sep. 1963, *H. L. Tsiung 35504* (IBK); **Lushui**, Daheba, 890 m, 10 Oct. 1991, *Exped. Gaoligongshan 7331* (KUN); Daxindi, 930 m, 14 Oct. 2002, *Gaoligong Shan Biodiversity Survey 15945* (CAS, KUN); Luzhang, 13 Sep. 2013, *Y. F. Deng et al. 24887* (IBSC); **Miyi**, s. loc., 17 Oct. 1982, *D. Liu 23965* (CDBI); s. loc., 1200 m, 20 Oct. 1985, *H. Li 097* (KUN); Puwei, 1200 m, 12 Jul. 1983, *Qinghai-Xizang Exped. 11969* (KUN, PE); **Muli**, Eya, 2000 m, 7 Aug. 1978, *Q. S. Zhao et al. 7932* (CDBI, SZ); Erya, 1900 m, 28 Sep. 1983, *Qinghai-Xizang Exped. 14849* (KUN, PE); s. loc., 1800 m, 19 Nov. 1959, *S. K. Wu 3688*

(KUN); **Ninglang**, Yunchou, 1250 m, 21 Sep. 1938, *T. T. Yü 17660* (KUN, PE); Tuodian, Bank of Jinshajiang, 1600 m, *S. K. Wu 2942* (CDBI, KUN); Lobin Badahe, 800 m, 22 Nov. 1984, *H. Sun 0423* (KUN); **Ningnan**, Hulukou, 750 m, 1 Sep. 2012, *Y. F. Deng 23354* (IBSC); Baihetan, 11 Oct. 2014, *Y. F. Deng 25871* (IBSC); **Panzhihua**, Dukou, Mar. 1984, *G. D. Tao 14564* (HITBC); Minzheng, Baguanhe, 1400 m, 17 Jun. 1983, *Qinghai-Xizang Exped. 11205* (KUN); **Puge**, Jiaojihe, 2200 m, 10 Aug. 1960, *Sichuan Yaoyuan Pucha Dui 25655* (NAS); **Shangri-la** (Zhongdian), Jinsha Jiang Gorge on Jinsha Jiang, E of Hutiaoxia, 1585 m, 20 Jul. 2006, *D. E. Boufford et al. 35053* (A); Zhongxia of Hutiaoxia (Tiger Leaping Gorge) on Jinsha Jiang, 2000–2100 m, 25 Jul. 2006, *D. E. Boufford et al. 35211* (A); **Shimian**, Sanba, Jiangbian, 1700–1800 m, 2010, *Kham Expedition 10-2194* (PE); bank of Daduhe, 1000 m, 3 Aug. 2006, *Y. F. Deng 19069* (IBSC); bank of Daduhe, 11 Sep. 1981, *Sorvig 31* (K); **Tianquan**, Erlangshan, 9 Sep. 1953, *H. L. Tsiung 35504* (IBSC, SZ); **Xichang**, on the way from Mopan to Jinhe Yanyuan, 1300 m, 2015, *FLPH Sichuan Exped. 152261* (PE); **Yanbian**, Yumen, Guanyinyan, 1250 m, 24 Sep. 2002, *X. F. Gao et al. 3878* (CDBI, PE); **Yangbi**, Yang-pe, Jul. 1929, *G. Forrest 27914* (PE); **Yanyuan**, Dajinhe, 1270 m, 29 Oct. 1982, *Vegetation Group 30171* (CDBI); between Meidsepu and LuMapu, 1500–1650 m, 13 Oct. 1914, *H. Handel-Mazzetti 5586* (WU); Datiaoku, 1200–1400 m, 24 Sep. 1914, *H. Handel-Mazzetti 5303* (WU); **Yongsheng**, Tze-li, 15 Sep. 1939, *R. C. Ching 21613* (A, KUN, PE); Zhili, 1400 m, 1 Dec. 1999, *Y. F. Deng 14093* (IBSC); Pianjiao, 1200 m, *Y. F. Deng 23840* (IBSC); **Yulong**, Daju, Hutiaoxia, 16 Sep. 2013, *Y. F. Deng et al. 25077* (IBSC); Daju, Hutiaoxia, 1700 m, 3 Jun. 1985, *Q. Lin 771623* (KUN); Duinaoko, 1500–2000 m, 25 Oct. 1916, *H. Handel-Mazzetti 12996* (WU); Baoshan, Shitoucheng, 1700 m, 14 Sep. 2012, *Y. F. Deng 25279* (IBSC); Fengko, 1750 m, 22 Dec. 1937, *T. T. Yü 14894* (KUN, PE); Tongshan, 2700 m, 1913, *G. Forrest 11385* (PE). **U IRRAWADDY: Tengchong**, Pingpo to Youngchang, Oct. 1922, *J. F. Rock 6984* (A); Lujiangba, 850 m, 26 Dec. 1958, *X. Zhou 218* (KUN).

5. *Barleria integrisepala* H. P. Tsui, Acta Bot. Yunnan. 12: 269. 1990; Tsui in Hu (ed.), Fl. Reipubl. Popularis Sin. 70: 66. 2002; Hu, Deng & Daniel in Wu, Raven & Hong (eds.), Fl. China 19: 469. 2011. Type: [S Hengduan] China, Sichuan, Muli, Kala, 2000 m, 25 Aug. 1983, *Exped. Pl. Xizang 13398* [holotype: PE (PE01558197); isotype: PE (PE01558198)].

Subshrubs, 0.5–1 m high; stems sparsely lenticellate, strigose, soon glabrescent. Leaves clustered at shoot apices; petiole 3–5 cm long, strigose; blade lanceolate, 3–4.5 cm long, 0.7–1.8 cm broad, both surfaces pilose and strigose along veins, lateral veins 4–6 pairs, base cuneate and decurrent onto petiole, margin serrate, apex acute. Flowers axillary, solitary. Bracteoles linear, 1.2–1.4 cm long, 1–1.3 mm broad, sparsely pilose. Outer calyx lobes elliptic, 1.1–1.3 cm long, 0.6–0.7 cm broad, abaxially sparsely strigose, inside pilose, base cuneate, margin ciliate, apex acute; inner calyx lobes lanceolate, 4–5 mm long, 1–1.2 mm broad, margin ciliate, apex acuminate. Corolla bluish purple, funnel-shape, 3.2–3.5 cm long; tube base cylindric, 6–7 mm long, throat 1.2–1.5 cm long; lobes suborbicular, 1.3–1.5 cm long, 1.2–1.4 cm broad, subequal. Stamens 4; filaments basally pilose, longer pair 1.4–1.7 cm long, shorter pair 3–4 mm long; anthers of longer pair 3–4 mm long, shorter pair 1.5–2 mm long; staminode 1, 3–4 mm long. Ovary subellipsoid, 3–4 mm long, glabrous; stigma enlarged. Capsules ellipsoid, 1.5–1.7 cm long, with a beak towards apex, 4-seeded. Seeds ovate in outline, compressed, 4–5 mm long, 4–5 mm broad. Figure 46-22.

Figure 46-22: **Barleria integrisepala** H. P. Tsui: **a**, portion of plant; **b**, corolla, stamens and pistil; **c**, bracts and young fruit; **d**, capsule; **e**, seed and retinaculum. Drawn by Han-Ping YU; reproduced from *Fl. Reipubl. Popularis Sin.* 70: 64, pl. 7: 7–11. 2002.

Phenology: flowering from August to September; fruiting from November to December.

Habitats: thickets along streams; 1900–2000 m.

Distribution: endemic; S Hengduan (Map 46-38).

S HENGDUAN: Muli, bank of Yalong Jiang, 2100 m, 8 Jul. 1937, *T. T. Yü 6957* (GH, KUN, PE); Kala, 2000 m, 25 Aug. 1983, *Exped. Pl. Xizang 13398* (PE).

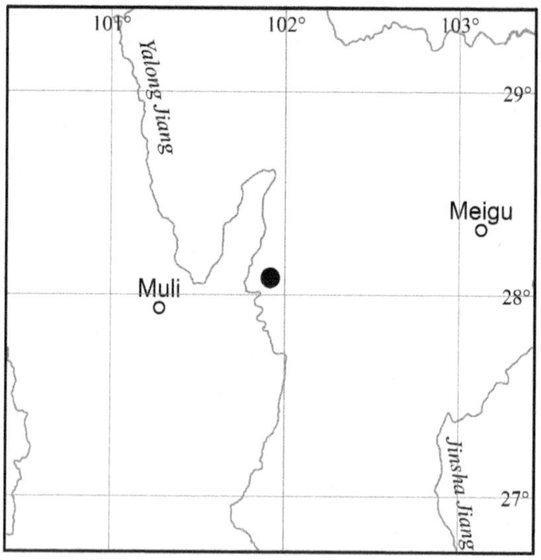

Map 46-38: **Barleria integrisepala** H. P. Tsui

7. LEPIDAGATHIS Willd.

Willdenow, Sp. Pl. 3: 400. 1800; Nees in de Candolle (ed.), Prodr. 11: 249. 1847; Bentham & Hooker, Gen. Pl. 2: 1101. 1876; Clarke in Hooker (ed.), Fl. Brit. India 4: 515. 1885; Lindau in Engler & Prantl (eds.), Pflanzenfam. 4(3b): 312. 1895; Nasir & Ali in Malik & Ghafoor (eds.), Fl. Pakistan 188: 9. 1988; Wood in Grierson & Long (eds.), Fl. Bhutan 2(3): 1285. 2001; Hu, Fl. Reipubl. Popularis Sin. 70: 196. 2002; Hu, Deng & Daniel in Wu, Raven & Hong (eds.), Fl. China 19: 469. 2011. Type: *Lepidagathis cristata* Willd.

Herbs or rarely shrubs, often anisophyllous; with cystoliths. Leaves opposite; leaf blade entire or crenate-serrate. Inflorescences dense spikes or cymes, condensed in to globose heads. Bracts and bracteoles present. Calyx 5-lobed; lobes heteromorphic, posterior one usually largest, lateral 2 smallest, anterior 2 usually connate to a greater degree than other lobes. Corolla small; tube usually expanded into a distinct throat near the middle; limb 2-lipped, lower lip 3-lobed, upper lip entire to 2-lobed; lobes quincuncial in bud. Stamens 4, didynamous, usually include; filaments short; anthers bithecous or sometimes shorter pair monothecous; thecae oblong, parallel, or one lower and somewhat divergent, base muticous. Ovary with 2 ovules per locule; stigma subcapitate, entire or slightly 2-cleft. Capsules subellipsoid to oblong to linear-conical, 1 4-seeded; retinacula present. Seeds discoid, pubescent with hygroscopic trichomes. **Chromosomes** 2n = 18, 20, 22, 24, 40, 42, 44, 84.

About 100 species: tropical to subtropical regions worldwide; four species in the Pan-Himalaya.

1a. Leaves sessile; blade linear to lanceolate, less than 5 mm broad ······························ 1. *L. trinervis*
1a. Leaves distinctly petiolate; blade ovate to oblong-lanceolate, broader than 7 mm.
 2a. Leaves on lateral twigs much smaller than those on main axis; bracts and bracteoles cuspidate or spine-tipped; spikes 4-fariously; seeds 2–2.5 mm long....................................... 2. *L. cuspidata*
 2b. Leaves on lateral twigs similar to those on main axis; bracts and bracteoles not cuspidate; spikes unilateral; seeds 1–1.5 mm long.

3a. Spikes usually terminal; bracts not striate, densely ciliate with mostly eglandular hairs; corolla spotted on both lips .... 3. *L. incurva*
3b. Spikes mostly borning from leaf axils; bracts striate, sparsely ciliate with mostly glandular hairs; corolla spotted on the middle lobe of lower lip only .... 4. *L. purpuricaulis*

1. Lepidagathis trinervis Nees in Wallich (ed.), Pl. Asiat. Rar. 3: 96. 1832; Nees in de Candolle (ed.), Prodr. 11: 254. 1847; Clarke in Hooker (ed.), Fl. Brit India 4: 517. 1885; Nasir & Ali in Malik & Ghafoor (eds.), Fl. Pakistan 188: 13. 1988. ≡ *Barleria trinervis* Wall., Numer. List n. 2500. 1830, nom. invalid. Type: India, 6 Jan. 1813, *F. Buchanan-Hamilton in Wallich Cat. 2500a* [lectotype **here designated**: K-W (K001116220); isolectotypes: GZU (GZU000251623), K]

Lepidagathis prostrata Dalzell, Hooker's J. Bot. Kew Gard. Misc. 2: 138. 1850, **syn. nov.**; Clarke in Hooker (ed.), Fl. Brit India 4: 518. 1885. Type: India, s. d., *Herb. Dalzell s. n.* [holotype: K (K000950053)].

Subshrubs, 20–30 cm high, with rootstock; stems prostrate to suberect, quardrangular, puberulent or glabrous. Leaves sessile; blade linear to narrowly lanceolate, 2–4.5 mm long, 3–5 mm broad, glabrous, palmately 3-veined, base cuneate, margin entire, apex acute, mucronulate. Spikes axillary, solitary or fascicled near rootstock, 1–2(–2.5) cm long, oblong-ovoid, one-sided. Bracts ovate or elliptic-ovate, 4–5 mm long, 2.5–3 mm broad, minutely puberulous, apex with a recurved spine for 3.5–4.5 mm long, margin ciliate toward apex; bracteoles 3.5–4.5 mm long, outer ones oblong-lanceolate, apex acuminate-aristate, inner ones eiform. Calyx lobes 7–8 mm long, apex acute, adaxially hairy, abaxially glabrescent. Corolla light pink, spotted inside, 5–6 mm long, densely long silky hairy in bud, upper lip emarginate, lower lip patent, 3-lobed; lobes ovate, 1.2–1.5 mm long, 1–1.2 mm broad, apex obtuse. Capsules ovoid-lanceolate, 5–6 mm long, glabrous, 2-seeded. Seeds oblong, 2–3 mm long, with long hygroscopic hairs.

Phenology: flowering from September to February of next year; fruiting November to March of next year.

Habitats: roadside; 300–1200 m.

Distribution: U Ganga & Indus, Sikkim & Darjiling (Map 46-39); also in other parts of Pakistan and India.

U GANGA & INDUS: Kangra, 300–1200 m, s. d., *C. B. Clarke s. n.* (K). **SIKKIM &**

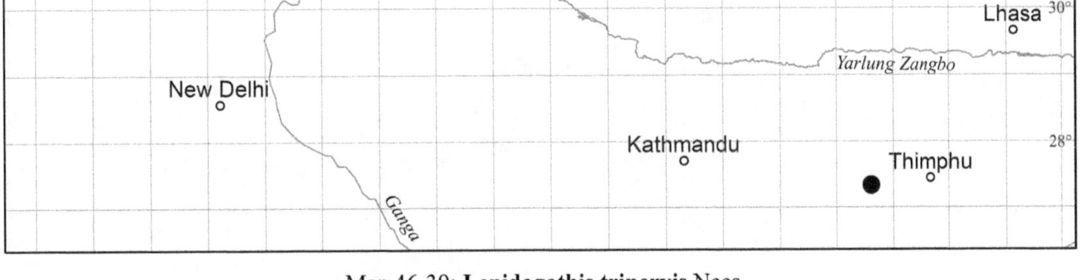

Map 46-39: **Lepidagathis trinervis** Nees

DARJILING: Sikkim, s. d., *Herb. Griffith 6142* (A, K).

2. **Lepidagathis cuspidata** Nees in Wallich (ed.), Pl. Asiat. Rar. 3: 97. 1832; Nees in de Candolle (ed.), Prodr. 11: 258. 1847; Anderson, J. Linn. Soc., Bot. 9: 495. 1865; Clarke in Hooker (ed.), Fl. Brit. Ind. 4: 519. 1884; Nasir & Ali in Malik & Ghafoor (eds.), Fl. Pakistan 188: 11. 1988. ≡ *Ruellia cuspidata* Wall., Numer. List n. 2405a, b. 1830, nom. invalid. Type: [U Ganga & Indus] India, Hurdwar, 1825, *Wallich Cat. 2405b* [lectotype **here designated**: K-W (K001115871); isolectotype: GZU (GZU000251519)].

Herbs or subshrubs, 0.4–1 m high; stems erect or decumbent, quadrangular, densely glandular-pubescent to subglabrous. Leaves on lateral twigs much smaller than those on main axis; petiole (0–)0.5–2.5 cm long; blade oblong-lanceolate to elliptic-lanceolate, 1.5–10(–12) cm long, 0.7–4 cm broad, glabrous or glandular-pubescent, base attenuate, tapering onto petiole, margin entire, apex acute or acuminate, mucronate or cuspidate. Spikes terminal, 5–10 cm long. Bracts elliptic-oblong or obovate, 9–12 mm long, 4–6 mm broad, apex acute, spine-tipped, glandular; bracteoles lanceolate, 6–8 mm long, 2–3 mm broad, apex spine-tipped, glandular-pubescent. Calyx 1–1.2 cm long, unequally 5-lobed to base, abaxially glandular-pubescent; lobes unequal, larger two lanceolate, 9–11 mm long, 3–4 mm broad, spine-tipped, smaller ones linear, 8–9 mm long, 1–2 mm broad, apex acute. Corolla white, purple-streaked on upper lip, 1.2–1.6 cm long, sparsely pubescent outside; tube 5–6 mm long; upper lip 4–5 mm long, ovate, apex emarginate or slightly 2-lobed, lower lip 4–5 mm long, 3-lobed; lobes orbicular-oblong, 3–4 mm long, 2–2.5 mm broad. Filaments glabrous, longer pair 5–5.5 mm long, shorter pair 2.5–3 mm long; anthers bearded, thecae 1.2–1.5 mm long. Ovary glabrous; style 7–8 mm long, hairy; stigma slightly 2-cleft. Capsules ovoid-lanceolate, 6–9 mm long, glabrous, 4-seeded. Seeds ovate in outline, compressed, 2–2.5 mm long, flattened. **Chromosomes** 2n = 22.

Phenology: flowering from April to May.

Habitats: roadside, slopes, thickets, dry places; 900–1600 m.

Distribution: Jammu & Kashmir, U Ganga & Indus, W Nepal, C Nepal (Map 46-40); also in other parts of Pakistan and India.

JAMMU & KASHMIR: Jammu, Adh Kanwari, 1600 m, 1933, *S. K. Kapoor 14455* (RRLH). **U GANGA & INDUS: Almora**, 1350 m, s. d., *R. Strachey & J. E. Winterbottom 1* (A); below Ranikhet, 1050 m, 19 Jun. 1923, *R. N. Parker 2007* (A); **Chamba**, 1800 m, *Herb. G. Watt s. n.* (E); Choane, 900 m, 22 May 1878, *Herb.*

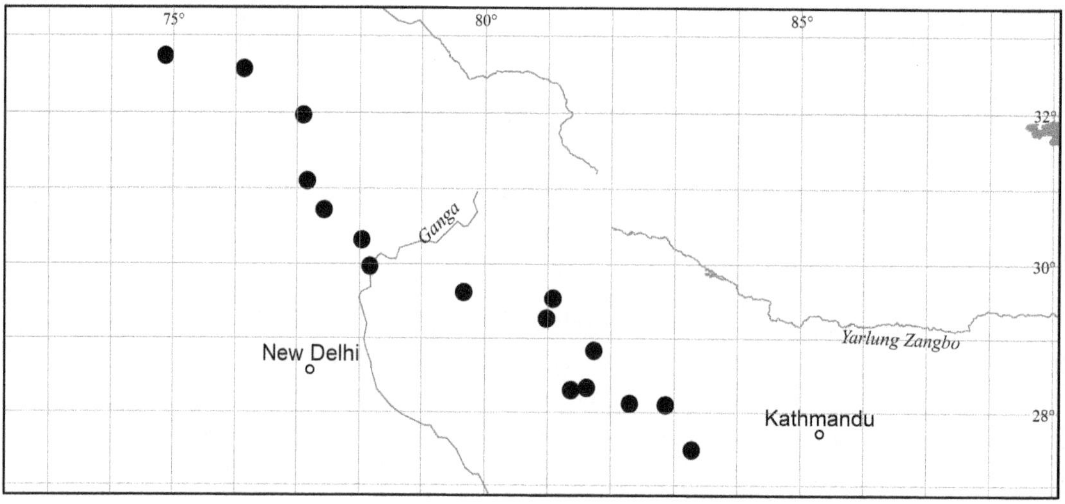

Map 46-40: **Lepidagathis cuspidata** Nees

G. Watt 505 (E); **Dehradun**, s. d., *A. Smythies 142* (E); 1928, *U Singh 189* (A); **Huridwar**, s. d., *Wallich Cat. 2405a, b* (GZU, K-W); **Kulu**, Kulu Nuddi, s. d., *D. Ritchie 1214* (E); **Mussorie**, s. d., *R. R. Stewart 16795* (DD, fide Raizada & Saxena, Flora of Mussoorie: 541. 1978); **Simla**, *G. Watt 4216* (E); Simla to Changdigarh, 10 May 1970, *S. Kurosawa S. K. 287* (TI). **W NEPAL: Bheri**, Banke, Napelganj, 1974, *J. F. Dobremez N2404* (BM); Banke, Napelganj, 250 m, 12 Apr. 1974, *J. F. Dobremez 2606 (N.P. Nanauthal 74-33)* (BM, E, KATH); Bardiya, Thari Dammar, 700 m, 1979, *K. R. Rajbhandari & B. Roy 2634* (KATH); Dailekh, Chhipa Chhipe Dani, 1090 m, 25 Feb. 1991, *N. P. Manandhar 395-91* (KATH); Dailekh, Lohari, 790 m, 27 Feb. 1991, *NN. P. Manandhar 490-91* (KATH); Dailekh, 1400 m, 1974, *K. R. Rajbhandari & B. Roy 2808* (KATH); **Rapti**, Dang, Koilabas, Pathraiya Khola, 180 m, 2 Apr. 1976, *M. K. Ghimire 734* (KATH); Dang, Ghorahi, 650 m, 1976, *N. P. Manandhar & P. M. Regmi 356* (KATH); Dang, Koilabas, 1973, *N. P. Manandhar 10152* (KATH); Pyuthan, Bigjwar, 780 m, 5 Mar. 1976, *N. P. Manandhar & P. M. Regmi 192* (KATH); **Seti**, Bajhang, S of Raskot, 700 m, 19 Mar. 1968, *J. D. A. Stainton 6155* (BM); Doti, Mahapani, *J. F. Dobremez 1866* (BM, E); Doti, 1430 m, 5 Apr. 1984, *P. Pradhan et al. 1162* (KATH). **C NEPAL: Lumbini**, Rupandehi, Butwal, 300 m, *P. R. Shakya 2260* (BM).

3. Lepidagathis incurva Buch.-Ham. ex D. Don, Prodr. Fl. Nepal.: 119. 1825; Nasir & Ali in Malik & Ghafoor (eds.), Fl. Pakistan 188: 12. 1988; Wood in Grierson & Long (eds.), Fl. Bhutan 2(3): 1286. 2001; Hu, Fl. Reipubl. Popularis Sin. 70: 200. 2002; Hu, Deng & Daniel in Wu, Raven & Hong (eds.), Fl. China 19: 471. 2011. Type: [C] Nepal, Bagmati, Kathmandu, Narainhetty, 27 Jan. 1803, *F. Buchamann-Hamilton. s. n.* [holotype: BM (BM001124550)].

Lepidagathis hyalina Nees, Pl. Asiat. Rar. 3: 95. 1832; Clarke in Fl. Brit. India 4: 521. 1885. Type: Myanmar, Mt. Prome, 1826, *F. Buchanan-Hamilton in Wallich Cat. 2365h* [lectotype **here designated**: K-W (K001115748); isolectotype: C].

Lepidagathis ustulata Nees, Pl. Asiat. Rar. 3: 95.1832. ≡ *Lepidagathis hyalina* var. *ustulata* (Nees) C. B. Clarke in Hooker (ed.), Fl. Brit. India. 4: 521. 1885. ≡ *Lepidagathis incurva* var. *ustulata* (Nees) R. R. Stewart, [Annot. Cat. Vasc. Pl. West Pakistan Kashmir 677. 1972, nom. invalid. (Art. 41.5)] Kew Bull. 29: 433. 1974; Nasir & Ali in Malik & Ghafoor (eds.), Fl. Pakistan 188: 12. 1988. ≡ *Ruellia ustata* Wall., Numer. List n. 2366b. 1830, nom. invalid. Type: India,"Circa Gurrukpur et Sukanaghar", 13 Apr. 1814, *F. Buchanan-Hamilton s. n. in Wallich Cat. 2366b* [lectotype **here designated**: K-W (K001115764); isolectotypes: BM, K-W (K001115765)].

Lepidagathis semiherbacea Nees, Pl. Asiat. Rar. 3: 96. 1832. ≡ *Lepidagathis hyalina* var. *semiherbacea* (Nees) C. B. Clarke in Hooker (ed.), Fl. Brit. India 4: 521. 1885. Type: Myanmar, Ripa Attran, 1827, *Wallich Cat. 7164a* [lectotype **here designated**: K-W (K001126809)].

Lepidagathis hyaline Nees var. *mollis* C. B. Clarke in Hooker (ed.), Fl. Brit. India 4: 521. 1885. ≡ *Lepidagathis incurva* var. *mollis* (C. B. Clarke) Raizada & Saxena, Ind. Forester 92: 315. 1966. Type: India, *s. coll. 320* [holotype: K (K000950070)].

Herbs 30–100 cm high; stems quadrangular, sulcate, pubescent or soon glabrescent. Petiole 0.5–3.5 cm long; leaf blade ovate to narrowly elliptic, 2.5–11 cm long, 1–4.5 cm broad, both surfaces pubescent, or sparsely pubescent along veins, lateral veins (4–)7–9 pairs, base cuneate and conspicuously decurrent onto petiole, margin entire and slightly sinuate, apex acute to shortly acuminate. Spikes (0.5–)1–3.8 cm long, usually terminal. Bracts oblong-lanceolate, 6–10 mm long, 1.5–2 mm broad, abaxially pubescent with or sometimes without glandular- or eglandular-pubescent, 1-veined, margin usually ciliate, apex acuminate; bracteoles similar to bracts except usually smaller. Calyx 8–10 mm long, pubescent with or sometimes without glandular- or eglandular-pubescent, or glabrescent; posterior lobe oblong-lanceolate, 8–9 mm long, 1.5–1.7mm broad, 3-veined; lateral lobes lanceolate, 6–7.5 mm long, 0.7–0.8 mm broad; anterior lobes connate at base for 1/3–1/2 of their length. Corolla white streaked with purple potted on both lip, 8–10 mm long, outside sparsely pubescent; upper lip emarginate; lower lip 2.5–3 mm long, 3-lobed; lobes ovate, 1.5–3 mm long, 1.3–1.8 mm broad. Filaments 2.5–3 mm long, glabrous; anther-thecae ovate, 1–1.2 mm long. Ovary 1–1.5 mm long, hispid towards apex; style 7–9 mm long, subglabrous; stigma bifid. Capsules ovate-oblong, 5–6 mm long, pilose towards apex, 4-seeded. Seeds subcircular in outline, 1.5–1.8 mm in diameter, pubescent. **Chromosomes** 2n = 20, 44, 60, 84. Figure 46-23.

Phenology: flowering and fruiting from October to March.

Habitats: grasslands, thickets, roadsides, hedgerows, stream banks; 100–2200m.

Distribution: N Pakistan, Jammu & Kashmir, U Yarlung Zangbo, W Nepal, C Nepal, E Nepal, Sikkim & Darjiling, Bhutan (Map 46-41); also in Bangladesh, Laos, Malaysia, Thailand, Vietnam, and other parts of Pakistan, India, Myanmar and China (S Yunnan, Guangxi, Guangdong, Hainan).

Uses: medicine.

N PAKISTAN: Azad Kashmir, Kotli, Saidpur, 600 m, May 29 1930, *R. R. Stewart s. n.* (RAW, fide Malik & Ghafoor, 1988); Mirpur, 900 m, Apr. 1954, *R. R. Stewart 27245* (RAW, fide Malik & Ghafoor, 1988). **Hazara**, Murree, Murree Hills, 600 m, *R. R. Stewart 29037*(RAW, fide Malik & Ghafoor, 1988); Margalla hills, Mar. 1938, *Dwarkanath 369* (RAW, fide Malik & Ghafoor, 1988). **JAMMU & KASHMIR: Jammu**, Tope Village, 300 m, 1969, *B. M. Sharma 13* (RRLH). **U GANGA & INDUS: Dehradun**, Lachiwarla, 19 Feb. 1931, *E. K. Kinglanan 10* (A); **Kumaon**, outer hills of Kumaon, *R. Strachey & J. E. Winterbottom 2* (K); Kumaon, 1845, *s. coll. 1965* (K); **Mussorie**, *M. B. Saxena 206* (DD, fide Raizada & Saxana, Flora of Mussoorie 1: 542. 1978). **W NEPAL: Bheri,** Banke, Tulsipur to Phalwang, 730–1840 m, 1982, *s. coll. W144(9)/82* (KATH); Bardiya, Ganga Patuwa, 180 m, 1983, *R. N. Shukla 5165* (KATH); Bardiya, Kohalpur, 1980, *T. K. Bhattachaya 391* (KATH); Bardiya, Khapare Khola, near Babai, 1976, *K. J. Malla et al. 899* (KATH); Dailekh, Sattala, 1050 m, 1985, *K. J. Malla 1657* (KATH); Dailekh, Chhip Chhipe Dani, 1090 m, 1991, *N. P. Manandhar 436-91* (KATH); Jajarkot, Kuiyapani, 720 m, 1991, *N. P. Manandhar 671-91* (KATH); Surkhet, Birandra Nagar, 720 m, 1983, *P. P. Kurmi & D. K. Shrestha 4992* (KATH); Surkhet, Ranimatta, 1200, 1986, *K. J. Malla & P. R. Sharma 1555* (KATH); Surkhet, Kalyan, 870 m, 1991, *N. P. Manandhar 224-91* (KATH); **Karnali**, Jumla, Chutiaghati, 750 m, 19 Mar. 1966, *Herbarium Banerji s. n.* (A); **Mahakali**, Pyuthan, Bijwas, 660 m, 1983, *Pandyal 5322* (KATH). **Rapti,** Dang, Koilabas, Pedikhola; 300 m, 1982, *K. J. Malla 251* (KATH); Dang, Ghorahi, 1974, *K. J. Malla 14524* (KATH); Dang, Dang Valley, 300 m, 1979, *T. K. Bhattacharya 503* (KATH); Dang, Dang Valley, 300 m, 1979, *T. K. Bhattacharya 506* (KATH); Rolpa, Nimrigaon, 1080–1880 m, 1982, *s. coll. W333/82* (KATH); Salyan, Phalabang, 1350 m, 27 Mar. 1952, *O. V. Polunin, W. R. Sykes & L. H. J. Williams 655* (A, BM); Salyan, Marma Khola, 900 m, 31 Mar. 1952, *O. V. Polunin, W. R. Sykes & L. H. J. Williams 715* (A, BM, E). **C NEPAL: Bagmati,** Kathamandu, Swayambhu, 1300 m, 7 Mar. 1970, *N. P. Manandhar 9409* (KATH); Kathamandu, Nagarjun, 1970, *H. Kanai & P. R. Shakya 1685* (KATH); Kathamandu, Nagarjung, 1969, *H. Kanai 9718* (KATH); Kavre Palanchok, Swoyamkhoo, 1400 m, 1986, *P. R. Shakya et al. 8830* (KATH); Latitpur, Godamchaurl, 480 m, 30 Nov. 2001, *T. Watanabe et al. NPL-SP011213(126)* (MBK); **Dhawalagiri**, Baglung, 900 m, 20 Apr. 1954, *J. D. A. Stainton, W. R. Sykes & L. H. J. Williams 56* (BM, E); Baglung, Kali Gandaki, 840 m, 20 Apr. 1954, *J. D. A. Stainton, W. R. Sykes & L. H. J. Williams 92* (BM, E); Nuwakot, Trisuli, Bhainse to Betrabati, 650 m, 15 Feb. 1975, *M. M. Amatya & K. R. Rajbhandari 48/75* (KATH); Parbat, Limithana, 850 m, 3 Jun. 1992, *N. P. Manandhar & S. K. Acharya 646-92* (KATH); Rusuwa, Haku, 2060 m, 3 Feb. 1983, *S. B. Rajbhandary & N. P. Manandhar 9358* (KATH); **Gandaki,** Gorkha, 1100 m, 12 May 1987, *N. P. Manandhar & L. P. Katel 11722* (KATH); Gorkha, Arunghat, 9 miles above Arughat Bazar, 610 m, *P. C. Gardner 282* (BM); Kaski, Kuibang, 1300 m, 2003, *P.*

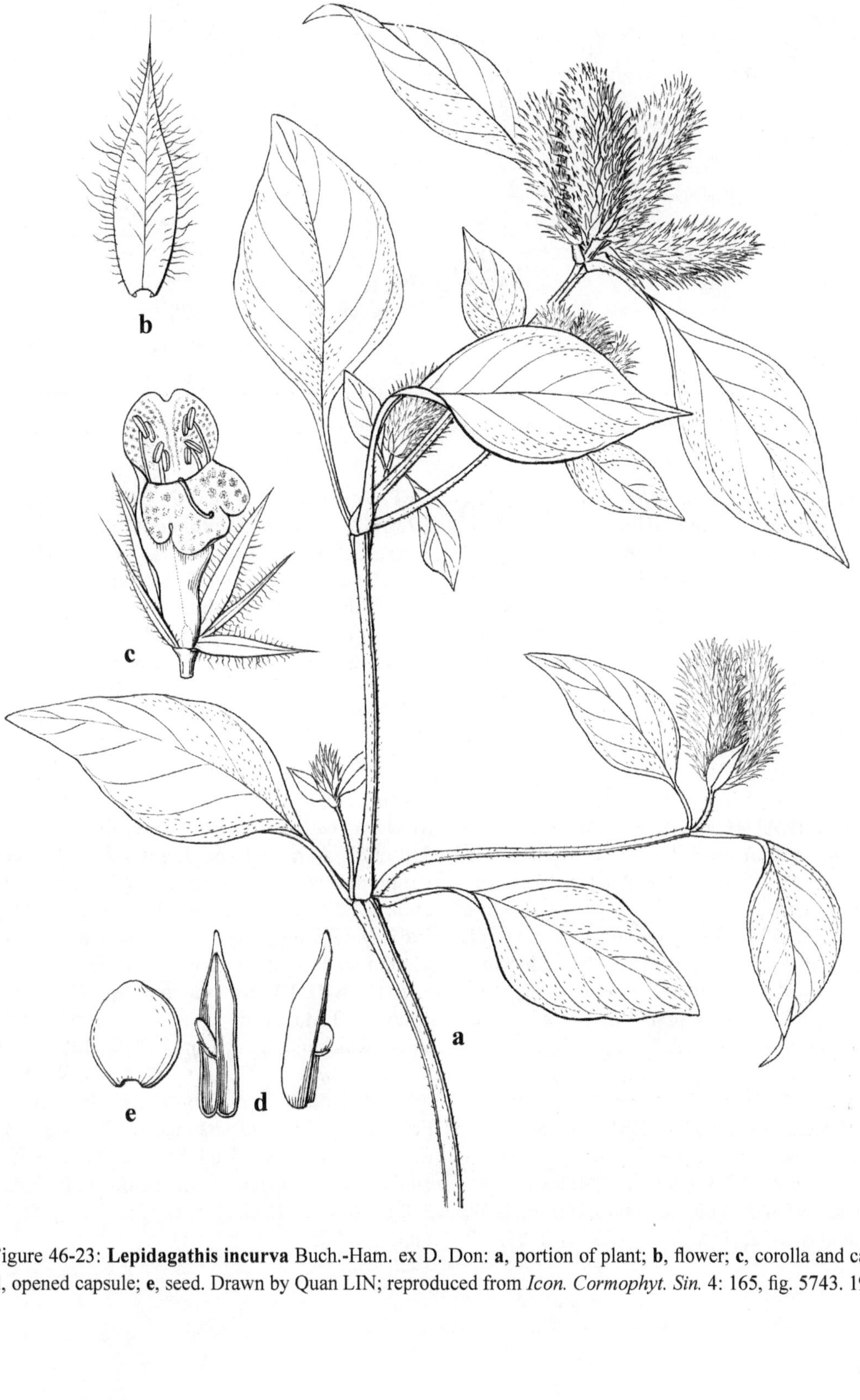

Figure 46-23: **Lepidagathis incurva** Buch.-Ham. ex D. Don: **a**, portion of plant; **b**, flower; **c**, corolla and calyx; **d**, opened capsule; **e**, seed. Drawn by Quan LIN; reproduced from *Icon. Cormophyt. Sin.* 4: 165, fig. 5743. 1975.

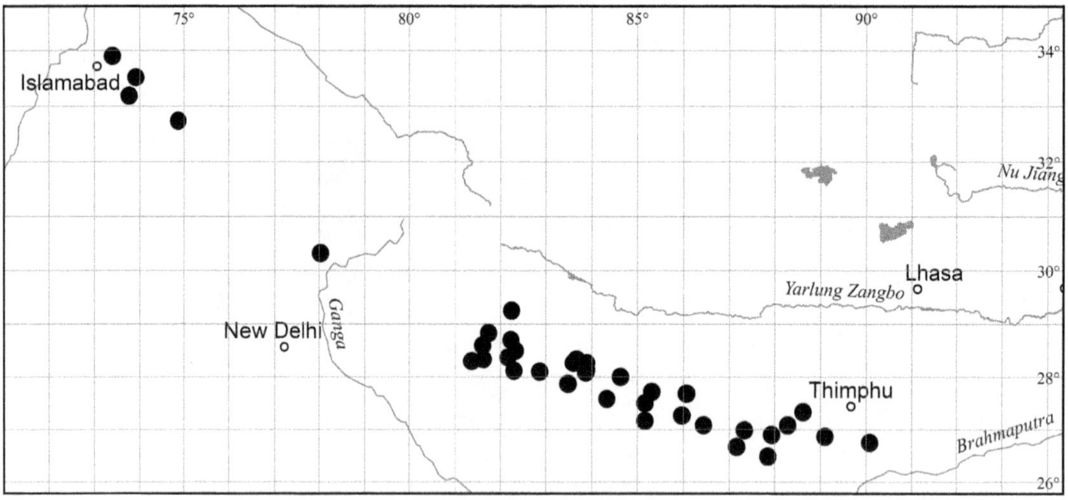

Map 46-41: **Lepidagathis incurva** Buch.-Ham. ex D. Don

R. Shakya & D. B. Shrestha 10635 (KATH); Syangja andhi Khola, Bhorle Panchayat, 600 m, 1974, *D. P. Joshi & M. M. Amatya 74/1480* (KATH); Syangja, Raja Swanra, Najbe Kateri Panchayat, 1650 m, 16 Mar. 1974, *D. P. Joshi & M. M. Amatya 74/1587* (KATH); **Janakpur**, Dolakha, Pandrang, 1100 m, 1987, *N. P. Manandhar & L. P. Katel 11722* (KATH); Sindhuli, Dhungre Bhanjyang, 1200 m, 3 Jun. 1979, *N. P. Manandhar & M. K. Adhikari 2046* (KATH); **Lumbini,** Palpa, Tansen, Basani Khola, 1680 m, 1 Mar. 1976, *N. P. Manandhar & P. M. Regmi 39* (KATH); Palpa, Hathikot Base, 650 m, 1974, *D. P. Joshi & M. M. Amatya 74/1265* (KATH); Palpa, Sattewati, Koldanda Panchayat, 1350 m, 1974, *D. P. Joshi & M. M. Amatya 74/1352* (KATH); **Narayani,** Bara, Tamagardi, 1993, *L.P. Kaltel 273/93* (KATH); Chitwan, 1996, *M. Mikage et al. 9614147* (E, TI); Chitwan, Sauraha to Padampur, 17 Jan. 1996, *K. Yoda et al. 9614115* (E, KATH, TI); Chitwan, N. of Tikoli, 270 m, 12 Mar. 1966, *Herbarium Banerji 1726* (A); Chitwan, Ram Nagar, 500 m, Jun. 1990, *K. Wesche 154* (BM); Makawanpur, Rapti Valley, Hetaura, 450 m, 9 Mar. 1966, *J. D. A. Stainton 5219* (BM, E); Makawanpur, Mokwanpur Garhi to Pass, 1000–1200 m, 1969, *H. Kanai & M. S. Bista 11838* (KATH, KYO). **E NEPAL: Koshi**, Dhankuta, Pokhare to Dharpani, 1984, *T. K. Bhathacharya 5785* (KATH); Sunsari, Chatra to Bararkshatra, 150 m, 1982, *K. R. Rajbhandari et al. 6052* (KATH, TI); **Mechi**, Ilam, Ranga Pani to Ghorsa, 9 Dec. 1963, *H. Hara et al. 6303607* (KYO); Ilam, near Fikkal, 900 m, 20 Apr. 1962, *K. Nishioka 579* (KYO); Ilam, Dhara Pani to Illam, 5 Dec. 1963, *H. Hara et al. 6303610* (KYO); Illam, Illam to Ranga, 8 Dec. 1963, *H. Hara et al. 6303611* (BM, KYO); Jhapa, Jal-Thol Forests near Prithwinagar, 100 m, 2003, *N. Thapa et al. 2072* (KATH). **SIKKIM & DARJILING: Darjiling**, Rayong, 250 m, 21 Apr. 1960, *M. Togashi 5837* (KYO); Bijanbari, 900 m, 23 Sep. 1977, *H. Ohashi et al. 774361* (E); **Sikkim**, Damthang, 17 May 1967, *N. C. Majumdar & R. M. Datta 381* (MO); Munoong, 1140 m, 25 Nov. 1908, *W. G. Craib 361* (E). **BHUTAN: Phuntsholong**, Kharbandi, 1000m, 19 Mar. 1988, *J. R. I. Wood 6193* (E); above Rinchending, 660 m, 16 Feb. 1982, *A. J. C. Grierson & D. G. Long 2955* (E); W slope of Raidak valley, S of Tabji Khola, 1400 m, 23 Feb. 1986, *A. J. C. Grierson & D. G. Long 3167* (E); **Samchi**, Samchi High School, 400 m, 7 Jan. 1989, *J. R. I. Wood 6857* (E); **Sankosh**, Pinkhua, 310 m, 16 Mar. 1982, *A. J. C. Grierson & D. G. Long 3751* (E).

4. Lepidagathis purpuricaulis Nees in Wallich (ed.), Pl. Asiat. Rar. 3: 97. 1832; Nees in de Candolle (ed.), Prodr. 11: 257. 1847; Clarke in Hooker (ed.), Fl. Brit. India 4: 519. 1885. Type: India, "in Burhi Simar", 8 Mar. 1811, *F. Buchanan-Hamilton in Wallich Cat. 2366C* [lectotype **here designated**: GZU (GZU000251525); isolectotype: K-W (K001115762)].

Barleria pungens Wall., Numer. List n. 2498. 1830, nom.nud.

Ruellia mucronata Wall., Numer. List n. 2366. 1830, nom.nud.

Herbs, 0.3–0.6 m high; stems decumbent, glabrous. Leaves unequal on each pair; petiole (0–)0.5–1 cm long; blade lanceolate to oblong-lanceolate, (2.3–)6.5–13 cm long, (0.7–)1.7–3 cm broad, slightly pubescent on both surfaces when younng, then glabrescent, lateral veins 5–7 pairs, base attenuate, margin entire, apex acute. Spikes mostly borne from leaf axils, 1.2–3 cm long, 3–11-flowered. Bracts lanceolate, 6–8 mm long, 2–2.5 mm broad, glandular-pilose, margin sparsely ciliate, apex aristate; bracteoles 2, linear-lanceolate, 6–9 mm long, 0.8–1 mm broad, glandular-pubescent, 1-veined, margin ciliate. Calyx 1–1.2 cm long, glandular-pilose; lobes unequal, elliptic or oblong, larger ones 8–10 mm long, 2.3–3 mm broad, 5-veined, smaller ones subequal, 6–9 mm long, 0.5–1 mm broad, 1-veined. Corolla 8–10 mm long, white or pale mauve with darker mauves pots on mid-lobe of lower lip; tube 6–7 mm long; upper lip 5–6 mm long, apex emarginate or slightly 2-lobed, lower lip 5–7 mm long, 6–7 mm broad, 3-lobed; lobes orbicular or oblong, 3–3.5 mm long, 1.5–2 mm broad. Filaments of longer pair 3.5–4 mm long, shorter pair 1.6–2 mm long; anthers glabrous, 1.2–1.5 mm long. Ovary glabrous; style 7–7.5 mm long; stigma slightly 2-cleft. Capsules linear-ellipsoid, 6–8 mm long, glabrous. Seeds ovoid, 0.9–1.2 mm long. **Chromosomes** 2n = 22, 24, 84.

Phenology: flowering from March to April; fruiting from April to May.

Habitats: roadsides, open forests; 300–1600 m.

Distribution: U Ganga & Indus, W Nepal, C Nepal, E Nepal (Map 46-42); also in other parts of India and Myanmar.

U GANGA & INDUS: Kumaon, *R. Strachey & J. E. Winterbottom 3* (A, K); **Huridwar**, *Wallich Cat. 2366f* (K-W); **Mussorie**, *M. B. Saxena 290* (DD, fide Raizada & Saxena, 1978). **W NEPAL:** Katai, 27 Mar. 1929, *Bis Ram 143* (BM); **Bheri**, Banke, Kohalpur, Napelganj, 1973, *N. P. Manandhar & Malla 17150* (KATH); **Jajarkot**, Kuepani, Bheri River, 910 m, 1 Apr. 1952, *O. V. Polunin, W. R. Sykes & L. H. J.*

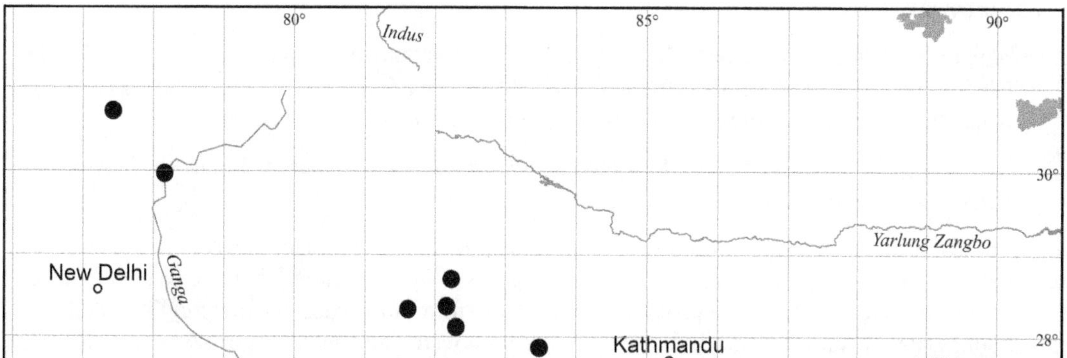

Map 46-42: **Lepidagathis purpuricaulis** Nees

Williams 3799 (A, BM, E). **Rapti,** Dang, Harnok, 1300 m, 1976, *N. P. Manandhar & P. M. Regmi 237* (KATH); Salyan, Jagyana, Salleyana, 850 m, 1976, *N. P. Manandhar & P. M. Regmi 410* (KATH). **C NEPAL: Janakpur,** Sindhuli, Mulkot, 1450 m, 21 Nov. 1988, *N. P. Manandhar 12329* (KATH); Ramechhap, Banchare, 1150 m, 5 Mar. 1976, *N. P. Manandhar & P. M. Regmi 179* (KATH); **Lumbini**, Palpa, Chaune Bhanjyang, 1370 m, 1976, *N. P. Manandhar & P. M. Regmi 35* (KATH). **E NEPAL: Mechi**, Ilam, Ranga Pani to Ghowa, 9 Dec. 1963, *H. Hara et al. s. n.* (TI).

Tribe 3. Ruellieae Dumort., Anal. Fam. Pl.: 23. 1829. Type: *Ruellia* L.

Plant with cystoliths. Calyx 5-lobed. Corolla usually contorted or rarely slightly bilabiate. Stamens usually 4, rarely 2, anthers bithecous. Capsules 2–4- or many-seeded.

Recent references: Bremekamp & Nannenga-Bremekamp, 1948. A preliminary survey of the Ruelliinae (Acanthaceae) of the Malay Archipelago and New Guinea. *Verh. Kon. Ned. Akad. Wetensch., Afd. Natuurk., Sect. 2* 45(1): 1–39. Tripp et al., 2013. Phylogenetic relationships within Ruellieae (Acanthaceae) and a revised classification. *Intern. J. Pl. Sci.* 174: 97–137; Tripp & Fekadu, 2014. Comparative leaf and stem anatomy in selected species of Ruellieae (Acanthaceae) representative of all major lineages. *Kew Bull.* 69: 1–8.

About 30 genera and 2500 species: worldwide; eight genera and 105 (35 endemic) species in the Pan-Himalaya.

8. ERANTHEMUM L.

Linnaeus, Sp. Pl. 1: 9. 1753; Nees in de Candolle (ed.), Prodr. 11: 445. 1847; Benth & Hooker, Gen. Pl. 2: 1097. 1876; Lindau in Engler & Prantl (eds.), Pflanzenfam. 4(3b): 311. 1895; Nasir & Ali in Malik & Ghafoor (eds.), Fl. Pakistan 188: 75. 1988; Wood in Grierson & Long (eds.), Fl. Bhutan 2(3): 1275. 2001; Tsui in Hu (ed.), Fl. Reipubl. Popularis Sin. 70: 58. 2002; Hu, Deng & Daniel in Wu, Raven & Hong (eds.), Fl. China 19: 432. 2011. ≡ *Pigafetta* Adanson, Fam. Pl. 2: 223. 1763, non *Pigafetta* (Blume) Becc., Malesia 1: 89. 1877, nom. orth. & type cons. Type: *Eranthemum capensis* L.

Daedalacanthus T. Anderson, Enum. Pl. Zeyl.: 229. 1864; Bentham & Hooker, Gen. Pl. 2: 1097. 1876; Clarke in Hooker (ed.), Fl. Brit. India 4: 417. 1885. Type: *Daedalacanthus montanus* (Roxb.) T. Anderson (≡ *Eranthemum montanum* Roxb. = *Eranthemum capensis* L.).

Upudalia Raf., Fl. Tellur. 4: 66. 1838. Type: *Upudalia pulchella* (Andrews) Raf. (≡ *Eranthemum pulchellum* Andrews).

Shrubs or perennial herbs; with cystoliths. Leaves petiolate; leaf blade entire or crenate. Spikes terminal or rarely axillary, sometimes in panicles. Bracts sometimes colored; bracteoles small, narrow. Calyx 5-lobed; lobes narrow, subequal. Corolla subsalverform; tube base narrowly cylindric, slender, throat sometimes inconspicuous; limb 5-lobed; lobes obovate, subequal, contorted in bud. Stamens 2, inserted below throat; anthers bithecous; thecae parallel, base muticous; staminodes 2, clavate or filiform. Ovary with 2 ovules per locule; style filiform, glabrous or pubescent; stigma unequally 2-lobed. Capsules clavate, with a solid stalk at base, 4-seeded; retinacula present. Seeds discoid, compressed, pubescent with hygroscopic trichomes. **Chromosomes** 2n = 24, 30, 40, 42, 84.

Recent references: Fosberg & Sachet, 1953. Reinstatement of *Eranthemum pulchellum* Andr. *Taxon* 2: 135–136; Wood, 1994. Notes relating to the Flora of Bhutan: XXIX Acanthaceae, with special reference to *Strobilanthes*. *Edinburgh J. Bot.* 51(2): 175–273.

About 40 species: tropical and subtropical Asia; six species (two endemic) in the Pan-Himalaya.

1a. Corolla pink or red; tube ventricose from the middle.
  2a. Corolla pink; bracts and leaves almost glabrous .... 1. *E. griffithii*
  2b. Corolla crimson to orange-red; bracts and leaves distinctly pubescent at least on veins.
    3a. Corolla tube less than 3 cm long; bracts obovate, abruptly contracted to point .... 2. *E. splendens*
    3b. Corolla tube 3–3.4 cm long; bracts elliptic to broadly ovate, gradually narrowed to point .... 3. *E. tubiflorum*
1a. Corolla blue; tube narrowly cylindric for almost all its length.
  4a. Bracts glabrous to pubescent, apex mucronate; inflorescences of paniculate spikes .... 4. *E. pulchellum*
  4b. Bracts ciliate, apex acuminate but not mucronate; inflorescences of solitary spikes.
    5a. Spikes 3–11 cm long, elongated, 5–several-flowered .... 5. *E. ciliatum*
    5b. Spikes 1–2 cm long, subcapitate, 3–5-flowered .... 6. *E. wardii*

1. Eranthemum griffithii (T. Anderson) Bremek. & Nann.-Bremek., Verh. Kon. Ned. Akad. Wetensch., Afd. Natuurk. 45(1): 35. 1948; Wood in Grierson & Long (eds.), Fl. Bhutan 2(3): 1276. 2001. ≡ *Daedalacanthus griffithii* T. Anderson, J. Linn. Soc., Bot. 9: 486. 1867; Clarke in Hooker (ed.), Fl. Brit. India 4: 418. 1884. Type: India, Assam, Ghoorgong, 2 Jan. 1838, *W. Griffith 1437* (lectotype designated by Wood, Edinburgh J. Bot. 51: 180. 1994: K).

Subshrubs, 1–2.5 m high; stems tetragonous, glabrous except for few hairs near nodes. Petiole 0.5–1.5 cm long, subglabrous; leaf blade lanceolate to narrowly elliptic, 5–26 cm long, 2–8 cm broad, adaxially glabrous, abaxially glabrous or puberulent, lateral veins 8–10 pairs, base attenuate, margin entire, apex acute or slightly falcate. Spikes terminal and axillary, branched, 4–15 cm long; peduncles 0–3 cm long. Bracts elliptic, 2–3 cm long, 1.1–1.5 cm broad, glabrous, variegated green and yellow, apex acuminate; bracteoles linear, 5–6 mm long. Calyx 6–8 mm long, sparsely glandular-pubescent, 5-lobed to the middle; lobes linear, apex acuminate. Corolla pink, curved, sparsely glandular-hairy; tube 2–3 cm long, cylindric, gradually widened to ca. 8 mm broad at mouth; lobes ovate, 4–6 mm long, 3–4 mm broad. Anthers 4–5 mm long. Capsules ellipsoid, 1.6–2 cm long, glabrous. Seeds ovate in outline, 0.8–1 mm long and broad. Figure 46-24.

Phenology: flowering from December to February.

Habitats: forests; 200–900 m.

Distribution: Bhutan (Map 46-43); also in other parts of N India.

BHUTAN: Deothang, s. d., *W. Griffith s. n.* (BM); **Phuntsholing**, Torsa River, 200 m, 11 Feb. 1982, *A. J. C. Grierson & D. G. Long 2860* (E, K); between Kharbandi (Richending) and Sorchen, 700 m, 18 Mar. 1990, *J. R. I. Wood 7238* (E); between Kharbandi (Richending) and Sorchen,700 m, 26 Nov. 1991, *J. R. I. Wood 7522* (E); W side of Torsa River, 400 m, 19 Jan. 1992, *J. R. I. Wood 7530* (E); **Sankosh**, Sarbhang, Pinkhua, 310 m, 16 Mar. 1982, *A. J. C. Grierson & D. G. Long 3752* (E, K); between Sarbhang and Damphu, 700m, 19 Feb. 1988, *J. R. I. Wood 6149* (E); Sarbhang River, towards Damphu, 500m, 18 Feb. 1990, *J. R. I. Wood 7233* (E).

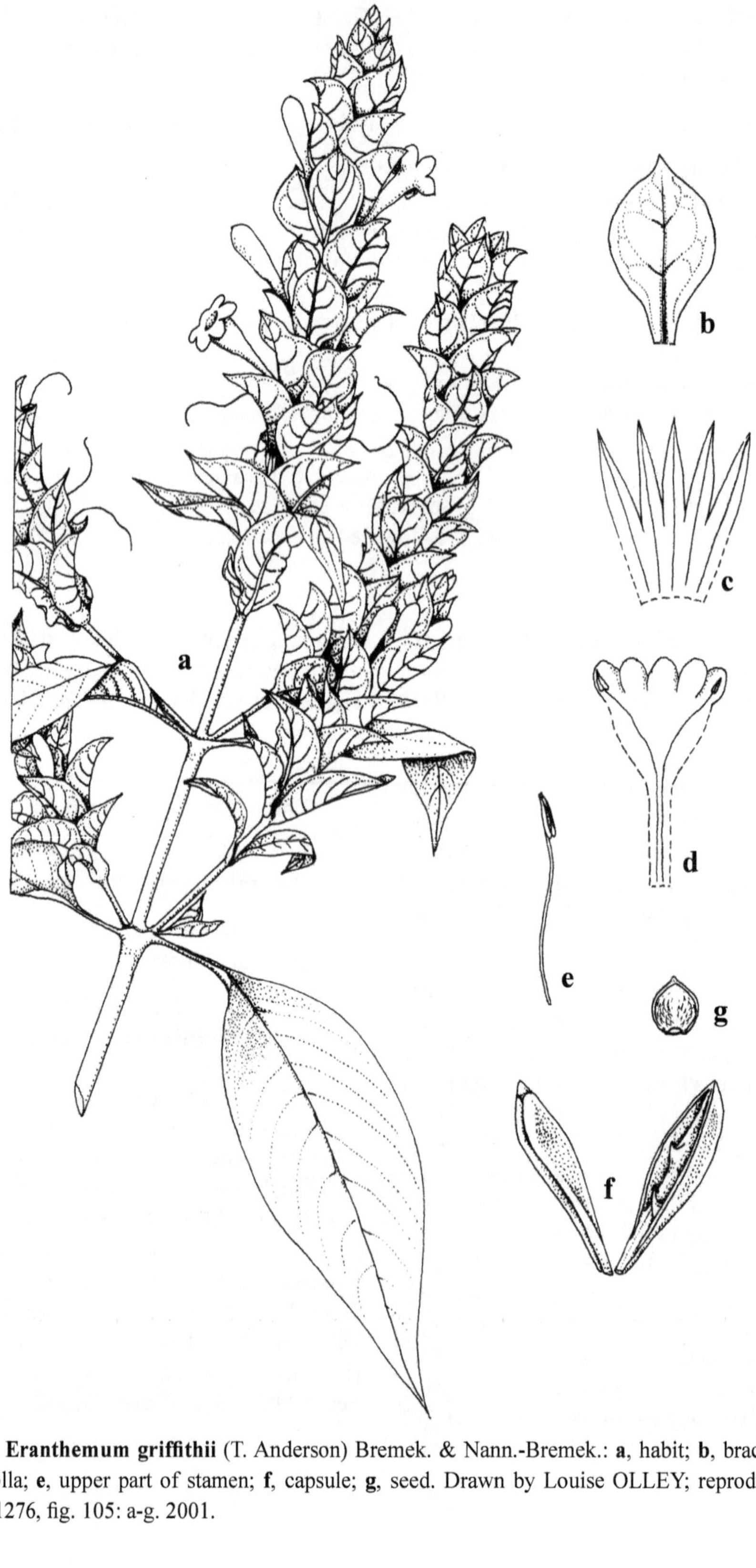

Figure 46-24: **Eranthemum griffithii** (T. Anderson) Bremek. & Nann.-Bremek.: **a**, habit; **b**, bract; **c**, calyx; **d**, disserted corolla; **e**, upper part of stamen; **f**, capsule; **g**, seed. Drawn by Louise OLLEY; reproduced from *Fl. Bhutan* 2(3): 1276, fig. 105: a-g. 2001.

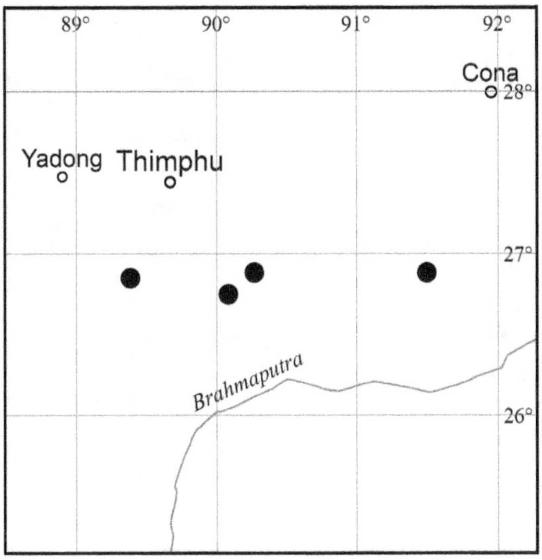

Map 46-43: **Eranthemum griffithii** (T. Anderson) Bremek. & Nann.-Bremek.

2. Eranthemum splendens (T. Anderson) Hort. ex Siebert & Voss, Vilm. Blumengartn., ed. 3, 1: 806. 1894; Wood in Grierson & Long (eds.), Fl. Bhutan 2(3): 1278. 2001. ≡ *Daedalacanthus splendens* T. Anderson, J. Linn. Soc., Bot. 9: 486. 1867; Clarke in Hooker (ed.), Fl. Brit. India 4: 418. 1884. Type: [Sikkim & Darjiling] India, Darjiling, *Griffith Kew Distrib. No. 6129* [lectotype designated by Wood, Edinburgh J. Bot. 51: 181. 1994: K (K000882643); isolectotypes: C (C10004857, C10004858)].

Shrubs, to 1.5 m high; stems pubescent upwards. Petiole 1–2.5 cm long, pubescent; leaf blade ovate to elliptic, 5–20 cm long, 3–8 cm broad, sparsely pilose, lateral veins 7–10 pairs, base cuneate or attenuate, decurrent onto petiole, margin entire, apex shortly acuminate. Spikes 5–35 cm long, branched; peduncle 2–5 cm. Bracts obovate, 1.5–2.5(–4.5) cm long, 0.8–1.7(–2.1) cm broad, glandular-pubescent, strongly varigated-reticulate with green veins, apex mucronate; bracteoles linear-lanceolate, 5–12 mm long, glandular-pubescent. Calyx 7–9 mm long, glandular-puberulent; lobes linear-lanceolate, 1–1.2 mm broad. Corolla orange-red, 4.2–5 cm long, slightly curved, pubescent; tube 2.3–2.8 cm long, 6–7 mm broad at mouth; lobes elliptic, 5–6 cm long, 3–4 mm broad, patent or recurved. Filaments glabrous, 2–2.5 cm long; anthers oblong, 2.5–3 mm long. Ovary 2.5–3 mm long; style glabrous. Capsules clavate, 1.5–1.8 cm long, glandular-pilose. Seeds ovate in outline, compressed, 2.5–3 mm in diameter, glandular-pilose. **Chromosomes** 2n = 24.

Phenology: flowering from December to January; fruiting from March to April.

Habitats: stream banks, in forests; 200–1700 m.

Distribution: E Nepal, Sikkim & Darjiling, Bhutan (Map 46-44); also in Bangladesh, and other parts of India.

E NEPAL: Mechi, Ilam, Ilam to Ranga Pani, 700–1200 m, 8 Dec. 1963, *H. Hara et al. 3829* (TI); Ilam, Ranga Pani to Ghorwa, 300–700 m, 9 Dece. 1963, *H. Hara et al. 4230* (TI); Ilam, Garuwa, Ghorwa to Sanichare, 200–300 m, 10 Dec. 1963, *H. Hara et al. 4256* (BM, TI). **SIKKIM & DARJILING: Darjiling**, Sivoke Hills, 600m, *J. S. Gamble 7562* (K); s. loc., 300–900 m, *J. D. Hooker Acanth. 23* (BM, C, CGE, K); Shekhin, *J. S. Gamble 2420* (K); Panchkilla, 300 m, *C. B. Clarke 27079* (K, BM); Panchkilla, 300 m, 1876, *C. B. Clarke 31712* (CAL, K); **Sikkim**, Kurseong, 1857, *Thomson s. n.* (CAL);

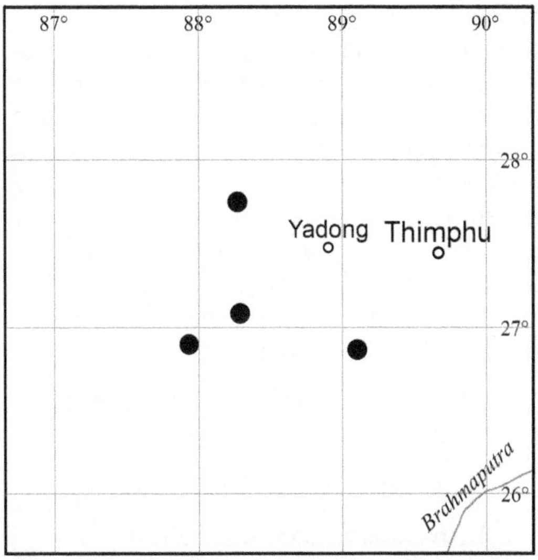

Map 46-44: **Eranthemum splendens** (T. Anderson) Hort. ex Siebert & Voss

Pankabary, 600 m, 1873, *J. S. Gamble 3537* (CAL, K). **BHUTAN: Sanchi**, Kharpaga Hill, 390 m, 1963, *K. Thothathri 10347* (CAL).

3. **Eranthemum tubiflorum** (T. Anderson) Radlk. ex Lindau in Engler & Prantl (eds.), Nat. Pflanzenfam. 4(3b): 311. 1895; Wood in Grierson & Long (eds.), Fl. Bhutan 2(3): 1276. 2001. ≡ *Daedalacanthus tubiflorus* T. Anderson, J. Linn. Soc., Bot. 9: 485. 1867; Clarke in Hooker (ed.), Fl. Brit. India 4: 417. 1884. Type: Yarlung Zangbo-Brahmaputra, Mishmi, *Griffith ex Kew Distrib. No. 6122* [lectotype designated by Wood, Edinburgh J. Bot. 51: 180. 1994: K (K000882640)].

Daedalacanthus roseus var. *neesii* C. B. Clarke in Hooker (ed.), Fl. Brit. India 4: 420. 1884. ≡ *Eranthemum roseum* var. *neesii* (C. B. Clarke) Karthik & Moorthy, Fl. Pl. India 1: 16. 2009. Type: India, Assam, *Griffith s. n.* (holotype: K).

Subshrubs, 0.6–1.2 m high; stems subquadrangular, bifariously pubescent, then glabrescent. Petiole 1.5–2 cm long, pubescent; leaf blade elliptic, 9–23 cm long, 5–8 cm broad, adaxially glabrous, abaxially sparsely pubescent, especially along veins, lateral veins 10–13 pairs, base attenuate, margin undulate or obscurely crenulate, apex acute. Spikes axillary, solitary, 3–6 cm long; peduncles 0.8–1.5 cm long. Bracts elliptic to broadly ovate, 1.4–2.2(–2.8) cm long, 0.9–1.4 cm broad, densely pubescent on veins, white with green reticulate veins, margin entire and ciliate, apex tapering mucronate; bracteoles sublate or lanceolate, 4–5 mm long, scarious, apex acuminate. Calyx 0.8–1.0 cm long; lobes lanceolate, puberulent, apex acuminate. Corolla crimson, sparsely glandular-pubescent; tube 3.0–3.4 cm long; lobes oblong, 6–9 mm long, 4–6 mm broad. Filaments glabrous; anthers ellipsoid, 4–5 mm long. Capsules unknown.

Phenology: flowering from December to February.

Habitats: stream banks; ca. 300 m.

Distribution: endemic; Bhutan, Yarlung Zangbo-Brahmaputra (Map 46-45).

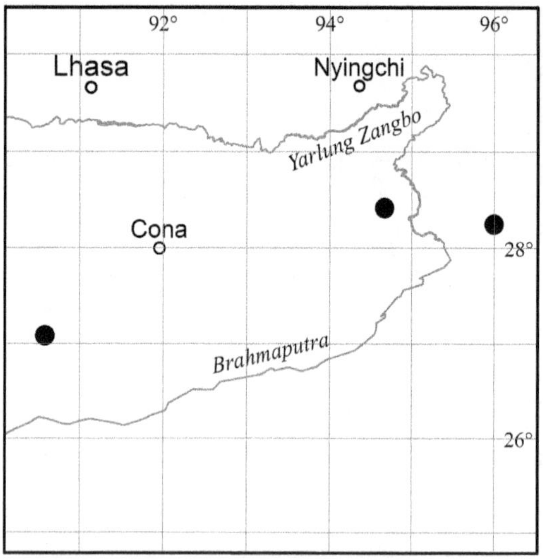

Map 46-45: **Eranthemum tubiflorum** (T. Anderson) Radlk. ex Lindau

BHUTAN: Gaylegphug, 1963, *S. Sen Gupta 847* (CAL). **YARLUNG ZANGBO-BRAHMAPUTRA: Abor Hills**, 300 m, *Kingdon-Ward 8753* (K); **Mishmi Hills**, 1 Dec. 1836, *Griffith, Kew Distrib. No. 6122* (K); the bank of the Brahmaputra River, s. d., *Lowther s. n.* (CAL).

4. **Eranthemum pulchellum** Andrews, Bot. Repos. t. 88. 1798; Nasir & Ali in Malik & Ghafoor (eds.), Fl. Pakistan 188: 76. 1988; Wood in Grierson & Long (eds.), Fl. Bhutan 2(3): 1278. 2001; Tsui in Hu (ed.), Fl. Reipubl. Popularis Sin. 70: 59. 2002; Hu, Deng, & Daniel in Wu, Raven & Hong (eds.), Fl. China 19: 432. 2011. ≡ *Justicia pulchella* (Andrews) Roxb., Pl. Corom. 2: 41. t. 177. 1798. ≡ *Ruellia varians* Vent., Descr. Pl. Nouv. Hort. Cels. t. 46. 1801, nom. illeg. (Art. 52.1 & 52.2). ≡ *Upudalia pulchella* (Andrews) Raf., Fl. Tellur. 4: 66. 1838 "1836". ≡ *Pseuderanthemum pulchellum* (Vahl) Merr., Philipp. J. Sci., C 7: 248. 1912. Type: [*icon*] Andrews, Bot. Repos. 2: pl. 88. 1798 (lectotype designated by Nasir & Ali, 1988).

Justicia nervosa Vahl, Enum. Pl. 1: 164.1804. ≡ *Eranthemum nervosum* (Vahl) R. Br., Prodr.

1: 477. 1810; Nees in de Candolle (ed.), Prodr. 11: 465. 1847. ≡ *Daedalacanthus nervosus* (Vahl) T. Anderson, J. Linn. Soc., Bot. 11: 487. 1876; Clarke in Hooker (ed.), Fl. Brit. India 4: 418. 1884. Type: India, "Habitat in India orientali, 7 Jan. 1799, *J. P. Rottller464* [lectotype **here designated**: C (C10004854); isolectotypes: C (C10004855), M (M0186318)].

Eranthemum nervosum (Vahl) R. Br. ß *scabrum* Nees, Pl. Asiat. Rar. 3: 106. 1832. Type: [U Ganga & Indus] India, Srinagur, Rungpur, *Kamrup in Wallich Cat. 2493g* [lectotype **here designated**: K-W (K001116198); isolectotype: K (K001116197)].

Eranthemum edgeworthianum Nees in de Candolle (ed.), Prodr. 11: 44. 1847. Type: India, Mar Bar, 2000–3000 ft., Sep. 1840, *M. P. Edgeworth 246* [lectotype **here designated**: GZU (GZU000249935)].

Herbs or subshrubs; 1.2–1.5 m high; stem glabrous. Petiole 0.2–1.5 cm long, glabrous; leaf blade ovate to elliptic, 2–6 cm long, 2–4 cm broad, glabrous, lateral veins 3–7 pairs, base cuneate, margin entire, apex acute to caudate-acuminate. Panicles terminal or axillary, 3–5 cm long; peduncles 1.5–3 cm long. Bracts foliaceous, elliptic, 1–2 cm long, 0.5–1 cm broad, glabrous to pubescent; bracteoles lanceolate, 4–8 mm long, glabrous, margin ciliate, apex acuminate. Calyx 0.6–1.2 cm long; lobes lanceolate, 0.5–1 cm long, 1–1.2 mm broad. Corolla blue, 2.5–3 cm long, curved, outside puberulent, inside bifariously pubescent; tube base cylindric, 2–3 cm long, 1.5–2 mm broad, abruptly widened to 1.5–1.8 cm broad at mouth; lobes oblong or obovate, 1–1.5 cm long, 1–1.4 cm broad. Filaments 0.5–1 cm long, glabrous; anther-thecae oblong, 2–4 mm long. Staminodes 2, 1.5–2 mm long. Ovary ellipsoid, 3–5 mm long, glabrous; style 3–4 cm long, pubescent; stigma 2-lobed. Capsules oblong, 1.5 cm long, glabrous. Seeds orbicular, 4–5 mm in diameter, compressed, hairy. **Chromosomes** 2n = 42, 44, 54. Figure 46-25.

Phenology: flowering and fruiting from November to April of next year.

Habitats: moist gullies, stream banks, dry forests or cultivated; 100–2300 m.

Distribution: N Pakistan, Jammu & Kashmir, U Ganga & Indus, W Nepal, C Nepal, E NEPAL, Sikkim & Darjiling, Bhutan (Map 46-46); also in Bangladesh, and other parts of India and Pakistan.

N PAKISTAN: Hazara, Murree, Chattar Bagh, Murree Road, 750 m, 1954, *R. R. Stewart 25730* (A). **JAMMU & KASHMIR: Jammu**, Dewan Garden, 5 Mar. 1943, *B. M. Sharma 450* (RRLH). **U GANGA & INDUS: Dehradun**, 690 m, 21 Apr. 1975, *D. Chambulain 19* (E); **Haridwar**, 1825, *Wallich Cat. 2493f* (K-W); **Kangra**, 900 m, 6 Apr. 1900, *G. S. Hart 43* (E); **Mussorie**, *M. B. Saxena 418* (DD, fide Raizada & Saxena, 1978); **Simla**, 1887, *J. R. Drummond 20940* (K); **Srinagur**, Rungpur, s. d., *Wallich Cat. 2493g* (K-W). **W NEPAL: Bheri**, Banke, Amila Nara, 250 m, 11 Apr. 1974, *J. F. Dobremez 2389 (N. P. Ranaudha 74-18)* (E, KATH); Dailekh, 1976, *L. R. Sharma 1583* (KATH); Dailekh, Dullu, 1210 m, 7 Mar. 1977, *K. J. Malla 1674* (KATH); **Mahakali**, Kanchanpur, Shuklaphata, 180 m, 1980, *T. K. Bhattacharya 3686* (KATH); **Rapti**, Dang, Rapti Nadi, 300 m, 24 Mar. 1952, *O. V. Polunin, W. R. Sykes & L. H. J. Williams 3618* (A, BM); Dang, Rapti Valley, Hetaura, 480 m, 10 Mar. 1966, *J. D. A. Stainton 5224* (BM, KATH); Dang, Chorkholi, Koilabas, 240 m, 20 Mar. 1982, *K. J. Malla 225* (KATH); Pyuthan, Bijuwar, 780 m, 6 Mar. 1976, *N. P. Manandhar & P. M. Regmi 197* (KATH); Salyan, Sarda Khola Valley, Dharkhani Village, 1150 m, 30 Mar. 1952, *O. V. Polunin, W. R. Sykes & L. H. J. Williams 1805* (BM, E); **Seti**, Bajura, Kandra, 180 m, 1975, *L. R. Sharma 839* (KATH); Kailali, Burtola to Chisapami, 28 Apr. 1971, *P. R. Shakya & D. P. Joshi 377* (KATH); Kailali, Dhangarhi, 1 Apr. 1974, *N. P. Manandhar et al. 33* (KATH). **C NEPAL: Bagmati**, Dailekh, Kharguar, 1979, *S. Dayal 2379* (KATH); Kavre Palanchok, Phulchoki, Thulo Pahara, 1210 m, 30 Mar. 1967, *R. K. Manandhar et al. 6845* (KATH); Lalitpur, Godavari, 1500 m, 10 Apr. 1973, *AMA 729* (E); Lalitpur, Godawati, 1670 m, 18 Mar. 1966, *P. Pradhan 4310* (KATH); **Dhawalagiri**, Baglung, Burtibang, east of Kutharpekot, 1200 m, 28 Apr. 1954, *J. D. A. Stainton, W. R. Sykes &*

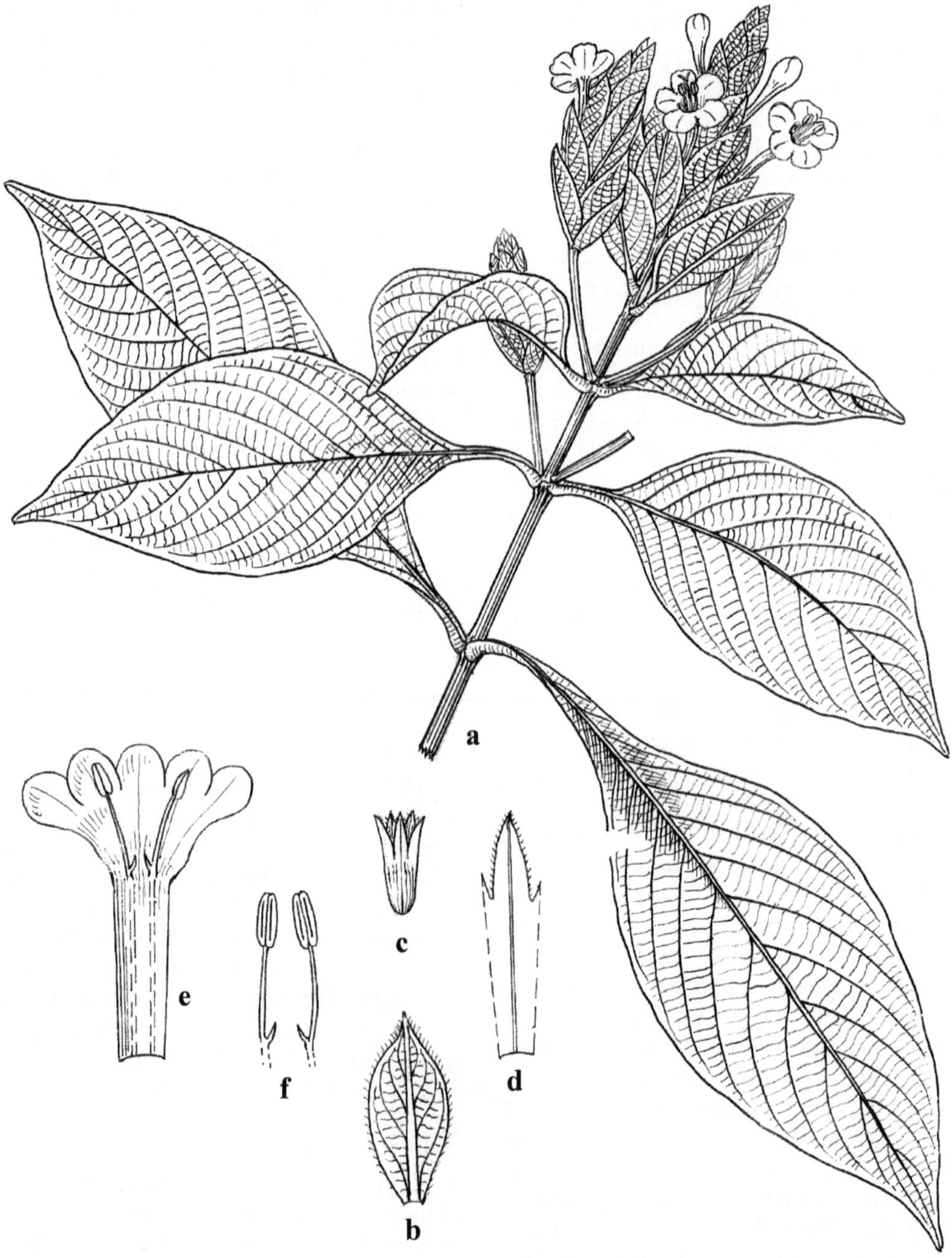

Figure 46-25: **Eranthemum pulchellum** Andrews: **a**, portion of plant; **b**, bract; **c**, calyx; **d**, calyx lobe; **e**, corolla opened showing androecium; **f**, stamens. Drawn by Han-Ping YU; reproduced from *Fl. Reipubl. Popularis Sin.* 70: 57, pl. 6: 6–12. 2002.

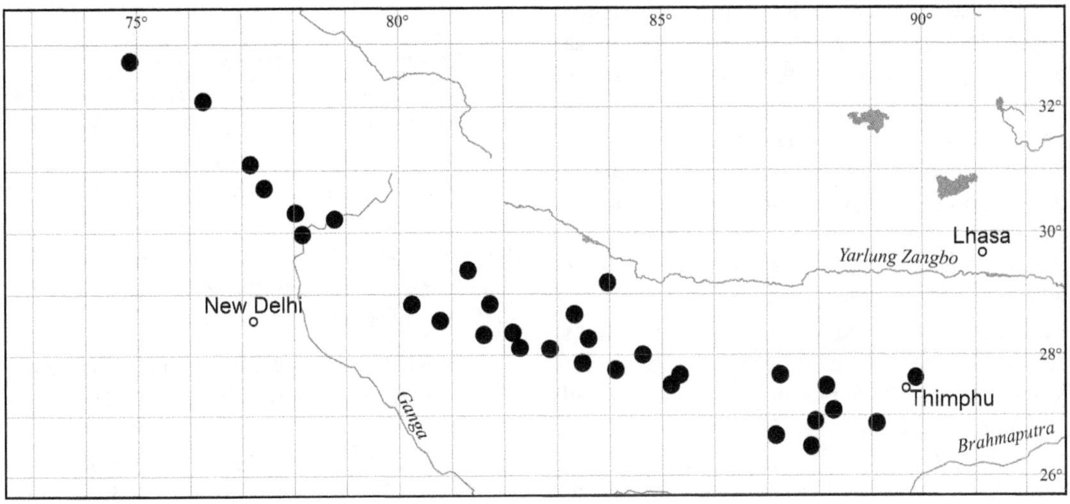

Map 46-46: **Eranthemum pulchellum** Andrews

L. H. J. Williams 297 (BM, E); Mustang, Lete, 2300 m, 23 Mar. 1978, *P.M. Amatya & B. Roy 407/78* (KATH); Myagdi, Darbang, 300 m, 1979, *T. K. Bhattachaya & K. J. Malla 308* (KATH); **Gandaki**, Gorkha, Budhi Gandaki River, 9 miles above Arughat Bazar, 610 m, 11 Apr. 1953, *P. C. Gardner 279* (E); **Karnali**, Jumla, Jaspal to Gothdanda, 1969, *H. Kanai & M. S. Bista 12209* (KATH, TI); **Lumbini**, Nawalparasi, Barghat, Nawal Parasi, 150m, 20 Mar. 1971, *J. Makin 99* (BM); Palpa, Basani Khola, 1680 m, 1 Mar. 1976, *N. P. Manandhar & P. M. Regmi 43* (KATH); Rapandehi, Butwal, 300 m, 6 Mar. 1966, *P. R. Shakya 2249* (KATH); **Narayani**, Bara, Adhabar, Parsa Wildlife Reserve, 235 m, 8 Oct. 1995, *M. Mikage et al. 9552763* (TI); Chitwan, between Kashara and Dhoba, 15 Dec. 1976, *H. Tabata et al. 7618* (KYO); Chitwan, Sauraha to Tal, 18 Jan. 1996, *M. Mikage et al. 9611793* (TI); Makwanpur, Churia to Mai Area, 610 m, 19 Mar. 1966, *T. B. Shrestha 3934* (KATH); Makwanpur, Hetauda, 650 m, 26 May 1986, *N. K. Bhattarai 86/106* (KATH). **E NEPAL: Koshi**, Sankhuwasabha, Sedua, 1800 m, 15 Sep. 1971, *M. Ohsawa & P. R. Shakya 1805* (TI); Sankhuwasabha, Mangsewa Phedi, 600 m, 20 Mar. 1989, *T. B. Shrestha & P. R. Shakya 9303* (KATH); Sunsari, Sanguri, 19 Mar. 1978, *S. Bhattarai et al. 7856* (KATH); **Mechi**, Ilam, Mai Khola, 300 m, 25 Feb. 1969, *J. D. A. Staiinton 6413* (BM); Ilam, on the way to Soktim, 300 m, 2 Apr. 1967, *J. B. Malla 7799* (KATH); Jhapa, KankaniRiver, 250 m, 1967, *D. H. Nicolson 3109* (BM); Tapjung, Tuwa to Taplethok, 5 Nov. 1963, *H. Hara et al. s. n.* (TI); **Sagarmatha**, Udayapur, Simule, 210 m, 25 Oct. 1995, *M. Mikage et al. 9558071* (TI). **SIKKIM & DARJILING: Darjiling**, above Tista Bazar, 500 m, 13 Aug. 1963, *H. Hara et al. 6300280* (KATH); Tista, 350 m, 20 Jun. 1960, *H. Kanai et al. 5900* (A, KYO); **Sikkim**, Sikkim Tarai, Jan. 1890, *H. H. Heinis 2429* (K). **BHUTAN: Phuntsholing**, 11 Dec. 1963, *G. V. Subla Rao 58* (A); **Punakha**, ca. 30 km south of Wangdi towards Chirang, 1100 m, 22 Apr. 1990, *J. R. I. Wood 7246* (E); Sankosh Valley ca. 20 km S of Wangdi Phodrang, 900 m, 1 Apr. 1989, *J. R. I. Wood 6882* (E); **Samchi**, Dain Khola, 360 m, 1 Mar. 1982, *A. J. C. Grierson & D. G. Long 3335* (A, E, K).

5. Eranthemum ciliatum (Craib) Benoist, Fl. Indo-Chine 4: 661. 1935; Wood in Grierson & Long (eds.), Fl. Bhutan 2(3): 1279. 2001; Hu, Deng & Daniel in Wu, Raven & Hong (eds.), Fl. China 19: 432. 2011. ≡ *Daedalacanthus ciliatus* Craib, Bull. Misc. Inform. Kew 1913: 202. 1913. Type: Thailand, Nan, Lao Cha Hawn, by edge of stream in evergreen jungle, 195 m, 21 Feb. 1912, *A. F. G. Kerr 2398* [lectotype **here designated**: K (K000882623); isolectotypes: BM (BM000950036), E, K (K000882624)].

Herbs to 50 cm high; stems quadrangular, pubescent. Petiole 0.7–2.5 cm long, pilose; leaf blade elliptic, ovate or elliptic-ovate, 2–10 cm long, 1–4.5 cm broad, abaxially pubescent along veins, adaxially glabrous, lateral veins 4–6 pairs, base cuneate to subrounded, usually decurrent onto petiole, margin entire, apex acute to acuminate. Spikes 3–11 cm long; peduncle 1.5–3 cm long, densely villous. Bracts yellowish white with green along veins, ovate to elliptic, 1.2–2 cm long, 0.4–0.8 cm broad, abaxially villous, adaxially glabrous, margin ciliate; bracteoles triangular-ovate, 5–6 mm long, 1–1.5 mm broad, margin ciliate, apex acute. Calyx 5–7 mm long, abaxially pilose, 5-lobed to the middle; lobes lanceolate, 0.8–1 mm broad. Corolla bluish-purple, 3.5–4 cm long, glabrous or pubescent; tube base cylindric, 2.5–3 cm long, throat 3.5–4 mm long; lobes obovate, 6–8 mm long, 5–8 mm broad, apex 2-lobed. Filaments 3–4 mm long. Ovary glabrous or pubescent; style pilose. Capsules ellipsoid, 1–1.5 cm long, glabrous or pubescent. Seeds ovate in outline, black, 3–4 mm long, 2.5–3 mm broad.

Phenology: flowering from November to January of next year.

Habitats: forests, thickets in ravines; below 1700 m.

Distribution: W Nepal, C Nepal, E Nepal, Bhutan (Map 46-47); also in Myanmar, Laos, and Thailand.

Notes: The species was previously misidentified as *Eranthemum strictum* Colebr. ex Roxb. by Yamazaki (in Bull. Univ. Museum, Univ. Tokyo 2: 1–393. 1966) or *E. purpurascens* Nees by Press et al. (Annotated Checklist of the Flowering Plants of Nepal: 1. 2000). *E. purpurascens* Nees is easily distinguished from *E. ciliatum* by bracts ovate, obtuse and tipped at apex and spikes often long-pedunculate.

W NEPAL: Bheri; Banke, Shame Shergunj, 6 Nov. 1980, *R. N. Shukla 3388* (KATH); Banke, Napelganj, 1977, *R. Rana 1858* (KATH); Bardiya, Babai, 9 Dec. 1983, *T. K. Bhattachaya & P. P. Kurmi 2540* (KATH); Bardiya, Kohalpur Dudwa, 1973, *N. P. Manandhar & K. J. Malla 17626* (KATH); Bardiya, Royal Bardia National Park, Tiger Top, 3 Oct. 1996, *M. Mikage et al. 9689132* (TI); Surkhet, Kutke Danda, 1330 m, 18 Oct. 1976, *L. R. Sharma 1512* (KATH); Surkhet, Katuka to Rata Nangla, 1470 m, 23 Jul. 1985, *P. R. Shakya et al. 8366* (KATH); Surkhet, Ramghat, 1982, *S. R. Adhikary & N. P. Manandhar 9168* (KATH); **Karnali**, Mugu, Padikhola, 1980, *T. K. Batthacharya 175* (KATH); **Rapti,** Dang, Harnok, 1520 m, 22 Sep. 1984, *P. M. Regmmi 2852* (KATH); Dang, Kurpani, 1670 m, 3 Sep. 1952, *O. V. Polunin, W. R. Sykes & L. H. J. Williams 1326* (BM); Dang, Ranitakura, 1200 m, 24 Oct. 1952, *O. V. Polunin, W. R. Sykes & L. H. J. Williams 5878* (BM, E);

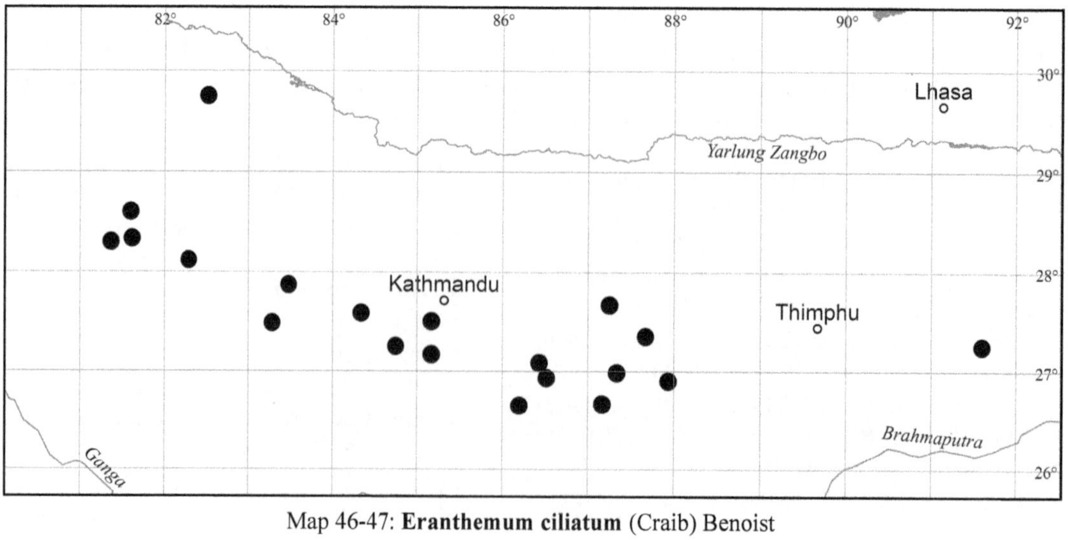

Map 46-47: **Eranthemum ciliatum** (Craib) Benoist

Seti, Kailali, outer Siwalik, 220 m, 6 Dec. 1966, *D. H. Nicolson 2841* (BM, KATH); Pyuthan, Bardara, 1979, *T. K. Bhattacharya & S. B. Malla 2856* (KATH). **C NEPAL: Bagmati**, Kathmandu, Naubise–Mugling, 5 Nov. 2000, *T. Watanabe et al. s. n.* (MBK); Rasuwa, Bagdeo to Saraswati, 120 m, 4 Nov. 1970, *H. Kanai & P. R. Shakya 674762* (KATH, TI); Rasuwa, Lenti Khola, 360 m, 1981, *P. R. Shakya 6469* (KATH); **Lumbini**, Nawalparasi, Daunne, Daunne Hill, 420 m, 8 Nov. 2012, *B. Adhikari & G. D. Bhatt 22* (E); Palpa, Ramdighat, Koli Gandaki, 610 m, 3 Oct. 1954, *J. D. A. Stainton, W. R. Sykes & L. H. J. Williams 8734* (BM); Palpa, between Rumre and Bodegaum, 700 m, 18 Sep. 1953, *T. Fujimura 20* (KYO); Rupandehi, Butwal to Nawakot, 220–840 m, 1973, *D. P. Joshi & M. M. Amatya 73/880* (KATH); **Narayani**, Bara, on south side of main Terai highway east from Pathlaiya, 100 m, 26 Nov. 2004, *C. A. Pendry et al. (Second Darwin Nepal Fieldwork Training Expedition) A114* (E); Bara, Pathlaiya, 160 m, 19 Nov. 2012, *B. Adhikari & D. R. Kandel 59* (E); Chitawan, Tiges Top, 200 m, 1975, *D. P. Joshi et al. 75/3507* (KATH); Makwanpur, Bhainse, 520 m, 6 Nov. 2012, *B. Adhikari & G. D. Bhatt 4* (E); Makwanpur, Nayabasi, Manahari, 4 km W Lothar, 270 m, 8 Nov. 2012, *B. Adhikari & G. D. Bhatt 18* (E); **Parsa**, Parsa, Wildlife Reserve, along Mahadev Khola, 190–200 m, 9 Oct. 1995, *M. Mikage et al. 9552775* (E, TI); **E NEPAL: Koshi**, Dhankuta, 400m, 11 Oct. 1971, *J. F. Dobremez 1378* (BM, E); Dhankuta, Mohamabedi Khola to Dhara Pani, 700 m, 9 Jul. 1972, *H. Kanai et al. 721219* (TI); Dhankuta, Teku Nala to Dhara Pani, 400 m, 4 Sep. 1977, *H. Ohashi et al. 771333* (KATH, TI); Samkhuwasabha, Arun Valley, Tumlingtar, West of Chainpur, 300 m, 28 Sep. 1956, *J. D. A. Stainton 1786* (BM, E); Samkhuwasabha, E bank of Arun River, Piluwa Khola, 310 m, 24 Sep. 2001, *D. G. Long et al. ENEP46* (E, TI); Sunsari, Chatara, 120 m, 25 Nov. 2012, *B. Adhikari & D. R. Kandel 89* (E); Sunsari, Dharan, 390 m, 22 Sep. 1967, *L. H. J. Williams & J. D. A. Stainton 8552* (BM, K, TI); Sunsari, Mulghat, 200 m, 1977, *S. K. Dhungna 980* (KATH); **Mechi**, Ilam, Ghorwa to Sanichare, 10 Dec. 1963, *H. Hara et al. 6303681* (KYO, TI); Ilam, Dhara Pani to Phankuta, 11 Oct. 1963, *H. Hara et al. 6303682* (BM, E, K, KATH, KYO, TI); Taplejung, Nessum, 200 m, 1977, *S. K. Dhungana 904* (KATH); **Sagarmatha**, Siraha, d'Aridungaa Belsot, 320 m, 8 Nov. 1954, *A. Zimmermann 2117* (E, K); Udayapur, Simule–Dhaplang to Chamling, 200 m, 25 Oct. 1995, *M. Mikage et al. 9554136* (E, TI); Udayapur, Simule, 1 km S of Chamling, 170 m, 25 Oct. 1995, *M. Mikage et al. 9558071* (KATH); Trijuga Khla, 300 m, 5 Nov. 1969, *J. D. A. Stainton 6647* (BM, TI). **BHUTAN: Mongar**, near Lingmethang, 120 km W of Tashigang, 850 m, 1991, *C. Parker 4916* (E).

6. **Eranthemum wardii** (W. W. Sm.) N. P. Balakr., J. Bombay Nat. Hist. Soc. 67: 63. 1970. ≡ *Daedalacanthus wardii* W. W. Sm., Notes Roy. Bot. Gard. Edinburgh 10: 174. 1918. Type: [U Irrawaddy] Myanmar, Kachin State, Putao to Myitkyina, 19 Dec. 1914, *F. Kingdon-Ward 2003* [holotype: E (E00273510)].

Shrubs, to 1 m high; stems yellowish hirsute. Petiole 1–1.5 cm long, densely yellowish hirsute;

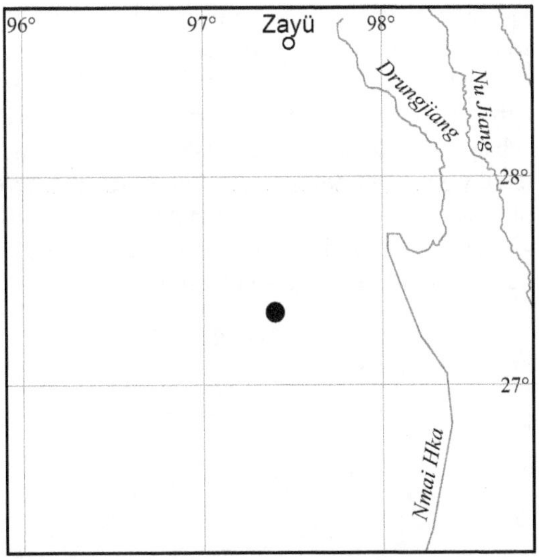

Map 46-48: **Eranthemum wardii** (W. W. Sm.) N. P. Balakr.

leaf blade elliptic-lanceolate or subelliptic, 5–8 cm long, 2–3.5 cm broad, yellowish pubescent along veins on both surfaces, base broadly cuneate or subrounded, margin undulate, apex shortly acuminate. Spikes subcapitate, 1–2 cm long, 3–5-flowered; peduncle 0.8–1.2 cm long, densely yellowish hirsute. Bracts lanceolate, 1.5–2 cm long, yellow hirsute, margin long ciliate, apex acuminate; bracteoles linear-lanceolate, 2–3 mm long. Calyx 2–3 mm long; lobes linear-lanceolate, pubescent. Corolla 2.5–3 cm long, violate, pubescent; tube curved, limb 0.8–1 cm broad at mouth; lobed oblong. Ovary and style sparsely hirsute. Capsules unseen.

Phenology: flowering from December to January of next year.

Habitats: forests.

Distribution: endemic; U Irrawaddy (Map 46-48).

Notes: This species is similar to *Eranthemum ciliatum* (Craib) Benoist and differs only by broader leaves and much shorter and few-flowered spikes. The further studies is necessary.

U IRRAWADDY: Putao, way to Myitkyina, 19 Dec. 1914, *F. Kingdon-Ward 2003* (E).

9. PETALIDIUM Nees

Nees in Wall., Pl. Asiat. Rar. 3: 75. 1832; Nees in de Candolle (ed.), Prodr. 11: 114. 1847; Bentham & Hooker, Gen. Pl. 2: 1081. 1876; Clarke in Hooker (ed.), Fl. Brit. India 4: 416. 1885; Lindau in Engler & Prantl (eds.), Nat. Pflanzenfam. 4(3b): 299. 1895; Nasir & Ali in Malik & Ghafoor (eds.), Fl. Pakistan 188: 68. 1988. Type: *Petalidium barlerioides* (Roth) Nees (≡ *Ruellia barlerioides* Roth).

Subshrubs. Leaves opposite; blade ovate or narrowly oblong, margin entire or toothed. Flowers solitary, usually crowded on lateral branches, sessile or subsessile. Bracts absent or small and narrow; bracteoles 2, ovate, reticulately veined and ultimately scarious. Calyx unequally 5-lobed to base; lobes narrowly linear-lanceolate. Corolla straight or curved; tube dilated upwards or infundibuliform; limb subequally 5-lobed; lobes twisted to left in bud. Stamens 4, didynamous; filaments monadelphous basally via a sheath; anthers bithecous, thecae oblong, sagittate and bristle-tipped at bases. Ovary bilocular, 2 ovules per locule; stigma unequally 2-lobed. Capsules compressed, slightly stipitate, 2- or 4-seeded; septa with retinacula separating elastically and rising from inner capsule wall at dehiscence. Seeds discoid, covered with hygroscopic hair, borne on retinacula.

About 35 species: mainly in tropical and S Africa and Indo-Pakistan subcontinent; one species in the Pan-Himalaya.

1. **Petalidium barlerioides** (Roth) Nees, Pl. Asiat. Rar. 3: 82. 1832; Clarke in Hooker (ed.), Fl. Brit. India 4: 416. 1885; Nasir & Ali in Malik & Ghafoor (eds.), Fl. Pakistan 188: 68. 1988. ≡ *Ruellia barlerioides* Roth, Nov. Pl. Sp.: 310. 1821. Type: India, *Heyne s. n.* (not located).

Ruellia bracteata Roxb., Fl. Ind. ed. 2. 3: 47. 1832. Type: [U Ganga & Indus] India, Deyra Dhun, 1825, *Roxburgh in Wallich Cat. 2389c* [lectotype **here designated:** K-W (K001115837); isolectotypes: CAL, K].

Subshrubs, 1–3 m high; stems subquadrangular, red glandular-hairy when young, then glabrescent. Leaves subequal or rarely unequal in each pair; petiole 4–10 mm long; blade elliptic-ovate to narrowly oblong, 5–15 cm long, 2–8 cm broad, grey puberulent along veins on both surfaces, base cuneate to slightly decurrent onto petiole, margin entire or crenulate-serrate, apex acute or aculminate. Flowers axillary, solitary or in small clusters, sessile or subsessile. Bracteoles ovate, 2–2.5 cm long, 1–1.5 cm broad, minutely white pubescent, 4–6-veined, base cuneate margin entire, apex acuminate. Calyx 1–1.3 cm long; lobes subequal, one slightly longer than others, linear-lanceolate or subulate, 0.8–1.2 cm long, 2–5 mm broad, glandular-pubescent abaxially, minutely pubescent adaxially, 1-veined, margin ciliate, apex acute. Corolla funnel-shaped, 3–3.5 cm long, pale blue or white, yellow-tinged with purple streaks, outside pubescent, inside glabrous except for a stout pilose retaining style; tube cylindric, 0.8–1 cm long, throat 1–1.2 cm long, gradually widened to 1.3–1.5 cm at mouth; lobes obovate, 1.3–1.5 cm long, 1–1.1 cm broad, margin minutely crenulate, apex obtuse. Stamens 4, didynamous; filaments sparsely hirsute, longer pair exserted, 12.–1.5 cm long, shorter pair subexserted, 5–6 mm long; anther 3.5–5 mm long, 2–3 mm broad. Ovary ovoid, 3–4 mm long, 1.8–2 mm broad, glandular-pubescent above; style 2–2.2 cm long, exserted, shortly hairy. Capsules clavate, 1–2 cm long, shortly stipitate, glabrous, dehiscing elastically from base, usually 2-seeded. Seeds discoid, 4–5 mm long and broad, with shaggy hygroscopic hairs. **Chromosomes** 2n = 32. Figure 46-26.

Phenology: flowering from January to April.

Habitats: open, dry areas, forests, thickets; 100–500 m.

Distribution: U Ganga & Indus, W Nepal, C Nepal (Map 46-49); also in other parts of Pakistan and India.

U GANGA & INDUS: Dehradun, *C. R. Babu 35203* (CAL); 1825, *Wallich Cat. 2389b* (K-W); **Solan**, Kasauli, 1885, *J. R. Drummond 26729* (E). **W NEPAL: Bheri**, Banke, Napelganj, 450 m, 12 Apr. 1974, *J. F. Dobremez 2403 (N. P. Manauthar 74-32)* (BM, E, KATH); Bardiya, Babai Khola, 460 m, 10 Apr. 1968, *J. D. A. Stainton 6222* (BM, K); Surkhet, Imlibas, 300 m, 25Aug. 1980, *T. K. Bhathacharya 3075* (KATH); **Rapti**, Dang, Koilabas, 21 Mar. 1973, *N. P. Manandhar et al. 10153* (KATH); Dang, Koilabas, Kabaikhda, 180 m, 28 Mar. 1982, *K. J. Malla 266* (KATH); **Seti**, Kailali, Golmura, 380 m, 19 Aug. 1965, *J. D. A. Stainton 4803* (BM, KATH); Kailali, Golmura (Terai), 380 m, 19 Aug. 1965, *J. D. A. Stainton 4804* (BM). **C NEPAL: Bagmati**, Dhading, Chorkholi, 300 m, 5 Mar. 1977, *T. K. Bhathacharya & K. J. Malla 310* (KATH); Dhading, Chorkholi, 26 Mar. 1976, *S. A. Siddqui 1576* (KATH); Dhading, Daokhuri, 26 Mar. 1976, *S. A. Siddqui 1583* (KATH).

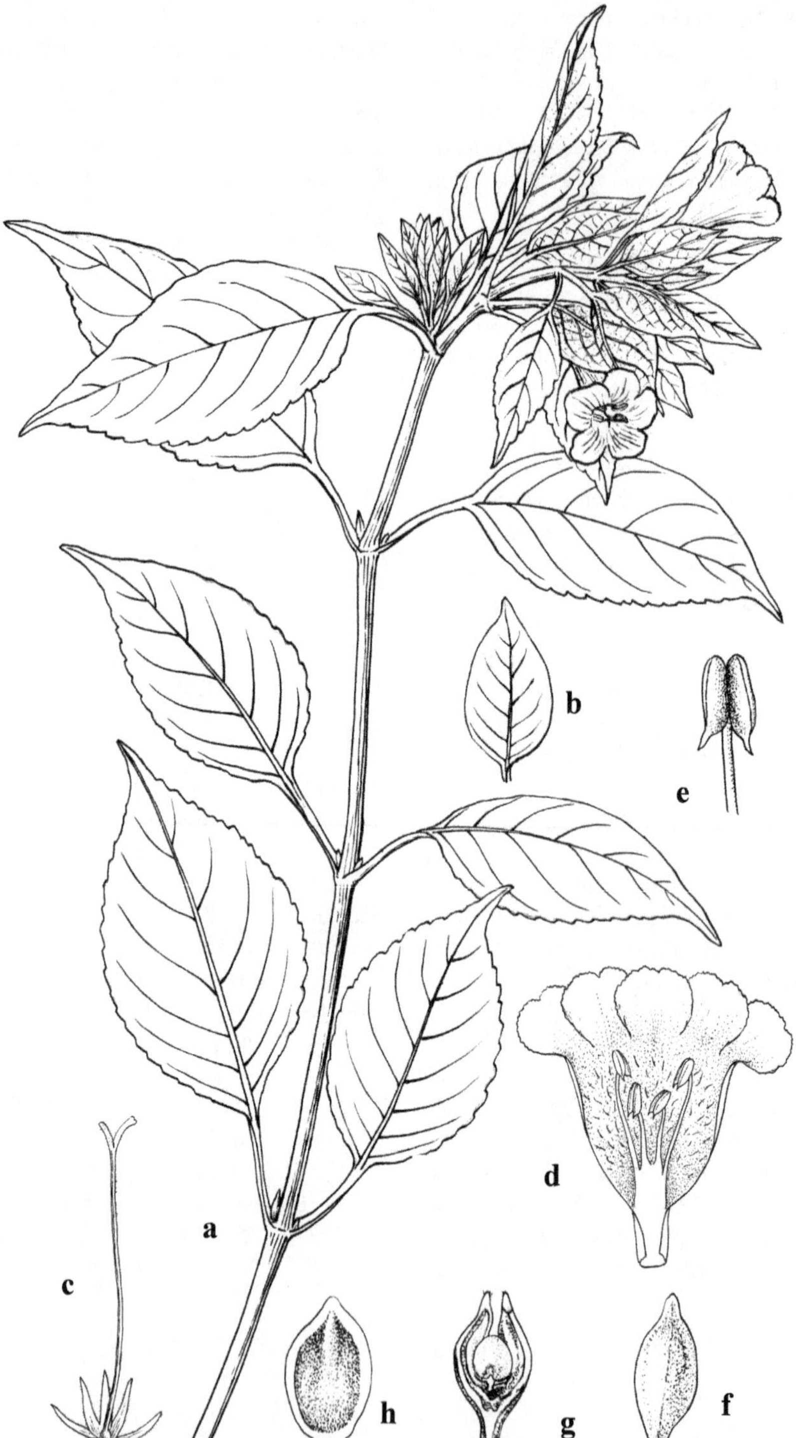

Figure 46-26: **Petalidium barlerioides** (Roth) Nees: **a**, flowering branch; **b**, abaxial surface of bracteole; **c**, calyx, ovary and style; **d**, opened corolla showing stamens; **e**, anther-thecae; **f**, capsule; **g**, opened capsule; **h**, seed. Drawn by Yun-Xiao LIU.

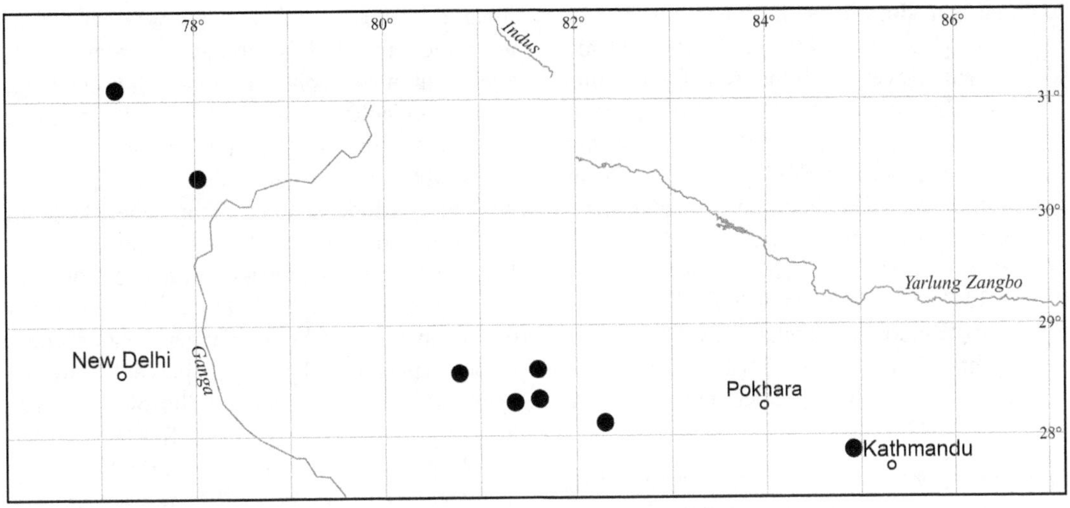

Map 46-49: **Petalidium barlerioides** (Roth) Nees

10. PHAULOPSIS Willd.

Willdnow, Sp. Pl. 3: 4, 342. 1800, "*Phaylopsis*", nom. et orth. cons.; Wood in Grierson & Long (eds.), Fl. Bhutan 2(3): 1274. 2001; Hu, Fl. Reipubl. Popularis Sin. 70: 76. 2002; Hu, Deng & Daniel in Wu, Raven, Hong (eds.), Fl. China 19: 429. 2011. Type: *Phaulopsis parviflora* Willd., nom. illeg. [= *Phaulopsis dorsiflora* (Retz.) Santapau].

Micranthus J. C. Wendl., Bot. Beob.: 38. 1798, nom. rej., vs. *Micranthus* (Pers.) Ecklon, Topogr. Verzeichniss Pflanzensamml.: 43. 1827, nom. cons.; Lindau in Engler & Prantl (eds.), Nat. Pflanzenfam. 4(3b): 298. 1895. Type: *Micranthus oppositifolius* J. C. Wendl.

Herbs, perennial or annual; with cystoliths. Leaves long petiolate; blade margin entire or crenate. Inflorescences terminal or axillary, secund strobilate spikes. Bracts orbicular or reniform, imbricate; bracteoles absent. Calyx 5-lobed almost to base; lobes unequal, posterior one larger than others. Corolla small; tube cylindrical; limb 2-lipped, lower lip 3-lobed, upper lip 2-lobed; lobes contorted in bud. Stamens 4, didynamous, included; filaments each two coherent at base; anthers bithecous; thecae parallel, minutely spurred at base. Ovary with 2 ovules per locule; style hairy; stigma unequally 2-lobed. Capsules shortly stipitate, 4-seeded; retinacula present; septa with attached retinacula separating from inner wall of mature capsule. Seeds discoid, densely covered with hygroscopic trichomes.

Recent references: Bennet, 1978. Nomenclature of few flowering plants. *Indian J. Forest.* 1: 22–24; Manktelow, 1996. *Phaulopsis* (Acanthaceae)–a monograph. *Symb. Bot. Upsal.* 31(2): 1–183.

Twenty-two species: tropical regions of Africa and Asia; one species in the Pan-Himalaya.

1. Phaulopsis dorsiflora (Retz.) Santapau, Kew Bull. [3]: 276. 1948; Hu, Deng & Daniel in Wu, Raven & Hong (eds.), Fl. China 19: 429. 2011. ≡ *Ruellia dorsiflora* Retz., Observ. Bot. 6: 31. 1791. ≡ *Aetheilema reniforme* Nees in Wallich (ed.), Pl. Asiat. Rar. 3: 94. 1832, nom. superfl. illeg. (Art. 52). ≡ *Micranthus dorsiflorus* (Retz.) C. E. C. Fisch., Bull. Misc. Inform. Kew 1932(2): 63. 1932. Type: India, s. d., *Konig s. n.* [lectotype designated by Manktelow, Symb. Bot. Upsal. 31(2): 159. 1996: LD (LD1745105); isolectotypes: C, M (M0186725)].

Micranthus oppositifolius J. C. Wendl., Bot. Beob.: 39. 1798; Hu, Fl. Reipubl. Popularis Sin. 70: 76. 2002. ≡ *Phaulopsis oppositifolia* (J. C. Wendl.) Lindau in Engler & Prantl (eds.), Nat. Pflanzenfam. 1: 305. 1897; Hu, Fl. Reipubl. Popularis Sin. 70: 76. 2002. ≡ *Phaulopsis parviflora* Willd., Sp. Pl. 3: 342. 1800, nom. superfl. illeg. (Art. 52); Clarke in Hooker (ed.), Fl. Brit. India 4: 417. 1884. ≡ *Aetheilema parviflorum* Spreng., Syst. Veg. 2: 826. 1825, nom. superfl. illeg. (Art. 52). Type: *Wendland s. n.* (lectotype designated by Manktelow, Symb. Bot. Upsal. 31(2): 159. 1996: GOET; isolectotype: B-W).

Blechum anisophyllum Juss., Ann. Mus. Hist. Nat. 9: 270. 1807. ≡ *Aetheilema anisophyllum* (Juss.) E. Meyer ex Nees, Prodr. 11: 262. 1847, nom. illeg., non *Aetheilema anisophyllum* R. Br., Prodr.: 478. 1810. Type: Réunion, *P. Commerson s. n.* [holotype: P-JU; isotypes: G (G00085138), P (P00435950)].

Phaulopsis imbriacta auct. non (Forssk.) Sweet: Wood in Grierson & Long (eds.), Fl. Bhutan 2(3): 1275. 2011.

Herbs 30–50 cm high; stems ascending, quadrangular, brown retrorsely pubescent. Leaves unequal in each pair; petiole 4–6 cm long; blade ovate to elliptic, 7.5–11 cm long, 3.5–5cm broad, abaxially pubescent along veins, adaxially hirsute, lateral veins 5–7 pairs, base cuneate to attenuate and slightly oblique, margin entire, apex acuminate. Spikes terminal, 3–6 cm long. Bracts orbicular to reniform, 1–1.1 cm long, 1.3–1.4 cm broad, glandular-pubescent. Calyx lobes abaxially glandular-pubescent, posterior one ovate-elliptic, 8–10 mm long, others linear to subulate, 5–6 mm long. Corolla white, 5–6 mm long, sparsely pilose; lobes ovate to ovate-oblong, 1–2.4 mm long, 0.6–1.2 mm broad. Filaments 0.7–1 cm long, glabrous; anther-thecae oblong. Ovary sparsely glandular-pubescent towards apex; style pilose. Capsules ellipsoid, 5–7 mm long, apex glandular-pubescent. Seeds orbicular, 1.5–2 mm long, 1.3–1.7 mm broad, densely hygroscopically pubescent. **Chromosomes** 2n = 56, ca. 64, 70. Figure 46-27.

Phenology: flowering from November to May; fruiting from February to July.

Habitats: thickets, roadsides, stream banks; 200–800 m.

Distribution: E Nepal, Sikkim & Darjiling, Bhutan, Yarlung Zangbo-Brahmaputra, U Irrawaddy (Map 46-50); also in Bangladesh, Laos, Thailand, Vietnam, South Africa, Indian Ocean islands (Réunion), Madagascar, and other parts of India, Myanmar and China (S Yunnan).

E NEPAL: Koshi, Sunsari, Phursi to Sanguari Bhanjang to Dharm Pani, 16 Oct. 1963, *H. Hara 6303684* (KATH, KYO, TI). **SIKKIM & DARJILING: Darjiling**, *J. M. Cowan s .n.* (E). **BHUTAN: Gaylegphung**, above Aiebridge, 500 m, 22 Mar. 1982, *A. J. C. Grierson & D. G. Long 3925* (E, K); Gaylegphug, 300 m, 29 Oct. 1988, *J. R. I. Wood 6829* (E); **Phuntsholing**, 250 m, 15 Feb. 1982, *A. J. C. Grierson & D. G. Long 2929* (E); **Samchi**, 500 m, 28 Feb. 1982, *A. J. C. Grierson & D. G. Long 3270* (E, K); Samchi High School, 400 m, 7 Jan. 1989, *J. R. I. Wood 6862* (E). **YARLUNG ZANGBO-BRAHMAPUTRA:** Papum Pare, 300–400 m, 10 Oct. 2015, *A. Bawri 401* (RGU). **U IRRAWADDY:** Mali Kha Valley, 600 m, 1 Dec. 1930, *F. Kingdon-Ward 9006* (BM).

Figure 46-27: **Phaulopsis dorsiflora** (Retz.) Santapau: **a**, portion of plant; **b**, bract and inflorescence; **c**, bracteole and flower; **d**, corolla spreading out showing stamens; **e**, dehisced capsule; **f**, seed. Drawn by Han-Ping YU; reproduced from *Fl. Reipubl. Popularis Sin.* 70: 77, pl. 9: 1–6. 2012.

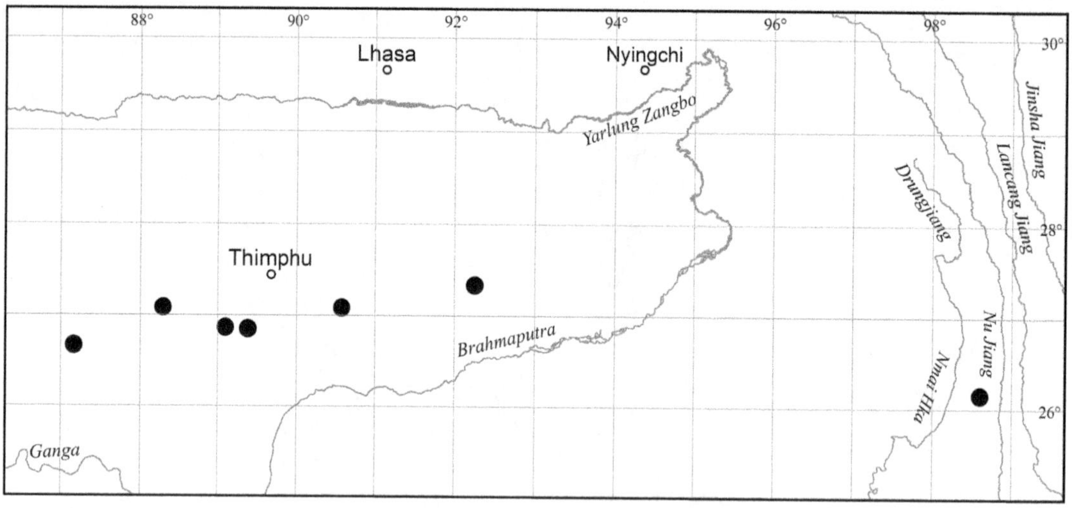

Map 46-50: **Phaulopsis dorsiflora** (Retz.) Santapau

11. STROBILANTHES Blume

Blume, Bijdr. Fl. Ned. Ind.: 781. 1826; Nees in de Candolle (ed.), Prodr. 11: 177. 1847; Benthan & Hooker, Gen. Pl. 2: 1086. 1876; Clarke in Hooker (ed.), Fl. Brit. India 4: 429. 1884; Lindau in Engler & Prantl (eds.), Nat. Pflanzenfam. 4(3b): 304 1895; Nasir & Ali in Malik & Ghafoor (eds.), Fl. Pakistan 188: 51. 1988; Wood in Grierson & Long (eds.), Fl. Bhutan 2(3): 1255. 2001; Hu, Fl. Reipubl. Popularis Sin. 70: 178. 2002; Hu, Deng & Wood in Wu, Raven & Hong (eds.), Fl. China 19: 381. 2011. Type [designated by Bremekamp, Verh. Kon. Ned. Akad. Wetensch., Afd. Natuurk., Sect. 2 41(1): 39. 1944]: *Strobilanthes cernua* Blume.

Phlebophyllum Nees in Wallich (ed.), Pl. Asiat. Rar. 3: 75, 83. 1832. Type: *Phlebophyllum kunthianum* Nees [≡ *Strobilanthes kunthianum* (Nees) T. Anderson ex Benth.].

Aechmanthera Nees, Pl. Asiat. Rar. 3: 75, 87. 1832; Nees in de Candolle (ed.), Prodr.: 170. 1847; Bentham & Hooker, Gen. Pl. 2: 1088. 1876; Clarke in Hooker (ed.), Fl. Brit. India 4: 428. 1884; Lindau in Engler & Prantl, Pflanzenfam. 4(3b): 303. 1897; Nasir & Ali in Malik & Ghafoor (eds.), Fl. Pakistan 188: 69. 1988; Wood in Grierson & Long (eds.), Fl. Bhutan 2(3): 1253. 2001; Tsui in Hu (ed.), Fl. Reipubl. Popularis Sin. 70: 87. 2002. Type: *Aechmanthera tomentosa* Nees [≡ *Strobilanthes tomentosa* (Nees) Terao ex J. R. I. Wood].

Endopogon Nees in Wallich (ed.), Pl. Asiat. Rar. 3: 76, 98. 1832, nom. illeg., non *Endopogon* Raf., Amer. Monthly Mag. & Crit. Rev. 3: 101. 1818. Type: *Endopogon consanguineus* Nees [≡ *Strobilanthes consanuinea* (Nees) C. B. Clarke].

Goldfussia Nees, Pl. Asiat. Rar. 3: 75, 87. 1832; Nees in de Candolle (ed.), Prodr. 11: 101, 147. 1847; Tsui & Hu in Hu (ed.), Fl. Reipubl. Popularis Sin. 70: 159. 2002. ≡ *Strobilanthes* sect. *Goldfussia* (Nees) T. Anderson, J. Linn. Soc., Bot. 9: 475. 1867. Type [designated by Bremekamp, Verh. Kon. Ned. Akad. Wetensch., Afd. Natuurk., Sect. 2 41(1): 226. 1944]: *Goldfussia capitata* Nees [≡ *Strobilanthes capitata* (Nees) T. Anderson].

Strobilanthes subg. *Pteracanthus* Nees, Pl. Asiat. Rar. 3: 87. 1832. ≡ *Strobilanthes* ser. *Pteracanthi* (Nees) Nees in de Candolle (ed.), Prodr. 11: 192. 1847. ≡ *Pteracanthus* (Nees) Bremek., Verh.

Kon. Ned. Akad. Wetensch., Afd. Natuurk., Sect. 2 41(1): 198. 1944; Tsui & Hu in Hu (ed.), Fl. Reipubl. Popularis Sin. 70: 123. 2002. Type: *Strobilanthes wallichii* Nees [≡ *Pteracanthus alatus* (Wall. ex Nees) Bremek. = *Strobilanthes atropurpurea* Nees].

Hemigraphis Nees, Prodr. 11: 722. 1847; Clarke in Hooker (ed.), Fl. Brit. India 4: 66. 1884; Lindau in Engler & Prantl (eds.), Nat. Pflanzenfam. 4(3b): 303 1895; Nasir & Ali in Malik & Ghafoor (eds.), Fl. Pakistan 188: 66. 1988; Wood in Grierson & Long (eds.), Fl. Bhutan 2(3): 1255. 2001; Tsui in Hu (ed.), Fl. Reipubl. Popularis Sin. 70: 86. 2002. Type (designated by Pferiffer, Nom. 1: 1602. 1874): *Hemigraphis elegans* Nees [≡ *Ruellia elegans* Hook., 1835, non *Ruellia elegans* Poir., 1816 ≡ *Strobilanthes pavala* (Roxb.) J. R. I. Wood].

Gutzlaffia Hance, Hooker's J. Bot. Kew Gard. Misc. 1: 142. 1849; Tsui in Hu (ed.), Fl. Reipubl. Popularis Sin. 70: 100. 2002. Type: *Gutzlaffia aprica* Hance [≡ *Strobilanthes aprica* (Hance) T. Anderson].

Baphicacanthus Bremek., Verh. Kon. Ned. Akad. Wetensch., Afd. Natuurk., Sect. 2 41(1): 190. 1944; Hu, Fl. Reipubl. Popularis Sin. 70: 113. 2002. Type: *Baphicacanthus cusia* (Nees) Bremek. [≡ *Goldfussia cusia* Nees ≡ *Strobilanthes cusia* (Nees) Kuntze].

Championella Bremek., Verh. Kon. Ned. Akad. Wetensch., Afd. Natuurk., Sect. 2 41(1): 150. 1944; Hu, Fl. Reipubl. Popularis Sin. 70: 89. 2002. Type: *Championella tetrasperma* (Champ. ex Benth.) Bremek. [≡ *Ruellia tetrasperma* Champ. ex Benth. ≡ *Strobilanthes tetrasperma* (Champ. ex Benth.) Druce].

Diflugossa Bremek., Verh. Kon. Ned. Akad. Wetensch., Afd. Natuurk., Sect. 2 41(1): 235. 1944; Tsui & Hu in Hu (ed.), Fl. Reipubl. Popularis Sin. 70: 170. 2002. Type: *Diflugossa colorata* (Nees) Bremek. [≡ *Goldfussia colorata* Nees ≡ *Strobilanthes colorata* (Nees) T. Anderson = *Strobilanthes hamiltoniana* (Steud.) Bosser & Heine].

Dossifluga Bremek., Verh. Kon. Ned. Akad. Wetensch., Afd. Natuurk., Sect 2 41(1): 61, 234. 1944. Type: *Dossifluga suborbicularis* (J. B. Imlay) Bremek. (≡ *Strobilanthes suborbicularis* J. B. Imlay).

Listrobanthes Bremek., Verh. Kon. Ned. Akad. Wetensch., Afd. Natuurk., Sect. 2 41(1): 253.1944. Type: *Listrobanthes khasiana* (Nees) Bremek. [≡ *Strobilanthes khasyana* (Nees) T. Anderson].

Perilepta Bremek., Verh. Kon. Ned. Akad. Wetensch., Afd. Natuurk., Sect. 2 41(1): 191. 1944; Hu, Fl. Reipubl. Popularis Sin. 70: 115. 2002. Type: *Perilepta auriculata* (Nees) Bremek. (≡ *Strobilanthes auriculata* Nees).

Pseudaechmanthera Bremek., Verh. Kon. Ned. Akad. Wetensch., Afd. Natuurk., Sect. 2 41(1): 188. 1944; Hu, Fl. Reipubl. Popularis Sin. 70: 111. 2002. Type: *Pseudaechmanthera glutinosa* (Nees) Bremek. (≡ *Strobilanthes glutinosa* Nees).

Tarphochlamys Bremek., Verh. Kon. Ned. Akad. Wetensch., Afd. Natuurk., Sect. 2 41(1): 156. 1944; Hu, Fl. Reipubl. Popularis Sin. 70: 105. 2002. Type: *Tarphochlamys affinis* (Griff.) Bremek. [≡ *Strobilanthes affinis* (Griff.) Terao ex J. R. I. Wood & J. Benn.].

Tetraglochidium Bremek., Verh. Kon. Ned. Akad. Wetensch., Afd. Natuurk., Sect. 2 41(1): 214. 1944; Hu, Fl. Reipubl. Popularis Sin. 70: 156. 2002. Type: *Tetraglochidium thunbergiiflorum* (S. Moore) Bremek. (= *Strobilanthes echinata* Nees).

Tetragoga Bremek., Verh. Kon. Ned. Akad. Wetensch., Afd. Natuurk., Sect. 2 41(1): 299. 1944; Hu, Fl. Reipubl. Popularis Sin. 70: 190. 2002. Type: *Tetragoga nagaensis* Bremek. [= *Strobilanthes cruciata* (Bremek.) Terao].

Paragutzlaffia H. P. Tsui, Acta Bot. Yunnan. 12: 273. 1990; Tsui in Hu (ed.), Fl. Reipubl. Popularis Sin. 70: 103. 2002. Type: *Paragutzlaffia henryi* (Hemsl.) H. P. Tsui (≡ *Strobilanthes henryi* Hemsl.).

Clarkeasia J. R. I. Wood, Edinburgh J. Bot. 51: 187. 1994; Wood in Grierson & Long (eds.), Fl. Bhutan 2(3): 1254. 2001. Type: *Clarkeasia parviflora* (T. Anderson) J. R. I. Wood (≡ *Strobilanthes parviflora* T. Anderson).

Herbs, subshrubs, or rarely small trees, isophyllous or weakly to strongly anisophyllous; with cystoliths; stems and branches usually quadrangular, often sulcate. Leaves opposite, petiolate or sessile; leaf margin variously dentate, serrate, crenate, undulate or entire. Inflorescences axillary and/or terminal, bracteate heads, head-like clusters, spikes, sometimes panicles. Bracts persistent or caducous at anthesis, very variable in size and shape; bracteoles 2 per flower, usually small, sometimes absent. Calyx usually 5-lobed to base, commonly accrescent in fruit; lobes equal or one distinctly longer than others, sometimes partially fused and forming bipartite or tripartite calyx. Corolla tubular or funnel-shaped, inside glabrous apart from trichomes retaining style or rarely pubescent; tube straight or curved, campanulate or gibbous; limb 5-lobed; lobes contorted in bud. Stamens usually 4 and didynamous, rarely 2; filaments basally monadelphous forming a sheath; anthers included or exserted, bithecous; thecae oblong or subspherical, erect or incurved, muticous at base, rarely with connective extended to a mucronate tip; staminodes 1 or 2. Ovary oblong to obovoid, ovules 2–8 per locule; style filiform, retained in place by trichomes on one side of corolla tube; stigma 2-cleft with one longer than another. Capsules oblong to narrowly obovoid, sometimes fusiform to narrowly ellipsoid, (2–)4(–16)-seeded; retinacula present. Seeds usually ovate or orbicular in outline, usually pubescent with appressed mucilaginous trichomes.

Recent references: Bremekamp, 1944. Materials for a monograph of the Strobilanthinae (Acanthaceae). *Verh. Kon. Ned. Akad. Wetensch., Afd. Natuurk., Sect. 2* 41(1): 1–305; Shakya, 1975. Three new species of Acanthaceae from Nepal. *Jap. J. Bot.* 50: 97–103; Terao, 1982. Two new species of *Strobilanthes* from Burma and S. W. China. *Notes Roy. Bot. Gard. Edinburgh* 40: 153–155; Wood, 1994. Notes relating to the Flora of Bhutan: XXIX Acanthaceae, with special reference to *Strobilanthes*. *Edinburgh J. Bot.* 51: 175–273; Carine & Scotland, 1998. Pollen morphology of *Strobilanthes* Blume (Acanthaceae) from southern India and Sri Lanka. *Rev. Palaeobot. Palynol.* 103: 143–165; Carine & Scotland, 2002. Classification of Strobilanthinae (Acanthaceae): trying to classify the unclassifiable. *Taxon* 51: 259–279; Wang & Blackmore, 2003. Pollen morphology of *Strobilanthes* Blume (Acanthaceae) in China and its taxonomic implications. *Grana* 42: 82–87; Wood et al., 2003. Notes on *Strobilanthes*: the *Sympagis* group. *Kew Bull.* 58: 131–173; Wood & Scotland, 2003a. The 2-lipped species of *Strobilanthes* (Acanthaceae). *Kew Bull.* 58: 83–129; Wood & Scotland, 2003b. *Strobilanthes*: panicled species from east Asia. *Kew Bull.* 58: 679–702; Moylan et al., 2004. Phylogenetic relationships among *Strobilanthes s. l.* (Acanthaceae): evidence from ITS nrDNA, *trnL-F* cpDNA and morphology. *Amer. J. Bot.* 91: 724–735; Moylan et al., 2005. Comparative floral anatomy of Strobilanthinae (Acanthaceae), with particular reference to internal partitioning of the flower. *Pl. Syt. Evol.* 249: 77–98; Deng et al., 2006. New and reassessed species of *Strobilanthes* (Acanthaceae) in the flora of China. *Bot. J. Linn. Soc.* 150: 369–390; Wood & Scotland, 2006. Some notes on the *Strobilanthes pentstemonoides* (Acanthaceae) complex in China, India and South East Asia. *Kew Bull.* 61: 5–16; Bennett et al., 2008. Uncorrelated variation in widespread species: species delimitation in *Strobilanthes echinata* Nees (Acanthaceae). *Bot J. Linn. Soc.* 156: 131–141; Wood & Scotland, 2009. New and little-known species of *Strobilanthes* (Acanthaceae) from India and South East Asia. *Kew Bull.* 64: 3–47; Deng et al., 2010a. New combinations and new species of *Strobilanthes* Blume (Acanthaceae) from China. *J. Trop. Subtrop. Bot.* 18: 469–484; Deng et al., 2010b. *Strobilanthes ovata* (Acanthaceae), a new species from Gaoligong Shan, in Yunnan, China. *Novon* 20: 143–146. Albertson & Wood, 2012. Forgotten types of *Strobilanthes* (Acanthaceae) in the Central National Herbarium,

Kolkata, India. *Phyototaxa* 43: 49–60; Wood, 2012. *Strobilanthes lachenensis*. *Curtis's Bot. Mag.* 29: 34–43; Wood, 2014. *Strobilanthes attenuata*. *Curtis's Bot. Mag.* 31: 154–167; Wood & Adhikari, 2014. *Strobilanthes nutans*. *Curtis's Bot. Mag.* 31: 168–179; Callmander et al., 2015. The transfer of the Malagasy species of *Strobilanthes* Blume to *Acanthopale* C. B. Clarke (Acanthaceae). *Candollea* 70: 253–256; Wood et al., 2019. *Strobilanthes twangensis* (Acanthaceae), a new species from the East Himalayas. *Kew Bull.* 74: 41.

About 400 species: tropical Asia and few species extending to Pacific Islands and N Australia; 88 species (32 endemic) in the Pan-Himalaya.

Notes: The genus was reported from Madagascar by some authors. However, the Malagasy species are quite different from Asian plants and have been transferred to *Acanthopale* C. B. Clarke (Callmander et al., 2015).

1a. Fertile stamens 2, staminodes 2.
  2a. Spikes abbreviated to capitula, 1–2.5 cm; calyx tripartite, 3 lobes partially fused, other 2 lobes free to base .... 1. *S. aprica*
  2b. Spikes elongate, longer than 3 cm; calyx subequally 5-lobed to base.
    3a. Flowers distant on spikes at least on lower portion; stamens exserted; capsules glabrous except for a few glandular hairs towards apex.
      4a. Plants subglabrous or grey pubescent; spikes pedunculate, simple, borne on leafy branchlets; bracts linear-spatulate .... 2. *S. henryi*
      4b. Plants densely covered with brown hairs; spikes subsessile, usually several clustered on leaf axils; bracts elliptic or oblanceolate-spatulate .... 3. *S. khasyana*
    3b. Flowers imbricate on rachis; stamens included; capsules pubescent.
      5a. Bracts obovate, apex emarginate .... 4. *S. tubiflos*
      5b. Bracts oblong to ovate, apex acute .... 5. *S. kachinensis*
1b. Fertile stamens 4, staminodes 1 or absent.
  6a. Ovules 3–8 per locule; capsules 6–16-seeded.
    7a. Herbs; corolla less than 1.5 cm.
      8a. Petiole 0–5 mm long; leaf blade elliptic or obovate, 0.7–2.5 cm long, 0.4–1.8 cm broad .... 6. *S. hirta*
      8b. Petiole 1.5–3.5 cm long; leaf blade ovate, 2.5–7.5 cm long, 1.5–3.5 cm broad .... 7. *S. pavala*
    7b. Subshrubs; corolla longer than 2 cm long.
      9a. Flowers in 1-sided axillary panicles of spikes; capsules 12–16-seeded .... 8. *S. violifolia*
      9b. Flowers in small clusters along axis forming an open panicle of capitula; capsules 6–8-seeded .... 9. *S. tomentosa*
  6b. Ovules 2 per locule; capsules 4-seeded.
    10a. Inflorescences an open panicle with capillary branches; flowers not arranged in opposite pairs.
      11a. Inflorescences axillary, shorter than or subequaling to leaves .... 10. *S. panichanga*
      11b. Inflorescences terminal or rarely axillary, longer than leaves.
        12a. Calyx lobes long acuminate; inflorescences covered with long setae .... 11. *S. setosa*
        12b. Calyx lobes obtuse; inflorescences without setae but sometimes glandular-pilose.
          13a. Corolla arcuate, 2.2–3.5 cm long; rachis with glandular trichomes .... 12. *S. euantha*
          13b. Corolla straight, 3.5–4 cm long; rachis glabrous or with glandular-pilose when fruiting .... 13. *S. hamiltoniana*
    10b. Inflorescences spicate, elongate, subcapitate or capitellate, sometimes aggregated into a panicle; flowers in opposite pairs, sometimes only one flower fertile in each pair.
    14a. Filaments of longer pair usually unequal, shorter pair reflexed; anther-thecae incurve,

subspherical, as broad as long.

15a. Plants isophyllous; flowers arranged in subsessile or shortly pedunculate axillary head-like spikes; bracts usually present.

16a. Inflorescence pendulous; bracts caducous .... 14. *S. nutans*

16b. Inflorescence erect; bracts present.

17a. Calyx lobes mucronate, mucro accrescent into long awn in fruit... 15. *S. accrescens*

17b. Calyx lobes acute, not accrescent in fruit.

18a. Bracts brownish pilose; bracts entire on margin; corolla cream .... 16. *S. jennyae*

18b. Bracts white hairy; bracts serrate, repand or dentate on margin.

19a. Leaves petiolate; petiole 2.5–5.5 cm long; bracts foliose, broadly oblong-elliptic, 1–1.2 cm broad, margin repand to undulate .... 17. *S. twangensis*

19b. Leaves sessile or shortly petiolate; petiole shorter than 5 mm; bracts spatulate or broadly elliptic to narrowly obovate, margin dentate.

20a. Leaves shortly petiolate, base cuneate; bracts spatulate, 3-lobed towards apex, apex retuse .... 18. *S. cuneata*

20b. Leaves subsessile, base broadly cuneate to rounded; bracts broadly elliptic to narrowly obovate, margin serrate...................................... 19. *S. bheriensis*

15b. Plants anisophyllous; flowers in long pedunculate heads or spikes.

21a. Inflorescences spicate; flowers opposite, distinct on rachis, or sometimes uppermost ones aggregated in head.

22a. Bracts mostly 1.5–2.5 cm long, relatively persistent.

23a. Plants isophyllous; leaves equal in each pair; corolla pubescent...... 20. *S. lamiifolia*

23b. Plants strongly anisophyllous; leaves unequal in each pair; corolla glabrous .... 21. *S. angustifrons*

22b. Bracts less than 1.5 cm long, caducous.

24a. Leaf blade ovate, base rounded to subcordate; anthers slightly exserted .... 22. *S. medogensis*

24a. Leaf blade lanceolate to ovate-elliptic, base rounded to cuneate; anthers included.

25a. Spikes simple, terminal to borne from upper leaf axil; corolla pubescent outside.

26a. Leaf blade margin entire to sinuate; calyx 1.6–2 cm long..............23. *S. secunda*

26b. Leaf blade margin serrate; calyx less than 1.3 cm long.

27a. Spikes terminal or borne from upper leaf axil; bracteoles absent; corolla glandular-pubescent on lobes ....24. *S. oligocephala*

27b. Spikes axillary; bracteoles present; corolla eglandular pubescent .... 25. *S. thomsonii*

25b. Spikes usually forming axillary or terminal panicle, rarely simple; corolla glabrous.

28a. Stems usually zigzag; leaves strongly unequal in each pair .... 26. *S. divaricata*

28b. Stems not zigzag; leaves subequal or unequal in each pair.

29a. Leaf blade narrowly lanceolate, broad less than 4 cm; inflorescences densely glandular-pilose... 27. *S. tenax*

29b. Leaves ovate to broadly oblong-elliptic to ovate, broad more than 5 cm; inflorescences sparsely glandular-pilose.

30a. Spikes simple not forming panicle; bracteoles absent .... 28. *S. multidens*

30b. Spikes usually forming a panicle; bracteoles present.

31a. Basal leaves petiolate, apical ones sessile; leaf blade ovate to ovate-lanceolate, base rounded to broadly cuneate; calyx-lobes acute at apex........ .... 29. *S. mastersii*

31b. Leaves all petiolate; blade narrowly elliptic, base narrowly cuneate to attenuate; calyx-lobe subobtuse at apex .... 30. *S. laevigata*

21b. Inflorescences capitate, head-like, occasionally becoming subcapitate in fruit; flowers few densely clustered in groups of 4 or more at apex of rachis.

32a. Bracts (0.9–)1.5–3 cm long, persistent, outermost commonly sterile.

33a. Inflorescences covered with purple multicellular hairs or glandular-hairs; leaves strongly unequal in each pair .... 31. *S. glomerata*

33b. Inflorescences covered with soft whitish hairs; leaves slightly unequal in each pair.

34a. Outermost bracts lanceolate to narrowly oblong-elliptic, margin often dentate, apex obtuse; isolated flowers sometimes present below main head ....32. *S. kingdonii*

34b. Outermost bracts ovate, margin entire, apex acute or acuminate; inflorescences strictly capitate ....33. *S. capitata*

32b. Bracts less than 1 cm long, all normally flower-bearing, often scarious and early caducous.

35a. Corolla outside glabrous even on lobes and in bud.

36a. Leaf blade abaxially whitish; calyx in fruit with lobes subequal ....34. *S. discolor*

36b. Leaf blade abaxially paler but not whitish; calyx in fruit with one lobe longer than others.

37a. Bracts suborbicular, apex rounded or obtuse .... 35. *S. pentstemonoides*

37b. Bracts ovate, apex acute to acuminate.

38a. Stems somewhat zigzag; all leaves petiolate; all stamens fully developed ....36. *S. dimorphotricha*

38b. Stems not zigzag; upper leaves ovate, sessile; shorter pair of stamens not fully developed..37. *S. straminea*

35b. Corolla pubescent or glandular-pubescent at least on lobes or in bud.

39a. Leaf blade linear-lanceolate, 5–9 mm broad......................................38. *S. arenicola*

39b. Leaf blade ovate to elliptic, wider than 1.5 cm broad.

40a. Bract orbicular to suborbicular, apex rounded .... 39. *S. pubiflora*

40b. Bracts ovate to elliptic, apex acute or obtuse.

41a. Leaves ovate-lanceolate or oblong-lanceolate, less than 2 cm broad, apex long acuminate to caudate; bracts elliptic, apex obtuse .... 40. *S. persicifolia*

41b. Leaves ovate-elliptic to elliptic, always broader than 3 cm broad, apex acute or acuminate; bracts ovate, apex acute.

42a. Leaves pale green beneath; bracteoles caducous...................... 41. *S. bracteata*

42b. Leaves whitish green beneath, whitish when dry; bracteoles present.

43a. Leaves petiolate; leaf blade apex acuminate; corolla glandular-pubescent..... ....42. *S. oxycalycina*

43b. Leaves sessile; leaf blade apex acute; corolla pilose on lobes.

44a. Corolla glandular on lips; leaves rhombida-elliptic, base cuneate............. ....43. *S. rhombifolia*

44b. Corolla glabrous; leaves ovate-elliptic or ovate, base rounded to cordate at base ....44. *S. ramulosa*

14b. Filaments of longer pair equal, shorter pair erect; anther-thecae oblong, ca. 2 times as long as broad.

45a. Low herbs, sometimes trailing, always isophyllous; corolla less than 2.5 cm long and often less than 2 cm broad, not obviously curved.

46a. Inflorescences of solitary flowers in leaf axils; leaf margin entire ..... 45. *S. szechuanica*

46b. Inflorescences of terminal or sometimes axillary heads or spikes; leaf margin serrate or crenulate.

47a. Petiole coarsely hirsute with large-celled hairs; bracts oblanceolate to obovate-spatulate; corolla less than 1 cm long .... 46. *S. wilsonii*

47b. Glabrous or nearly so; bracts oblanceolate to oblong-lanceolate; corolla 1.5–2 cm long .... 47. *S. japonica*

45b. Herbs or subshrubs, never trailing, usually greater than 50 cm high, anisophyllous or isophyllous; corolla longer than 2 cm, often strongly curved.

48a. Inflorescences subcapitate or capitellate, with flowers densely clustered in groups of 4 or more, rarely solitary in leaf axil, heads occasionally becoming subspicate in fruit.

49a. Flowers arranged in small head-like spikes along an axis forming a panicle .... 48. *S. congesta*

49b. Flowers arranged in axillary or terminal capitula or rarely solitary in leaf axil.

50a. Bracts ovate, or pandurate in outline with a broad petiolar base, longer than 1.8 cm; corolla straight, white.

51a. Outer bracts sessile, ovate.

52a. Bracts dentate; corolla glabrous .... 49. *S. echinata*

52b. Bracts entire; corolla hirsute .... 50. *S. wardiana*

51b. Outer bracts with a broad petiolar base, pandurate in outline.

53a. Petiolar base of bracts forming a rigid involucre, apical portion spreading and ovate .... 51. *S. cruciata*

53b. Petiolar base of bracts not forming an involucre, apical portion narrowly lanceolate .... 52. *S. simonsii*

50b. Bracts obovate, oblong-elliptic to lanceolate, shorter than 1.2 cm; corolla curved, purplish-blue.

54a. Bracts obovate, broader near apex, apex rounded .... 53. *S. lihengiae*

54b. Bracts oblong-elliptic to lanceolate, broader below the middle, apex acute.

55a. Plants anisophyllous; corolla bent for 90° above tube base cylindric .... 54. *S. cyphantha*

55b. Plants isophyllous; corolla straight or slightly curved .... 55. *S. yunnanensis*

48b. Inflorescences spicate or flowers in leaf axils or arranged singly or in opposite pairs in a panicle.

56a. Inflorescences of axillary spikes, sometimes reduced to solitary in leaf axils, with basal persistent expanded leaf-like bracts; spikes simple; flowers usually distant on rachis.

57a. Rachis of inflorescences strongly winged.

58a. Calyx shorter than 2 cm long; capsules glabrous .... 56. *S. tibetica*

58b. Calyx 2.5–3.6 cm long; capsules comose .... 57. *S. pterygorrhachis*

57b. Rachis of inflorescences not winged.

59a. Leaf blade usually blackish when dry; calyx with one lobe much longer than others; corolla straight or slightly curved .... 58. *S. cusia*

59b. Leaf blade usually green when dry; calyx lobe equal or subequal; corolla bent.

60a. Corolla glabrous outside; calyx 1–1.5 cm long .... 59. *S. atropurpurea*

60b. Corolla pubescent outside; calyx up to 1.4 cm long.

61a. Stems, leaves and bracts glabrous; bracts subentire, apex obtuse .... 60. *S. urophylla*

61b. Stems, leaves and bracts grey-pubescent; bracts margin dentate, apex acuminate .... 61. *S. inflata*

56b. Inflorescences of terminal or axillary spikes, without basal leaf-like bracts; spikes aggregated into a paniclke; flowers imbricate or not.

62a. Bracts imbricate so flowers densely arranged on spike rachis.

63a. Leaves sessile; bracteoles absent .... 62. *S. auriculata*

63b. Leaves petiolate; bracteoles present.

64a. Corolla pilose or pubescent outside.

65a. Bracts longer than or subequaling to calyx; stamens strongly exserted from corolla .... 63. *S. affinis*

65b. Bracts shorter than calyx; stamens included....................... 64. *S. himalayana*

64b. Corolla glabrous or pubescent outside; stamens included in corolla.

66a. Bracts elliptic-obovate, margin shouldered, with one distinct marginal tooth on each side .... 65. *S. frondosa*

66b. Bracts ovate, oblong-elliptic, margin entire.

67a. Bracts oblong to elliptic, 9–13 mm long, broadest above the middle.......... .... 66. *S. stolonifera*

67b. Bracts ovate, 7–8 mm long, broadest at base....................... 67. *S. glutinosa*

62b. Bracts not imbricate so flowers clearly separate apart from basalmost.

68a. Flowers in axillary leafless spikes, that sometimes from apicalmost axils; bracts distinct from leaves, often inconspicuous and merging into bracteoles and calyx lobes; spikes not aggregated into a leafless panicle; corolla outside glabrous.

69a. Calyx glabrous or with glandular-pilose, margin never ciliate or ciliolate.

70a. Inflorescences with flowers solitary on one side of spike; bracts linear, persistent.. 68. *S. helicta*

70b. Inflorescences with flowers paired and on two sides of spike; bracts ovate, somewhat caducous .... 69. *S. rubescens*

69b. Calyx lobes sometimes also with glandular-pilose, margin ciliate or ciliolate.

71a. Leaves petiolate; bracts obovate .... 70. *S. fimbriata*

71b. Leaves sessile; bracts lanceolate .... 71. *S. nobilis*

68b. Inflorescences terminal or terminal and axillary, sometimes forming a panicle of spikes, branches bearing leaves below spikes, if apparently axillary then corolla outside pubescent or bracts expanded and clearly different from calyx.

72a. Corolla outside glabrous.

73a. Calyx subequally 5-lobed to base.

74a. Spikes axillary or terminal, many-flowered, 5–10 cm long; bracts persistent.. 72. *S. pinetorum*

74b. Spikes axillary, few-flowered, 2–3.5 cm long; bracts caducous.

75a. Leaf blade 1–3.5 cm long, 1–1.8 cm broad, margin undulate to slightly crenate; bracts and calyx glandular-pubescent .... 73. *S. parvifolia*

75b. Leaf blade 4–13 cm long, 1.5–4.5 cm broad, margin serrate; bracts and calyx glabrous.. 74. *S. candida*

73b. Calyx 2-lipped, 3 upper lobes partially fused.

76a. Upper leaves sessile, leaf blade ovate or ovate-elliptic, base cordate or broadly cuneate... 75. *S. ovata*

76b. All leaves petiolate, leaf blade ovate to elliptic, base narrowly cuneate.

77a. Spikes with 1–3 flower pairs; bracts 9–14 mm long .... 76. *S. atroviridis*

77b. Spikes with 1–7 flower pairs; bracts less than 8 mm long.

78a. Bracts linear-lanceolate; rachis wingless; plant drying blackish green .... 77. *S. longispica*

78b. Bracts ovate-oblong; rachis winged; plant not drying blackish green .... 78. *S. dryadum*

72b. Corolla outside pubescent at least in bud, on lobes or tube.

79a. Leaves sessile.

80a. Leaf blade ovate to suborbicular, abaxially whitish; bracts orbicular to obovate, 5–8 mm long .... 79. *S. tamburensis*

80b. Leaves ovate to rarely ovate-oblong, abaxially green; bracts lanceolate, 7–10 mm long .... 80. *S. forrestii*

79b. Leaves petiolate.

81a. Corolla ventricose and strongly bent just below mouth, pubescent over its whole length.

82a. Corolla inside pubescent especially on lobes; spikes often secund .. .. 81. *S. oresbia*

82b. Corolla inside glabrous even on lobes; spikes 2-sided..................... ....82. *S. versicolor*

81b. Corolla gradually widened above base, curved but not very strongly so, pubescent on lips only.

83a. Bracts obovate-suborbicular or ovate, nearly as broad as long.

84a. Leaf blade glabrous; bracts glandular-pilose; calyx lobes linear-oblong; corolla ventricose .... 83. *S. sabiniana*

84b. Stem and leaves hirsute; bracts glabrous; calyx lobes inear-oblanceolate or spaulate; corolla strongly saccate .....84. *S. saccata*

83b. Bracts linear to lanceolate, length/broadth ratio > 2.

85a. Spikes extremely lax, with apical flower pairs commonly separated by greater than 1 cm along rachis; bracts ovate.

86a. Leaf base cordate, truncate or rounded; calyx lobes lanceolate... ....85. *S. attenuata*

86b. Leaf base cuneate; calyx lobes conspicuously spatulate ....86. *S. extensa*

85b. Spikes apically relatively dense, with flower pairs becoming imbricate along rachis; bracts linear to lanceolate; calyx lobes not usually conspicuously spatulate.

87a. Leaf blade (1.5–)3–6.5cm broad, pilose on both surfaces, equal .... 87. *S. lachenensis*

87b. Leaf blade 0.8–2 cm broad, glabrous or sparsely pilose on both surfaces, often unequal...................................... 88. *S. subnudata*

1. Strobilanthes aprica (Hance) T. Anderson in Bentham, Fl. Hongk.: 262. 1861. ≡ *Gutzlaffia aprica* Hance, Hooker's J. Bot. Kew Gard. Misc. 1: 143. 1849; Bremekamp, Verh. Kon. Ned. Akad. Wetensch., Afd. Natuurk., Sect. 2 41(1): 154. 1944; Tsui in Hu (ed.), Fl. Reipubl. Popularis Sin. 70: 101. 2002; Hu, Deng & Wood in Wu, Raven & Hong (eds.), Fl. China 19: 400. 2011. ≡ *Phlebophyllum apricum* (Hance) Benth., Hooker's J. Bot. Kew Gard. Misc. 5: 131. 1853. Type: China, Hong Kong, *H. F. Hance 536* [lectotype designated by Albertson & Wood, Phytotaxa 43: 50. 2012: CAL (CAL0000019794); isolectotypes: BM, GH (GH00387581), P (P00719255, P00719256)].

Strobilanthes dielsiana W. W. Sm., Notes Roy. Bot. Gard. Edinburgh 8: 207. 1914. ≡ *Gutzlaffia dielsiana* (W. W. Sm.) S. Moore, J. Bot. 63: 239. 1925; Bremekamp, Verh. Kon. Ned. Akad. Wetensch., Afd. Natuurk., Sect. 2 41(1): 155. 1944. Type: [S Hengduan] China, Yunnan, western flank of Lichiang Range, 27°25′N, 10,000 ft., Aug. 1910, *G. Forrest 6264* [holotype: E (E00133365); isotypes: BM (BM000796850), K (K000882742), PE (PE01545751)].

Strobilanthes mairei H. Lév., Cat. Fl. Yunnan: 6. 1915. Type: China, Yunnan, Dongchuan, "Pâtures des montagnes à Tong-Tchouan", 2600 m, Jun. 1912, *E. E. Maire s. n.* [holotype:

E (E000265788); isotypes: E (E000265789), P].

Shrubs or perennial herbs, 30–70 cm high, isophyllous; stems quadrangular, sulcate, retrorsely white hirsute. Petiole 5–10 mm long, shortly white pubescent; leaf blade elliptic to oblong-elliptic, 2–7 cm long, 1–3 cm broad, both surfaces usually scabrous-hispid but abaxially sometimes whitish silky tomentose and occasionally glabrous except for pubescent midvein, lateral veins 5–7 pairs, base broadly cuneate, margin entire and scabrid to sparsely ciliate, apex acute. Spikes axillary or terminal,1 –2.5 cm long shortly subcapitate; peduncles (0–) 0.5–3(–7) cm long. Bracts ovate-lanceolate, 1.3–1.5 cm long, 3–5 mm broad, abaxially covered white multicellular trichomes, adaxially glabrous, apex long acuminate; bracteoles linear, 1–1.1 cm long, 0.8–1 mm broad, abaxially sparsely pilose with multicellular trichomes, adaxially glabrous. Calyx 1–1.3 cm long, tripartite; lobes linear, lower 2 free nearly to base, upper 3 connate to the middle, abaxially pubescent to subglabrous, adaxially glabrous, margin pilose. Corolla lilac with a white tube, funnel-shaped, 2.5–3.5 cm long, outside glabrous, inside glabrous except for trichomes retaining style; tube base cylindric, 1.3–1.5 cm long, 0.8–1 mm broad, upwards abruptly inflated to 0.8–1 cm broad at mouth; lobes suborbicular, 4–5 mm in diameter, subequal, apex obtuse. Stamens 2, exserted; filaments 0.8–1 cm long, glabrous; anther-thecae ellipsoid, 2–2.5 mm long. Ovary narrowly oblong, pilose towards apex; style 3.2–3.5 cm long, pilose. Capsules oblong, 1–1.3 cm long, glabrous except for a few trichomes towards apex, 4-seeded. Seeds ovate in outline, compressed, 1.5–2 mm long and broad, convex on both surfaces, pubescent. Figure 46-28.

Phenology: flowering from August to January of next year; fruiting from September to February of next year.

Habitats: dry forests, montane thickets; 1500–3900 m.

Distribution: S Hengduan, U Irrawaddy (Map 46-51); also in Cambodia, Laos, Thailand, Vietnam, and other parts of Myanmar and China (C & E Yunnan, S Sichuan, Guizhou, Guangxi, Guangdong, Jiangxi).

S HENGDUAN: Dali, Taipingpu, 2350 m, 23 Dec. 1958, *X. Zhou 214* (KUN); W flank of Tali Range, 3000 m, 1913, *G. Forrest 11587* (A, BM, E, IBSC, K, N, PE, TI); **Dechang**, 2100 m, 8 Aug. 1932, *T. T. Yü 1255* (A, IBSC, PE); **Dêqên**, Gujiagou, 3900 m, 29 Nov. 1937, *T. T. Yü 10892* (KUN); **Heqing**, Baiya, 2150 m, 8 Sep. 1940, *R. C. Ching 24495* (KUN, PE); **Huili**, s. loc., 1932, *T. T. Yü 1365* (A, IBSC, PE); s. loc., 1900 m, 31 Aug. 1932, *T. T. Yü 1391* (A, IBSC, PE); s. loc., 1932, *T. T. Yü 1394* (IBSC, PE); between Hokou and Fongsaying, 1800–2000 m, 13 Sep. 1914, *H. Handel-Mazzetti 5104*(WU); **Jianchuan**, Haihong, 2160 m, 4 Aug. 1984, *H. R. Wang 754* (YUKU); Cuifeng, Mingxing dadui, Dahu Pool, 7 Aug. 1975, *P. Y. Chiu 60396* (KUN); **Longyang**, Wayao, 1950 m, 2 Jan. 1959, *X. Zhou 227* (KUN); **Luding**, Dewei, Kuiwu dadui, Yaoping, 1500 m, 6 Jun. 1982, G. H. Xu *41070* (CDBI); **Ningnang**, S. E. of Yungning, Sep. 1922, *G. Forrest 22418* (CAL, E, K); **Panzhihua**, Renhe, Datian, Yinlu, Baoxingshan, 1900 m, 3 Aug. 2003, *Y. L. Peng 6273* (CDBI); **Puge**, Daogou, 2500 m, 8 Aug. 1960, *Sichuan Medicine Exped. 25254* (CNBI, IBSC, NAS, PE); **Shangri-la** (Zhongdian), Haba, Enhe, 2500 m, 19 Jun. 1962, *Exped. Zhongdian 2167* (KUN, PE); Sape et Haba, 2400–3000 m, 2 Aug. 1914, *H. Handel-Mazzetti 4409* (WU); **Xichang**, Lushan, 1700 m, 3 Sep. 1963, *L. N. Zhao 2007* (KUN, PE); 2100 m, 8 Aug. 1932, *T. T. Yü 1255* (PE); **Yanbian**, Heai, Lanba, 1990 m, 6 Aug. 2003, *Y. L. Peng 6537* (CDBI); **Yangbi**, from Cangshanxi to Taiping, 1700 m, 3 Sep. 2013, *Y. F. Deng 24196* (IBSC, KUN); **Yongsheng**, N Yung-peh, 3000–3300 m, Aug. 1922, *G. Forrest 22055* (BM, CAL, E, K, N, PE); between Yungpeh and Taoming, 2 Jun. 1914, *C. Schneider 1715* (K); Qina, 1720 m, 19 Aug. 1981, *1977 Jinjiang Exped. 77312* (YUKU); **Yulong** (Lijiang): Likiang Range, 3000 m, Aug. 1910, *G. Forrest 6264* (BM, E, K, PE); W slopes of Likiang Snow Range, 30 May–6 Jun. 1922, *J. F. Rock 5414* (A, E, N); Likiang Range, 3 Jul. 1914, *H. Handel-Mazzetti 3425* (WU). **U IRRAWADDY: Tengchong:** NW Tengyueh, Aug. 1912, *G. Forrest 8397* (E); Tengyueh, Jul. 1912, *G. Forrest 8797* (E); W Tengyueh, 1800–2100 m, Jul. 1912,

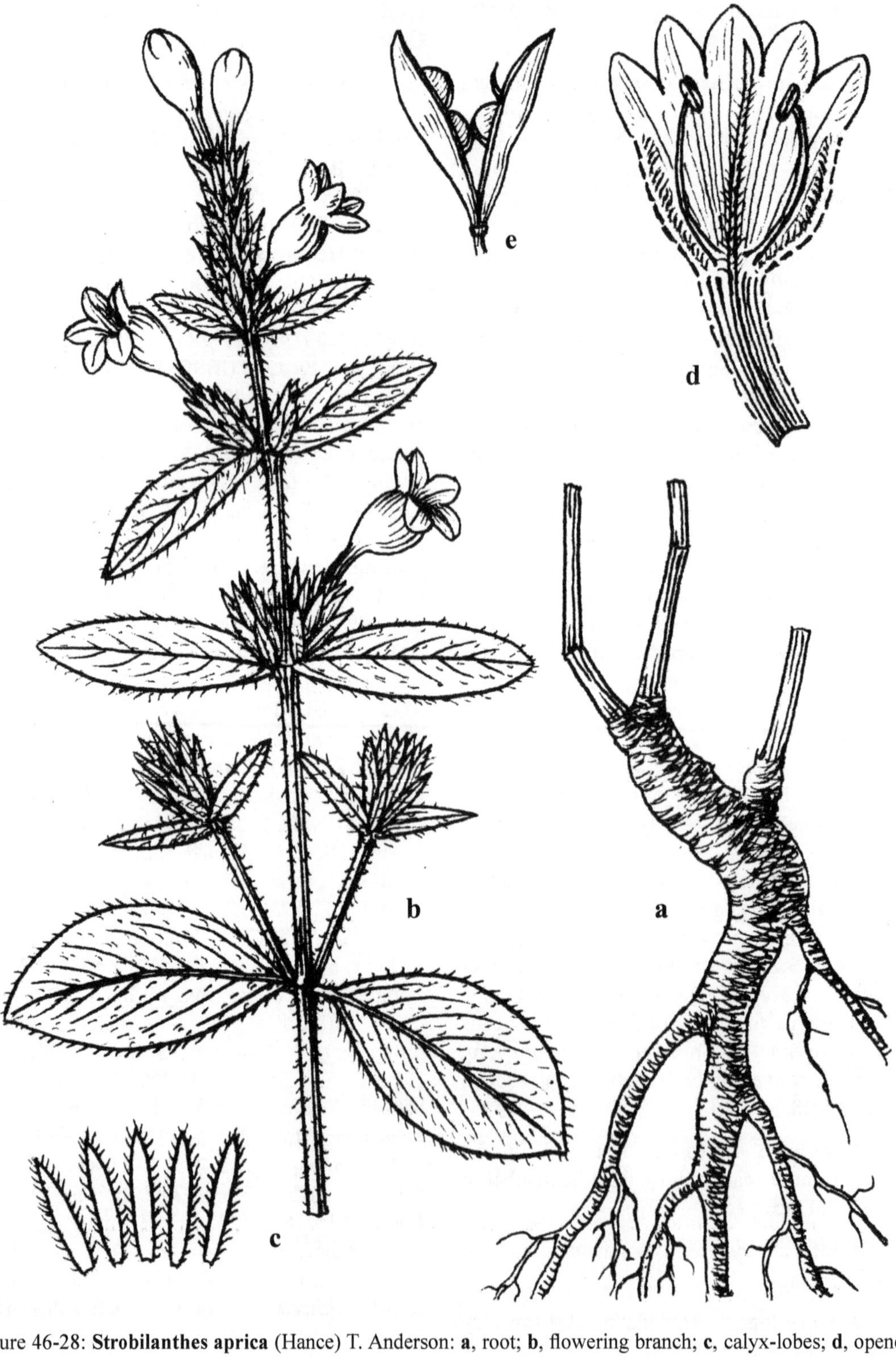

Figure 46-28: **Strobilanthes aprica** (Hance) T. Anderson: **a**, root; **b**, flowering branch; **c**, calyx-lobes; **d**, opened corolla showing stamens and style; **e**, capsule. Drawn by Han-Ping YU; reproduced from *Fl. China Illustr.* 19: fig. 294: 1–5. 2012.

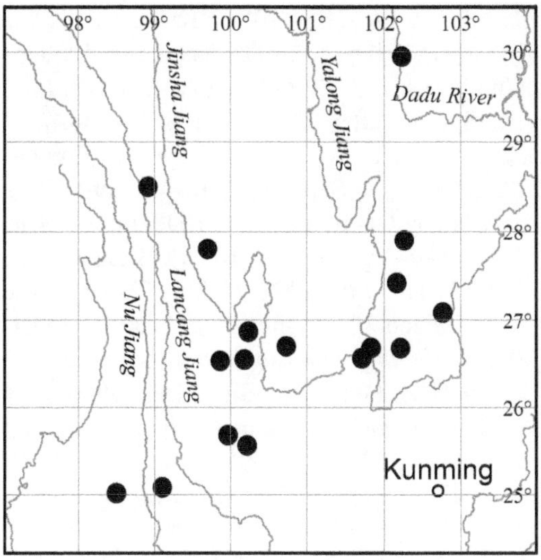

Map 46-51: **Strobilanthes aprica** (Hance) T. Anderson

G. Forrest 9083 (E, IBSC, K).

2. Strobilanthes henryi Hemsl., J. Linn. Soc., Bot. 26: 240. 1890; Hu, Deng & Wood in Wu, Raven & Hong (eds.), Fl. China 19: 398. 2011. ≡ *Gutzlaffia henryi* (Hemsl.) C. B. Clarke ex S. Moore, J. Bot. 63: 167. 1925. ≡ *Paragutzlaffia henryi* (Hemsl.) H. P. Tsui, Acta Bot. Yunnan 12(3): 274. 1990. Type: China, Hubei, "Hupeh", Ichang and immediate neighbourhood, May 1888, *A. Henry 4269* [lectotype **here designated:** K (K000882711); isolectotypes: BM (BM000906336), GH (GH00249040), K (K000882711, K000882712), P (P00719342), US].

Ruellia lyi H. Lév., Fedde, Rep. Sp. Nov. 12: 2. 1913. ≡ *Gutzlaffia lyi* (H. Lév.) E. Hossain, Not. Bot. Gard. Edinburgh 32(2): 407. 1972. ≡ *Paragutzlaffia lyi* (H. Lév.) H. P. Tsui, Acta Bot. Yunnan. 12(3): 274. 1990. Type: China, Guizhou, "Kouy-Tcheou, rochers sousbois", a Tin-Fan, Nov. 1904, *J. Cavaleri 1888* [lectotype **here designated:** E (E00133593); isolectotypes: E (E00133594), K (K000882708)].

Strobilanthes anisandrus Benoist, Bull. Mus. Hist. Nat. Paris 28: 190. 1922. ≡ *Gutzlaffia anisandra* (Benoist) Hand.-Mazz., Symb. Sin. 7: 891, 1936; Bremekamp, Verh. Kon. Ned. Akad. Wetensch., Afd. Natuurk., Sect. 2 41(1): 154. 1944. Type: China, Yunnan, environs de Paeulgay (prefecture de Tchaotong), Oct. 1904, *M. Mey in herb. Ducloux 2902* [lectotype **here designated**: P (P00719249)].

Gutzlaffia forrestii S. Moore, J. Bot. 63: 167. 1925. Type: [S Hengduan] China, Yunnan, Lijiang, margin of thickets on Tongshan in Yangtze bend, 2700m, 15 Aug. 1913, *G. Forrest 10966* [holotype: BM (BM000810977); isotypes: E (E00133364), K (K000882706), PE (PE01558422)].

Gutzlaffia multiramosa Hand.-Mazz., Symb. Sin. 7: 891. 1936. Type: [S Hengduan] China, Yunnan, Hanged et rockenenstr, St. am Mekong unter Lota, 27°50–55′N, 1950 m, Monszwischen Sitjitong und Tjionatongo ber Tschamutong am Salwin, 5 Oct. 1915, *H. Handel-Mazetti 8463* [holotype: WU (WU 0035625); isotype: W (W1934-0005483)].

Gutzlaffia anisandra var. *drosothyrsa* Hand.-Mazz., Symb. Sin. 7: 891. 1936. Type: [S Hengduan] China, Yunnan, Dschungdien, Tonschiefer, 2400 m, 25 Aug. 1915, *H. Handel-Mazzeti 7780* [holotype: W (W 1940-0001573); isotypes: E, WU (WU0035611, WU0035611)].

Subshrubs to 70 cm high, weakly anisophyllous; stems slender, erect or ascending, quadrangular, villous to subglabrous. Petiole 0.5–3 cm long; leaf blade ovate, ovate-elliptic or subrhomboidal, 2–9 cm long, 1.5–4.5 cm broad, both surfaces sparsely to densely pilose, lateral veins 5 or 6 pairs, base broadly cuneate, slightly oblique and usually decurrent onto petiole, margin crenulate, apex acute. Spikes borne on the terminal of leafy branchlets, 3–8 cm long, simple. Peduncles 1.5-4 cm long. Bracts oblong to spatulate, 6–7 mm long, 1–2 mm broad, persistent, abaxially glandular-pubescent to subglabrous; bracteoles linear, 6–7 mm long, 0.6–0.8 mm broad, abaxially glandular-pubescent to subglabrous. Flowers distant basally but apically imbricate on rachis. Calyx 8–11 mm long, subequally 5-lobed almost to base; lobes linear, 6–7 mm long, outside glandular-pubescent to subglabrous. Corolla light purple, blue or white, funnel-shaped, 1.7–2.3 cm long, outside pubescent, inside glabrous except for trichomes retaining style; tube base cylindric, 3.5–6 mm long, upwards bent and gradually widened to 0.8–1 cm broad at mouth; lobes orbicular, 2–3 mm in diameter, spreading. Stamens 2; filaments 3–3.5 mm long, pubescent; anther-thecae ovoid-oblong, exserted. Staminodes 2. Ovary oblong, glabrous but with a few glands towards apex; style 1.8–2 cm long, glabrous. Capsules oblong, 7–10 mm long, glabrous apart from a few apical glands, 4-seeded. Seeds suborbicular in outline, 1.5–2 mm in diameter, densely villous; areole small. Figure 46-29.

Phenology: flowering and fruiting from July to January.

Habitats: mountain slopes; 500–3000 m.

Distribution: N Hengduan, S Hengduan (Map 46-52); also in other parts of China (C & NE Yunnan, S Sichuan, Guizhou, Hubei, Hunan).

N HENGDUAN: Baoxing, 1700 m, 5 Sep. 1936, *K. L. Chü 3796* (BM, E, IBSC, PE, SZ); the city to Yemaoping, 1150 m, 29 Aug. 1938, *C. Pei 8093* (PE); Fuchuanyu, Yemaoping, 1150 m, *C. Pei 8097* (PE); 1000 m, 27 Aug. 1963, *K. C. Kuan & W. T. Wang 2752* (HIB, PE); 1000 m, 27 Aug. 1963, *K. C. Kuan & W. T. Wang 2863* (HIB). **S HENGDUAN: Bingchuan**, Waxi to Dayingjie, 15 Oct. 1946 , *T. N. Liou 21591* (IBSC, PE); Pingchuan, 1800 m, Aug. 1965, *W. M. Zhu & J. L. Wu 2555* (YUKU); **Butuo**, Jiaojihe, Datiba, 1150m, 20 Aug. 1959, *Chuan Jing Liang 5606* (KUN, PE); Niujiaowan, Xixihe, 680 m, 11 Oct. 2014, *Y. F. Deng 25862* (IBSC); **Cawarong (Tsawarong)**, Salwin-Kiukiang divide, Jul. 1919, *G. Forrest 19325* (E, K); **Dali**, Xiaguan, 8 Sep. 1924, *K. K. Tsoong 2379* (PE, PEY, SYS), Xiaguan to Dali, 26 Nov. 1946, *T. N. Liou 17388* (IBSC, KUN, PE); Tali Range, 3000 m, Aug. 1913, *G. Forrest 11512* (E, K, PE); Xiaguan, Xieyangfeng, 2000 m, 5 Aug. 1961, *K. J. Luo et al. 614265* (YIM); **Dêqên**, from Yanmen to Yanling, 16 Mar. 2009, *S. S. Zhou 5345* (HITBC, QBG); **Eryuan**, Zhujiaying, Dawan, 20 Jul. 1929, *R. C. Ching 23306* (KUN, PE); **Gongshan**, Soo-roo-la, Cham-pu-tung, 2000 m, Sep. 1935, *C. W. Wang 66570* (A, HIB, IBSC, KUN, LE, NAS, PE); Salwin Valley, NW of Champutong, 1750 m, 25 Jun. 1938, *T. T. Yü 19161* (GH, E, KUN, PE); Cikai, Salvin Valley, 1600 m, 20 Sep. 1938, *T. T. Yü 23010* (A, E, KUN, PE); Dayon to Lumba, 2200–2500 m, 1 Sep. 1940, *K. M. Feng 7224* (KUN, PE); Gaoligongshan, 1000 m, 24 Apr. 1979, *Exped. Nujiang 218* (KUN); Binzhongluo, Bibili, 1570 m, 5 Jul. 2000, *H. Li 12021* (A, CAS, KUN); Binzhongluo, Qiunatong, 1700 m, 2 Oct. 2002, *Gaoligong Shan Biodiversity Survey 15661* (CAS, KUN); Pengdang, Bibili He, 1500 m, 28 Sep. 2002, *Gaoligong Shan Biodiversity Survey 15558* (CAS, KUN); Bingzhongluo, 1750 m, 8 Oct. 2002, *Gaoligong Shan Biodiversity Survey 17062* (CAS, KUN); Puladi, Qida, 1420 m, 16 Aug. 2006, *Gaoligong Shan Biodiversity Survey 33426* (KUN); Shi Ji Tong north to bridge over Salween, 1660 m, 13 Sep. 1997, *Gaoligong Shan Expedition (1997) 8981* (E, KUN, MO); Bibili, 8 Nov. 1959, *s. coll. 24431* (KUN); Bingzhongluo to Nidadang, 1650 m, 23 Jun. 1982, *Exped. Qinghai-Xizang 7370* (KUN, PE); Chamutung, 2 Jan. 1914, *F. Kingdon-Ward s. n.* (E); Mekong Valley, 2700 m, Aug. 1914, *G. Forrest 13151* (E, IBSC, PE, SYS), Bingzhongluo, Sijitong, 1500 m, 2008, *X. H. Jin et al. DLJ-2677* (PE); Bingzhongluo, Sijitong, 1780 m, 2008, *X. H. Jin et al. DLJ-ET 2720* (PE); **Lanping,** 2800 m, 13 Oct. 1933, *H. T. Tsai 56228* (A, IBSC, KUN, LBG, NAS, PE, SZ); **Luding**, Detuo, Daduhe, 1000 m, 2 Sep. 1981, G. H. Xu *25886* (CDBI); Qiaotou, Guanyinpo, 1400 m, 31 Oct. 1960, *Sichuan Medicine Pl. Exped. 16016* (NAS); **Panzhihua**, Dukou, 1000 m, Mar. 1984, *G. D. Tao 14545* (HITBC); **Shangri-la** (Zhongdian), Hutiaoxia, 2000–2100 m, 25 Jul. 2006, *D. E. Boufford et al. 35223* (A); **Shimian**, Nanya, 850 m, 27 Sep. 1993, *Z. C. Zhao 1213* (PE); Xiyoufang, Hepinggou, 1800 m, 7 Oct. 1993, *Z. C. Zhao 2762* (PE); Xiyoufang, Hepinggou, 1700 m, 7 Oct. 1993, *Z. C. Zhao 2772* (PE); Xiyoufang, Bailu, 1700 m, 7 Oct. 1993, *Z. C. Zhao 2852* (PE); **Weixi**, Kangpu to Badi, 14 Mar. 2009, *S. S. Zhou 5303* (HITBC, QBG); Lota, 5 Oct. 1915, *H. Handel-Mazzetti 8463* (WU); Kangpu to Badi, 2009, *S. S. Zhou 5303* (QBG); **Yanyuan**, 1700 m, 20 Oct. 1932, *T. T. Yü 1745* (IBSC, K); **Yuexi**, s. loc., 1932, *T. T. Yü 1784* (PE); 1500 m, 9 Nov. 1932, *T. T. Yü*

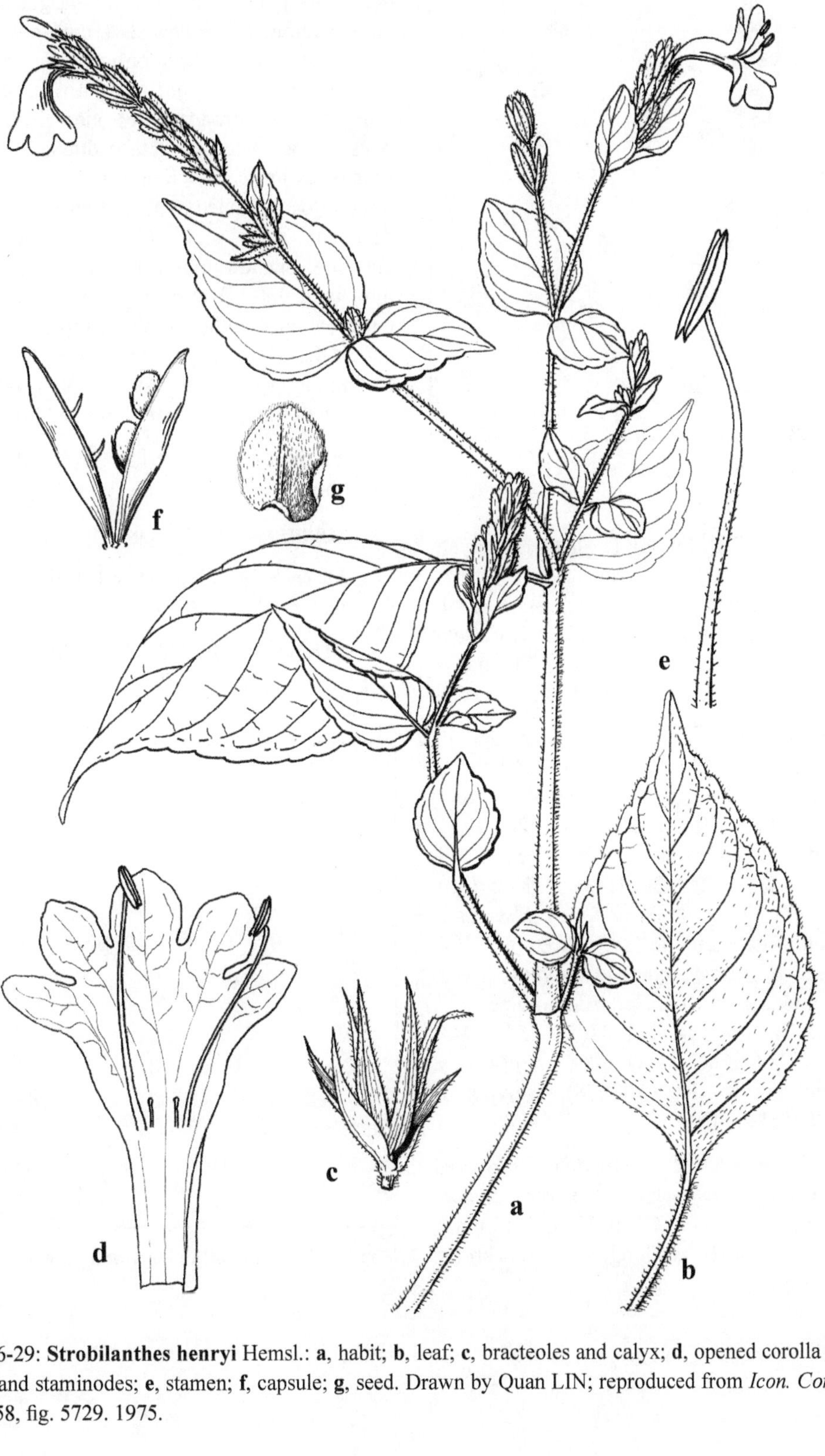

Figure 46-29: **Strobilanthes henryi** Hemsl.: **a**, habit; **b**, leaf; **c**, bracteoles and calyx; **d**, opened corolla showing stamens and staminodes; **e**, stamen; **f**, capsule; **g**, seed. Drawn by Quan LIN; reproduced from *Icon. Cormophyt. Sin.* 4: 158, fig. 5729. 1975.

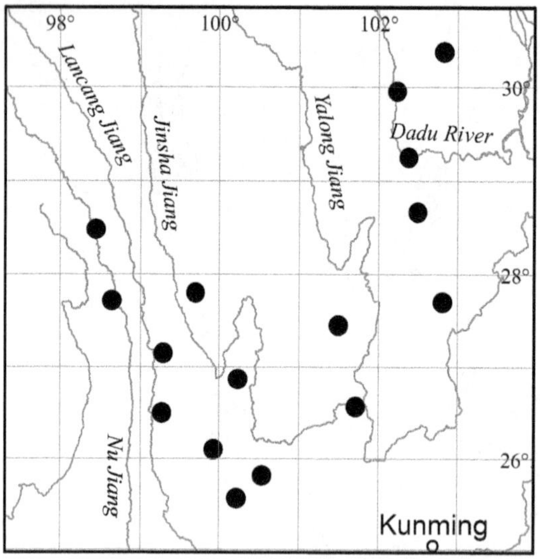

Map 46-52: **Strobilanthes henryi** Hemsl.

1787 (A, IBSC, PE); **Yulong** (Lijiang), Longpan, Lunan, 24 Aug. 2007, *Y. F. Deng 19635* (IBSC); Longpan, Hutiaoxia, 15 Sep. 2013, *Y. F. Deng 25044* (IBSC); Dadong, 16 Sep. 2013, *Y. F. Deng 25048* (IBSC); Gucheng, Dadong, 1740 m, 16 Sep. 2013, *Y. F. Deng 25281* (IBSC).

3. Strobilanthes khasyana (Nees) T. Anderson, J. Linn. Soc., Bot. 9: 471. 1867; Clarke in Hooker (ed.), Fl. Brit. India 4: 436. 1884; Wood, Edinburgh J. Bot. 51: 220. 1994; Wood in Grierson & Long (eds.), Fl. Bhutan 2(3): 1259. 2001; Wood, Kew Bull. 58: 171. 2003. ≡ *Endopogon khasyanus* Nees in de Candolle (ed.), Prodr. 11: 104. 1847. ≡ *Listrobanthes khasyana* (Nees) Bremek., Verh. Kon. Ned. Akad. Wetensch., Afd. Natuurk., Sect. 2 41(1): 254. 1944. Type: India, Meghalaya, *W. Griffith s. n.* [holotype: K (K000882908); isotype: K (K000882912)].

Subshrubs, 0.5–1.2 m high, densely covered with spreading brown hairs. Leaves unequal in each pair; petiole 1–6 cm long; leaf blade narrowly ovate, 4–10 cm long, 2.5–5 cm broad, densely brown pilose on both surfaces, lateral veins 7–9 pairs, base cuneate, margin crenulate, apex acuminate. Spikes 2–8 cm long, sessile, usually several clustered on leaf axils; flowers in opposite pairs, 3–5 mm apart. Bracts 4–11 mm long, 2–7 mm broad, lower ones elliptic, upper ones oblanceolate-spatulate, caducous at anthesis; bracteoles linear, 5–7 mm long. Calyx 0.6–1 cm long, pilose, 5-lobed almost to base; lobes linear, subequal, apex acute. Corolla white, 1.2–1.5 cm long, outside pubescent; tube cylindric, 4–6 mm long, 0.8–1.2 mm broad, widened to 0.8–1 cm broad at mouth; lobes spreading. Stamens 2; anther-thecae exserted; staminodes 2. Capsules 6–8 mm long, sparsely pilose towards apex. Seeds ovate in outline, 1–1.2 mm long and broad.

Phenology: flowering from August to November.

Habitats: stream banks.

Distribution: Sikkim & Darjiling (Map 46-53); also in other parts of India.

SIKKIM & DARJILING: **Darjiling**, s. d., *Griffith s. n.* (K).

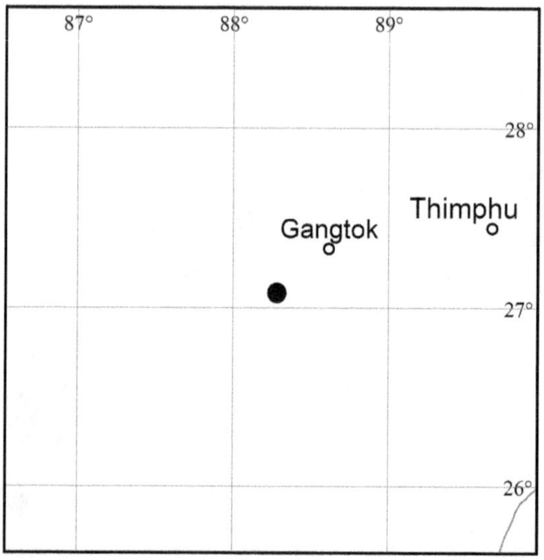

Map 46-53: **Strobilanthes khasyana** (Nees) T. Anderson

4. Strobilanthes tubiflos (C. B. Clarke) J. R. I. Wood, Edinburgh J. Bot. 51: 264. 1994; Hu, Deng & Wood in Wu, Raven & Hong (eds.), Fl. China 19: 398. 2011. ≡ *Strobilanthes petiolaris* var. *tubiflos* C. B. Clarke in Hooker (ed.), Fl. Brit. India 4: 458. 1885. Type: Yarlung Zangbo-Brahmaputra, Mishmi Hills, *W. Griffith Kew District 6095* [lectotype (first step) designated by Wood, Edinburgh J. *Bot.* 51: 264. 1994, second step **here designated**: K (K000883127); isolectotypes: BM, GH (GH00295528), K (K0008831278), P].

Subshrubs; stems quadrangular to subterete, sulcate, sparsely pilose in grooves. Petiole 5–10 mm long; leaf blade ovate-elliptic, 6.5–16 cm long, 3.5–8 cm broad, abaxially pubescent along midvein, adaxially glabrous, lateral veins 4–7 pairs, base cuneate and decurrent onto petiole, margin dentate, apex acuminate. Spikes terminal, branched, 2–4 cm long; rachis glandular-pilose. Bracts obovate, 5–6 mm long, 2.5–3.5 mm broad, densely yellowish glandular-pubescent, apex usually emarginate; bracteoles oblanceolate, 4–5 mm long, 0.8–1 mm broad, abaxially glandular-pubescent especially apically, apex obtuse. Calyx 7–8 mm long, subequally 5-lobed almost to base; lobes oblong, 0.8–1 mm broad, abaxially ciliate and glandular-pilose toward apex, adaxially glabrous. Corolla blue, 2–3 cm long, curved, outside pilose, inside glabrous except for trichomes retaining style; tube base cylindric, 2–3 mm long, 1–1.3 mm broad, upwards gradually widened to 0.8–1 cm long at mouth; lobes broadly ovate, 1.5–2 mm long, 2.5–3 mm broad, apex rounded. Stamens 2, included; filaments glabrous, 3–4 mm long; anther-thecae oblong, erect, 1.5–2 mm long, 0.6–0.8 mm broad. Ovary oblong-ovate, 1–1.5 mm long, apically pilose; style 1.6–2 cm long, glabrous. Capsules oblong, 5–6 mm long, 2–1.5 mm broad, pubescent, 4-seeded. Seeds suborbicular in outline, 1–1.5 mm long, 0.8–1.1 mm broad, pubescent; areole prominent.

Phenology: flowering from August to September.

Habitats: valleys, stream banks; 200–300 m.

Distribution: endemic; Yarlung Zangbo-Brahmaputra (Map 46-54).

YARLUNG ZANGBO-BRAHMAPUTRA: **Mishmi Hills**, s. d., *Griffith K. D. 6095* (K); Mar. 1949, *F. Kingdon-Ward 18409* (BM, NY).

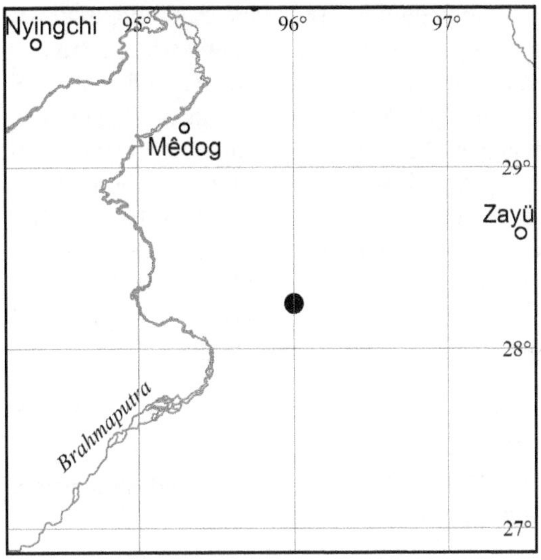

Map 46-54: **Strobilanthes tubiflos** (C. B. Clarke) J. R. I. Wood

5. Strobilanthes kachinensis J. R. I. Wood & J. R. Benn., Kew Bull. 58: 170. 2003. Type: Myanmar, Kachin Hills, *S. M. Toppin 4040* [holotype: E (E00133385)].

Perennial herbs, 1–2 m high; stems sulcate, glabrous, decumbent and rooting at lower nodes, upwards ascending or erect. Leaves equal or slightly unequal in each pair; petiole 0.3–2.7 cm long, sparsely pubescent to subglabrous; blade ovate, 2.2–18.5 cm long, 1–8.5 cm broad, adaxially glabrous, or sparsely appressed hairy, green variegated with white patches, abaxially puberulent along veins, uniformly violet, base attenuate and decurrent onto petiole, margin coarsely dentate, apex acuminate to slightly falcate. Spikes 2–10 cm long, axillary, simple or commonly in panicles; peduncles 2–11 cm long, subglabrous to puberulent; rachis densely glandular-pilose; flowers in opposite pairs, imbricate. Bracts oblong to obovate, 6–11 mm long, 3–5.5 mm broad, diminishing in size upwards, densely glandular-villous, apex acute, persistent; bracteoles oblong to obovate, 3–5 mm long, 1–1.5 mm broad. Calyx 5–7 mm long, subequally 5-lobed to base; lobes oblong, densely glandular-villous towards apex on both surfaces. Corolla 2.7–3.2 cm long, blue, ventricose, outside sparsely to densely pilose, inside glabrous except for hairs retaining style; tube cylindric, 6–8 mm long, upwards gently curved and widened to 8–9 mm broad at mouth; lobes ovate, 4–5 mm long, 4–5 mm broad, apex obtuse. Stamens 2, included; filaments 3.5–4 mm long, glabrous, erect; anthers 1.8–2 mm long, glabrous. Ovary villous; style 2–2.2 cm long, glabrous. Capsules oblong-ellipsoid, 6–8 mm long, 2–2.5 mm broad, silky-pubescent, 4-seeded. Seeds lenticular, 1.5–2 mm long, 1–1.5 mm broad, pubescent with appressed mucilaginous hairs. Figure 46-30.

Phenology: flowering from December to January of next year.

Habitats: shaded gullies and streamside; 100–2400 m.

Distribution: endemic; U Irrawaddy (Map 46-55).

U IRRAWADDY: **Mahtum**, Sensi Gorge to Sadon, 2400 m, Sep. 1912, *G. Forrest 9097* (E, K); **Putao**, Kachin Hills, 950 m, 8 Dec. 1930, *F. Kingdon-Ward 9027* (BM); Sumpra Brum, 1600 m, 8 Jan. 1953, *F. Kingdon-Ward 20358* (BM).

6. Strobilanthes hirta (Vahl) Blume, Bijdr. Fl. Ned. Ind. 14: 797. 1826. ≡ *Ruellia hirta* Vahl, Symb. Bot. 3: 84. t. 67. 1794. ≡ *Hemigraphis hirta* (Vahl) T. Anderson, J. Linn. Soc., Bot. 9: 462. 1867; Clarke in Hooker (ed.), Fl. Brit. India 4: 422. 1884. Type: India Orientali, *O. R. Sehestedt s. n.* [lectotype **here designated**: C (C10004868)].

Ruellia sarmentosa Nees, Pl. Asiat. Rar. 3: 83. 1832. Type: *Herbs Roxburgh in Wallich Cat. no. 2382a* [lectotype **here designated**: K-W (K001115810)].

Herbs, 25–30 cm high; stems prostrate or ascending, densely white pubescent. Leaves subsessile or shortly petiolate; petiole 0–5 mm long; blade broadly elliptic or obovate, 0.7–2.5 cm long, 0.4–1.8 cm broad, densely pubescent on both surfaces, base cuneate, margin crenate, apex obtuse. Flowers in dense heads, terminal on main stems and on short branchlets from upper leaf axil; peduncles 0–1 cm long. Bracts elliptic-obovate, 1.1–1.5 cm long, 1–2 mm broad, pilose, apex obtuse to acute; bracteoles absent. Calyx 0.9–1.2 cm long, base connate to 3–5 mm; lobes subequal, one longer than others, linear-spatulate, 6–7 mm long, 0.7–1 mm broad, margin ciliate, apex acute. Corolla 1.2–1.5 cm long, pale blue or violet, funnel-shaped, straight, pubescent outside; tube base cylindric, 4–5 mm long, 0.8–1.2 mm broad, upwards gradually widened to 6–7 mm broad at mouth; lobes oblong to ovate, 2.5–3 mm long and broad. Stamens 4, included; filaments of longer pair 1.8–2.2 mm long, hirsute toward base, shorter pair 1–1.2 mm long, glabrous; anthers 0.8–1 mm long, connective not produced. Ovary comose; ovules 5–7 per locule; style 8–9 mm long, pubescent; stigma slightly 2-cleft. Capsules 0.9–1.0 cm long, 1–2 mm broad, minutely glandular-puberulent, 10–12-seeded. Seeds ovate in outline, 1–1.2 mm long and broad, compressed, pubescent with appressed mucilaginous hairs.

Phenology: flowering from February to May; fruiting from April to July.

Habitats: weeds in grassy places in plains; below 400 m.

Distribution: W Nepal, C Nepal, E Nepal (Map 46-56); also in other parts of India.

W NEPAL: **Bheri**, Banke, Napelganj, 180 m, 1973, *N. P. Manandhar 12298* (KATH); Banke, Nepalgungj, 180 m, 16 Oct. 2001, *T. Watanabe et al. s. n.* (MBK); Bardiya, 180 m, 20 Jun. 1983, *P. P. Kurmi et al. 5416* (KATH); **Karnali**, Mugu, Banya bhar, 1977, *Ramyar 2610* (KATH); Mugu, Khata, 180 m, 1984, *L. R. Sharma & J. K. Bhattacharya 2725* (KATH); **Mahakali**, Kanchanpur, Gobraiya, 1982, *L. P. Kattel 719* (KATH); **Rapti**, Dang, Gadhawa,

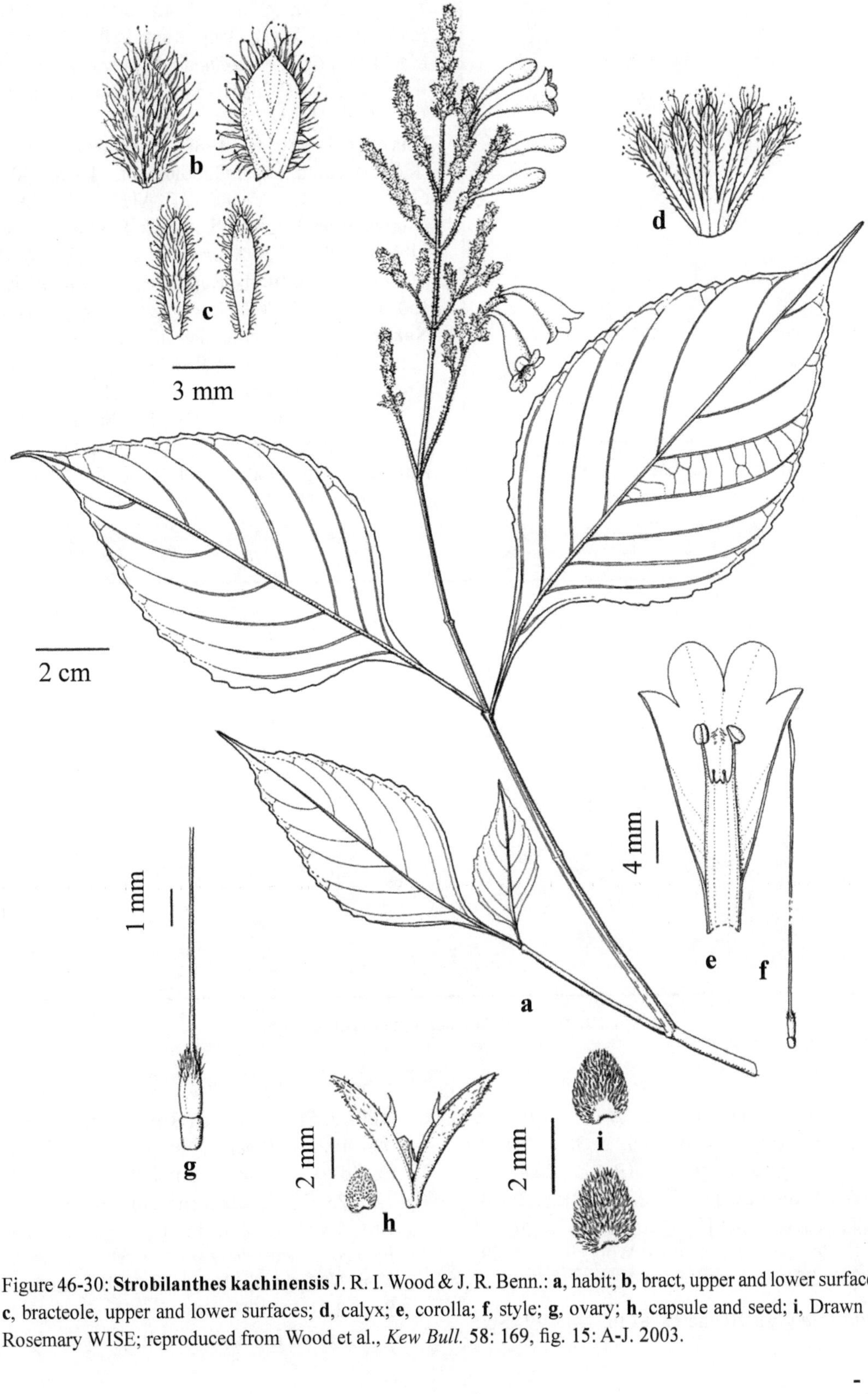

Figure 46-30: **Strobilanthes kachinensis** J. R. I. Wood & J. R. Benn.: **a**, habit; **b**, bract, upper and lower surfaces; **c**, bracteole, upper and lower surfaces; **d**, calyx; **e**, corolla; **f**, style; **g**, ovary; **h**, capsule and seed; **i**, Drawn by Rosemary WISE; reproduced from Wood et al., *Kew Bull.* 58: 169, fig. 15: A-J. 2003.

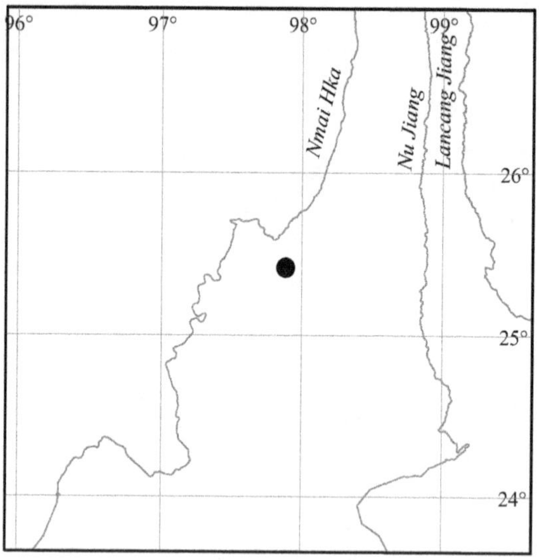

Map 46-55: **Strobilanthes kachinensis** J. R. I. Wood & J. R. Benn.

280 m, 11 Mar. 1976, *N. P. Manandhar & P. M. Regmi 299* (KATH); Dang, Rapti Valley, 30 m, 24 Mar. 1952, *O. V. Polunin, W. R. Sykes & L. H. J. Williams 3604* (BM, E); **Seti**, Kailali, Ataria, 13 Jun. 1976, *H. Tabata et al. 387* (KYO); Kailali, **Dhangarhi**, 180 m, 1983, *P. P. Kurmi 4784* (KATH); Kailali, near Ghodaghodi, 210 m, 2003, *G. D. Bhatta & S. Khatri 54* (KATH). **C NEPAL**: **Dhawalagiri**, Baglung, Dundwa, Bankuj, 180 m, 21 May 1983, *P. R. Sharma & Awamidyal 4539* (KATH); **Lumbini**, Rupandehi, Bhairahawa, 150 m, 2006, *M. Ghimine et al. 0629* (KATH); **Narayani,** Bara, Nijgadh, 19 Apr. 1969, *H. Kamai & M. S. Bista 11493* (KATH). **E NEPAL**: **Koshi**, Morang, Biratnagar, 80 m, 18 May 1969, *L. H. J. Williams 49* (BM, E); **Sagarmatha**, Saptari, Koshi Tappu Wildlife Reserve, from Pathari to Trijuga, 28 May 1993, *P. R. Shakya et al. 178* (KATH).

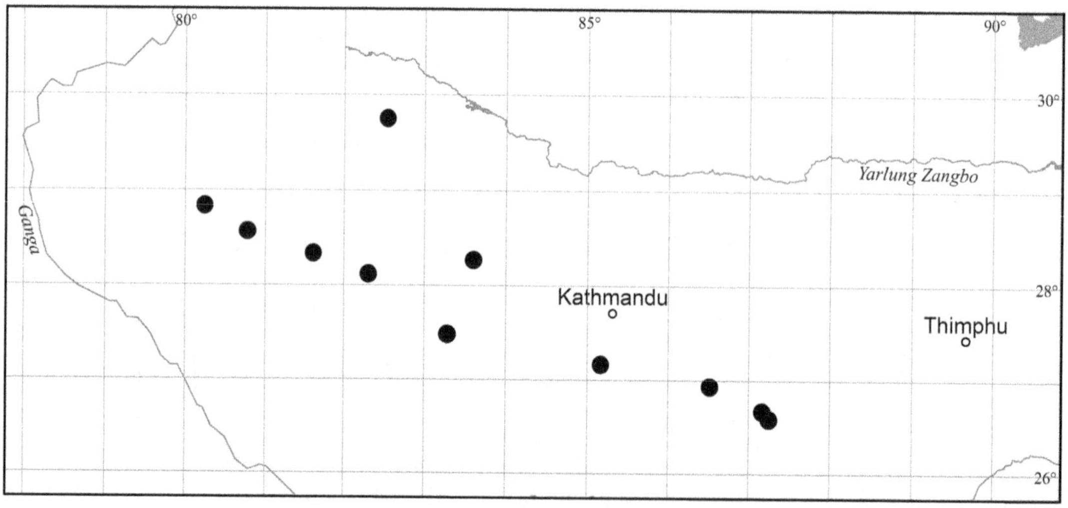

Map 46-56: **Strobilanthes hirta** (Vahl) Blume

7. Strobilanthes pavala (Roxb.) J. R. I. Wood, Novon 23: 392. 2014. ≡ *Ruellia pavala* Roxb., Fl. Ind. ed. 2, 3: 47. 1832. ≡ *Hemigraphis pavala* (Roxb.) Kurz, J. Asiat. Soc. Bengal, Pt. 2, Nat. Hist. 39(2): 78. 1870. Type: [*icon*] *Icon. Roxburgh 327* (lectotype designated by Wood, Novon 29: 392. 2014: K).

Ruellia latebrosa Heyne ex Roth, Nov. Pl. Sp.: 307. 1821, non *Strobilanthes latebrosa* Ridl., J. Straits Branch Roy. Asiat. Soc. 86: 304. 1922. ≡ *Hemigraphis latebrosa* (Heyne ex Roth) Nees in de Candolle (ed.), Prodr. 11: 723. 1847; Clarke in Hooker (ed.), Fl. Brit. India 4: 423. 1884. ≡ *Hemigraphis latebrosa* var. *heyneana* Bremek., Verh. Kon. Ned. Akad. Wetensch., Afd. Natuurk., Sect. 2 41(1): 139. 1944, nom. invalid. (Art. 24.3). Type: India (or Pakistan), Islamnagar,

26 Dec. 1811, *F. Buchanan-Hamilton s. n. ex Wallich Cat. no. 2395a* [neotype designated by Wood, Novon 29: 392. 2014: K-W (K001115850, p. p. plant at bottom of sheet)].

Ruellia diffusa Wall. ex Nees, Pl. Asiat. Rar. 3: 83. 1832, non *Ruellia diffusa* Vell., Fl. Flumin. 266. 1829. ≡ *Crossandra diffusa* Wall., Numer. List n. 2383a. 1830, nom. invalid. Type: India, Chunar, 1 May 1813, *Hamilton in Wallich Cat. 2383a* [lectotype **here designated**: K-W (K0011115819); isolectotypes: K-W (K001115817, K001115820, K001115821)].

Ruellia elegans Hook., Curtis's Bot. Mag., 62: tab. 3389. 1835, nom illeg., non *Ruellia elegans* Poir., Encyc. Suppl. 4. 727. 1816. ≡ *Hemigraphis elegans* Nees in de Candolle (ed.), Prodr. 11: 722. 1847. Type: [India.] Plant cultivated at Glasgow from the "East Indies", specimen ex Herb. Hooker, labeled on left side of sheet "*Ruellia* Bot Mag. Hort Gl. ex Ind. Or."and below "*Ruellia diffusa* N ab E"in Nees's hand, crossed out and then in another hand, "*Hemigraphis elegans* Nees" (lectotype designated by Wood, Novon 29: 392. 2014: K).

Ruellia crenata Benth. ex Hohen., Flora 32: 558. 1849, **syn. nov.** ≡ *Hemigraphis elegans* var. *crenata* (Benth. ex Hohen.) C. B. Clarke in Hooker (ed.), Fl. Brit. India 4: 425. 1884. Type: India, Karnataka, *R. F. Hohenacker 737* [lectotype **here designated**: K (K000882590); isolectotypes: BM, K (K000882589), LE, TUB].

Ruellia ebracteata Dalzell, Hooker's J. Bot. Kew Gard. Misc. 2: 342. 1850. ≡ *Hemigraphis ebracteata* (Dalzell) C. B. Clarke in Hooker (ed.), Fl. Brit. India 4: 424. 1884, "*ebracteolata*". ≡ *Hemigraphis latebrosa* var. *ebracteata* (Dalzell) Cooke, Fl. Bombay 2: 353. 1905, "*ebracteolata*". Type: India. "Crescit raro in Concano septentrionale," s. d., *Herb. Dalzell s. n.* [lectotype designated by Wood, Novon 29: 392. 2014: K (K001096869)].

Ruellia satpoorensis Wawra, Oesterr. Bot. Z. 31: 281. 1881. Type: India. Satpoora, s. d., *Wawra I/1450* (holotype: W).

Hemigraphis latebrosa var. *rupestris* C. B. Clarke in Hooker (ed.), Fl. Brit. India 4: 424. 1884; Malik & Ghafoor in Nasir & Ali (eds.), Fl. Pakistan 188: 66. 1988. Type: India. "Specimina accepti a veneratiss. Beni. Hevne ex India Orientali", 27 Sep. 1798, *Heyne in [J. P.] Rottler s. n.* [lectotype designated by Wood, Novon 29: 392. 2014: K (fertile piece lowermost on sheet at right)].

Hemigraphis latebrosa var. *beddomei* C. B. Clarke in Hooker (ed.), Fl. Brit. India 4: 424. 1884. Type: India. Bellary, s. d., *R. H. Beddome 5780* [holotype: K (K000882594)].

Hemigraphis latebrosa var. *incana* Gamble, Fl. Madras 2: 1019. 1921. Type: India. s. d., *Herb. Wight 2272* [holotype: K (K000882593)].

Diffuse herbs, up to 80 cm high; stems quadrangular, viscidly white pubescent. Petiole 1.5–3.5 cm long, pubescent; leaf blade ovate, 2.5–7.5 cm long, 1.5–3.5 cm broad, densely pubescent, often viscid aromatic, lateral veins 5 pairs, base cuneate, margin crenate-serrate, apex acuminate. Flowers in dense, axillary and terminal heads, blue. Bracts ovate or elliptic, 6–10 mm long, 2–3 mm broad, margin entire, ciliate, apex acute, outer ones longer than calyx, pubescent; bracteoles absent. Calyx 8–9 mm long; lobes linear, one larger than others, whitish with green tips, margin ciliate. Corolla purplish-blue, 1.2–1.5 cm long, funnel-shaped, glabrous outside; lobes obovate, 3–4 mm long, 3–4 mm broad, apex obtuse. Stamens 4; filaments of shorter pair glabrous, longer pair beset with long white hairs; anther-thecae oblong. Ovary pilose; style 1–1.2 cm long, hirsute. Capsules 6–8 mm long, minutely pubescent, 6–8-seeded. Seeds suborbicular or ellipsoid, 1.5–2 mm in diameter, yellowish-brown, pubescent with appressed mucilaginous hairs.

Phenology: flowering from October to February of next year.

Habitats: hillslopes, waste lands and forests; 600–1700 m.

Distribution: Jammu & Kashmir, U Ganga & Indus (Map 46-57); also in other parts of India.

JAMMU & KASHMIR: Jammu, Pouni Chak, 1969, *B. M. Sharma 8* (RRLH); **Rajouri**, Drhal, 1650 m, 2010, *Akhtar 1042* (KASH). **U GANGA & INDUS**: **Chamba**, ca. 600 m, *J. R. Drummond 1722* (K); **Haridwar**, 1825, *Wallich 2383b* (K-W); **Kangra**, s. d., *J. R. Drmmond 26209* (E, K); **Kumaon**, Decan to Birmdeo, 25 Dec. 1848, *R. Strachey & J. E. Winterbottom 263* (K); Outer Hills of Kumaon, 600–1500 m, *R. Strachey & J. E. Winterbottom s.n.* (BM, K); **Mussorie**, *M. B. Saxena 370* (DD, fide Raizada & Saxena, 1978); **Simla**, 1956, *J. W. Chiddell 99* (BM).

8. **Strobilanthes violifolia** T. Anderson, J. Linn. Soc. Bot. 9: 485. 1867; Clarke in Hooker (ed.), Fl. Brit. India 4: 477. 1884. ≡ *Pteracanthus violifolius* (T. Anderson) Bremek., Verh. Kon. Ned. Akad. Wetensch. Afd. Natuurk., Sect. 2 41(1): 200. 1944. Type: [Sikkim & Darjiling] INDIA. Sikkim, Ratong to Yoksum, 2500–5000 ft., *T. Anderson 1051* [lectotype designated by Albertson & Wood, Phytotaxa 43: 58. 2012: CAL (CAL0000019841)].

Echinacanthus parviflorus T. Anderson, J. Linn. Soc. Bot. 9: 459. 1867, non *Strobilanthes parviflora* Beddome, Icon. Pl. Ind. Or.: 45. 1868; Clarke in Hooker (ed.), Fl. Brit. India 4: 415. 1884. ≡ *Clarkeasia parvifora* (T. Anderson) J. R. I. Wood, Edinburgh J. Bot. 51: 189. 1994; Wood in Grierson & Long (eds.), Fl. Bhutan 2(3): 1254. 2001. Type: India, Meghalaya (Khasia), *W. Griffith s. n.* [lectotype designated by Wood, Edinburgh J. Bot. 51: 189. 1994: K (K000882651); isolectotypes: CAL].

Echinacanthus andersonii C. B. Clarke in Hooker (ed.), Fl. Brit. India 4: 415. 1884. Type: [Sikkim & Darjiling] India, Darjiling, 31 Oct. 1870, *C. B. Clarke 13633* [lectotype **here designated**: K (K000882661); isolectotypes: BM (BM000950014), K (K000882660)].

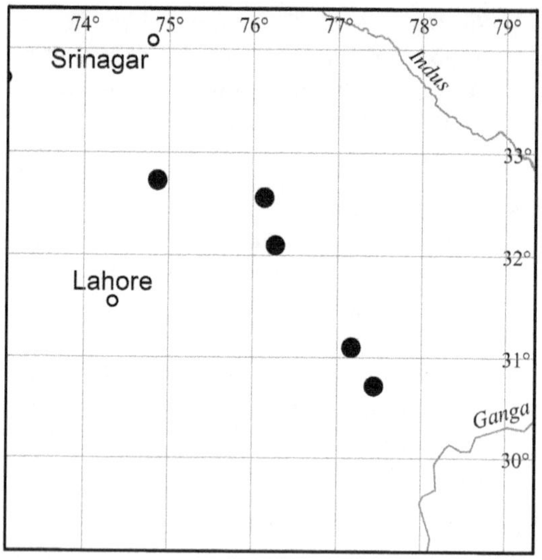

Map 46-57: **Strobilanthes pavala** (Roxb.) J. R. I. Wood

Echinacanthus andersonii var. *viscosa* C. B. Clarke in Hooker (ed.), Fl. Brit. India 4: 415. 1884. Type: [Sikkim & Darjiling] India, Sikkim, 21 Oct. 1875, *C. B. Clarke 25468* [lectotype **here designated**: K (K000882657); isolectotypes: BM, K (K000882658, K0008826609)].

Echinacanthus longistylus C. B. Clarke in Hooker (ed.), Fl. Brit. India 4: 415. 1884. Type: [Sikkim & Darjiling] India, Darjiling, 15 Nov. 1870, *C. B. Clarke 13845* [lectotype **here designated**: K (K000882655); isotype: K (K000882654)].

Echinacanthus pumilio C. B. Clarke in Hooker (ed.), Fl. Brit. India 4: 415. 1885. Type: India, Meghalaya, Khasia, Cohra Coal Hill, 4500 ft., 3 Nov. 1872, *C. B. Clarke 17801* [lectotype **here designated**: K (K000882652); isotypes: BM (BM000950012), K (K000882653)].

Clarkeasia parvifora (Anderson) J. R. I. Wood var. *albescens* J. R. I. Wood, Edinburgh J. Bot. 51: 191. 1994, **syn. nov.**; Wood in Grierson & Long (eds.), Fl. Bhutan 2(3): 1254. 2001. Type: India, West Bengal, Western Dicas, 1879, *D. Brandis s. n.* [holotype: K (K000882662)].

Subshrubs, 0.5–2 m high; stems striate, glabrous or sparsely pubescent. Leaves unequal in each pair, smaller one 1/2–2/3 of larger one in size; petiole 1–2 cm long, pilose; blade decreasing in size upwards, lower ones ovate-elliptic, 5–12 cm long, 1.5–6.5 cm broad, uppermost ones linear-lanceolate, 1.2–2.5 cm long, 2–5 mm broad, glabrous or pubescent along veins, base usually asymmetric, cuneate, rounded or cordate, margin sinuate, apex shortly acuminate. Spikes axillary, 1-sided, usually several in panicles, 2–9 cm long; flowers solitary or paired, somewhat remote, 5–15 mm apart; rachis often zigzag. Bracts linear, 2.5–10 mm long, 0.6–1 mm broad, glandular-pilose, apex obtuse; bracteoles linear, 2–3 mm long, glandular-pilose. Calyx 9–15 mm long, glabrous, strigose or glandular-pilose, united for 2.5–3 mm long above base; lobes linear, apex acuminate. Corolla blue, 2.5–3 cm long, ventricose, sparsely pubescent; tube base cylindric, 4–5 mm long, 1–1.2 mm broad, upwards gradually widened to 0.8–1 cm broad at mouth; lobes ovate, 3–4 mm long, 2.5–3 mm broad, apex obtuse. Stamens 4, included; filaments of longer pair 6–8 mm long, hirsute, shorter pair 5–6 mm long, glabrous; anther-thecae oblong, 2–2.5 mm long, 1–1.5 mm broad. Ovary glabrous; style 1.6–2 cm long, hirsute; stigma unequally 2-lobed. Capsules 1–1.2 cm long, glabrous or with few glandular hairs near tip, 16-seeded. Seeds ovate in outline, compressed, 1–1.5 mm long, 0.8–1 mm broad, appressed pubescent. Figure 46-31.

Phenology: flowering and fruiting from August to January of next year.

Habitats: rocks and cliffs beside rivers in subtropical forests; 100–2300 m.

Distribution: U Ganga & Indus, W Nepal, C Nepal, E Nepal, Sikkim & Darjiling, Bhutan (Map 46-58); also in Thailand, and other parts of India.

U GANGA & INDUS: **Kumaon**, descent to Bel Khate, s. d., *J. E. Winterbottom s. n.* (K). **W NEPAL**: **Bheri**, Dailekh, Seri-9, Guranse, 2150 m, 6 Sep. 2014, *B. Adhikari & R. Basnet 46* (E); **Seti**, Kailali, Godavari Kailali, 750 m, 10 Nov. 2012, *B. Adhikari & G. D. Bhatt 33* (E); Kailali, Budhetola, 1140 m, 10 Nov. 2012, *B. Adhikari & G. D. Bhatt 41* (E). **C NEPAL**: **Bagmati**, Laliptur, Lele, 1540 m, 1990, *I. Shrestha & N. Pradhan 138* (KATH); Laliptur, Chappa Khodaka, 2250 m, 1963, *s. coll. 8301* (KATH); **Gandaki**, Gorkha, Dobhan to Jagat, 1290 m, 27 Jul. 2008, *Manaslu 2008 Expedtion (H. Ikeda et al.) 20816039* (TI); Kaski, Lamjung Himal, 17 Sep. 1969, *T. Wraber 269* (BM); Lamjung, below Seilung, along Midam Khola, 17 Sep. 1969, *T. Wraber 36638* (BM); **Lumbini**, Nawalparasi, Daunne, Daunne Danda, 1200 m, 18 Dec. 1975, *D. P. Joshi et al. 75/3588* (KATH); Palpa, Masure, Bhairabsthan, 1280 m, 15 Nov. 2012, *B. Adhikari 50* (E, KATH); Rupandehi, Marek, near Butwal, 610 m, 1954, *J. D. A. Stainton, W. R. Sykes & L. H. J. Williams 8813* (BM); **Narayani**, Bara, on the road from Hetauda to Birgunj, 480 m, 26 Nov. 2004, *C. A. Pendry et al. A129* (E, KATH); Chitwan, Jarwang, 1050 m, 16 Nov. 1989, *N. P. Manandhar 13933* (KATH); Makwanpur, Bara Badhikim Bridge, on the road from Hetauda to Birgunj, 480 m, 26 Nov. 2004, *C. A. Pendry et al. DNEP2 A129* (E). **E NEPAL**: **Koshi**, Sunsari, Sapt Kosi Gorge, 200 m, 8 Nov. 1969, *J. D. A. Stainton 6658* (BM, KATH, TI); Sunsari, Phusri–Sanguri Bhanjang–Dhari Pani, 16 Oct. 1963, *H. Hara et al. 4907* (BM, KATH, KYO, TI); Sunsari, Baraha Chetra, 160 m, 25 Nov. 2012, *B. Adhikari & D. R. Kandel 91* (E, KATH); Sunsari, Baraha Chetra, 160 m, 25 Nov. 2012, *B. Adhikari & D. R. Kandel 92* (E, KATH); Sankhuwasabha, near Obak, E side of Arun Valley, 1260m, 15 Oct. 1991, *D. G. Long et al. 834* (E, KATH); Sankhuwasabha, Merak, near Butwal, 610 m, 6 Oct. 1954, *J. D. A. Stainton, W. R. Syke & L. H. J. Williams 8813* (BM); **Mechi**, Ilam, Selpur, Goduk-1, 790 m, 25 Nov. 2012, *B. Adhikari & D. R. Kandel 75* (E, KATH). **SIKKIM & DARJILING**: **Darjiling**, Pomong, 900 m, 31 Oct. 1870, *C. B. Clarke 13633* (K, BM); Punkabari, 15 Nov. 1870, *C. B. Clarke 13812* (BM, K); Kurseong, 15 Nov. 1875, *C. B. Clarke 13845* (K); Punkabari, 80 m, 23 Jun. 1871, *C. B. Clarke 13897* (BM); Panchkilla, 300 m, 29 Feb. 1876, *C. B. Clarke 27079* (BM); Selim, 1100 m, 12 Oct. 1884, *C. B. Clarke 36738* (BM, K); Kurseongroad, 1000 m, *J. Gamble 155A* (K); Tista Valley, 300 m, 7 Jan. 1873 m, *J. Gamble 3565* (K); Kurseong, 1400 m, s. d., *J. Gamble 3574* (K); Pankaban, 750 m, Nov.

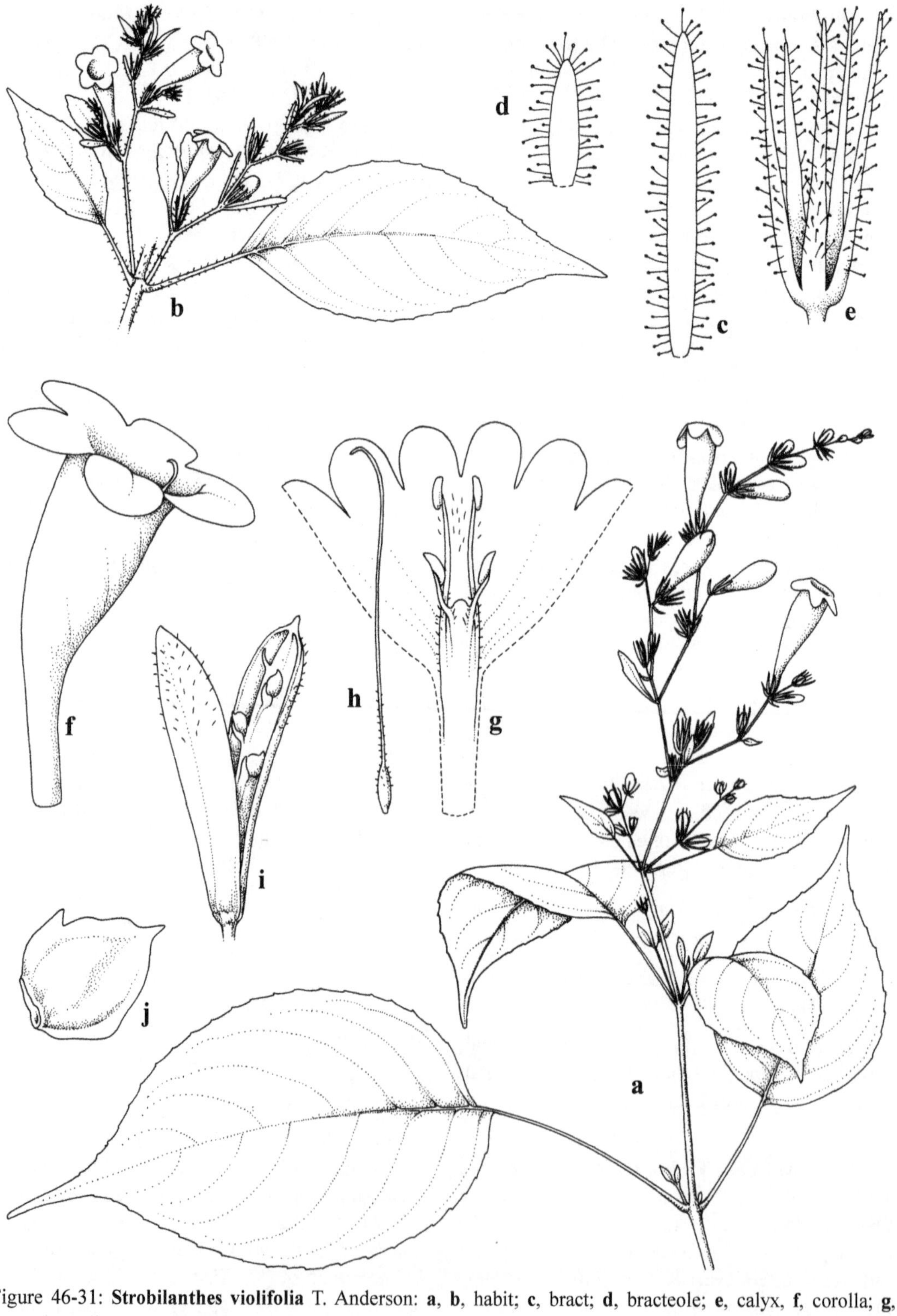

Figure 46-31: **Strobilanthes violifolia** T. Anderson: **a**, **b**, habit; **c**, bract; **d**, bracteole; **e**, calyx, **f**, corolla; **g**, opened corolla showing androecium; **h**, pistil; **i**, capsule; **j**, seed. Drawn by Mary BATES; reproduced from Wood, *Edinburgh J. Bot.* 51: 189, fig. 3. 1994.

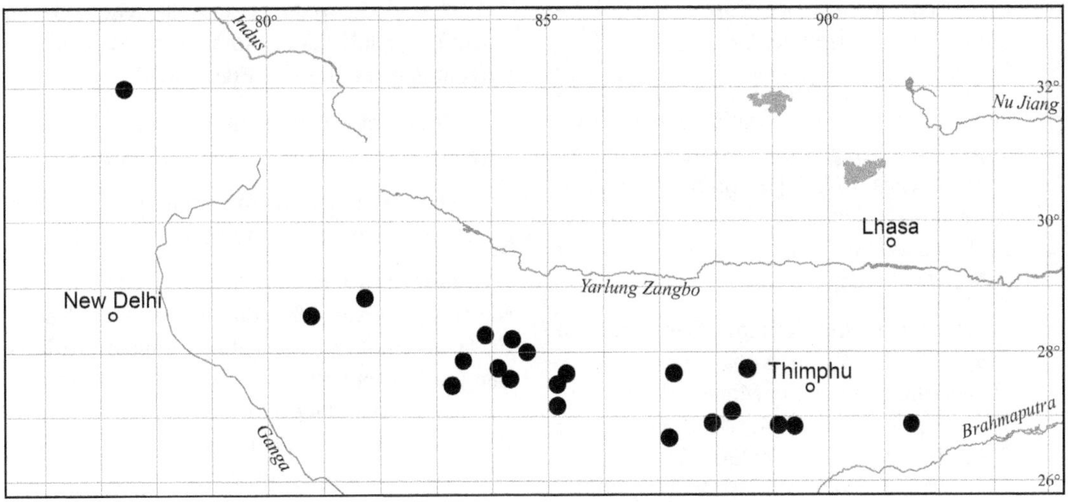

Map 46-58: **Strobilanthes violifolia** T. Anderson

1879, *J. Gamble 7382* (K); **Sikkim**, Lingcham, 1200 m, 21 Oct. 1875, *C. B. Clarke 25468* (BM, K); s. loc., 1878, *G. King s. n.* (K). **BHUTAN**: **Deothang**, N Deothang, 1400 m, 18 Oct. 1987, *J. R. I. Wood 5970* (E, K, THIMPHU); **Samchi**, W bank of Torsa River, 200 m, 17 Feb. 1982, *A. J. C. Griersan & D. G. Long 2987* (E); Charmurchi River, 800 m, 20 Nov. 1991, *J. R. I. Wood 7514* (E, K); **Phuntsholing**, Torsa River, 300 m, 19 Jan. 1992, *J. R. I. Wood 7533* (E).

9. Strobilanthes tomentosa (Nees) Terao ex J. R. I. Wood, Kew Bull. 64: 16. 2009; Hu, Deng & Wood in Wu, Raven & Hong (eds.), Fl. China 19: 389. 2011. ≡ *Aechmanthera tomentosa* Nees, Pl. Asiat. Rar. 3: 87. 1832; de Candolle, Prodr. 11: 171. 1847; Clarke in Hooker (ed.), Fl. Brit. India 4: 428. 1884; Nasir & Ali in Malik & Ghafoor (eds.), Fl. Pakistan 188: 70. 1988; Hu, Fl. Reipblic. Popularis Sin. 70: 191. 2002. ≡ *Aechmanthera wallichii* var. *tomentosa* (Nees) Nees in de Candolle (ed.), Prodr. 11: 171. 1847. ≡ *Ruellia tomentosa* Wall., Numer. List n. 2333. 1830, nom. invalid. Type: [U GANGA & INDUS] India, Kumaon, *Wallich Cat. 2333b* [lectotype designated by Wood & Scotland, Kew Bull. 64: 16. 2009: K-W (K001115679); isolectotypes: E (E00273437), CGE].

Ruellia gossypina Wall., Pl. Asiat. Rar. 1: 38. t. 42. 1830, non *Strobilanthes gossypina* T. Anderson, J. Linn. Soc., Bot. 9: 466. 1867. ≡ *Aechmanthera gossypina* (Wall.) Nees, Pl. Asiat. Rar. 3: 87. 1832; Bremekamp, Verh. Kon. Ned. Akad. Wetensch., Afd. Natuurk., Sect. 2 41(1): 149. 1944; Wood in Grierson & Long (eds.), Fl. Bhutan 2(3): 1253. 2001; Hu, Fl. Reipubl. Popularis Sin. 70: 191. 2002. ≡ *Aechmanthera wallichii* Nees var. *gossypina* (Wall.) Nees in de Candolle (ed.), Prodr. 11: 171. 1847; T. Anderson, J. Linn. Soc. 9: 485. 1867. ≡ *Aechmanthera wallichii* Nees in de Candolle (ed.), Prodr. 11: 171. 1847, nom. illeg. (Art. 52.1 & 52.2). ≡ *Aechmanthera tomentosa* var. *wallichii* C. B. Clarke in Hooker (ed.), Fl. Brit. India 4: 428. 1884, nom. illeg. (Art. 52.1 & 52.2). ≡ *Aechmanthera wallichii* Nees f. *fomentaria* Nees in de Candolle (ed.), Prodr. 11: 171. 1847. ≡ *Ruellia gossypina* Wall., Numer. List n. 2332. 1830, nom. invalid. Type: [U Ganga & Indus] Deyra Dhun, *Wallich Cat. 2332b* [lectotype designated by Wood & Scotland, Kew Bull. 64: 16. 2009: K-W (K001115678); isolectotypes: E (E00273434, E00273435)].

Aechmanthera wallichii Nees [subvar.] β *leuconeura* Nees in de Candolle (ed.), Prodr. 11: 170. 1847. Type: India, specimen annotated *Ruellia leuconeuron, Royle s. n.* (holotype: LIV; isotype: LE).

Aechmanthera leiosperma C. B. Clarke in Hooker (ed.), Fl. Brit. India 4: 429. 1884. Type: India, Meghalaya (Mungot, Jaintea), *C. B. Clarke 14825* [lectotype designated by Wood & Scotland, Kew Bull. 64: 16. 2009: K (K000885421); isolectotype: BM (BM001191495)].

Aechmanthera claudiae Bernardi, Candollea 18: 243 1963. Type: [E] Nepal, Mechi Zone, Taplejunng District, Mane Bhanjyang, 1580 m, 5 Nov. 1954, *A. Zimmermann 2064* [holotype: G (G00236119); isotypes: A (A00093575), BM (BM000950011), K (K000885421), LE (LE00017228), U (U0000002)].

Subshrubs, 0.5–1 m high, isophyllous; stems usually gray tomentose, sometimes sparsely pilose. Leaves equal in each pair; petiole 1–3 cm long, yellow multicellular glandular-pubescent; blade elliptic-ovate to elliptic-oblong, 3–13 cm long, 2–6.5 cm broad, abaxially sericeous, adaxially sparsely pilose, densely covered with linear cystoliths, lateral veins 7–10 pairs, base subrounded, cuneate or cordate, sometimes decurrent onto petiole, margin crenate, apex acute. Inflorescences usually paniculate, 10–30 cm long, sometimes 3–8 flowers clustered on axillary branches in leaf axils; branches 3–8 cm long. Bracts linear-ligulate, 0.8–1.2 cm long, abaxially glandular-pubescent, margin glandular-ciliate, apex obtuse; bracteoles narrowly linear, 4–5.5 mm long. Calyx 0.8–1.2 cm long, 5-lobed almost to base; lobes linear, unequal with 2 longer and 3 shorter, sparsely glandular-pilose, apex obtuse. Corolla blue, 2.5–3.5 cm long, funnel-shaped, straight, ventricose, outside glabrous, inside glabrous except for trichomes retaining style; tube base cylindric, 4–8 mm long, 0.8–1 mm broad, upwards gradually widened to 0.8–1 cm broad at mouth; lobes oblong, 4–5 mm long, 4–5 mm broad, subequal, apex obtuse. Stamens 4, included; filaments of longer pair 1–1.2 cm long, pilose, shorter pair 5–6 mm long, glabrous; anther-thecae oblong; connective minutely excurrent for a mucro. Ovary comose, 3 or 4 ovules per locule; style sparsely hairy. Capsules oblong, 7–10 mm long, pubescent, 6–8-seeded. Seeds discoid, densely hairy, sometimes glabrescent. **Chromosomes** 2n = 30. Figure 46-32.

Phenology: flowering from June to December; fruiting from August to December.

Habitats: grassy mountain slopes, open forests in relatively dry areas; 600–2500 m.

Distribution: N Pakistan, Jammu & Kashmir, U Ganga & Indus, W Nepal, C Nepal, E Nepal, Sikkim & Darjiling, Bhutan, Yarlung Zangbo-Brahmaputra, S Hengduan (Map 46-59); also in Bangladesh, Laos, and other parts of Pakistan, India, Myanmar and China (C, S & E Yunnan, Guizhou, Guangxi).

N PAKISTAN: Azad Kashmir, Nakial, 29 Sep. 1953, *A. Rashid, 3E. Nasir & R. R. Stewart s. n.* (RAW, Malik & Ghafoor, 1988). **JAMMU & KASHMIR**: Kashmir, Sradabad, Oct. 1880, *Young s. n.* (BM); **Jammu**, Nandani, 1969, *B. M. Sharma 63* (RRLH); **Rajouri**, Dudang, 1000 m, 2010, *K. Akhtar 1058* (KATH). **U GANGA & INDUS**: **Almora**, Cheelkote, 2340 m, 1950, *F. C. Thanas 20880* (CAS); **Chamba**, Masrund, 21 Jun. 1916, *R. R. Stewart 2318* (K); **Dehradun**, Sahanidara, 950 m, Oct. 1890, *J. S. Gamble 22304* (K); Mohand Pass, 900 m, 22 Sep. 1969, *U. C. Bhattacharrya 39188* (KATH); **Kumaon**, *Blinkworth in Wallich Cat. 2333b* (E, K-W); 1200 m, *R. Strachey & J. E. Winterbottom s. n.* (BM, K); **Lahul & Spiti**, Mila, 1900 m, 1941, *N. L. Bor 15841* (K); **Mussorie**, *D. Brandis 1619* (DD, fide Raizada & Saxena, 1978); *R. R. Stewart 73268* (DD, fide Raizada & Saxena, 1978); **Simla**, Sahanidara, 950 m, *J. S. Gamble 22304* (K); **Solan**, Sabathu, 11 Oct. 1831, *Lady Dalhousie s. n.* (K); **Tehre Garhwal**, near Ghausali, 1954, *K. C. Sahui 21606* (BO); Garhwal, *S. Eardley-Wilmot 80* (FHO). **W NEPAL**: **Bheri**, Dailekh, Rati Khola, 1700 m, 1975, *T. B. Shrestha & N. P. Manandhar 517* (KATH); Jajarkot, Gothe Hill (Phul.), 1820 m, 30 Aug. 1967, *N. P. Manandhar et al. 7356* (BM); Jajarkot, Lan, Thuli, Bheri River, 2420 m, 10 Sep. 1952, *O. V. Polunin, W. R. Sykes & L. H. J. Williams 3304* (BM, E, GH, KATH, TI); Surkhet, 1520 m, 22 Nov. 1964, *J. J. Burnet 10* (BM); Surkhet, Rani Matta, 2900 m, 1975, *T. B. Shrestha & N. P. Manandhar 539* (KATH);

Figure 46-32: **Strobilanthes tomentosa** (Nees) Terao ex J. R. I. Wood: **a**, habit; **b**, opened corolla; **c**, anther; **d**, capsule. Drawn by Quan LIN; reproduced from *Icon. Cormophyt. Sin.* 4: 158, fig. 5730. 1975.

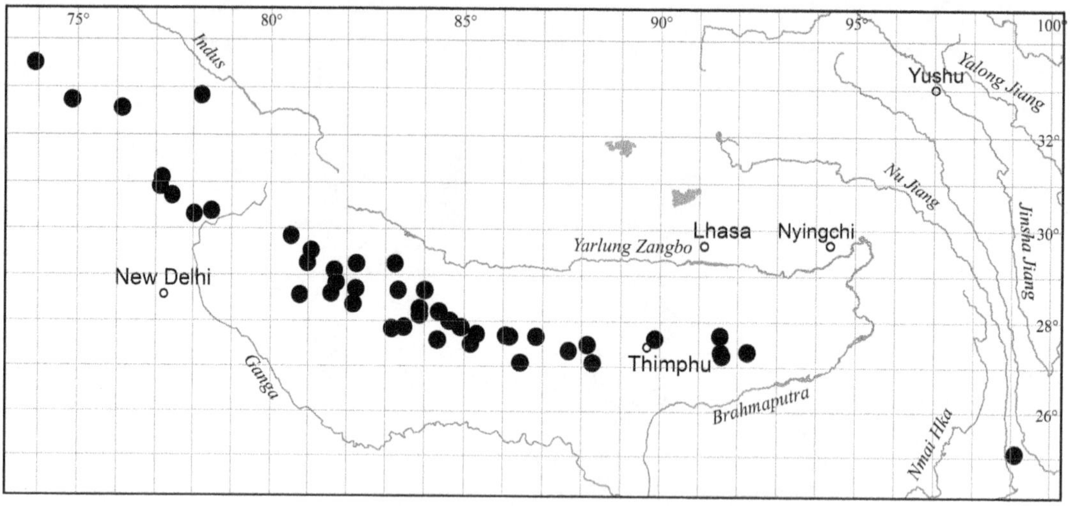

Map 46-59: **Strobilanthes tomentosa** (Nees) Terao ex J. R. I. Wood

Surkhet, Rot, 1250 m, 1982, *S. R. Adhikari & N. P. Manandhar 9185* (KATH); **Karnali**, Dolpa, Narku, between Wahu and Dla, Bheri River, 1970 m, 10 Sep. 1952, *O. V. Polunin, W. R. Sykes & L. H. J. Williams 3207* (BM, E, KATH, TI); Dolpa, Liwange Genl, 950 m, 1982, *N. P. Manandhar & N. R. Bhattara 8707* (KATH); Humla, Toli, 23 Aug. 1976, *H. Tabata et al. 2857* (KYO); Jumla, Katiyar Khola, 2440 m, 1975, *T. B. Shrestha & N. P. Manandhar 306* (KATH); Kalikort, 1430 m, 26 Sep. 1982, *N. P. Mannadhar & N. R. Bhattarai 9102* (KATH); **Mahakali**, Dadeldhura, Lipha, 1050 m, 1983, *K. R. Rajbhandari et al. 7033* (KATH); Darchula, Maiekholi-Sipti, 1660–1810 m, 1982, *M. M. Amatya & P. M. Regmi W559/82* (KATH); **Rapti**, Phulgam, 1670 m, 25 Aug. 1952, *O. V. Polunin, W. R. Sykes & L. H. J. Williams 592* (BM, KATH, TI); Salyan, Marma Khola, 1050 m, 24 Aug. 1952, *O. V. Polunin, W. R. Sykes & L. H. J. Williams 559* (BM); **Seti**, Bajhang, 1950 m, 23 Aug. 1991, *M. Suzuki et al. 9192519* (A); Agar to Kinara, 1400–2070 m, 23 Aug. 1991, *M. Suzuki et al. 9160970* (A, TI); Doti, Ghanteswor, 1680 m, 11 Nov. 2012, *B. Adhikari & G. D. Bhatt 45* (E); Kailali, Budhetola, 1140 m, 2012, *B Adhikari & G. D. Bhatt 39* (E). **C NEPAL**: **Bagmati**, Dhading, Todake, 1150 m, 1989, *N. P. Manandhar 13207* (KATH); Dhading, Todake, 1150 m, 1989, *N. P. Manandhar 13167* (KATH); Kathmandu, hills around Nepal Valley, 1520 m, Oct. 1953, *D. Pround 193* (BM); Kabhre, Khardarpati, 1550 m, 1988, *N. P. Manandhar 12141* (KATH); Nuwakot, Trisuli Sazrar, 1210 m, 10 Oct. 1974, *J. D. A. Stainton 7304* (BM); Rasuwa, Bhargu, 1850 m, 5 Sep. 1971, *J. F. Dobremez 953* (BM, E, KATH); Rasuwa, Rasua Gadhi to Thangjet, 1500–1800 m, 12 Jul. 1949, *O. V. Polunin 976* (BM, KATH); Rasuwa, Chilime Kharka to Chilimegoan, 2120 m, 26–28 Jul. 1949, *O. V. Polunin 1447* (BM, KATH); Rasuwa, Manigaon, 1180 m, 1983, *S. B. Rajbhandari & N. P. Manandhar 9274* (KATH); **Dhawalagiri,** Mustang, Marengley, 2100 m, 2 Nov. 1931, *K. N. Sharma E181* (BM); Myagdi, Mayangdi Khola, Arman, 1820 m, 23 Apr. 1954, *J. D. A. Stainton, W. R. Sykes & L. H. J. Williams 163* (BM); Myagdi, above Mayangdi Khola, 1670 m, 26 Aug. 1984, *J. D. A. Stainton, W. R. Sykes & L. H. J. Williams 4041* (BM, KATH); Myagdi, Tatopani, S of Dana, 1360 m, 4 Sep. 1954, *J. D. A. Stainton, W. R. Sykes & L. H. J. Williams 7682* (BM, KATH, TI); Myagdi, near Mayangdi Khola, 1360 m, 28 Oct. 1954, *J. D. A. Stainton, W. R. Sykes & L. H. J. Williams 9181* (BM, E); **Gandaki**, Gorkha, Gorkhali gaun to Ghat Khola, 2100 m, 1963, *K. Itoh & K. R. Rajbhandari 1256* (KATH, TI); Gorkha, Buri Gandaki, Khorlak, 1820 m, 5 Oct. 1967, *J. D. A. Stainton 6009* (BM, KATH); Kaski, Pokhara, 1050 m, 15 Aug. 1954, *J. D. A. Stainton, W. R. Sykes & L. H. J. Williams 6895* (BM, E, GH, KATH, TI); Kaski, Poklarachaur, 1520 m, 14 Oct. 1957, *J. D. A. Stainton, W. R. Sykes & L. H. J. Williams 8937* (BM); Lamjung, Jagat, 1969 m, 1 Nov. 1969, *T. Wraber 604* (BM); Manang, Marsyandi Valley, Jogat to Bahundara,

1969, *T. Wraber 36835* (BM); **Janakpur**, Dolakha, de Pasa Nangon a Charikot, 1700 m, 10 Sep. 1954, *A. Zimmermann 1155* (E); Dolakha, Pheda Charikot, 900 m, *Herbariium Banerji 983* (GH); Kirantichhap, 910 m, 22 Nov. 1971, *P. R. Shakya & M. Ohsawa 1048* (KYO, TI); **Lumbini**, Arghakhachi, Narpani, 1740 m, 15 Jul. 2016, *B. Adhikari & G. Parmar BG59* (E); Palpa, Plaha, near Tansing, 1520 m, 10 Oct. 1957, *J. D. A. Stainton, W. R. Sykes & L. H. J. Williams 8864* (BM, KATH, TI); **Narayani**, Chitwan, Syandala, 1650 m, 1989, *N. P. Manandhar 13996* (KATH); Makwanpur, about 10 km N of Bimphede, 1700 m, 29 Nov. 2004, *C. Pendry et al. B198* (E, KATH, TI); Makwanpur, Pairang, 1690 m, 1988, *N. P. Manandhar 12704* (KATH). **E NEPAL**: **Koshi**, Dhankuta, 1100 m, 11 Oct. 1971, *J. F. Dobremez 1369* (TI); Sunsari, Phusri to Dhar Pani, 16 Oct. 1963, *H. Hara et al. 6303719* (BM, K, KATH, KYO, TI); **Mechi**, Taplejung, Khebang, 24 Nov. 1963, *H. Hara et al. 6304739* (BM); **Sargamatha,** Okhadhunga, Mane Bhanjyang, 1580 m, 5 Nov. 1954, *A. Zimmermann 2064* (A, BM,G, K, U, LE); Solukhumbu, Bhoti Kosi Valley, 1250 m, 15 Oct. 1975, *A. D. Schilling et al. 2117* (K); Solukhumbu, Dudh Kosi, Waspa Khani, 1950 m, 26 Oct. 1969, *J. D. A. Stainton 6605* (BM, KATH, TI). **SIKKIM & DARJILING**: **Darjiling**, Rangit, 650 m, 13 Oct. 1869, *C. B. Clarke 9449* (BM); **Sikkim**, 1200 m, J. D. Hooker *& T. Thomson s. n.* (GH). **BHUTAN**: Samtigong, Augduphnong, 7 Sep. 1914, *R. E. Cooper 2007* (BM, E); Samtigong, 1800 m, 10 Oct. 1914, *R. E. Cooper 2336* (E); Dainertsi, Kuru Chu Valley, 25 Aug. 1915, *R. E. Cooper 4572* (BM, E); **Lhuntse**, 1300 m, *F. Ludlow & G. Sherriff 18729* (BM); **Mongar**, Yadi, 1600 m, 22 Oct. 1987, *J. R. I. Wood 5993* (E); 1300 m, 24 Oct. 1990, *J. R. I. Wood 7362* (E); Yadi, 1500 m, Nov. 1991, *J. R. I. Wood 7491* (E); **Punakha**, Naptaika Punakha, 1500 m, 2 Sep. 1914, *R. E. Cooper 3110* (BM, E); between Chuzomsa and Samtengang, 1300 m, 2 Oct. 1987, *J. R. I. Wood 5900* (E); **Tashigang**, 1300 m, 15 Oct. 1934, *F. Ludlow & G. Sherriff 1061* (BM, E). **YARLUNG ZANGBO-BRAHMAPUTRA**: **Dirang**, 6 Sep. 1938, *F. Kingdon-Ward 14204* (BM, TI); **Nyam Jang Chu**, Gyipu, 1500 m, 6 Sep. 1938, *F. Ludlow, G. Sherriff & G. Taylor 6717* (BM, E, TI); near Beting, 2300 m, 5 Nov. 1938, *F. Ludlow, G. Sherriff & G. Taylor 7080* (BM, E). **S HENGDUAN**: **Longyang**, Mangkuan, Baihualing, Dajie Shan, 1530 m, 10 Sep. 2003, *Gaoligong Shan Biodiversity Survey 19080* (CAS, KUN).

10. Strobilanthes panichanga (Nees) T. Anderson, J. Linn. Soc., Bot. 9: 478. 1867; Clarke in Hooker, Fl. Brit. India 4: 473. 1884. ≡ *Asystasia panichanga* Nees in Wallich, Pl. Asiat. Rar. 3: 90. 1832. ≡ *Ruellia panichanga* Wall., Numer. List n. 2387. 1830, nom. invalid. Type: India, Meghalaya, "Silhet", *F. Buchanan-Hamilton in Wallich Cat. 2387a* [lectotype designated by Wood, Kew Bull. 68: 682. 2003: K-W (K001115831); isolectotype: S (S09-2695)].

Subshrubs; stems quadrangular, densely pubescent. Leaves equal in each pair; petiole 1.5–2 cm long, brown pubescent; blade ovate, 7–14 cm long, 4–8 cm broad, pubescent on both surfaces, lateral veins 5–7 pairs, base cuneate, margin serrate, apex acuminate. Panicles axillary, 10–15 cm long, shorter than or subequaling to leaves; branches densely pubescent; flowers distant on rachis. Bracts triangular-subulate, 2.5–3 mm long, densely pubescent, persistent; bracteoles triangular-linear, 4–5 mm long, 1.6–2 mm broad, densely pubescent, apex acute. Calyx 0.8–1 cm long, densely pubescent, 5-lobed almost to base; lobes lanceolate, 2.5–3 mm broad, margin ciliate, apex acute. Corolla white with purple on lobes, 3.7–4.2 cm long, ventricose, outside densely pubescent; tube base cylindrical, 1.2–1.5 cm long, upwards gradually widened to 2.2–2.5 cm broad at mouth; lobes oblong, 4–5 mm long, 4–5 mm wide, apex retuse. Stamens 4, included; filaments glabrous; anther-thecae oblong, glabrous. Ovary pilose; 2 ovules per locule; style 3.2–3.5 cm long. Capsules ellipsoid, 1.8–2 cm long. Seeds ovate in outline, compressed, 1.8–2.2 mm long, 1.6–1.8 mm broad, pubescent.

Phenology: flowering from November to January of next year.

Habitats: shaded places, often secondary bushland; 300–1300 m.

Distribution: U Irrawaddy (Map 46-60); also in other parts of India.

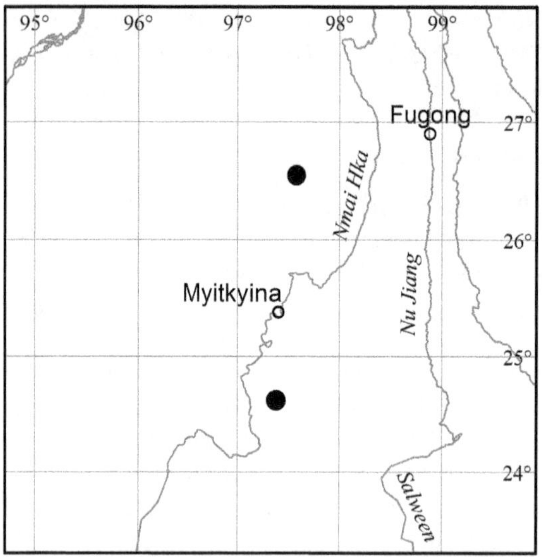

Map 46-60: **Strobilanthes panichanga** (Nees) T. Anderson

U IRRAWADDY: **Putao,** Kachin Hills, 500 m, 10 Dec. 1930, *F. Kingdon-Ward 9034* (BM, NY); SE of Sumprabum, 300–650 m, Dec. 1953, *F. Kingdon-Ward 21704* (BM, TI).

11. Strobilanthes setosa J. R. I. Wood, Edinburgh J. Bot. 51: 259. 1994. Type: [U Irrawaddy] Myanmar, Hills E of Putao, 460 m, 10 Dec. 1937, *F. Kingdon-Ward 13572* [holotype: BM (BM000793212); isotype: TI].

Subshrubs, 30–60 cm high; stems glabrous, sometimes glandular-setose. Leaves unequal in each pair, smaller one 3/5–2/3 of larger one in size; petiole 0.3–1 cm long; blade of larger ones elliptic, 6–13.5 cm long, 2–5 cm broad, apex acuminate, smaller ones ovate, 2.5–4 cm long, 1.5–3 cm broad, sparsely setose on both surfaces, adaxially dark green and with numerous cystoliths, abaxially paler, base cuneate, margin serrulate and setose, apex acute. Inflorescences of branched, axillary spikes bearing a pair of reduced leaves at branching point and forming a lax terminal panicle, 15-25 cm long, much longer than leaves; rachis flexuose zigzag, covered with setae; flowers 4–8 cm long apart on rachis. Bracts linear, 2–6 mm long, 0.4–0.5 mm broad, glabrous or sparsely setose, persistent; bracteoles linear, 1.7–2 mm long, caducous at anthesis. Calyx 1.2–1.3 cm long, glabrous or sparsely setose, 5-lobed almost to base; lobes subequal, linear, apex long acuminate. Corolla 3.5–4.5 cm long, pendulous, pale yellow, straight, outside glabrous, inside glabrous; tube base cylindric, 2 mm broad, upwards gradually widened to 1.0–1.3 mm broad at mouth; lobes orbicular, 5–6 mm long and broad. Stamens 4; filaments glabrous, shorter pair 3–4 mm long, longer pair 6–8 mm long; anthers ellipsoid, 2.5–3 mm long. Ovary with 2 ovules per locule; style 3.2–3.6 cm long, glabrous. Capsules narrowly oblong, 1.2–1.5 cm long, apex acuminate, 4-seeded. Seeds ovate in outline, 2.5–3 mm long, 2.2–2.5 mm broad, villous. Figure 46-33.

Phenology: flowering from December to January of next year.

Habitats: forests; ca. 500 m.

Distribution: endemic; U Irrawaddy (Map 46-61).

U IRRAWADDY: **Putao**, Hills E of Putao, 460 m, 10 Dec. 1937, *F. Kingdon-Ward 13572* (BM); without locality or date, *S. M. Toppin 3226* (E).

12. Strobilanthes euantha J. R. I. Wood, Kew Bull. 58: 697. 2003; Hu, Deng & Wood in Wu, Raven & Hong (eds.), Fl. China 19: 428. 2011. Type: [S Hengduan] China, Yunnan, Lushui, Gulangha, along Gulang Jiang, between Pianma and Gangfang, N. of Pianma on W side of Gaoligong Shan, 26°05.28′N, 98°35.5′E, 1560 m, 13 Oct. 1998, *H. Li, B. Bartholomew & Z. L. Dao 10336* [holotype: CAS (CAS435547); isotype: KUN].

Goldfussia tengyuehensis C. Y. Wu, Ind. Fl. Yunnan. 2: 1675. 1984, nom. invalid. (Art. 38.1).

Herbs 0.5–1.5 m high, slightly anisophyllous; stems quadrangular, sulcate, glabrous. Petiole 0–2.6 cm long, glabrous; leaf blade ovate to ovate-elliptic, 4–20 cm long, 3–12 cm broad, abaxially sparsely pilose along veins, adaxially pilose, lateral veins 5–8 pairs, base cordate to attenuate, somewhat oblique, margin serrate, apex acuminate. Panicles terminal, 8–15 cm long; peduncle and rachis 1–3 cm

Figure 46-33: **Strobilanthes setosa** J. R. I. Wood: **a**, habit; **b**, calyx. Drawn by Mary BATES; reproduce from Wood, *Edinburgh J. Bot.* 64: 263, fig. 13: d-e. 1994.

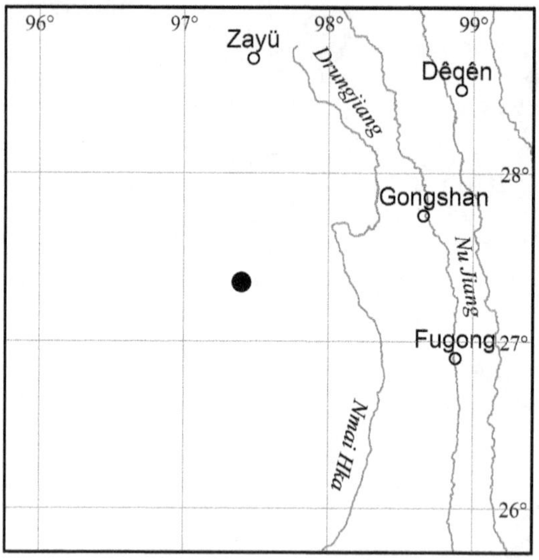

Map 46-61: **Strobilanthes setosa** J. R. I. Wood

long, glandular-pilose; flowers distant on rachis, usually 1 per node. Bracts obovate, 1–2 mm long, caducous at anthesis, apex rounded; bracteoles similar to bracts. Pedicel 1–3 mm long, glandular-pilose. Calyx 3–4 mm long, accrescent to 6–7 mm in fruit, glabrous, 5-lobed to 1 mm above base; lobes oblong to oblanceolate, apex truncate to emarginate. Corolla pale lavender to pale pink, 2.2–3.5 cm long, curved, glabrous; tube base cylindric, 0.6–1.9 cm long, upwards ventricose and widened to 8–10 mm broad at mouth; lobes ovate, 3–4 mm long, 3–4 mm broad, apex acute. Stamens 4; filaments glabrous, longer pair 1–1.2 cm long, erect, slightly exserted, shorter pair 2–4 mm long, incurved; anther-thecae spherical, 1–1.2 mm in diameter. Ovary glabrous; 2 ovules per locule; style 2.2–2.7 cm long, glabrous. Capsules narrowly oblong-obovoid, 1.2–1.4 cm long, glabrous, 4-seeded. Seeds ovoid-suborbicular in outline, 2–2.5 mm long, 1.7–2 mm broad, pubescent. Figure 46-34.

Phenology: flowering from September to October; fruiting from November to December.

Habitats: forests; 900–2300 m.

Distribution: S Hengduan, U Irrawaddy (Map 46-62); also in other parts of China (C Yunnan).

SHENGDUAN: **Longyang**, Wama, Shweili-Salwine divide, 2100 m, 1924, *G. Forrest 25332* (E); **Lushui**, Xiapianma, 1800 m, 22 Sep. 1980, *Exped. W Yunnan 11110* (YUKU); Gulangha, between Pianma and Gangfang, along Gulang Jiang, 1560 m, 13 Oct. 1998, *H. Li 10336* (CAS, KUN). **U IRRAWADDY**: **Chaungmaw Kyaun**, Hkinlum, North Triangle, 1350 m, 25 Aug. 1953, *F. Kingdon-Ward 21279* (BM, TI); **Laukhaung**, 950–1900 m, 24 Nov. 1938, *F. Kingdon-Ward 24* (A, NY); **Mahtum**, 1450 m, 2 Sep. 1939, *Kaulback 375* (BM); **Tengchong**, Pingdi, 2000 m, 12 Nov. 1958, *C. Chen 294* (KUN); Qushi, Jiangju, 1500–1830 m, 12 Sep. 1960, *W. Q. Yin 60-1037* (KUN); Sanyun, Datianpo, 2010 m, 5 Nov. 1998, *H. Li 11534* (CAS, KUN); Wuhe, 1760 m, 25 May 2005, *Gaoligong Shan Biodiversity Survey 24920* (KUN); Xinhua, Longjin, vicinity of Xiaoheiqin, 1850 m, 14 May 2006, *Gaoligong Shan Biodiversity Survey 29636* (KUN, MO).

13. Strobilanthes hamiltoniana (Steud.) Bosser & Heine, Bull. Mus. Natl. Hist. Nat., B, Adansonia 10: 148. 1988; Wood in Grierson & Long (eds.), Fl. Bhutan 2(3): 1273. 2001; Hu, Deng & Wood in Wu, Raven & Hong (eds.), Fl. China 19: 427. 2011. ≡ *Goldfussia colorata* Nees in Wallich, Pl. Asiat. Rar. 3: 89. 1832, non *Ruellia colorata* Blume, Bijdr. Fl. Ned. Ind. 14: 795. 1826; Nees in de Candolle (ed.), Prodr. 11: 176. 1847. ≡ *Ruellia hamiltoniana* Steud., Nomencl. Bot., ed. 2: 481. 1840. ≡ *Strobilanthes colorata* (Nees) T. Anderson, J. Linn. Soc., Bot. 9: 481. 1867, nom. illeg., non *Strobilanthes colorata* Nees, Companion Bot. Mag. 2: 312. 1857; Clarke in Hooker (ed.), Fl. Brit. India 4: 473. 1884. ≡ *Diflugossa colorata* (Nees) Bremek., Verh. Verh. Kon. Ned. Akad. Wetensch., Afd. Natuurk., Sect. 2 41(1): 237. 1944; Tsui & Hu in Hu (ed.), Fl. Reipubl. Popularis Sin. 70: 171. 2002. ≡ *Ruellia colorata* Wall., Numer. List n. 2388. 1830, nom. invalid. Type: India, Borjora, 29 Nov. 1808, *F. Buchanan-Hamiltonin in Wallich Cat. 2388* [lectotype designated by Bosser & Heine, Bull. Mus. Natl. Hist. Nat., B, Adansonia 10: 148. 1988: K-W (K001115833); isolectotypes: CAL, E, GZU, K].

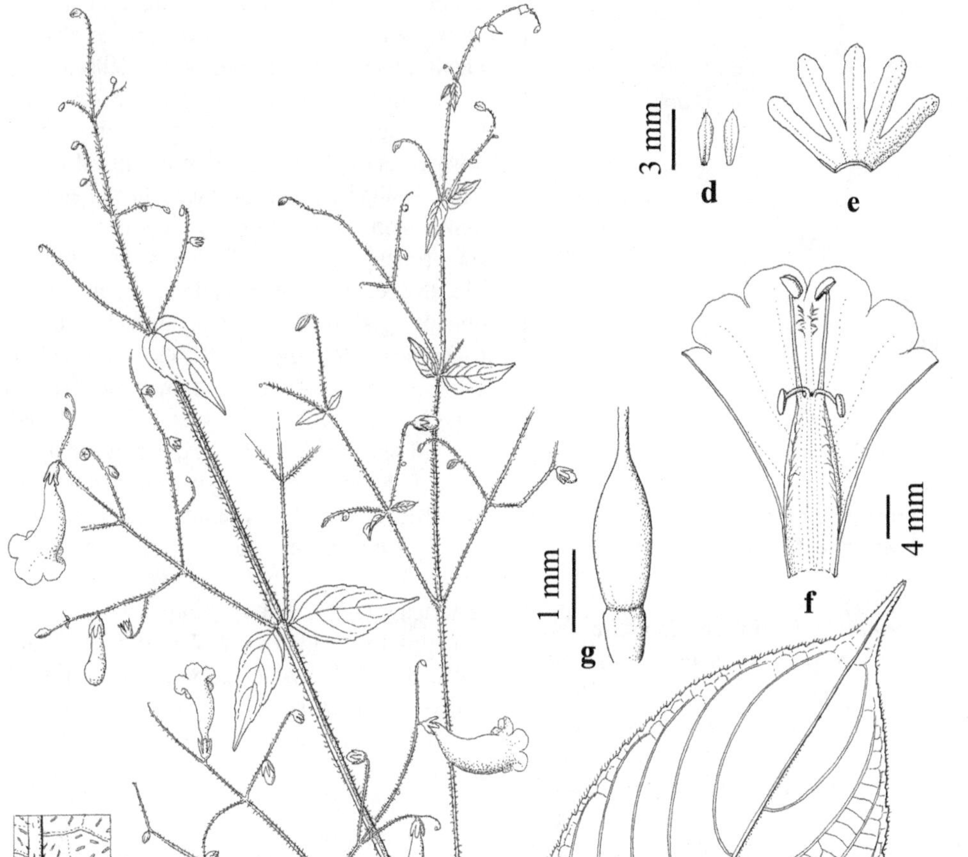

Figure 46-34: **Strobilanthes euantha** J. R. I. Wood: **a**, portion of plant; **b**, portion of adaxial leaf surfaces; **c**, portion of abaxial leaf surface; **d**, bract, upper and lower surfaces; **e**, calyx; **f**, corolla; **g**, ovary; **h**, capsule; **i**, seeds. Drawn by Rosemary WISE; reproduced from *Fl. China Illustr.* 19: fig. 337. 2012 (orginal from Wood & Scotland, *Kew Bull.* 58: 698, fig. 5. 2003).

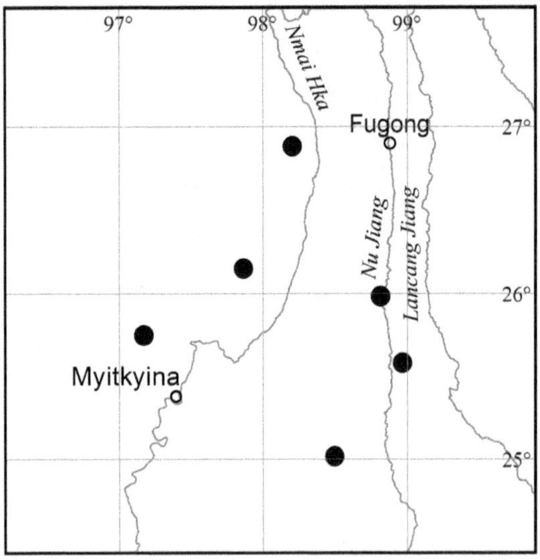

Map 46-62: **Strobilanthes euantha** J. R. I. Wood

Goldfussia crinita Nees in de Candolle (ed.), Prodr. 11: 176. 1847. ≡ *Strobilanthes crinita* (Nees) T. Anderson, J. Linn. Soc., Bot. 9: 481. 1867. ≡ *Strobilanthes colorata* var. *crinita* (Nees) C. B. Clarke in Hooker (ed.), Fl. Brit. India 4: 473. 1884. ≡ *Diflugossa crinita* (Nees) Bremek., Verh. Kon. Ned. Akad. Wetensch., Afd. Natuurk., Sect. 2 41(1): 246. 1944. Type: India, Assam, *W. Griffith s. n.* [lectotype **here designated**: K (K000883196); isolectotype: K (K000883195)].

Shrubs 0.5–1.5 m high, anisophyllous; stems quadrangular, sulcate, glabrous. Leaves unequal in each pair, petiolate or apical ones sessile or subsessile; petiole (0–)0.8–5(–8) cm long, glabrous; blade lanceolate to ovate, smaller pair at least 2/3 size of larger one, 5–19 cm long, 2–8.5 cm broad, glabrous, densely covered with linear cystoliths, lateral veins 4–7 pairs, base oblique, attenuate, or subcordate, margin serrulate, apex acuminate. Panicles axillary or terminal, 15–30 cm long, much branched; rachis angled at about 60°, glabrous or rarely pubescent, gland-tipped in fruit. Bracts broadly obovate, 3–5 mm long, 1.5–2 mm broad, concave, caducous, glabrous, becoming gland-tipped in fruit, margin entire, apex retuse; bracteoles oblanceolate to narrowly elliptic, 3–4 mm long, 0.8–1 mm broad, glabrous, concave, caducous, margin entire, apex retuse. Flowers solitary on each node, distant on rachis. Pedicel 1.5–5 mm long, glabrous. Calyx 8–10 mm long, accrescent to 1.1–1.3 cm long in fruit, glabrous, sometimes glandular-pubescent towards apex, becoming densely gland-tipped in fruit, 5-lobed almost to base; lobes unequal, one longer than others, narrowly elliptic, apex acute to sometimes retuse. Corolla 3.5–4 cm long, white, lilac or deep pink with a white tube, subventricose, outside glabrous, inside glabrous except for trichomes retaining style; tube base cylindric, 1.1–1.3 cm long, 1.6–2 mm broad, upwards slightly ventricose and gradually widened to 1.4–1.8 cm broad at mouth; lobes ovate, 4–5 mm long, 5–6 mm broad, apex rounded to retuse. Stamens 4, included; filaments glabrous, longer pair 7–9 mm broad, unequal, erect, shorter pair equal, 1–1.5 mm long, recurved; anther-thecae spherical, 1–1.2 mm in diameter. Ovary 2.2–2.5 mm long, glabrous; 2 ovules per locule; style 3.4–3.8 cm long, sparsely pubescent. Capsules fusiform, 1–1.5 cm long, glabrous or glandular-pilose, 4-seeded. Seeds ovate in outline, compressed, 2.5–3 mm long and broad, pubescent with long mucilaginous trichomes; areole small. Figure 46-35.

Phenology: flowering from December to January of next year.

Habitats: forests on mountain slopes; 100–2000 m.

Distribution: U Ganga & Indus, C Nepal, E Nepal, Bhutan, Sikkim & Darjiling, Yarlung Zangbo-Brahmaputra, U Irrawaddy (Map 46-63); also in other parts of India and Myanmar, widely cultivated or naturalized in tropical regions.

U GANGA & INDUS: **Deheradun**, cult. in Fr. Sch. Garden, Sep. 1891, *J. S. Gamble s. n.* (K). **C NEPAL: Bagmati**, Dhading, Malekhu, 370 m, 2012, *B. Adhikari & D. R. Kandel 57* (E); **Gandaki**, Kaski, between Pokhara and Naudanda, 1120 m, 4 Oct. 1976, *H. Tabata et al. 6394* (A, KATH, KYO); Kaski, Hyanja, between Hyanja and Naudanda, 1090 m, 4 Oct. 1976, *H. Tabata et al. 6435* (BM, KATH, KYO); Kaski, Pokhara, 1200 m, 7 Sep. 1954, *J. D. A. Stainton, W. R. Sykes & L. H. J. Williams 7141* (A, BM, KATH, TI);

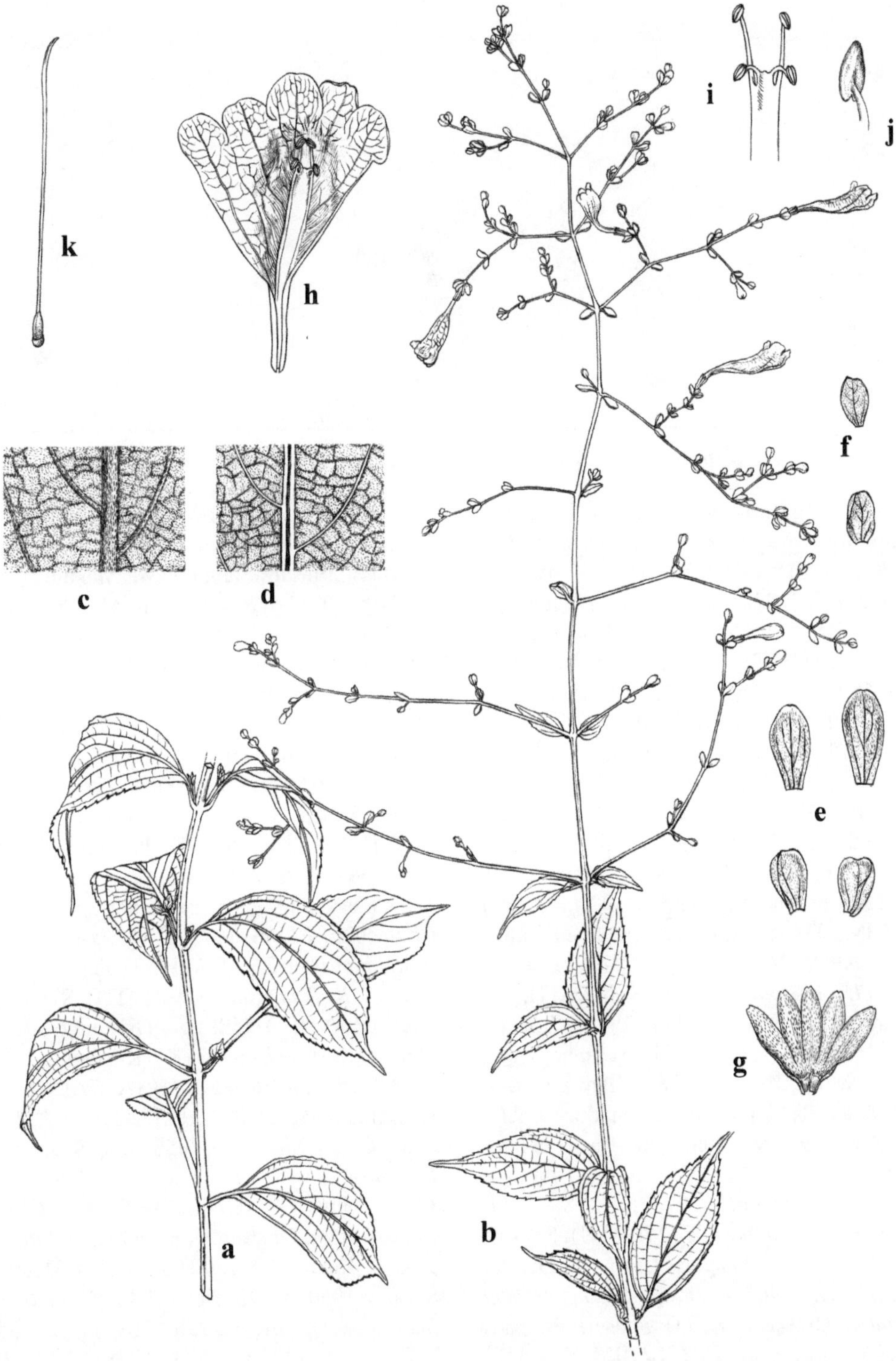

Figure 46-35: **Strobilanthes hamiltoniana** (Steud.) Bosser & Heine: **a**, branch; **b**. flowering branch; **c**, abaxial leaf surface; **d**, adaxial leaf surface; **e**, bracts; **f**, bracteoles; **g**, calyx; **h**, opened corolla; **i**, stamens; **j**, anther; **k**, pistil. Dawn by Yun-Xiao LIU.

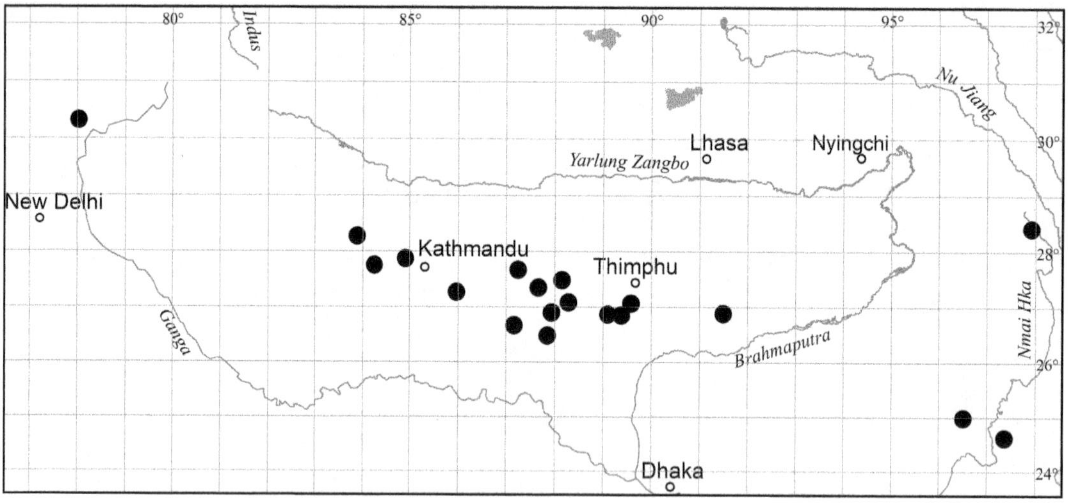

Map 46-63: **Strobilanthes hamiltoniana** (Steud.) Bosser & Heine

Kaski, Ghabung Khola, 910 m, 4 Nov. 1954, *J. D. A. Stainton, W. R. Sykes & L. H. J. Williams 9280* (BM, KATH, TI); Kaski, Beynas, 790 m, *B. K. Basnet et al. 2017211* (KATH); Kaski, Pokhara to Naudanda Phedi, 1150 m, 1988, *M. Suzuki et al. 8800049* (KATH, TI); Tanahu, Sisaghat, Mardi Khola, 500 m, 18 Oct. 1967, *J. D. A. Stainton 6067* (BM, TI); **Janakpur**, Sindhui, Sukhchauri, 1000 m, 7 Nov. 1954, *A. Zimmermann 2098* (E, G, K). **E NEPAL**: **Koshi**, Sankhuwasabha, Num, Arun Valley, between Num and Arun River, 1000 m, 22 Sep. 1991, *D. G. Long et al. EMAK164* (E, TI); Sankhuwasabha, Arun Valley, Num, 1520 m, 30 Aug. 1956, *J. D. A. Stainton 1464* (BM, TI); Sankhuwasabha, Arun Valley, Sabhaya Khola, N of Chainpur, 1210 m, 6 Sep. 1956, *J. D. A. Stainton 1548* (BM, KATH, TI); Sankhuwasabha, Sedua to Num, 1988, *M. Suzuki et al. 8821190* (KATH); Sankhuwasabha, Gadhi Danda to Mude, 880 m, 1998, *S. Noshiro et al. 9830182* (KATH, TI, TNS); Sankhuwasabha, Arun River, Below Num, 780 m, 1981, *P. R. Shakya 6975* (KATH); Sankhuwasabha, Hedangna to Pheksinda, 970 m, 1986, *T. B. Shrestha & P. R. Shakya 9240* (KATH); Sunsari, Sanguribhangyang, Sanguri Lekh, N Dharan, 1210 m, 3 Sep. 1967, *L. H. J. Williams & J. D. A. Stainton 8349* (BM, K, TI); Sunsari, Phusri to Pani, 1963, *Hara et al. 6303719* (BM, K, KATH, TI); Sunsari, N Dharan, 1200 m, 1980, *A. D. Schilling et al. 2452* (BM); Sunsari, Sanguri, 1230 m, 1981, *P. R. Shakya 6458* (KATH); Sunsari, Sangoori to Dharapani, 1200 m, 1972, *T. B. Shrestha & T. K. Bhattacharya 72-292* (KATH); Sunsari, Sangure, Mt. Makalu, 1200 m, 27 Aug. 1971, *P. R. Shakya & M. Ohsawa 1083* (KATH, KYO, TI); Sunsari, Sanguri, 1230 m, 1981, *P. R. Shakya 6458* (KATH); **Mechi**, Ilam, Dhara Pani to Dharan Bazar, 1000 m, 1977, *H. Ohashi et al. 771385* (BM, KYO, TI); Ilam, Dhara Pani to Dharan Bazar, 5 Sep. 1977, *H. Ohashi et al. 775601* (TI); Ilam, Dhara Pani to Dgaran Bazar, 5 Sep. 1977, *H. Ohashi et al. 774219* (KATH, TI); Jhapa, N of Chula Chuli, 1520 m, 15 Sep. 1967, *L. H. J. Williams & J. D. A. Stainton 8524* (BM, K); Jhapa, Bisnetar to Kawal Dhor, Chulachuli, 270 m, 1972, *T. B. Shrestha & T. K. Battacharya 72-221* (KATH); Taplejung, 1400 m, 1977, *S. K. Dhungana 905* (KATH). **SIKKIM & DARJILING**: **Darjiling**, Dumping, 700 m, 1869, *C. B. Clarke 9967* (K); Kurseong, 1300 m, 26 Oct. 1939, *K. Biswas 3749* (A, CAL); **Sikkim**, Kumai to Rongu, 840 m, 14 Oct. 1949, *K. Biswas 9882* (CAL); Munypoo, 1050 m, 25 Sep. 1908, *W. W. Smith 108* (CAL); Munypoo, 1050 m, 1908, *W. W. Smith 132* (CAL). **BHUTAN: Chukka**, Mahadeothau Lavrang, N Kalikola, 700 m, 1988, *J. R. I. Wood 6789* (E, FHO, THIM); **Deothang**, Satsalor, 950 m, 13 Nov. 1938, *F. Ludlow, G. Sherriff & G. Taylor 7256* (BM, TI); ca. 15 km N Deothang, 1400 m, 18 Oct. 1987, *J. R. I. Wood 5969* (E); Radeng, 13 Oct. 1974, *R. Nawawn & S. Shering s. n.* (E); **Manas**, between Panbang and Goshing, 300–400 m, Aug. 1991, *L. Reader s. n.*

(E); **Phuntsholing**, above Rinchending, 950 m, 1984, *I. W. J. Sinclair & D. G. Long 5775* (E); above Phuntsholing, 1200 m, 12 Oct. 1988, *J. R. I. Wood 6799* (E); above Rinchending, 950 m, 21 Oct. 1984, *I. W. J. Sinclair & D. G. Long 5775* (E); **Samchi**, Tamangdhanra Forests, 750 m, 5 Mar. 1982, *A. J. C. Grierson & D. G. Long 3495* (K). **YARLUNG ZANGBO-BRAHMAPUTRA**: **Daffla** Hill, Feb. 1890, *B. Khan 133* (CAL); 1890, *King 62* (CAL); Nacyklow, Oct. 1891, *H. Collett s. n.* (CAL); 600–950 m, 3 Dec. 1931, *F. Kingdon-Ward 10196* (BM); **Zayü**, Zhuwagen, 600–950 m, 3 Dec. 1931, *F. Kingdon-Ward 10196* (BM). **U IRRAWADDY**: **Putao**, valley of the Nmai Hka, 600 m, 17 Nov. 1919, *F. Kingdon-Ward 3727* (E).

14. Strobilanthes nutans (Nees) T. Anderson, J. Linn. Soc., Bot. 9: 475. 1867; Clarke in Hooker (ed.), Fl. Brit. India 4: 454. 1884; Wood & Adhikari, Curtis's Bot. Mag. 31: 177. 2014. ≡ *Goldfussia nutans* Nees, Pl. Asiat. Rar. 3: 88. 1832. Type: Nepal, 1821, *Wallich Cat. 2362* [lectotype designated by Wood & Adhikari, Curtis's Bot. Mag. 31: 177. 2014: K-W (K001115732); isolectotypes: BM, CGE, E (E00273506, E00273507), K].

Ruellia strobilina Wall., Numer. List n. 2362. 1830, nom. invalid.

Perennial herbs, 30–50 cm high, isophyllous; stems decumbent and rooting at nodes on lower portion, then ascending upwards, deflexed pilose. Leaves equal in each pair; petiole 0.5–5.5 cm long, pilose; blade ovate-elliptic, 5–14 cm long, 3–7 cm broad, densely pilose on both surfaces, lateral veins 7–9 pairs, base narrowly attenuate, margin serrate, apex acuminate. Spikes axillary, 3–5 cm long, cone-like, pendulous; peduncles (1–)3–5.5 cm long, pubescent, deflexed. Bracts oblong to obovate, 1.0–1.5 cm long, 0.6–1.0 cm broad, glabrous, concave, apex retuse and minutely mucronate, caducous ; bracteoles obovate to oblanceolate, 1.0–1.4 cm long, 0.4–0.5 cm broad, apex retuse. Calyx 8–9 mm long, accrescent to 15 mm long in fruit, glabrous, 5-lobed to 1.5–2 mm above base; lobes unequal, one longer than others, lanceolate, 1.2–1.5 mm broad, glandular-pilose, apex acute. Corolla white, 3.2–3.5 cm long, straight, outside pilose on tube, inside glabrous except for hairs retaining style; tube base cylindric, 6–7 mm long, 2–2.5 mm broad, upwards somewhat ventricose and widened to 1–1.1 cm broad at mouth; lobes ovate-deltoid, 5–6 mm long, 5–6 mm broad, glabrous, apex obtuse. Stamens 4, included; filaments of longer pair slightly unequal, shorter one 5–6 mm long, longer one 6–7 mm long, pilose, shorter pair 1–1.5 mm long, recurved 180°, glabrous; anther-thecae oblong, 0.75–1 mm long, 1–1.2 mm broad. Ovary oblong, comose towards apex; 2 ovules per locule; style glabrous; stigma curved. Capsules oblong, 1.2–1.4 cm long, 2–3 mm broad, sparsely glandular towards apex. Seeds lenticular, 2–2.5 mm long, 1.8–2 mm broad, densely covered with thin hygroscopic trichomes; areole small.

Phenology: flowering from August to October.

Distribution: endemic; C Nepal (Map 46-64).

Habitats: moist evergreen forests; 2100–2700 m.

C NEPAL: **Bagmati**, Kathmandu, Shivapuri, 2400 m, Aug. 1962, *T. B. Shrestha & A. V. Upadhyay 532* (KATH); Kathmandu, Forests office to Shiwapuri Summit, 2300 m, 13 Sep. 1988, *M. Suzuki et al. 8811389* (KATH); Kathmandu, Forests Office to Shiwapuri summit, 13 Sep. 1988, *M. Suzuki et al. 8881846* (BM,

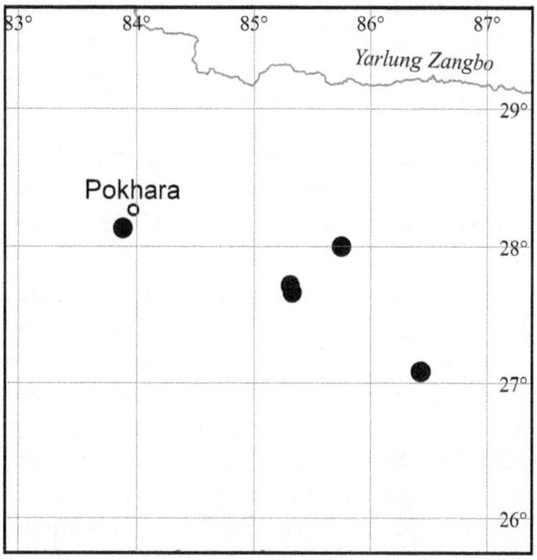

Map 46-64: **Strobilanthes nutans** (Nees) T. Anderson

TI); Kathmandu, Forests Office to Shiwapuri summit, 2400 m, 13 Sep. 1988, *M. Suzuki 8881809-1* (KATH, TI); Kathmandu, Shiwapuri, 2290 m, 24 Aug. 1990, *M. Minaki et al. 9070226* (TI); Kathmandu, Nepal Valley, 2120 m, Oct.–Nov. 1953, *D. Pround 137* (BM); Kathmandu, Mulkaharka to Chipu Danda, 2200 m, 20 Aug. 1969, *S. B. Malla & H. Kanai 673504* (KATH, KYO, TI); Kathmandu, Sundarijal, near Mulakharka, 2200 m, 20 Aug. 1969, *S. B. Malla 16015* (KATH); Kathmandu, Bagdwar, 2320 m, 2 Aug. 1966, *S. B. Malla 4864* (CAL, KATH); Lalitpur, Godawari to Camp site (N ridge of Mt. Phulchoki), 2130 m, 3 Oct. 1995, *M. Mikage et al. 9552659* (KATH, TI); Nuwakot, Pati Bhanjyang, 2270 m, 9 Sep. 1974, *J. D. Stainton 7196* (BM); Nuwakot, Pati Bhanjyang, 2400 m, 6 Aug. 1974, *J. H. de Haas 2061* (BM); Nuwakot, Mangen to Khodang Danda, 2600 m, 28 Aug. 1972, *H. Kanai et al. 722028* (KATH, KYO, TI); Nuwakot, Khodang Danda to Gut Bhanjyaug, 2400 m, 29 Aug. 1972, *H. Kanai et al. 722044* (A, BM, KUN, TI); **Rasuwa**, Chipling, 2160 m, 21 Aug. 1969, *H. Kanai & S. B. Malla 673580* (KATH, TI); **Sindupalchok**, Malemchigaon, 2250 m, 10 Sep. 1970, *J. D. A. Stainton 6664* (A, BM, KATH, TI).

15. Strobilanthes accrescens J. R. I. Wood, Edinburgh J. Bot. 51: 225. 1994; Wood in Grierson & Long (eds.), Fl. Bhutan 2(3): 1264. 2001. Type: Bhutan, between Chimakothi and Chukka, 1600 m, 29 Sep. 1991, *J. R. I. Wood 7464* [holotype: E (E00273420); isotypes: E (E00273421), K (K000883178), THIMPHU].

Shrubs, 1.5–2 m high, plietesial, isophyllous; stems weakly quadrangular, sulcate, glabrous. Leaves subequal in each pair; petiole 1–7 cm long; blade broadly elliptic or ovate, 5–25 cm long, 2–11 cm broad, diminishing in size upwards, glabrous on both surfaces, base attenuate and decurrent onto petiole, margin serrate, apex shortly acuminate. Spikes head-like, axillary, solitary or several aggregated on branchlets forming panicles 10–20 cm long; peduncles glabrous, 0.5–2 cm long; heads 1–1.5 cm long, 1–2 cm broad. Bracts heteromorphic, outer ones sterile, strongly concave, ovate or broadly elliptic, 1.3–1.8 mm long, 0.7–0.8 mm broad, glabrous, margin entire or crenulate, inner ones fertile, concave, broadly elliptic, 1.2–1.4 cm long, 0.5–0.6 cm broad, pale green with black sessile glands near tip, margin ciliate above the middle, apex acute; bracteoles concave, oblong-elliptic, 1–1.2 cm long, 0.3–0.4 cm broad, abaxially dotted with black glands and minutely puberulent, apex obtuse. Calyx 1–1.2 cm long, pale green, adaxially glabrous, 5-lobed almost to base; lobes subequal, linear-oblanceolate, dotted with black glands, margin ciliate in upper half, apex mucronate; mucro accrescent to 2.5–3 cm in fruit, curved, sticky glandular-pilose. Corolla bluish-purple, 3.3–3.7 cm long, curved, pilose; tube base cylindric and dirty-white, 0.8–1 cm long, 1.3–1.5 mm broad, upwards gradually widened to 1–1.2 cm broad at mouth; lobes ovate, 5–6 mm long, 5–6 mm broad, apex rounded. Stamens 4, included; filaments, sparsely pilose near base, longer pair 7–8 mm long, shorter pair 1.5–2 mm long; anther-thecae oblong, 2–2.5 mm long. Ovary sparsely glandular-hairy; 2 ovules per locule; style 2.5–2.7 cm long, sparsely glandular-pilose basally; stigma simple, curved. Capsules oblong, 1.2–1.3 cm long, apiculate, glabrous except for few glandular hairs near tip, 4-seeded. Seeds ovate in outline, compressed, hairy; areole absent. Figure 46-36.

Distribution: endemic; Bhutan (Map 46-65).

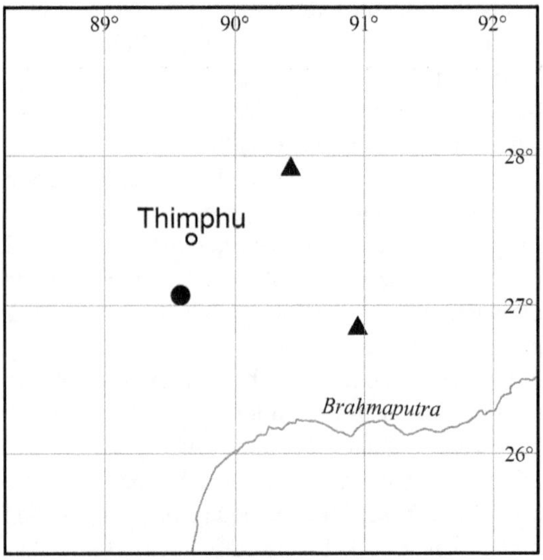

Map 46-65: **Strobilanthes accrescens** J. R. I. Wood (●) subsp. **accrescens**, (▲) subsp. **teraoi**

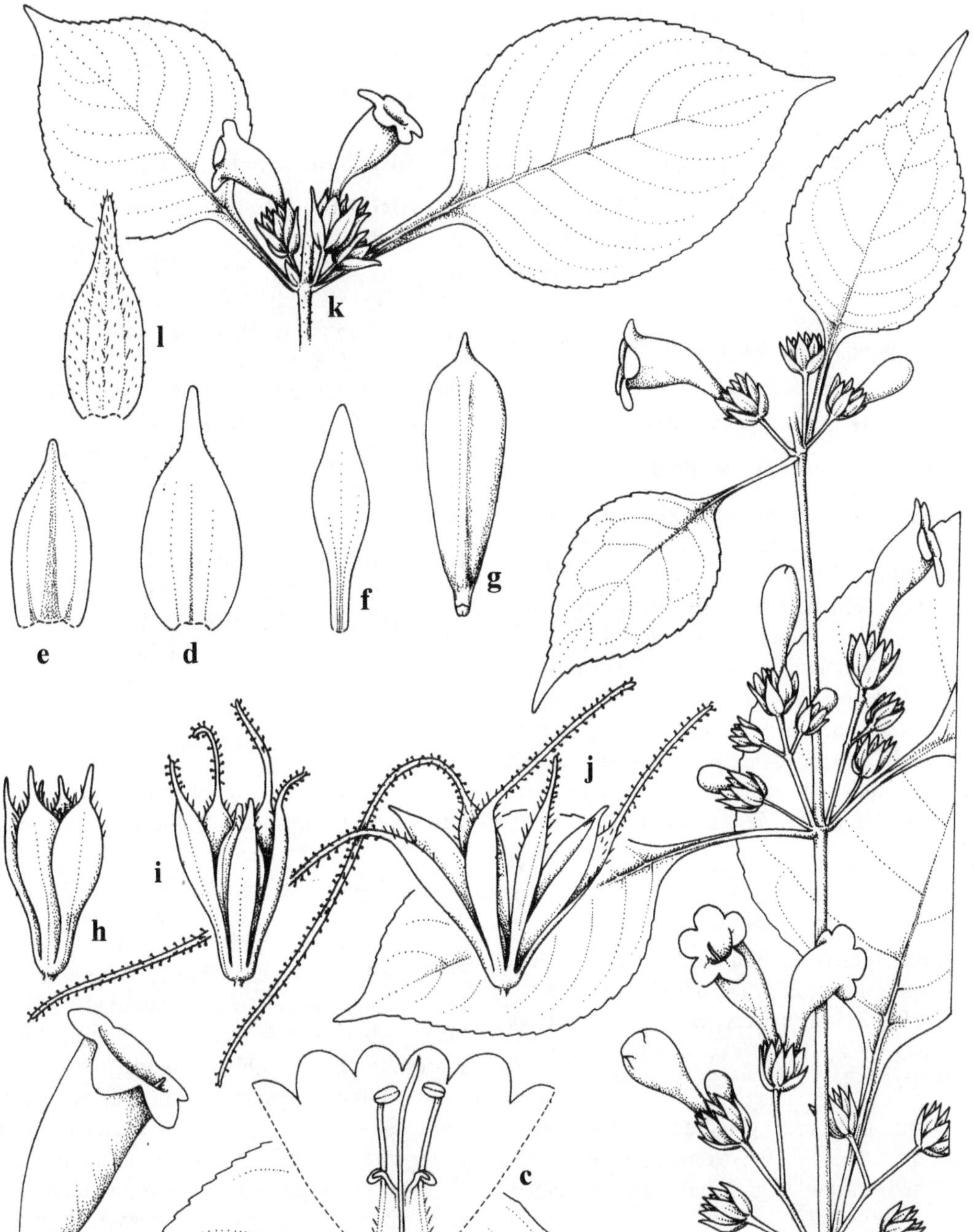

Figure 46-36: **Strobilanthes accrescens** J. R. I. Wood subsp. **accrescens**: **a**, habit; **b**, corolla; **c**, corolla opened up; **d**, outer bract; **e**, inner bract; **f**, bracteole; **g**, capsule; **h**, young calyx; **i**, calyx; **j**, calyx in fruiting. **Strobilanthes accrescens** J. R. I. Wood subsp. **teraoi** J. R. I. Wood: **k**, habit; **l**, outer bract. Drawn by Mary BATES; reproduced from *Edinburgh J. Bot.* 51(2): 226, fig. 7. 1994.

1a. Stem, leaves, and outer bracts glabrous .... 15a. subsp. *accrescens*
1b. Stem, leaves, and outer bracts pubescent .... 15b. subsp. *teraoi*

15a. Strobilanthes accrescens subsp. **accrescens**

Stem, glabrous. outer bracts glabrous. Leaf blade broadly elliptic, glabrous, base gradually narrowed. Capsules sparsely glandular-hairy towards apex.

Phenology: flowering from September to October.

Habitats: moist forests; 1500–2200m.

Distribution: **endemic**; Bhutan.

BHUTAN: **Chukka**, Chukka, 1550 m, 18 Nov. 1914, *R. E. Cooper* 3612 (BM, E); below Chimakothi, 1800 m, 29 Sep. 1991, *J. R. I. Wood 7462* (E, FHO, K); between Chimakothi and Chukka, 1600 m, 29 Sep. 1991, *J. R. I.Wood 7464* (E, K); between Chukka and Taktichu (Awakha), 1700 m, 29 Sep. *J. R. I. Wood 7469* (E, FHO, K); Lobneka, between Bunakha and Chapcha, 2200 m, 20 Sep. *J. R. I. Wood 7576* (E, K, TI); Asinabari, between Gedu and Chukka, 1900 m, 20 Sep. *J. R. I. Wood 7581* (E, FHO, K).

15b. Strobilanthes accrescens J. R. I. Wood subsp. **teraoi** J. R. I. Wood, Edinburgh J. Bot. 51: 227. 1994; Wood in Grierson & Long (eds.), Fl. Bhutan 2(3): 1264. 2001. Type: Bhutan, *J. R. I. Wood* 7472 [holotype: E (E00273422); isotypes: E (E00273423), FHO (FHO00057879), K (K000883179), THIMPHU].

Stems pubescent, gland-dotted. Leaf blade pubescent adaxially, grey-tomentose abaxially. Bracts and bracteoles grey pubescent. Capsules pilose on upper half.

Phenology: flowering from September to October.

Habitats: rough matshy scrub and in clearings in dry broad leaved forests; 1700–2700m.

Distribution: endemic; Bhutan.

BHUTAN: Punakha, Mo Chu, 1750 m, 7 Oct. 1967, *B. Lyon 5769* (BM); **Tongsa**, Tashiling, 2700 m, 10 Sep. 1914, *R. E. Cooper 2062* (E); Tashiling, 2300 m, 7 Oct. 1914, *R. E. Cooper* 2325 (E, BM); W Tashiling, 2500 m, 18 Oct. 1991, *J. R. I. Wood* 7472 (E, K); Tashiling, 2500 m, 18 Oct. 1991, *J. R. I. Wood 7474* (E, FHO); W Tongsa, 2100 m, 19 Oct. 1991, *J. R. I. Wood 7475* (E, FHO); **Mangde Chu** gorge below Tongsa, 1800 m, 19 Oct. 1991, *J. R. I. Wood 7478* (E, FHO).

16. Strobilanthes jennyae J. R. I. Wood, Edinburgh J. Bot. 51: 243. 1994; Wood in Grierson & Long (eds.), Fl. Bhutan 2(3): 1262. 2001. Type: Bhutan, Tashigang, between Tashiyangtse and Bumdeling, 2100 m, 8 Nov. 1991, *J. R. I. Wood 7499* [holotype: E (E00273493)].

Perennial herbs, 10–30 cm high, isophyllous; stems decumbent, densely pilose. Leaves equal in each pair, petiolate below, subsessile above; petiole 0–2.5 cm long; leaf blade ovate-elliptic, 3.5–6 cm long, 1.5–5 cm broad, both surfaces dark green and roughly pilose, base gradually or abruptly narrowed cuneate, margin crenate, apex acute. Spikes pseudo-terminal, 2–3 cm long; peduncles 0.5–1 cm long, pilose; flowers subsessile. Bracts oblong-elliptic, 0.9–1.5 cm long, 0.3–0.4 cm broad, brownish pilose, margin entire, apex acute, caducous; bracteoles linear, 6–9 mm long, white pilose and sparsely purple glandular-hairy towards tip, margin ciliate, apex acute. Calyx 5–7 mm long in flower, accrescent to 8–10 mm in fruit, 5-lobed almost to base; lobes equal, linear, margin ciliate, apex acute. Corolla cream, 2.8–3 cm long, straight, outside densely glandular-hairy; tube base cylindric, 1.2–1.5 mm broad, upwards gradually widened to 1–1.2 cm broad at mouth; lobes ovate, 3–3.5 mm long and broad, apex obtuse. Stamens 4, included; filaments glabrous, longer pair 3–4 mm long, shorter pair 1–2 mm long; anthers-thecae ellipsoid, 1–1.2 mm long, nodding. Ovary pilose towards tip; 2 ovules per locule; style 1.5–1.7 cm long, pilose, reddish; stigma curved. Capsules unknown. Figure 46-37.

Phenology: flowering from October to December.

Habitats: thickets; 2100–2400 m.

Distribution: endemic; Bhutan (Map 46-66).

BHUTAN: **Tashigang**, Lumphi, E Tashiyangtse, 2350 m, 14 Oct. 1915, *R. E. Cooper 4446* (BM, E); between Tashiyangtse and Bumdeling, 2100 m, 8 Nov. 1991, *J. R. I. Wood 7499* (E).

17. Strobilanthes twangensis J. R. I. Wood & D. Borah, Kew Bull. 74: 41. 2019. Type: Yarlung Zangbo-Brahmaputra, Tawang, Jang, 2524 m, 1 Sep. 2018, D. Borah *L. Yama & M. Puranjoy 098* (holotype: CAL; isotypes: ARUN, ASSAM).

Weakly anisophyllous perennial herbs. Stems decumbent, rooting at nodes, eventually ascending to 50 cm high, glabrous. Leaves petiolate, slightly unequal in each pair; petiole 2.5–5.5 cm long, diminishing in size upwards, glabrous; blade ovate or elliptic, 6.5–12 cm long, 4.5–6.5 cm broad, glabrous on both surfaces, lateral veins 5–6 pairs, base asymmetric, rounded to cuneate, shortly decurrent onto petiole, margin crenate, ciliate with brownish hairy, apex shortly acuminate. Spikes axillary, capitate, 2.5–3 cm long; peduncles (0.5–)1.7–4 cm long, spreading or somewhat pendulous, glabrous; flowers distant for 3–4 mm on rachis. Bracts diminishing in size, outer ones foliose, broadly oblong-elliptic, 1.8–2.1 cm long, 1–1.2 cm broad, margin repand to undulate, sparsely ciliate with scattered white hairs, apex rounded to obtuse, inner ones ovate, 1.3–1.5 cm long, 0.4–0.5 cm broad, margin ciliate, apex obtuse; bracteoles oblong to oblong-pandurate, 8–9 mm long, 1.5–2 mm broad, weakly keeled, margin sparsely ciliate, apex obtuse to rounded. Calyx 1.3–1.5 cm long, subequally 5-lobed to base; lobes linear-oblong, 2–2.2 mm broad, margin ciliate, apex narrowed to subacute or obtuse. Corolla 3.2–3.4 cm long, white with dark pink veins, campanulate, straight, outside pilose, inside glabrous except for trichomes retaining style; lobes, broadly ovate, 5–6 mm long, 4.5–5 mm broad, apex rounded. Stamens 4; filaments sparsely pilose towards base, shorter pair 9–10 mm long, longer pair slightly unequal, 14–15 mm long; anthers broadly ellipsoid, 0.7–1.2 mm long, 0.5–0.7 mm broad, incurved; connective shortly excurrent. Ovary ellipsoid, comose; style 2.7–3 cm long, glabrous. Capsules unseen.

Phenology: flowering from September to October.

Habitats: slopes, thickets; 2500–2600 m.

Distribution: endemic; Yarlung Zangbo-Brahmaputra (Map 46-67).

YARLUNG ZANGBO-BRAHMAPUTRA: **Nyam Jang Chu**, Moindawang, Jang, 2520 m, 1 Sep. 2018, *D. Borah et al. 098* (ASSAM, CAL, fide Wood et al., 2019).

18. Strobilanthes cuneata (Shakya) J. R. I. Wood, Edinburgh J. Bot. 51(2): 218. 1994; Hu, Deng & Wood in Wu, Raven & Hong (eds.), Fl. China 19: 420. 2011. ≡ *Dossifluga cuneata* P. R. Shakya, Jap. J. Bot. 50: 99. 1975. Type: [C] Nepel, Dhaulagiri, Mustang, Lete, South of Tukuche, Kali Gandhaki Valley, 2420 m, 27 Aug. 1954, *J. D. A. Stainton, W. R. Sykes & L. H. J. Williams 7498* [holotype: BM (BM000906295); isotypes: E (E00273503), KATH (KATH001118), TI (TI10010687)].

Herbs, to 30 cm high, isophyllous; stems quadrangular, sulcate, glabrous except for pubescent on nodes. Leaves equal in each pair, shortly petiolate; petiole 1–3 mm long,

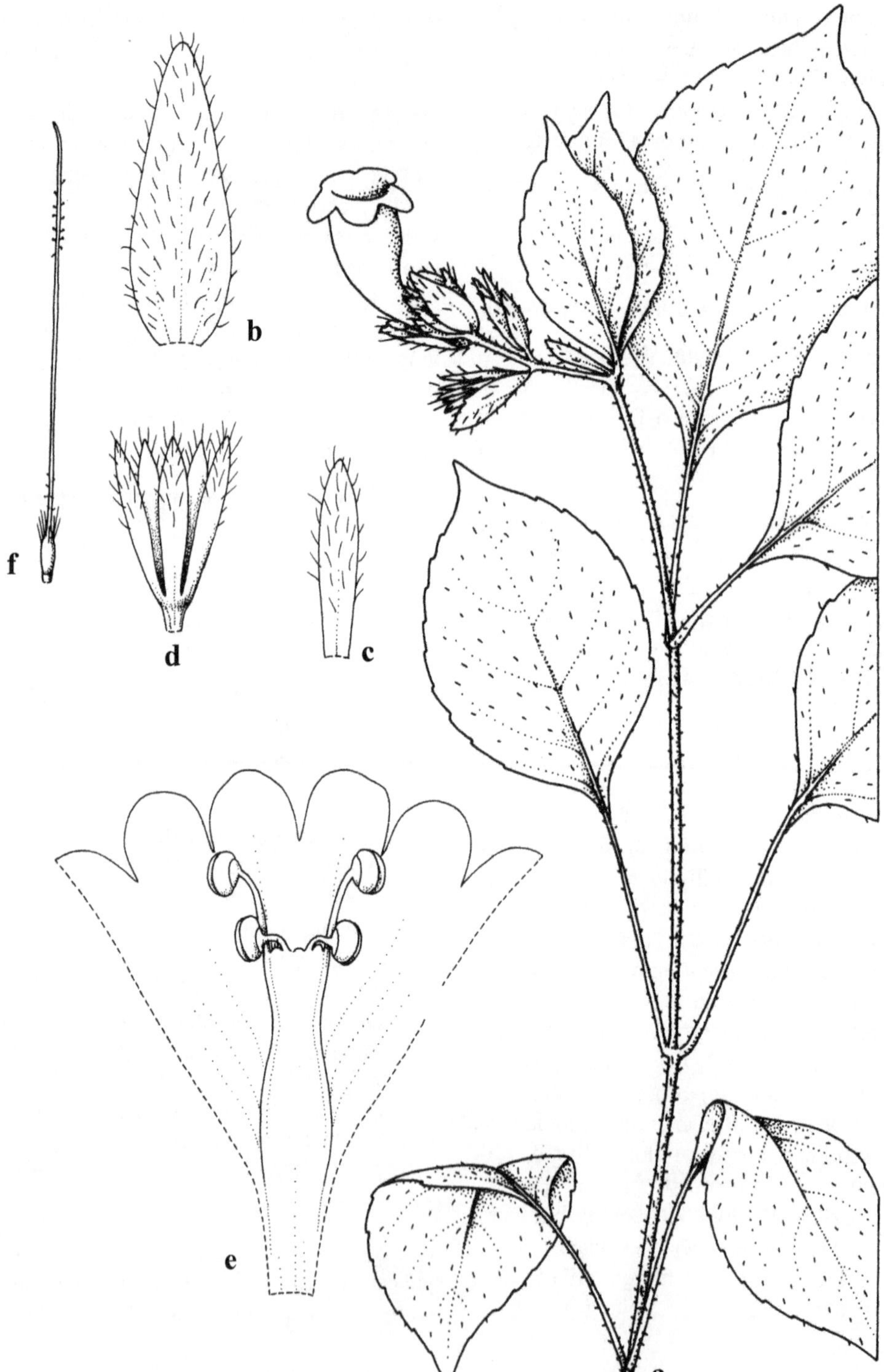

Figure 46-37: **Strobilanthes jennyae** J. R. I. Wood: **a**, habit; **b**, bract; **c**, bracteole; **d**, calyx; **e**, corolla opened out to show stamens; **f**, ovary and style. Drawn by Mary BATES; reproduced from Wood, *Edinburgh J. Bot.* 51: 244, fig. 10. 1994.

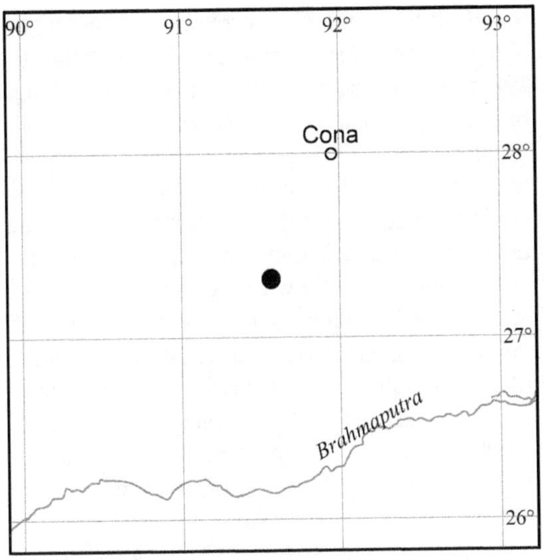

Map 46-66: **Strobilanthes jennyae** J. R. I. Wood

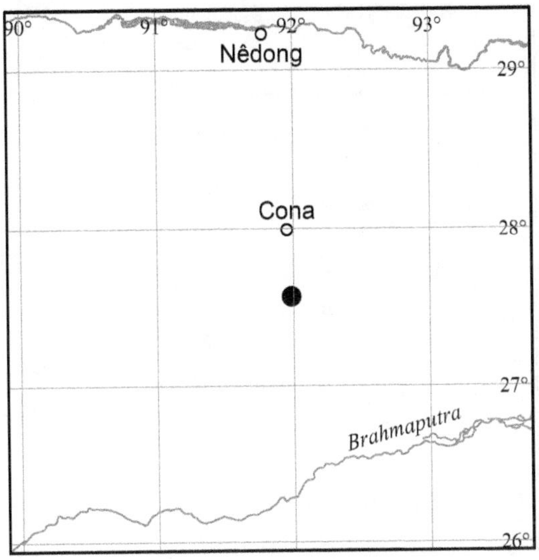

Map 46-67: **Strobilanthes twangensis** J. R. I. Wood & D. Borah

subglabrous; blade ovate or broadly elliptic, 2–3.5 cm long, 5–10 cm broad, adaxially glabrous, abaxially sparsely pilose along veins, lateral veins 6 or 7 pairs, base cuneate, margin serrate and ciliate, apex acuminate. Flowers usually 3 arranged in axillary spikes; peduncles 0–5 mm long. Bracts spatulate, 0.7–1.2 cm long, 2.5–4 mm broad, pilose, margin 3-lobed torwads apex, apex obtuse, caducous; bracteoles linear-oblong, 2.5–5 mm long, 1.2–1.5 mm broad, sparsely pilose. Calyx 1.2–1.5 cm long, 5-lobed almost to base; lobes subequal, linear-lanceolate, 1–1.5 mm broad, pilose toward apex, margin ciliate. Corolla 3.5–4 cm long, pale blue, campanulate; tube base cylindric, 4–5 mm long, 0.8–1.2 mm broad, upwards bent and gradually widened to 1.5 cm broad at mouth; lobes oblong, 6–8 mm long, 6–7 mm broad, apex retuse. Stamens 4, included; filaments glabrous, longer pair unequal, 4–5 mm long, erect, shorter pair equal, 1–1.5 mm long, curved; anthers-thecae oblong, 1–1.2 mm long. Ovary glabrous; 2 ovule per locule; style 2–2.3 cm long, sparsely glandular-pilose. Capsules unseen.

Phenology: flowering from July to August.

Habitats: slopes, thickets; 1700–2800 m.

Distribution: endemic; C Nepal (Map 46-68).

C NEPAL: **Bagamati**, Rasuwa, Langtang Khola, 2400 m, 1962, *J. D. A. Stainton 4033* (KATH); **Dhawalagiri**, Mustang, Lete, 2420 m, 27 Aug. 1954, *J. D. A. Stainton, W. R. Sykes & L. H. J. Williams 7498* (BM, E, KATH, TI); Mustang, Titigaon to Ghasa, 2410 m, 6 Aug. 1996, *T. Hoshino et al. 9666190* (A, CBM, E, KATH, TI); Myagdi, Kabhre to Kalopani, 1800 m, 28 Aug. 1988, *M. Suzuki et al. 8811144* (BM, TI); Myagdi, Kabhre (Kabre), 1720 m, 28 Aug. 1988, *M. Suzuki et al. 8881524* (BM, E, KUN, L, MO, TI).

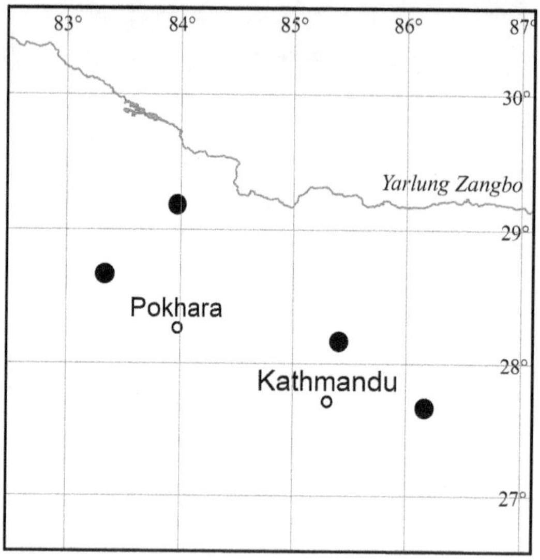

Map 46-68: **Strobilanthes cuneata** (Shakya) J. R. I. Wood

19. Strobilanthes bheriensis (Shakya) J. R. I. Wood, Edinburgh J. Bot. 51(2): 218. 1994. ≡ *Dossifluga bheriensis* Shakya, Jap. J. Bot. 50: 97. 1975. Type: [W] Nepal, Karnali Zone, Dolpa District, between Hurta and Narku, Bheri river, 2134 m, 6 Sep. 1952, *O. V. Polunin, Syes & L. H. J. Williams 3233* [holotype: BM (BM000906318); isotypes: KATH (KATH001112), TI (TI10010685)].

Perennial herbs, 15–30 cm high, isophyllous; stems quadrangular, sulcate, densely pilose. Leaves sessile; blade broadly elliptic, ovate or obovate, 1.5–3.5 cm long, 1–3 cm broad, adaxially densely pilose, abaxially sparsely pilose along veins, base rounded, margin serrate, apex acute or obtuse. Spike axiallary and terminal, 1–3 cm long, 2–4-flowered; peduncles 1–1.5 cm long; flowers imbricate. Bracts broadly elliptic to narrowly obovate, 1–1.5 cm long, 0.7–1 cm broad, pilose, margin serrate, apex acute or obtuse; bracteoles oblong, 2.5–3 mm long, 1–1.4 mm long, pilose. Calyx 1.1–1.5 cm long, pilose, 5-lobed almost to base; lobes unequal, usually two slightly longer than others, linear-lanceolate, 0.8–1 mm broad. Corolla 3.2–3.6 cm long, mauve, ventricose, pilose outside; tube base cylindric, 4–5 mm long, upwards slightly curved and gradually widened to ca. 1 cm broad at mouth; lobes equal, oblong, 4–5 cm long, 4–5 mm broad, apex retuse. Stamens 4, didynamous; filaments glabrous, longer pairs unequal, longer one 5–6 mm long, shorter one 4–5 mm long, shorter pair 1.5–2 mm long, curved; anther bithecous, thecae 1.2–1.5 mm long, connective not produced. Ovary pilose towards apex; 2 ovules per locule; style 2.5–3 cm long, sparse pilose. Capsules unknown.

Phenology: flowering from July to September.

Habitats: thickets; 1400–3500 m.

Distribution: endemic; W Nepal, C Nepal (Map 46-69).

W NEPAL: Dhawalagiri, Myagdi, Boghara to Jugepani (Naura), 1470–2260 m, 11 Sep. 1996, *M. Mikage et al. 9685335* (TI, TNS); Myagdi, 1820–2360 m, 7 Sep. 1996, *M. Mikage et al. 9682322* (TI); Myagdi, valley of Konaban Khola, 2360–3300 m, 8 Sep. 1996, *M. Mikage et al. 9681282* (TI); **Karnali**, Dolpa, Hurta, 2120 m, 6 Sep. 1952, *O. V. Polunin, W. R. Sykes & L. H. J. Williams 3233* (BM, KATH, TI); **Rapti**, Rukum, Ranmamikot, 2880 m, 1 Jul. 1954, *J. D. A. Stainton, W. R. Sykes & L. H. J. Williams 3325* (BM, E, TI). **C NEPAL: Bagmati**, Sindupalchok, Lamjure Peak, 3420 m, 1966, *M. L. Banerji & P. R. Shakya 5529* (BM, KATH).

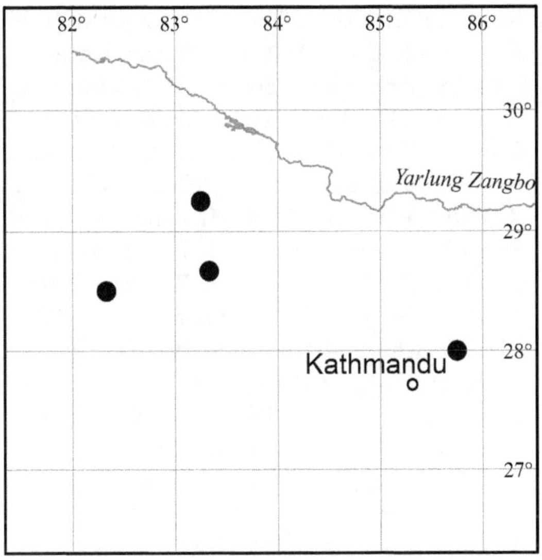

Map 46-69: **Strobilanthes bheriensis** (Shakya) J. R. I. Wood

20. Strobilanthes lamiifolia (Nees) T. Anderson, J. Linn. Soc., Bot. 9: 476. 1867; Clarke in Hooker (ed.), Fl. Brit. India 4: 463. 1884, "*lamiifolius*"; Wood in Grierson & Long (eds.), Fl. Bhutan 2(3): 1266. 2001; Hu, Deng & Wood in Wu, Raven & Hong (eds.), Fl. China 19: 422. 2011. ≡ *Goldfussia lamiifolia* Nees, Pl. Asiat. Rar. 3: 88. 1832; de Candolle, Prodr. 11: 171. 1847. ≡ *Ruellia lamiifolia* Wall., Numer. List n. 2347. 1830, nom. invalid. Type: Nepal, 1821, *Wallich Cat. 2347* [lectotype designated by Deng et al., Bot. J. Linn. Soc. 150: 377. 2006: K-W (K001115704); isolectotypes: BM (BM000906307), CGE, K (K000883041, K000883042, K000883043), LE, WU].

Ruellia rotundifolia D. Don, Prodr. Fl. Nepal: 120. 1825, non *Strobilanthes rotundifolia* Benoist, Bull. Mus. Natl. Hist. Nat. 28: 97. 1922. ≡ *Pteracanthus rotundifolius* (D. Don) Bremek., Verh. Kon. Ned. Akad. Wetensch., Afd. Natuurk., Sect. 2 41(1): 199. 1944; Tsui & Hu in Hu (ed.), Fl. Reipubl. Popularis Sin. 70: 147. 2002. Type: Nepal, *N. Wallich s. n.* (holotype: BM).

Strobilanthes austinii C. B. Clarke ex W. W. Sm., Notes Bot. Gard. Edinburgh 10: 190. 1918. ≡ *Goldfussia austinii* (C. B. Clarke ex W. W. Sm.) Bremek., Verh. Kon. Ned. Akad. Wetensch., Afd. Natuurk., Sect. 2 41(1): 231. 1944; Hu, Fl. Reipulic. Popularis. Sin. 70: 161. 2002. Type: China, Yunnan, s. d., *A. Henry 13382* [lectotype designated by Deng et al., Bot. J. Linn. Soc. 150: 377. 2006: K (K000882694)].

Strobilanthes feddei H. Lév., Repert. Spec. Nov. Regni Veg. 12: 20. 1913; Bremekamp, Verh. Kon. Ned. Akad. Wetensch., Afd. Natuurk., Sect. 2 41(1): 274. 1944. ≡ *Goldfussia feddei* (H. Lév.) E. Hossain, Notes Roy. Bot. Gard. Edinburgh 32: 407. 1973; Hu, Fl. Reipubl. Popularis Sin. 70: 163. 2002. Type: China, Yunnan, Route de Mong-Tse à Yunnan-sen, Pieddesrochers au borddelaroute, aucol de Koan-Yu-Chan, 27 Oct. 1896, *E. Bodinier s. n.* [holotype: E (E00047324)].

Strobilanthes hancockii C. B. Clarke ex W. W. Sm., Notes Roy. Bot. Gard. Edinburgh 10: 193. 1918. ≡ *Goldfussia hancockii* (C. B. Clarke ex W. W. Sm.) Bremek., Verh. Kon. Ned. Akad. Wetensch., Afd. Natuurk., Sect. 2 41(1): 231, 234. 1944. Type: China, Yunnan, Mengzi, Mengtsz mountains, Sep. 1893, *Hancock 104* [lectotype designated by Deng et al., Bot. J. Linn. Soc. 150: 377. 2006: K (K000882692); isolectotype: NY].

Strobilanthes mahongensis H. Lév., Cat. Pl. Yunnan: 6. 1915, "*makongensis*"; Bremekamp, Verh. Kon. Ned. Akad. Wetensch., Afd. Natuurk., Sect. 2 41(1): 280. 1944. ≡ *Goldfussia mahongensis* (H. Lév.) E. Hossain, Notes Roy. Bot. Gard. Edinburgh 32: 407. 1973. Type: China. Yunnan, pâturages à Mahong, 2800 m, Aug.1912, *E. E. Maire s. n.* [lectotype designated by Deng et al., Bot. J. Linn. Soc. 150: 377. 2006: E (E00047324); isotype: BM (BM000906313)].

Herbs, 30–60 cm high, isophyllous; stems decumbent and often rooting at lower nodes, ascending upwards, bifariously pilose or glabrescent. Leaves equal in each pair, petiolate or uppermost ones sessile; petiole 0–3 cm long, densely yellowish-brown pilose; leaf blade elliptic, ovate, obovate or ovate-lanceolate,

1–7(–11) cm long, 1–4(–6) cm broad, abaxially glabrous or pilose, adaxially glabrous or sparsely pilose, lateral veins 4–7 pairs, base cuneate and often decurrent onto petiole, margin serrulate to crenulate, apex acute. Spikes terminal or axillary, 2–5 cm long; peduncle 0–6 cm long, pubescent. Bracts oblong to ovate-elliptic, 1.1–2.5 cm long, concave, abaxially pubescent and covered with cystoliths, caduceus or somewhat persistent at anthesis, margin crenulate; bracteoles narrowly oblong-elliptic, 6–7 mm long, 1–2 mm broad, scarious, caducous, abaxially puberulent. Calyx 0.8–1.4 cm long, abaxially glandular-ciliate, 5-lobed almost to base; lobes unequal, one distinctly longer than others, linear-oblong apex acute. Corolla purplish red to blue, 3.5–5 cm long, ventricose, outside pubescent; tube base cylindric, 1–1.2 cm long, 1.5–2 mm broad, upwards slightly curved and gradually widened to 1–1.2 cm broad at mouth; lobes orbicular, 5–6 mm in diameter. Stamens 4, included; filaments subglabrous, longer pair 5–7 mm long, erect, shorter pair 2–3 mm long, strongly reflexed; anther-thecae subspherical, 1–1.2 mm long. Ovary comose; 2 ovules per locule; style 3.2–3.5 cm long. Capsules fusiform, 1–1.2 cm long, sparsely pilose towards apex, 4-seeded. Seeds ovate in outline, 2–2.5 mm long, 1.5–2 mm broad, light brown, densely pubescent; areole small, glabrous. Figure 46-38.

Phenology: flowering from August to November.

Habitats: dry grasslands and open forests; 1000–2800 m.

Distribution: U Yarlung Zangbo, C Nepal, Bhutan, Yarlung Zangbo-Brahmaputra, S Hengduan, U Irrawaddy (Map 46-70); also in other parts of India and China (C Yunnan, S Sichuan).

U Yarlung Zangbo: Nylam, Zhangmu, Lixin, 2490 m, 2010, *PE-Expedition to Xizang 882* (PE). **C NEPAL**: **Bagmati**, Kathmandu, Sundarijal, between Sundarijal and Pati-Banjang, 1900 m, 4 Oct. 1974, *J. H. de Haas 2614* (BM); Kathmandu, Nagarjong Danda, Nagurjong Hill, 2000 m, 23 Oct. 1966, *D. H. Nicolson 2698* (BM, BO, LE); Kathmandu, 1550–1860 m, 22 Sep. 1993, *Himalaustria Project Team 242*(K); Kathmandu, Nagarjun, 2020 m, 1969, *T. B. Shrestha & M. S. Bista 12442* (KATH); Kathmandu, Nagarjun, northern nope, 1500–1800 m, *M. S. Bista et al. 8890* (KATH); Kathmandu, Tareveer, Seopuri, 1520 m, 1970, *M. M. Amatya 12608* (KATH); Nuwakot, Pati Bhanjhang, 2300 m, 1974, *J. D. A. Stainton 7197* (BM, KATH); Nuwakot, Latsu–Pati Bhanjyang–Bhorlang, 1972, *H. Kanai et al. 4901* (KATH, KYO, TI); Rasuwa, Syabru, 2100 m, 1975, *J. D. A. Stainton 7484* (BM); Sindhupalchok, above Thulo Pakhar, 1950 m, 1975, *Schilling 2114* (K); **Gandaki**, Lamjung, near Thange, 2200 m, 1979, *P. R. Shakya et al. 5305* (KATH); Manang, Dona Khola to Thonje, 16 Nov. 1963, *T. Namba 1116004* (KYO); Manang, Dharapani, 2000 m, 1983, *N. P. Manandhar 9708* (KATH); **Janakpur**, Dolakha, Bhumludandi to Risingo, 2000 m, 1964, *M. L. Banerji 1329* (A); Dolakha, Suri Doban, 1040 m, 14 Sep. 2007, *S. Noshiro et al. 20710044* (KATH, TI); **Narayani**, Makwanpur, old road from Hetauda to Kathmandu via Bimphede, 1090 m, 29 Nov. 2004, *T. Miller et al. A208* (E, KATH). **BHUTAN**: Hing Lai La, 2700 m, 24 Aug. 1949, *F. Ludlow & G. Sherriff & Hicks 19639* (A, BM, TI); **Bumthang**, Burdoh School, from Trissa to Shiugtilic, 1600 m, Aug. 1991, *Lesley Reader s. n.* (E); **Lhuntse**, Junior High School, 1600 m, 16 Oct. 1990, *J. R. I. Wood 7354* (E); **Mansa**, between Khimsaer and Zilingbi, 800–1200 m, Aug. 1991, *Lesley Reader s. n.* (E); **Mongar**, Limethang, 1500 m, 11 Oct. 1990, *J. R. I. Wood 7351* (E); around Mongar, 1700 m, 5 Nov. 1991, *J. R I. Wood 7490* (E, FHO, TI); **Punkha**, E Wangdi, 1200 m, 13 Aug. 1989, *King & Rose 1045* (E); Hing Lai La, 2800 m, 24 Aug. 1949, *Ludlow, Sherriff & Hicks 19639* (BM); Wangdi Phodrang, 1800 m, 11 Sep. 1967, *Bowes Lyon 5002* (BM); near Lomitsawa, 2100 m, 2 Oct. 1987, *J. R. I. Wood 5887* (E, FHO); above Thinleygang, 2000 m, 2 Sep. 1988, *J. R. I. Wood 6627* (E); **Tashigang**, around Tashiyangtse, 1900–2100 m, 8 Nov. 1991, *J. R. I. Wood 7502* (E). **YARLUNG ZANGBO-BRAHMAPUTRA**: **Dirang**, 1500–1800 m, 13 Oct. 1935, *F. Kingdon-Ward 12421* (BM, TI); **Nyam Jang Chu**, Pangchen, Shati, 2150 m, 3 Nov. 1938, *F. Ludlow, G. Sherriff & G. Taylor 7033* (BM, TI); Pangchen, Shati, Gyipo-Karteng,

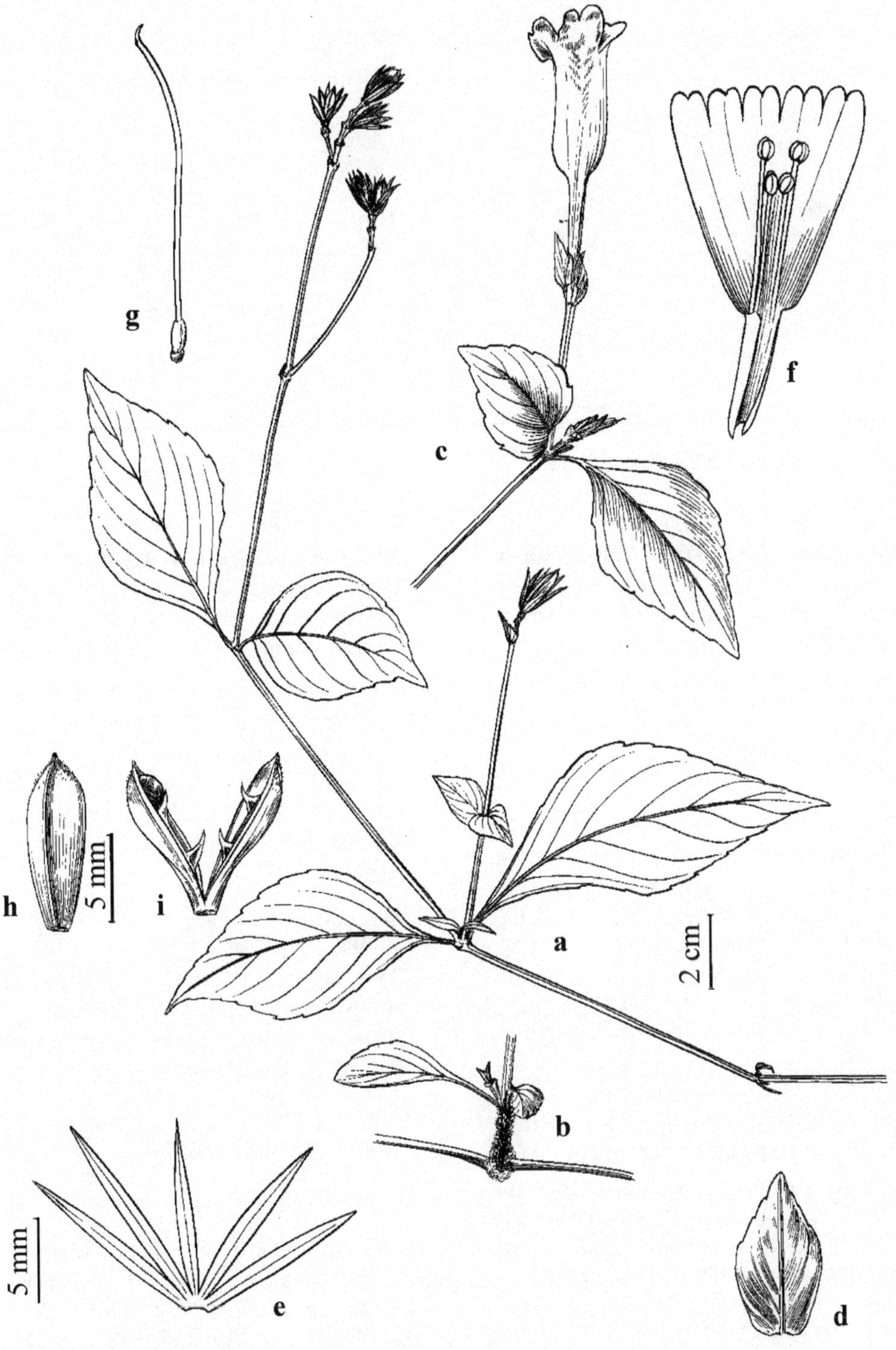

Figure 46-38: **Strobilanthes lamiifolia** (Nees) T. Anderson: **a**, habit; **b**, stem; **c**, flowering branch; **d**, bract; **e**, calyx; **f**, opened corolla showing stamens; **g**, pistil; **h**, capsule; **i**, opened capsule showing seeds. Drawn by Ding-Han CUI.

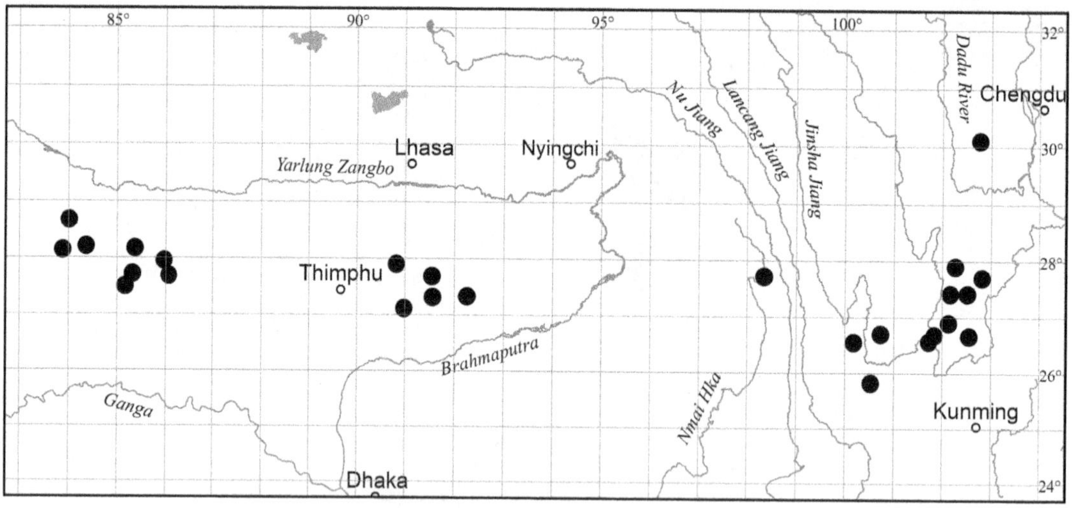

Map 46-70: **Strobilanthes lamiifolia** (Nees) T. Anderson

2150 m, 4 Nov. 1938, *F. Ludlow, G. Sherriff & G. Taylor 7073* (BM). **S HENGDUAN**: **Bingchuan**, Yangkoupingzi, 2000 m, *W. M. Zhu & J. L. Wu 2533* (YUKU); **Butuo,** Jiaojihe, Lianhe, way to Sanchahe, 1400 m, 22 Aug. 1959, *Chuan Jing Liang 5727* (CDBI, KUN, PE); **Dali**, Pingpo, 2000 m, 8 Aug. 1966, *K. J. Luo et al. 613487* (KUN); **Dechang**, Longwo Qu, 2200 m, 5 Sep. 1959, *S. F. Zhu 20199* (KUN, PE); Longwo Qu, 2100 m, 4 Sep. 1959, *S. F. Zhu 20185* (KUN, PE); **Huidong**, 2500 m, 14 Sep. 1983, *D. M. Liu 107* (KUN). **Heqing**, Ta-Pin-Tze, Aug. 1882, *M. Delavay 242* (K, P, US), Hechuang, 6 Oct. 1958, *W. T. Wang Dali-381* (KUN); **Miyi,** Malong, 2300 m, 1958, *S. Y. Chen et al. 11133* (NAS, SZ); **Panzhihua**, Panzhihua *Cycas* Natural Reserve, 2151 m, 13 Sep. 2010, *Exped. to Cycas Panzhihuaensis (B. Liu & J. F. Ye) 448* (IBSC, PE); **Puge**,Tuodang, 2600 m, 10 Aug. 1960, *Exped. to Sichuan Medic. 25898* (IBSC, NAS); **Tianquan**, 900 m, 14 Oct. 1936, *K. L. Chü 3989* (E, IBSC, PE, SZ); **Xichang**, 1932, *T. T. Yü 1318* (PE); **Yanbian**, Zemulong, Dahuoshan, 2000 m, 16 Sep. 2002, *X. F. Gao et al. 3538* (CDBI, PE); **Yongsheng**, 2 Jul. 1906, *F. Ducloux 4421* (P); Yunpe, 1920, *S. Ten 137* (E). **U IRRAWADDY**: **Drungjiang**, 2200 m, 1914, *C. Schneider 2566* (A, K, US).

21. Strobilanthes angustifrons C. B. Clarke in Hooker (ed.), Fl. Brit. India 4: 466. 1885. ≡ *Pteracanthus angustifrons* (C. B. Clarke) Bremek., Verh. Kon. Ned. Akad. Wetensch., Afd. Natuurk., Sect. 2 41(1): 200. 1944. Type: [U Ganga & Indus] India, Dehradun, *T. Thomson s. n.* (lectotype designated by Wood, Edinburgh J. Bot. 51: 228. 1994: K).

Goldfussia edgeworthiana Nees in de Candolle (ed.), Prodr. 11: 173. 1847, non *Strobilanthes edgeworthiana* Nees, Prodr. 11: 190. 1847. Type: [U Ganga & Indus] India, Upper Ganges, s. d., *M. P. Edgeworth* (lectotype designated by Wood, Edinburgh J. Bot. 51: 228. 1994: GZU).

Goldfussia acuminata P. R. Shakya, J. Jap. Bot. 50: 99. 1976. Type: [C] Nepal, Dhaulagiri Zone, Baglung District, Bhuji Khola, 2120 m, 16 Oct. 1954, *J. D. A. Stainton, W. R. Sykes & L. H. J. Williams 9059* [holotype: BM (BM000521513); isotypes: KATH, TI (TI10011304)].

Pteracanthus agrestis auct. non (C. B. Clarke) Bremek.: Hara, Enum. Fl. Plants Nepel. 3: 143 (1982), p. p.

Subshrubs, 0.5–1 m high, anisophyllous; stems quadrangularis, glabrous. Leaves unequal in each pair, smaller one 1/5–1/3 of larger one in size; petiole 1–3.5 cm long, glabrous; blade elliptic, larger ones 15–30 cm long, 6–10 cm broad, smaller ones 3–10 cm long, 1.5–4 cm broad, glabrous, base cuneate, margin entire, apex long acuminate. Inflorescences of panicle of head-like spikes, axillary or terminal. Bracts ovate, 1.2–1.5 cm long, 0.4–0.5 cm broad, glabrous, margin ciliate, apex long acuminate, persistent; bracteoles elliptic or lanceolate, 9–11 mm long, 1.5–2 mm long, margin ciliate. Calyx 1–1.3 cm long, abaxially pilose, 5-lobed almost to base; lobes linear-lanceolate, 0.5–1 mm broad. Corolla 4–4.5 cm long, campanulate, ventricose, outside glabrous, inside pilose; tube cylindric, 0.8–1 cm long, 1.2–1.5 mm broad, upwards gradually widened to ca. 1 cm broad at mouth; lobes oblong, 5–6 mm long, 4–5 mm broad, apex retuse. Stamens 4, included; filaments sparsely pilose, longer pair unequal, 4–5 mm long, erect, short pair subequal, 1–1.5 mm long, curved; anther-thecae oblong, 1–1.2 mm long, connective not produced. Ovary glabrous; 2 ovules per locule; style 2.8–3.2 cm long. Capsules oblong, 1–1.2 cm long, 4-seeded. Seeds ovoid, 2–2.5 mm long, 1.6–2 mm broad, densely villose.

Phenology: flowering and fruiting from September to December.

Habitats: thickets, stream banks; 600–2800 m.

Distribution: N Pakistan, U Ganga & Indus, W Nepal, C Nepal, E Nepal, Sikkim & Darjiling (Map 46-71); also in other parts of Pakistan and India.

N PAKISTAN: **Hazara**, **Abbottabad**, Laswani Khala, Malkot, 1370 m, Dec. 1898, *Gamble 27499* (K). **U GANGA & INDUS**: **Dehradun**, 600–760 m, 1896, *J. S. Gamble 22315* (K); Rajpur, 910 m, Oct. 1891, *J. S. Gamble 23164* (K); Robbers Cave, 760 m, *J. S. Gamble 24580* (K); 600–760 m, 1896, *J. S. Gamble 24581* (K); Nalapani, 910 m, *J. S. Gamble 24602* (K); Rajpur, 910 m, Sep. 1889, *J. S. Gamble 27230* (K); Thamo, 910 m, Dec. 1898, *J. S. Gamble 27506* (K); **Kangra**, 910 m, 1874, *C. B. Clarke 23633* (K); **Mussorie**, 1970, *V. I. J. 53* (AAU); **Tehri Gahrwal**, W Sangui, 1220 m, Sep. 1898, *J. S. Gamble 27204* (K). **W NEPAL**: **Bheri**, Surkhet, 1520 m, 29 Nov. 1964, *J. J. Burnet 38* (BM); Surkhet, Kot, 1250 m, 1982, *S. R. Adhikari & N. P. Manandhar 9195* (KATH); **Karnali**, Dolpa, Belor Lunh, Bheri River, 1670 m, 8 Oct. 1952, *O. V. Polunin, W. R. Sykes & L. H. J. Williams 3286* (BM, KATH). **C NEPAL**: **Bagmati**, Kathmandu, Shioupuri, 2700 m, 14 &15 1969, *H. Ohashi 69113* (TI); **Dhawalagiri**, Baglung, Bhuji Khola, 2120 m, 1954, 16 Oct. *J. D. A. Stainton, W. R. Sykes & L. H. J. Williams 9059* (BM, KATH, TI); Baglung, 910 m, 31 Oct. 1954, *J. D. A. Stainton, W. R. Sykes & L. H. J. Williams 9213* (BM, TI); Myagdi, Tatopani to Beni, 820–1240 m, 29 Sep. 1995, *M. Mikage et al. 9550516* (TI). **E NEPAL**: **Koshi**, Dhankuta, Tute to Tinjure Phedi, 14 Jul. 1991, *H. Ohba et al.*

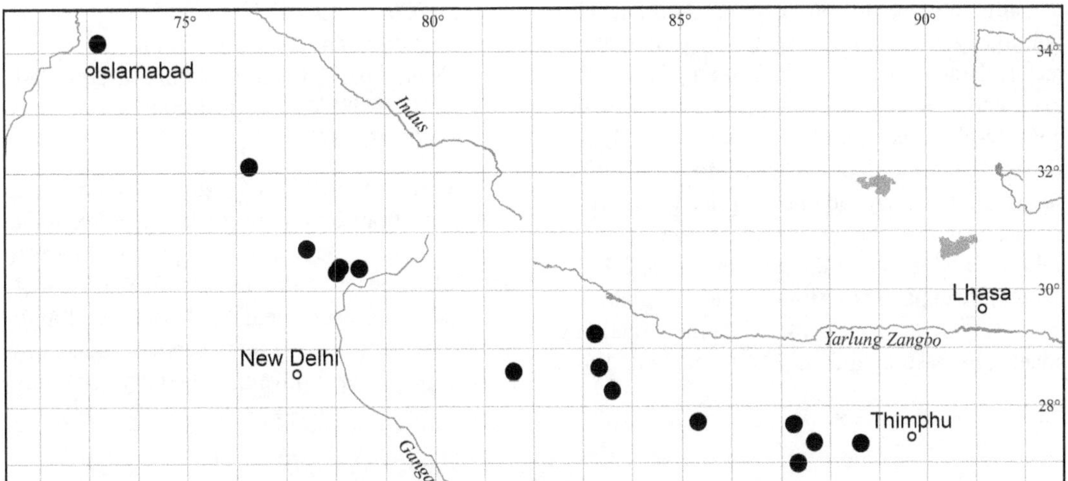

Map 46-71: **Strobilanthes angustifrons** C. B. Clarke

9154047 (TI); **Sankhuwasabha**, Milke Danda, 2740 m, 1980, *Schilling 2459* (K). **SIKKIM & DARJILING**: **Sikkim**, Rangbee, 17 Sep. 1867, *T. Anderson s. n.* (BO).

22. Strobilanthes medogensis (H. W. Li) J. R. I. Wood & Y. F. Deng, Bot. J. Linn. Soc. 150: 383. 2006; Hu, Deng & Wood in Wu, Raven & Hong (eds.), Fl. China 19: 426. 2011. ≡ *Goldfussia medogensis* H. W. Li in Wu, Fl. Xizang. 4: 413. 1985. Type: [Yarlung Zangbo-Brahmaputra] China, Tibet, Mêdog xian, Gedang, 2400m, 27 Aug. 1974, *Quinghai-Xixang Complex Exped. 74-4911* [holotype: KUN (KUN484455); isotypes: PE (PE01649727, PE01649727, PE01649727)].

Strobilanthes campaniformis J. R. I. Wood, Edinburgh J. Bot. 51 (2): 235 (1994). Type: [Yarlung Zangbo-Brahmaputra] China, Xizang, Lung, Chayul Chu, Chaynechu, 9000 ft., 2750 m, 10 Jul. 1936, *F. Ludlow & G. Sherriff 2336* [holotype: BM (BM000898032); isotype: E (E00133439)].

Subshrubs 2–3 m high, anisophyllous; stems quadrangular, slightly sulcate, bifariously puberulent. Leaves unequal in each pair, smaller of pair about 2/3 of larger one in size, subsessile or petiolate; petiole 0–5 mm long; blade ovate, 3–13 cm long, 2–5.5 cm broad, glabrous, base rounded to subcordate, margin serrulate, apex acuminate and falcate. Spikes axillary, 3–6 cm long, simple or bifurcate; peduncle 2–5cm long; rachis bifariously pubescent; flowers 3–11 mm apart on rachis. Bracts obovate-oblanceolate, 2.5–3 mm long, herbaceous, caducous, glabrous; bracteoles obovate-oblanceolate, 1.2–1.5 mm long, glabrous. Calyx 1–1.2 cm long, 5-lobed to base; lobes subequal, narrowly oblong-elliptic, glabrous or glandular-pubescent, apex acuminate. Corolla yellowish-white but dull purple on lobes, 2.2–2.5 cm long, campanulate, straight, outside glabrous, inside glabrous except for trichomes retaining style; tube base cylindric, 0.8–1 cm long, 1.5–2 mm broad, upwards gradually widened to 1–1.5 cm broad at mouth; lobes broadly ovate, 3–5 mm long, 6–8 mm broad, apex rounded. Stamens 4, more or less exserted; filaments pilose towards base, longer pair 9–10 mm long, shorter pair 7–8 mm long; anther-thecae subspherical, 1–1.2 mm long, incurved. Ovary glabrous; 2 ovules per locule; style 2.7–2.8 cm long, glabrous. Capsules unknown. Figure 46-39.

Phenology: flowering from July to August.

Habitats: forests; 1100–2800 m.

Distribution: endemic; L Yarlung Zangbo, Yarlung Zangbo-Brahmaputra (Map 46-72).

L YARLUNG ZANGBO: **Lhünzê**, Chayne chu, 2750 m, 10 Jul. 1936, *F. Ludlow & G. Sherriff 2336* (BM, E); **Nyingchi**, Pailong, 2175 m, 8 Sep. 2006, *Tiber-MacArthur 261* (A); **Tangmai**, road to Tongmai, 2100 m, 23 Jul. 1973, *R. F. Huang 966* (HNWP); Tongmai Road, 2100 m, 23 Jul. 1973, *R. F. Huang 967* (HNWP). **YARLUNG ZANGBO-BRAHMAPUTRA**: **Bangxin**, Pangxin, 1100 m, 1 Dec. 1982, *S. Z. Cheng & B. S. Li 3238* (PE); **Gutang**, 2400 m, 27 Aug. 1974, *Quinghai-Xixang Complex Exped. 74-4911* (KUN, PE); 2450 m, 11 Sep. 1982, *B. S. Li & S. Z. Cheng 814* (PE); 1940 m, 13 Oct. 1982, *S. Z. Cheng & B. S. Li 1457* (PE).

23. Strobilanthes secunda T. Anderson, J. Linn. Soc., Bot. 9: 480. 1867; Hu, Deng & Wood in Wu, Raven & Hong (eds.), Fl. China 19: 427. 2011. Type: Yarlung Zangbo-Brahmaputra, Mishmi Hills, *W. Griffith 6122* [lectotype **here designated**: K (K000883231), isolectotypes: C, K (K000886132), LE].

Shrubs 0.6–1.5 m high, anisophyllous; stems and branches often zigzag, quadrangular, sulcate, rufous pubescent in grooves. Leaves unequal in each pair; petiole 3–8 cm long, appressed pubescent; leaf blade oblong-elliptic, 4–18 cm long, 1.5–7.5 cm broad, abaxially pubescent along midvein, adaxially glabrous,

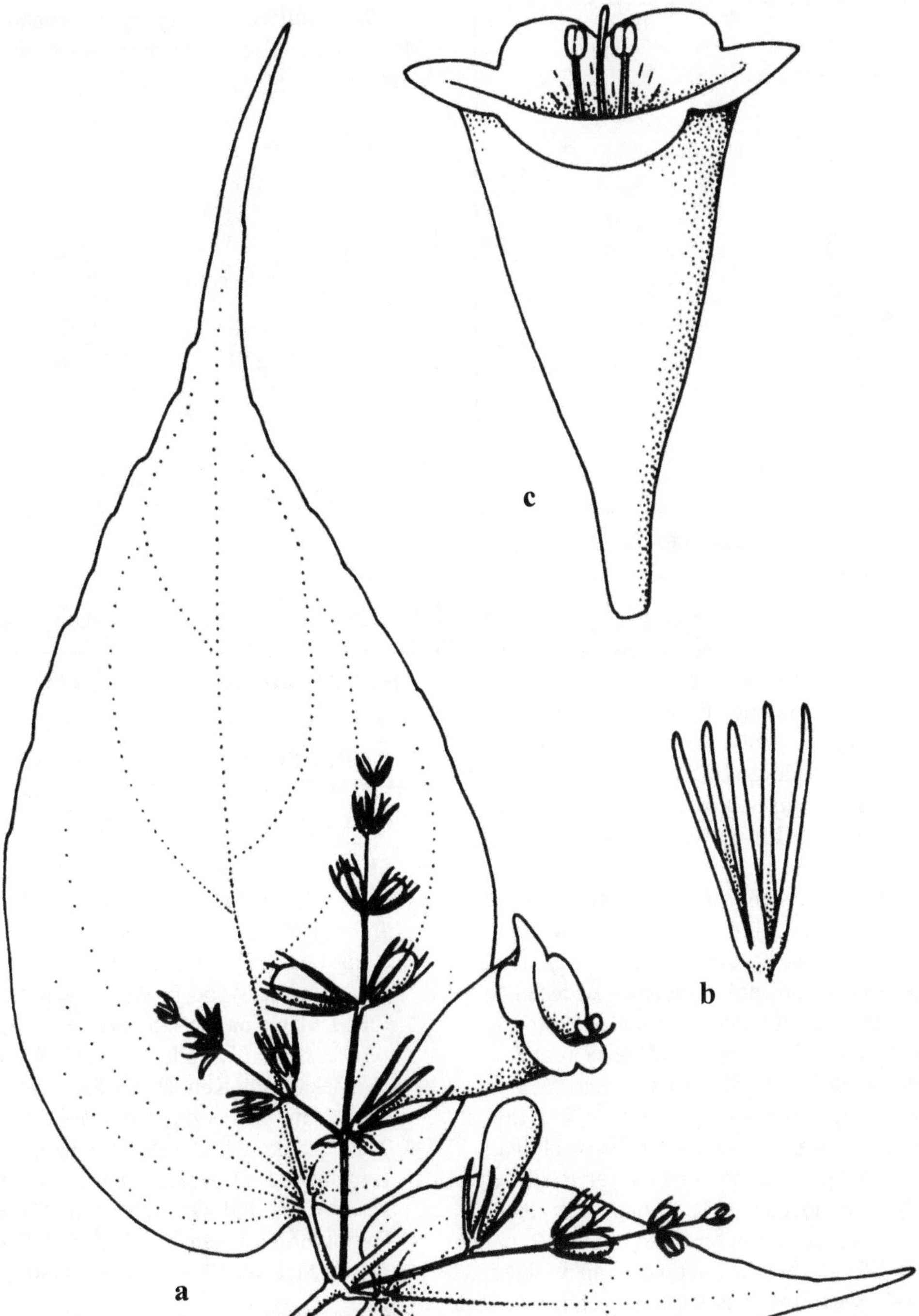

Figure 46-39: **Strobilanthes medogensis** (H. W. Li) J. R. I. Wood & Y. F. Deng: **a**, portion of plant; **b**, calyx; **c**, corolla. Drawn by Mary NATES; reproduced from Wu, Raven & Hong, *Fl. China Illustr.* 19: fig. 331: 5–7. 2012 (original from *Edinburgh J. Bot.* 51: 234, fig. 8: C–D, 1994).

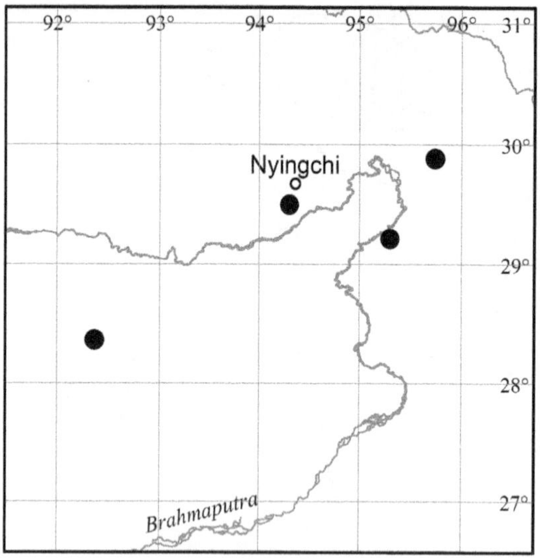

Map 46-72: **Strobilanthes medogensis** (H. W. Li) J. R. I. Wood & Y. F. Deng

lateral veins 6–9 pairs, base narrowly cuneate, margin entire to sinuate, apex acuminate. Spikes axillary, 3–18 cm long; flowers solitary or in opposite pairs on rachis, 1.5–3 cm apart. Bracts and bracteoles caducous. Calyx 1.6–2 cm long, subequally lobed to 3–4 mm above base; lobes linear-lanceolate, 1.2–1.6 cm long, glabrous, apex finely acuminate. Corolla yellow, 2.6–3 cm long, curved and somewhat ventricose, outside pubescent, inside glabrous except for trichomes retaining style; tube base cylindric, 6–7 mm long, 2–3 mm broad, upwards gradually widened to 1.2–1.5 cm broad at mouth; lobes broadly ovate, 8–9 mm broad, 4–5 mm broad, apex rounded. Stamens 4, included; filaments of shorter pair 2–3 mm long, glabrous, longer pair 4–5 mm long, villous towards base; anther-thecae oblong, 1.8–2 mm long, 1–1.2 mm broad. Ovary comose; 2 ovules per locule; style 2.4–2.6 cm long, pilose. Capsules narrowly oblong, 1.8–2.2 cm long, 2–2.5 mm broad, pubescent, apex acute, 4-seeded. Seeds ovate in outline, compressed, 3.2–3.5 mm long, 2.4–2.6 mm broad, pilose; areole densely pilose.

Phenology: flowering from December to January of next year; fruiting from January to April.

Habitats: forests.

Distribution: Yarlung Zangbo-Brahmaputra (Map 46-73), also in other parts of India and Myanmar.

Map 46-73: **Strobilanthes secunda** T. Anderson

YARLUNG ZANGBO-BRAHMAPUTRA: Akha Hills, Jan. 1890, *King's Collector 64* (BM); **Mishmi Hills**, *W. Griffith 6122* (C, K, LE).

24. Strobilanthes oligocephala T. Anderson ex C. B. Clarke in Hooker (ed.), Fl. Brit. India 4: 461. 1885; Wood in Grierson & Long (eds.), Fl. Bhutan 2(3): 1265. 2001; Hu, Deng & Wood in Wu, Raven & Hong (eds.), Fl. China 19: 425. 2011. ≡ *Goldfussia oligocephala* (T. Anderson ex C. B. Clarke) Bremek., Verh. Kon. Ned. Akad. Wetensch., Afd. Natuurk., Sect. 2 41(1): 234. 1944. Type: [Sikkim & Darjiling] Indian, Sikkim/ Darjiling, *J. D. Hooker s. n.*, "*Goldfussia* sp. 24, herb. Ind. Or. Hook. f. & Thomson" [lectotype designated by Wood, Edinburgh J. Bot. 51: 248. 1994: K (K000883062)].

Goldfussia thomsonii Hook., Bot. Mag. 75: t. 5119. 1859, non *Strobilanthes thomsonii* T. Anderson, J. Linn. Soc., Bot. 9: 478. 1867. Type: [*icon*] Hooker, Bot. Mag. 75: t. 5119. 1859 (lectotype designated by Wood, Edinburg J. Bot. 51: 249. 1994: K).

Strobilanthes oligocephala var. *treutleri* C. B.

Clarke in Hooker (ed.), Fl. Brit. India 4: 461. 1885. Type: [Sikkim & Darjiling] India, Sikkim, Tonglo and Sundukphoo, *D. Treutler s. n.* [holotype: K (K000883063)].

Strobilanthes paupera C. B. Clarke in Hooker (ed.), Fl. Brit. India 4: 463. 1885. ≡ *Diflugossa paupera* (C. B. Clarke) Bremek., Verh. Kon. Ned. Akad. Wetensch., Afd. Natuurk., Sect. 2 41(1): 246. 1944. Type: India, Meghalaya (Khasia), *J. D. Hooker & T. Thomson s. n.* ("*Goldfussia* sp. 26, Herb. Ind. Or. Hook. f. & Thomson") [lectotype **here designated**: K (K000883064); isolectotypes: BM, K (K000883066)].

Herbs 40–50 cm high, isophyllous; stems quadrangular, sulcate, sparsely and retrorsely white hairy. Leaves equal in each pair; petiole 3.5–4 cm long, sparsely pilose; leaf blade broadly ovate to elliptic, 4–7(–10) cm long, 2–4 cm broad, abaxially sparsely pilose along veins, adaxially glabrous to sparsely pubescent, lateral veins 4–6 pairs, base broadly cuneate, margin remotely serrate, apex acuminate. Spikes terminal or borne from upper leaf axils, 2–3 cm long, elongating to 3–4 cm in fruit, sometimes reduced to a single flower pair. Flowers imbricate on rachis. Bracts diminishing in size upwards, outer ones 1.3–1.5 cm long, inner ones smaller, white pubescent; bracteoles absent. Calyx 7–9 mm long, subequally 5-lobed almost to base; lobes linear, adaxially pubescent toward apex. Corolla bluish purple, 3–3.5 cm long, usually bent, outside sparsely glandular-pubescent on lobes, inside glabrous except for trichomes retaining style; tube base cylindric, 1.2–1.5 cm long, 1.5–2 mm broad, upwards gradually widened to 1.1–1.3 cm broad at mouth; lobes suborbicular, 4.5–5 mm long and broad. Stamens 4, included; filaments of shorter pair 2–3 mm long, glabrous, longer pair 5–6 mm long, pilose; anther-thecae oblong, 1.8–2 mm long, 1–1.1 mm broad. Ovary pubescent towards apex; 2 ovules per locule; style 3–3.2 cm long, pilose. Capsules narrowly obovoid, 9–11 mm long, pubescent towards apex, 4-seeded. Seeds ovate in outline, compressed, 2–2.5 mm long, 1.5–2 mm broad, pubescent. **Chromosomes** 2n = 60. Figure 46-40.

Phenology: flowering from July to September; fruiting from October to November.

Habitats: forests, shaded moist grasslands; 1000–3700 m.

Distribution: U Irrawaddy, C Nepal, E Nepal, Sikkim & Darjiling, Bhutan, Yarlung Zangbo-Brahmaputra (Map 46-74); also in other parts of India.

Notes: Wood (1994) designated *Clarke 12959* as the lectotype for the name *Strobilanthes oligocephala* var. *treutleri* C. B. Clarke. However, Clarke only mentioned Treutler's collection in the protologue.

U YARLUNG ZANGBO: **Nylam**: Zhangmu, 2200 m, *L. M. Gao et al. GLM-081659* (KUN). **C NEPAL**: **Bagmati**, Dhading, Pati Kharka to Pabil Kharka, 2920 m, 5 Aug. 1994, *F. Miyamoto et al. 9430093* (E, KATH, TI, TNS); Kathmandu, near the Pashupati Nath temple, 12 Nov. 1972, *T. Wraber 500* (BM); Rasuwa, Parbati Kund (near Gatlang) to Yure Kharka, 2590 m, 25 Jul. 1994, *F. Miyamoto et al. 9420029* (E, TI, TNS); **Gandaki**, Gorkha, Ghap to Namrung, 2190 m, 31 Jul. 2008, *H. Ikeda et al. 20818082* (TI); Kaski, Bhainsi Kharka to Banthanti, 2500 m, 23 Aug. 1988, *M. Suzuki et al. 8881247* (KATH, TI); Kaski, Bhainsi Kharka to Banthati, 2500 m, 23 Aug. 1988, *M. Suzuki et al. 8810923* (KATH, TI); **Dhawalagiri**, Myagdi, Maraini to Jalja La, 2900 m, 15 Sep. 1996, *M. Mikage et al. 9682576* (TI); Myagdi, Ritu Kharka to Lumsung, 2920 m, 21 Sep. 1996, *M. Mikage et al. 9682963* (TI); **Janakpur**, Ramechhap, Neju to Choarma, 2760–3650 m, 1985, *H. Ohba et al. 8540210* (E, KATH); **Narayani**, Makwanpur, from Hetauda to Kathmandu via Bimphede, 1090 m, 2004, *T. Miller et al. A203* (E, KATH). **E NEPAL**: **Koshi**, Sankhuwasabha, below Chamtang to Chyangrima, 2415 m, 20 Aug. 1998, *S. Noshiro et al. 9830079* (E, KATH, TI, TNS); Sankhuwasabha, Arun valley, Kasowa Khola, W of Nom, 3030 m, 12 Sep. 1956, *J. D. A. Stainton 1640* (BM); Sankhuwasabha, Thudam, between Tudam and Ritak, 11 Nov. 1971, *L. W. Beer et al. 12202* (BM); Sankhuwasabha, Redak to Chhurchathanga, 9 Aug. 1977, *H. Ohashi et al. 770600* (TI); Sankhuwasabha, Redak to Chhurchathanga, 2600 m, 9 Aug. 1977, *H.*

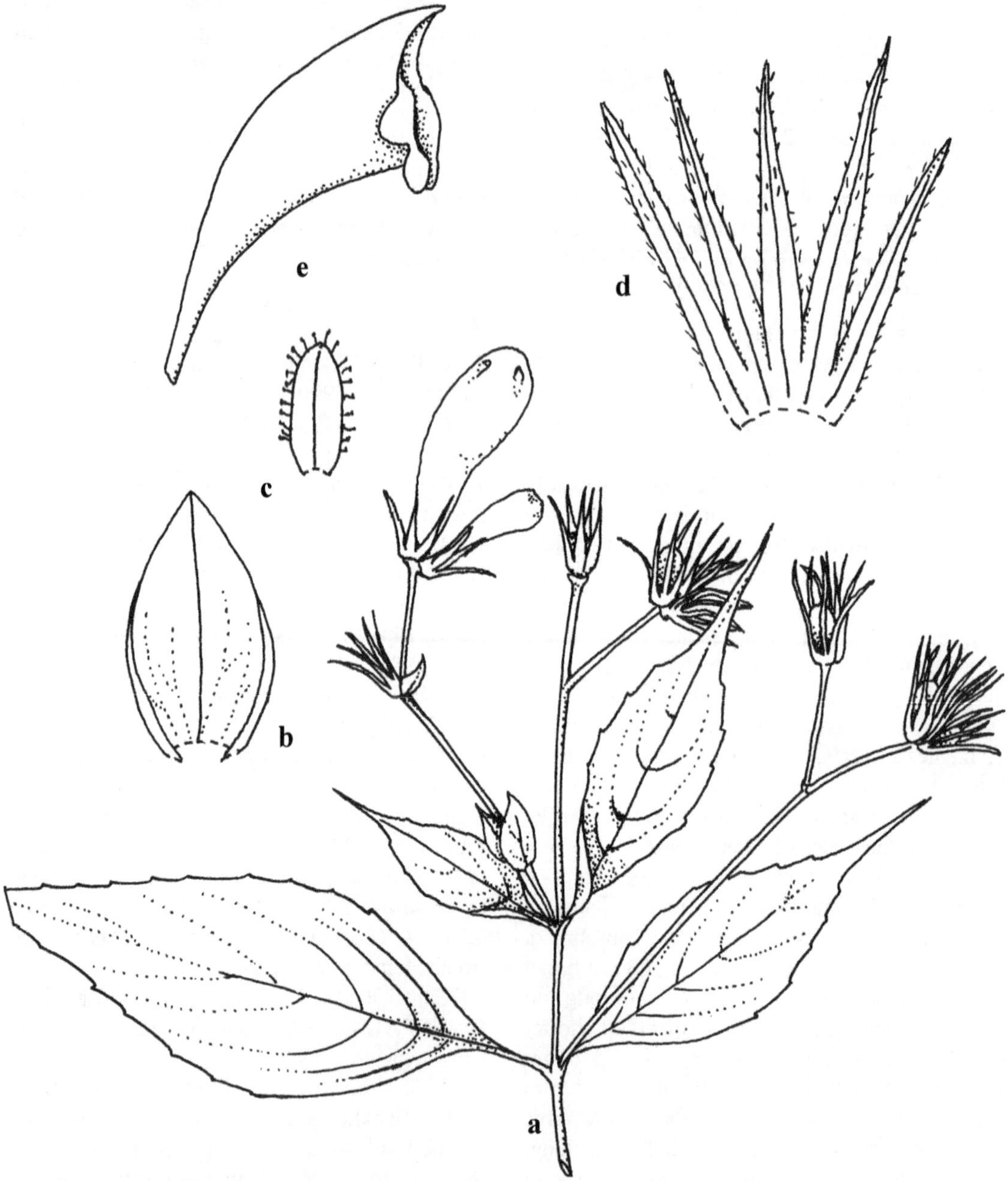

Figure 46-40: **Strobilanthes oligocephala** T. Anderson ex C. B. Clarke: **a**, portion of plant; **b**, outer bract; **c**, inner bract; **d**, calyx; **e**, corolla. Drawn by Louise OLLEY; reproduced from From *Fl. China* 19: fig. 335: 1–5. 2012 [original from Wood in Grierson & Long (eds.), *Fl. Bhutan* 2(3): 1266, fig. 103: a-e. 2001].

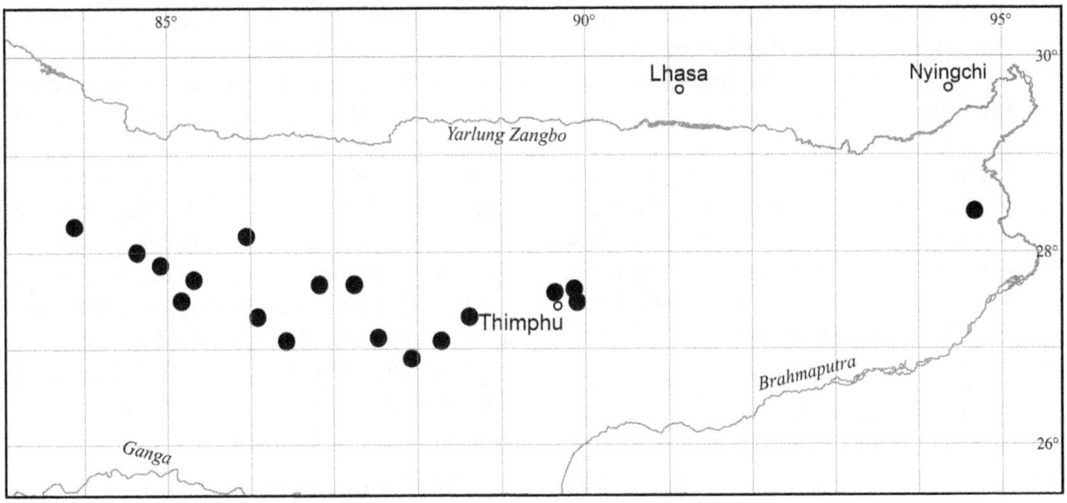

Map 46-74: **Strobilanthes oligocephala** T. Anderson ex C. B. Clarke

Ohashi et al. 773607 (TI); Terethum, Chauke to Tute, 2300–2700 m, 1 Sep. 1977, *H. Ohashi et al. 773183* (BM, TI); **Mechi**, Ilam, Mani Bhanjyang to Tonglu, 2000–3000 m, 1972, *H. Kanai et al. 72730* (BM, TI); Ilam, Tonglu to Sandakphu, 3000–3600 m, 1972, *H. Kanai et al. 72731* (BM, KATH, TI); Taplejung, Thakma Khola to Diorali Bhanjang, 14 Nov. 1963, *H. Hara et al. 3633* (BM, KYO, TI); **Sagarmatha**, Solukhumbu, W Lukla, Dukh Kosi valley, 2725 m, 1974, *J. H. de Haas 2887* (BM, L). **SIKKIM & DARJILING: Darjiling**, Tingh, 2700 m, 13 Sep. 1875, *C. B. Clarke 27412* (K); Tingb, 1000 m, 13 Sep. 1875, *C. B. Clarke 27538* (K); Tumbok, 2700 m, 11 Oct. 1870, *C. B. Clarke 12982* (K); Tonglu, 2400 m, 16 Aug. 1904, *I. H. Burkill 25123* (CAL); Tongloo, 3000 m, Aug. 1874, *J. S. Gamble 3562* (K); Tongloo, 3020 m, Nov. 1874, *J. S. Gamble 3552* (K); Tongloo, 3000 m, Sep. 1876, *J. S. Gamble 1611* (K); Tonglo, 2700 m, Sep. 1880, *J. S. Gamble 8412* (CAL, E, K); **Sikkim**: Tambok, 2700 m, 11 Oct. 1870, *C. B. Clarke 12959* (K); Dumbok, 2700 m, 11 Oct. 1870, *C. B. Clarke 13425* (BM); Neebay, 1950 m, 11 Oct. 1875, *C. B. Clarke 25249* (K, LE); Neebay, 2100 m, 16 Oct. 1875, *C. B. Clarke 25320* (K); Yoksna, 1950 m, 17 Oct. 1875, *C. B. Clarke 25379* (BM); Yokswa, 1950 m, 17 Oct. 19875, *C. B. Clarke 25394* (K); Phullort, 22 Aug. 1857, *T. Thomson 58* (CAL); Phullort, 22 Aug. 1857, *T. Thomson 59* (CAL); 3000 m, 12 Sep. 1874, *Treuteler 750* (K); Below Tongloo, 2700 m, 5 Oct. 1875, *J. S. Gamble 4* (K). **BHUTAN: Sankhuwasabha**, Bakang, above Sekidim, ca. 2600 m, 15 Aug. 1981, *C. Grey-Wilson et al. 4033* (K); **Punakha**, between Thinlaygang and Lomitsawa, 2000 m, 8 Nov. 1987, *J. R. I. Wood 6045* (E); above Thinleygang, 2100 m, 20 Oct. 1991, *J. R. I. Wood 7487* (E); **Thimphu**, Hulé La, 2700 m, 13 Oct. 1914, *R. E. Cooper 2379* (BM, E); near Thimphu River above Dotena, towards Barshong, 2500 m, 11 Sep. 1988, *J. R. I. Wood 6661* (E); above Thimphu River, above Dotena, towards Barshong, 2500 m, 11 Sep. 1988, *J. R. I. Wood 6662* (E, FHO); above the Thimphu River, Dotena, 2500 m, 11 Sep. 1988, *J. R. I. Wood 6676* (E, FHO); above the Thimphu River, Dotena, 2500 m, 11 Sep. 1988, *J. R. I. Wood 6677* (E, FHO); Dodena, 2500 m, 25 Aug. 1990, *J. R. I. Wood 7321* (E); above Dotena, upper Thimphu Valley, 2600 m, 22 Sep. 1992, *J. R. I. Wood 7583* (E, FHO); Thimphu Chu above Dotena, 2560 m, 5 Sep. 1984, *I. W. J. Sinclair & D. G. Long 4875* (E); Thimphu Chu above Dotena, 2560 m, 5 Sep. 1984, *I. W. J. Sinclair & D. G. Long 4877* (E). **YARLUNG ZANGBO-BRAHMAPUTRA**: **Dirang**, Senge Dzong, 2100–2400 m, 26 Aug. 1938, *F. Kingdon-Ward 14164* (BM, TI).

25. Strobilanthes thomsonii T. Anderson, J. Linn. Soc., Bot. 9: 478. 1867, "*thomsoni*"; Clarke in Hooker (ed.), Fl. Brit. India 4: 467. 1886; Wood, Edinburgh J. Bot. 51(2): 223. 1994; Wood in Grierson & Long (eds.), Fl. Bhutan 2(3): 1268. 2001; Hu, Deng & Wood in Wu, Raven & Hong (eds.), Fl. China 19: 426. 2011. Type: [Sikkim & Darjiling] India, Sikkim, 10,000 ft., *J. D. Hooker & T. Thomson "Goldfussia no. 30"* (lectotype **here designated**: K (K000883031)].

Subshrubs 1–2 m high, anisophyllous; stems and branches quadrangular, sulcate, glabrous or sparsely pubescent. Leaves unequal in pair, sessile or petiolate; petiole (0–)2–14 mm long, glabrous; leaf blade ovate to narrowly ovate-elliptic, 12–20 cm long, 4–7 cm broad, glabrous, lateral veins 5–7 pairs, base cuneate and decurrent onto petiole, margin serrate, apex long acuminate and usually falcate. Spikes axillary, simple or branched, 3–9 cm long. Flowers imbricate on rachis or to ca. 1 cm apart. Bracts variable in shape, ovate, elliptic, oblanceolate or spatulate, 4–6 mm long, 2–3 mm broad, glabrous or glandular-pubescent, caducous; bracteoles linear, 2.2–2.5 mm long, 0.8–1 mm broad, caducous. Calyx 5–9 mm long, accrescent to 1.1–1.3 cm in fruit, subequally 5-lobed almost to base; lobes linear, glabrous or pilose, apex acute. Corolla purple, funnel-shaped, 2.5–3.2 cm long, curved, outside pubescent, inside glabrous except for trichomes retaining style; tube base cylindric, 0.8–1 cm long, 1.6–2 mm broad, upwards curved and widened to 1–1.3 cm broad at mouth; lobes ovate, 4–4.5 mm long, 4–5 mm broad, apex rounded. Stamens 4; filaments glabrous, longer pair 3.5–4 mm long, erect, shorter pair 1.5–2 mm long, strongly incurved; anther-thecae subspherical, 0.8–1 mm in diameter, incurved. Ovary pubescent; 2 ovules per locule; style 2.5–3 cm long, sparsely pilose towards base. Capsules narrowly obovoid, 1–1.2 cm long, glandular-pubescent, 4-seeded. Seeds orbicular in outline, 1.5–2 mm in diameter, densely covered with appressed trichomes.

Phenology: flowering from September to November.

Habitats: dry forests; 100–2800 m.

Distribution: W Nepal, C Nepal, E Nepal, Bhutan, Sikkim & Darjiling, Yarlung Zangbo-Brahmaputra (Map 46-75); also in other parts of India.

W NEPAL: Dhawalagiri, Baglung, Suprang, 2175 m, 21 Aug. 2014, *B002* (E, TI). **C NEPAL: Bagmati**, Syabru, 1520 m, 23 Sep. 1974, *J. D. A. Stainton 3337* (BM, KATH); **Janakpur**, Ramechhap, Naju to Choarma, 2760 m, 2 Aug. 1985, *H. Ohba et al. 8571062* (TI). **E NEPAL: Koshi**, Sankhuwasbha, Arun Valley, at mouth of

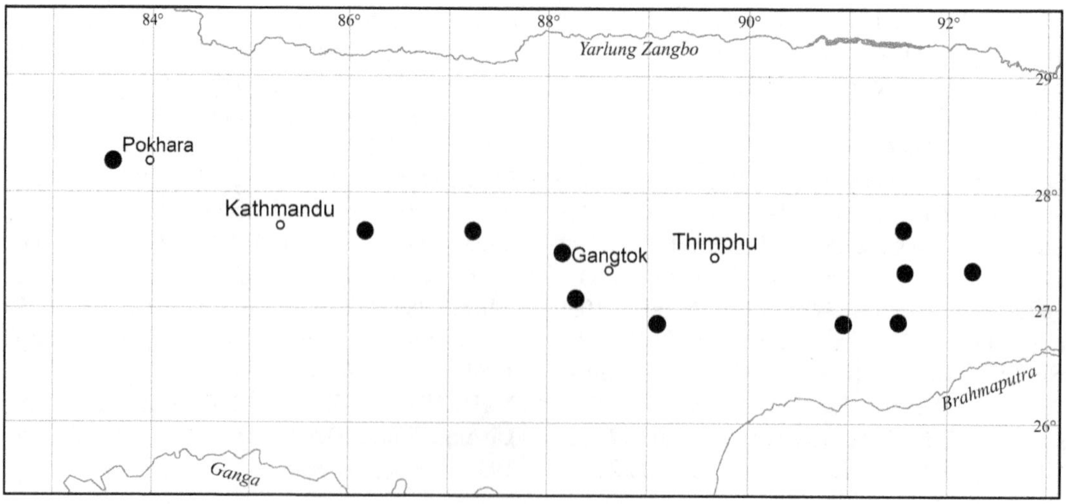

Map 46-75: **Strobilanthes thomsonii** T. Anderson

the Choyang Khola into Arun, 900–1000 m, 1972, *T. Wraber 34774* (BM); Sankhuwasbha, Arun Valley, 900 m, 3 Nov. 1972, *T. Wraber 467* (BM); **Mechi**, Taplejung, Baroya Khimty to Thakma Khola, 16 Nov. 1963, *H. Kanai et al. 3627* (TI); Taplejung, Thakma Khola to Yamphodin, 14 Nov. 1963, *H. Hara et al. 3734* (TI). **SIKKIM & DARJILING**: **Darjiling**, Saigtaca, 900 m, Nov. 1881, *J. S. Gamble 10017* (CAL); Sindul, 2700 m, 27 Aug. 1869, *C. B. Clarke 8855* (CAL); **Sikkim**, Dikchu, 700 m, 1 Nov. 1961, *G. S. Gapta 200* (CAL); Kalimpoong, Bhot, 900 m, 20 Sep. 1869, *C. B. Clarke 9351* (BM, CAL); Tungleng, 1200 m, Oct. 1877, *C. B. Clarke 25431* (CAL); Lingchaw, 1500 m, 22 Oct. 1875, *C. B. Clarke 25498* (BM); Ratong to Yoksum, 750–1500 m, 4 Oct. 1862, *Herb. Sikkim. T. Anderson 1042* (BM, CAL); Pimenbong, 180–360 m, 30 Sep. 1862, *Herb. Sikkim. T. Anderson 1050* (CAL). **BHUTAN**: **Deothang**, N Deothang, 1400 m, 18 Oct. 1987, *J. R. I. Wood 5971* (E); **Samchi**, path from Samchi to Dorokha, 1000 m, 23 Nov. 1991, *J. R. I. Wood 7518* (E); **Tashigang**, between Pam and Kanglung, 1100 m, 6 Nov. 1991, *J. R. I. Wood 7495* (E, TI); S Tashiyangtse to Tashigang, 1700 m, 7 Nov.1991, *J. R. I. Wood 7496* (E); **Tongsa**, below Kuenga, Rapten to Shemgang, 1400–1900 m, 28 Oct. 1988, *J. R. I. Wood 6824* (E, FHO, TI); **Wangdi**, ca. 10 km S of Wangdiphodrang in the Sankosh valley, 1100 m, 8 Oct. 1989, *J. R. I. Wood 7193* (E); ca. 20 km S of Wangdiphodrang, 1000 m, 8 Oct. 1989, *J. R. I. Wood 7200* (E); Sarbhang, ca. 10 km north of Sarbhang towards Chirang, 800 m, 29 Oct. 1988, *J. R. I. Wood 6838* (E). **YARLUNG ZANGBO-BRAHMAPUTRA**: **Dirang**, 1500–1800 m, 13 Sep. 1938, *F. Kingdon-Ward 14427* (BM); **Nyam Jang Chu**, Karteng,16 Nov. 1938, *F. Ludlow, G. Sherriff & G. Taylor 6723* (BM, E); Karteng, Tsang Po Valley, 1200–1500 m, 16 Nov. 1938, *F. Ludlow, G. Sherriff & G. Taylor 5743* (E, TI).

26. Strobilanthes divaricata (Nees) T. Anderson, J. Linn. Soc. 9: 478. 1867; Clarke in Hooker (ed.), Fl. Brit. India 4: 468. 1884; Wood in Grierson & Long (eds.), Fl. Bhutan 2(3): 1273. 2001. ≡ *Goldfussia divaricata* Nees, Pl. Asiat. Rar. 3: 89. 1832. ≡ *Diflugossa divaricata* (Nees) Bremek., Verh. Kon. Ned. Akad. Wetensch., Afd. Natuurk., Sect. 2 41(1): 246. 1944. ≡ *Acanthopale divaricata* (Nees) C. B. Clarke, J. Asiat. Soc. Bengal 74(2): 659. 1907. ≡ *Ruellia divaricata* Wall., Numer. List n. 2372. 1830, nom. invalid. Type: Nepal, Higher parts of Nepal towards the Himalaya, 1810, *Wallich Cat. 2372* [lectotype **here designated**: K-W (K001115782); isolectotypes: BM, CGE, GZU (GZU000251597), K].

Subshrubs, 10–50 cm high, strongly anisophyllous; stems glabrous, often zigzag above. Leaves very unequal in each pair, smaller one about one third of larger one in size, larger one shortly petiolate, smaller one subsessile; petiole 0–3 mm long; blade of smaller one ovate, 1.5–6 cm long, 0.5–2 cm broad, larger one broadly lanceolate to lanceolate-elliptic, 7–18 cm long, 2–5.5 cm broad, glabrous, base cuneate or rounded, margin obscurely serrate, apex acute to acuminate. Spikes axillary, several in lax panicles; flowers opposite or sometimes solitary on rachis; peduncles 1–3 cm long, glabrous to sparsely pubescent. Bracts elliptic, 4–7 mm long, 1–2 mm broad, caducous; bracteoles obovate, 1–2 mm long, caducous. Calyx 8–15 mm long, glabrous when young, 5-lobed almost to base; lobes linear-lanceolate, subequal, pilose towards apex, apex acuminate. Corolla deep purple, 3–3.6 cm long, ventricose, glabrous; tube base 9–10 mm long, upwards slightly curved and gradually widened to ca. 1 cm broad at mouth; lobes oblong, 5–5.5 mm long, 5–6 mm broad. Stamens 4, included; filaments of longer pair unequal, longer one 5–6 mm long, shorter one 4–5 mm long, shorter pair equal, 1–1.5 mm long, curved; anthes subspherical, 1–1.2 mm in diameter. Ovary glabrous; 2 ovules per locule; style 2.2–2.5 cm long. Capsules 1.2–1.3 cm long, glabrous, 4-seeded. Seeds ovate in outline, compressed, 2–2.5 mm long and broad, densely pubescent. Figure 46-41.

Phenology: flowering from July to November.

Habitats: moist broad-leaved hill forests; 700–3000 m.

Distribution: C Nepal, E Nepal, Sikkim & Darjiling, Bhutan, Yarlung Zangbo-Brahmaputra (Map 46-76); also in other parts of NE India.

Figure 46-41: **Strobilanthes divaricata** (Nees) T. Anderson: **a**, habit; **b**, bract; **c**, bracteole; **d**, calyx; **e**, corolla. Drawn by Louise OLLEY; reproduced from Wood, *Fl. Bhutan* 2(3): 1269, fig. 104: l-p. 2001.

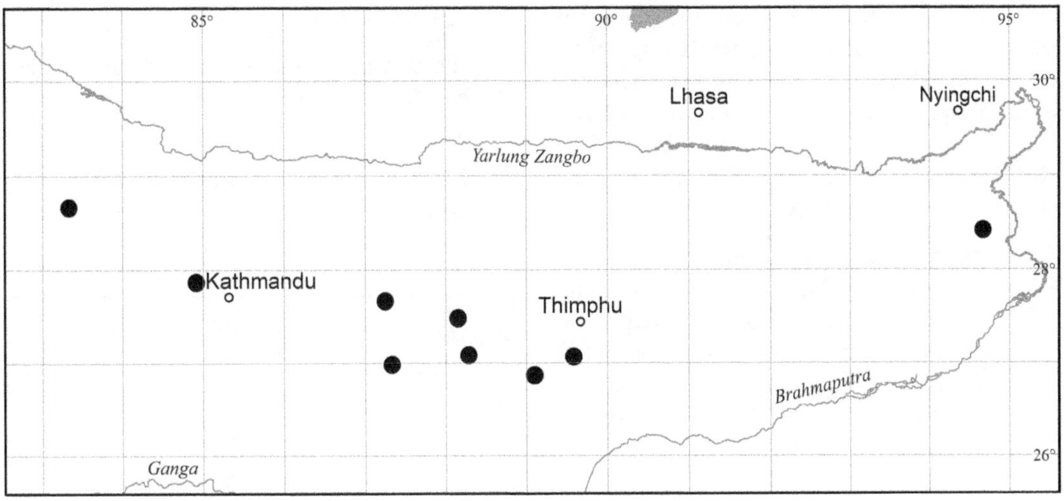

Map 46-76: **Strobilanthes divaricata** (Nees) T. Anderson

C NEPAL: **Bagmati**, between Dhaibunget and Ramche, 1500 m, 15 Sep. 1966, *D. H. Nicholson 2349* (BM); **Dhawalagiri**, Myagdi, Dobang–lapche Kharka–Lipshe–Boghara, 1820–2360 m, 10 Sep. 1996, *M. Mikage et al. 9681329* (A, BM); Myagdi, near Lumsum, 2120 m, 1954, *J. D. A. Stainton, W. R. Sykes & L. H. J. Williams 4274* (BM, KATH, TI); **Gandaki**, Kaski, between Pokhara and Lamjung, 18 Nov. 1953, *J. Fujiamura s.n.* (KYO). **E NEPAL**: **Koshi,** Dhankuta Diatrict,way to Chitray, 2100 m, 1962, *T. B. Shrestha 603* (KATH); Sankhuwasabha, Arun Valley, 2000 m, 29 Aug. 1972, *T. Wraber 142* (BM); Sankhuwasabha, Arun Valley, between Bhotebas and Chichila, 1940 m, 20 Sep. 1991, *D. G. Long et al. EMAK81* (E); Sankhuwasabha, Simbung to Hatiya, 2075 m, 1998, *S. Noshiro et al. 9830041* (KATH); Sankhuwasabha, Simbung to Hatiya, 2020 m, 16 Aug. 1998, *S. Noshiro et al. 9840046* (A, BM, E, KATH); Sankhuwasabha, Manebanjang to Chichila, 1930 m, 28 Jul. 1990, *M. Minaki et al. 9010053* (TI); Sankhuwasabha, Bhotebas, 1710 m, 1981, *P. R. Shakya 6610* (KATH); **Mechi**, Taplejung, between Wallunchun Gola and Pembu Tsoki, 2400 m, 19 Aug. 1962, *K. Nishioka 788* (KYO). **SIKKIM & DARJILING: Darjiling**, s. d., *W. Griffith 6124* (CAL, K, LE); *N. L. Bor 6191* (CAL); Rimbik, 25 Sep. 1962, *B. Safui 1860* (CAL); Kaseong, May 1905, *E. A. C. Uodde 37* (CAL); 1800 m, 30 Aug. 1868, *C. B. Clarke 8893* (CAL, LE); 2100 m, 21 Aug. 1870, *C. B. Clarke 12424*(K); 17 Aug. 1875, *C. B. Clarke 27037* (CAL, K); 1800 m, Sep. 1881, *J. S. Gamble 9860* (K); 2100 m, Jul. 1882, *J. S. Gamble 10571* (CAL, K); Tongboo, 2400–3000 m, 3 Aug. 1862, *T. Anderson 1034* (E); 2100 m, 27 Aug. 1862, *T. Anderson 1035* (E, PE); Scuchal, 2100 m, Sep. 1902, *J. H. Lace 2377* (E, K); 2100 m, 25 Aug. 1874, *D. Treutler C. N. 536* (E, LE); **Sikkim**, Ratong to Yoksum, 750–1500 m, 4 Oct. 1862, *T. Anderson 1061* (E); Khuping, 1650 m, 26 Sep. 1884, *C. B. Clarke 35904* (LE); Khuping, 1650 m, 26 Sep. 1884, *C. B. Clarke 35920* (LE); Goong ridge, 28 Sep. 1857, *T. Thomson s. n.* (BO, CAL, LE); Gangtok, 1350 m, 20 Oct. 1952, *K. Biswas 9765* (CAL); Gangtok, 1500 m, 20 Oct. 1952, *K. Biswas 9766* (CAL); Damthang to Namchi, 2100 m, 21 Sep. 1968, *N. C. Majumdar & R. N. Baneyu 599* (CAL); Saceil, 1950 m, Sep. 1879, *J. S. Gamble 7063*(K); Tonglo, 2700 m, 8 Oct. 1874, *D. Treutler* 847 (K); Lemp, 1200 m, *J. D. Hooker s. n.* (K); Tormpehar Forests, 2100 m, 1891, *H. H. Haines 2750* (K); Tompuhar, 1891, *H. H. Haines 2750* (K). **BHUTAN**: **Chukka**, S Gedu, 2000 m, 11 Nov. 1987, *J. R. I. Wood 6071* (E, FHO); Chapcha River Gorge, 2300 m, 20 Sep. 1992, *J. R. I. Wood 7574* (E, FHO, TI); between Chapcha and Chimakothi, 2200 m, 20

Sep. 1992, *J. R. I. Wood 7577* (E, FHO); between Takti Chu and Asinabari, 1900 m, 20 Sep. 1992, *J. R. I. Wood 7582* (E); **Samchi**, near Sengden Scholl above Samchi, Doya, 1700 m, 1991, *J. R. I. Wood 7516* (E). **YARLUNG ZANGBO-BRAHMAPUTRA**: Tutung, 700 m, 6 Nov. 1958, *R. S. Rao 17351* (CAL).

27. Strobilanthes tenax Dunn, Bull. Misc. Inform., Kew 1920: 208. 1920; Hu, Deng & Wood in Wu, Raven & Hong (eds.), Fl. China 19: 427. 2011. Type: Yarlung Zangbo-Brahmaputra, Abor Hills, Kobo, 140 m, Mar. 1912, *I. H. Burkill 36785* [holotype: K (K000883132); isotype: CAL].

Subshrubs, 0.6–1 m high; stems quadrangular, sulcate, glandular-pubescent. Leaves petiolate or uppermost one sessile; petiole (0–)2–5 mm long; blade lanceolate, 8–10 cm long, 2.5–4 cm broad, glabrous, lateral veins 4–6 pairs, base cuneate, margin serrate, apex acuminate. Panicle of branched spikes, terminal, 3–12 cm long; rachis glandular-pilose. Flowers sessile, opposite, pairs 1–2.5 cm apart on rachis. Bracts variable in shape and size, outer ones sterile, leaf-like, oblong-lanceolate to ovate, 8–12 mm long, 3–5 mm broad, inner ones fertile, oblong-elliptic, 5–7 mm long, 1.8–2 mm broad, scarious, sparsely glandular-pilose, apex obtuse, caducous; bracteoles absent. Calyx 9–11 mm long, abaxially glandular-pubescent, 5-lobed almost to base; lobes linear, apex acute. Corolla 3–3.2 cm long, straight, outside glabrous, inside glabrous except for trichomes retaining style; tube base cylindric, 7–8 mm long, 1–2 mm broad, upwards abruptly widened to ca. 1 cm broad at mouth; lobes broadly ovate, 5–6 mm long, 3–4 mm broad, apex rounded. Stamens 4, included; filaments glabrous, longer pair 3–4 mm long, shorter pair 1–2 mm long; anther-thecae obliquely ellipsoid, 1.2–1.5 mm long, incurved. Ovary obovoid, 2–2.5 mm long, glandular pilose towards apex; 2 ovules per locule; style 1.9–2 cm long, sparsely pilose. Capsules narrowly obovoid, 1.3–1.5 cm long, glandular-pubescent, 4-seeded.

Phenology: flowering from February to March.

Habitats: forests; below 300 m.

Distribution: endemic; Yarlung Zangbo-Brahmaputra (Map 46-77).

YARLUNGZANGBO-BRAHMAPUTRA: Abor Hills, Kobo, 140 m, Mar. 1912, *I. H. Burkill 36785* (K).

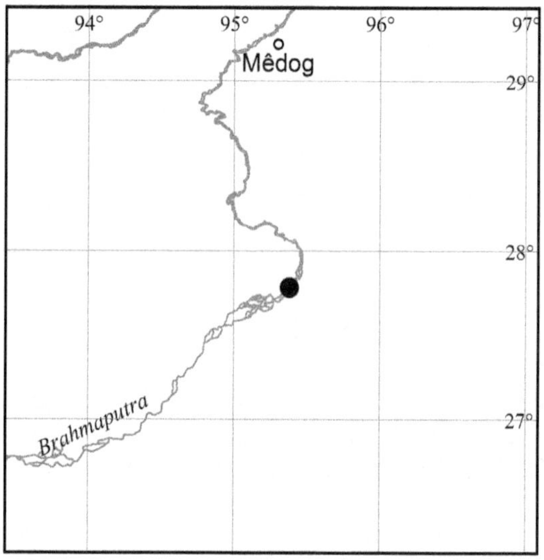

Map 46-77: **Strobilanthes tenax** Dunn

28. Strobilanthes multidens C. B. Clarke in Hooker (ed.), Fl. Brit. India 4: 461. 1895; Wood in Grierson & Long (eds.), Fl. Bhutan 2(3): 1266. 2001; Hu, Deng & Wood in Wu, Raven & Hong (eds.), Fl. China 19: 425. 2011. ≡ *Goldfussia multidens* (C. B. Clarke) Bremek., Verh. Kon. Ned. Akad. Wetensch., Afd. Natuurk., Sect. 2 41(1): 234. 1944. ≡ *Strobilanthes pentstemonoides* (Nees) T. Anderson var. *multidens* (C. B. Clarke) Kuntze, Revis. Gen. Pl. 2: 499. 1891. Type: Bhutan, *W. Griffith 2398* [lectotype designated by Wood, Edinburgh J. Bot. 51: 247. 1994: BM (BM000793161); isotype: K (K000883071)].

Strobilanthes agrestis C. B. Clarke in Hooker (ed.), Fl. Brit. India 4: 466. 1885. ≡ *Pteracanthus agrestis* (C. B. Clarke) Bremek., Verh. Kon. Ned. Akad. Wetensch., Afd. Natuurk., Sect. 2 41(1): 200.1944.

Type: [Sikkim & Darjiling] India, Darjiling, Rishap, 3 Mar. 1871, *C. B. Clarke 13911* [lectotype designated by Wood, Edinburgh J. Bot. 51: 247. 1994: K (K000883070); isolectotype: US].

Strobilanthes agrestis var. *hemiotis* C. B. Clarke in Hooker (ed.), Fl. Brit. India 4: 466. 1885. Type: India, Meghalaya, Khasia, *G. Watt s. n.* [lectotype designated by Wood, Edinburgh J. Bot. 51: 247. 1994: E (E00273490)].

Subshrubs, 1–2 m high, strongly anisophyllous. Leaves unequal in each pair; petiole 0–5 mm long; blade ovate-elliptic, 2–20 m long, 1–9.5 cm long, usually pubescent at least along veins, base attenuate and oblique, margin serrate, apex shortly acuminate and sometimes falcate. Spikes axillary, simple, head-like, trifurcate, sometimes with scattered flower pairs below heads; peduncles glandular-pilose. Bracts ovate-suborbicular, 3–6 mm long, accrescent to 1.3–1.6 cm long in fruit, glandular-pubescent; lobes linear, one lobe 2–3 mm longer than others, apex acute. Corolla blue, 3.5–4.8 cm long, outside glabrous, inside glabrous except for trichomes retaining style; tube base cylindric, 1.6–2 mm broad, upwards widened to 1.2–1.5 cm broad at mouth. Stamens 4; filaments glabrous, longer pair slightly unequal, 5–6 mm long, erect, shorter pair equal, 1–2 mm long, recurved; anther-thecae subspherical, 1–1.2 mm in diameter. Ovary 2–2.5 mm long, glandular-pubescent; 2 ovules per locule; style 2.2–2.5 cm long, sparsely pilose toward base. Capsules oblong, 1.3–1.8 cm long, 2.5–3 mm broad, apically glandular-pilose, 4-seeded. Seeds ovate in outline, compressed, 2.5–3 mm long, 2–2.5 mm broad, appressed pubescent.

Phenology: flowering from September to December.

Habitats: stream banks; 900–2800 m.

Distribution: C Nepal, E Nepal, Sikkim & Darjiling, Bhutan, Yarlung Zangbo-Brahmaputra (Map 46-78); also in other parts of India.

C NEPAL: Gandaki, Kaski, en route from Khare to Birathanti, 5 Oct. 1976, *H. Tabata et al. 6012* (KATH, KYO). **E NEPAL: Koshi**: Sankhuwasabha, Mt. Makalu, above Nhorbu Gaon, 2160 m, 23 Oct. 1971, *M. Ohsawa & P. R. Shakya 2470* (KYO); Sankhuwasabha, Bhainsi Kharta–Danda Kharka to Khongma, 3000 m, 4 Aug. 1990, *M. Minaki et al. 9020615* (TI); Sankhuwasabha, Sedua to Tashi Gaun, 1965 m, 2 Aug. 1990, *M. Minaki et al. s. n.* (TI); **Mechi**, Taplejung, Mul Pokhari to Damhan, 1963, *H. Hara et al. 6303700* (KYO, TI). **SIKKIM & DARJILING**: **Darjiling**, Rungbee, 1500 m, 31 Oct. 1869, *C. B. Clarke 10311* (K); Rarsiong, 1200 m, 29 Feb. 1876, *C. B. Clarke 27050* (K); Jakdah, Black V (Hoom), 1560 m, 5 Dec. 1899, *C. G. Rogers 1* (CAL). **Sikkim**, Sitong, 1500 m, Feb. 1882, *J. S. Gamble 10262* (K); Rishap, 3 Mar. 1871, *C. B. Clarke 13911* (K, US), Champoong, 1800 m, 7 Oct. 1875, *C. B. Clarke 25090* (K). **BHUTAN**: **Chukka**, ridge S of Tala, 1950 m, 22 Feb. 1982, *A. J. C. Grierson & D. G. Long 3141* (K); **Galegphug**, between Chabley Khola and Tama, 1880 m, 30 Mar. 1982, *A. J. C. Grierson & D. G. Long 4140* (K); **Tsongsa**, Tashiling to Tsongsa, 16 Apr, 1967, *H. Kanai et al. s. n.* (KYO). **YARLUNG ZANGBO-BRAHMAPUTRA**: **Dirang**, Rupa to Bomdila, 1700–2710 m, 12 Apr., 1957, *G. Panigrahi 6773* (CAL); from Rupa to Bomdila, 1800 m, 11 May 1957, *G. Panigrahi 7293* (CAL).

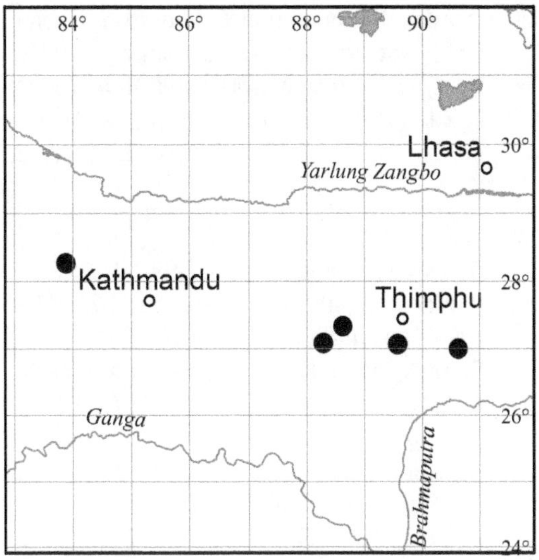

Map 46-78: **Strobilanthes multidens** C. B. Clarke

29. Strobilanthes mastersii T. Anderson, J. Linn. Soc., Bot. 9: 481. 1867, "*mastersi*"; Hu, Deng & Wood in Wu, Raven & Hong (eds.), Fl. China 19: 427. 2011. Type: India, Assam, *Masters s. n.* (lectotype **here designated**: K).

Strobilanthes scoriarum W. W. Sm., Notes Roy. Bot. Gard. Edinburgh 10: 199. 1918; J. R. I. Wood, Edinburgh J. Bot. 51: 223, 258. 1994. ≡ *Goldfussia scoriarum* (W. W. Sm.) Bremek., Verh. Kon. Ned. Akad. Wetensch., Afd. Natuurk., Sect. 2 41(1): 237. 1944. ≡ *Diflugossa scoriarum* (W. W. Sm.) E. Hossain, Notes Roy. Bot. Gard. Edinburgh 32: 406. 1973; Tsui & Hu in Hu (ed.), Fl. Reipubl. Popularis Sin. 70: 172. 2002. Type: [U Irrawaddy] China, Yunnan, Lavabed west of Tengyueh, lat. 25°N, alt. 5000 ft., Aug. 1912, *G. Forests 8862* [holotype: E (E00047320); isotype: PE (PE01620735)].

Strobilanthes schweliensis W. W. Sm., Notes Roy. Bot. Gard. Edinburgh 12: 224. 1920. ≡ *Diflugossa schweliensis* (W. W. Sm.) E. Hossain, Notes Roy. Bot. Gard. Edinburgh 32: 406. 1973. Type: China, Yunnan, Shweili-Salween Divide, 3000 m, lat. 25°20′N, Sep. 1917, *G. Forrest 16107* [lectotype **here designated**: E (E00133609); isotypes: E (E00133608, E00133610), K (K000882688)].

Strobilanthes aborensis Dunn, Bull. Misc. Inform., Kew 1920: 208. 1920. Type: Yarlung Zangbo-Brahmaputra, Outer Abor Hills, Makum, 140 m, *I. K. Burkill 35757* [holotype: K (K000883152); isotype: CAL].

Herbs, 1–1.5 m high, anisophyllous; stems quadrangular, sulcate, glabrous to glandular-pilose. Leaves unequal in each pair, basal ones petiolate or uppermost ones sessile; petiole 0–2 cm long, subglabrous; leaf blade ovate to ovate-lanceolate, 10–15 cm long, 4–7 cm broad, glabrous, lateral veins 5–10 pairs, base broadly cuneate on basal leaves and rounded on apical leaves, margin serrate and ciliate, apex acuminate. Panicles of spikes terminal, narrowly ovoid, 15–30 cm long; peduncle and rachis glabrous when young, densely glandular-pilose at maturity; flowers opposite pairs or solitary through abortion. Bracts elliptic, 5–6 mm long, 2–3 mm broad, caducous, apex obtuse; bracteoles oblong, 2.5–3 mm long, 0.8–1 mm broad, abaxially glabrous, caducous. Calyx 0.8–1.2 cm long, subglabrous with sessile glands when young, densely glandular-pilose and accrescent in fruit, 5-lobed almost to base; lobes lanceolate, unequal, 3 lobes slightly longer than other 2, apex acuminate. Corolla purple, 2.5–3 cm long, straight, ventricose, outside glabrous, inside glabrous except for trichomes retaining style; tube base cylindric, 1.6–2 mm broad, 1–1.3 cm long, upwards widened to 1–1.2 cm broad at mouth; lobes orbicular, 4.5–5 mm in diameter. Stamens 4, included; filaments glabrous, longer pair 5–6 mm long, shorter pair 2–3 mm long; anther-thecae subspherical, 0.8–1 mm long in diameter, incurved. Ovary slightly pilose towards apex; 2 ovules per locule; style 2–2.2 cm long, sparsely pilose towards base. Capsules clavate, 1.2–1.5 cm long, glandular-pilose toward apex, 4-seeded. Seeds brown, ovate in outline, 2–2.5 mm long, 1.5–2 mm broad; areole small, verrucose.

Phenology: flowering from September to January of next year.

Habitats: forests on mountain slopes; 100–3000 m.

Distribution: Yarlung Zangbo-Brahmaputra, S Hengduan, U Irrawaddy (Map 46-79); also in other parts of India, Myanmar and China (SW Yunnan).

YARLUNG ZANGBO-BRAHMAPUTRA: Abor Hills, Makum, 140 m, 21 Nov. 1911, *Burkill 35757* (K); Dihang Valley, 350–700 m, 12 Feb. 1928, *F. Kingdon-Ward 7855* (K); **Baibung**, 950 m, *H. Sun et al. ETM-1060* (KUN); **Mêdog**, near Mêdog, 1100 m, *H. Sun et al. 2910* (KUN); near Mêdog, 1400 m, 17 Jan. 1997, *H. Sun et al. ETM-2985* (KUN); Dangmu to Miri, 1200 m, 3 Sep. 1974, *Exped. Qinghai-Xizang 74-5022* (KUN, PE); Damor, 1300 m, 11 Aug. 2011, *South Tibet Expedition Team STET2538* (PE). **S HENGDUAN**: **Lushui**, Pianma, 1750 m, *Exped. Gaoligongshan 7237* (KUN); **Weixi**, 2800 m, *C. W. Wang 67606* (PE). **U IRRAWADDY**: **Tengchong**, W of Mekong, Pingpo to Youngchang and Tengyueh, Oct. 1922, *J. F. Rock 7044* (E, US); W Tengyueh, 1600 m,

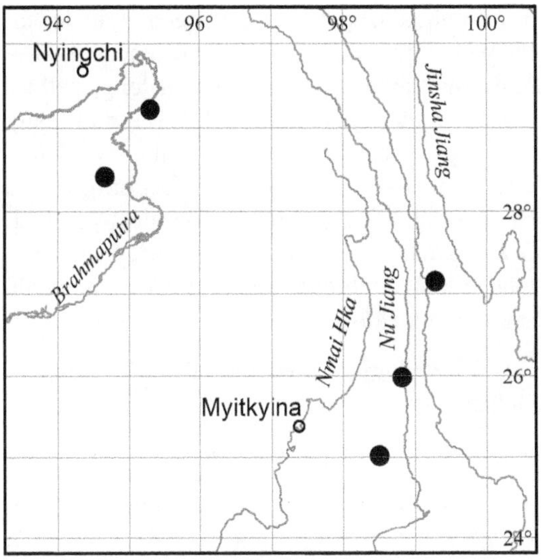

Map 46-79: **Strobilanthes mastersii** T. Anderson

G. Forrest 8862 (E, PE); Shweli-Salween Divide, 3000 m, Sep. 1917, *G. Forrest 16107* (E, K).

30. Strobilanthes laevigata C. B. Clarke in Hooker (ed.), Fl. Brit. India 4: 467. 1884. Type: [Sikkim & Darjiling] India, Sikkim, *J. D. Hooker s. n.* [lectotype **here designated**: K (K000883235); isolectotype: K (K000883234)].

Shrubs, anisophyllous; stems quadrangular, glabrous. Leaves unequal in each pair; petiole 1–2 cm long, glabrous; blade narrowly elliptic, 13–18 cm long, 4–6 cm broad, glabrous, lateral veins 6–8 pairs, base narrowly cuneate to attenuate, margin serrate, apex acuminate. Panicles terminal and axillary. Bracts and bracteoles caducous, unknown. Calyx 0.8–10 mm long, sparsely pilose with scattered stalked glands, 5-lobed almost to base; lobes subequal, somewhat enlarged in fruit, narrowly lanceolate, apex subobtuse. Corolla 3.5–4 cm long, outside glabrous, inside glabrous except for hairs retaining style; tube base cylindric, 1.2–1.5 cm long, 1.5–2 mm broad, upwards gradually widened to 1–1.2 cm broad at mouth; lobes ovate, 4–5 mm long, 4–5 mm broad, apex emarginate. Stamens 4, included; filaments of longer pair subunequal, 6–8 mm long, shorter pairs 2–3 mm long, recurved; anther-thecae subspherical, 1–1.3 mm in diameter. Ovary comose towards apex; 2 ovules per locule; style 2.8–3.2 cm long, hirsute at base. Capsules 1.2–1.5 cm long, glandular-pubescent towards apex.

Phenology: unknown.

Habitats: forests.

Distribution: endemic; Sikkim & Darjiling (Map 46-80).

Notes: Both Bremekamp (1944) and Wood (1994) have mistakenly treated *Strobilanthes laevigata* as a synonym of *S. hamiltoniana*. However, it is quite different from *S. hamiltoniana* in the flowers opposite on rachis. It is similar to *S. angustifrons*, but differs by caducous bracts and bracteoles.

SIKKIM & DARJILING: **Sikkim**, *J. D. Hooker s. n.* (K).

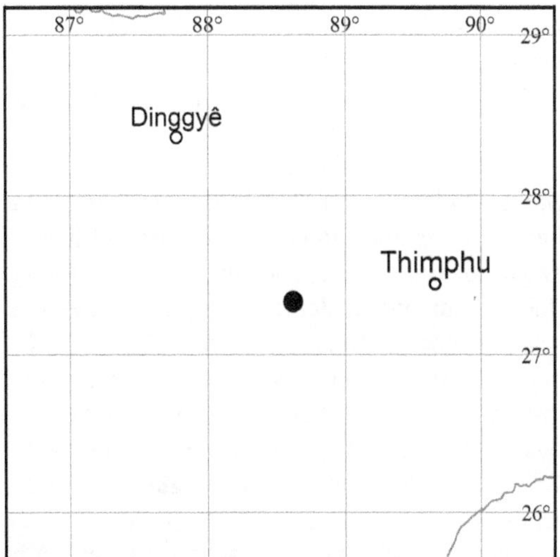

Map 46-80: **Strobilanthes laevigata** C. B. Clarke

31. Strobilanthes glomerata (Nees) T. Anderson, J. Linn. Soc., Bot. 9: 475. 1867; Clarke in Hooker (ed.), Fl. Brit. India 4: 448. 1884; Hu, Deng & Wood in Wu, Raven & Hong (eds.), Fl. China 19: 420. 2011. ≡ *Goldfussia glomerata* Nees, Pl. Asiat. Rar. 3: 88. 1832; Tsui & Hu in Hu (ed.), Fl. Reipubl. Popularis Sin. 70: 164. 2002. ≡ *Ruellia glomerata* Wall., Numer. List n. 2361. 1830, nom. invalid. Type: Bangladesh, Mt. Syllet, *F. De Silva in Wallich Cat. 2361* [lectotype designated by Bennett & Scotland, Kew Bull. 58: 43. 2003: K-W (K001115731); isolectotype: GZU (GZU000251600)].

Goldfussia glomerata Nees var. *speciosa* Hook., Bot. Mag. 80: t. 4767. 1854; Bremekamp, Verh. Kon. Ned. Akad. Wetensch., Afd. Natuurk., Sect 2 41(1): 230. 1944. Type: [*icon*] Hooker, Curtis's Bot. Mag. 80: t. 4767. 1854 (lectotype **here designated**).

Subshrubs, 0.6–1 m high, strongly anisophyllous; stems and branches quadrangular to subterete, setose. Leaves very unequal in each pair, smaller one about 1/5 of larger one in size; petiole 0.6–3 cm long, densely pubescent; blade elliptic, broadly elliptic or narrowly ovate, 5.5–19 cm long, 2.7–10 cm broad, both surfaces densely purplish pubescent, lateral veins 5–7 pairs, base oblique, rounded and decurrent onto petiole, margin serrate and reddish ciliate, apex acuminate. Spikes capitate, covered with purple multicellular hairs or glandular hairs. Bracts variable in size and shape, outer ones lanceolate, 2.5–3 cm long, 0.5–0.7 cm broad, densely purplish pubescent, margin serrate, inner ones lanceolate, 1.6–1.8 cm long, 0.4–0.5 cm broad, margin entire and ciliate, apex acuminate; bracteoles lanceolate, 8–12 mm long, 1.8–2 mm broad, pubescent. Calyx 1.2–1.5 cm long, 5-lobed almost to base; lobes 10–13 mm long, 1–1.5 mm broad, slightly unequal, abaxially purplish pubescent especially towards apex, adaxially glabrous. Corolla purple, 4–6 cm long, subventricose, outside pubescent with glandular-pilose, inside glabrous except for trichomes retaining style; tube base cylindric, 0.6–1.6 cm long, 1.5–2 mm broad, upwards bent to about 90° and gradually widened to 1.5–1.7 cm broad at mouth; lobes ovate, 4.5–5.5 mm long, 6–7 mm broad, apex retuse. Stamens 4, included; filaments glabrous, longer pair unequal, 7–9 mm long, shorter pair 1.5–2 mm long; anther-thecae obliquely ellipsoid, 1.2–1.5 mm long. Ovary glabrous; 2 ovules per locule; style 2.6–3 cm long, glabrous. Capsules fusiform, 1–1.2 cm long, 2.5–3 mm broad, glabrous, 4-seeded. Seeds ovate in outline, compressed, 2–2.2 mm long, 1.8–2 mm broad, densely pubescent; areole small. Figure 46-42.

Phenology: flowering from August to October.

Habitats: forests; ca. 300 m.

Distribution: Sikkim & Darjiling, Yarlung Zangbo-Brahmaputra (Map 46-81); also in Bangladesh, and other parts of India.

SIKKIM & DARJILING: **Sikkim**, Tista Valley, 21 Jan., 1908, *Riba 401* (CAL). **YARLUNG ZANGBO-BRAHMAPUTRA**: **Arbor Hills**, 300 m, *F. Kingdom-Ward 8751* (K); Kobo, 1 Dec. 1911, *I. H. Burkill 35953* (CAL).

32. Strobilanthes kingdonii J. R. I. Wood, *Edinburgh J. Bot.* 51: 244. 1994; Hu, Deng & Wood in Wu, Raven & Hong (eds.), Fl. China 19: 421. 2011. Type: Yarlung Zangbo-Brahmaputra, Lohit Valley, Kahao, 1220–1525 m, 16 Nov. 1926, *F. Kingdon-Ward 7654* (holotype: K).

Subshrubs, 0.5–3 m high, branched, slightly anisophyllous; stems quadrangular, bifariously pilose, apically glabrescent. Petiole 0.5–2 cm long, pubescent; leaf blade ovate, 5–12 cm long, 1.5–5 cm broad, both surfaces pubescent, lateral veins 5–8 pairs, base attenuate, margin crenate to serrate, apex acute to acuminate and slightly falcate. Spikes axillary, head-like, usually simple or (2- or) 3-furcate, isolated flowers sometimes present below main head; peduncle 1–3 cm long, pilose; rachis sparsely pilose. Bracts variable in size and shape, outer ones lanceolate to narrowly oblong-elliptic, 1.3–2.7 cm long, 0.3–0.8 cm broad, pubescent, margin dentate, inner ones ovate, ovate-elliptic or subrhomboidal, 7–10 mm long, 2–4 mm broad, persistent, abaxially ciliate and eglandular white pubescent, apex obtuse; bracteoles linear-oblanceolate, 4–8 mm long, 1.5–2 mm broad, margin ciliate and eglandular

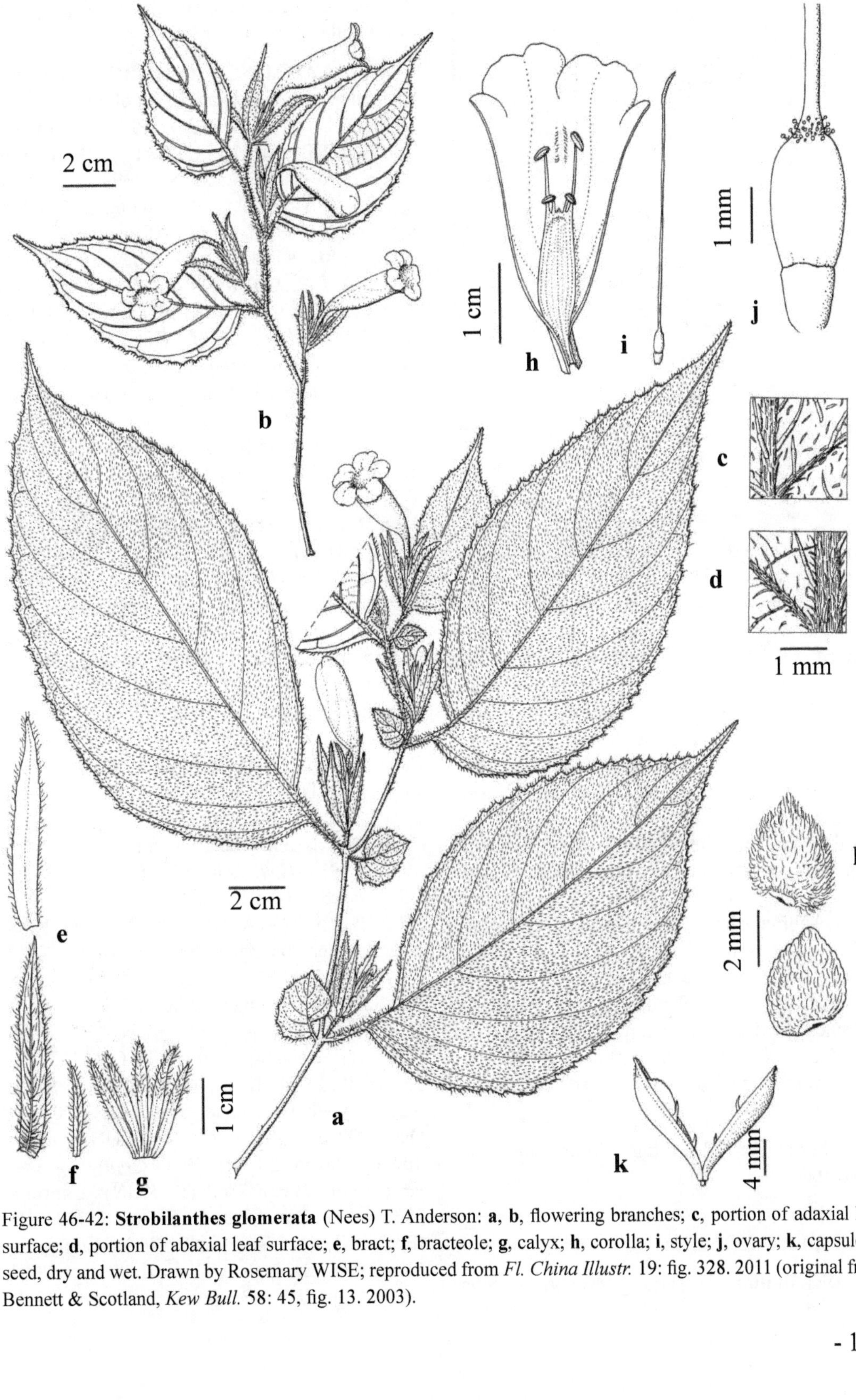

Figure 46-42: **Strobilanthes glomerata** (Nees) T. Anderson: **a**, **b**, flowering branches; **c**, portion of adaxial leaf surface; **d**, portion of abaxial leaf surface; **e**, bract; **f**, bracteole; **g**, calyx; **h**, corolla; **i**, style; **j**, ovary; **k**, capsule; **l**, seed, dry and wet. Drawn by Rosemary WISE; reproduced from *Fl. China Illustr.* 19: fig. 328. 2011 (original from Bennett & Scotland, *Kew Bull.* 58: 45, fig. 13. 2003).

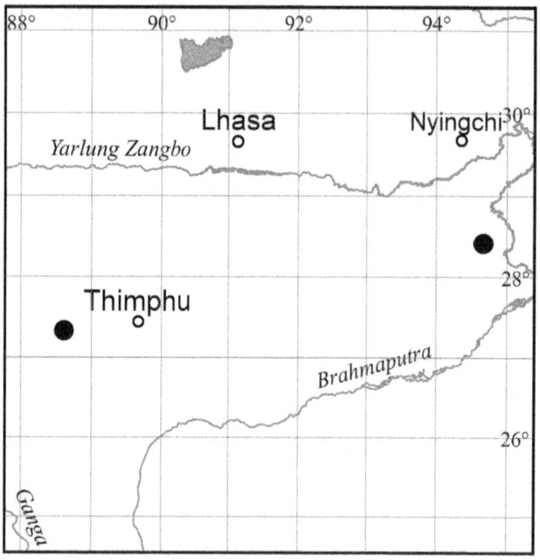

Map 46-81: **Strobilanthes glomerata** (Nees) T. Anderson

white pubescent or sometimes with stalked glands toward apex, apex obtuse to acute. Calyx 0.8–1.0 cm long, abaxially densely white eglandular and glandular pilose, adaxially glabrous, subequally 5-lobed to base; lobes narrowly oblong to lanceolate,7–9 mm long, 1.2–1.5 mm broad, apex acute. Corolla purple, blue, or rose, 3.5–4.2 cm long, straight, outside pilose, inside glabrous except for trichomes retaining style; tube base cylindric, 8–10 mm long, 1.5–2 mm broad, upwards ventricose and widened to 1–1.2 cm broad at mouth; lobes broadly ovate, 6–7 mm long, 5–6 mm broad, apex rounded. Stamens 4, included; filaments of longer pair 6–8 mm long, pilose, shorter pair 1.5–3.5 mm long, glabrous; anther-thecae ellipsoid, 1–1.2 mm long, incurved. Ovary glandular-pilose towards apex; 2 ovules per locule; style 1.7–1.9 cm long, sparsely pilose. Capsules oblong, 1–1.2 cm long, glandular-pilose, 4-seeded. Seeds ovate in outline, compressed, 2.5–3 mm long, 1.8–2 mm broad, pubescent. Figure 46-43.

Phenology: flowering from July to November.

Habitats: evergreen broad-leaved forests by streams; 1200–2800 m.

Distribution: U Yarlung Zangbo, Yarlung Zangbo-Brahmaputra, S Hengduan (Map 46-82); also in other parts of N Myanmar.

U YARLUNG ZANGBO: Gyirong, Zhaxiangresuo to Lexianglabi, 2600 m, 31 Aug. 1990 *B. S. Li et al. 13317* (PE). **YARLUNG ZANGBO-BRAHMAPUTRA: Bamgxing**, Pangxin, Gemendeng, 26 Nov. 1982, *s. coll. s. n.* (PE); **Dexing**, Dege, Wunagongba, 160–800 m, 6 Feb. 1983, *B. S. Li & S. Z. Cheng 2763* (PE); **Luhit**, Kahao, Zayul Valley, 1525 m, 20 Jul. 1926, *F. Kingdon-Ward 7158* (K); Lohit Valley, Kahao, 1220–1525 m, 16 Nov. 1926, *F. Kingdon-Ward 7654* (K); Lohit Valley, 610 m, 18 Mar. 1933, *F. Kingdon-Ward 10313* (BM); **Mêdog**, Jialasha, 1200 m, 19 Feb. 1993, *H. Sun et al. ETM-3537* (KUN); 1600 m, 5 Sep. 1990, *G. Yao et al. 3903* (NAS); near Jialasa, 1250 m, 13 Nov. 1982, *S. Z. Cheng & B. S. Li 1767* (PE); Darmo, 1200 m, 14 Nov. 2019, *Y. F. Deng 28930* (IBSC); Darmo, 880 m, 14 Nov. 2019, *Y. F. Deng 28953* (IBSC); **Xiazayü**: Cibagou, 1500 m, 15 Sep. 2012, *FLPH Tibet Expedition 12-1242* (PE); way to Shage, 1500 m, 29 Aug. 1983, *B. S. Li et al. 6965* (PE); near Xiachayu Farm, 4 Sep. 1962, *C. Y. Wu 5485* (KUN); **Zayü**: Shama, 1700 m, 22 Jul. 1973, *J. W. Zhang 0874* (PE); Cibagou Natural Reserve, *Southeast Tibet Expedition Team SET-ET 757* (IBSC); Gedui, 1700 m, 2 Sep. 2003, *X. F. Gao et al. 6844* (CDBI); Gedui, 1700 m, 2 Sep. 2003, *X. F. Gao et al. 6845* (CDBI); Shage, 1650 m, 28 Aug. 1983, *B. S. Li et al. 6924* (PE). **S HENGDUAN**: **Fugong**, 1500 m, 10 Sep. 1934, *H. T. Tsai 54221* (A, IBSC, KUN, LBG, NAS, PE, SZ); 2800 m, 20 Oct. 1934, *H. T. Tsai 58806* (A, IBSC, KUN, LBG, NAS, PE, SZ); near Shangpa, 2000 m, 23 Oct. 1934, *H. T. Tsai 58909* (A, IBSC, KUN, LBG, NAS, PE, SZ); near Shangpa, 2000 m, 24 Oct. 1934, *H. T. Tsai 58956* (A, IBSC, KUN, LBG, NAS, SZ); near Shangpa, 2000 m, 31 Oct. 1934, *H. T. Tsai 59125* (A, IBSC, KUN, LBG, NAS, PE, SZ); Shiyueliang, Meluo, Shiluowa, 1400 m, 6 Oct. 2007, *H. Li & R. Li 1150* (KUN); Fugong Town, 1300 m, 20 Oct. 1996, *Gaoligong Shan Expedition (1996) 7862* (E, KUN); **Lanping**, 2800 m, 13 Oct. 1933, *H. T. Tsai 56229* (GH, IBSC, KUN, LBG, NAS, PE); **Longyang**, Mangkuan, Baihualing, Jiujiezi, 8 Sep. 2003, *Gaoligong Shan Biodiversity Survey 18885*

Figure 46-43: **Strobilanthes kingdonii** J. R. I. Wood: **a**, habit; **b**, bract; **c**, bracteole; **d**, calyx; **e**, opened corolla showing androecium and pistil. Drawn by Tai-Li ZHANG; reproduced from *Fl. Reipubl. Popularis Sin.* 70: 175, pl. 26. 2002.

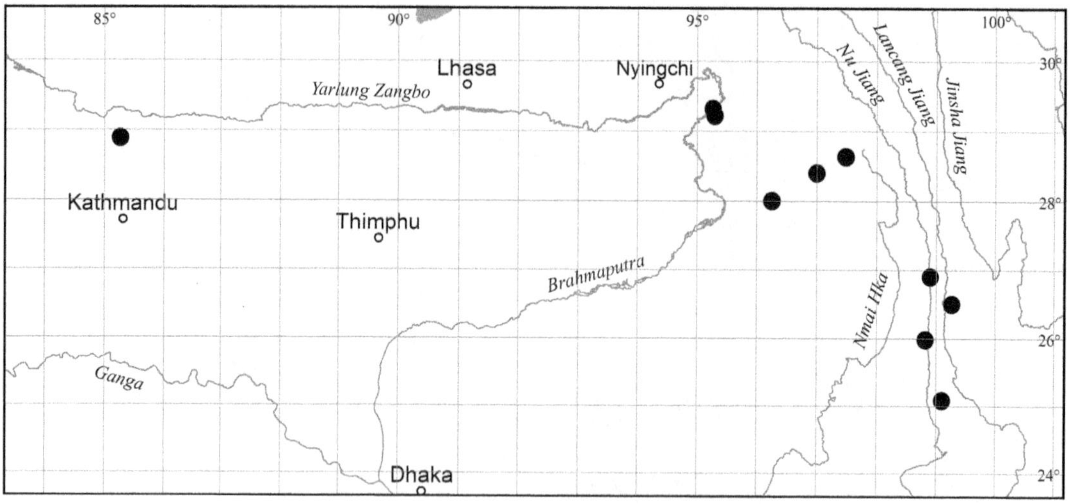

Map 46-82: **Strobilanthes kingdonii** J. R. I. Wood

(KUN); Baihualing, 24 Jan. 2008, *Y. F. Deng 20488* (IBSC); Baihualing, 12 Sep. 2013, *Y. F. Deng 24736* (IBSC); **Lushui**, Luzhangsha, 1950 m, 13 Sep. 1996, *Exped. Gaoligongshan 7075* (KUN); Luzhangshan, 1750 m, 20 Oct. 1996, *Exped. Gaoligongshan 7106* (KUN); Yaojiaping, 2180 m, 13 Sep. 1996, *Exped. Gaoligongshan 7208* (KUN); Luzhang, 1950 m, 13 Sep. 1996, *Exped. Gaoligongshan 7087* (KUN); Luzhang, 13 Sep. 2013, *Y. F. Deng 24843* (IBSC).

33. Strobilanthes capitata (Nees) T. Anderson, J. Linn. Soc., Bot. 9: 475. 1867; Clarke in Hooker (ed.), Fl. Brit. India 4: 459. 1884; Wood in Grierson & Long (eds.), Fl. Bhutan 2(3): 1263. 2001; Hu, Deng & Wood in Wu, Raven & Hong (eds.), Fl. China 19: 420. 2011. ≡ *Goldfussia capitata* Nees, Pl. Asiat. Rar. 3: 88. 1832; Tsui & Hu in Hu (ed.), Fl. Reipubl. Popularis Sin. 70: 162. 2002. ≡ *Ruellia capitata* Wall., Numer. List n. 2351. 1830, nom. invalid. Type: Nepal, 1821, *N. Wallich 2351* [lectotype **here designated**: K-W (K001115708); isolectotypes: BM (BM000793162), GZU (GZU000251594, GZU000251595), K (K000883084), LE].

Goldfussia biceps Nees, Pl. Asiat. Rar. 3: 88. 1832. ≡ *Strobilanthes biceps* (Nees) Miq., Fl. Ned. Ind. 2: 804. 1858. Type: Myanmar, Taong Dong, 27 Nov. 1826, *Wallich Cat. 7161* [lectotype **here designated**: K-W (K001126806); isolectotype: GZU (GZU000250395)].

Strobilanthes capitata var. *nitida* C. B. Clarke in Hooker (ed.), Fl. Brit. India 4: 459. 1884. Type: India, Meghalaya, Morsto, Khasia, 2500 ft., 10 Nov. 1877, *C. B. Clarke 18153* (holotype: K; isotype: CAL).

Subshrubs or perennial herbs, 0.5–1 m high, anisophyllous; stems decumbent at lower portion, ascending upwards, pilose or glabrescent. Leaves unequal in each pair; petiole 1–3 cm long; blade lanceolate, elliptic-ovate, or elliptic, 4–14 cm long, 1.5–9 cm broad, abaxially pubescent along veins, adaxially sparsely pilose, lateral veins 6–8 pairs, base attenuate and oblique, margin serrate, apex acuminate. Panicles of pedunculated head-like spikes axillary, heads 1–3, ovoid to ellipsoid, 1.5–2 cm long, 1–3 cm broad; peduncle 1–6 cm long, glabrous or pubescent. Bracts heteromorphic, outer ones sterile, ovate, 1.5–2.3 cm long, not overtopping capitula, concave, caducous at anthesis, pubescent or glabrous, margin crenate, apex obtuse, acute, or acuminate; inner ones fertile, oblong-elliptic, 1.3–1.6 cm long, margin entire or with 1 or 2 teeth; bracteoles oblong-elliptic, 7–11 mm long, pilose, caducous. Calyx 1–1.2 cm long, accrescent to 1.5–2 cm long in fruit, abaxially pubescent, 5-lobed almost to base; lobes linear-oblanceolate, equal, margin ciliate. Corolla blue, 4.3–5.5 cm long, straight, outside sparsely pilose, inside glabrous except for trichomes retaining style; tube base cylindric, 1.2–1.5 cm long, 1.5–2 mm broad, upwards gradually widened to 1.2–1.5 cm broad at mouth; lobes oblong, 5–6 mm long, 4–5 mm broad. Stamens 4, included; filaments glabrous, longer pair 7–8 mm long, erect, unequal, shorter pair 1.5–2 mm long, nodding; anther-thecae spherical, 1–1.2 mm in diameter, incurved. Ovary glandular-pilose; 2 ovules per locule; style 3.2–3.5 cm long, sparsely pilose. Capsules oblong, 1.4–2 cm long, apical half glandular-pilose, 4-seeded. Seeds ovate in outline, 2.5–3 mm long, 2.2–2.5 mm broad, pubescent; areole small. Figure 46-44.

Phenology: flowering from September to February.

Habitats: forests; 100–3400 m.

Distribution: U Ganga & Indus, W Nepal, C Nepal, E Nepal, Sikkim & Darjiling, Bhutan, Yarlung Zangbo-Brahmaputra (Map 46-83); also in other parts of India and Myanmar.

U GANGA & INDUS: Nainital, above Kathgodam, 900 m, 5 Oct. 1920, *W. Dudgeon & L. A. Kenoyer 492* (MO). **W NEPAL: Rapti**, Dang, Koilabas, 760 m, 20 Nov. 2012, *B. Adhikari & D. R. Kandel 66* (E); **Seti**, Bajhang, Agar, 1950 m, 23 Aug. 1991, *M. Suzuki et al. 9192519* (E, TI); Bajhang, Agar, 1390 m, 23 Aug. 1991, *M. Suzuki et al. 9160970* (E). **C NEPAL**: **Bagmati**, Kathmandu, Naubise to Mugling, 5 Nov. 2000, *T. Watanabe et al. NPL-SP001105 (048)* (MBK); **Gandaki**, Ghorka, above Jagate, 1360 m, 13 Oct. 2006, *K. Adhikari 490* (TUCH); Gorka, Ngyak to Den, 5 Oct. 1963, *T. Namba 1005030* (KYO); Kaski, en route from Kurdi to Chhumro, 2200–3050 m, 3 Oct. 1976, *Y. Suehiro 1461* (KYO); Kaski, en route from Kurdi to Chhumro, 2200–3050 m, 3 Oct. 1976, *Y. Suehiro 2577* (KYO); Kaski, Pokhara, 800 m, 10 Nov. 2008, *A. Sukhorukov 69* (E); Kaski, Mardi Khola, 1670 m, 16 Sep. 1954, *J. D. A. Stainton, W. R. Sykes & L. H. J. Williams 8453* (BM, KATH, TI); Kaski, Begnas Tal, 910 m, 23 Oct. 1981, *J. D. A. Stainton 8484* (E); Kaski, between Pisapur and Tapura, 1600 m, 27 Sep. 1953, *T. Fujimura 80* (KYO); Manang, Marsyangdi River, Between Bahundara and Khudi, 2 Nov. 1969, *T. Wraber 612* (BM); Syangja, Kali Gandaki, 500 m, below Andhi Khola, 500 m, 5 Oct. 1991, *K. A. Lye 17539* (E); Tanahu, Naya Pul, 410 m, 27 Nov. 2012, *B. Adhikari & D. R. Kandel 98* (E); Tanahu, Ghansi Kuwa, 580 m, 27 Nov. 2012, *B. Adhikari & D. R. Kandel 99* (E); Tanahu, Ghansi Kuwa, Palim, 530 m, 28 Nov. 2012, *B. Adhikari & D. R. Kandel 103* (E); **Janakpur**, Sindhuli, Napani, 1240 m, 27 Nov. 1988, *N. P. Manandhar 12611* (KATH); **Lumbini**, Nawalparasi, Daunne Danda, 400 m, 18 Dec. 1975, *D. P. Joshi et al. 75/3586* (KATH); Palpa, Masure, Bhairabsthan, 1276 m, 13 Nov. 2012, *B. Adhikar 51* (E); Palpa, Marek, near Butwal, 610 m, 6 Oct. 1954, *J. D. A. Stainton, W. R. Sykes & L. H. J. Williams 8769* (BM, E); **Narayani**, Bara, Near Badahikim Bridge, 670 m, 27 Nov. 2004, *T. MillerA169* (E, KATH); Chitwan, Sauraha, Sauraha to Devi Tal, 130–160 m, 21 Jan. 1996, *M. Migage et al. 9614248* (E, KATH, TI); Chitwan, Tiger Top, Chitwan National Park, 300 m, 12 Dec. 1975, *D. P. Joshi et al. 75/3500* (KATH); Makwanpur, Jyamire, Bhainse, from Hetauda to Daman, 1060 m, 28 Nov. 2004, *T. Miller et al. A194* (E, KATH); Makwanpur, Saune Khola, Bhainse to Daman, 770 m, 7 Nov. 2012, *B. Adhikari & G. D. Bhatt 8* (E); Makwanpur, between Birganj and Hettoura, Bichiakoh pass, 30 Nov. 1907, *I. H. Burkill 29515* (CAL); Makwanpur, near Hettounda, 30 Nov. 1907, *I. H. Burkill 29530* (CAL); Suparifar, 1 Dec. 1907, *I. H. Burkill 29564* (CAL); Makwanpur, Suparifar to Pancan, 1 Dec. 1907, *I. H. Burkill 29565* (CAL). **E NEPAL**: **Koshi**, Dhankuta, 1200 m, 19 Oct. 1963, *H. Hara et al. 6303703* (BM, KATH, KYO, TI); Dhankuta, 2100 m, 21 Oct. 1963, *H. Hara et al. 6303711* (A, KATH, KYO, TI); Dhankuta, 1200 m, 20 Oct. 1963, *H. Hara et al. 6303714* (TI); Dhankuta, Pakhribas to Diyale, 1390 m, 16 Sep. 1981, *P. R. Shakya 6522* (KATH); Sankhuwasabha, Tumlingtar to Khandbari, 660 m, 21 Sep. 1981, *P. R. Sharya 6569* (KATH); Sankhuwasabha, Below Khandbari, 900 m, 22 Sep. 1981, *P. R. Shakya 6582* (KATH); Sankhuwasabha, Hedangna, 1120 m, 11 Oct. 1981, *P. R. Shakya 7012* (KATH); Sankhuwasabha, below Jimigaon, 750 m, 8 Nov. 1981, *P. R. Shakya 7357* (KATH); Sankhuwasabha, below Khandbari, 1060 m, 29 Nov. 1990, *P. R. Shakya 9369* (KATH); Sankhuwasabha, Bareni, 480 m, 1 Dec.

Figure 46-44: **Strobilanthes capitata** (Nees) T. Anderson: **a**, portion of plant; **b**, outermost bract; **c**, outer bract; **d**, inner bract; **e**, bracteole; **f**, calyx; **g**, dissected corolla. Drawn by Louise OLLEY; reproduced from Wu, Raven & Hong (eds.), *Fl. China Illustr.* 19: fig. 335: 6–12. 2012 (original from Wood in Grierson & Long (eds.), *Fl. Bhutan* 2(3): 1267, fig. 103: f–l. 2001).

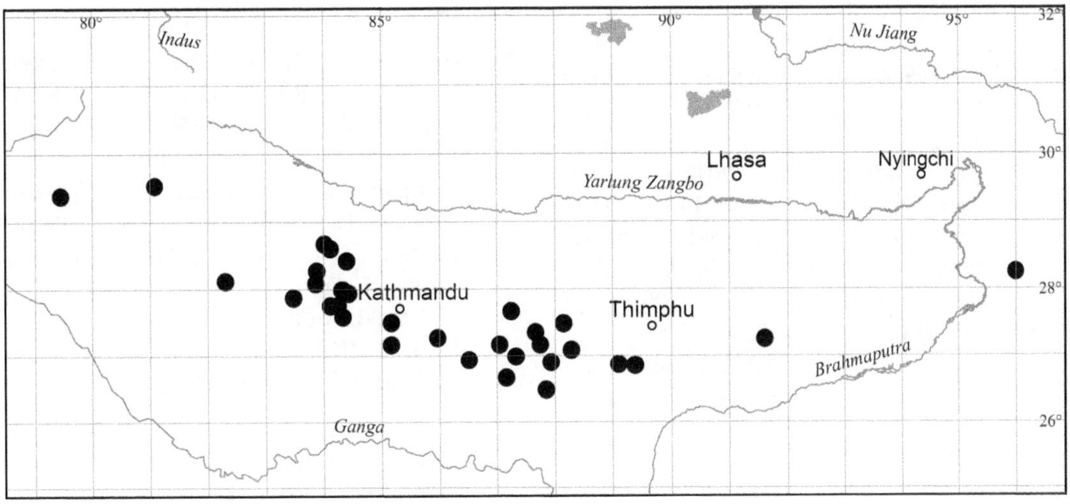

Map 46-83: **Strobilanthes capitata** (Nees) T. Anderson

1990, *P. R. Shakya 9376* (KATH); Sankhuwasabha, Dhyaksinda, 860 m, 21 Dec. 1991, *P. R. Shakya 9535* (KATH); Sankhuwasabha, Arun Valley, between Khandbari and Bhotebas, 1140 m, 18 Sep. 1991, *D. G. Long et al. EMAK11* (E); Sankhuwasabha, Mani Bhanjyang, 1260 m, 26 Sep. 2001, *D. G. Long et al. 68* (E, TI); Sankhuwasabha, between Manebhanjyang and Bhote Bas, 1260 m, 26 Sep. 2001, *D. G. Long ENEP68* (TI); Sankhuwasabha, Tumlingtar, R. Arun, 540 m, 15 Dec. 1961, *A. H. Norkett 8072* (BM, TI); Sankhuwasabha, Tumlingtar, R. Arun, 540 m, 25 Dec. 1961, *A. H. Norkett 8134* (BM, KATH, TI); Sansari, 1410 m, 3 Dec. 1990, *P. R. Shakya 9402* (KATH); **Mechi**, Ilam, Ranga Pani to Ghorwa, 9 Dec. 1963, *H. Hara et al. 6303705* (A, BM, E, KATH, KYO, TI); Ilam, Jog Mai to Rang Pank, 8 Dec. 1963, *H. Hara et al. 6303706* (BM, E, KATH, KYO, TI); Ilam, Dhara Pani, 5 Dec. 1963, *H. Hara et al. 6303708* (BM, E, KATH, KYO, TI); Ilam, 1200 m, 6 Dec. 1963, *H. Hara et al. 6303709* (BM, K, KYO, TI); Ilam, Ghorwa to Sanichare, 10 Dec. 1963, *H. Hara et al. 6303710* (KATH, KYO, TI); Ilam, Jog Mai to Ranga Pana, 8 Dec. 1963, *H. Hara et al. 6304215* (KYO); Ilam, Chulachuli, 540 m, 22 Sep. 1972, *T. B. Shrestha & T. K. Bhattacharya 72-97* (KATH); Ilam, Chulachuli, 360 m, 22 Sep. 1972, *T. B. Shrestha & T. K. Bhattacharya 72-121* (KATH); Jhapa, Chulachuli, N of Chulachuli, 1210 m, 15 Sep. 1967, *L. H. J. Williams & Staintan 8520* (BM, K, TI); Panchthar, Memeng, 2390 m, Narelung Community Forests, 29 Sep. 2007, *s. coll. 287* (E); Panchthar, Tharbu to near Chyangthaphu, 26 Nov. 1963, *H. Hara et al. 6303707* (A, BM, E, KATH, KYO, TI); Panchthar, Bharomdin to Tharpu, 25 Nov. 1963, *H. Hara et al. 6303713* (TI); Taplejung, Bagang Khola, 3330 m, 11 Nov. 1971, *L. W. Beer et al. 10746* (BM); Taplejung, Dumhan, Tamur River, 700 m, 31 Oct. 1963, *G. Murata et al. 6303701* (KATH, KYO, TI, TNS); Taplejung, Taplejung to Garhi Danra, 3 Nov. 1963, *H. Hara et al. 6303702* (KYO, TI); Taplejung, Dumhan to Taplejung, 1 Nov. 1963, *H. Kanai et al. 6303704* (KYO, TI); Taplejung, Khebang to Bharomdin, 24 Nov. 1963, *H. Hara et al. 6303712* (A, BM, E, K, KATH, KYO, TI); **Sagarmatha**, Bhojpate, Between Dingla and Baulia, 1210 m, 20 Dec. 1971, *Croucher 24* (BM); Bhojpate, between Dingla and Baulia, 1210 m, 20 Dec. 1971, *Bowring 24* (BM); Bhojpate, Bungling to Irkhuwa Khola, 300 m, 11 Nov. 1981, *P. K. Shakya 7400* (KATH); Bhojpate, Simule to Chamling, 150 m, 25 Oct. 1995, *M. Mikage et al. 9554124* (A, TI); Khotang, Khani Danda to Dorpa Churi Danda, 1440 m, 29 Oct. 1995, *M. Mitage et al. 9554200* (TI); Udayapur, Bastim to Pharkushe, 1480 m, 2 Nov. 1995, *M. Mikage et al. 9554262* (E, TI, TNS); Udyapur, Udaipur Garhi, 600 m, 3 Nov. 1969, *J. D. A. Stainton 6634* (BM, TI). **SIKKIM & DARJILING**: **Darjiling**, Choonbultee, 750 m,

14 Dec. 1876, *C. B. Clarke 31719* (K, LE); Pashok, 750 m, 17 Oct. 1902, *J. H. Lace 2427* (CAL,E); Mahanauda Wildlife Sanctuary, Sukna, 4 Mar. 1995, *K. H. Maity 8938* (CAL); Teesta Bridge, 150 m, 23 Nov. 1962, *J. K. Maheshwari 5025* (CAL); **Sikkim**, *Griffith 2400* (BM); *Griffith 2401*(BM); Balaum Valley, 600 m, 14 Nov. 1874, *D. Treutler 1061* (E, L); Nov. 1875, *O. Kuntze 6566* (K); Tarai, 17 Jan. 1890, *H. H. Haines 2751* (K); Khursiong, 1200 m, 25 Sep. 1884, *C. B. Clarke 3836* (LE); Rummar, 750 m, 3 Oct. 1875, *C. B. Clarke 24908* (CAL, K); Khubiong, 1200 m, 25 Sep. 1884, *C. B. Clarke 35846* (BM, CAL, K); Tharia, K & J. Hills, 12 Dec. 1939, *G. K. Deka 19309* (CAL); 750 m, *G. King 262* (CAL); Sitong, 11 Dec. 1908, *W. C. Craib 614* (CAL); Titi-guglati, N., Kodahari, Nov. 1879, *J. S. Gamb le 7381* (CAL); Mungpoo, 1060 m, 23 Sep. 1908, *W. W. Smith 97* (CAL); Tista Valley, Rungho to Tista Bridge, 240–360 m, 4 Nov. 1909, *I. H. Burkill 34068* (CAL). **BHUTAN**: **Mongar**, W Mongar, 1700 m, 10 Nov. 1991, *J. R. I. Wood 7507* (E); between Mongar and Lingmethang, 1300 m, 10 Nov. 1991, *J. R. I. Wood 7508* (E); between Lingmethang and Namning, 1500 m, 10 Nov. 1991, *J. R. I. Wood 7509* (E); between Lingmethang and Wamning, 1200 m, 24 Oct. 1990, *J. R. I. Wood 7364* (E); **Phuntsholing**, above Rinchending, 780 m, 16 Feb. 1982, *A. J. C. Grierson & D. G. Long 2963* (E); above Kharbandi, 1000 m, 12 Nov. 1987, *J. R. I. Wood 6087* (E, TI); between Kharbandi and Sorchen, 800 m, 26 Nov. 1991, *J. R. I. Wood 7523* (E); **Samchi**, Khagra Valley near Gokti, 420 m, 2 Mar. 1982, *A. J. C. Grierson & D. G. Long 3383* (E, K); N of Samchi Bazaar, 400 m, 1 Jan. 1989, *J. R. I. Wood 6856* (E); Samchi to Dorokha, 700 m, 23 Nov. 1991, *J. R. I. Wood 7519* (E); **Sarpang**, Manas, Lungdurbi, 500 m, 1991, *L. Reader s. n.* (E); Manas, between Nyakar and Buli, 1100 m, 1991, *L. Reader s. n.* (E); Manas, between Mamung and Gomphu, 300–1390 m, 1991, *L. Reader s. n.* (E); Manas, between Lungdurbi and Sarphey La, 790–1760 m, 1991, *L. Reader s. n.* (E); **Shamgong**, Aug.–Sep. 1985, *I. Broad s. n.* (E). **YARLUNG ZANGBO-BRAHMAPUTRA**: **Lhünzê**, Tsangpo Gorge, Gompo He, 1500 m, 12 Oct. 1924, *F. Kingdon-Ward 6379* (E, K); **Mishmi Hills**, Lakhuni, Naga Hills, 450–600 m, 2 Jan. 1949, *F. Kingdon-Ward 18379* (BM); 900 m, 29 Feb. 1876, *C. B. Clarke 27071* (LE).

34. Strobilanthes discolor (Nees) T. Anderson, J. Linn. Soc., Bot. 9: 477. 1867; Clarke in Hooker (ed.), Fl. Brit. India 4: 462. 1844; Hu, Deng & Wood in Wu, Raven & Hong (eds.), Fl. China 19: 425. 2011. ≡ *Goldfussia discolor* Nees in de Candolle (ed.), Prodr. 11: 172. 1847. Type: India, Meghalaya, Khasia, *Griffith s. n.* [lectotype designated by Wood, Edinburgh J. Bot. 51: 237. 1994: K (K000883054); isolectotypes: K (K000883055, K000883056)].

Strobilanthes discolor var. *nudicalyx* C. B. Clarke in Hooker (ed.), Fl. Brit. India 4: 462. 1884. Type: India, Meghalaya (Khasia), 9 Nov. 1850, J. D. Hooker *& T. Thomson, Goldfussia 25* [lectotype **here designated**: K (K000883051); isolectotypes: BM (BM000906329), C, K (000883052)].

Diflugossa nagaensis Bremek., Verh. Kon. Ned. Akad. Wetensch. Afd. Natuurk. Sect. 2 41(1): 237. 1944. ≡ *Strobilanthes nagaensis* (Bremek.) W. J. Cress & DeFilipps, Contr. U.S. Natl. Herb. 45: 483 2003. Type: India, Nagaland, Dec. 1907, *Meebold 7381* [holotype: B (B_10_0272654); isotype: K (K000883057)].

Subshrubs, 1–2 m high, strongly anisophyllous; stems and branches quadrangular to subterete, sulcate, glabrous, apically somewhat zigzag. Leaves very unequal in each pair; petiole 0.5–2.5 cm long, sulcate, glabrous; blade elliptic to ovate-elliptic, 3–14 cm long, 2–6.5 cm broad, glabrous, abaxially whitish, lateral veins 6–8 pairs, base usually slightly oblique, attenuate, shortly decurrent onto petiole, margin serrate, apex acuminate to caudate and often falcate. Panicles of pedunculate head-like spikes, axillary. Bracts oblanceolate to spatulate, 3–5 mm long, 1–2 mm broad, glabrous, caducous; bracteoles oblanceolate, caducous. Calyx 6–8 mm long, 5-lobed almost to base; lobes subequal. linear-oblong, 5–7 mm long, 0.8–1 mm broad, subequal, pilose towards apex, apex obtuse, accrescent to 0.8–1 cm long in fruit, becoming glandular-pilose. Corolla violet, 3.2–3.8 cm long, outside glabrous, inside glabrous except for trichomes retaining style; tube base cylindric, 8–10 mm long, 1.6–

2 mm broad, upwards gradually widened to ca. 1 cm broad at mouth; lobes ovate, 5–6 mm long, 6–7 mm broad, apex rounded. Stamens 4; filament of longer pair 7–8 mm long, glabrous or sparsely pilose, shorter pair 1.5–2 mm long, strongly reflexed, glabrous; anther-thecae ellipsoid, 1–1.2 mm long, incurved. Ovary glabrous; 2 ovules per locule; style 2.3–2.8 cm long, hirsute. Capsules narrowly obovoid, 9–11 mm long, 2–2.5 mm broad, sparsely glandular-pilose towards apex, 4-seeded. Seeds ovate in outline, compressed, 2.5–mm long, 1.6–2 mm broad, pilose. **Chromosomes** 2n = 22. Figure 46-45.

Phenology: flowering from November to February of next year.

Habitats: forests; 1500–2100 m.

Distribution: U Ganga & Indus, Yarlung Zangbo-Brahmaputra (Map 46-84); also in other parts of India.

U GANGA & INDUS: Stich, Chagorrie, *Duthie 504* (CAL); Sama, 10 Oct. 1920, *S. H. Howard 6* (CAL); **Chamba**, Kalel Khad, 1500–2100 m, 28 Sep. 1896, *J. H. Lace 1512* (CAL). **YARLUNG ZANGBO-BRAHMAPUTRA**: **Dirang**, 1550–1860 m, 5 Sep. 1938, *F. Kingdon-Ward 14202* (BM).

35. Strobilanthes pentstemonoides (Nees) T. Anderson, J. Linn. Soc., Bot. 9: 477. 1867; Clarke in Hooker (ed.), Fl. Brit. India 4: 460. 1884; Wood in Grierson & Long (eds.), Fl. Bhutan 2(3): 1265. 2001; Hu, Deng & Wood in Wu, Raven & Hong (eds.), Fl. China 19: 422. 2011. ≡ *Goldfussia pentstemonoides* Nees, Pl. Asiat. Rar. 3: 88. 1832; de Candolle, Prodr. 11: 175. 1847. ≡ *Ruellia pentstemonodes* Wall., Numer. List n. 2340. 1832, nom. invalid. Type: Nepal, 1821, *Wallich Cat. 2340* [lectotype **here designated**: K-W (K001115689); isolectotypes: BM, CGE, E (E00273483), K (K000883081, K000883082), K-W (K001115692), P (P00719400)].

Goldfussia flexuosa Nees, Pl. Asiat. Rar. 3: 88. 1832. ≡ *Ruellia flexuosa* Wall., Numer. List n. 2412. 1830, nom. invalid. Type: Nepal, *Herb Wight in Wallich Cat. 2412* [lectotype **here designated**: K-W (K001115881); isolectotypes: E, GZU, K].

Ruellia capitata Buch.-Ham. ex D. Don, Prodr. Fl. Nepal.: 120. 1825, non *Goldfussia capitata* Nees in Wallich (ed.), Pl. Asiat. Rar. 3: 88. 1832, nec *Strobilanthes capitata* (Nees) T. Anderson, J. Linn. Soc., Bot. 9: 475. 1867. Type: Nepal?, 24 Oct. 1802, *F. Buchanan-Hamilton s. n.* [holotype: BM (BM000884894)].

Ruellia cephalotes Wall., Numer. List n. 2352. 1832, nom. invalid.

Herbs, 0.5–1 m high, anisophyllous; stems quadrangular, sulcate, subglabrous. Leaves unequal in each pair, subsessile; blade elliptic to elliptic-lanceolate, 4–15 cm long, 1–4 cm broad, abaxially pubescent along veins, adaxially glabrous, lateral veins 5–7 pairs, base cuneate, margin serrulate, apex acute. Spikes terminal, abbreviated head-like, simple or 3-furcately branched, sometimes with 1 or a few distant flowers; peduncle glandular-pilose. Bracts suborbicular, 6–9 mm long and broad, unequal in size, scarious, caducous, glabrous, apex rounded or obtuse; bracteoles absent. Calyx 0.5–1.2 cm long, accrescent to 1.8–2 cm in fruit, 5-lobed almost to base; lobes linear, unequal, one slightly longer than others, abaxially glandular-pilose, apex obtuse. Corolla violet, 3.2–3.5 cm long, slightly curved, outside glabrous, inside glabrous except for trichomes retaining style; tube base cylindric, 7–8 mm long, 1.5–2 mm broad, upwards gradually widened to 1–1.2 cm broad at mouth; lobes ovate, 3.5–4 mm long and broad. Stamens 4, included; filaments subglabrous, longer pair 7–8 mm long, shorter pair 3–4 mm long, incurved; anther-thecae spherical, 1–1.3 mm in diameter, incurved. Ovary pilose towards apex; 2 ovules per locule; style 2.5–2.8 cm long, sparsely hirsute. Capsules clavate, 1–1.2 cm long, glandular-pubescent, 4-seeded. Seeds suborbicular in outline, 2–2.5 mm in diameter, pubescent.

Distribution: U Ganga & Indus, U Yarlung Zangbo, W Nepal, C Nepal, E Nepal, Sikkim & Darjiling, Bhutan, Yarlung Zangbo-Brahmaputra (Map 46-85); also in other parts of India.

1a. Stem and leaves subglabrous; leaf blade

Figure 46-45: **Strobilanthes discolor** (Nees) T. Anderson: **a,** portion of plant; **b**, bract; **c**, bracteole; **d**, young calyx; **e**, mature calyx. Drawn by Mary BATES; reproduced from Wu, Raven & Hong (eds.), *Fl. China Illustr.* 19: fig. 336: 2–6. 2012 (original from Wood, *Edinburgh J. Bot.* 51: 257, fig. 12: G–K. 1994).

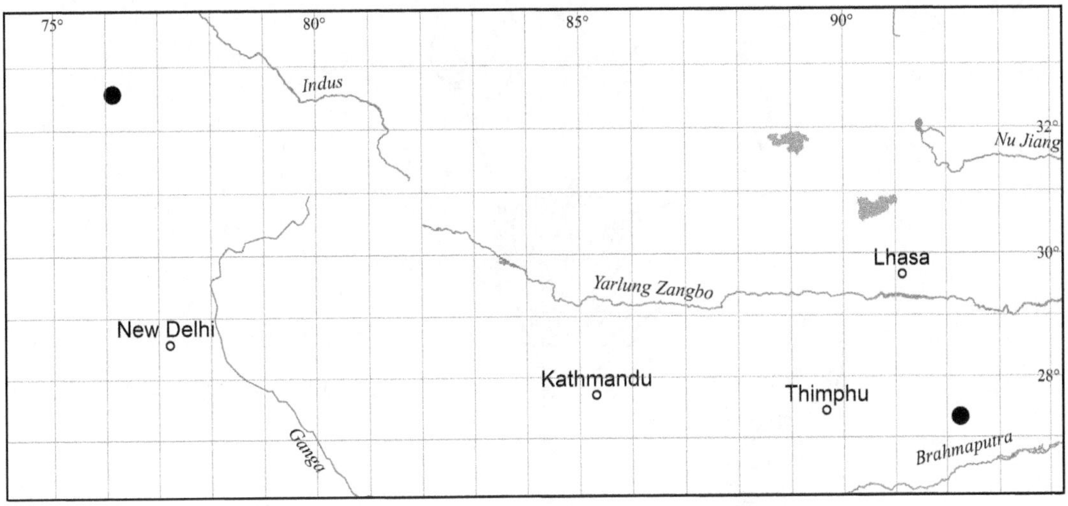

Map 46-84: **Strobilanthes discolor** (Nees) T. Anderson

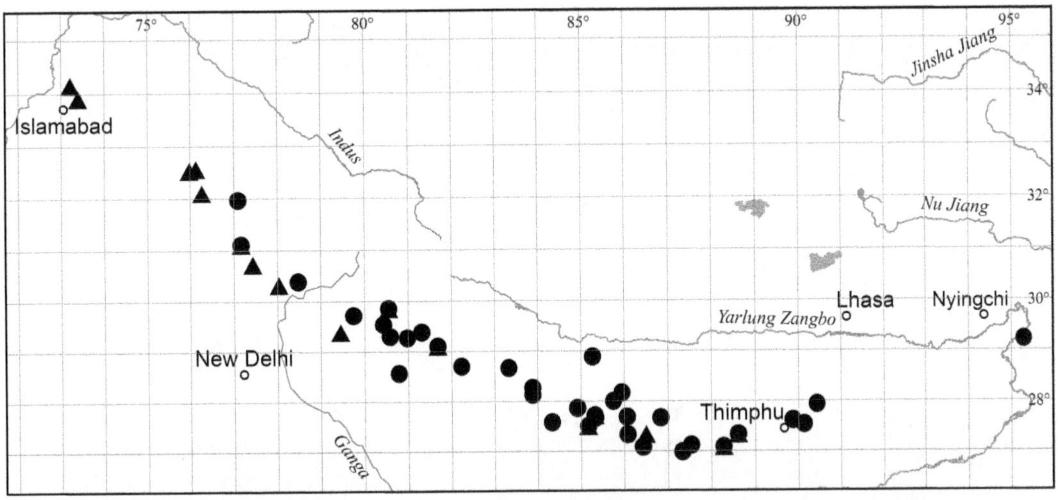

Map 46-85: **Strobilanthes pentstemonoides** (Nees) T. Anderson var. **pentstemonoides (●)**, var. **dalhousieana(▲)**

elliptic to elliptic-lanceolate, 4–15 cm long, 1–4 cm broad .... 35a. *S. pentstemonoides* var. *pentstemonoides*

1b. Stems and leaves hirsute; leaf blade elliptic, 10–20 cm long, 4–6 cm broad .... 35b. *S. pentstemonoides* var. *dalhousieana*

35a. Strobilanthes pentstemonoides var. **pentstemonoldes**

Stem and leaves subglabrous; leaf blade elliptic to elliptic-lanceolate, 4–15 cm long, 1–4 cm broad. Figure 46-46.

Phenology: flowering from August to November; fruiting from September to December.

Habitats: mountain slopes; 1100–3400 m.

Distribution: N Pakistan, Jammu &

Figure 46-46: **Strobilanthes penstemonoides** (Nees) T. Anderson: **a**, portion of plant; **b**, corolla; **c**, capsule with fruiting calyx. Drawn by Mary BATES; reproduced from *Fl. China Illustr.* 19: fig. 331. 2011 (original from *Edinburgh J. Bot.* 51: 252, fig. 11: H–J. 1994).

Kashmir, U Ganga & Indus, U Yarlung Zangbo, W Nepal, C Nepal, E Nepal, Sikkim & Darjiling, Bhutan, Yarlung Zangbo-Brahmaputra; also in other parts of India.

U GANGA & INDUS: **Binsar**, 2250 m, 23 Oct. 1960, *H. M. Heybroek 170* (L); **Kulu**, Plach Kulu, 1800 m, 2 Oct. 1888, *G. Watt 9601* (E); Kulu Valley, Malana Valley, 2000 m, 21 Oct. 1962, *J. Tanaka 26* (KYO); **Kumaon**, Jul. 1969, 1200 m, *P. N. Mehra 9* (K); **Simla**, 24 Sep. 1831, *Herb. Royle 10023* (LIV); Jakho, 2400 m, Oct. 1907, *A. Meebold 5047* (CAL); **Tehui Garhwal**, Dhanaulti, 2100 m, Aug. 1944, *R. R. Stewart s. n.* (K). **U YARLUNG ZANGBO**: **Nyalam**, Zhangmu, Youyiqiao, 18 Nov. 2011, *Y. S. Chen et al. 637* (PE); Zhangmu, Lixin, 2490 m, 14 Aug. 2010, *PE Xizang Expedition 976* (PE); **Gyirong**, Gyirong to Mangmu, 2100 m, *Z. C. Ni et al. 2315* (PE); Mangmu, 2300 m, 4 Nov. 1967, *S. Jiang & C. F. Zhao 351* (PE). **W NEPAL**: **Bheri**, Dailekh, Ranimatta to Dungesar, 31 Jul. 1991, *M. Suzuki et al. 9191054* (TI); Jajarkot, Garanchi, 1600 m, 1 Aug. 1981, *N. P. Manandhar 6454* (KATH); **Karnali**, Kalikot, Panipokhari, 2020 m, 4 Aug. 1991, *M. Suzuki et al. 9160237* (E, TI); Kalokot, Chaukebada to Badarigaon, 5 Aug. 1991, *M. Suzuki et al. 9191144* (TI); **Mahakali**, Baitadi, Patan to Gwani, 1200–1900 m, 20 Sep. 1981, *I. Sharma et al. 601* (KATH); Dadeldhura, Gogan to Niglad, 1420–1720 m, 15 Sep. 1981, *I. Sharma et al. 427* (KATH); Dadeldhura, Niglad, 1700 m, 1 Sep. 1982, *L. P. Kattel 998* (KATH); Dadeldhura, Chiri Kitta, 2000 m, 1 Sep. 1982, *L. P. Kattel 1032* (KATH); Darchula, Maie Kholi, 1810 m, 29 Aug. 1982, *M. M. Amatya & P. M. RegmiW553/82* (KATH); Darchula, Maie Kholi to Sipti, 30 Aug. 1982, *M. M. Amatya & P. M. RegmiW559/82* (KATH); Darchula, Maie Kholi to Sipti. 1810–1860 m, 30 Aug. 1982, *M. M. Amatya & P. M. Regmi W598/82* (KATH); Darchula, Makari ged to Thumti, 1330–2060 m, *M. M. Amatya & P. M. Regmi W682/82* (KATH); Darchula, Chamelia Valley, near Ghusa Village, Ghusa, 2100 m, 15 Oct. 2014, *S. K. Ghimire et al. ANCA 541* (TUCH); Darchula, between Naaliand Sithi, Ghusa, 2300 m, 16 Oct. 2014, *S. K. Ghimire et al. ANCA 620* (TUCH); **Seti**, Bajhang, Dantola, 12 Aug. 1976, *H. Tabata et al. 2441* (KYO); Bajura, Rajatoli, 18 Aug. 1976, *H. Tabata et al. 2582* (KYO); Bajura, Pandaseni, 17 Aug. 1976, *H. Tabata et al. 2656* (BKF, KYO, TNS); Toli, 23 Aug. 1976, *H. Tabata et al. 2859* (BKF, KYO, TI); Doti, Budar, Budar to Dadelhura, 1409 m, 11 Nov. 2012, *B. Adhikari & G. D. Bhatt 44* (E); Doti, Ghanteswor, 1900 m, 16 Aug. 1980, *K. R. Rajbhandari et al. 5410* (KATH); Kailali, Budhetoda, 1140 m, 11 Nov. 2012, *B. Adhikari & G. D. Bhatt 37* (E). **C NEPAL**: **Bagmati**, Dhading, Thambu, 1200 m, 4 Nov. 1989, *N. P. Manandhar 12892* (KATH); Birjet, 1620 m, 4 Nov. 1989, *N. P. Manandhar 12933* (KATH); Dhading, Tharlang, 1550 m, *N. P. Manandhar 13000* (KATH); Kathmandu, Kathmandu Valley, 1500–2500 m, 12 Nov. 2000, *T. Watanabe NPL-SP001112(048)* (MBK); Kathmandu, Halchok, 18 Feb. 1969, *H. Kanai 15086* (KATH); Kathmandu, Swyambhu, 1300 m, 21 Jun. 1970, *N. P. Manandhar 9420* (KATH); Kathmandu, Bhid Nilkantha to Banjang, N of Kathmandu, 28 Mar. 1969, *H. Kanai 11132* (KATH); Kathmandu, between Sundarrjal and Pati, Banjang, 2500 m, 5 Sep. 1974, *J. H. de Haas 2636* (BM); Kathmandu, above the road to Pharping, 1370 m, 1 Dec. 2005, *C. Pendry et al. B6* (E); Kathmandu, Gokarna, 1330 m, 2 Nov. 1966, *P. Pradhan & R. Thapa 4564* (KATH); Kathmandu, Nagajung Danda, 1520 m, 24 Oct. 1920, *s. coll. 8803*(BM); Kathmandu, Airport, 13 Oct. 1963, *H. Hara et al. 6303697* (KYO, TI); Kathmandu, Chhap to Shiwapuri, 1800 m, 13 Sep. 1988, *M. Suzuki et al. 8811541* (KATH, TI); Kathmandu, Shiwapuri Summit to Borlang Bhanjyang, 2500 m, 14 Sep. 1988, *M. Suzuki et al. 8881973* (KATH, TI); Kathmandu, Chandragiri, SW Kathmandu, 2100 m, 12 Jan. 1975, *D. P. Joshi et al. 75/308* (KATH); Kathmandu, Chandragiri Lehk, S W of Kathmandu, 2100 m, 12 Juanuary 1975, *D. P. Joshi 75/298* (KATH); Kathmandu, Shivapuri, 1800 m, Jan. 1968, *M. M. Amatya 12834* (KATH); Kathmandu, Simpani Danda, SW of Kathmandu, 2350 m, 9 Jan. 1975, *D. P. Joshi 75/187* (KATH); Kavre Palanchok, Khardar Pati, 1550 m, 19 Nov. 1988, *N. P. Manandhar 12186* (KATH); Lilatpur, Phulchoki, 2400–2700 m, 19 Sep. 1970, *H. Kanai & C. Chuma 674751* (TI); Lalitpur, Phulchoki, 1900 m, 19 Sep. 1970, *H. Kanai & C. Chuma 674780* (TI); Lalitpur, Phulchoki, 2100 m, 6 Sep. 1972, *H. Kanai 722131* (TI); Lalitpur,Godavari,

S of Kathmandu, 1600 m, *Schilling 677* (K); Lalitpur, Godavari to Mt. Pulchoki, 2200 m, 3 Sep. 1966, *D. H. Nicholson 2248* (BM, BO, E, KATH); Lalitpur, Phulckoki above Godawari, 2200 m, 20 Oct. 1991, *K. A. Lye 17692* (E); Lalitpur, Phulchoki, 2400–2700 m, 28 Nov. 1968, *Malla & I. Shrestha 10749* (KATH); Lalitpur, Phulchok, 2200–2770 m, 9 Aug. 1969, *H. Kanai 15084* (KATH); Lalitpur, Godavari, 2200–2800 m, 27 Nov. 1948, *B. D. Pande 49* (BM); Lalitpur, Godawan, 2130 m, 3 Oct. 1995, *M. Mikage et al. 9552659* (E); Lalitpur, Godavari, near Kathandu, 1500 m, 18 Sep. 1963, *H. Hara et al. 6303698* (TI); Lalitpur, Tikabhairab, 1520 m, 14 Oct. 1971, *P. Pradhan & S. Gurung 16986* (KATH); Lalitpur, 1540 m, 4 Oct. 1989, *I. Shrestha & N. Pradhan 7* (KATH); Lalitpur, Hattiban Village, 1600 m, 6 Jan. 1975, *D. P. Joshi et al. 75/73* (KATH); Lalitpur, Godavari, 1600 m, 24 Nov. 1974, *I. Sharma & M. Sharma 74/23* (KATH); Lalitpur, Lele, 1540 m, 4 Oct. 1989, *I. Shrestha & N. Pradhan 7* (KATH); Nuwakot, near Kakani, 14–15 Dec. 1952, *S. Makao s. n.* (KYO); Nuwakot, Latsu to Borlang, 1700–2400m, 30 Aug. 1972, *H. Kanai et al. 72830* (BM, KYO, TI); Rasuwa, between Dhunche and Dhimsa, 2000 m, 5 Oct. 2001, *D. G. Long et al. ENEP143* (E, TI); Rasuwa, Sharpagoan, 2750 m, 23–31 Aug. 1949, *O. V. Polunin 1843* (BM, TI); Rasuwa, Dunche to Singum Gompa, 2800 m, 22 Aug. 1972, *H. Kanai et al. 10011531* (TI); Rasuwa, Khodang Danda to Latsu, 2100–2500 m, 29 Aug. 1972, *H. Hara et al. s. n.* (TI); Rasuwa, Dunche to Singum Gompa, 1900–3200 m, *H. Hara et al. 72822* (TI); Rasuwa, Sing Gomba to Dhunche, 2000 m, 31 Jul. 1995, *T. Hoshino et al. 9535329* (TI); Shidhupalchok, Helumbu, Malenchi, 2420 m, 18 Sep. 1974, *J. D. A. Stainton 7215* (BM, TI); Shidhupalchok, Kalinchok, 2240 m, 10 Sep. 1970, *H. Kanai et al. 674766* (KATH, TI); Shidhupalchok, Tashi Thong, S of Tharke Ghyong, 2550 m, 7 Sep. 1974, *J. H. de Haas* 2679 (BM, L); **Dhawalagiri**, Myagdi, near Lumsum, 2400 m, 17 Jul. 1954, *J. D. A. Stainton, W. R. Sykes & L. H. J. Williams 3522* (BM, E, KATH, TI); Myagdi, Dana, 1670 m, 30 Nov. 1954, *J. D. A. Stainton, W. R. Sykes & L. H. J. Williams 8283* (BM, KATH, TI); Myagdi, Shika to Narchang, 1800 m, 26 Aug. 1988, *M. Suzuki 8881440* (BM, KATH, TI); Myagdi, Boghara to Dobang, 7 Sep. 1996, *M. Mikage et al. 9681231* (TI); Myagdi, Ritu Kharka to Lulang, 2880 m, 21 Sep. 1996, *M. Mikage et al. 9682965* (TI); Myagdi, Dharapani to Muri, 4 Sep. 1996, *M. Mikage et al. 9685154* (TI); Myagdi, Boghara to Dobang, 7 Sep. 1996, *M. Mikage et al. 9687197* (TI); **Gandaki**, Kaski, en route from Kare to Chandrakot, 1350–1400 m, 22 Sep. 1976, *Y. Suehiro 274* (KYO); Kaski, Pothana–Bhichuk–Tolka, 2040 m, 20 Aug. 1988, *M. Suzuki et al. 8800198* (TI); Kaski, Pothana to Tolka (Thorka), 1800 m, 20 Aug. 1988, *M. Suzuki 8810827* (BM, KATH, TI); Kaski, Pothana to Tolka, 1800 m, 20 Aug. 1988, *M. Suzuki 8830147* (TI); Kaski, Pothana to Tolka, 1970 m, 20 Aug. 1988, *M. Suzuki 8860511* (TI); Kaski, Ghandruk to Bhainsi Kharka, 2100 m, 22 Aug. 1988, *M. Suzuki et al. 8881170* (TI); Kaski, Ulleri, 2030 m, 16 Sep. 1995, *M. Kikage et al. 9552153* (E, TI); Kaski, Tirkhedhunga to Ghorepani, 16 Sep. 1995, *M. Mikage et al. 9550070* (TI); Kaski, vicility of Tamage, 2090 m, 8 Aug. 1999, *M. Mikage et al. 9966513* (TI); Kaski, Banjan–Mt. Panchase–Tamage, 10 Aug. 1999, *M. Mikage et al. 9967065* (KATH, TI); Kaski, Pamdur, near Pokhara, 1500 m, 31 Oct. 1954, *J. D. A. Stainton, W. R. Sykes & L. H. J. Williams 8314* (BM, KATH, TI); Kaski, Dhumpur Lumle, 1900 m, 1 Oct. 1969, *H. Flatt 104* (BM); Kaski, Lumle, 1700 m, 4 Nov. 1974, *G. Dawson* 244 (BM); **Janakpur**, Dolakha, above Charikot (now Bhimeshwar), 2120 m, 15 Sep. 1964, *M. L. Banerji et al. 2763* (BM, KATH); Dolakha, Khare Khola, Phedi Khola, 1700 m, 14 Sep. 1983, *H. Ohba et al. 8332090* (BM, E, KATH, TI); Dolakha, Pheti Kharka to Khanigaon, 14 Sep. 1983, *H. Ohba et al. 8351515* (TI); Dolakha, Thulnai, 1900 m, 1 Oct. 1970, *H. Kanai et al. 674708* (KYO); Ramechhap, Thosey, 2800–3400 m, 1930, *L. Dhwoj 351* (BM, E); Ramechhap, Bhandar to Shivalaya, 6 Aug. 1985, *H. Ohba et al. 8571395* (BM, KATH, KYO, TI); Ramechhap, Shivalaya to Bhandar, 1800–2700 m, 17 Aug. 1985, *H. Ohba et al. 8571690* (TI); Ramechhap, Shivalaya to Bhandar, 1800–2300 m, 17 Sep. 1985, *H. Ohba et al. 8530830* (E, TI); Ramechhap, Shivalaya to Bhandar, 1800–2300 m, 17 Sep. 1985, *H. Ohba et al. 8571689* (TI); Ramechhap, Kalingchok, Thala to Tale Bisauna, 2050 m, 10 Sep. 1970, *H. Kanai et al. 672791* (KATH, TI);

Ramechhap, Bhandar to Shivalaya, 2360 m, 3 Sep. 1997, *M. Wakabayashi et al. 9720377* (TI); **Narayani**, Chitwan, Kot, 1400 m, 17 Nov. 1989, *N. P. Manandhar 14109* (KATH); Makwanpur, Namtar, between Aghor and Lamidanda, 1620 m, 7 Nov. 2012, *B. Adhikari & G. D. Bhatt 17* (E). **E NEPAL**: **Koshi**, Dankhuta, Lobuche Khola Valley, Manmaya to Mangmaya Khola, 2130 m, 27 Sep. 1981, *P. R. Shakya 6726* (KATH); Dankhuta, Mangsing Danda, 2100 m, 3 Oct. 1981, *P. R. Shakya 6857* (KATH); Dankhuta, Mangmaya Danda, 2460 m, 6 Oct. 1981, *P. R. Shakya 6907* (KATH); Tehrathum, 1550 m, 16 Nov. 1978, *P. Pradhan et al. 901* (KATH); Sankhuwasabha, Milke Danda, 2000 m, 16 Sep. 1993, *P. Sondergaard P1991-5197* (C); Sankhuwasabha, Nurbu Gaon, 2140 m, 20 Aug. 1997, *S. Noshiro et al. 9760229* (TI); **Mechi**, Taplejung, Shewaden to Papung, 2000–2600 m, 22 Aug. 1977, *H. Ohashi et al. 772677* (TI); Taplejung, Shewaden to Papung, 2000–2600 m, 22 Aug. 1977, *H. Ohashi et al. 773987* (TI); Taplejung, Bhandukay to Ghatte, 16Nov. 1963, *H. Hara et al. 10011331* (TI); **Sagarmatha**, Solukhumbu, Namikhil to Bhandar, 1550–2300 m, 10 Sep. 1985, *H. Ohba et al. 8572546* (TI); Solukhumbu, Eswa Valley, 2510 m, 1 Nov. 1981, *P. R. Shakya 7299* (KATH). **SIKKIM & DARJILING**: **Darjiling**, Ryang, 11 Nov. 1870, *C. B. Clarke 13624* (LE); Ralinpoony, 1200 m, 20 Sep. 1869, *C. B. Clarke 9178* (LE); Dikeeling, 1350 m, 13 Oct. 1869, *C. B. Clarke 9719* (LE); **Sikkim**, Burting, 1200 m, 5 Oct. 1875, *C. B. Clarke 25057* (LE); Runghee, 17 Feb. 1867, *T. Anderson s. n.* (L). **BHUTAN**: Gosa Dzong, 2700 m, 1 Sep. 1958, *S. Nakao 386* (KYO); between Gasa and Tamji, 2200 m, 2 Sep. 1958, *S. Nakao 435* (KYO); **Upper Mo Chu**, Between Goen Gaza and Gaza Zam, 2450m, 13 Sep. 1984, *I. W. J. Sinclair & D. G. Long 4982* (E); **Punakha**, 10 km below Nobding, 2190 m, 11 Oct. 1984, *I. W. J. Sinclair & D. G. Long 5633* (E). **YARLUNG ZANGBO-BRAHMAPUTRA**: **Mêdog**, Renqinpeng, 1500 m, *G. Yao et al. 3648* (NAS).

35b. Strobilanthes pentstemonoides var. **dalhousieana** (Nees) Kuntze, Revis. Gen. Pl. 2: 499. 1891; Wood, Edinburgh J. Bot. 51: 253. 1994. ≡ *Goldfussia dalhousieana* Nees, Prodr. 11: 174. 1847, "*dalhousiana*". ≡ *Strobilanthes dalhousieana* (Nees) C. B. Clarke in Hooker (ed.), Fl. Brit. India 4: 460. 1884; Nasir & Ali in Malik & Ghafoor (eds.), Fl. Pakistan 188: 55. 1988. Type: [U Ganga & Indus] India, Simla, "Ad Simla, Indiae Orientalis", Sep. 1831, *Lady Dalhousie s. n.* [holotype: K (K000883073); isotypes: E, K].

Stems and leaves hirsute; leaf blade elliptic, 10–20 cm long, 4–6 cm broad.

Phenology: flowering from July to October.

Habitats: slopes, thickets; 300–2600 m.

Distribution: N Pakistan, Jammu & Kashmir, U Ganga & Indus, W Nepal, C Nepal, E Nepal, Sikkim & Darjiling; also in other parts of Pakistan and India.

N PAKISTAN: **Hazara**: Kalapani, 2100 m, 20 Aug. 1956, *R. R. Stewart s. n.* (RAW, fide Malik & Ghafoor, 1988); **Muree Hills**, Brewery woods, 2 Sep. 1915, *E. M. Sounders s. n.* (K); Potha, 1350 m, 3 Oct. 1949, *R. R. Stewart & Gorrie 23503* (K). **JAMMU & KASHMIR**: Basaoli, 1370 m, 26 Sep. 1876, *C. B. Clarke 31566* (K); **Dharamshala**; 1460 m, 27 Sep. 1974, *D. R. Sharma 13322* (RRLH), 2130 m, 17 Oct. 1874, *C. B. Clarke* 23997 (K); 1350 m, 1 Dec. 1932, *R. N. Parker 3249* (A, K). **U GANGA & INDUS**: Perganun Chuari, 2100 m, 27 Nov. 1919, *R. N. Parker 5554* (K); Chatri, 2100 m, 15 Oct. 1987, *McBeath 1974* (E); **Chamba**, 16 km to Dalhousie, 1111 m, 12 Oct. 1977, *van der Maesen 2895* (K); Kilor, Bhaudal Valley, 2100 m, 30 Sep. 1919, *R. N. Parker s. n.*(A); Kainthli Reservor, Dalhousie Range, 2050 m, 27 Nov. 1919, *R. N. Parker s. n.* (A); Kalelkhed to Chanan, 1500–2100 m, 28 Sep. 1896, *J. H. Lace 1512* (E); **Dalhousie**, 2130 m, 10 Sep. 1874, *C. B. Clarke 22030* (K); **Dehradun**:

Jaunsar, Komain, Desbau, 2400 m, Sep. 1890, *K. Naud 305* (E); Kanasar, Jaunsar, 1950 m, 17 Sep. 1936, *M. B. Raizada 7183* (E); Jaunsar, Chahrata, 2100 m, *J. S. Gamble 7253*(K); Jaunsar, Tauncar, 2100 m, Oct. 1898, *J. S. Gamble 27382* (K); Jaunsar, Jakolani, 2400 m, Sep. 1898, *J. S. Gamble 27271* (K); **Kangra**, Mandi, Ghogar Dhar, 1800 m, 16 Oct. 1985, *R. J. D. McBeath 1678* (E); **Mussorie**, Gracemount, Oct. 1917, *A. Anderson s. n.* (E); 1820–2130 m, Jul. 1938, *R. R. Stewart 1673A* (A); **Nainital**, 2134m, *R. Strachey & J. E. Winterbottom* (K); **Shimla**, 1830 m, Oct. 1885, *Collett* 870 (K); *Lady Dalhousie 264* (E); 1800 m, Sep. 1885, *H. Collett 670* (BM); Elipani Hills, 13 Sep. 1876, *J. S. Gamble 991* (K); Jakko, 2400 m, Sep. 1877, *J. S. Gamble 5465* (K); Theog, 2100 m, 8 Sep. 1878, *J. S. Gamble 6573* (K); 1500–2100 m, Oct. 1887, *G. Watt 8854* (CAL, E, K); Mashvbra, 2100 m, Oct. 1887, *G. Watt 8855* (E, K); 1885, *J. R. Drummond 26213* (E, K); 2100 m, Aug. 1938, *P. C. Jocki s. n.* (A); Jakks, 16 Sep. 1915, *H.H. Rich 17* (K); 11 Nov. 1948, *E. K. Janaki Ammal 1410* (K). **W NEPAL**: **Karnali**, Kalikot, Chaukebada to Badarigaon, 900–2340 m, 5 Aug. 1991, *M. Suzuki et al. 9160297* (TI); **Mahakali**, Darchula, Chamelia Valley, Janjari, Ghusa, 2130 m, 22 Oct. 2013, *S. K. Ghimire et al. ANCA 007* (TUCH). **C NEPAL**: **Bagmati**, Kathmandu, Pulchoki, 20 Sep. 1993, *Pollenpproparat 235* (K); **Narayani**, Makwanpur, Daman, 2100 m, Oct. 1962, *P. N. Suwal et al. 904* (BM, KATH); Makwanpur, below Aghari, on Hetauda to Kathamandu road, 2000 m, *C. Pendry et al. B228* (E, KATH); Makwanpur, Shimbhanjyang, 2400 m, s. d., *S. B. Shah 166* (KATH). **E NEPAL**: **Sagarmatha**, Okhaldunga, Tozke to Okhaldunga, 1990m, 2 Nov. 1954, *A. Zimmermann 1990* (K, G). **SIKKIM & DARJILING**: **Darjiling**, *Griffth 3 (Herbs T. Anderson 2398)* (CAL); **Sikkim**, Kulhaut, 2100 m, 11 Oct. 1870, *C. B. Clarke 12990* (CAL); 300 m, 23 Sep. 1874, *D. Treutler 750* (LE).

36. Strobilanthes dimorphotricha Hance, J. Bot. 21: 355. 1883, "*dimorphotrichus*"; Hu, Deng & Wood in Wu, Raven & Hong (eds.), Fl. China 19: 423. 2011. ≡ *Goldfussia dimorphotricha* (Hance) Bremek., Verh. Kon. Ned. Akad. Wetensch., Afd. Natuurk., Sect. 2 41(1): 231, 274. 1944. Type: China, Guangdong, Lienchau, in silvula umida ad Fuk-shan-man, 300 m, 25 Oct. 1881, *B. C. Henry ex herb. Hance no. 22110* [holotype: BM (BM000884978)].

Strobilanthes gracilis T. Anderson, J. Linn. Soc. 9: 474. 1867, nom. illeg., non *Strobilanthes gracilis* Beddome, J. Linn. Soc., Bot. 9: 474. 1864. ≡ *Strobilanthes geniculata* C. B. Clarke in Hooker (ed.), Fl. Brit. India 4: 459. 1885; Wood, Edinburgh J. Bot. 51: 240. 1994. ≡ *Goldfussia geniculata* (C. B. Clarke) Bremek., Verh. Kon. Ned. Akad. Wetensch., Afd. Natuurk., Sect. 2 41(1): 233. 1944. Type: [Yarlung Zangbo-Brahmaputra], Mishmi, *Griffith K. D. 6116* [lectotype designated by Wood, Edinburgh J. Bot. 51: 240. 1994: K (K000883090)].

Strobilanthes geniculata var. *integra* C. B. Clarke in Hooker (ed.), Fl. Brit. India: 459. 1885. Type: India, Meghalaya (Khasia), 1873, *C. B. Clarke 21919* [lectotype designated by Wood, Edinburgh J. Bot. 51: 240. 1994: K (K000883089); isolectotype: K (K000883088)].

Strobilanthes burkillii Dunn, Bull. Misc. Inform., Kew 1920: 208, 1920. Type: Yarlung Zangbo-Brahmaputra, Outer Abor Hills, between Kobo and Pasighat, 12 Dec. 1911, *I. H. Burkill 37107* [holotype: K (K000883086); isotype: CAL].

Perennial herbs, 0.40–1.5 m high, anisophyllous; stems somewhat zigzag upwards, glabrous or yellowish pubescent. Leaves unequal in each pair, smaller of pair 1/3–2/3 of larger one in size; petiole 0.5–1.5 cm long, sparsely pubescent or glabrous; blade elliptic, elliptic-lanceolate, ovate, oblong, or oblong-elliptic, 1.5–15 cm long, 0.5–4.5 cm broad, adaxially white appressed pubescent, abaxially glabrous or whitish pubescent along veins, lateral veins 5–7 pairs, base cuneate to attenuate, margin serrulate, apex acuminate, caudate, or acute. Panicles of head-like spikes, axillary or terminal, 1–3-forked; heads subspherical, very compact, 2- or 3-flowered; peduncle 1.5–2.5 cm long, glabrous or pubescent. Bracts ovate-elliptic,

1–1.3 cm long, glabrous, apex acute to acuminate, caducous; bracteoles lanceolate, 4–5 mm long, caducous. Calyx 7–9 mm long, accrescent to 1.5–1.7 cm long in fruit, 5-lobed almost to base; lobes unequal, one lobe slightly longer than others, lanceolate, glandular-pubescent, apex acuminate. Corolla violet, 3–4 cm long, slightly curved, outside subglabrous to sparsely pilose, inside glabrous except for trichomes retaining style; tube base cylindric, 5–7 mm long, 1.5–2 mm broad, upwards gradually widened to 1.2–1.5 cm broad at mouth; lobes suborbicular, 5–6 mm long and broad, apex emarginate. Stamens 4, included; filaments glabrous, longer pair 5–7 mm long, unequal, erect, shorter pair 2–2.5 mm long, equal, incurved; anther-thecae spherical, 1–1.2 mm in diameter. Ovary pilose; 2 ovules per locule; style 3.2–3.5 cm long, pilose. Capsules oblong-clavate, 1.4–1.8 cm long, glandular-pubescent, 4-seeded. Seeds ovate in outline, compressed, 2–2.5 mm long and broad, lenticular, pubescent. Figure 46-47.

Phenology: flowering from August to February of next year.

Habitats: thickets on limestone hills, stream banks; 200–2800 m.

Distribution: Yarlung Zangbo-Brahmaputra, S Hengduan, U Irrawaddy (Map 46-86); also in Vietnam, and other parts of India and China (C, E, NE & S Yunnan, S, C, E Sichuan, Chongqing, Hubei, Hunan, guizhou, Guangxi, Hainan, Guangdong, Jiangxi, Fujian, Zhejiang, Taiwan).

YARLUNG ZANGBO-BRAHMAPUTRA: **Abor Hills**, Pilung between Kobo and Pasighat, 12 Dec. 1911, *L. H. Burkill 37107* (K); **Lohit**, foothills of Badusu, *R. S. Rao 10308* (CAL); **Mishmi Hills**, *W. Griffith, Kew Distr. No. 6116* (K). **S HENGDUAN**: **Dali**, 13 Jun. 1949, *T. N. Liou 22903* (IBSC, KUN); Tchen-fong-chan, 650 m, *E. E. Maire s. n.* (IBSC); Cangshan, 1 Aug. 2019, *Y. F. Deng 28634* (IBSC); **Fugong**, Lumadeng, Yaping, 2040 m, 21 Aug. 2005, *Gaoligong Shan Biodiversity Survey 28801* (KUN); way to Lanping, Xishaofang, near Daluo, *S. K. Wu 8627* (HIB, KUN, LBG); Labuluohe, 2000 m, 11 Jul. 1978, *Exped. Bijiang 1020* (KUN, YUKU); Kongdong, Labujing, 2800 m, *Exped. Nujiang 942* (KUN, YUKU); **Gongshan**, Qiqi to Dongshao Ffang, 2050–2200 m, 19 Sep. 1997, *Gaoligong Shan Exped. (1997) 9334* (E); Cikai, 1550 m, 25 Sep. 2002, *Gaoligong Shan Biodiversity Survey 15456* (CAS, KUN); Pengdang, 1540 m, 29 Sep. 2002, *Gaoligong Shan Biodiversity Survey 15586* (CAS, KUN, TI); Cikai, Heiwadi, 1900 m, 4 Oct. 2002, *Gaoligong Shan Biodiversity Survey 15739* (CAS, KUN); Cikai, Shigu, 1500 m, 9 Oct. 2002, *Gaoligong Shan Biodiversity Survey 15894* (CAS, KUN, TI); Cikai, Gongshan to Qiqi, 1550 m, 25 Sep. 2002, *Gaoligong Shan Biodiversity Survey 16584* (CAS, KUN, TI); Cikai, Dangdang Park, 1890 m, 26 Sep. 2002, *Gaoligong Shan Biodiversity Survey 16634* (CAS, KUN); Lishadi, Yaduo, below Shibali, 2170 m, 10 Aug. 2005, *Gaoligong Shan Biodiveristy Survey 26937* (CAS, KUN, MO); Cikai, Yimaluo, 1520 m, 14 Aug. 2006, *Gaoligong Shan Biodiversity Survey 33298* (KUN); Cikai, Yimaluo, 1520 m, 14 Aug. 2006, *Gaoligong Shan Biodiversity Survey 33311* (KUN); South to Changputong, Shuanglanghe, 1700–1800 m, 28 Sep. 1940, *K. M. Feng 8067* (KUN, PE); Mekong-Salwin divide, Bahanlo, 2400 m, 15 Sep. 1938, *T. T. Yü 22696* (A, E, KUN, PE); Salwin Valley, Pengda to Baihanluo, 2000 m, 23 Sep. 1938, *T. T. Yü 23088* (A, E, KUN, PE); Urus Valley, 16 Jul. 2008, *X. H. Jin et al. DLJ-ET 2999* (PE); **Longyang**: Mankuan, Baihualing, 1580 m, 23 Nov. 2000, *H. Li 13405* (CAS, KUN); Mankuan, Baihualing, 1700 m, 28 Nov. 2000, *H. Li 13481* (KUN, MBK), Wayao, 1850 m, 2 Jan. 1959, *X. Zhou 228* (KUN); Wayao to **Yongping**, 551 km, marker of Yunnan-Burma Rd., 1900 m, 2 Jan. 1959, *C. Chen 934* (KUN); **Lushui**, Yaojiaping to Luzhang, 2440 m, 25 Oct. 1996, *Gaoligong Shan Expedition (1996) 8055* (E); **Miyi**, Rongdong, 1600–1800 m, 12 Sep. 2002, *G. M. Zhang 470* (PE); **Yangbi**, Pingpo, 5 Sep. 2013, *Y. F. Deng et al.* 24180 (IBSC); Halamo to Dacun, 13 Nov. 1946, *T. N. Liou 32903* (PE); **Yongping**, 16 km west of the city, 1950 m, 2 Jan. 1959, *X. Zhou 234* (KUN). **U IRRAWADDY**: **Bahang**, Salween Valley, 300 m, 16 Oct. 1922, *F. Kingdon-Ward 5419* (E); **Drungjiang**, Shanglawa Village, 1650 m, 9 Jul. 2009, *X. H. Jin et al. DLJ-ET 2362*

Figure 46-47: **Strobilanthes dimorphotricha** Hance: **a**, habit; **b**, outer bract; **c**, inner bract; **d**, calyx; **e**, opened corolla; **f**, capsule; **g**, seed. Drawn by Han-Ping YU; reproduced from *Icon. Cormophyt. Sin.* 4: 162, fig. 5738. 1975.

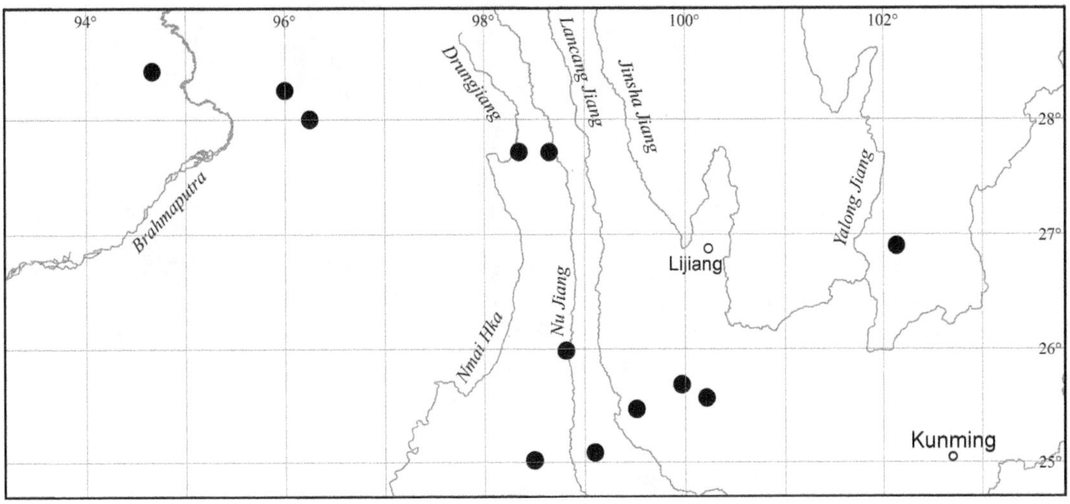

Map 46-86: **Strobilanthes dimorphotricha** Hance

(PE); **Tengchong**, Sanyun, Chuanlong, 1530 m, 3 Nov. 1998, *H. Li 11426* (CAS, KUN, MBK); Sanyun, Chuanlong, 1980 m, 4 Nov. 1998, *H. Li 11494* (CAS, KUN).

37. Strobilanthes straminea W. W. Sm., Notes Roy. Bot. Gard. Edinburgh 10: 200. 1918; Wood, Edinburgh J. Bot. 51: 260. 1994. ≡ *Goldfussia straminea* (W. W. Sm.) C. Y. Wu & C. C. Hu in Hu (ed.), Fl. Reipubl. Popularis Sin. 70: 169. 2002. Type: [U Irrawaddy] Myanmar, 40 km S of Putao, 600 m, Dec. 1914, *F. Kingdon-Ward 1986* [holotype: E (E00273477)].

Herbs, 50–80 cm high, anisophyllous; stems not zigzag, glabrous or glabrescent. Leaves unequal in each pair, subsessile or shortly petiolate; petiole (0–)2–5 mm long; blade broadly ovate to narrowly elliptic, 8–11 cm long, 5–6.5 cm broad, glabrous, lateral veins 6–8 pairs, base cuneate or rounded, margin serrate, apex acute or acuminate. Panicles of trifurcate head-like spikes, 12–18 cm long; spikes usually with a single distant flower below main head; rachis glabrous or slightly pilose. Bracts oblong-obovate, scarious, caducous, glabrous, margin glandular-ciliate, apex obtuse; bracteoles similar to bracts, but narrower. Calyx 1–1.2 cm long, 5-lobed almost to base; lobes unequal, one slightly longer than others, linear-lanceolate, glabrous, margin ciliate towards apex. Corolla 2.6–3 cm long, white with faint purple lines, curved, outside glabrous, inside pilose on throat; tube base cylindric, upwards abruptly widened; lobes oblong, 5–6 mm long and broad. Stamens 4, filaments glabrous, shorter pairs usually not developed. Ovary white pilose; 2 ovules per locule; style 2–2.3 cm long. Capsules clavate, 0.8–1 cm long, glandular-pilose, 4-seeded. Seeds ovate in outline, compressed, 1.6–2 mm long and broad, pubescent. Figure 46-48.

Phenology: flowering and fruiting from November to February of next year.

Habitats: forests; 100–1800 m.

Distribution: endemic; U Irrawaddy (Map 46-87).

U IRRAWADDY: **Nam Tamai Valley**, 1200–1500m, 6 Nov. 1922, *F. Kingdon-Ward 5506* (E); **Putao**, between Mache Ga and Tsuptang, 1200–1800 m, 26 Dec. 1961, *Keenan et al. 3041* (K); S Putao, 600 m, 22 Dec. 1914, *F. Kingdon-Ward 1986* (E); Hills S of Fort Hertz, 600–900 m, 22 Dec. 1931, *F. Kingdon-Ward 10223* (BM); 1100 m, 31 Dec. 2015, *S. S. Zhou 235* (HITBC); Upper Shankaung, 450 m, 12 Oct. 2014, *Putao Expedition Team (X. H. Jin et al.) PT-ET 111* (PE); Chaungsone Camp to Ziyadam, 1050–1520 m, 19 Oct. 2014, *Putao Expedition Team (X. H. Jin et al.) PT-ET 682* (PE); Machanbaw, 420 m, 25 Oct. 2014, *Putao Expedition Team (X. H. Jin*

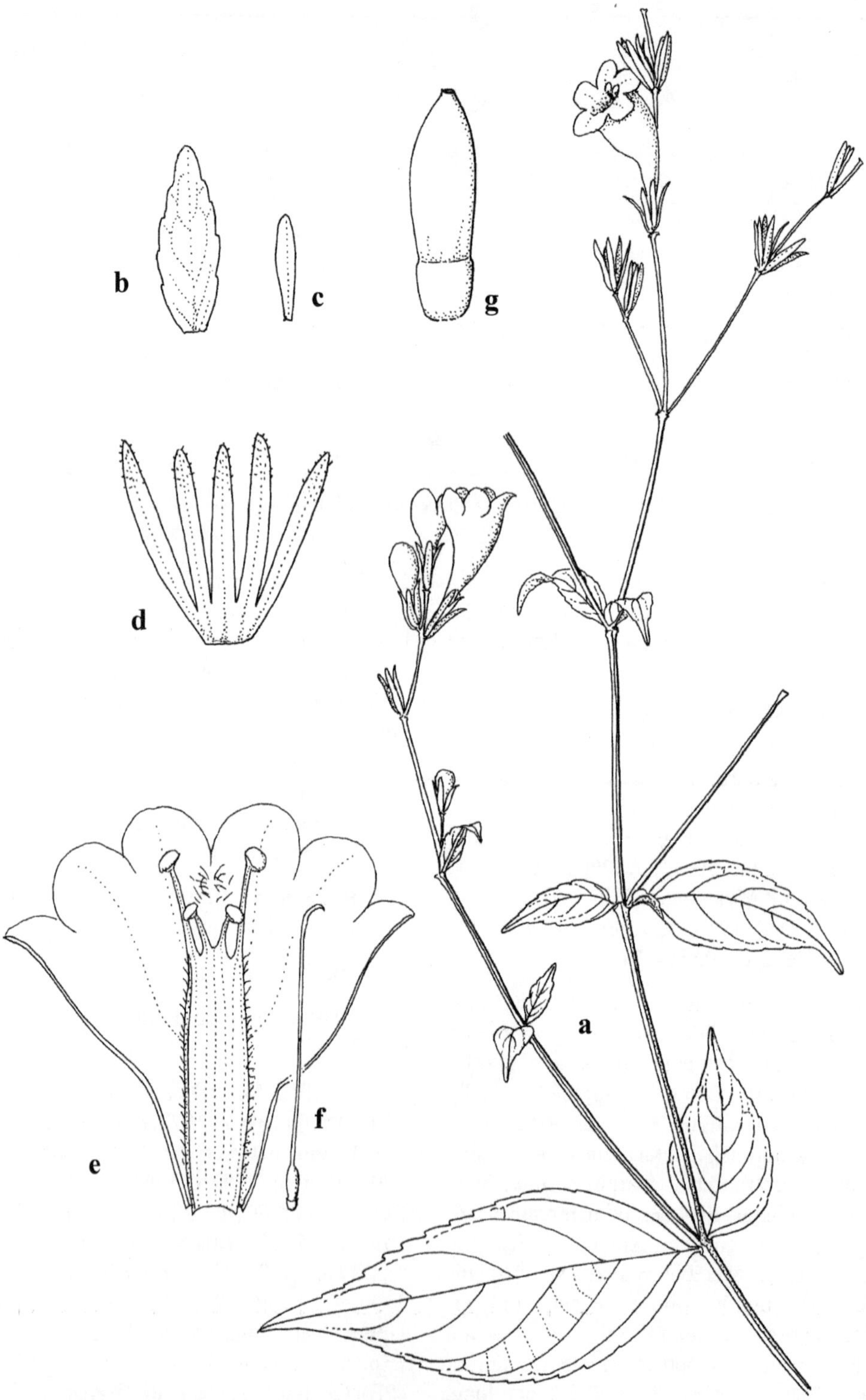

Figure 46-48: **Strobilanthes straminea** W. W. Sm.: **a**, habit; **b**, bract; **c**, bracteole; **d**, calyx; **e**, corolla opened out to show stamens; **f**, pistil; **g**, capsule. Drawn by Rosemary WISE; reproduced from *Kew Bull.* 64: 45, fig. 22: G-P. 2009.

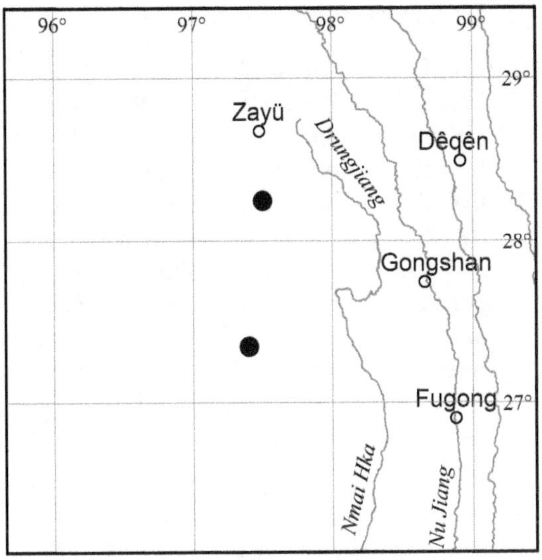

Map 46-87: **Strobilanthes straminea** W. W. Sm.

et al.) PT-ET 1259 (PE).

38. Strobilanthes arenicola W. W. Sm., Notes Roy. Bot. Gard. Edinburgh 10: 190. 1918. Type: Myanmar, Bed of Mali Kha, Insop, 600–1000 ft., 18 Dec. 1914, *F. Kingdon-Ward 2000* [lectotype **here designated**: E (E00133443)].

Shrubs, 30–45 cm high; stems quadrangular, pubescent, then glabrescent. Petiole 1–3 mm long, glabrous; leaf blade linear-lanceolate, 3–5 cm long, 0.5–0.9 cm broad, adaxially glabrous, abaxially sparsely pilose or glabrescent, lateral veins 5 pairs, base cuneate, margin subremotely serrulate, apex acute or subacuminate. Panicles of 1–3-furcate spikes, terminal, 5–8 cm long. Bracts linear-oblong, 3–4 mm long, glabrous, caducous; bracteoles similar to bracts, but smaller. Calyx 6–8 mm long, glandular-hirsute, equally 5-lobed to base; lobes linear-oblong, apex acute. Corolla 3–3.5 cm. long, violet, ventricose, outside sparsely puberulent to subglabrous, inside sparsely white villose; tube base cylindric, 1.5–1.7 cm long, 1.2–1.4 mm broad; lobes suborbicular, 5–7 mm long and broad. Stamens 4, included; filaments sparely pilose toward base. Ovary pilose; 2 ovules per locule; style glabrous. Capsules 1–1.2 cm long, sparsely pilose.

Phenology: flowering from November to January of next year.

Habitats: stream banks, forests; below 500 m.

Distribution: **endemic**; U Irrawaddy (Map 46-88).

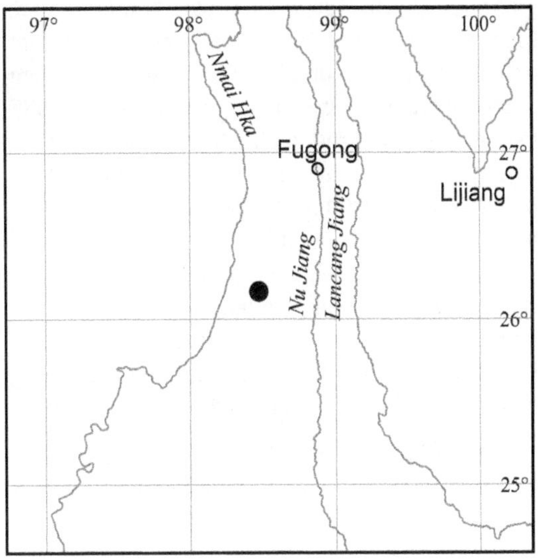

Map 46-88: **Strobilanthes arenicola** W. W. Sm.

U IRRAWADDY: Mali Hka Valley, 50 m, 300–450 m, 5 Jan. 1953, *F. Kingdon-Ward 20341* (BM); SE Sumprabum, 360 m, 15 Dec. 1953, *F. Kingdon-Ward 21693* (BM); Tamu, 300 m, 21 Nov. 1938, *F. Kingdon-Ward 18* (BM).

39. Strobilanthes pubiflora J. R. I. Wood, Edinburgh J. Bot. 51: 255. 1994; Wood in Grierson & Long (eds.), Fl. Bhutan 2(3): 1265. 2001. Type: Bhutan, Phuntsoling, between Kharbandi (Rinchending) and Sorchen, 800 m, 26 Nov. 1991, *J. R. I. Wood 7524* [holotype: E (E00273482); isotypes: K, THIMPHU].

Subshrubs, 0.5–1 m high, anisophyllous; stems glabrous. Leaves unequal in each pair; petiole 0.3–2 cm long; blade broadly oblong-elliptic, 6–20 cm long, 2–7 cm broad, glabrous on both surfaces, base attenuate, margin serrate, apex shortly acuminate, slightly falcate. Panicles of head-like spikes, axillary; peduncles 1–2.5 cm long, glabrous. Bracts suborbicular to orbicular,

2–9 mm long and broad, glabrous, apex rounded, concave, scarious, caducous at anthesis; bracteoles absent. Calyx 5–7 mm long in flower, accrescent to 1.6–2.0 cm in fruit, abaxially glandular-pubescent, adaxially glabrous, 5-lobed to base; lobes linear, one 1.5–2 mm longer than others, apex acute. Corolla 3.2–5 cm long, blue, slightly curved, outside pubescent, inside glabrous except for trichomes retaining style; tube base cylindric, 1–1.2 cm long, 1.5–2 mm broad, upwards gradually widened to 1.2–1.6 cm broad at mouth; lobes ovate, 4–5 mm long, 6–7 mm broad, apex rounded. Stamens 4, included; filaments pilose, longer pair 5–6 mm long, shorter pair 2–3 mm long; anther-thecae ellipsoid, 1–1.2 mm long. Ovary glandular-puberulent; 2 ovules per locule; style 2.5–3 cm long, glabrous, reddish; stigma curved. Capsules oblong, sparsely glandular-pubescent, 1.4–1.5 cm long, 4-seeded. Seeds ovate in outline, compressed, 1.6–2 mm long and broad, pubescent with mucilaginous hairs. Figure 46-49.

Phenology: flowering from September to October; fruiting from November to January of next year.

Habitats: woodlands; 600–1700 m.

Distribution: endemic; Sikkim & Darjiling, Bhutan (Map 46-89).

SIKKIM & DARJILING: Darjiling, Dumsong Forests, 900 m, 8 Sep. 1923, *J. M. Cowan s. n.*(E); 600–1500 m, *J. D. Hooker s. n.* (BM, K); Kalimpong, 1070 m, 15 Nov. 1875, *C. B. Clarke 26364* (BM, K); Kurseong, 1370 m, 27 Sep. 1884, *C. B. Clarke 35943* (BM); Tinkdah (Takdah), 1520 m, 21 Nov. 1875, *J. Gamble 305* (K); Surreil, 1540 m, *J. Gamble 9904* (K). **BHUTAN**: **Chukka**, escarpment above Kharbandi (Rinchending), 1100 m, 12 Nov. 1987, *J. R. I. Wood 6089* (E, FHO); **Phuntsoling**, between Kharbandi (Rinchending) and Sorchen, 800 m, 26 Nov. 1991, *J. R. I. Wood 7524* (E, K); near Kamjee, between Gedu and Phuntsholing, 1700 m, 18 Jan. 1992, *J. R. I. Wood 7527* (E); **Samchi**, in valley forests, 1 km S of Dorokha, 1000 m, 22 Nov. 1991, *J. R. I. Wood 7517* (E, FHO, K); **Sarbhang** towards Chirang, 1100 m, 19 Nov. 1988, *J. R. I. Wood 6154* (E); **Tongsa**, between Dakpai and Tintibi, 1000 m, Nov. 1989, *J. R. I. Wood 7227* (E).

40. Strobilanthes persicifolia (Lindl.) J. R. I. Wood, Kew Bull. 64: 27. 2009. ≡ *Ruellia persicifolia* Lindl., Bot. Reg. 11: pl. 955 (1 Feb. 1826). Type: Plant grown by Mr Colvill's nursery from seeds introduced by Mrs Fairlie, presumably from Calcutta, specimens in Herb. Lindley, unlabelled portion on top left of sheet labelled *Ruellia anisophylla*, annotated "1" and cross referenced below to H. H. S. (lectotype designated by Wood & Scotland, Kew Bull. 64: 27. 2009: CGE).

Ruellia anisophylla G. Lodd., Bot. Cab. 11(7): pl. 1070. 1826. ≡ *Goldfussia anisophylla* (G. Lodd.) Nees, Pl. Asiat. Rar. 3: 88. 1832. ≡ *Strobilanthes anisophylla* (G. Lodd.) T. Anderson, Cat. Pl. Calcutta: 43: 1865; Anderson, J. Linn. Soc. Bot. 9: 478. 1867; Clarke in Hooker (ed.), F. Brit. India 4: 462. 1884; Wood in Grierson & Long (eds.), Fl. Bhutan 2(3): 1273. 2001. Type: [*icon*] Lodd, Bot. Cab.11: pl. 1070. 1826 (lectotype designated by Wood & Scotland, Kew Bull. 64: 28. 2009).

Subshrubs, 30–60 cm high, anisophyllous; stems sometimes zigzag, glabrous. Leaves very unequal, smaller very small and soon caducous or equal; petiole (0–)3–9 mm long; blade ovate-lanceolate or oblong-lanceolate, 1.5–9 cm long, 0.4–1.3 cm broad, glabrous, apex acuminate, margin obscurely serrulate, base cuneate. Panicles of head-like spikes, axillary; peduncles 2–12 mm long. Bracts elliptic, 6–7 mm long, apex obtuse, caducous; bracteoles oblong-lanceolate, 3–4 mm. Calyx 6–9 mm long, reddish glandular-pubescent; lobes linear, subequal, apex acute. Corolla pale blue or white, 2.5–2.8 cm long, outside glandular-pubescent; tube slightly curved, upwards gradually widened to ca. 1 cm broad at mouth. Ovary pilose; 2 ovules per locule; style 2–2.3 cm long. Capsules 0.9–1.0 cm long, glandular-pubescent. **Chromosomes** 2n = 20, 26.

Phenology: flowering from October to December.

Habitats: rocks and cliffs beside rivers in subtropical forests; 300–1000 m.

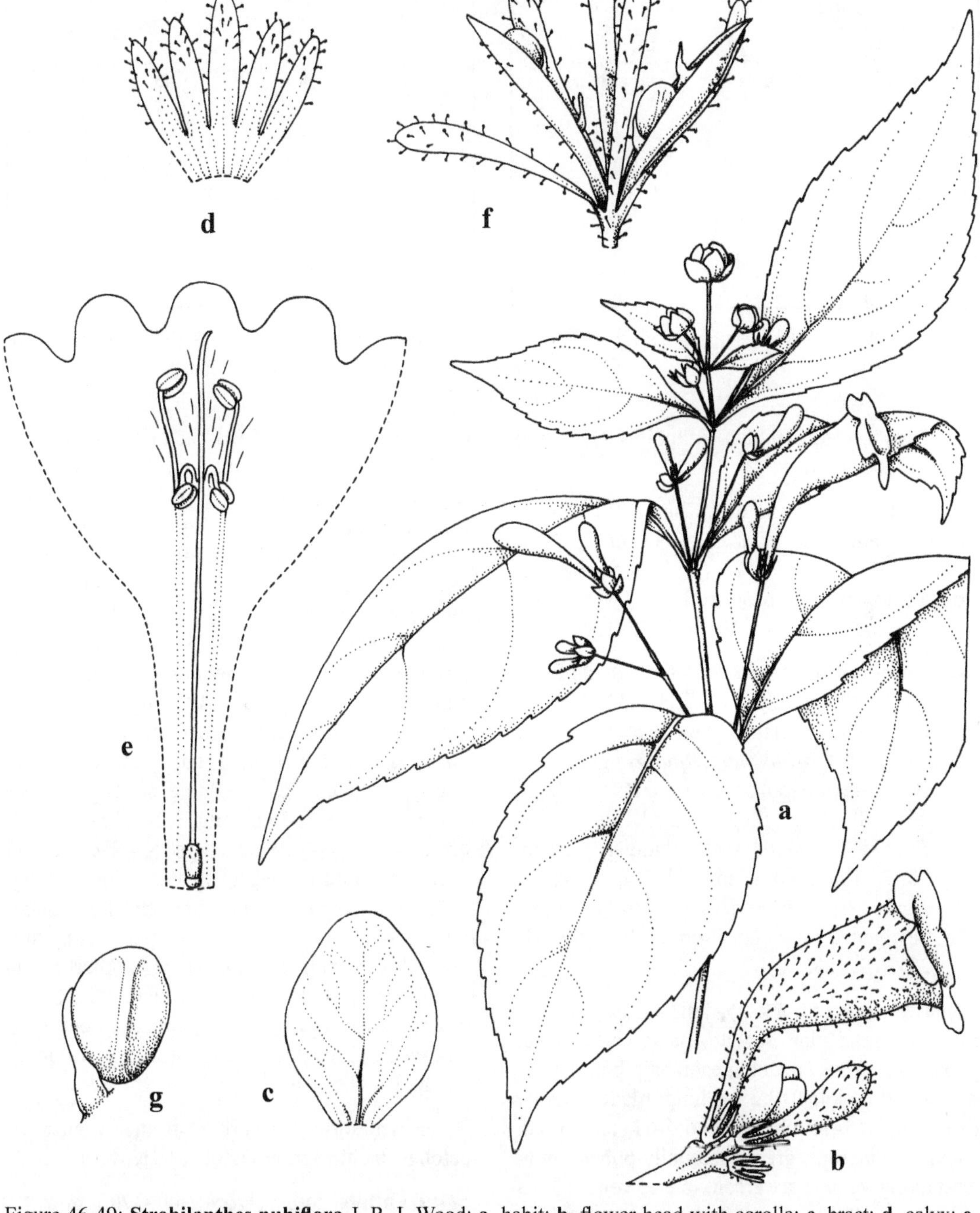

Figure 46-49: **Strobilanthes pubiflora** J. R. I. Wood: **a**, habit; **b**, flower-head with corolla; **c**, bract; **d**, calyx; **e**, corolla opened showing stamens and pistil; **f**, capsule with fruiting calyx; **g**, seed. From *Edinburgh J. Bot.* 51: 252, fig. 11: A-G. 1994.

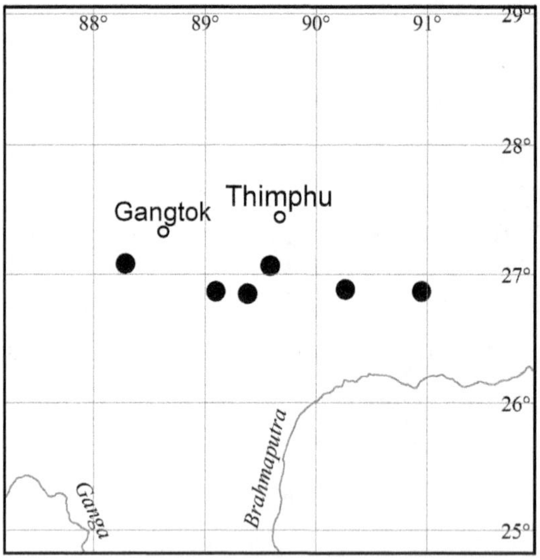

Map 46-89: **Strobilanthes pubiflora** J. R. I. Wood

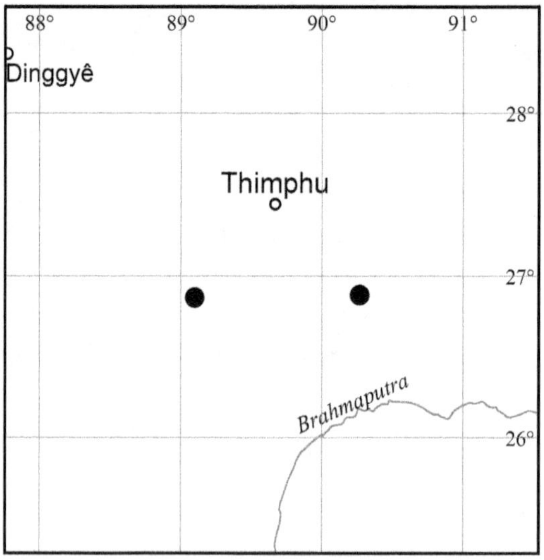

Map 46-90: **Strobilanthes persicifolia** (Lindl.) J. R. I. Wood

Distribution: Bhutan (Map 46-90); also in other parts of India.

BHUTAN: **Samchi**, Chamurchi River, between Samchi and Dorokha, 600 m, 23 Nov. 1991, *J. R. I. Wood 7520* (E, TI); **Sarbhang**, Longa Khola near Phipsoo, 300 m, 17 Mar. 1982, *A. J. C. Grierson & D. G. Long 3795* (E).

41. Strobilanthes bracteata (Nees) J. R. I. Wood, Edinburgh J. Bot. 51: 232. 1994. ≡ *Goldfussia bracteata* Nees, Pl. Asiat. Rar. 3: 88. 1832. ≡ *Strobilanthes quadrangularis* C. B. Clarke in Hooker (ed.), Fl. Brit. India 4: 460. 1885, nom. illeg. (Art. 52.1 & 52.2). ≡ *Ruellia bracteata* Wall., Numer. List n. 2357. 1830, nom. invalid. Type: Nepal, *Herb. Heyne in Wallich 2357b* (lectotype designated by Wood, Edingurgh J. Bot. 51: 232. 1994: K-W).

Herbs or subshrubs, 0.5–1 m high, isophyllous; stems quadrangular, sulcate, narrowly winged, pubescent. Leaves equal in each pair, petiolate or subsessile; petiole 0–2.5 cm long; blade obovate to elliptic, 15–30 cm long, 6.5–13 cm broad, adaxially glabrous, green, abaxially pubescent at least along veins, pale green, lateral veins 11–13 pairs, base attenuate and decurrent onto petiole, margin serrate, apex shortly acuminate. Panicles of trifurcate head-like spikes, umbellate, axillary; peduncles 2.5–7.5 cm long, puberulent. Bracts broadly ovate, 1–1.5 cm long, 1–1.3 cm broad, pubescent, apex acute, caducous; bracteoles oblong-ovate, 2–3 mm long, 1–1.2 mm broad, pubescent, caducous. Calyx 0.7–1.4 cm long, glandular-pubescent abaxially, 5-lobed to 2–3 mm above base; lobes unequal, one lobe longer than others, oblong-spatulate, longer one 9–10 mm long, 2–2.5 mm broad, smaller ones 5–7 mm long, 1.8–2 mm broad, apex obtuse. Corolla blue, 3.5–4.5 cm long, outside pubescent; tube base cylindric, 0.5–1 cm long, 1.2–1.5 mm broad, upwards gradually widened to 1–1.2 cm broad at mouth; lobes oblong to ovate, 4–5 mm long, 5–7 mm broad, apex retuse. Stamens 4, included; filaments of longer pair unequal, 6–8 mm long, shorter pair 1–2 mm long; anther-thecae subspherical, 1–1.2 mm in diameter. Ovary pilose; 2 ovules per locule; style 2.5–3 cm long, pubescent. Capsules oblong, 1.2–1.5 cm long, glandular-pubescent.

Phenology: flowering from August to October; fruiting from October to November.

Habitats: forest slopes; 900–3700 m.

Distribution: U Ganga & Indus, W Nepal, C Nepal, E Nepal (Map 46-91); also in other parts of India.

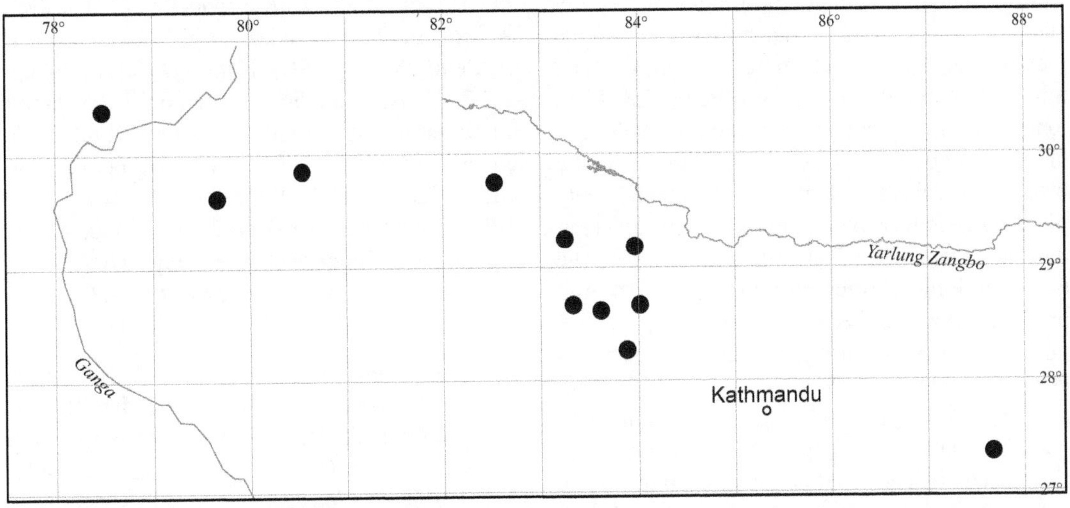

Map 46-91: **Strobilanthes bracteata** (Nees) J. R. I. Wood

U GANGA & INDUS: Almora, Maya-Ka-Danda Reserve, 30 Sep. 1932, *A. E. Osmaston 1507* (A); **Tehri Garhwal**, near Kuliya Ghat, Kaleemondee, Garhwal, 2440 m, *Madden s. n.* (E). **W NEPAL: Karnali**, Mugu, between Gum and Lumsa, Karnali Valley, 1980 m, 14 Aug. 1952, *O. V. Polunin, W. R. Sykes & L. H. J. Williams 5204* (BM, E); Dolpa, Rachi to Ankhe, 2760 m, 9 Oct. 1991, *M. Minaki et al. 9104450* (KATH, TI); Dolpa, Suligad, 3 Sep. 2014, *s. coll. A229* (E); **Mahakali**, Darchula, Sunchera, 1800 m, 25 Aug. 1980, *K. R. Rajbhandari & K. J. Malla 5571* (KATH). **C NEPAL: Dhawalagiri**, Mustang, Marpha to Titigaon, 2570 m, 5 Aug. 1996, *T. Hoshino et al. 9666171* (A, CBM, E, KATH, TI); Mustang, Lete, 2210 m, 3 Nov. 2014, *K. R. Rajbhandari et al. 201411049* (KATH); Myagdi, Lumsum, 2120 m, 9 Sep. 1954, *O. V. Polunin, W. R. Sykes & L. H. J. Williams 4302* (BM, E, TI); **Gandaki**, Kaski, Lumle, 1700 m, 13 Oct. 1976, *G. Dawson 553* (BM); Kaski, Dhampus, 1880 m, 1 Oct. 1969, *H. Flatt 104a* (BM); Kaski, Bhainsi Kharka to Banthanti, 2500 m, 23 Aug. 1988, *M. Suzuki et al. 8810935* (KATH, TI); Manang, Tilche, 2160 m, 13 Aug. 1994, *M. Suzuki et al. 9485401* (A, BM, E, TI); Manang, Bimtang, near Gho village, 2480 m, 14 Aug. 2008, *s. coll. 20818153* (E). **E NEPAL: Mechi,** Taplejung, Doban to Nessum, 900 m, 28 Aug. 1977, *H. Ohashi et al. 772755* (KATH).

42. Strobilanthes oxycalycina J. R. I. Wood, Edinburgh J. Bot. 51: 250. 1994; Hu, Deng & Wood in Wu, Raven & Hong (eds.), Fl. China 19: 424. 2011. Type: Yarlung Zangbo-Brahmaputra, Dambuk, 305 m, 2 Dec. 1928, *F. Kingdon-Ward 8748* [holotype: K (K000883148); isotype: K (K000883148)].

Subshrubs, 1–2.5 m high, isophyllous or slightly anisophyllous; stems somewhat sulcate, glabrous. Leaves equal or slightly unequal in each pair; petiole 0.5–1.7 cm long, glabrous; blade broadly elliptic to oblong-elliptic, 5–14 cm long, 2–7 cm broad, both surfaces glabrous, lateral veins 6–9 pairs, base cuneate, margin serrate, apex acuminate. Spikes terminal, head-like, simple or 3-furcate branched, often with a few flowers some distant below head; peduncles 2–4 cm long, glabrous. Bracts ovate, 6–7 mm long, 2.5–3 mm broad, caducous, sparsely pubescent, apex shortly acuminate; bracteoles oblong-obovate, 2.5–3 mm long, 1–2 mm broad, apex acute. Calyx 8–9 mm long, accrescent to 1.1–1.2 cm long in fruit,

5-lobed almost to base; lobes linear to narrowly oblong-elliptic, equal, abaxially pubescent with glandular-pilose, adaxially glabrous, apex acute. Corolla blue, 3–4 cm long, straight or slightly curved, outside glandular-pubescent, inside glabrous except for trichomes retaining style; tube base cylindric, 1–1.2 mm broad, 0.8–1 cm long, upwards gradually widened to 8–9 mm broad at mouth; lobes ovate, 4–5 mm long and broad, apex obtuse. Stamens 4; filaments finely pubescent, longer pair slightly exserted, erect, unequal, longer one 7–8 mm long, shorter one 6–7 mm long, shorter pair included, 3–4 mm long, recurved; anther-thecae spherical, 1.1–1.3 mm in diameter. Ovary glandular-pilose towards apex; 2 ovules per locule; style 2.2–2.5 cm long, pilose. Capsules oblong, 1.1–1.3 cm long, glandular-pubescent on upper half especially near tip, 4-seeded. Seeds ovate in outline, compressed, 2.5–3 mm long and broad, densely covered with appressed trichomes. Figure 46-50.

Phenology: flowering from December to February.

Habitats: forests; 300–2200 m.

Distribution: endemic; U Yarlung Zangbo, Yarlung Zangbo-Brahmaputra (Map 46-92).

U YARLUNG ZANGBO: **Gyirong**, Gyirong to Mangmu, 2100 m, 16 Sep. 1981, *Z. C. Ni et al. 2315* (PE). **YARLUNG ZANGBO-BRAHMAPUTRA**: **Baibung**, 1st Bridge to Hanmi, 800–1010 m, 25 Sep. 2003, *J. T. Yin et al. 511* (HITBC); Baibung to Gelin, 1100 m, 12 Dec. 2019, 1100 m, 12 Dec. 2019, *Y. F. Deng 28896* (IBSC); Hanmi to No. 1 Bridge, 2000 m, 23 Sep. 2003, *X. F. Gao et al. 7851* (CDBI); Renqinpeng, 2100 m, 23 Apr. 1993, *H. Sun et al. ETM-5231* (KUN); Hanmi to Maniweng, 1000 m, 6 Aug. 1974, *Exped. Qinghai-Xizang 74-4156* (KUN, PE); **Bangxing**, Gemendeng, 1400 m, *B. S. Li & S. Z. Cheng 3103* (PE); **Dexing**, 730 m, 13 Nov. 2019, *Y. F. Deng 28918* (IBSC); **Gutang**, Sangzhenka, 1754 m, 14 Nov. 2019, *Y. F. Deng 28937* (IBSC); Gedang to Xingkai, 2200 m, 31 Aug. 1974, *Exped. Qinghai-Xizang 74-4967* (KUN, PE); Gutang to Xingkai, 2100 m, 28 Sep. 1982, *B. S. Li & S. Z. Cheng 1107* (PE); Gutang, 1750 m, 1 Sep. 1980, *W. L. Chen 14578* (PE); Gutang, 1750 m, *W. L. Chen 16578* (PE); Damor to Gutang, 1605 m, 13 Nov. 2016, *Y. J. Ge et al. 16CS14419* (IBSC); **Lohit Valley**, 600–900 m, 3 Dec. 1926, *F. Kingdon-Ward 7690* (K); **Mêdog**, on the way to Baibung, 13 Nov. 2019, *Y. F. Deng 28905* (IBSC); Hanmi, 2100 m, 5 Aug. 1973, *Xizang Expedition 1742* (HNWP); Damor to Daguoji, 1400–1600 m, 29 Oct. 1982, *B. S. Li & S. Z. Cheng 1543* (PE); Nigulong, 1050 m, 30 Jan. 1983, *B. S. Li & S. Z. Cheng 3535* (PE); Bangushan, 1900 m, 27 Jul. 1980, *Alpine Group from Department of Ecology 11373* (PE); Buqiong Lake, 1400 m, *H. Sun et al. ETM-1315* (KUN); *H. Sun et al. ETM-3103* (KUN).

43. Strobilanthes rhombifolia C. B. Clarke in Hooker (ed.), Fl. Brit. India 4: 461. 1885; Hu, Deng & Wood in Wu, Raven & Hong (eds.), Fl. China 19: 425. 2011. ≡ *Goldfussia sessilis* Nees in de Candolle (ed.), Prodr. 11: 172. 1847, non *Strobilanthes sessilis* Nees in Wallich (ed.), Pl. Asiat. Rar. 3: 85. 1832. Type: Yarlung Zangbo-Brahmaputra, "Assam", *Griffith s. n.* [holotype: K (K000883068); isotypes: GZU (GZU000250034), K].

Herbs or subshrubs, strongly anisophyllous; stems quadrangular, glabrous. Leaves unequal in each pair, lower ones petiolate, apical ones subsessile; petiole 0–10 mm long, glabrous; blade rhomboidal-elliptic, 2–15 cm long, 1.5–10 cm broad, glabrous, lateral veins 5–7 pairs, base cuneate, margin serrate, apex acute. Spikes head-like, axillary, simple or 2- (or 3-) furcately branched, 2–10 cm long; peduncle sulcate, bifariously pubescent along sulci. Bracts variable in shape and size, out ones sterile, ovate, 5–6 mm long, 4–5 mm broad, relatively persistent; inner ones fertile, obovate to oblong-obovate, 5–6 mm long, 1.2–1.5 mm broad, soon scarious and caducous; bracteoles absent. Calyx subequally 5-lobed to base; lobes linear-elliptic, 7–12 mm long, 1–1.2 mm broad, glandular and eglandular pubescent. Corolla blue, 4–4.5 cm long, outside sparsely glandular-pilose in bud, then glabrescent, inside glabrous except for trichomes retaining style; tube base cylindric, 1.5–2 mm broad, 0.8–1 cm long, upwards widened to 1.2–1.7 cm broad at mouth; lobes suborbicular, 5–6 mm long,

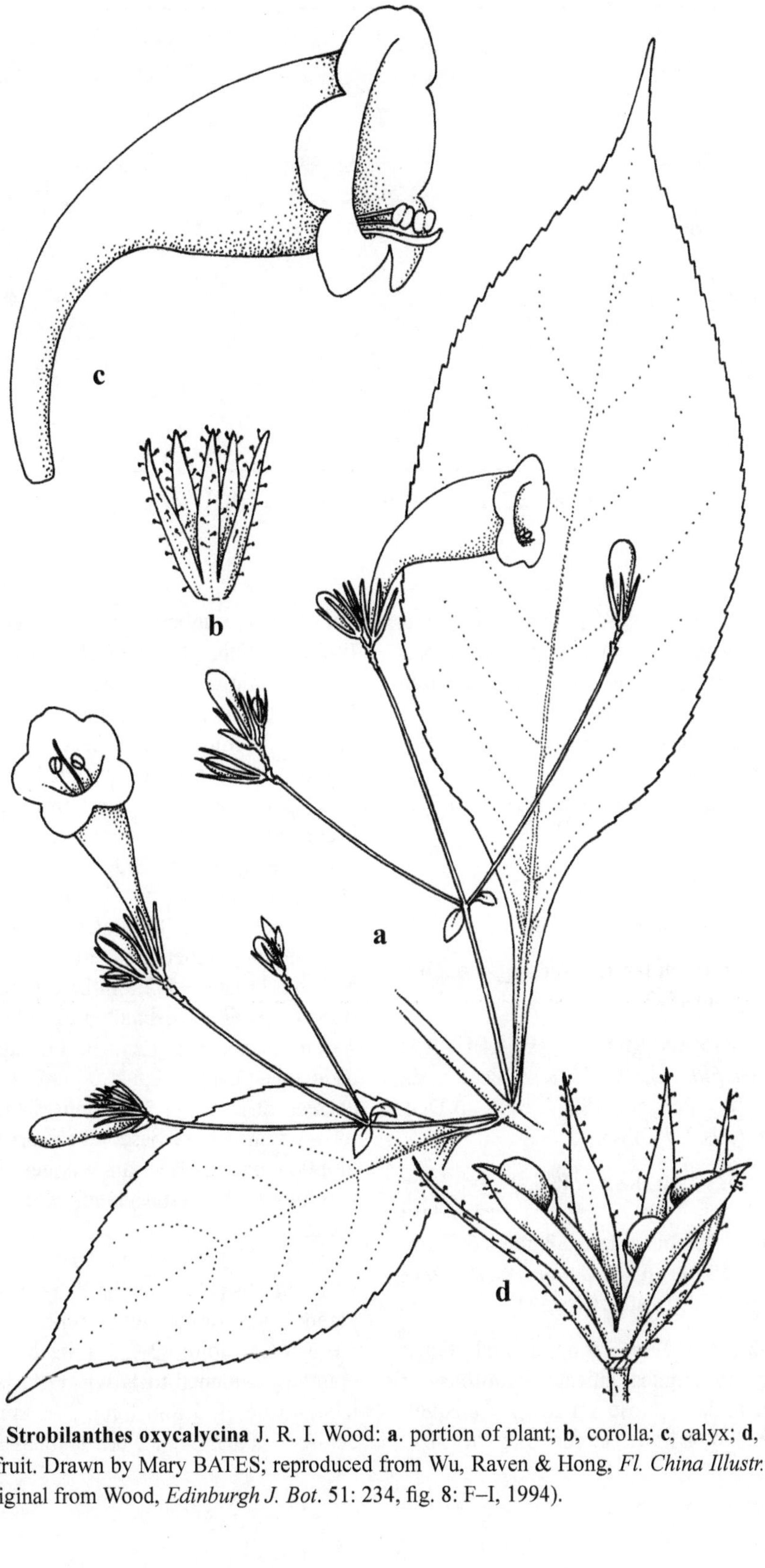

Figure 46-50: **Strobilanthes oxycalycina** J. R. I. Wood: **a**. portion of plant; **b**, corolla; **c**, calyx; **d**, mature calyx with opened fruit. Drawn by Mary BATES; reproduced from Wu, Raven & Hong, *Fl. China Illustr.* 19: fig. 332: 1–4. 2012 (original from Wood, *Edinburgh J. Bot.* 51: 234, fig. 8: F–I, 1994).

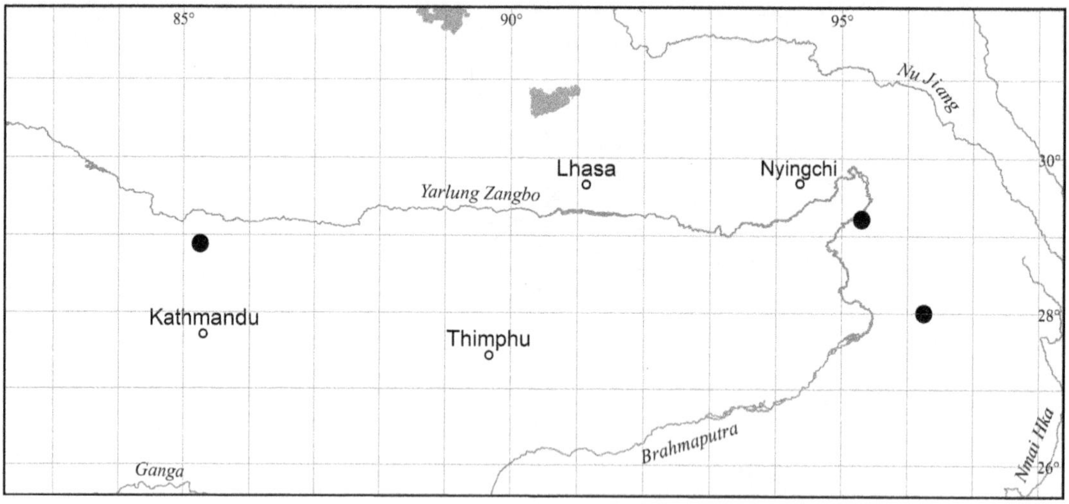

Map 46-92: **Strobilanthes oxycalycina** J. R. I. Wood

6–7 mm broad. Stamens 4, included; filaments glabrous, longer pair 5–6 mm long, shorter pair 2–3 mm long; anther-thecae oblong, 1.2–1.5 mm long, 1–1.2 mm broad, incurved. Ovary narrowly ovoid, glandular pilose towards apex; 2 ovules per locule; style 2.5–2.7 cm long, sparsely pilose towards base. Capsules unknown. Figure 46-51.

Phenology: flowering from November to January of next year.

Habitats: forests; below 400 m.

Distribution: endemic; Yarlung Zangbo-Brahmaputra (Map 46-93).

YARLUNG ZANGBO-BRAHMAPUTRA: s. loc., s. d., *Griffith s. n.* (GZU, K); s. loc., s. d., *Jemkens s. n.* (K); Pengare, 180–230 m, 26 Dec. 1935, *Barnard AHP31* (BM).

44. Strobilanthes ramulosa J. R. I. Wood, Kew Bull. 64: 43. 2009. Type: [U Irrawaddy] Myanmar, Kachin State, Putao, 500 m, 9 Dec. 1930, *F. Kingdon-Ward 9031* [holotype: BM (BM001046227)].

Subshrubs, 0.6–1 m high, anisophyllous; stems terete, sometimes sulcate, glabrous or sparsely glandular pilose. Leaves unequal, smaller one 3/4–4/5 of larger one in size, petiolate or uppermost ones sessile; petiole 0–7 mm long, glabrous; blade ovate-elliptic or ovate, diminishing in size upwards, 2.5–14 cm long, 2–6 cm cm broad, glabrous, base rounded, slightly asymmetric and decurrent onto petiole, margin serrate, apex acuminate. Panicles of head-like spikes, axillary and terminal, trifurcately branched 2–3 times; spikes 1–4-flowered; peduncles glabrous. Bracts variable in side and shape, diminishing in size upwards, glabrous, outer ones sterile, lanceolate, 1.0–2.0 cm long, 3–6 mm long, inner ones fertile, elliptic-obovate, 3–4 mm broad, 1.5–2 mm broad, apex obtuse to truncate, caducous; bracteoles oblong-spatulate, 4–5 mm long, 0.8–1 mm broad, apex rounded, caducous. Calyx 0.8–1.0 cm long, weakly accrescent to 1.1–1.3 cm long in fruit, 5-lobed to 1–1.5 mm above base; lobes narrowly oblong-elliptic, one 1–1.5 mm longer than others, glabrous with sessile glands when young, then glandular and eglandular hirsute, apex obtuse. Corolla 3.2–3.8 cm long, pale violet or white with purplish lines, slightly curved, sparsely glandular-pilose on tube; tube base cylindric, 1.0–1.5 cm long, 2–2.5 mm broad, upwards gradually widened to 1–1.1 cm broad at mouth; lobes ovate, 4–5 mm long, 5–6 mm long, apex obtuse. Stamens 4, included; filaments glabrous,

Figure 46-51: **Strobilanthes rhombifolia** C. B. Clarke: **a**, portion of plant. Drawn by Mary BATES; reproduced from *Fl. China Illustr.* 19: figure 336: 1. 2012 (original from Wood, *Edinburgh J. Bot.* 51: 257, fig. 12: E. 1994).

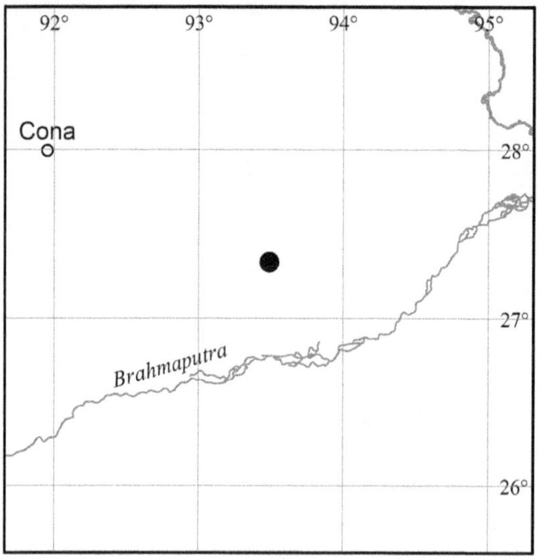

Map 46-93: **Strobilanthes rhombifolia** C. B. Clarke

longer pair 5–6 mm long, shorter pair 2–2.5 mm long; anther-thecae subspherical, 1–1.2 mm in diameter, nodding. Ovary glabrous; 2 ovules per locule; style 2.3–2.5 cm long, glabrous. Capsules unknown. Figure 46-52.

Phenology: flowering from December to January of next year.

Habitats: stream banks, shaded thickets; 100–600 m.

Distribution: endemic; U Irrawaddy (Map 46-94).

U IRRAWADDY: **Putao**, 500 m, 9 Dec. 1930, *F. Kingdon-Ward 9031* (BM); between Fort Hertz (Putao) and Sumprabum, 22 Dec. 1931, *F. Kingdon-Ward 10223* (BM).

45. Strobilanthes szechuanica (Batalin) J. R. I. Wood & Y. F. Deng, Bot. J. Linn. Soc. 150: 375. 2006; Hu, Deng & Wood in Wu, Raven & Hong (eds.), Fl. China 19: 391. 2011. ≡ *Hemigraphis szechuanica* Batalin, Trudy Imp. S.-Peterburgsk. Bot. Sada 13: 384. 1894. Type: [N Hengduan] China, Sichuan, inter Ta shui wan et Shi tsüan, 1 Sep. 1893, *G. N. Potanin s. n.* [holotype: LE; isotypes: K (K000190211, K000190211)].

Herbs to 20 cm high, isophyllous; stems decumbent, quadrangular, subglabrous, rooting at basal nodes. Leaves equal in each pair; petiole 3–5 mm long, puberulent; blade ovate to obovate, or suborbicular, 1–2 cm long, 0.8–1 cm broad, glabrous, lateral veins 3 or 4 pairs, base broadly cuneate and decurrent onto petiole, margin entire, sparsely ciliate, apex obtuse to rounded. Flowers axillary, solitary; bracteoles oblanceolate, 5–7 mm long, 1–1.2 mm broad, pilose, margin sparsely ciliate. Calyx 8–10 mm long, 5-lobed almost to base; lobes linear to linear-lanceolate, margin ciliate, apex obtuse. Corolla pale lilac, funnel-shaped, 1.8–2.3 cm long, outside pubescent; tube base cylindric, 4–5 mm long, 0.8–1 mm broad, upwards gradually widened to 7–8 mm broad at mouth; lobes suborbicular, 2–3 mm long and broad, margin ciliate. Stamens 4, included; filaments of longer pair 5–6 mm long, pilose, shorter pair 2–3 mm long, glabrous; anther-thecae oblong, 1.5–2 mm long, 1–1.2 mm broad. Ovary oblong, comose; 2 ovules per locule; style 1.4–1.8 cm long, pilose. Capsules 0.8–1 cm long, pilose towards apex, 4-seeded. Seeds ovate in outline, compressed, 1.5–2 mm in diameter, pilose.

Phenology: flowering from July to September.

Habitats: stream banks, on rocks; 700–1500 m.

Distribution: endemic; N Hengduan (Map 46-95).

N HENGDUAN: **Maoxian**, 1450 m, 28 Jul. 1930, *F. T. Wang 21867* (PE); Tumen, Lushun, Ranggangou, 1250 m, 8 Aug. 1975, *s. coll. 8618* (CDBI, PE).

Figure 46-52: **Strobilanthes ramulosa** J. R. I. Wood: **a**, habit; **b**, lower leaf surface; **c**, bract; **d**, bracteole; **e**, calyx; **f**, corolla opened out to show stamens; **g**, pistil; **h**, capsule; **i**, seeds. Drawn by Rosemary WISE; reproduced from Wood, *Kew Bull.* 64: 45, fig. 22: A–F. 2009.

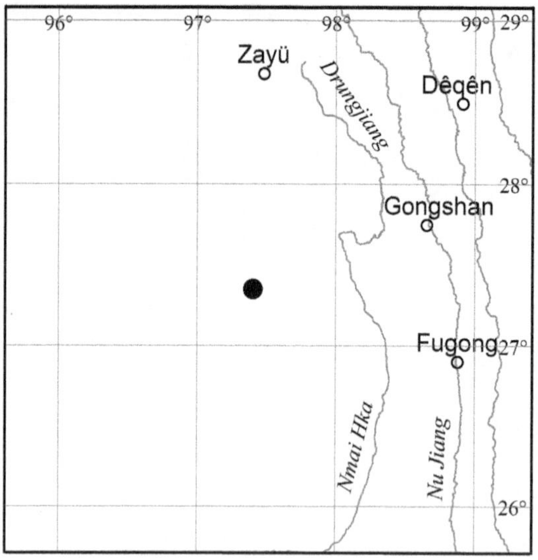

Map 46-94: **Strobilanthes ramulosa** J. R. I. Wood

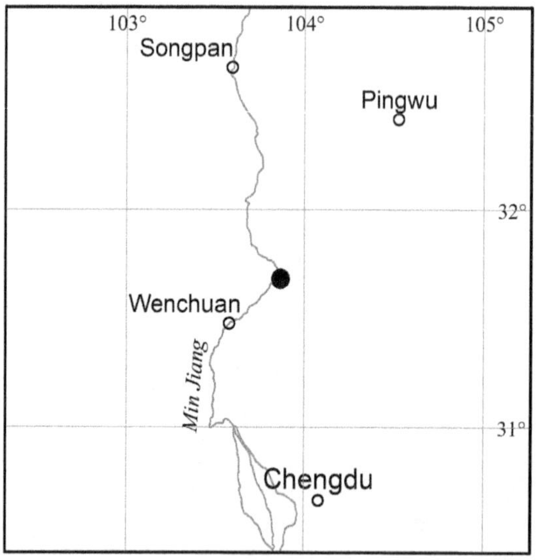

Map 46-95: **Strobilanthes szechuanica** (Batalin) J. R. I. Wood & Y. F. Deng

46. Strobilanthes wilsonii J. R. I. Wood & Y. F. Deng, Bot. J. Linn. Soc. 150: 371. 2006; Hu, Deng & Wood in Wu, Raven & Hong (eds.), Fl. China 19: 392. 2011. Type: China, Sichuan, Leshan, "Kurting", 3 Oct. 1903, *E. H. Wilson 4301* [holotype: K (K000882829); isotypes: K (K000882828), BM (BM000906323), HBG (HBG510923)].

Herbs, 30–60 cm high, isophyllous; stems procumbent, sulcate, glabrous or sparsely bifariously pilose, rooting at lower nodes. Leaves equal in each pair; petiole 0.5–2 cm long, bifariously hirsute; blade narrowly elliptic, broadly elliptic, or obovate-spatulate, 1.5–3 cm long, 1–1.5 cm broad, sparsely pilose towards base, lateral veins 3–5 pairs, base attenuate, margin serrate, apex shortly acuminate to obtuse. Spikes axillary or terminal, head-like, 1–1.5 cm long, few-flowered; peduncle 0.2–1.8 cm long, pilose. Bracts diminishing in size upwards, obovate-spatulate, 4–10 mm long, 2–6 mm broad, abaxially glabrous, adaxially sparsely pilose toward base, margin crenulate above the middle, ciliate, apex obtuse; bracteoles oblanceolate to obovate-spatulate, 2–3 mm long, 0.4–0.5 mm broad, margin ciliate, apex rounded. Calyx 4–5 mm long, subequally 5-lobed almost to base; lobes linear-oblong, 3–3.2 mm long, 0.8–1 mm broad, pilose, apex obtuse. Corolla blue, 7–8 mm long, funnel-shaped, outside pilose; tube base cylindric, 2–4 mm long, 0.8–1 mm broad, upwards gradually widened to 4–5 mm broad at mouth; lobes ovate, 1.5–2 mm long and broad, apex retuse. Stamens 4; filaments glabrous, longer pair 4–6 mm long, slightly exserted, shorter pair 2–3 mm long, included; anther-thecae oblong, 0.5–1 mm long, 0.2–0.3 mm broad; staminode 0.7–1 mm long. Ovary pilose; 2 ovules per locule; style 6–8 mm long. Capsules 2.5–4 mm long, pilose, 4-seeded. Seeds ovate in outline, compressed, lenticular, 1.5–2 mm long and broad, densely covered with appressed trichomes; areole small, glabrous. Figure 46-53.

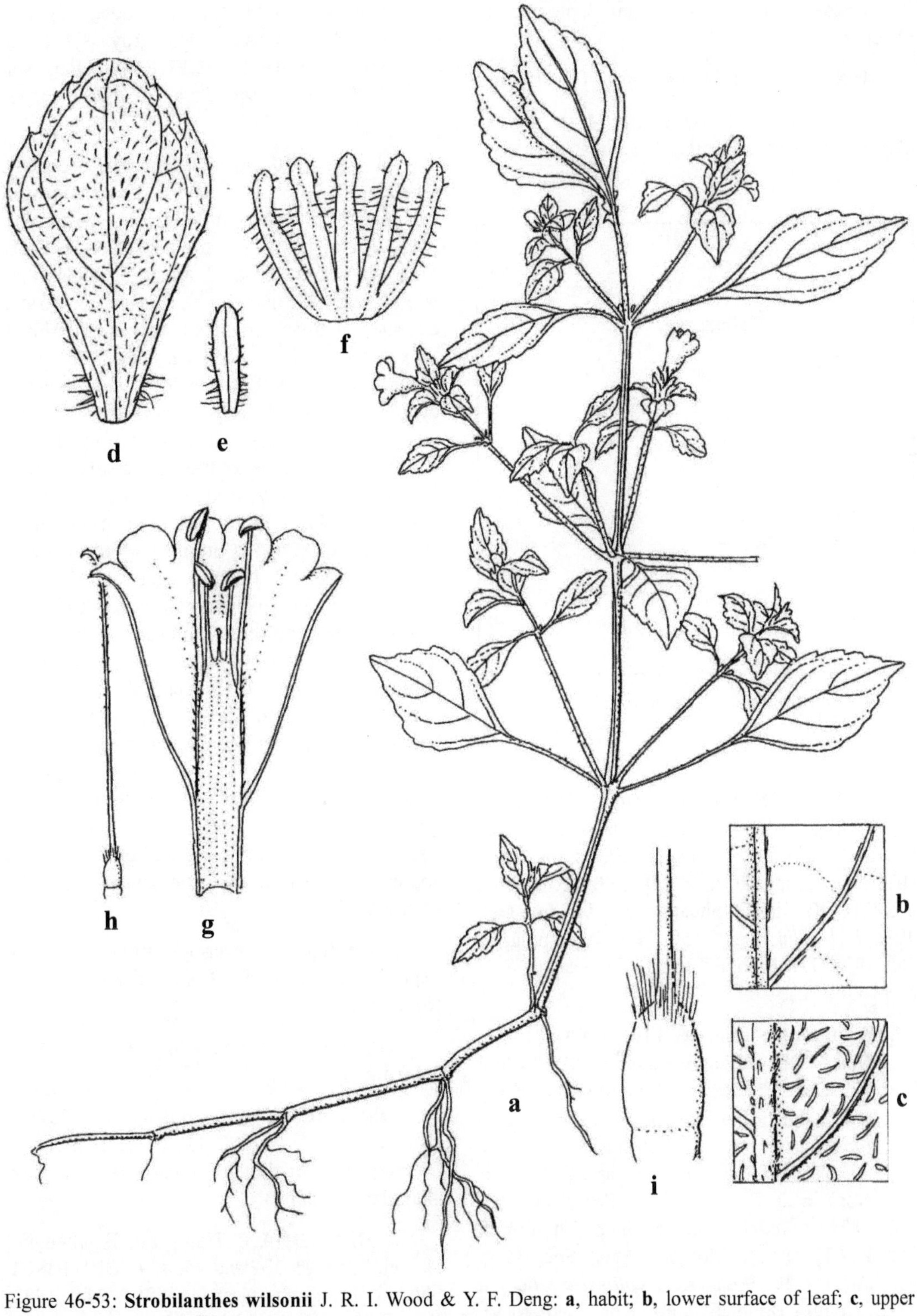

Figure 46-53: **Strobilanthes wilsonii** J. R. I. Wood & Y. F. Deng: **a**, habit; **b**, lower surface of leaf; **c**, upper surface of leaf; **d**, bract; **e**, bracteole; **f**, calyx; **g**, corolla, opened out; **h**, style; **i**, ovary. Drawn by Rosemary WISE; reproduced from Deng et al., *Bot. J. Linn. Soc.* 150: 374, fig. 9. 2006.

Phenology: flowering from September to October.

Habitats: moist shaded places; 700–1100 m.

Distribution: S Hengduan (Map 46-96); also in other parts of China (C Sichuan).

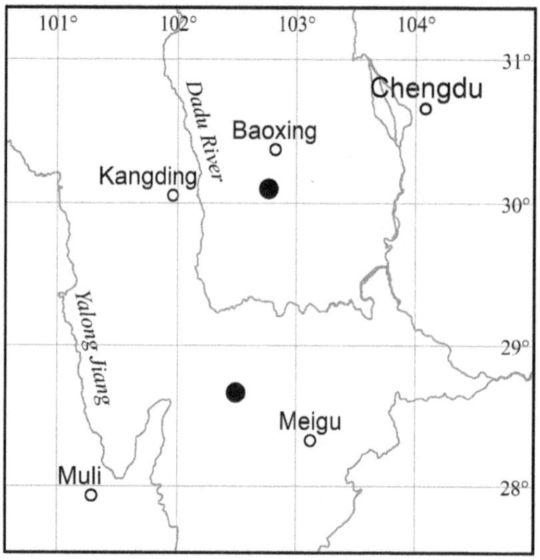

Map 46-96: **Strobilanthes wilsonii** J. R. I. Wood & Y. F. Deng

S HENGDUAN: **Tianquan**, Wenbifeng Park, 5 Aug. 2006, *Y. F. Deng 19079* (IBSC); 780 m, 16 Oct. 2008, *Y. F. Deng & H. J. Xin 20906* (IBSC); Yuyang, 870 m, 4 Jul. 1982, *D. Y. Peng 45347* (CDBI, IBSC); Shuangchang, 900 m, 8 Oct. 1936, *K. L. Chü 3983* (IBSC, PE); Kugaoping, 18 Oct. 1935, *Y. Y. Ho 6861* (NAS); **Yuexi**, s. d., *S. L. Sun 1337* (PE).

47. Strobilanthes japonica (Thunb.) Miq., Ann. Mus. Bot. Lugduno-Batavi 2: 124. 1866; Hu, Deng & Wood in Wu, Raven & Hong (eds.), Fl. China 19: 392. 2011. ≡ *Ruellia japonica* Thunb. in Murray, Syst. Veg., ed. 14: 576. 1784; Thunb., Fl. Jap.: 254. 1784; Nees in de Candolle (ed.), Prodr. 11: 153. 1874. ≡ *Acanthopale japonica* (Thunb.) C. B. Clarke ex S. Moore, J. Asiat. Soc. Beng. 74(2): 659. 1907. ≡ *Championella japonica* (Thunb.) Bremek., Verh. Kon. Ned. Akad. Wetensch., Afd. Natuurk., Sect. 2 41(1): 150. 1944; Hossain, Not. Roy. Bot. Gard. 38(3): 475, 1980; Hu, Fl. Reipub. Pop. Sin. 70: 90. 2002. Type: Japan, *C. P. Thunberg s. n.* [holotype: UPS [UPS-THUNB 14525)].

Herbs 20–50 cm high, isophyllous; stems quadrangular, slender, glabrous. Leaves equal in each pair; petiole 2–5 cm long, white pubescent; blade narrowly elliptic to lanceolate, 2–5 cm long, 0.5–1.8 cm broad, both surfaces glabrous, lateral veins 3–5 pairs, base cuneate to broadly cuneate, margin crenulate, apex acuminate. Spikes terminal, 1–3(–5) cm long, sometimes reduced to a single flower pair. Bracts imbricate, oblanceolate to oblong-lanceolate, 7–10 mm long, 2–3 mm broad, pubescence toward base, margin entire, apex obtuse; bracteoles linear, 4–5 mm long. Calyx equally 5-lobed almost to base; lobes linear, 6–7 mm long, margin ciliate, apex rounded. Corolla purplish-white or white, funnel-shaped, 1.2–1.5 cm long, straight, outside pubescent, inside glabrous except for trichomes retaining style; tube base cylindric, 5–6 mm long; lobes oblong-elliptic, 5–6 mm long. Stamens 4, included; filaments of longer pair 0.9–1.2 cm long, shorter pair 6–8 mm long; anther-thecae oblong, 1.2–1.5 mm long. Ovary glabrous; 2 ovules per locule; style 1–1.2 cm long, pilose; stigma 2-cleft. Capsules oblong, 6–8 mm long, glabrous, 4-seeded. Seeds ovate in outline, compressed, 2.5–3 mm long and broad, densely appressed pubescent. **Chromosomes** 2n = 30. Figure 46-54.

Phenology: flowering from June to September; fruiting from September to November.

Habitats: stream banks or wet places; 500–1100 m.

Distribution: S Hengduan (Map 46-97); also in other parts of China (C & E Sichuan, Chongqing, Guizhou, Hunan, Hubei), and naturalized in Japan.

S HENGDUAN: **Tianquan**, Beichengjie, 18 Jun. 1982, *D. Y. Peng 45241* (CDBI, IBSC); Shiyang, Duogong, Qieshan, 770 m, 3 Jul. 1982, *D. Y. Peng 45312* (CDBI, IBSC).

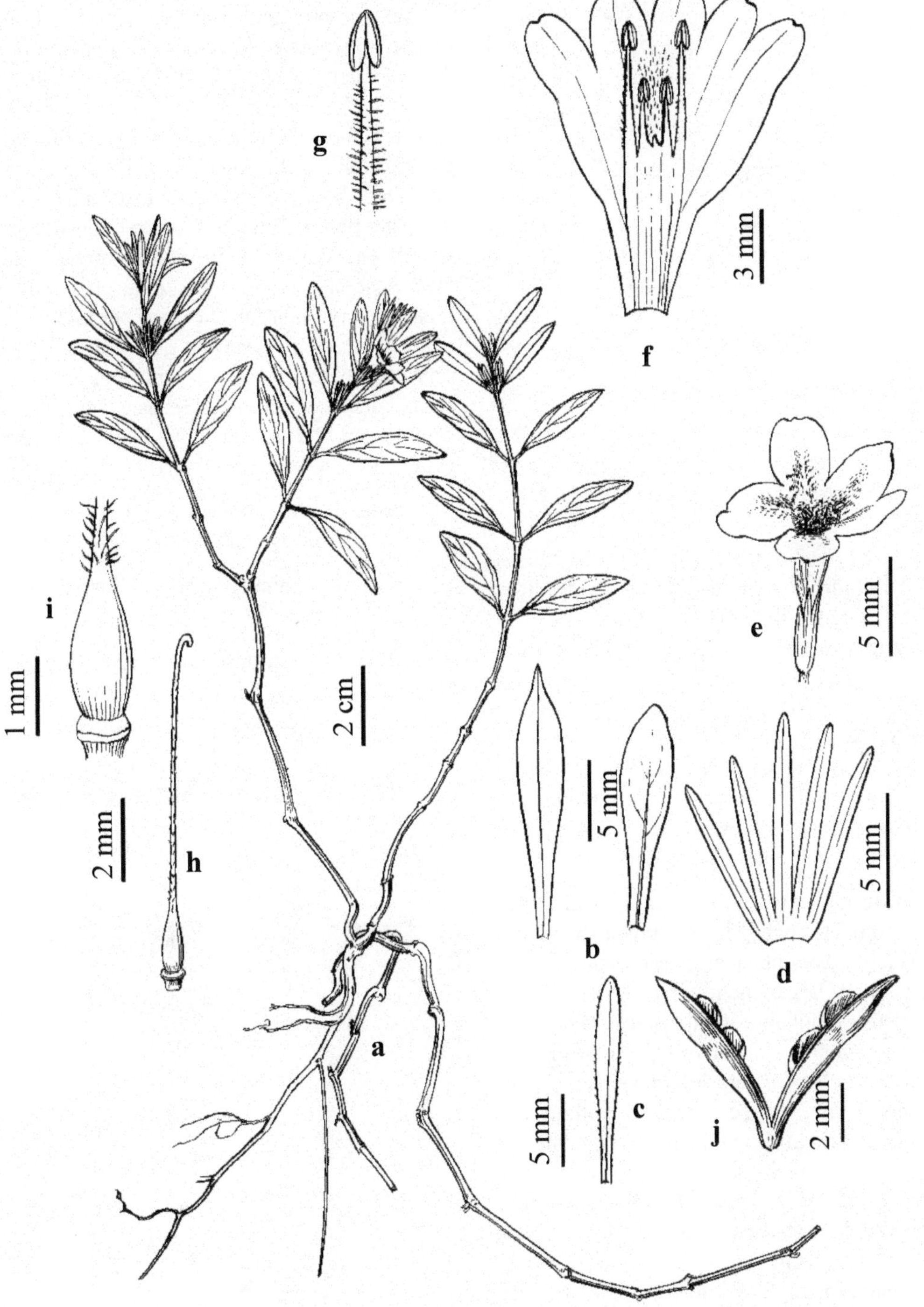

Figure 46-54: **Strobilanthes japonica** (Thunb.) Miq.: **a**, habit; **b**, bracts; **c**, bracteole; **d**, calyx; **e**, corolla; **f**, opened corolla showing stamens; **g**, stamen; **h**, pistil; **i**, ovary; **j**, opened capsule showing seeds. Drawn by Ding-Han CUI.

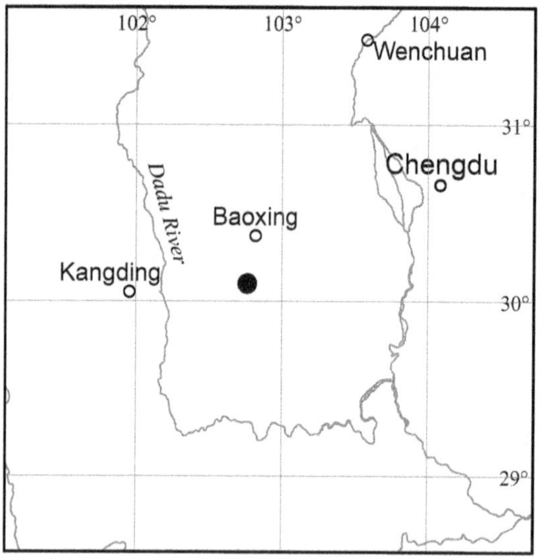

Map 46-97: **Strobilanthes japonica** (Thunb.) Miq.

48. Strobilanthes congesta Terao, Notes Roy. Bot. Gard. Edinburgh 40: 153. 1982; Hu, Deng & Wood in Wu, Raven & Hong (eds.), Fl. China 19: 411. 2011. ≡ *Pteracanthus congestus* (Terao) C. Y. Wu & C. C. Hu in Hu (ed.), Fl. Reipubl. Popularis Sin. 70: 133. 2002. Type: [U Irrawaddy] Myanmar, Hpawte valley, descending towards the Ngaw Chang valley, 22 Aug. 1919, *R. J. Farrer 1263* [holotype: E (E00133515)].

Subshrubs, 0.6–1.8 m high, anisophyllous; stems subterete, whitish to brownish yellow pubescent. Leaves unequal in each pair, smaller one 3/5–2/3 of larger one in size; petiole 1–7 mm long, densely hirsute; blade ovate-lanceolate, 2–9 cm long, 0.8–3.5 cm broad, both surfaces brownish to yellow hirsute, lateral veins 4–7 pairs and inconspicuous, base attenuate, oblique and shortly decurrent onto petiole, margin crenate and ciliate, apex acuminate and often falcate. Panicles of spikes, axillary or terminal; spikes head-like, 2–8-flowered, 1–4-congested; rachis 3–7 cm long, densely white or yellow pilose. Bracts linear-oblanceolate, 3–8 mm long, 0.8–1.5 mm broad, pilose, apex obtuse; bracteoles linear, 5–6 mm long, 0.8–1 mm broad, white pilose. Calyx 1.1–1.4 cm long, abaxially pilose, adaxially pilose except for glabrous towards base, 5-lobed almost to base; lobes ovate to lanceolate, 1.5–5 mm broad, unequal, one 1.5–2 mm longer than others, margin abruptly narrowed at the middle, apex obtuse, mucronate. Corolla pale whitish-purple, 2.2–2.5 cm long, curved, outside pubescent; tube base cylindric, 5–6 mm long, 1.2–1.5 mm broad, upwards abruptly inflated and widened to 1–1.2 cm broad at mouth; lobes oblong, 4–5 mm long, 3.5–4 mm broad. Stamens 4, included; filaments of longer pair 5–6 mm long, pilose towards base, shorter pair 2.5–3.5 mm long, glabrous; anther-thecae oblong, 1.6–2 mm long, 0.4–0.5 mm broad. Ovary glabrous; 2 ovules per locule; style 1.6–1.8 cm long, pilose and glandular-pubescent. Capsules oblong, 0.8–1.0 cm long, 2–3 mm broad, glabrous, 4-seeded. Seeds ovate in outline, compressed, 2.5–3 mm long, 2–2.5 mm broad, pubescent.

Phenology: flowering from July to September; fruiting from November to December.

Habitats: thickets, tropical forests; 1500–2200 m.

Distribution: endemic; S Hengduan and U Irrawaddy (Map 46-98).

S HENGDUAN: **Fugong**, Shang-Pa, 1500 m, 24 Sep. 1933, *H. T. Tsai 54653* (A, BO, IBSC, NAS, PE, SZ). **U IRRAWADDY**: **Chawngmaw Kyaun**, 2120 m, 10 Aug. 1919, *F. Kingdon-Ward 3481* (E); **Drungjiang**, N Bapo, 1510 m, 16 Jul.

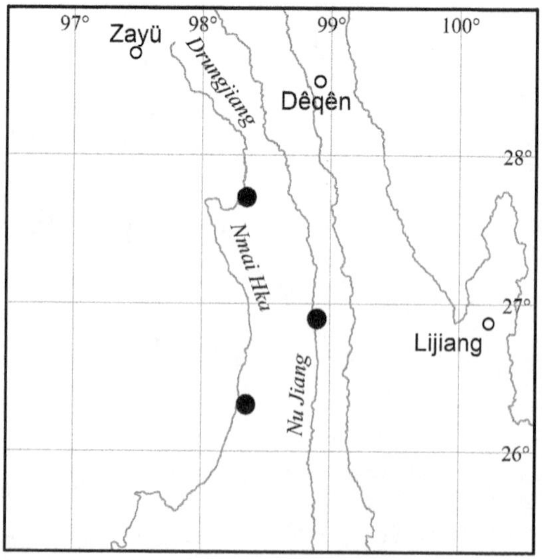

Map 46-98: **Strobilanthes congesta** Terao

2000, *H. Li 12897* (E, KUN); Kiukiang Valley, Taron, Chingen, 1600 m, 20 Aug. 1938, *T. T. Yü 19923* (A, E, PE); **Kang Fang**, Hpawte valley, descending towards the Ngaw Chang valley, 22 Aug. 1919, *R. J. Farrer 1263* (E).

49. Strobilanthes echinata Nees, Pl. Asiat. Rar. 3: 85. 1832; Wood in Grierson & Long (eds.), Fl. Bhutan 2(3): 1262. 2001; Hu, Deng & Wood in Wu, Raven & Hong (eds.), Fl. China 19: 420. 2011. ≡ *Strobilanthes pectinata* T. Anderson, J. Linn. Soc., Bot. 9: 474. 1867, nom. superfl. illeg. (Art. 52.1 & 52.2). ≡ *Goldfussia echinata* (Nees) N. P. Balakr., Fl. Jowai, 2: 355. 1983. ≡ *Ruellia pectinata* Wall., Numer. List n. 2356. 1830, nom. invalid. Type: India, Meghalaya, *F. de Silva* in *Wallich Cat. 2356* [lectotype designated by Bennett et al., Bot. J. Linn. Soc. 156: 133. 2008: K-W (K001115724); isolectotype: BM (BM000884989)].

Strobilanthes pectinata var. *daltonii* C. B. Clarke in Hooker (ed.), Fl. Brit. India 4: 447. 1884. ≡ *Strobilanthes echinata* var. *daltonii* (C. B. Clarke) Bremek., Verh. Kon. Ned. Akad. Wetensch., Afd. Natuurk., Sect. 2 41(1): 274. 1944. ≡ *Goldfussia echinata* var. *daltonii* (C. B. Clarke) V. Singh & P. Singh, J. Econ. Taxon. Bot. 15(3): 705. 1992. Type: [Sikkim & Darjiling] India, Sikkim, *J. D. Hooker s. n.* [lectotype designated by Bennett et al., Bot. J. Linn. Soc. 156: 133. 2008: K (sheet annotated by Anderson and Clarke)].

Strobilanthes calvata J. R. I. Wood, Edinburgh J. Bot. 51(2): 233. 1994. Type: [U Irrawaddy] Myanmar, Chawngmaw, 10 Aug. 1919, *F. Kingdon-Ward 3482* [holotype: E (E00133440)].

Herbs, 1–2 m tall, anisophyllous; stems subterete to quadrangular, sulcate, brownish pubescent or glabrescent. Leaves unequal in each pair, smaller one 1/2–2/3 of larger one in size, petiolate or uppermost ones subsessile; petiole (0–)0.5–4 cm long, slightly winged, pubescent or glabrous; blade ovate, elliptic, or ovate-lanceolate, 4–20 cm long, 1.5–7.5 cm broad, both surfaces pilose to subglabrous, lateral veins 7–11 pairs, base cuneate and decurrent onto petiole, margin crenate-serrate, apex acuminate. Spikes capitate, solitary, 4–8-flowered. Bracts variable in shape and size, outer ones sterile, sessile, ovate, ovate-oblong, or oblong-ovate, 3–5 cm long, 2–2.5 cm broad, equal or unequal, persistent, margin dentate, apex obtuse, rounded, or fimbriate, inner ones fertile, obovate to spatulate, 2–3 cm long, 0.7–0.8 cm broad, margin ciliate; bracteoles linear to oblanceo-late, 1.3–2.3 cm long, 1.5–3 mm broad, pubescent or glabrous, margin entire, sometimes dentate or fimbriate toward apex, apex acuminate. Calyx 1.5–2 cm long, pubescent, 5-lobed almost to base; lobes linear, 1–2 mm broad, subequal, abaxially pubescent towards apex. Corolla white, 4.5–5.5 cm long, straight, outside glabrous, inside glabrous except for trichomes retaining style; tube base cylindric, 1.5–1.8 cm long, 2–3 mm broad, upwards ventricose and widened to 1.2–1.5 cm broad at mouth; lobes ovate, 8–9 mm long, 8–10 mm broad, apex rounded. Stamens 4, included; filaments of longer pair 5–7 mm long, pilose, shorter pair 3–5 mm long, glabrous; anther-thecae oblong, 2–3 mm long, 1.7–2 mm broad; connective excurrent; staminode small or inconspicuous. Ovary oblong, 2–3 mm long, glabrous; 2 ovules per locule; ovules 2 per locular; style 2.3–2.5 cm long, hirsute. Capsules fusiform, 1.2–1.8 cm long, glabrous, 4-seeded. Seeds brown, ovate in outline, 3–5 mm long, 2.5–4 mm broad, pilose with appressed mucilaginous hairs; areole small. Figure 46-55.

Phenology: flowering and fruiting from May to September.

Habitats: moist forests; 600–2700 m.

Distribution: Sikkim & Darjiling, Bhutan, U Irrawaddy (Map 46-99); also in Cambodia, Indonesia, Laos, Malaysia, Singapore, Thailand, Vietnam, and other parts of India, Myanmar and China (S Yunnan, Guangxi, Guangdong).

SIKKIM & DARJILING: **Darjiling**, 2190 m, *C. B. Clarke 27024* (LE); 2190 m, 17 Aug. 1875, *C. B. Clarke 27029* (K); 2100 m, 8 Jul. 1862, *T. Anderson 1059* (BM); May 1874, *J. S. Gamble 3572* (K); 22 Jun. 1874, *J. S. Gamble 3573* (K); Goompahar Ramp, Nagu Spa, 27 May 1902, *J. H.*

Figure 46-55: **Strobilanthes echinata** Nees: **a**, portion of branch; **b**, bracts; **c**, bracteole and calyx; **d**. corolla. Drawn by Ai-Li LI; reproduced from *Fl. Reipubl. Popularis Sin.* 70: 158, pl. 23. 2002.

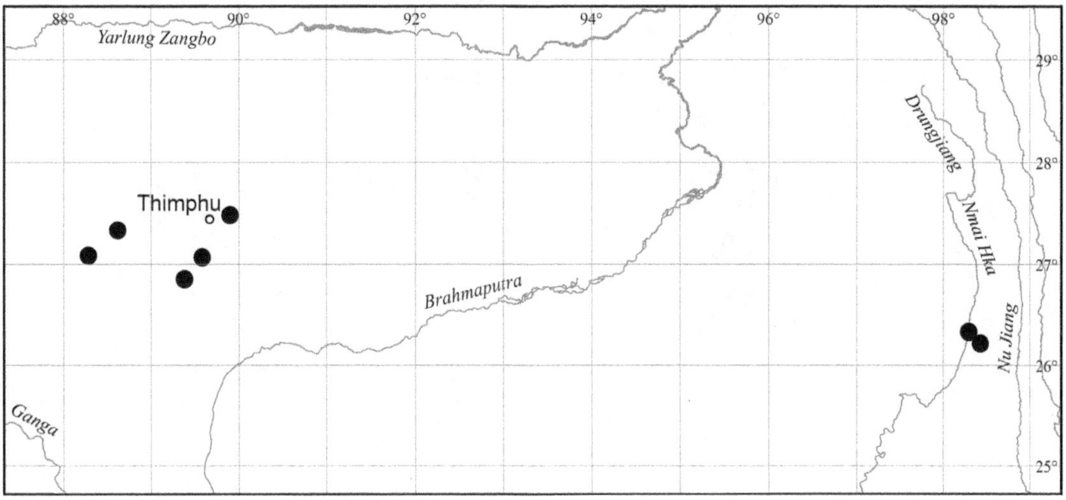

Map 46-99: **Strobilanthes echinata** Nees

Lace 2237 (E); Valley of the Nam Tisang, 600 m, *F. Kingdom-Ward 7374* (K); **Sikkim**, Rungbee, 13 Aug. 1869, *C. B. Clarke 8572* (K, LE); Rungbee, 1650 m, Sep. 1879, *J. S. Gamble 7062* (K); 2100 m, 9 Jun. 1874, *D. Treutler 178* (E, K, LE); Sinchal, 2400 m, Jul. 1862, *T. Anderson 1059* (A). **BHUTAN**: **Chukha**, on the summit atne Gedu, 2100 m, 23 Jun. 1990, *J. R. I. Wood 7271* (E); **Phuntsholing**, between Ganglakha and Sorchen, above Sintilakha, 1800 m, 9 Jun. 1991, *J. R. I. Wood 7380* (E); **Thimphu**, Mirichama, 1200–1500 m, 3 Jul. 1917, *R. E. Cooper 1168* (BM, E); **Tsongsa** to Tsanka, 2160–2250 m, 28 Jun. 1938, *B. J. Gould 627* (K). **U IRRAWADDY**: **Chaungmaw Kyaun**, Laktang, Trimple area, 1500 m, 3 Aug. 1919, *F. Kingdon-Ward 3458* (E); 1800–2150 m, 10 Aug. 1919, *F. Kingdon-Ward 3482* (E); **Putao**, hills E of Mali Hka, 600–900 m, 29 Jul. 1937, *F. Kingdon-Ward 12842* (BM).

50. Strobilanthes wardiana J. R. I. Wood, Kew Bull. 64: 28. 2009. Type: [U Irrawaddy] Myanmar, Nam Ti, 650 m, 18 Apr. 1926, *F. Kingdon-Ward 6632* [holotype: K (K000882830); isotype: K (K000882831)].

Subshrubs, 1.5–2 m high, isophyllous; stems quadrangular, sulcate, sparsely puberulent. Leaves equal in each pair, petiolate; petiole 4–7 cm long, densely puberulent; blade elliptic, 9–25 cm long, 4.5–12 cm broad, adaxially glabrous, abaxially sparsely puberulent, base attenuate, margin undulate, apex caudate. Spikes axillary, capitate, dense-flowered. Bracts ovate or ovate-elliptic, 1.9–2.8 cm long, 1.1–1.3 cm broad, pubescent, margin entire, brown ciliate, apex obtuse; bracteoles narrowly elliptic, 1–1.1 cm long, 2.5–3 mm broad, margin ciliate, apex obtuse. Calyx 8–9 mm long, subequally 5-lobed to basse; lobes lanceolate, 7–8 mm long, 1–1.5 mm broad, adaxially sericeous, margin brown ciliate, apex acuminate. Corolla 4–4.2 cm long, white with purplish veins, outside hirsute, inside glabrous except hairs retaining style; tube base subcylindric, 1–1.2 cm long, 2–3 mm broad, upwards gently curved 90° and gradually widened to 1.2–1.5 cm broad at mouth; lobes broadly ovate, 6–7 mm long, 3–4 mm broad, apex rounded. Stamens 4, included; filaments glabrous, longer pair 5–6 mm long, shorter pair 2–3 mm long; anther-thecae oblong, 3–4 mm long, 1–1.2 mm broad, connective sometimes mucronate at apex. Ovary comose towards apex; 2 ovules per locule; style 3.2–3.5 cm long, glabrous. Capsules unknown. Figure 46-56.

Phenology: flowering from March to April.

Habitats: rain forests; 600–700 m.

Distribution: **endemic**; U Irrawaddy (Map 46-100).

U IRRAWADDY: **Nam Tamai Valley**, Nam Ti, 650 m, 18 Apr. 1926, *F. Kingdon-Ward 6632* (K).

Figure 46-56: **Strobilanthes wardiana** J. R. I. Wood: **a**, habit and lower leaf; **b**, leaf; **c**, adaxial leaf surface; **d**, abaxial leaf surface; **e**, bract; **f**, bracteole; **g**, calyx; **h**, corolla opened out to show stamens; **i**, ovary. Drawn by Rosemary WISE; reproduced from *Kew Bull.* 64: 30, fig. 13. 2009.

Map 46-100: **Strobilanthes wardiana** J. R. I. Wood

51. Strobilanthes cruciata (Bremek.) Terao, Acta Phytotax. Geobot. 31: 59. 1980; Hu, Deng & Wood in Wu, Raven & Hong (eds.), Fl. China 19: 419. 2011. ≡ *Tetragoga cruciata* Bremek., Verh. Kon. Ned. Akad. Wetensch., Afd. Natuurk., Sect. 2 41(1): 300. 1944. Type: Indonesia, Sumatra, East Coast Govt, between Bandarr Baru and Dolok Baru, 900 m, 1 May 1918, *J. A. Lörzing 5668* [holotype: L (L0002848); isotypes: GH (GH00295522), SING (SING0045507)].

Tetragoga nagaensis Bremek., Verh. Kon. Ned. Akad. Wetensch., Afd. Natuurk., Sect. 2 41(1): 299. 1944; Hu, Fl. Reipubil. Popularis Sin. 70: 191. 2002. Type: India, Nagaland, Naga Hill, 1500 m, *Meebold 4891* (holotype: B).

Subshrubs, 1–2 m high, anisophyllous; stems quadrangular, sulcate, rust-colored hirsute or glabrous. Leaves unequal in each pair, smaller one 1/4–1/3 of larger one in size; petiole 1–1.5 cm long, pubescent; leaf blade ovate to ovate-lanceolate, 4–13 cm long, 1.5–5.5 cm broad, abaxially pale white tomentose especially along midvein, adaxially sparsely tomentose, lateral veins 7–9 pairs, base cuneate, margin serrate and ciliate, apex acuminate. Spikes axillary or terminal, subcapitate; peduncle 0–2.5 cm long, glandular-pilose. Bracts variable in shape and size, outer ones leaf-like, petiolar base forming a rigid involucre around capitulum, 1–2 cm long, 0.5–1 cm broad, abaxially densely brownish glandular-pilose to subglabrous, adaxially glabrous, apical leaf blader portion ovate, 2–4 cm long, 2–3 cm broad, erect, inner ones linear-pandurate, 1.2–1.6 cm long, 3–5 mm broad, abaxially pilose and glandular-pubescent towards apex, adaxially glabrous, margin entire, apex acute; bracteoles narrowly linear, 1.2–1.3 cm long, 1–1.2 mm broad. Calyx 1.1–1.2 cm long, abaxially pilose except for glabrous toward base, adaxially silky villous towards apex, margin ciliate towards apex. Corolla white, 2.5–2.8 cm long, outside glabrous, inside glabrous except for trichomes retaining style; tube base cylindric, 1–1.2 cm long, 2–3 mm broad, upwards ventricose and gradually widened to 1–1.2 cm broad at mouth; lobes ovate, 4–5 mm long and broad, apex obtuse. Stamens 4, included; filaments of longer pair 5–6 mm long, hirsute, shorter pair 2–3 mm long, glabrous; anther-thecae oblong, 2–3 mm long, 1–1.2 mm broad; connective shortly excurrent. Ovary oblong, glabrous; 2 ovules per locule; style 2.2–2.5 cm long, glandular-pilose. Capsules fusiform, 1.2–1.5 cm long, glabrous, 4-seeded. Seeds brown, ovate in outline, compressed, 3–4 mm long, 2.5–3 mm broad, densely appressed pubescent; areole small. Figure 46-57.

Phenology: flowering and fruiting from December to July.

Habitats: dense moist mixed forests; 300–1500 m.

Distribution: U Irrawaddy (Map 46-101); also in Indonesia, Thailand, and other parts of India, Myanmar and China (Hainan).

U IRRAWADDY: **Putao**, Fort Hertz (Putau), Hkamti Plain, 370 m, 19 Dec. 1930, *F. Kingdon-Ward 9052* (BM).

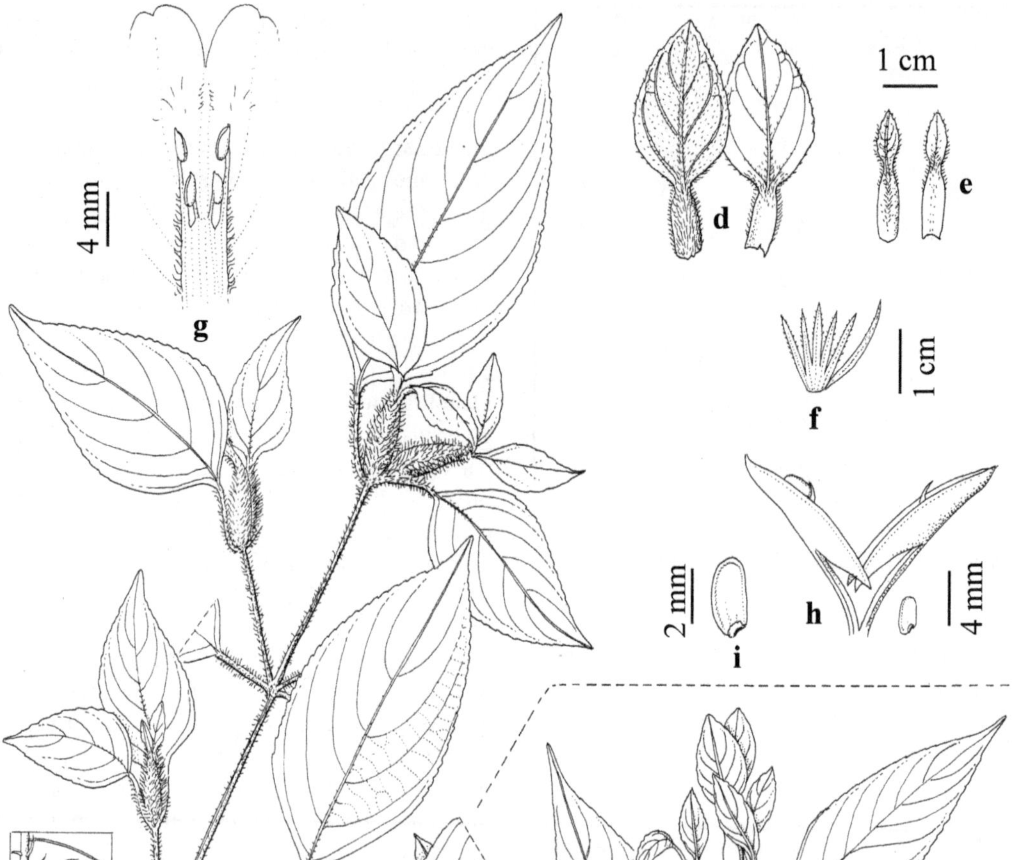

Figure 46-57: **Strobilanthes cruciata** (Bremek.) Terao: **a**, portion of plant; **b**, portion of adaxial leaf surface; **c**, portion of abaxial leaf surface; **d**, outer bracts; **e**, inner bracts; **f**, calyx and bracteole; **g**, corolla spread out showing stamens; **h**, capsule; **i**, seed; **j**, portion of near glabrous. Drawn by Rosemary WISE; reproduced from Wu, Raven & Hong, *Fl. China Illustr.* 19: fig. 326. 2012 (original from Wood & Scotland, *Kew Bull.* 64: 22, fig. 9. 2009).

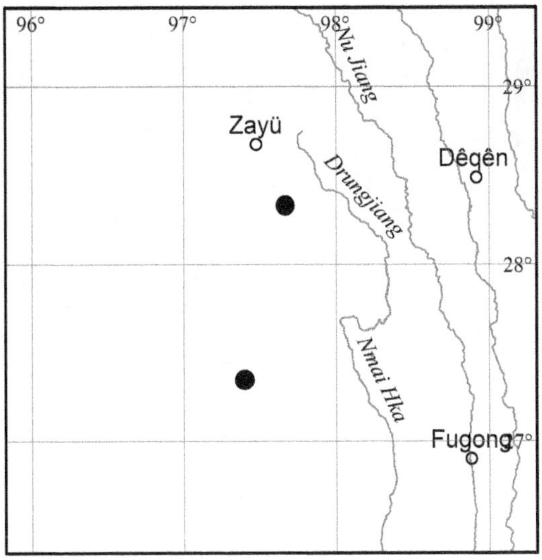

Map 46-101: **Strobilanthes cruciata** (Bremek.) Terao

52. Strobilanthes simonsii T. Anderson, J. Linn. Soc., Bot. 9: 474. 1867; Wood in Grierson & Long (eds.), Fl. Bhutan 2(3): 1263. 2001; Hu, Deng & Wood in Wu, Raven & Hong (eds.), Fl. China 19: 419. 2011. Type: Bhutan, Deothang ("Dewangari"), *Simons s. n.* [lectotype designated by Wood, Kew Bull. 64: 23. 2009: K (K000882999)].

Subshrubs, 0.5–1 m high, anisophyllous; stems and branches quadrangular, sulcate, tomentose. Leaves unequal in each pair; petiole 2–10 mm long, sulcate, tomentose; blade ovate-elliptic, smaller one 1.2–4 cm long, 0.6–1.4 cm broad, larger one 3.5–18 cm long, 1.6–6.5 cm broad, both surfaces tomentose, lateral veins 5–7 pairs, base cuneate and oblique, margin crenulate-serrate, apex acuminate and falcate. Spikes axillary, capitate, few-flowered; peduncle 0.2–1 cm long, pubescent. Bracts variable in size and shape, outermost ones leaf-like, pandurate with a constricted middle and broad petiolar base, 3–4 cm long, 4–5 mm broad, basal portion not forming an involucre, apical portion lanceolate inner ones lanceolate, 1.3–2.0 cm long, 2.5–3 mm broad, abaxially tomentose, apex acute; bracteoles linear, 2–3 cm long, 2.5–cm broad, abaxially pubescent towards apex. Calyx 1.3–1.5 cm long, subequally 5-lobed almost to base; lobes linear, outside pilose towards apex, apex acute. Corolla white, 3.5–4 cm long, straight, outside glabrous, inside glabrous except for trichomes retaining style; tube base cylindric 2–2.5 mm broad for 6–8 mm long, upwards gradually widened to 1.6–1.8 cm broad at mouth; lobes oblong, 5–6 mm long, 4–5 mm broad, apex rounded. Stamens 4, included; filaments of longer pair 0.7–1 cm long, hirsute, shorter pair 2–3 mm long, glabrous; anther-thecae oblong, 1.5–2 mm long, 0.8–1 mm broad, base divaricate, apex apiculate; connective shortly excurrent. Ovary glabrous; 2 ovules per locule; style 2.8–3.2 cm long, hirsute. Capsules clavate, 1.8–2 cm long, glabrous, 4-seeded. Seeds ovate in outline, compressed, 2–3 mm long, 2 mm broad, densely pubescent. Figure 46-58.

Phenology: flowering from March to August.

Habitats: forests; 1300–1700 m.

Distribution: endemic; Bhutan, Yarlung Zangbo-Brahmaputra (Map 46-102).

BHUTAN: **Deothang**, Dewangari, s. d., *Simons s. n.* (K). **YARLUNG ZANGBO-BRAHMAPUTRA**: **Baibung**, Gelin, Ni Legu, Shang Yuan, 1700 m, 24 May 1983, *B. S. Li et al. 03773* (PE); **Mizoram**, Tuisen, 1300 m, Apr. 1927, *A. D. Parry 203* (K).

53. Strobilanthes lihengiae Y. F. Deng & J. R. I. Wood, Bot. J. Linn. Soc. 150: 379. 2006; Hu, Deng & Wood in Wu, Raven & Hong (eds.), Fl. China 19: 394. 2011. Type: [U Irrawaddy] China, Yunnan, Tengchong Xian, Jietou Xiang, Shaba, community forests of Lidazhai, 2000 m, 23 Dec. 2000, *H. Li, Z. L. Dao & L. W. Yin 13630* (holotype: CAS; isotypes: KUN, MBK).

Herbs to 40 cm high, perennial, isophyllous; stems terete to slightly quadrangular, sulcate, bifariously pubescent to subglabrous. Leaves equal in each pair; petiole 0.2–4 cm long, pilose; leaf blade ovate to ovate-elliptic, 1.5–11 cm long, 0.5–6 cm broad, both surfaces pilose especially along veins, lateral veins 4–6 pairs, base cuneate, margin dentate to crenate, apex shortly acuminate. Spikes terminal or axillary, head-like, very compact. Bracts obovate, 8–11 mm long, 6–9 mm broad, broader next apex, shortly and

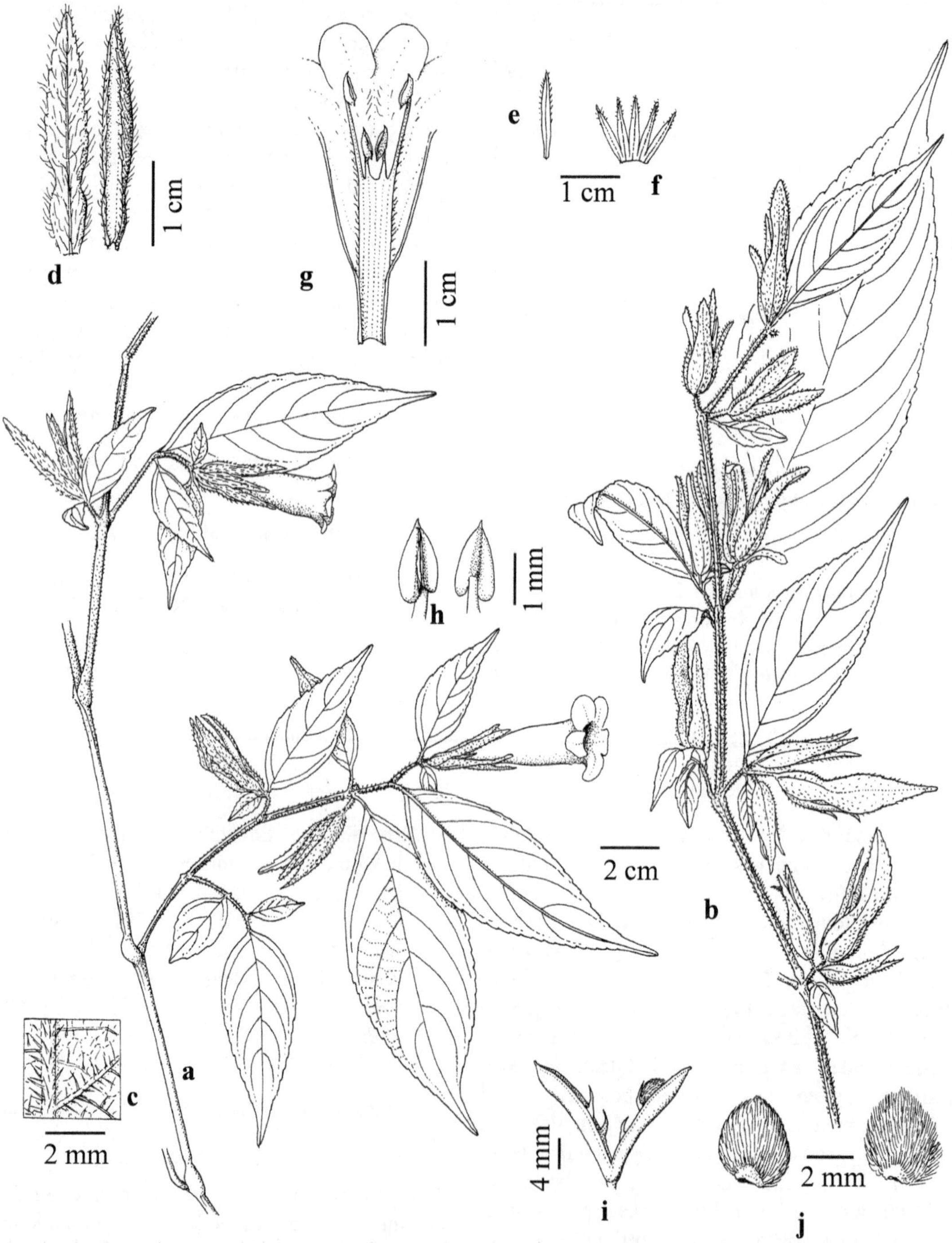

Figure 46-58: **Strobilanthes simonsii** T. Anderson: **a**, portion of plant; **b**, habit in fruit; **c**, portion of abaxial leaf surface; **d**, bracts, outer and inner surfaces; **e**, bracteole; **f**, calyx; **g**, corolla spread out showing stamens; **h**, anther; **i**, capsule; **j**, seeds. Drawn by Rosemary WISE; reproduced from Wu, Raven & Hong, *Fl. China Illustr.* 19: fig. 324. 2011 (original from Wood & Scotland, *Kew Bull.* 64: 24, fig. 10. 2009).

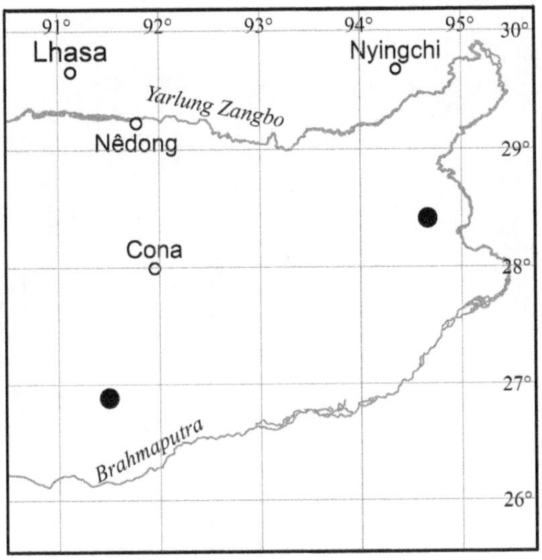

Map 46-102: **Strobilanthes simonsii** T. Anderson

densely tomentose, margin entire, apex rounded; bracteoles oblong-oblanceolate, 5–7 mm long, 1.5–2 mm broad, margin long ciliate. Calyx 8–9 mm long, subequally 5-lobed to base; lobes green-tipped, lanceolate-oblong, 1.1–1.3 mm broad, margin long ciliate, apex obtuse. Corolla blue, 3.2–4 cm long, ventricose, sparsely pilose on lobes; tube base cylindric, 1–1.2 mm broad, 8–10 mm long, upwards bent and expanded gradually to 1.2–1.5 cm broad at mouth; lobes ovate, 4–5 mm long and broad, apex acute. Stamens 4, included; filaments coarsely pilose, longer pair 5–6 mm long, shorter pair 2–3 mm long; anther-thecae narrowly oblong, 2.5–3 mm long, 0.5–0.7 mm broad. Ovary villous; 2 ovules per locule; style 2.5–3 cm long, basally sparsely pilose towards base. Capsules cylindric, 1–1.2 cm long, 2.5–3 mm broad, comose, 4-seeded. Seeds ovoid-elliptic in outline, 2–2.5 mm long, 1.5–2 mm broad, pilose. Figure 46-59.

Phenology: flowering from November to December.

Habitats: disturbed broad-leaved evergreen forests; 1700–2000 m.

Distribution: S Hengduan, U Irrawaddy (Map 46-103); also in other parts of China (C Yunnan).

S HENGDUAN: **Dali**, Taipingpu, 23 Dec. 1958, *X. Zhou 213* (KUN); 9 Jun. 1940, *T. N. Liou 17021* (KUN). **Yangbi**: Ertachai, *R. C. Ching 25309* (KUN, PE). **U IRRAWADDY**: **Drungjiang**, Dulongjiang valley, from Gongshan to Kongdang, 1 Sep. 2006, *Gaoligong Shan Biodiversity Survey 34410* (KUN); Drungjiang, Xianjiudang, 1700 m, 24 Aug. 1982, *Qinghai-Xizang Exped. 9601* (KUN); Puer, 1909 m, 2014, *Y. J. Guo et al. 14CS9920* (KUN); **Tengchong**, Jietou, Shaba, Lidazhai, 2000 m, 23 Dec. 2000, *H. Li et al. 13630* (CAS, KUN, MBK).

54. Strobilanthes cyphantha Diels, Notes Roy. Bot. Gard. Edinburgh 5: 162. 1912; Hu, Deng & Wood in Wu, Raven & Hong (eds.), Fl. China 19: 397. 2011. ≡ *Pteracanthus cyphanthus* (Diels) C. Y. Wu & C. C. Hu in Hu (ed.), Fl. Reipulic. Popularis Sin. 70: 134. 2002. Type: [S Hengduan] China, Yunnan, E plank of the Tali Range, 9000–11,000 ft., Jun. 1906, *G. Forrest 4549* [holotype: E (E00047309); isotype: P (P00719291)].

Subshrubs, 45–60 cm high, anisophyllous; stems quadrangular, sulcate, pubescent. Petiole 2–5 cm long, pilose; leaf blade ovate, 3–7 cm long, 1.5–4.5 cm broad, both surfaces pilose, lateral veins 6–8 pairs, base cuneate, decurrent onto petiole, margin serrate, apex acute. spikes terminal or axillary, subcapitate, dense-flowered, sessile or pedunculate; peduncles (0–)2–5 mm long, white pilose. Bracts variable in shape and size, outer ones leaf-like, inner ones elliptic, 0.8–1.2 cm long, 4–5 mm broad, white pilose, margin dentate and ciliate; bracteoles linear, 1.4–1.6 cm long, 1–2 mm broad, sparsely pilose. Calyx 1–1.3 cm long, sparsely pilose, 5-lobed almost to base; lobes linear, 1–1.2 mm broad, subequal. Corolla blue, 3–4 cm long, strongly bent and ventricose, outside sparsely pilose, inside pubescent on lobes and glabrous on tube except for trichomes retaining style; tube base cylindric, 1–1.2 cm long, 2–3 mm broad, upwards bent 90° at the middle and widened to 1–1.2 cm broad at mouth; lobes oblong, 4–5 mm long, 3.5–4 mm broad. Stamens 4, included; filament of longer pair 3–4 mm long, sparsely pilose towards base, shorter pair 2–2.5 mm long, glabrous; anther-

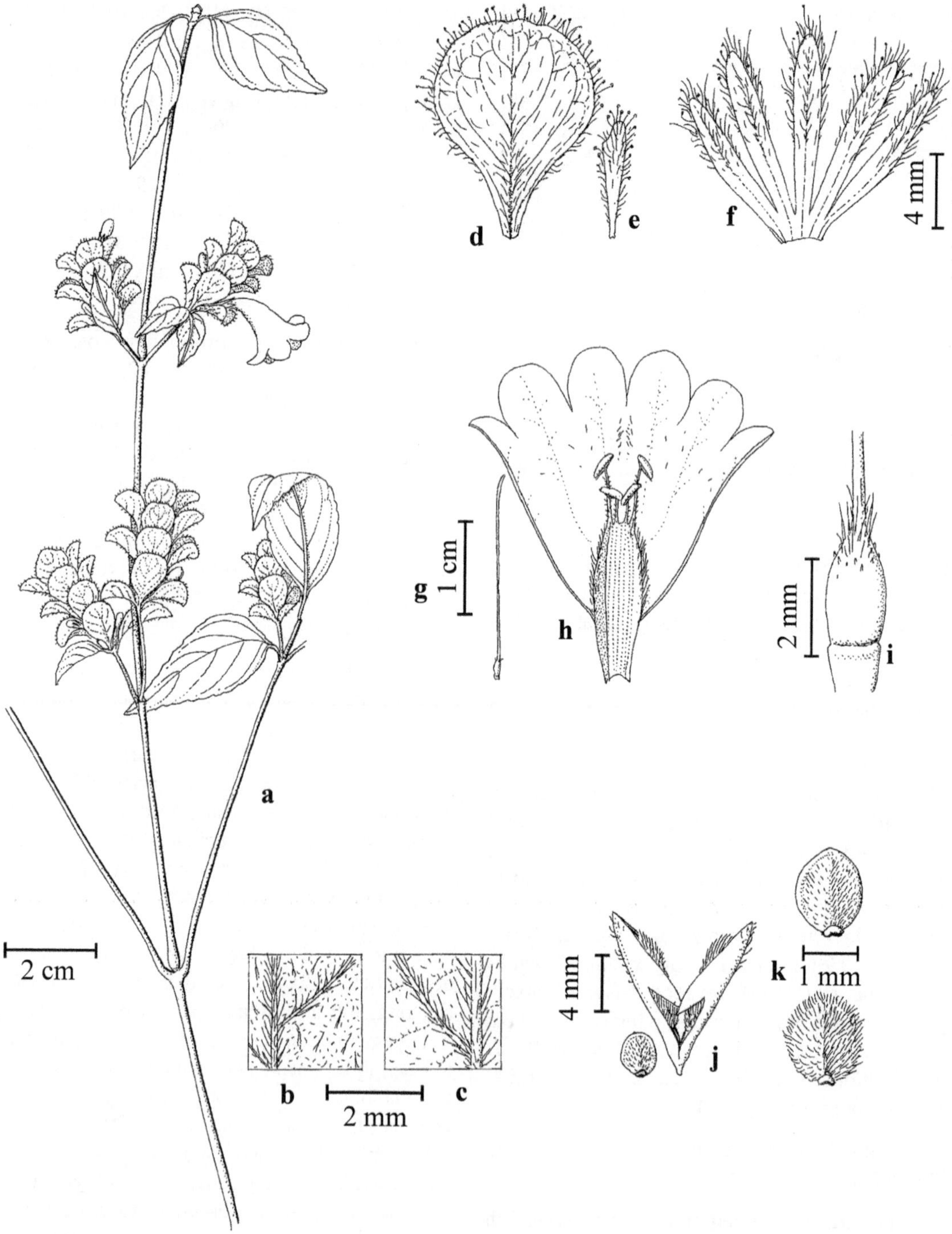

Figure 46-59: **Strobilanthes lihengiae** Y. F. Deng & J. R. I. Wood: **a**, habit; **b**, lower surface of leaf; **c**, upper surface of leaf; **d**, bract; **e**, bracteole; **f**, calyx; **g**, style; **h**, corolla opened showing androecium; **i**, ovary; **j**, capsule; **k**, seeds. Drawn by Rosemary WISE; reproduced from Deng, Wood & Scotland, *Bot. J. Linn. Soc.* 150: 381, fig. 13. 2006.

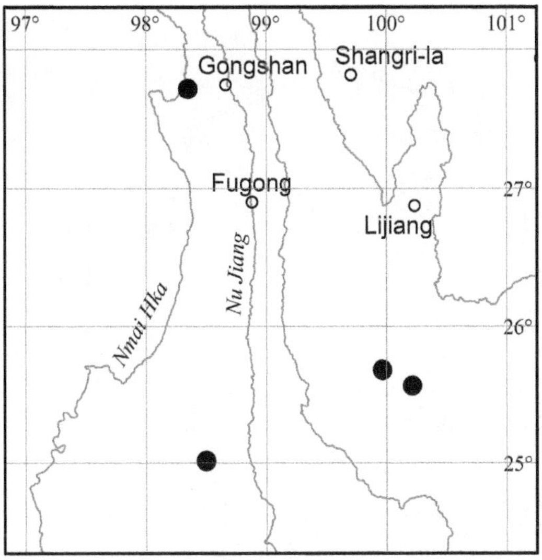

Map 46-103: **Strobilanthes lihengiae** Y. F. Deng & J. R. I. Wood

thecae oblong, 2–2.5 mm long, 1–1.5 mm broad. Ovary comose; 2 ovules per locule; style 2.7–3.2 cm long, pubescent. Capsules 1.2–1.4 cm long, pilose towards apex, 4-seeded. Seeds ovate in outline, compressed, 2–2.5 mm long, 1.5–2 mm broad, densely covered with appressed trichomes. Figure 46-60.

Phenology: flowering from July to October; fruiting from November to December.

Habitats: mixed forests; 1200–3600 m.

Distribution: Yarlung Zangbo-Brahmaputra, N Hengduan, S Hengduan, U Irrawaddy (Map 46-104); also in other parts of China (NE Yunnan, SE Sichuan, Guizhou, Hunan).

YARLUNG ZANGBO-BRAHMAPUTRA: **Baibung**, Lage to Hanmi, 22 Sep. 2003, *X. F. Gao et al. 7692* (CDBI); Nage, Duoxiongla Mountain, 2700 m, 24 Oct. 1992, *H. Sun et al. ETM-732* (KUN); **Lohit Valley**, 1950, *F. Kingdon-Ward 20221* (BM). **N HENGDUAN**: **Wenxian**, Bikou, Xuanmaguan to Xigouyan, 1700 m, 14 Jul. 2006, *Exped. Baishuijiang 2007* (PE). **S HENGDUAN**: **Baoxing**, Xinglong, Shanmugou, 1560 m, 17 Jun. 1982, *Z. B. Cheng et al. 116539* (SZ); s. loc., 17 Jul. 1936, *K. L. Chü 3137* (E, PE, SZ); Dengchigou, 1700 m, *T. H. Tu 4891* (PE); **Dali**, Huangjianpu to Beidou, 1200 m, 24 Dec. 1958, *X. Zhou 216* (KUN); E flank of Tali range, 2400–2700 m, Jul.–Aug. 1906, *G. Forrest 4521* (CAL, E, K); E flank of the Tali Range, 2700–3000 m, Jun. 1906, *A. Henry 4549* (E); E Cangshan, Baijiangou, 22 Jun. 1940, *R. C. Ching 22998* (KUN, PE); Wuliqiaoqing, 5 Jun. 1942, *H. C. Wang 2264* (IBSC, KUN, PE); **Fugong**, Shangpa, 2800 m, 20 Oct. 1934, *H. T. Tsai 58812* (A, IBSC, KUN, NAS, SZ); **Gongshan**, Mekong-Salwin Divide, Londjrela, 3600 m, 29 Sep. 1938, *T. T. Yü 23163* (E, KUN, PE); Maikha-Salvin divide, 1924-1925, *G. Forrest 25044* (PE); Lanngyyin to Lou Lan, Sep. 1906, *F. Ducloux 161* (E); Mount Kenyichunpo and region of Champutong, Bayau longba, 2700–3000 m, Oct. 1923, *J. F. Rock 11648* (A, US); **Hanyuan**, Lingshan, 3200 m, 14 Aug. 1936, *T. P. Wang 8749* (PE); **Heqing**, Lianping, Jiguiguang, 2850 m, 20 Aug. 1940, *R. C. Ching 23785* (KUN, PE); **Longyang**, Wayao, 1950 m, 2 Jan. 1959, *X. Zhou 226* (KUN); Wayao, 1850 m, 2 Jan. 1959, *X. Zhou 230* (KUN); **Tianquan**, Yinchangping, 2350 m, *D. Y. Peng 45906* (IBSC); Erlangshan, Longdanxi, 2210 m, *D. Y. Peng 46456* (IBSC); Erlangshan, Longdanxi, 2130m, *D. Y. Peng 46534* (IBSC); Wajinyan, 2140 m, 2 Oct. 1923, *T. H. Tu 4670* (PE); **Yongping**, 1210 m, 3 Jan. 1959, *X. Zhou 237* (KUN). **U IRRAWADDY**: **Drungjiang**, 3000 m, *Exped. Qinghai-Xizang 8642* (KUN, PE); Xiongdang to Longdegeng, 1900–2000 m, 3 Sep. 1982, *Exped. Qinghai-Xizang 9833* (KUN, PE).

55. Strobilanthes yunnanensis Diels, Notes Roy. Bot. Gard. Edinburgh 5: 164. 1912; Hu, Deng & Wood in Wu, Raven & Hong (eds.), Fl. China 19: 396. 2011. ≡ *Goldfussia yunnanensis* (Diels) H. P. Tsui, Act. Bot. Yunnan 12: 275. 1990. ≡ *Pteracanthus yunnanensis* (Diels) C. Y. Wu & C. C. Hu in Hu (ed.), Fl. Reipblic. Popularis Sin. 70: 150. 2002. Type: [S Hengduan] China, Yunnan, E flank of the Lichiang Range, 27°12'N, 9,000-10,000 ft., Jun. 1906, *G. Forrest 2361* [lectotype **here designated**: E (E00047318); isolectotypes: BM (BM000810989), GH (GH00063067), K (K000882826), P (P00719465)].

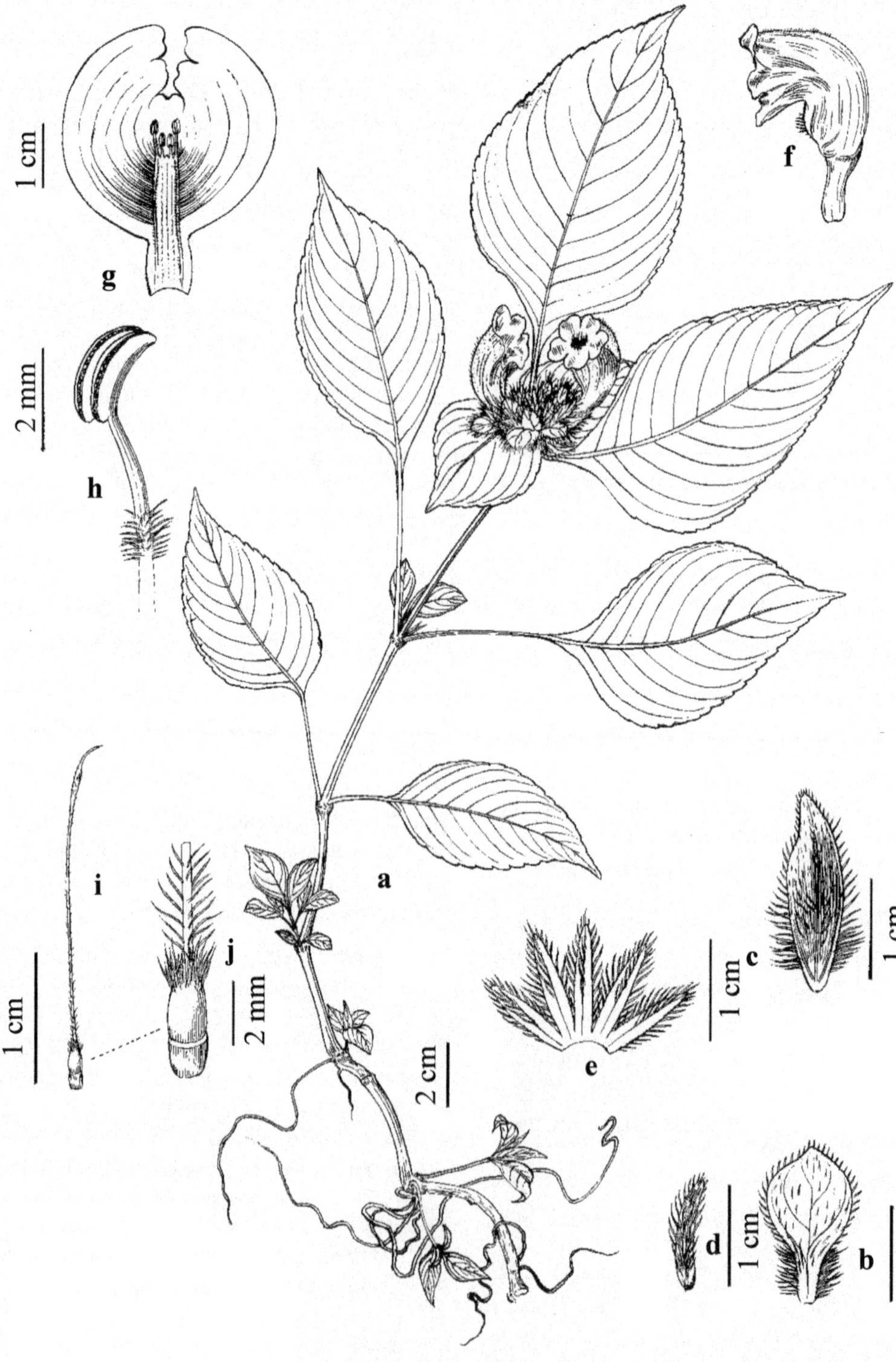

Figure 46-60: **Strobilanthes cyphantha** Diels: **a**, habit; **b**, **c**, bracts; **d**, bracteole; **e**, calyx; **f**, corolla; **g**, opened corolla showing androecium; **h**, stamen; **i**, pistil; **j**, ovary. Drawn by Ding-Han CUI.

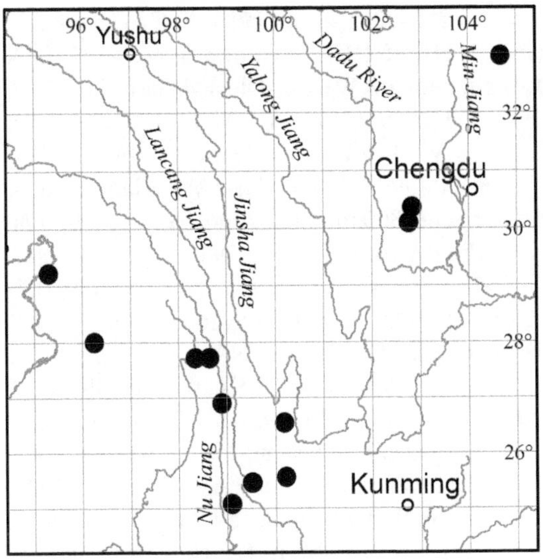

Map 46-104: **Strobilanthes cyphantha** Diels

Strobilanthes limprichtii Diels, Fedde Repert. 12: 488. 1922; Bremekamp, Verh. Kon. Ned. Akad. Wetensch., Afd. Natuurk., Sect. 2 41(1): 279. 1944; Hu, Fl. Reipblic. Popularis Sin. 70: 186. 2002. Type: [S Hengduan] China, Sichuan, Yachoufu, Thaldes Yaho, zw. Tienchuanchou, Chushikping, 1200 m, 7 Jun. 1914, *H. W. Limpricht 1545* [holotype: WRSL; isotypes: A (A00063103), WU (WU0035624)].

Strobilanthes mekongensis W. W. Sm., Notes Roy. Bot. Gard. Edinburgh 10: 195. 1918; Bremek., Verh. Kon. Ned. Akad. Wetensch., Afd. Natuurk., Sect. 2 41(1): 280. 1944. ≡ *Pteracanthus mekongensis* (W. W. Sm.) C. Y. Wu & C. C. Hu, Fl. Reipblic. Popularis Sin. 70: 144. 2002. Type: [S Hengduan] China, Yunnan, Mekong valley, lat. 27°40′N, alt. 6000 ft., Jul. 1914, *G. Forrest 12882* [holotype: E (E00045284); isotypes: IBSC (IBSC0561893), PE (PE01432629)].

Strobilanthes hygrophiloides C. B. Clarke ex W. W. Sm., Notes Roy. Bot. Gard. Edinburgh 10: 194. 1918. ≡ *Pteracanthus hygrophiloides* (C. B. Clarke ex W. W. Sm.) H. W. Li [in Wu (ed.), Index. Fl. Yunnan. 2: 1682, 1984, nom. invalid. (Art. 41.5)] in Wu (ed.), Fl. Xizang 4: 415. 1985; Tsui & Hu in Hu (ed.), Fl. Reipblic. Popularis Sin. 70: 142. 2002. Type: [S Hengduan] China, Sichuan, Kangding, "Thibet Oriental. Ta-tsien-lou (Principauté de Kiala)", 1893, *Soulié 591* [lectotype **here designated**: K (K000882825); isolectotypes: MPU (MPU018235), P (P00719350)].

Strobilanthes hygropiloides var. *brachytrichus* Hand.-Mazz., Symb. Sin. 7: 895. 1936. ≡ *Pteracanthus hygrophiloides* var *brachytrichus* (Hand.-Mazz.) C. Y. Wu, Index Fl. Yunnan. 2: 1268. 1984, nom. invalid. (Art. 41.5). Type: [S Hengduan] China, Yunnan, Shangri-la, Bodo, 4 Aug. 1914, *H. Handel-Mazzetti 4463* [holotype: WU (WU0035622); isotypes: E, W (W 1935-0000116)].

Strobilanthes hygrophiloides var. *subnudus* Benoist, Null. Mus. Hist. Naurk. Pais. 28: 189. 1922. Type: China, Yunnan, Ta tchay, près Kiao Kia 3 jours au nord-est et à 3 lieues du Fleuve Bleu, 11 Jul. 1908, *S. Ten in herb. Ducloux 5503* [lectotype **here designated**: P (P00719349); isolectotype: P (P00719352)].

Subshrubs, 0.5–2.5 m high, isophyllous; stems quadrangular, bifariously whitish pubescent, sometimes glabrescent. Leaves equal in each pair; petiole 2–4 cm long, whitish pubescent; blade ovate, elliptic, or elliptic-ovate, 3–11 cm long, 2–5 cm broad, abaxially sparsely pubescent, adaxially pubescent, lateral veins 4–7 pairs, base narrowly cuneate and decurrent onto petiole, margin crenate-serrate, apex acuminate and sometimes falcate. Spikes axillary or terminal, subcapitate, rarely reduced to a single flower; peduncle 0–4 mm long, densely white pubescent. Bracts variable in shape and size, hirsute, outermost ones sterile, leaf-like, lanceolate to ovate-lanceolate, 1.5–2.5 cm long, 3–4 cm broad, petiolate, inner ones fertile, linear-oblong, 1.2–1.6 cm long; bracteoles linear to oblanceolate, 9–10 mm long, 1.5–2 mm broad, pilose, apex obtuse. Calyx 1.2–1.6 cm long, 5-lobed almost to base; lobes linear-lanceolate, 1.1–1.5 cm long, 1–1.5 mm broad, subhyaline, pilose especially along midvein, margin ciliate, apex acuminate. Corolla pale blue to pinkish,

3.5–4.5 cm long, straight, outside densely white pilose, inside glabrous except for trichomes retaining style; tube base cylindric, 1–1.5 cm long, 8–9 mm broad, upwards gradually widened to 1.6–2 cm broad at mouth; lobes orbicular, 6–8 mm in diameter. Stamens 4, included; filaments pilose, longer pair 4–7 mm long, shorter pair 2–3 mm long; anther-thecae oblong, 3–4 mm long, 1–1.5 mm broad. Ovary whitish comose; 2 ovules per locule; style 2.8–3.4 cm long, sparsely pilose. Capsules oblong, 1.1–1.3 cm long, 2–2.5 mm broad, sparsely glandular-pubescent towards apex, 4-seeded. Seeds ovate in outline, compressed, 2.5–3.5 mm long, 2–2.5 mm broad, densely appressed pilose. Figure 46-61.

Phenology: flowering from July to September; fruiting from October to November.

Habitats: moist open shady places in thickets; 800–3400 m.

Distribution: L Yarlung Zangbo, Yarlung Zangbo-Brahmaputra, N Hengduan, S Hengduan, U Irrawaddy (Map 46-105); also in other parts of China (C & NE Yunnan, S Sichuan).

L YARLUNGZANGBO: **Yi'ong**, 2400m, *Alpine Group from Department of Ecology 15465* (PE); near Yi'ong, 2300–2500 m, 10 Sep. 1980, *Z. C. Ni et al. 1585* (PE); **Nyingchi**, Pailong, near old bridge, 2175 m, 8 Sep. 2006, *Tibet-McArthur 232* (US). **YARLUNG ZANGBO-BRAHMAPUTRA**: **Cona**, Lexiang to Xilianhua Temple, 4 Sep. 2012, *FLPH Tibet Expedition 12-0991-b* (PE); **Gutang**, Gedang to Xingkai, 2300 m, 31 Aug. 1974, *Exped. Qinghai-Xizang 74-4969* (KUN, PE); **Mêdog**, *S. Z. Cheng & B. S. Li 3045* (PE); **Shangzayü**, 2100 m, 20 Jun. 1979, *R. F. Huang 0109* (HNWP); near Renbuzong, 24 Jul. 1980, *Z. C. Ni et al. 718* (PE); Rongyu, 1978 m, 23 Aug. 2009, *Southeast Tibet Expedition Team SET-ET 653* (IBSC); **Tawang**, Tawang Chu, 27 Jul. 2002, *Bowes Lyon 15067* (E); Tongkyuk, 2700–3000 m, 29 June 1935, *F. Kingdon-Ward 12081* (BM, IBSC, TI); **Xiazayü**, Gayao, 2413 m, 20 Aug. 2009, *Southeast Tibet Expedition Team SET-ET 514* (IBSC); **Zayü**, Putsang river, Rang To valley, 1800 m, 30 Oct. 1933, *F. Kingdon-Ward 10934* (BM); Dongqiong to Baan, 1600 m, 17 Jul. 1973, *Exped. Qinghai-Xizang 73-769* (KUN, PE); near Baantong, 1500 m, 8 Jul. 1980, *Z. C. Ni et al. 531* (PE). **N HENGDUAN**: **Danba**, *s. coll. 391* (SZ); **Litang**, between Wa-Erh-Dje and Garu, Jul.–Aug. 1928, *J. F. Rock 16896* (A, BM, E, IBSC, PE, US); between Wa-Erh-Dje and Garu, Jul.–Aug. 1928, *J. F. Rock 16902* (E, GH, US); **Wenxian**, Bikoou, Datuanyuhe, 800 m, 8 Jul. 2006, *Exped. Baishuihe 1461* (PE); **Yajiang**, 30 Jul. 1940, *K. Huang et al. 515* (PE, SZ); *K. Huang et al. 585* (KUN, PE). **S HENGDUAN**: **Baoxing**, way to Linguan, 19 Sep. 1939, *C. Pei 8254* (IBK, NAS, PE, SZ); 1954, *T. P. Soong 38241* (IBSC, KUN, NAS, PE, SZ); Puxi, 2375 m, Jul. 26, 1954, *T. P. Tsoong 39239* (SZ); **Bingchuan**, Jizushan, 2500–2800 m, 25 Jul. 1965, *Muben Youliao Diaochazu 65-0618* (KUN, LBG); **Dali**, W flank of the Tali Range, Aug. 1917, *G. Forrest 15471* (BM, E, K); Huadianba, 1 Aug. 1942, *H. C. Wang 2634* (IBSC); 6 Sep. 1942, *H. C. Wang 2867* (IBSC); **Eryuan**, Xiaoluoping, 16 Jun. 1942, *H. C. Wang 2368* (IBSC, KUN, PE); Liantie, Luopingshan, 7 Jul. 1940, *R. C. Ching 23173* (KUN, PE); Sanying, Dongshan, 23 Sep. 1975, *P. Y. Qiu 60985* (CDBI, HIB, KUN); **Fugong**, Lumadeng, Yaping above old Shibali, 2040 m, 21 Aug. 2005, *Gaoligong Shan Biodiversity Survey 28797* (CAS, KUN, MO); **Gongshan**, Baihanluo, 20 Nov. 1959, *K. M. Feng 24625* (KUN); **Heqing**, Baiya, 2600 m, s. d., *R. C. Ching 24341* (KUN, PE); **Jiulong**, 9 Jun. 1930, *K. Huang et al. 325* (KUN); Naiqu, 2300 m, 9 Aug. 1979, *Q. Q. Wang 21415* (CDBI); **Kangding**, between Batang and Tachienlu, Sep.–Oct. 1904, *Kew Herb. N. H2001/00972 39* (K); Ta-tsien-lou, 1893, *J. A. Soulie 591* (K); Ta Tsien Lou, s. d., *R. P. Soulie 819* (A, E, K, P); **Mianning**, Fubao, 2700 m, 27 Jul. 1959, *S. K. Wu 2319* (KUN, PE); **Muli**, Xiamaidi, 1 Aug. 2006, *Y. F. Deng 19038* (IBSC); Boya, Niwa to Dawanzi, 1800–3400 m, 7 Jul. 1960, *J. S. Ying 4459* (PE); 3000 m, *F. Kingdon-Ward 4568* (E); Laoyingyan, 2200–2300 m, *S. K. Wu 2462* (KUN); Lwa-lang-kang-din, 2800 m, 30 Jul. 1937, *T. T. Yü 7477* (A, KUN, PE); Chutouwan, 2600 m, 13 Jun. 1978, *Y. B. Yang 4782* (CDBI, SZ); Maidi to Mianyashankou, 2800 m, 5 Aug. 1978, *K. H. Mou & Y. B. Yang 6992* (CDBI); Bahe, 2300 m, 19 Jul. 1978, *Q. S. Zhao 6064* (CDBI, SZ); Taoba, 2200 m, 30 Aug. 1978, *Q. S. Zhao et al. 8189* (CDBI, SZ); Zhutouwan, 2520 m, 13 June 1978, *Y. B. Yang 4790* (CDBI, SZ); Taoba, 2620 m, 19 Jul.

Figure 46-61: **Strobilanthes yunnanensis** Diels: **a**, stem; **b**, flowering branch; **c**, bract, bracteoles and calyx; **d**, opened corolla showing androecium; **e**, pistil. Drawn by Quan LIN; reproduced from *Icon. Cormophyt. Sin.* 4: 164, fig. 5741. 1975.

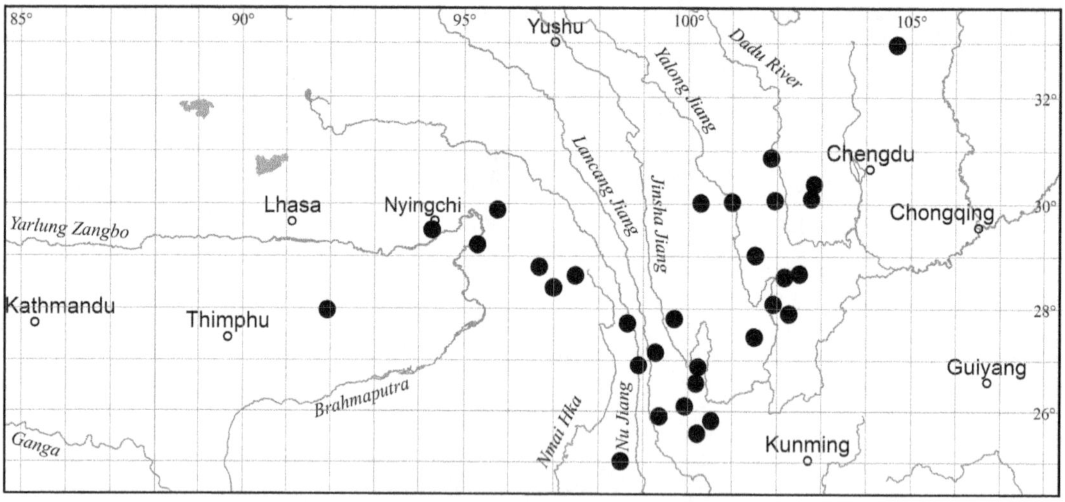

Map 46-105: **Strobilanthes yunnanensis** Diels

1978, *J. S. Liu 7605* (CDBI, SZ); Kangwuliangzi, 2500 m, 14 Oct. 1982, *F. T. Pu et al. 115* (CDBI, NAS); Xiahetaowan, 2350 m, 23 Aug. 1978, *Q. S. Zhao et al. 8259* (CDBI, SZ); Hetaowan, 2300–2500 m, 23 Oct. 1982, *F. T. Pu & K. Yao 421* (CDBI, NAS); Xiamaidi to Mianbuyakou, 2368 m, 3 Aug. 2010, *E. D. Liu et al. 2570* (KUN); Xiamaidi to Mianbuyakou, 2368 m, 3 Aug. 2010, *E. D. Liu et al. 2571* (K, KUN); **Shangri-la** (Zhongdian), Hutiaoxia, nearby Luoshihe Power Station, 24 Sep. 2007, *H. Peng et al. 6667* (IBSC, KUN); Bodo, 2500–2600 m, 4 Aug. 1914, *H. Handel-Mazzetti 4463* (E, WU), near Baidi, 2400 m, 18 Aug. 1962, *Exped. Zhongdian 1040* (YUKU), near Baidi, 2400 m, *Exped. Zhongdian 1046* (KUN, PE); between Tung-pa-tze and Bodo, 10 Sep. 1939, *K. M. Feng 2343* (GH, KUN, PE); Kung-shing-shu, Snow Mt., on the way to Kai-lou-wei (Yangtze bank), 2800 m, 25 Jul. 1939, *K. M. Feng 1844* (GH, KUN); **Tianquan**, 13 Sep. 1939, *F. C. Tai & C. M. Teng 5198* (SZ); Xichuanheba, 1000 m, Nov. 1928, *Y. Chen 5683* (NAS); Xingou, 2400 m, 1 Jul. 2010, *X. F. Gao et al. 11995* (CDBI); **Weixi**, 1800 m, Jul. 1914, *G. Forrest 12882* (E, PE); **Xichang**, 2150 m, 24 Jun. 1932, *T. T. Yü 1087* (PE); **Yanbian**, Guosheng, Bailinshan, Yiwanshui, 2600 m, Sep. 11, 2002, *X. H. Jin et al. 138* (CDBI, PE); Zemulong, Shaohuizuizui, 2600–2650 m, *X. F. Gao et Y. L. Peng 3336* (CDBI); Chushikping, 1200 m, 7 Jun. 1914, *H. W. Limpricht 1545* (WU); **Yanyuan**, near Guanbie, 2200 m, 28 Jul. 1983, *Exped. Qinghai-Xizang 12606* (KUN, PE); Yuanbao, Lingbinggou, 2800–3000 m, 30 Jul. 1983, *Exped. Qinghai-Xizang 12634* (KUN, PE); **Yuexi**, 2600 m, 14 Jun. 1932, *T. T. Yü 1043* (PE); **Yulong** (Lijiang), Ciju, 2600 m, *Exped. N. W. Yunnan 9750* (KUN, PE); Lijue to Tongdian, 2800 m, 7 Jul. 1960, *Nan Shui Bei Diao (N. W. Yunnan) 9759* (YUKU), E flank of the Likiang Range, 2700–3000 m, Jun. 1906, *G. Forrest 2361* (A, BM, E, K); Xuesong, Longhuangmiao, 2800 m, 9 Aug. 1942, *K. M. Feng 8880* (PE); Xuesong, Longhuangmiao, 2800 m, 9 Aug. 1942, *K. M. Feng 8881* (KUN, PE); Yülung-schan, Jun. 1914, *H. Handel-Mazzetti 3742* (WU); Yushuizhai, 17 Sep. 2013, *Y. F. Deng et al. 25137* (IBSC); **Yunlong**, Tianchi Natural Reserve, 2500 m, 27 Jun. 1984, *Xiangliao Kaochazu 104* (KUN). **U IRRAWADDY**: **Tengchong**, E Tengyueh, 1800 m, May 1912, *G. Forrest 7783* (E, K); S. Tengyueh, 1800–2100 m, Jun. 1912, *G. Forrest 8123* (E, IBSC, K); S Tengyueh, Jun. 1912, *G. Forrest 8386* (E); Tengyueh, 1911, *E. B. Howell 302* (E).

56. Strobilanthes tibetica J. R. I. Wood, Edinburgh J. Bot. 51: 262. 1994; Hu, Deng & Wood in Wu, Raven & Hong (eds.), Fl. China 19: 409. 2011. ≡ *Pteracanthus tibeticus* (J. R. I. Wood) C. Y. Wu & C. C. Hu in Hu (ed.), Fl. Reipubl. Popularis Sin. 70: 132. 2002. Type: [Yarlung Zangbo-Brahmaputra] China, Xizang, Zayül,

Putsang River, Rong To Valley, 8000 ft., 2 Nov. 1933, *F. Kingdon-Ward 10942* [holotype: BM (BM000906335), isotypes: E (E00265800), TI].

Subshrubs, 0.8–1.5 m high, isophyllous; stems slightly quadrangular, glabrous or obscurely bifariously pubescent. Leaves equal in each pair; petiole 1–3 cm long, sparsely pilose; blade ovate to elliptic, 7–14 cm long, 3–9 cm broad, abaxially sparsely pilose towards base, adaxially sparsely pubescent, lateral veins 5 or 6 pairs, base narrowly cuneate, margin serrate, apex acuminate and slightly falcate. Spikes axillary, 4–12 cm long, secund, sometimes reduced to single flowers; rachis finely puberulent to glandular-pilose, straight or obscurely zigzag, winged, wings 0.2–1 mm broad; flowers 1–1.5 cm apart on rachis. Bracts glabrous, variable in shape and size, basal ones leaf-like, with a petiole to 4 mm long, ovate, 1.5–2.5 cm long, apex acuminate, apical ones sessile, narrowly oblong-elliptic, 1–1.2 cm long, apex acuminate; bracteoles oblong, 2.5–4 mm long, apex obtuse, glabrous. Calyx 0.9–2 cm long, subequally 5-lobed to 1–1.5 mm above base; lobes almost colorless, linear, glabrous or sparsely glandular-pilose, apex obtuse. Corolla violet, 3.2–4 cm long, outside pubescent but silky gray in bud, inside glabrous; tube strongly inflated from near base, upwards widened to 1.2–1.5 cm broad at the middle and bent to ca. 90°; lobes ovate, 4–5 mm long, 5–7 mm broad, apex rounded to slightly emarginate. Stamens 4, included; filaments equal, 2.5–3.5 mm long, glabrous; anther-thecae oblong, 1.5–2 mm long. Ovary glabrous; 2 ovules per locule; style 2–2.3 cm long, glabrous or sparsely hirsute. Capsules oblong, 1.1–1.2 cm long, glabrous, 4-seeded. Seeds oblong in outline, 2–3 mm long, 1.5–2 mm broad, covered with stiff white trichomes. Figure 46-62.

Phenology: flowering from August to September; fruiting from October to November.

Habitats: forests; 2300–2800 m.

Distribution: endemic; U Yarlung Zangbo, L Yarlung Zangbo, Yarlung Zangbo-Brahmaputra (Map 46-106).

U YARLUNG ZANGBO: **Nyalam**, Zhangmu to Kangbajiba, 2400 m, 22 Jun. 1966, *Y. T. Zhang & K. Y. Lang 4350* (PE). **L YARLUNG ZANGBO**: **Yi'ong**, Yi'ong Farm, 2300–2500 m, *B. S. Li et al. 6751* (PE). **YARLUNG ZANGBO-BRAHMAPUTRA**: **Baibung**, Hanmi to Bangge, 2500 m, 23 Sep. 1980, *W. L. Chen 15136* (PE); **Delei Valley**, 2340–2740 m, 26 Sep. 1928, *F. Kingdon-Ward 8674* (K); **Gutang**, Gedang to Bengben, 2650 m, 13 Oct. 1982, *B. S. Li & S. Z. Cheng 1287* (PE); 68 km south of Galonglashan, 2650 m, *B. S. Li & S. Z. Cheng 67* (PE); 68 km south of Galonglashan, 2500 m, 15 Aug. 1982, *B. S. Li & S. Z. Cheng 84* (PE); **Zayü**: Putsang River in Rong To Valley, 2340 m, 2 Nov. 1933, *F. Kingdon-Ward 10942* (BM, E, TI).

57. Strobilanthes pterygorrhachis C. B. Clarke, J. Linn. Soc., Bot. 25: 54. 1889; Hu, Deng & Wood in Wu, Raven & Hong (eds.), Fl. China 19: 410. 2011. Type: India, Kegwima, Naga Hill, 5500 ft., 11 Nov. 1885, *C. B. Clarke 41919* [lectotype **here designated**: K (K000883142); isolectotypes: BM (BM000884902), K (K000883141, K000883143)].

Subshrubs, 1.5–3.5 m high, subglabrous; stems and branches weakly quadrangular. Petiole 2–9 cm long, glabrous; leaf blade ovate, 12–16 cm long; 6–7.5 cm broad, glabrous, base cuneate, margin serrate, apex acute to shortly acuminate. Spikes terminal or axillary, 3–25 cm long, secund; rachis slightly zigzag, strongly winged, glabrous; flowers mostly paired, 3–4 cm apart on rachis. Bracts leaf-like, petiolate, lanceolate, diminishing in size, 0.6–2 cm long, 3–7 mm broad; bracteoles linear-oblanceolate, 5–10 mm long. Calyx 2.5–3.6 cm long, subequally 5-lobed almost to base; lobes narrowly lanceolate, 2.2–3.2 cm long, 2–3 mm broad, margin minutely ciliate, apex acuminate or mucronate. Corolla yellow, 3.2–3.5 cm long, outside pubescent, inside glabrous except for trichomes retaining style; tube base shortly cylindric, upwards abruptly widened and bent just below mouth. Stamens 4, included; filaments glabrous, longer pair 5–6 mm long, shorter pair 3–4 mm; anther-thecae oblong, 3–3.5 mm long, 1–1.2 mm broad. Ovary comose; 2 ovules per locule; style 3–3.2 cm long, glabrous. Capsules comose.

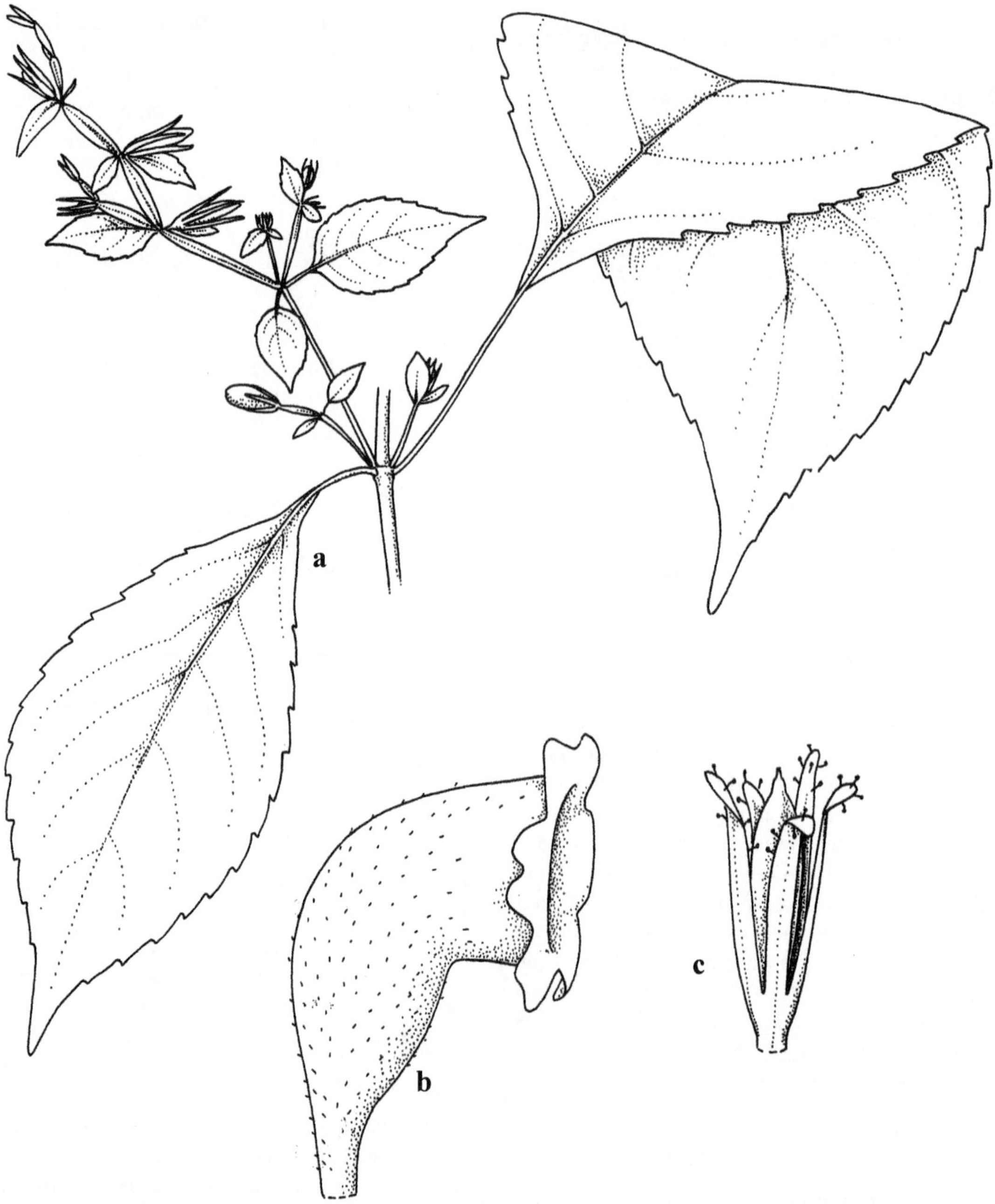

Figure 46-62: **Strobilanthes tibetica** J. R. I. Wood: **a**, Portion of plant; **b**, corolla; **c**, capsule and calyx. Drawn by Mary BATES; reproduced from Wu, Raven & Hong, *Fl. China Illustr.* 19: fig. 310: 1–3. 2012 (original from *Edinburgh J. Bot.* 51: 263, fig. 13: A–C. 1994).

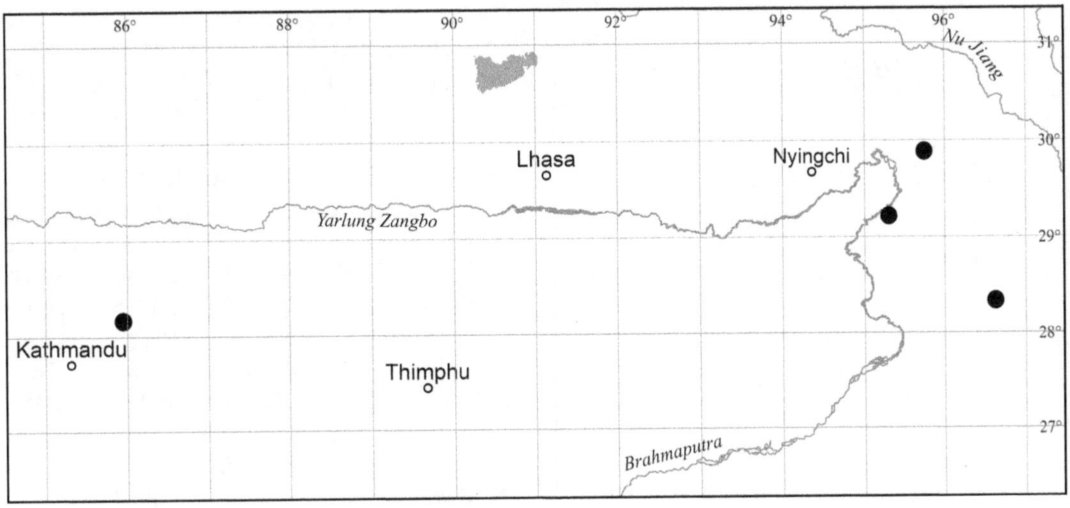

Map 46-106: **Strobilanthes tibetica** J. R. I. Wood

Phenology: flowering from August to October.

Habitats: moist places in forests; ca. 1800 m.

Distribution: Yarlung Zangbo-Brahmaputra (Map 46-107); also in other parts of India.

YARLUNG ZANGBO-BRAHMAPUTRA: s. loc., s. d., *R. S. Rao 10497* (CAL).

58. Strobilanthes cusia (Nees) Kuntze, Revis. Gen. Pl. 2: 499. 1891; Wood in Grierson & Long (eds.), Fl. Bhutan 2(3): 1274. 2001; Hu, Deng & Wood in Wu, Raven & Hong (eds.), Fl. China 19: 407. 2011. ≡ *Goldfussia cusia* Nees, Pl. Asiat. Rar. 3: 88. 1932; de Candolle, Prodr. 11: 175. 1847. ≡ *Baphicacanthus cusia* (Nees) Bremek., Verh. Kon. Ned. Akad. Wetensch., Afd. Natuurk., Sect. 2 41(1): 59, 190. 1944; Hu, Fl. Reipubl. Popularis Sin. 70: 113. 2002. ≡ *Ruellia cusia* Wall., Numer. List n. 2386. 1830, nom. invalid. Type: India, Gongachora, 29 May 1809, *F. Buchanan-Hamilton in Wallich Cat. 2386* [lectotype **here designated**: K-W (K001115830); isolectotype: GZU].

Ruellia indigofera Griff., J. Traval. Ind.: 327. 1847. Type: not designated.

Strobilanthes flaccidifolius Nees in de Candolle (ed.), Prodr. 11: 194. 1847; Clarke in Hooker (ed.), Fl. Brit. India 4: 468. 1886. Type: India, Assam, *Griffith s. n.* in Herb. Hook. [holotype: K (K000883229)].

Ruellia indigotica Fortune, Risid in China: 158. 1857. Type: not designated.

Strobilanthes flaccida G. Mann, Assam For. Rep.: 135. 1876-1877. Type: not designated.

Herbs, 0.5–1.5 m high, isophyllous to weakly anisophyllous; stems glabrous or minutely brown puberulent. Leaves equal or unequal in each pair; petiole 0.5–7 cm long; leaf blade elliptic to ovate, 4–20 cm long, 2–9 cm broad, usually blackish when dry, both surfaces glabrous or abaxially minutely puberulent along veins, lateral veins 7–9 pairs, base attenuate, margin serrate, apex acute. Spikes terminal or axillary, 1–6 cm long, usually several aggregated in a leafy branched panicle; peduncle 1–12 cm long. Bracts petiolate, oblanceolate, obovate, or spatulate, 1.2–2.5 cm long; bracteoles linear-oblanceolate, 2–3 mm long, caducous. Calyx 0.8–1.5 mm long, accrescent to 2.2–2.5 cm long in fruit, minutely puberulent, 5-lobed almost to base; lobes unequal, one much longer than others, longer one oblanceolate, shorter ones linear-lanceolate, apex acute to obtuse. Corolla blue, 3.5–5 cm long, straight to slightly bent, outside glabrous; tube base cylindric, 2–3 mm broad, 1–1.5 cm long, upwards slightly curved and gradually widened to 1.2–1.5 cm broad at mouth; lobes oblong, 8–9 mm long, subequal.

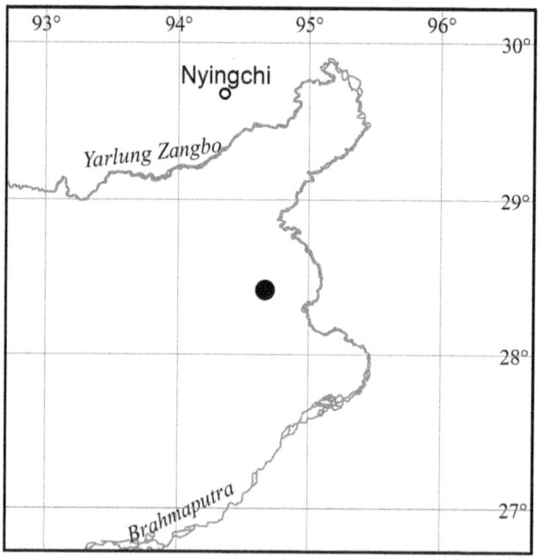

Map 46-107: **Strobilanthes pterygorrhachis** C. B. Clarke

Stamens 4, included; filaments glabrous, longer pair 6–7 mm long, shorter pair 2–3 mm long; anther-thecae oblong, 2.5–3 mm long. Ovary oblong, puberulent; 2 ovules per locule; style 3–3.2 cm long, glabrous. Capsules 1.5–2.2 cm long, glabrous, 4-seeded. Seeds ovate in outline, compressed, 3–3.5 mm long and broad, covered with appressed trichomes; areole small. **Chromosomes** 2n = 16, 32. Figure 46-63.

Phenology: flowering from July to February; fruiting from December to February of next year.

Habitats: moist wooded places, sometimes cultivated; below 2100 m.

Distribution: Bhutan, Yarlung Zangbo-Brahmaputra, S Hengduan, U Irrawaddy (Map 46-108); also in Bangladesh, Laos, Thailand, Vietnam, and other parts of India, Myanmar and China (E & S Yunnan, SE Sichuan, Guizhou, Guangxi, Hainan, Hunan, Hubei, Jiangxi, Fujian, Zhejiang, Taiwan).

Uses: medicinal, dyer.

BHUTAN: **Deothang**, Mikuri to N Decheling, N Nanglan, 1990, *R. Reynolds s. n.* (E); **Punakha**, Mendegang between Dochong La and Lobesa, 1620 m, 10 Oct. 1984, *I. W. J. Sinclair & D. G. Long 5582* (E). **YARLUNG ZANGBO-BRAHMAPUTRA**: **Mishmi Hills**, Lakhuni, Nagahills, 300–450 m, 2 Jan. 1949, *F. Kingdon-Ward 18377* (A, BM); upper part of Brahmaputra (Tsangpo), June-Sep. 1913, *C. Bailey s. n.* (E). **S HENGDUAN**: **Dali**, Tali valley, 2010 m, Aug. 1906, *G. Forrest 4685* (E); 1800–2100 m, *G. Forrest 9383* (E, K); 2100 m, 20 Oct. 1915, *H. Handel-Mazzetti 8548* (WU); **Longyang**: Baihualing, 1560 m, 17 Mar. 2005, *S. Z. Zhang et al. 14884* (SZBG); Mankuan, Baihualing, Wangjiawa, 1550 m, 9 Dec. 2000, *H. Li 13520* (KUN). **U IRRAWADDY**: **Chaungmaw Kyaun**, Hkinlum, North Triangle, 1200 m, 21 Nov. 1953, *F. Kingdon-Ward 21622* (A, BM); **Putao**, Kachi Hills, 300–450 m, 12 Dec. 1930, *F. Kingdon-Ward 9036* (BM); Haka, 1860 m, Nov. 1937, *F. G. Dickason 7730* (A).

59. Strobilanthes atropurpurea Nees, Pl. Asiat. Rar. 3: 86. 1832; Nees in de Candolle (ed.), Prodr. 11: 195. 1847; Clarke in Hooker (ed.), Fl. Brit. India 4: 472. 1884; Nasir & Ali in Malik & Ghafoor (eds.), Fl. Pakistan 188: 54. 1988; Hu, Deng & Wood in Wu, Raven & Hong (eds.), Fl. China 19: 408. 2011. ≡ *Ruellia atropurpurea* Wall., Numer. List n. 2398. 1830, nom. invalid. Type: Nepal (?), 1821, *Wallich Cat. 2398* [lectotype designated by Malik & Ghafoor in Malik & Ghafoor (eds.), Fl. Pakistan 188: 54. 1988: K-W (K001115854); isolectotypes: BM, LE].

Ruellia alata Wall., Pl. Asiat. Rar. 1: 26. 1830, non *Strobilanthes alata* Blume, Bijdr. Fl. Ned. Ind. 14: 798. 1826. ≡ *Strobilanthes wallichii* Nees, Pl. Asiat. Rar. 3: 87. 1832; Nees in de Candolle (ed.), Prodr. 11: 193. 1847; Clarke in Hooker (ed.), Fl. Brit. India 4: 471. 1884; Nasir & Ali in Malik & Ghafoor (eds.), Fl. Pakistan 188: 55. 1988; Wood in Grierson & Long (eds.), Fl. Bhutan 2(3): 1271. 2001. ≡ *Pteracanthus alatus* (Wall.) Bremek., Verh. Kon. Ned. Akad. Wetensch., Afd. Natuurk., Sect. 2 41(1): 199. 1944; Hu, Fl. Reipubl. Popularis Sin. 70: 129. 2002. Type: Nepal, *Wallich Cat. 2371a* [lectotype designated by Malik & Ghafoor, 1988: K-W (K001115780); isolectotypes:

Figure 46-63: **Strobilanthes cusia** (Nees) Kuntze: **a**, habit; **b**, calyx; **c**, opened corolla showing androecium; **d**. capsule. Drawn by Quan LIN; reproduced from *Icon. Cormophyt. Sin.* 4: 160, fig. 5733. 1975.

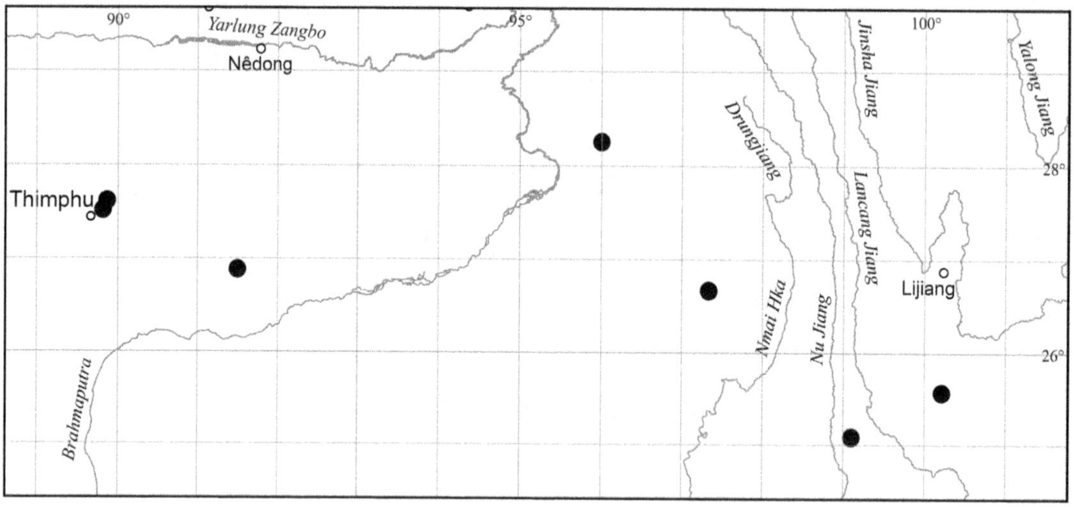

Map 46-108: **Strobilanthes cusia** (Nees) Kuntze

BM (BM000793209), CGE, E (E00273463, E00273464), K (000883216), LE, P (P0071946)].

Strobilanthes wallichii var. *microphylla* Nees in de Candolle (ed.), Prodr. 11: 193. 1847. Type: [U Ganga & Indus] India, Kumaon, *Wallich Cat. 2371b* [lectotype **here designated**: K-W (K001115781); isolectotypes: BM (BM000793211), E (E00273463, E00273464), CGE, K].

Strobilanthes atropurpurea var. *berevifolia* Kuntze, Revis. Gen. Pl. 2: 499. 1891, **syn. nov.** Type: [Sikkim & Darjiling] India, Sikkim, 2700 m, Nov. 1875, *C. E. O. Kuntze 6888* [holotype: NY (NY00179308)].

Herbs, 30–50(–100) cm high, isophyllous or slightly anisophyllous; stems slender, quadrangular, sulcate, rooting at nodes, glabrous or pubescent. Leaves equal or unequal in each pairs, basal ones petiolate, apical ones sessile; petiole 0–2 cm long, winged, glabrous or pubescent; blade ovate, elliptic or narrowly lanceolate, 1–6(–16) cm long, 0.5–3.5(–6.5) cm broad, puberulent or glabrous, lateral veins 5 or 6 pairs, base rounded to attenuate, decurrent onto petiole, margin serrate to coarsely crenate, apex acute to acuminate. Spikes axillary, 2–7 cm long, secund, few-flowered, usually reduced to flowers solitary or paired; rachis zigzag. Bracts ovate to subcordate, 8–10 cm long, 2–3 mm broad, glabrous; bracteoles linear-oblong, 4–5 mm long, 1.5–2 mm broad, glabrous. Calyx 1–1.5 cm long, accrescent to 2–2.5 cm long in fruit, 5-lobed almost to base; lobes unequal, one slightly larger than others, linear, 1–1.5 mm broad, glabrous, pilose, or rarely glandular-pubescent. Corolla pale purple, bluish purple, or rarely white, 2.5–3.5 cm long, outside glabrous, inside glabrous except for trichomes retaining style; tube base cylindric, 1.6–2 mm broad, 1–1.2 cm long, upwards inflated, bent and widened to 1.5–1.8 cm broad at mouth; lobes oblong, 3–4 mm long, 4–4.5 mm broad, apex obtuse. Stamens 4, included; filaments glabrous, longer pair 6–8 mm long, shorter pair 3–4 mm long; anther-thecae oblong, 2–2.3 mm long. Ovary glabrous; 2 ovules per locule; style 2.2–2.5 cm long, glabrous; stigma acute. Capsules fusiform, 1.2–1.8 cm long, glabrous, 4-seeded. Seeds ovate in outline, compressed, 2–3 mm long and broad, pubescent with appressed trichomes; areole 0.2–0.3 mm long. Figure 46-64.

Phenology: flowering from June to December; fruiting from October to February of next year.

Habitats: moist places on mountain slopes, forests, river banks; 900–4000 m.

Figure 46-64: **Strobilanthes atropurpurea** Nees: **a**, flowering branch; **b**. bract and calyx; **c**, opened corolla swing androecium; **d**, capsule. Drawn by Quan LIN; reproduced from *Icon. Cormophyt. Sin.* 4: 163, fig. 5740. 1975.

Distribution: N Pakistan, Jammu & Kashmir, U Ganga & Indus, W Nepal, U Yarlung Zangbo, C Nepal, E Nepal, Sikkim & Darjiling, M Yarlung Zangbo, Bhutan, Yarlung Zangbo-Brahmaputra, S Hengduan, U Irrawaddy (Map 46-109); also in Vietnam, and other parts of Pakistan, India, Myanmar and China (C, E & S Yunnan, S Sichuan, Chongqing, Guizhou, Hainan, Guangxi, Guangdong, Jiangxi, Hunan, Hubei, Fujian, Taiwan, Zhejiang).

Notes: *Goldfussia thomsonii* Hook. was cited as the synonym of *Strobilanthes wallichii* Nees by Yamazaki (Fl. East Himal.: 304. 1966) and Tsui & Hu [in Hu (ed.), Fl. Reipubl. Popularis Sin. 70: 129. 2002]. However, it is a synonym of *S. lamiifolia* (Nees) T. Anderson.

N PAKISTAN: **Dir**, s. d., *Salim 145* (K); **Hazara**, Bhurj, KaganValley, 28 Sep. 1897, *Luayat 22078* (CAL); Mansehra, Dargun, Kagan, *Luayat 22079* (CAL, K). **JAMMU & KASHMIR**: Kashmir, Jul. 1956, *T. A. Rao 308* (CAL); Premula Valley, 3000 m, Jun. 29, 1951, *T. C. Thomas 20782* (CAS); Muhasoo to Kumaon, 2700–3300 m, *Madden 365* (E); **Kishtwar**, Banihal Pass, *R. R. Stewart 14687* (RAW, fide Malik & Ghafoor, 1988); **Sonamarg**, *R. R. Stewart 6393* (K, RAW). **U GANGA & INDUS**: **Chamba**, Kajiar, 2100 m, 2 Oct. 1874, *C. B. Clarke 24071* (BM); Kaletop forests, 2400 m, 6 Jul. 1898, *J. H. Lace 1723* (CAL, E); **Dalhousie**, 2400 m, *C. B. Clarke 22633* (CAL); **Drehradun**, Thauo, 900 m, *J. S. Gamble 27506* (CAL); Jaunsar., Decban, 2700 m, Oct. 1894, *J. S. Gamble 25229* (CAL); Jaunsar, Sep. 1890, *K. Nand 305* (FHO); Pandawa Forets, Jaunsar, 2400 m, Aug. 1898, *B. B. Osmaston 27151* (CAL); Kanasar to Konain Road, Jaunsar, 2400 m, 18 Sep. 1936, *M. B. Raizada 7238* (E); **Kulu**, Manali, Kothi, 3000 m, 7 Jul. 1986, *Kirkpatrick & McBeath 1797* (E); Pulgn, Parbatti valley, 2790 m, 7 June 1934, *S. E. Parkinson 4075* (E); Parbitti Valley, 2700 m, 19 Jun. 1934, *S. E. Parkinson 4124* (E); Solang, 3000 m, 24 Sep. 1916, *R. E. Cooper 5684* (E); **Pithoragarh**, Dhauli valley, 3000–3300 m, 6 Aug. 1886, *J. F. Duthie 5877* (BM, CAL, LE); Palang Gadh, Byans, 3000–3300 m,19 Jul. 1886, *J. F. Duthie 5878* (BM, CAL); **Tehri Garhwal**, Nag Tiba, 2400 m, 7 Aug. 1898, *P. W. Machinnon s. n.* (CAL); **Simla**, Kufri, 2400 m, 4 Aug. 1918, *R. N. Parker 19618* (BM); Kufri, 2400 m, 29 Sep. 1918, *R. N. Parker s. n.* (BM); Pangi, Bushahr, 14 Jul. 1934, *N. Parmanand 499* (CAL, E). **U YARLUNG ZANGBO**: **Dinggyê**, Kamahe, 3000 m, 21 Jun. 1960 m, *s. coll. 473* (PE); **Gyirong**, Gyirong, 3000 m, 2 Jun. 1972, *Exped. Xizang Chinese Herds 88* (PE); Jiaji line, 3380 m, 9 Aug. 2010, *PE-Expedition to Xizang 620* (PE); Gyirong Valley, 3140 m, 24 Jul. 2013, *PE Expedition to Xizang 4149* (PE); **Nyalam**, Quxiang, 3200 m, 28 Aug. 1972, *Xizang Chinese Herd Exped. 1587* (PE); Zhangmu, 3300 m, 15 Aug. 1972, *Xizang*

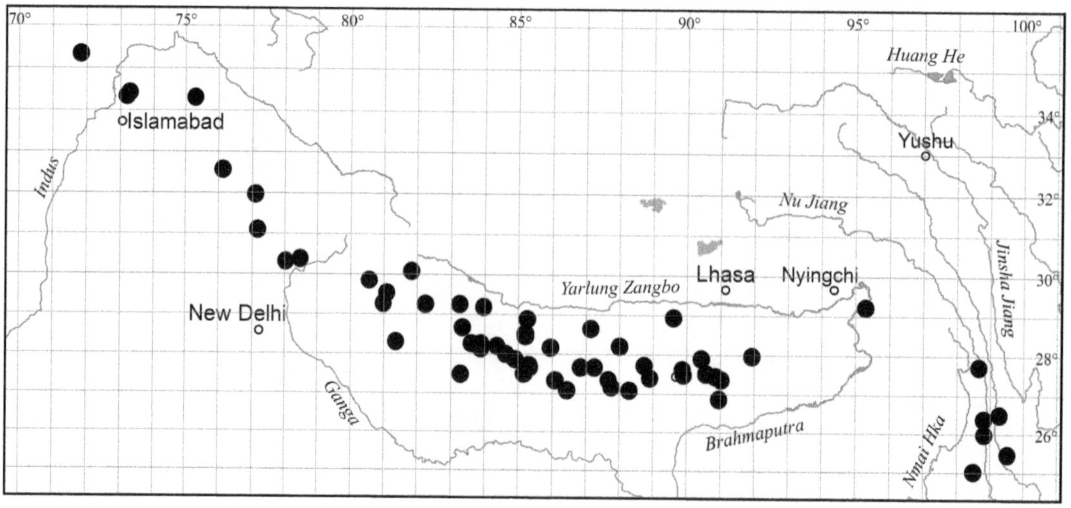

Map 46-109: **Strobilanthes atropurpurea** Nees

Chinese Herd Exped. 1199 (PE); Zhangmu, Lixin to Songsong, 2600 m, 8 Oct. 1990, *B. S. Li & H. Li 14050* (PE); Zhangmu, 3300 m,15 Aug. 1972, *Exped. Xizang Chinise Herds 1199* (PE); Zhangmu, 2950 m, 29 Jun. 1975, *Exped. Qinghai-Xizang 6071* (KUN, PE); Zhangmu, Lixin to Xuebugang, 2800 m, 29 Jun. 1975, *Exped. Qinghai-Xizang 6620* (KUN, PE); Zhangmu, 3205 m, 28 Jul. 2013, *PE-Expedition to Xizang 4363* (PE); **Tingri**, Mount Everest, Puket Ahumus inside Agage, 3300 m, 27 Jun. 1924, *R. W. G. Hinston 387* (K). **W NEPAL**: **Bheri**, Bardiya, Topke Gola to Shewaden, *H. Kanai et al. 720840* (A, TI); Karnali, Dolpa, Tarakot, 2730 m, 8 Jul. 1952, *O. V. Polunin, W. R. Sykes & L. H. J. Williams 2402* (BM, TI); Dolpa, near Tarakot, Bheri river, 3030 m, 10 Jul. 1952, *O. V. Polunin, W. R. Sykes & L. H. J. Williams 2425* (A, BM, K, KATH, TI); Dolpa; Tarakot, 3500 m, 19 Jul. 1973, *S. Einarsson et al. 2458* (BM); Jumla, Maharigaon, 3480 m, 16 Jul. 1952, *O. V. Polunin, W. R. Sykes & L. H. J. Williams 204* (BM); Jumla, Chorta, 3180 m, 29 Jul. 1952, *O. V. Polunin, W. R. Sykes & L. H. J. Williams 4936* (BM); Jumla, between Jumla and Garjigoth, 3330 m, 8 Aug. 1952, *O. V. Polunin, W. R. Sykes & L. H. J. Williams 5033* (A, BM, KATH, TI); Humla, Marghore, 3640 m, 21 Jul. 1968, *S. B. Malla 14125* (BM, TI); **Mahakali,** Darchula, Simaar, 2780 m, 19 Jul. 2015, *S. K. Ghimire et al. ANCA1129* (TUCH); Darchula, Upper Chameliya Valley, Dhauley Oodar, 3400 m, 19 Jul. 2015, *S. K. Ghimire et al. ANCA1186* (TUCH); **Seti**, Bajhang, Ghodi lekh, 13 Aug. 1976, *H. Tabata et al. 2430* (BM, KYO); Bajura, Pategaon to Badigaon, 16 Aug. 1991, *M. Suzuki et al. 9160665* (TI); Doti, Khaptad, 3000 m, 3 Aug. 1972, *M. S. Bista & D. P. Joshi 229* (KATH). **C NEPAL**: **Bagmati**, Dhading, Dhading, 1500 m, 20 Jun. 1986, *N. P. Manandhar 11999* (KATH); Dhading, Ankhu Khola, 3640 m, 10 Jul. 1962, *J. D. A. Stainton 3974* (BM, KATH); Kathemandu, opposite Budhi village, 2700–3000 m, *J. F. Duthie 5877* (CAL); Kathemandu, Mani Bhanjyang to Tonglu, 30 Jul. 1972, *H. Kanai et al. 71730* (KUN); Lalitpur, Godawari, 1500 m, 14 Oct. 1991, *K. A. Lye 17650* (E); Nuwakot, Mangen to Khodang Danda, 28 Aug. 1972, *H. Kanai et al. 722010* (KATH, KYO, TI); Rasuwa, Mailung Khola, 3180 m, 26 Jun. 1975, *J. D. A. Stainton 7388* (E); Rasuwa, Parbati Kund, 3005 m, 25 Jul. 1994, *F. Miyamoto et al. 9420023* (E, KATH); Rasuwa, Parbati Kund, 3005 m, 25 Jul. 1994, *F. Miyamoto et al. 9430017* (E, KATH, TI, TNS); Rasuwa, Gosainkund to Syabrubensi, 3300 m, 16 Aug. 1974, *J. H. de Haas 2308* (BM, L); Rasuwa, Gosainkund, Syabrubensi, 3200 m, 17 Aug. 1974, *J. H. de Haas 2348* (BM, L); Rasuwa, Parbati Kunda, near Gatlang, 2590 m, 25 Jul. 1994, *F. Miyamoto et al. 30017* (KATH); Rasuwa, Gutlang, 3300 m, 25 Jul. 1994, *F. Miyamoto et al. 9420023* (KATH, TI, TNS); Sindupalchok, Saddle Kharka, 2962 m, 12 Sep. 2011, *M. F. Watson et al. EKSIN 101* (E, MBK); Sindupalchok, Singalila, Sandapku, 3800–3970 m, 16 Jul. 1969, *H. Hara et al. 4907* (BM, KUN, TI); Sindupalchok, Marghore, 3600 m, 21 Jul. 1968, *S. B. Malla 14125* (A, TI); Sindupalchok, Thari, 3600 m, Apr. 1888, *King s. n.* (CAL); Sindupalchok, Pongsing, 2730–3940 m, 1929, *L. Dhwoj 119* (BM); Sindupalchok, Tama Khani, 3640 m, 1928, *L. Dhwoj 236* (E); Sindupalchok, Baraha Pokhri ridge, 3030 m, 18 Sep. 1950, *D. G. Lowndes 1522* (BM); **Dhawalagiri**, Mustang, Lete, 3030 m, 6 Jun. 1954, *J. D. A. Stainton, W. R. Sykes & L. H. J. Williams966* (BM, KATH); Mustang, Tukuche, 3030 m, 17 Jun. 1954, *J. D. A. Stainton, W. R. Sykes & L. H. J. Williams 1198* (A, BM, KATH, TI); Mustang, Taglung, 3030 m, 11 Jul. 1954, *J. D. A. Stainton, W. R. Sykes & L. H. J. Williams 1695* (BM, KATH, TI); Mustang, Larjung, 3030 m, 23 Jul. 1954, *J. D. A. Stainton, W. R. Sykes & L. H. J. Williams1972* (BM, K, KATH, TI); Mustang, Lete, 3030 m, 4 Jun. 1954, *J. D. A. Stainton, W. R. Sykes & L. H. J. Williams5563* (BM, KATH); Myagdi, Ghorepani, 2500 m, 21 Jul. 1973, *C. Grey-Wilson & B. Phillips 330* (K); Myagdi, Gurjakhani, 3030 m, 7 Jun. 1954, *J. D. A. Stainton, W. R. Sykes & L. H. J. Williams 3044* (BM, TI); Myagdi, Lumsum, 1910 m, 22 Sep. 1996, *M. Mikage et al. 9683009* (A, TI); **Gandaki**, Gorkha, Sama Gaon, 3640 m, 4 Jul. 1953, *P. C. Gardner 1131* (BM); Kaski, Siklis, Above Siklis, 2700 m, 28 Aug. 1976, *R. G. Troth 956* (BM);

Kaski, Mardi Khola, 3940 m, 19 Sep. 1954, *J. D. A. Stainton, W. R. Sykes & L. H. J. Williams 8489* (BM); Lamjung, between Kuti and Lamjung, 1000 m, 16 Nov. 1953, *J. Jujimura s. n.* (KYO); Lamjung, Rambrong Pass, 3480 m, 29 Jun. 1954, *J. D. A. Stainton, W. R. Sykes & L. H. J. Williams 6013* (A, BM, KATH, TI); **Janakpur**, Dolakha, Jiri to Shivalaya, 1740–2000 m, 16 Aug. 1985, *M. Suzuki et al. 8571636* (TI); Ramechhap, Deurali, 2700–3400 m, 6 Jul. 1985, *H. Ohba et al. 8530128* (E, KYO, TI); Ramechhap, Deurali to Serdingma, 2700–3400 m, 6 Jul. 1985, *H. Ohba et al. 8570196* (BM, KATH, TI); Ramechhap, Shivalaya to Bhandar, 1800–2700 m, 17 Aug. 1985, *H. Ohba et al. 8571698* (TI); Ramechhap, Deorali to Serdingma, 6 Jul. 1985, *H. Ohba et al. 8580160* (KATH, TI); **Lumbini,** Kapilvastu, Budhi village, 2700–3000 m, 18 Jul. 1886, *J. F. Duthie 5877* (CAL); **Narayani**, Makwanpur, Daman, Rikheshor temple, 2306 m, 7 Nov. 2012, *B. Adhikari & G. D. Bhatt 15* (E). **E NEPAL**: **Koshi**, Sankhuwasabha, Arun Valley, Kasuwa Khola, N Nom, 3180 m, 10 Jun. 1956, *J. D. A. Stainton 594* (BM, KATH, TI); Sankhuwasabha, Maghang Khola (Arun Valley), 3030 m, 1 Jul. 1956, *J. D. A. Stainton 811* (A, BM, K, KATH, TI); Sankhuwasabha, Redak to Chhurchathanga, 2400–3600 m, 9 Aug. 1977, *H. Ohashi et al. 770598* (TI); Sankhuwasabha, Khongma to Tashi Gaun, 2160–3150 m, 31 Jul. 1988, *M. Suzuki et al. 8821107* (TI); Sankhuwasabha, Danda Kharka to Unshisa Kharka, 2820 m, 4 Aug. 1988, *M. Suzuki et al. 8850358* (TI); Sankhuwasabha, around Yangri Kharka, 3750 m, 19 Jul. 1988, *M. Suzuki 8880639* (TI); Sankhuwasabha, below Chamtang, 2380 m, 20 Aug. 1998, *S. Noshiro et al. 9830077* (A, BM, L, MBK, MO); Sankhuwasabha, Chamtang, 2380 m, 20 Aug. 1998, *S. Noshiro et al. 9830077* (BM, E); Sankhuwasabha, Deurali, 3230 m, 22 Aug. 1998, *S. Noshiro et al. 9830116* (A, E, TI); Sankhuwasabha, Nehe Kharka, S Barun Khola, 3700 m, 29 Sep. 1991, *D. G. Long et al. 413* (E, KATH); Sankhuwasabha, Obak Khola, 3330 m, 26 Jul. 1971, *L. W. Beer et al. 9463* (BM, TI); **Mechi**, Panchthar, Phedung Danda, 3400 m, 20 Jun. 1992, *S. Noshiro et al. 9240912* (BM); Taplejung, Sele La, 3200 m, 27 Jul. 1958, *S. Makao 139* (KYO); Taplejung, Topke Gola to Shewaden, 3300 m, 28 Jun. 1972, *H. Kanai et al. 720840* (A, BM, TI); Taplejung, Topke to Shewaden, 2600–3600 m, 21 Aug. 1977, *H. Ohashi et al. 770979* (TI); Taplejung, Topke to Shewaden, 2600–3600 m, 21 Aug. 1977, *H. Ohashi et al. 772657* (TI); Taplejung, Topke to Shewaden, 2600–3600 m, 21 Aug. 1977, *H. Ohashi et al. 773888* (TI); Taplejung, Mane Bhanjyang to Jaljale Pokhari, 3500 m, 8 Aug. 1999, *Y. Omori et al. 9950030* (A, E, KATH, TI); Taplejung, Chandanbari, 2700 m, Jan. 1975, *H. K. Sainju & P. M. Regmi 762* (KATH). **Sagarmatha**: Solukhumbu, Pike, near Okhardhunga, 3000–3500 m, 9–10 Sep. 1976, *K. Itoh 56* (KYO); Solukhumbu, Nashing Dinga to Chalem Kharka, 2820 m, 4 Aug. 1997, *M. Wakabayashi et al. 9715072* (TI); Solukhumbu, Pike Dongshar to Namikhil, 9 Sep. 1985, *H. Ohba et al. 8572495* (KATH, TI); Solukhumbu, Nashing Dingma, 2700 m, 4 Aug. 1997, *M. Wakabayashi et al. 9710130* (TI). **SIKKIM & DARJILING**: **Darjiling**, Sindul, 2700 m, 27 Aug. 1869, *C. B. Clarke 8855* (CAL, LE); 2700 m, 29 Aug. 1869, *C. B. Clarke 8880* (CAL); Ralimpoong, 20 Sep. 1869, *C. B. Clarke 9351* (CAL); Singaleleh, 3300 m, 8 Oct. 1870, *C. B. Clarke 13431* (CAL); 2700 m, 3 Sep. 1875, *C. B. Clarke 27372* (BM); Inelo, 2400 m, 14 Sep. 1875, *C. B. Clarke 27436* (CAL, LE); Sandakphu, 16 Sep. 1962, *B. Safrei 1730* (CAL); Kurseong, 26 Oct. 1939, *K. Biswas 3726* (CAL); Danglu, 3000 m, 17 Oct. 1904, *I. H. Burkill 25150* (CAL); between Phalluk & Sandakpu, 3300 m, 22 Oct. 1904, *I. H. Burkill 25365* (CAL); Sonada, 1800 m, Nov. 1911, *A. Meebold 16005* (CAL); 1950 m, 1862, *T. Anderson 1046* (CAL); 1800–2400 m, 1862, *Herb. Anderson 1045* (CAL); Sanduhpho, 3600 m, Jul. 1881, *J. S. Gamble 9452* (CAL); **Sikkim**, near Karponang, Gangtok Road, 2400 m, 3 Mar. 1949, *F. Ludlow, G. Sherriff & J.H. Hicks 17564A* (BM); Tonglo, 3000 m, 5 Aug. 1874, *D. Treutler C. N. 475* (E); north, Glacial valley above Yakche, N of Lachung, 3180 m, 15 Jul. 1996, *Edinburgh*

Expedition to Northern Sikkim (1996) 219 (E); 2100–3000 m, *J. D. Hooker s. n.* (A, BM, CAL, CGE, L); 2100 m, 27 Jul., 1874, *D. Treutler 420* (LE); Mungpoo Bhulkiya, 900 m, 22 May, 1914, *C. W. Cousins 159* (CAL); Linchal, 2400 m, Jul. 1862, *Herb. T. Anderson 1071* (CAL); Linchal, 2400 m, 8 Sep. 1868, *Herb. Anderson 1044* (CAL); Lachung, 2700 m, 2 Sep. 1892, *G. A. Gammie 1101* (CAL); Tongsi to Aloktang, 9 Oct. 1862, *Herb. T. Anderson 1058* (CAL); Zamu Valley, 2850 m, 9 Jul., 1909, *Smith & Cave 1032* (CAL); Kujiar, Chunba, 2100 m, 2 Oct. 1874, *C. B. Clarke 24071* (CAL); Sitong, 1500 m, Feb. 1882, *J. S. Gamble 10262* (CAL); Linchal, 2 Feb. 1877, *S. Kurz 4080* (CAL); Celow Tangloo, 5 Oct. 1875, *J. S. Gamble 1104A* (CAL); Layloo, 2700 m, Aug. 1879, *J. S. Gamble 3558C* (CAL); Gangtok, 1500 m, 5 Oct. 1983, *Sangumdur 16329* (CAL); Rithabu Rigo, 2280 m, Sep. 1879, *J. S. Gamble 7062* (CAL); Karponang, 3200–3500 m, 3 Nov. 1961, *G. Sen Gupta 246* (CAL). **M YARLUNG ZANGBO**: **Gyangzê**, Lelai, 3200–3300 m, 12 Jun. 1975, *Supp. Qinghai-Xizang Exped. 750316* (KUN); **Yadong**, Chumbi Valley, 27 Jul. 1952, *P. C. Tsoong 5959* (PE); Chumbi, Rinclunginy, 22 Jul. 1877, *Dungboo 4309* (CAL, E, K); Champitang to Yatung, 3000–3600 m, 2 Aug. 1936, *F. S. Chapman 681* (K); Xiayadong, near Biwutang, 3300 m, 12 Sep. 1974, *Exped. Qinghai-Xizang 74-2323* (KUN, PE); Xiayadong, 23 Jul., 2011, *X. F. Gao et al. 14360* (CDBI); Tanggapu Valley, 3000–3800 m, 8 Jul. 2013, *PE Expedition to Xizang 2897* (PE); Asang, 3000 m, 8 Jul. 2013, *PE Expedition to Xizang 2934* (PE). **BHUTAN**: Donga La, 3000 m, 27 Jul. 1933, *F. Ludlow& G. Sherriff 330* (BM); Nahi to Hinglai-La, 1950–3300 m, 6 Jul. 1938, *B. J. Gould 845* (E); **Bumthang**, Yuto La, 9 Jul. 1933, *F. Ludlow & G. Sherriff 245* (BM); W side of Yuto La, 3330 m, 9 Jul. 1979, *A. J. C. Grierson & D. G. Long 2617* (BM, E); Dhur Valley, 3300–3900 m, 22 Jul. 1949, *F. Ludlow & G. Sherriff 16930* (BM, TI); **Byakar**, W side of Ura La, SE Byakar, 3500 m, 13 Jun. 1979, *A. J. C. Grierson & D. G. Long 1868* (E); **Mangde Chu**, 10 Jun. 1966, *S. B. Lyon* 3435 (BM); **Punakha**, Duke La, 2700 m, 17 May 1914, *R. E. Cooper 2671* (E); Dochong La Ridge, 3300 m, 16 Jul. 2000, *S. B. Lyon 13002* (E); **Sengor**, below Sengor, 3000 m, 6 Jul. 1979, *A. J. Grierson & D. G. Long 2522* (CAS, E, K, TI); **Thimphu**, above Talukah Monastery, 3700 m, 19 Jun. 1988, *J. R. I. Wood 6411* (E); west of Hongtsu, 3100 m, 10 Jul. 1988, *J. R. I. Wood 6456* (E, FHO); above Phajoding Monastery, 3500 m, 17 Jul. 1988, *J. R. I. Wood 6464* (E, FHO); Phajoding Monastery, 3000 m, 17 Jul. 1988, *J. R. I. Wood 6481* (E, FHO); E Thimphu, 3400 m, 24 Jul. 1988, *J. R. I. Wood 6495* (E); above Motithang, 3200 m, 9 Aug. 1987, *Bigger 2960* (E); **Tsampa**, Shimitang, 3000 m, 1 Jul. 1949, *F. Ludlow, G. Shrrifd & Hicks 19274* (BM, E, TI); **Upper Kuru Chu**, Jalu, Khoma Chu, 3000–3300 m, 3 Aug. 1915, *R. E. Cooper 4318* (BM, E); below Singhi Dzong, 2400 m, 21 Jul. 1949, *F. Ludlow, G. Sherriff & Hicks 21307* (BM, E, TI); **Upper Mangde Chu**, Saga La, 3900 m, 15 Jul. 1949, *F. Ludlow, G. Shrrifd & Hicks 16845* (A, BM, E, TI). **YARLUNG ZANGBO-BRAHMAPUTRA**: **Baibung**, Gelin, 1600 m, Dec. 27, 1992, *H. Sun, Z. K. Zhou & H. Y. Yu ETM-2768* (KUN); **Cona**, Mama, 2900 m, 3 Sep. 1975, *Qinghai-Xizang Exped. 751828* (KUN, PE). **Mêdog,** Rengqingpeng, 1860 m, 19 Jan. 1983, *B. S. Li & S. Z. Cheng 3499* (PE). **S HENGDUAN**: **Gongshan**, Shweli-Salwin divide, Aug. 1918, *G. Forrest 17620* (BM, E, K); Upper Kiukiang valley (Clulung), Letahca, 1950 m, 31 Jul. 1938, *T. T. Yü 19536* (A, E, KUN, PE); **Lanping**, Xiaoyanjing, 3100 m, 29 Jun. 1981, *Exped Hengduan Shan 915* (PE); **Lushui**, Shangpianma to Yakou, bank of Pianmahe, 2700 m, 17 Jul. 1964, *S. K. Wu 8032* (HITBC, KUN); Pianma, Yakou to Pianma, Xinlu, 3400 m, 18 Aug. 1964, *S. K. Wu 8472* (HITB, KUN, LBG); **Yongping**: Jinguangsi, *s. coll. 162* (SWFC). **U IRRAWADDY**: **Chimili Valley**, 3000 m, 21 Aug. 1919, *R. Farrer 1247* (E); **Tengchong**: Qushi, Daba to Zhaigongfang, 2500 m, 7 Sep. 1995, *X. C. Shi & S. X. Yang 788* (KUN); s. loc., 3400 m, 29 Oct. 1965, *Tengchong Gongzuozhan 653253* (YIM).

60. Strobilanthes urophylla (Nees) Nees in de Candolle (ed.), Prodr. 11: 192, 1847; Clarke in Hooker (ed.), Fl. Brit. India 4: 472. 1885, "*urophyllus*"; Wood in Grierson & Long (eds.), Fl. Bhutan 2(3): 1271. 2001; Hu, Deng & Wood in Wu, Raven & Hong (eds.), Fl. China 19: 410. 2011. ≡ *Dipteracanthus urophyllus* Nees in Wallich Pl. Asiat. Rar. 3: 82. 1832. ≡ *Pteracanthus urophyllus* (Nees) Bremek., Verh. Kon. Ned. Akad. Wetensch., Afd. Natuurk., Sect. 2 41(1): 199. 1944; Tsui & Hu in Hu (ed.), Fl. Reipblic. Popularis Sin. 70: 148. 2002. ≡ *Ruellia urophylla* Wall., Numer. List n. 2344. 1830, nom. invalid. Type: India, Meghalaya, Pundua, *F. de Silva* in *Wallich Cat. 2344* [lectotype **here designated**: K-W (K001115699); isolectotypes: BM, K (K000883203, K000883205)].

Strobilanthes urophylla var. *sikkimensis* C. B. Clarke in Hooker (ed.), Fl. Brit. India 4: 472. 1884. Type: [Sikkim & Darjiling] India, Darjiling, *W. Griffith K. D. 6125* [lectotype **here designated**: K (K000883202); isolectotypes: BM, GH, P (P00719462), W].

Subshrubs, 30–60 cm high, anisophyllous; stems quadrangular, glabrous. Leaves unequal in each pair; petiole 0.8–1.6 cm long; blade narrowly ovate-elliptic, 5–14 cm long, 1.5–5.5 cm broad, abaxially pubescent along veins, adaxially glabrous, lateral veins 6–9 pairs, base cuneate to broadly cuneate, decurrent onto petiole, margin remotely serrate, apex shortly acuminate to caudate. Spikes axillary; flowers in opposite pairs, 1–1.5 cm apart on rachis. Bracts lovate-elliptic, 0.8–4.5 cm long, glabrous, pinnativeined, margin subentire, apex obtuse; bracteoles obovate-spatulate, 3–4 mm long, subglabrous, 3-veined. Calyx 7–8 mm long, pubescent, 5-lobed almost to base; lobes unequal, one longer than others, lanceolate, apex acuminate. Corolla pale purple, 2–2.5 cm long, ventricose, outside pubescent, inside subglabrous except for trichomes retaining style; tube base cylindric, 1.6–2 mm broad, 4–5 mm long, upwards slightly bent and gradually widened to 1–1.2 cm broad at mouth; lobes oblong, 4–5 mm long, 3–4 mm broad, apex obtuse. Stamens 4, included; filaments hirsute, longer pair 5–6 mm long, shorter pair 3–4 mm long; anther-thecae oblong, 2–2.5 mm long, 1–1.2 mm broad. Ovary comose; 2 ovules per locule; style 1.5–1.7 cm long, hirsute. Capsules 1.2–1.4 cm long, glabrous, 4-seeded. Seeds ovate in outline, compressed, 2–2.5 mm long, 1.5–2 mm broad, pubescent; areole small.

Phenology: flowering from May to June; fruiting from July to August.

Habitats: evergreen broad-leaved forests; 1000–2200 m.

Distribution: Sikkim & Darjiling, Yarlung Zangbo-Brahmaputra (Map 46-110); also in other parts of India.

SIKKIM & DARJILING: **Darjiling**, East Birch Hill Road, 2200 m, 5 Jul. 1969, *H. Hara et al. 4904* (KYO); 1800 m, 16 Sep. 1874, *J. S. Gamble 3559* (CAL); Nagri Forests, 1800 m, 17 May 1902, *J. H. Lace 2238* (CAL, E); Boyapani, 1050 m, 29 Oct. 1871, *C. B. Clarke 16362* (CAL, E); **Sikkim**, Keochiperi Lake, 22 Jun., 1993, *S. Pradham 15068* (CAL); 1980 m, *J. K. Maheshwari 4962* (CAL); *G. King 891* (CAL); below Chuttakpore, 1800 m, 19 May 1862, *Herb. Sikkim T. Anderson 249* (CAL). **YARLUNG ZANGBO-BRAHMAPUTRA**: **Baibung**, Gelin, Wumingshan, 1800–2000 m, 20 May 1983, *B. S. Li et al. 3654* (PE); Gelin, nilequ, Shangyuan, 1700 m, 24 May 1983, *B. S. Li et al. 3773* (PE); Gelin, 1580 m, 12 Nov. 2019, *Y. F. Deng 28893* (IBSC); **Gutung**, Damu to Gedang, 14 Nov. 2016, *Y. J. Guo et al. 16CS14439* (IBSC); **Lhünzê**, Zhari to Cuogahu, 8 Aug. 2013, *Y. S. Chen et al. 13-0771* (PE); **Luhit (Lohit)**, Dreyi to Shoeliang, 1200–1600 m, 13 Nov. 1957, *R. S. Rao 10539* (CAL); **Mêdog**: Aniqiao, 1200 m, 30 Jun. 1980, *W. L. Chen 10839* (PE).

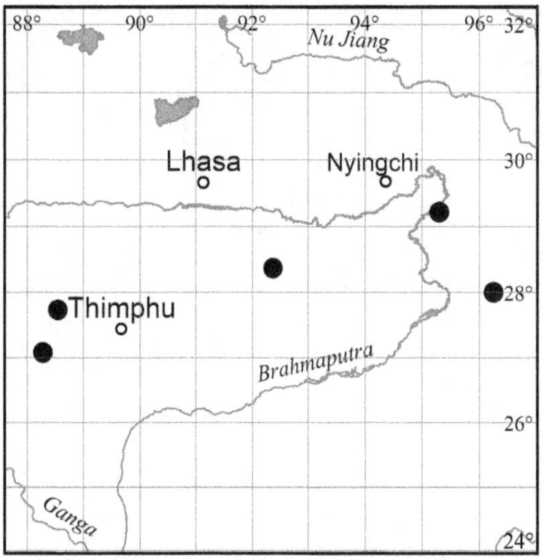

Map 46-110: **Strobilanthes urophylla** (Nees) Nees

61. Strobilanthes inflata T. Anderson, J. Linn. Soc., Bot. 9: 476. 1867, "inflatus"; Clarke in Hooker (ed.), Fl. Brit. India 4: 471. 1885, "*inflatus*"; Wood in Grierson & Long (eds.), Fl. Bhutan 2(3): 1272. 2001; Hu, Deng & Wood in Wu, Raven & Hong (eds.), Fl. China 19: 410. 2011. ≡ *Pteracanthus inflatus* (T. Anderson) Bremek., Verh. Kon. Ned. Akad. Wetensch., Afd. Natuurk., Sect. 2 41(1): 278. 1944; Tsui & Hu in Hu (ed.), Fl. Reipblic. Popularis Sin. 70: 142. 2002. Type: [Sikkim & Darjiling] India, Sikkim, 5000–8000 ft., J. D. Hooker *& Thomson s. n. "28 Goldfussia"* [lectotype **here designated**: K (K000883209); isolectotypes: K (K000883210), BM, C (C10005202), E (E00273496), MPU (MPU018237), NY, P (P00719353, P00719354)].

Strobilanthes wardii W. W. Sm., Notes Roy Bot. Gard. Edinburgh 10: 201. 1921. Type: Valley of Baw Chaung, 9,000-10,000 ft., 28 Aug. 1914, *F. Kingdon-Ward 1912* [lectotype **here designated**: E (E00273495); isolectotype: E (E00273494)].

Herbs, 50–100 cm high, isophyllous; stems straight or somewhat zigzag, slightly quadrangular, sulcate, pubescent. Leaves equal in each pair; petiole 3–3.5(–7.5) cm long, subglabrous, pubescent, or densely yellowish brown hirsute; blade oblong-ovate, elliptic or ovate, 5–20 cm long, 1–10 cm broad, abaxially hirsute, pubescent, or subglabrous and with conspicuous cystoliths, adaxially pubescent, brownish and whitish tomentose along veins, sparsely setose, or densely yellowish brown hirsute on midvein, lateral veins 6–8 pairs, base oblique and broadly rounded to cuneate, decurrent onto petiole, margin serrate to crenate, apex acuminate to caudate. Spikes terminal or axillary, 2.5–12(–21) cm long, secund, sometimes reduced to 1 or 2 flower pairs; peduncle 1.5–3 cm long; rachis usually geniculate at first node then straight or somewhat zigzag; flowers 1–2 cm apart on rachis. Bracts variable in size and shape, lowermost ones leaf-like, with a petiole for 3–4 mm long, ovate to narrowly ovate, 2–3 cm long, 0.8–1.3 cm broad, margin serrate, upper ones obovate, elliptic, or ovate-spatulate, 0.6–3 cm long, 0.3–1.2 cm broad, pubescent, margin usually with 1 or 2 teeth, apex acuminate; bracteoles oblong to linear, 2–3.5 mm long, apex rounded. Calyx 1.1–1.4 cm long, white glandular or eglandular pilose, 5-lobed almost to base; lobes unequal, linear, reflexed, glabrous, sometimes glandular-pubescent in fruit, apex acute. Corolla pinkish blue to blue, 3.5–4 cm long, outside pubescent, inside glabrous except for trichomes retaining style; tube base cylindric, 1.6–2 mm broad, 4–5 mm long, upwards bent to ca. 90° and abruptly widened to 1–1.3 cm broad at mouth; lobes oblong, 4–5 mm long, 5–6 mm broad. Stamens 4, included; filaments glabrous, longer pair 5–6 mm long, erect, shorter pair 3–4 mm long, incurved; anther-thecae oblong, 1.5–2 mm long, 1–1.2 mm broad. Ovary comose; 2 ovules per locule; style 1.6–1.8 cm long, sparsely glandular-pubescent towards base. Capsules brownish yellow, narrowly ellipsoid, 1.6–2 cm long, glabrous, 4-seeded. Seeds orbicular in outline, compressed, 2–3 mm in diameter, villous; areole inconspicuous. **Chromosomes** 2n = 32.

Phenology: flowering from June to November.

Habitats: stream banks in shady places, forests, forest margins; 1200–3800 m.

Distribution: Yarlung Zangbo-Brahmaputra, S Hengduan, U Irrawaddy (Map 46-111); also in Indonesia, and other parts of India, Myanmar and China (S Yunnan).

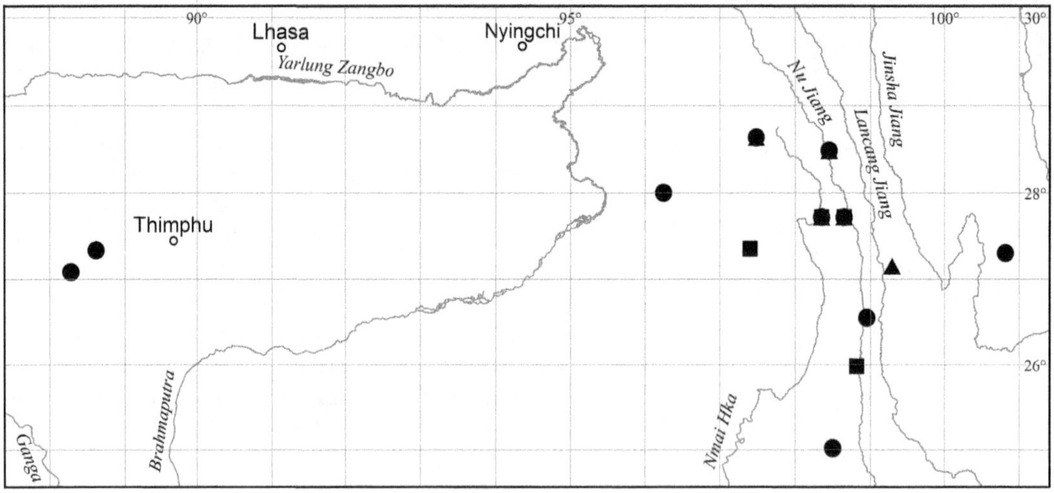

Map 46-111: **Strobilanthes inflata** T. Anderson var. **inflata** (●), var. **aenobarba** (▲), var. **gongshanensis** (■)

1a. Stems, petiole and inflorescences densely covered with long yellowish brown trichomes; leaf blade broadly ovate, 12–20 cm long, 6–10 cm broad .... 61c. var. *aenobarba*
1b. Stems, petiole and inflorescences subglabrous or white pubescent; leaf blade ovate-elliptic, less than 12 cm long and 6 cm broad.
  2a. Leaf blade white pubescent .... 61a. var. *inflata*
  2b. Leaf blade subglabrous .... 61b. var. *gongshanensis*

61a. Strobilanthes inflata var. **inflata**

Stems, petiole and inflorescences white pubescent. Leaf blade elliptic to ovate, 5–8 cm long, 3–6 cm broad, white pubescent.

Phenology: flowering from June to November.

Habitats: forests; 1200–3800 m.

Distribution: Sikkim & Darjiling, Yarlung Zangbo-Brahmaputra, S Hengduan, U Irrawaddy; also in other parts of India, Myanmar and China (S Yunnan).

SIKKIM & DARJILING: **Darjiling**, Jorpoku, 2250 m, 4 Sep. 1902, *J. H. Lace 2367* (CAL); Dloyd Botanical Garden, 1950 m, 4 Jul. 1956, *D. Chatterjee 10* (CAL); Nov. 1996, *J. T. K. Bhattachaya et al. 24335* (CAL); **Sikkim**, Manipo, 1950 m, *C. B. Clarke 36029* (K, LE); 1500–2400 m, *J. D. Hooker 28* (A, L, P, S); Rungbee, 17 Feb. 1867, *T. Anderson s. n.* (CAL, L); Chiwa Bhaujiang, Oct. 1908, *Ribo 345* (CAL); Gompoha Forests, 1800–2100 m, 1891, *H. H. Haines 2752* (K); Rungbool, Oct. 1872, *J. S. Gamble 3570* (K); Sonadah, 1800 m, Jul. 1879, *J. S. Gamble 3571* (K). **YARLUNG ZANGBO-BRAHMAPUTRA**: **Lohit,** Dreyi to Shoeliang, 1200–1625 m, 13 Nov. 1957, *Rolla & Rao 10491* (CAL); **Zayü**, Aug. 2009, *T. Chen 0908305* (SZBG); 20 Jul. 2008, *X. H. Jin et al. DLJ-ET 3404* (PE). **S HENGDUAN**: **Bijiang**, Lahe valley, 2700 m, 16 Jul. 1978, *Exped. Bijiang 1299* (KUN, YUKU); **Cawalong** (Tsawalung), 3750 m, May–Jun. 1932, *J. F. Rock 23248* (BM, E, K, KUN, SYS, US); **Gongshan**, Cikai, 2700 m, 11 Oct. 2004, *Gaoligong Shan Biodiversity Survey 22364* (CAS, KUN); way to Dulong, 3000 m, 26 Jul. 1982, *Exped. Qinghai-Xizang 8642* (PE); Cikai to Drungjiang, Heipu Tunnel, 2 Sep. 2011, *S. X. Yu & W. Wang 6526* (PE); Quanmugulu to Aludeng, 3100–3600 m, 2 Aug. 1979, *Lin Qin 791708* (KUN); **Ninglang**, s. loc., s. d., *H. T. Tsai 57389* (IBSC, KUN, PE); s. loc., 1933–

1934, *H. T. Tsai 57433* (A, BO, IBSC, KUN, LBG, NAS, PE, SZ); Yunlong, Tianchi, 2550 m, 3 Aug. 2010, *H. Peng et al. H-Dianyunlong-Z 4214* (KUN); Tianchi Nature Reserve, 2581 m, 29 Jul. 2010, T. *Zhang et al. 10CS2125* (K). **U IRRAWADDY**: **Drungjiang**, Moqiuwang, 2240 m, 1 Sep. 2006, *Gaoligong Shan Biodiversity Survey 34431* (KUN); Moqiuwang, 2240 m, 1 Sep. 2006, *Gaoligong Shan Biodiversity Survey 34448* (KUN); Moqiuwang, 2240 m, 1 Sep. 2006, *Gaoligong Shan Biodiversity Survey34470* (KUN); Dizheng Dang, 2500 m, 4 Jul. 2008, *X. H. Jin et al. DLJ-ET 1435* (PE); **Tengchong**, Wuhe, Xiaodifang, 2211 m, 22 Aug. 2003, *Gaoligong Shan Biodiversity Survey 17411* (CAS, KUN); Sanyun, 2300 m, 4 Sep. 2003, *Gaoligong Shan Biodiversity Survey 18696* (CAS, KUN); Muchengzhai, 23 Nov. 1954, *J. S. Yang 1389* (KUN); Mingguang, Zizhi, 2750–2850 m, 14 May 2006, *Gaoligong Shan Biodiversity Survey 29238* (KUN); Mackha-Salwin Divide, Aug. 1919, *G. Forrest 18305* (BM, CAL, E, K).

61b. Strobilanthes inflata var. **gongshanensis** (H. P. Tsui) J. R. I. Wood & Y. F. Deng, Bot. J. Linn. Soc. 150: 387. 2006; Hu, Deng & Wood in Wu, Raven & Hong (eds.), Fl. China 19: 411. 2011. ≡ *Pteracanthus gongshanensis* H. P. Tsui, Acta Bot. Yunnan. 12: 277. 1990; Tsui & Hu in Hu (ed.), Fl. Reipubl. Popularis Sin. 70: 137. 2002. Type: [U Irrawaddy] China, Yunnan, Gongshan, Drungjiang to Jiku, 1900–2200m, 6 Aug. 1982, *Exped. Qinghai-Xizang 8953* [holotype: PE (PE01558622); isotypes: KUN, PE (PE01558620, PE015586201)].

Strobilanthes unilateralis J. R. I. Wood, Edinburgh J. Bot. 51: 264. 1994. Type: [U Irrawaddy] Myanmar, North Triangle, Wring Bum, above Ahkail, 2338 m, 11 Nov. 1953, *F. Kingdom Ward 21576* [holotype: BM (BM000884990); isotypes: A (A00423037), TI].

Stems, petiole and inflorescences subglabrous. Leaf blade ovate-elliptic, 5–12 cm long, 1–4.5 cm broad, subglabrous. Figure 46-65.

Phenology: flowering from August to September; fruiting from November to December.

Habitats: evergreen broad-leaved forests; 1900–3300 m.

Distribution: **endemic**; S Hengduan, U Irrawaddy.

S HENGDUAN: **Gongshan**, *T. T. Yü 103* (YUKU); Gaoligongshan, 2000 m, 26 Jul. 1982, *Exped. Qinghai-Xizang 8825* (KUN, PE); **Lushui**, Luobenzhuo, Ega, 3300 m, 7 Aug. 2005, *Gaoligong Shan Biodiversity Survey 25698* (CAS, KUN, MO); Luobenzhuo, Ega, 2920 m, 8 Aug. 2005, *Gaoligong Shan Biodiversity Survey 25772* (CAS, KUN). **U IRRAWADDY**: **Chaungmaw Kyaun**, Hkinlum, 1070 m, 25 Aug. 1953, *F. Kingdon-Ward 21280* (BM, TI); North Triangle, Wring Bum, above Ahkail, 2338 m, 11 Nov. 1953, *F. Kingdom Ward 21576* (A, BM, TI); **Drungjiang**, between Shigong Qiao and Panjiasheng, 2100 m, 30 Oct. 2004, *Gaoligong Shan Biodiversity Survey 21996* (CAS, KUN); W side of the Dulong Jiang valley, NW Maku, 2068 m, 17 Aug. 2006, *Gaoligong Shan Biodiversity Survey 32376* (KUN, MO); Drungjiang to Jiku, 1900–2200 m, 6 Aug. 1982, *Exped. Qinghai-Xizang 8953* (KUN, PE); **Putao**, Wring Bum, above Ahkail, 2338 m, 11 Nov. 1953, *F. Kingdon-Ward 21516* (BM).

61c. Strobilanthes inflata var. **aenobarba** (W. W. Sm.) J. R. Wood & Y. F. Deng, Bot. J. Linn. Soc. 150: 386. 2006; Hu, Deng & Wood in Wu, Raven & Hong (eds.), Fl. China 19: 411. 2011. ≡ *Strobilanthes aenobarba* W. W. Sm., Notes Roy. Bot. Gard. Edinburgh 13: 185. 1921. ≡ *Pteracanthus aenobarbus* (W. W. Sm.) C. Y. Wu & C. C. Hu, Fl. Reip. Popularis Sin. 70: 128. 2002. Type: [S Hengduan] China, Xizang, Salween-Kiukiang divide, Tsarong, Aug. 1914, *G. Forrest 19238* [lectotype **here designated**: E (E00047316); isotypes: E (E00047316), K (K000882703)].

Stems, petiole, blade and inflorescences densely covered with long yellowish brown trichomes. Leaf blade broadly ovate, 12–20 cm long, 6–10 cm broad.

Phenology: flowering from August to October.

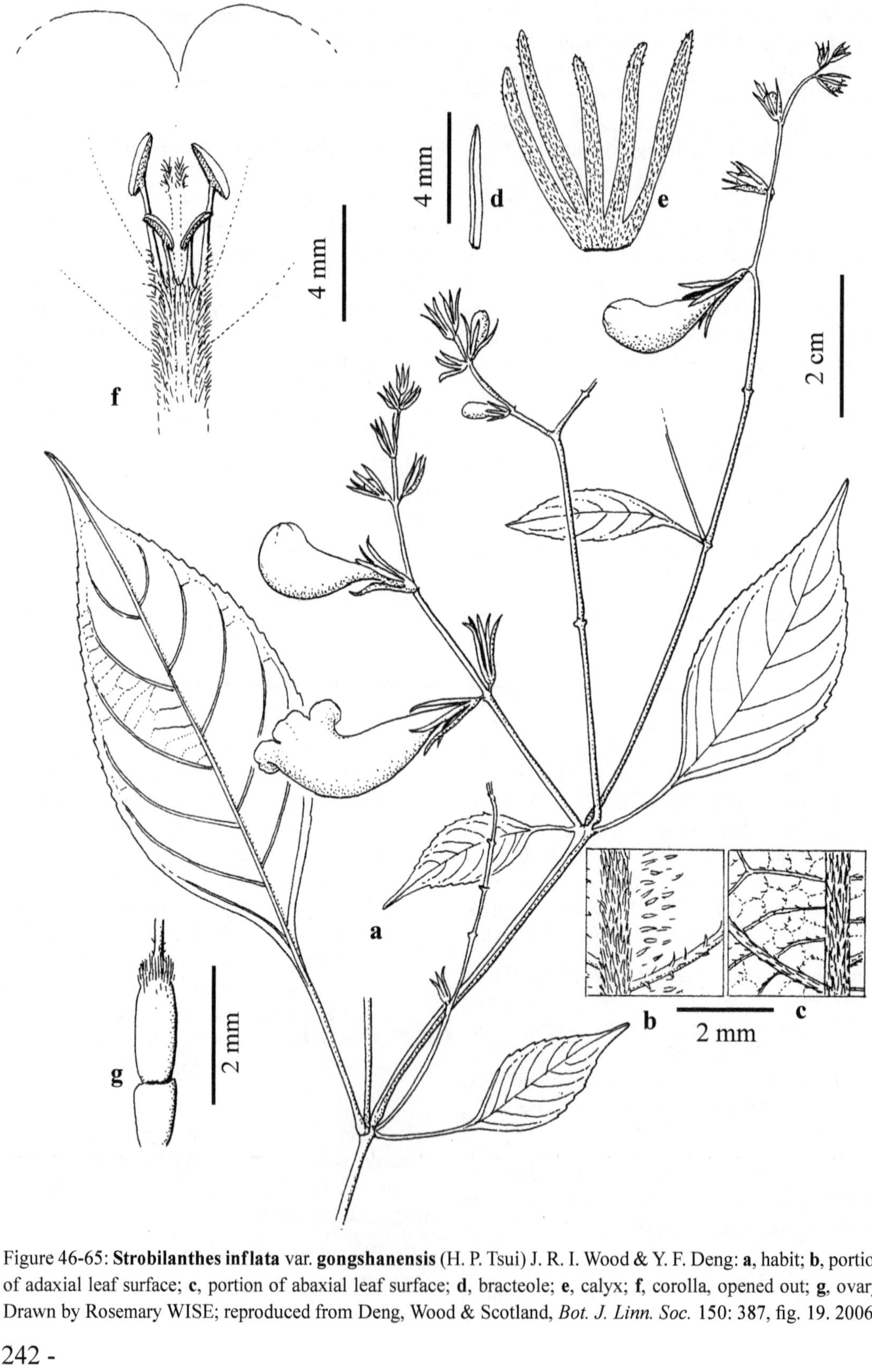

Figure 46-65: **Strobilanthes inflata** var. **gongshanensis** (H. P. Tsui) J. R. I. Wood & Y. F. Deng: **a**, habit; **b**, portion of adaxial leaf surface; **c**, portion of abaxial leaf surface; **d**, bracteole; **e**, calyx; **f**, corolla, opened out; **g**, ovary. Drawn by Rosemary WISE; reproduced from Deng, Wood & Scotland, *Bot. J. Linn. Soc.* 150: 387, fig. 19. 2006.

Habitats: stream banks in shady places, forest margins; 1900–3200 m.

Distribution: S Hengduan, U Irrawaddy; also in Indonesia, and other parts of Myanmar and China (SW Yunnan).

S HENGDUAN: **Cawalong** (Tsawarong), Salwin-Kiu-chiang divide, Guoda, Aug. 1919, *G. Forrest 19238* (E, K); **Gongshan**, Divide of Lancangjiang and Nujiang, Dayon-lumba, 3000–3200 m, 25 Aug. 1940, *K. M. Feng 6904* (KUN, PE); Cikai, Heipushan, Oct. 6, 1940, *K. M. Feng 8200* (KUN, PE); **Weixi**: Baimaluo, 2300–2500 m, 4 Jun. 1940, *K. M. Feng 4461* (KUN, PE). **U IRRAWADDY**: **Drungjiang**, Dizhengdang, 1950 m, 28 Nov. 1990, *Dulong Jiang Investigation Team 2213* (CAS, KUN); Sansuo, 22 Nov. 1990, *Dulong Jiang Investigation Team 771* (KUN, PE).

62. Strobilanthes auriculata Nees in Wallich (ed.), Pl. Asiat. Rar. 3: 86. 1832; de Candolle, Prodr. 11: 191. 1847; Clarker in Hooker (ed.), Fl. Brit. India 4: 453. 1884; Wood in Grierson & Long (eds.), Fl. Bhutan 2(3): 1260. 2001; Hu, Deng & Wood in Wu, Raven & Hong (eds.), Fl. China 19: 399. 2011. ≡ *Perilepta auriculata* (Nees) Bremek., Verh. Kon. Ned. Akad. Wetensch., Afd. Natuurk., Sect. 2 41(1): 194. 1944; Hu, Fl. Reipubl. Popularis Sin. 70: 116. 2002. ≡ *Ruellia auriculata* Wall., Numer. List n. 2341. 1830, nom. invalid. Type: Nepal, *F. Buchanan-Hamilton in Wallich Cat. 2341a* [lectotype **here designated**: K-W (K001115690); isolectotypes: BM (BM000906333), GZU (GZU000249669), K, K-W (K001115691)].

Strobilanthes amplectens Nees, Pl. Asiat. Rar. 3: 86. 1832. ≡ *Perilepta amplectens* (Nees) Bremek., Verh. Kon. Ned. Akad. Wetensch., Afd. Natuurk., Sect. 2 41(1): 195 & 269. 1944. Type: Myanmar, Taong Dong, 27 Nov. 1826, *Wallich Cat. 7158* [lectotype **here designated**: K-W (K001126803)].

Strobilanthes plumulosa Nees, Pl. Asiat. Rar. 3: 86. 1832. ≡ *Strobilanthes auriculata* var. *plumulosa* C. B. Clarke in Hooker (ed.), Fl. Brit. India 4: 453. 1884. ≡ *Perilepta plumulosa* (Nees) Bremek., Verh. Kon. Ned. Akad. Wetensch., Afd. Natuurk., Sect. 2 41(1): 195 & 283. 1944. Type: Myanmar, Prome Mountains, 1826, *Prome in Wallich Cat. 7157* [lectotype **here designated**: K-W (K001126802); isotypes: BM (BM000906337), E (E00273411), K (K000882957, K000882958)].

Strobilanthes edgeworthiana Nees in de Candolle (ed.), Prodr. 11: 190. 1847. ≡ *Strobilanthes auriculata* var. *edgeworthiana* (Nees) C. B. Clarke in Hooker (ed.), Fl. Brit. India 4: 453. 1884. ≡ *Perilepta edgeworthiana* (Nees) Bremek.,Verh. Kon. Ned. Akad. Wetensch., Afd. Natuurk., Sect. 2 41(1): 195 & 274. 1944; Hu, Fl. Reipubl. Popularis Sin. 70: 117. 2002. Type: [U Ganga & Indus] India, Dehradun, *M. P. Edgeworth s. n.* [lectotype designated by Wood & Scotland, Kew Bull. 58: 90. 2003: K (K000882956)].

Strobilanthes maclellandii var. *latipes* C. B. Clarke in Hooker (ed.), Fl. Brit. India 4: 454. 1884. Type: Myanmar, Tenasserim, or "Pegu Yoma", Nov. 1879, *R. H. Beddome 97* (lectotype **here designated**: K (K000882953); isolectotyp: K (K000882954)).

Strobilanthes auriculata var. *bracteolata* C. B. Clarke in Hooker (ed.), Fl. Brit. India 4: 453. 1884. Type: India, Meghalaya, Khasi Hills, *J. D. Hooker & G. Thomson s. n.* [lectotype designated by Wood & Scotland, Kew Bull. 58: 89. 2003: K (K000882965)].

Strobilanthes dyeriana Mast., Gard. Chron. 1: 442. 1893. ≡ *Strobilanthes auriculata* var. *dyeriana* (Mast.) J. R. I. Wood, Kew Bull. 58: 92. 2003. ≡ *Perilepta dyeriana* (Mast.) Bremek., Verh. Kon. Ned. Akad. Wetensch., Afd. Natuurk., Sect. 2 41(1): 195 & 274. 1944; Hu, Fl. Reipubl. Popularis Sin. 70: 117. 2002. Type: a specimen cultivated at Kew in Feb. 1916 (neotype designated by Wood & Scotland, Kew Bull. 58: 92. 2003: K).

Strobilanthes trichophora C. E. C. Fisch., Bull. Misc. Inform., Kew 1932: 202. 1932. Type:

India, Meghalaya, Garo Hills, Nov. 1929, *N. E. Parry 1076* [holotype: K (K000882960, K000882961)].

Subshrubs, 0.5–2 m high, anisophyllous; stems quadrangular, sometimes zigzag, glabrous or sparsely hirsute. Leaves unequal, sessile; leaf blade narrowly to broadly oblong-elliptic, or oblong-oblanceolate, rarely ovate, 2–20 cm long, 1.5–6 cm broad, both surfaces sparsely pilose, lateral veins 12–15 pairs, base attenuate, rounded, or auriculate, margin serrate to undulate, apex acuminate. Spikes axillary or terminal, 3–10 cm long, sometimes branched; peduncle 2–4 cm long, quadrangular, sulcate. Bracts variable in shape and size, lower ones sterile, ovate, 0.8–2 cm long, 1.1–1.5 cm broad, upper ones fertile, imbricate, broadly ovate to obovate-spatulate, 4–9 mm long, densely villous and ciliate, apex usually apiculate, sometimes rounded, or emarginate, recurved; bracteoles absent. Calyx 6–12 mm long, glandular-pilose, 5-lobed almost to base; lobes linear, unequal, 2 slightly shorter than others, margin ciliate, apex acute. Corolla pale purple to violet, funnel-shaped, 2.5–3 cm long, curved, outside glandular-pilose on lobes, inside glabrous; tube base cylindric, 4–5 mm long, upwards bent, abruptly inflated and widened to 2.2–2.5 cm broad at mouth; lobes oblong, 2–3 mm long and broad, apex emarginate. Stamens 4, included; filaments glabrous, longer pair 4–5 mm long, shorter pair 2–3 mm long; anther-thecae oblong, 1.2–1.5 mm long, 0.8–1 mm broad. Ovary glabrous; 2 ovules per locule; style 3–3.2 cm long. Capsules narrowly oblong-obovoid, 8–10 mm long, 2–3 mm broad, glabrous, apex apiculate, 4-seeded. Seeds ovate-suborbicular in outline, 1.5–2 mm long, 1.5–1.8 mm broad, pubescent; areole very small. Figure 46-66.

Phenology: flowering from September to January; fruiting from January to February of next year.

Habitats: valleys; 200–2100 m.

Distribution: U Ganga & Indus, W Nepal, C Nepal, E Nepal, Sikkim & Darjiling (Map 46-112); also in Bangladesh, Malaysia, Thailand, and other parts of Pakistan, India, Myanmar and China (S Yunnan).

Notes: The cultivated plants with reddish leaf blade were sometimes treated as an independent species *Strobilanthes dyeriana* or a variety of *Strobilanthes auriculata* var. *dyeriana*, which is not recognized here.

U GANGA & INDUS: Kumaon, Sitaban, 600 m, *R. Strachey & J. E. Winterbottom 1* (K); Kumaon, Sitaban, 600 m, *R. Strachey & J. E. Winterbottom 2* (K); **Dehera Dun**, Rajpur, 2050 m, 20 Jan. 1962, *S. K. Sharma 19754* (RRLH); **Kangra**, Bhali Forests, 4 Apr. 1901, *G. S. Hast 520* (E); **Musoorie**, *M. B. Saxena 1584* (DD, fide Raizada & Saxena, 1978). **W NEPAL**: **Bheri**, Jajarkot, 1060 m, 17 Oct. 1952, *O. V. Polunin, W. R. Sykes & L. H. J. Williams 5798* (BM, E); Surkhet, Kalyan, 24 Feb. 1991, *N. P. Manandhar 328-91* (KATH); **Rapti**, Dang, Koilabas, 210 m, 30 Dec. 1977, *K. J. Malla 1465* (KATH); Dang, Koilabas, 300 m, 25 Nov. 1992, *P. P. Kurmi KB378* (KATH); **Seti**, Doti, Budar, approx. 3 km from Budar to Dadeldhura, 1410 m, 11 Nov. 2012, *B. Adhikari & G. D. Bhatt 43* (E); Kailali, Dhangarhi, 240 m, 24 Mar. 1981, *T. K. Bhattachaya 3612* (KATH); Kailali, Ghante, 300 m, 1976, *H. Saiju et al. 259* (KATH). **C NEPAL**: **Narayani**, Makwanpur, Chuniya to Purano Gaon, 1220 m, Hetauda to Daman, 28 Nov. 2004, *T. Miller et al. A185* (E, KATH); Makwanpur, Jyamire, Bhainse, on the road from Hetauda to Daman, 1060 m, 28 Nov. 2004, *T. Miller et al. A193* (E, KATH); **Chitwan**, Sauraha to Nandan Tal, 18 Jan. 1996, *M. Mikage et al. 961456* (TI). **E NEPAL**: **Koshi**, Dhankuta, 1200 m, 21 Oct. 1963, *H. Hara et al. 6303690* (A, BM, K, KATH, KYO, TI); **Mechi**, Panchthar, Tharpu to near Chyangthaphu, 26 Nov. 1963, *H. Kanai et al. 6303683* (KATH, KYO, TI). **SIKKIM & DARJILING**: **Darjiling**, Teran, Oct. 1877, *C. B. Clarke 28066* (K); **Sikkim**, Parthoh 450 m, Nov. 1779, *J. S. Gamble 7355* (K).

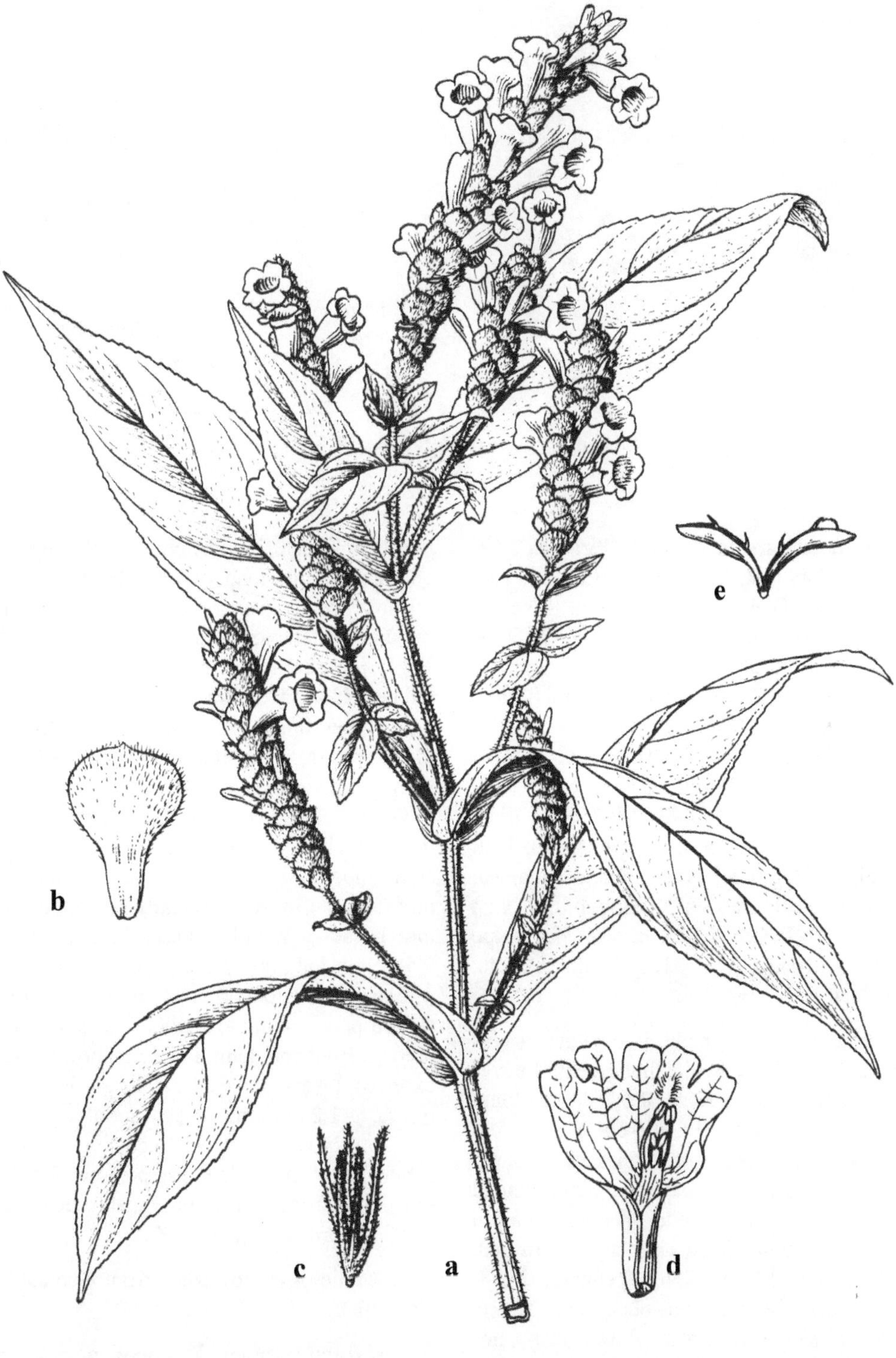

Figure 46-66: **Strobilanthes auriculata** Nees: **a**, flowering branch; **b**, bract; **c**, calyx; **d**, corolla; **e**, capsule. Drawn by Quan LIN; reproduced from *Icon. Cormophyt. Sin.* 74: 159, fig. 5732. 1975.

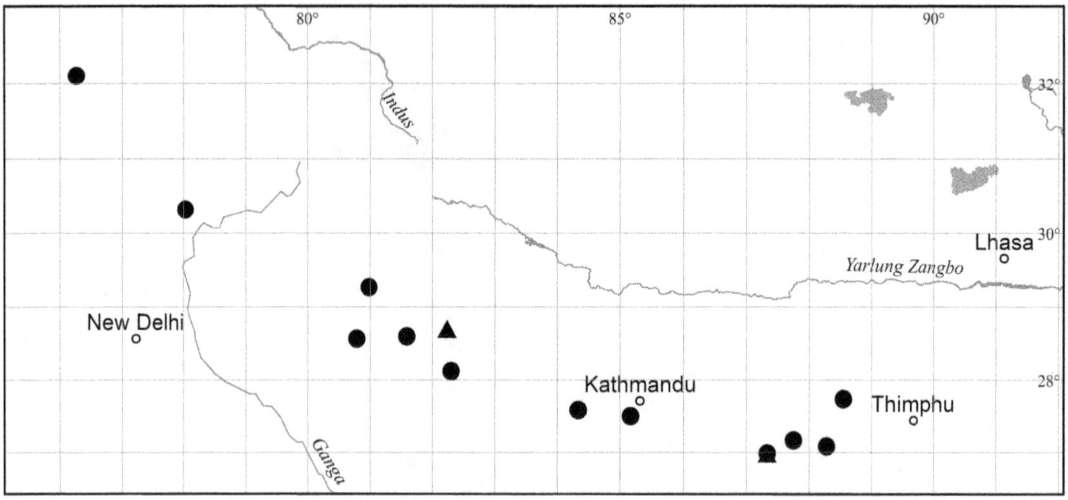

Map 46-112: **Strobilanthes auriculata** Nees (●), var. **dyeriana** (▲)

63. Strobilanthes affinis (Griff.) Terao ex J. R. I. Wood & J. R. Benn., Kew Bull. 58: 134. 2003; Hu, Deng & Wood in Wu, Raven & Hong (eds.), Fl. China 19: 399. 2011. ≡ *Adenosma affinis* Griff., Not. Pl. Asiat. 4: 133. 1854. ≡ *Tarphochlamys affinis* (Griff.) Bremek., Verh. Kon. Ned. Akad. Wetensch., Afd. Natuurk., Sect. 2 41(1): 156. 1944. ≡ *Strobilanthes acrocephalus* T. Anderson, J. Linn. Soc., Bot. 9: 473. 1867 nom. illeg. (Art. 52); Clarke in Hooker (ed.), Fl. Brit. India 4: 454. 1884, Type: India, Meghalaya, Khasia on the descent of Bogupin, 7 Nov. 1835, *W. Griffith K. D. No. 6098* [lectotype **here designated**: K (K000883121); isolectotype: CAL].

Herbs, 30–60 cm high, isophyllous; stems quadrangular, sulcate, white pubescent. Leaves equal in each pair; petiole 0.5–1.5 cm long, pubescent; blade ovate to ovate-elliptic, 3–8 cm long, 2–4.5 cm broad, both surfaces sparsely strigose, lateral veins 5–7 pairs, base subrounded to cuneate, decurrent onto petiole, margin crenulate, apex acute. Spikes axillary or terminal, dense-flowered, 2–6 cm long; peduncle 0.5–2 cm long, densely glandular-pubescent. Bracts variable in shape and size, lowermost ones sometimes sterile, suborbicular to ovate, 5–10 mm long, 5–10 mm broad, strigose, upper ones fertile, imbricate, broadly obovate, rhomboidal, or elliptic, 6–7 mm long, 3–4 mm broad, glandular-pilose, 5-veined, margin ciliate, apex sometimes reflexed; bracteoles oblong-linear, 4–5 mm long, 1–1.2 mm broad, glandular-pubescent, margin ciliate. Calyx 5–7 mm long, 5-lobed almost to base; lobes linear, unequal, one slightly longer than others, glandular-pubescent, 1-veined, margin ciliate, apex obtuse. Corolla purplish-blue to violet, 1.2–1.5 cm long, curved, outside pilose, inside glabrous except for trichomes retaining style; tube base cylindric, 1–1.2 mm broad, 6–7 mm long, upwards gradually widened to 6–8 mm broad at mouth; lobes unequal, orbicular, 2.5–3 mm long and broad, apex retuse. Stamens 4; filaments of longer pair 5–6 cm long, hirsute, shorter pair 3–4 mm long, glabrous; anther-thecae oblong, 1.6–2 mm long, 1–1.2 mm broad, strongly exserted. Ovary pilose; 2 ovules per locule; style pilose, 1–1.2 cm long. Capsules oblong, 7–9 mm long, pubescent, 4-seeded. Seeds brown, ovate to suborbicular in outline, compressed, 1.5–2 mm long and broad, pubescent, glabrescent; areole small, glabrous. Figure 46-67.

Phenology: flowering from September to November.

Habitats: mountain slopes, thickets; 600–2400 m.

Figure 46-67: **Strobilanthes affinis** (Griff.) Terao ex J. R. I. Wood & J. R. Benn.: **a**, habit; **b**, calyx lobes; **c**, corolla opened showing stamens and pistil; **d**, capsule; **e**, seed. Drawn by Quan LIN; reproduced from *Icon. Cormophyt. Sin.* 4: 159, fig. 5732. 1975.

Distribution: Yarlung Zangbo-Brahmaputra, S Hengduan, U Irrawaddy (Map 46-113); also in N Vietnam, and other parts of India, Myanmar and China (E & S Yunnan, Guizhou, Guangxi, Hunan).

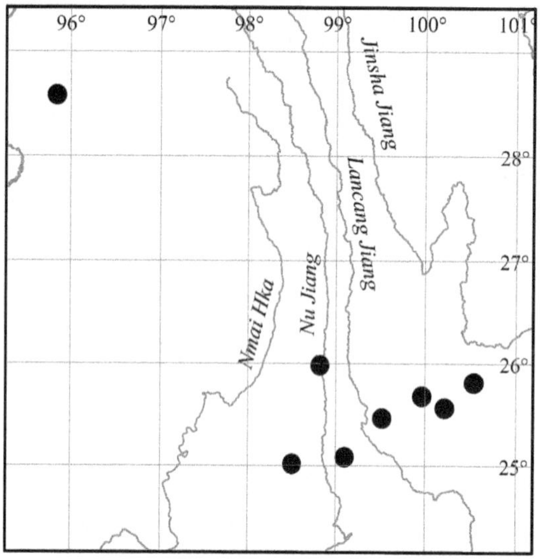

Map 46-113: **Strobilanthes affinis** (Griff.) Terao ex J. R. I. Wood & J. R. Benn.

YARLUNG ZANGBO-BRAHMAPUTRA: s. loc., *A. A. Mao 101917* (ASSAM). **S HENGDUAN: Bingchuan**, Xianyang to Bangying, 14 Oct. 1946, *T. N. Liou 21564* (IBSC, PE); **Dali**, 5 Oct. 1908, *K. K. Tsoong 2533* (SYS); **Longyang**, Yiluochi, 15 Nov. 1959, *J. S. Yang 1350* (KUN); Mangkuan, Laomiancheng, 1000 m, 18 Oct. 1998, *H. Li 10584* (CAS, E, KUN, MBK); Mangkuan, Baihuiling, Dajie Shan, 1525 m, 10 Sep. 2003, *Gaoligong Shan Biodiversity Survey 19079* (CAS, KUN); Mangkuan, Baihuiling, Dajie Shan, 1525 m, 10 Sep. 2003, *Gaoligong Shan Biodiversity Survey 19103* (CAS, KUN); Mangkuan, Baihuiling, Dajie Shan, 1525 m, 10 Sep. 2003, *Gaoligong Shan Biodiversity Survey 19126* (CAS, KUN); Jinji, 6 Nov. 2014, *Y. P. Chen EM155* (IBSC, KUN); **Lushui**, S of Liuku 900 m, 17 Oct. 1998, *H. Li 10534* (CAS, E, KUN, MBK). **Yangbi**, *s. coll. 2533* (PE); between **Yangbi** and **Yongping**, 2100–2400 m, Sep. 1900, G. *Forrest 7920* (E, K). **U IRRAWADDY**: **Tengchong**, Dieshuihe, 1500 m, 15 Oct. 1977, *P. H. Yu & Y. H. Li 20219* (HITBC).

64. Strobilanthes himalayana J. R. I. Wood, Edinburgh J. Bot. 51: 241. 1994; Wood in Grierson & Long (eds.), Fl. Bhutan 2(3): 1260. 2001. Type: [Sikkim & Darjiling] India, Sikkim, 4000 ft., 1875, *O. Kuntze 6661* [holotype: E (E00273498); isotypes: K (K000883101), NY (NY01163549, NY01163550, NY01163552)].

Strobilanthes petiolaris auct. non Nees: Clarke in Hooker (ed.), Fl. Brit. India. 4: 458. 1885.

Undershrubs, 0.5–1 m high; stems pilose at nodes. Leaves unequal in each pair; petiole 0.3–3 cm long; blade broadly elliptic, 1.5–12 cm long, 1–5 cm broad, adaxially often black-spotted, abaxially pilose along veins, base cuneate and often decurrent onto petiole, margin serrate, apex acute. Spikes terminal or axillary, 3-forked, dense-flowered; spikes 2–7 cm long, 0.7–1 cm broad; peduncles 0–2 cm long, glandular-pubescent. Bracts obovate, 4–6 mm long, glandular-pubescent, apex rounded, persistent; bracteoles linear, 3–4 mm long, apex obtuse. Calyx 7–8 mm long, glandular-pubescent; lobes linear, subequal, apex acute. Corolla blue, 2.5–3 cm long, straight, outside pilose, inside glabrous except for few trichomes retaining style; tube base cylindric, 4–5 mm long, 1.6–2 mm broad, upwards gradually widened to 7–8 mm broad at mouth; lobes oblong, 3–4 mm long and broad. Stamens 4, included. Ovary comose; 2 ovules per locule; style 2–2.2 cm long. Capsules oblong, 7–8 mm long, pilose. Seeds with large areole.

Phenology: flowering and fruiting from June to October.

Habitats: forests; 900–3100 m.

Distribution: **endemic**; Sikkim & Darjiling, Bhutan (Map 46-114).

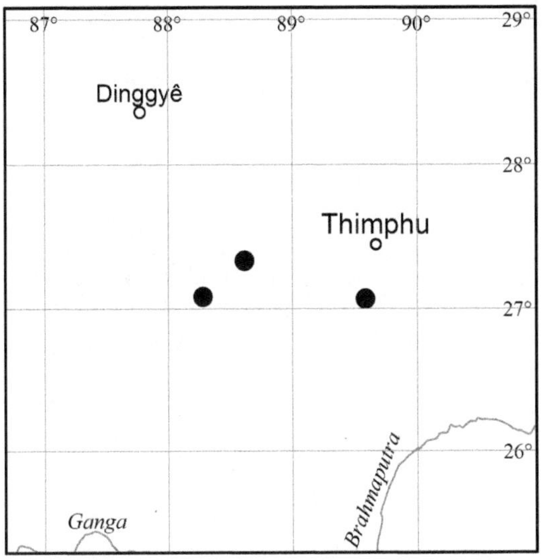

Map 46-114: **Strobilanthes himalayana** J. R. I. Wood

SIKKIM & DARJILING: **Darjiling**, Rishap, 1200 m, 16 Sep. 1860, *C. B. Clarke 9133* (K); s. loc., 1050 m, *C. B. Clarke 9657* (CAL); Rishap, 910 m, 5 June 1870, *C. B. Clarke 11949* (K); Mongpo, 910 m, 24 Sep. 1875, *C. B. Clarke 24778* (BM, K); Kahmpong, Yakla, 1220 m, *C. B. Clarke 26366* (BM, K, LE); Rungpo valley, 1200 m, 19 Sep. 1875, *J. S. Gamble 3556* (K); Kurseong, Oct. 1875, *J. S. Gamble 3557* (CAL, K); 1219 m, Nov. 1875, *O. Kuntzc 6661* (E, K); Mongpo, 1067 m, Oct. 1881, *J. S. Gamble* 9903 (K); **Sikkim**, Surail, 1200–1800 m, *R. H. Beddome 221* (BM); Yakla, 3000 m, 15 Oct. 1869, *C. B. Clarke 9825* (K); Sitong, 1500 m, Feb., 1882, *J. S. Gamble 10242* (CAL); on the road to Mungpoo below Sureil, 6 Nov. 1894, *R. Prondtock s. n.* (CAL); Mongpu, 1901, *Prain 153* (CAL); s. loc., 1200 m, Nov. 1875, *O. Kuntze 6661* (K). **BHUTAN**: s. loc., 1837, *W. Griffith 2403* (BM, K).

65. Strobilanthes frondosa J. R. I. Wood, Edinburgh J. Bot. 51: 239. 1994; Wood in Grierson & Long (eds.), Fl. Bhutan 2(3): 1260. 2001. Type: Bhutan, Chukka, between Chukka and Chimakothi, 1600 m, 29 Sep. 1991, *J. R. I. Wood 7465* [holotype: E (E00273500); isotypes: FHO, K, THIMPHU].

Herbs, 30–50 cm high, anisophyllous; stems sulcate, pilose, decumbent at base, rooting at nodes on lower portion, often nodding near tips. Leaves unequal in each pair, lower ones petiolate, upper ones sessile; petiole 0–4 cm long, pilose; leaf blade diminishing in size upwards, ovate to elliptic, (1–)6–12.5 cm long, (0.8–)2–6 cm broad, pilose on both surfaces, base cuneate, margin serrate, apex acuminate. Spikes terminal or axillary, subcapitate, 1–2 cm long, nodding; peduncles pilose, 0.8–2.5 cm long. Bracts imbricate, broadly elliptic-obovate, 7–8 mm long, 3–4 mm broad, pilose and sparsely orange gland-dotted, apex subacute, margin shouldered, with one distinct marginal tooth on either side; bracteoles linear, 5–8 mm long, pilose, apex subacute. Calyx 6–7 mm long, pilose outside, gland-dotted, 5-lobed to base; lobes subequal, linear, apex obtuse. Corolla 1.5–1.8 cm long, pale blue on lobes, white on tube, straight, outside glabrous on lobes, inside glabrous except for trichomes retaining style; tube base cylindric, 5–6 mm long, 1.3–1.5 mm broad, upwards gradually widened to 0.5–0.6 cm broad at mouth; lobes ovate, 4–5 mm long and broad, apex obtuse. Stamens 4, included or slightly exserted; filaments glabrous, longer pair 3–4 mm long, shorter pair 1.5–2.5 mm long; anther-thecae oblong-ellipsoid, 1–1.2 mm long. Ovary pilose; 2 ovules per locule; style 1.2–1.4 cm long, sparsely pilose; stigma simple, curved. Capsules oblong-elliptic, 5–7 mm long, pilose towards apex, 4-seeded. Seeds ovate in outline, compressed, 1.5–2 mm long and broad, pubescent, with a small areole. Figure 46-68.

Phenology: flowering from September to October.

Habitats: shaded stream banks in moist hill forests; 1600–1700 m.

Distribution: Bhutan (Map 46-115); also in other part s of India and Myanmar.

BHUTAN: Chukka, between Chukka and Chimakothi, 1600 m, 29 Sep. 1991, *J. R. I. Wood 7465* (E, K, THIMPHU), between Awakha and Takti Chu, 1600 m, 20 Sep. 1992, *J. R. I. Wood 7580* (E, K, THIMPHU, TI).

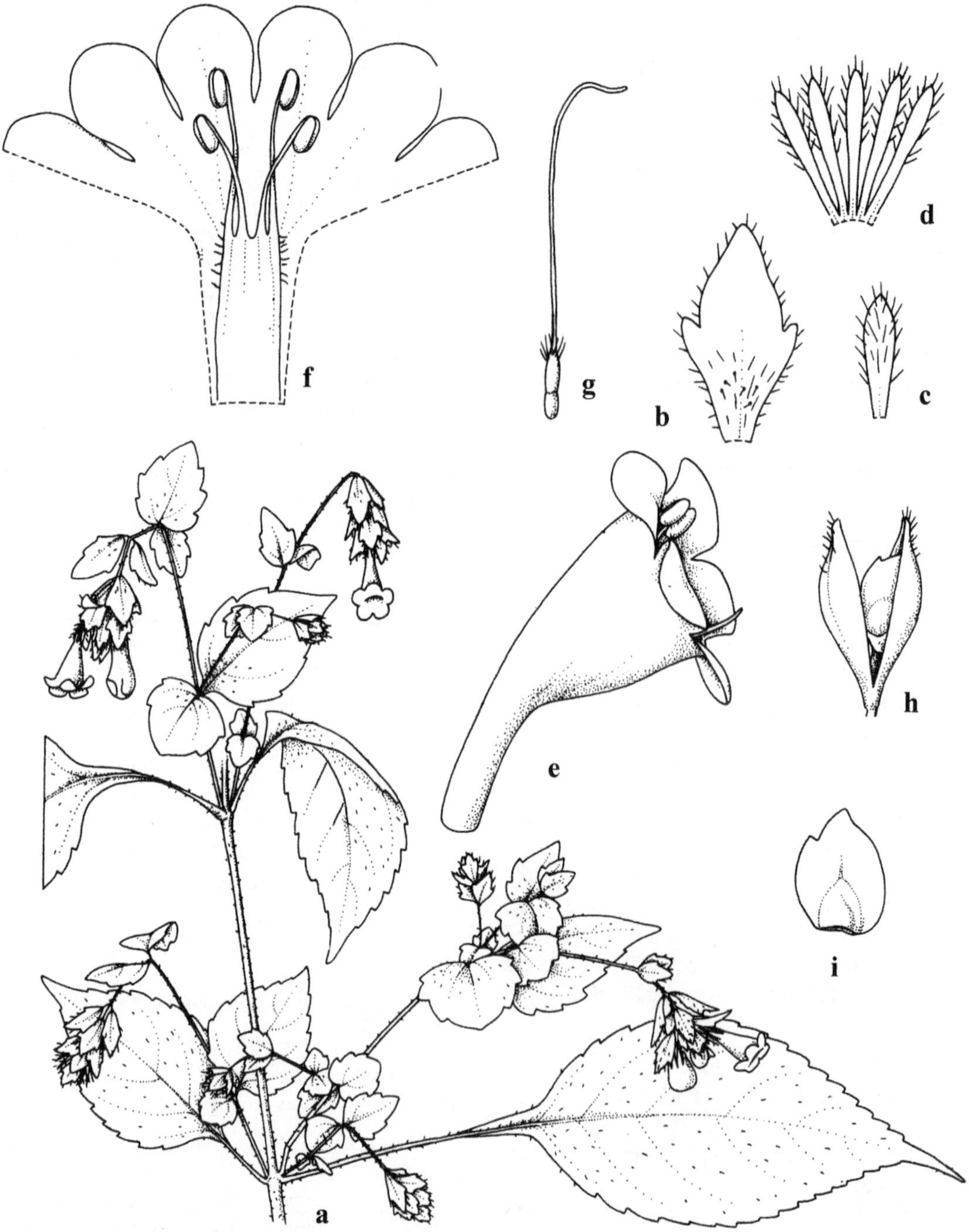

Figure 46-68: **Strobilanthes frondosa** J. R. I. Wood: **a**, habit; **b**, bract; **c**, bracteole; **d**, calyx; **e**, corolla; **f**, corolla opened out to show stamens; **g**, ovary and style; **h**, capsule; **i**, seed. Drawn by Mary BATES; reproduced from Wood, *Edinburgh J. Bot.* 51: 239, fig. 9. 1994.

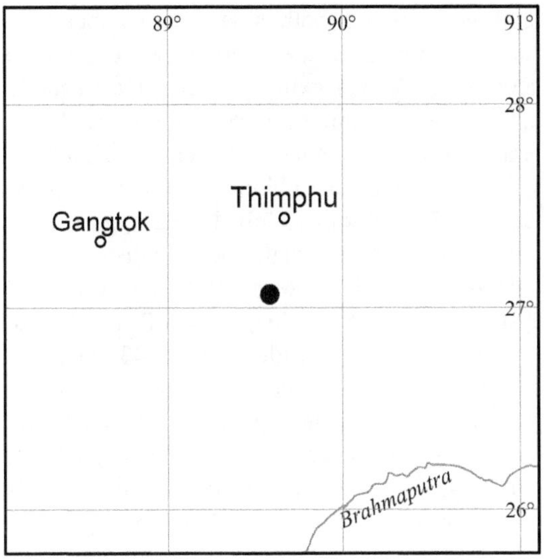

Map 46-115: **Strobilanthes frondosa** J. R. I. Wood

66. Strobilanthes stolonifera Benoist, Bull. Mus. Hist. Nat. Paris 28: 98. 1922; Bremekamp, Verh. Kon. Ned. Akad. Wetensch., Afd. Natuurk., Sect. 2 41(1): 286. 1944; Hu, Fl. Reipubl. Popularis Sin. 70: 189. 2002; Hu, Deng & Wood in Wu, Raven & Hong (eds.), Fl. China 19: 406. 2011. Type: [S Hengduan] China, Yunnan, Heqing, Bais à Tapintse, 22 Jun. 1888, *M. Delavay 2382* [lectotype **here designated**: P (P00719444); isolectotypes: BM (BM000957909), GH (GH00295523), K (K000882686), P (P00719443, P00719445, P00719446), NY (NY00278314), US (US00588605)].

Subshrubs, 20–60 cm high, isophyllous; stems quadrangular, sulcate, usually rooting at lower nodes, glabrous. Leaves equal in each pair; petiole 3–15(–20) mm long, glabrous; blade elliptic, (3–)5–10 cm long, (1–)2–2.5 cm broad, abaxially sparsely pubescent, adaxially pubescent but soon glabrescent, lateral veins 5–7 pairs, base decurrent onto petiole, margin crenate to crenate-serrate, apex acuminate. Spikes terminal, 1–2 cm long; rachis pilose. Bracts oblong to elliptic, 9–13 cm long, 2–3 mm broad, pubescent; bracteoles oblong, 7–9 mm long, 1–1.5 mm broad, pilose. Calyx 1.2–1.4 cm long; lobes linear, 0.8–1 mm broad, glandular-pubescent and eglandular-pubescent. Corolla violet, 3–3.5 cm long, subventricose, outside glabrous, inside glabrous except for trichomes retaining style; tube base cylindric, 0.8–1 cm long, 1.6–2 mm wide, upwards slightly curved and gradually widened to 0.8–1 cm broad at mouth; lobes ovate, 3.5–4.5 mm long and broad. Stamens 4, included; filaments pilose, longer pair 5–6 mm long, shorter pair 2–3 mm long; anther-thecae oblong, 2.5–3 mm long, 0.8–1 mm broad. Ovary oblong, 3–4 mm long, puberulent towards apex with some glandular-pilose; 2 ovules per locule; style 2.6–3 cm long, villous. Capsules narrowly oblong, 1–1.3 cm long, 2–3 mm broad, pilose, 4-seeded. Seeds orbicular in outline, 2–3 mm long, 2–2.5 mm broad, villous.

Phenology: flowering from June to October; fruiting from September to November.

Habitats: dry slopes, thickets; 1800–2400 m.

Distribution: endemic; S Hengduan (Map 46-116).

S HENGDUAN: **Binchuan**, Jizushan, near Zhushengsi, 2100 m, *Y. F. Deng & H. Peng 23866* (IBSC); Jizusi, Biyun Temple, 2335 m, 16 Apr. 2011, *E. D. Liu FW11021* (IBSC, KUN); **Dali**, between Hsingai and Tietsun, 1650–1800 m, 16 May 1915, *H. Handel-Mazzetti 6347* (W, WU); **Heqing**, Tapintse, 22 Jun. 1888, *M. Delavay 2382* (BM, GH, K, NY, P, US); Ta Pin Tze, s. d., *M. Delavay s. n.* (A); **Yongsheng**, Ta pin siou, près de Kieou ya pin, Sep. 1906, *M. Martin in herb. Ducloux 4420* (P); **Yulong** (Lijiang), Dongshan, Baishuihe, *K. M. Feng 2596* (A, KUN, PE); Daju, Hutiaoxia, 16 Sep. 2013, *Y. F. Deng et al. 25076* (IBSC); Baoshan, Shitoucheng, 1900 m, *Y. F. Deng 25275* (IBSC); Baoshan, Shitoucheng, 1500–1700 m, 14 Jan. 2014, *Y. F. Deng 25281* (IBSC); Daju, 1800 m, 14 Jan. 2014, *Y. F. Deng 25282* (IBSC); Baoshan, Shitoucheng, 1690 m, 19 Apr. 2014, *Y. F. Deng 25288* (IBSC).

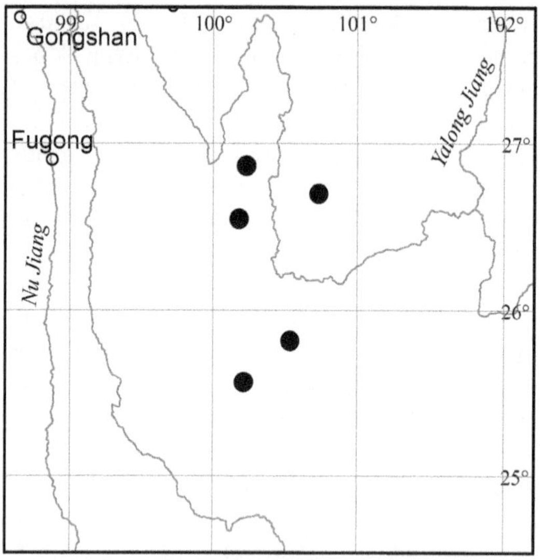

Map 46-116: **Strobilanthes stolonifera** Benoist

67. Strobilanthes glutinosa Nees, Pl. Asiat. Rar. 3: 86. 1832; C. B. Clarke, Fl, Brit. India 4: 458. 1884; Nasir & Ali in Malik & Ghafoor (eds.), Fl. Pakistan 188: 53. 1988. ≡ *Pseudaechmanthera glutinosa* (Nees) Bremek., Verh. Kon. Ned. Akad. Wetensch., Afd. Natuurk., Sect. 2 41(1): 188. 1944. ≡ *Ruellia glutinosa* Wall., Numer. List n. 2350. 1830, nom. invalid. Type: Nepal, 1821, *Wallich Cat. 2350* [lectotype designated by Malik & Ghafoor, 1988: K-W (K001115707); isotypes: BM (BM000793207), K (K000883093, K000883094)].

Ruellia hirta D. Don, Prodr. Fl. Nepal.: 119. 1825, non *Ruellia hirta* Vahl, Symb. Bot. 3: 84, t. 67. 1879. Type: Nepal, s. d., *N. Wallich s. n.* [holotype: BM (BM001190990)].

Ruellia jacquemontiana Nees, Prodr. 11: 145. 1847. Type: India, Caschemyr, *V. V. Jacquemont s. n. in Herb. Nees 148* [holotype: GZU (GZU000250301)].

Shrubs, 0.6–1.5 m high; stems terete, glutinous hairy. Petiole 6–15 mm long; leaf blade elliptic-ovate, 5–8 cm long, 1.5–3 cm broad, densely hairy on both sides, base cuneate to rounded, margin crenate-serrate, apex acute to acuminate. Spikes axillary or terminal, usually capitate or sometimes reduced to flower solitary. Bracts ovate, 7–8 mm long, densely glutinose-villose; bracteoles oblong, densely villose. Calyx 8–13 mm long, 5-lobed to base, glandular-pubescent; lobes unequal, linear, obtuse. Corolla pale blue, 4.5–5 cm long, glabrous; tube base cylindric, 2.2–2.5 cm long, 1.5–1.8 mm broad, upwards curved and widened to 2.5–3 cm broad at mouth; lobes patent. Stamens 4, filaments glabrous. Ovary glandular-pilose; 2 ovules per locule; stigma reflexed. Capsules oblong, 1.5–1.8 cm long, glandular-pubescent, 4-seeded. Seeds ovate in outline, compressed, 3–4 mm long, hairy.

Phenology: flowering from November to March of next year.

Habitats: forests; 600–2300 m.

Distribution: N Pakistan, U Ganga & Indus, W Nepal, C Nepal, Yarlung Zangbo Brahmaputra (Map 46-117); also in other parts of Pakistan and India.

N PAKISTAN: Azad Kashmir, Nawal Nadi, *R. R. Stewart & E. Nasir 23760* (RAW, fide Malik & Ghafoor, 1988); Muzaffarabad, Domel, Jhelum Valley, 670 m, 17 Feb. 1940, *F. Ludlow & G. Sherriff 7567* (BM, KYO, TI); **Hazara**, Abbottabd, hill S. E. of Town, 1380 m, 28 Mar. 1958, *B. L. Burtt B486* (E); Abbottabd, *M. Nath 99* (RAW, fide Malik & Ghafoor, 1988); Murree, Tret-Murree Road, 1000 m, *R. R. Stewart 7115* (RAW, fide Malik & Ghafoor, 1988). **U GANGA & INDUS: Bageshwar**, Kapkot, 1050 m, *R. Strachey & J. E. Winterbottom 3* (A, K, LE); **Chamba**, Chamba to Bathie, 23 Feb. 1896, *J. H. Lace 6. c.* (E); **Dehradun**, 1200 m, Feb. 1890, *J. S. Gamble 23443* (K); s. loc., Feb. 1894, *J. S. Gamble 24541* (K); Malkot Hills, 1350 m, Feb. 1895, *J. S. Gamble 25285* (K); Paltharkhan, 1500 m, *J. S. Gamble 27501* (K); **Kangra**, Hissar, Baijrath, 15 Jan. 1933, *W. Koelz 4025* (L); Hissar, Baijrath, 15 Jan., 1933, *W. Koelz 4027a* (L); **Mussoorie**, Nov. 1938, *R. R. Stewart 17194* (A); 1800–2100 m, Jan. 1935, *R. R. Stewart 14741A* (A); **Smila**, 1500 m, *H. Collett 411*(K); The Glen,

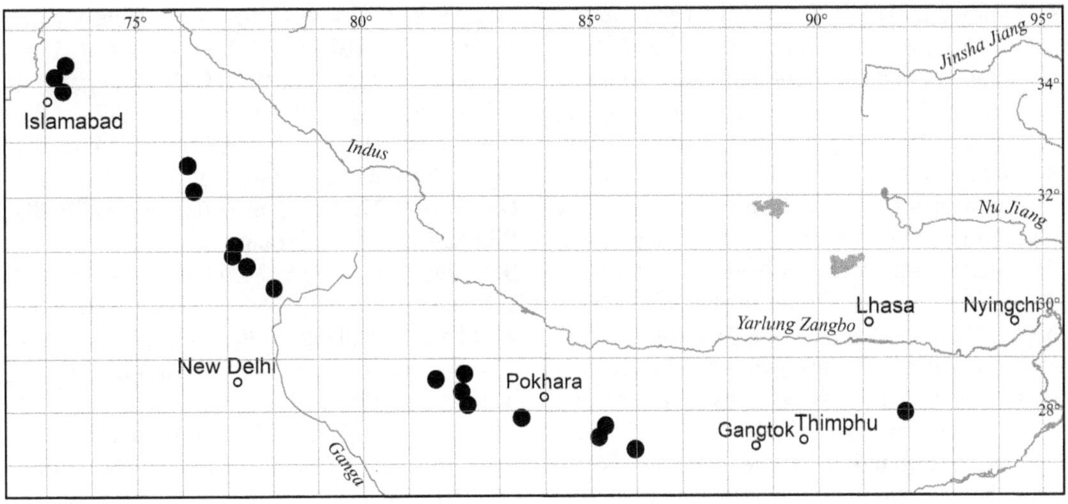

Map 46-117: **Strobilanthes glutinosa** Nees

1800 m, 8 Mar. 1890, *G. Watt 10031* (E); **Solam**, Kasanli, 1885, *J. R. Drummond 26735* (E); Kasanli, 1860 m, 14 Nov. 1890, *G. Watt 13680* (E). **W NEPAL**: **Bheri**, Jajarkot, Jhapra, 1500 m, 3 Mar. 1991, *N. P. Manandhar 617-91* (KATH); Surkhet, Ranimatha, 1800 m, Jan. 1977, *K. J. Malla 1563* (KATH); **Rapti,** Dang, Newalapur, 1380 m, 14 Mar. 1996, *N. P. Manandhar & P. M. Regmi 372* (KATH); Salyan, Phalabang, 1750 m, 27 Mar. 1952, *O. V. Polunin, W. R. Sykes & L. H. J. Williams 648* (BM, KATH, TI). **C NEPAL**: **Bagmati**, Kathmandu, hills around Nepal Valley, 1500–2300 m, 1953, *D. Pround 188* (BM, TI); **Janakpur**, Sindhuli, near Sindhuli Garhi, 1200 m, 1 Jan. 1977, *P. R. Shakya 7 M.B. Roychhetri 3870* (KATH); **Lumbini**, Palpa, Sattewati, Koldanda Pachya, 1350 m, 1 Mar. 1974, *D. P. Joshi & M. M. Amatya 74/1350* (KATH); **Narayani**, Makwanpur, Between Hettauda and the valley of Nepal, Cheesapani, 1800 m, 2 Dec. 1907, *I. H. Burkill 29592* (CAL); Makwanpur, Dunge Garhi, 1000 m, 16 Nov. 1970, *J. Kanai 674979* (TI). **YARLUNG ZANGBO-BRAHMAPUTRA**, Kuaon, Jeolikote, 1200 m, 18 Dec. 1909, *N. Gill 389* (CAL); Tehu, 1200 m, Apr. 1895, *J. S. Gamble 25351* (CAL).

68. Strobilanthes helicta T. Anderson, J. Linn. Soc., Bot. 9: 479. 1867, "*helictus*"; Clarke in Hooker (ed.), Fl. Brit. India 4: 469. 1885; Wood in Grierson & Long (eds.), Fl. Bhutan 2(3): 1272. 2001; Hu, Deng & Wood in Wu, Raven & Hong (eds.), Fl. China 19: 405. 2011. ≡ *Asystasia calycina* Nees in Wallich, Pl. As. Rar. 3: 90. 1832, non *Strobilanthes calycina* Nees, Companion Bot. Mag. 2: 312. 1836. ≡ *Echinacanthus calycinus* (Nees) Nees in de Candolle (ed.), Prodr. 11: 168. 1847. ≡ *Pteracanthus calycinus* (Nees) Bremek., Verh. Kon. Ned. Akad. Wetensch., Afd. Natuurk., Sect. 2 41(1): 199. 1944; Hu, Fl. Reipblic. Popularis Sin. 70: 132. 2002. ≡ *Ruellia calycina* Wall., Numer. List n. 2417. 1830, nom. invalid. Type: Bangladesh, Mount Sillet, *F. da Silva in Wallich Cat. 2417* [lectotype **here designated**: K-W (K001115887); isolectotype: GZU].

Subshrubs, 50–100 cm high, isophyllous; stems quadrangular, sulcate, brown pubescent when young, then glabrescent. Leaves equal in each pair; petiole 0.2–5 cm long, bifariously pubescent; blade ovate-lanceolate to elliptic,

5–12 cm long, 1–6 cm broad, pubescence nearby margin towards base, lateral veins 6 or 7 pairs, base attenuate and decurrent onto petiole, margin serrate to rarely subentire, apex acuminate. Spikes axillary, 2–12 cm long, secund, sometimes few-flowered; rachis usually strongly zigzag; flowers 6–10 mm apart on rachis, only one sterile per node. Bracts linear, 5–13 mm long, glabrous or glandular-pilose; bracteoles linear, 2–3 mm long, glabrous or glandular-pilose. Calyx 1.3–2.5 cm long, glabrous or glandular-pubescent, 5-lobed almost to base; lobes unequal, one shorter than others, linear often keeled, apex acuminate. Corolla white or flushed palepurple, 3.5–4 cm long, outside glabrous, inside glabrous except for trichomes retaining style; tube base cylindric, 5–7 mm long, 1.8–2 mm broad, upwards abruptly widened to 1.6–1.8 cm broad at mouth, finally narrowed and bent to ca. 90° near mouth; lobes unequal, oblong, 4–5 mm long, 3–4 mm broad, apex emarginate. Stamens 4, included; filaments glabrous, longer pair 6–8 mm long, shorter pair 3–4 mm long; anther-thecae oblong, 2–3 mm long, 1–1.2 mm broad. Ovary glabrous; 2 ovules per locule; style 2.8–3.2 cm long, apically hirsute. Capsules oblong, 1.8–2 cm long, glabrous, 4-seeded, apex apiculate. Seeds ovoid to obovate in outline, 3.5–4 mm long, 2.5–3 mm broad, appressed pubescent. Figure 46-69.

Phenology: flowering from September to November; fruiting form November to January.

Habitats: evergreen broad-leaved forests; 600–3300 m.

Distribution: E Nepal, Sikkim & Darjiling, Bhutan, L Yarlung Zangbo, Yarlung Zangbo-Brahmaputra, S Hengduan, U Irrawaddy (Map 46-118); also in other parts of India, Myanmar and China (SW Yunnan).

E NEPAL: Mechi, Ilam, Mabu, Dobate to Hangetham, 2656 m, 5 Sep. 2007, *s. coll. 191* (E). **SIKKIM & DARJILING**: **Darjiling**, Debepani, 1950 m, *J. H. Lace 2369* (CAL); Jorpokri, 2250 m, *J. H. Lace 2368* (CAL); Jorpokie, 2250 m, 4 Sep. 1902, *B. B. Osmaston 2368* (E); Debrepani, 1950 m, 4 Sep. 1902, *B. B. Osmaston 2369* (E); **Sikkim**, between Yoktun and Necbu, Ratong, 13 Oct. 1862, *T. Thomson 1061* (CAL); 1800 m, 15 Oct. 1874, *D. Treutler 905* (K); Rungbee, *R. H. Beddome 239* (BM); 750–1500 m, 4 Oct. 1862, *T. Anderson 1061* (K). **BHUTAN**: **Mongar**, just below Namning, 2000 m, 22 Oct. 1987, *J. R. I. Wood 6002* (E, TI). **L YARLUNGZANGBO**: **Nyingchi**, Pailong, 2200 m, 9 Sep. 2003, *X. F. Gao et al. 7202* (CDBI). **YARLUNG ZANGBO-BRAHMAPUTRA**: **Dirang**, Sessa, 1050 m, 12 Sep. 1964, *J. Joseph 39860* (CAL); Bomdila, 2780 m, *J. Joseph 39936* (CAL); **Bangxing**, west bank of Yajiang, 1700 m, 8 Dec. 1982, *S. Z. Cheng & B. S. Li 3315* (PE); **Gutang**, Gedang, Xiabolong, 2000–2200 m, 15 Sep. 1982, *B. S. Li & S. Z. Cheng 927* (PE); **Zayü**, Putsang river, Raug To valley, 1800 m, *F. Kingdon-Ward 10933* (BM, TI); 2500 m, *Forests Bureau of Xizang 1502* (SWFC), Shaqiong, 2000 m, 3 Jul. 1980, *Z. C. Ni et al. 0366* (PE). **S HENGDUAN**: **Longyang**, from Heishuitang to Silk-ancient-road, *Y. J. Guo et al. 12CS5066* (IBSC); Nankang Botany garden (Lihuipo), 2150 m, *H. Li 13257* (CAS, KUN). **U IRRAWADDY**: **Chaungmaw Kyaun**, North Triangle, Tama Bum, 2100 m, 5 Jul. 1953, *F. Kingdon-Ward 21109* (A); **Drungjiang**, Drungjiang to Maku, 2200 m, *Exped. Qinghai-Xizang 8954* (KUN, PE); Drungjiang to Maku, 2200 m, *Exped. Qinghai-Xizang 8955* (KUN, PE); **Putao**, Maikha-Salwin divide, 3000–3300 m, Aug. 1924, *G. Forrest 24866* (E, K, PE); **Tengchong**, Tengyueh, *E. B. Howell 104* (E).

69. Strobilanthes rubescens T. Anderson, J. Linn. Soc., Bot. 9: 479. 1867; Clarke in Hooker (ed.), Fl. Brit. India 4: 469. 1884; Wood in Grierson & Long (eds.), Fl. Bhutan 2(3): 1262. 2001; Hu, Deng & Wood in Wu, Raven & Hong (eds.), Fl. China 19: 405. 2011. ≡ *Pteracanthus rubescens* (T. Anderson) Bremek., Verh. Kon. Ned. Akad. Wetensch., Afd. Natuurk., Sect. 2 41(1): 199. 1944. Type: India, Meghalaya, *W. Griffith 309/K. D. 6108* [holotype: K (K000883220); isotype: GH].

Strobilanthes boerhaavioides T. Anderson, J. Linn. Soc., Bot. 9: 479. 1867. ≡ *Pteracanthus boerhaavioides* (T. Anderson) Bremek., Verh. Kon. Ned. Akad. Wetensch., Afd. Natuurk., Sect. 2 41(1): 199. 1944. Type: India, Meghalaya, *J. D. Hooker & T.*

Figure 46-69: **Strobilanthes helicta** T. Anderson: **a**, portion of plant; **b**, bract; **c**, bracteole; **d**, calyx; **e**, corolla. Drawn by Louise OLLEY; reproduced from *Fl. China Illustr.* 19: fig. 335: 13–17. 2012 (original from Wood in Grierson & Long (eds.), *Fl. Bhutan* 2(3): 1266, fig. 103: m–p. 2001).

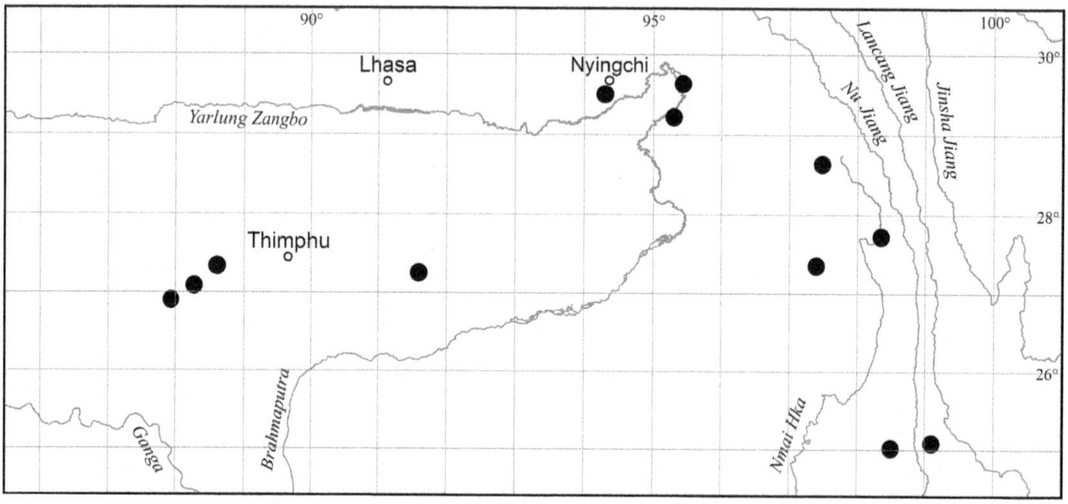

Map 46-118: **Strobilanthes helicta** T. Anderson

Thomson 27 [lectotype **here designated**: K; isolectotypes: BM (BM000884903), CGE, E (E00273480), P (P00719429, P00719430, P00719431, P00719432)].

Strobilanthes rubescens var. *microsperma* C. B. Clarke in Hooker (ed.), Fl. Brit. India 4: 469. 1884. Type: India, Meghalaya, *J. D. Hooker & T. Thomson s. n.* (lectotype **here designated**: K).

Subshrubs, 0.5–2 m high, anisophyllous; stems somewhat sulcate, pilose, then glabrescent, basally woody. Leaves equal in each pair, smaller of pair 3/5–2/3 of larger one in size, petiolate or uppermost ones sessile; petiole 0–2 cm long, pilose; blade ovate to suborbicular, 4–10 cm long, 2–5 cm broad, abaxially sparsely pilose along veins, adaxially sparsely hispid along midvein, base rounded, cordate, or cuneate, decurrent onto petiole, margin crenulate-dentate, apex acute to shortly acuminate, falcate. Spikes axillary, 2–5 cm long, secund, flowers paired, sometimes reduced to solitary flowers; rachis straight or obscurely zigzag, pilose; flowers 0.8–2 cm apart on rachis. Bracts variable in shape and size, lowermost ones ovate, 1–4 cm long, 0.7–2.2 cm broad, glabrous, apex acuminate, upper ones fertile, ovate, 2–3 mm long, 1–1.5 mm broad, glabrous, apex acuminate; bracteoles oblong-lanceolate, 1.5–3 mm long, 0.5–1 mm broad, caducous, glabrous. Calyx 0.9–1.6 cm long, subequally 5-lobed to base; lobes almost colorless, linear, 1–2 mm broad, glabrous, apex acute. Corolla white, 2–2.5 cm long, straight, outside glabrous, inside glabrous except for trichomes retaining style; tube base cylindric, 4–5 mm long, 1.5–1.8 mm broad, upwards gradually widened to 1–1.2 cm broad at mouth; lobes ovate, 4–6 mm long and broad, apex rounded. Stamens 4, included; filaments glabrous, longer pair 5–6 mm long, shorter pair 2–3 mm long; anther-thecae spherical, 1–1.2 mm in diameter, glabrous. Ovary glabrous; 2 ovules per locule; style 2–2.3 cm long, glabrous. Capsules 1–1.2 cm long, sparsely pilose towards apex, 4-seeded. Seeds ovate in outline, compressed, 2–2.4 mm long, 1.7–2 mm broad, hairy.

Phenology: flowering and fruiting from October to May of next year.

Habitats: limestone hills; 700–2000 m.

Distribution: Sikkim & Darjiling, Bhutan (Map 46-119); also in other parts of India and China (S Yunnan).

SIKKIM & DARJILING: **Darjiling**, Rungbee, *R. H. Beddome 238* (BM); Darjiling Pass, *R. H. Beddome 237* (BM). **BHUTAN**: **Deothang**, between Deothang and Wamrun, 1800 m, 18 Oct. 1987, *J. R. I. Wood 5979* (E, FHO); **Gaylegphung**, "Chablikhola" above Shamkhara on Gaylephung-Shamgang road, 2000 m, 13

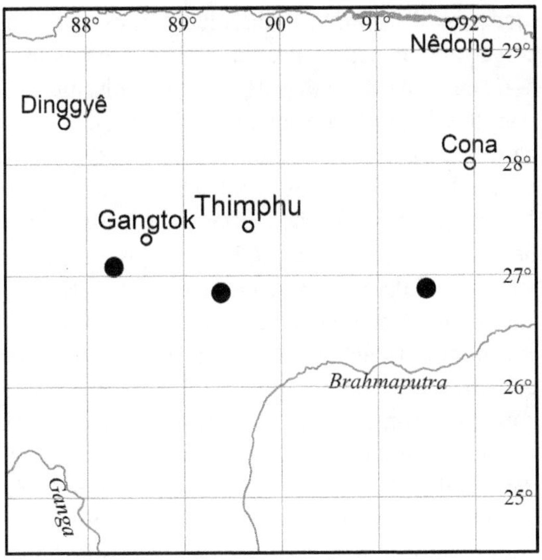

Map 46-119: **Strobilanthes rubescens** T. Anderson

May 1987, *J. R. I. Wood 5464* (E); **Phuntsholing**, hillside above Rinchending, Phuntsholing, 780 m, 16 Feb. 1982, *A. J. C. Grierson & D. G. Long 2965* (E, K).

70. Strobilanthes fimbriata Nees, Pl. Asiat. Rar. 3: 85. 1832; Clarke in Hooker (ed.), Fl. Brit. India 4: 447. 1884; Hu, Deng & Wood in Wu, Raven & Hong (eds.), Fl. China 19: 417. 2011. ≡ *Ruellia fimbriata* Wall., Numer. List n. 2363. 1830, nom. invald. Type: Bangladesh, Mount Sillet, *de Silva & G. Gomez in Waliich Cat. 2363* [lectotype **here designated**: K (K001115733); isotypes: BM (BM000810983), E (E00273501), GZU (GZU000249624), K (K000883002), M (M0186783), NY (NY00278317), P (P00719322)].

Endopogon macrostegius Nees in de Candolle (ed.), Prodr. 11: 104. 1847. ≡ *Strobilanthes macrostegia* (Nees) C. B. Clarke in Hooker (ed.), Fl. Brit. India 4(11): 456. 1884. Type: India, Assam, *W. Griffith s. n.* [lectotype **here designated**: K (K000883109); isolectotypes: GZU (GZU000280522), K (K000883109)].

Strobilanthes neesii Kurz, J. Asiat. Soc. Bengal, Pt. 2, Nat. Hist. 42(2): 93. 1873. Type: Myanmar, Karen Country, Toukyeghat, Nakawa Chg, Jul. 1891, *W. S. Kurz 835* [holotype: CAL; isotype: M (M0186771)].

Strobilanthes fimbriata var. *majuscula* W. W. Sm., Notes Roy. Bot. Gard. Edinburgh 10: 196. 1918. Type: Myanmar, Kachin Hills, N. Myitkina, on the road to Kanpti Long via Malihka branch, 1500 ft., Dec., 1911, *S. Toppin 4244* [lectotype **here designated**: E (E00133507); isolectotype: CAL].

Shrubs, 0.6–1.5 m high. Branches terete, sparsely brown pubescent, then glabrescent. Leaves equal in each pair; petiole 0.5–2.5 cm long; leaf blade oblong-elliptic to elliptic, 13–18 cm long, 4.5–7 cm broad, abaxially sparsely brown pubescent along veins, adaxially glabrous, lateral veins 8–12 pairs, base cuneate, margin subentire, apex acuminate. Spikes axillary, 3–6 cm long, with 6–8 flower pairs; peduncle 0–3 mm long, glabrous; flowers imbricate on rachis. Bracts obovate, 1–1.3 cm long, 3.5–5 mm broad, both surfaces densely covered with cystoliths, margin ciliate with rigid brown trichomes towards apex, apex abruptly narrowed to a mucro; bracteoles oblanceolate, 2–3 mm long, 0.8–1 mm broad, abaxially fulvous pubescent towards apex, margin with a distinct shoulder, apex cuspidate. Calyx 1–1.3 cm long, accrescent to 1.6–2 cm long in fruit; lobes linear-lanceolate, 0.8–1 mm broad, apex long acuminate. Corolla violet, 4–4.5 cm long, outside glabrous, inside glabrous except for trichomes retaining style; tube base cylindric, 6–8 mm long, upwards abruptly widened to 1.5–1.8 cm broad at mouth; lobes broadly ovate, 5–6 mm long, 6–7 mm broad, apex rounded. Stamens 4, included; filaments of longer pair 6–8 mm long, pilose, shorter pair 2–3 mm long, glabrous; anther-thecae oblong, 2.5–3 mm long, 1.2–1.5 mm broad. Ovary comose towards apex; 2 ovules per locule; style 3.5–4 cm long, pilose. Capsules oblong, 1–2 cm long, comose, 4-seeded. Seeds ovate in outline, compressed, 3–4 mm long, 2–3 mm broad, densely appressed pilose.

Phenology: flowering from December to January of next year.

Habitats: moist places in forests; 300–600 m.

Distribution: U Irrawaddy (Map 46-120); also in Bangladesh, and other parts of India and Myanmar.

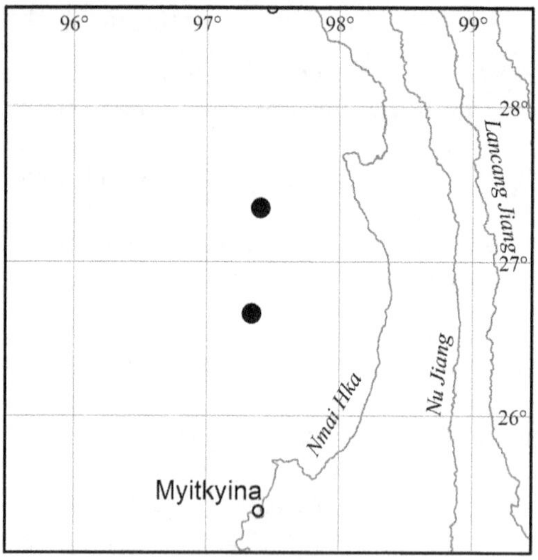

Map 46-120: **Strobilanthes fimbriata** Nees

U IRRAWADDY: Putao, Hills east of Putao, 300–600 m, 10 Dec. 1937, *F. Kingdon-Ward 13571* (BM, TI).

71. Strobilanthes nobilis C. B. Clarke in Hooker (ed.), Fl. Brit. India 4: 471. 1884; Hu, Deng & Wood in Wu, Raven & Hong (eds.), Fl. China 19: 418. 2011. ≡ *Pteracanthus nobilis* (C. B. Clarke) Bremek., Verh. Kon. Ned. Akad. Wetensch., Afd. Natuurk., Sect. 2 41(1): 199. 1944. Type: India, Meghalaya Jaintea Hills, Jarain, alt. 3500 ft., 21 Nov. 1872, *C. B. Clarke 17869* [lecotype **here designated**: K (K000883211); isolectotypes: BM (BM001190986), K].

Isophyllous herbs, 50–100 cm high; stems quadrangular, sulcate, reddish-brown pubescent. Leaves sessile; blade ovate-lanceolate to oblong-elliptic, sometimes pandurate, 7–17(–25) cm long, 3.5–8 cm broad, abaxially glabrous or sparsely pilose especially along veins, lateral veins 6–8 pairs and somewhat parallel, base cuneate to rounded, auriculate, margin crenate, apex shortly acuminate. Spikes axillary or terminal, simple or basally branched, 5–20 cm long, many-flowered; rachis reddish-brown puberulent; flowers 1–1.5 cm apart. Bracts lanceolate, 1.5–2 cm long, 2–4 cm broad, glabrous or glandular-pilose, margin ciliolate, apex long acuminate; bracteoles narrowly linear-lanceolate, 1.5–1.8 cm long, 0.8–1 mm broad, glabrous, apex acuminate. Calyx 1–2.2 cm long, 5-lobed almost to base; lobes linear, 1.4–2.0 cm long, 0.7–1 mm broad, equal, margin ciliolate, apex acuminate. Corolla blue, 2.5–3 cm long, outside glabrous, inside glabrous except for trichomes retaining style; tube base cylindric, 6–7 mm long, 2–2.5 mm broad, upwards strongly bent and abruptly widened to 1–1.4 cm broad at mouth; lobes ovate, 4–5 mm long and broad, apex emarginate. Stamens 4, included; filaments of longer pair 6–7 mm long, pilose towards base, shorter pair 2–3 mm long, glabrous; anther-thecae oblong, 2–3 mm long, 1–1.2 mm. Ovary glabrous; 2 ovules per locule; style 2.2–2.5 cm long, pilose. Capsules clavate, 1.2–2.2 cm long, glabrous, 4-seeded. Seeds ovate in outline, compressed, 2.4–3 mm long, 2–2.5 mm broad, densely appressed pubescent; areole small. Figure 46-70.

Phenology: flowering from September to November; fruiting from December to March.

Habitats: stream banks, forests; 1100–1800 m.

Distribution: U Irrawaddy (Map 46-121); also in N Vietnam, and other parts of India Myanmar and China (S Yunnan).

U IRRAWADDY: **Drungjiang**, Kongdang, *Dulong Jiang Investigation Team 3025* (KUN); Bapo, 1350 m, 14 Nov. 1990, *Dulong Jiang Investigation Team 278* (CAS, KUN); Jimudeng, 1400 m, *Dulong Jiang Investigation Team 874* (CAS, KUN); Dandanghe, 1400 m, 17 Jan. 1991, *Dulong Jiang Investigation Team 3239* (KUN); along Dulongjiang River, 1400–1500 m, 24 Jul. 2013, *X. H. Jin et al. ST0930* (PE); **Putao**, banks of the Hpuginhka river, 1600 m, Mar. 1962, *J. Keenan et al. 3927* (A).

Figure 46-70: **Strobilanthes nobilis** C. B. Clarke: **a**, flowering branch; **b**, fruiting branch; **c**, basal leaf; **d**, portion of abaxial leaf surface; **e**, bract; **f**, bracteole; **g**, calyx spread out; **h**, corolla spread out showing androecium; **i**, ovary and base of style; **j**, capsule; **k**, seeds. Drawn by Rosemary WISE; reproduced from Wu, Raven & Hong, *Fl. China Illustr.* 19: fig. 323. 2011.

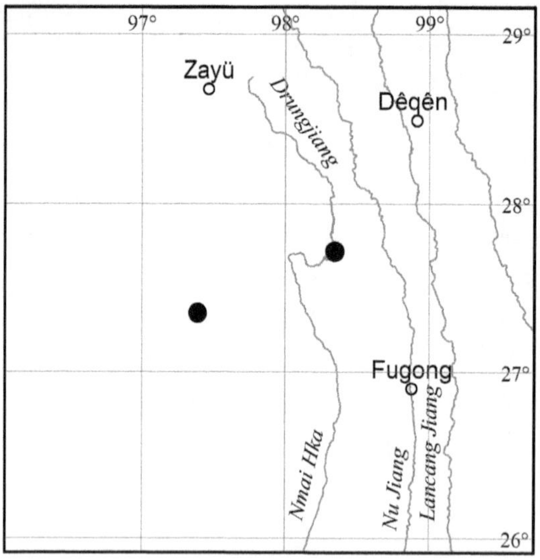

Map 46-121: **Strobilanthes nobilis** C. B. Clarke

72. Strobilanthes pinetorum W. W. Sm., Notes Roy. Bot. Gard. Edinburgh 10: 197. 1918; Bremek., Verh. Kon. Ned. Akad. Wetensch., Afd. Natuurk., Sect. 2 41(1): 283. 1944; Hu, Deng & Wood in Wu, Raven & Hong (eds.), Fl. China 19: 405. 2011. ≡ *Diflugossa pinetorum* (W. W. Sm.) C. Y. Wu & C. C. Hu in Hu (ed.), Fl. Reipubl. Popularis Sin. 70: 174. 2002. Type: [U Irrawaddy] China, Yunnan, hills to the South of Tengyueh, lat. 25° N, alt. 6000 ft., 15 Jun. 1912, *G. Forrest 8385* [holotype: E (E00045282 & E00045283); isotype: PE (PE01432631)].

Subshrubs, 0.5–1 m high, anisophyllous; stems glabrous or subglabrous. Leaves unequal in each pair; petiole (0–)1–2 cm long, subglabrous, winged; blade broadly ovate to suborbicular, smaller one 2–3 cm long, 1–1.5 cm broad, larger one 6–10 cm long, 3.5–5 cm broad, abaxially glabrous, adaxially sparsely pilose, soon glabrescent, lateral veins 5 or 6 pairs, base subrounded to broadly cuneate and decurrent onto petiole, or cordate for uppermost ones, margin serrate, apex acuminate. Spikes axillary or terminal, simple or branched, 5–10 cm long. Bracts variable in shape and size, lowermost ones ovate, 1–5 cm long, 0.7–3.5 cm broad, base cordate, apex shortly acuminate, upper ones ovate-oblong, 7–11 mm long, 2–3 mm broad; bracteoles linear-lanceolate, 5–6 mm long, 1–1.2 mm broad, glabrous, apex obtuse. Calyx 5–7 mm long, glabrous, 5-lobed almost to base; lobes linear-oblong, 4–6 mm long, 0.7–1 mm broad, apex obtuse. Corolla whitish-purple, 3.5–4 cm long, curved, ventricose, outside glabrous, inside glabrous except for trichomes retaining style; tube base cylindric, 1.1–1.3 cm long, 1.7–2 mm broad, upwards curved and gradually widened to 1–1.2 cm broad at mouth; lobes oblong, 3–4 mm long and broad. Stamens 4, included; filaments subglabrous, longer pair 5–7 mm long, shorter pair 3–4 mm long; anther-thecae oblong, 2–3 mm long, 1.2–1.5 mm broad. Ovary pilose towards apex; 2 ovules per locule; style 2.8–3.2 cm long, sparsely hirsute towards apex. Capsules unknown.

Phenology: flowering from June to September.

Habitats: forests; 1800–2000 m.

Distribution: endemic; U Irrawaddy (Map 46-122).

U IRRAWADDY: Tengchong, S Tengyueh, 1800 m, Jun. 1912, *G. Forrest 8385* (E, PE).

73. Strobilanthes parvifolia J. R. I. Wood, Kew Bull. 64: 25. 2009; Hu, Deng & Wood in Wu, Raven & Hong (eds.), Fl. China 19: 428. 2011. Type: Yarlung Zangbo-Brahmaputra, Lohit Valley, 1600 m, 31 Oct. 1928, *F. Kingdon-Ward 8718* (holotype: K).

Herbs, anisophyllous; stems somewhat sulcate, glabrous, basally geniculate. Leaves unequal in each pair, smaller of pair 1/4–1/3 size of larger one, petiolate or rarely sessile; petiole 0–4 mm long, glabrous; blade lanceolate, ovate or ovate-elliptic, 1–3.5 cm long, 1–1.8 cm broad, adaxially glabrous, abaxially glabrous or rarely sparsely glandular-pilose along veins, base broadly to narrowly cuneate and decurrent onto petiole, margin undulate to slightly crenate, apex acute. Spike terminal, 3-forked; flowers in opposite pairs. Bracts oblong, 4–5 mm long, 1–1.2 mm broad, glandular-pilose, caducous; bracteoles oblong, 1.8–2.2 mm long, 0.8–1 mm broad. Calyx 0.9–1.1 cm long, accrescent to 1.7–2 cm long in fruit, 5-lobed almost to base; lobes narrowly oblong-lanceolate, 1–1.5 mm broad, subequal, densely glandular-pilose, apex obtuse. Corolla white, with violet stripes, 2.5–3

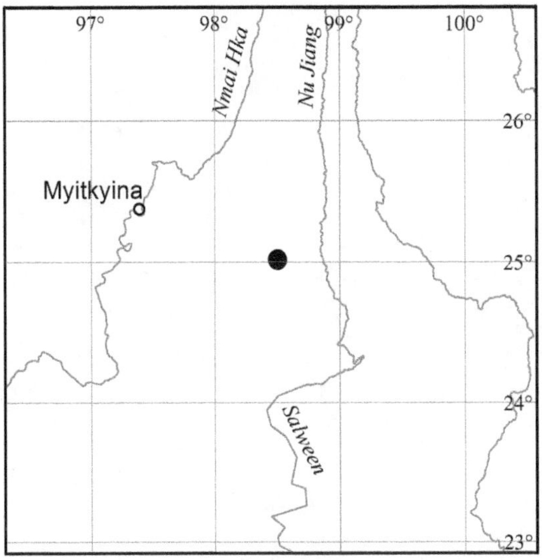

Map 46-122: **Strobilanthes pinetorum** W. W. Sm.

cm long, outside glabrous, inside pubescent; tube base cylindric, 0.8–1 cm long, 2–3 mm broad, upwards bent to ca. 90° and gradually widened to 1.2–1.5 cm broad at mouth; lobes broadly ovate, 5–6 mm long and broad, apex obtuse. Stamens 4, included; filaments glabrous, longer pair 5–6mm long, shorter pair 1.5–2 mm long; anther-thecae oblong, 2–2.2 mm long, 0.5–0.6 mm broad. Ovary glabrous; 2 ovules per locule; style 1.2–1.5 cm long, glabrous. Capsules narrowly clavate, 2–2.2 cm long, glabrous, 4-seeded. Mature seeds unknown. Figure 46-71.

Phenology: flowering and fruiting from September to October.

Habitats: forest margins on hills; ca. 1600 m.

Distribution: endemic; Yarlung Zangbo-Brahmaputra (Map 46-123).

YARLUNG ZANGBO-BRAHMAPUTRA: **Lohit Valley**, 1600 m, 31 Oct. 1928, *F. Kingdon-Ward 8718* (K).

74. Strobilanthes candida J. R. I. Wood, Edinburgh J. Bot. 51: 236. 1994. Type: Myanmar, Kachin State, Putao, Mountains E of Fort Hertz, 915 m, 3 Sep. 1926, *F. Kingdon-Ward 42006* [holotype: K (K000883170 & K000883171)].

Subshrubs, 0.5–6 m high; stems woody below, glabrous or obscurely bifariously pubescent. Leaves somewhat unequal especially on inflorescence; petiole 0.3–3.5 cm long, glabrous; leaf blade elliptic or oblong-elliptic, 4–13 cm long, 1.5–4.5 cm broad, glabrous, base cuneate, margin serrate, apex acuminate. Spikes axillary, 2–3.5 cm long, 2–6-flowered; flowers ca. 1 cm apart on rachis. Bracts obovate, 1.5–2.5mm long, glabrous; bracteoles oblanceolate, 2–3 mm long, glabrous. Calyx 6–9 mm long, accrescent to 1.1–1.2 cm long in fruit, glabrous, gland-dotted; lobes linear, subequal, apex acuminate. Corolla white, 2.5–3 cm long, curved, outside glabrous, inside pilose; tube base cylindric, 1.8–2 mm broad, 5–6 mm long, upwards gradually expanded to 1.2–1.5 cm broad at mouth; lobes broadly oblong, 3–4 mm long, 7–9 mm broad, apex rounded. Stamens 4, included; filaments glabrous, longer pair 5–6 mm long, shorter pair 1–2 mm long; anthers conical, 1.6–1.8 mm long, 0.8–1 mm broad. Ovary with numerous, small sessile glands; 2 ovules per locule; style 2–2.2 cm long, glabrous. Capsules unknown. Figure 46-72.

Phenology: flowering from August to October.

Habitats: stream banks, moist banks in forests; 900–1100 m.

Distribution: endemic; U Irrawaddy (Map 46-124).

U IRRAWADDY: **Nam Tamai valley**, 1070 m, 27 Aug. 1931, *F. Kingdon-Ward 13063* (BM); **Putao**, Mountains E of Fort Hertz, 920 m, 3 Sep. 1926, *F. Kingdon-Ward 7352* (K); Mountains E of Fort Hertz, 915 m, 3 Sep. 1926, *F. Kingdon-Ward 42006* (K).

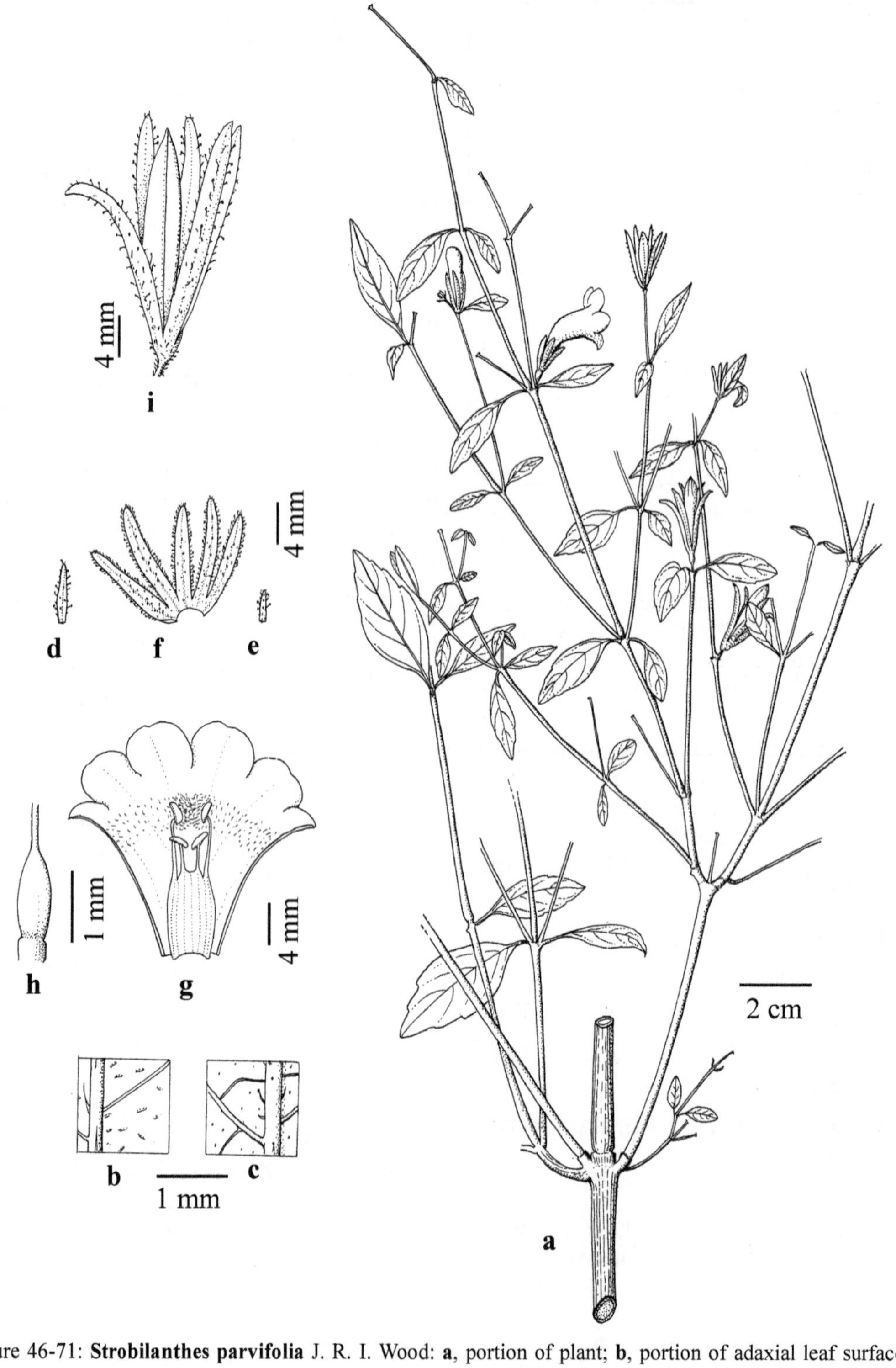

Figure 46-71: **Strobilanthes parvifolia** J. R. I. Wood: **a**, portion of plant; **b**, portion of adaxial leaf surface; **c**, portion of abaxial leaf surface; **d**, bract; **e**, bracteole; **f**, calyx; **g**, corolla opened out; **h**, ovary; **i**, accrescent fruiting calyx and capsule. Drawn by Rosemary WISE; reproduced from Wu, Raven & Hong, *Fl. China Ill.* 19: fig. 276. 2012 (original from Wood & Scotland, *Kew Bull.* 64: 26, fig. 11. 2009).

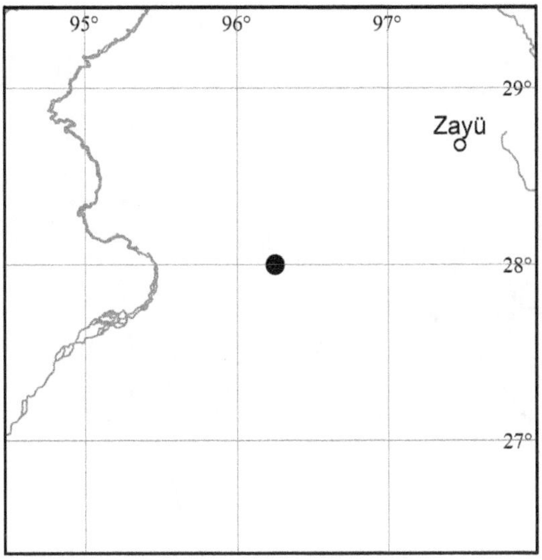

Map 46-123: **Strobilanthes parvifolia** J. R. I. Wood

75. Strobilanthes ovata Y. F. Deng & J. R. I. Wood, Novon 20: 143. 2010; Hu, Deng & Wood in Wu, Raven & Hong (eds.), Fl. China 19: 404. 2011. Type: [S Hengduan] China, Yunnan, Baoshan, Nankang Botany Garden, Lihuipo, 2100 m, 17 Nov. 2000, *H. Li with Z.L. Dao & L.W. Yin 13171* (holotype: KUN; isotype: CAS).

Herbs, 30–40 cm high; stems quadrangular, sulcate, glabrous, narrowly winged, wing 0.6–0.8 mm broad. Leaves petiolate and uppermost ones sessile; petiole (0–)3–5 mm long, glabrous; blade ovate to ovate-elliptic, 2–5 cm long, 1–2.3 cm broad, glabrous, lateral veins 2–4 pairs, base cordate or cuneate and decurrent onto petiole, margin crenate to crenate-serrate, apex acuminate. Spikes terminal or axillary, 1–7 cm long; rachis narrowly winged, glandular-pilose or eglandular-pilose. Bracts ovate, oblong-ovate, or rounded, 3–7 cm long, 1–5 mm broad, pubescent; bracteoles oblanceolate, 4–5 mm long, 0.8–1 mm broad, subglabrous, eglandular-pilose or glandular-pilose, apex rounded. Calyx subglabrous, slightly 2-lipped, upper lip 8–10 mm long, 3-lobed to 2–3 mm from base, lobes triangular, 5–7 mm long, margin glandular-ciliate, lower lip 2-lobed almost to base, lobes linear-oblanceolate to subspatulate, 9–10 mm long, apex obtuse. Corolla light blue, 3–3.5 cm long, gently curved, subventricose, outside glabrous, inside glabrous except for trichomes retaining style; tube base cylindric, 0.8–1 cm long, 1.8–2 mm broad, upwards gradually widened to 1.1–1.3 cm broad at mouth; lobes ovate, 3–4 mm long and broad, subequal. Stamens 4, included; filament of longer pair 5–6 mm long, pilose towards base, shorter pair 2–3 mm long, glabrous; anther-thecae oblong, 1.7–2 mm long, 1–1.2 mm broad. Ovary oblong, 1.6–2 mm long, comose towards apex; 2 ovules per locule; style 2–2.2 cm long, sparsely pilose. Capsules narrowly oblong-ellipsoid, 1–1.2 cm long, 1.7–2 mm broad, glabrous except for glandular-pubescent towards apex, 4-seeded. Seeds ovoid to suborbicular in outline, 2–3 mm long, 1.5–2 mm broad, pilose. Figure 46-73.

Phenology: flowering from October to February; fruiting from December to March of next year.

Habitats: moist forests; 1600–2200 m.

Distribution: S Hengduan, U Irrawaddy (Map 46-125); also in other parts of China (C & S Yunnan).

S HENGDUAN: **Longyang**, Nankang Botany Garden (Lihuipo), 2050 m, 16 Nov. 2000, *H. Li 13101* (CAS, KUN, MBK); Nankang Botany Garden (Lihuipo), 2100 m, 17 Nov. 2000, *H. Li 13171* (CAS, KUN); Lihuipo, *H. J. Xin 39* (IBSC); Lihuipo, 2100 m, 23 Jan. 2008, *Y. F. Deng 20458* (IBSC); Lihuipo, 2200 m, 11 Sep. 2013, *Y. F. Deng 24729* (IBSC). **U IRRAWADDY**: **Tengchong**, 1959, *X. Zhou 553* (KUN).

76. Strobilanthes atroviridis Y. F. Deng & J. R. I. Wood, J. Trop. Subtrop. Bot. 18: 480. 2010; Hu, Deng & Wood in Wu, Raven & Hong (eds.), Fl. China 19: 403. 2011. Type: China, Yunnan, Jingdong, His-So-Tou-Shan, 2400 m, 26 Nov. 1939, *M. K. Li 2086* (holotype: IBSC; isotypes: HITBC, KUN).

Subshrubs to 3 m high, isophyllous or slightly anisophyllous; stems quadrangular, sulcate, glabrous; nodes pustulose, pubescent. Leaves equal or slightly unequal in each pair; petiole 2–5 cm long, bifariously pubescent; blade ovate-elliptic, 9–22 cm long, 5.5–14 cm broad, abaxially glabrous, adaxially, sparsely

Figure 46-72: **Strobilanthes candida** J. R. I. Wood: **a**, habit. Drawn by Mary BATES; reproduced from *Edinburgh J. Bot.* 51: 234, fig. 8: a. 1994.

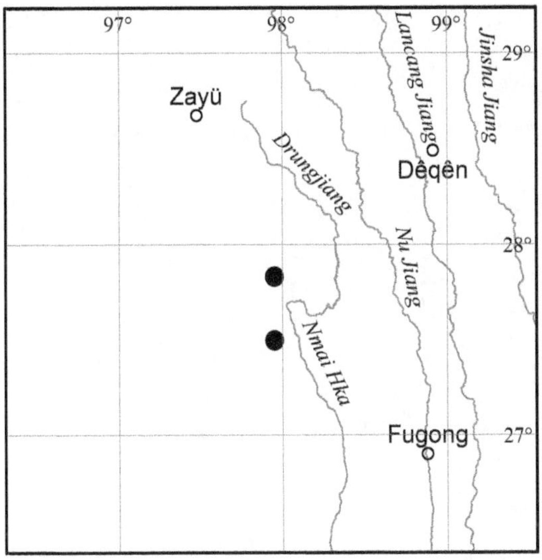

Map 46-124: **Strobilanthes candida** J. R. I. Wood

pubescent, lateral veins 6–10 pairs, base narrowly cuneate and decurrent onto petiole, margin crenulate, apex acute to long acuminate. Spikes axillary, 1–3(–6) cm long, 2(–8)-flowered; rachis glandular-pubescent. Bracts linear-lanceolate, 9–14 mm long, 2–3 mm broad, abaxially glandular-pubescent, adaxially glabrous; bracteoles linear-lanceolate, 9–13 mm long, 2–2.5 mm broad, abaxially glandular-pubescent, adaxially glabrous. Calyx 1–1.5 cm long, abaxially glandular-pubescent, adaxially glabrous, 2-lipped, lower lip 2-lobed almost to base, lobes linear-lanceolate, 2–3 mm broad, apex acute, upper lip 3-lobed to 4–7 mm above base, lobes lanceolate, 5–7 mm long, 2–3 mm broad. Corolla purplish-blue, 3–4 cm long, straight, outside glabrous, inside glabrous except for trichomes retaining style; tube base cylindric, 6–8 mm long, 1.7–2 mm broad, upwards gradually widened to 1.8–2 cm broad at mouth; lobes ovate, 6–9 mm long and broad, apex emarginate. Stamens 4, included; filaments pubescent towards base, longer pair 7–8 mm long, shorter pair 5–6 mm; anther-thecae oblong. Ovary with 2 ovules per locule; style 2–2.2 cm long, glabrous. Capsules unknown. Figure 46-74.

Phenology: flowering from November to April of next year.

Habitats: moist places in forests; 2100–2400 m.

Distribution: S Hengduan (Map 46-126); also in other parts of China (C Yunnan).

S HENGDUAN: Yongping, Jinguangsi, 18 Mar. 1989, *Du & Chen 2033* (SWFC).

77. Strobilanthes longispica (H. P. Tsui) J. R. I. Wood & Y. F. Deng, Bot. J. Linn. Soc. 150: 383. 2006; Hu, Deng & Wood in Wu, Raven & Hong (eds.), Fl. China 19: 402. 2011. ≡ *Adenacathus longispicus* H. P. Tsui in Hu (ed.), Fl. Reipubl. Popularis Sin. 70: 123, 348. 2002. Type: [S Hengduan] China, Yunnan, Gongshan, 1380 m, Jan. 1991, *Exped. Pl. Drung Jiang 3793* [holotype: KUN (KUN1263782); isotype: KUN].

Strobilanthes tripartita J. R. I. Wood, Kew Bull. 58(1): 108. 2003. Type: [U Irrawaddy] Myanmar, "Burma-Tibet frontier", Adung River., 1300–1600 m, 2 Feb. 1931, *F. Kingdon-Ward 9184* [holotype: BM (BM000906299); isotypes: E (E00133491), KUN].

Herbs, 0.7–1.2 m high, isophyllous; stems quadrangular, sulcate, bifariously pubescent. Leaves equal in each pair; petiole (0.2–)1–2.8 cm long, hirsute; leaf elliptic to ovate, 4.5–13 cm long, 1.4–5 cm brad, both surfaces slightly strigose, lateral veins 7 or 8 pairs, base attenuate, margin serrate, apex acuminate. Spikes axillary or terminal, 3–20 cm long, with 2–7 flower pairs; peduncle 0.8–1 cm long, pubescent; rachis pubescent and glandular-pilose. Bracts linear-lanceolate, 6–8 mm long, 1–2 mm broad, abaxially pubescent with glandular-pilose, 1-veined, apex acute; bracteoles linear, 4–6 mm long, 0.8–1 mm broad, pubescent with glandular-pilose, apex acute; flowers 1.5–3.5 cm apart on rachis, usually only one sterile per node. Calyx 1.2–1.8 cm long, abaxially pilose with glandular-pilose, 2-lipped, lower lip 2-lobed almost to base, lobes lanceolate,1.0–1.6 cm long, 1.2–1.5 mm broad, upper lip 3-lobed, lobes triangular, 4–6 mm long, 1.2–1.5 mm broad. Corolla violet blue, 3.2–3.5 cm long, slightly curved, outside glabrous, inside glabrous except for trichomes retaining style; tube base cylindric, 6–7 mm long, 1.7–2 mm broad, upwards gradually widened to 0.8–1 cm

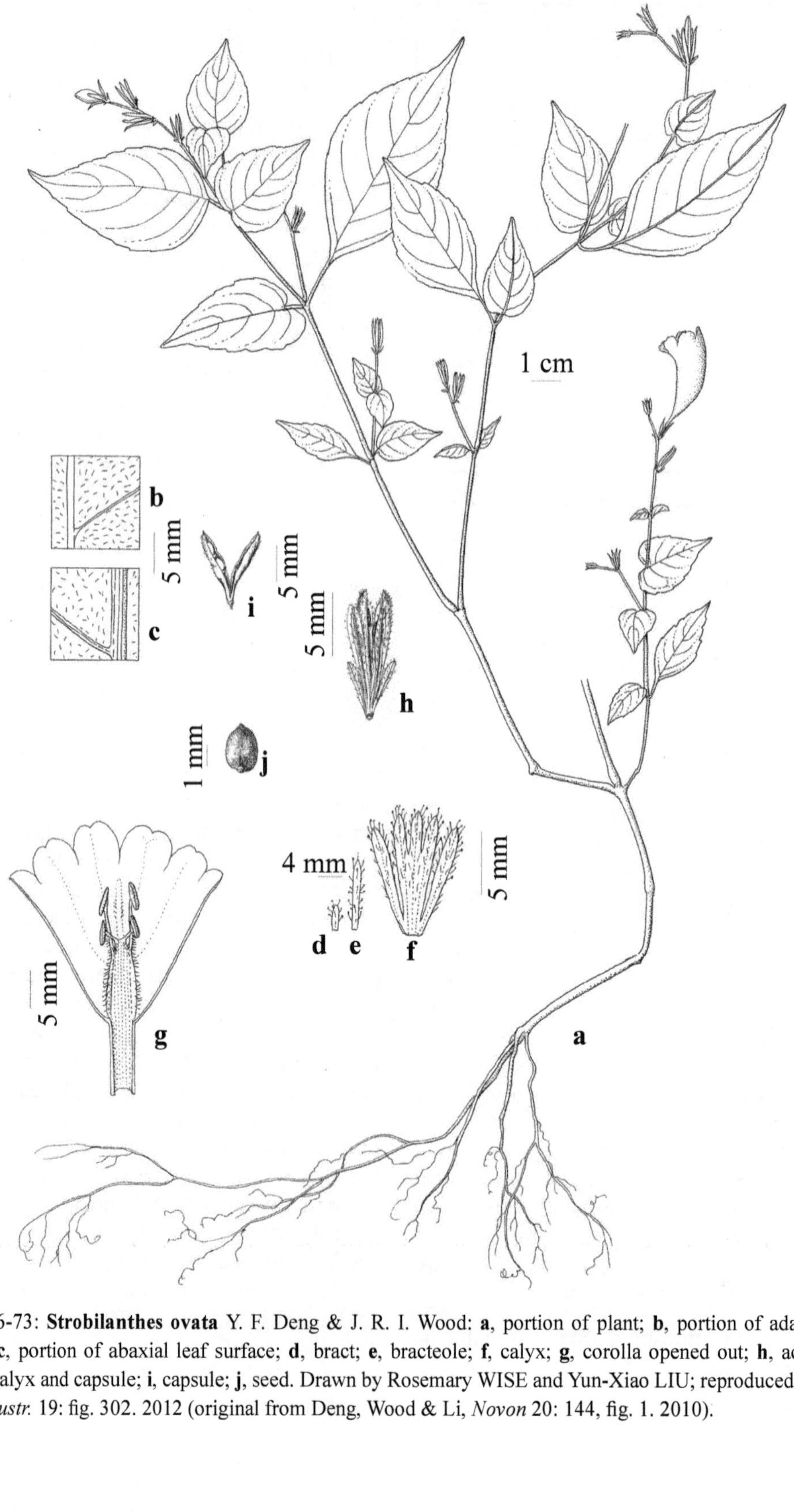

Figure 46-73: **Strobilanthes ovata** Y. F. Deng & J. R. I. Wood: **a**, portion of plant; **b**, portion of adaxial leaf surface; **c**, portion of abaxial leaf surface; **d**, bract; **e**, bracteole; **f**, calyx; **g**, corolla opened out; **h**, accrescent fruiting calyx and capsule; **i**, capsule; **j**, seed. Drawn by Rosemary WISE and Yun-Xiao LIU; reproduced from *Fl. China Illustr.* 19: fig. 302. 2012 (original from Deng, Wood & Li, *Novon* 20: 144, fig. 1. 2010).

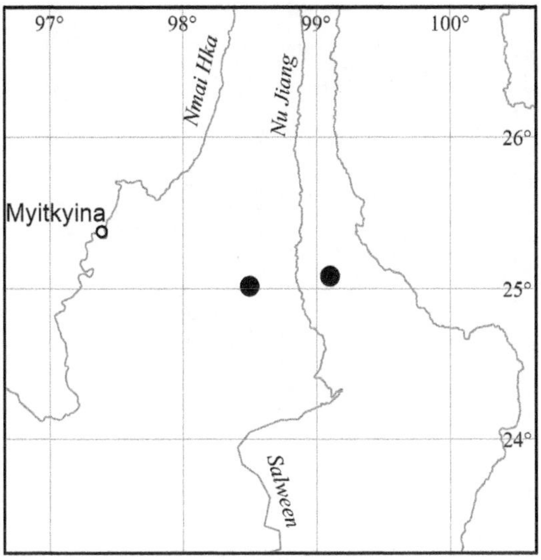

Map 46-125: **Strobilanthes ovata** Y. F. Deng & J. R. I. Wood

broad at mouth; lobes oblong-ovate, 6–7 mm long, 5–6 mm broad, apex emarginate. Stamens 4, included; filaments hirsute, longer pair 5–6 mm long, shorter pair 2–2.5 mm long; anther-thecae oblong, 2–2.2 mm long. Ovary ellipsoid, 2–3 mm long, glabrous; 2 ovules per locule; style 1.5–1.8 cm long, sparsely glandular-pubescent. Capsules oblong, 1.5–1.8 cm long, glandular-pubescent towards apex, 4-seeded. Seeds ovate in outline, compressed, 2–3 mm long, 1.5–2 mm broad, convex, densely pilose; areole small, glabrous. Figure 46-75.

Phenology: flowering from October to January of next year.

Habitats: thickets on mountain slopes; 1300–2400 m.

Distribution: endemic; S Hengduan and U Irrawaddy (Map 46-127).

S HENGDUAN: Lushui, Gutanhe, 2400 m, 3 Jun. 1981, *Exped. Hengduanshan 414* (PE). **U IRRAWADDY**: **Drungjiang**, E Drung Jiang, Bapo, Langwangduo, 1570 m, 1 Nov. 2004, *Gaoligong Shan Biodiversity Survey 22148* (CAS, KUN); Langwangduo, SSE of Bapo, 1380 m, Jan. 1991, *Dulong Jiang Investigation Team 3793* (CAS, KUN); **Putao**, Adung R., 1300–1600 m, 2 Feb. 1931, *F. Kingdon-Ward 9184* (BM, E, KUN).

78. Strobilanthes dryadum Benoist, Bull. Mus. Natl. Hist. Nat. 28: 94. 1922; Hu, Deng & Wood in Wu, Raven & Hong (eds.), Fl. China 19: 402. 2011. ≡ *Pteracanthus dryadum* (Benoist) C. Y. Wu & C. C. Hu in Hu (ed.), Fl. Reipubl. Popularis Sin. 70: 135. 2002. Type: [S Hengduan] China, Yunnan, Heqing, "Les bois de Ta long tan, près de Ta-Pin-Tze", 12 Jan. 1889, *M. Delavay 4142* [holotype: P (P00719310); isotypes: A (A00387584), BM (BM000796833), E (E00047308), GH (GH00286754), K (K000882730), NY (NY00278305), P (P00719311, P00719312), US (US00588597)].

Subshrubs, 1 m tall, isophyllous; stems quadrangular, sulcate, bifariously pubescent. Leaves equal in each pair; petiole 1–3 cm long, sulcate, bifariously pubescent; blade ovate to lanceolate, 15–25 cm long, 4.5–8 cm broad, glabrous, lateral veins 6 or 7 pairs, base attenuate and decurrent onto petiole forming wings, margin crenate-dentate, apex acuminate. Spikes terminal, 3–6 cm long, with 2–6 flower pairs; rachis pubescent with glandular-pilose, winged, wing 0.5–1 mm. Bracts ovate-oblong, 4–7 mm long, 1–2 mm broad, glandular-pubescent, apex obtuse; bracteoles oblong, 3–4 mm long, 0.5–0.6 mm broad, glandular-pubescent; flowers 1–2 cm distant on rachis, usually one sterile per node. Calyx 1–1.5 cm long, accrescent in fruit, abaxially glandular-pubescent, 2-lipped, lower lip 2-lobed almost to base, upper lip 3-lobed to below the middle, lobes linear-oblong, 1.1–1.5 cm long, 1–2 mm broad. Corolla blue, 3–3.4 cm long, outside glabrous, inside glabrous except for trichomes retaining style; tube base cylindric, 6–8 mm long, 1.6–2 mm broad, upwards gradually widened to 0.8–1 cm broad at mouth; lobes oblong, 6–8 mm long, 6–8 mm broad. Stamens 4, included; filaments hirsute, shorter pair 3–4 mm long, longer pair 5–7 mm; anther-thecae oblong, 1.6–2 mm. Ovary puberulent towards apex; 2 ovules per locule; style 2.2–2.4 cm long, sparsely pilose. Capsules narrowly oblong-obovoid, 1–1.2 cm long, apically glandular-pubescent, 4-seeded. Seeds ovate in outline, compressed, 2–2.5 mm long, 1.6–2.2 mm broad, pilose; areole small. Figure 46-76.

Figure 46-74: **Strobilanthes atroviridis** Y. F. Deng & J. R. I. Wood: **a**, habit; **b**, bract; **c**, bracteole; **d**, calyx; **e**, corolla opened out; **f**, ovary and style. Drawn by Ling WANG; reproduced from Deng et al., *J. Trop. Subtrop. Bot.* 18: 481, fig. 7. 2010.

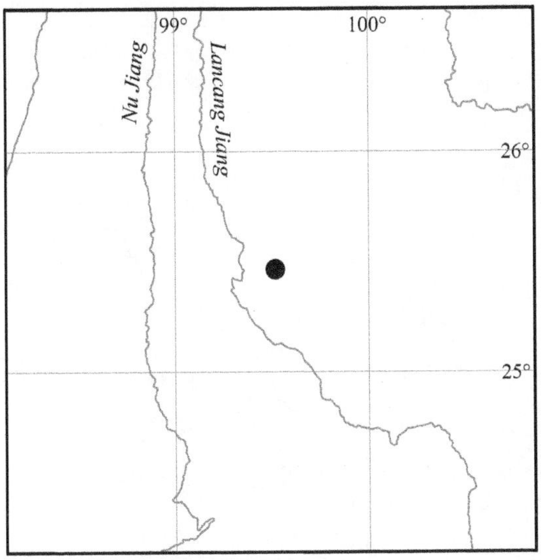

Map 46-126: **Strobilanthes atroviridis** Y. F. Deng & J. R. I. Wood

Phenology: flowering from September to November; fruiting from January to March.

Habitats: forests slope; 2000–2500 m.

Distribution: S Hengduan, U Irrawaddy (Map 46-128); also in other parts of China (C Yunnan).

S HENGDUAN: **Heqing**, Ta long tan, Ta Pin Tze, 12 Jan. 1889, *M. Delavay 4142* (A, BM, E, GH, K, P, NY, US); **Yangbi**, Haba, 2200–2300 m, 1984, *Sino-German Exped. 1055* (KUN); **Yunlong**, Tianchi, 2500 m, in 1985, *J. Peng et al. 19280* (YUKU). **U IRRAWADDY**: Kambaiti, 73 km E of Myitkyina, 2100 m, 1934, *R. Malaise 226* (S).

79. Strobilanthes tamburensis C. B. Clarke in Hooker (ed.), Fl. Brit. India 4: 454. 1884; Wood in Grierson & Long (eds.), Fl. Bhutan 2(3): 1261. 2001; Hu, Deng & Wood in Wu, Raven & Hong (eds.), Fl. China 19: 404. 2011. Type: [E] Nepal, Mechi Zone, Taplejung, the Tambur River, 1210 m, *J. D. Hooker s. n.* [holotype: K (K000882949)].

Herbs, 30–50 cm high, slightly anisophyllous; stems decumbent and rooting at nodes on lower portion, then ascending, quadrangular, sulcate, bifariously pilose. Leaves unequal in each pair, smaller one about two third of larger one in size, usually sessile or shortly petiolate; petiole 0–3 mm long; blade ovate to suborbicular, 8–45 cm long, 8–45 cm broad, both surfaces glabrous or pilose, abaxially whitish, adaxially green, lateral veins 5–7 pairs, base rounded to subcordate, margin entire or obscurely crenulate-serrate, apex acute to obtuse. Spikes terminal or axillary, simple, 3–8 cm long; peduncle 0–3.5 cm long, pilose; rachis pilose; flowers 2–2.5 cm apart basally but imbricate apically on rachis. Bracts orbicular to obovate, 5–8 mm long, 5–8 mm broad, persistent until flowers falling, pilose; bracteoles obovate, 2–4 mm long, 1–2 mm broad, pilose. Calyx 6–10 mm long, 5-lobed almost to base; lobes linear-spatulate, margin ciliate, apex acute. Corolla blue, 2.5–3 cm long, outside pilose, inside glabrous except for trichomes retaining style; tube base cylindric, 2.5–3 mm broad for 1–1.2 cm long, upwards bent to ca. 90° and abruptly widened to 8–10 mm broad at mouth; lobes broadly ovate, 5–6 mm long, 6–7 mm broad. Stamens 4, included; filaments subglabrous, longer pair 6–7 mm long, shorter pair 3–4 mm long; anther-thecae oblong, 1.8–2 mm long, 0.8–1 mm broad. Ovary gland-tipped; 2 ovules per locule; style 2.2–2.5 cm long, sparsely pilose, apically glabrous. Capsules 1–1.2 cm long, glandular-pilose, 4-seeded. Seeds orbicular in outline, 2–3 mm long and broad, densely pilose; areole absent.

Phenology: flowering from August to November.

Habitats: wet places in forests; 1200–2200 m.

Distribution: endemic; E Nepal, Sikkim & Darjiling, Bhutan, Yarlung Zangbo-Brahmaputra (Map 46-129).

E NEPAL: **Koshi**, Sankhuwasabha, Chichila, 1780 m, 27 Sep. 2001, *D. G. Long et al. 95* (E, TI); Sankhuwasabha, Arun Valley, between Chichila and Kuwapani, 1910 m, 21 Sep. 1991, *D. G. Long et al. 139* (E, KATH); Khandbari, 2120 m, 31 Aug. 1956, *J. D. A. Stainton 1482* (BM, KATH, TI); **Mechi**, Taplejung, Tambur River, 1210 m, J. D. Hooker *s. n.* (K). **SIKKIM & DARJILING: Sikkim**, Sitong, 1500 m, Feb. 1882, *J. S. Gamble 10242* (K). **BHUTAN**:

Figure 46-75: **Strobilanthes longispica** (H. P. Tsui) J. R. I. Wood & Y. F. Deng: **a**, portion of plant; **b**, bract, bracteole and calyx; **c**, corolla opened showing stamens. Drawn by Ai-Li LI; reproduced from *Fl. Reipubl. Popularis Sin.* 70: 124, pl. 18. 2002.

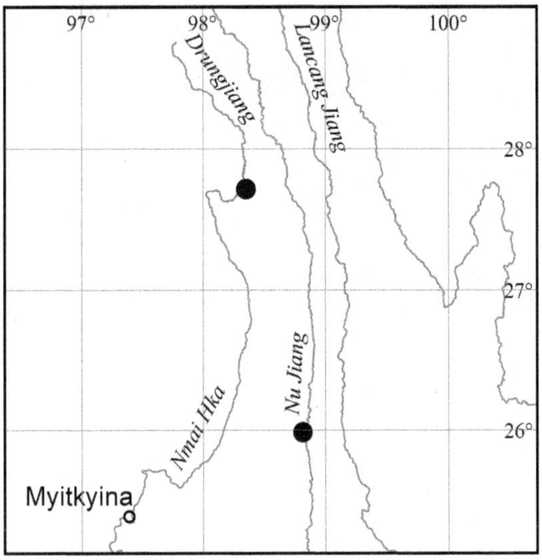

Map 46-127: **Strobilanthes longispica** (H. P. Tsui) J. R. I. Wood & Y. F. Deng

Chukka, ca. 3 km S of Chimakothi, 11 Nov. 1987, *J. R. I. Wood 6065* (E, FHO); **Tashigang**, between Tashiyangtse and Bumdeling, 2100 m, 8 Nov. 1991, *J. R. I. Wood 7503* (E, FHO, TI). **YARLUNG ZANGBO-BRAHMAPUTRA:** Senge Dzong, 1800–2100 m, 1938, *F. Kingdon-Ward 14186* (BM).

80. Strobilanthes forrestii Diels, Notes Roy. Bot. Gard. Edinburgh 5: 162. 1912; Bremekamp, Verh. Kon. Ned. Akad. Wetensch., Afd. Natuurk., Sect. 2 41(1): 275. 1944; Wood, Edinburgh J. Bot. 51(2): 238. 1994; Hu, Deng & Wood in Wu, Raven & Hong (eds.), Fl. China 19: 412. 2011. ≡ *Pteracanthus forrestii* (Diles) C. Y. Wu [Ind. Fl. Yunnan. 2: 1682. 1984, nom. invalid., Art. 41.5] ex H. P. Tsui in Wang (ed.), Vascul. Pl. Hengduan Mount. 2: 1873. 1994; Tsui & Hu in Hu (ed.), Fl. Reipubl. Popularis Sin. 70: 137. 2002; Hu & Tsui in Wu (ed.), Fl. Yunnan 16: 696. 2006. Type: [S Hengduan] China, Yunnan, E flank of the Lichiang Range, 10,000 ft., Aug. 1906, *G. Forrest 2760* [lectotype designated by Wood, Edinburgh J. Bot. 51: 238. 1994: E (E00045291); isolectotypes: BM (BM000884900), IBSC (IBSC 0285827)].

Strobilanthes panpienkaiensis H. Lév., Cat. Pl. Yunnan: 6. 1915; Bremekamp, Verh. Kon. Ned. Akad. Wetensch., Afd. Natuurk., Sect. 2 41(1): 275. 1944. Type: China, Yunnan, Vallée de Pan-Pien-Kai, 2500 m, Jul. 1912, *E. E. Maire s. n.* [holotype: E (E00045289)].

Strobilanthes rotundifolia Benoist, Bull. Mus. Hist. Nat. Paris 28: 97. 1922; Bremekamp, Verh. Kon. Ned. Akad. Wetensch., Afd. Natuurk., Sect. 2 41(1): 284. 1944. ≡ *Pteracanthus rotundifolius* (Benoist) Bremek., Verh. Kon. Ned. Akad. Wetensch., Afd. Natuurk., Sect. 2 41(1): 199. 1944; Tsui & Hu in Hu (ed.), Fl. Reipubl. Popularis Sin. 70: 147. 2002; Hu & Tsui in Wu, Fl. Yunnan 16: 703. 2006. Type: China, Yunnan, Songming, Yo li chan, 1910, *F. Ducloux 7495* [holotype: P (P00228389); isotypes: K (K000882716), P].

Diflugossa muliensis H. P. Tsui, Acta Bot. Yunnan. 12: 275. 1990. Type: [S Hengduan] China, Sichuan, Muli, Erya, 2300–2400 m, 1 Oct. 1983, *Exped. Qinghai-Xizang 14898* [holotype: PE (PE01545539); isotypes: KUN (KUN427915, KUN1255287), PE (PE01545540)].

Herbs, 30–75 cm high, isophyllous; stems quadrangular, sparsely eglandular-pilose or glandular-pilose, glabrescent. Leaves equal in each pair, subsessile; blade ovate to rarely ovate-oblong, 2–5 cm long, 1.2–3 cm broad, both surfaces sparsely pilose, lateral veins 4–6 pairs, base rounded or cordate, margin crenate to subentire, apex acute to obtuse. Spikes terminal or rarely axillary, 5–15 cm long; rachis pilose with glandular-pilose; flowers 1–2.5 cm apart on rachis. Bracts variable in shape and size, diminishing upwards, glandular- and eglandular-pilose, lower ones ovate-elliptic, 1–3 cm long, 1–2 cm broad, margin obscurely crenate, upper ones lanceolate, 7–10 mm long, 2–3 mm broad, margin entire; bracteoles linear to narrowly oblong, 6–11 mm long, 1–1.2 mm broad, eglandular and glandular-pilose. Calyx 0.8–1.2 cm long, 5-lobed almost to base; lobes linear, unequal, one 1.5–2 mm longer than others, eglandular- and glandular-pilose. Corolla blue to purple, 3.2–4 cm long, curved, outside

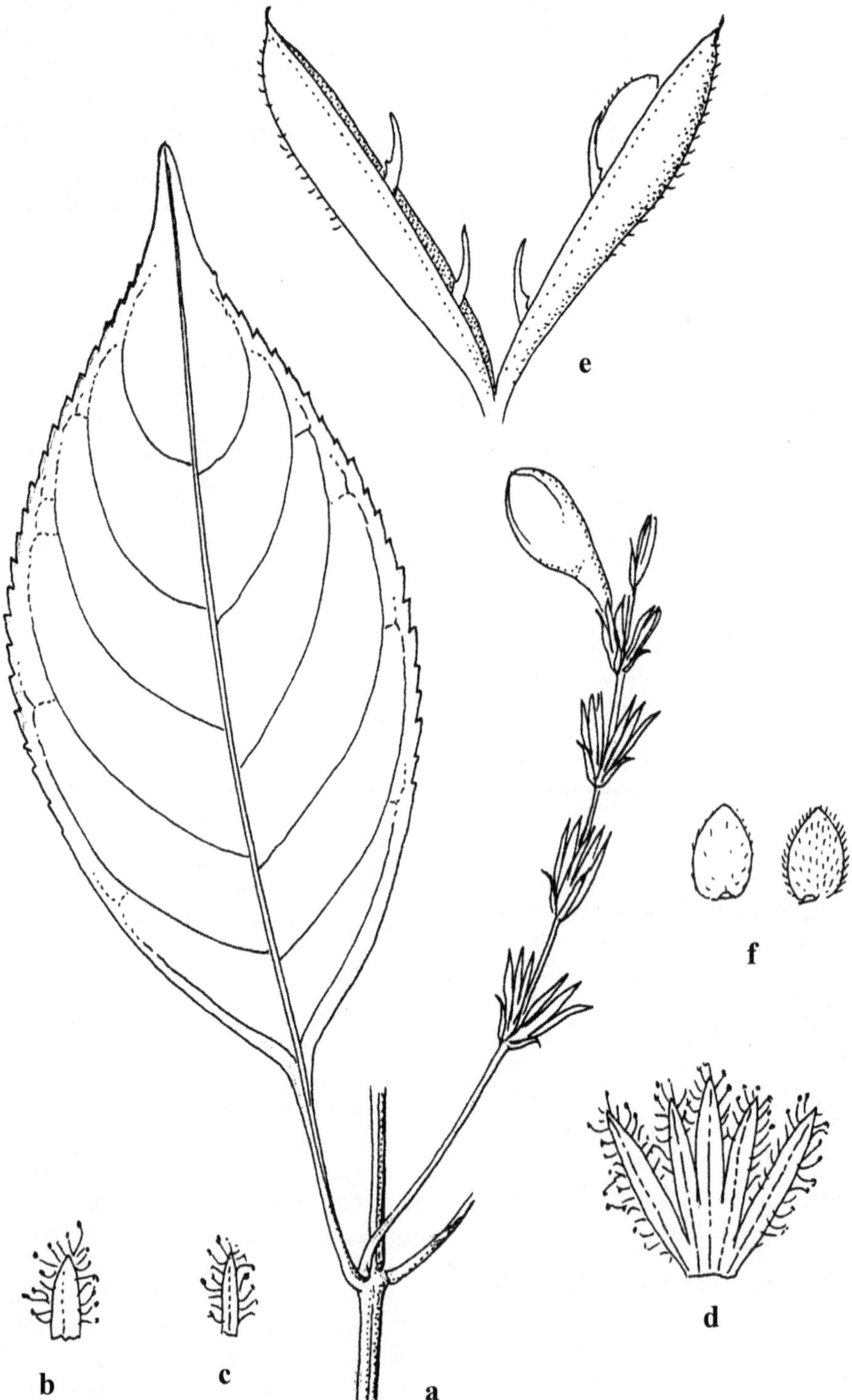

Figure 46-76: **Strobilanthes dryadum** Benoist: **a**, portion of inflorescence with leaf; **b**, bract; **c**, bracteole; **d**, calyx; **e**, capsule; **f**, seeds. Drawn by Rosemary WISE; reproduced from Wu, Raven & Hong, *Fl. China Illustr.* 19: fig. 301: 9–13. 2012 (original from Wood, Bennett & Scotland, *Kew Bull.* 58: 111, fig. 7: J–N. 2003).

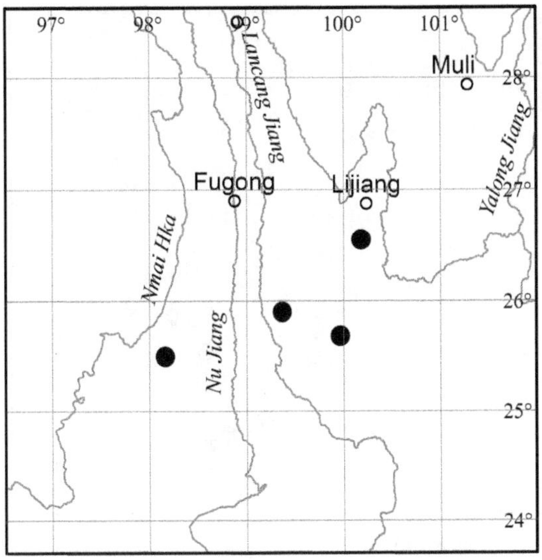

Map 46-128: **Strobilanthes dryadum** Benoist

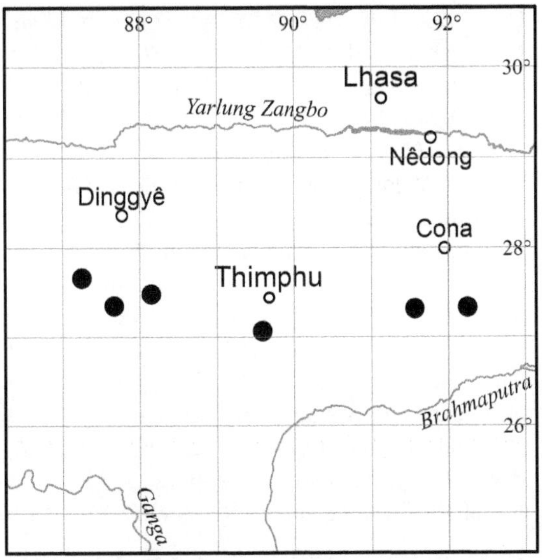

Map 46-129: **Strobilanthes tamburensis** C. B. Clarke

pilose, inside glabrous except for trichomes retaining style; tube base cylindric, 5–6 mm long, 1.8–2 mm broad, upwards bent to ca. 90° and widened to 1.2–1.5 cm broad at mouth; lobes ovate, 5–6 mm long, 6–7 mm broad, apex obtuse. Stamens 4, included; filaments of longer pair 6–8 mm long, pubescent, shorter pair 2–3 mm long; anther-thecae oblong, 1.8–2 mm long, 0.6–0.8 mm broad. Ovary glandular-puberulent towards apex; 2 ovules per locule; style 2.5–2.8 cm long. Capsules oblong, 1–1.2 cm long, glandular-pilose towards apex, 4-seeded. Seeds ovate in outline, 2–3 mm long and broad, pubescent; areole white, small.

Phenology: flowering from July to August.

Habitats: forests, grass slopes; 1900–4000 m.

Distribution: U Yarlung Zangbo, E Nepal, M Yarlung Zangbo, S Hengduan (Map 46-130); also in other parts of China (C Yunnan).

U YARLUNG ZANGBO: **Gyirong**, Gyirong Zhen, Zhaxiang, 3353 m, *Z. Zhou et al. ZJW-3842* (KUN). **E NEPAL: Mechi**, Taplejung, Iladanda to Helok, 13 Nov. 1963, *H. Kanai et al. 6303693* (KATH, TI). **M YARLUNG ZANGBO**: **Yadong**, Asang, 3000 m, 8 Jul. 2013, *PE Expedition to Xizang 2934* (PE). **S HENGDUAN**: **Dali**, E flank of Tali Range, 3000–3300 m, Jul. 1906, *G. Forrest 4553* (BM, E); E flank of the Tali Range, 2700–3000 m, Aug. 1906, *G. Forrest 4551* (BM, E); Xiaoguan to Dacang, 2400–2500 m, 9 Jul. 1984, *1984 Sino-Amer. Bot. Exped. 936* (A, KUN); Huadianba, 2900–3300 m, 18 Jul. 1984, *1984 Sino-Amer. Bot. Exped. 1146* (KUN); Diancang Shan, Xiaohuadianba, 2900–3200 m, 19 Jul. 1984, *1984 Sino-Amer. Bot. Exped. 1164* (A, CAS, E, TI, US); Xiaguan, 1960 m, 5 Aug. 1962, *K. J. Luo et al. 614266* (YIM); Lutaoping, Cangshan, 2900 m, 25 Jun. 1964, *X. Zhou 1276* (KUN); Cangshan, Huadianba to Xiaohuadianba, 3100 m, 21 Jul. 2009, *Z. J. Yin & H. J. Dong 1476* (KUN); 19 Sep. 1919, *K. K. Tsoong 7413* (PEY); **Eryuan**, Liantie to Fengxiang, 9 Jul. 1940, *R. C. Ching 23189* (KUN, PE); Fengyu to Liantie, 2946 m, 18 Aug. 2010, *E. D. Liu et al. 2749* (KUN); **Jiulong**, Wuxugou, 3340 m, 29 Jun. 1960, *Z. T. Guan 35-0139* (KUN); **Luding**, Erlangshan, Ganhaizi to Yazikou, 7 Jul. 1951, *W. K. Hu & C. He 10270* (PE, SZ); Erlangshan, Machanggou, 2400 m, 17 June 1953, *H. L. Tsiung & T. H. Tsiung 34484* (IBSC, PE); **Muli**, Wochio, 2800 m, 5 Aug. 1937, *T. T. Yü 7595* (KUN, PE); Shuzhupingzi, 2750 m, Jun. 24, 1978, *Y. B. Yang 7043* (SZ); Zhutoutang, 2700 m, 28 Jun. 1978, *Y. B. Yang 7108* (CDBI, SZ); Maidi, way to Mianyashankou, 2750 m, 5 Aug. 1978, *Q. S. Zhao et al. 6994* (SZ); Tangjian, Poer, 2600 m, 19 Aug. 1978, *Q. S. Zhao 8580* (CDBI, SZ); Taoba,

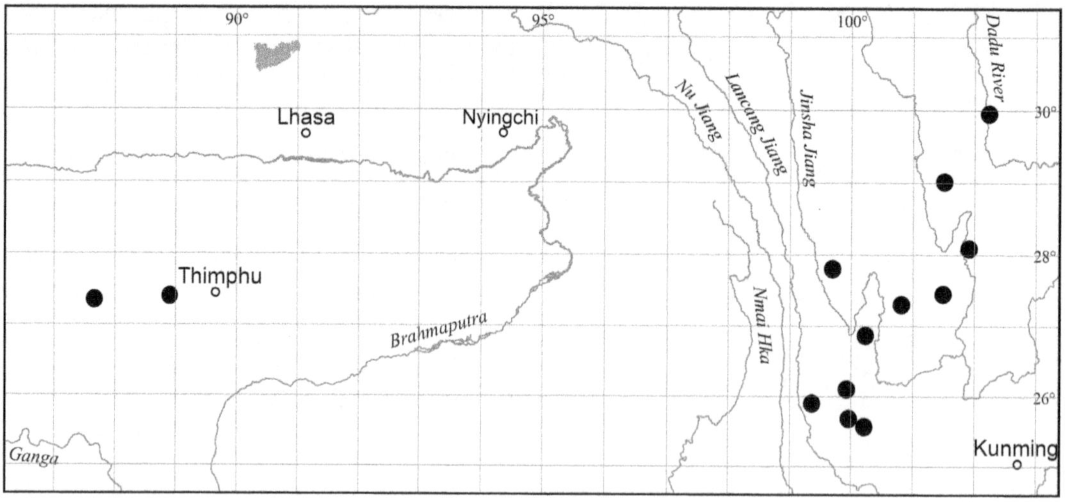

Map 46-130: **Strobilanthes forrestii** Diels

2550 m, 20 Jul., 1978, *S. S. Liu 7694* (CDBI); s. loc., Jul. 1930, *G. Forrest 28435* (E, MO, PE); s. loc., 2700 m, 28 Jul. 1978, *K. H. Mou & Y. B. Yang 7388* (SZ); S Muli, 3000 m, 21 Aug. 1924, *J. Kingdon-Ward 4850* (E); NE Yangtze band, 3000 m, Sep. 1913, *G. Forrest 11223* (BM, E, K, PE); **Ninglang**, between Yongning and Mutichiu, 23 Jun. 1916, *C. Schneider 1624* (E, K); **Shangri-la** (Zhongdian), Haba, Aug. 1962, *A. L. Zhang 100952* (KUN); Haba, Longwang to Shaba, 2800 m, 27 Aug. 1962, *Exped. Zhongdian 1488* (KUN, PE, YUKU), Haba, Baiziduo, 2900 m, 17 Sep. 1962, *Exped. Zhongdian 2135* (KUN, PE); Geza, Wengshui dadui, 27 Jun. 1975, *J. S. Yang 7659* (KUN); Gezan, 3550 m, 25 Aug. 1983, *J. S. Yang 83-286* (KUN); Gezan, 4000 m, 4 Sep. 2010, *Kham Expedition 10-3298* (PE); **Xiangcheng**, Maanshan, 20 Jul. 2010, *Kham Expedition 10-0252* (PE); **Yangbi**, Meiweng, Zhongshan to Sanchahe, 2400 m, 24 Nov. 1964, *Z. H. Yang 101884* (PE); **Yanyuan**, Yenyuan, 11 Jun. 1914, *C. Schneider 1482* (E); **Yulong** (Lijiang), Snow range, Luomeiluo, 2800–3000 m, 10 Aug. 1942, *K. M. Feng 8907*(KUN, PE); Xueshan, Liuougu, 3600 m, 15 Aug. 1942, *K. M. Feng 9042* (KUN, PE); E flank of the Lichiang Range, 3000 m, Aug. 1906, *G. Forrest 2760* (BM, E, IBSC); E flank of the Lichiang Range, Aug. 1910, *G. Forrest 6329* (E); Lichiang Rang, 3300 m, Jul. 1914, *G. Forrest 12700* (E); W slopes of Likiang Snow Range, 30 May–6 Jun. 1922, *J. F. Rock 4263* (A, N, UC); E slopes of Likiang Snow Range, May-Oct. 1922, *J. F. Rock 4855* (A); Yulongxueshan, Muzhuchao, 3 Jul. 1962, *A. L. Zhang 100325* (KUN); Yulongshan, Labahe, 3000 m, 7 Jul. 1961, *R. L. Xiong 712086* (YIM), Ganhaizi, 3100 m, 3 Aug. 1961, *R. L. Xiong & Y. F. Qi 612532* (YIM), Yulongshan, near Botanical Garden, 2800 m, 21 Sep. 1963, *S. W. Yu 63012* (KUN); Yulongshan, Sandaoting, 2200 m, 6 Jun. 1985, *Exped. KUN & E 85-593* (KUN); Yulongshan, Muzhugou, 28 June 1959, *Kunming Working Station 22323* (KUN); Baisha, Yuhu, 2800 m, 9 Aug. 2005, *E. D. Liu 1250* (IBSC, KUN); 2600 m, streamside, 8 Aug. 1962, *Exped. Zhongdian 603* (KUN, PE, YUKU); **Yunlong**, Bi Jiang River valley, Yunlong to Lanping, 3035 m, 30 Jul. 2010, *T. Zhang et al. 10CS2149* (K).

81. Strobilanthes oresbia W. W. Sm., Notes Roy. Bot. Gard. Edinburgh 10: 196. 1918; Hu, Deng & Wood in Wu, Raven & Hong (eds.), Fl. China 19: 411. 2011. ≡ *Pteracanthus oresbius* (W. W. Sm.) C. Y. Wu & C. C. Hu in Hu (ed.), Fl. Reipubl. Popularis Sin. 70: 142. 2002. Type: [S Irrawaddy] Myanmar, mountains above Lajing Kha, Valley of Naw Chuang, 11,000–12,000 ft., 28 Aug. 1914, *F. Kindon-Ward 1911* [holotype: E (E00045278)].

Goldfussia grandissima H. P. Tsui, Acta Bot. Yunnan. 12: 276. 1990. ≡ *Strobilanthes grandissima* (H. P. Tsui) J. R. I. Wood, Edinburgh J. Bot. 51: 207. 1994. ≡ *Pteracanthus grandissimus* (H. P. Tsui) C. Y. Wu & C. C. Hu in Hu (ed.), Fl. Reipubl. Popularis Sin. 70: 140. 2002. Type: [S Hengduan] China, Yunnan, Gongshsan, Salwin-Kiukiang Divide, Neewahlung, Danglahtung, 3100–3400 m, 11 Jul. 1937, *T. T. Yü 19329* [holotype: P (PE01545731); isotypes: KUN, PE (PE01545732)].

Strobilanthes larium Hand.-Mazz., Symb. Sin. 7: 893, 1936; Hu, Fl. Reipubl. Popularis Sin. 70: 186. 2002; Hu, Deng & Wood in Wu, Raven & Hong (eds.), Fl. China 19: 411. 2011. Type: [S Hengduan] China, Yunnan, Doker-la, 2700–3500 m, 16 Sep. 1915, *H. Handel-Mazzetti 8043* [holotype: WU (WU 0035623); isotype: W (W1940-0001850)].

Herbs, to 2 m high, isophyllous; stems quadrangular, densely white pubescent with glandular-pilose, glabrescent. Leaves equal in each pair; petiole 2–13 cm long, glandular-pubescent, glabrescent, apically winged; blade ovate to orbicular, 7–22 cm long, 4–18 cm broad, pubescent especially along veins, lateral veins 9–11 pairs, base cuneate, rounded, or sometimes subcordate and decurrent onto petiole, margin deeply dentate or serrate, apex acute to acuminate. Spikes axillary, 8–15 cm long, secund, sometimes reduced to single flowers; peduncle 0.5–7 cm long, densely glandular-pubescent; rachis somewhat zigzag; flowers usually 1 fertile on each node, 0.8–2.5 cm apart on rachis. Bracts variable in shape and size, diminishing upwards, lowermost ones ovate, upper ones lanceolate, 1–4 cm long, 1–2 cm broad, glandular-pubescent, base cuneate, margin serrate or entire, apex obtuse; bracteoles linear-oblanceolate, 5–7 mm long, 0.8–1.1 mm broad, glandular-pubescent, apex obtuse. Calyx 1.2–2.1 cm long, densely glandular-pubescent, 5-lobed almost to base; lobes unequal, one 1.5–2 mm longer than others, linear-oblanceolate, apex obtuse. Corolla purplish-blue, 4–5 cm long, ventricose, outside sparsely white pubescent, inside pubescent; tube base cylindric, 1.8–2.2 mm broad, 5–6 mm long, strongly ventricose, upwards bent to ca. 90° and widened to 1.2–1.5 cm broad at mouth; lobes oblong, 5–6 mm long, 4–5 mm broad, apex obtuse. Stamens 4; filaments glabrous, longer pair 4–5 mm long, shorter pair 2–2.5 mm long; anther-thecae oblong, 2–2.5 mm long, 1–1.2 mm broad. Ovary glabrous; 2 ovules per locule; style 2.2–2.5 cm long pilose. Capsules oblong, 1.2–1.5 cm long, 2–3 mm broad, glandular-pilose towards apex, 4-seeded. Seeds blackish-brown, ovate in outline, 3–3.5 mm long, 2–2.5 mm broad, densely pilose; areole very small. Figure 46-77.

Phenology: flowering from July to September; fruiting from October to December.

Habitats: moist alpine meadows and forests; 1600–4100 m.

Distribution: U Yarlung Zangbo, Yarlung Zangbo-Brahmaputra, S Hengduan, U Irrawaddy (Map 46-131); also in other parts of India, Myanmar and China (NE Yunnan, S Sichuan, Chongqing).

Notes: Yu et al. (in Taxon 60: 1168. 2011) incorrectly considered the name *Goldfussia grandissima* to be not validly published by Tsui (in Acta Bot. Yunnan. 12: 276. 1990) required by Art. 37.3 of Vienna Code (now Art. 40.7 of Shenzhen Code). However, they overlooked that Tsui clearly indicated in the footnote on page 270 that all specimens cited in that paper are kept in the herbarium of Institute of Botany, Chinese Academy (PE) otherwise the herbarium is indicated. Therefore, the name *Goldfussia grandisimma* has been validly published by Tsui.

U YARLUNG ZANGBO: **Gyirong**, Gyirong, near Zhacun, 2800–3970 m, 23 Jul. 2013, *PE Xizang Exped. 4096* (PE). **YARLUNG ZANGBO-BRAHMAPUTRA**: Lo La, Pashakshiri, 3 Jul. 1936, 3300 m, *F. Ludlow & G. Sherriff 1905* (BM, E); Tha Chu, 3000–3600 m, 15 Jul. 1950, *F. Kingdon-Ward 19640* (BM, TI); **Mainling**, Doshong La, Tsangpo Valley, 3750 m, 14 Jul. 1936, *F. Ludlow, G. Sherriff & G. Taylor 5238* (BM); Tsangpo Valley, 2750 m, 14 Jul. 1936, *F. Ludlow, G. Sherriff & G. Taylor 5238a* (BM); **Zayü**, Putsang River, Rong To Valley, 3000 m, 5 Nov. 1933, *F. Kingdon-Ward 10953* (BM, TI); **Mêdog**, 4 Sep. 1990, *G. Yao et al. 3680* (NAS); **Xiazayü**, 11 Jul. 2008, *X. H. Jin et al. DLJ-ET 1840* (PE). **S HENGDUAN**: **Cawalong** (Tsawarong), 16 Sep. 1915, *H. Handel-Mazzetti 8043* (WU);

Figure 46-77: **Strobilanthes oresbia** W. W. Sm.: **a**, portion of plant; **b**, leaf. Drawn by Han-Ping YU; reproduced from *Fl. Reipubl. Popularis Sin.* 70: 167, pl. 24: 5–8. 2002.

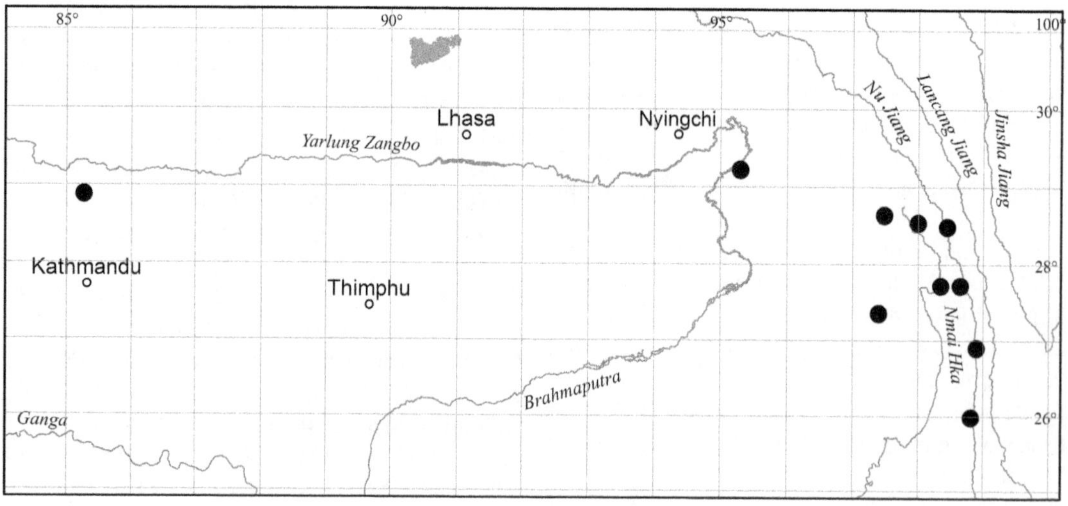

Map 46-131: **Strobilanthes oresbia** W. W. Sm.

Dokerla, 3200–3500 m, 16 Sep. 1915, *H. Handel-Mazzetti 8043* (A, E, WU); Qinnatong, 3000 m, Aug. 1935, *C. W. Wang 65195* (PE); **Fugong**, Shala Valley, 2800 m, 16 Jul. 1978, *Exped. Bijiang 1185* (KUN); Lishadi, Yaduo, 3040 m, 8 Aug. 2005, *Gaoligong Shan Biodiversity Survey 26740* (CAS, KUN, MO); Lishadi, Yaduo, 3120 m, 9 Aug. 2005, *Gaoligong Shan Biodiversity Survey 26857* (CAS, GH, KUN, MO); Lishadi, Yaduo, 3120 m, 9 Aug. 2005, *Gaoligong Shan Biodiversity Survey 26861* (CAS, KUN, MO); Lumadeng, Yaping, 3240m, 13 Aug. 2005, *Gaoligong Shan Biodiversity Survey 27140* (CAS, KUN); Lumadeng, Yaping, 3120 m, 13 Aug. 2005, *Gaoligong Shan Biodiversity Survey 27162* (CAS, GH, KUN, MO); Lumadeng, Yaping, 2700 m, 21 Aug. 2005, *Gaoligong Shan Biodiversity Survey 28830* (CAS, KUN); **Gongshan**, Bingzhongluo, 3450–4100 m, 3 Aug. 1979, *X. F. Deng 791475* (KUN); Bingzhongluo, Shi-ge-tung, Champutung, 2800 m, Oct. 1935, *C. W. Wang 67148* (A, KUN, NAS, PE); Bingzhongluo, Shi-gi-tung,Champutung, 2700 m, Oct. 1935, *C. W. Wang 67234* (A, KUN, NAS, PE); Bingzhongluo, Shi-gi-tung, Champutung, 3000 m, Oct. 1935, *C. W. Wang 67238* (KUN, PE); Cikai, Heipushan, 12 Oct. 1940, *K. M. Feng 8400* (KUN, PE); Cikai, Dabaidi, 3120 m, 12 Nov. 2004, *Gaoligong Shan Biodiversity Survey 22530* (KUN); Cikai, Cikeluo Qiao, 3030 m, 21 Aug. 2006, *Gaoligong Shan Biodiversity Survey 33814* (KUN); Cikai, Cikeluo Qiao, 3030 m, 21 Aug. 2006, *Gaoligong Shan Biodiversity Survey 33871* (KUN); Cikai, Cikeluo Qiao, 3030 m, 21 Aug. 2006, *Gaoligong Shan Biodiversity Survey 33890* (KUN); Cikai, Danzhu, along the Damawadi He, 3220 m, 26 Aug. 2006, *Gaoligong Shan Biodiversity Survey 34156* (KUN); Cikai, Danzhu, along the Damawadi He, 3220 m, 26 Aug. 2006, *Gaoligong Shan Biodiversity Survey 34164* (KUN); Cikai, Danghatu, 3280 m, 2 Sep. 2006, *Gaoligong Shan Biodiversity Survey 34504* (KUN); way to Dulong, Dongshaofang, 3000 m, 25 Jul. 1982, *Qinghai-Xizang Exped. 8582* (KUN); Cikai, Danzhu valley, Danzhupass, 3340 m, 3 Oct. 2007, *H. Li & R. Li 1028* (KUN); Cikai, Danzhu Valley, Danzhu Pass, 3340 m, 3 Oct. 2007, *H. Li & R. Li 1030* (KUN); Salwin-Kiukiang Divide Neewahlung, Danglahtung, 3100–3400 m, 11 Jul. 1937, *T. T. Yü 19329* (IBSC, KUN, PE); Taron-Taru Divide, Valley of Bucahwang, 2500–2800 m, 3 Sep. 1938, *T. T. Yü 22074* (E, KUN, PE); Salwin-Kiukiang Divide, Swangchiang, 2800 m, 13 Jul. 1938, *T. T. Yü 22081* (E, GH, KUN, PE); Cikai, 2650m, 1 Jul. 2000, *H. Li 11837* (KUN); Cikai to Drungjiang, Heipu Channel, 2 Sep. 2011, *S. X. Yu & W. Wang 6540* (PE); Tjiontson-lumba, Tschamutong, 3250 m, 3 Jul. 1916, *H. Handel-Mzzetti 9253* (E, WU); **Lushui**, Luobenzhuo, Ega, 3000 m, 6 Aug. 2005, *Gaoligong Shan Biodiversity Survey 25656* (KUN); Luobenzhuo, Ega, 2920 m, 8 Aug. 2005, *Gaoligong Shan Biodiversity Survey 25781* (CAS, KUN); Luzhang, Pianma Yakou, 3250 m, 15 Oct. 2002, *Gaoligong Shan Biodiversity Survey 15946* (CAS, KUN). **U IRRAWADDY**: **Drungjiang**, Gongshan Yakou to Kongdang, 12 Jul., 2011, *Y. H. Tong 11071212* (IBSC); Dulong, W Gaoligongshan, 2600–2900 m, 26 Jul. 1982, *Exped. Qinghai-Xizang 8807* (KUN, PE); Drungjiang to Gongshan, 1979, *Q. Lin & X. F. Deng 790680* (KUN); Sandui, from Xishaofang to Bapo, along Gamolai He, 2570 m, 17 Jul. 2002, *H. Li 15040* (KUN); Heiwadi, along the new road from Kongdang to Gongshan County, 2400 m, 26 Jul. 2002, *H. Li 15311* (KUN); near Yakou of Kongdang, 3400 m, *X. H. Jin et al. DLJ-ET 2264* (PE); **Putao**, W flank of N'Maikha-Salween divide, 3600–3900 m, 25 Sep. 1925, *G. Forrest 27249* (E, K, US); **Ridong**, Shiladu, 3000–3100 m, 7 Sep. 1982, *Exped. Qinghai-Xizang 10030* (KUN, PE); **Seuighku Valley**, 3000 m, 26 Jul. 1930, *F. Kingdom-Ward 7225* (K); 3000–3300 m, *F. Kingdon-Ward 7516* (K).

82. Strobilanthes versicolor Diels, Notes Roy. Bot. Gard. Edinburgh 5: 163. 1912; Hu, Deng & Wood in Wu, Raven & Hong (eds.), Fl. China 19: 412. 2011. ≡ *Pteracanthus versicolor* (Diels) H. W. Li in Wu, Fl. Xizang 4: 415. 1985; Tsui & Hu in Hu (ed.), Fl. Reipubl. Popularis Sin. 70: 149. 2002. Type: [S Hengduan] China, Yunnan, eastern flank of the Likiang range, 11,000 ft., *G. Forrest 2308* [holotype: E (E00045280); isotypes: BM, K (K000882832), P (P00719461)].

Strobilanthes flexa Benoist, Bull. Mus. Natl. Hist. Nat. 28: 186. 1922, "*flexus*", **syn. nov.**; Bremek., Verh.

Kon. Ned. Akad. Wetensch., Afd. Natuurk., Sect. 2 41(1): 275. 1944; Hu, Deng & Wood in Wu, Raven & Hong (eds.), Fl. China 19: 412. 2011. ≡ *Pteracanthus flexus* (Benoist) C. Y. Wu & C. C. Hu, Fl. Reipubl. Popularis Sin. 70: 137. 2002. Type: China, Chongqing, Chengkou, "Héoupin près de Tchen-keou-tin", 2000 m, 30 Jun. 1895, *R. P. Farges 1349* [lectotype **here designated**: P (P00719323); isolectotypes: A (A00295526), BM (BM000906284), CAS (CAS425118, CAS425119, CAS425120), GH (GH00295525), IBSC (IBSC0285824, IBSC02858245), K (K000882722, K000882726), P (P00719325, P00719326), PE (PE01620740, PE016207401, PE016207402), NY (NY00278306), US].

Herbs, 0.5–0.8(–1) m high, isophyllous; stems quadrangular, sparsely to densely pubescent, glabrescent. Leaves equal in equal; petiole (0–)1–5 cm long, sparsely pubescent; blade obovate, ovate, elliptic, or suborbicular, 2.5–15 cm long, 1.5–7 cm broad, adaxially glabrous, abaxially pilose on midvein towards base, lateral veins 4–8 pairs, base broadly cuneate and decurrent onto petiole, margin coarsely dentate or serrate, apex acute. Spikes axillary or terminal, 4–15 cm long, lax, simple or sparingly branched, sometimes reduced to a single flower; rachis glandular-pilose; flowers 1–2.5 cm apart on rachis. Bracts diminishing in size upward, oblong-oblanceolate, 1.2–3.5 cm long, 0.2–1 cm broad, glandular-pubescent, apex acute to acuminate, persistent, basal ones petiolate, suborbicular to oblong-elliptic, margin dentate, upper ones sessile, narrowly oblong-elliptic, margin entire; bracteoles linear-oblanceolate to oblong, 4–8 mm long, 1–2 mm broad, glandular-pubescent, apex obtuse. Calyx 0.8–1.5 cm long, glandular-pubescent, 5-lobed almost to base; lobes subequal, linear-oblong, 7–11 mm long, 0.8–1.2 mm broad, apex obtuse. Corolla purplish-blue, 3.5–4 cm long, outside pilose, inside glabrous except for trichomes retaining style; tube base cylindric, 1–1.5 cm long, 2–2.5 mm broad, upwards abruptly inflated, bent to ca. 90° above the middle and finally widened to 1.2–1.5 cm broad at mouth; lobes ovate, 4–5 mm long and broad, apex rounded. Stamens 4, included; filaments of longer pair 5–7 mm long, pilose, shorter pair 2–4 mm long, glabrous; anther-thecae oblong, 2–2.5 mm long, 1–1.2 mm broad, erect. Ovary comose; 2 ovules per locule; style 2.8–3.2 cm long, sparsely pilose. Capsules oblong, 1.5–1.8 cm long, apically glandular-pubescent, 4-seeded. Seeds ovate in outline, compressed, 2–3 mm long, 2–3 mm broad, appressed pubescent. **Chromosomes** 2n = 16. Figure 46-78.

Phenology: flowering from June to September; fruiting from August to December.

Habitats: grasslands at forest margins; 1400–4300 m.

Distribution: U Yarlung Zangbo, N Hengduan, S Hengduan (Map 46-132); also in other parts of China (NE Yunnan, S Sichuan, Chongqing, Guizhou, Hubei).

U YARLUNG ZANGBO: **Gyirong**, Tuodang, 3 Jul. 1975, *C. Y. Wu et al. 75-525* (PE); Gyirong to Sale, 2900 m, 28 Jul. 2015, *L. Wei & J. C. Hao 15386* (BJNU). **N HENGDUAN**: **Maoxian**, Nanxin, Chashan, 2720 m, *X. F. Gao 2260* (CDBI). **S HENGDUAN**: **Baoxing**, Dengchigou, 2880 m, 26 Jun. 1933, *T. H. Tu 4309* (IBK, PE); s. loc., 1954, *T. P. Soong 38335* (IBSC, KUN, PE); s. loc., *T. P. Soong 39117* (KUN, NAS, SZ); **Cawarong** (Tsawarong), 3500 m, 13 Jul. 2010, *South Tibet Expedition Team (X. H. Jin et al.) STET 0270* (PE); 3500 m, 13 Jul. 2010, *South Tibet Expedition Team (X. H. Jin et al.) STET 0287* (PE); **Dali**, Tali Range, Sep. 1929, *G. Forrest 27954* (BM, E); Dali to Zhongdian, Jiaochuang, 2350–2450 m, *Alpine Garden Society Expedition to China 1994 ACE 36* (E, K); Hsingai to Tietsun, 1650–1800 m, 16 May 1915, *H. Handel-Mazzetti 6357* (W), Zhongsi to Zhonghefeng, 3200 m, 15 Jul. 1960, *Exped. N. W. Yunnan 10516* (KUN); Diangcangshan mountain range, en route from Huadianba to Xizhou, 2850 m, 20 Jul. 1984, *1984 Sino-Amer. Bot. Exped. 1217* (A, CAS, E, KUN, US); Santaipi, 2600m, 11 Jun. 1986, *H. Li et al. 1078* (KUN); Lanfengshan, 16 Jun. 1946, *H. C. Wang 4757* (IBSC, KUN, PE); Huadianba, 2700 m, 20 Jun. 2004, *Y. F. Ma & D. H. Fu 20040055* (YIM); Tsang (Cang) Shan, Huadianba, 2920 m, 17 Oct. 1990, *Chungtien-Lijanng-Dali Expedition (CLD-90) 1457* (E); **Dêqên**, Tehching, Muyung, 3200m, 27 Sep. 1937, *T. T. Yü 10490* (A, KUN, PE); **Dêqên**, Yunlin, Xidang, Yubeng, way to Qishe Waterfall,

Figure 46-78: **Strobilanthes versicolor** Diels: **a**, habit; **b**, stem; **c**, calyx; **d**, corolla opened showing stamens; **e**, capsule. Drawn by Quan LIN; reproduced from *Icon. Cormophyt. Sin.* 4: 161, fig. 5736. 1975.

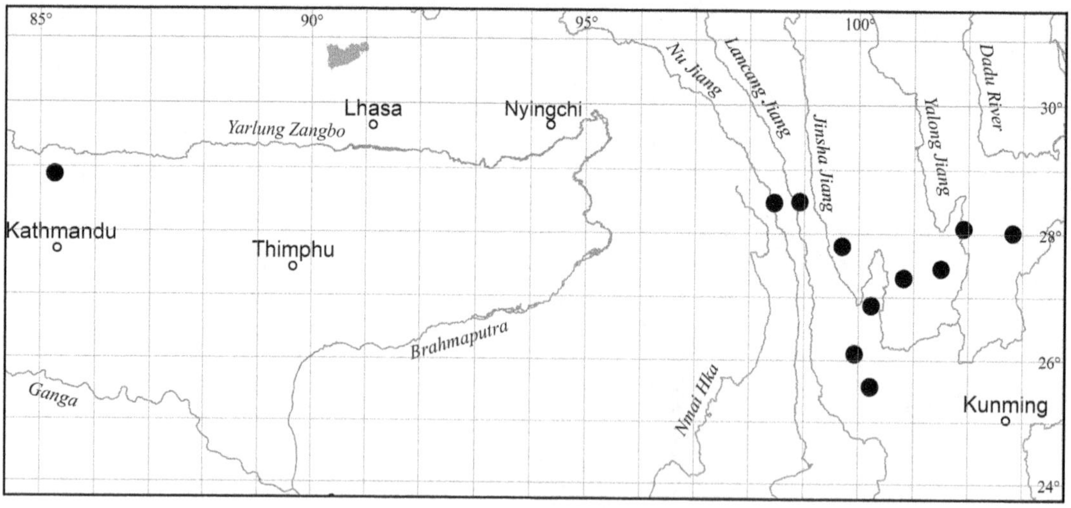

Map 46-132: **Strobilanthes versicolor** Diels

3400 m, 4 Oct. 1984, *J. S. Yang 84-1021* (KUN); Meilixueshan, 4210 m, 24 Aug. 1986, *W. M. Zhu et al. 20868* (YUKU); **Eryuan**, Biaocao, Huapingzi, 17 Jul. 1940, *R. C. Ching 23265* (KUN,PE); **Kangding**, Tachienlu, 2700–4000 m, *A. E. Partt 1* (BM, CAL, E, K); **Muli**, Mt. Gibboh, S Muli Gomb, 3450–3600 m, Apr.–May 1932, *J. F. Rock 24180* (A, E); Litang River Divide, 3150 m, 3 Jun. 1921, *F. Kingdon-Ward 4146* (E); 3000 m, 1 Aug. 1921, *J. Kingdon-Ward 4568* (E); Litang River, 3000 m, 21 May 1922, *F. Kingdom-Ward 5173* (E); Relading Shan, 4000m, 27 Jul. 1978, *Q. S. Zhao 6263* (SZ); Chutouwan, 2600 m, 28 Jun. 1978, *Y. B. Yang 7104* (CDBI, SZ); Muli Range, near Yungning, 3000 m, 16 Jun. 1922, *F. Kingdon-Ward 5234* (E); **Ninglang**, Yungning, 3150 m, 23 Jun. 1914, *H. Handel-Mazzetti 3161* (WU); **Shangri-la** (Zhongdian), NE Yangtze band, 3000 m, Sep. 1913, *G. Forrest 11223* (BM, E, PE); Haba, Enluoyoubuyan, 3350 m, *Exped. Zhongdian 2138* (HIB, KUN, PE); Tehgoh, 3400 m, 27 Jul. 1937, *T. T. Yü 12469* (KUN, PE); Pica, 3000 m, 5 Jul. 1937, *T. T. Yü 11901* (A, KUN, PE); Kongsinshu, 3200 m, 25 Jun. 1937, *T. T. Yü 11774* (A, KUN, PE); Haba, Lihai, 29 Jul. 1962, *A. L. Zhang 100683* (KUN); Baidi, Baishuitai, 2600 m, 26 Jun. 1963, *Exped. N. W. Yunnan 4464* (KUN, PE, YUKU); Haba, Sinuo, 2800 m, 7 Jul. 1963, *Exped. Zhongdian 63-3240* (KUN); Kung-shing-shu, 3600 m, 23 Jul. 1939, *K. M. Feng 1779* (A, KUN); south of the town, 3232 m, 3 Jul. 1994, *Alpine Garden Society Expedition to China 1994 ACE 906* (E, K); Road to Haba Shan, 3328 m, 14 Jun. 1994, *Alpine Garden Society Expedition to China 1994 ACE 231* (E, K); Habashan, 3300 m, 14 Sep. 1962, *Exped. Zhongdian 1079* (KUN, PE, YUKU); Habashan, 3300 m, 22 Aug. 1962, *Exped. Zhongdian 1279* (KUN, PE); Haba, Baidi, Baishuitaidi, 2570 m, 10 Jul. 1963, *Exped. Zhongdian 63-3304* (KUN); Haba, Sinuo, 2800 m, 7 Jul. 1963, *Exped. Zhongdian 63-3223* (KUN); Zhongdian to San-ba, 3000 m, 7 Jul. 1963, *Alpine Garden Society Expedition to China ACE 1942* (E); **Weixi**, Kazi Pass, Mekong-Yangtze Divide, 3000–3300 m, 31 Jul. 1921, *F. Kingdon-Ward 5305* (E); **Yanyuan**, Wolo-ho, 3300 m, 15 Jun. 1914, *H. Handel-Mazzetti 3069* (E, WU); Tzemananish, inter Woloho & Chora, 2900 m, 15 Jun. 1914, *C. Schneider 1571* (E, K, US); Jisha Linchang, 3300 m, 21 Aug. 2010, *Kham Expedition 10-2562* (PE); **Yulong** (Lijiang), Yulongshan, Ameiluo, 2800 m, 17 Nov. 1962, *A. L. Zhang & S. W. Yu 100181* (KUN); Yulongshan, Yunshanping, 3240 m, 11 Jul. 1962, *A. L. Zhang & S. W. Yu 100946* (KUN); Lichiang Range, Oct. 1913, *G. Forrest 11448* (BM, E, K); Xuesong, 2800 m, 5 Sep. 1937, *T. T. Yü 15531* (A, KUN, PE); eastern flank of the Likiang Range, 3300 m, Jun. 1906, *G. Forrest 2456* (BM, E, K); Eastern plank of Likiang Range, Aug. 1910, *G. Forrest 6344* (BM, E, K, PE); Ganhaitze, SW of Likiang Snow Range, 14 Jul. 1939, *R. C. Ching 20987*

(GH), W slope of Likiang Snow Range, May 30–Jun. 6, 1922, *J. F. Rock 4223* (SYS, US); E slopes of Likiang Snow Range, May–Oct. 1922, *J. F. Rock 4349* (A, UC); E slope of Likiang Snow Range, above Mahong Patze, 390 m, 24 Jun. 1922, *J. F. Rock 4674* (US); E slopes of Likiang Snow Range, May–Oct. 1922, *J. F. Rock 4689* (A, E, US); Yulongshan, Mahuangba, 3500 m, 6 Jun. 1962, *Kunming Working Station 100107* (KUN); Yulongshan, 2800 m, Jun. 1957, *Yunnan Daxue Shengwuxi 1021* (YUKU); Yulongshan, Yezhulin, 3500 m, 22 Jul. 1961, *R. L. Xiong 612396* (YIM); Shitougou, 3000 m, 31 May 1986, *H. Li et al. 749* (KUN); Yulong-Shan, Sandawan, 6 Jun. 1985, *Kunming Edinbourgh Yulong Shan Expedtion (1985) 593* (E); Xuesong, Heixueshan, 28 Jun. 1939, *X. G. Zhao 30311* (KUN, PE); Yulong Shan, below Mahuangba, 3250 m, 2 Jun. 1987, *Sino-British Expedition to Lijiang Co. (China, W Yunnan)-1987 SBL 342* (E); Yuhu, 2650 m, 4 Sep. 1959, *Kunming Working Station 22682* (KUN, PE); Yulongshan, *Yunnan Daxue Shengwuxi 1021* (YUKU); Yulongshan, 28 Aug. 1957, *Yunnan Daxue Shengwuxi 1940* (YUKU); 2800 m, Jul. 1935, *C. W. Wang 70913* (KUN, PE); Ma Huang Ba, Sep. 1987, *B & L 12048* (E); Canghaitze, SW of Likiang Snow Range, 14 Jul. 1939, *R. C. Ching 20997* (KUN); Lichiang, 3400 m, 1914, *C. Schneider 3658* (A); Yülung-schan, Jun. 1914, *H. Handel-Mazzetti 3743* (WU); Likiang Range, 3000–3300 m, June 1906, *G. Forrest 2308* (E, K); Ganhaizi, 14 Jul. 1939, *H. C. He 21002* (KUN); W slope of Likiang Snow Range, 30 May–6 Jun. 1922, *J. F. Rock 4148* (US); W slope of Likiang Snow Range, May 1922, *J. F. Rock 3599* (US); Yulogshan, 24 June 1962, *Lijiang Botanical Garden 100315* (KUN); Yulongxueshan, 3200 m, 4 Aug. 2019, *Y. F. Deng 28646* (IBSC); **Zhaojue**, Sikai, 2500 m, 30 Jul. 1976, *s. coll. 12825* (PE).

83. Strobilanthes sabiniana (Wall. ex Lindl.) Nees in Wallich (ed.), Pl. Asiat. Rar. 3: 86. 1832; Wood in Long & Grierson (eds.), Fl. Bhutan 2(3): 1261. 2001. ≡ *Ruellia sabiniana* Wall. ex Lindl., Bot. Reg. 15: t. 1238. 1829. Type: Nepal, Pundua, 1821, *F. de Silva in Wallich Cat. 2338a* [lectotype **designated here**: K-W (K001115693); isolectotypes: BM (BM000884899), GZU (GZU000249663), LE].

Subshrubs, 0.5–2 m, strongly anisophyllous; stems quadrangular, sulcate, glabrous except for pubescent at nodes, reddish-brown. Leaves strongly unequal in each pair, smaller one less than half of larger one in size; petiole 0–5 mm long; blade ovate, elliptic, or subrhomboidal, large one 6–15 cm long, 1.8–6 cm broad, smaller one 1.5–6 cm long, 0.8–3.5 cm broad, glabrous, base rounded or cuneate and decurrent, margin entire, serrate, or obscurely crenate, apex acute. Spikes axillary, 3–8 cm long, simple or sometimes 1–2-branched; peduncles 1–1.5 cm long, glabrous; rachis glandular-pubescent. Bracts obovate-suborbicular, 2.5–6 mm long, 2.5–5 mm broad, glandular-pubescent, apex rounded, sometimes lightly retuse; bracteoles linear, 3–4 mm long, 0.5–0.8 mm broad, glandular-pubescent. Calyx 4–8 mm long, glandular-pubescent, 5-lobed almost to base; lobes subequal, linear-oblong, apex obtuse. Corolla pale blue, 2.5–3 cm long, outside pubescent on lobes; tube base cylindric, 4–5 mm long, 1.5–1.8 mm broad, upwards bent 90° and abruptly widened to 0.8–1 cm broad at mouth; lobes oblong, 4.5–5 mm long and broad, apex obtuse or shallowly retuse. Stamens 4; filaments of longer pair 1.4–1.6 cm long, exserted, shorter pair 4–5 mm long, included; anther-thecae oblong, 1.8–2 mm long, connective not produced. Ovary pilose; 2 ovules per locule; style 2–2.5 cm long, glabrous. Capsules oblong, 6–7 mm long, glandular-pubescent, 4-seeded. Seeds ovate in outline, compressed, 2–2.3 mm long and broad, pubescent; areole ca. 0.5 mm long. Figure 46-79.

Phenology: flowering from November to January of next year; fruiting from December to March of next year.

Habitats: forests; 400–2000 m.

Distribution: C Nepal, Sikkim & Darjiling, Bhutan (Map 46-133); also in other parts of India.

C NEPAL: **Gandaki**, Kaski, Pothana to Tolka, 1750–1970 m, *M. Suzuki et al. 8881102* (TI); **Narayani**, Makwanpur, Saune Khola, Bhainse to Daman, 768 m, 7 Nov. 2012, *B. Adhikari & G. D. Bhatt 6* (E, KATH). **SIKKIM & DARJILING**: **Darjiling**, Pashetang, 1800 m, Dec. 1879, *J. S. Gamble 7533* (K). **BHUTAN**: **Gaylegphung**,

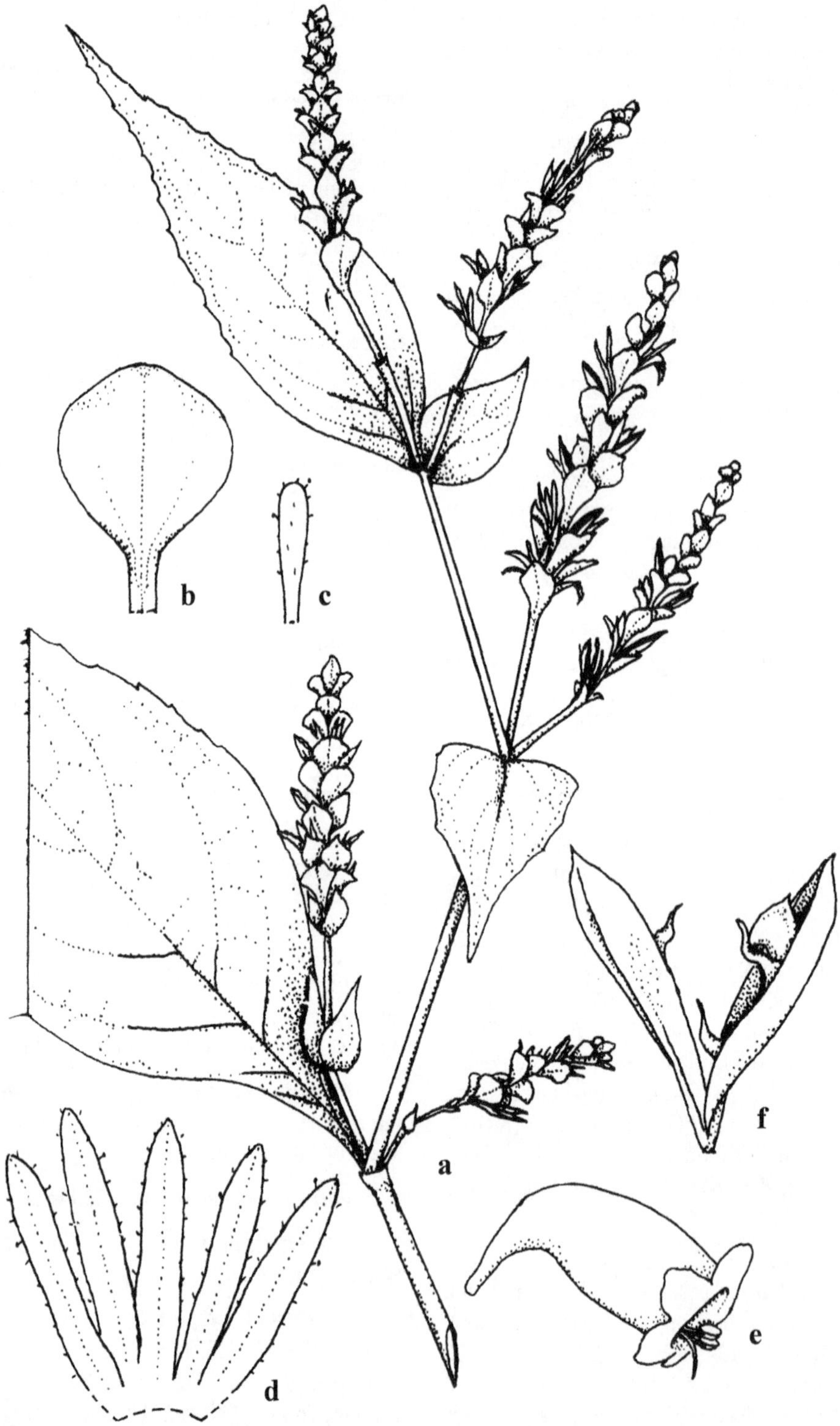

Figure 46-79: **Strobilanthes sabiniana** (Wall. ex Lindl.) Nees: **a**, habit; **b**, bract; **c**, bracteole; **d**, calyx; **e**, corolla; **f**, dehiscent capsule. Drawn by Louise OLLEY; reproduced from Wood in Long & Grierson (eds.), *Fl. Bhutan* 2(3): 1269, fig. 104, a–f. 2001.

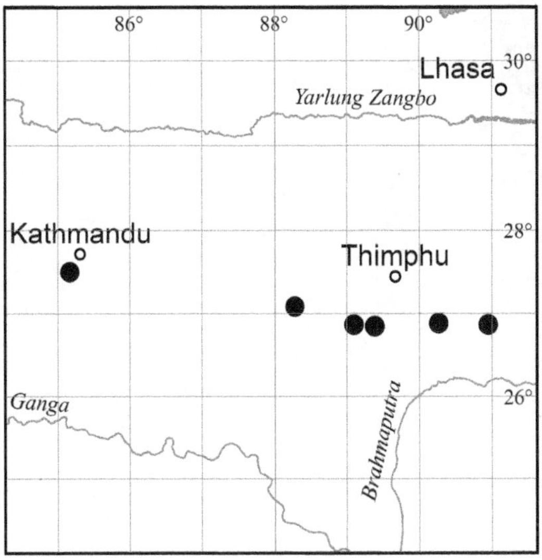

Map 46-133: **Strobilanthes sabiniana** (Wall. ex Lindl.) Nees

near Tatopani, 700 m, 11 Feb. 1990, *J. R. I. Wood 7231* (E); **Phuntsholing**, between Kharbandi and Sorchen, 700 m, 18 Mar. 1990, *J. R. I. Wood 7237* (E); **Samchi**, Tamangdhanra Forests, 700 m, 28 Feb. 1982, *A. J. C. Grierson & D. G. Long 3308* (A, E, K); **Sarbhang**, Kami Khola, 420 m, 10 Mar. 1982, *A. J. C. Grierson & D. G. Long 3634* (E, K); between Sarbhang Bazaar and the High School, 600 m, 7 Feb. 1988, *J. R. I. Wood 6134* (E); **Tongsa**, between Pertimi and Tintibi Bridge, Mangse Chu, 1030 m, 5 Apr. 1982, *A. J. C. Grierson & D. G. Long 4348* (E).

84. Strobilanthes saccata J. R. I. Wood, Edinburgh J. Bot. 51: 256. 1994. Type: [C] Nepal, Gandaki, Kaski, Dhumpus/ Lumle, 1930 m, 1 Oct.1969, *H. Flatt 106* [holotype: BM (BM000810980); isotype: TI (10011535)].

Strobilanthes sabiniana var. *argentea* Nees in Wallich (ed.), Pl. Asiat. Rar. 3: 86. 1832, **syn. nov.** ≡ *Ruellia argentea* Wall., Numer. List n. 2339. 1830, nom. invalid. Type: Nepal, 1830, *Wallich Cat. 2339* [lectotype **here designated**: K (K000882951); isolectotypes: E, (K000882950), LE].

Herbs, 30–40 cm high, anisophyllous; stems decumbent or ascending, slightly zigzag upwards, pilose. Leaves unequal in each pair, smaller 1/2–2/3 of larger one in size, subsessile or petiolate; petiole 0–2 cm long, diminishing in length upwards, pilose; blade ovate, 2–8 cm long, 1–3.5 cm broad, adaxially sparsely pilose, glabrescent, abaxially sparsely pilose on veins, base rounded or broadly cuneate, decurrent onto petiole, margin entire or obscurely crenulate, apex shortly acuminate. Spikes axillary and terminal, 4–9 cm long, simple; peduncles 0–3 cm long, sparsely pilose; rachis sparsely bifariously pubescent; flowers in opposite pairs, 5–10 mm apart on rachis. Bracts variable in shape and size, lowermost ones ovate, 0.8–3.2 cm long, 0.9–2.5 cm broad, apex acute, upper ones obovate, 2.5–5 mm long, glabrous or with subsessile glands, apex rounded, often submucronate; bracteoles oblong-oblanceolate, 2–2.5 mm long, glabrous or sparsely pilose. Calyx 5–6 mm long, accrescent to 8–9 mm long in fruit, 5-lobed almost to base, glabrous or pilose when young, then glandular-pubescent; lobes subequal, two slightly longer than other three, linear-oblanceolate or spatulate, margin slightly ciliate. Corolla blue, 2.6–3 cm long, outside sparsely pilose on lobs; tube base cylindric, 2–5 mm long, 1–1.4 mm broad, upwards bent 90°, strongly inflated and widened to 1.2–1.5 cm long; lobes broadly ovate, 3–4 mm long, 5–6 mm broad, apex shallowly emarginate. Stamens 4, included; filaments glabrous, shorter pair 2–3 mm long, longer pair 7–8 mm long; anther-thecae narrowly ellipsoid, 2–2.5 mm long. Ovary pilose towards apex; 2 ovules per locule; style 2.2–2.5 cm long, pilose. Capsules oblong, 8–10 mm long, pubescent towards apex, 4-seeded; seeds oblong ovate, 1.5–2 mm long and broad, pilose with mucilaginous hairs, areole small. Figure 46-80.

Phenology: flowering from August to October.

Habitats: forests; 1500–2000 m.

Distribution: endemic; C Nepal, E Nepal (Map 46-134).

C NEPAL: **Bagmati**, Lalitpur, Phulchoki, Botton of Phulchoko Hill, 1650 m, 24 Aug. 2014, *B. Adhikari et al. 1* (E); Lalitpur, Godawari, near Naudhara, 1570 m, 16 Sep. 2014, *B. Adhikari & G. D. Bhatt 59* (E); Sindupalchok, beside Malemchi

Figure 46-80: **Strobilanthes saccata** J. R. I. Wood: **a**. habit. Drawn by Mary BATES; reproduced from Wood, *Edinburgh J. Bot.* 51: 263, fig. 13: I. 1994.

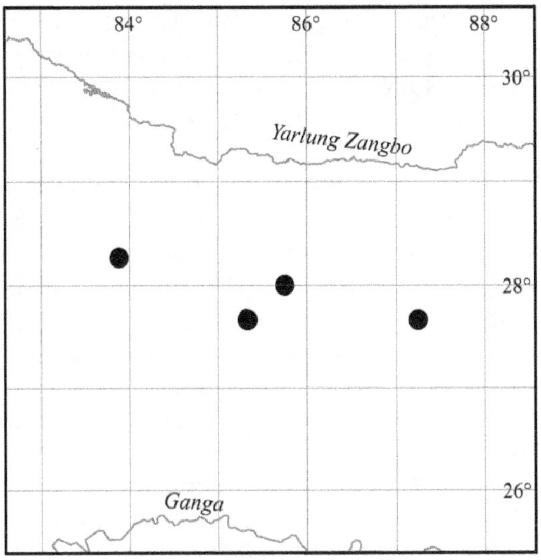

Map 46-134: **Strobilanthes saccata** J. R. I. Wood

Khola near bridge between Tarke Ghuang and Malemchigaon, 2000 m, 17 Sep. 1974, *J. H. de Haas 2773* (BM, L); **Gandaki**, Kaski, Dhumpus/ Lumle, 1930m, 1 Oct. 1969, *H. Flatt 106* (BM, TI). **E NEPAL**: **Koshi**, Sankhuwasabha, Chichila, below school, 1780 m, 27 Sep. 2001, *D. G. Long et al. ENEP95* (E, TI).

85. Strobilanthes attenuata (Nees) Nees in de Candolle (ed.), Prodr. 11: 193. 1847. ≡ *Ruellia attenuata* Nees in Wallich (ed.), Pl. Asiat. Rar. 3: 83. 1832. ≡ *Pteracanthus attenuatus* (Nees) Bremek., Verh. Kon. Ned. Akad. Wetensch., Afd. Natuurk., Sect. 2 41(1): 199. 1944. Type: Nepal, 2000–3500m, *Wallich Cat. 2346 (as "2345")* [lectotype designated by Wood, Edinburgh J. Bot. 51: 230. 1994: K-W (K001115703); isolectotype: LIV].

Strobilanthes reflexa Nees in de Candolle (ed.), Prodr. 11: 194. 1847. ≡ *Pteracanthus reflexus* (Nees) Bremek., Verh. Kon. Ned. Akad. Wetensch. Afd. Natuurk., Sect. 2 41(1): 199. 1944. Type: India, *Royle s. n.* (holotype: LIV).

Strobilanthes newii Beddome ex C. B. Clarke in Hooker (ed.), Fl. Brit. India 4: 464. 1885. ≡ *Mackenziea newii* (Beddome ex C. B. Clarke) Bremek., Verh. Kon. Ned. Akad. Wetensch., Afd. Natuurk., Sect. 2 41(1): 183. 1944. Type: India, Karnataka, Western Mysore, *R. H. Beddome 236* [holotype: K (K000260015)].

Strobilanthes alata Nees in de Candolle (ed.), Prodr. 11: 194. 1847, nom. illeg., non *Strobilanthes alata* Blume, Bijdr. Fl. Ned. Ind. 14: 798. 1826; Clarke in Hooker (ed.), Fl. Brit. India 4: 464. 1884. ≡ *Strobilanthes urticifolia* Wall. ex Kuntze, Revis. Gen. Pl. 2: 499. 1891; Nasir & Ali in Malik & Ghafoor (eds.), Fl. Pakistan 188: 57. 1988. ≡ *Pteracanthus urticifolius* (Kuntze) Bremek., Verh. Kon. Ned. Akad. Wetensch., Afd. Natuurk., Sect. 2 41(1): 199. 1944; Tsui & Hu in Hu (ed.), Fl. Reipubl. Popularis Sin. 70: 149. 2002. ≡ *Ruellia urticifolia* Wall., Numer. List n. 2346. 1830, nom. invalid. Type: India, Meghalaya, Pundua, *Wallich Cat. 2346* [lectotype designated by Nasir & Ali in Malik & Ghafoor (eds.), Fl. Pakistan 188: 58. 1988: K-W (K001115703)].

Perennial herbs or subshrubs, 0.6–1.2 m high, isophyllous; stems quadrangular, sparsely pubescent. Leaves equal in each pair; petiole 3–7.5 cm long, winged; blade ovate, 8–16 cm long, 4.5–8 cm broad, sparsely pubescent on both surfaces, base cordate, truncate or rounded, margin strongly serrate, apex acuminate. Panicles of spikes, terminal, 10–25 cm long; rachis glandular-filose; flowers in distant opposite pairs, mostly 6–13 mm apart. Bracts variable in shape and size, diminishing upwards, lower ones ovate, base rounded, margin serrate, apex acuminate, uppermost ones lanceolate to oblong-elliptic, 2–5(–10) mm long, 0.5–2 mm broad, eglandular pubescent, caducous; bracteoles linear, 2–2.5 mm long, 0.5–0.6 mm broad, glandular-filose, caducous. Calyx 6–7 mm long, accrescent to 1.2–1.5 cm long in fruit, 5-lobed almost to base; lobes unequal, one 2–3 mm longer than others, lanceolate, glandular-pilose, apex obtuse. Corolla 3.2–3.5 cm long, dark blue, rarely pale blue, pinkish, or white, curved, outside pubescent on lobes; tube base cylindric, 1–1.2 cm long, 2–3 mm broad; lobes ovate, 4–4.5 mm long, 4.5–5 mm broad. Stamens 4, didynamous, filaments glabrous, longer pair 7–8 mm long, shorter pair 3–4 mm long, anther-thecae ellipsoid. Ovary pilose; style sparsely pilose; 2 ovules per locule.

Capsules narrowly elliptic-oblong, 1.3–1.5 cm long, 2.5–3 mm broad, sparsely glandular-pilose to glabrescent, 4-seeded. Seeds ovate in outline, 2.5–3 mm long, 2–1.5 mm broad, lenticular, appressed pilose with hygroscopic hairs, which spread when moistened. Figure 46-81.

Phenology: flowering from July to October; fruiting from September to November.

Habitats: moist places in forests; 1200–4000 m.

Distribution: N Pakistan, Jammu & Kashmir, U Ganga & Indus, W Nepal, Sikkim & Darjiling (Map 46-135); also in other parts of Afghanistan, Pakistan and India.

N PAKISTAN: **Azad Kashmir**, Besan, 4 Oct. 1977, *Shahzad & Ayaz 1269* (MO); Hattian, 1885, *J. R. Drummond 26211* (K); Hattian, Karauli Forests, near Rampur, Jhelum Valley, 1800 m, 5 Jul. 1940, *F. Ludlow & G. Sherriff 7723* (BM, TI); Muzaffarabad, Leepa Valley, *M. Qaiser & Azmat Ali 5024* (KUH, fide Malik & Ghafoor, 1988); **Hazara**, Cachemire, 3200–4000 m, 28 Jul. 1953, *F. Schimid 539* (BM); Richmosi, Panjul Saran Range, 22 Aug. 1899, *J. F. Duthie s. n.* (K); Abbottabad, Dunga Gali, 1800 m, 29 Sep. 1977, *M. N. Chaudhri, Faiqah 2* (MO); Babusar Pass, Jul.–Sep. 1954, *John Abei 153* (BM); 30 Aug. 1896, *Luayat 20036* (CAL, K); Kagan, Nuri, 4 Aug. 1899, *J. E. Duthie s. n.* (K); Kagan, Bhonja Kagan, 31 Jul. 1897, *Luayat 22077* (CAL, K); Kagan Valley, Uahandri, 21 Sep. 1927, *s. coll. s. n.* (LE); Mansehra, from Kwai to Shogran, 2000 m, *S. Abedin & M. Qaiser 8779* (KUH, fide Malik & Ghafoor, 1988); Murree Hills, *E. M. Sauders s. n.* (K); Murree Hills, 2100 m, Sep. 1885, *J. R. Drummond 1777* (K); Murree Hills, Ghikka Gali–Ghatcal Road, 1950 m, Aug. 1918, *T. A. Sprague 150* (K); Murree Hills, 1800–2100 m, 13 Oct. 1944, *M. Nath 21456* (K); **Swat**, Kalam, 5 Sep. 1966, *Sadruddin s. n.* (KUH, fide Malik & Ghafoor, 1988); Shang-la, Shangla hills, 1600 m, *M. Qaiser & A. Ghafoor 4731* (KUH, fide Malik & Ghafoor, 1988). **JAMMU & KASHMIR**: **Anantnag**, Bringhi River, 2100 m, 26 Aug. 1939, *F. Ludlow & G. Sherriff 7561* (BM, TI); Pahalgam, 2200 m, 23 Sep. 1961, *B. M. Wadhura & J. N. Volrra 168* (CAL); Pahalgam, Sallar, 1950 m, *J. R. Drummond 14316* (K); **Bandipur**, 1800 m, 4 Aug. 1928, *B. B. Osmaston 6* (K); **Ganderbal**, Naca Nag, Wangar Valley, 12 Aug. 1940, *P. W. Pinfold 258* (BM, TI); **Jammu**, Chhanji Chhat, Katra, 13 Jan. 1960, *A. K. Dutt 6887* (RRLH); **Kishtvar**, Kishtvar to the Pir Patsaski or Kishtvar Pass, 1–4 Aug. 1856, *Herbarium Schlagintweit 3802* (E); **Ladak**, in montibus Karakorum, inter Gund et Karghan, 13 Sep. 1933, *N. Ambolt 6373* (S); **Liddar Valley**, 20 Aug. 1902, *Luayat 25736* (K); **Srinagar**, Gurais and Pir Panjal Mountains, 1902, *E. F. Noel 455* (LIV). **U GANGA & INDUS**: **Almora**, Ranikhet, 1800 m, 22 Aug. 1885, *J. F. Duthie 4279* (K); **Chamba**, Aug. 1880, *R. Ellis 465* (CAL); Kala to Reserve, Pergauna Bathri, 1919, *R. Rang s. n.* (A); Pangi, 2550–2700 m, 25 Jan. 1878, *G. Watt 87B* (E); Pangi, Killar to Phindra, 2400 m, 1 Aug. 1899, *Harsukh s. n.* (K); Kalatop Forests, 2400 m, 5 Sep. 1895, *J. H. Lace 779* (CAL, E); **Dalhousie**, Kaibthli Reserve, 1890 m, *Herbs Schlagintweit 3307* (LE); 2100 m, 11 Sep. 1874, *C. B. Clarke 22208* (LE); 2100 m, 14 Sep. 1874, *C. B. Clarke 22400* (CAL); 1800 m, 28 Sep. 1874, *C. B. Clarke 23206* (CAL, LE); **Dehradun**, Snow Viea, 24 Jul. 1881, *J. S. Gamble s. n.* (K); Paled, 1200 m, Feb. 1889, *J. S. Gamble 24594* (CAL, K); Jaunsar, Chamata, Oct. 1894, *J. S. Gamble 25226* (K); Jaunsar, Mundali, 2400 m, Sep. 1898, *J. S. Gamble 27310* (K); Kanasar-Konain Road, Jaunsar, 2400 m, 18 Sep. 1936, *M. B. Raizada 7151* (E); **Deoban**, Jansar, 2700 m, Aug. 1890, *K. Nand 184* (E); Jawala Mukhi, Purti, 2700 m, 18 Aug. 1879, *Rithies 465* (K); **Kanaur**, Jaunsar, 1950 m, 17 Sep. 1936, *M. B. Raizada 7184* (E); **Kulu**, Banjar–Jibhi, 27 Jul. 1888, *J. R. Drummond 23271* (K); Manali, 2300 m, Aug.–Sep. 1989, *C. A. Chadwell 347*(A); Aug. 1864, *D. Brandis s. n.* (CAL); Oct. 1877, *C. B. Clarke s. n.* (LE); Kulu Division, 10 Oct. 1936, *N. Singh 2* (KEP); **Kumaon**, Muktesar, 5 Sep. 1938, *R. L. Badhwar 5050* (RRLH); **Lahul & Spiti**, Khoksar, 3 Jul. 1938, *N. L. Bor s. n.* (CAL); 2400 m, 13 Aug. 1941, *N. L. Bor 15325* (K); 2700 m, 2 Aug. 1941, *N. L. Bor 15589* (K); Manali, 1950 m, 8 Aug. 1891, *N. L. Bor 15306* (K); **Mussorie**, Toward Kempti Falls, 1500–2100 m, 13 Aug. 1944, *R. R. Stewart 21129* (K); s. loc., *J. Marten s. n.*(BM); s. loc., *Roban 34* (CAL); **Nainital**, Kumaon, 2100 m, *R. Strachey*

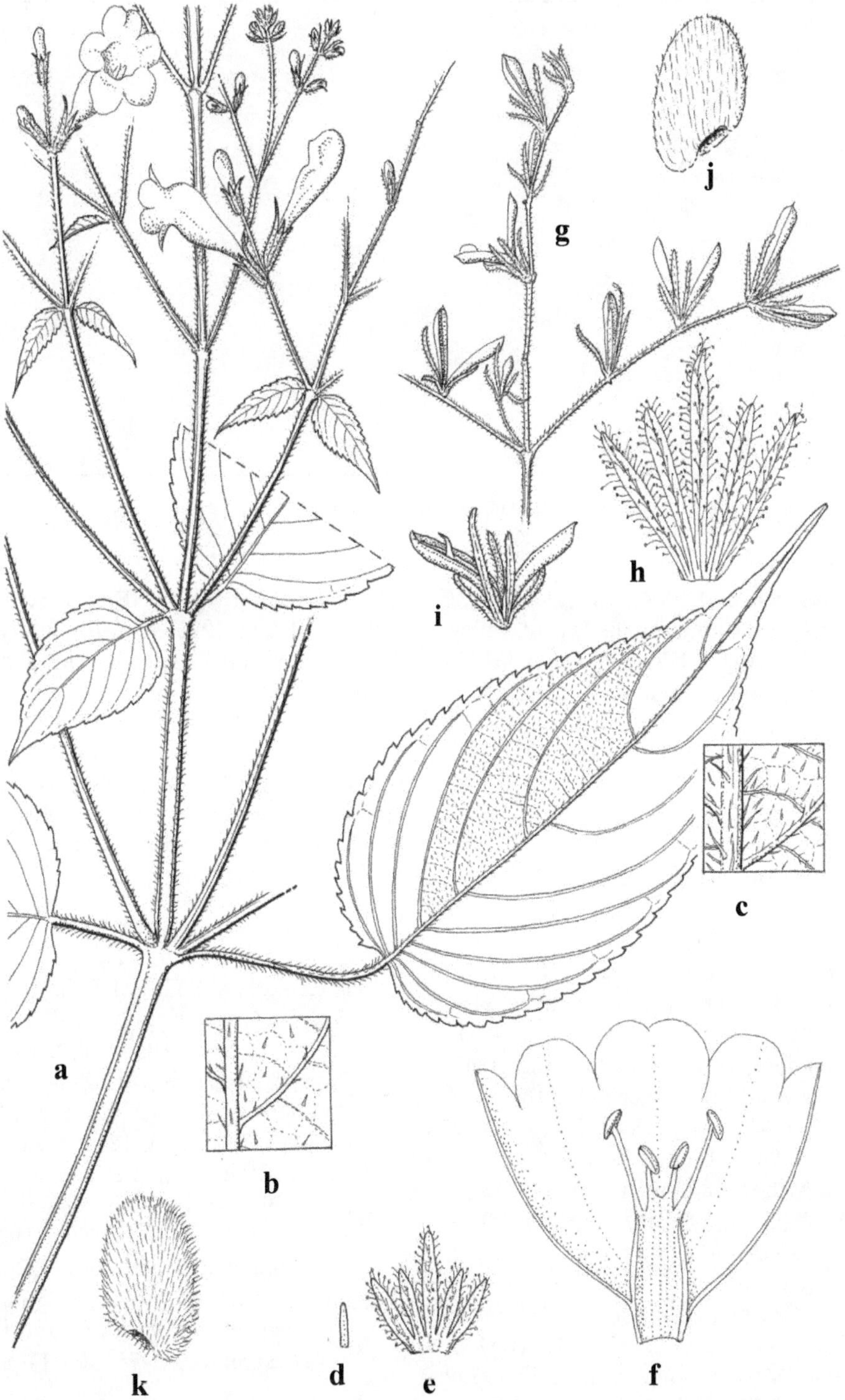

Figure 46-81: **Strobilanthes attenuata** (Nees) Nees: **a**, habit; **b**, upper surface of leaf; **c**, lower surface of leaf; **d**, bracteole; **e**, calyx at anthesis; **f**, corolla opened out to show stamens; **g**, fruiting inflorescence; **h**, calyx in fruit3; **i**, capsule and calyx; **j**, seed when dry; **k**, seed when wetted. Drawn by Rosemary WISE; reproduced from Wood, *Curtis's Bot. Mag.* 31: 154–167. 2014.

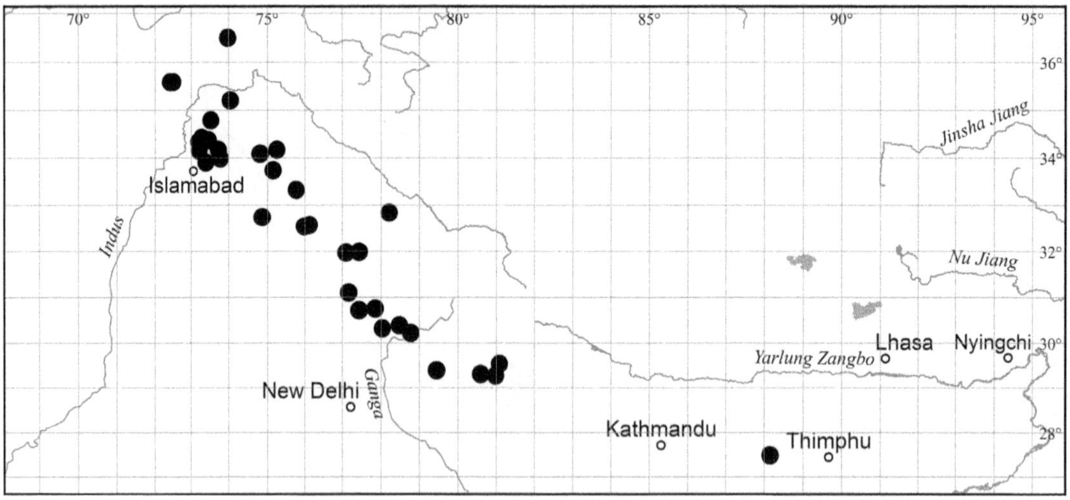

Map 46-135: **Strobilanthes attenuata** (Nees) Nees

& *J. E. Winterbottom 5* (BM); Oct. 1949, *N. N. Sen s. n.* (SING); **Pithoragarh**, Dhauli valley, 2700–3000 m, 7 Aug. 1887, *J. F. Duthie 5881a* (K); near Sosa, 14 Jul. 1886, *J. F. Duthie 5881* (CAL); **Shimla**, Bashahr, Baturang, 19 Aug. 1891, *J. H. Lace 354* (CAL, E); Bashahr, 1884, *Lobha Rain ex herb. J. R. Drummond 1887* (E, K); near Rampur, 1350 m, 27 Aug. 1919, *R. R. Sterwat 5115* (K); Naskanda, 2850 m, 6 Aug. 1964, *R. A. Geesteranus 14052* (L); 1884, *J. R. Drummond 1888* (K); Mahasa, 2400 m, Aug. 1878, *J. S. Gamble 6410* (K); Jakko, 15 Sep. 1915, *H. H. Rich 11* (K); The Glea, 4 Aug. 1917, *H.H. Rich 675* (K); Simla to Mynee Valley, 2100 m, *Madden 367*(E); 2120 m, Jul.–Aug. 1847, *L. R. Maclagan 500* (E); 2120 m, Jul.–Aug. 1847, *L. R. Maclagan 502* (E); Theog, 1950 m, *J. S. Gamble 658* (CAL); Aug. 1849, *Fleming s. n.* (E); Mahasa, 2400 m, 11 Aug. 1877, *J. S. Gamble 689* (CAL); Meflea, 1800 m, 1 Sep. 1878, *J. S. Gamble 6478* (CAL); Makasu woods, 2400 m, Oct. 1885, *H. Collett 865* (K); Shabi, 2100 m, *H. Collett s. n.* (K); 2400 m, Oct. 1907, *A. Meebold 5077* (CAL); Saharan, Sulej Valley, 2100 m, 11 Aug. 1939, *F. Ludlow & G. Sherriff 7531* (BM, E, TI); The Glar, 1800 m, 1 Aug. 1888, *G. Watt 9399* (E); **Srinagar to Kashwir**, Dachigam, 1950 m, 13 Oct. 1970, *V. L. Shrestha 166* (KATH); **Tehri Garhwal**, 2400 m, Aug. 1898, *B. B. Osmarta 27178* (CAL, K); Ganda Valley, 2100–2400 m, *D. J. Duthie 1887* (BM). **W NEPAL: Seti**, Bajhang, 1500 m, 18 Aug. 1972, *M. S. Bista & D. P. Joshi 564* (KATH); Doti, Salimgado–Choyakhatneodar, 2850 m, 2 Aug. 1972, *M. S. Bista & D. J. Joshi 195* (KATH); **Mahakali**, Dadeldhura, Niglad, 1700 m, 1 Sep. 1982, *L. P. Kattel 998* (KATH). **SIKKIM & DARJILING**: **Sikkim**, 2100–3000 m, *J. D. Hooker s. n.* (A); Sikkim border, Singilala Ridge, near Phalat, 22 Aug. 1991, *Kew Accession Number 324.85.03716* (K).

86. Strobilanthes extensa (Nees) Nees in de Candolle (ed.), Prodr. 11: 191. 1847; Clarke in Hooker (ed.), Fl. Brit. India 4: 463. 1884; Wood in Grierson & Long (eds.), Fl. Bhutan 2(3): 1270. 2001; Hu, Deng & Wood in Wu, Raven & Hong (eds.), Fl. China 19: 413. 2011. ≡ *Goldfussia extensa* Nees in Wallich (ed.), Pl. Asiat. Rar. 3: 88. 1832. ≡ *Pteracanthus extensus* (Nees) Bremek., Verh. Kon. Ned. Akad. Wetensch., Afd. Natuurk., Sect. 2 41(1): 199. 1944; Tsui & Hu in Hu (ed.), Fl. Reipubl. Popularis Sin. 70: 137. 2002. Type: India, "montibus provinciae Sillet", 1830, *W. Gomez* in *Wallich Cat. 7160* [lectotype **here designated**: K-W (K001126805)].

Strobilanthes claviculata C. B. Clarke ex W. W. Sm., Notes Roy. Bot. Gard. Edinburgh 10: 191. 1918, "*claviculatus*"; Wood in Grierson & Long (eds.), Fl. Bhutan 2(3):

1270. 2001. ≡ *Pteracanthus claviculatus* (C. B. Clarke ex W. W. Sm.) C. Y. Wu [Ind. Yunnan. Fl. 2: 1682, 1984, nom. invalid., Art. 41.5] ex H. P. Tsui in Wang (ed.), Vasc. Pl. Hengduan Moumt. 2: 1874. 1994; Tsui & Hu in Hu (ed.), Fl. Reipubl. Popularis Sin. 70: 132. 2002. Type: China, Yunnan, Mi-lê, in mountain wood, *A. Henry 10319* [holotype: K (K000882729); isotype: NY (NY278302)].

Strobilanthes duclouxii C. B. Clarke ex Benoist, Bull. Mus. Hist. Nat. Paris 28: 96. 1922. ≡ *Pteracanthus duclouxii* (C. B. Clarke ex Benoist) C. Y. Wu & C. C. Hu in Hu (ed.), Fl. Reipblic. Popularis Sin. 70: 135. 2002; Hu & Tsui in Wu, Fl. Yunnan 16: 695. 2006. Type: China, Yunnan, environs de Lan ngy tsin, 9 Aug. 1904, *P. Py in herh. Ducloux 2603* [lectotype **here designated**: P (P00719313); isolectotypes: P (P00719314, P00719315)].

Subshrubs, 0.5–2 m high, anisophyllous; stems glabrous or pilose. Leaves unequal in each pair, sessile or petiolate; petiole (0–)5–10 mm long, narrowly winged, sparsely pubescent; blade ovate, lanceolate, or oblong-lanceolate, 4–15 cm long, 1–5 cm broad, abaxially subglabrous to pilose, adaxially glabrous or sparsely pilose, lateral veins 5 or 6 pairs, base cuneate, margin remotely serrate, apex acuminate and falcate. Spikes axillary or terminal, 4–8 cm long, simple, 2–6-flowered; peduncle and rachis densely hirsute; flowers usually paired, 0.5– 4 cm apart on rachis. Bracts variable in shape and size, lowermost ones ovate to lanceolate, 2.5–6 cm long, 1–2.5 cm broad, glabrous or pubescent, base cuneate to cordate, upper ones ovate to linear, 3–6 mm long, 1.5–4 mm broad, glandular-pilose; bracteoles oblong to elliptic, 2–4 mm long, 1–3 mm broad, glandular-pilose. Calyx 1–2 cm long, strongly accrescent in fruit, glandular-pilose, 5-lobed almost to base; lobes unequal, one 2–3 mm longer than others, oblong-oblanceolate, apex spatulate and darker green. Corolla pale whitish purple to violet, 3.2–4 cm long, ventricose, outside pubescent only on lobes, inside glabrous except for trichomes retaining style; tube base cylindric, 1–1.2 cm long, 1.8–2 mm broad, upwards gradually widened to 1.2–1.5 cm broad and bent at mouth; lobes orbicular, 4–5 mm in diameter. Stamens 4, included; filaments glabrous, longer pair 7–8 mm long, shorter pair 3–4 mm long; anther-thecae oblong, 2–2.5 mm long, 1–1.2 mm broad, erect. Ovary sparsely glandular-pilose; style 3–3.2 cm long, sparsely pilose; 2 ovules per locule. Capsules clavate, 1.0–1.6 cm long, 2–2.5 mm broad, sparsely glandular-pilose towards apex, 4-seeded. Seeds ovate in outline, compressed, 2–3 mm long, 2–2.5 mm broad, pubescent; areole small. Figure 46-82.

Phenology: flowering from August to October; fruiting from October to December.

Habitats: mixed forests; 900–3400 m.

Distribution: U Yarlung Zangbo, C Nepal, E Nepal, Sikkim & Darjiling, Bhutan, Yarlung Zangbo-Brahmaputra, S Hengduan, U Irrawaddy (Map 46-136); also in other parts of India and China (C & NE Yunnan, S Sichuan).

U YARLUNG ZANGBO: **Nylam**, near Zhangmu, 2500 m, *Y. T. Zhang & K. Y. Lang 3307* (PE); **Gyirong**, Gyirong, 2895 m, 7 Aug. 2010, *PE Xizang Exped. 295* (PE); Langjiu to Chongse, 2700 m, 11 Nov. 2011, *Y. S. Chen et al. 371* (PE). **C NEPAL: Bagmati**, Katmandu, Thankot, 1950 m, *K. B. Codrington 223* (BM); Lalitpur, Kitini, 1500 m, 6 Nov. 2001, *T. Watanabe et al. NPL-SP011213(033)* (MBK); Lalitpur, Godavari, 10 miles SE of Kathmandu, 1 Sep. 1966, *D. H. Nicolson 2189* (KATH, LE); Lalitpur, Educational Garden, 1515 m, 16 Aug. 2002, *N. Joshi & B. Silwal 400* (KATH); Rasuwa, Syaphru, 2250 m, 11 Jul. 1992, *H. Takayama et al. 9239103* (E); **Dhawalagiri**, Myagdi, above Chitre, 2570 m, 3 Oct. 2006, *C. A. Pendry et al. 80* (E); Gandaki, Gorkha, Lho, 3340 m, 2 Aug. 2008, *s. coll. 20818099* (E). **E NEPAL: Koshi**, Solukhumbu, Dudh Kosi Valley, ca. 1 km W of Lukla, 2750 m, 1 Oct. 1974, *J. H. de Haas 2885* (BM, L); Solukhumbu, Mopung to Thosa Kharka, 21 Aug. 1985, *H. Ohba et al. 8571891* (KYO); **Mechi**, Taplejung, between Wallunchun Gola and Pembu Tsoki, 2600 m, 19 Aug. 1962, *K. Nishioka 641* (KYO). **SIKKIM & DARJILING: Sikkim**, *Falconer 734* (CAL, LE). **BHUTAN**: **Chukka**, south side of Chapcha

Figure 46-82: **Strobilanthes extensa** (Nees) Nees: **a**, portion of plant; **b**, lower bract; **c**, bract from top of inflorescence; **d**, calyx; **e**, corolla. Drawn by Louise OLLEY; reproduced from Wood, *Fl. Bhutan* 2(3): 1269, fig. 104: g–k. 2001.

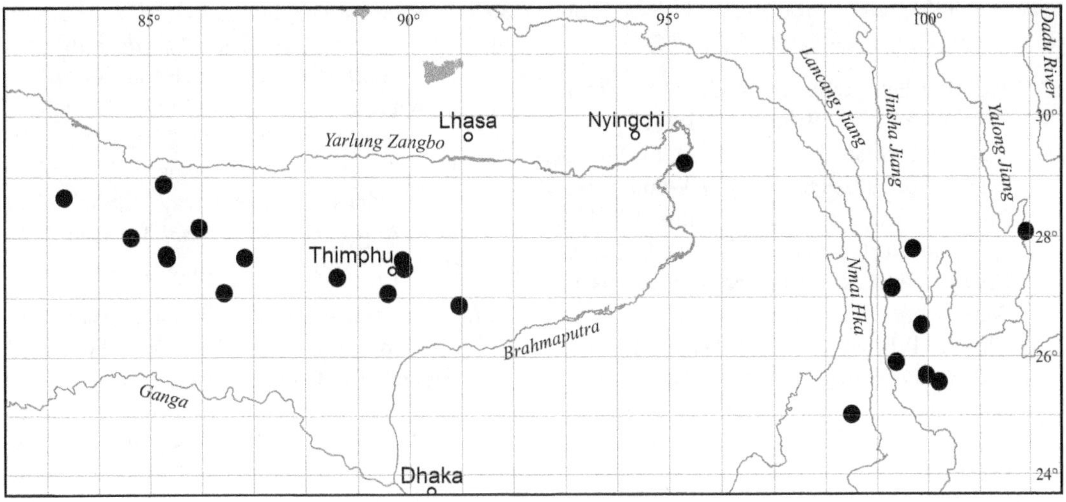

Map 46-136: **Strobilanthes extensa** (Nees) Nees

Gorge, 2300 m, 20 Sep. 1992, *J. R. I. Wood 7575* (FHO); **Punakha**, Tilagong Punakha, 2100 m, 18 Sep. 1914, *R. E. Cooper 3338* (BM, E); between Ranchu and Nobding, 2400 m, 18 Oct. 1991, *J. R. I. Wood 7470* (E, FHO); **Thimphu**, 2400 m, 15 Jul. 1914, *R. E. Cooper 1542* (BM, E); Dotwa, Valley Ginekah Village, 2500 m, 18 Sep. 1990, *J. R. I. Wood 7341* (E); Gidakom Valley, 900 m, 24 Aug. 1991, *J. R. I. Wood 7439* (E); Gidakom Valley, 2400 m, 14 Sep. 1991, *J. R. I. Wood 7451* (FHO, TI); around Ginekah Village, 2600–2900 m, 19 Sep. 1991, *J. R. I. Wood 7454* (E, FHO); N Chapcha, 2500 m, 28 Sep. 1991, *J. R. I. Wood 7461* (FHO); around Dorji Drah Gompha, 2900 m, 16 Aug. 1992, *J. R. I. Wood 7566* (E, FHO); **Tongsa**, Dorji Gompah, 2600 m, 10 Oct. 1990, *J. R. I. Wood 7347* (E); between Chendebji and Tongsa, 2200 m, 9 Oct. 1990, *J. R. I. Wood 7345* (E). **YARLUNG ZANGBO-BRAHMAPUTRA**: **Mêdog**, Jialasa, Gudeng, Wenquan, 1850 m, *B. S. Li & S. Z. Cheng 2156* (PE). **S HENGDUAN**: **Dali**, Cangshan, 20 Jul. 2006, *Y. F. Deng 19100* (IBSC); Mount Tsang, Sep. 1914, C. Schneider *2864* (A); Tsang (Cang) Shan, Shu Ching Qan valley, NW Huadianba, 2885 m, 17 Oct. 1990, *Chungtien-Lijanng-Dali Expedition (CLD-90) 1460* (E, K); Tchangchan, 23 Aug. 1905, *F. Ducloux 148* (E); Cangshan, Xiyangfeng, 2550 m, 14 Oct. 1965, *W. L. Shi 652126* (YIM), Tali, Pe Yun Mt., *McLarsen 148B* (BM, E, K, P); Qingbixi, 2050–2200 m, 21 Jul. 1988, *Sino-Germ. Exped. 88-016* (KUN); Xiaoguan to Dacang, 2400–2500 m, 9 Jul. 1984, *1984 Sino-Amer. Bot. Exped. 909* (A, CAS, KUN); **Jianchuan**, Madeng, Maliqing, 2410 m, 15 Aug. 2010, *E. D. Liu et al. 2720* (KUN); **Meigu**, Houbuliutuo, Yisuojue, Gaikelamu, 20 Aug. 1959, *Chuan Jiang Liang 1814* (KUN, PE); **Muli**, Eya, Sanjiangkou, 2300–2400 m, 1 Oct. 1983, *Qinghai-Xizang Exped. 14898* (KUN, PE); **Shangri-la** (Zhongdian), Wu-tso, near Longhuangmiao, bank of Jinshajiang, 2200 m, 9 Nov. 1939, *K. M. Feng 3265* (GH, KUN, PE); Sanba, Haba, ca. 125 km at East Ring, 2853 m, 13 Aug. 2010, *E. D. Liu et al. 2713* (KUN); Sanba, Jiangbian, Mianshaba, 2850 m, 28 Aug. 2010, *Kham Expedition 10-2981* (PE); **Weixi**, Wei-si Hsien, 2800 m, Aug.–Sep. 1935, *C. W. Wang 67606* (A, KUN, NAS); **Yangbi**, Weihsi-Djientschwan,Yenaping, 2850–3200 m, 22 Sep. 1916, *H. Handel-Mazzetti 10057* (WU); Pingpo, 7 Nov. 1946, *T. N. Liou 22609* (IBSC, PE); **Yunlong**, Bi Jiang River valley, Yakou, Yunlong to Lanping, 2955 m, 30 Jul. 2010, *T. Zhang et al. 10CS2138* (K). **U IRRAWADDY**: **Tengchong**, Tengyueh, *E. B. Howell 68* (E, IBSC, K); Jietou, Datang, 2410 m, 14 May 2006, *Gaoligong Shan Biodiversity Survey 30185* (KUN); Mazhang, 25 Oct. 1998, *H. Li* 10922 (CAS, KUN); Dading, 1000 m, 9 Sep. 1960, *Z. Y. Wang & R. W. Zhang*

6006002 (YIM), Jiangnan, 1800 m, 11 Sep. 1960, *R. W. Zhang & Z. Y. Wang 6006172* (YIM).

87. Strobilanthes lachenensis C. B. Clarke in Hooker (ed.), Fl. Brit. India 4: 465. 1884; Wood in Grierson & Long (eds.), Fl. Bhutan 2(3): 1270. 2001; Hu, Deng & Wood in Wu, Raven & Hong (eds.), Fl. China 19: 413. 2011. ≡ *Pteracanthus lachenensis* (C. B. Clarke) Bremek., Verh. Kon. Ned. Akad. Wetensch., Afd. Natuurk., Sect. 2 41(1): 200. 1944. Type: [Sikkim & Darjiling] India, Sikkim, Lachen, Medous, 31 Jul. 1849, *J. D. Hooker s. n.* [lectotype designated by Wood, Edinburgh J. Bot. 51: 245. 1994: K (K000883039)].

Strobilanthes xanthantha Diels, Notes Roy. Bot. Gard. Edinburgh 5: 163. 1912. ≡ *Championella xanthantha* (Diels) Bremek., Verh. Kon. Ned. Akad. Wetensch., Afd. Natuurk., Sect. 2 41(1): 151. 1944; Hu, Fl. Reipubl. Popularis Sin. 70: 96. 2002. Type: [S Hengduan] China, Yunnan, Tali Range, 25°40′N, 8,000-9,000 ft., Jul.–Aug. 1906, *G. Forrest 4550* [lectotype **here designated**: E (E00133387); isotypes: BM (BM000906296), E (E00133388)].

Strobilanthes leucotricha Benoist, Bull. Mus. Hist. Nat., Paris 28: 96. 1922; Tsui & Hu in Hu (ed.), Fl. Reipubl. Popularis Sin. 70: 137. 2002. Type: China, Yunnan, Torrents sous bois à Tien Sin, 2800 m, *E. Maire s. n.* [holotype: P (P00218436)].

Strobilanthes nemorosa Benoist, Bull. Mus. Natl. Hist. Nat. 28: 97. 1922; Bremekamp, Verh. Kon. Ned. Akad. Wetensch., Afd. Natuurk., Sect. 2 41(1): 281.1944; Hu, Deng & Wood in Wu, Raven & Hong (eds.), Fl. China 19: 413. 2011. ≡ *Pteracanthus nemorosus* (Benoist) C. Y. Wu & C. C. Hu in Hu (ed.), Fl. Reipubl. Popularis Sin. 70: 144. 2002. Type: China, Chongqing, "Su-tchuenoriental", Tchen-keou-tin, 10 Aug. 1897, *R. P. Farges 1408* [holotype: P (P00719381); isotypes: BM (BM000884929), CAS (CAS0005644, CAS0033156, CAS0033157), E (E00133598, E00133599), GH (GH00094308, GH00094309), K, NY (NY00278311), P (P00719383, P00719384, P00719386), PE].

Strobilanthes pandurata Hand.-Mazz., Symb. Sin. 7: 893. 1936, "*panduratus*". ≡ *Pteracanthus pandurata* (Hand.-Mazz.) C. Y. Wu & C. C. Hu in Hu (ed.), Fl. Reipubl. Popularis Sin. 70: 146. 2002. Type: [S Hengduan] China, Yunnan, Yungning, "inter vicos Hoso et Hwayi", 2900–3200 m, 7 Aug. 1915, *H. Handel-Mazzetti 7518* [holotype: WU (WU0024088)].

Diflugossa muliensis H. P. Tsui, Acta Bot. Yunnan. 12(3): 275. 1990. Type: [S Hengduan] China, Sichuan, Muli xian, Eya, Sanjiangkou, 2300–2400 m, 1 Oct. 1983, *Exped. Qinghai-Xizang 14898* [holotype: PE (PE01545539); isotypes: PE (PE01545540), KUN (KUN427915, KUN1255287)].

Strobilanthes attenuata subsp. *nepalensis* J. R. I. Wood, Edinburgh J. Bot. 51(2): 232.1994. Type: [C] Nepal, Dhaulagiri, Myagdi, near Gurjakhani, 28°36′N, 83°13′E, 11 Sep. 1954, *J. D. A. Stainton, W. R. Sykes & L. H. J. Williams 4353* [holotype: BM (BM000906302); isotype: E].

Herbs, 30–100 cm high, isophyllous; stems quadrangular, sulcate, pubescent, then glabrescent. Leaves equal in each pair, sessile or petiolate; petiole (0–)0.5–3.5 cm long, sparsely pubescent to glabrescent; blade ovate, obovate, lanceolate to oblong-elliptic, or broadly elliptic, 3–20 cm long, 1.5–6.5 cm broad, both surfaces sparsely white pubescent especially along veins, lateral veins 5–12 pairs, base cuneate and often decurrent onto petiole, occasionally pandurate, margin serrate to crenate, apex acute to shortly acuminate. Spikes axillary or terminal, 3–10 cm long, usually several in panicles; rachis glandular-pilose; flowers basally to 2.5 cm apart on rachis but apically congested. Bracts variable in size and shape, lower ones ovate, 2–5 cm long, 0.5–2 cm broad, persistent, margin usually toothed, upper ones oblong to sometimes elliptic, 5–12 mm long, 1–2 mm broad; bracteoles linear, 4–5 mm long, 1–1.2 mm broad, white glandular-pilose. Calyx 0.8–1.6 cm long, accrescent in fruit, glandular-pilose, 5-lobed almost to base; lobes unequal, one longer than others, linear or linear-spatulate, 7–12 mm long, 0.8–1 mm broad.

Corolla blue, 3–3.6 cm long, curved, outside pubescent on lobes, inside glabrous except for trichomes retaining style; tube base cylindric, 7–8 mm long, 1.8–2 mm broad, slightly bent at mouth and gradually widened to 1–1.2 cm broad at mouth; lobes ovate, 4–4.5 mm long and broad. Stamens 4, included; filaments of longer pair 7–8 mm long, sparsely; pilose shorter pair 2–3 mm long, glabrous; anther-thecae oblong, 2–2.2 mm long, 1–1.3 mm broad, erect. Ovary comose; 2 ovules per locule; style 2.7–3.2 cm long, sparsely glandular-pilose. Capsules clavate, 1.1–1.5 cm long, glandular-pilose towards apex, 4-seeded. Seeds ovate in outline, compressed, 2–3 mm broad, 1.5–2 mm broad, densely appressed pubescent. Figure 46-83.

Phenology: flowering from June to October.

Habitats: stream banks, forests, thickets; 900–4000 m.

Distribution: U Ganga & Indus, U Yarlung Zangbo, W Nepal, C Nepal, E Nepal, Sikkim & Darjiling, Bhutan, M Yarlung Zangbo, Yarlung Zangbo-Brahmaputra, S Hengduan (Map 46-137); also in other parts of India and China (C Yunnan, S Sichuan).

U GANGA & INDUS: Dachegam, 1950 m, 13 Oct. 1970, *V. L. Shrestha 170* (KATH); **Kulu**, Kullu Valley, Malana Valley, 16 Sep. 1962, *J. Tanaka 7* (KYO). **U YARLUNG ZANGBO**: **Dinggyê**, Chentang, Nadang, Chentanggou, 2470 m, 12 Jul. 2013, *PE-Xizang Expedition 3046* (PE); Chentang, Nadang, Chentanggou, 2470 m, 12 Jul. 2013, *PE-Xizang Expedition 3124* (PE); **Gyirong**, Tuodang, 3300m, 3 Jul. 1975, *C. Y. Wu et al. 75-525* (KUN); Magagou, 26 Jun. 1975, *Anonymous 75-356* (PE); Gyirong, Zhacun, 3150 m, *Exped. Qinghai-Xizang 6012* (PE); Gyirong, Tuocun, 3150m, 19 Jul. 1975, *Qinghai-Xizang Exped. 7012* (KUN, PE); near Rucheng, 3100 m, 15 Jul. 1975, *Exped. Qinghai-Xizang 6362* (KUN, PE); Zhaxianchongse to Jiangcun, 2600–2100 m, *B. S. Li & H. Li 13169* (PE); Rongqu, 2900–3100 m, *B. S. Li et al. 13629* (PE); Gyirong, 3100 m, 9 Jun. 1972, *Xizang Chinese Herd Exped. 194* (PE); **Nyalam**, Zhangmu to Kangbajiba, 2400 m, 22 Jun. 1966, *Y. T. Zhang & K. Y. Lang 4350a* (PE herb. no. 932698); **Tingri**, Rongshar Valley, 3355 m, 21 Jul. 1924, *Hingston 309* (K). **W NEPAL**: Thongnada, 2270 m, 7 Jul. 1981, *P. R. Shakya et al. 6341* (KATH); **Bheri**, Dailekh, Ranimatta, 31 Jul. 1991, *M. Suzuki 9160125* (E); Dailekh, Ranimatta, 2115 m, 31 Jul. 1991, *M. Szuzuki et al. 9193062* (E); Dailekh, Jajarkot, Samaila, 2100 m, 1 Aug. 1981, *N. P. Manandhar 6434* (KATH); **Karnali**, Dolpa, near Balangoa Pass, 3700 m, 19 Jul. 1952, *O. V. Polunin, W. R. Sykes & L. H. J. Williams 2510* (BM, KATH); Dolpa, Haharigaon to Chote, 2850 m, 20 Aug. 1966, *T. B. Shrestha 5184* (KATH); Jumla, Ghurchi Lekh between Lumra and Muma, 11 Aug. 1952, *O. V. Polunin, W. R. Sykes & L. H. J. Williams 5126* (BM, KATH); Mugu, Chuchi Lekh, betweeen Lumsa and Murma, 3000 m, 11 Aug. 1952, *O. Polunin, W. R. Sykes & L. H. J. Williams 5126* (KATH); **Mahakali**, Darchula, between Naali and Sithi, Ghusa, 2300 m, 16 Oct. 2014, *S. K. Ghimire et al. ANCA 603* (TUCH); **Seti**, Bajhang, Lokundo, 8 Aug. 1976, *H. Tabata et al. 2279* (KYO); Bajhang, 2481 m, 11 Jul. 2009, *H. Ikeda et al. 20913095* (TI). **C NEPAL**: **Bagmati**, Kathmandu, Phulchoki, 2200 m, 9 Aug. 1969, *H. Kanai 673380* (KATH, KYO, TI); Kathmandu, Phulchoki, 1500–2200 m, 6 Sep. 1972, *H. Kanai 722166* (KATH, TI); Kathmandu, Thankot, 1670 m, Aug. 1956, *K. D. B. Codrington 247* (BM); Kathmandu, Thankot, 1980 m, 1956, *K. D. B. Codrington 272* (BM, TI); Kathmandu, Phulchoke, SW of Kathmandu, 2340m, 27 May 1965, *A. D. Schilling 605* (K, TI); Kathmandu, Phulchoki, 2400–2700 m, 19 Sep. 1970, *H. Kanai & C. Chuma 674781* (KATH); Lalitpur, Godawari to Phuckoki, 1540–2760 m, 29 Jul. 2000, *T. Watanabe et al. NPL-SP00079(005)* (MBK); Lalitpur, Phulchoki, near Falamkhani, 2320 m, 16 Sep. 2014, *B. Adhikari & G. D. Bhatt 60* (E); Lalitpur, Lele, 1 Jan. 1991, *I. Shrestha & N. Pradhan 227* (KATH); Lalitpur, Godawari, 1500 m, 5 Nov. 1969, *N. P. Manandhar et al. 12959* (KATH, TI); Lalitpur, Godavari, 1600 m, 24 Nov. 1974, *Indira et al. 74/32* (KATH); Lalitpur, Royal Botanical Garden, 21 Sep. 1974, *S. Bhahrai & M. Gorkhali 229* (KATH); Lalitpur, Phulchoki, Phulchoki hills, 2700 m, 5 Dec. 1986, *L. P. Kattel 101* (KATH); Rasuwa, Dunche to Tricali Khola to Singum Gompah, 2600 m, 22

Figure 46-83: **Strobilanthes lachenensis** C. B. Clarke: **a**, flowering branch; **b**, calyx; **c**, opened corolla showing stamens. Drawn by Xi-Chou LI; reproduced from Wu, *Fl. Xizang.* 4: 414, fig. 179. 1985.

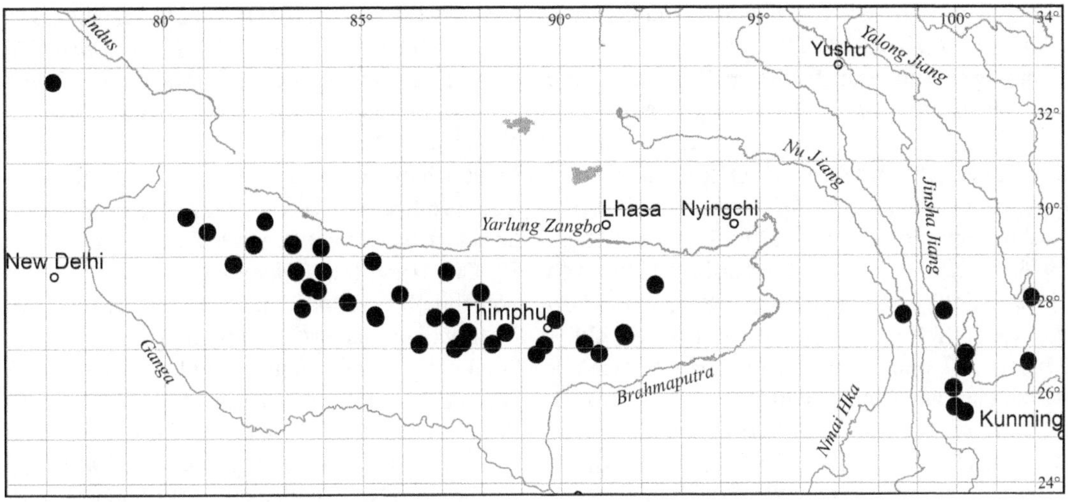

Map 46-137: **Strobilanthes lachenensis** C. B. Clarke

Aug. 1972, *H. Kanai et al. 751805* (BM, KUN, KYO, TI); Rasuwa, Langtang, 3490 m, 22 Jun. 1949, *O. V. Polunin 522* (BM); Rasuwa, near Langtang River, 3350 m, 25 June 1949, *O. V. Polunin 587* (BM); Rasuwa, Langtang, 3350 m, Jun. 1949, *O. V. Polunin 602* (BM); Rasuwa, Lende Khola, 2450–2900 m, 9–11 Jul. 1949, *O. V. Polunin 869* (BM, KATH, TI); Rasuwa, Langtang, Langtang forests area, 2880 m, 1 Aug. 1949, *O. V. Polunin 1640* (BM); Rasuwa, Lende Khola, 2400–2850 m, 11 Jul. 1949, *O. V. Polunin 851* (BM); Rasuwa, Langtang, Ghosa Tabela, 3000–3600 m, 26 Jul. 1976, *Durham University Expedition D2* (BM); Rasuwa, Pati Kharka, 2800 m, 5 Aug. 1994, *F. Miyamoto et al. 9420187* (E, KATH, TI, TNS); Rasuwa, Pati Kharka to Kharka, 3200 m, 5 Aug. 1994, *F. Miyamoto et al. 9440093* (CBM, E, KATH, TI, TNS); Rasuwa, Ghoratabala, 3000 m, 14 Jul. 1984, *H. K. Saiju & B. Roy 55* (KATH); **Dhawalagiri**, Myagdi, Ghorepani to Tatopani, 2550m, 17 Sep. 1995, *M. Mikage et al. 9552216* (A, BM, E, L, MO, TI, TNS); Myagdi, Boghara to Dobang, 1920 m, 7 Sep. 1996, *M. Mikage et al. 9687189* (TI); Myagdi, Ghorapani to Tatopani, 2420 m, 27 Jul. 1996, *T. Hoshino et al. 9666045* (CBM, TI); Myagdi, near Gurjakhani, 2660 m, 21 Jul. 1954, *J. D. A. Stainton, W. R. Sykes & L. H. J. Williams 3557* (BM, KATH); Myagdi, near Gurjakhani, 2550 m, 21 Jul. 1954, *J. D. A. Stainton, W. R. Sykes & L. H. J. Williams 3561* (BM, KATH); Myagdi, near Gurjakhani, 2440m, 11 Sep. 1954, *J. D. A. Stainton, W. R. Sykes & L. H. J. Williams 4353* (BM, E, KATH); Mustang, Ghasa, S of Tukuchu, Kali Gandaki, 3050 m, 4 Jul. 1954, *J. D. A. Stainton, W. R. Sykes & L. H. J. Williams 1523* (BM, KATH); Mustang, Ghasa, S Tukuchu, Kali Gandaki, 3200 m, 4 Jul. 1954, *J. D. A. Stainton, W. R. Sykes & L. H. J. Williams 1603* (BM, KATH); Mustang, Larjung, S Tukuchu, 15 Sep. 1954, *J. D. A. Stainton, W. R. Sykes & L. H. J. Williams 7852* (BM); Mustang, Larjung, S Tukuchu, 3000 m, 17 Oct. 1954, *J. D. A. Stainton, W. R. Sykes & L. H. J. Williams 8179* (BM); Mustang, Dhunche to Triculi Khola to Singum, 2600 m, 22 Aug. 1972, *H. Kanai et al. 721805* (BM, KATH); Mustang, Tukuche, 3030 m, 17 Jul. 1982, *J. D. A. Stainton 8563* (BM); Mustang, Banthnti to Deurali. 2660 m, 24 Aug. 1988, *M. Suzuki 8860646* (KATH); Mustang, Dhampus to Kokethanti, 2480 m, 29 Aug. 1988, *M. Suzuki 8860811* (KATH); Mustang, Ghasa, 2260 m, 16 Aug. 1999, *M. Mikage et al. 9962920* (KATH, TI); Mustang, Ghasa to Lefa, 2100 m, 22 Jul. 1979, *P. R. Shakya et al. 5074* (KATH); Mustang, Lete, 2400 m, 24 Jul. 1974, *D. P. Joshi & T. K. Bhattacharya 74/1969* (KATH); Parbat, Ulleri, Mayathanti, 2500 m, 21 Jul. 1974, *D. P. Joshi & T. K. Bhattacharya 74/1832* (KATII); **Gandaki**, Gorkha, en route Dovan to Hinko, 2700–3500 m, 28 Sep. 1976, *Y. Suehiro 1116* (KYO); Gorkha, Thorang Khola, 2300 m, 4 Aug. 1994, *M. Suzuki et al. 9485270* (E); Gorkha,

Namrung to Lho, near Barchham, 2610 m, 1 Aug. 2008, *H. Ikeda et al. 20818089* (E); Gorkha, Namrung to Lho, near Barchham, 2610 m, 1 Aug. 2008, *H. Ikeda et al. 20818092* (E); Gorkha, Lho to Shyala, near Lho, 3170 m, 2 Aug. 2008, *H. Ikeda et al. 20818096* (E, KATH); Gorkha, 3550 m, 2 Aug. 2008, *H. Ikeda et al. 20818103* (E); Kaski, Chhumro to Dovan, 2200–2700 m, 27 Sep. 1976, *Y. Suehiro 974* (KYO); Kaski, Chhumro to Dovan, 2200–2700 m, 27 Sep. 1976, *Y. Suehiro 985* (KYO); Kaski, Chhumro to Dovan, 2200–2700 m, 27 Sep. 1976, *Y. Suehiro 989* (KYO); Kaski, Lamjung Himal, 3030 m, 17 Jul. 1954, *J. D. A. Stainton, W. R. Sykes & L. H. J. Williams 6416* (BM, KATH); Kaski, Chiyanglung, Seti Khola, 2300 m, 11 Sep. 1954, *J. D. A. Stainton, W. R. Sykes & L. H. J. Williams 8553* (BM, KATH); Kaski, Bhainsi Kharka to Banthanti, 2450 m, 23 Aug. 1988, *M. Suzuki et al. 8810940* (KATH); Manang, Bimtang to Gho, near Gho village, 2480 m, 14 Aug. 2008, *H. Ikeda et al. 20818153* (E); Manang, Pisang to Chame, 6 Aug. 1979, *P. R. Shakya et al. 5301* (KATH); **Lumbini**, Palpa, Satyawatital, 2120 m, 12 Oct. 1954, *J. D. A. Stainton, W. R. Sykes & L. H. J. Williams 8898* (BM, KATH, TI). **E NEPAL: Koshi**, Dhankuta, Tute, 2340 m, 14 Jul. 1991, *H. Ohba et al. 9110046* (E); Dhankuta, Basantapur, Bilbatay, Bhanjang-Hatisar, 2200–2700 m, 27 Oct. 1983, *H. Hara et al. 6303695* (BM, K, KYO, TI); Sankhuwasabha, Kauma, 2160 m, 31 Jul. 1988, *M. Suzuki et al. 8821099* (A, BM, KATH, TI); Sankhuwasabha, Chamtang, 2380 m, 20 Aug. 1998, *S. Noshiro et al. 9830078* (A, BM, E, TNS); Sankhuwasabha, Hongaon to Chantang, 7 Aug. 1977, *H. Ohashii 774947* (BM, KYO, TI); Sankhuwasabha, Tinjure Ridge, between Door Pani and Chanki, 2900 m, *Kew, Edinburgh, Kathmandu Expedition KEKE 55* (E, K, KATH); Sankhuwasabha, Hati Sar to Minchin Dhap, 28 Oct. 1963, *H. Hara et al. 603694* (KYO); Terathum, Tinjure Darda, 2750 m, 4 Jul. 1969, *L. H. J. Williams 1118* (BM); Terathum, Tinjure Danda Ridge between Door Pani and Chauki, 2900 m, 29 Aug. 1989, *C. Grey-Wilson et al. 55* (KATH); Terathum, Barantpur, 13 Aug. 1976, *N. P. Manandhar 345* (KATH); **Mechi**, Taplejung, Iladanda to Helok, 13 Nov. 1963, *H. Kanai et al. 6303693* (KATH, TI); Taplejung, Hellok to Iladanda, 8 Nov. 1963, *H. Kanai et al. 6303699* (K, KATH, KYO); Taplejung, Ghunsa to Gola, 3000 m, 28 Jun. 1962, *K. Nishioka 186* (KYO); Taplejung, between Mallunchun Gola and Pembu Tsoki, 3200 m, 19 Aug. 1962, *K. Nishioka 636* (KYO); Taplejung, between Mallunchun Gola and Pembu Tsoki, 2600 m, 19 Aug. 1962, *K. Nishioka 641* (KYO); **Sagarmatha**, Solukhumbu, Namche Bazaar to Dudh Kosi, 2750 m, 17 Oct. 1954, *A. Zimmermann 1743* (G, K); Solukhumbu, 2840 m, 1 Aug. 1997, *M. Wakabayashi et al. 9715043* (CBM, TNS); Solukhumbu, Takisindu to Junbesi, 2850 m, 1 Sep. 1995, *F. Miyamoto et al. 9584321* (CBM, E, TI); Solukhumbu, Namkhli to Gori to Gnaula, 3050 m, 21 Jul. 1995, *F. Miyamoto et al. 9584021* (CBM, E, TI); Solukhumbu, Pike Dongshar to Namikhil, 9 Sep. 1985, *H. Ohba 8572489* (A, BM, KYO, TI); Solukhumbu, Mopung to Thosa Kharka, 21 Aug. 1985, *H. Ohba et al. 8581060* (KATH, TI); Solukhumbu, Toketey, 3940 m, 1930, *L. Dhwoj 594* (BM, E). **SIKKIM & DARJILING**: **Darjiling**, Jakdah, Black V (Hoom), 1560 m, 5 Dec. 1899, *C. G. Rogers 1* (CAL); Rungbee, 1500 m, 31 Oct. 1869, *C. B. Clarke 10311* (BM, K); Rarsiong, 1200 m, 29 Feb. 1876, *C. B. Clarke 27050* (K); 11 Mar. 1876, *C. B. Clarke 27243* (K); **Sikkim**, Surail, *R. H. Beddome 234* (BM); Sitong, 1500 m, Feb. 1882, *J. S. Gamble 10262* (K); *V. Naratanaswami 1363* (CAL); Champoong, 1800 m, 7 Oct. 1875, *C. B. Clarke 25090* (BM, K). **BHUTAN**: **Chukka,** between Gedu and Phunsholing, 1700 m, 12 Oct. 1988, *J. R. I. Wood 6801* (E); Chukka, 1500 m, 29 Sep. 1991, *J. R. I. Wood 7467* (E, FHO); near Awakha, between Chukka and Taktichu, 1700 m, 17 Nov. 1991, *J. R. I. Wood 7511* (E); between Gedu and Takhtichu, 1800 m, 26 Nov. 1991, *J. R. I. Wood 7526* (E, FHO); ridge S of Tala, 1950 m, 22 Feb. 1982, *A. J. C. Grierson & D. G. Long 3141* (E, K, TI); **Gaylegphug** (Sarbhang), between Chabley Khola and Tama, 1880 m, 30 Mar. 1982, *A. J. C. Grierson & D. G. Long 4140* (E, K); **Manas**, way to Goozhang, Kheng County, 1400–2400 m, 1991, *L. Reader s. n.* (E); **Mongar**, above Yadi, 1800 m, 5 Nov. 1991, *J. R. I. Wood 7492* (E, FHO); **Phuntsholing**, between Gedu and Kharbandi, 1700 m, 12 Nov. 1987, *J. R. I. Wood 6094* (E); between Phuntsholing and Gedu, 1600 m, Oct. 1989, *J. R. I. Wood 7215* (E, FHO);

between Sorchen and Gedu, 1500–2000 m, 18 Jan. 1992, *J. R. I. Wood 7528* (E, TI); **Punakha**, W side of Pele La, 3000–3200m, 19 Oct. 1991, *J. R. I. Wood* 7480 (E, K, THIMPHU); **Samdrop Jongkar**, N Deothang, 18 Oct. 1987, *J. R. I. Wood 5976* (E); **Shamgang**, to Mangde Chu, 1600 m, 26 Feb. 1988, *J. R. I. Wood 6175* (E, FHO); **Tashigang**, near Rongthong, between Tashigang and Kanglung, 1800 m, 6 Nov. 1991, *J. R. I. Wood 7494* (E, FHO); **Tongsa**, between Pertimi and Tintibi Bridge Mangde Chu, 980 m, 5 Apr. 1982, *A. J. C. Grierson & D. G. Long 4353* (BM, E); Tashiling to Tongsa, 16 Apr. 1967, *H. Kanai et al. 5178* (BM, TI); near Tongsa, 2200 m, 19 Oct. 1991, *J. R. I. Wood 7476* (E); between Khorila and Shemgang, 1400–1800 m, 15 Mar. 1992, *J. R. I. Wood 7541* (E). **M YARLUNG ZANGBO**: **Dinggyê**, Chentangqu, Kamahe valley, 2500 m, 7 Jun. 1975, *Qinghai-Xizang Exped. 5619* (KUN, PE). **YARLUNG ZANGBO-BRAHMAPUTRA**: **Dirang**, Senge Dzong, 2740 m, Aug. 1938, *Kingdon-Ward 14095* (BM, TI); **Lhünzê**, Zhari to Majia, 2780 m, 7 Aug. 2013, *Y. S. Chen et al. 13-0658* (PE). **S HENGDUAN**: **Dali**, E flank of the Tali Range, 2400–2700 m, *G. Forrest 4550* (BM, E); Changshan, from Shi Jo to Huadianba, 2750 m, *Chungtien, Lijiang & Dali Expedition (1990) 1392* (E); Yinglefeng, *H. C. Wang 1115* (KUN, PE); Zhonghe Peak, Cangshan Mountain, 14 Jul. 2013, *X. H. Jin et al. ST0049* (PE); **Eryuan**, Xiaoluoping, *H. C. Wang 2372* (IBSC, KUN, PE); **Gongshan**, Mekong-Salwin Devide, Londjrela, 3600 m, *T. T. Yü 23158* (A, E, KUN, PE); Cikai, 3350 m, *Gaoligong Shan Biodiversity Survey 32144* (KUN, MO); **Heqing**, Songgui, Lianping, Maershan, 3400 m, *R. C. Ching 24083* (PE); **Meigu**, Wahei to Shuwo, 1800 m, 28 Jul. 1976, *s. coll. 13600* (CDBI, PE); **Muli**, Liziping, 2500 m, 29 Jul. 2006, *Y. F. Deng 18921* (IBSC); Hoso et Hwayi, 2900–3200 m, 1915, *H. Handel-Mazzetti 7518* (WU); Eya, Sanjiangkou, 2300–2400 m, *Exped. Qinghai-Xizang 14898* (PE); **Shangri-la** (Zhongdian), S of the town, 3 Jul. 1994, *Alpine Garden Scociety ACE906* (LIV); **Weixi**, Kari Pass, Mekong-Yangtze divide, 22 Jul. 1923, *F. Kingdon-Ward 5306* (E); **Yanbian**, Zemulong, Zhongcao, 2650 m, 11 Sep. 2002, *X. F. Gao et al. 3347* (CDBI); **Yangbi**, Mubanzuzu, *R. C. Ching 25336* (KUN, PE); Maidi, *H. C. Wang 2387* (IBSC, KUN, PE); **Yongsheng**, Yangping, 31 Oct. 2014, *Y. P. Chen & Q. R. Zhao EM136* (IBSC); **Yulong** (Lijiang), Yulongshan, Dahuoshan, banli, 3000 m, *R. L. Xiong & Y. F. Qi 612680* (KUN); Ludian, to Lidiping, 3104 m, 14 Aug. 2010, *E. D. Liu et al. 2718* (KUN); Yulongshan, Laomeiluo, 3240 m, *Yunnan Univ. 661* (IBSC, SWFC, YUKU).

88. Strobilanthes subnudata C. B. Clarke in Hooker (ed.), Fl. Brit. India 4: 472. 1885; Wood in Grierson & Long (eds.), Fl. Bhutan 2(3): 1271. 2001. ≡ *Pteracanthus subnudatus* (C. B. Clarke) Bremek., Verh. Kon. Ned. Akad. Wetensch., Afd. Natuurk., Sect. 241(1): 200. 1944. Type: [Sikkim & Darjiling] India, Sikkim, Lachong, 16 Aug. 1949, *J. D. Hooker 29* [lectotype designated by Wood, Edinburgh J. Bot. 51: 261. 1994: K (K000883208)].

Subshrubs, 20–50 cm high, anisophyllous; stems quadrangular, sulcate, glabrous, reddish-brown. Leaves very unequal in size; blade ovate, elliptic or subrhomboidal, 1.5–6 cm long, 0.8–2 cm broad, glabrous, base rounded or cuneate and decurrent, apex acute, margin entire or obscurely crenate. Spikes axillary, 3–8 cm long, usually solitary, simple, sometimes 1–2-branched; peduncles 1–1.5 cm long, glabrous; rachis stickly glandular-pubescent. Bracts obovate-suborbicular, 5–9 mm long, 3–5 mm broad, glandular-pubescent, apex rounded; bracteoles linear-spatulate, 4–5 mm long, 1–1.2 mm broad, pilose. Calyx 9–13 mm long, glandular-pubescent; lobes unequal, one longer than others, linear, apex obtuse. Corolla pale blue, 2.8–3.5 cm long, pubescent on lobes; tube base cylindric, 5–6 mm long, upwards bent 90° and abruptly widened to 0.8–1 cm broad at mouth. Ovary pilose; 2 ovules per locule; style 1–2.2 cm long, sparsely pilose. Capsules clavate, 6–7 mm long, glandular-pubescent toward apex, 4-seeded. Seeds ovate in outline, compressed, 2–3 mm long and broad, densely adpressed pilose.

Phenology: flowering from January to April.

Habitats: moist forests; 1700–3700 m.

Distribution: C Nepal, E Nepal, Sikkim & Darjiling (Map 46-138); also in other parts of India.

C NEPAL: **Bagmati**, Rasuwa, Chimang Village, 2700 m, Jan. 1974, *T. B. Shrestha &*

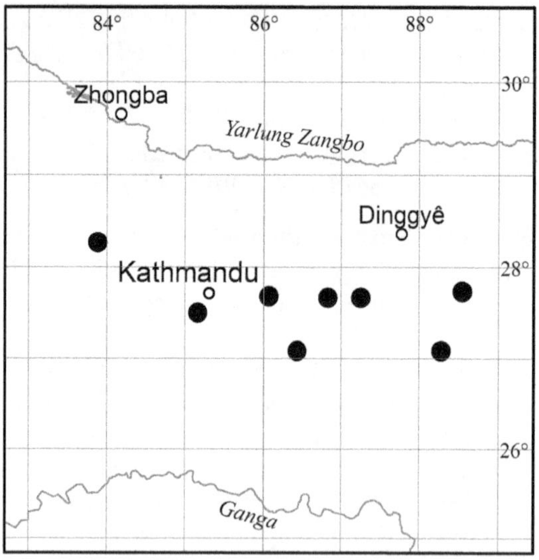

Map 46-138: **Strobilanthes subnudata** C. B. Clarke

M. S. Bista 2162 (KATH); **Gandaki**, Kaski, hill above Lumle, 2100 m, 11 Dec. 1974, *G. Dawson 279* (BM); Kaski, en route from Ghandrung to Chhumro, 2070–2200 m, 25 Sep. 1976, *Y. Suehiro 680* (KYO); Kaski, en route from Ghandrung to Chhumro, 2070–2200 m, 25 Sep. 1976, *Y. Suehiro 707* (KYO); Pokhara, Machhapuchhare Camp Base, 3700 m, 1 Oct. 1976, *Y. Suehiro 2434* (KYO); above Siklis School, 2500 m, 29 Aug. 1976, *R. G. Troth 1004* (BM); Ulleri Pass, 2300 m, 28 Aug. 1967, *Lange* 13 (K); **Janakpur**, Dolakha, Jiri-Those, 1820 m, 23 Sep. 1964, *M. J. Banerji el al. 2885* (BM, KATH); **Narayani**, Makwanpur, Simhanjhyang near Daman on Hetauda to Kathmandu road, 2420 m, 20 Nov. 2004, *T. Miller et al. A236* (E, KATH); Makwanpur, Daman, 2310 m, 18 Aug. 1964, *S. B. Rajbhandary & A. V. Upadhyay 2241* (BM, KATH). **E NEPAL**: **Koshi**, Sankhuwasabha, below Tashigaon, 1900 m, 23 Sep. 1991, *D. G. Long et al. EMAK 179* (BM, E, KATH); **Sagarmatha**, Solukhumbu, Dudh Kosi Valley, 1 km W of Lukla, 2750 m, 1 Oct. 1974, *J. H. de Haas 2885* (BM, KATH, L). **SIKKIM & DARJILING**: **Darjiling**, Patasi below Tanglu, 1830 m, 1 Apr. 1975, *Chamberlain 44* (E); **Sikkim**, Lachung, 2740 m,16 Aug. 1849, *J. D. Hooker 29* (K).

12. HYGROPHILA R. Br.

Brown, Prodr: 479. 1810; Bentham & Hooker, Gen. Pl. 2: 1075. 1876; Clarker in Hooker (ed.), Fl. Brit. India 4: 406. 1884; Nasir & Ali in Malik & Ghafoor (eds.), Fl. Pakistan 188: 183. 1988; Wood, in Grierson & Long (eds.), Fl. Bhutan 2(3): 1251. 2001; Hu, Fl. Reipubl. Popularis Sin. 70: 68. 2002; Hu, Deng & Daniel in Wu, Raven & Hong (eds.), Fl. China 19: 430. 2011. Type: *Hygrophila angustifolia* R. Br. [= *Hygrophila ringens* (L.) R. Br. ex Steud.].

Asteracantha Nees in Wallich, Pl. As. Rar. 3: 75, 81. 1832; Lindau in Engler & Prantl (eds.), Nat. Pflanzenfam. 4(3b): 297. 1895. ≡ *Hygrophila* subg. *Asteracantha* (Nees) C. B. Clarke in Hooker (ed.), Fl. Brit. India 4: 408. 1884. Type: *Asteracantha longifolia* (L.) Nees.

Adenosma Nees in Wallich (ed.), Pl. Asiat. Rar. 3: 75, 79. 1832, nom. illeg., non *Adenosma* R. Br., Prodr. Fl. Nov. Holland. 442. 1810; Nees in de Candolle (ed.), Prodr. 11: 67. 1847. Type **here designated**: *Adenosma biplicata* Nees [≡ *Hygrophila biplicata* (Nees) Sreem.].

Hemiadelphis Nees in Wall., Pl. Asiat. Rar. 3: 75 & 80. 1832. ≡ *Hygrophila* sect. *Hemiadelphis* (Nees) Lindau in Engler & Prantl (eds.), Nat. Pflanzenfam. 4(3b): 297. 1895. Type: *Hemiadelphis polysperma* (Roxb.) T. Anderson [≡ *Hygrophila polysperma* (Roxb.) T. Anderson].

Synnema Benth. in de Cadolle (ed.), Prodr. 10: 527, 538. 1846. ≡ *Cardanthera* Buch.-Ham. ex Benth. & Hook. f., Gen. Pl. 2: 1074. 1876, nom. illeg. (Art. 52). Type: *Synnema avanum* (Wall. ex Benth.) Benth. [≡ *Pedicularis avana* Wall. ex Benth. ≡ *Cardanthera avana* (Wall. ex Benth.) Benth. ex C. B. Clarke ≡ *Hygrophila avana* (Wall. ex Benth.) E. Hossain].

Herbs, annual or perennial, often aquatic or hygrophilous, sometimes with axillary spines, cystoliths present. Leaves whorled or opposite, sessile or shortly petiolate; leaf blade entire, crenulate, or sometimes undulate. Flowers sessile, in terminal spikes or several in axillary clusters; bracteoles present or absent. Calyx 5-lobed, rarely 4-lobed. Corolla ventricose, limb 2-lipped, upper lip apex shallowly dentate, lower lip 3-lobed; lobes contorted in bud. Stamens 4, didynamous, or 2; anthers bithecous; thecae equally inserted; staminodes 2 or absent. Ovary with 4–16 ovules per locule; stigma entire or 2-dentate. Capsules linear-ellipsoid to narrowly oblong, (4–)8–30-seeded; retinacula present. Seeds discoid, covered with long mucilaginous trichomes.

Recent references: Heine, 1962. Notes on some West African Acanthaceae: The reduction of the genus *Asterucuntha* Ness to *Hygrophila* R. Br. *Kew Bull.* 16: 171–173; Sreemadhavan, 1968. The taxonomic status of *Cardanthera* (Acanthaceae). *Bull. Bot. Surv. India* 10: 22–233; Heine, 1971. Une nouvelle espéce hétérophylle du genre *Hygrophila* R. Br. *Adansonia* 13: 652–659; Hossain, 1974. Taxonomic and nomenclatural notes on *Synnema* Benth. (Acanthaceae). *Bangladesh J. Bot.* 3: 45–49; Fosberg & Sachet, 1981. Clarification of *Hygrophila triflora. Baileya* 21: 145–148; Fosberg & Sachet, 1984. Identify and typification of *Ruellia triflora* Roxb. and *H. difformis* L. f. *Baileya* 22: 138–140; Almeida & Almeida, 1987. Identity of "*Bahel schulli*" of *Hortus Malabaricus*. *J. Bombay Nat. Hist. Soc.* 83(Suppl.): 221–222; Cramer, 1989. The *Hygrophila* complex (Acanthaceae) in India and Ceylon. *Nordic J. Bot.* 9: 261–263; Cramer, 1992. Name changes in the Acanthaceae of India and Ceylon (Sri Lanka). *J. Nat. Sci. Counc. Sri Lanka*, 20: 59–69; Furness, 1994. The pollen morphology of *Hygrophila* and *Brillantaisia* (Acanthaceae: Ruellieae). *Acta Bot. Gallic*a. 141. 267–278; Angerstein & Lemke, 1994. First records of the aquatic weed *Hygrophila polysperma* (Acanthaceae) from Texas. *Sida* 16: 365–371; Chou & Lo, 1996. A revision of the Nantian Xianzi specific names of *Hygrophila. Guihaia* 16: 373–374.

About 100 species: widely distributed in tropical and subtropical regions; five species in the Pan-Himalaya.

1a. Stems armed with axillary spines; leaves whorled; calyx 4-lobed; seeds less than 10 ········ 1. *H. schulli*

1b. Stems unarmed; leaves opposite; calyx 5-lobed; seeds more than 12.
  2a. Flowers in terminal bracteolate spikes; bracts imbricate.
    3a. Corolla pubescent; stamens 2 ········ 2. *H. polysperma*
    3b. Corolla glabrous or obscurely puberulent; stamens 4 ········ 3. *H. difformis*
  2b. Flowers solitary or clustered in leaf axils; bracts distinct from one another.
    4a. Leaf blade narrowly lanceolate to oblanceolate, glabrous or slightly puberulent; calyx sparsely hispid-ciliate ········ 4. *H. ringens*
    4b. Leaf blade elliptic, obovate or oblong, densely strigose or hirsute on both surfaces; calyx densely hispid-ciliate ········ 5. *H. phlomoides*

1. **Hygrophila schulli** (Buch-Ham.) M. R. Almeida & S. M. Almeida, J. Bombay Nat. Hist. Soc. 83(Suppl.): 221. 1987. ≡ *Bahel schulli* Buch-Ham., Trans. Linn. Soc. Lond. 14: 288. 1824. Type: "India", [*icon*] Hortus Malabaricus 2: p. 87. fig. 45. 1679 (holotype).

Barleria longifolia L., Cent. Pl. II.: 22. 1756; Amoen. Acad. 4: 320. 1759. ≡ *Ruellia longifolia* (L.) Roxb., Fl. India (ed. 2) 3: 50. 1832. ≡ *Asteracantha longifolia* (L.) Nees in Wallich (ed.), Pl. Asiat. Rar. 3: 90. 1832; Nees in de Candolle (ed.), Prodr. 11: 247. 1847. ≡ *Hygrophila spinosa* T. Anderson in Thwaites, Enum. Pl. Zeyl.: 225. 1860; Anderson, J. Linn. Soc., Bot. 7: 22. 1863; Clarke in Hooker (ed.), Fl. Brit. India 4: 408. 1884. ≡ *Hygrophila longifolia* (L.) Kurz, J. Asiat Soc. Bengal 39(2): 78. 1870, nom. illeg., non *Hygrophila longifolia* Nees in Martius (ed.), Fl. Bras. 9: 21. 1847. Type: not designated.

Barleria auriculata Schumach., Beskr. Guin. Pl.: 285. 1827. ≡ *Hygrophila auriculata* (Schumach.) Heine, Kew Bull. 16: 172. 1962; Wood in Grierson & Long (eds.), Fl. Bhutan 2(3): 1252. 2001. ≡ *Asteracantha auriculata* (Schumach.) Nees in de Candolle (ed.), Prodr. 11: 248. 1847. Type: Ghana, *Thonning 185* [holotype: C (C10003397)].

Barleria hexacantha Moris, Mem. Reale Accad. Sci. Torino 36: 200. t. 7. 1833. Type: [*icon*] Moris, Mem. Reale Accad. Sci. Torino 36: 200. t. 7. 1833 (lectotype **here designated**).

Perennial herbs, 0.5–1.5 m high; stems quadrangular to terete, sparsely to densely hispid, prominent axillary spines. Leaves in whorls of 6, two large, others reduced in size, sessile; blade lanceolate or oblong-lanceolate, rarely hastate, 3–15 cm long, 0.5–4 cm broad, subglabrous to white hispid, base attenuate to subcordate, margin entire, apex subacute to acuminate. Flowers in dense axillary whorls, each whorl usually with 6 spines; spines 1–3.5 cm long, yellowish to brownish. Bracts linear-lanceolate, 2–3 cm long, hispid. Calyx 0.8–2.0 cm long, 4-lobed; lobes unequal, one broader than others, linear-lanceolate 1.2–1.5 cm long, margin coarsely ciliate. Corolla pale bluish-purple, 3–3.5 cm long, glabrous or puberulent; tube 1.4–1.7 cm long, dilated above, upper lip ovate to elliptic, 2-lobed, lobes 2–5 mm long, apex rounded, lower lip broadly ovate, 3-lobed, lobes oblong, 4–7 mm long, apex rounded or emarginate. Stamens 4; filaments 0.5–15 cm long, glabrous or slightly pubescent near base; anther 2–3 mm long. Ovary oblong, glabrous; style 2.5–3 cm long, pubescent; stigma 2-cleft, lobes 1.7–2 mm long. Capsules oblong, 9–11 mm long, glabrous, 4–8-seeded; retinacula prominent, distinctly hooked. Seeds orbicular, 1.5–2 mm long.

Phenology: flowering from January to October.

Habitats: wet places of river bank, river beds and waste rice fields.

Distribution: U Ganga & Indus, W Nepal, C Nepal, E Nepal, Sikkim & Darjiling (Map 46-139); widespread in tropical regions of Africa and S Asia.

U GANGA & INDUS: **Simla**, Kaupa Valley, 600 m, 20 Oct. 1886, *H. Collett s. n.* (K); **Kumaon**, Bel Khet, 1200 m, 24 Dec. 1848, *R. Strachey & J. E. Winterbottom s. n.* (K). **W NEPAL: Bheri**, Banke, Salyani Bagia, 180 m, 6 Jun. 1975, *L. R. Sharma 277* (KATH); Banke, Napelganj, Sallyan Bagh, 20 Oct. 1975, *K. J. Malla & P. R. Sharma 761* (KATH); Banke, Napelganj, 180 m, 2 Dec. 1979, *T. K. Bhattachaya 1956* (KATH); Banke, Chisa Pani, 240 m, 24 Sep. 1983, *P. P. Kurmi & D. K. Shrestha 4747* (KATH); Banke, Puraina, 180 m, 20 Oct. 1981, *R. M. Sharma 5049* (KATH); Banke, Napelganj, 150 m, 8 Mar. 1968, *J. D. A. Stainton 6112* (BM); Banke, Napelganj, 26 Oct. 1972, *N. P. Manandhar 10461* (KATH); Banke, Napelganj, 600 m, Apr. 1979, *T. K. Bhathacharya T19* (KATH); Surkhet, Solughopte, 700 m, 21 Feb. 1991, *N. P. Manandhar et al. 17/91* (KATH); Rapti, Dang, 8 Mar. 1976, *N. P. Manandhar & P. M. Regmi 263* (KATH); Dang, Garhara, 11 Mar. 1976, *N. P. Manandhar & P. M. Regmi 303* (KATH); Dang, Dang Valley, 300 m, 20 Mar. 1979, *T. K. Bhattachaya & K. J. Malla 350* (KATH); Dang, Gardwa, 28 Aug. 1982, *N. P. Manandhar & N. K. Bhattarai 8515* (KATH); Dang, Tulspur,

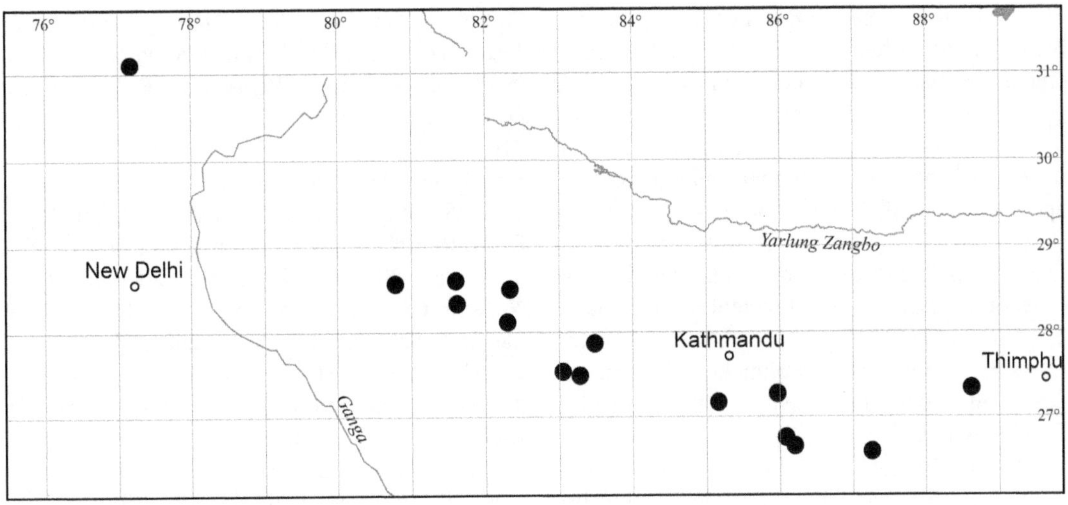

Map 46-139: **Hygrophila schulli** (Buch-Ham.) M. R. Almeida & S. M. Almeida

2000 ft., 24 Oct. 1952, *O. V. Polunin, W. R. Sykes & L. H. J. Williams 5877* (BM, E); Dang, Utter Amrai, Southeast of Bijauri, 610 m, *O. V. Polunin, W. R. Sykes & L. H. J. Williams 3635* (A, E); Dang, Koilabas, Mar. 1976, *N. K. Calimire 227* (KATH); Rolpa, Baijapur, 4 Nov. 1980, *R. N. Shukla 3316* (KATH); **Seti**, Kailali, 5 miles NE of Dhangarhi, 200 m, 5 Dec. 1966, *D. H. Nicolson 2836* (BM, KATH). **C NEPAL**: **Janakpur**, Dhanusa, Barkurwa, 150 m, 18 Feb. 1971, *N. P. Manandhar 17036* (KATH); Dhanusa, Dhalkebar, between Dhalkebar and Janakpur, 59 m, 26 Nov. 2012, *B. Adhikari & D. R. Kandel 95* (E); Sindhui, Napani, 1240 m, 27 Nov. 1988, *N. P. Manandhar 12530* (KATH); **Lumbini**, Kapilvastu, Karma, 200 m, 7 Nov. 1992, *P. P. Kurmi KB481* (KATH); Palpa, Shrinagar, 12 Oct. 1979, *Joziram & R. K. Pol 2096* (KATH); Rupandehi, Bhairahawa, 150 m, 20 Nov. 1973, *D. P. Joshi & M. M. Amatya 73/855* (KATH); Rupandehi, Tilorakot, 25 miles west of Butwal, Banks of Barn Gunga, 6 Mar. 1956, *K. B. Codrington 367* (BM); **Narayani**, Bara, Pragati Nagar, 120 m, 16 Feb. 1993, *L. P. Kattel 193/93* (KATH). **E NEPAL: Mechi**, Morang, Gauriganja-Rangali, 13 Dec. 1963, *H. Hara et al. 6303679* (KATH); Morang, Kathgara, Gauri Ganj, 14 Dec. 1963, *H. Hara et al. 6303680* (A, BM, K, KYO); **Sagarmatha**, Siraha, Pryagpur Chock, 118 m, 19 Nov. 2012, *B. Adhikari & D. R. Kandel 60* (E). **SIKKIM & DARJILING: Sikkim**, 11 Jan. 1875, *D. Treutler 1216* (K).

2. Hygrophila polysperma (Roxb.) T. Anderson, J. Linn. Soc., Bot. 9: 456. 1867; Clarke in Hooker (ed.), Fl. Brit. India 4: 406. 1884; Nasir & Ali in Malik & Ghafoor (eds.), Fl. Pakistan 188: 19. 1988; Wood in Grierson & Long (eds.), Fl. Bhutan 2(3): 1252. 2001; Hu, Fl. Reipubl. Popularis Sin. 70: 74. 2002; Hu, Deng & Daniel in Wu, Raven & Hong (eds.), Fl. China. 19: 430. 2011. ≡ *Justicia polysperma* Roxb., Fl. India 1: 120. 1820. ≡ *Adenosma polysperma* (Roxb.) Spreng., Syst. Veg. 829. 1825. ≡ *Hemiadelphis polysperma* (Roxb.) Nees in Wallich, Pl. As. Rar. 3: 80. 1832. Type: "Bengal", [*icon*] *Roxburgh Icon. 1016* (lectotype designated by Nasir & Ali in Malik & Ghafoor (eds.), Fl. Pakistan 188: 21. 1988: K).

Hygrophila polysperma f. *albiflora* Kuntze, Revis. Gen. Pl. 2: 490. 1891, **syn. nov**. Type: [Sikkim & Darjiling] India, Sikkim, Terrai, 1 Dec. 1875, *C. E. O. Kuntze 7205* [holotype: NY (NY00312057)].

Hemiadelphis polysperma var. *joshianus* P. Rao & S. N. Biswas, Indian Forester 94: 657. 1968. Type: India, Chandmani near Babaras Hindu University, 72 m, 11 Jan. 1967, *S. P. Singh 2558-A* (holotype: CAL).

Annuals or perennials herbs, 10–20 cm high; stems often prostrate, quadrangular, slightly swollen above nodes, puberulent, soon glabrescent. Petiole 1–5 mm; leaf blade oblong-lanceolate to ovate, 2–3.5 cm long, 0.6–1.3 cm broad, glabrous, base attenuate, margin entire, apex obtuse. Spikes terminal, 1.3–5 cm long. Bracts ovate, elliptic or obovate, 5–10 mm long, 1.5–3.5 mm broad, both surfaces densely hirtellous, apex obtuse; bracteoles lanceolate, 3.5–5 mm long, 0.5–1 mm broad, margin ciliate. Calyx 3.5–6 mm long; lobes unequal, lanceolate, 2.5–3 mm long, pubescent, apex acuminate. Corolla white, 5–6 mm long, pubescent; tube 3–3.5 mm long, lower lip 1.5–2 mm long, 3-lobed, lobes orbicular and subequal, upper lip oblong-ovate, 0.8–1 mm long, 2-lobed. Stamens 2, slightly exserted; filaments 1–1.5 mm long; staminodes 2, bristle-like, short. Ovary pilose; style 4–5 mm long; stigma capitate. Capsules linear-oblong, 5.5–8 mm long, pubescent, 20–30-seeded. Seeds orbicular-oblong, 0.7–1 mm long, 0.4–0.5 mm broad. **Chromosomes** 2n = 32. Figure 46-84.

Phenology: flowering from September to November; fruiting from October to January of next year.

Habitats: stream banks; below 1000 m.

Distribution: U Ganga & Indus, W Nepal, C Nepal, E Nepal, Sikkim & Darjiling, Bhutan, U Irrawaddy (Map 46-140); also in Sri Lanka, Vietnam, and other parts of India, Myanmar and China (S Yunnan, Guangxi, Guangdong, Jiangxi, Taiwan); naturalized in tropical and temperate regions.

U GANGA & INDUS: Laley Ka Bagh, 12 Aug. 1897, *B. M. Sharma 18524* (RRLH); **Kangra**, 600 m, 17 Apr. 1938, *R. N. Parker 3423* (K); 900 m, Jun. 1911, *Druii s. n.* (CAL). **W NEPAL: Bheri**, Banke, Napelganj, 180 m, 5 Jan. 1993, *P. P. Kurmi KB505* (KATH); Surkhet, Bheri, 600 m, 14 Apr. 1973, *J. F. Dobremez 2512 (N. P. Manandhar 74-141)* (E, KATH); Surkhet, Solighopte, 700 m, 21 Feb. 1991, *N. P. Manandhar 29/91* (KATH); Surkhet, Kalyan, 870 m, 23 Feb. 1991, *N. P. Manandhar 208/91* (KATH); **Rapti**, Pyuthan, Pyuthan, Bigjwar, 780 m, 6 Mar. 1926, *N. P. Manandhar & P. M. Regmi 201* (KATH); Salyan, Marma Khola, 910 m, 1 Apr. 1952, *O. V. Polunin, W. R. Sykes & L. H. J. Williams 1835* (BM, E); **Seti**, Doti, Kaptar, 210 m, 24 Apr. 1973, *J. F. Dobremez 1956* (E); Kailali, Ghoda Ghodi Tal, 205 m, 24 Mar. 2003, *G. D. Bhatt & S. Khatri 37* (KATH). **C NEPAL: Dhawalagiri**, Baglung, 100 m, Range Danda, 100 m, 24 Jan. 2003, *N. Trapa et al. 2030* (KATH); **Narayani**, Chitwan, Chitwan National Park, Hatti Sar Sauraha, 150 m, 21 Nov. 2004, *T. Miller et al. (DNEP) A53* (E, KATH); Chitwan, Sauraha to Kasara, 160–1800 m, 13 Jan. 1996, *M. Suzuki et al. 9614037* (TI); Chitwan, Sauraha–Padampur, 16 Jan. 1996, *M. Suzuki et al. 9614085* (A, BM, E, TI); Chitwan, Sauraha, riverside of the Rapti River, 12 Dec. 1976, *H. Tabata et al. 7529* (A, KYO). **E NEPAL: Koshi**, Sunsari, Barampur, near Dharan, 230 m, 19 May 1969, *L. H. J. Williams 79* (BM); Sunsari, Barampur, 230 m, 19 May 1969, *L. H. J. Williams 79* (BM); Sunsari, Tarahara, 220 m, 19 May 1969, *s. coll. 12329* (KATH); **Mechi**, Jhapa, Sibganja to Gauriganja, 12 Dec. 1963, *H. Hara et al. 6303672* (BM, KYO); Jhapa, Sibganja to Mahara Bahara, 200 m, 12 Dec. 1963, *G. Murata & T. Togashi 6303673* (BM, K, KATH, KYO); **Sagarmatha**, Saptari, Rajbiraj, 13 Feb. 1971, *N. P. Manandhar 17087* (KATH); Siraha, Lahan, 150 m, 15 Feb. 1971, *N. P. Manandhar 17098* (KATH). **SIKKIM & DARJILING: Darjiling**, *Griffith s. n.* (CAL, K); **Sikkim**, Baguakoti, 22 Feb. 1911, *R. Rhomoo 4953* (CAL); Sowka, 5 Jan. 1910, *R. Rhomoo 2317* (CAL); Tista, 22 Feb. 1910, *R. Rhomoo 3691* (CAL). **BHUTAN**: s. loc., 300 m, *J. D. Hooker s. n.* (K). **U IRRAWADDY**: **Putao**, Hkamti Plain, 27 Dec. 1930, *F. Kingdon-Ward 9063* (BM, E).

3. Hygrophila difformis (L. f.) Blume, Bijdr. Fl. Ned. Ind. 14: 804. 1826. ≡ *Ruellia difformis* L. f., Suppl. Pl.: 289. 1782, "1781". ≡ *Cardanthera difformis* (L. f.) Druce, Rep. Bot. Soc. Exch. Club Brit. Isles 4 (Suppl. 2): 612. 1917. Type: India, 1774, *J. G. Koenig s. n.* [lectotype designated by Fosberg & Sachet, Baileya 21: 147. 1981: LINN (LINN802.20); isolectotype: C (C10004892)].

Adenosma triflora Roxb. ex Nees in Wallich (ed.), Pl. Asiat. Rar. 3: 89. 1832; Nees in

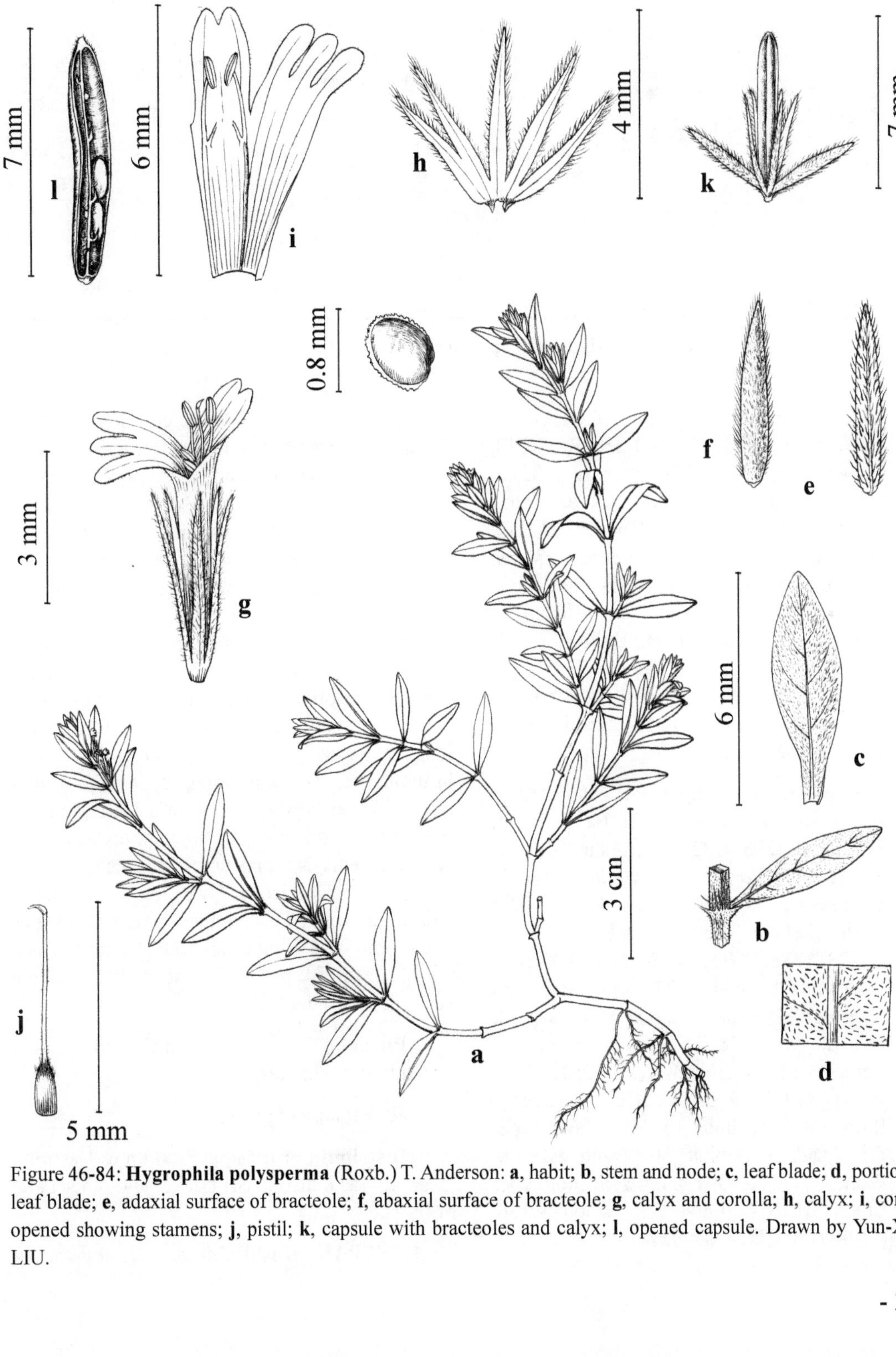

Figure 46-84: **Hygrophila polysperma** (Roxb.) T. Anderson: **a**, habit; **b**, stem and node; **c**, leaf blade; **d**, portion of leaf blade; **e**, adaxial surface of bracteole; **f**, abaxial surface of bracteole; **g**, calyx and corolla; **h**, calyx; **i**, corolla opened showing stamens; **j**, pistil; **k**, capsule with bracteoles and calyx; **l**, opened capsule. Drawn by Yun-Xiao LIU.

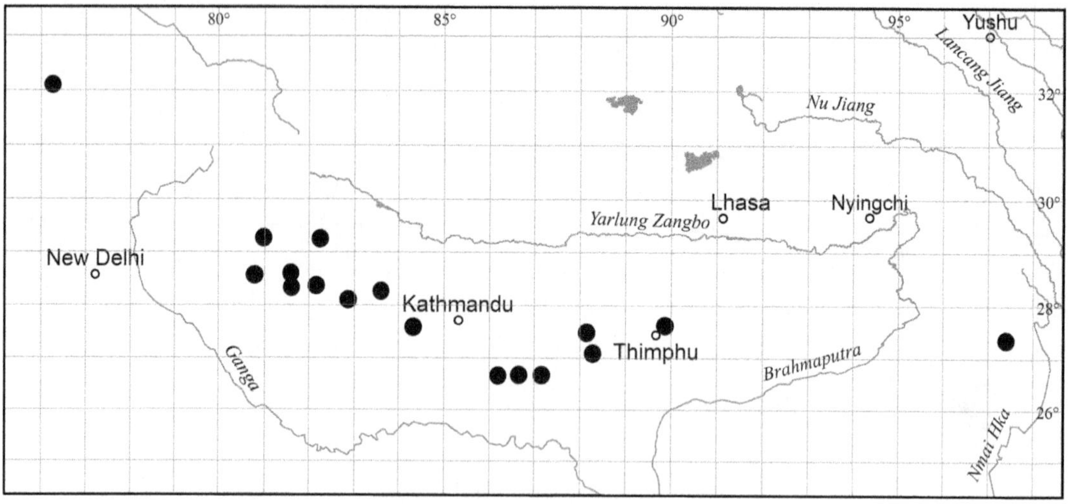

Map 46-140: **Hygrophila polysperma** (Roxb.) T. Anderson

de Candolle (ed.), Prodr. 11: 68. 1847. ≡ *Ruellia triflora* (Roxb. ex Nees) Roxb., Fl. India (ed. 2) 3: 52. 1832 (Oct.–Nov. 1832). ≡ *Cardanthera triflora* (Roxb. ex Nees) Ham.-Buch. ex C. B. Clarke in Hooker (ed.), Fl. Brit. India 4: 405. 1884. ≡ *Synnema triflorum* (Roxb. ex Nees) Kuntze, Revis. Gen. Pl. 2: 500. 1891. ≡ *Hygrophila triflora* (Roxb. ex Nees) Fosberg & Sachet, Baileya 21(3): 147. 1981. Type: India, *Wallich Cat. 7147a* [lectotype designated by Fosberg & Sachet, Baileya 21: 147. 1981: K-W (K001126784); isolectotypes: C, K].

Ruellia uliginosa L. f., Suppl. Pl.: 290. 1782. ≡ *Adenosma uliginosa* (L. f.) Nees, Verm. Schrifr. 3: 298. 1827; Pl. Asiat. Rar. 3: 79. 1832. ≡ *Cardanthera uliginosa* (L. f.) Buch.-Ham. ex Benth., Gen. Pl. 2: 1074. 1876; Clarke in Hooker (ed.), Fl. Brit. India 4: 403. 1885. ≡ *Hygrophila helodes* Heine, Adansonia 11: 658. 1971, nom. illeg., non *Hygrophila helodes* Nees in Martius (ed.), Fl. Bras. 9: 23. 1847; Wood in Grierson & Long (eds.), Fl. Bhutan 2(3): 1251. 2001. ≡ *Hygrophila uliginosa* (L. f.) Sreem., Bull. Bot. Suw. India 10: 233. 1968. Type: Sri Lanka, "Habitat in Tranquebar & Prope", 1777, *J. G. Koenig s. n.* [holotype: LINN (LINN-HL804-22); isotypes: C (C10004904, C10004905)].

Annual herbs, 30–40 cm high; stems decumbent or ascending from a fibrous rootstock, sparsely pubescent, then glabrescent. Leaves subsessile; blade oblong to subovate, 1–3.5 cm long, 0.4–1 cm broad, usually glabrous, base cuneate, margin entire or crenulate, apex acute. Spikes terminal, 1–1.3 cm long. Bracts broadly elliptic, 8–10 mm long, 7–8 mm road, apex abtuse; bracteoles obovate, 3.5–4.5 mm long, 2.5–3 mm broad. Calyx 1–1.2 cm long, pubescent; lobes linear-oblong, 8–9 mm long, 0.8–1 mm broad. Corolla blue with paler tube, 1.1–1.3 cm long, glabrous or obscurely puberulent; tube infundibular, 5–7 mm long, upper lip 4 mm long, 2-lobed, lower lip 5–6 mm long, 3-lobed. Stamens 4, posterior pair smaller than anterior pair; filaments 2–5 mm long, pubescent; anther-thecae linear, muticous. Ovary narrowly linear-elliptic, 2–3 mm long; ovules several. Capsules linear-oblong, 6–8 mm long, 0.8–1.1 mm broad, pilose towards apex, many-seeded; retinacula very short, not hooked.

Phenology: flowering and fruiting from November to February.

Habitats: wet places.

Distribution: E Nepal, Sikkim & Darjiling (Map 46-141); also in Sri Lanka, and other parts of India.

E NEPAL: **Koshi**, Morang, Kathgara, 14

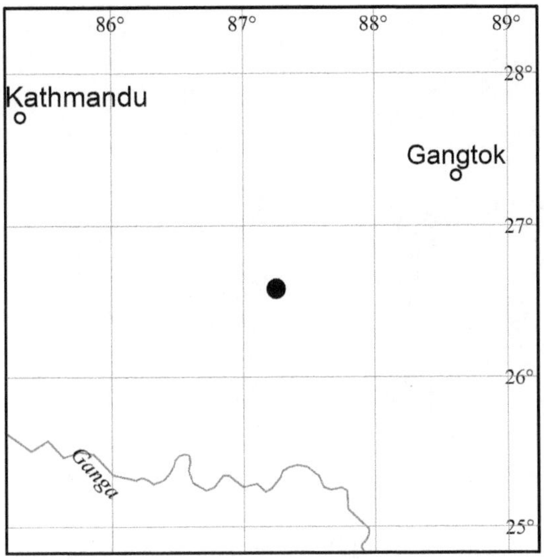

Map 46-141: **Hygrophila difformis** (L. f.) Blume

Dec. 1963, *H. Hara et al. 6303627* (KATH, KYO); **Mechi**, Jhapa, Sibganja to Mahara Bahara, 12 Dec. 1963, *G. Murata & M. Togashi 6303626* (KATH, KYO). **SIKKIM & DARJILING**: **Darjiling**, s. d., *Griffith s. n.* (K).

4. Hygrophila ringens (L.) R. Br. ex Spreng., Syst. Veg. 2: 828. 1825; Hu, Deng & Daniel in Wu, Raven & Hong (eds.), Fl. China, 19: 431. 2011. ≡ *Ruellia ringens* L., Sp. Pl. 2: 635. 1753. ≡ *Dipteracanthus ringens* (L.) Abeyw., Ceylon J. Sci., Biol. Sci. 2(1): 84. 1959. Type: India, *Herb. Linn. No. 804.13* (lectotype designated by Hansen, in Nicolson et al., Interpret. Van Rheede's Hort. Malab. 40. 1988: LINN).

Ruellia salicifolia Vahl, Symb. Bot. 3: 84. 1794. ≡ *Hygrophila salicifolia* (Vahl) Nees, Pl. Asiat. Rar. 3: 81. 1832; Nees in de Candolle (ed.), Prodr. 11: 92. 1847; Clarke in Hooker (ed.), Fl. Brit. India 4: 407. 1884; Wood in Grierson & Long (eds.), Fl. Bhutan 2(3): 1252. 2001; Hu, Fl. Reipubl. Popularis Sin. 69. 2002. Type: India, *J. König s. n.* [holotype: C (C10004916)].

Ruellia undulata Vahl, Symb. Bot. 3: 82. 1794. Type: India Orient., *Koenig s.n.* [holotype: C (C10004917)].

Ruellia quadrivalvis Buch.-Ham., Trans. Linn. Soc. London 14: 291. 1825. ≡ *Hygrophila quardrivalvis*. Buch.-Ham, Trans. Linn. Soc. London 14: 291. 1825; Nees in Wallich, Pl. As. Rar. 3: 81. 1832; Clarke in Hooker (ed.), Fl. Brit. India 4: 408. 1884. ≡ *Hygrophila angustifolia* var. *quadrivalvis* (Buch.-Ham.) Williams, Bull. Herb. Boiss. Ser. 2, 5: 435. 1905. ≡ *Hygrophila salicifolia* var. *quadrivalvis* (Buch.- Ham.) J. B. Imlay, Bull. Misc. Inform. Kew 1939: 111. 1939. Type: [*icon*] Hortus Malabaricus 2: 89, fig. 46. 1679 (holotype).

Perennials, 30–80 cm high; stems quadrangular, striate, slightly pubescent. Petiole 0.5–1 cm long, glabrous; leaf blade narrowly lanceolate to oblanceolate, 3–10 cm long, 0.5–1.5 cm broad, both surfaces glabrous or slightly pubescent, lateral veins 8–11 pairs, base attenuate and decurrent onto petiole, margin entire or slightly undulate, apex acute to obtuse. Flowers solitary or 2–10 clustered in leaf axils, sessile; bracteoles narrowly ovate, 3–5 mm long, margin pubescent, apex obtuse. Calyx narrowly campanulate, 5–6 cm long, 5-lobed to the middle; lobes linear-lanceolate, grayish pubescent or fulvous strigose, apex acuminate. Corolla pale purple, 1–2.5 cm long; tube 6–7 mm long,1.5–2 mm broad, glabrous, limb 2-lipped, lower lip obovate, 2–3 mm long, 3-lobed to the middle, lobes ovate, apex obtuse, upper lip elliptic, 2–3 mm long, apex shallowly 2-lobed. Stamens 4, included; filaments glabrous, longer pair 4–5 mm long, shorter pair 2–3 mm long; anther-thecae 1–2 mm long. Ovary glabrous; style 6–8 mm long, included; stigma unequally 2-lobed. Capsules narrowly oblong, 0.8–2.2 cm long, 1.2–1.5 mm broad, glabrous, 12–18-seeded. Seeds ovoid in outline 0.8–1 mm long and broad, pubescent. **Chromosomes** 2n = 44. Figure 46-85.

Phenology: flowering from August to October; fruiting from December to February of next year.

Habitats: stream banks, wet places; below 1000 m.

Distribution: W Nepal, C Nepal, E Nepal, Sikkim & Darjiling, Bhutan (Map 46-142); widespread in tropical and subtropical region of Asia.

Figure 46-85: **Hygrophila ringens** (L.) R. Br. ex Spreng.: **a**, flowering branch; **b**, fruiting branch; **c**, flower; **d**, opened corolla showing stamens; **e**, stamens; **f**, ovary and style; **g**, capsule; **h**, opened capsule showing seeds. Drawn by Quan LIN; reproduced from *Icon. Cormophyt. Sin.* 4: 154, fig. 5722. 1975.

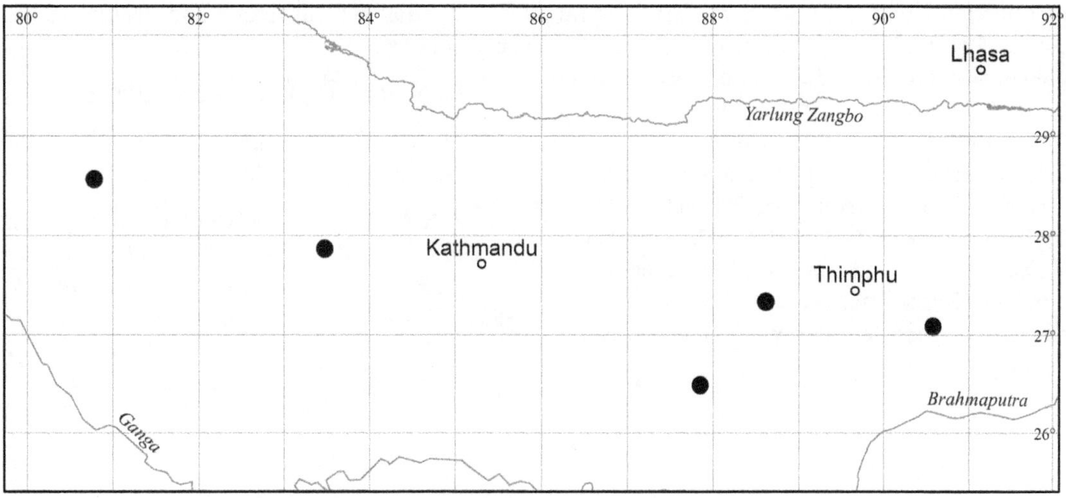

Map 46-142: **Hygrophila ringens** (L.) R. Br. ex Spreng.

W NEPAL: Mahakal, Royal Suklaphanta Woldlife Reserve, Jhilmila, Shuklaphat, 200 m, 8 Oct. 1996, *M. Mikage et al. 9689144* (TI); **Seti**, Kailali, Balmi, Beladevipur, 180 m, 10 Nov. 2012, *B. Adhikari & G. D. Bhatt 32* (E). **C NEPAL: Lumbini**, Palpa, Tansen, 910 m, 5 Oct. 1954, *J. D. A. Stainton, W. R. Sykes & L. H. J. Williams 8751* (BM, E); **Narayani**, Chitwan, Sauraha to Kasara, 13 Jan. 1996, *M. Suzuki et al. 9614284* (TI). **E NEPAL: Mechi**, Jhapa, Gauri Ganj, 200 m, 13 Dec. 1963, *H. Kanai et al. 6303671* (KATH). **SIKKIM & DARJILING: Sikkim**, Dalkajhar, 150 m, 15 Oct. 1884, *C. B. Clarke 36914* (K). **BHUTAN: Gaylephug**, Gaylegphug, 300 m, 29 Oct. 1988, *J. R. I. Wood 6831* (E).

5. Hygrophila phlomoides Nees in Wallich (ed.), Pl. Asiat. Rar. 3: 80. 1832; Nees in de Candolle (ed.), Prodr. 11: 90. 1847; Clarke in Hooker (ed.), Fl. Brit. India 4: 408. 1884; Wood in Grierson & Long (eds.), Fl. Bhutan 2(3): 1252. 2001; Hu, Fl. Reipubl. Popularis Sin. 70: 72. 2002; Hu, Deng & Daniel in Wu, Raven & Hong (eds.), Fl. China 19: 431. 2011. Type: Bangladesh, Silhet, *Wallich Cat. 2376a* [lectotype **here designated**: K-W (K001115798); isolectotypes: C (C10004911), K (K000882379), K-W (K001115799)].

Perennial herbs, 30–80 cm high; stems quadrangular, brown strigose. Petiole 0–3 mm long, hirsute; leaf blade elliptic, obovate, or oblong, 2–9 cm long, 1–3 cm broad, papery, both surfaces hirsute, lateral veins 8–15 pairs, base usually attenuate and decurrent onto petiole, margin entire or undulate, apex acute

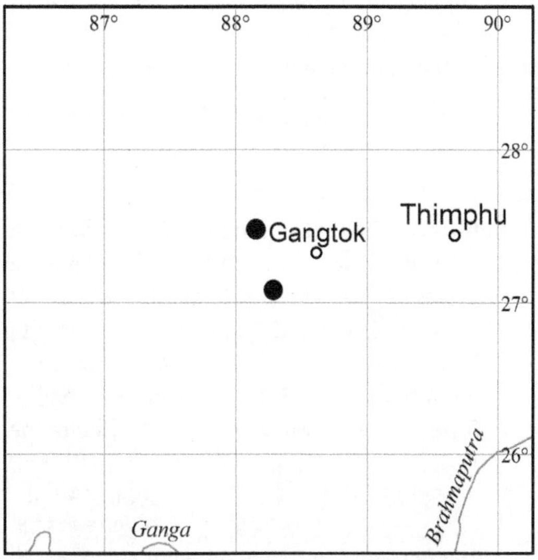

Map 46-143: **Hygrophila phlomoides** Nees

or sometimes obtuse. Flowers axillary, several clustered or in whorls upwards; bracteoles linear-oblong, 4–5 mm long, 1.5–2 mm broad, hirsute. Calyx 0.9–1.1 cm long, white hirsute, 5-lobed to the middle; lobes linear. Corolla 1.8–2.2 cm long, pubescent, lower lip oblong, sparely pilose, 3-lobed, upper lip triangular, 2-lobed. Stamens 4; filaments glabrous, longer pair 4–5 mm long, shorter pair 2–3 mm long; anther-thecae 1.7–2 mm long. Ovary glabrous; style 1.5–1.8 cm long, pubescent. Capsules narrowly oblong, 1.2–1.5 cm long.

Phenology: flowering and fruiting from July to December.

Habitats: wet places; below 1200 m.

Distribution: Sikkim & Darjiling (Map 46-143); also in tropical region of Asia.

SIKKIM & DARJILING: **Darjiling**, *J. M. Cowan s. n.* (E, K); **Sikkim**, 900 m, *J. D. Hooker s. n.* (A, BM, CGE, K); Dalkajhar, 160 m, 14 Oct. 1884, *C. B. Clarke 36945* (BM, K).

13. ECHINACANTHUS Nees

Nees in Wallich (ed.), Pl. Asiat. Rar. 3: 75, 90. 1832; Nees in de Candolle (ed.), Prodr. 11: 168. 1847; Bentham & Hooker, Gen. Pl. 2: 1064, 1080. 1865; Clarke in Hooker (ed.), Fl. Brit. India 4: 414. 1884; Lindau in Engler & Prant., Nat. Pflanzenf. 4(3B): 301. 1895; Wood in Grierson & Long (eds.), Fl. Bhutan 2(3): 1253. 2001; Hu, Fl. Reipubl. Popularis 70: 81. 2002; Hu, Deng & Daniel in Wu, Raven & Hong (eds.), Fl. China 19: 433. 2011. Type: *Echinacanthus attenuatus* Nees.

Perennial herbs; with cystoliths. Leaves opposite; leaf blade entire. Flowers in terminal branched thyrses. Bracts narrow; bracteoles inconspicuous. Calyx often glandular-hairy, deeply 5-lobed; lobes equal or subequal. Corolla purple or yellow, funnel-shaped to campanulate; tube ventricose, limb 5-lobed; lobes subequal, contorted in bud. Stamens 4, didynamous; filaments connate at base into 2 pairs; anthers bithecous; thecae parallel, hirsute, spurred at base. Ovary with 4–8 ovules per locule; stigma 2-lobed, one lobe reduced. Capsules linear-cylindric to linear-oblong, 8–16-seeded; retinacula present. Seeds subovate in outline, compressed, covered with hygroscopic trichomes.

One species: endemic to the Pan-Himalaya.

1. Echinacanthus attenuatus Nees in Wallich (ed.), Pl. Asiat. Rar. 3: 90. 1832; Clarke in Hooker (ed.), Fl. Brit. India 4: 414. 1884; Wood in Grierson & Long (eds.), Fl. Bhutan 2(3): 1253. 2001. ≡ *Ruellia attenuata* Wall., Numer. List n. 2345. 1830, nom. invalid. Type: Nepal, *Wallich Cat. 2345* [lectotype designated by Tripp et al., Intern. J. Pl. Sci. 174: 130. 2013: GZU (GZU000249910); isolectotypes: BM, CGE, E (E00273504, E00273505), K (K000882663), K-W (K001115700, K001115701, K001115702)].

Perennial herbs, 30–60 cm high. Petiole 0.3–2 cm long; blade elliptic to oblanceolate, 5–18 cm long, 2–7 cm broad, glabrous or sparsely pubescent especially along veins, lateral veins 6–9 pairs, base attenuate, margin entire, apex acute. Thyrses terminal, 6–20 cm long; cymes 3–6 cm long, usually 3-flowered; rachis glandular-pubescent. Bracts narrowly oblong, 1–4 mm long, 0.2–0.6 mm broad, margin glandular-pilose; bracteoles linear, 8–15 mm long, glandular-pubescent. Calyx 1.5–1.8 cm long, glandular-pilose; lobes linear. Corolla purplish-blue, straight, 3.2–4.2 cm long, pubescent outside, basal tube cylindric, 0.9–1.2 cm long, 1–1.3 mm broad, upwards gradually widened to 8–10 mm broad at mouth; lobes ovate, 4–5 mm long and broad, apex rounded. Stamens 4, didynamous; filaments of longer pair 1–2 cm long, shorter pair 6–8 mm long; anthers bithecous, thecae oblong, 2–2.2 mm long, 1–1.2 mm broad, spurred at base. Ovary glabrous; style 2–2.3 cm long, glabrous, stigma unequally 2-cleft. Capsules cylindric clavate, 1.5–1.8 cm long, glabrous. Seeds ovate in outline, compressed, 1.8–2 mm long and broad, pubescent. Figure 46-86.

Phenology: flowering from September to October.

Habitats: thickets; 100–1700 m.

Distribution: endemic; U Ganga & Indus, W Nepal, C Nepal, E Nepal, Sikkim & Darjiling (Map 46-144).

U GANGA & INDUS: Almora, East Almora, Toti Valley, 1060 m, 5 Jan. 1932, *A. E. Armsaton 1472* (K); **Nainital**, Ramnagar Forests, *H. H. Haines 3947* (K). **W NEPAL**: **Mahakali**, Dadeldhura, Okhaldhunga to Chyanum, 1680 m, Sep. 1964, *M. L. Banerj et al. 3037* (KATH); **Rapti**, Dang, Rapti Nadi, Hetaura, 450 m, 9 Mar. 1966, *J. D. A. Stainton 5221* (BM, TI). **C NEPAL**: **Bagmati**, Kathamandu, Kuwapani & Banepa, 1370 m, 5 Oct. 1956, *Herbarium Banerji 1108* (A, CAL); Kavrepalanchok, Dolalghat, Sunkosi River, 700 m, 21 Oct. 1969, *H. Kanai 16604* (KATH); **Janakpur**, Dolakha, Khanigaon to Suri Dhoban, 1000–1700 m, *H. Ohba et al. 8332114* (TI); Sindhuli, Hariharpur Gadhi, 460 m, 18 Dec. 1980, *J. D. A. Stainton 8232* (E); Dolakha, Kiratichhap, 1210 m, 25 Oct. 1978, *A. D. Schilling et al. 2433A* (BM); **Narayani**, Makwanpur, Makwanpur gadi, 900 m, 16 Nov. 1970, *H. Kanai & P. R. Shakya 674761* (KATH, KYO, TI); Makwanpur, near Kuro-Nadi, near Hettaunda, between Birganj & Hettoura, *I. H. Burkill 29527* (CAL); Makwanpur, Brindaban Herbal Farm, 600 m, Oct. 1964, *N. S. Bista 3279* (KATH); Chitwan, Between Kashara and Dhoba, *H. Tabata et al. 7597* (A, KYO). **E NEPAL**: **Kosi**, Bhojpur, Bastin to Pharkushe, 300 m, 2 Nov. 1995, *M. Mikage et al. 9554274* (E, TI); Dhankuta, Dharapani, 12 Nov. 1978, *P. Pradhan et al. 746* (KATH); Dhankuta, 1200 m, 19 Oct. 1963, *H. Hara et al. 6303689* (BM, KATH, KYO, TI); Sankhuwasabha, Khahare to Tumlingtar, 360 m, 26 Jul. 1990, *M. Minaki et al. 9077011* (TI); Sankhuwasabha, Tumlingtar, 470 m, 8 Nov. 1972, *T. Wraber 490* (BM); Sankhuwasabha, Bareni, 480 m, 30 Nov. 1990, *P. R. Shakya 9373* (KATH); Sankhuwasabha, Tumlingtar, 550 m, 13 Dec. 1961, *A. H. Norkett 8049* (BM, KATH); Sunsari, Baraha Chetra, Near Baraha Chetra, 160 m, 25 Nov. 2012, *B. Adhikari & D. R. Kandel 90* (E); Terathum, Kyaureni, 1400 m, 17 Nov. 1978, *P. Pradhan et al. 955* (KATH); Terathum, Arun valley, Tumlingtar, 470 m, 8 Nov. 1972, *T. Wraber 34797* (BM); **Mechi**, Ilam, Rangapani–Chisa Pani to Ghorwa, 9 Dec. 1963, *H. Hara et al. 6303687* (A, BM, E, KATH, KYO); Ilam, Dhara Pani, Dhara Pani to Dhankuta, 17 Oct. 1963, *H. Hara et al. 6303688* (BM, K, KYO); Jhapa, Satasidham temple to Satasidham temple, 245 m, 23 Nov. 2012, *B. Adhikari & D. R. Kandel 84* (E); Jhapa, Chulachuli, Chola Choli, 150 m, 3 Feb. 1985, *J. D. A. Stainton 8907* (E); Panchthar, Phidim, 1270 m, 22 Nov. 2012, *B. Adhikari & D. R. Kandel 79* (E); Jhapa, Sibganja to Gauriganja, 12 Dec. 1963, *H. Hara et al. 6303685* (KYO, TI); Sunsari, Phusre to Sanguri Bhanjang to Dhara Pani, 16 Oct. 1963, *H. Hara et al. 6303686* (A, BM, E, KATH, KYO, TI); Taplejung, near Hellock, Nov. 1963, *A. V. Upadhyay 1528* (KATH); Taplejung, Yagu to Bungling, 480 m, 10 Nov. 1981, *P. R. Shakya 7395* (KATH); **Sargamatha**, Khotang, Dihigaon to Regmitar, 200 m, 27 Oct. 1995, *M. Mikage et al. 9554166* (TI); Udayapur, Panchawati, Udaipur Garhi, 1060 m, 1 Nov. 1969, *J. D. A. Stainton 6626* (BM, TI); Udayapur, Trijuga Khola, 230 m, 7 Nov. 1969, *J. D. A. Stainton 6655* (BM); Udayapur, between Chisapani and Katari, 31 Dec. 1976, *H. Tabata et al. 7685* (KYO); Udayapur, Trijuga Khola, 225 m, 7 Oct. 1969, *J. D. A. Stainton 9956* (TI); Okhaldhunga, 1700 m, 5 Oct. 1964, *M. L. Banerjee et al. 3037* (KATH); Khotang, Regmitar-Khani Danda, 380 m, 28 Oct. 1995, *M. Mikage et al. 9554187* (E, TI). **SIKKIM & DARJILING**: **Darjiling**, Nihal, 600–1200 m, *J. D. Hooker s. n.* (A); **Sikkim**, Raging, 760 m, 6 Oct., 1875, *C. B. Clarke 25079* (K); 1200 m, 3 Nov. 1874, *D. Treutler 1016* (K).

Figure 46-86: **Echinacanthus attenuatus** Nees: **a**, habit; **b**, calyx, ovary and style; **c**, calyx and corolla; **d**, opened corolla showing stamens; **e**, anther. Drawn by Yun-Xiao LIU.

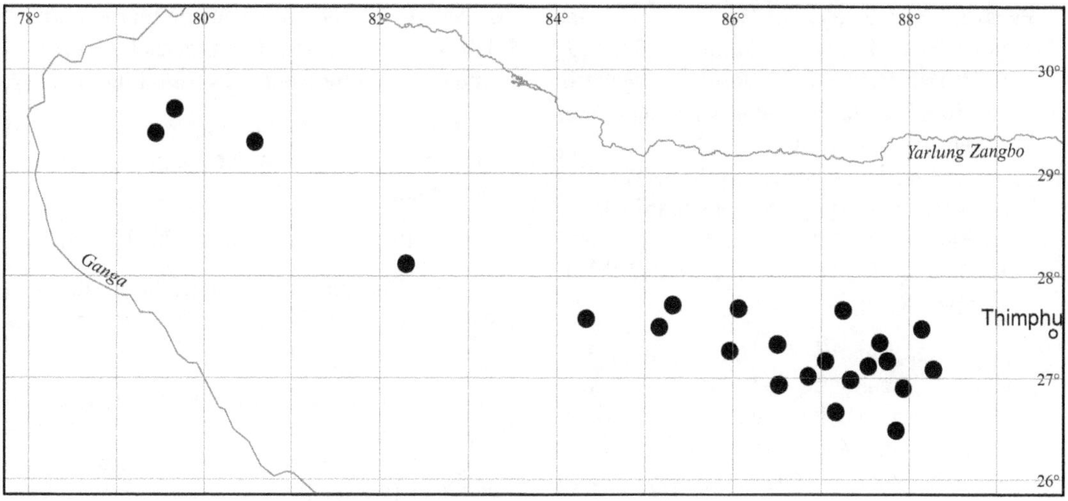

Map 46-144: **Echinacanthus attenuatus** Nees

14. PARARUELLIA Bremek. & Nann.-Bremek.

Bremekamp & Nannenga-Bremekamp, Verh. Kon. Ned. Akad. Wetensch., Afd. Natuurk., Sect. 2 45(1): 25. 1948; Tsui in Hu (ed.), Fl. Reipubl. Popularis Sin. 70: 52. 2002; Hu, Deng & Daniel in Wu, Raven & Hong (eds.), Fl. China 19: 434. 2011. Type: *Pararuellia sumatrensis* (C. B. Clarke) Bremek.

Ruellia sect. *Schizothecium* Baill., Bull. Mens. Soc. Linn. Paris 107: 852. 1890 & Hist. Pl. 10: 408. 1891. Type: *Ruellia delavayana* Baill. [≡ *Pararuellia delavayana* (Baill.) E. Hossain].

Herbs, perennial; stems short; with cystoliths. Leaves opposite, rosulate, petiolate; blade undulate, crenulate, or rarely subentire. Inflorescences terminal, spikes or thyrses. Bracts usually subleaf-like; bracteoles present. Flowers sessile. Calyx subequally to equally 5-lobed. Corolla white, bluish, or pink; tube base cylindric and slender, apically expanded into a throat, limb 5-lobed; lobes subequal, contorted in bud, apex rounded or obtuse. Stamens 4, didynamous, inserted at base of throat; filaments glabrous; anthers bithecous; thecae usually widely separated by a broad connective, more or less parallel; staminodes absent. Ovary glabrous, with 4–8 ovules per locule; style pubescent; stigma pubescent, 2-cleft, upper lobe very short or reduced. Capsules cylindric, not stipitate, 8–16-seeded; retinacula present. Seeds lenticular, pubescent with hygroscopic hairs.

About ten species: China, Vietnam, Malaysia, Indonesia and the Philippines; one species in the Pan-Himalaya.

1. **Pararuellia delavayana** (Baill.) E. Hossain, Notes Roy. Bot. Gard. Edinburgh 32: 409. 1973; Tsui in Hu (ed.), Fl. Reipubl. Popularis Sin. 70: 55. 2002; Hu, Deng & Daniel in Wu, Raven & Hong (eds.), Fl. China 19: 435. 2011. ≡ *Ruellia delavayana* Baill., Hist. Pl. 10: 408. 1891. Type: [S Hengduan] China, Yunnan, Heqing, Tapintze, 16 Aug. 1888, *M. Delavay 174* [holotype: P (P00650208); isotypes: K, P].

Hemigraphis drymophila Diels, Notes Roy. Bot. Gard. Edinburgh 5: 161. 1912. ≡ *Ruellia drymophila* (Diels) Hand.-Mazz., Akad. Wiss. Wien, Math.-Naturwiss. Kl., Anz. 61: 169. 1924. ≡ *Pararuellia drymophila* (Diels) C. Y. Wu & H. S. Lo, Fl. Hainan. 3: 593. 1974. Type: [S Hengduan] China, Yunnan, Lijiang Range eastern flank, 8500-10,000 ft., Aug. 1906, *G. Forrest 2774* [holotype: E (E00095245); isotype: P (P 00650209)].

Ruellia arcuata Lingelsh. & Borza, Repert. Spec. Nov. Regni Veg. 13: 390. 1914. Type: China, Yunnan, Lufeng, 27 Aug. 1913, *H. W. Limpricht 868* (holotype: WRSL).

Herbs; stems 1–2 cm long. Leaves rosette; petiole 0.5–2 cm long, pubescent; blade oblong to oblong-elliptic, sometimes obovate, elliptic, or lanceolate, 4–12 cm long, 2.5–4 cm broad, abaxially pubescent, adaxially strigose, lateral veins 7–9 pairs, base cuneate and decurrent onto petiole, margin undulate to crenate, apex acute, obtuse, or rounded. Spikes 3–6 cm long, with 1–3 nodes. Bracts elliptic to ovate, 1–4.7 cm long, 0.5–2.5 cm broad, reduced in size upwards on rachis, both surfaces sparsely strigose, base rounded, margin ciliate, apex acute to obtuse; bracteoles linear, 5–7 mm long, 0.8–1.2 mm broad. Calyx 8–9 mm long; lobes triangular-lanceolate, 0.8–1 mm broad, margin ciliate towards apex. Corolla white, light blue, or pink, 1.1–1.3 cm long; tube base narrowly cylindric, 3–4 mm long, throat 4–5 mm long; lobes orbicular, 3–4 mm long and broad, subequal, apex emarginate. Stamen 4; filaments of longer pair 3–3.5 mm long, shorter pair 1–1.5 mm long; anther-thecae oblong, 1–1.2 mm long. Ovary ellipsoid; style 1.2–1.4 cm long, hirsute. Capsules clavate, 1.6–2 cm long, 16-seeded. Seeds ovoid to subcircular in outline, 1.1–1.4 mm long and broad, densely appressed trichomes. **Chromosomes** 2n = 28. Figure 46-87.

Phenology: flowering from July to September; fruiting from November to February of next year.

Habitats: forests, slopes; 700–3000 m.

Distribution: S Hengduan (Map 46-145); also in other parts of China (C & S Sichuan, C & E Yunnan).

S HENGDUAN: Bingchuan, Xiayang to Bangying, 14 Oct. 1946, *T. N. Liou 21572* (PE); Dongsheng 2000 m, 4 Jul. 1966, *Y. Y. Hu & Y. F. Qi 662029* (YIM); **Dali**, *K. K. Tsoong 2981* (SYS); *K. K. Tsoong 2421* (SYS); Pingpo, Shangpingpo, 2000 m, *K. J. Luo et al. 613440* (YIM), Pingpo, Shangpingpo, 2800 m, 7 Aug. 1961, *K. J. Luo et al. 613435* (KUN, YIM), Haidong to Heqing, 2100 m, 9 Aug. 2013, *K. Wu & J. C. Hao YN157* (BNU); **Eryuan**, from Changyi to Maidi, 1658 m, 18 Aug. 2010, *E. D. Liu et al. 2754* (KUN); **Heqing**, Tapintze, 16 Aug. 1888, *M. Delavay 4578* (K); Tapintze, Aug. 1882, *M. Delavay 174* (P); Duomei, Shanghechuan, 1850 m, 18 Sep. 1988, *Y. H. Li et al. 573* (KUN); Jiangying, 2000 m, 3 Sep. 1940, *R. C. Ching 24614* (KUN, PE); Jiangyingba, 1600 m, 2 Aug. 1963, *Exped. N.W. Yunnan Jinshajiang 5392* (KUN); **Huili**, 20 km N of the city, 1700 m, 21 Aug. 2001, *C. M. Hu 9087* (IBSC); s. loc., 2000 m, 31 Aug. 1932, *T. T. Yü 1395* (A, PE); **Jianchuan**, Diannan, 15 Sep. 2013, *Y. F. Deng 25024* (IBSC); **Longyang**, Mangkuang, Baihualing, Yutang, 1520 m, 11 Oct. 2007, *H. Li 1295* (KUN); Mangkuan, Baihualing, Dajieshan, 1525 m, 11 Oct. 2007, *Gaoligong Shan Biodiversity Survey 19054* (CAS, KUN); Mangkuan, Baihualing, Dajieshan, 1525 m, 10 Sep. 2003, *Gaoligong Shan Biodiversity Survey 19073* (CAS, KUN); Mangkuan, Baihualing, Dajieshan, 1525 m, 10 Sep. 2003, *Gaoligong Shan Biodiversity Survey 19100* (CAS, KUN); Baihualing, 1100 m, 28 Aug. 1995, *X. C. Shi & S. X. Yang 631* (KUN); **Lushui**, near Liuku, 740 m, 21 Aug. 1964, *S. K. Wu 8507* (KUN); Luzhang, 6 Jul. 1975, *J. S. Yang 7697* (KUN); Liuku, 14 Sep. 2013, *Y. F. Deng 24920* (IBSC); **Miyi**, s. loc., 29 Oct. 1963, *s. coll. 281* (CDBI); Xinhe,

Figure 46-87: **Pararuellia delavayana** (Baill.) E. Hossain: **a**, portion of plant; **b**, flower; **c**, opened corolla showing stamens; **d**, stamen; **e**, capsule. Reproduced from *Icon. Cormophyt. Sin.* 4: 155, fig. 5724. 1975.

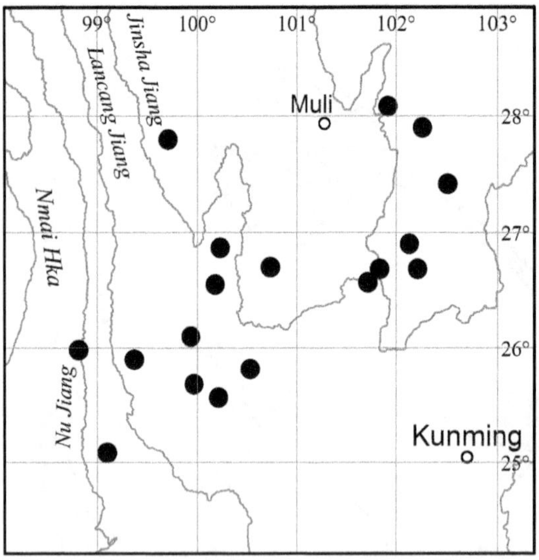

Map 46-145: **Pararuellia delavayana** (Baill.) E. Hossain

1100 m, 1 Aug. 1963, *Miyi Working Station 164* (CDBI); 29 Oct. 1963, *Miyi Working Station 281* (CDBI); Puwei, 1800 m, *s. coll. 11243* (SZ); **Muli**, Mosuogou to Qingao, 1900–2280 m, 6 Sep. 1959, *S. K. Wu 2399* (KUN); Szechuan-ping on Yangtze near Fongkou, 2300 m, 14 Sep. 1939, *K. M. Feng 2950* (A, KUN); Xiamaidi, 2600 m, 3 Aug. 1978, *Q. S. Zhao et al. 6985* (CDBI); **Panzhihua**, Renhe, Datian, Baoxingshan, 1700 m, 27 Sep. 2002, *X. F. Gao et al. 4096* (CDBI); Renhe, Datian, Yinlu, Baoxingshan, 1700 m, 3 Aug. 2003, *Y. L. Peng & H. L. Zhao 6324* (CDBI); Renhe qu, Zhongba, 1620 m, 4 Aug. 2003, *Y. L. Peng 6347* (CDBI); Cycas Natural Reserve, 1613 m, 11 Jul. 2010, *B. Liu & J. F. Ye 50* (PE); Cacas Natural Reserve, Fengjialiangzi, 2218 m, 11 Jul. 2010, *B. Liu & J. F. Ye 250* (PE); **Puge**, Luowugou, 1500 m, 4 Aug. 1976, *s. coll. 13927* (CDBI); **Shangri-la** (Zhongdian), Bödö (Peti), Waschwa, 2500 m, 4 Aug. 1914, *H. Handel-Mazzetti 4462* (WU), s. loc., Apr.–May 1922, *J. F. Rock 24693* (BM); Taku, 3000 m, Aug. 1914, *C. Schneider 2156* (A, K); NE of the Yangtze bend, 2700–3000 m, Aug. 1913, *G. Gorrest 10710* (PE); Chungtien Plateau, 2400 m, Aug. 1913, *G. Forrest 10890* (K, PE); between Bodo and Haba, 17 Aug. 1939, *K. M. Feng 2068* (A, KUN); Boo to Washwa, 17 Aug. 1939, *K. M. Feng 2069* (KUN); between Haba to Tung-pa-tze near Anangu, 2 Sep. 1939, *K. M. Feng 2254* (A, KUN); Lochlho, 1300 m, 20 Aug. 1939, *K. M. Feng 2711* (KUN); Haba, 3000 m, 20 Sep. 1937, *T. T. Yü 13570* (A, KUN); **Xichang**, Lushan, 1700 m, 29 Jul. 1964, *L. N. Zhao 2253* (KUN, SZ); **Yanbian**, Huimin, Deyang, 1450 m, 23 Sep. 2002, *X. F. Gao et al. 3845* (CDBI); Yongxing, Tianshengqiao, 1280 m, 29 Jul. 2003, *X. F. Gao et al. 6105* (CDBI); Qinghe, Toudihe, 29 Ju. 2003, *X. F. Gao et al. 6142* (CDBI); Yankou to Qinghe, 23 Sep. 2002, *X. F. Gao et al. 6199* (CDBI); Xinjiu, Lanjiahe, 1600 m, 6 Aug. 2003, *H. L. Zhao 6400* (CDBI); **Yangbi**, city to Meixi, 1900 m, 3 Aug. 1963, *Exped. Zhongdian 63-4011* (KUN, LBG), Cangshanxi, on the way from Cangshanxi Zhen to Taiping, ca. 1300 m, 13 Sep. 2013, *Y. F. Deng et al. 24188* (IBSC, KUN); **Yongsheng**, 2050 m, 4 Aug. 1976, *Wu 4025* (KUN); Renli, 1250 m, *Wu 4061* (KUN); Pianjiao, near Jinshajiang, 1200 m, 14 Sep. 2012, *Y. F. Deng & H. Peng 23842* (IBSC); Pianjiao, 1470 m, 14 Sep. 2012, *Y. F. Deng & H. Peng 23829* (IBSC); Yungbei, vic. Dahwaschu, 1800–2000 m, 2 Jul. 1914, *H. Handel-Mazzetti 3388* (K); Yung-pe mountains, 2700 m, Sep. 1913, *G. Forrest 11038* (K, PE); **Yulong** (Lijiang), Heilongtan, 10 Oct. 1970, *s. coll. 7419* (HITBC); Heilongtan, 2400 m, 10 Aug. 1962, *Exped. Zhongdian 669* (KUN, PE); Heilongtan, 2400 m, 4 Aug. 2019, *Y. F. Deng 28642* (IBSC); s. loc., 2800 m, 10 Aug. 1914, *C. Schneider 2223* (A, K, US); near Shangfengke, 2500 m, 3 Oct. 1983, *Exped. Qinghai-Xizang 14925* (KUN, PE); W slopes of Likiang Snow Range, Jul. 1922, *J. F. Rock 5411* (A); Lichiang Range, 1933, *H. D. McLaren 41* (BM); **Yunlong**, Jiuzhoubi, Cijiang, 4 Aug. 2010, *H. Peng et al. H-Dianyunlong-Z-4303* (KUN); Changxin, Songnong, 14 Sep. 2013, *Y. F. Deng 24957* (IBSC).

15. RUELLIA L.

Linnaeus, Sp. Pl. 2: 634. 1753; Bentham & Hooker, Gen. Pl. 2: 1077. 1876; Clarke in Hooker (ed.), Fl. Brit. India 4: 411. 1884; Lindau in Engler & Prantl (eds.), Nat. Pflanzenfam. 4(3b): 308. 1895; Nasir & Ali in Malik & Ghafoor (eds.), Fl. Pakistan 188: 58. 1988; Hu, Deng & Daniel in Wu, Raven & Hong (eds.), Fl. China 19: 435. 2011. Lectotype (designated Britton & Brown, Ill. Fl. N.U.S. ed. 2. 3: 241. 7 Jun. 1913; affirmed by Green, Prop. Brit. Bot. 169. Aug. 1929): *Ruellia tuberosa* L.

Cryphiacanthus Nees, Delect. Sem. Horto Bot. Vratisl. 1841: [3]. 1841. Lectotype (designated by Bremekamp, ING card 00272. 1955): *Cryphiacanthus barbadensis* Nees, nom. illeg. (Art. 52) (≡ *Ruellia tuberosa* L.).

Dipteracanthus Nees in Wallich (ed.), Pl. Asiat. Rar. 3: 75, 81. 1832; Tsui in Hu (ed.), Fl. Reipubl. Popularis Sin. 70: 50. 2002. ≡ *Ruellia* sect. *Dipteracanthus* (Nees) Lindau in Engler & Prantl (eds.), Nat. Pflanzenfam. 4(3b): 309. 1895. Type (lectotype designated by Bremekamp, Verh. Kon. Ned. Akad. Wetensch., Afd. Natuurk., Sect. 2 45(1): 15. 1948): *Dipteracanthus prostratus* (Poiret) Nees (≡ *Ruellia prostrata* Poiret).

Herbs or shrubs, perennial, prostrate to erect; with cystoliths. Leaves sessile or petiolate; blade entire, crenulate, or dentate. Inflorescences axillary or terminal, reduced or expanded dichasia, sometimes forming dichasiate spikes, thyrses, or panicles, rarely reduced to a solitary flower. Bracts opposite, usually green, margin entire; bracteoles 2 or absent. Flowers sessile or subsessile to pedicellate. Calyx deeply 5-lobed; lobes equal or subequal. Corolla funnel-shaped; tube base with a narrowly cylindric portion, usually expanded apically into a distinct throat; limb 5-lobed; lobes usually ovate to orbicular, equal to unequal in size, contorted in bud. Stamens 4, didynamous, usually included; filaments sometimes connate in pairs at base; anthers bithecous; thecae equal, parallel to subsagittate, muticous at base; staminode 1 or absent. Ovary with 4–10 ovules per locule; style usually included in corolla tube or slightly exserted; stigma 2-lobed, lobes equal or unequal. Capsules stipitate or not stipitate, 12–26-seeded; retinacula present; septa with attached retinacula remaining attached. Seeds discoid, usually pubescent with hygroscopic trichomes.

Recent references: Tripp, 2007. Evolutionary relationships within the species-rich genus *Ruellia* (Acanthaceae). *Syst. Bot.* 32: 628–649; Tripp et al., 2009. New molecular and morphological insights prompt transfer of *Blechum* to *Ruellia* (Acanthaceae). *Taxon* 58: 893–906.

About 250 species: tropical to temperate regions worldwide; two species in the Pan-Himalaya.

1a. Flowers clustered in leaf axil; bracteoles linear; seeds white-margined .... 1. *R. beddomei*
1b. Flowers several in terminal or axillary cymes; bracteoles linear-oblong; seeds not white-margined .... 2. *R. tuberosa*

1. Ruellia beddomei C. B. Clarke in Hooker (ed.), Fl. Brit. India 4: 506. 1884. ≡ *Dipteracanthus beddomei* (C. B. Clarke) Santapau, Indian Forester 79: 322. 1953. Type: India, Madras, Kurneol, Nallaymallays, 2000 ft., *R. H. Beddome 60* [holotype: BM (BM000950024)].

Herbs, 0.4–1 m high; stems borne from a slender woody stock, scarcely suffruticose, obtusely tetragonous and grooved, smooth or minutely strigose, sparsely hairy upwards, rooting at lower nodes. Leaves petiolate; petiole 1–3 cm long; blade elliptic to lanceolate, 5–13 cm long, 2.5–5 cm broad, strigose, lateral veins 5–7 pairs, base suddenly narrowed and somewhat decurrent onto petiole, margin undulate-crenate, ciliate, apex acuminate. Flowers axillary, several clustered, sessile. Bract elliptic to ovate, 1.5–2 cm long, 0.5–1.5 cm broad, strigillose, apex acute or acuminate; bracteoles linear. Calyx 1.2–1.5 cm long; lobes linear-lacneolate, 7–8 mm long, abaxially adpressed hairy, margin ciliate. Corolla 7–8 cm long, light purple blue on lobes, outside pubescent; tube white, cylindric, 2.5–3 cm long, upwards gradually widened to 1.3–1.6 cm at mouth; lobes orbicular-oblong, 1.5–2 cm long, 1.2–2 cm broad. Stamens 4; filaments of longer pair 8–9 mm long, shorter pair 3–4 mm long; anther 2.5–3 mm long. Ovary glabrous; style 5–6 mm long; stigma 2-cleft. Capsules clavate, 1.5–2.5 cm long, glabrous. Seeds orbicular in outline, strongly compressed, white-margined with dense flattened hairs.

Phenology: flowering from August to December; fruiting from November to December.

Habitats: forests; 100–2200 m.

Distribution: W Nepal, C Nepal, E Nepal (Map 46-146); also in other parts of India.

W NEPAL: **Bheri**, Banke, Napelganj, Deurali, 6 Oct. 1977, *R. Rana 1865* (KATH); **Mahakali**, Kanchanpur, Betkot to Lipna, 12 Aug. 1980, *K. R. Rajbgandarii 5225* (KATH); **Rapti,** Dang, Ambas, 1800 m, 29 Aug. 1979, *T. K. Bhattachaya & K. J. Malla 2910* (KATH); Dang, Tulsipur to Ramri, 2220 m, 8 Sep. 2014, *B. Adhikari 51* (E, KATH); Dang, Sun Dhabari, 316 m, 16 Jul. 2016, *B. Adhikari & G. Parmar BG65* (E, KATH); Dang, between Tulsipur and Amiliya, 685 m, 16 Jul. 2016, *B. Adhikari & G. Parmar GB68* (E, KATH); Salyan, Sitalpati, 1060 m, 27 Aug. 1952, *O. V. Polunin, W. R. Sykes & L. H. J. Williams 1213* (BM, KATH). **C NEPAL**: **Bagmati**, Nuwakot, Trisuli, 600 m, 14 Sep. 1966, *D. H. Nicolson 2324* (KATH); **Lumbini**, Palpa, Rumre, 750 m, 18 Sep. 1953, *T. Fujimura 27* (KYO); Rupandehi, Butwal, 150 m, 7 Oct. 1954, *J. D. A. Stainton, W. R. Sykes & L. H. J. Williams 8829* (BM, KATH); **Narayani**, Makwanpur, Hetauda to Amlekhjanj, 460 m, 21 Aug. 1967, *P.*

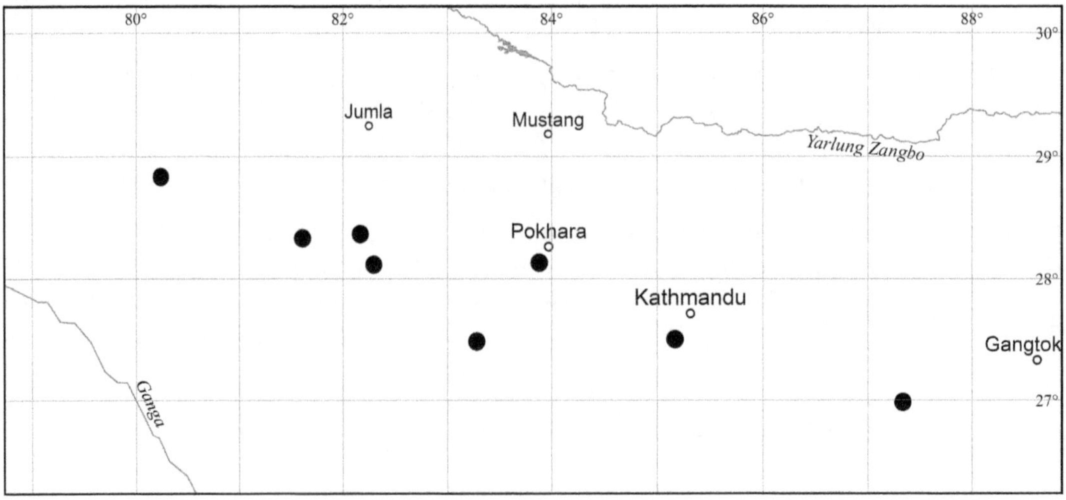

Map 46-146: **Ruellia beddomei** C. B. Clarke

R. Shakya 9336 (KATH); Makwanpur, Hetaura, Rapti Valley, 22 Aug. 1967, *L. H. J. Williams & J. D. A. Stainton 8194* (BM, KATH). **E NEPAL**: **Koshi**, Dhankuta, Diyale Bhaluse to Hangmaya, 460 m, 25 Aug. 1986, *T. B. Shrestha & P. R. Shakya 8884* (KATH).

2. Ruellia tuberosa L., Sp. Pl. 2: 635. 1753; Nasir & Ali in Malik & Ghafoor (eds.), Fl. Pakistan 188: 63. 1988; Hu, Deng & Daniel in Wu, Raven & Hong (eds.), Fl. China: 437. 2011. ≡ *Cryphiacanthus barbadensis* Nees, Prodr. 11: 197. 1847, nom. illeg. (Art. 52). Type: "Habitats in Jamaica", *Herb. Linn. 804.8* [lectotype designated by Howard, Fl. Lesser Antilles 6: 380. 1989: LINN (LINN 804.8)].

Herbs, 20–45 cm high; roots with elongate tuber-like swellings; stems quadrangular, strigillose on angles. Petiole 0.5–1 cm long, glabrous; leaf blade oblong-obovate, 4–8 cm long, 1.5–4.2 cm broad, both surfaces glabrous, base cuneate and tapering onto petiole, margin undulate to subentire, apex acute. Flowers in axillary or terminal lax dichotomous cymes; peduncle 1–3 cm long. Bracts oblong-lanceolate, 3–9 mm long, 1–3 mm broad; bracteoles linear-oblong, 1.2–1.5 mm long. Pedicel 6–10 mm long, minutely pilose. Calyx tube 5–6 mm long, abaxially tomentose; lobes linear-lanceolate, 0.9–2 cm long. Corolla pale blue, 2.2–5.5 cm long, outside hirtellous; tube cylindric, 0.8–1 cm long, upwards widened to 1–1.3 cm at mouth; lobes suborbicular, 1.3–1.6 cm long, 1.2–1.5 cm broad. Stamens 4; filaments glabrous, longer pair 6–8 mm long, shorter pair 3–4 mm; anther-thecae white, hairy on dorsal side. Ovary sparsely glandular-pilosetowards apex; style 2.1–2.3 cm long, sparsely hirsute; stigma 2-lamellate. Capsules oblong, linear-ellipsoid, 1.8–2.5 cm long, 3–4 mm broad, glandular-pilose towards apex, 20–26-seeded. Seeds discoid, 1.6–2 mm long and broad, covered with appressed hygroscopic trichomes. **Chromosomes** 2n = 32, 34. Figure 46-88.

Phenology: flowering from December to January of next year; fruiting from May to July.

Habitats: roadsides; below 600 m.

Distribution: W Nepal, C Nepal, E Nepal (Map 46-147); native to tropical America, introduced and naturalized in tropical regions.

W NEPAL: Bheri, Banke, Napelganj, 200 m, 16 Jun. 2001, *S. K. Acharya & B. Adhikari 2500* (TUCH); Banker, Nepalgungj, 180 m, 16 Oct. 2001, *T. Watababe et al. NPL-SP011213(007)* (MBK). **C NEPAL: Bagmati**, Lalitpur, Botanical Garden, 15 Jun. 1968, *R. Mapa s. n.* (KATH); **Lumbini**, Nawalparasi, Parasi, 350 m, 1 Jul. 1993, *P. P. Kurmi KB868* (KATH). **E NEPAL: Koshi**, Morang, Biratnagar, 120 m, 19 Jul. 1969, *T. K. Bhattachatya 7555* (KATH).

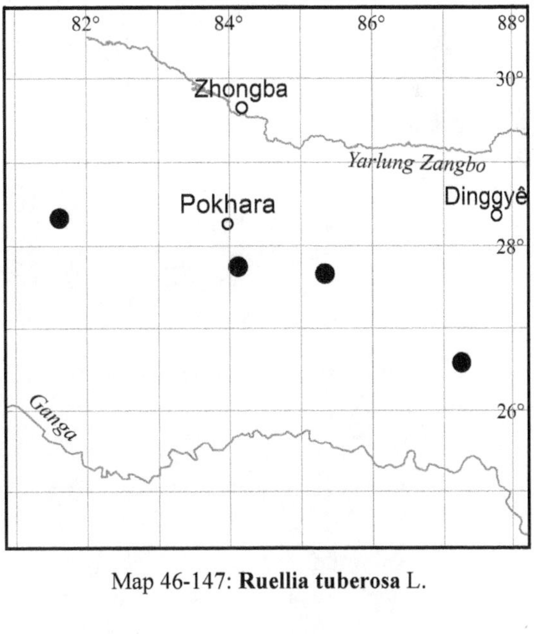

Map 46-147: **Ruellia tuberosa** L.

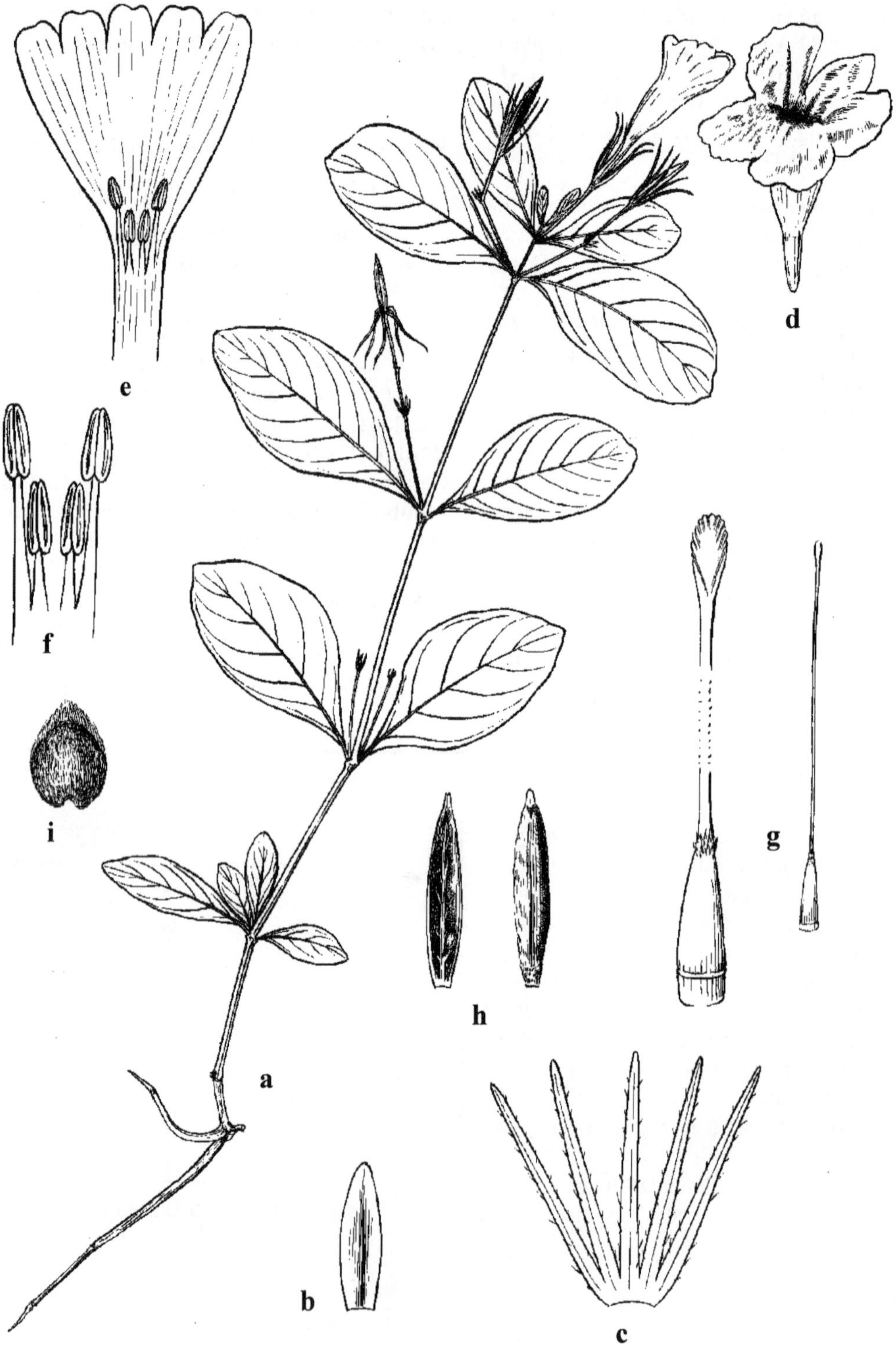

Figure 46-88: **Ruellia tuberosa** L.: **a**, habit; **b**, bracteole; **c**, calyx; **d**, corolla; **e**, opened corolla showing stamens; **f**, stamens; **g**, pistil; **h**, capsule; **i**, seed. Drawn by Ding-Han CUI.

Tribe 4. Andrographideae Endl., Gen. Pl.: 707. 1839. Type: *Andrographis* Nees.

Plants with cystoliths. Calyx 5-lobed. Corolla bilabiate, upper lip 2-lobed, lower lip 3-lobed. Stamens 2, anthers bithecous or rarely reduced to monothecous. Ovules several per locule. Capsules clavate, many-seeded, seeds borne on hook-like retinacula.

Six genera and about 100 species in Asia; three genera and 12 (one endemic) species in the Pan-Himalaya.

Recent references: Scotland, 1992. Pollen morphology of Andrographideae (Acanthaceae). *Rev. Palaeobot. Palynol.* 72: 229–243.

16. PHLOGACANTHUS Nees

Nees in Wallich (ed.), Pl. Asiat. Rar. 3: 76, 99. 1832; Nees in de Candolle (ed.), Prodr. 11: 320. 1847; Clarke in Hooker (ed.), Fl. Brit. India 4: 510. 1885; Lindau in Engler & Prantl (eds.), Nat. Pflanzenfam. 4(3b): 323. 1895; Wood in Grierson & Long (eds.), Fl. Bhutan 2(3): 1284. 2001; Hu, Fl. Reipubl. Popularis Sin. 70: 207. 2002; Hu, Deng & Daniel in Wu, Raven & Hong (eds.), Fl. China 19: 474. 2011. Type: *Phlogacanthus curviflorus* Nees, **here designated.**

Cystacanthus T. Anderson, J. Linn. Soc., Bot. 9: 457. 1867; Clarke in Hooker (ed.), Fl. Brit. India 4: 513. 1885; Hu, Fl. Reipubl. Popularis Sin. 70: 212. 2002; Hu, Deng & Daniel in Wu, Raven & Hong (eds.), Fl. China 19: 476. 2011. Type: *Cystacanthus paniculatus* T. Anderson, **here designated**.

Loxanthus Nees in Wallich, Pl. Asiat. Rar. 3: 76, 122. 1832. Type: *Loxanthus gomezii* Nees.

Janasia Raf., Fl. Tellur. 4: 63. 1838 [1836]. Type: *Janasia rosea* Raf., nom. illeg. (Art. 52) [≡ *Justicia quadriangularis* Sims = *Phlogacanthus vitellinus* (Roxb.) T. Anderson].

Herbs, shrubs, or small trees, with cystoliths. Leaves opposite; blade margin entire or obscurely crenate. Flowers in terminal thyrses or axillary cymes. Bracts small; bracteoles small or absent. Calyx deeply 5-lobed; lobes equal to unequal. Corolla tubular, slightly curved, limb subequally 5-lobed or 2-lipped, lower lip 3-lobed; lobes ascending cochlear in bud, upper lip 2-cleft. Stamens 2, inserted at the middle or basal portion of tube, included or sometimes slightly exserted beyond corolla lip; anthers bithecous; thecae equal, base divaricate and muticous; staminodes 2, small. Ovary usually glabrous; stigma subentire. Capsules linear, quadrangular, basal portion solid but lacking a conspicuous basal stalk; retinacula present, 8–16-seeded. Seeds lenticular, glabrous or pubescent.

Recent references: Xia & Deng, 2013. *Phlogacanthus yangtsekiangensis*, a new combination in Chinese Acanthaceae. *Phytotaxa*, 104: 58–60; Turner, 2015: The botanical legacy of Thomas Hardwicke's journey to Srinagar in 1796. *Europ. J. Taxon.* 108: 1–25.

About 35 species: S to SE Asia; nine (one endemic) species in the Pan-Himalaya.

Notes: The genus *Cystacanthus* T. Anderson was established by Anderson (in J. Linn. Soc., Bot. 9: 457. 1867) and recognized as an independent genus by many authors (Scotland, Rev. Palaeobot. Palynol. 72: 229–243. 1992; Scotland & Vollesen, Kew Bull. 55: 580. 2000; Hu, Fl. Reipubl. Popularis Sin. 70: 212. 2002; Hu & Fu, *Higher Pl. China* 10: 388. 2005; Wu et al., *The families and genera of angiosperms in China*: 1093. 2003, *The areal-types of seed plants and their origin and differentiation*: 224. 2006; Hu et al., Fl. China 19: 476. 2011). It differs from *Phlogacanthus* Nees only in the corolla shape and indumentum and thus, many authors merged *Cystacanthus* with *Phlogacanthus* (Benoist, *Fl. Gén. Indochine* 4: 703. 1935; Imlay, *Bull. Misc. Inform. Kew* 1939: 109-150. 1939; Brummit, *Vasc. Pl. Fam. Gen.* 1992; Mabberley, Mabberley's Pl.-Book 271. 2017). Our recently unpublished molecular analysis, however, showed that species of *Cystacanthus* are nested within *Phlogacanthus* and they might be treated as a single genus, *Phlogacanthus*.

1a. Corolla tube abruptly inflated near the middle and usually bent ca. 90° (sect. I. **Cystacanthus**).
  2a. Corolla outside pubescent with glandular-pilose.. 1. *P. yangtsekiangensis*
  2b. Corolla outside eglandular pubescent ....2. *P. vittelinus*
1b. Corolla tube cylindric and slightly curved.
  3a. Inflorescences axillary cymes or racemes; stamens exserted (Sect. II. **Pubinervia**).
    4a. Leaves sparsely pubescent; bracts, bracteoles and calyx covered with eglandular pubescent .... 3. *P. pubinervius*
    4b. Leaves densely softly hairy on both surfaces; bracts, bracteoles and calyx densely glandular-pubescent...4. *P. lambertii*
  3b. Inflorescences terminal thyrses, or sometimes with several borne on upper leaf axils; stamens included (sect. III. **Phlogacanthus**).
    5a. Leaves densely softly villous beneath ....5. *P. tubiflorus*
    5b. Leaves glabrous or sparsely pubescent along veins beneath.
      6a. Flowers arranged in lax narrow dichotomously branched panicle; pedicels 2–5 mm long.. ....6. *P. parviflorus*
      6b. Flowers arranged in densely thyrses; pedicels usually less than 2 mm long.
        7a. Corolla 4.5–5 cm long...7. *P. curviflorus*
        7b. Corolla less than 3 cm long.
          8a. Thyrses dense, not interrupted; corolla orange-red; anthers 3–4 mm .... 8. *P. thyrsiformis*
          8b. Thyrses interrupted below with internodes up to 2 cm; corolla orang or orange-yellow; anthers 2–3 mm .... 9. *P. guttatus*

I. Phlogacanthus sect. **Cystacanthus** (T. Anderson) Y. F. Deng, st. nov. Basionym: *Cystacanthus* T. Anderson, J. Linn. Soc., Bot. 9: 457. 1867; Clarke in Hooker (ed.), Fl. Brit. India 4: 513. 1885; Hu, Fl. Reipubl. Popularis Sin. 70: 212. 2002; Hu, Deng & Daniel in Wu, Raven & Hong (eds.), Fl. China 19: 476. 2011. Type: *Phlogacanthus paniculatus* (T. Anderson) J. B. Imlay (≡ *Cystacanthus paniculatus* T. Anderson).

Thyrses or racemes terminal or axillary. Corolla tube abruptly inflated near the middle and usually bent ca. 90°. Stamens included.

Twenty-six species in S and SE Asia; two species in the Pan-Himalaya.

1. **Phlogacanthus yangtsekiangensis** (H. Lév.) C. Xia & Y. F. Deng, Phytotaxa 104: 68. 2013. ≡ *Strobilanthes yangtsekiangensis* H. Lév., Cat. Pl. Yun-Nan, 6. 1915. ≡ *Cystacanthus yangtsekiangensis* (H. Lév.) Rehder, J. Arnold Arbor. 16: 315. 1935; Hu, Fl. Reipubl. Popularis 70: 213. 2002; Hu, Deng & Daniel in Wu, Raven & Hong (eds.), Fl. China 19: 476. 2011. Type: [S Hengduan] China Yunnan, Dêqên Xian, Rives du fleuve Bleu à Ta-Tchai, 400–450 m, May–June 1912, *E. E. Maire s. n.* [holotype: E (E00095241); isotypes: A (A00063102, A00063097), E (E00095242)].

Cystacanthus affinis W. W. Sm., Notes Roy. Bot. Gard. Edinburgh 9: 103. 1906; Hu, Fl. Reipubl. Popularis 70: 212. 2002; Hu, Deng & Daniel in Wu, Raven & Hong (eds.), Fl. China 19: 476. 2011. Type: [S Hengduan] China, Yunnan, Lijiang, Fengkow Valley, Lat. 27°40′N", 8000 ft., Jun. 1914, *G. Forrest 12493* [holotype: E (E00095255); isotypes: A, IBSC (IBSC0004957), K (K000885626)].

Cystacanthus yunnanensis W. W. Sm., Notes Roy. Bot. Gard. Edinburgh 9: 104. 1916; Hu, Fl. Reipubl. Popularis 70: 214. 2002; Hu, Deng & Daniel in Wu, Raven & Hong (eds.), Fl. China 19: 476. 2011. Type: China, Yunnan, Dengchuan Xian, "Tengchwan Valley", Lat. 26°5′N, 7000 ft., May 1913, *G. Forrest 10007* [holotype: E (E00095254); isotypes: A (A00093824), K (K000885625)].

Shrubs, 1–2 m high; stems subterete or slightly quadrangular, striate, densely whitish or yellowish pubescent. Petiole 0.5–1.5 cm long, pubescent; leaf blade ovate to ovate-lanceolate, (2.5–)5–10 cm long, (1–)3.5–4 cm broad, both surfaces pubescent and glaucous, lateral veins 10–12 pairs, base cuneate, margin entire, apex acute. Thyrses terminal or axillary, racemose, secund, 5–10 cm long, densely glandular-hairy. Bracts linear-lanceolate, 5–8 mm long, densely glandular-villous; bracteoles similar to bracts. Pedicel 2–4 mm long, glandular-pubescent. Calyx 7–10 mm long; lobes linear, densely yellowish pubescent. Corolla blue or red, campanulate, 2.2–2.5 cm long, ventricose and curved, pubescent with glandular-pilose outside; tube base cylindric, 1–1.2 cm long, 4–5 mm broad, upwards bent to 90° and widened to 1.2–1.3 cm broad; lobes broadly triangular, apex obtuse. Stamens 2, included; filaments 5–7 mm long, barbate; anther-thecae ovoid, connective barbate with yellowish brown trichomes; staminodes 2–2.5 mm long. Ovary pubescent; style 2–2.3 cm long, glabrous; stigma slightly 2-cleft. Capsules 3–3.5 cm long, 4–5 mm broad, glandular-villous. Seeds disciform, 2–3 mm long and broad, puberulent. Figure 46-89.

Phenology: flowering from March to May.

Habitats: thickets in valley; 400–2400 m.

Distribution: U Yarlung Zangbo, S Hengduan (Map 46-148); also in other parts of China (C & NE Yunnan, SE Sichuan).

U YARLUNG ZANGBO: **Nylam**, Zangmu, Youyiqiao, 11 May 1966, *Y. T. Zhang & K. Y. Lang 3277* (KUN). **S HENGDUAN**: **Bingchuan**, Hetaoping, 30 Mar. 1957, *Y. C. Xu s. n.* (SWFC); **Dali**, N of the Tali valley, 2010 m, May 1922, *G. Forrest 21148* (A, CAL, E, K); **Miyi**, Huanhai, Dakaiwan, 17 Jun. 1958, *S. Y. Chen et al. 10234* (NAS, SZ); Binggu, Anshouhe, 1100 m, 11 Apr. 1963, *Mao (Cymbopogon) Working Group CMY 31* (CDBI); Binggu, 1020 m, 10 Mar. 1964, *s. coll. 391* (CDBI); **Ningnan**, Hulukou, 750 m, 1 Sep. 2012, *Y. F. Deng 23363* (IBSC); **Panzhihua**, Dukoui, 1000 m, Mar. 1984, *G. D. Tao 38990* (HITBC); Dukou, 1000 m, Mar. 1984, *G. D. Tao 40157* (HITBC); near Ertan, 1300 m, 2 May 1980, *Z. G. Liu & Q. Q. Wang 21196* (CDBI); **Xichang**, Lushan, 1800–2000 m, 8 Apr. 1959, *Chuan Jing Xi 18* (CDBI, KUN, SZ); s. loc., 15 Mar. 1964, *J. Y. He & Z. L. Wang 394* (CDBI); **Yanbian**, Qinghe, 1400 m, 3 May 2003, *X. F. Gao et al. 4676* (CDBI); Qinghe, 1400 m, 3 May 2003, *X. F. Gao 4677* (CDBI); **Yongsheng**, Yongpeh, 18 Apr. 1939, *R. C. Ching 20052* (KUN); Yongbei, 18 Apr. 1939, *R. Q. Feng 20052* (KUN); valley south of Yung-pei, 1500 m, 18 May 1921, *F. Kingdon-Ward 3871* (E); **Yunlong** (Lijiang), Feng-Kou, the flank of Yangtze, 2400 m, 15 Apr. 1922, *F. Kingdon-Ward 5042* (E); Likiang Snow Range, 6-11 May 1922, *J. F. Rock 3185* (A);

Figure 46-89: **Phlogacanthus yangtsekiangensis** (H. Lév.) C. Xia & Y. F. Deng: **a**, flowering branch; **b**, calyx; **c**, corolla opened showing androecium; **d**, stamen; **e**, pistil. Drawn by Quan LIN; reproduced from *Icon. Cormophyt. Sin.* 4: 168, fig. 5750. 1975.

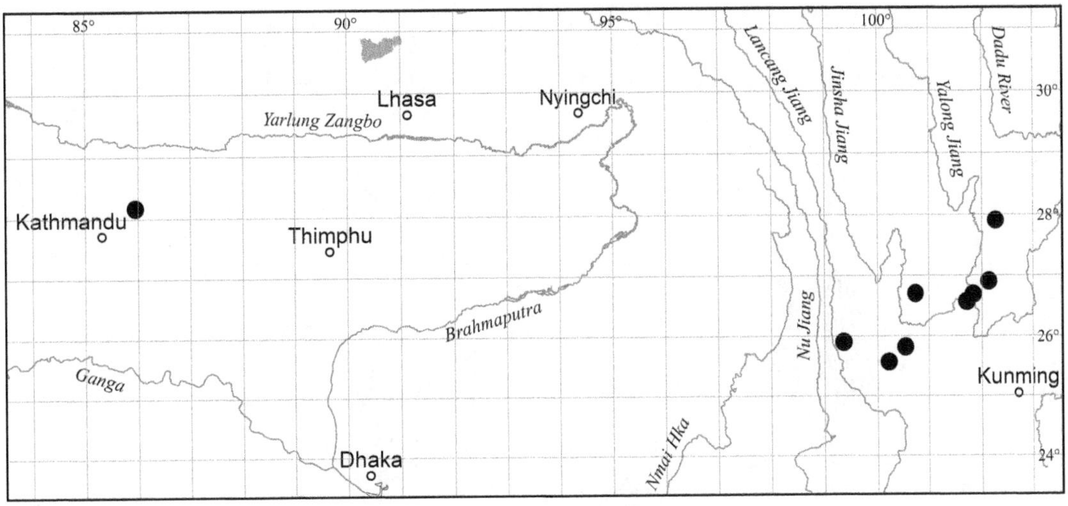

Map 46-148: **Phlogacanthus yangtsekiangensis** (H. Lév.) C. Xia & Y. F. Deng

Daju, Hutiaoxia, 1700 m, 3 Jun. 1985, *Q. Lin et al. 771619* (KUN).

2. Phlogacanthus vitellinus (Roxb.) T. Anderson, J. Linn. Soc., Bot. 9: 507. 1867; Wood in Grierson & Long (eds.), Fl. Bhutan 2(3): 1285. 2001; Hu, Fl. Reipubl. Popularis Sin. 70: 209. 2002. ≡ *Justicia vitellina* Roxb., Fl. Ind. 1: 117. 1820. ≡ *Endopogon vitellinus* (Roxb.) Nees, Prodr. 11: 723. 1847. ≡ *Cystacanthus vitellinus* (Roxb.) Y. F. Deng, Fl. China 19: 477. 2011. Type: [*icon*] *Roxburgh Icon. 2166* (lectotype **here designated**: K).

Justicia quadrangularis Hook., Bot. Mag. 55: t. 2845. 1828; Bot. Reg. 16: tab. 1340. 1830. ≡ *Phlogacanthus quadrangularis* (Hook.) Heine, Bull. Mus. Natl. Hist. Nat., B, Adansonia Sér. 4, 10: 148. 1988. Type: [*icon*] Bot. Mag. 55: t. 2845. 1828 (lectotype **here designated**).

Justicia asperula Lodd., Bot. Cab. 17: tab. 1681. 1831. ≡ *Phlogacanthus asperulus* (Lodd.) Nees, Pl. Asiat. Rar. 3: 99. 1832; Nees in de Candolle (ed.), Prodr. 11: 321. 1847; Clarke in Hooker (ed.), Fl. Brit. India 4: 511. 1884; Lindau in Engler & Prantl (eds.), Nat. Pflanzenfam. 4(3b): 323. 1895. Type: India, *Wallich Cat. 7171* [lectotype **here designated**: K-W (K001126820); isotype: BM].

Phlogacanthus pulcherrimus T. Anderson, J. Linn. Soc., Bot. 9: 507. 1867. ≡ *Cystacanthus pulcherrimus* (T. Anderson) C. B. Clarke, Fl. Brit. India 4: 514. 1884. Type: Myanmar, Tavoy Island, Moulmein, in 1862, *Parish 416* [lectotype **here designated**: K (K000885620)].

Shrubs, 1–1.5 m high; stems terete, glabrous. Petiole 1–5 cm long; leaf blade broadly ovate to oblong-lanceolate, 10–30 cm long, 6–11 cm broad, abaxially slightly pubescent or glabrescent, adaxially glabrous or slightly pubescent, lateral veins 9–11 pairs, base attenuate and decurrent onto petiole, margin slightly undulate, apex acute to acuminate. Thyrses 10–20 cm long. Bracts minute. eglandular-pubescent; bracteoles linear, 1–1.5 mm long, 0.4–0.5 mm broad, eglandular pubescent. Pedicel 1–3 mm long. Calyx 5–6 mm long, pubescent; lobes linear-lanceolate. Corolla purplish- red or white, 1.2–1.8 cm long, outside pubescent, bent to 90° and abruptly inflated at the middle. Stamens included; filaments 0.7–1 cm long, hirsute towards base; anther-thecae oblong, 2–2.3 mm long, connective glabrous; staminodes very small. Ovary glabrous; style 1–1.5 cm long, glabrous; stigma slightly 2-cleft. Capsules clavate, 2–2.3 cm long, 8-seeded. Seeds ovate in outline, compressed, 2.5–3 mm long, 2–2.5 mm broad, puberulent. Figure 46-90.

Figure 46-90: **Phlogacanthus vitellinus** (Roxb.) T. Anderson: **a**, flowering branch; **b**, leaf; **c**, flower; **d**, anther; **e**, capsule. Drawn by Quan LIN; reproduced from *Icon. Cormophyt. Sin.* 4: 167, fig. 5747. 1975.

Phenology: flowering and fruiting from November to March of next year.

Habitats: forests; 100–1800 m.

Distribution: Yarlung Zangbo-Brahmaputra, U Irrawaddy (Map 46-149); also in Bangladesh, Thailand, and other parts of India, Myanmar and China (S Yunnan).

YARLUNG ZANGBO-BRAHMAPUTRA: Lohit, 4 Dec., 1969, *J. Joseph 48479* (CAL); Pasighat, 5 Mar., 1912, *I. H. Burkill 36863* (CAL); *I. H. Burkill 38159* (CAL). **U IRRAWADDY: Putao**, Sumprabun, from Sunprabum to Kumon Range, between Hpuginhku and N'Dum Zup, 150–1800 m, 15 Jan. 1962, *J. Keenan et al. 3252* (E).

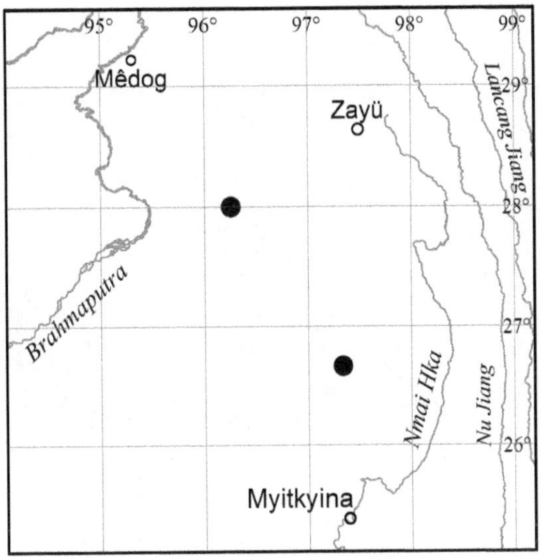

Map 46-149: **Phlogacanthus vitellinus** (Roxb.) T. Anderson

II. Phlogacanthus sect. Pubinervia Y. F. Deng, sect. nov. Type: *Phlogacanthus pubinervius* T. Anderson.

Inflorescences axillary cymes or racemes. Corolla tube cylindric and slightly curved. Stamens exserted.

Three species: Myanmar, China, India, Nepal, Bhutan; two species in the Pan-Himalaya.

3. Phlogacanthus pubinervius T. Anderson, J. Linn. Soc., Bot. 9: 508. 1867; Clarke in Hooker (ed.), Fl. Brit. India 4: 513. 1885; Wood in Grierson & Long (eds.), Fl. Bhutan 2(3): 1285. 2001; Hu, Fl. Reipubl. Popularis Sin. 70: 210. 2002. Type: [Sikkim & Darjiling] INDIA, Sikkim, 5000 ft., *J. D. Hooker s. n.* [lectotype **here designated**: K (K000950012); isolectotype: C (C10005087)].

Shrubs or small trees, 2–5 m high. Petiole 0.5–2 cm long; leaf blade elliptic-oblong to oblong, (5–)8–18 cm long, (1–)3.5–5 cm broad, abaxially puberulent along veins, adaxially scabrous, lateral veins 5–7 pairs, base attenuate, margin undulate, apex acuminate to long acuminate. Cymes axillary, 1–3 cm long, 1–4-flowered; peduncle 0.8–1.6 cm long, quadrangular, pubescent. Bracts small, caducous; bracteoles subulate, 2.5–3 mm long, 0.8–1 mm broad, caducous. Pedicel 0.8–1.5 cm long. Calyx 7–8 mm long, abaxially pubescent, adaxially gray tomentose; lobes linear-lanceolate, unequal. Corolla orange, 1.6–1.8 cm long, outside pubescent; tube 1.1–1.3 cm long, slightly curved, lower lip deeply 3-lobed, lobes ovate, upper lip 2-cleft. Stamens 2, exserted; filaments 2–2.5 cm long, glabrous; anther-thecae oblong, parallel; staminodes 2, small. Ovary glabrous; style 1.8–2 cm long, glabrous. Capsules 2.5–3 cm long, glabrous, 8-seeded. Seeds ovate in outline, compressed, 2–2.5 mm long, 1.8–2.2 mm broad, puberulent. Figure 46-91.

Phenology: flowering from November to January of next year; fruiting from January to March.

Habitats: forests, thickets; 100–2100 m.

Distribution: W Nepal, C Nepal, E Nepal, Sikkim & Darjiling, Bhutan, S Hengduan, U Irrawaddy (Map 46-150); also in other parts of India, Myanmar and China (C, S, SE & SW Yunnan, SE Sichuan, Guangxi, Guizhou).

Figure 46-91: **Phlogacanthus pubinervius** T. Anderson: **a**, portion of plant; **b**, corolla; **c**, stamen; **d**, capsule. Drawn by Quan LIN; reproduced from *Icon. Cormophyt. Sin.* 4: 168, fig. 5749. 1975.

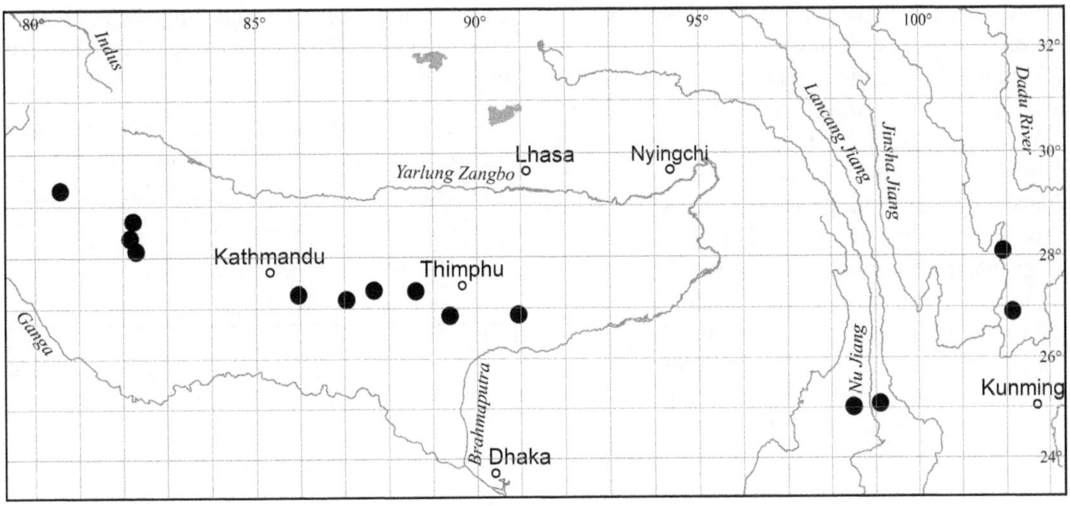

Map 46-150: **Phlogacanthus pubinervius** T. Anderson

W NEPAL: **Bheri**, Jajarkot, Dar, north of Jajakot, 1210 m, 3 Apr. 1952, *O. V. Polunin, W. R. Sykes & L. H. J. Williams 3816* (A, BM); Jajarkot, Khurpa, 1670 m, 5 Apr. 1968, *J. D. A. Stainton 6208* (BM, KATH); **Mahakali**, Dadeldhura, Kamala Khola, 300 m, 10 Feb. 1982, *J. D. A. Stainton 8496* (E); **Rapti**, Dang, Jaldhung to Haripur, 10 Feb. 1967, *M. L. Banerji s. n.* (KATH); Salyan, Jugyana, Salleyana, 850 m, 18 Mar. 1976, *N. P. Manandhar & P. M. Regmi 412* (KATH). **C NEPAL**: **Gandaki**, Lamjung, Hile, 1400 m, 13 Jan. 1983, *T. Kadota s. n.* (KYO); **Janakpur**, Sindhuli, Hariharpur Garhi, 460 m, 15 Feb. 1967, *J. D. A. Stainton 5637* (BM, E, KATH); Sindhuli, Gwang Khola, 600 m, 31 Dec. 1976, *P. R. Shakya & M. B. Raychhetri 3867* (KATH). **E NEPAL**: **Koshi**, Bhojpur, between Zonkedebhan and Phedi, 23 Jan. 1977, *H. Tabata et al. 7817* (KATH); **Mechi**, Taplejung, Tuwa to Taplethok, 5 Nov. 1963, *H. Hara et al. s. n.* (KYO); Taplejung, Sanghu, 1760 m, 9 Jan. 1962, *A. H. Norkett 8630* (BM, E). **SIKKIM & DARJILING**: **Sikkim**, 1500–2100 m, *J. D. Hooker s. n.* (A, BM, CGE), Silong, 1500 m, Feb. 1882, *J. S. Gamble 10266* (K); Shittang, towards Lalpauchor, 1650 m, 1 Jan. 1871, *J. S. Gamble 3575A* (K). **BHUTAN**: **Phuntsholing**, Sorchen (between Phuntsholing and Gedu), 1200 m, 18 Jan. 1992, *J. R. I. Wood 7529* (E); Chima Khoti, 2 Jun. 1967, *H. Hara et al. 57* (TI); **Shamgang**, N Shamgang, 1100 m, 16 Mar. 1992, *J. R. I. Wood 7553* (E); **Tongsa**, near Pertimi, 1400 m, 3 Apr. 1982, *A. J. C. Grierson & D. G. Long 4277* (A, E, K). **S HENGDUAN**: Salwin Valley, 900 m, Apr. 1906, *G. Forrest 5011* (A, E, LE); **Longyang**, Mangkuan, Baihualing, 1600 m, 20 Dec. 2000, *H. Li 13593* (KUN); Baihualing, 1600 m, 24 Jan. 2008, *Y. F. Deng 20485* (IBSC); **Lushui**, Liuku, Yuejinqiao, 13 Sep. 2013, *Y. F. Deng et al. 24904* (IBSC); **Miyi**, Dakaiwan, 17 June, 1958, *S. Y. Chen et al. 10234* (NAS); **Muli**, Tsungchi, 2200 m, 18 Jun. 1937, *T. T. Yü 6393* (PE); **Yulong** (Lijiang), 17 Sep. 2013, *Y. F. Deng et al. 25181* (IBSC). **U IRRAWADDY**: **Putao**, Sumprabum, 900 m, 17 Jan. 1953, *F. Kingdon-Ward 20395* (E); **Tengchong**, south of Tengyueh, 1800 m, Feb. 1925, *G. Forrest 26203* (A, AU, BM, E, IBSC, PE); Machang to Mangbang, 2013, *Y. F. Deng 24698* (IBSC).

4. **Phlogacanthus lambertii** Raizada, Indian Forester 60: 535. 1934. Type: [U GANGA & INDUS] INDIA, Uttarakhand, Nainital, Danda, Haldwani Div., 2800 ft., 22 Jan. 1927, *A. E. Osmaston 1327* [holotype: DD; isotype: K (K000950008)].

Shrubs, 1–3 m high; stems pubescent. Petiole 1–3 cm long, villous; leaf blade elliptc, 13–20 cm long, 5–10 cm broad, both surfaces densely pubescent when young, then glabrescent, lateral veins 7–9 pairs, base attenuate, margin entire, apex acuminate. Cymes axillary, 1–1.5 cm long, (1–)2–4-flowered. Pedicels 3–6 mm long, pubescent. Bracts subulate, 1.7–2 mm long, densely glandular-pubescent; bracteoles subulate, 2–3 mm long, 0.8–1 mm broad, densely glandular-pubescent. Calyx densely glandular-pubescent, 5-lobed almost to base; lobes narrowly oblong, 7–10 mm long, apex acuminate. Corolla 1.1–1.3 cm long, orange with dark purple veins, tubular-inflated from base, slightly curved, outside densely puberulent and glandular, inside pubescent; lobes ovate. Stamens 2; filaments glabrous, curved, 1.8–2.5 cm long. Ovary glandular-hairy; style 1.5–1.8 cm long. Capsules oblong, elongate, 2–2.8 cm long, glabrous, 8-seeded. Seeds ovate in outline, compressed, 2–2.5 mm long, 1.8–2.2 mm broad, pubescent.

Phenology: flowering from January to February; fruiting from March to April.

Habitats: forests; 700–1400 m.

Distribution: endemic; U Ganga & Indus (Map 46-151).

Notes: This species differs from *P. pubinervius* T. Anderson by possessing larger leaves and densely glandular-pubescent bracts, bracteoles and caylx while eglandular-pubescent in *P. pubinervius*.

U GANGA & INDUS: **Almora**, near Baram, Gori valley, *W. J. Lambert s. n.* (DD, fide Raizada & Saxena, 1978); **Nainital**, Danda, Haldwani, 840 m, 22 Jan. 1927, *A. E. Osmaston 1327* (K).

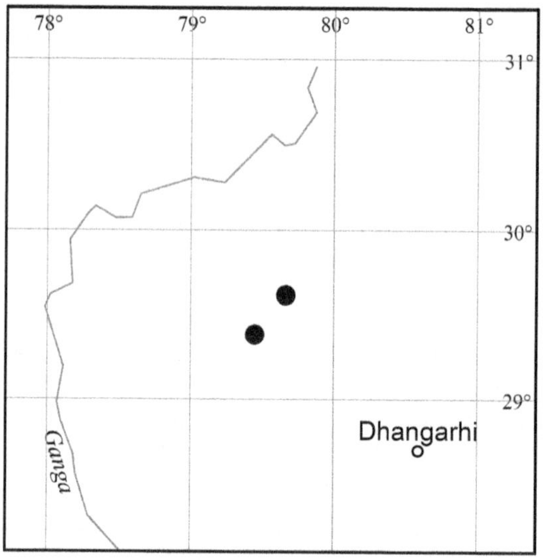

Map 46-151: **Phlogacanthus lambertii** Raizada

III. Phlogacanthus sect. **Phlogacanthus**. Type: *Phlogacanthus curviflorus* Nees.

Inflorescences terminal thyrses, or sometimes with several borne on upper leaf axils. Corolla tube cylindric and slightly curved. Stamens included.

About ten species from Himalaya to SE Asia through Indochina; five species in the Pan-Himalaya.

5. Phlogacanthus tubiflorus Nees, Pl. Asiat. Rar. 3: 99. 1832; Nees in de Candolle (ed.), Prodr. 11: 321. 1847; Clarke in Hooker (ed.), Fl. Brit. India 4: 511. 1885. ≡ *Justicia tubiflora* Buch.-Ham. ex Wall., Numer. List n. 2428. 1830, nom. invalid. Type: India, "In montibus Camrupae ad Goalparam", 12 Dec. 1808, *Hamilton s. n. in Wallich Cat. 2428* [lectotype **here designated**: K-W (K001115915); isolectotype: GZU (GZU000251442)].

Phlogacanthus tubiflorus var. *angustifolius* Nees in de Candolle (ed.), Prodr. 11: 321. 1847, **syn. nov**. Type: India, Assam, *W. Griffith s. n.* [holotype: K "herb. Hook."; isotype: GZU (GZU000251447)].

Shrubs, 1–3 m high. Petiole 3–5 cm long; leaf blade 20–30 cm long, 10–13 cm broad, elliptic or ovate, adaxially glabrous, abaxially pubescent especially along veins; lateral veins 8–10 pairs, base cuneate or decurrent onto petiole, margin entire, subentire, or obscurely crenulate, apex acuminate. Thyrses terminal, 10–15 cm long, somewhat hairy. Pedicels 2–4 mm long, pubescent. Bracts 2.5–3 mm long, triangular, pubescent; bracteoles subulate, 0.8–1 mm long.

Calyx lobes triangular, lanceolate, 4–6 mm long. Corolla 2.5–3 cm long, orange-red, slightly curved, pubescent, upper lip 2-lobed, lobes ovate, 3.5–4 mm long, 2.5–3 mm broad, lower lip 3-lobed, lobes triangular-ovate, 1.8–2 mm long and broad. Stamens 2; filaments 2.5–3 cm long, glabrous; anthers 3–3.5 mm long, arcuate. Ovary glabrous; style 2.5–3 cm long. Capsules clavate, 3–3.5 cm long, 3–3.5 mm broad, glabrous, 12-seeded. Seeds ovate in outline, compressed, 4–5 mm long, 3.5–4 mm broad, glabrous, compressed, blackish-brown.

Phenology: flowering and fruiting from December to March of next year.

Habitats: hill forest floors.

Distribution: U Irrawaddy (Map 46-152); also in Bangladesh, and other parts of India.

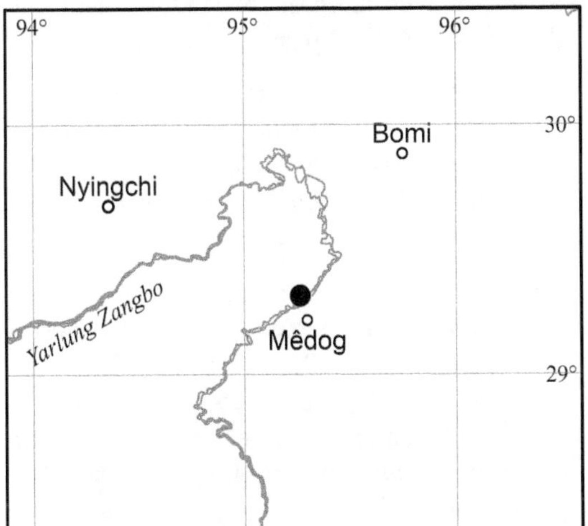

Map 46-152: **Phlogacanthus tubiflorus** Nees

U IRRAWADDY: **Chaungmaw Kyaun**, North Triangle, Hkrang Hka valley, 900 m, 30 Mar. 1953, *F. Kingdon-Ward 21649* (A, BM); **Putao**, Pinbow, 29 Nov. 1926, *C. E. Parkinson 366* (FHO); Supra Burma, 1200 m, 27 Jan. 1953, *F. Kingdon-Ward 20440* (A, E); SE of Sumprabum 300–450 m, 19 Dec. 1953, *F. Kingdon-Ward 21714* (A, BM); Hot Springs 7 miles from Hsenwi on Lashlo road, 760 m, 7 Jan. 1958, *H. S. McKee 6005* (K); Wasi, 180 m, 12 Dec. 1911, *S. M. Toppin 4084* (CAL, E); Nearr Rat ula Uray wyo, 1050 m, 29 Oct. 1916, *A. Rodger 656* (NAS); upper Burma, Jan. 1901, *D. Brandis s. n.* (K); Shwebo, Dec. 1914, *H. H. Heines s. n.* (K).

6. Phlogacanthus parviflorus T. Anderson, J. Linn. Soc., Bot. 9: 506. 1867; Clarke in Hooker (ed.), Fl. Brit. India 4: 512. 1885. Type: Yarlung Zangbo-Brahmaputra, Mishmee, *Herb. Griffith 6155* [lectotype **here designated**: (K000950021); isolectotypes: K (K000950017), P].

Shrubs, 2–4 m high; stems quadrangular, sulcate, sparsely pubescent. Petiole 2–4 cm long; leaf blade lanceolate, 9–15 cm long, 2.5–8 cm broad, glabrous, glabrous, lateral veins 6–10 pairs, base narrowly cuneate and decurrent onto petiole, margin entire, apex acute. Thyrses terminal or axillary, dichotomously branched, 5–15 cm long; rachis pubescent. Pedicels 2–5 mm long, pubescent. Bracts subulate, 1.8–2 mm long, margin ciliate; bracteoles subulate, 1.3–1.5 mm long, margin ciliate. Calyx 3–4 mm long, 5-lobed to base; lobes linear-lanceolate, apex acuminate, margin ciliate. Corolla 2.5–3 cm long, slightly curved, pubescent outside. Ovary 2–3 mm long, hairy; style 1.8–2.3 cm long, hirsute. Capsules clavate, 2.5–3 cm long.

Phenology: flowering from October to February of next year; fruiting from February to March.

Habitats: forests; 900–2500 m.

Distribution: Yarlung Zangbo-Brahmaputra, U Irrawaddy (Map 46-153); also in other parts of N Myanmar and NE India.

YARLUNG ZANGBO-BRAHMAPUTRA: Luhit Valley, Theronliang Tidding Valley, 1200–1800 m, 6 Mar. 1928, *F. Kingdom-Ward 7921* (K); Tidding Saddle, 900–1500 m, 7 Feb. 1950, *F. Kingdon-Ward 19125* (BM); Mishmee, *Herbs Griffith 6154* (K); *Herbs Griffith 6155* (K). **U IRRAWADDY**: **Putao**, Hills east of Lort Herty, 1200 m, 29 Dec. 1930, *F. Kingdon-Ward 9076* (BM); Hkamdaukmyit Camp, 2457 m, 21 Oct. 2014, *Putao Expedition Team (X. H. Jin et al.) PT-ET 1058* (E).

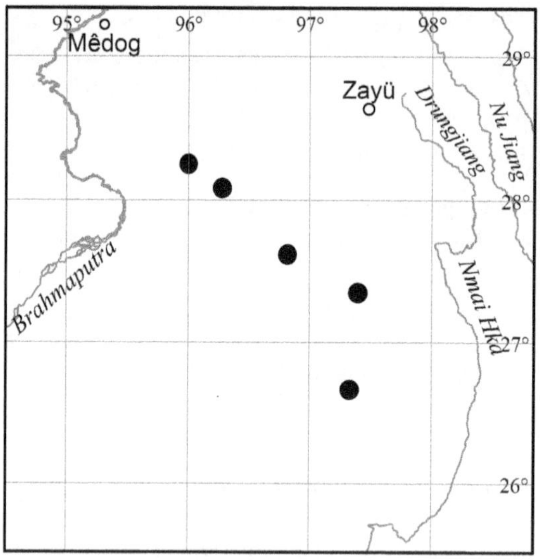

Map 46-153: **Phlogacanthus parviflorus** T. Anderson

7. Phlogacanthus curviflorus (Wall.) Nees in Wallich (ed.), Pl. Asiat. Rar. 3: 99. 1832; Nees in de Candolle (ed.), Prodr. 11: 320. 1847; Clarke in Hooker (ed.), Fl. Brit. India 4: 511. 1885; Hu, Fl. Reipubl. Popularis 70: 210. 2002; Hu, Deng & Daniel in Wu, Raven & Hong (eds.), Fl. China 19: 475. 2011. ≡ *Justicia curviflora* Wall., Pl. Asiat. Rar. 2: 9. t. 114. 1831. Type: Bangladesh, e mount Sillet, *Wallich Cat. 2429a* [lectotype **here designated**: K-W (K001115916); isolectotypes: BM, E (E00273556), K (K00950024)].

Phlogacanthus curviflorus var. *brevicalyx* C. B. Clarke, Fl. Brit. India 4: 511. 1885. Type: Myanmar, Pegu, 15 Jul. 1872, *S. Kurz 2163* [lectotype **here designated**: K (K000950026)].

Shrubs, 3–6 m high. Petiole 1.5–5 cm long; leaf blade elliptic to oblong, 12–35 cm long, 9–18 cm broad, abaxially puberulent along veins, then glabrescent, adaxially glabrous, lateral veins 12–17 pairs, base broadly cuneate and decurrent onto petiole, margin entire to subcrenate, apex acute to acuminate. Thyrses terminal, 14–18 cm long, 6–8 cm broad; cymes (1–)3–5-flowered. Pedicel 5–12 mm long, densely tomentose. Bracts subulate to lanceolate, 3–7 mm long; bracteoles subulate 2–3 mm long. Calyx 6–11 mm long; lobes slightly unequal, subulate to lanceolate, subequal, outside sparsely pubescent, inside densely brown tomentose. Corolla purplish-red, 4.5–5 cm long, outside densely brownish pubescent and sometimes with glandular-pilose; tube 4–4.2 cm long, lower lip 3-lobed, lobes oblong 1.8–2 mm long, 1.4–1.6 mm broad; upper lip 2-cleft. Stamens slightly exserted; filaments glabrous; anther-thecae ellipsoid, 5–5.5 mm long; staminodes 2, small. Ovary and style glabrous. Capsules clavate, 3.5–5 cm long, 8–10-seeded. Seeds 5.5–6.3 mm long, 4.2–4.5 mm broad, pubescent. Figure 46-92.

Phenology: flowering from October to February; fruiting from February to May.

Habitats: thickets, forest margins, ravines; 100–600 m.

Distribution: U Irrawaddy (Map 46-154); also in Bangladesh, Laos, Thailand, Vietnam, and other parts of India, Myanmar and China.

U IRRAWADDY: **Putao**, Upper Shankaung, 450 m, 12 Oct. 2014, *Putao Expedition Team (X. H. Jin et al.) PT-ET 79* (PE); Kaungmulon, 410 m, 26 Oct. 2014, *Putao Expedtion Team (X. H. Jin et al.) PT-ET 1316* (PE).

8. Phlogacanthus thyrsiformis (Roxb. ex Hardw.) Mabb. in Manila (ed.), Bot. Hist. Hortus Malabar.: 83. 1980; Mabberley, Taxon 29: 606. 1980; Wood in Grierson & Long (eds.), Fl. Bhutan 2(3): 1284. 2001. ≡ *Justicia thyrsiformis* Roxb. ex Hardw., Asiat. Res. 6: 349 (1799). Type: [*icon*] *Hardwicke Drawing* No. 16 (lectotype designated by Turner, Europ. J. Tax. 108: 3. 2015: BM).

Justicia thyrsiflora Roxb., [Hort. Bengal. 3. 1814, nom. invalid. (Art. 38.1)] Fl. Ind. (ed. Carey & Wallich) 1: 116. 1820; Fl. Ind. (ed. Carey) 1: 114. 1832. ≡ *Phlogacanthus thyrsiflora* (Roxb.) Nees in Wallich (ed.), Pl. Asiat. Rar. 3: 99. 1832; Nees in de Candolle (ed.), Prodr. 11: 321. 1847; Anderson, J. Linn. Soc. 9: 506. 1867; Clarke in Hooker (ed.), Fl. Brit. India 4: 512, 1884. Type: *Herb. Roxb. in Wallich Cat. 2430a* [lectotype **here designated**: K-W (K001115918); isolectotyes: BM, CGE, K (K000950020), K-W (K001115924)].

Figure 46-92: **Phlogacanthus curviflorus** (Wall.) Nees: **a**, leaf; **b**, flowering branch; **c**, flower; **d**, opened corolla showing stamens; **e**, capsule. Drawn by Quan LIN; reproduced from *Icon. Cormophyt. Sin.* 4: 167, fig. 5748. 1975.

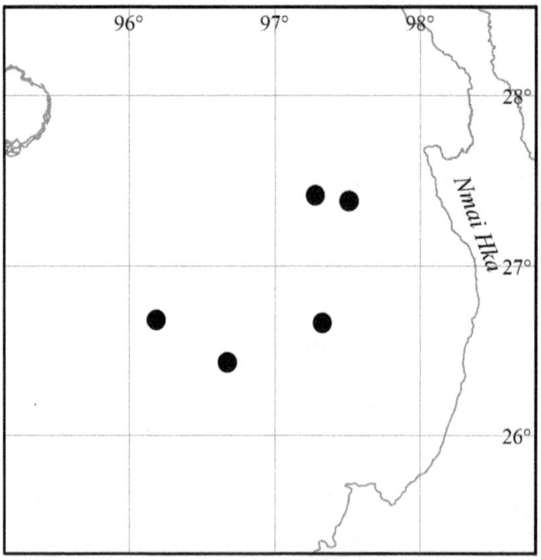

Map 46-154: **Phlogacanthus curviflorus** (Wall.) Nees

Evergreen shrubs, to 3 m high; stems yellowish-brown, quadrangular, striate, glabrous. Leaves often crowded near branch tips; petiole 1–3 cm long; blade oblanceolate or elliptic-oblong, (6–)15–20 cm long, (3–)5.0–6 cm broad, adaxially glabrous, abaxially glabrescent, lateral veins 8–12 pairs, arcuate, base cuneate, decurrent onto petiole, margin entire, apex acute or acuminate, usually caudate. Thyrses terminal, 8–23 cm long, usually solitary, sometimes 2–4 together; peduncle 10–15 cm long, quadrangular, pubescent; rachis pubescent; pedicels 1.5–2.5 mm long, purplish-brown pubescent. Bracts linear, 1–2 cm long, 0.5–0.6 cm broad, densely pubescent, caducous; bracteoles absent. Calyx tube 2.5–3 mm long; lobes linear-lanceolate, abaxially densely pubescent, 1-veined. Corolla orange-red, 2.5–3.0 cm long, tubular, outside pubescent; tube broad from base, slightly curved, upper lip suberect, 6–7 mm long, 2-lobed, lobes ovate, 3–4 mm long, 2–3 mm broad, lower lip patent, 3-lobed, lobes ovate, 1.5–2 mm long, 1.5–2 mm broad. Stamens 2; filaments 1.2–1.3 cm long, yellow, hirsute at base, anthers 3–4 mm long, purple. Ovary 2–2.2 mm long, hairy, 10–14-ovules per locule; style 1.6–1.7 cm long, glabrous. Capsules narrowly clavate, 2–3 cm long. Seeds orbicular, 2–3 mm long and broad, much compressed, brown.

Phenology: flowering from February to March; fruiting from April to June.

Habitats: forests, scrubs, gullies and stream banks; below 2400 m.

Distribution: U Ganga & Indus, W Nepal, C Nepal, E Nepal, Sikkim & Darjiling, Bhutan, Yarlung Zangbo-Brahmaputra (Map 46-155); also in Bangladesh, and other parts of India.

U GANGA & INDUS: **Dehradun**, *A. G. Hobart-Hampden 76A* (FHO); Mar. 1893, *J. S.*

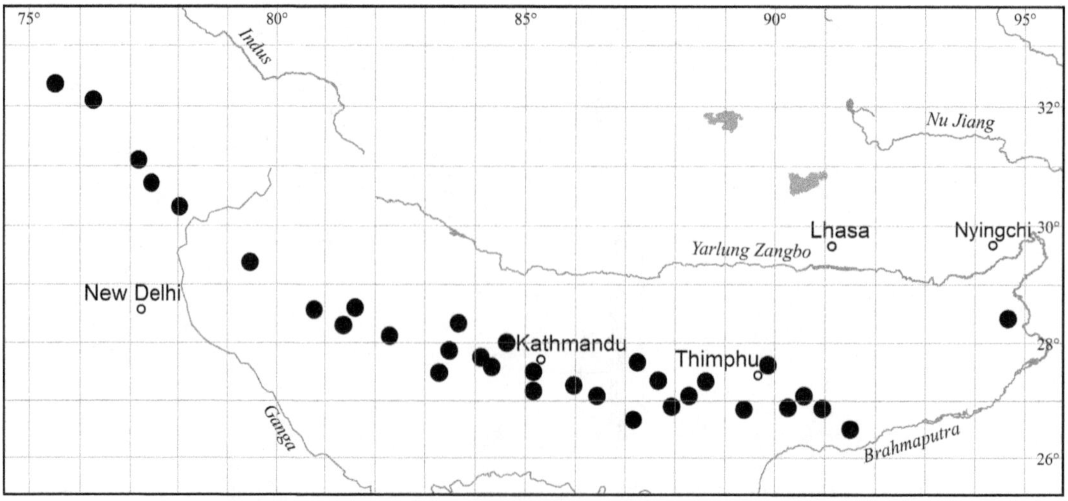

Map 46-155: **Phlogacanthus thyrsiformis** (Roxb. ex Hardw.) Mabb.

Gamble s. n. (FHO); *A. Smythies 718* (FHO); Apr. 1930, *M. R. Raizada 54* (A); Siwalik & Jausar Division, Kalsi, 16 Apr. 1918, *K. N. Deyr 124* (FHO); Kalsi, 750 m, 5 May 1913, *L. S. Murti 74* (CGE), 19 Dec. 1891, *G. Watt 114* (E); Kurwapari, 750 m, 28 Feb. 1894, *T. W. Forster 81* (CAL); Siwalik & Jaunsar Divsions, Kalsi, *M. Singh 22* (CAL); Laehiwala, Siwalik & Jaunsar division, *K. R. Choudhury 89* (A); **Kangra**, Bhali Forests, 900 m, 4 Apr. 1901, *G. S. Hart 521* (E); **Kathua**, Kadu Hill, Mohungin, 450 m, 22 Feb. 1910, *J. H. Lace 5218* (E); **Kumaon**, near Bel Khet, 1350 m, *R. Strachey & J. E. Winterbottom s. n.* (A, LE); Butu hill, 600 m, *R. Strachey & J. E. Winterbottom s. n.* (A, K); **Mussorie**, 1800–2200 m, 9 Mar. 1961, *J. C. Sen Gupta et al. 14011* (LE); **Nainital**, 1200 m, 31 Jan. 1910, *N. Gill 421* (BO, E); Jeolikot, 1350 m, 22 Nov. 1909, *N. Gill 342* (CAL); Jeolikot, 1350 m, 18 Apr., 1913, *N. Gill 523* (CAL); **Shimla**, Forests Office, *Herb. George Watt s. n.* (E). **W NEPAL**: **Bheri**, Bardiya, Babai, Ambash, 23 Mar. 1976, *K. J. Malla et al. 885* (KATH); Bardiya, Babai, Ambash, 23 Mar. 1976, *K. J. Malla et al. 900* (KATH); Surkhet, Huke, 780 m, 22 Feb. 1991, *N. P. Manandhar 82-91* (KATH); Surkhet, Kalyan, 870 m, 24 Feb. 1991, *N. P. Manandhar 302-91* (KATH); **Rapti**, Dang, Koulabas, 22 Jan. 1974, *J. Bahadur et al. 14514* (KATH); Dang, Deupur, Beldamar (Deokhuri), 26 Feb. 1974, *K. J. Malla 14583* (KATH); Dang, Deokhuri, 8 Apr. 1978, *S. A. Siddqui 1582* (KATH); **Seti**, Kailali, Golmura, Terai, 600 m, 22 Mar. 1965, *J. D. A. Stainton 4805* (BM, KATH); Kailali, Kachli Ghat, 200 m, 14 Apr. 1976, *L. A. Sharma 1014* (KATH). **C NEPAL**: **Bagmati**, Rasuwa, Sherpagaon, 2600 m, 24 Sep. 1986, *N. K. Bhattarai 86/970* (KATH); **Dhawalagiri**, Parbat, Dharmpani, 10 Mar. 1974, *D. P. Joshi & M. M. Amatya 74/1510* (KATH); Parbat, Laguwa, 1300 m, 27 Nov. 1973, *D. P. Joshi & M. M. Amatya 73-1048* (KATH); Parbat, Amarai, 1300 m, 27 Nov. 1973, *D. P. Joshi & M. M. Amatya 73-1102* (KATH); **Gandaki**, Gorkha, Arughat, Buri Gandaki River, 9 miles above Arughat Bazzar, 610 m, 11 Apr. 1953, *P. C. Gardner 275* (BM); Kaski, between Chalegaon and Pokhara, 750 m, 18 Nov. 1853, *J. Fujimura s. n.* (KYO); **Janakpur**, Sindhuli, Saraswati, 600 m, 19 Apr. 1969, *H. Kanai & M. S. Bista 11500* (KATH); **Lumbini**, Nawalparasi, Dauney Hills, 200 m, Mar. 1971, *J. Makin 118* (BM); Nawalparasi, Nawal Parasi, 200 m, Apr. 1972, *J. Makin 250* (BM); Palpa, Thungse Khola, 580 m, 28 Feb. 1974, *D. P. Joshi & M. M. Amatya 74/1311* (KATH); Palpa, Dowmme Danda, 400 m, 18 Dec. 1975, *D. P. Joshi et al. 75/3572* (KATH); Palpa, Tansen, Chame Bhanjhyang, 1370 m, 1 Mar. 1976, *N. P. Manandhar & P. M. Regmi 9* (KATH); Rupandehi, Butwal, 300 m, 6 Mar. 1966, *P. R. Shakya 2261* (KATH); **Narayani**, Bara, Lower Nijgarh Forests, 140 m, 14 Feb. 1993, *L. P. Kattel 147/93* (KATH); Bara, Tamagadi, 130 m, 20 Feb. 1993, *L. P. Kattel 275/93* (KATH); Chitwan, Rapti Valley, 380 m, 18 Mar. 1966, *J. D. A. Stainton 5287* (BM); Chitwan, Sauraha to Padampur, 17 Jan. 1996, *K. Yoda et al. 9614102* (E, KATH); Makwanpur, Hetauda, 520 m, 10 Mar. 1966, *T. B. Shrestha 3917* (KATH); Makwanpur, Rapti Valley, Hetauda, 460 m, 10 Mar. 1966, *J. D. A. Stainton 5227* (BM, KATH); Makwanpur, Makaldamar, 500 m, 30 Nov. 1988, *N. P. Manandhar 12580* (KATH). **E NEPAL**: **Koshi**, Sankhuwasabha, Mangleura Phedi, 570 m, 20 Mar. 1989, *T. B. Shrestha & P. R. Shakya 9298* (KATH); Sankhuwasabha, near Waleng, 25 Jan. 1977, *H. Tabata et al. 7825* (KATH, KYO); Sunsarit, Chota to Barakshstra, 150 m, 22 Jan. 1982, *K. R. Rajbhandari 6048* (KATH); Sunsarit, Phusre, 450 m, 25 Jan. 1982, *K. R. Rajbhandari et al. 6069* (KATH); **Mechi**, Ilam, near Chisapani, 19 Jan. 1982, *M. N. Subedi et al. 9* (KATH); Ilam, Garuwa, Khebang to Bharomdin, 24 Nov. 1963, *H. Hara et al. 6303603* (A, BM, K, KYO, TNS); Ilam, Mai Khola, 300 m, 25 Feb. 1969, *J. D. A. Stainton 6418* (BM); Ilam, Ratua Khola, 270 m, 31 May 1969, *L. H. J. Williams 313* (BM); Ilam, Godak, 650 m, 15 Apr. 1979, *P. Pradhan et al. 724* (KATH); Taplejung, Khebang to Bharomdin, 24 Nov. 1963, *H. Haraet al. 6303605* (BM). **SIKKIM & DARJILING**: **Darjiling**, 450 m, *W. Schlich 396* (FHO); 1800–2400 m, Jun.–Jul. 1855, *Gen. No. of Catalogue 12313* (A); Sukha, Siliguri, 15 Apr. 1961, *J. C. Sen Gupta et al. 75* (LE); Sukha, Tepan, Siliguri, 15 Apr. 1961, *J. C. Sen Gupta et al. 89* (LE); **Sikkim**, between Badantam and the Ranjit, 600 m, 9 Apr. 1913, *Herbs Lacaita HIV512* (BM); Punkaburi, 80 m, 23 Feb. 1871, *C. B. Clarke 13898* (BM, LE); Rishi, 510 m, 10 Dec. 1980, *P. Charrabaty 1102* (CAL); Singtam, 600 m, 14 May 1967, *N. C. Majumdar*

& R. M. Datta 420 (MO); 300 m, 30 Dec. 1874, *D. Treutler 1209* (LE). **BHUTAN**: **Gaylegphug**, between Gaylegphug and Shemgang, near Surey, 1100 m, 26 Feb. 1988, *J. R. I. Wood 6168* (E); **Phuntsholing**, 900 m, 3 Mar. 1967, *H. Hara et al. 1809* (A, BM, KYO, TNS); 900 m, 3 Apr. 1967, *H. Hara et al. 1957* (BM, KYO); Torsa River, 200 m, 2 May 1979, *A. J. C. Grierson & D. G. Long 782* (E); N Phuntsholing, 220 m, 13 Feb. 1982, *A. J. C. Grierson & D. G. Long 2903* (E, K); above Rinchending, 600 m, 16 Feb. 1982, *A. J. C. Grierson & D. G. Long 2946* (E, K) between Sorchen and Kharbandi, 800 m, 16 Mar. 1991, *J. R. I. Wood 7372* (E); **Punakha**, S Wangdiphodrang towards Chirang, 900 m, 24 Feb. 1989, *J. R. I. Wood 6877* (E); **Sarbhang**, valley above Sarbhang High School, 600 m, 7 Feb. 1988, *J. R. I. Wood 6132* (E); Yaylegphug, 300 m, 14 Dec. 1963, *G. Sen Gupta 824* (CAL); W Singi Khola, 260 m, 20 Mar. 1982, *A. J. C. Grierson & D. G. Long 3870* (A, E, K); **Tashigang**, 1200 m, 19 Feb. 1925, *F. Kingdom-Ward 6434* (E, K); **Tongsa**, between Tintibi and Dakpai, 1000 m, 15 Mar. 1992, *J. R. I. Wood 7546* (E). **YARLUNG ZANGBO-BRAHMAPUTRA**: Kobo, 26 Feb. 1912, *I. H. Burkill 38110* (CAL).

9. **Phlogacanthus guttatus** (Nees) Nees in Wallich (ed.), Pl. Asiat. Rar. 3: 99. 1832; Nees in de Candolle (ed.), Prodr. 11: 321. 1847; Clarke in Hooker (ed.), Fl. Brit. India 4: 512. 1885; Wood in Grierson & Long (eds.), Fl. Bhutan 2(3): 1285. 2001. ≡ *Justicia guttata* Wall., Pl. Asiat. Rar. 1: 24. 1830. Type: Bangladesh, Shilhet, *Wallich Cat. 2431* [lectotype **here designated**: K-W (K001115926); isolectotypes: BM, E, K (K000950010), K-W (K001115926)].

Herbs or shrubs, 3–5 m high; stems quadrangular, glabrous. Leaf blade oblong or elliptic, 11–20 cm long, 4–7 cm broad, glabrous, base narrowly cuneate, margin undulate or crenulate, apex acuminate. Thyrses 7–14 cm long, 3–3.5 cm broad, lax and interrupted especially below with internodes up to 2 cm long. Bracts and bracteoles sublate. Calyx 7–9 mm long. Corolla orange or orange-yellow, 1.5–1.8 cm long, 5–6 mm broad at mouth, densely glandular-pubescent, 2-lipped to the middle of corolla, lower lip strongly bent downwards, slightly shorter than upper lip. Stamens 2, included; filaments 8–10 mm long; anthers 2–3 mm long. Ovary glabrous; style 2.2–2.5 cm long, glabrous to pubescent. Capsules clavate, 2.5–3 cm long, 12-seeded. Seeds ovate in outline, compressed, 2–2.5 mm long, 1.8–2.2 mm broad.

Phenology: flowering and fruiting from November to March of next year.

Habitats: moist, shady places of forests; 300–1200 m.

Distribution: Bhutan, Yarlung Zangbo-Brahmaputra (Map 46-156); also in other parts of India and Myanmar.

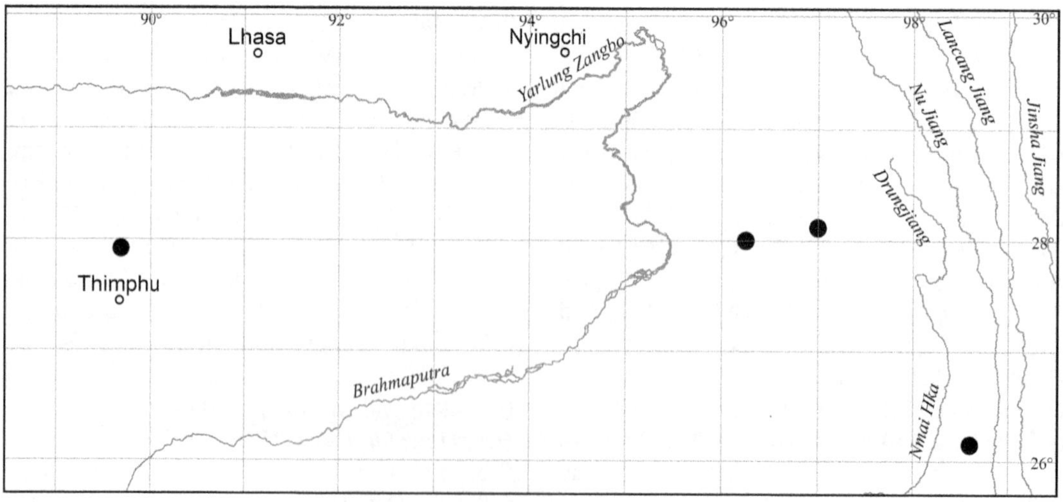

Map 46-156: **Phlogacanthus guttatus** (Nees) Nees

Notes: *Phlogacanthus guttatus* can be easily distinguished from other specis by its corolla 2-lipped to half of length, lower lip patent with purple dots in side.

BHUTAN: s. loc., *S. coll. 1260* (CAL). **YARLUNG ZANGBO-BRAHMAPUTRA**: **Lohit Valley**, Mar.–May, 1916, *E. Alezander 25* (CAL); **Walung**, 900 m, 8 Nov. 1872, *C. B. Clarke 17820* (CAL).

17. ANDROGRAPHIS Wall. ex Nees

Nees in Wallich (ed.), Pl. Asiat. Rar. 3: 77, 116. 1832; Nees in de Candolle (ed.), Prodr. 11: 515. 1847; Clarke in Hooker (ed.), Fl. Brit. India 4: 501. 1885; Lindau in Engler & Prantl (eds.), Pflanzenfam. 4(3b): 323. 1895; Hu, Fl. Reipubl. Popularis Sin. 70: 204. 2002; Hu, Deng & Daniel in Wu, Raven & Hong (eds.), Fl. China 19: 473. 2011. Type: *Andrographis paniculata* (Burm. f.) Wall. ex Nees (≡ *Justicia paniculata* Burm. f.).

Neesiella Sreem., Phytologia 15: 270. 1967, non *Neesiella* Schiffn., Hepaticae: 32. 1893. ≡ *Indoneesiella* Sreem., Phytologia 16: 466. 1968. ≡ *Andrographis* subg. *Indoneesiella* (Sreem.) L. H. Cramer, Kew Bull. 51: 555. 1996. Type: *Andrographis echinoides* (L.) Nees (≡ *Justicia echioindes* L.).

Herbs or subshrubs; with cystoliths. Leaf blade entire. Inflorescences panicles, racemes, or sometimes spikes or dense clusters, terminal or axillary. Bracts present; bracteoles small or absent. Calyx deeply 5-lobed; lobes narrow, equal or subequal. Corolla tubular to funnelform at base; tube straight, limb 2-lipped, lower lip 3-lobed, upper lip entire or 2-lobed; lobes ascending cochlear in bud. Stamens 2, exserted; filaments sometimes pubescent, dilated at attachment; anthers bithecous; thecae parallel, equal or subequal, base muticous and conspicuously pubescent; connective pilose or woolly at base. Ovary bilocular, 6–8 ovules per locule. Capsules linear-oblong or ellipsoid, compressed perpendicular to septum, 10–14-seeded; retinacula present. Seeds subglobose, blocky, or more or less compressed, surfaces smooth or ornamented, lacking trichomes.

Recent references: Sreemadhavan, 1967. *Neesiella*, a new genus of Acanthaceae. *Phytologia* 15: 270–271; Sreemadhavan, 1968. A substitute name in Acanthaceae. *Phytologia* 16: 466; Hansen, 1985. Notes on *Andrographis* and *Gymnostachyum* (Acanthaceae). *Nordic J. Bot.* 5: 353–356; Chu, 1991. A revision of the *Andrographis* (Acanthaceae) of China. *Bull. Bot. Res., Harbin* 11(1): 45–48; Cramer, 1996. Notes on Sri Lankan Acanthaceae. *Kew Bull.* 51: 553–556; Gnanasekaran & Murthy, 2015. Lectotypification of four names in *Andrographis* (Acanthaceae). *Phytotaxa* 227: 175–181.

About 30 species: tropical and subtropical Asia; two species in the Pan-Himalaya.

1a. Panicles terminal; ovules 3–6 per locule; capsule linear-oblong (subg. I. **Andrographis**) .... 1. *A. paniculata*

1b. Raceme axillary; ovules 2 per locule; capsule ovoid oblong (subg. II. **Indoneesiella**) .... 2. *A. echinoides*

I. Andrographis subg. **Andrographis**. Type: *Andrographis paniculata* (Burm. f.) Wall. ex Nees.

Panicles terminal; ovules 3–6 per locule; capsule linear-oblong.

About 25 species: India and Sri Lanka; one species in the Pan-Himalaya.

1. **Andrographis paniculata** (Burm. f.) Wall. ex Nees in Wallich (ed.), Pl. Asiat. Rar. 3: 116. 1832; Nees in de Candolle (ed.), Prodr. 11: 515. 1847; Clarker in Hooker (ed.), Fl. Brit. India 4: 501. 1884; Hu, Fl. Reipubl. Popularis Sin. 70: 207. 2002; Hu, Deng & Daniel in Wu, Raven & Hong (eds.), Fl. China 19: 474. 2011. ≡ *Justicia paniculata* Burm. f., Fl. Ind.: 9. 1768. Type: [*icon*] Rheede, Hort. Malab. 9: 109, t. 56. 1689 (lectotype designated by Gnanasekaran & Murthy, Phytotaxa 227: 180. 2015).

Andrographis subspathulata C. B. Clarke, Fl. Brit. India 4: 502. 1884. Type: India, Kurnool at the foot of the Nellymallays, *R. H. Beddome s. n.* [holotype: K (K000545977)].

Herbs, 30–50 cm high; stems quadrangular, glabrous. Petiole 0.3–1 cm long; leaf blade ovate-lanceolate, lanceolate, or narrowly elliptic, 1.5–7 cm long, 1–2.5 cm broad, both surfaces glabrous, lateral veins 3–5 pairs, base attenuate and decurrent onto petiole, margin entire, apex acute to shortly acuminate. Panicles or rarely racemes terminal; rachis glabrous to sparsely pubescent. Bracts triangular to subulate, 1–1.5 mm long; bracteoles linear to subulate, 1–1.5 mm long. Pedicel 2–9 mm long, sparsely pubescent with glandular and eglandular trichomes. Calyx 2.5–3 mm long, abaxially glabrous or glandular-pubescent; lobes subulate. Corolla white, 0.9–1.5 cm long, outside glandular-pubescent; tube base funnel-form, 4–8 mm long, upper lip 5–7 mm long, reflexed, 2-lobed, lobes 1–1.2 mm long and broad, lower lip with purple dots, 5–7 mm long, erect, lobes 2.5–3 mm long. Style 6–10 mm long, sparsely pilose toward base. Capsules linear-oblong, compressed, 1.5–2 cm long, 3–4 cm broad, glabrous, or sparsely pubescent with glandular-pilose, 12-seeded. Seeds 1.7–2 mm long, 1.3–1.5 mm broad, rugose. **Chromosomes** 2n = 50. Figure 46-93.

Phenology: flowering and fruiting throughout year.

Habitats: roadsides, thickets, waste places or cultivated.

Distribution: Jammu & Kashmir, W Nepal, C Nepal (Map 46-157); native to Sri Lanka, and other parts of India; widely cultivated or naturalized in tropical and subtropical regions of the world.

Uses: medicine.

JAMMU & KASHMIR: Jammu, Chatha Farm, 300 m, 5 Apr. 1998, *T. N. Srivastava & B. K. Kapahi 17646* (RRLH). **W NEPAL**: **Bheri**,

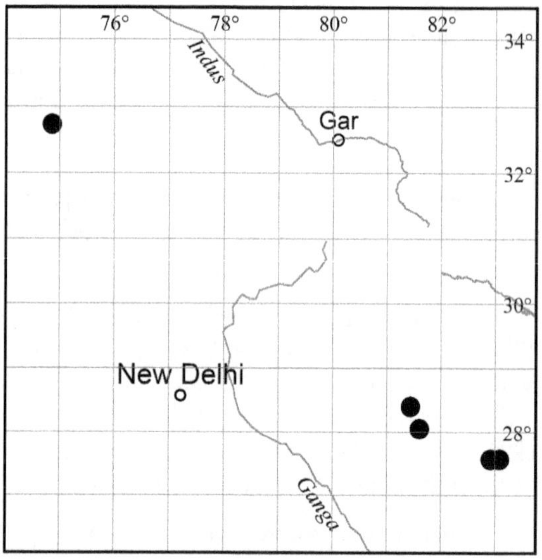

Map 46-157: **Andrographis paniculata** (Burm. f.) Wall. ex Nees

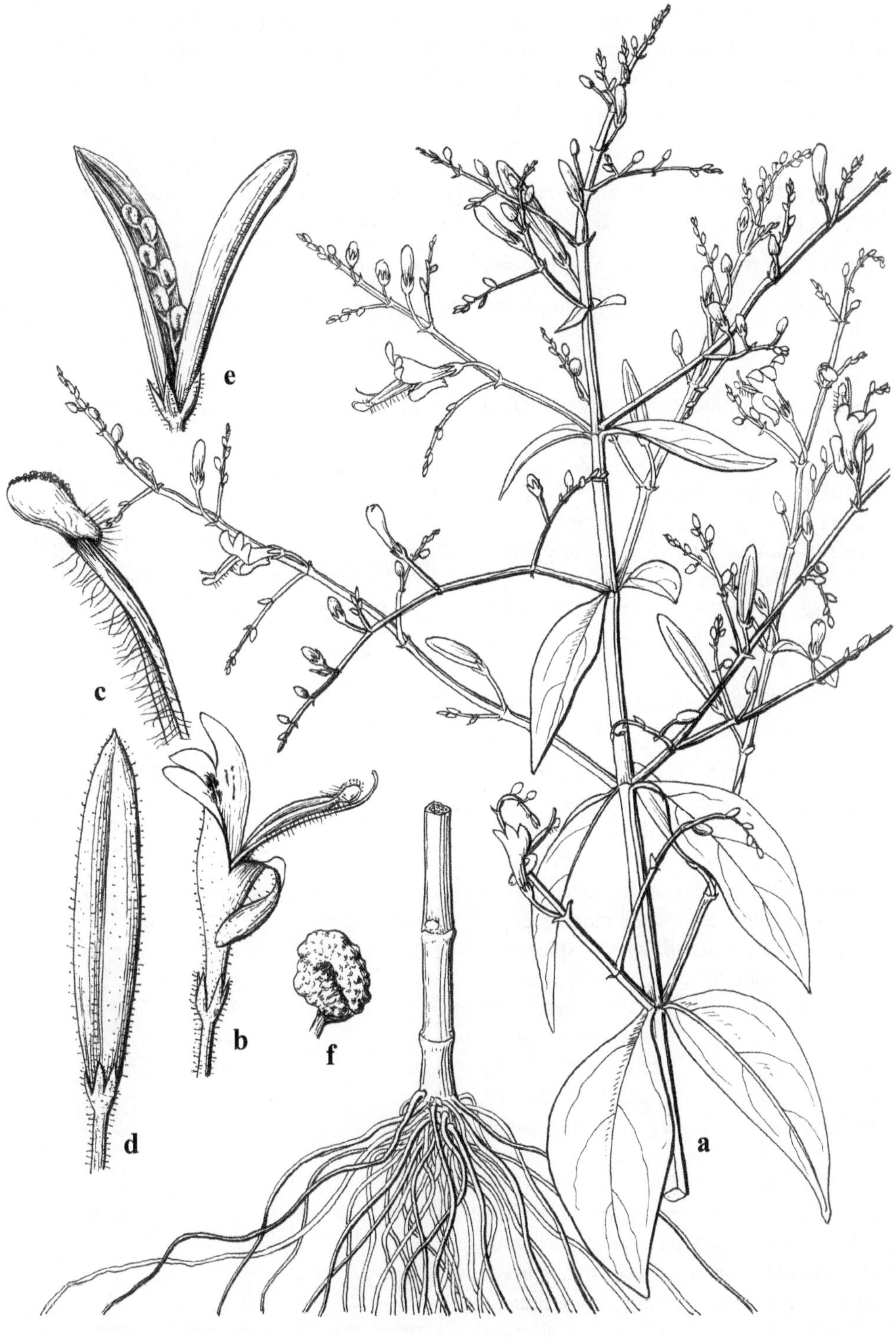

Figure 46-93: **Andrographis paniculata** (Burm. f.) Wall. ex Nees: **a**, habit; **b**, flower; **c**, stamen; **d**, capsule; **e**, opened capsule showing seeds; **f**, seed. Drawn by Quan LIN; reproduced from *Icon. Cormophyt. Sin.* 4: 166, fig. 5746. 1975.

Bardiya, Bandarbania, 180 m, 19 Feb. 1993, *P. P. Kurmi KB615* (KATH); Banke, Napelganj, 20 Nov. 1972, *N. P. Manandhar 6992* (KATH); Banke, Khajura, 180 m, 6 Nov. 1980, *T. K. Bhattacarya 3564* (KATH); Banke, Napelganj, 180 m, 12 Jan. 1993, *P. P. Kurmi KB575* (KATH). **C NEPAL: Lumbini**, Kapilvastu, Tilaurakot, 85 m, 6 Apr. 2007, *P. P. Kurmi 10123* (KATH); Baraipur, 180 m, 18 Oct. 1995, *P. P. Kurmi 10082* (KATH).

II. Andrographis subg. **Indoneesiella** (Sreem.) L. H. Cramer, Kew Bull. 51: 555. 1996. ≡ *Indoneesiella* Sreem., Phytologia 16 (6): 466. 1968. Type: *Andrographis echinoides* (L.) Nees.

Raceme axillary; ovules 2 per locule; capsule ovoid oblong.

Two species in India; one species in the Pan-Himalaya.

2. Andrographis echinoides (L.) Nees in Wallich (ed.), Pl. Asiat. Rar. 3: 117. 1832; Nees in de Candolle (ed.), Prodr. 11: 518. 1847; Clarke in Hooker (ed.), Fl. Brit. India 4: 505. 1885. ≡ *Justicia echinoides* L., Sp. Pl. 16: 1753. ≡ *Neesiella echinoides* (L.) Sreem., Phytologia 15: 271. 1967. ≡ *Indoneesiella echinoides* (L.) Sreem., Phytologia 16(6): 466. 1968. Type: "Habit in India", *Herb. Hermann 4: fol. 40* (lectotype designated by Stearn ex Cramer in Dassanayake & Clayton, Revised Handb. Fl. Ceylon 12: 100. 1982: BM).

Annual herbs, erect, 20–40 cm high; stem quadrangular, glandular-hirsute. Leaves sessile or subsessile; leaf blade oblong or oblong-spatulate, 4–7.5 cm long, 1.5–2.4 cm broad, villous with white hairs on both surfaces, lateral veins 3–7 pairs, base attenuate, margin entire and ciliolate, apex acute. Racemes spiciform, 2–7 cm long, unilateral; peduncles densely glandular-hirsutulous. Bracts and bracteoles filiform, 2–3 mm long. Flowers erect. Calyx-tube 1.5–2 mm long, limb pinkish-white; lobes linear-filiform, 5–8 mm long, abaxially densely glandular-hirsute, apex acute. Corolla tubular; tube 4 mm long, faintly inflated at base, outside sparsely glandular-hirsutillous, limb white, upper lip oblong, 4.5–5.5 mm long, 1.7–2 mm broad, 2-lobed above, faintly hooded, lower lip to 7 mm long; lobes oblong-lanceolate, marked with purple blotches. Filaments flattened, sparsely bristly, anther-thecae bearded at base. Capsules broadly ovoid, 9–10 mm long, 2.5–3.5 mm broad, sparsely glandular-bristly towards apex; retinacula lanceolate, hooked. Seeds oblong cylindric, 1.3–1.5 mm long, tuberculate. **Chromosomes** 2n = 38, 48.

Phenology: flowering from March to June; fruiting from October to December.

Habitats: dry places.

Distribution: Jammu & Kashmir (Map 46-158); also in Bangladesh, Sri Lanka, and other parts of India.

JAMMU & KASHMIR: **Jammu**, Ram Nagar, 15 Sep. 1969, *B. M. Sharma 156* (RRLH).

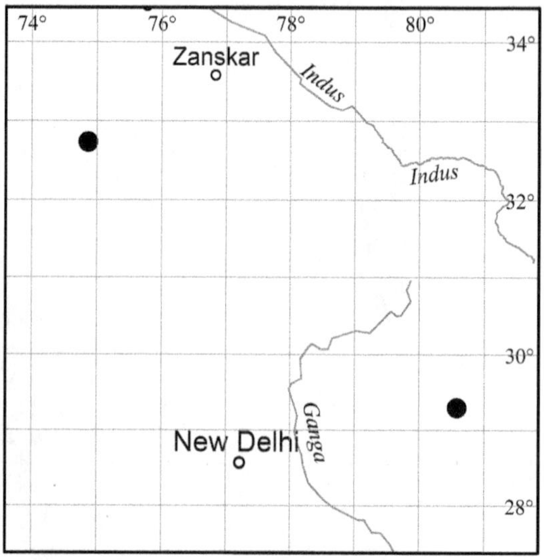

Map 46-158: **Andrographis echinoides** (L.) Nees

18. HAPLANTHUS Nees

Nees in Wallich, Pl. Asiat. Rar. 3: 77, 115. 1832, non *Haplanthus* T. Anderson, J. Linn. Soc., Bot. 9: 445. 1866. Type: *Haplanthus tener* Nees [= *Haplanthus laxiflorus* (Blume) Gnanasek., G. V. S. Murthy & Y. F. Deng].

Herbs perennial, with cystoliths; stems subterete to quadrangular, glabrous to glandular-pubescent, swollen at nodes towards base of plant; rooting at lower nodes. Leaf blade ovate-elliptic, lanceolate, minute entire or undulate. Inflorescences racemose, axillary and terminal, forming a terminal lax or reduced panicle, sometimes flowers almost glomerulate in leaf axils. Bracts lanceolate, minute; bracteoles 2, linear to lanceolate. Calyx subequal 5-lobed almost to base. Corolla subactinomorphic; tube curved, inconspicuously ventricose, limb slightly 2-lipped, upper lip deeply 2-lobed, lower lip 3-lobed. Stamens 2, included, adnate to base of ventricose portion of corolla tube; filaments 3–4 mm long, pouched at apex, filiform at point of attachment; anthers bithecous, thecae oblong; connectives woolly dorsally. Ovary oblong, 2-locular; ovules 6–8 per locule; style antrorsely bristled; stigma linear, green. Capsules linear-oblanceolate, compressed at right angles to septum with a median longitudinal groove, glandular-hairy, 8–16-seeded. Seeds compressed, verrucose, brownish.

Recent references: Sreemadhavan, 1971. A nomenclatural note on *Andrographis laxiflora* (Acanthaceae). *Bull. Bot. Surv. India* 11(1–2): 183; Gnanasekaran et al., 2016. Resurrection of the genus *Haplanthus* (Acanthaceae: Andrographinae). *Blumea* 61: 165–169.

About five species: E. Himalaya to SE Asia through Indochina; one species in the Pan-Himalaya.

1. Haplanthus laxiflorus (Blume) Gnanasek., G. V. S. Murthy & Y. F. Deng, Blumea 61(3): 168. 2016. ≡ *Justicia laxiflora* Blume, Bijdr. 14: 789. 1826. ≡ *Andrographis laxiflora* (Blume) Lindau in Engler & Prantl (eds.), Nat. Pflanzenfam. 4(3b): 323. 1895; Hu, Fl. Reipubl. Popularis Sin. 70: 204. 2002; Hu, Deng & Daniel in Wu, Raven & Hong (eds.), Fl. China 19: 474. 2011. Type: Indonesia, Java, s. d., *C. Blume s. n.* [holo L (L0003148)].

Haplanthus tener Nees in Wallich (ed.), Pl. Asiat. Rar. 3: 115. 1832. ≡ *Andrographis tenera* (Nees) Kuntze, Revis. Gen. Pl. 2: 382. 1891. ≡ *Andrographis tenuiflora* T. Anderson, J. Linn. Soc., Bot. 9: 502. 1867, nom. illeg. superfl. (Art. 52.1). Type: Myanmar, Tanintharyi, Tavoy, 2 Oct. 1827, *W. Gomez in Wallich Cat. 7185a* [lectotype **here designated**: K-W (K000014472)].

Herbs, 20–50 cm tall, isophyllous or anisophyllous; stems quadrangular, glabrous or pubescent. Petiole 3–10 mm; leaf blade ovate, 1.5–9 cm long, 1.5–3(–5.5) cm broad, both surfaces glabrous, lateral veins 5 or 6 pairs, base slightly attenuate and decurrent onto petiole, margin entire, apex acuminate to shortly acuminate. Racemes or panicles of racemes axillary or terminal, sometimes flowers clustered in leaf axils; rachis glabrous or glandular- or eglandular-pubescent. Bracts linear, subulate, or lanceolate, 1–3 mm long, glabrous or pubescent; bracteoles subulate, glabrous or pubescent. Pedicel 0.5–2 mm long, glabrous or pubescent. Calyx 2–4 mm long, abaxially glabrous or pubescent; lobes subulate. Corolla white with purplish or pinkish coloration on lips, 0.8–1.2 cm long, outside pubescent with gland-pilose; tube 6–9 mm long, base curved, dorsally gibbous; lower lip 1.5–4 mm long, lobes 0.5–2 mm long, upper lip 1.5–4 mm long, erect (± parallel to lower lip), apex entire to 2-lobed. Stamens included. Capsules linear-cylindric, 10–20 cm long, 1.8–2 mm broad, glabrous or pubescent with glandular-pilose, 16–20-seeded. Seeds 0.8–1 mm long, 0.8–0.8 mm broad, rugose. Figure 46-94.

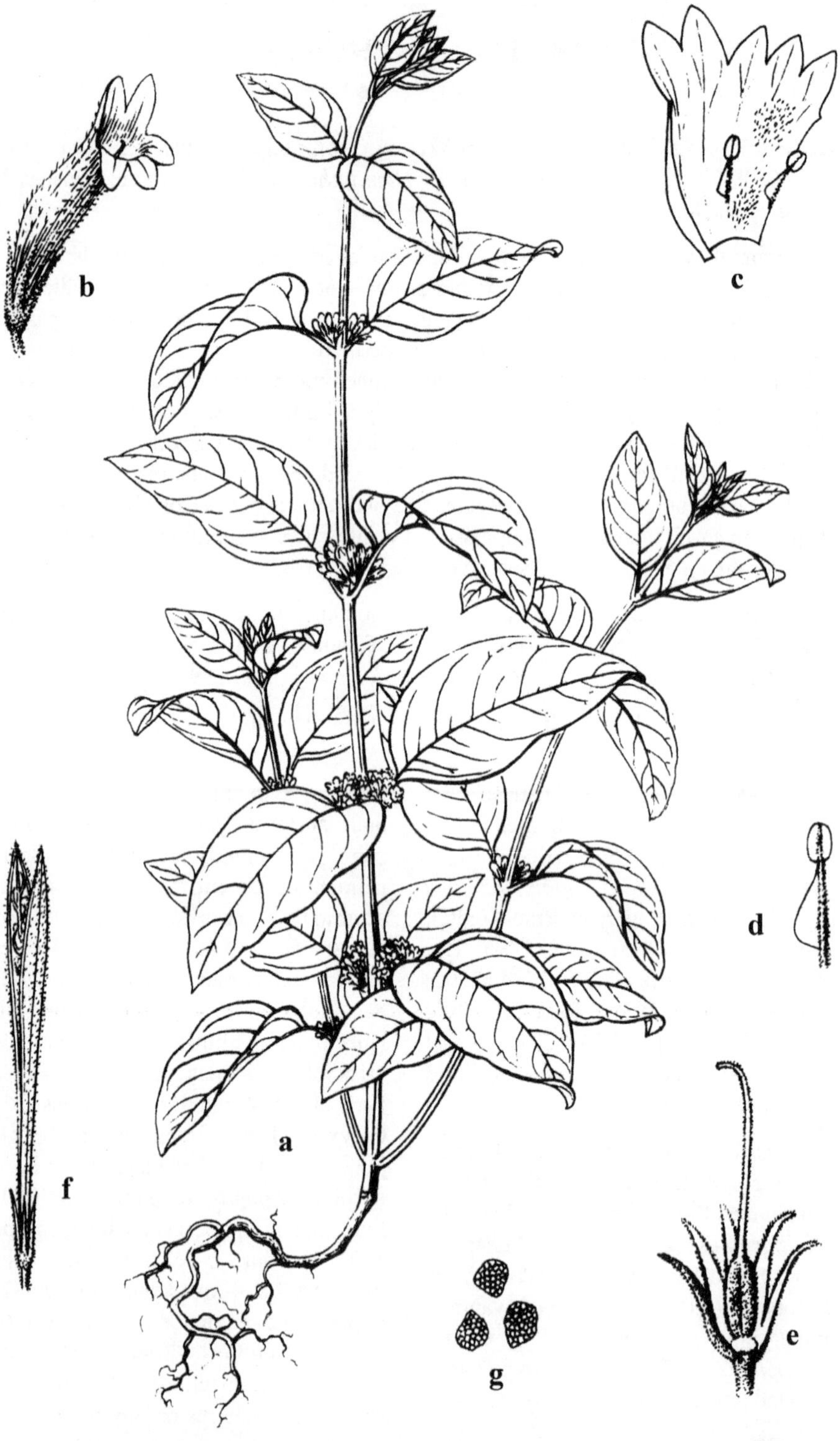

Figure 46-94: **Haplanthus laxiflorus** (Blume) Gnanasek., G. V. S. Murthy & Y. F. Deng: **a**, habit; **b**, corolla; **c**, corolla opened showing stamens; **d**, stamen; **e**, calyx and pistil; **f**, capsule; **g**, seeds. Drawn by Shu-Qin CAI; reproduced from Ying & Zhang, *Chinese End. Gen. Seed Pl.*: 51. 1994.

Phenology: flowering from August to November; fruiting from November to March of next year.

Habitats: thickets; 600–800 m.

Distribution: S Hengduan (Map 46-159); also in Thailand, Laos, Vietnam, Cambodia, Malaysia, Indonesia, and other parts of India (Assam, Manipur), Myanmar and China (S Yunnan).

S HENGDUAN: **Lushui**, Liuku, 14 Sep. 2013, *Y. F. Deng 24906* (IBSC, KUN).

Tribe 5. **Justicieae** Dumort., Anal. Fam. Pl.: 23. 1829. Type: *Justicia* L.

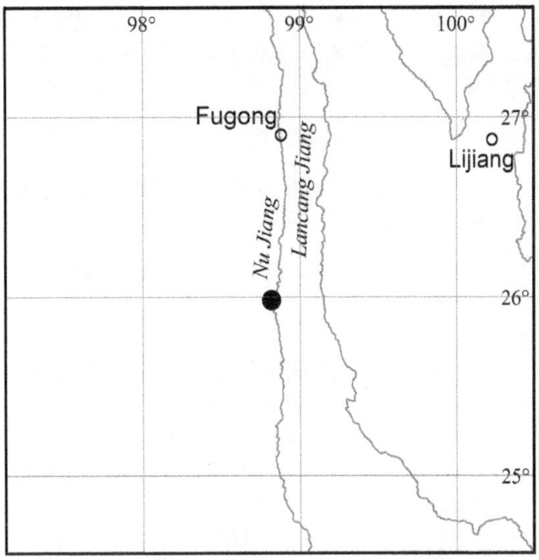

Map 46-159: **Haplanthus laxiflorus** (Blume) Gnanasek., G. V. S. Murthy & Y. F. Deng

Stamens 2, rarely 4; staminodes 2 or absent. Ovary 2 per locule. Capsules 4-seeded.

About 100 genera and 2000 species, pantropic; 15 genera and 49 (four endemic) species in the Pan-Himalaya.

19. ASYSTASIA Blume

Blume, Bijdr.: 796. 1826; Nees in de Candolle (ed.), Prodr. 11: 100, 163. 1847; Clarke in Hooker (ed.), Fl. Brit. India 4: 492. 1885; Lindau in Engler & Prantl (eds.), Pflanzenfam. 4(3b): 325. 1895; Nasir & Ali in Malik & Ghafoor (eds.), Fl. Pakistan 188: 67. 1988; Wood in Grierson & Long (eds.), Fl. Bhutan 2(3): 1282. 2001; Hu, Fl. Reipubl. Popularis Sin. 70: 215. 2002; Hu, Deng & Daniel in Wu, Raven & Hong (eds.), Fl. China 19: 437. 2011. ≡ *Intrusaria* Raf., Fl. Tell. 4: 66. 1838 ('1836'), nom. illeg. (Art. 52.1). Type: *Asystasia intrusa* (Forssk.) Blume (≡ *Ruellia intrusa* Forssk. ≡ *Intrusaria bicolor* Raf.).

Asystasiella Lindau in Engler & Prantl (eds.), Nat. Pflanzenfam. 4(3b): 326. 1895; Hu, Fl. Reipubl. Popularis Sin. 70: 2002. Lectotype (designated by Ensermu et al., Kew Bull. 47: 670. 1992): *Asystasiella atroviridis* (T. Anderson) Lindau (≡ *Asystasia atroviridis* T. Anderson).

Herbs or subshrubs; with cystoliths. Leaves petiolate; leaf blade entire to crenate. Inflorescences terminal or axillary, unilateral spikes or racemes, or panicles of these. Bracts small, shorter than calyx; bracteoles similar to bracts, or sometimes absent. Calyx 5-lobed; lobes equal to subequal. Corolla campanulate; tube expanded distally into a throat, or subsalver form with cylindric basal portion of tube, limb slightly 2-lipped, often appearing subactinomorphic with 5 equal lobes; lobes ascending cochlear in bud. Stamens 4, included in or partially exserted from corolla tube; anthers bithecous, thecae parallel to sagittate to subperpendicular, equally to unequally inserted, base muticous; staminodes absent. Ovary with 2 ovules per locule; stigma capitate, 2-lobed or 2-dentate. Capsules with a solid stalk at base, clavate, up to 4-seeded; retinacula present. Seeds lenticular, lacking trichomes.

Recent references: Ensermu et al., 1992. A reconsideration of *Asystasiella* Lindau (Acanthaceae). *Kew Bull.* 47: 669–675; Kiew & Vollesen, 1997. *Asystasia* (Acanthaceae) in Malaysia. *Kew Bull.* 52(4): 965–971.

About 40 species: tropical and subtropical regions of Old World; two species in the Pan-Himalaya.

1a. Corolla campanulate with cylindric basal portion of tube less than 1.5 times as long as throat and limb combined .... 1. *A. macrocarpa*

1b. Corolla subsalverform with cylindric basal portion of tube at least 2 times as long as throat and limb combined .... 2. *A. neesiana*

1. Asystasia macrocarpa Nees in Wallich (ed.), Pl. Asiat. Rar. 3: 89. 1832; Nees in de Candolle (ed.), Prodr. 11: 163. 1847; Clarke in Hooker (ed.), Fl. Brit. India 4: 495. 1884; Wood in Grierson & Long (eds.), Fl. Bhutan 2(3): 1282. 2001. ≡ *Mackaya macrocarpa* (Nees) Das, Fl. Assam 3: 447. 1939. ≡ *Ruellia macrocarpa* Wall., Numer. List n. 3348. 1830, nom. invalid. Type: Nepal, 1821, *Wallich Cat. 2348* [lectolotype **here designated**: K-W (K001115705)].

Decumbent or ascending subshrubs, 1–2 m high; stems quadrangular, sulcate, sparsely pilose when young, then glabrescent. Petiole 0–1.5 cm long, pilose; leaf blade lanceolate, ovate, or oblong, 2–13 cm long, 1–4.5 cm broad, adaxially sparsely pilose, abaxially sparsely pilose especially along veins, base usually attenuate, rarely cuneate, margin entire, apex acuminate. Racemes, terminal, 2–6 cm long; rachis pilose; flowers in opposite pairs, internode 0.8–2 cm long below; pedicels 5–6 mm long. Bracts linear, 2–3 mm long, pilose; bracteoles linear, 1–2 mm. Calyx 5–7 mm long, glandular-pubescent; lobes lanceolate, apex acuminate. Corolla campanulate with cylindrical portion of tube less than 1.5 times as long as throat and limb combined, pale purple, 2.8–3.8 cm long, sparsely glandular-pubescent, ventricose; tube base cylindric, 0.8–1.2 cm long, 1.5–2 mm broad, upwards gradually widened to 1–1.2 cm broad at mouth; lobes elliptic to ovate, 0.6–1 cm long and broad. Stamens 4, included; filaments of longer pair 8–10 mm long, shorter pair 4–5 mm long. Ovary comose; style 3.2–4 cm long. Capsules 2.8–3.6 cm long, cylindric, glandular-pubescent, base solidly stalked, sterile portion 1.5–2 cm long. Seeds suborbicular in outline, much compressed, 5–6 mm in diameter, tubercular, rugose, glabrous.

Phenology: flowering and fruiting from November to February.

Habitats: moist places in forests; 0–2100 m.

Distribution: U Ganga & Indus, Nepal, W Nepal, C Nepal, E Nepal, Sikkim & Darjiling, Bhutan (Map 46-160); also in other parts of India and Myanmar.

U GANGA & INDUS: **Kumaon**, Sitaban, 600 m, *R. Trachey & J. E. Winterbotom s. n.* (K); **Mussorie**, *Raizada s. n.* (DD, fide Raizada & Saxena, 1978). **W NEPAL**: **Bheri**, 500 m, 15 Apr. 1974, *J. F. Dobremez 2527 (N. P. Nanandhar 74-156)* (E); **Rapti**, Dang, Phedi, Dang to Deokhuri, 440 m, 27 Aug. 1982, *N. P. Manandhar 8471* (KATH); Puythan, Asurkot, 1000 m, 5 Mar. 1976, *N. P. Manandher & P. M. Regmi 168* (KATH). **C NEPAL**: **Bagmati**, Kathmandu, Dokobhanjyang, N of Kathmandu, 2300 m, 16 Mar. 1975, *D. P. Joshi & K. R. Rajbhandari 75/664* (KATH); Lalitpur, 23 Mar. 2003, *T. Watanabe & K. J. Malla NPL-SP030323 (001)* (MBK); Nuwakot, Kakani, 1969, *O. V. Polunin 9561* (BM); **Gandaki**, Gorkha, Arughat, 610 m, 9 Apr. 1953, *P. C. Gardner 224* (BM); Kaski, Tirkhedhungo, Kali Gandaki, before Tirkhedhungo, 1520 m, 5 Apr. 1978, *C. G. Rice 878* (BM); Manang, Thonje, 1800 m, 28 Apr. 1978, *J. F. Dobremez 7815* (E); **Lumbini**, Gulmi, Ridhi, 450 m, 2 Mar. 1976, *N. P. Manandher & P. M. Regmi 74* (KATH); Palpa, Tansen, Channe, Bhanjyang, 1370 m, 1 Mar. 1976, *N. P. Manandhar & P. M. Regmi 31* (KATH); Palpa, Taldhunga, 750 m, 15 Oct. 1977, *H. Saiju et al. 229* (KATH); Rypandehi, Butawl, 610 m, 6 Oct. 1954, *J. D. A. Stainton, W. R. Sykes & L. H. J. Williams 8805* (BM); **Narayani**, Chitwan, Tikoli,

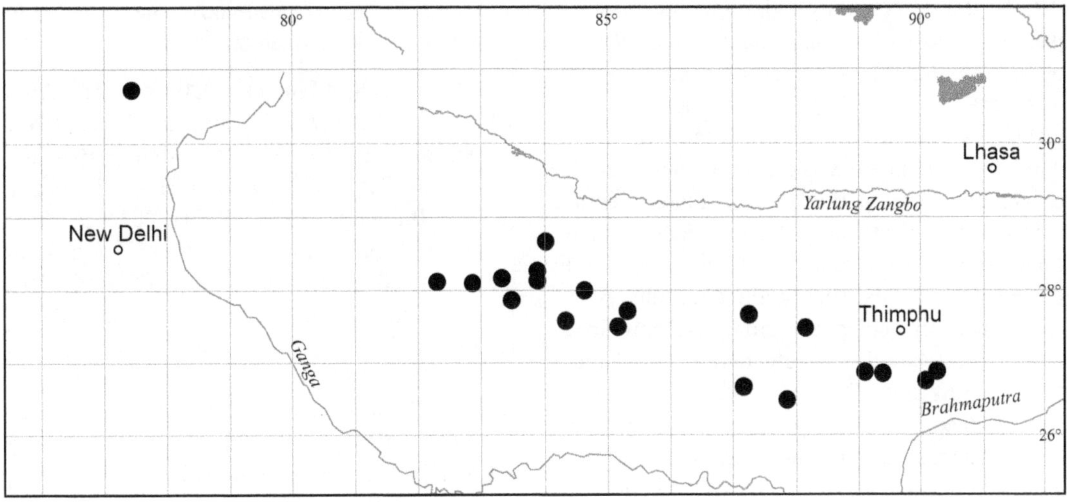

Map 46-160: **Asystasia macrocarpa** Nees

270 m, 12 Mar. 1966, *T. B. Shrestha 3983* (BM, KATH); Chitwan, 250 m, 22 Mar. 1976, *R. G. Troth 696* (BM); Chitwan, Tikoli, 300 m, 12 Mar. 1966, *J. D. A. Stainton 5249* (BM); Makwanpur, Bhimphedi, 31 Mar.–3 Apr. 1953, *P. C. Gardner 1620* (BM); Makwanpur, Bhimphedi, 31 Mar.–3 Apr. 1953, *P. C. Gardner 1620* (BM). **E NEPAL**: **Koshi**, Sankhuwasabha, Mangteura Phedi, 570 m, 20 Mar. 1989, *T. B. Shrestha & P. R. Shakya 9301* (KATH); Sankhuwasabha, Chirkhu Khola to Tumlingtar, 270 m, 12 Nov. 1981, *P. R. Shakya 7415* (KATH); Sunsari, Dharan, 380 m, 16 Apr. 1956, *J. D. A. Stainton 27* (BM); Sunsari, Chatara, Chatara Ghat (Kosi), 28 Mar. 1978, *B. Bhattarai et al. 78105* (KATH); Sunsari, Dharan, E of Dharan, 150 m, 3 Feb. 1985, *J. D. A. Stainton 8905* (E); **Mechi**, Jhapa, Hokse, 300 m, 5 Mar. 1969, *J. D. A. Stainton 6426* (BM). **SIKKIM & DARJILING**: **Sikkim**, Khaibari Forests, Sikkim Tarai, 17 *Jan.* 1890, *H. H. Haines 164* (K); Punkabari, 500 ft., 15 Nov. 1870, *C. B. Clarke 13834* (K); Punkaburi, 80 m, 23 Feb. 1871, *C. B. Clarke 13900* (K); Mar. 1871, *C. B. Clarke 16778* (K). **BHUTAN**: **Phuntsholing**, N Phuntsholing, 230 m, *A. J. C. Grierson & D. G. Long 2900* (E, K); **Samchi**, W bank of Torsa River, Phuntsholing, 200 m, *A. J. C. Grierson & D. G. Long 2988* (E, K); **Sankosh**, 2 km W of Pinkhua, 340 m, 16 Mar. 1982, *A. J. C. Grierson & D. G. Long 3772* (E, K); **Sarbhang**, Singi Khola 10 Km SW of Sarbhang, 390 m, 9 Mar. 1982, *A. J. C. Grierson & D. G. Long 3619* (E, K).

2. Asystasia neesiana (Wall.) Nees in Wallich (ed), Pl. Asiat. Rar. 3: 89. 1832; Nees in de Candolle (ed.), Prodr. 11: 494. 1947; Clarke in Hooker (ed.), Fl. Brit. India 4: 496. 1885; Hu, Deng & Daniel in Wu, Raven & Hong (eds.), Fl. China 19: 437. 2011. ≡ *Ruellia neesiana* Wall., Pl. Asiat. Rar. 1: 73. 1831. ≡ *Asystasiella neesiana* (Wall.) Lindau in Engler & Prantl (eds.), Nat. Pflanzenfam. 4(3b): 326. 1893; Hu, Fl. Reipubl. Popularis Sin. 70: 218. 2002. ≡ *Mackaya neesiana* (Wall.) Das, Fl. Assam 3: 447. 1939. Type: Bangladesh, Sylhet, *F. de Silva in Wallich Cat. 2342* [lectotype **here designated**: K-W (K001115697); isolectotype: K].

Asystasia chinensis S. Moore, J. Bot. 13: 228. 1875. ≡ *Asystasiella chinensis* (S. Moore) E. Hossain, Notes Roy. Bot. Gard. Edinburgh 32: 405. 1973; Hu, Fl. Reipubl. Popularis Sin. 70: 218. 2002. Type: China, Jiangxi, Jiujiang, "Kew Kiang", 1873, *Shearea s. n.* [holotype: K (K000885425)].

Ruellia auriculata Wall., Numer. List. 2341. 1830, nom. invalid.

Herbs, 0.4–1 m high; stems quadrangular, sulcate, bifariously puberulent to subglabrous.

Petiole 0.5–5 cm long, glabrous or sparsely pubescent; leaf blade lanceolate, ovate, elliptic, oblong, or uppermost sometimes ovate-cordate, 4–25 cm long, 2–11 cm broad, both surfaces glabrous or slightly pubescent, lateral veins 6–8 pairs, base attenuate and decurrent onto petiole, or cordate for uppermost ones, margin entire, sinuate-crenate, or subdentate, apex acuminate. Spikes or racemes terminal, 6–15 cm long, simple or branched; rachis usually glandular-pubescent. Bracts and bracteoles triangular, 1–2 mm long, 0.8–1.2 mm broad, sparsely pubescent. Pedicel (0–)1–3 mm long. Calyx 4–6.5 mm long; lobes linear-lanceolate, usually glandular-pubescent. Corolla subsalverform, pink to purplish blue, 4–6 cm long, outside glandular-pubescent; tube base cylindric, elongate, 3–4 cm long, at least 2 times as long as throat and limb combined; lower lip 3-lobed, lobes ovate, 5–6 mm long, 7–8 mm broad, upper lip 2-lobed, lobes suborbicular, 6–7 mm long and broad. Stamens 4, didynamous; filaments of longer pair 4–5 mm long, shorter pair 1.5–2 mm long; anther-thecae subequally inserted. Style 3.6–5.2 cm long, glabrous. Capsules 1.8–2.8 cm long, glandular-pubescent, 4-seeded, basal stalk 1–1.5 cm long. Seeds ovate in outline, compressed, 3.5–4 long, 2.5–3 mm broad. Figure 46-95.

Phenology: flowering from July to December; fruiting from October to January of next year.

Habitats: forests, stream banks, ditches, trails; 100–2100 m.

Distribution: Bhutan, Yarlung Zangbo-Brahmaputra, N Hengduan, S Hengduan, U Irrawaddy (Map 46-161); also in Indonesia, Laos, Malaysia, Thailand, Vietnam, and other parts of India, Myanmar and China (S Yunnan, SE Sichuan, Guizhou, Guangxi, Guangdong, Hubei, Hunan, Jiangxi, Fujian, Taiwan, Zhejiang, Anhui, Jiangsu).

Notes: *Asystasia neesiana* is sometimes treated in *Asystasiella* Lindau, a genus of two species with spheroidal, 4- or 5-porate and spinulate (vs. prolate-spheroidal to subprolate, 3-colporate and reticulate) pollen (Ensermu et al., 1992). The generic position of *Asystasia neesiana* and some of its relatives have yet not to be thoroughly investigated.

BHUTAN: **Tsonga**, 900 m, 28 Aug. 1915, *R. E. Cooper 4728* (BM, E). **YARLUNG ZANGBO-BRAHMAPUTRA**: **Mêdog**, Baibung, 950 m, 8 Oct. 1992, *H. Sun et al. ETM-151* (KUN); Baibung, 750 m, 19 Oct. 2015, *J. Hu et al. HJ05176* (CDBI, NAS); Miri to 113 km, 1300 m, 1 Oct. 2003, *X. F. Gao et al. 8070* (CDBI); Hanmi to No. 1 Bridge, 1300 m, 23 Sep. 2003, *X. F. Gao et al. 7766* (CDBI); Kabu to Damu, 1464 m, 15 Nov. 2016, *Y. J. Guo et al. 16CS14449* (IBSC); Damu, Bolonggong to Daguoqiao, 1420 m, 7 Oct. 2017, *J. Luo 20171004-05* (XZE). **N HENGDUAN**: **Wenxian**, Bikou, Dadaoling, 1200 m, 23 Oct. 1958, *Z. P. Wei 3332* (WUK); Bikou, Dadaoling, 1000 m, 23 Oct. 1958, *Y. C. Ho 1455* (WUK); Bikou, Zhuyuan, 800 m, 8 Sep. 1959, *Z. Y. Zhang 14687* (WUK); Bikou, Zhuyuan, 800 m, 9 Sep. 1959, *Z. Y. Zhang 14712* (WUK); Bikou, Zhuyuan to Caoba, 800 m, 9 Sep. 1959, *Z. Y. Zhang 14811* (PE, WUK); Bikou, Fanba, 1000 m, 2 Aug. 1964, *Q. X. Li & X. C. Zhao 2471* (PE); Bikou, Shilonggou, 1000 m, 22 Aug. 2011, *Z. M. Jiang et al. 3226* (WUK). **S HENGDUAN**: **Baoxing**, Bao-Hsin, Ma-Lu-shan, 900 m, Jul.–Aug. 1939, *S. Y. Hu 1553* (A); **Dali**, Tali Range, *G. Forrest 11717* (BM, E, K); W flank of Tali Range, *G. Forrest 15566* (E, K); **Heqing**, Ta-long-tan, Ta-pin-tze, 21 Jun. 1886, *M. Delavay 2386* (K); Ta-long-tan, Ta-pin-tze, 17 Sep. 1887, *M. Delavay s. n.* (BM, E, K); **Gongshan**, Maku, 2000 m, 7 Nov. 2008, *X. H. Jin 9570* (PE); Maku, 2000 m, 7 Nov. 2008, *X. H. Jin 9579* (PE); **Longyang**, Mangkuan, Baihualing, Hannong, 1540 m, 10 Oct. 2007, *H. Li 1229* (KUN); Mangkuan, Baihualing, 1600 m, 12 Oct. 2007, *H. Li 1341* (KUN); Nujiangba, Qianjin, 680 m, 27 Sep. 1979, *Y. H. Li & P. H. Yu 20461* (HITBC), Baihualing, Laomanggou, 1500 m, 27 Aug. 1935, *X. C. Shi & S. X. Yang 622* (KUN); Mangkuan, 1740 m, 18 Oct. 1998, *H. Li 10565* (KUN, MBK); Mankuan, Baihualing, Hanglong, 1730 m, 28 Nov. 2000, *H. Li 13482* (CAS, KUN, MBK); Mangkuan, Baihualing, Zaotang, 1590 m, 9 Sep. 2003, *Gaoligong Shan Biodiversity Survey 18950* (CAS, KUN); **Puge**, Qiaowo, Dacaohe, 1600 m, 7 Aug. 1976, *s. coll. 14212* (CDBI, PE); **Tianquan**, Wenbifeng Park, 780 m, 16 Oct.

Figure 46-95: **Asystasia neesiana** (Wall.) Nees: **a**, portion of plant; **b**, calyx lobes and pistil; **c**, corolla spread out showing stamens; **d**, stamens; **e**, capsule; **f**, seed. Drawn by Yun-Xi ZHU; reproduced from *Fl. China Ill.* 19: 346, fig. 346. 2011.

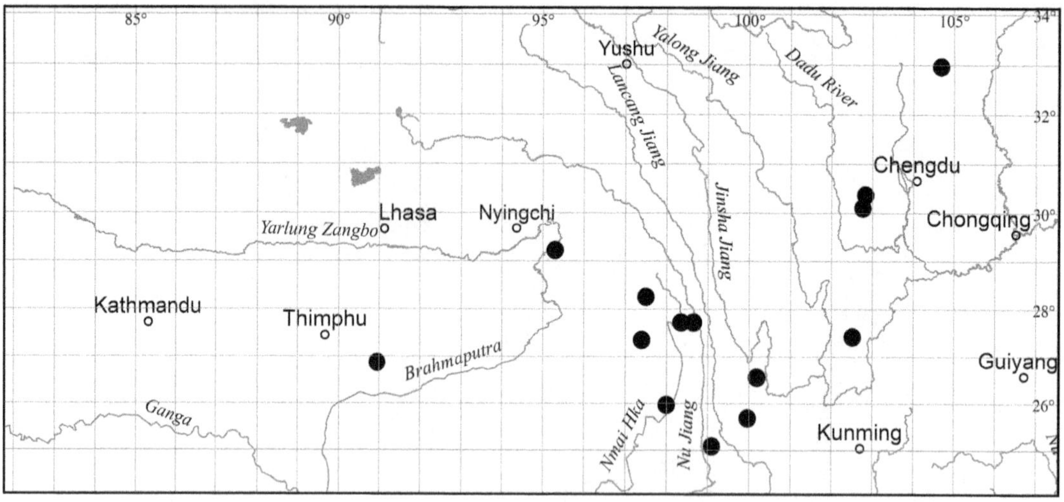

Map 46-161: **Asystasia neesiana** (Wall.) Nees

2006, *Y. F. Deng & H. J. Xin 20903* (IBSC); Shuitaping, 4 Aug. 1953, *H. L. Tsiang 35142* (IBK, IBSC, PE, SZ); 10 Aug. 1939, *F. C. Tai & C. M. Teng 5087* (SZ); 850 m, Sep. 13, 1963, *Exped. W. Sichuan (K. J. Guan & W. T. Wang) 3441* (HIB, PE); Kugaoping, 18 Oct. 1935, *Y. Y. Ho 6863* (NAS); **Yangbi**, Pingpo, in the thickets, 5 Sep. 2013, *Y. F. Deng et al. 24163* (IBSC, KUN); Shimenguan, 1750–2100 m, 31 Jul. 2013, *Q. W. Lin 2013-283* (PE). **U IRRAWADDY**: **Chaungmaw Kyaun**, North Triangle, 1200 m, 30 Aug. 1953, *F. Kingdon-Ward 21297* (A, BM, TI); **Drungjiang**, 2 km S of Bapo near the confluence of the Dulong Jiang and the Teliwang He, 1410 m, 28 Oct. 2004, *Gaoligong Shan Biodiversity Survey 20684* (CAS, HUNN, KUN, MO); ca. 2 km S of Bapo near the confluence of the Dulong Jiang and the Teliwang He, 1410 m, 28 Oct. 2004, *Gaoligong Shan Biodiversity Survey 20699* (CAS, KUN, MO); Gamolai He, Bapo, 1460 m, 30 Oct. 2004, *Gaoligong Shan Biodiversity Survey 21158* (CAS, KUN); Kiukiang Valley, Mongting, 1400 m, 20 Sep. 1938, *T. T. Yu 20375* (E); **Mahtum**, Hkawng Gaw, 750 m, 10 Sep. 1939, *R. Kaulback 388* (BM); **Nam Tamai Valley**, 1500 m, 9 Sep. 1938, *R. Kaulback 120* (BM); **Putao**, Chaunsone Camp to Ziyadan, 1050–1516 m, 19 Oct. 2014, *Putao Expedition Team (X. H. Jin et al.) PT-ET 688* (PE); **Tengchong**, Shweili-Salwin Divide, *G. Forrest 16050* (E).

20. PHIALACANTHUS Benth.

Bentham, Gen. Pl. 2(2): 1102. 1876; Clarke in Hooker (ed.), Fl. Brit. India 4: 523. 1885; Lindau in Engler & Prantl (eds.), Nat. Pflanzenfam. 4(3B): 335. 1895. Type: *Phialacanthus griffithii* Benth.

Herbs or small shrubs, with cystoliths. Leaves opposite, subequal, petiolate, entire. Inflorescences of terminal panicles, or frequently reduced nearly to a raceme or spike, glabrous or very nearly so; pedicels spreading, with nodding flowers. Bracts small, inconspicuous. Corolla-tube linear or cylindric, conspicuously 2-lipped, upper lip erect, subentire, lower lip 3-lobed; lobes imbricate in bud. Calyx-lobes united into a tube at base. Stamens 4; anthers bithecous, thecae oblong, base muticous, one placed half-way below another, or nearly at equal height, base muticous. Ovary glabrous, or hairy on lower portion; ovules 2 in each locule; style shortly 2-fid, or with one branch obsolete. Capsules stalked, with 4 smooth, flat seeds in clavate upper part.

Four species in Himalaya and Malaysia; one species in the Pan-Himalaya.

1. Phialacanthus griffithii Benth., Gen. Pl. 2(2): 1102. 1876; Clarke in Hooker (ed.), Fl. Brit. India 4: 523. 1885. Type: Yarlung Zangbo-Brahmaputra, Mishme, *Herbs Griffith s. n.* [holotype: K (K000950125); isotype: K (K000950124)].

Shrubs, erect; stems glabrous. Petiole 2–4 cm long; leaf blade elliptic, 15–20 cm long, 5–6.5 cm broad, glabrous, apex acuminate, margin entire, base broadly cuneate. Cymes terminal, 4–7.5 cm long; pedicels slender, 3–5 mm long. Bracts minute. Calyx tubular, campanulate; lobes minute, triangular. Corolla 3.5–4 cm long, orange with yellow mouth, ventricose, 2-lipped, upper lip erect, subentire, lower lip 3-lobed, lobes subequal. Stamens 4; filaments glabrous; anthers pubescent. Ovary glabrous; 2 ovules per locule; style filiform, glabrous; stigma minutely 2-lobed. Capsules oblong, 5–6 mm long, glabrous, shortly stalked, 4-seeded. Figure 46-96.

Phenology: unknown.

Habitats: thickets.

Distribution: Yarlung Zangbo-Brahmaputra (Map 46-162); also in other parts of NE India.

YARLUNG ZANGBO-BRAHMAPUTRA: Mishme, *Herb. Griffith s. n.* (K).

21. MACKAYA Harv.

Harvey, Thes. Cap. 1: 8. 1859, nom. cons., non *Mackaya* Arnott, Mag. Zool. Bot. 2: 550. 1838; Lindau in Engler & Prantl (eds.), Nat. Pflanzenfam. 4(3B): 336. 1895; Wood in Grierson & Long (eds.), Fl. Bhutan 2(3): 1283. 2001; Hu, Deng & Daniel in Wu, Raven & Hong (eds.), Fl. China 19: 441. 2011. Type: *Mackaya bella* Harv.

Odontonemella Lindau, Bot. Jahrb. Syst. 18: 56. 1893; Lindau in Engler & Prantl (eds.), Nat. Pflanzenfam. 4(3B): 335. 1895. Type: *Odontonemella indica* (Nees) Lindau (≡ *Thyrsacanthus indicus* Nees).

Herbs, perennial, with cystoliths. Leaves petiolate. Inflorescences axillary or terminal, secund racemes. Bracts linear-lanceolate or narrowly deltoid; bracteoles 2, at base of pedicel. Calyx much shorter than corolla tube, 5-lobed almost to base; lobes linear-lanceolate. Corolla subactinomorphic; tube base cylindric, campanulate toward apex, limb 2-lipped, lower lip 3-lobed, upper lip 2-lobed; lobes suborbicular, ascending cochlear in bud. Stamens 2, inserted at base of campanulate portion of corolla tube, included; anthers bithecous; thecae oblong, dorsally pubescent; staminodes 2. Ovary ellipsoid; 2 ovules per locule; style subglabrous or pubescent near base; stigma 2-lobed. Capsules stipitate, 2- or 4-seeded; retinacula present. Seeds somewhat rugose.

Recent references: Ensermu et al., 1992. A reconsideration of *Asystasiella* Lindau (Acanthaceae). *Kew Bull.* 47: 669–675; Deng & Wu, 2009. A new combination in *Mackaya* (Acanthaceae), with lectotypification for *M. tapingensis*. *Novon* 19: 307–309.

Two species: disjunctly distributed in Himalayan region and South Africa; one species in the Pan-Himalaya.

Figure 46-96: **Phialacanthus griffithii** Benth.: portion of plant. Drawn by Yun-Xiao LIU.

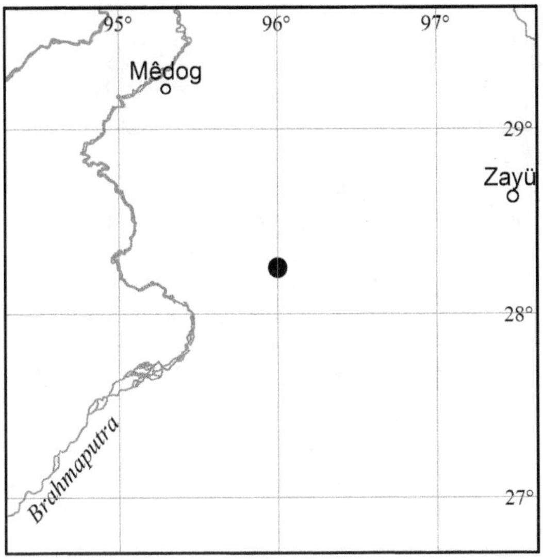

Map 46-162: **Phialacanthus griffithii** Benth.

1. **Mackaya indica** (Nees) Ensermu Kelbessa in Kew Bull. 47(4): 673. 1992; Wood in Grierson & Long (eds.), Fl. Bhutan 2(3): 1283. 2001. ≡ *Thyrsacanthus indicus* Nees in de Candolle (ed.), Prodr. 11: 325. 1847. ≡ *Odontonemella indica* (Nees) Lindau, Bot. Jahrb. 18: 56. 1894; Lindau in Engler & Prantl (eds.), Naturl. Pflanzenfam. 4(3b): (1995) 336. ≡ *Eranthemum indicum* (Nees) C. B. Clarke in Hooker (ed.), Fl. Brit. India 4: 497. 1884. ≡ *Pseuderanthemum indicum* (Nees) A. M. & J. M. Cowan, Trees Bengal: 98. 1929. Type: India, Assam, *Mack s. n.* [lectotype designated by Ensermu et al., Kew Bull. 47: 673. 1992: K (K000885509)].

Eranthemum tapingense W. W. Sm., Notes Roy. Bot. Gard. Edinburgh 10: 177. 1918, **syn. nov**. ≡ *Pseuderanthemum tapingense* (W. W. Sm.) C. Y. Hu & H. S. Lo, Fl. Hainan 4: 558. 1975; Hu, Fl. Reipubl. Popularis Sin. 70: 226. 2002. ≡ *Mackaya tapingense* (W. W. Sm.) Y. F. Deng & C. Y. Wu, Novon 19: 308. 2009; Hu, Deng & Daniel in Wu, Raven & Hong (eds.), Fl. China 19: 442. 2011. Type: Myanmar, Kachin State, Taping Valley, lat. 24°20′N, 2000 ft., Feb. 1913, *G. Forrest 9655* [lectotype designated by Deng & Wu, Novon 19: 308. 2009: E (E00198372); isotypes: A (A00273528), E (E00198371), IBSC (IBSC0306451), K (K000885507)].

Herbs or shrubs, 1–2.5 m high; stems erect, subterete, brown, glabrous or subglabrous. Petiole 0.5–3.5 cm long, sparsely puberulent or glabrescent; leaf blade elliptic, lanceolate to oblong-lanceolate, 3–21 cm long, 1.5–8 cm broad, glabrous, membranous, both surfaces glabrous or glabrescent, abaxially paler, adaxially dark green, lateral veins 8–12 pairs, base cuneate, margin entire to undulate, apex shortly acuminate. Inflorescences of lax terminal and axillary racemes, those in leaf axils shorter and always solitary; racemes 5–19 cm long, flowers usually in opposite pairs, internodes up to 1.5 cm long below, less above; rachis glabrous or puberulent; pedicel 0–5 mm long, sparsely pilose or glabrescent. Bracts linear triangular, 2–3 mm long, pilose or glabrescent; bracteoles linear, ca. 1 mm long. Calyx 5–8 mm long, puberulent; lobes lanceolate, 3–4 mm long, 1.5–2 mm broad, glabrous, apex acuminate. Corolla white or pink with darker red vines, 2.5–3.5 cm long, finely glandular pubescent; tube base cylindric, 0.8–1 cm long, weakly 2-lipped, upper lip 2-lobed, lower lip trilobed; lobes suborbicular, 7–12 mm long and broad, subequal, margin dentate. Stamens 2, included; filaments 6–12 mm long, pilose; anther-thecae 4–5 mm long, muticous at base; staminodes 1–4 mm long. Ovary glabrous; style persistent long after corolla fall. Capsules 2.5–3 cm long, glabrous, stipitate at base. Figure 46-97.

Phenology: flowering from December to April.

Habitats: moist forests especially in wood gullies, 400–4500 m.

Distribution: C Nepal, E Nepal, Sikkim & Darjiling, Bhutan, Yarlung Zangbo-Brahmaputra (Map 46-163); also in other parts of India, Myanmar and China (SW Yunnan).

C NEPAL: **Bagmati**, Kathamandu, Thankot, 1800 m, 12 Jan. 1975, *D. P. Joshi & K. R. Rajbhandari 75-276* (KATH); Kathamandu, Shivapuri, Okhareni, 1820 m, 18 Mar. 1970, *T. B. Shrestha & P. R. Shakya 11657* (KATH); Kathamandu, Shivapuri, Budhanikantha-Banjyang, 26 Mar. 1969, *H. Kanai 15075* (KATH, KYO);

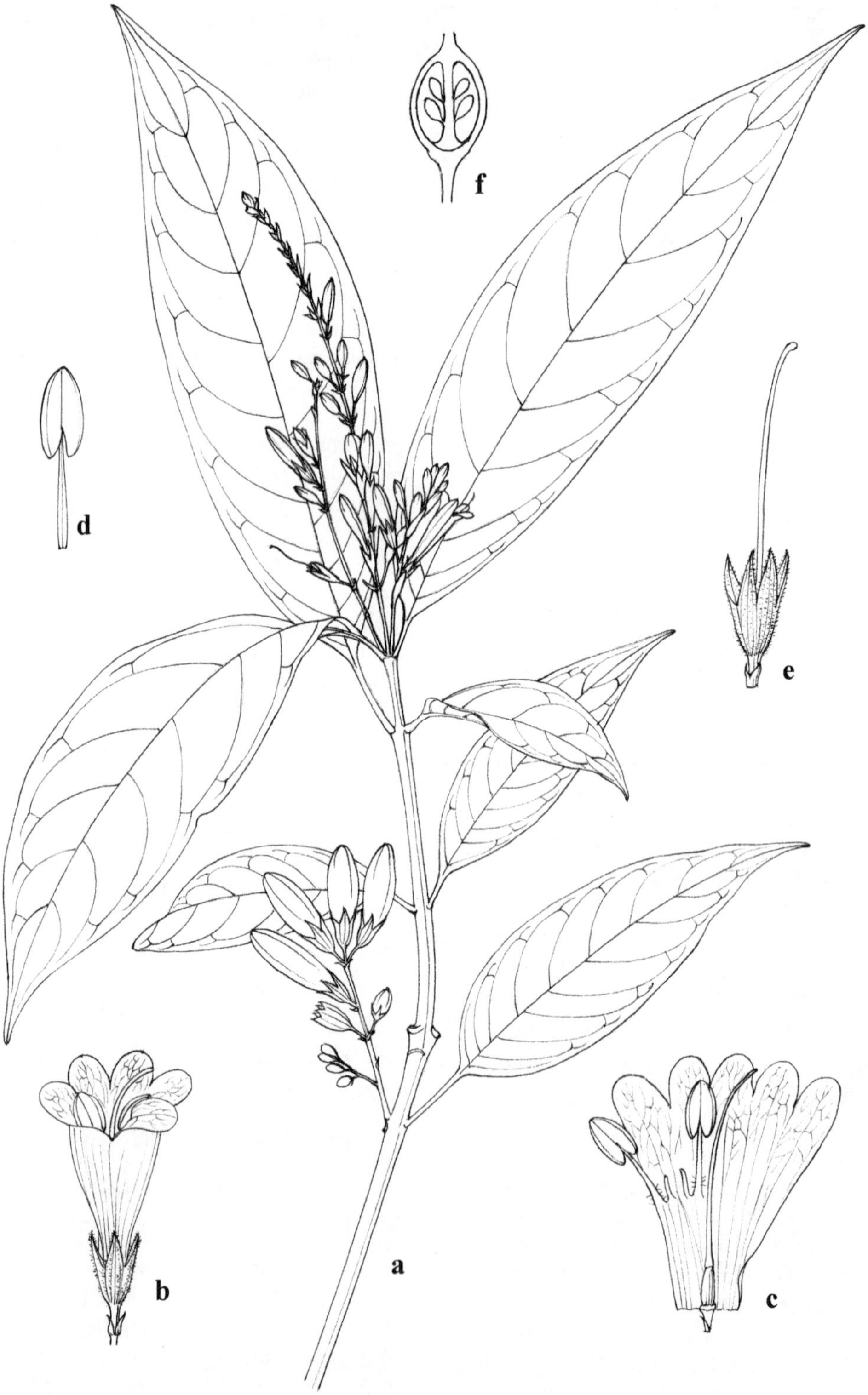

Figure 46-97: **Mackaya indica** (Nees) Ensermu Kelbessa: **a**, portion of plant; **b**, flower; **c**, opened corolla showing stamens; **d**, stamen; **e**, calyx and pistil; **f**, opened ovary showing ovules. Drawn by Xi-Lin WU.

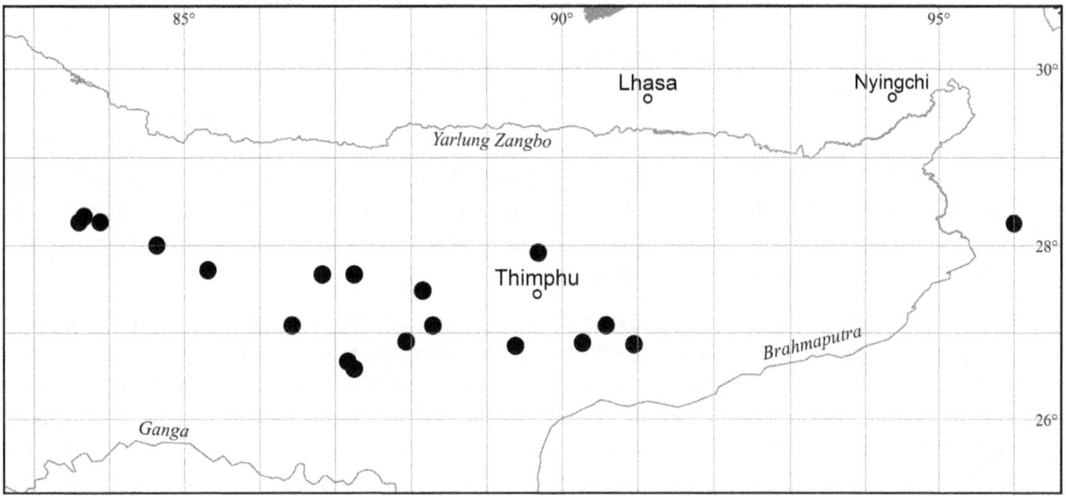

Map 46-163: **Mackaya indica** (Nees) Ensermu Kelbessa

Rusuwa, Trisuli Valley, Bokajhunda, 600 m, 14 Apr. 1962, *S. Bowes Lyon 18* (BM, E); **Dhawalagiri**, Baglung, Tarakholagaon, north of Beni, 2120 m, 24 Apr. 1954, *J. D. A. Stainton, W. R. Sykes & L. H. J. Williams 172* (BM, E); Parbat, Bhoking, 1900 m, 11 Mar. 1974, *D. P. Joshi & M. M. Amatya 74/1531* (KATH); **Gandaki**, Gorkha, N. of Gorkha, Gorkha Himalaya, 1290 m, 31 Mar. 1983, *G. Miehe 2661* (BM); Gorkha, Philim, Buri Gandaki, above Philim, 1820 m, 14 Apr. 1953, *P. C. Gardner 387* (BM); Kaski, Hills north of Pokhara, 2120 m, 17 Apr. 1954, *J. D. A. Stainton, W. R. Sykes & L. H. J. Williams 4845* (BM, E). **E NEPAL: Koshi**, Sankhuwasabha, Absuwa Khola, 1800 m, 28 Mar. 1989, *P. R. Shakya 9331* (KATH); Sunsari, Dharah, Cochuhola, 460 m, *J. D. A. Stainton 5678* (BM); Morang, Meklajam, N. E. of Dharah, 1820 m, *J. D. A. Stainton 6380* (BM); **Mechi**, Hams Pokhari, W. of Mechi Khola, *J. D. A. Stainton 6448* (BM); Ilam, Mai Pokhari, 2100 m, 6 Apr. 1967, *D. H. Dicolson 3153* (BM, KATH); **Sagarmatha**, Solukhumbu, between Kharikhola and Jubing, *H. Tabata et al. 7768* (A, KYO). **SIKKIM & DARJILING**: **Darjiling**, Lecbry, 1500 m, *C. B. Clarke* 27233 (K, BM); below Darjiling, 1500–1950 m, Apr. 1881, *G. Watt 7019* (E); Jalvar, 1800 m, *C. B. Clarke* 15815 (BM); **Sikkim**, Rishap, 900 m, *C. B. Clarke 13914* (LE); Bvich Hills, 2100 m, *J. S. Gamble 6879A* (K); Punkabari, 600 m, *C. B. Clarke 13884* (BM); Around lake Kheriperi, 1800 m, *S. Z. Lucksom 2802* (CAL); Shillong, Shillong Peak, 1800 m, *N. L. Bor 19* (L). **BHUTAN: Gasa**, Laya Gewog, North Triangle (Laja), 1200 m, *F. Kingdon-Ward 21640* (A, TI); **Gaylegphug**, W. bank of Chabley Khola, 55 km along Tongsa Road, 1720 m, 27 Mar. 1982, *A. J. C. Grierson & D. G. Long 4034* (A, E, K); Gal Chu Valley, 1650 m, *F. Ludlow & G. Sherriff 2951* (BM); **Phuntsholing**, between Ganglakha and Sintilakha, 1700 m, 3 Apr. 1992, *J. R. I. Wood 7554* (E); hillside above Rinchending, Phuntscholing, ca. 780 m, 16 Feb. 1982, *A. J. C. Grierson & D. G. Long 2967* (E); **Sarbhang**, Valley behind Sarbhang School, 600 m, 7 Feb. 1988, *J. R. I. Wood 6131* (E); near Tori Bari 18 km along Chirang road, ca. 1020 m, 11 Mar. 1982, *A. J. C. Grierson & D. G. Long 3658* (A, E, K); ca. 8 km from Sarbhang towards Dauplu, ca. 800 m, 19 Feb. 1988, *J. R. I. Wood 6152* (E); Sarpang, Gelephu (Surelakha), 12 m, N. of Hatisar, 1200 m, *F. Ludlow, G. Sherriff & J. H. Hicks 18511* (BM); **Tashigang**, betwee Pumakna and Gara, 1500 m, 10 Mar. 1991, *J. R. I. Wood 7369* (E); **Tongsa**, SW of Mangde Chu crossing below Shemgang, 1000 m, 2 Feb. 1988, *J. R. I. Wood 6130* (E); Between Shemgang and Dakpai, 1700 m, 15 Mar. 1992, *J. R. I. Wood 7543* (E). **YARLUNG ZANGBO-BRAHMAPUTRA: Dirang**, *G. Panigrahi 6059* (CAL); **Mishmi Hills**, Kamlang river, 4500 m, *F. Kingdon-Ward 18427* (BM).

22. CODONACANTHUS Nees

Nees in de Candolle (ed.), Prodr. 11: 103. 1847; Clarke in Hooker (ed.), Fl. Brit. India 4: 500. 1884; Lindau in Engler & Prantl (eds.), Nat. Pflanzenfam. 4(3B): 330. 1895; Wood in Grierson & Long (eds.), Fl. Bhutan 2(3): 1283. 2001; Hu, Fl. Reipubl. Popularis Sin. 70: 227. 2002; Hu, Deng & Daniel in Wu, Raven & Hong (eds.), Fl. China 19: 448. 2011. Type: *Codonacanthus pauciflorus* (Nees) Nees (≡ *Asystasia pauciflora* Nees).

Herbs, perennial; with cystoliths. Leaves petiolate; leaf blade entire to subsinuate. Inflorescences of axillary or terminal racemes, sometimes branched and forming a panicle, pedicellate. Bracts and bracteoles minute. Flowers 1(–3 or more) on each node. Calyx deeply 5-lobed; lobes subequal. Corolla white with colored markings on lower lip, broadly campanulate; tube short; limb slightly 2-lipped, lower lip 3-lobed, upper lip 2-lobed; lobes ascending cochlear in bud. Stamens 2, included in to slightly exerted from corolla tube; anthers bithecous; thecae unequal, subparallel, equally to subequally inserted; staminodes 2, short. Ovary with 2 ovules per locule; stigma capitate to shallowly 2-cleft. Capsules with a solid stalk at base, fertile portion ellipsoid, usually with a medial constriction, 4-seeded; retinacula present. Seeds slightly discoid, surfaces and margin smooth to rugose.

Two species: E and S Asia; one species in the Pan-Himalaya.

1. Codonacanthus pauciflorus (Nees) Nees in de Candolle (ed.), Prodr. 11: 103. 1847; Clarke in Hooker (ed.), Fl. Brit. India 4: 500. 1884; Wood in Grierson & Long (eds.), Fl. Bhutan 2(3): 1284. 2001; Hu, Fl. Reipubl. Popularis Sin. 70: 228. 2002; Hu, Deng & Daniel in Wu, Raven & Hong (eds.), Fl. China 19: 448. 2011. ≡ *Asystasia pauciflora* Nees in Wallich (ed.), Pl. Asiat. Rar. 3: 90. 1832. ≡ *Ruellia pauciflora* Wall., Numer. List no. 2369, 1830, nom. invalid. Type: Bangladesh, Silhet, *F. D. in Wallich Cat. 2369* [holotype: K-W (K001115768); isotypes: CGE, E (E00273509), GZU (GZU000280513, GZU000280514), K (K000885627), LE].

Codonacanthus acuminatus Nees, Prodr. 11: 103. 1847. Type: India, Assam, *Griffith s. n.* [holotype: K ("herb. Hook."); isotype: GZU (GZU000280512)].

Herbs, 20–60 cm high; stems quadrangular, sulcate, bifariously pubescent. Petiole 3–15 mm long, puberulent or glabrous; leaf blade elliptic, ovate, or lanceolate, (2–)4–12 cm long, (0.5–)1.5–4.5 cm broad, abaxially glabrous or sparsely pubescent along veins, adaxially glabrous or with few multicellular hairs, lateral veins 5–7 pairs, base rounded, cuneate, or attenuate and decurrent onto petiole, apex acute or acuminate. Racemes simple or terminal ones usually forming a panicle, 5–15 cm long; rachis puberulent. Bracts lanceolate to linear, 1–2.5 mm long, puberulent; bracteoles lanceolate, 1–2 mm long, puberulent. Pedicel 1–7 mm long, puberulent. Calyx 2.5–4 mm long; lobes lanceolate, puberulent. Corolla white with purple spots at base of lower lip, 6.5–10 mm long, glabrous; tube base cylindric, 1–2 mm long, 1–1.5 mm broad, upwards abruptly widened to 3.5–6.5 mm broad at mouth; lobes 3.5–5.5 mm long, 2–3.5 mm broad. Stamens 2; filaments 2–2.5 mm long, glabrous; anther-thecae 1.2–1.4 mm long. Staminodes 2, minute. Ovary glabrous; style 4–6.5 mm long. Capsules 1–1.9 cm long, glabrous. Seeds ovate in outline, compressed, 2–3 mm long and broad. Figure 46-98.

Phenology: flowering and fruiting from August to April of next year.

Habitats: evergreen broad-leaved forests, wet places in ravines; below 1700 m.

Distribution: Bhutan, Yarlung Zangbo-Brahmaputra, U Irrawaddy (Map 46-164); also in Bangladesh, Cambodia, Japan, Thailand, Vietnam, and other parts of India, Myanmar and

China (S Yunnan, Guizhou, Guangxi, Guangdong, Hainan, Hunan, Jiangxi, Fujian, Taiwan).

BHUTAN: **Gaylephug**, Taklai Khola E of Gaylegphug, 370 m, 23 Mar. 1982, *A. J. C. Grierson & D. G. Long 3973* (E); **Sarbhang**, valley above Sarbhang High School, 600 m, 7 Feb. 1988, *J. R. I. Wood 6133* (E, TI); Lam Pati 10 km E of Sarbhang, 410 m, 7 Mar. 1982, *A. J. C. Grierson & D. G. Long 3525* (E, K). **YARLUNG ZANGBO-BRAHMAPUTRA:** Pasighat, Dihang Valley, 150–300 m, 9 Feb. 1928, *F. Kingdon-Ward 7817* (ABD, K); Makum Dihang River, 16 Mar. 1894, *G. A. Gammie 20* (CAL); Sher, SW of the river Janakmakh, *I. H. Burkill 36461* (CAL); Edhe of Serpo valley near Rengging Camp, 630 m, *I. H. Burkill 36311* (CAL); Yambung to Sissin, 300 m, *I. H. Burkill 36017* (CAL); Akha Hills, *Badul Khan 146* (CAL); **Aprunyi**, Kambong to Pakam, *G. K. Murthy 13033* (CAL); **Minhmi**, *Master 119* (CAL); **Riga**, Kaleng Gate, *S. Panigrahi 5906* (CAL). **U IRRAWADDY**: **Putao**, 350 m, *S. M. Toppin 5045* (CAL, E).

23. PSEUDERANTHEMUM Radlk.

Radlkofer, Sitzungsber. Math.-Phys. Cl. Königl. Bayer. Akad. Wiss. München 13: 282. 1883; Lindau in Engler & Prantl (eds.), Nat. Pflanzenfam. 4(3B): 330. 1895; Wood in Grierson & Long (eds.), Fl. Bhutan 2(3): 1282. 2001; Hu, Fl. Reipubl. Popularis Sin. 70: 220. 2002; Hu, Deng & Daniel in Wu, Raven & Hong (eds.), Fl. China 19: 439. 2011. Lectotype (designanted by Leonard, *Contr. U.S. Natl. Herb.* 31: 292. 1958): *Pseuderanthemum bicolor* (Schrank) Lindau (≡ *Eranthemum bicolor* Schrank).

Herbs or shrubs; with cystoliths. Leaves sessile or petiolate; leaf blade entire or subentire or lobed. Flowers in axillary or terminal thyrses, racemes, or spikes. Bracts and bracteoles present, much shorter than calyx. Calyx deeply 5-lobed; lobes equal to subequal. Corolla salverform; tube slender, not or scarcely apically enlarged; tube usually long cylindric and straight, or rarely short and curved; limb 2-lipped; upper lip 2-lobed, lower lip 3-lobed; lobes ascending cochlear in bud. Stamens 2, included in corolla tube or exserted from corolla mouth; anthers bithecous; thecae equal to subequal, parallel, inserted at same height, base muticous; staminodes 2. Style included or exserted; stigma 2-lobed; lobes equal. Capsules clavate, base with a long solid stalk, 4-seeded; retinacula present. Seeds lenticular, ornamented or smooth, lacking trichomes.

Recent references: Fosberg & Sachet, 1985. History of the generic name *Pseuderanthemum* Radlkofer (Acanthaceae). *Baileya* 22(4): 178–179; Hansen, 1989. Notes on SE Asian Acanthaceae 1. *Nordic J. Bot.* 9: 209–215; Choopan et al., 2018. An annotated checklist of the genus *Pseuderanthemum* Radlk. (Acanthaceae) in Thailand. *Thai Forests Bull. (Bot.)* 46: 90–111. 2018.

About 50 species: pantropical; three species in the Pan-Himalaya.

1a. Corolla tube curved, swollen at base, constricted above and narrowed to mouth ... 1. *P. leptanthus*
1b. Corolla salverform with long cylindric tube, straight, slender, not constricted above and narrowed to mouth.
  2a. Herbs; flowers in racemes; bracts distant on rachis..2. *P. latifolium*
  2b. Shrub; flowers in thyrses; bracts imbricate on rachis...................................... 3. *P. polyanthum*

Figure 46-98: **Codonacanthus pauciflorus** (Nees) Nees: **a**, portion of plant; **b**, inflorescence; **c**, opened corolla showing stamens; **d**, stamen; **e**, calyx and pistil; **f**, capsule. Drawn by Quan LIN; reproduced from *Icon. Cormophyt. Sin.* 4: 170, fig. 5754. 1975.

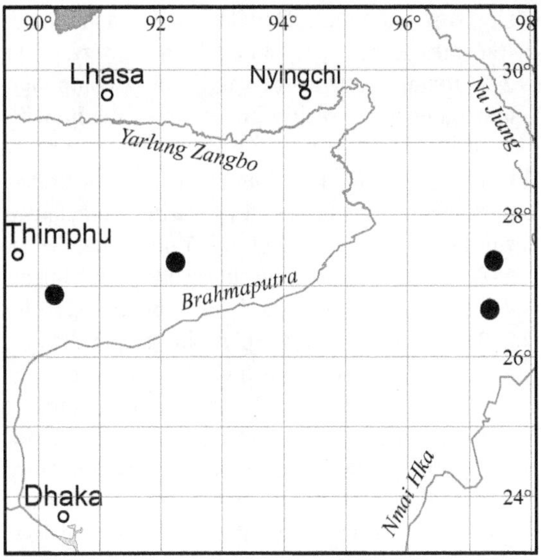

Map 46-164: **Codonacanthus pauciflorus** (Nees) Nees

1. Pseuderanthemum leptanthus (C. B. Clarke) Lindau in Engler & Prantl (eds.), Nat. Pflanzenfam. 4(3b): 330. 1895. ≡ *Eranthemum leptanthus* C. B. Clarke in Hooker (ed.), Fl. Brit. India 4: 500. 1884. Type: Yarlung Zangbo-Brahmaputra, Mishime, *Herbs Griffith 6181* [lectotype **here designated**: K (K000885485); isolectotype: K (K000885486)].

Sphinctacanthus parkinsonii C. E. C. Fisch., Bull. Misc. Inform. Kew 1927: 210. 1927. Type: Myanmar, Mytkyina, Nawraw Chaung, 10 Dec. 1925, *C. E. Parkinson 390* [holotype: K (K000885487)].

Shrubs; stems glabrous. Petiole 1.5–2.2 cm long, glabrous; leaf blade narrowly elliptic to oblanceolate, 12–22 cm long, 3.5–9 cm broad, glabrous, lateral veins 8–13 pairs, base narrowed onto petiole, margin subundulate, apex acute or acuminate. Spikes terminal, many-flowered, 8–15 cm long; peduncle 4–5 cm long, sparsely puberulous. Bracts minute, linear. Pedicels slender, glandular-puberulent. Calyx 3.5–4 mm long, sparsely puberulous without, 5-lobed; lobes equal or anterior one shorter, linear-lanceolate, glabrous, margin ciliate, apex acute or subacute. Corolla 1.8–2 cm long, mauve, glabrous; tube curved, swollen at base, constricted above and narrowed to mouth, limb 8–9 mm long, upper lip 2-lobed to the middle, lobes oblong, rounded, lower lip deeply 3-lobed, lobes narrower, rounded. Stamens 2, exserted; filaments flat, broad, glabrous, 1.5–2.5 mm long, base united by a narrow ligament; anthers bithecous, thecae separate, suborbicular or reniform, one slightly below another, base muticous. Staminode 2, ligulate. Ovary ellipsoid, glabrous, 2–3 mm long; style filiform, glabrous, shortly exserted; stigma slightly 2-lobed. Capsules unseen.

Phenology: flowering from December to February of next year.

Habitats: thickets; 300–3000 m.

Distribution: Yarlung Zangbo-Brahmaputra, U Irrawaddy (Map 46-165); also in other parts of India.

YARLUNG ZANGBO-BRAHMAPUTRA: Rumi Dam Bang, 16 *Jan.* 1912, *I. H. Burkill 36408* (K); **Abor Hills**, Nyamghat, 300 m, 28 Aug. 1928, *F. Kingdon-Ward 8733* (K); outer Abor Hills, 23 Jan. 1912, 1650 m, *I. H. Burkill 36211* (K); **Lohit Valley**, 1200 m, 18 Dec. 1926, *F. Kingdom Ward 7699* (K); **Mishime**, *Herbs Griffith 6181* (K). **U IRRAWADDY**: **Putao**, Upper Shanlaung to Wasadam, 1000–3000 m, 14 Oct. 2013, *Putao Expedition Team (X. H. Jin et al.) PT-ET 225* (PE); Hkamdaukmyit

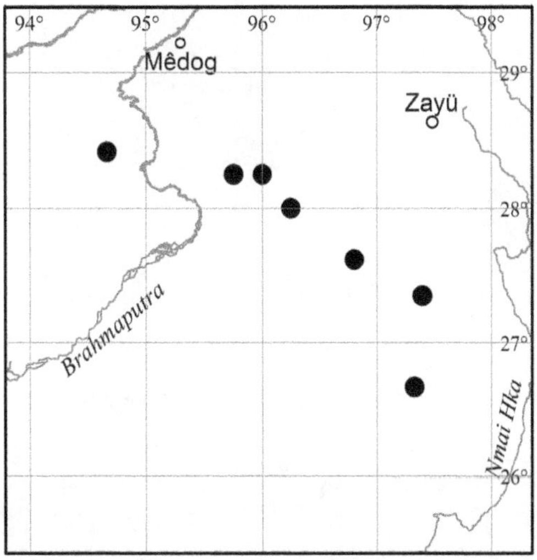

Map 46-165: **Pseuderanthemum leptanthus** (C. B. Clarke) Lindau

Camp, 2457 m, 21 Oct. 2014, *Putao Expedition Team (X. H. Jin et al.) PT-ET 1046* (PE).

2. Pseuderanthemum latifolium (Vahl) B. Hansen, Nordic J. Bot. 9: 213. 1989; Hu, Fl. Reipubl. Popularis Sin. 70: 225. 2002; Hu, Deng & Daniel in Wu, Raven & Hong (eds.), Fl. China 19: 440. 2011. ≡ *Justicia latifolia* Vahl, Symb. Bot. 2: 4. 1791. Type: India, *Koenig s. n.* [holotype: C (C10005094)].

Justicia palatifera Wall., Pl. Asiat. Rar. 1: 80. t. 92. 1830. ≡ *Eranthemum palatiferum* (Wall.) Nees in Wallich, Pl. As. Rar. 3: 108. 1832; Nees in de Candolle (ed.), Prodr. 11: 457. 1847; Clarke in Hooker (ed.), Fl. Brit. India 4: 498. 1884; Wood in Grierson & Long (eds.), Fl. Bhutan 2(3): 1282. 2001. ≡ *Pseuderanthemum palatiferum* (Wall.) Radlk. ex Lindau in Engl. & Prantl., Pflanzenfam. 4(3b): 330. 1895. Type: [*icon*] Wallich, Pl. Asiat. Rar. 1: 80, t. 92, f. 1–3. 1830 (lectotype designated by Hansen, Nordic J. Bot. 9: 213. 1989).

Herbs, 0.4–1 m high, perennial; stems quadrangular, pubescent when young, then glabrescent. Petiole 0.5–2.5 mm; leaf blade elliptic to elliptic-obovate, 5–20 cm long, 1.5–8 cm broad, abaxially glabrous except villous on midvein, adaxially glabrous, lateral veins 5 or 6 pairs, base cuneate, margin subentire to slightly undulate-crenulate, apex acuminate. Racemes terminal, 15–30 cm long, solitary, often 2–5 arising from tip of stem, rarely compound and paniculate; rachis pubescent, usually with 3-flowered clusters at nodes. Bracts linear-triangular, 4–6 mm long, 1.6–2 mm broad, pubescent, distant on the rachis; bracteoles similar to bracts. Pedicel 0.5–5 mm long. Calyx 5–7 mm long, adaxially pubescent; lobes linear, 4.5–5.5 mm long, ca. 1 mm broad. Corolla 2–3.5 cm long, light purple with purple dots on lower lip, pubescent on tube; tube cylindric for 1.5–2.5 cm long, 1.7–2 mm broad at mouth; lobes elliptic, 3–4 mm long, 2.5–3 mm broad, 2 slightly smaller than other 3 giving a slightly 2-lipped appearance, mauve with a yellow spot, glabrous. Anthers acute at base. Ovary comose; style basally pubescent. Capsules 2–2.5 cm long, pubescent. Seeds ovate in outline, compressed, 3–4 mm long and broad, rugose.

Phenology: flowering from April to June; fruiting from August to September.

Habitats: forests; 100–1600 m.

Distribution: E Nepal, Sikkim & Darjiling, Bhutan, Yarlung Zangbo-Brahmaputra (Map 46-166); also in Cambodia, Laos, Malaysia, Thailand, Vietnam, and other parts of India, Myanmar and China (Yunnan, Guizhou, Guangxi, Hainan).

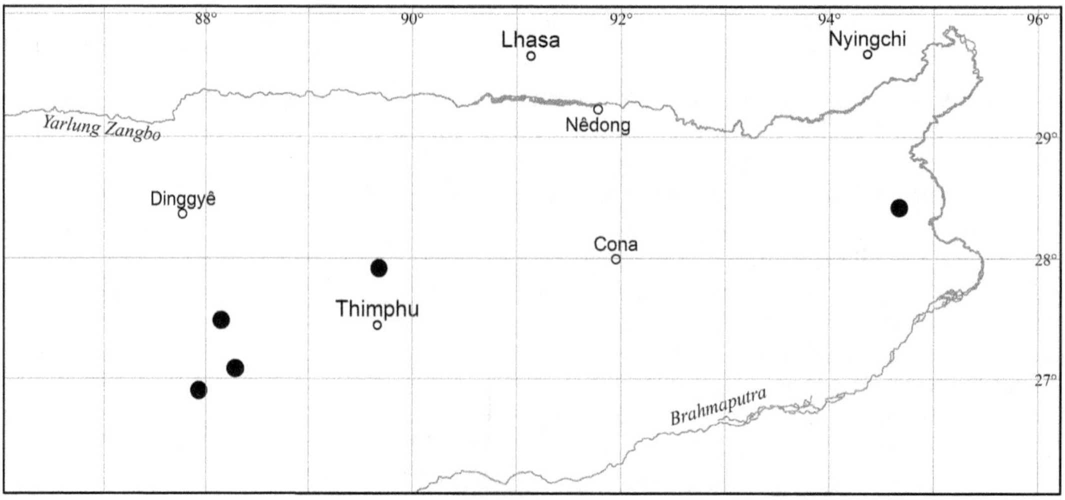

Map 46-166: **Pseuderanthemum latifolium** (Vahl) B. Hansen

E NEPAL: **Mechi**, Ilam, Soktim Tea Estate, Mai Khola, 460 m, 17 Apr. 1971, *J. D. A. Stainton 6760* (KATH). **SIKKIM & DARJILING**: **Darjiling**, Ducha Thar, Apr. 1876, *J. S. Gamble 494C* (K); Ryang, 900 m, 11 Nov. 1870, *C. B. Clarke 13709* (K); **Sikkim**, 300 m, 29 Apr. 18274, *D. Treutler 40* (K); 1500 m, *J. D. Hooker s. n.* part B (K). **BHUTAN**: Hatisan, 300 m, 24 Apr. 1937, *F. Ludnow & G. Sherriff 2922* (BM). **YARLUNG ZANGBO-BRAHMAPUTRA**: Outer **Abor Hills**, *I. H. Burkill 36819* (K).

3. Pseuderanthemum polyanthum (C. B. Clarke ex Oliv.) Merr., Brittonia 4: 175. 1941; Hu, Fl. Reipubl. Popularis Sin. 70: 225. 2002; Hu, Deng & Daniel in Wu, Raven & Hong (eds.), Fl. China 19: 439. 2011. ≡ *Eranthemum polyanthum* C. B. Clarke ex Oliv., Hooker's Icon. Pl. 20: t. 2000. 1891. Type: India, Nempean in the Patkye Mountains, between Assam and Burma, *W. Griffith s. n.* (lectotype **here designated**: K).

Shrubs; stems grayish, subterete, glabrescent. Petiole 1–2.5 cm long; leaf blade broadly ovate to oblong, 7–17 cm long, (3–)4–9 cm broad, both surfaces glabrous, lateral veins 7–9 pairs, base cuneate and decurrent onto petiole, margin entire, apex acute. Thyrses 5–12 cm long. Bracts triangular, 3.5–4 mm long, 1.2–1.5 mm broad, imbricate on the rachis; bracteoles 1.8–2 mm long, 0.4–0.5 mm broad. Pedicel 1.5–2.5 mm long. Calyx 1–1.1 cm long; lobes lanceolate. Corolla bluish purple; tube 3–3.5 cm long, lower lip 3-lobed; lobes oblong, 1.2–1.5 mm long, 0.5–0.6 cm broad, upper lip 2-lobed; lobes 1–1.1 cm long, 2–3 mm broad. Stamens 2; filaments short, distinct, inserted at throat. Ovary pilose; style 2.5–2.8 cm long, pilose towards base; stigma spherical. Capsules clavate, 2.2–2.5 cm long, pubescent. Seeds ovate in outline, compressed, 2–3 mm long and broad, verrucose, glabrous. Figure 46-99.

Phenology: flowering from April to June; fruiting from July to September.

Habitats: forests, thickets; 300–1600 m.

Distribution: U Irrawaddy (Map 46-167); also in Malaysia, Thailand, Vietnam, and other parts of India, Myanmar and China (S Yunnan).

U IRRAWADDY: **Putao**, 1931, *G. Forrest 29392* (PE).

24. WUACANTHUS Y. F. Deng, N. H. Xia & H. Peng

Deng et al., Pl. Divers. 38: 318. 2016. Type: *Wuacanthus microdonta* (W. W. Sm.) Y. F. Deng, N. H. Xia & H. Peng (≡ *Justicia microdonta* W. W. Sm.).

Shrubs. Leaves opposite, petiolate, blade margin entire, penninerved. Inflorescence terminal thyrses terminal. Bracts linear-lanceolate; bracteoles 2, similar to bracts. Calyx 5-lobed almost to base; lobes linear-lanceolate. Corolla tube short; limb longer than tube, 2-lipped, upper lip 2-lobed, lower lip 3-lobed. Stamens 2, included; filaments pilose; anthers bithecous, thecae oblong, each with two white spurs at base. Staminodes 2, minute. Ovary glabrous, 2 ovules per locule; style densely pilose. Capsules 4-seeded, stalked at base. Seeds reddish brown, ovoid.

Recent references: Deng et al., 2016. *Wuacanthus* (Acanthaceae), a new Chinese endemic genus segregated from *Justicia* (Acanthaceae). *Pl. Divers.* 38: 312–321.

Monospecific genus, endemic to China, occuring in Jinshajiang Valley; also in the Pan-Himalaya.

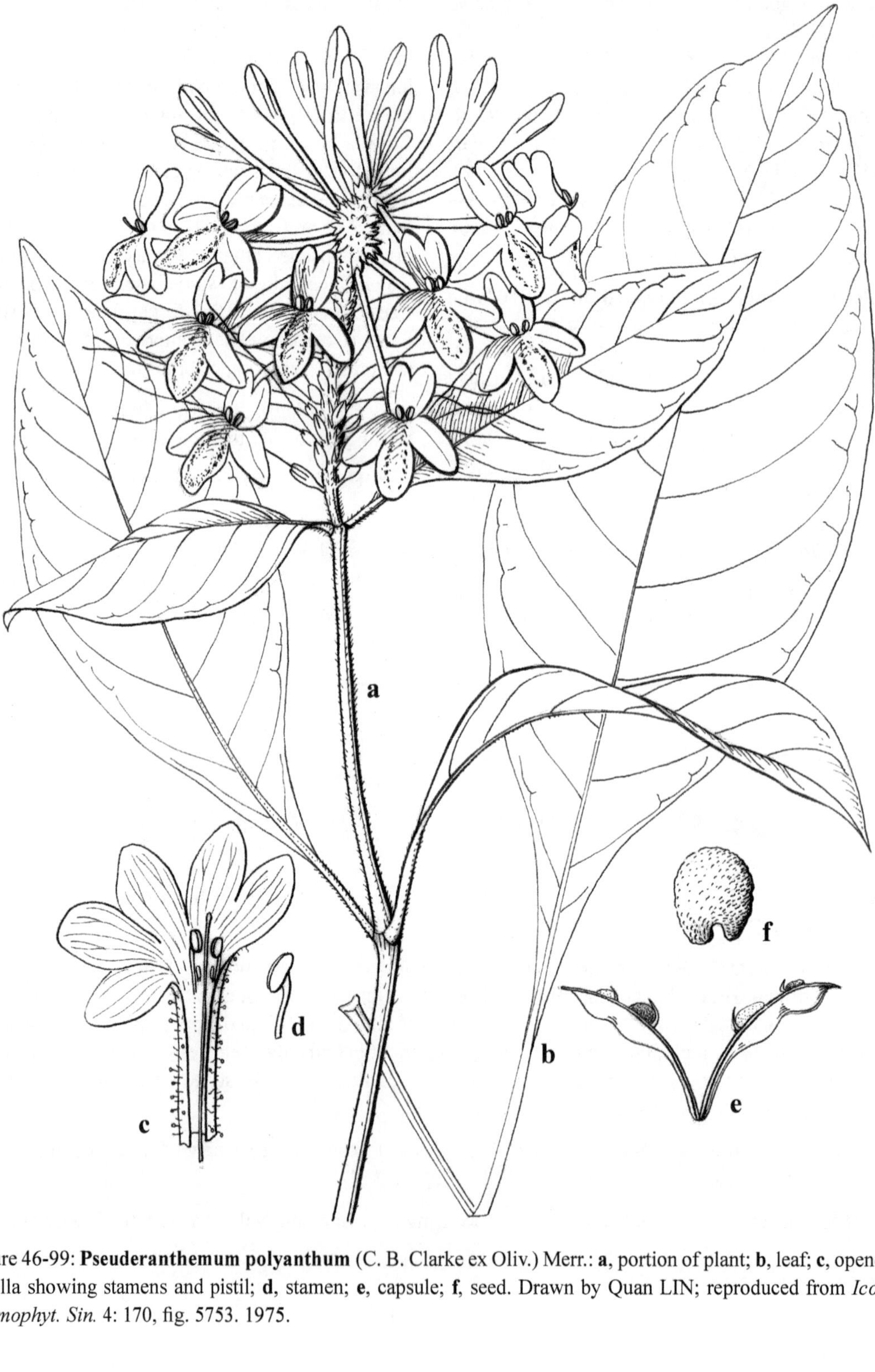

Figure 46-99: **Pseuderanthemum polyanthum** (C. B. Clarke ex Oliv.) Merr.: **a**, portion of plant; **b**, leaf; **c**, opened corolla showing stamens and pistil; **d**, stamen; **e**, capsule; **f**, seed. Drawn by Quan LIN; reproduced from *Icon. Cormophyt. Sin.* 4: 170, fig. 5753. 1975.

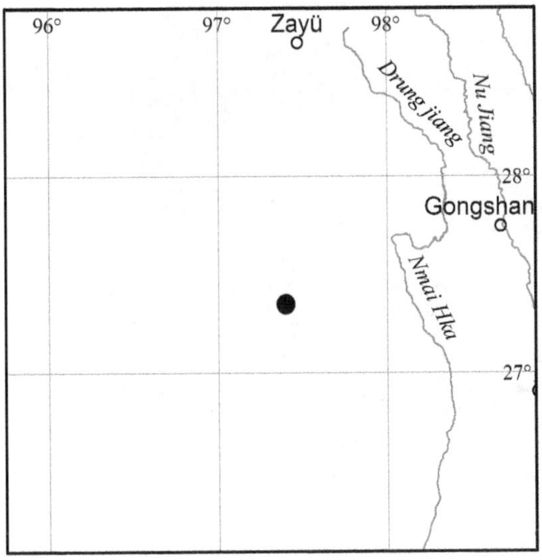

Map 46-167: **Pseuderanthemum polyanthum** (C. B. Clarke ex Oliv.) Merr.

1. **Wuacanthus microdonta** (W. W. Sm.) Y. F. Deng, N. H. Xia & H. Peng, Pl. Divers. 38: 320. 2016. ≡ *Justicia microdonta* W. W. Sm., Notes Roy. Bot. Gard. Edinburgh 10: 183. 1918; Hu, Deng & Daniel in Wu, Raven & Hong (eds.), Fl. China 19: 459. 2011. ≡ *Mananthes microdonta* (W. W. Sm.) C. Y. Wu & C. C. Hu in Hu (ed.), Fl. Reipubl. Popularis Sin. 70: 296. 2002. Type: [S Hengduan] China, Yunnan: mountains of the Chungtien plateau, Jul. 1914, *G. Forrest 12822* [lectotype designated by Deng et al., Pl. Divers. 38: 320. 2016: E (E00265762); isolectotypes: BM (BM001191509), K (K000884033)].

Shrubs, 0.6–2 m high, much branched. Young branches reddish when dry, quadrangular, sparsely pilose. Petiole 1–2 cm; leaf blade ovate, (2–)3–5.5 cm long, 1.5–2.5 cm broad, abaxially pilose along midvein, adaxially pilose especially along veins, base broadly cuneate and decurrent onto petiole, margin entire, apex shortly acuminate to acute. Thyrses terminal, 5–7(–10) cm long, nodes distant; cymes 2–4-flowered; peduncle short; rachis densely pilose. Bracts linear-lanceolate, 3–5 cm long; bracteoles similar to bracts. Calyx 5–7 mm long, 5-lobed almost to base; lobes linear-lanceolate, 5–6 mm long, pilose along midvein. Corolla white, 1–1.3 cm long, outside somewhat pubescent to glabrescent; tube short, limb longer than tube; lower lip 4–5 mm long, 3-lobed, lobes pilose inside and ciliate on margin, upper lip 2-lobed. Stamens 2; filaments pilose; anther-thecae parallel, superposed, with two white spur at base. Ovary glabrous; style densely pilose. Capsules 1.8–2 cm long, 4-seeded. Seeds reddish-brown, circular in outline, 3–4 mm in diameter, rugose-verruculose. Figure 46-100.

Phenology: flowering from August to October; fruiting from November to January of next year.

Habitats: thickets; 600–2100 m.

Distribution: S Hengduan (Map 46-168); also in other parts of China (S Sichuan).

S HENGDUAN: **Butuo**, Jiaojihe, Dadiba, *Chuan Jing Liang 5636* (KUN, PE); Niujiaowan, bank of Xixihe, 690 m, 11 Oct. 2014, *Y. F. Deng 25860* (IBSC, KUN); **Dêqên**, Yangtze Valley at Paung-tzu-la, 2100 m, Aug. 1914, *G. Forrest 13162* (E, K); **Ganluo**, near Suxiong, 800 m, Sep. 24, 1976, *s. coll. 14291* (CDBI, PE); **Shangri-la** (Zhongdian), mountains of the Chungtien plateau, Jul. 1914, *G. Forrest 12822* (E); **Yulong** (Lijiang), Baoshan, Shitoucheng, Oct. 2014, *Z. K. Wu s. n.* (IBSC).

25. HYPOESTES Sol. ex R. Br.

Brown, Prodr.: 474. 1810; Nees in de Candolle (ed.), Prodr. 11: 464. 1847; Clarke in Hooker (ed.), Fl. Brit. India 4: 557. 1885; Lindau in Engler & Prantl (eds.), Nat. Pflanzenfam. 4(3B): 333. 1895; Wood in Grierson & Long (eds.), Fl. Bhutan 2(3): 1293. 2001; Hu, Fl. Reipubl. Popularis Sin. 70: 248. 2002; Hu, Deng & Daniel in Wu, Raven & Hong (eds.), Fl. China 19: 467. 2011. Type: *Hypoestes floribunda* R. Br.

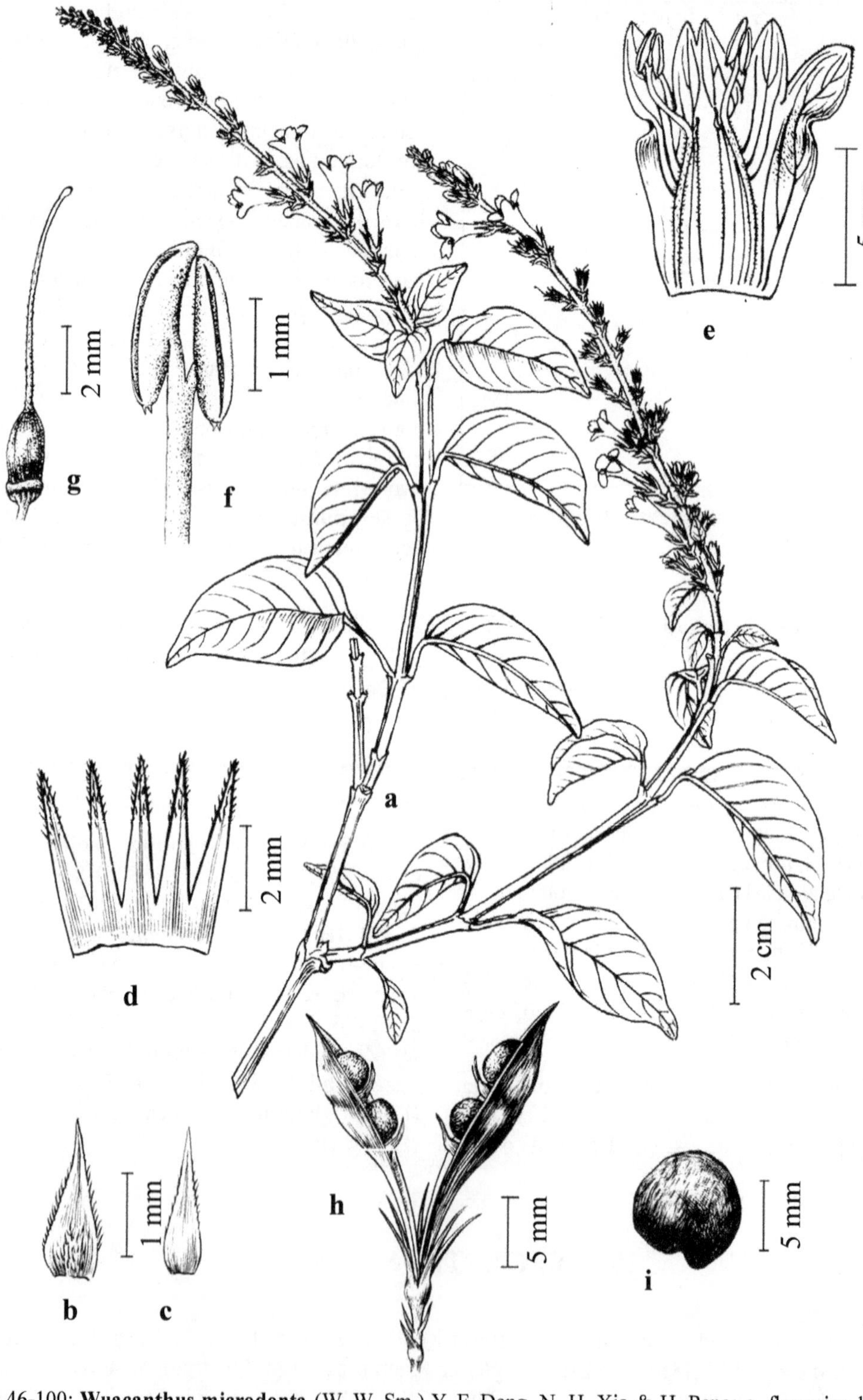

Figure 46-100: **Wuacanthus microdonta** (W. W. Sm.) Y. F. Deng, N. H. Xia & H. Peng: **a**, flowering branch; **b**, bract; **c**, bracteole; **d**, spreading calyx; **e**, opened corolla showing androecium; **f**, anther; **g**, pistil; **h**, capsule; **i**, seed. Drawn by Ding-Han CUI; reproduced from *Pl. Divers.* 38: 319, fig. 4. 2016.

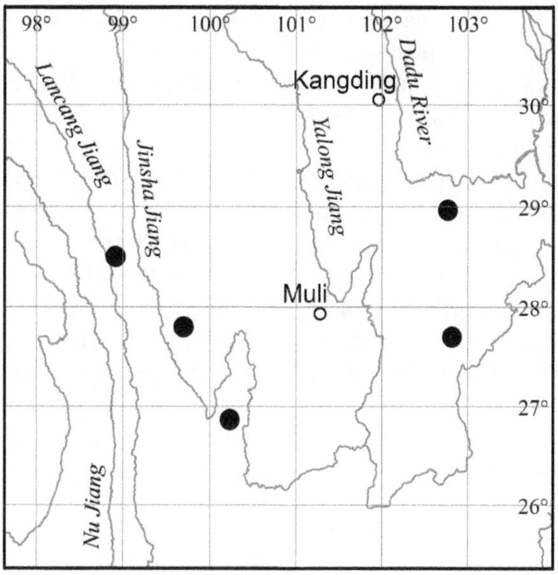

Map 46-168: **Wuacanthus microdonta** (W. W. Sm.) Y. F. Deng, N. H. Xia & H. Peng

Shrubs or perennial herbs, decumbent to erect; with cystoliths. Leaves opposite, petiolate; leaf blade entire or dentate. Inflorescences of axillary or terminal bracteate spikes to panicles of spikes or of variously disposed cymes or sometimes clustered in leaf axils; bracteoles usually 4, in 2 to rarely more pairs, distinct or connate, inner pair sometimes partially adnate to outer pair, forming an involucre enclosing 1 or sometimes more flowers. Calyx small, usually obscured by bracteoles, scarious or membranous, 5-lobed; lobes equal or subequal in length. Corolla rose-pink, purplish, or whitish, resupinate 180° by torsion of tube; tube subcylindric to expanded distally into a narrow throat, limb 2-lipped; lower lip in upper position shallowly 3-lobed; upper lip in lower position subentire or shallowly 2-lobed; lobes ascending cochlear in bud. Stamens 2, inserted near apex of corolla tube, exserted from mouth of corolla; anthers monothecous; anther-thecae oblong, base muticous. Ovary with 2 ovules per locule; style exserted from mouth of corolla; stigma 2-lobed, lobes equal or unequal. Capsules stalked at base, 2–4-seeded; retinacula present. Seeds smooth or ornamented.

About 150 species: Old World; one species in Pan-Himalaya.

1. Hypoestes triflora (Forssk.) Roem & Schult., Syst. Veg. 1: 141. 1817; Nees in de Candolle (ed.), Prodr. 11: 506. 1847; Clarke in Hooker (ed.), Fl. Brit. India 4: 557. 1885; Wood in Grierson & Long (eds.), Fl. Bhutan 2(3): 1293. 2001; Hu, Fl. Reipubl. Popularis Sin. 70: 251. 2002; Hu, Deng & Daniel in Wu, Raven & Hong (eds.), Fl. China 19: 468. 2011. ≡ *Justicia triflora* Forssk., Fl. Aegypt.-Arab. 4. 1775. Type: Yemen, Al Hadiyah, 1763, *P. Forsskal 1382* [lectotype **here designated**: C (C10002468); isolectotype: LD].

Hypoestes wallichii Nees, Pl. Asiat. Rar. 3: 114. 1832. Type: Nepal, 1820, *Wallich Cat. 2466e* [lectotype **here designated**: K (K001116131); isolectotypes: GZU (GZU000249482, GZU000249483)].

Dicliptera riparia var. *yunnanensis* Hand.-Mazz., Symb. Sin. 7(4): 898. 1936. Type: [S Hengduan] China, Yunnan, "Hedso bei Gwangdung", 6 Sep. 1914, *H. Handel-Mazzetti 4885* (holotype: W).

Herbs, 30–60 cm high, decumbent, much branched; stems quadrangular, sulcate, strigose.

Petiole 1–3 cm long, pubescent; leaf blade ovate, ovate-elliptic, or oblong, (1.3–)3–10 cm long, (1–) 2–4 cm broad, both surfaces sparsely pubescent, lateral veins 4–7 pairs, base cuneate and decurrent onto petiole, margin minutely crenate to denticulate, apex acuminate. Cymes axillary, simple or in a terminal cluster, (1–)3(–5)-flowered. Outer pair of bracteoles elliptic, obovate or oblanceolate, 0.7–1.7 cm long, 0.4–0.6 cm broad, abaxially pilose, base cuneate, apex rounded to acute; inner pair of bracteoles linear-lanceolate to oblanceolate, smaller, basally connate, apex acute. Calyx 5–6 mm long, 5-lobed almost to base; lobes linear-lanceolate, 4–5 mm long. Corolla white to pink with maroon markings, 1.2–1.5 cm long, outside pubescent; tube base cylindric, 1–1.3 mm broad, 0.8–1 cm long, up lip in lower position lanceolate, 4–5 mm long, 1.2–1.5 mm broad, apex slightly 2-lobed, lower lip in upper position oblong, 4–5 mm long, 2.5–3 mm broad, 3-lobed. Ovary pubescent; style 1.2–1.4 cm long, glabrous; stigma unequally 2-lobed. Capsules 7–9 mm long, pubescent. Seeds black, ovate in outline, 1.5–2 mm long and broad, verrucose. **Chromosomes** 2n = 30. Figure 46-101.

Phenology: flowering from July to September; fruiting September to January of next year.

Habitats: trailsides, forests; 300–3000 m.

Distribution: U Yarlung Zangbo, W Nepal, C Nepal, E Nepal, Sikkim & Darjiling, Bhutan, Yarlung Zangbo-Brahmaputra, S Hengduan, U Irrawaddy (Map 46-169); also in Thailand, Africa, and other parts of India, Myanmar and China (S Yunnan).

U YARLUNG ZANGBO: **Nyalam**, Zhangmu, Youyiqiao, 18 Nov. 2011, *Y. S. Chen et al. 654* (PE). **W NEPAL: Bheri**, Jajarkot, Above Maina, 2120 m, 12 Oct. 1952, *O. V. Polunin, W. R. Sykes & L. H. J. Williams 5573* (A, BM, E). **C NEPAL: Bagmati**, Dhading, Jharlang, 1550 m, 5 Nov. 1989, *N. P. Manandhar 13040* (KATH); Kathmandu, Swayambhu Nath, 23 Sep. 1963, *H. Hara et al. 6303717* (KYO, TI); Kathmandu, 90 km NE Kathm, zw Brabise und Kodari, 1650–1750 m, 22 Oct. 1993, *Projektteam 561-1* (K); Kathmandu, Shiwapuri, 2100 m, 13 Sep. 1988, *M. Suzuki et al. 8881854* (A, BM, E, TI); Kathmandu, Chandragiri, 2250 m, 4 Sep. 1970, *H. Kanai & T. B. Shrestha 672762* (KATH, KYO, TI); Kathmandu, Swayambha, 1300 m, 26 Sep. 1970, *N. P. Manandhar 9434* (KATH); Kathmandu, Balaju, 1210 m, 26 Nov. 1948, *B. D. Pande 19* (BM); Kathmandu, Champadevi, 3 Sep. 2007, *P. Maharjan 71* (TUCH); Lalitpur, Godavari, 1500 m, 5 Nov. 1969, *N. P. Manandhar et al. 12958* (KATH, TI); Lalitpur, Godawari, S of Kathmandu, 1 Nov. 1969, *H. Kanai 674474* (KATH, KYO, TI); Lalitpur, Godawari, 1590 m, 24 Nov. 1974, *Murari et al. 74-50* (KATH); Lalitpur, Godawari, 450 m, 9 Dec. 1966, *C. Chuma 2871* (TI); Lalitpur, Godawari, 1500 m, 8 Sep. 1970, *C. Chuma 674783* (KATH, TI); Lalitpur, Naudhara, Godawari, 1590 m, 24 Nov. 1974, *Murari et al. 74/50* (KATH); Lalitpur, Lele Bhanjyang, 1540 m, 1 Jan. 1991, *I. Shrestha & N. Pradhan 249* (KATH); Lalitpur, Godawari, 1500 m, 5 Nov. 1992, *N. Joshi & I. Shrestha 258* (KATH); Lalitpur, Lele, 1540 m, 4 Oct. 1989, *I. Shrestha & N. Pradhad 44* (KATH); Lalitpur, Education Garden, 1515 m, 12 Aug. 2002, *M. Joshi & R. K. Bista 368* (KATH); Lalitpur, Bajrabarahi, 12 Oct. 1972, *P. Pradhan & S. Guryng 8765* (KATH); Lalitpur, Tikabhairab, 1490 m, 11 Oct. 1971, *P. Pradhan & S. Gurung 16910* (KATH); Nuwakot, Kakani, 1830 m, 24 Sep. 1993, *Projektteam 306* (K); Nuwakot, Kakani Hill, 24 Sep. 1963, *H. Hara et al. 6303716* (TI); Nuwakot, Salme, 1850 m, 19 Nov. 1979, *N. P. Manandhar 3274* (KATH); Sindhupalchok, Tingoang to Khosori Khabre, 2000 m, 12 Sep. 1970, *H. Kanai et al. s. n.* (TI); Rasuwa, Lama hotel to bridge over Langtang Khola, near Langtang view Hotel, 2340 m, 16 Oct. 2001, *D. G. Long et al. ENEP512* (E, TI). **Dhawalagiri**, Myagdi, Tatopani, 1420 m, 18 Sep. 1995, *M. Mikage et al. 9552254* (A, BM, E, TI); Myagdi, Boghara–Lipshe–Lapche Kharka–Dobang, 7 Sep. 1996, *M. Mikage 9557182* (TI); Myagdi, Boghara–Lipshe–Lapche Kharka–Dobang, 7 Sep. 1996, *M. Mikage 9681232* (TI); Myagdi, Lipshe to Lapche Kharka, 1920 m, 7 Sep. 1996, *M. Mikage et al. 9684074* (TI); Myagdi, Boghara–Lipshe–Lapche Kharka–Dobang, 7 Sep. 1996, *M. Mikage 9685236* (TI); Myagdi, Gurjakhani, 2580 m, 11 Sep. 1954, *J. D. A. Stainton, W. R. Sykes & L. H. J. Williams 4361* (A, BM, E); Mustang,

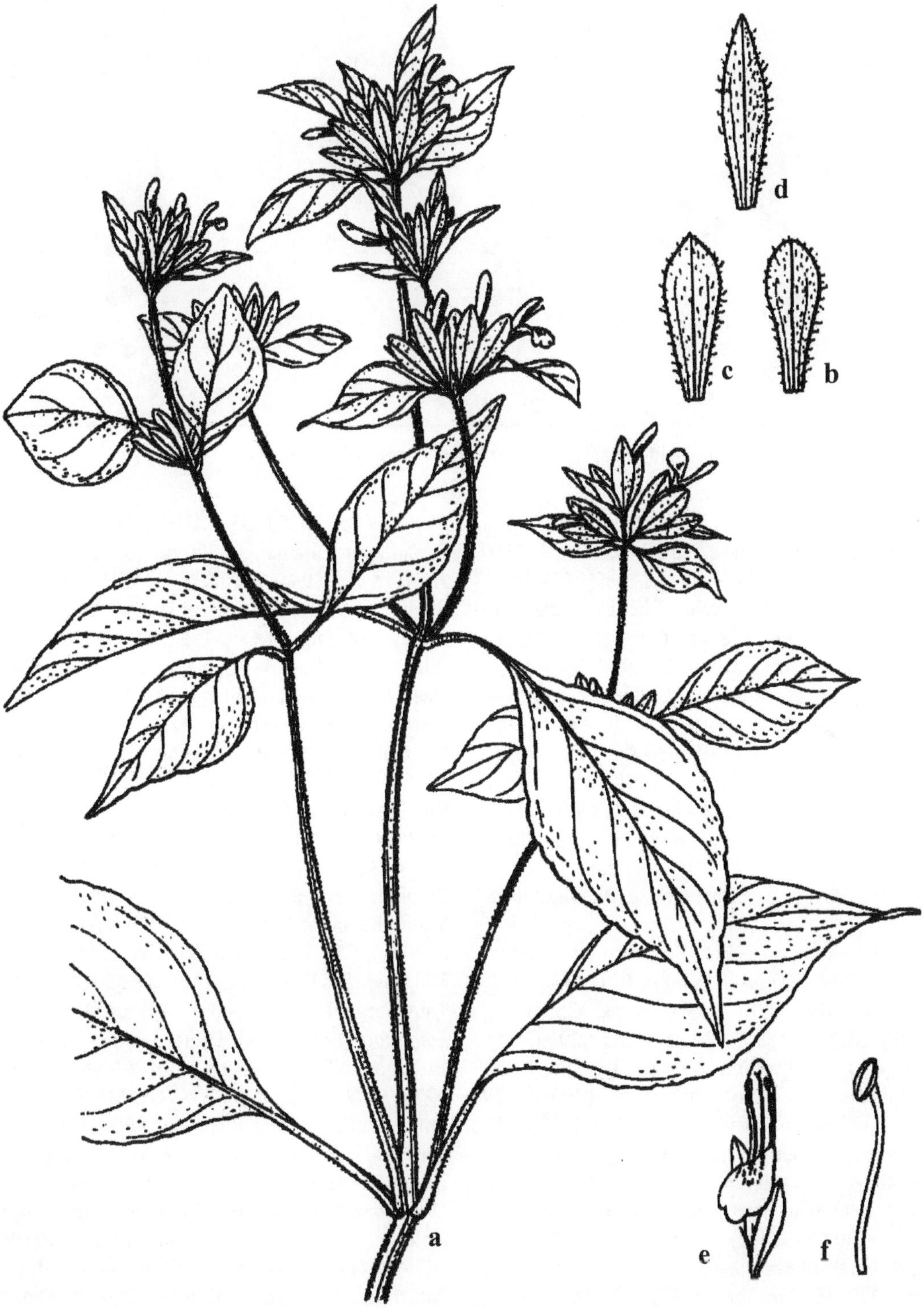

Figure 46-101: **Hypoestes triflora** (Forssk.) Roem & Schult.: **a**, portion of plant; **b**, outer bract; **c**, inner bract; **d**, calyx lobe; **e**, corolla; **f**, stamen. Drawn by Quan LIN; reproduced from *Icon. Cormophyt. Sin.* 4: 173, fig. 5759. 1975.

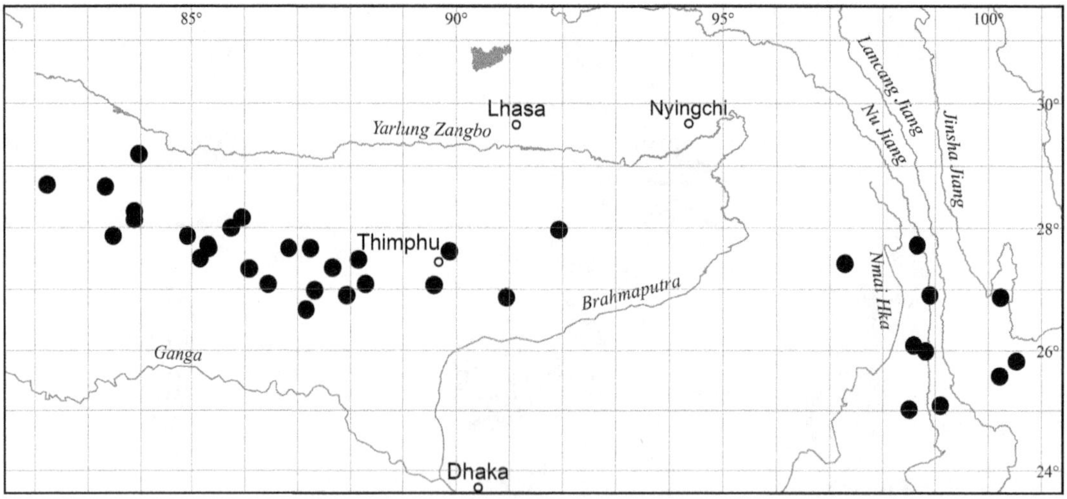

Map 46-169: **Hypoestes triflora** (Forssk.) Roem & Schult.

Ghasa, 16 Oct. 1976, *H. Tabata et al. 6534* (A, KATH, KYO); **Gandaki**, Kaski, Banthanti, 2240 m, 26 Jul. 1996, *T. Hoshino et al. 9666017* (CBM, E); **Janakpur**, Dolakha, Phedi Kharka to Khanigaon, 14 Sep. 1983, *H. Ohba et al. 8332110* (TI); Dolakha, Shivalaya to Jiri, 1740–2000 m, 12 Sep. 1985, *N. Kurosaki & S. K. Wu 8572613* (TI); Dolakha, Shivalaya–Those–Kattike–Kune–Jiri, 1740–2000 m, 12 Sep. 1985, *N. Kurosaki & S. K. Wu 8572615* (TI); Ramechhap, Bhandar to Sete, 1650–2550 m, 18 Aug. 1985, *H. Ohba et al. 8530866* (TI); Ramechhap, Bhandar to Shivalaya, 2360 m, 3 Sep. 1997, *M. Wakabayashi et al. 9720376* (KATH, TI); Ramechhap, Bhandarkhola, 1320 m, 10 Jun. 1962, *I. Shrestha et al. 675* (KATH); **Lumbini**, Palpa, Satyawatital, 1520 m, 12 Oct. 1954, *J. D. A. Staointon, W. R. Sykes & L. H. J. Williams 8891* (BM); Palpa, Chilauney, 2420 m, 8 Nov. 1831, *K. N. Sharma E200* (BM); **Narayani**, Makwanpur, about 10 km N of Bimphede on road to Kathmandu, 1700 m, 29 Nov. 2004, *C. Pendry et al. B221* (E, KATH); Makwanpur, Deurali, 1933 m, 6 Nov. 2012, *B. Adhikari & G. D. Bhatt 2* (E). **E NEPAL**: **Koshi**, Bhojpur, Dorpa Churi Danda to Deurali Danda, 1440 m, 30 Oct. 1995, *M. Mikage et al. 9554221* (E, TI); Dhankuta, Chauke to Tute, 2600 m, 1 Sep. 1977, *H. Ohashi et al. 771250* (BM, KATH, TI); Dhankuta, Hilay to Murhay, 23 Oct. 1963, *H. Hara et al. 6303664* (TI); Dhankuta, Dhara Pani to Tambur Bridge to Dhankuta, 17 Oct. 1963, *H. Hara et al. 6303665* (KYO, TI); Dhankuta, Basantapur, Bhanjang, 2200 m, 26 Oct. 1963, *H. Hara et al. 6303669* (BM, K, KATH, KYO, TI); Sankhuwasabha, slopes near Hatiya, Upper Arun Valley, 1600 m, 13 Oct. 1991, *D. G. Long et al. EMAK 745* (E, KATH); Sankhuwasabha, Bhote Bash to Chichila, 1860 m, 24 Sep. 1981, *P. R. Shakya 6649* (KATH); Sunsari, Phusri to Sanguri Bhanjang to Dhara Pani, 16 Oct. 1963, *H. Hara et al. 6303670* (A, KATH, KYO, Ti); **Mechi**, Ilam, Mai Majuwa to Dhara Pani, 4 Dec. 1963, *H. Hara et al. 6303965* (TI); Ilam, Mahabharat Lekh, 1820 m, 14 Sep. 1967, *L. H. J. Williams & J. D. A. Stainton 8486* (BM, K, KATH); Ilam, Mainajhuwa, 1820 m, 27 Aug. 2007, *N. B. Khatri et al. C009* (KATH); Panchthar, Yektin to Akasay, 30 Nov. 1963, *H. Hara et al. 6303668* (TI); Taplejung, Nessum, 1300 m, 28 Aug. 1977, *S. K. Dhaungana 901* (KATH); Taplejung, Papung to Bir Gaon to Sangrati Pati, 1050–2000 m, 26 Aug. 1977, *H. Ohashi et al. 772738* (A, TI); Taplejung, Surke Pati to Gupha Pokhari, 2100–2800 m, 30 Aug. 1977, *H. Ohba et al. 775446* (TI); Taplejung, Deorali Forests, 2800 m, 30 Aug. 1977, *S. K. Dhungana 938* (KATH); Taplejung, Iladanda to Helok, 13 Nov. 1973, *H. Kanai et al. 6303667* (KYO, TI); **Sagarmatha**, Solukhumbu, Surke Pati to Gupha Pokhari, 30 Aug. 1977, *H. Ohashi et al. 774126* (KYO); Solukhumbu, Namikhil–Chamare–Likhu–Bhandar, 1550–2300 m, 10 Sep. 1985, *H. Ohba et al. 8581496* (TI); Solukhumbu,

Sete to Kenja, 2500 m, 2 Sep. 1997, *M. Wakabayashi et al. 9715310* (CBM, KATH, TI); Solukhumbu, Namikhil, 10 Sep. 1985, *H. Ohba et al. 8581496* (KATH). **SIKKIM & DARJILING**: **Darjiling**, Takuda, 1600 m, 19 Apr. 1960, *H. Hara et al. 2104* (BM); Three Miles to Six Miles, 24 Sep. 1977, *H. Ohashi et al. 774399* (E); **Sikkim**, Ratong loyoksum, 750–1500 m, 4 Oct. 1862, *T. Anderson 1063* (A); Neebay, 2100 m, 16 Oct. 1875, *C. B. Clarke 25333* (BM, K). **BHUTAN**: **Punkha**, near Lomitsawa, 2100 m, 2 Oct. 1987, *J. R. I. Wood 5890* (E); Thimpu, 6000 ft., 18 Nov. 1914, *R. E. Cooper 3610* (BM, E); **Tongsa**, 2100 m, 6 Oct. 1914, *R. E. Cooper 2297* (E). **YARLUNG ZANGBO-BRAHMAPUTRA**: **Cona**, Lai to Marmang, 2445 m, 11 Sep. 2017, *PE-Xizang Expedition PE6326* (PE); Lai, Demang, 2488–2994 m, 10 Sep. 2017, *PE-Xizang Expedition PE6233* (PE). **S HENGDUAN**: **Binchuan**, Jizushan, near Zhushengsi, 2100 m, *Y. F. Deng & H. Peng 23863* (IBSC); **Dali**, Talifu, 2250 m, *H. Handel-Mazzetti 8566* (WU); Gwangdung, 1830 m, *H. Handel-Mazzetti 4885* (E, IBSC, WU); 2250 m, *G. F. Handel-Mazzetti 1619* (A); Sep. 31 1914, *C. Schneider 2715* (A, K); Caojian, 14 Sep. 2013, *Y. F. Deng et al. 24943* (IBSC); **Fugong**, Shangpa, 1300 m, *Gaoligong Shan Expedition 7870* (E, KUN); 1500 m, *H. T. Tsai 54641* (IBSC, LBG, NAS, SZ); 1500 m, *H. T. Tsai 54958* (A, IBSC, LBG, SZ); Shangpa, 2000 m, *H. T. Tsai 58877* (A); **Gongshan**, Pengdang, Jimudeng, 1550 m, *Gaoligong Shan Biodiversity Survey 15547* (KUN); Pengdang, from Yonglaga to Di, 1640 m, *Gaoligong Shan Biodiversity Survey 15852* (CAS, KUN); Cikai, Danzhu, 1400 m, *Gaoligong Shan Biodiversity Survey 22293* (CAS, KUN); Cikai, Danzhu, 1430 m, *Gaoligong Shan Biodiversity Survey 23192* (CAS, KUN); Cikai, Danzhu, 1410 m, *Gaoligong Shan Biodiversity Survey 22385* (CAS, KUN); Cikai, Dangdan Park, 1600 m, *Dulong Jiang Investigation Team 182* (CAS); **Jianchuan**, Diannan, 15 Sep. 2013, *Y. F. Deng et al. 25025* (IBSC); **Longyang**, Lihuipo, 2100 m, 23 Jan. 2008, *Y. F. Deng 20478* (IBSC); Mangkuang, Baihualing, *H. Li 1346* (KUN); Wayao, 1220 m, *Gaoligong Shan Expedition 7323* (E, KUN); Mangkuan, Sanba Gou, 1000 m, *H. Li 10655* (CAS, KUN, MBK); Lujiang, Nankang, 2200 m, *H. Li 10675* (KUN, MBK); Mangkuan, Baihualing, Hanlong, 1580 m, *H. Li 13408* (CAS, KUN); **Lushui**, Xingangfang, along the Gulang Jiang, between Pianma and Gangfang, 1640 m, *H. Li 10341* (KUN, MBK); **Yangbi**, W side of Cangshan, 1500 m, 26 Feb. 2009, *Z. J. Yin & H. J. Dong 009* (MBK); **Yulong** (Lijiang), Yulongshan, Ganhaizi, *H. C. Wang 2947* (IBSC); Longpan, 15 Sep. 2013, *Y. F. Deng et al. 25033* (IBSC). **U IRRAWADDY: Chaungmaw Kyaun**, North Triangle, Hkinlum, 1200 m, 30 Aug. 1953, *F. Kingdon-Ward 21294* (A, BM); **Hpimaw**, Hpimaw Hills, 1500-2100 m, 5 Oct. 1919, *s. coll. 1387* (E); **Putao**, Upper Shankaung to Wasadam, 13 Oct. 2014, *Putao Expedition Team (X. H. Jin et al.) PT-ET 174* (PE); **Tengchong**, Heshun, 1650 m, *Y. H. Li 20257* (HITBC), Ruidian, 1800 m, *Y. H. Li 20377* (HITBC); Tengyueh, *G. Forrest 1024* (E); Tengyueh, *G. Forrest 613* (E); Fuyu, *R.L. Xiong 580984* (YIM); Guanpo, 2050 m, *780-Group 438* (HITBC); Laifengshan, 10 Sep. 2013, *Y. F. Deng 24584* (IBSC).

26. DICLIPTERA Juss.

Jussieu, Ann. Mus. Natl. Hist. Nat. 9: 267. 1807, nom. cons.; Nees in de Candolle (ed.), Prodr. 11: 478. 1847; Clarke in Hooker (ed.), Fl. Brit. India 4: 550. 1885; Lindau in Engler & Prantl (eds.), Nat. Pflanzenfam. 4(3B): 332. 1895; Nasir & Ali in Malik & Ghafoor (eds.), Fl. Pakistan 188: 22. 1988; Wood in Grierson & Long (eds.), Fl. Bhutan 2(3): 1291. 2001; Hu, Fl. Reipubl. Popularis Sin. 70: 234. 2002; Hu, Deng & Daniel in Wu, Raven & Hong (eds.), Fl. China 19: 462. 2011. ≡ *Diapedium* K. D. Koenig, Ann. Bot. 2: 189. 1805, nom. rej. Type: *Dicliptera chinensis* (L.) Juss. (≡ *Justicia chinensis* L.).

Herbs, diffuse, annual, biennial, or perennial; with cystoliths. Leaves petiolate; blade entire or subsinuate. Inflorescences of cymes in leaf axils or in terminal spike-like thyrses or panicles of thyrses. Flowers subtended by an involucre of several pairs of bracteoles. Outermost pair of bracteoles equal or unequal in size, usually concealing inner pairs, calyx and base of corolla. Calyx deeply 5-lobed; lobes narrow, equal, apex acute. Corolla pink, resupinate 180° by torsion of tube; tube slender, upwards widened; limb 2-lipped; lower lip in upper position 3-lobed, upper lip in lower position entire or emarginate; lobes ascending cochlear in bud. Stamens 2, exserted from mouth of corolla; anthers bithecous; thecae unequally inserted to superposed, base muticous; staminodes absent. Stigma 2-cleft. Capsules substipitate or stipitate, 2–4-seeded; retinacula present; septa with retinacula separating elastically and rising from inner capsule wall at dehiscence. Seeds lenticular, papillate, lacking trichomes.

Recent references: Hara, 1980. Comments on the East Asiatic plants (8). *J. Jap. Bot.* 55: 321–327; Panigrahi & Dubey, 1983. *Dicliptera bupleuroides* var. *roxburghiana*, a new name for *D. roxburghiana* C. B. Clarke, not Nees (Acanthaceae). *Taxon* 32(2): 286–288; Hara, 1984. Comment on species of *Dicliptera*. *Taxon* 33(2): 296.

About 100 species: tropical and temperate regions worldwide; two species in the Pan-Himalaya.

1a. Outer bracteoles oblong-lanceolate to oblanceolate, greater than 2 times as long as broad .... 1. *D. bupleuroides*
1b. Outer bracteoles subcircular, or obovate, less than 2 times as long as broad .... 2. *D. elegans*

1. Dicliptera bupleuroides Nees in Wallich (ed.), Pl. Asiat. Rar. 3: 111. 1832; Nees in de Candolle (ed.), Prodr. 11: 485. 1847; Nasir & Ali in Malik & Ghafoor (eds.), Fl. Pakistan 188: 23. 1988; Wood in Grierson & Long (eds.), Fl. Bhutan 2(3): 1292. 2001; Hu, Fl. Reipubl. Popularis Sin. 70: 235. 2002; Hu, Deng & Daniel in Wu, Raven & Hong (eds.), Fl. China 19: 462. 2011. ≡ *Dicliptera roxburghiana* var. *bupleuroides* (Nees) C. B. Clarke in Hooker (ed.), Fl. Brit. India 4: 554. 1885. Type: [C] Nepal, *Wallich Cat. 2423* [lectotype designated by Hara, Jap. J. Bot. 55: 324. 1980: K-W (K001115893); isolectotypes: BM, C (C10004810, C10004811), GZU (GZU000249825)].

Dicliptera bupleuroides var. *sessilis* Nees in de Candolle (ed.), Prodr. 11: 485. 1847, **syn. nov.** Type: India, Assam, 1830, *W. Griffith s. n.* [lectotype **here designated**: K ("Herb. Hook."); isolectotype: GZU].

Dicliptera bupleuroides var. *nazimii* Malik & Ghafoor in Nasir & Ali (eds.), Fl. Pakistan 188: 25. 1988, **syn. nov**. Type: Pakistan, Rawalpindi, Ayub National Park, *S. Nazimuddin & S. Abedin 1207* (holotype: KUH).

Dicliptera bupleuroides var. *ciliata* Malik & Ghafoor in Nasir & Ali (eds.), Fl. Pakistan 188: 25. 1988, **syn. nov.** Type: Pakistan: Karachi, Malir, 20 Apr. 1971, *T. Hussain s. n.* (holotype: KUH).

Dicliptera bupleuroides var. *qaiseri* Malik & Ghafoor in Nasir & Ali (eds.), Fl. Pakistan 188: 26. 1988, **syn. nov**. Type: Pakistan, 6 miles from Garhi Habibulllah on way to Balakot, *K. A. Malik & M. Qaiser 197* (holotype: KUH).

Dicliptera bupleuroides var. *roxburghiana* Panigrahi & Dubey, Taxon 32: 287. 1983. Type: Bangladesh, Sylhet, *F. de Silva in Wallich Cat. 2466D* [holotype: K-W (K001116135); isolectotype: E (E00273541)].

Dicliptera roxburghiana auct. non Nees: Anderson, J. Linn. Soc., Bot. 9: 519. 1867; Clarke in Hooker (ed.), Fl. Brit. India 4: 553. 1885.

Diffuse herbs, 30–80 cm high, with a woody

root-stock; stems subhexagonal, shallowly sulcate, often turgid at nodes, pubescent. Petiole 0.5–2.5 cm long; leaf blade elliptic, rhomboidal, or lanceolate, 2–6 cm long, 1–2.5 cm broad, adaxially glabrescent, abaxially pubescent, or with 2- celled trichomes, lateral veins 4 or 5 pairs, base broadly cuneate, margin subentire or obscurely undulate, apex shortly acuminate to acute. Flowers in axillary and subsessile or sometimes terminal clusters often in umbel-like cymes, 1-to several-flowered, pinkish; peduncle 2–5 mm long. Outer bracteoles obong-lanceolate to oblanceolate, unequal, larger one 6–12 mm long, 3–7 mm broad, smaller one 5–9 mm long, 2.5–3 mm broad, abaxially pubescent, base cuneate, margin entire and ciliate, apex mucronate; inner bracteoles 2, lanceolate, 4–5 mm long, 1.2–1.5 mm broad, green with yellowish margin, adaxially pubescent. Calyx 3.5–4 mm long, pilose, 5-lobed near to base; lobes subulate. Corolla light purplish-red, 1.2–1.6 cm long, outside pubescent; upper lip sub-orbicular, 1.2–1.5 mm long, 1.2–1.5 mm broad, lower lip oblong, 1.6–2 mm long, 0.8–1 mm broad, 3-lobed, lobes ovate, 0.2–0.3 mm long and broad. Stamens 2; filaments 0.8–1 mm long, hairy; anther-thecae spherical, superposed. Ovary pilose towards apex; ovules 2 per locule; style filiform, 1–1.2 cm long, sparsely pubescent; stigma bifid. Capsules clavate, 3–4 mm long. Seeds ovate or suborbicular in outline, 1–1.2 mm long, papillate. **Chromosomes** 2n = 26.

Phenology: flowering from November to May; fruiting from December to June.

Habitats: roadsides, forests, shady places; 100-4500 m.

Distribution: N Pakistan, Jammu & Kashmir, U Ganga & Indus, W Nepal, C Nepal, E Nepal, Sikkim & Darjiling, Bhutan, U Irrawaddy (Map 46-170); also in Cambodia, Laos, Thailand, Vietnam, and other parts of Afghanistan, Pakistan, India, Myanmar and China (S Yunnan).

Notes: The species *Dicliptera riparia* was reported from E Nepal by Hara (An Enumeration of the Flowering Plants of Nepal. 3: 139. 1982. London: British Museum (Natural History).). It is possible to be misidentified for this species. *Dicliptera riparia* differs from this species by its retuse bracts.

N PAKISTAN: Azad Kashmir, Mirpur, Potha, Murree Hills, 1320 m, 3 Oct. 1949, *R. R. Steward 23504* (K); **Chitral**, Loari Pass, *N. Ali 10267* (RAW, fide Malik & Ghafoor, 1988); Dir, 2 miles from Dir on way to Tamirgrah, *S. Abedin 8179* (KUH, fide Malik & Ghafoor, 1988); Rani village on way to Dir, *M. Qaiser & A. Ghafoor 1723* (KUH, fide Malik & Ghafoor, 1988); **Hazara**, Pakii, 30 Nov. 1976, *Shankat & Nisar 2564* (A); Abbottabad, *S. Abedin 2996* (KUH, fide Malik & Ghafoor, 1988); Murree Hill, 27 May 1955, *S. Nakao s. n.* (KYO); **Swat**, between

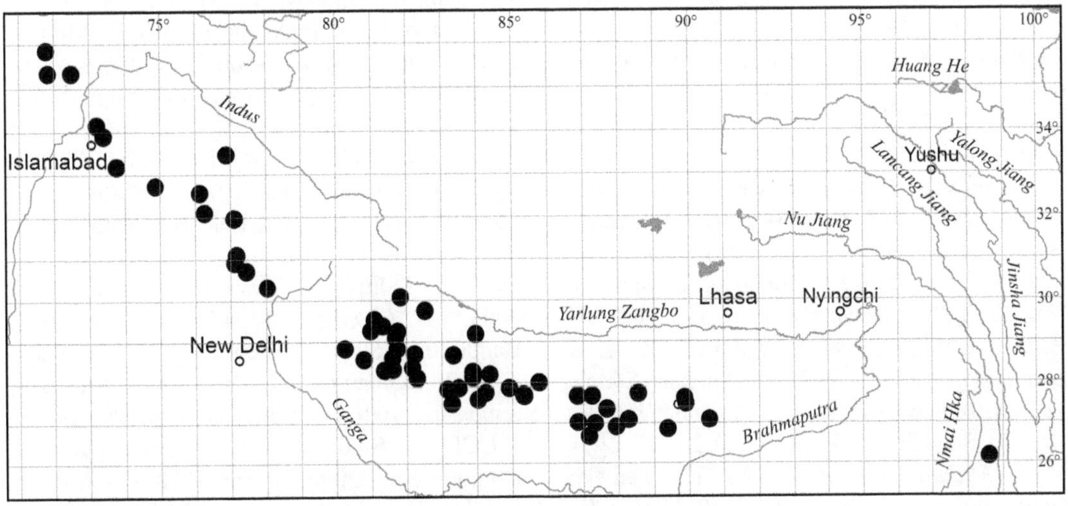

Map 46-170: **Dicliptera bupleuroides** Nees

Madyan andalam, 1700 m, 22 Aug. 1962, *K. H. Rechinger 19383* (E); near Saidu Sharif village, ca. 1050 m, 10 Aug. 1952, *R. J. Rodin 5405* (K). **JAMMU & KASHMIR**: **Jammu**, Ramnagar, 6 Sep. 1970, *B. M. Sharma 712* (RRLH); **Padum**, Naoshahra, 18 Jan. 1902, *J. R. Drummond 13870* (K). **U GANGA & INDUS**: **Chamba**, Bassa & Bakam, 1350 m, 13 Jun. 1896, *J. H. Lace 1426* (E); Pass Chamba, 1800–2100 m, 23 May 1878, *Herb. G. Watt 542* (E); 900 m, 22 May 1878, *Herb. G. Watt 495* (E); **Chatri**, 2100 m, 15 Oct. 1987, *R. McBeath 1971* (E); **Dehradun**, Jaunsar, Binal God, 1200 m, Jun. 1890, *K. Nand 267* (FHO); Jaunsar, Binak Gaa, 1200 m, *K. Nand 267* (E); Gritgadh, Jaunsar, 2100 m, May 1895, *J. S. Gamble 25507* (K); **Kangra**, 900m, 22 Nov. 1936, *W. Koelz 10315* (A); **Kullu**, Naggar, 27 Oct. 1936, *W. Koelz 10261* (A); outer hill of **Kumaon**, 1200 m, *R. Strachey & J. E. Winterbottom s. n.* (P); **Mussorie**, the Park, 1950 m, Oct. 1915, *A. Anderson s. n.* (E); Dick Road, 1950 m, Nov. 1914, *A. Anderson s. n.* (E); **Simla**, Kumsu, Bushahr, 14 Jun. 1934, *N. Parmanand 88* (E); Tapru, Bushahr, 11 Jul. 1934, *N. Parmanand 440* (E); Rampur, Bushahr, 2 Jul. 1934, *N. Parmanand 343* (E); below Jara Devi, 1500–1800 m, 6 Oct. 1889, *Herb. G. Watt 8856* (E); *G. Watt 4204* (E); 2100 m, May 1882, *W. Schlich 398* (FHO); 1200 m, Giri, Sep. 1884, *J. R. Drummond 1891* (E, K); Koti Rd., N. of Simla, 1950 m, 25 May 1877, *J. S. Gamble 4285* (K); Takko, 22 Sep. 1915, *H. H. Rich 38* (K); 1500 m, Nov. 1885, *H. Collett 412* (K); 2100 m, Sep. 1885, *H. Collett 869* (K); Vinaut Hill, 11 Nov. 1948, *E. R. Jauok Amal 1403* (K); Kálka via Kassáuli to Sabathu, 600–1380 m, *Gen No of Catalogue 7775* (A); 2020 m, 30 Jun. 1877, *J. S. Gamble 4472* (K); Shimla,1800–2190 m, *Sclagintweit Gen. No of catalogue 4887* (P); **Solan**, Kasanli, 1886, *J. R. Drummond 26673* (K). **W NEPAL**: **Bheri**, Banke, Babai, 24 Mar. 1976, *K. J. Malla et al. 898* (KATH); Banke, Napelganj, Puraina, 30 Nov. 1972, *N. P. Manandhar et al. 8977* (KATH); Babdia, Kohalpur, 180 m, 9 Apr. 1979, *T. K. Bhattachaya 3683* (KATH); Dailekh, Sattale, 1120 m, 28 Feb. 1977, *K. J. Malla 1666* (KATH); Dailekh, Sanodhungeshwar, 560 m, 24 Feb. 1991, *N. P. Manandhar 266-91* (KATH); Dailekh, Gairai, 900 m, 25 Feb. 1991, *N. P. Manandhar 380-91* (KATH); Jajarkot, Khiyapani, 720 m, 3 Mar. 1991, *N. P. Manandhar 679/91* (KATH); Surkhet, 1250 m, 8 Dec. 1982, *S. R. Adhikari & N. P. Manandhar 9194* (KATH); **Karnali**, Humla, Just below Darma, 2020 m, 14 Jun. 2008, *C. A. Pendry et al. JRS A136* (E, KATH); Humla, Syamme Khola, 2120 m, 2 Nov. 1985, *M. N. Subedi 19/85* (KATH); Kalikot, Khallagad to Talasero gaon, 1500 m, 10 Jul. 1979, *K. R. Rajbhandari & B. Roy 3140* (KATH); Mugu, Karnali Valley, between Gum and Lumsa, 6500 ft., *O. V. Polunin, W. R. Sykes & L. H. J. Williams 5205* (BM, P); Mugu, Naina, 6500 ft., 12 Oct. 1952, *O. V. Polunin, W. R. Sykes & L. H. J. Williams 5581* (A, BM); **Mahakali**, Kanchanpur, Mahendra Nagar, 15 miles NE from Mahendranagar, 450 m, 9 Dec. 1966, *D. H. Nicolson 2871* (BM, KATH); **Rapti**, Dang, Hapur Kola, Southeast of Bijanri, 610 m, 25 Mar. 1952, *O. Polunin, W. R. Sykes & L. H. J. Williams 3657* (BM); Salyan, Phalabang, 1500m, 2 Apr. 1992, *P. P. Kurmi KB143* (KATH); Salyan, Chorkholi, 300 m, 5 Dec. 1978, *T. K. Batthacharya & S. B. Malla 307* (KATH); **Seti**, Bajhang, Trail from Tameil to Chainpur, 1142 m, *Bajhang 2009 Expedition 20913063* (E); Bajura, Baudi Khola, S of Kuru, 1768 m, 21 Aug. 2017, *B159* (E); Doti, Bahela, on the way to Doti, 1290 m, 23 Apr. 1965, *T. B. Shrestha 4002* (KATH); Doti, to Siligarhi, 4500 m, *N. Ecker-Racz 19* (MO); Kailali, Dhangarhi, 180 m, 24 Mar. 1981, *T. K. Bhattachaya 3604* (KATH). **C NEPAL**: **Bagmati**, Dhading, Tharlang, 1550 m, 5 Nov. 1989, *N. P. Manandhar 13054* (KATH); Kathmandu, Jawal Tole, 1300 m, 22 Dec. 1966, *D. H. Nicolson 2880* (BM, KATH); Kathmandu, Tika to Ghairab to Champi, 1380–1570 m, 16 Mar. 1973, *M. Amatya & N. P. Manandhar 73-51* (KATH); Kathmandu, Dolhu, SW of Kathmandu, 7 Jan. 1975, *D. P. Joshi et al. 75/80* (KATH); Lalitpur, Lele Bhanjyang, 1540 m, 1 Jan. 1991, *I. Shrestha & N. Pradhan 240* (KATH); Lalitpur, Godawari, 1500 m, 9 May 2003, *A. Bhaukaje 34* (TUCH); Nuwakot, Banderithan, 800 m, 19 Feb. 1975, *M. M. Amatya & K. R. Rajbhandari 121/75* (KATH); Nuwakot, Sundarijal, 1600 m, 10 Feb. 2004, *S. Shrestha 51* (TUCH); Sindupalchok, Lamdsangu to Dhulopakhra, 720 m, 18 Apr. 1978, *I. Sharma et al. 78/111* (KATH); **Dhawalagiri**, Bajhang,

Karnali Valley, N of Raskot, 1200 m, 22 Mar. 1968, *J. D. A. Stainton 6171* (E, KATH); Mustang, Lower Lete to Tatopani, 1240–2380 m, 28 Sep. 1995, *M. Mikage et al. 9550497* (A, TI); Mustang, Lower Lete to Tatopani, 28 Sep. 1995, *M. Mikage et al. 9550508* (TI); Mustang, Kali Gandaki Valley, S of Gharsa, 2000 m, 10 Sep. 1973, *C. Grey-Wilson & B. Phillips 865A* (K); Myagdi, Mayagdi Khola, Beni, 970 m, 22 Apr. 1954, *J. D. A. Stainton, W. R. Sykes & L. H. J. Williams 96* (BM, E); Myagdi, Takam, near Takumsibang, 1670 m, 6 Sep. 1954, *J. D. A. Stainton, W. R. Sykes & L. H. J. Williams 4251* (BM); Myagdi, Dana, 1520 m, 5 Sep. 1954, *J. D. A. Stainton, W. R. Sykes & L. H. J. Williams 7688* (BM); Myagdi, Muna, Terrace banks, 1820 m, 27 Oct. 1954, *J. D. A. Stainton, W. R. Sykes & LH.J. Williams 9157* (BM); Myagdi, below Dhar, 1790 m, 13 Sep. 1996, *M. Mikage et al. 9682485* (TI); Myagdi, Tatopani to Beni, 29 Sep. 1995, *M. Mikage et al. 9550550* (TI). **Gandaki**, Kaski, Tirkhedhungo, 1080 m, 7 Sep. 1988, *M. Suzuki et al. 8881777* (BM, TI); Lamjung, Tarku, 500 m, 25 Nov. 1970, *J. F. Boremez 644* (BM, E, KATH); Tanahu, Aanbu Khaireni, 323m, 27 Nov. 2012, *B. Adhikari & D. R. Kandel 97* (E); **Lumbini**, Arghakhachi, Matipur, 900 m, 3 Mar. 1976, *N. P. Manandhar & P. M. Regmi 123* (KATH); Papa, Channe, Bhanjyang, 1370 m, 1 Mar. 1976, *N. P. Manandhar & P. M. Regmi 32* (KATH); Papa, Tansen, Basani Khola, 1680 m, 1 Mar. 1976, *N. P. Manandhar & P. M. Regmi 34* (KATH); Papa, Basani Khola, 1680 m, 1 Mar. 1976, *N. P. Manandhar & P. M. Regmi 54* (KATH); Papa, Rupandehi, Butwal, near Butwal, 610 m, 6 Oct. 1954, *J. D. A. Stainton, W. R. Sykes & L. H. J. Williams 8984* (BM); Palpa, Masem, 1300 m, 25 Nov. 1973, *D. P. Joshi & M. M. Amatya 73/992* (KATH); Rupandehi, Butwal, 610 m, 6 Oct. 1954, *J. D. Staiton, W. R. Sykes & L. H. J. Williams 8771* (BM); **Narayani**, Chitwan, Sauraha, 160–200 m, 15 Jan. 1996, *Y. Koda & T. Kurosawa 9614074* (E, TI). **E NEPAL**: **Koshi**, Dhankuta, 1200 m, 21 Oct. 1963, *H. Hara et al. 6303654* (A, TI); Sankhuwasabha, Tumlingtar to Khandbari, 660 m, 21 Sep. 1981, *P. R. Shakya 6568* (KATH); Sankhuwasabha, Chhoyang to Yagu, 880 m, 9 Nov. 1981, *P. R. Shakya 7365* (KATH); Sunsari, Phusri to Sanguri Bhanjang to Dhara Pani, 16 Oct. 1963, *H. Hara et al. 6303659* (BM, E, KATH, TI); Sunsari, Tarahara, 12 Dec. 1977, *S. Bhaharai et al. 78/94* (KATH); Tarhara Forests, 300 m, 3 Sep. 1972, *T.K. Bhattacharya 17792* (KATH); Sunsari, Tarhara, 230 m, 23 Jan. 1974, *A. P. Singh & N. B. Raos Chetry 17931* (KATH); **Mechi**, Ilam, Kolbas, below Kolnas, 758 m, 20 Nov. 2012, *B. Adhikari & D. R. Kandel 64* (E); Ilam, 1200 m, 6 Dec. 1963, *H. Hara et al. 6303653* (K, TI); Ilam, 1200 m, 7 Dec. 1963, *H. Hara et al. 6303657* (TI); Ilam, Ranga Pani to Chorwa via Chisa Pani and Lochya Mai, 9 Dec. 1963, *H. Harar et al. 6303658* (KATH, TI); Ilam, Ranga Pani to Chorwa via Chisa Pani and Lochya Mai, 31 Oct. 1963, *H. Harar et al. 6303661* (TI); Ilam, near Glam, 1520 m, Nov. 1963, *A. V. Upadhyay 1533* (BM); Ilam, Godak, 1170 m, 3 Oct. 1977, *P. Pradhan et al. 214* (KATH); Taplejung, Mewa Khola, S side of Mewa Khola Valley, 820 m, 15 Dec. 1971, *Bowring 16* (BM); Taplejung, Tawa to Kiwa to Tpaplethok, 5 Nov. 1963, *H. Hara et al. 6303652* (BM, TI); Taplejung, Garhi Danra–Linkhim–Tuwa, 4 Nov. 1963, *H. Hara et al. 6303655* (TI); Taplejung, Dumhan to Taplejung, 1 Nov. 1963, *H. Kanai et al. 6303656* (TI); Taplejung, Dumhan, 700 m, 31 Oct. 1963, *H. Hara et al. 6303660* (KATH, TI); **Sagarmatha**, Khotang, Dorpa Churi Danda, 1440 m, 30 Oct. 1995, *M. Mikage et al. 9554222* (E, TI); Solukhumbu, Sanu Shree Nagar, 600 ft., 26 Oct. 1971, *Johirau & R. K. Prasal 2267* (KATH). **SIKKIM & DARJILING**: **Darjiling**, Panchkilla, 300 m, 14 Dec. 1816, *C. B. Clarke 31711* (K); **Sikkim**, Mingpoo, 900 m, 12 Nov. 1908, *W. G. Craib 18* (E); Kulkait Valley, 900 m, 16 Oct. 1874, *D. Treutler C. K. 926* (K). **BHUTAN**: **Gaylephung**, ca. 12 km north of Gaylephung, 800 m, 14 Feb. 1988, *J. R. I. Wood 6141* (E); **Phuntsholing**, on footpath in Tectona plantation, 200 m, 1 May 1979, *A. J. C. Grierson & D. G. Long 708* (E); hillside on N side of Phuntsholing, 230 m, 13 Feb. 1982, *A. J. C. Grierson & D. G. Long 2899* (E); escarpment above Kharbandi, 800–1000 m, 12 Nov. 1987, *J. R. I. Wood 6078* (E); **Punakha**, ca. 20 km S of Wangdi towards Chirang, 1100 m, 30 Oct. 1988, *J. R. I. Wood 6853* (E); **Samchi**, 600 m, 28 Feb. 1982, *A. J. C. Grierson & D. G. Long 3283* (E); **Wangdi**, 1000 m, 8 Oct. 1989, *J. R. I. Wood 7199* (E). **U IRRAWADDY**: **Kang Fang**, Bgawchang Valley, 1200 m, 29 Nov. 1938, *Vernay-Cutting Exped.*

(*Ward*) *50* (NY).

2. Dicliptera elegans W. W. Sm., Notes Roy. Bot. Gard. Edinburgh 10: 174. 1918; Hu, Fl. Reipubl. Popularis Sin. 70: 238. 2002; Hu, Deng & Daniel in Wu, Raven & Hong (eds.), Fl. China 19: 463. 2011. Type: [S Hengduan] China, Yunnan, "descent to the Yangtze from the eastern boundary of the Lichiang Valley", 27°15′N, 9000 ft., Jul. 1913, *G. Forrest 10697* [holotype: E (E00095244); isotypes: BM (BM000950075), GH (GH00063075), IBSC, K (K000884390)].

Dicliptera mairei Benoist, Bull. Mus. Natl. Hist. Nat. 2: 150. 1930. Type: China, Yunnan, cultives de Pa-eul-gou, 700 m, *E. E. Maire s. n.* [lectotype **here designated**: P (P00720022); isolectotype: P].

Herbs, 60–120 cm high; stems sulcate, sparsely pubescent. Petiole 0.5–1.5 cm long, sulcate, pubescent; leaf blade ovate to ovate-lanceolate, 4–8 cm long, 2–5 cm broad, abaxially sparsely pubescent, adaxially sparsely pubescent, lateral veins 4–6 pairs, base broadly cuneate and decurrent onto petiole, margin entire or sinuate, apex acuminate. Inflorescences axillary, each of 2 or 3 cymes; peduncle 4–7 cm long, white tomentose. Bracts linear, 7–10 mm long, 1–2 mm broad, puberulent; cymules 3–6 per cyme; outer bracteoles subcircular to obovate, 1.5–2 mm long, 1–1.5 cm broad, unequal, abaxially pilose, palmately 3- or 5-veined, base rounded, apex mucronate; inner bracteoles linear-lanceolate, 5–7 mm long, 0.8–1 mm broad, abaxially puberulent, apex acuminate. Pedicel 0.5–2 mm long. Calyx 6–7 mm long; lobes linear-lanceolate, adaxially puberulent, apex acuminate. Corolla pinkish-red, 1.5–1.8 cm long, outside pubescent. Stamens 2; filaments 4–5 mm long, glabrous; anther-thecae spherical, 1–1.2 mm long, inserted at slightly different levels. Ovary slightly pilose. Capsules 5–6 mm long, light yellow pilose. Seeds light brown, circular in outline, 1.2–1.5 mm long and broad, compressed, yellowish papillate. **Chromosomes** 2n = 20. Figure 46-102.

Phenology: flowering from August to October; fruiting from October to December.

Habitats: forest margins, thickets, roadsides; 600–3200 m.

Distribution: S Hengduan (Map 46-171); also in other parts of China (SW Sichuan).

S HENGDUAN: **Baidiao**, 2350 m, *Q. S. Zhao et al. 8459* (CDBI, SZ); Bank of Yalongjiang, 1830 m, 11 Sep. 1959, *S. K. Wu 2541* (CDBI, PE); **Bo'ao**, Baiwang, 1830 m, 11 Sep. 1959, *S. K. Wu 2541* (KUN); Baiwang, 2800 m, 24 Sep. 1953, *S. K. Wu 3224* (KUN); **Daocheng**, Dongyi Qu, Kasi, 3200 m, *Z. L. Wang Dao 0809* (CDBI); **Ganluo**, Suxiong, 800 m, 24 Sep. 1976, *s. coll. 14296* (CDBI); **Hanyuan**, Shunhe, 850 m, 14 Sep. 1993, *Z. C. Zhao 0257* (PE); **Jiulong**, Naiqu, Qingming, 2400 m, *Q. Q. Wang 21492* (CDBI, IBSC); between the city of Jiulong and the Yalong Jiang (River), 2250–2300 m, *D. E. Boufford et al. 32997* (A); Naiqu, Qiri, 2200 m, 30 Aug. 1979, *Jiulong Exped. 0430* (SM); Sanyanlong, 3000 m, 30 Aug. 1979, *Jiulong Exped. 0511* (SN); **Kangding**, Ta-Tsien-lou, *M. Bonvalot s. n.* (P); Jiju, 2460 m, 18 May 1960, *J. S. Ying 3444* (PE); **Luding**, Moxi, Gonghe, 1800 m, *Xu Guorong 23226* (CDBI); Detuo, Daduhe, 1000 m, *G. H. Xu 25883* (CDBI); Detuo, Daduhe, 1200 m, *G. H. Xu 26152* (CDBI); near Detuo, 1200 m, *X. H. Hu 29886* (CDBI); Detuo, Beitou, 1200 m, *Zhi Bei Zu 46044* (CDBI); Detuo, Tianchishan, Liangchagou, 9 Aug. 1983, *Zhi Bei Zu 46526* (CDBI); Miansha, 23 Sep. 1960, *Nan Shui Bei Diao Dui 4271* (KUN); 13 Oct. 1951, *C. Ho 11463* (PE, SZ); **Mianning**, Jinkuang, 1400 m, 23 Sep. 1960, *J. S. Ying 4270* (PE); **Muli,** Sihe, 2300 m, Jul. 20, 1978, *Q. S. Zhao 6123* (CDBI, SZ); Shuiluo, 31 Jul. 2006, *Y. F. Deng 18974* (IBSC); Dongzhi, 1400 m, *Q. S. Zhao 8677* (CDBI, SZ); Gangjian Linchang, Yalongjiang, 1900 m, *Y. L. Cao & G. Yao 356* (CDBI, NAS); Wu-Tsu-Bu, 3150 m, 3 Dec. 1937, *T. T. Yü 14815* (A, E, KUN, PE); Shuiluo, 2110 m, 24 Jan. 1959, *S. K. Wu 3491* (KUN); Kala, Tianzheng, 2000 m, 25 Aug. 1983, *Qinghai-Xizang Exped. 13347* (KUN, PE); **Shangri-la** (Zhongdian), Yangtze at Taku, 2800 m, Aug. 1914, *C. Schneider 2217* (A, K); **Shimian**, Ziliushan, 1600 m, *Z. C. Zhao 1020* (PE); Ziliushan, 1250 m, *Z. C. Zhao 1033* (PE); Chengguan, 850 m, *Z. C. Zhao 1098* (PE); Naya Dianzhan, 1400 m, *Z. C. Zhao 1232* (PE); Xiyoufang, Hepinggou, 1800 m, *Z.*

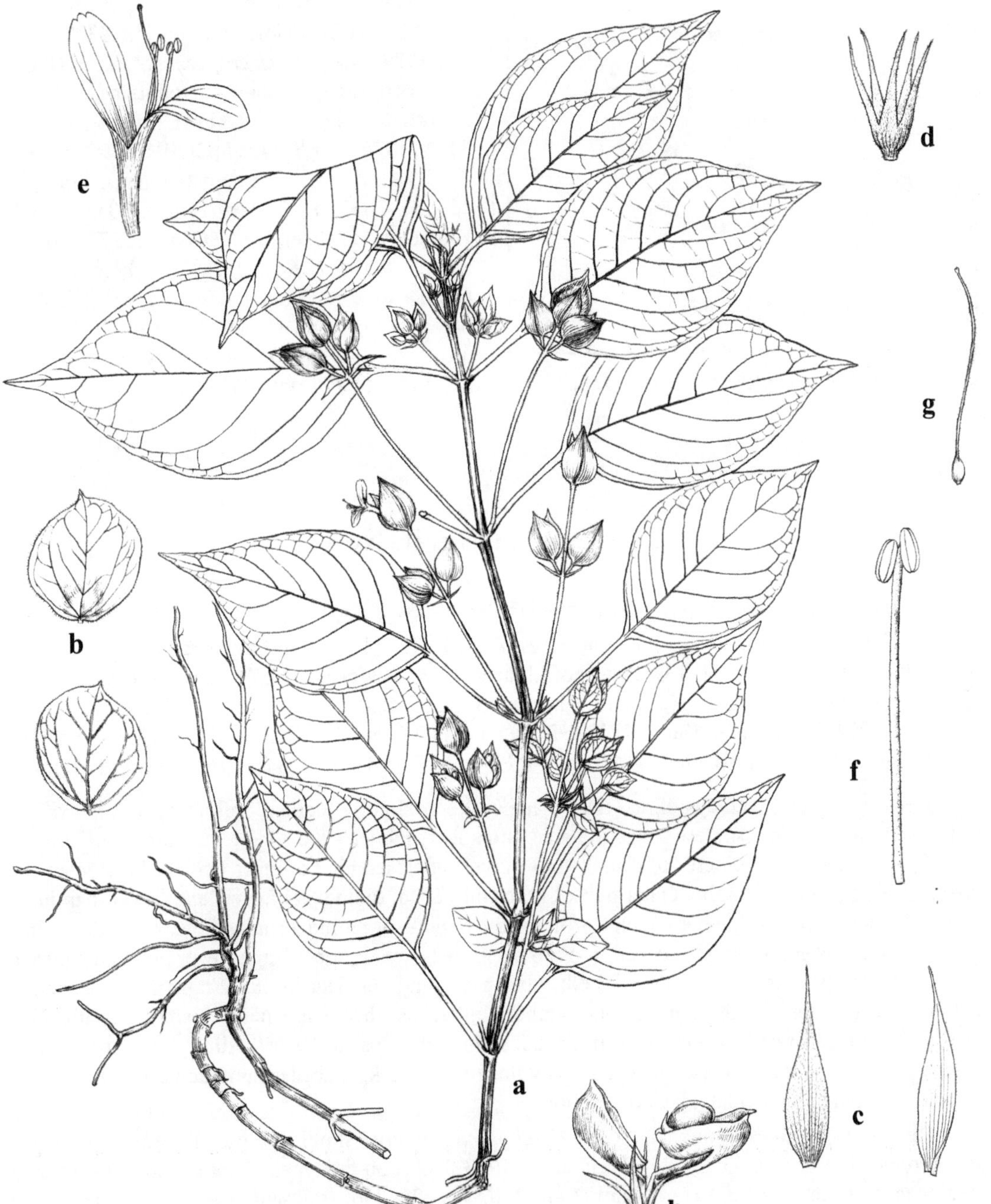

Figure 46-102: **Dicliptera elegans** W. W. Sm.: **a**, habit; **b**, outer bracteoles; **c**, inner bracteoles; **d**, calyx; **e**, corolla; **f**, stamen; **g**, pistil; **h**, capsule. Drawn by Yun-Xiao LIU.

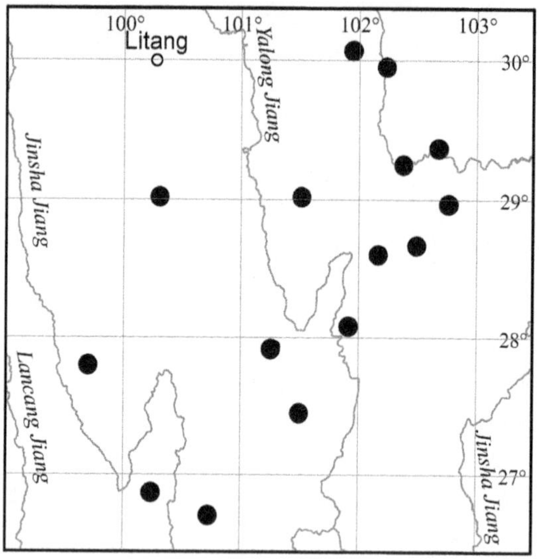

Map 46-171: **Dicliptera elegans** W. W. Sm.

C. Zhao 2717 (PE); Tianba, 1550 m, *Z. C. Zhao 2874* (PE); Wajiao, near Dagangshan Tunnel, 1170 m, 8 Sep. 2015, *FLPH Sichuan Expedtion 152444* (PE); **Yanyuan**, inter vicos Meidsepu et LuMapu, 1500–1650 m, *H. Handel-Mazzetti 5588* (E, WU), Guabie, Xiaojin He, 2300 m, 28 Jul. 1983, *Qinghai-Xizang Exped. 12566* (KUN, PE); **Yongsheng**, Yungpeh mountains, 1917, *G. Forrest 15182* (E, K, P); **Yuexi**, Shunhe, 1600 m, 3 Sep. 1976, *s. coll. 14251* (CDBI); 1500 m, 1932, *T. T. Yü 1786* (A); **Yulong** (Lijiang), Longpan, 15 Sep. 2013, *Y. F. Deng et al. 25036* (IBSC); Hutiaoxia, 1740 m, *Y. F. Deng 25277* (IBSC); near Ngaza, *J. F. Rock 10585* (E); between Bolo and Dsowa, *H. Handel-Mazzetti 1371* (IBSC); between Bolo and Dsowa, 2000 m, *H. Handel-Mazzetti 7579* (WU), Lichiang Valley, 2700 m, Jul. 1913, *G. Forrest 10697* (E, K).

27. PERISTROPHE Nees

Nees in Wallich (ed.), Pl. Asiat. Rar. 3: 77, 112. 1832; Nees in de Candolle (ed.), Prodr. 11: 492. 1847; Clarke in Hooker (ed.), Fl. Brit. India 4: 554. 1885; Lindau in Engler & Prantl (eds.), Nat. Pflanzenfam. 4(3B): 331. 1895; Nasir & Ali in Malik & Ghafoor (eds.), Fl. Pakistan 188: 27. 1988; Wood in Grierson & Long (eds.), Fl. Bhutan 2(3): 1292. 2001; Hu, Fl. Reipubl. Popularis Sin. 70: 240. 2002; Hu, Deng & Daniel in Wu, Raven & Hong (eds.), Fl. China 19: 464. 2011. Lectotype (designated by Phillips, Gen. S. African Fl. Pl. ed. 2: 710. 1951): *Peristrophe acuminata* Nees.

Herbs or shrubs; with cystoliths. Leaf blade entire or slightly toothed. Inflorescences in axillary or terminal cymes, these sometimes aggregated into a leafy panicle-like complex; cymes consisting of (1 or)2–4 involucres; involucres of usually 1–3 flowers enclosed by 2 or more pairs of bracteoles that conceal calyx and corolla tube. Outer pair of bracteoles equal to unequal in size and larger than inner pair. Calyx small, 5-lobed; lobes equal. Corolla white to pink, or purple, usually large, resupinate 180°; tube slender, twisted, apically widened into throat; limb 2-lipped; upper lip in lower position entire to 2-cleft, lower lip in upper position 3-lobed; lobes ascending cochlear in bud. Stamens 2, inserted at corolla throat, exserted from corolla tube; anthers bithecous; thecae unequally inserted, contiguous, or superposed, separated by gap; base muticous; staminodes absent. Ovary with 2 ovules per locule; stigma slightly enlarged or 2-cleft. Capsules with a solid stalk, 4-seeded; retinacula present. Seeds discoid, smooth to tuberculate, lacking trichomes.

Recent references: Yamazaki, 1970. *Dicliptera japonica* should be transferred to the genus *Peristrophe*. *Jap. J. Bot.* 45: 31–32; Wood et al., 1983. Notes on the types of some names of Arabian Acanthaceae in the Forsskal Herbarium. *Kew Bull.* 38: 429–456; Balkwill et al., 1986. Taxonomic studies in the Acanthaceae: tests microsculpturing in southern African species of *Peristrophe*. *S. African J. Bot.* 52: 513–520; Darbyshire & Vollesen, 2007. The transfer of the genus *Peristrophe* to *Dicliptera* (Acanthaceae), with a new species described from Eastern Africa. *Kew Bull.* 62: 119–128.

About 40 species: tropical and subtropical Africa, Asia and Madagascar, with one species possibly extending to Australia; six species in the Pan-Himalaya.

Notes: Recently, some authors (Darbyshire, 2007; Karthikeyan et al., 2009) merged *Peristrophe* with *Dicliptera*. The recent molecular studies (Kiel et al., 2017) showed that Asian *Peristrophe* excluding *P. paniculata* form the sister group with *Dicliptera* and *Peristrophe paniculata* complex. However, Asian *Peristrophe* excluding *P. paniculata* can be easily distinguished from *Dicliptera* by the linear anther, the septa not rising elastically from base of valves and pollen morphology. At present, I prefer to maintain *Peristrophe* as a distinct genus before further studies.

1a. Flowers in 3-dichotomous cymes, forming large, lax, divaricate panicles; corolla ca. 1 cm long; anther-thecae spherical, superposed (sect. I. **Paniculati**) ....1. *P. paniculata*
1b. Flowers in panicles clustered; corolla 1.8–5 cm long; anther-thecae oblong or linear, inserted at same level or at different levels but overlapping, touching each other.
2a. Corolla 1.8–6 cm long; anther-thecae oblong, inserted at same level (sect. II. **Yunnanenses**)...... ....2. *P. yunnanensis*
2b. Corolla longer than 3 cm; anther-thecae linear, inserted at different level (sect. III. **Peristrophe**).
3a. Outer involucral bracts abaxially pubescent with glandular-pilose; corolla ca. 4.5 cm long, pubescent with glandular- or eglandular-pubescent ....3. *P. lanceolaria*
3b. Outer involucral bracts abaxially glabrous or pubescent with eglandular trichomes.
4a. Outer involucral bracts linear-oblong or spatulate-oblanceolate, less than 3 mm broad ....4. *P. speciosa*
4b. Outer involucral bracts ovate to ovate-elliptic, 5–12 mm broad.
5a. Outer involucral bracts ovate to oblong, abaxially slightly villous, base subcordate .... 5. *P. fera*
5b. Outer involucral bracts ovate to elliptic, abaxially subglabrous to puberulent, base cuneate to truncate... 6. *P. japonica*

I. Peristrophe sect. **Paniculati** Y. F. Deng & Z. P. Hao, sect. nov. Type: *Peristrophe paniculata* (Forssk.) Brummitt.

Flowers in 3-dichotomous cymes, forming large, lax, divaricate panicles. Corolla less than 2 cm long. Anther-thecae spherical, superposed.

About twenty species: mainly in tropical Africa with two species in tropical Asia and Australia; one species in the Pan-Himalaya.

1. Peristrophe paniculata (Forssk.) Brummitt, Kew Bull. 38: 451. 1983; Nasir & Ali in Malik & Ghafoor (eds.), Fl. Pakistan 188: 27. 1988; Hu, Deng & Daniel in Wu, Raven & Hong (eds.), Fl. China 19: 464. 2011. ≡ *Dianthera paniculata* Forssk., Fl. Aegypt.-Arab.: 7. 1775. ≡ *Dicliptera paniculata* (Forssk.) I. Darbysh., Kew Bull. 62: 122. 2007. ≡ *Justicia ligulata* Lam, Encycl. 1: 632. 1785, nom. illeg. (Art. 52.1). Type: Yemen, Wadi Sordud, Jan. 1763, *Forsskal 385* [microfiche 38: III. 3-4] [lectotype designated by Wood et al., Kew Bull. 38: 451. 1983: C (C10002174)].

Dianthera bicalyculata Retz., Kongl. Vetensk. Acad. Handl. 297. 1775; Acta Holm. 1775: 279. 1776; Hu, Fl. Reipubl. Popularis Sin. 70: 243. 2002. ≡ *Justicia bicalyculata* (Retz.) Vahl, Symb. Bot. 2: 13. 1791. ≡ *Peristrophe bicalyculata* (Retz.) Nees in Wallich (ed.), Pl. Asiat. Rar. 3: 113. 1832; Nees in de Candolle (ed.), Prodr. 11: 496. 1847; Clarke in Hooker (ed.), Fl. Brit. India 4: 554. 1885. ≡ *Dianthera malabarica* L. f., Suppl. Pl.: 85. 1781 [1782], nom. illeg. (Art. 52.1). Type: India, s. d., *J. G. König s. n.* [lectotype **here designated**: C (C10005075); isolectotypes: C].

Peristrophe bicalyculata var. *rivinoides* C. B. Clarke in Hooker (ed.), Fl. Brit. India 4: 555. 1885. ≡ *Justicia rivinoides* Wall., Numer. List n. 2488. 1830, nom. invalid. Type: India, s. d., *Wallich Cat. 2488* [lectotype **here designated**: K-W (K001116175)].

Herbs, 0.5–1.2 m high, perennial; stems hirsute, usually 4- or 6-angled. Petiole 3–12 mm long, pubescent; leaf blade ovate to lanceolate, (1–) 3–4.5 cm long, (0.5–)1.5–2 cm broad, abaxially scabrid, adaxially slightly scabrid, base attenuate to broadly cuneate, margin remotely serrulate and slightly ciliate, apex acuminate, acute, or sometimes rounded. Cymes axillary, usually in panicles; rachis pubescent. Involucral bracts unequal, abaxially pubescent with glandular and eglandular trichomes, larger one linear, 9–10 mm long, 1.5–2 mm broad, 1-veined, apex acute and mucronate, smaller one subulate to lanceolate, 5–6 mm long, 0.1–0.3 mm broad; bracteoles narrowly linear to subulate, unequal in length. Calyx lobes lanceolate, 2.5–3 mm long, abaxially sparsely pubescent with eglandular trichomes, adaxially sparsely pubescent with glandular-pilose, margin membranous towards base, apex long acuminate. Corolla light to dark purple, 0.8–1 cm long, outside pubescent with eglandular trichomes; tube base cylindric, 4–5 mm long; lip in lower position narrowly elliptic, 5–7 mm long, 1–2 mm broad, apex emarginate; lip in upper position elliptic, 3–7 mm long, 2–3 mm broad, 3-lobed. Stamens 2; filaments 3.5–5 mm long, white hairy; anther-thecae oblong superposed, not touching. Ovary ovoid, pubescent and glandular; style glabrous; stigma 2-lobed. Capsules 0.9–1.3 cm long, pubescent with eglandular trichomes. Seeds 2.2–2.5 mm long, rough and tuberculate. **Chromosomes** 2n = 30.

Phenology: flowering from August to December; fruiting from October to February of next year.

Habitats: weedy areas, roadsides; 100–2200 m.

Distribution: U Ganga & Indus, W Nepal, C Nepal, E Nepal, Sikkim & Darjiling, S Hengduan (Map 46-172); also widespreading in tropical and subtropical region of Asia, Australia, and Africa.

JAMMU & KASHMIR: **Jammu**, Rihari, 300 m, 7 Jan. 1972, *B. M. Sharma 1002* (RRLH). **U GANGA & INDUS**: s. loc., 27 Mar. 1898, *Mayat 21737* (CAL); **Dehradun**, Hissar, 700 m, 31 Mar. 1904, *J. R. Drummond 15213* (K); Nak Raunda, 20 Feb. 1942, *M. B. Raizada 15665* (K); *J. R. Drummond 26671* (E, K); *J. R. Drummond 26672* (E, K); Trimor, in 1885, *J. R. Drummond 22487* (E, K); **Kumaun**, 19 Aug. 1886, *J. F. Duthie 5875* (CAL); **Mussoorie**, Oct., 1898, *P. W. Mackinnon s. n.* (CAL); Kubri Hill, 1950 m, Oct. 1916, *A. Anderson s. n.* (E); **Simla**, Simla Hills, 20 Oct. 1849, *Herbs Hooker s. n.* (K).

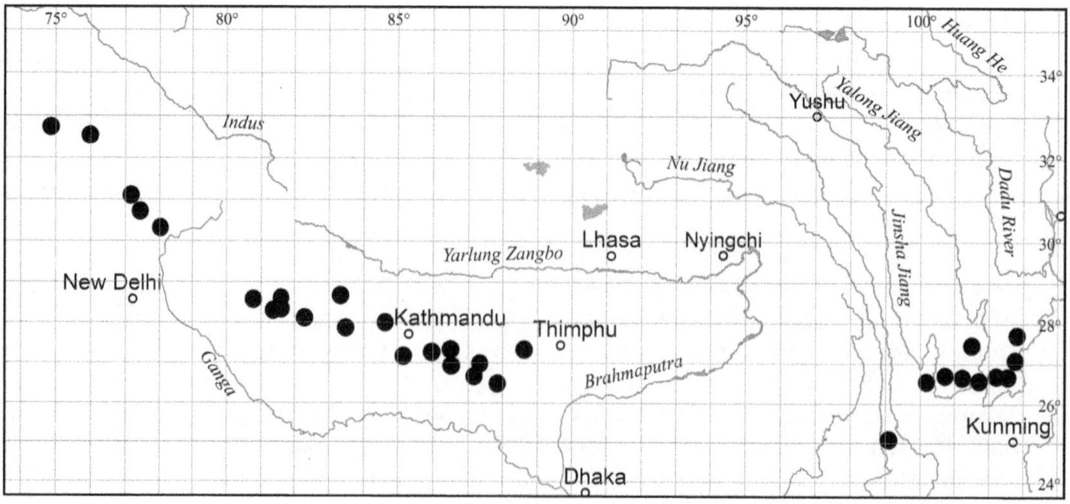

Map 46-172: **Peristrophe paniculata** (Forssk.) Brummitt

W NEPAL: **Bheri**, Banke, Napelganj, 200 m, 30 Jan. 1982, *D. K. Sharma & S. Dayal 4663* (KATH); Banke, Napelganj, 180 m, 16 Dec. 1979, *T. K. Bhatthacharya 1754* (KATH); Banke, Sallyani, Napelganj, 20 Oct. 1975, *K. J. Malla & P. R. Sharma 760* (KATH); Banke, Nepalganj, 24 Oct. 1972, *N. P. Manandhar 8035* (KATH); Banke, Napelganj, 180 m, 6 Jan. 1993, *P. P. Kurmi KB520* (KATH); Bardiya, Babai, 300 m, 9 Dec. 1983, *T. K. Bhattacharya 2524* (KATH); Bardiya, Gangapatia, 180 m, 16 Jan. 1985, *P. R. Sharma 3716* (KATH); Bardiya, Mau, 180 m, 3 Oct. 1981, *T. K. Bhattachaya 3787* (KATH); Bardiya, Basanta, 180 m, 16 Dec. 1982, *R. N. Shukla 5136* (KATH); Bardiya, Gulariya, 10 Feb. 1976, *L. R. Sharma 911* (KATH); Surkhet, Kalyan, 870 m, 23 Feb. 1991, *N. P. Manandhar 169-91* (KATH); **Rapti**, Dang, Koilabas, 230 m, 8 Aug. 1980, *T. K. Bhattachaya 3055* (KATH); Dang, Satbariya, Ward no. 4, Nimbukhutti, 228 m, 9 Nov. 2012, *B. Adhikari & G. D. Bhatt 28* (E); Dang, Ghorahi to Lape, 23 Dec. 1984, *T. K. Bhatthacharya 84.85/15* (KATH); **Seti**, Kailai, Dhangarhi, 200 m, 27 Nov. 1966, *D. H. Nicolson 2790* (A, BM, KATH, LE). **C NEPAL**: **Dhawalagiri**, Myagdi, Beni, 850 m, 30 Oct. 1954, *J. D. A. Stainton, W. R. Sykes & L. H. J. Williams 9202* (A, BM, E); **Gandaki**, Gorkha, Birendra Nagar, 720 m, 18 Jan. 1983, *P. P. Kurmi 4705* (KATH); **Janakpur**, Sindhuli, Mulkot, 1450 m, 21 Nov. 1988, *N. P. Manandhar 12328* (KATH); **Lumbini**, Palpa, Tansing, 300m, 5 Oct., 1954, *J. D. A. Stainton, W. R. Sykes & L. H. J. Williams 8756* (BM, CAL, P); Palpa, Tansen, 910 m, 5 Oct. 1953, *Staintan, W. R. Sykes & L. H. J. Williams 8947* (BM); Palpa, Ridhi, Gulmi, 450 m, 2 Mar. 1976, *N. P. Manandhar & P. M. Regmi 65* (KATH); **Narayani**, Bara, NW of Amlekganj, E of Bherah Khola, 330 m, 10 Oct. 1995, *M. Mikage et al. 9552817* (A, BM, E); Bara, Bara Badahikim Bridge, on the road from Hetauda to Birgunj, 480 m, 26 Nov. 2004, *C. A. Pendry et al. DENP2 A138* (E, KATH). **E NEPAL**: **Koshi**, Dankhuta, 13 Nov. 1978, *P. Pradhan et al. 764* (KATH); Sunsari, Phusri-Sanguri Bhanjang to Dhara Pani, 16 Oct. 1963, *H. Hara et al. 6303628* (BM, K, KATH); **Mechi**, Jhapa, Sunmai, Shantinagar VDG, 577 m, 20 Nov. 2012, *B. Adhikari & D. R. Kandel 62* (E); **Sagarmatha**, Okhaldhunga, Torke, 1590 m, 1 Dec. 1983, *N. P. Manandhar 12694* (KATH); Udayapur, 170 m, 25 Oct. 1995, *M. Mikage et al. 9558074* (KATH); Udayapur, Beltar–Kattike–Darula–Baplung–Simule, 420 m, 24 Oct. 1995, *M. Mikage et al. 9554115* (E); Udayapur, de Balaute a Sukhchauri, 400 m, 6 Nov. 1954, *A. Zimmermann 2082* (E, K). **SIKKIM & DARJILING**: **Sikkim**, Garudura, 17 Feb. 1911, *Ribu & Rhomoo s. n.* (CAL). **S HENGDUAN**: **Butuo**, Niujiaowan, Xixihe, 680 m, 11 Oct. 2014, *Y. F. Deng 25858* (IBSC); **Heqing**, Ta-pin-tze, 30 Jan. 1886, *M. Delavay s. n.* (P); Ta-pin-tze, 3 Nov. 1888, *M. Delavay 3563* (A, P, NY, UC); **Huaping**, Rongjiang, 1400 m, 15 Sep. 2012, *Y. F. Deng & H. Peng 23869* (IBSC); **Huidong**, Dacong, 11 Oct. 2014, *Y. F. Deng & H. Peng 25875* (IBSC); **Huili**, *H. Handel-Mazzetti 5666* (E, W); **Longyang**, Nujiangba, 66 km of Baoshan-Tengchong Highway, 850 m, *X. Zhou 217* (KUN); **Ningnan**, Hulukou, 750 m, 1 Sep. 2012, *Y. F. Deng 23362* (IBSC); Baihetan, 11 Oct. 2014, *Y. F. Deng & H. Peng 25872* (IBSC); **Panzhihua**, Dukou, Mar. 1984, *G. D. Tao 40137* (HITBC), Xinhua, 2100 m, *X. H. Hu et al. 23830* (CDBI); Geliping, 1200 m, *Y. F. Deng & H. Peng 23868* (IBSC); *Cycas* Natural Reserve, 1478 m, 16 Sep. 2010, *B. Liu & J. F. Ye 541* (PE); **Yanyuan**, Dayuhe, 2180 m, *Y. B. Yang et al. 57* (CDBI); **Yongsheng**, Pianjiao, near Jinshajiang, 1200 m, 14 Sep. 2012, *Y. F. Deng & H. Peng 23841* (IBSC); Pianjiao, 1470 m, 14 Sep. 2012, *Y. F. Deng & H. Peng 23837* (IBSC).

II. Peristrophe sect. **Yunnanenses** Y. F. Deng & Z. P. Hao, sect. nov. Type: *Peristrophe yunnanensis* W. W. Sm.

Flowers in panicles clustered. Anther-thecae oblong R, inserted at same level, touching each other.

One species, endemic to China; one species in the Pan-Himalaya.

2. Peristrophe yunnanensis W. W. Sm., Notes Roy. Bot. Gard. Edinburgh 10: 187. 1918; Hu, Fl. Reipubl. Popularis Sin. 70: 248. 2002; Hu, Deng & Daniel in Wu, Raven & Hong (eds.), Fl. China 19: 465. 2011. Type: [S Hengduan] China, Yunnan, Yangpi Valley, 15 Apr. 1914, *G. Forrest 12279* [holotype: E (E00706902); isotypes: IBSC (IBSC0560308, IBSC0560308), K (K000884389)].

Herbs, 1–2 m high; stems sulcate, densely grayish pubescent when young, later glabrescent. Petiole 1–10 mm long; leaf blade lanceolate to ovate, 2.5–3 cm long, 0.7–1 cm broad, abaxially pubescent, adaxially sparsely puberulent especially along veins, or glabrescent, base cuneate, margin entire, apex acuminate. Involucral bracts linear, 2–3 mm long; bracteoles linear, 2.5–6 mm long. Calyx 5–7 mm long, glabrous or subglabrous; lobes linear, apex acute. Corolla pink to violet, 1.6–1.8 cm long, outside slightly pubescent; tube base cylindric, 4–5 mm long, upper lip in lower position narrowly elliptic, 5–7 mm long, 1–2 mm broad, apex emarginate, lower lip in upper position elliptic, 3–7 mm long, 2–3 mm broad, 3-lobed, lobes 0.8–1.2 mm long. Stamens 2; filaments 1.5–1.8 cm long, pubescent towards base; anther-thecae oblong, 1.8–2 mm long, parallel, inserted at same level. Capsules 0.9–1.2 cm long, stipitate at base, pubescent with glandular-pilose. Seeds 2.5–3 mm long, 1.8–2.3 mm broad, smooth.

Phenology: flowering from December to February of next year.

Habitats: thickets; 1700–2200 m.

Distribution: S Hengduan (Map 46-173); also in other parts of China (C Yunnan).

S HENGDUAN: Binchuan, Jizushan, 2000 m, 6 Apr. 1982, *H. Li et al. 436* (HITBC, KUN); **Dali**, Qiliqiao, 10 May 1942, *H. C. Wang 2092* (PE); **Heqing**, between Chiang-Ing to Domei on Yangtze, 25 Aug. 1939, *K. M. Feng 816* (A, KUN); Xiyi, 23 Mar. 2009, *S. S. Zhou 5394* (HITBC, QBG); **Huili**, 1860 m, 26 Mar. 1914, *H. Handel-Mazzetti 219* (NAS); Laopaidse, ca. 1960 m, 26 April 1914, *H. Handel-Mazzetti 883* (A, E, IBSC, US); **Miyi**, Puwei, 2200 m, 23 Apr. 1959, *S. K. Wu 50* (KUN); **Yangbi**, Jinzhan, Diantou, 25 Apr. 1940, *R. C. Ching 22394* (KUN, PE); Yangtze Fahr La Ka tiang, *C. Schneider 4035* (K); Yangpi Valley, 15 Apr. 1914, *G. Forrest 12279* (E, K); s. loc., *H. D. McLaren 26* (E, K); **Yulong** (Lijiang), Baoshan, Shitoucheng, 1750 m, *Y. F. Deng 25276* (IBSC); Baoshan, Shitoucheng, 1750 m, 19 Apr. 2014, *Y. F. Deng 25286* (IBSC).

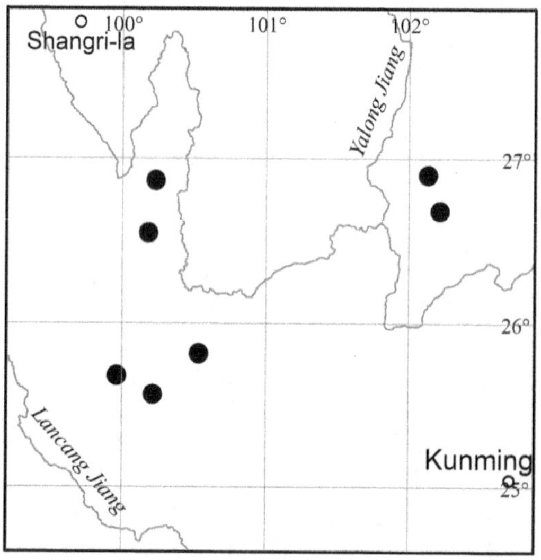

Map 46-173: **Peristrophe yunnanensis** W. W. Sm.

III. Peristrophe sect. **Peristrophe**. Type: *Peristrophe acuminata* Nees.

Flowers in panicles of clustered cymes; corolla 3–5 cm long; anther-thecae linear, inserted at different levels but overlapping, touching each other.

About 15 species: tropical and subtropical Asia; four species in the Pan-Himalaya.

3. Peristrophe lanceolaria (Roxb.) Nees in Wallich (ed.), Pl. Asiat. Rar. 3: 114. 1832; Nees in de Candolle (ed.), Prodr. 11: 496. 1847; Clarke in Hooker (ed.), Fl. Brit. India 4: 555. 1885; Hu, Fl. Reipubl. Popularis Sin. 70: 246. 2002; Hu, Deng & Daniel in Wu, Raven & Hong (eds.), Fl. China 19: 464. 2011. ≡ *Justicia lanceolaria* Roxb., Fl. Ind. 1: 122. 1820. Type: [*icon*] *Roxburgh Icon. 2415* (lectotype **here designated**: K).

Herbs, 0.4–1 m high. Branches densely covered with white retrorse trichomes or sometimes glabrescent. Petiole (0.5–)1.5–3.5 cm long; leaf blade lanceolate to ovate-lanceolate, (6–)10–17 cm long, (1–)3–6 cm broad, abaxially sparsely pubescent or glabrescent, adaxially pilose along midvein toward base, lateral veins 5–7 pairs, base narrowly cuneate and slightly oblique, margin entire, apex acuminate. Panicles of cymes axillary or terminal, 5–7 cm long. Involucral bracts linear-lanceolate, 1.0–1.8 cm long, 1.5–3 mm broad, glandular-pubescent; bracteoles similar to outer ones except smaller. Calyx 9–11 mm long; lobes linear, 1–1.2 mm broad. Corolla pink, 4–4.5 cm long, outside pubescent with glandular- or eglandular-pubescent; tube base cylindric, 2.2–2.5 cm long, 1.2–1.5 mm broad; upper lip in lower position obovate, 1.8–2 cm long, 6–8 nm broad, slightly 2-lobed; lower lip in upper position oblong, 1.7–2 cm long, 5–6 mm broad, 3-lobed, lobes ovate, 0.8–1 mm long and broad. Stamens 2; filaments 1.5–1.8 cm long, glabrous; anther-thecae oblong, 2.5–3 mm long, 0.5–0.6 mm broad, inserted at different height. Capsules 1.4–1.8 cm long, pubescent with glandular-pilose. Seeds 2–3 mm long, 1.5–2 mm broad, smooth.

Phenology: flowering from December to March of next year; fruiting from March to June.

Habitats: wet places; 400–1600 m.

Distribution: Sikkim & Darjiling, Bhutan (Map 46-174); also in Laos, Thailand, Vietnam, and other parts of India, Myanmar and China (S Yunnan).

SIKKIM & DARJILING: Darjiling, s. d., *J. S. Gamble 7571* (CAL); s. d., *C. B. Clarke 24796* (CAL); **Sikkim**, Mungpoo,750 m, *W. W. Smith 238* (CAL); Kangra, 900 m, *C. B. Clarke 29661*

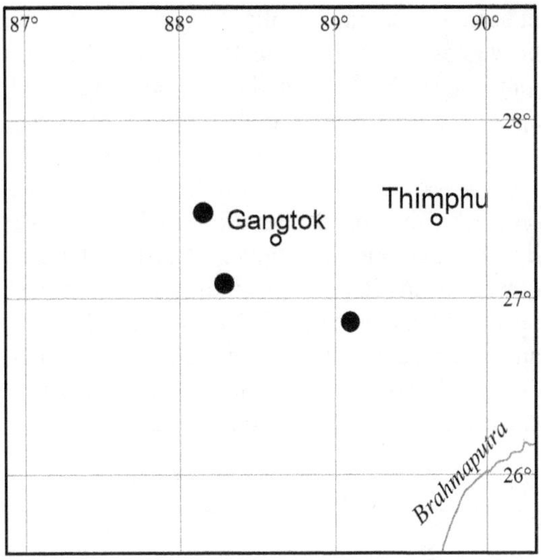

Map 46-174: **Peristrophe lanceolaria** (Roxb.) Nees

(LE); Selim, 450 m, *C. B. Clarke 36879* (LE); Rungbee, 600 m, *G. King 264* (CAL); Manpoo Peak, 900 m, 21 Sep. 1875, *Gamble 3582* (CAL, K). **BHUTAN**: **Shamgong**, between Shamgong and the Mangde river, 1520 m, *B. Bartholomew & D. E. Boufford 3826* (A); **Samchi**, Kharpaga Hill, 420 m, 13 Dec. 1963, *K. Thethathri 10343* (CAL).

4. Peristrophe speciosa (Roxb.) Nees in Wallich (ed.), Pl. Asiat. Rar. 3: 113. 1832; Nees in de Candolle (ed.), Prodr. 11: 495. 1847; Clarke in Hooker (ed.), Fl. Brit. India 4: 556. 1885; Wood in Grierson & Long (eds.), Fl. Bhutan 2(3): 1292. 2001. ≡ *Justicia speciosa* Roxb., [Hort. Bengal. 4. 1814, nom. invalid.]; Fl. Ind. 1: 123. 1820. ≡ *Dicliptera raui* Karthik. & Moorthy, Fl. Pl. India 12. 2009. Type: India, *Wallich Cat. 2468a* [lecotype **here designated**: K-W (K001116143)].

Perennial subshrubs, 1–2 m high; stems terete, scarcely striate, grey pubescent when young, then glabrescent. Petiole (0–)1–2.5 cm long, pubescent; leaf blade ovate-elliptic, 2–11 cm long, 1–5.5 cm broad, sparsely pubescent or glabrescent on both surfaces, base cuneate, margin crenulate, apex acute. Panicles of spikes terminal, 10–20 cm long. Rachis grey pubescent. Involucral bracts linear-oblong or spatulate-oblanceolate, 1.2–1.6 cm long, 1–3 mm broad, pubescent, margin usually ciliate, apex acute; bracteoles linear, 7–9 mm long. Calyx 5–6 mm long, puberulent; lobes linear, white-margined, apex acute. Corolla pink, 4–4.5 cm long, pubescent; tube slender, 2.3–2.7 cm long, upper lip in lower portion ovate, with a few purple spots on a whitish ground near mouth of tube, lower lip in upper portion elliptic, almost as long as tube, 3-lobed. Stamens protruding 1–1.2 cm long beyond tube; filaments 1.5–1.8 cm long, glabrous; anther-thecae linear, parallel, slightly unequal, 3–5 mm long, 0.5–0.8 mm broad. Capsules clavate, 1.5–1.8 mm long, pubescent. Seeds ovate in outline, compressed, 2–2.5 mm long, 1.5–2 mm broad. **Chromosomes** 2n = 60.

Phenology: flowering and fruiting from October to April of next year.

Habitats: rocky, scrub-covered slopes and gullies in river valleys; 400-2200 m.

Distribution: U Ganga & Indus, C Nepal, E Nepal, Sikkim & Darjiling, Bhutan, Yarlung Zangbo-Brahmaputra (Map 46-175); also in other parts of India and Pakistan; widely cultivated in the gardens in tropical regions.

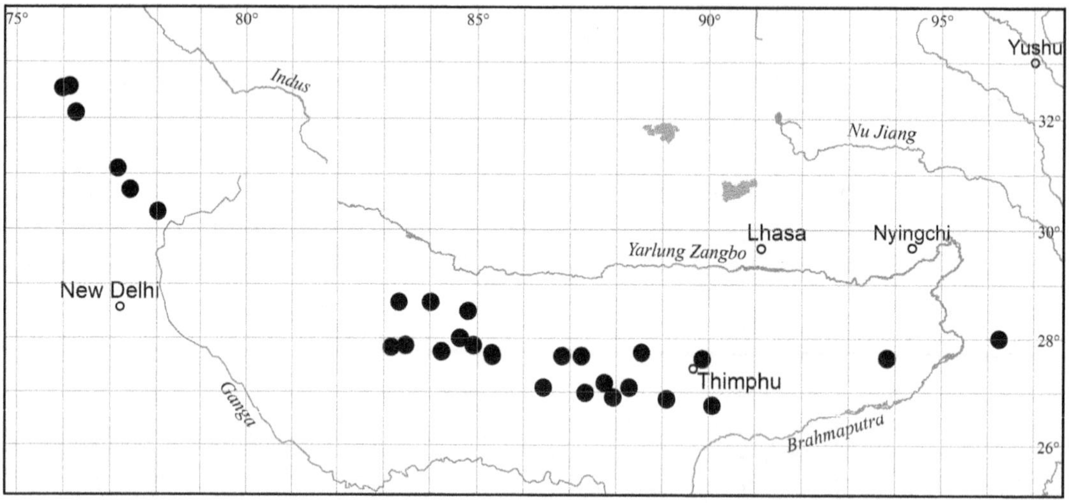

Map 46-175: **Peristrophe speciosa** (Roxb.) Nees

U GANGA & INDUS: **Chamba**, Chuari, Wazirat Bhatiyat, 960 m, 8 Jan. 1920, *R. N. Parker s. n.* (A, K); Chuari, Bhatigat, 960 m, 8 Jan. 1920, *R. N. Parker 5642* (K); **Dalhousie**, Donera, 900 m, 20 Feb. 1917, *R. R. Stewart 1238* (K, MO); **Dehradun**, 750 m, Jan. 1891, *J. S. Gamble 22535* (K); 600 m, Mar. 1894, *J. S. Gamble 24567* (CAL, K); Lachiwala, 17 Feb. 1942, *M. B. Rauzada 15650* (K); **Kangra**, Negrota, 900 m, 9 Mar. 1931, *W. Koelz 1703* (US); 900 m, 19 Oct., 1874, *C. B. Clarke 29661* (LE); **Kumaon**, Bagesar, 900 m, *R. Strachey & J. E. Winterbottom 12* (A); Chera Pani, 24 Dec. 1848, *R. Strachey & J. E. Winterbottom 2* (K, LE); **Mussoorei**, Rajpur, 25 Feb.–9 Mar. 1962, *H. J. M. Bowen 614* (K); **Simla**, Suni, 1800 m, Mar. 1889, *Herbs G. Watt s. n.* (E); Suni, 27 Jan. 1889, *G. Watt s. n.* (E). **NEPAL**: Kandrang Garhi, 1820 m, 30 Nov. 1935, *F. M. Beiley s. n.* (BM). **C NEPAL: Bagmati**, Dhading, Jharling, 1550 m, 5 Nov. 1989, *N. P. Manandhar 13054* (KATH); Kathmandu, Narainhetty, 1 Nov. 1802, *F. Buchanan-Hamilton s. n.* (BM); Lalitpur, Godavari, 10 m, 20 Dec. 1965, *A. D. Shilling 710* (K, KATH); alitpur, Godavari, 1500 m, 5 Nov. 1969, *N. P. Manandhar 12971* (KATH); Rasuwa, Bamboo hotel to Hot Springs, 1850 m, 17 Oct. 2001, *D. G. Long et al. ENEP524* (E); Rasuwa, between Syaphru and Langtang Khola, 1940 m, 1 Nov. 1976, *H. Tabata et al. 8692* (A, BM, KYO); Rasuwa, Langtang Valley, 1670 m, 23 Apr. 1963, *S. Bowes-Lyon 71* (BM, E); **Dhawalagiri**, Myagdi, Dana, Kali

Gandaki, 1650 m, 31 Oct. 1954, *J. D. A. Stainton, W. R. Sykes & L. H. J. Williams 8276* (A); **Gandaki,** Gorkha, Styar Khola, 2120 m, 14 Apr. 1953, *P. C. Gardner 353* (E); Gorkha, Lukuwa, 2200 m, 18 Dec. 1970, *J. F. Dobremez 724* (BM, E, KATH); Manag, Marsyangdi River, between Thonje and Jagat, 31 Sep. 1969, *T. Wraber 563* (BM); Tanahu, Aanbu Khaireni, Khani Khola, 361 m, 28 Nov. 2012, *B. Adhikari & D. R. Kandel 104* (E); **Lumbini**, Arghakhachi, Wangla, 1927, *S. Wigram 204* (CAL, K); Arghakhachi, Wangla, 1927, *C. Wigram 207* (E, K); Palpa, Ridi, 450 m, 2 Mar. 1976, *N. P. Manandhar & P. M. Regmi 76* (KATH). **E NEPAL**: **Koshi**, Dhankuta, Sangari to Dharapani, 1000 m, 25 Jan. 1982, *K. R. Rajbhandari et al. 6071* (KATH); Sankhuwasabha, Arun River near confluence of Barun Khloa, 1230 m, 14 Oct. 1991, *D. G. Long et al. EMAK796* (E, KATH); Sankhuwasabha, Hauling-Sedua, 1090 m, 12 Oct. 1981, *P. R. Shakya 7022* (KATH); Sankhuwasabha, Jaubari, 1420 m, 4 Dec. 1990, *P. R. Shakya 9404* (KATH); Sankhuwasabha, Apsuwa Doban to Pikhuwa, 610 m, 31 Aug. 1997, *S. Noshiro et al. 9760505* (TI); **Mechi**, Ilam, Puwa Khola, 1288 m, 22 Nov. 2012, *B. Adhikari & D. R. Kandel 82* (E); Panchthar, Yektin to Akasay, 30 Nov. 1963, *H. Hara et al. 6303630* (BM, K, KATH); **Sagarmatha**, Solu Khumbu, Dudh Kosi Valley, 1590 m, 1 Nov. 1978, *A. D. Schilling 2441A* (BM); Solu Khumbu, Jubhing, below Jubing, 1520 m, 26 Dec. 1971, *Bowring 44* (BM). **SIKKIM & DARJILING**: **Darjiling**, Mongpo, 600 m, 23 Sep. 1875, *C. B. Clarke 24796* (BM, CAL, K); Ryang, 450 m, 10 Nov. 1870, *C. B. Clarke 13621* (K, LE, US); Dec. 1879, *J. S. Gamble 7571* (K); **Sikkim**, Rungbee, 600 m, 16 Oct. 1878 *G. King 264* (CAL); Tingling, 1200 m, 19 Oct. 1875, *C. B. Clarke 25436* (K). **BHUTAN**: **Punakha**, ca. 20 km S of Wangdi Phodrang towards Chirang, 1100 m, 30 Oct. 1988, *J. R. I. Wood 6852* (E); **Samchi**, Kharpaga Hill, 420 m, 13 Dec. 1963, *K. Thethathri 10343* (CAL); by Chamurchi River on footpath from Samchi to Dorokha, 800 m, 20 Nov. 1991, *J. R. I. Wood 7512* (E); **Sankoshi**, between Dagapela and Daga Dzong, 1600 m, 23 Apr. 1990, *J. R. I. Wood 7241* (E); **YARLUNG ZANGBO-BRAHMAPUTRA**, Misik, 24 Oct. 1941, *K. Biswas 5884* (CAL); **Lohit Valley**, 600–900 m, 2 Dec. 1926, *F. Kingdon-Ward 7687* (K).

5. Peristrophe fera C. B. Clarke in Hooker (ed.), Fl. Brit. India 4: 556. 1885; Wood in Grierson & Long (eds.), Fl. Bhutan 2(3): 1293. 2001; Hu, Fl. Reipubl. Popularis Sin. 70: 243. 2002; Hu, Deng & Daniel in Wu, Raven & Hong (eds.), Fl. China 19: 466. 2011. ≡ *Dicliptera fera* (C. B. Clarke) Karthik. & Moorthy, Fl. Pl. India 10. 2009. Type: India, Khasia, Manla, 14 Nov. 1850, *J. D. Hooker & T. Thomson s. n.* [lectotype **here designated**: K (K000884472); isolectotype: K].

Herbs, 0.4–1 m high, erect; stems quadrangular, sulcate, pubescent. Petiole 0.5–1.0(–2.0) cm long; leaf blade ovate to ovate-lanceolate, 5–10 cm long, 2–4.3 cm broad, both surfaces pubescent, lateral veins 3–5 pairs, base broadly cuneate, margin subentire or obscurely dentate towards apex, apex acuminate. Cymes terminal or axillary, densely pubescent. Involucral bracts ovate to oblong, unequal in size, larger one 1.2–1.5 cm long, 0.8–1 cm broad, smaller one 0.9–1.2 cm long, 0.6–0.8 mm broad, slightly villous, base subcordate, margin ciliate, apex acuminate; bracteoles 2.2–2.5 mm long. Calyx 4–5 mm long, pubescent; lobes linear, apex acuminate. Corolla 3–4 cm long, outside pubescent; tube ca. 1.5 mm broad, upper lip in lower position broadly elliptic, apex emarginate, lower lip in upper position narrowly oblong, apex 3-lobed. Stamens 2, exerted; filaments 1.2–1.5 cm long, villous; anther-thecae linear, parallel, slightly unequally inserted at different height. Ovary pubescent; style hirsute towards base. Capsules 1.2–1.4 cm long, pubescent. Seeds ovate in outline, compressed, 1.5–2 mm long and broad.

Phenology: flowering from July to August.

Habitats: dense forests; 600–1800 m.

Distribution: Bhutan, Yarlung Zangbo-Brahmaputra, U Irrawaddy (Map 46-176); also in other parts of India, Myanmar and China (S Yunnan, Guangxi, Hainan).

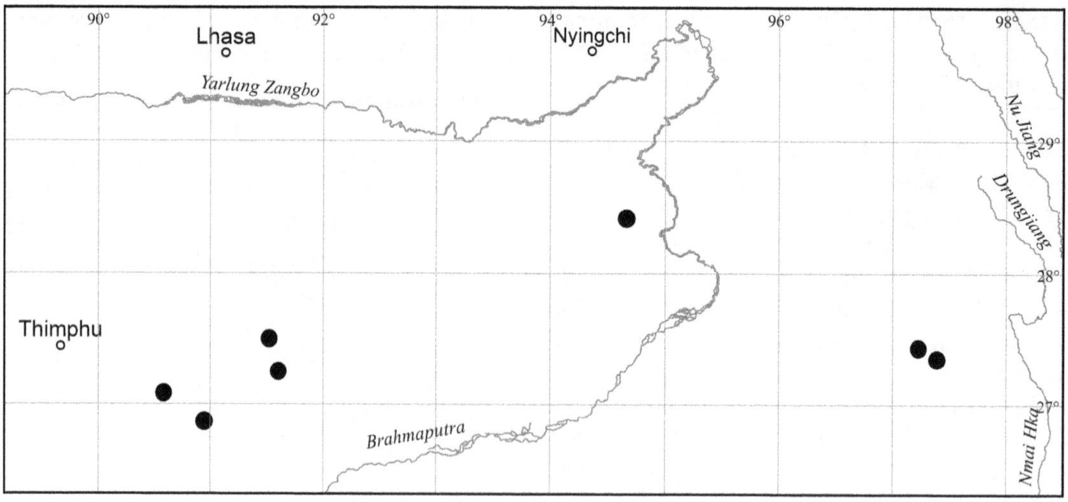

Map 46-176: **Peristrophe fera** C. B. Clarke

Notes: This species is very similar to *Peristrophe jalappifolia* (Vahl) Nees (≡ *Justicia jalappifolia* Vahl) from SE Asia and they may be merged into a single species. I do not treat them as such here due to few specimens examined.

BHUTAN: **Gaylegphug**, just west of Mangde Chu between Shemgang and Gaylephug, 1100 m, 26 Feb. 1988, *J. R. I. Wood 6172* (E); **Mongar**, between Namning and Lingmethang, 1800 m, 24 Apr. 1988, *J. R. I. Wood 6256* (E); **Shamgong**, between Shamgong and the Mangde Chu river, 1520 m, *B. Bartholomew & D. E. Boufford 3826* (A, CAS); **Tongsa**, above Tintibi, 1300 m, 3 Apr. 1989, *J. R. I. Wood 6892* (E); Mangde Chu Valley between Dakpai and Tintibi, 800 m, 15 Mar. 1992, *J. R. I. Wood 7549* (E); near Pertimi, 1400 m, 3 Apr. 1982, *A. J. C. Grierson & D. G. Long 4267* (E, K); **Trashiyangtse**, Nagton, 29 May 2000, *Y. Dorji, N. Pearce & P. Cribb 58* (E). **YARLUNG ZANGBO-BRAHMAPUTRA**: **Abor Hills**, Komsing, Dhang Valley, 670 m, 16 Feb. 1928, *F. Kingdon-Ward 7873* (ABD, K). **U IRRAWADDY**: **Putao**, Wasadam to Upper Shankaung, 23 Oct. 2014, *Putao Expedition Team (X. H. Jin et al.) 1177* (PE).

6. Peristrophe japonica (Thunb.) Bremek., Boissiera 7: 194. 1943; Hu, Fl. Reipubl. Popularis Sin. 70: 245. 2002; Hu, Deng & Daniel in Wu, Raven & Hong (eds.), Fl. China 19: 466. 2011. ≡ *Dianthera japonica* Thunb. in Murray, Syst. Veg., ed. 14: 64. 1784; Fl. Jap. 21. 1784. ≡ *Justicia japonica* (Thunb) Vahl, Symb. 2: 16. 1791, non *Justicia japonica* Thunb., Syst. Veg. (ed. 14): 63. 1784. ≡ *Dicliptera japonica* (Thunb.) Makino, Bot. Mag. (Tokyo) 17: 90. 1903. ≡ *Justicia crinita* Thunb., Trans. Linn. Soc. 2: 338. 1794, nom. illeg. (Art. 52.1 & 52.2). ≡ *Dicliptera crinita* Nees, Prodr. 11: 485. 1847, nom. illeg. (Art. 52.1 & 52.2). Type: Japan, *C. P. Thunberg s. n.* [holotype: UPS (UPS-THUNB 370)].

Dicliptera buergeriana Miq., Ann. Mus. Bot. Lugduno-Batavi 2: 125. 1865. Type: Japan, s. d., *H. Buerger s. n.* [lectotype **here designated**: L (L0367483); isotype: K].

Herbs perennial, 20–80 cm; stems quadrangular, lower portion prostrate and rooting at nodes, upwards ascending, bifariously pubescent. Petiole 0.5–1.5 cm long; leaf blade lanceolate to ovate-oblong, (3–)5–13 cm long, (1–)2.5–4.5 cm broad, both surfaces sparsely pubescent especially along midvein, lateral veins 3–5 pairs, base obtuse to cuneate, margin entire or slightly sinuate, apex acuminate to caudate. Panicles of cymes terminal or axillar; cymes 1–3-flowered; peduncle 0.7–5.5 cm long. Involucral bracts ovate to elliptic, subequal to unequal in size, larger one 1.1–2.5 cm long, 0.5–1.8 cm broad, smaller one 8–17 mm long, 3–9 mm broad, abaxially subglabrous to puberulent, base cuneate to truncate, margin entire, apex acute to obtuse; bracteoles linear, 2–3 mm long, abaxially puberulent, apex acuminate. Calyx 3.5–4 mm long; lobes subulate, pubescent. Corolla white, pale pink, or light purple, with pink specks or dark lines on limb, 2–3.4 cm long, outside pubescent with eglandular trichomes; tube base cylindric, 1.3–1.5 cm long, 1–1.2 mm broad, upper lip in lower position elliptic to subcircular, 8–15 mm long, 3–11 mm broad, apex entire or slightly emarginate, lower lip in upper position oblong, 9–15 mm long, 4–7 mm broad, 3-lobed. Stamens 2; filaments 5–7 mm long, pubescent; anther-thecae linear, 1.2–2 mm long, parallel, unequally inserted at different height. Ovary pilose; style 1.8–2 cm long; stigma 2-cleft. Capsules 0.8–1.2 cm long, pubescent with glandular- or eglandular-pubescent. Seeds ovate in outline, compressed, 1.6–2 mm in diameter, tuberculate. **Chromosomes** 2n = 48. Figure 46-103.

Phenology: flowering from July to February; fruiting from July to October.

Habitats: roadsides, grasslands, stream banks, montane slopes, trailsides, evergreen broad-leaved forests; 700–1800 m.

Distribution: Yarlung Zangbo-Brahmaputra, N Hengduan, S Hengduan (Map 46-177); also in Japan, Vietnam, Laos, and other parts of Myanmar and China (S of Changjiang Valley).

Uses: medicine.

YARLUNG ZANGBO-BRAHMAPUTRA: **Cona**, Lebugou, 2538 m, 10 Sep. 2015, *Z. Zhou et al. ZJW-3910* (HON). **N HENGDUAN**: **Pingwu**, Laohegou, 1300 m, 26 Oct. 2014, *W. G. Wang et al. 2396* (PE); **Wenxian**, Bikou, Bishangou to Hejiagou, 700 m, 31 Aug. 1967, *C. L. Tang 1745* (PE, WUK). **S HENGDUAN**: **Baoxing**, Donghe, Yanjinping, 1800 m, 10 Sep. 1936, *K.L. Chu 3827* (IBSC, PE); **Tianquan**, 8 Sep. 1928, *W. P. Fang 3451* (A, E, PE); 5 Sep. 1939, *P. C. Tai & C. M. Teng 4380* (KUN, SZ).

28. **RUNGIA** Nees

Nees in Wallich (ed.), Pl. Asiat. Rar. 3: 77, 109. 1832; Clarke in Hooker (ed.), Fl. Brit. India 4: 545: 1885; Lindau in Engler & Prantl (eds.), Nat. Pflanzenfam. 4(3B): 331. 1895; Wood in Grierson & Long (eds.), Fl. Bhutan 2(3): 1290. 2001; Fl. Reipubl. Popularis Sin. 70: 255. 2002; Hu, Deng & Daniel in Wu, Raven & Hong (eds.), Fl. China 19: 443. 2011. Lectotype (designated by Heine, Fl. Gabon 13: 204. 1966.): *Rungia repens* (L.) Nees (≡ *Justicia repens* L.).

Herbs, decumbent to erect; with cystoliths. Leaves petiolate; blade margin entire. Inflorescences axillary or terminal spikes, secund or 4-sided, sometimes reduced to a single flower. Bracts 2- or 4-ranked but usually only 2 ranks fertile, homomorphic or heteromorphic, usually imbricate, margin usually scarious or hyaline; bracteoles similar to bracts or narrower. Calyx deeply 5-lobed almost to base; lobes equal or subequal. Corolla small; tube base cylindric, distally expanded; limb 2-lipped, lower lip spreading, 3-lobed, upper lip erect, rugula (stylar furrow) present, margin entire to emarginate; lobes ascending cochlear in bud. Stamens 2, inserted in corolla throat; anthers bithecous; thecae subequal in size, superposed, lower one white spurred at base. Ovary with 2 ovules per locule; stigma entire or slightly 2-cleft. Capsules shortly stalked at base, ovoid or oblong, 4-seeded; retinacula present; septa with attached retinacula separating from inner wall of mature capsule. Seeds compressed, verrucose, lacking trichomes.

Recent references: Hansen, 1989. Notes on SE Asian Acanthaceae 1. *Nordic J. Bot.* 9(2): 209-215; Hansen, 1989. *Justicia* and *Rungia* (Acanthaceae) in the Indo-Chinese Peninsula. In: Holm-Nielsen et al. (eds.), Tropical forests and botanical diversity. Academic Press. pp. 201–210; Hedren et al., 1995. Relationships in the Acanthaceae and related families as suggested by cladistic analysis of rbcL nucleotide

Figure 46-103: **Peristrophe japonica** (Thunb.) Bremek.: **a**, portion of plant; **b**, stamen; **c**, capsule with seed; **d**, capsule. Drawn by Quan LIN; reproduced from *Icon. Cormophyt. Sin.* 4: 172, fig. 5758. 1975.

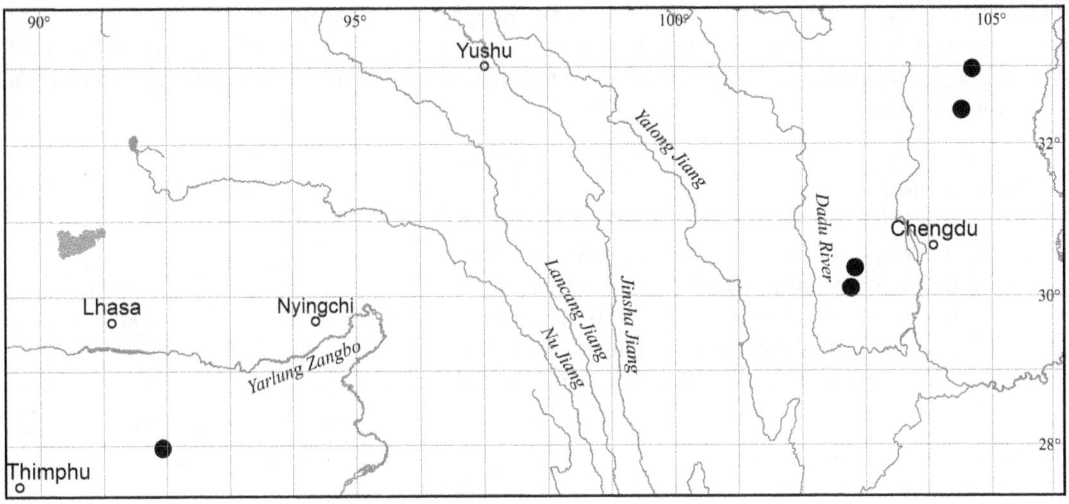

Map 46-177: **Peristrophe japonica** (Thunb.) Bremek.

sequences. *Plant Syst. Evol.* 194: 93–109; McDade et al., 2000. Phylogenetic relationships within the tribe Justicieae (Acanthaceae): Evidence from molecular sequences, morphology, and cytology. *Ann. MO Bot. Gard.* 87: 435–458; McDade & Moody. 1999. Phylogenetic relationships among Acanthaceae: Evidence from noncoding trnL-trnF chloroplast DNA sequences. *Amer. J. Bot.* 86: 70–80; Gao & Deng, 2007; Darbyshire et al., 2010. *Acanthaceae*, part 2. In: Beentje, H. (ed.), *Flora of tropical East Africa*. Richmond: Royal Botanic Gardens, Kew.A new section of *Rungia* Nees (Acanthaceae). *J. Trop. Subtrop. Bot.* 15: 549–560; Darbyshire & Vollesen, 2007. The transfer of the genus *Peristrophe* to *Dicliptera* (Acanthaceae), with a new species described from Eastern Africa. *Kew Bull.* 62: 119–128; Gao, 2010. *Phylogenetic relationships among Acanthaceae from China*. Ph. D. Thesis submitted to University of the Chinese Academy of Sciences. Guangzhou: South China Botanical Garden, Chinese Academy of Sciences; Wood, 2014. New names and combinations in Indian Acanthaceae. *Novon* 23: 385–395; Vollesen, 2015. Justicia. Pp. 162–224. In: Flora Zambesiaca vol. 8 part 6, Timberlake & Martins (eds.), Richmond: Royal Botanic Gardens, Kew.; Kiel et al., 2017. Unraveling relationships in the morphologically diverse and taxonomically challenging "justicioid" lineage (Acanthaceae: Justicieae). *Taxon* 66: 645–674; Lin & Deng, 2017. The identity of *Justicia burkilliana* (Acanthaceae). *Phytotaxa* 305: 293–298; Lin et al., 2020. Three new synonyms of *Rungia stolonifera* (Acanthaceae) from China and Vietnam. *Nordic J. Bot.* 38: e02649.

About 50 species: Old World tropics and subtropics; five species in the Pan-Himalaya.

Notes: The genus *Rungia* was originally placed in subtribe Diclipterinae and was considered to be related to *Dicliptera* as they have the same rising septa in ripe fruit (Nees, 1832, 1847; Bentham, 1876; Lindau, 1895; Hu, 2002). Based on pollen morphology, Raj (1961) suggested that *Rungia* should be included in Justicieae and some authors placed *Rungia* in subtribe Justiciinae and considered it to be close to *Justicia* (Anderson, 1867; Clarke, 1885; Bremekamp, 1965). Affinities between *Rungia* and *Justicia* was supported by recent studies of morphological and molecular evidence (Graham, 1988; McDade et al., 2000; Gao, 2010; Hu et al., 2011).

In some recent flora works, *Rungia* was merged with *Justicia* based mainly on morphological characters (Darbyshire & Vollesen, 2007; Darbyshire et al., 2010; Wood, 2014; Vollesen, 2015). However, recent molecular studies indicated that the genus *Justicia* is paraphyletic and *Rungia* is monophyletic nested with in broad *Justicia* (Hedren, 1995; McDade et al., 2000; Gao, 2010; Kiel et al., 2017). As stated by McDade & Moody (1999), once the overall diversity within *Justicia* is understood, it may be possible to recognize smaller monophyletic lineages that will help in organizing our knowledge of this unwieldy group. Therefore, before there is sufficient evidence to demonstrate that *Rungia* should be merged into *Justicia*, it seems more reasonable to treat *Rungia* as an independent genus for the present.

1a. Flowers solitary in apical leaf axils or in 2-sided spikes; bracts 2- or 4-ranked, all fertile (sect. I. **Stoloniferae**) .... 1. *R. stolonifera*
1b. Spikes 1-sided; bracts 4-ranked, each 2 fertile or sterile (Sect. II. **Rungia**).
  2a. Erect herbs; bracts acute at apex; corolla 3–8 mm long.
    3a. Bracts rhomboid or elliptic, margin entire .... 2. *R. burmanica*
    3b. Bracts broadly obovate, margin crenulate towards apex .... 3. *R. khasiana*
  2b. Prostrate or decumbent herbs, rooting at nodes; bracts rounded at apex; bracts 1.3–2 cm.
    4a. Spikes pedunculate; corolla pink or white, 1–1.2 cm long .... 4. *R. himalayensis*
    4b. Spikes sessile; corolla blue, 5–6 mm long .... 5. *R. pectinata*

Rungia sect. I. **Stoloniferae** C. M. Gao & Y. F. Deng, J. Trop. Subtrop. Bot. 15: 549. 2007. Type: *Rungia stolonifera* C. B. Clarke.

Flowers solitary in apical leaf axils or in 2-sided spikes. Bracts 2- or 4-ranked, all fertile.

Two species: Himalaya to Indochina; one species in the Pan-Himalaya.

1. Rungia stolonifera C. B. Clarke in Hooker (ed.), Fl. Brit. India 4: 547. 1885; Hu, Fl. Reipubl. Popularis Sin. 70: 265. 2002; Hu, Deng & Daniel in Wu, Raven & Hong (eds.), Fl. China 19: 445. 2011. ≡ *Diapedium stoloniferum* (C. B. Clarke) Kuntze, Revis. Gen. Pl. 2: 485. 1891. ≡ *Justicia stolonifera* (C. B. Clarke) H. Lév., Cat. PI. Yunnan.: 5. 1914–1915; Hansen, Nordic J. Bot. 9(2): 210. 1989, isonym. Type: India, Khasia, 24,000 ft., *J. D. Hooker & Thoms. s. n. "Diclipt-(14)"* [lectotype designated by Lin et al., Nodic J. Bot. 38: e02649. 2020: K (K000884202)].

Herbs, 30–60 cm high; stems rooting at basal nodes, quadrangular, sulcate, pubescent along groove. Petiole 5–10 mm long; leaf blade ovate, 2.5–7.5 cm long, 1.8–3.8 cm broad, subglabrous, lateral veins 5 or 6 pairs, base broadly cuneate, margin entire or undulate, apex acute. Spikes terminal or axillary. Bracts obovate to spatulate, 3–4 mm long, 2.5–3 mm broad, minutely pubescent, margin hyaline, apex rounded; bracteoles ovate-lanceolate, similar to bracts. Calyx 1–1.3 cm long, glabrous; lobes lanceolate, margin hyaline. Corolla 1.2–1.6 cm long; tube base cylindric, 2–3 mm broad, 6–8 mm long; limb 3.5–4 mm; upper lip triangular, erect, apex subentire, lower lip pendent, 3-lobed, lobes oblong and 1–1.2 mm long and broad. Stamens 2; filaments 4–5 mm long, glabrous; anther-thecae ovate, upper one muticous, lower one spurred at base. Ovary sparsely hirsute; style 1.5–1.8 cm long, exserted; stigma orbicular. Capsules 3–4 mm long, subglabrous. Seeds ovate in outline, compressed, 1.5–2 mm long, 1.2–1.5 mm broad, minutely glandular-papillose. Figure 46-104.

Phenology: flowering from September to October; fruiting from November to December.

Habitats: forests, trailsides; 1500–2000 m.

Distribution: S Hengduan (Map 46-178); also in Bangladesh, and other parts of India and China (E & NE Yunnan, Guizhou, Guangxi, Guangdong, Jiangxi, Fujian, Zhejiang).

Notes: Hansen (1989) incorrectly merged this species with *Rungia chinensis* Benth., which

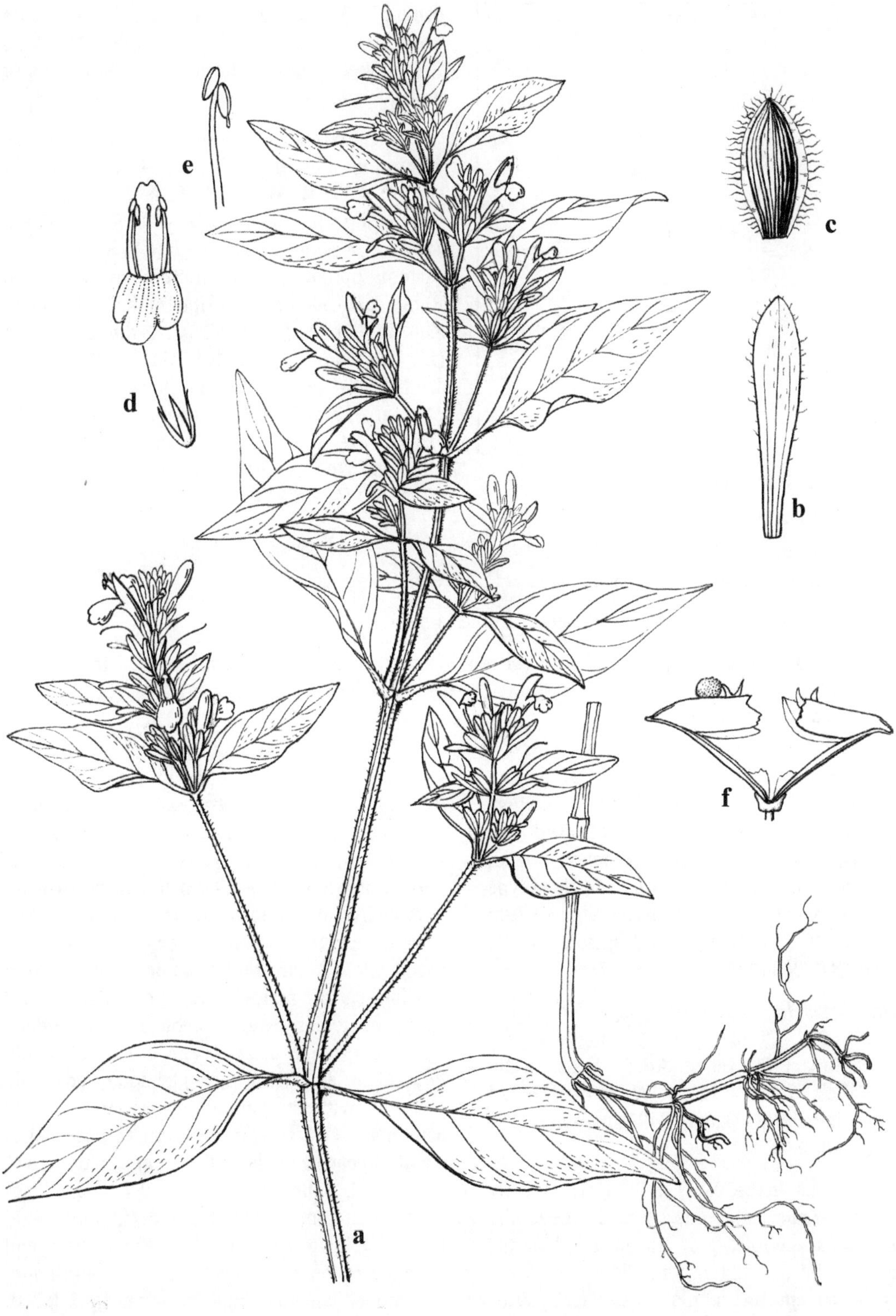

Figure 46-104: **Rungia stolonifera** C. B. Clarke: **a**, portion of plant; **b**, bract; **c**, bracteole; **d**, corolla with calyx; **e**, stamen; **f**, capsule. Drawn by Quan LIN; reproduced from *Icon. Cormophyt. Sin.* 4: 176, fig. 5766. 1975.

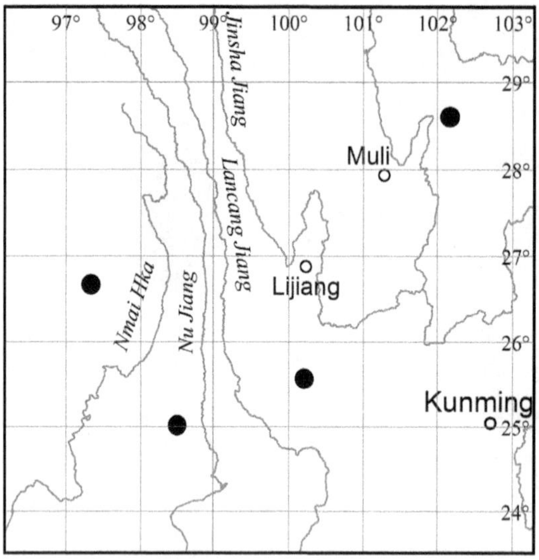

Map 46-178: **Rungia stolonifera** C. B. Clarke

occurs in S China and N Vietnam. This species can be easily distinguished from the latter by its four ranked bracts all fertile while two ranks of bracts fertile and two ranks sterile in the latter.

S HENGDUAN: Dali, *s. coll. 2606* (SYS); **Mianning**, Poshang, 1900 m, *T. T. Yü 18091* (KUN). **U IRRAWADDY: Putao**, Sumprabum, Ndum-Zup to Hpuginhku, 1800–1950 m, 30 Dec. 1961, *J. Keenan et al. 3093* (E); Sumprabum, eastern approaches from Sumprabum to Kumon Range, pathside between Hpuginhku and N'Dum Zup, 1500 m, 22 Feb. 1962, *J. Keenan et al. 3621* (E); **Tengchong**, Qushi, Baijiahe, 1510 m, *W. Q. Yin 60-1204* (KUN); Qushi, 10 Sep. 2013, *Y. F. Deng et al. 24682* (IBSC); Shweili-Salwin Divide, *G. Forrest 11787* (IBSC); Mazhan, Huoshan, 1885 m, 29 Oct. 2015, *Tengchong Expedition 520522151029859LY* (IMDY).

Rungia sect. II. **Rungia**. Type: *Rungia repens* (L.) Nees.

Spikes 1-sided. Bracts 4-ranked, each 2 fertile or sterile.

About 50 species: tropical and tropical regions of Asia and Africa; four species in the Pan-Himalaya.

2. Rungia burmanica (C. B. Clarke) B. Hansen, Nordic J. Bot. 9: 211. 1989. ≡ *Justicia burmanica* C. B. Clarke in Hooker (ed.), Fl. Brit. India 4: 527. 1885. Type: Myanmar. s. loc., s. d., *Herb. Griffith 6184* [lectotype designated by Lin & Deng, Phytotaxa 395: 293. 2017: K (K000884074); isolectotype: K (K000884073)].

Justicia burkilliana J. Bhattachaya, Kew Bull. 57: 231. 2002. Type: Yarlung Zangbo-Brahmaputra, outer Arbor Hills, Lalik Valley, 6 Jan. 1912, *I. H. Burkill 36725* [holotype: K (K000884076); isotype: K (K000884075)].

Perennial herbs, 0.5–1 m high; stems terete or subquadrate, glabrous. Leaves generally isophyllous; petiole 2–6 cm long, glabrous; blade elliptic to elliptic-rhomboidal, 10–20 cm long, 4–10 cm broad, glabrous, lateral veins 5–12 pairs, base cuneate to shortly attenuate, margin entire or subcrenulate, apex acute. Spikes terminal or axillary, 5–14 cm long; peduncle 0.5–1.5 cm long, glabrous; rachis quadrangular, glabrous or puberulent. Bracts 4–ranked, 2 sterile and 2 fertile; sterile bracts broadly obovate, 1.9–2.2 cm long, 1.7–1.9 cm broad, puberulent, without hyaline scarious margin, margin crenulate towards apex; fertile bracts obovate, 1.5–1.7 cm long, 0.9–1.2 cm broad, puberulent, without hyaline scarious margin, margin crenulate toward apex; bracteoles ovate-lanceolate to lanceolate, 1–1.6 cm long, 0.2–0.4 cm broad, abaxially puberulent and with tiny glandular, adaxially glabrous, without hyaline scarious margin, apex acuminate. Calyx 5-lobed almost to base; lobes equal, linear-lanceolate, 0.8–1 cm long, 0.8–1 mm broad, puberulent. Corolla 1.8–2 cm long, white with purple dotted stripes in throat and lower lip, glabrous; tube cylindric at base and enlarged in throat sharply, 1–1.2 cm long; upper lip unlobed, attenuate at apex, lower lip 3-lobed, lobes rounded at apex. Stamens 2; filaments 4–5 mm long, glabrous; anthers bithecous, thecae ovoid, superposed, upper one pubescent, lower one glabrous, spurred at base. Ovary glabrous, 1.2–1.5 mm long; 2 ovules per locule; style 0.8–1 cm long, slightly puberulent; stigma minutely 2-lobed. Capsules clavate, 1–1.2 cm long, 2.5–3 mm broad, glabrous, basal solid portion 5–7 mm long, stipitate, pubescent, apex obtuse. Seeds suborbicular, 2–3 mm long, 1.5–2 mm broad, compressed, dark brown, glabrous, surface verrucose. Figure 46-105.

Phenology: fruiting from November to January of next year.

Habitats: intermediate forests on rocky valley slopes; 700–1400 m.

Distribution: Yarlung Zangbo-Brahmaputra and U Irrawaddy (Map 46-179); also in other parts of Myanmar and China (S Yunnan).

YARLUNG ZANGBO-BRAHMAPUTRA: **Arbor Hills**, Lalik Valley, 6 Jan. 1912, *I. H. Burkill 36725* (K); s. loc., 22 Dec. 1911, *I. H. Burkill 37328* (CAL). **U IRRAWADDY**: **Putao**, Sumprabum, 25 Jul. 1953, *Tha Hla & Chit Ko Ko 4313* (K).

3. Rungia khasiana T. Anderson, J. Linn. Soc., Bot. 9: 518. 1867, non *Justicia khasiana* C. B. Clarke, Fl. Brit. India 4: 537. 1885; Clarke in Hooker (ed.), Fl. Brit. India 4: 548. 1885. ≡ *Diapedium khasiana* (T. Anderson) Kuntze, Revis. Gen. Pl. 2: 485. 1891. ≡ *Justicia meghalayana* J. R. I. Wood, Novon 23: 388. 2014. Type: India, s. d., *Griffith s. n.* [lectotype designated by Wood, Novon 23: 388. 2014: CAL (CAL0000020309)].

Herbs, 30–50 cm high; stems decumbent at base, upwards ascending, densely pilose. Leaves equal or unequal; petiole 0.5–2 cm long, pubescent; blade ovate, elliptic, or oblong, 2–10 cm long, 1–3 cm broad, pilose on both surfaces, lateral veins 4–6 pairs, base cuneate, margin entire, apex acute. Spikes terminal or axillary, 3–7 cm long; peduncles 0.5–2 cm long, pilose; rachis pilose. Bracts 4 ranks, 2 sterile and 2 fertile; steril bracts rhomboid, 8–10 mm long, 4–5 mm broad, densely pilose and glandular-pubescent, membranous margin ca. 1 mm broad, apex acute, margin entire and ciliate; fertile bracts elliptic, 8–9 mm long, 3–3.5 mm broad, densely pilose and glandular-pubescent, membranous margin 0.8–1 mm broad, margin entire and ciliate, apex acuminate; bracteoles lanceolate, 5–5.5 mm long, 1–1.2 mm broad, densely pilose and glandular-pubescent. Calyx densely pilose and glandular-pubescent, 5-lobed almost to base; lobes equal, linear, 4–5 mm long, 0.5–0.6 mm broad. Corolla 1.1–1.3 cm long, white with red stripes , outside pilose and glandular-pubescent; tube cylindric, 6–8 mm long; upper lip 2-lobed, lobes acute at apex, lower lip 3-lobed, lobes rounded. Stamens 2; filaments 2.5–3 mm long, glabrous, lower anther-thecae spurred. Ovary glabrous, 1.2–1.5 mm long; style 8–9 mm long, pilose on lower portion. Capsules ellipsoid, 5–7 mm long, stalked at base, densely pilose. Seeds suborbicular, 1.5–1.8 mm long and broad, rugose. **Chromosomes** 2n = 20.

Phenology: flowering from January to March; fruiting from February to April.

Habitats: moist places in the forests; 700–1800 m.

Distribution: U Irrawaddy (Map 46-180); also in other parts of India, Myanmar and China (S Yunnan, Hainan).

U IRRAWADDY: **Chaungmaw Kyaun**, North Triangle, Htangnam, 1050 m, 25 Mar. 1953, *F. Kingdon-Ward 22074* (A).

4. Rungia himalayensis C. B. Clarke in Hooker (ed.), Fl. Brit. India 4: 548. 1885; Wood in Grierson & Long (eds.), Fl. Bhutan 2(3): 1291. 2001. ≡ *Diapedium himalayense* (C. B. Clarke) Kuntze, Revis. Gen. Pl. 2: 485. 1891. ≡ *Justicia himalayensis* (C. B. Clarke) J. R. I. Wood, Novon 23: 387. 2014. Type: [Sikkim & Darjiling] India. "Sikkim, Darjiling", 3000–4000 ft., 22 Oct. 1869, *C. B. Clarke 9747* [lectotype designated by Wood, Novon 23: 387. 2014: K (K000884066)].

Perennial herbs; stems decumbent on lower portion, upwards ascending, rooting at lower nodes, sparsely puberulent. Petiole 3–14 mm long; leaf blade ovate or elliptic, often suffused with red, 1.0–6.0 cm long, 0.5–1.5 cm broad,

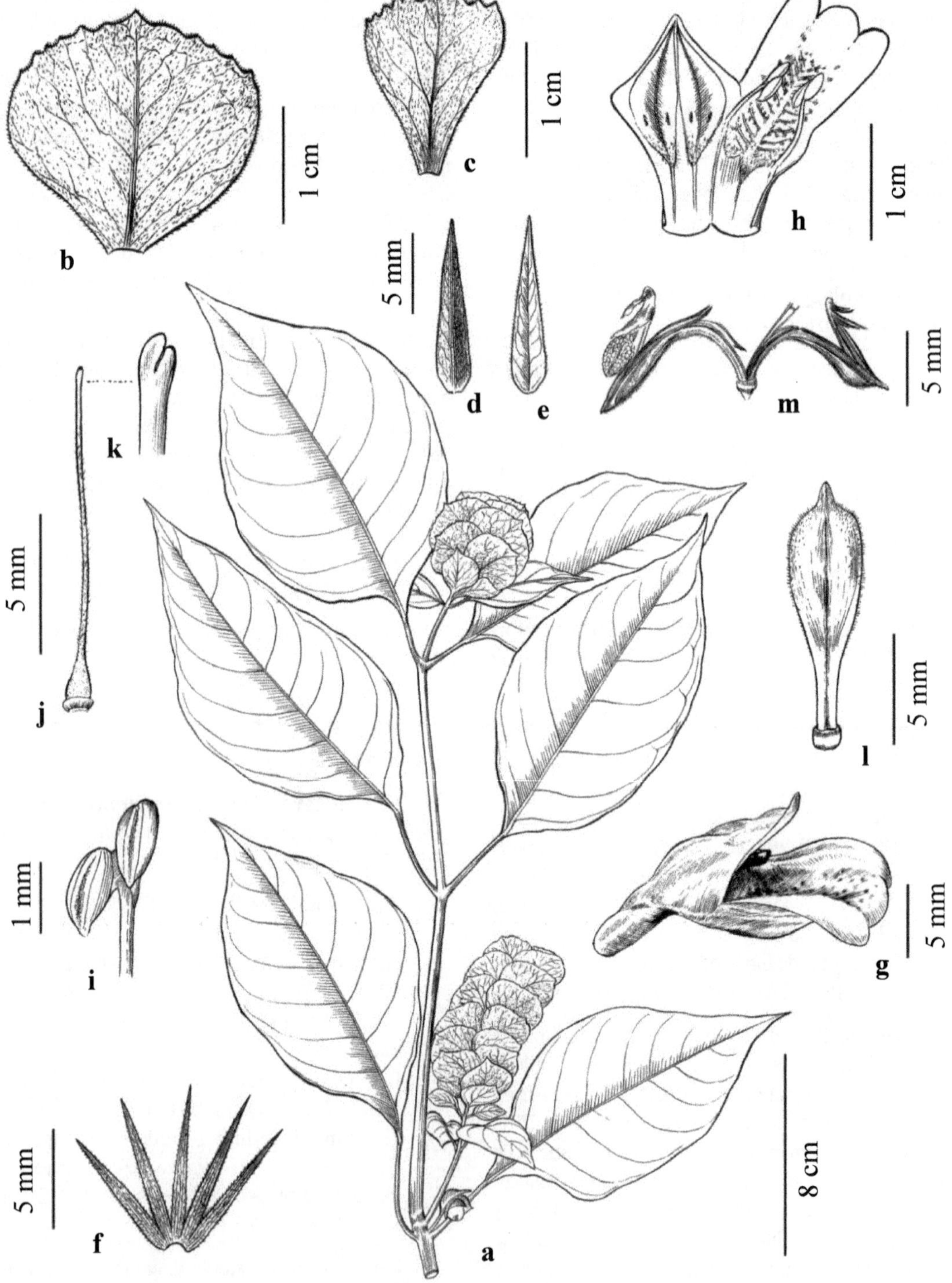

Figure 46-105: **Rungia burmanica** (C. B. Clarke) **B.** Hansen: **a**, flowering branch; **b**, sterile bract; **c**, fertile bract; **d**, abaxial view of bracteole; **e**, adaxial view of bracteole; **f**, calyx; **g**, corolla; **h**, corolla opened out; **i**, stamen; **j**, pistil; **k**, stigma; **l**, capsule; **m**, dehiscent capsule. Drawn by Ding-Han CUI; reproduced from Lin & Deng, *Phytotaxa* 305: 296, fig. 3. 2017.

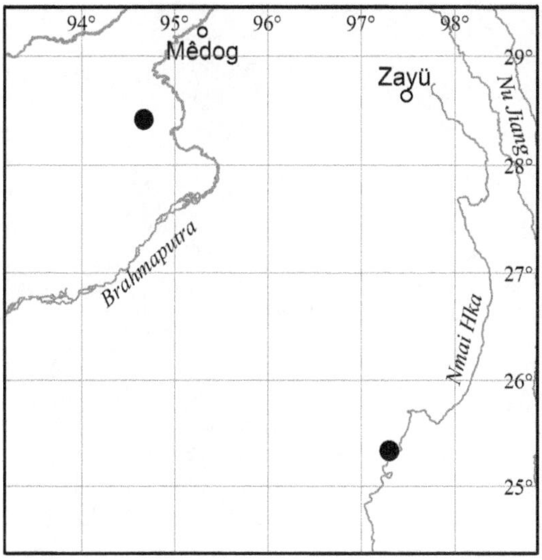

Map 46-179: **Rungia burmanica** (C. B. Clarke) B. Hansen

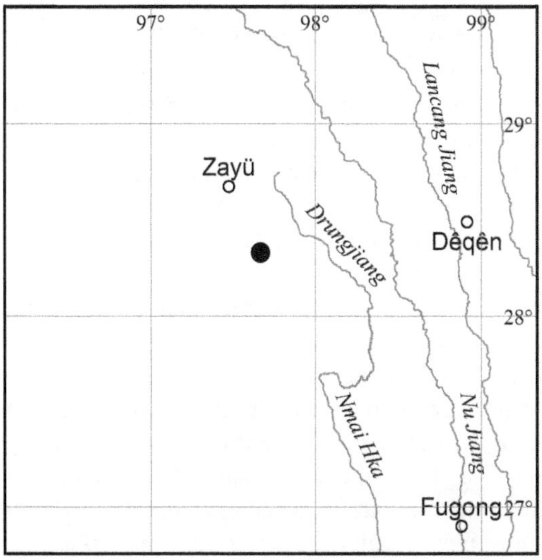

Map 46-180: **Rungia khasiana** T. Anderson

sparsely scabrid-pubescent, base shortly cuneate onto petiole, margin undulate or crenulate, apex acute. Spikes terminal and axillary spikes, 1–4 cm long, secund; peduncle (0–)1–3 cm long. Bracts all similar, ovate, 5–8 mm long, 2–4 mm broad, pubescent, membranous margin scarious, often pinkish, apex aristate; bracteoles linear, 4.5–5.5 mm long, 1–1.2 mm broad, pubescent, apex acuminate. Calyx 3.5–5 mm long, pubescent; lobes linear, margin ciliate, apex acute. Corolla white with pink spots on lower lip, 1–1.2 cm long, sparsely pubescent; tube 6–8 mm long; upper lip oblong, 2.5–3 mm long, 2-lobed, lobes oblong, lower lip 3–4 mm long, 3-lobed, lobes orbicular. Style 5–6 mm long. Capsules oblong, 3–4 mm long, pubescent. Seeds ovate in outline, compressed, 1.5–2 mm long and broad, verrucose, appearing punctate.

Phenology: flowering and fruiting from September to April of next year.

Habitats: open forests, shaded rocky slopes, roadsides; 100–1300 m.

Distribution: Jammu & Kashmir, U Ganga & Indus, W Nepal, E Nepal, Sikkim & Darjiling, Bhutan (Map 46-181); also in other parts of NE India.

JAMMU & KASHMIR: **Reasi**, 300 m, 28 Oct. 1960, *A. K. Dutt 6441* (RRLH). **U GANGA & INDUS**: **Chamba**, *J. F. Duthie 9680* (E). **W NEPAL**: **Bheri**, Banke, Baijapur, 180 m, *R. N. Shukla 3321* (KATH); Banke, Chatar, 30 Nov. 1973, *N. P. Manandhar & S. B. Malla 17147* (KATH); Banke, Napelganj, 162 m, 12 Nov. 2012, *B. Adhikari & G. D. Bhatt 49* (E, KATH); Bardiya, Dodhari, 180 m, 15 Jan. 1983, *R. N. Shukla 5133* (KATH); Bardiya, Zabdi, 180 m, 6 Oct. 1980, *T. K. Bhattachaya 3205* (KATH); Bardiya, Zaladi, 27 Jul. 1981, *R. J. Shah 4612* (KATH). **E NEPAL**: **Mechi**, Panchthar, Yektin-Akasay, Nov. 30, 1963, *H. Hara et al. 6303599* (A); Panchthar, Yektin-Akasay, 1100 m, 29 Nov. 1963, *H. Hara et al. 6303560* (K, KATH, KYO). **SIKKIM & DARJILING**: **Darjiling**, 900 m, 22 Oct. 1869, *C. B. Clarke 9661* (K); 1200 m, 22 Oct. 1869, *C. B. Clarke 9747* (K); **Sikkim**, 1200 m, *J. D. Hooker s. n.* (K). **BHUTAN**: **Gaylephug**, above Sarbhang High School, 7 Feb. 1988, *J. R. I. Wood 6133* (E); **Tongsa**, between Tintibi and Dakpai, 900 m, 15 Mar. 1992, *J. R. I. Wood 7548* (E); below Pertimi, 1300 m, 5 Apr. 1982, *Grierson & Long 4337* (E, K).

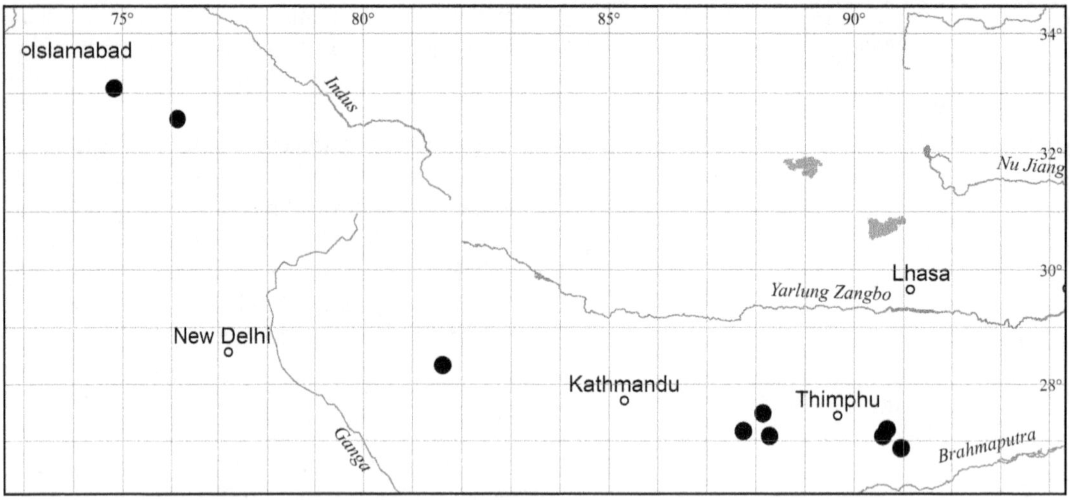

Map 46-181: **Rungia himalayensis** C. B. Clarke

5. **Rungia pectinata** (L.) Nees in A. Candolle, Prodr. 11: 470. 1847; Wood in Grierson & Long (eds.), Fl. Bhutan 2(3): 1291. 2001; Hu, Fl. Reipubl. Popularis Sin. 70: 262. 2002; Hu, Deng & Daniel in Wu, Raven & Hong (eds.), Fl. China 19: 445. 2011. ≡ *Justicia pectinata* L., Amoen. Acad. 4: 299. 1760. ≡ *Diapedium pectinatum* (L.) Kuntze, Revis Gen. Pl. 2: 484. 1891. ≡ *Rungia parviflora* subsp. *pectinata* (L.) L. H. Cramer, Revis. Handb. Fl. Ceylon 12: 105. 1998. ≡ *Rungia parviflora* var. *pectinata* (L.) C. B. Clarke in Hooker (ed.), Fl. Brit. India 4: 550. 1885; Trimen, Handb. Fl. Ceylon 3: 343. 1895. Type: *Herb. LINN No. 28.17* (lectotype designated by Cramer in Dassanayake & Clayton, Revised Handb. Fl. Ceylon 12: 105. 1998: LINN).

Justicia parviflora Retz., Obs. 5: 9. 1789. ≡ *Rungia parviflora* (Retz.) Nees in Wallich (ed.), Pl. Asiat. Rar. 3: 110. 1832; Nees in de Candolle (ed.), Prodr. 11: 469. 1847; Clarke in Hooker (ed.), Fl. Brit. India 4: 550. 1885. ≡ *Diapedium pectinata* var. *parviflora* (Retz.) Kuntze, Revis. Gen. Pl. 2: 484. 1891. Type: India, *Koenig s. n.* (microfiche folo Koenig 10 III 3-4) [lectotype **here designated**: C (C10005150)].

Justicia infracta Vahl, Enum. Pl. 1: 155. 1804. Type: *Herb. Vahl. s. n.* (microfiche photo Vahl nr. 41 I, 5-6) [holotype: C (C10005149)].

Herbs, annual or perennial, 20–50 cm high; stems basally prostrate and rooting at nodes, upwards ascending, sparsely pubescent. Petiole 0–7 mm long; leaf blade oblong-elliptic, 1–4 cm long, 0.4–1.4 cm broad, sparsely pilose along veins, lateral veins ca. 4 pairs, base cuneate and decurrent onto petiole, margin entire, apex acute. Spikes axillary or terminal, 0.5–2 cm long, 1-sided, solitary or sometimes 2 or 3 compound. Bracts dimorphic; sterile bracts green, elliptic, 3.5–4 mm long, 0.5–0.7 mm broad, glabrous, apex acute; fertile bracts circular to obovate, 4–5 mm long, 1.5–2 mm broad, pubescent, margin broadly hyaline, apex obtuse to rounded to emarginate and mucronulate; bracteoles elliptic, 2–3 mm long, margin ciliate, apex 2-cleft and subacute. Calyx 3–4 mm long, colorless, pubescent; lobes linear-lanceolate, 0.4–0.5 mm broad, margin narrowly hyaline, apex mucronulate. Corolla blue or white, 5–6 mm long, outside pubescent; upper lip ovate, 1–2 mm long, apex emarginate, lower lip 3-lobed, lobes triangular. Stamen 2; filaments glabrous. Ovary glabrous. Capsules ellipsoid, ca. 2.5 mm long, glabrous, 2–4-seeded, apex apiculate. Seeds orbicular in outline, 0.8–1 mm in diameter, minutely verrucose. **Chromosomes** 2n = 26. Figure 46-106.

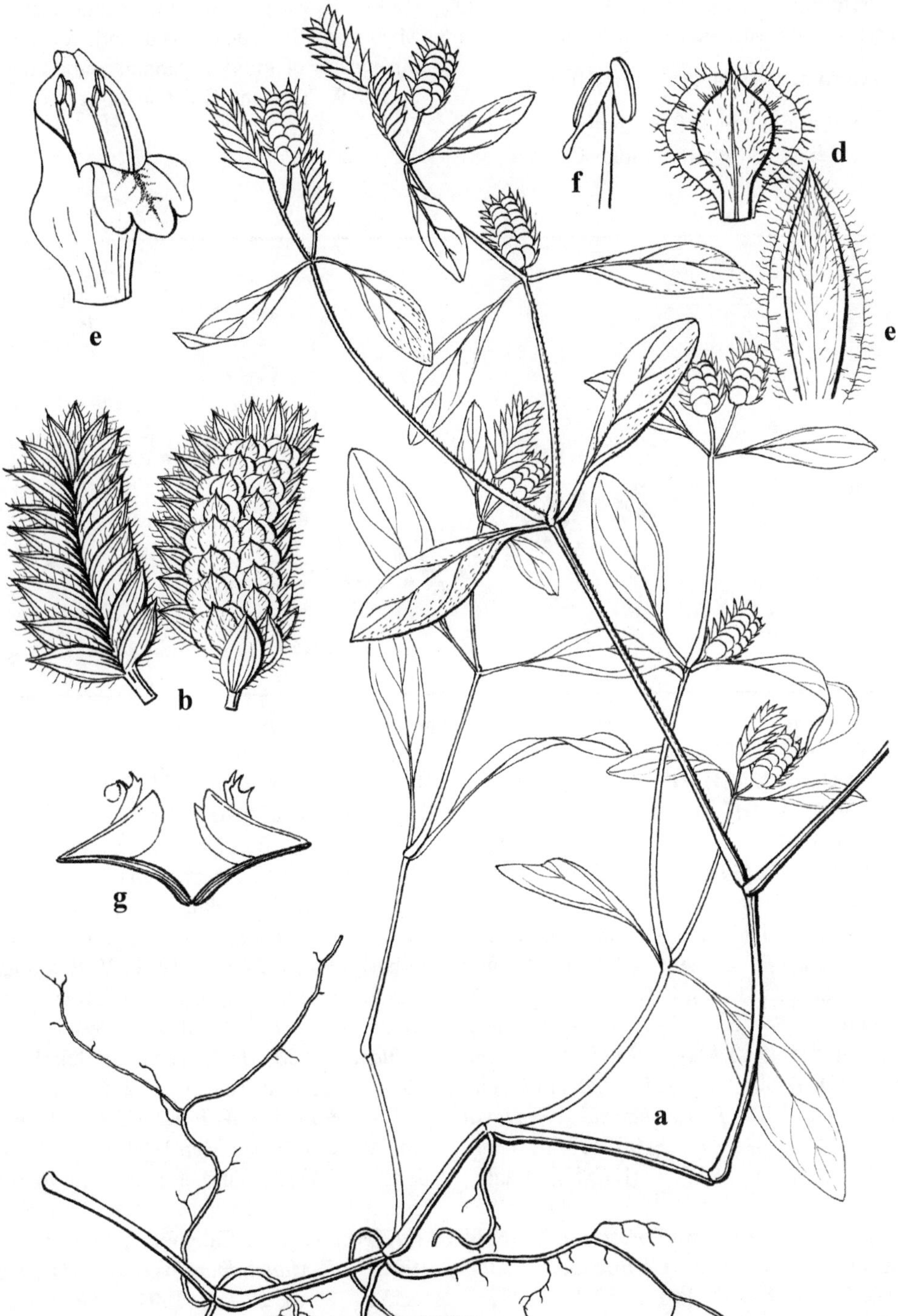

Figure 46-106: **Rungia pectinata** (L.) Nees: **a**, habit; **b**, spikes; **c**, sterile bract; **d**, fertile bract; **e**, corolla and stamens; **f**, stamen; **g**, dehisced capsule. Drawn by Quan LIN; reproduced from *Icon. Cormophyt. Sin.* 4: 177, fig. 5767. 1975.

Phenology: flowering from November to January; fruiting from January to April.

Habitats: wastelands; below 2100 m.

Distribution: ammu & Kashmir, U Ganga & Indus, C Nepal, E Nepal, Bhutan, U Irrawaddy (Map 46-182); also in Bangladesh, Laos, Myanmar, Sri Lanka, Thailand, Vietnam, and other parts of India, Myanmar and China (C, E, S & SW Yunnan, Guizhou, Guangxi, Hainan, Guangdong).

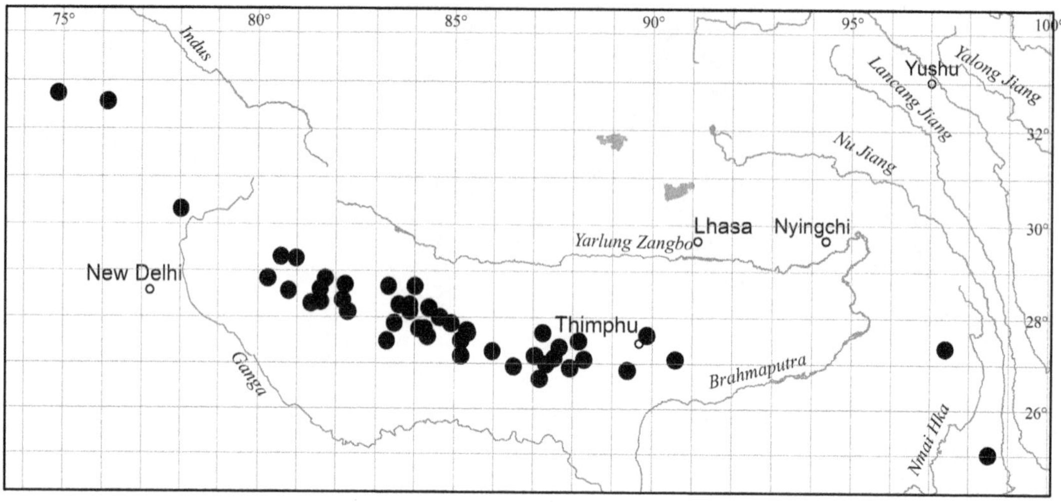

Map 46-182: **Rungia pectinata** (L.) Nees

JAMMU & KASHMIR: **Jammu**, Nandani, 1 Oct. 1969, *B. M. Sharma 56* (RRLH). **U GANGA & INDUS**: **Chamba**, 14 Dec. 1889, *J. F. Duthie 9682* (LE); **Dehradun**, Thano, 720 m, Jan. 1900, *Upendranath Kanjilal 804* (K); Sciwalihs, Nov. 1895, *J. S. Gamble 25593* (K); **Mussorie**, *M. B. Saxena 2338* (DD, fide Raizada & Saxena, 1978); **Sirmaur**, 1885, *J. R. Drummond 20804* (E, K). **W NEPAL**: **Bheri**, Banke, Napelganj, 18 May 1975, *L. R. Sharma 824* (KATH); Banke, Napelganj, Salyani Bagia, 180 m, 20 Jan. 1982, *D. K. Sharma & S. Dayal 4661* (KATH); Banke, Napelganj, 180 m, 6 Jan. 1993, *P. P. Kurmi KB513* (KATH); Banke, Napelganj, 30 Oct. 1972, *N. P. Manandhar 9864* (KATH); Banke, Mahadewa Forests, 16 km N of Napelganj, 4 Jan. 1973, *Chandrabali 10386* (KATH); Banke, Kamdi, 190 m, 12 Jan. 1982 *s. coll. W40/82* (KATH); Bardiya, Takia, 180 m, 3 Jan. 1978, *R. N. Shukla 5166* (KATH); Bardiya, Srinagar, 26 Feb. 1983, *J. Ram & R. K. Pradad 2253* (KATH); Bardiya, Bardiya National Park, 180 m, 29 Jan. 1995, *P. P. Kurmi & M. P. Pudaraini 10038* (KATH); Bardiya, Babai to Harre, 9 Dec. 1983, *T. K. Bhathacharya 2549* (KATH); Bardiya, Deorali to Babai, 20 Feb. 1980, *T. K. Bhathacharya 3412* (KATH); Bardiya, Ganga Patia, 180 m, 27 Nov. 1985, *T. K. Bhathacharya & P. P. Kurmi 2932* (KATH); Dailekh, Gauri, 900 m, 26 Feb. 1991, *N. P. Manandhar 465-91* (KATH); Dailekh, Lohari, 790 m, 27 Feb. 1991, *N. P. Manandhar 505-91* (KATH); Jajarkot, Bheri River, 760 m, 1 Apr. 1952, *O. V. Polunin, W. R. Sykes & L. H. J. Williams 1828* (BM, KATH); Jajarkot, Gad Rangchi, 1670 m, 16 Oct. 1952, *O. V. Polunin, W. R. Sykes & L. H. J. Williams 5734* (BM); Jajarkot, Khiyapani, 720 m, 3 Mar. 1991, *N. P. Manandhar 683-91* (KATH); Surkhet, Chen Chu, 640 m, 20 Dec. 1982, *P. P. Kurmi & D. K. Shrestha 4958* (KATH); Surkhet, Salighopte, 700 m, 21 Feb. 1991, *N. P. Manandhar 56-91* (KATH); Surkhet, Hunke, 780 m, 22 Feb. 1991, *N. P. Manandhar 96-91* (KATH); Surkhet, Kalyan, 870 m, 23 Feb. 1991, *N. P. Manandhar 241-91* (KATH); **Mahakali**, Dadeldhura, Runjatar to Okhaldhunga, 1400 m, 29 May 1979, *N. P. Manandhar & M. K. Adhikari 1954* (KATH); Kanchapur, Mahendra Nagar, 150 m, 14 Jan. 1983, *H.L. Rijal & B. R.*

Chhetri 1386 (KATH); **Rapti**, Dang, Tulsipur, 16 Apr. 1985, *P. M. Regmi 4325* (KATH); Dang, Rapti Nadi, 300 m, 24 Mar. 1952, *O. V. Poulnin, W. R. Sykes & L. H. J. Williams 3613* (BM); Dang, Koilabas, 230 m, 31 Oct. 1975, *M. K. Ghimire 112* (KATH); Dang, Koilabas, 23 Mar. 1973, *N. P. Manandhar 10302* (KATH); Dang, Koilabas, 180 m, 22 Feb. 1982, *K. J. Malla 233* (KATH); Salyan, Phalabang, 1210 m, 28 Mar. 1952, *O. V. Polunin, W. R. Sykes & L.H.J. 691* (BM); Salyan, Sarda Khola, 1060 m, 30 Mar. 1952, *O. V. Polunin, W. R. Sykes & L. H. J. 1812* (BM); **Seti**, Doti, Doti-Siligarhi, 1350 m, 27 Mar, 1967, N. Ecker-Racz 3 (LE); Kailali, Dhangarhi, 2 Apr. 1974, *N. P. Manandhar 47* (KATH); Kailali, Dhangarhi, 200 m, 27 Nov. 1966, *D. H. Nicolson 2792* (BM, KATH). **C NEPAL**: **Bagmati**, Dhading, Boksiling, 920 m, 3 Dec. 1988, *N. P. Manandhar 12806* (KATH); Dhading, Birjet, 1620 m, 4 Nov. 1989, *N. P. Manandhar 12929* (KATH); Kathmandu, Swayambhu, 1300 m, 7 Mar. 1970, *N. P. Manandhar 9410* (KATH); Kathmandu, Sundarijal Wajar Reservoir, 1463 m, 21 May 1976, *M. M. Amatya et al. 241/76* (KATH); Kathmandu, Gokarna, 14 Mar. 1966, *R. Mapa & P. Pradhan 4388* (KATH); Kathmandu, Nagarjun, 1500 m, 14 Jan. 1964, *Saman & Shrestha 661* (KATH); Lalitpur, Lele Bhanjyang, 1540 m, 28 December 1989, *I. Shrestha & N. Pradhan 63* (KATH); Lalitpur, Educational Garden, 1515 m, 12 Feb. 2002, *N. Joshi & R. K. Bista 430* (KATH); Lalitpur, Sanga, Bhanjyang, 1570 m, 4 Mar. 1975, *D. P. Joshi & K. R. Rajbhandari 75/342* (KATH); Lalitpur, Godavari, 1590 m, 24 Nov. 1974, *I. Sharma & M. Sharma 74/14* (KATH); Nuwakot, Samri Khola, 760m, 7 Apr. 1953, *P. C. Gaedner 163* (BM); Nuwakot, Bhaing to Pairegaon, Trisuli, 650–760 m, 15 Feb. 1975, *M. M. Amatya & K. R. Rajthandari 40/75* (KATH); Sindhupalchok, Lamosangu to Dhulopalahra, 720–1750 m, 18 Apr. 1978, *I. Sharma et al. 78/109* (KATH); **Dhawalagiri**, Baglung, 910 m, 20 Apr. 1954, *J. D. A. Stainton, W. R. Sykes & L. H. J. Williams 59* (BM, KATH); Myagdi, Myagdi Khola, 970 m, 22 Apr. 1954, *J. D. A. Stainton, W. R. Sykes & L. H. J. Williams 122* (BM); Myagdi, above Rithum Khola, 1810 m, 6 May 1954, *J. D. A. Stainton, W. R. Sykes & L. H. J. Williams 2731* (A, BM, KATH); Myagdi, Myagdi Khola, 1360 m, 28 Oct. 1954, *J. D. A. Stainton, W. R. Sykes & L. H. J. Williams 9179* (BM, KATH); Myagdi, Ranabang, 1000 m, 27 May 1992, *N. P. Manandhar & S. K. Acharya 363/92* (KATH); **Gandaki**, Gorkha, Taku Thele, 760 m, 11 May 1987, *N. P. Manandhar & L. P. Katel 11672* (KATH); Gorkha, Pandup Deorali, 1100 m, 12 May 1987, *N. P. Manandhar & L. P. Katel 11738B* (KATH); Gorkha, Dollu to Putalidanda, SW of Kathmmandu, 1580–1900 m, 7 Jan. 1975, *D. P. Joshi et al. 75-109* (KATH); Kaski, Pokhara, Pochbhaiya, Pupa Lake, 900 m, 17 Dec. 1973, *D. P. Joshi & M. M. Amatya 73/1223* (KATH); Kaski, Pokhara, on the bank of Phewa Tel, 15 Nov. 1985, *N. P. Manandhar 10449* (KATH); Kaski, Lumle, 1705 m, 20 Jan. 1970, *G. Dawson 303* (BM); Kaski, Dhampus, 2030 m, 22 Jan. 1969, *H. Flatt 2* (BM); Lamjung, Midim Khola, 4 Nov. 1969, *T. Wraber 629* (BM); Manang, Sabji Khola, 600 m, 14 Apr. 1974, *J. F. Dobremez 2492 (N. P. Nanandhar 74-121)* (BM, E, KATH); Tanahu, Sisaghat–Deorali-Deorali Phedi–Satmuhne, 14 Jan. 1970, *H. Kanai et al. 670640* (KATH); **Janakpur**, Dolakha, Kiratechap to Yarnsa, 25 Apr. 1963, *M. Numata 526* (KYO); Sindhuli, Malkot, 1450 m, 21 Nov. 1988, *N. P. Manandhar 12293* (KATH); Sindhuli, Khurkot, 1200 m, 22 Nov. 1988, *N. P. Manandhar 12381* (KATH); **Lumbini**, Nawalparasi, Tribeni, 20 Feb. 1956, *K. de B. Codrington 338* (BM); Nawalparasi, Daunne, 265 m, 8 Nov. 2012, *B. Adhikari & G. D. Bhatt 21* (E); Palpa, Masem, 25 Nov. 1973, *D. P. Joshi & M. M. Amatya 73-993* (KATH); Palpa, Basani Khola, 1680 m, 1 Mar. 1976, *N. P. Manandhar & P. M. Regmi 59* (KATH); Palpa. Dumre, 21 Apr. 1963, *N. Numata 239* (KYO); Rupandehi, Butwal, 460 m, 6 Oct. 1954, *J. D. A. Stainton, W. R. Sykes & L. H. J. Williams 8793* (A, BM, KATH); Rupandehi, Butwal, Charange Khola, 300 m, 26 Feb. 1974, *D. P. Joshi & M. M. Amatya 74/1253* (KATH); Rupandehi, Rummindel, SW Butwal, 1956, *K. de B. Codrington 363* (BM); Rupandehi, Tilorakot, 20 Feb. 1956, *K. de B. Codrington 371* (BM); Rupandehi, Butwal, 160 m, 20 Dec. 1975, *D. Joshi et al. 75/3631* (KATAH); **Narayani**, Bara, Badahikim Bridge, from Hetauda to Birgunj, 480 m, 26 Nov. 2004, *T. Miller et al. (DNEP2)*

A147 (E, KATH); Bara, near Badahikim Bridge, neat Badahikim Bridge, from Hetauda to Birgunj, 670 m, 27 Nov. 2004, *T. Miller et al. (DNEP2) A171* (E, KATH); Bara, Pragati Nagar, 16 Feb. 1993, *L. P. Kattel 178/93* (KATH); Bara, Gami, 130 m, 18 Feb. 1993, *L. P. Kattel 194/93* (KATH); Chitwan, Ram Nagar, 500 m, Jun. 1994, *K. Wesche 1170* (BM); Chitwan, Kashara, 190 m, 16 Dec. 1976, *H. Tabata et al. 7622* (KATH, KYO); Chitwan, Chitwan National Park, Tamor Tal, 140 m, 19 Nov. 2004, *C. Pendry et al. (DNEP2) B1* (E, KATH); Chitwan, Chitwan National Park, E Sukhibar Post, 140 m, 23 Nov. 2004, *C. Pendry et al. (DNEP2) B100* (E, KTH), Chitwan National Park, Lake on track to Bankadta from Sukibhar Post, 140 m, 23 Nov. 2004, *T. Miller et al. (DNEP2) A101* (E, KATH); Chitwan, Chitwan National Park Buffer Zone, Bis Hajar Tal (Twenty Southsan Lake), 180 m, 24 Nov. 2004, *T. Miller et al. A110* (E, KATH); Chitwan, Sauraha, 160 m, 13 Jan. 1996, *M. Suzuki et al. 9614025* (KATH); Chitwan, Sauraha to Padampur, 160 m, 16 *Jan.* 1996, *M. Suzuki et al. 9614078* (E, KATH); Chitwan, Tungati, 380 m, 15 Nov. 1989, *N. P. Manandhar 13875* (KATH); Chitwan, Tungati, 380 m, 15 Nov. 1989, *N. P. Manandhar 13894* (KATH); Chitwan, Jarwang, 1050 m, 16 Nov. 1989, *N. P. Manandhar 13922* (KATH); Chitwan, Saurah, 200 m, 6 Dec. 1975, *D.G. Joshi et al. 75/3344* (KATH); Makwanpur, Hetauda, 610 m, 21 Dec. 1963, *P. R. Shakya 1523* (BM, KATH); Makwanpur, Bhimphedi, 31 Mar.–3 Apr. 1953, *P. C. Gardner 1626* (BM); Makwanpur, Chanpakharka, 1450 m, 13 May 1981, *N. P. Manandhar 3472* (KATH); Makwanpur, Makaldamar, 500 m, 15 May 11987, *N. P. Manandhar 11852* (KATH); Makwanpur, Makaldamar, 500 m, 30 Nov. 1988, *N. P. Manandhar 12586* (KATH); Makwanpur, Brindavan Hesbal Farm, 600 m, 12 Jan. 1965, *s. coll. 3293* (KATH). **E NEPAL**: **Koshi**, Bhojpur, Irkhuwa Khola, 360 m, 11 Nov. 1981, *P. R. Shakya 7408* (KATH); Dhankuta, Dgara Pani to Tambur Bridge to Dhankuta, *H. Haraet al. 6303566* (A, KYO); Dhankuta, 1200 m, 20 Oct. 1963, *H. Hara et al. 6303567* (K); Sankhuwasabhu, Tumlingtar, 500 m, 27 Nov. 1990, *P. R. Shakya 9365* (KATH); Sankhuwasabhu, Tumlingtar, 550 m, 7 Dec. 1961, *A. H. Norkett 8004* (BM); Sankhuwasabhu, Arun Valley, 3 Nov. 1972, *T. Wraber 469* (BM); Sunsari, Dharan, Biratnagar, 380 m, 16 Apr. 1956, *J. D. A. Stainton 24* (BM, KATH); Sunsari, Tarahara, Biratnagar, 12 Feb. 1971, *N. P. Manandhar 17080* (KATH); Sunsari, Tarhara, 230 m, 23 Jan. 1974, *A. P. Single & M. B. Raue 17933* (KATH); Terathum, Kyaurehi, 1400 m, 16 Nov. 1978, *P. Pradhan et al. 938* (KATH); **Mechi**, Ilan, Ranga Pani to Chisa Pani to Ghorwa, 9 Dec. 1963, *H. Hara et al. 6303564* (KYO); Ilam, Phusri to Sanguri Bhanjang to Dhara Pani, 16 Oct. 1963, *H. Kanai et al. 6303600* (KATH, KYO); Ilam, Garuwa, 10 Dec. 1963, *H. Hara et al. 6303601* (BM, KATH, KYO); Taplejung, Sanghu, 1820 m, 7 Nov. 1961, *A. H. Norkett 6875* (BM); Taplejung, Dumhan, by the Tamur River, 700 m, 31 Oct. 1963, *G. Murata et al. 6303563* (KYO); Taplejung, Dumhan to Taplejung, 1 Nov. 1963, *H. Hara et al. 6303565* (KATH); **Sagarmatha**, Udayapur, Camp site to Dihigaon, 220 m, 26 Oct. 1995, *M. Mikage 9554146* (E); Udayapur, Pathari, Saptari, 28 May 1993, *P. R. Shakya et al. 182* (KATH). **SIKKIM & DARJILING: Darjiling**, Takdah to Rayang, 20 Apr. 1960, *H. Hara et al. 5840* (KYO); **Sikkim**, *D. Treutler 1132* (LE); 300–1200 m, *J. D. Hooker s. n.* (K). **BHUTAN**: **Gaylegphug**, forests slopes above Aie bridge, Gaylegphug, 500 m, 22 Mar. 1982, *Grierson & Long 3926* (E); **Phuntsoling**, hillside on N side of Phuntsholing, 220 m, 13 Feb. 1982, *Grierson & Long 2897* (E, K); Forests slopes above Phuntsholing, 350 m, A. J. C. Grierson & D. G. Long 761 (E); Phuntsoling, between Phuntsoling and Kharbandi, 400–800 m, Nov. 1987, *J. R. I. Wood 6075* (E); **Punakha**, Chuzomsa, 1100 m, 2 Nov. 1991, *J. R. I. Wood 7488* (E); **Samchi**, Samchi High School, 400 m, 7 *Jan.* 1989, *J. R. I. Wood 6858* (E); Samchi, 500 m, 28 Feb. 1982, *Grierson & Long 3272* (E, K). **U IRRAWADDY**: **Putao**, Putao to Myitkyina, 18 Dec. 1914, *F. Kingdon-Ward 2001* (E); **Tengchong**, the Taping Valley, *G. Forrest 12184* (IBSC); Wuhe, Sanjia Jie, 1200–1600 m, *W. Q. Yin 60-1482* (KUN).

29. JUSTICIA L.

Linnaeus, Sp. Pl. 1: 15. 1753; Nees in de Candolle (ed.), Prodr. 11: 426. 1847; Clarke in Hooker (ed.), Fl. Brit. India 4: 524. 1885; Lindau in Engler & Prantl (eds.), Nat. Pflanzenfam. 4(3B): 346. 1895; Nasir & Ali in Malik & Ghafoor (eds.), Fl. Pakistan 188: 29. 1988; Wood in Long & Grierson (eds.), Fl. Bhutan 2(3): 1286. 2001; Hu, Deng & Daniel in Wu, Raven & Hong (eds.), Fl. China 19: 449. 2011. ≡ *Ecbolium* Kuntze, Rev. Gen. 2: 486. 1891, nom. illeg., non *Ecbolium* Kurz, J. Asiat. Soc. Bengal, Pt. 2, Nat. Hist. 40(1): 75 1871. Type (designated by Hitchcock, Prop. Brit. Bot. 116. 1929): *Justicia hyssopifolia* L.

Adhatoda P. Mill., Gard. Dict. Abr. ed. 4. 1754; Hu, Fl. Reipubl. Popularis Sin. 70: 275. 2002. Type (designated by Heine, Fl. Gabon 13: 173. 1966): *Adhatoda vasica* Nees (≡ *Justicia adhatoda* L.).

Gendarussa Nees in Wallich (ed.), Pl. As. Rar. 3: 76, 103. 1832; Nees in de Candolle (ed.), Prodr. 11: 410. 1874; Hu, Fl. Reipubl. Popularis Sin. 70: 299. 2002. ≡ *Justicia* sect. *Gendarussa* (Nees) T. Anderson, J. Proc. Linn. Soc., Bot. 7: 41. 1863. ≡ *Justicia* subg. *Gendarussa* (Nees) Lindau in Engler & Prantl (eds.), Nat. Pflanzenfam. 4(3B): 347. 1895. Type: *Gendarussa vulgaris* Nees (≡ *Justicia gendarussa* Burm. f.).

Rostellaria Nees in Wallich (ed.), Pl. Asiat. Rar. 3: 76, 100. 1832, nom. illeg., non *Rostellaria* C. F. Gaertner, Suppl. Carp.: 135, t. 207: f. 1. 1807. ≡ *Justicia* sect. *Rostellaria* T. Anderson, J. Linn. Soc., Bot. 7: 38. 1863. ≡ *Justicia* subsect. *Rostellaria* (T. Anderson) V. A. W. Graham, Kew Bull. 43: 596. 1988. ≡ *Rostellularia* Reich., Handb. Nat. Pflr. Syst. 190. 1837; Nees in de Candolle (ed.), Prodr. 11: 370. 1847; Hu, Fl. Reipubl. Popularis Sin. 70: 302. 2002. ≡ *Justicia* sect. *Rostellularia* (Reich.) C. B. Clarke in Hooker (ed.), Fl. Brit. India 4: 536. 1885. Type (designated by Bremekamp, ING card 01967. 1956): *Rostellaria procumbens* (L.) Nees (≡ *Justicia procumbens* L.).

Rhaphidospora Nees in Wallich (ed.), Pl. Asiat. Rar. 3: 77, 115. 1832; de Candolle, Prodr. 11: 464, 499. 1847; Lindau in Engler & Prantl (eds.), Pflanzenfam. 4(3b): 329. 1895; Hu, Fl. Reipubl. Popularis Sin. 70: 253. 2002. ≡ *Rhaphidosperma* G. Don in Sweet, Hort. Brit. ed. 3: 559. 1839. ≡ *Justicia* sect. *Rhaphidospora* (Nees) T. Anderson, J. Proc. Linn. Soc., Bot. 7: 43. 1863. Type: *Rhaphidospora glabra* (Koenig ex Roxb.) Nees (≡ *Justicia glabra* Koenig ex Roxb.).

Hemichoriste Nees in Wallich (ed.), Pl. Asiat. Rar. 3: 76, 102. 1832. ≡ *Justicia* sect. *Hemichoriste* (Nees) Benth. & Hook. f., Gen. Pl. 2: 1109. 1876. Type: *Hemichoriste montana* Nees [≡ *Justicia montana* (Nees) Wall. ex T. Anderson].

Petalanthera Raf., Fl. Tellur. 2: 100 (1837), nom. illeg., non *Petalanthera* Nees, Laurin. Disp.: 15. 1833, nec *Petalanthera* Nutt., Nutt., J. Acad. Nat. Sci. Philadelphia 7(1): 107. 1834. Type: *Petalanthera punctata* Raf. (= *Justicia ventricosa* Hook.).

Nicoteba Lindau, Bot. Jahrb. Syst. 18: 56. 1893; Lindau in Engler & Prantl (eds.), Nat. Pflanzenfam. 4(3B): 329. 1895. Type (designated by Graham, Kew Bull. 43: 585. 1988): *Nicoteba betonica* (L.) Lindau (≡ *Justicia betonica* L.).

Calophanoides (C. B. Clarke) Ridl., Fl. Malay Penins. 2: 592. 1923; Hu, Fl. Reipubl. Popularis Sin. 70: 277. 2002. ≡ *Justicia* sect. *Calophanoides* C. B. Clarke in Hooker (ed.), Fl. Brit. India 4: 530 1885. Type (designated by Bremekamp, ING card 00178. 1955): *Calophanoides quadrifaria* (Nees) Ridl. [≡ *Gendarussa quadrifaria* Nees ≡ *Justicia quadrifaria* (Nees) T. Anderson].

Mananthes Bremek., Verh. Kon. Ned. Akad. Wetensch., Afd., Natuurk., Sect. 2. 45(2): 58. 1948; Hu, Fl. Reipubl. Popularis Sin. 70: 289. 2002. Type: *Mananthus sumatrana* (Miq.) Bremek. (≡ *Justicia sumatrana* Miq.).

Calymmostachya Bremek., Dansk Bot. Ark. 23: 217 (1966). Type: *Calymmostachya modesta* Bremek. [≡ *Justicia modesta* (Bremek.) V. A. W. Graham].

Herbs, subshrubs, or shrubs; cystoliths present. Leaves sessile or petiolate; leaf blade usually entire but sometimes sinuate or slightly serrate. Inflorescences dichasia in leaf axils, rarely deduced to a single flower, or axillary or terminal spikes, racemes, or thyrses, sometimes in panicles. Bracts variable in shape, sometimes prominent and/or brightly colored; bracteoles 2, similar to or smaller than bracts. Calyx deeply 4- or 5-lobed; lobes equal or subequal. Corolla tubular or funnel-shaped; limb strongly 2-lipped; upper lip entire to 2-lobed, internally rugulate, lower lip 3-lobed; lobes ascending cochlear in bud. Stamens 2; anthers bithecous; thecae equally or unequally inserted, parallel to perpendicular, one or both spurred at base or sometimes lacking basal appendages; staminodes absent. Ovary with 2 ovules per locule; stigma slightly 2-lobed. Capsules with a sterile basal stalk and a fertile head, 2–4-seeded; retinacula present. Seeds compressed.

Recent references: Ellis, 1979. Studies in *Justicia* Linn.: 1. *Bull. Bot. Surv. India* 20: 172; Ellis, 1982. Nomenclatural notes on *Justicia simplex* D. Don. *Bull. Bot. Surv. India* 22: 195–197; Graham, 1981. A new species of *Justicia* (Acanthaceae) from Nepal. *Jap. J. Bot.* 56(4): 117–119; Graham, 1988. Delimitation and infra-generic classification of *Justicia* (Acanthaceae). *Kew Bull.* 43: 551–624; Hansen, 1989. *Justicia* and *Rungia* (Acanthaceae) in the Indo-Chinese peninsula. In: Holm-Nielsen et al., Tropical forests and botanical diversity. Academic Press. pp. 201–210; Ellis, 1990. *Rostellularia diffusa* (Willd.) Nees (*Justicia diffusa* Willd.) complex (Acanthaceae). *Bull. Bot. Surv. India* 30: 126–133; Xia & Deng, 2005. Nomenclatural novelties in *Justicia* Linn. (Acanthaceae) from China. *J. Trop. Subtrop. Bot.* 13: 533–534; Soumya & Sunojkumar, 2017. *Justicia barapaniensis,* a new name for *Justicia salicifolia* T. Anderson (Acanthaceae). *Phytotaxa* 332: 98–100; Tong et al., 2019. *Justicia longipetiolata*, a new species of Acanthaceae from Yunnan, China. *Phytotaxa* 413: 281–288; Tong & Deng, 2019. The identity of *Dicliptera cyclostegia* (Acanthaceae). *Phytotaxa* 415: 137–144; Soumya & Sunojkumar, 2019. Notes on the identity of *Justicia diffusa* var. *hedyotidifolia* (Acanthaceae) with reduction of *Justicia gambleana. Phytotaxa* 415: 130–136.

About 700 species: tropical and temperate regions worldwide; 20 species (3 endemic) in the Pan-Himalaya.

Notes: Recent molecular studies indicated that the genus *Justicia* L. *s. l.* is paraphyletic and the further separation is necessary. But the further studied is still in need and therefore here I followed the treatment of Graham (1988) to maintain the broad sense of the genus.

1a. Calyx equally 4-lobed, or unequally 5-lobed with one lobe conspicuously smaller than others (sect. I. **Rostellaria**).
  2a. Bracts and bracteoles to ca. 2.5 mm long, shorter than calyx; capsules sub glabrous to puberulent .... 1. *J. diffusa*
  2b. Bracts and bracteoles to 6 mm long, equalling to calyx; capsules glabrous or hairy.
    3a. Leaves ovate to elliptic, the ratio of length and width < 3.
      4a. Leaves petiolate; petiole 3–8 mm; bracts ovate to elliptic-lanceolate, covered with non-glistening, unjointed hairs .... 2. *J. procumbens*
      4b. Leaves sessile or shortly petiolate; petiole 0–2 mm; bracts elliptic-ovate, covered with glistening, jointed hairs .... 3. *J. simplex*
    3b. Leaves linear or lanceolate to elliptic-lanceolate, the ratio of length and width > 3.
      5a. Petiole 8–10 mm long; bracts as long as or slightly longer than corollas; capsules

glabrous .... 4. *J. peploides*

5b. Petiole 1–4(–6)mm long; bracts nearly half as long as corollas; capsules hairy towards apex .... 5. *J. quinqueangularis*

1b. Calyx equally 5-lobed.

6a. Flowers in terminal or axillary elongated spikes, racemes or panicles.

7a. Bracts elliptic to ovate, broader than 5 mm, concealing calyx.

8a. Spikes with 1 flower per node; bracts obtuse, purple-margined, corolla to 2 cm long (sect. II ***Betonica***).. 6. *J. atkinsonii*

8b. Spikes with 2 flower per node; bracts acute, without distinct coloured margin; corolla ca. 3 cm long (sect. III. **Vasica**).. 7. *J. adhatoda*

7b. Bracts linear, to 3 mm broad, not concealing calyx (sect. IV. ***Rhaphidospora***).

9a. Flowers in axillary cymes forming a terminal lax leafy panicle............... 8. *J. pubigera*

9b. Flowers in terminal or axillary spikes or racemes.

10a. Leaf blade narrowly lanceolate, 1–1.5 cm broad; basal bracts longer than calyx upwards gradually smaller with apicalmost ones shorter than calyx .... 9. *J. gendarussa*

10b. Leaf blade ovate to elliptic, often broader than 3 cm; all bracts shorter than calyx.

11a. Leaf blade pubescent along veins abaxially; ovary and capsules pubescent...... .... 10. *J. virgata*

11b. Leaf blade glabrous; ovary and capsules glabrous .... 11. *J. vasculosa*

6b. Flowers in axillary abbreviated spikes, usually solitary or several clustered (sect. V. **Harnieria**).

12a. Leaves sessile; leaf base sub cordate .... 12. *J. tukuchensis*

12b. Leaves petiolate or subsessile; leaf base cuneate, rounded or truncate.

13a. Inflorescence bracts linear-lanceolate or ovate, broadest at base, 1-veined; calyx without membranous margin.

14a. Bracts 5–6 mm long; corolla 2–2.2 cm long .... 13. *J. wardii*

14b. Bracts 2–3 mm long; corolla 1.3–1.5 cm long................................ 14. *J. siccanea*

13b. Inflorescence bracts ovate to circular to elliptic to obovate to spatulate-obovate, broadest above base, penninerved; calyx with white membranous margin.

15a. Petioles 2–5.5 cm long; capsules ca. 2 times of calyx in length .... 15. *J. longipetiolata*

15b. Petioles 0.3–2(–2.5) cm long; capsules shorter than or subequal to calyx.

16a. Petioles (0.5–)1–2(–2.5) cm long; leaf blade usually longer than 3 cm.

17a. Leaf blade oblong, to lanceolate, or rarely orbicular, (2–)5.5–11.5 cm long, lateral veins 5–7 pairs .... 16. *J. quadrifaria*

17b. Leaf blade elliptic, 3–6.5 cm long, lateral veins 3 or 4 pairs .... 17. *J. cyclostegia*

16b. Petioles 3–5(–7) mm long; leaf blade usually shorter than 3 cm.

18a. Plants densely white villose for 3–5 mm long; ovary and capsules villose.. .... 18. *J. albovelata*

18b. Plants pilose or pubescent, trichomes usually shorter than 2 mm; ovary and capsules glabrous.

19a. Bracts lanceolate, apex acuminate; corolla whitish-yellow .... 19. *J. yunnanensis*

19b. Bracts elliptic, apex acute; corolla white ....20. *J. xerobatica*

I. **Justicia** sect. **Rostellaria** T. Anderson, J. Linn. Soc., Bot. 7: 511. 1863. ≡ *Rostellaria* Nees in Wallich (ed.), Pl. Asiat. Rar. 3: 76, 100. 1832, non *Rostellaria* C. F. Gaertner, Suppl. Carp. 135, t. 207: f. 1. 1807. ≡ *Justicia* subsect. *Rostellaria* (T. Anderson) V. A. W. Graham, Kew Bull. 43. 1988. ≡ *Rostellularia* Reich., Handb.: 190. 1837; Nees in de Candolle (ed.), Prodr. 11: 370. 1847; Hu,

Fl. Reipubl. Popularis Sin. 70: 302. 2002. ≡ *Justicia* sect. *Rostellularia* (Reich.) C. B. Clarke in Hooker (ed.), Fl. Brit. India 4: 538. 1885; Clarke, J. Asiat. Soc. Bengal 74: 682. 1905. Type (designated by Bremekamp, ING Card, 1956): *Justicia procumbens* L. [≡ *Rostellaria procumbens* (L.) Nees ≡ *Rostellularia procumbens* (L.) Nees].

Justicia subsect. *Diffusa* Ellis, Bull. Bot. Surv. India 30: 128. 1990. Type: *Justicia diffusa* Willd.

Herbaceous or rarely shrubby perennials or occasionally annuals. Inflorescences a simple spike with 2 flowers per node, axis eglandular. Bracts narrowly triangular, shorter than calyx, sometimes with a scarious margin. Calyx 4-lobed, or 5-lobed but one much reduced. Corolla white or lilac. Anther-thecae superposed, equal, parallel or slightly oblique, appendages large. Ovary glabrous. Capsules stalked at base, sterile portion about 1/3 of its length. Seeds scarcely compressed, rugulose.

About 30 species: Old World; five species in the Pan-Himalaya.

1. Justicia diffusa Willd., Sp. Pl. 1: 87. 1797; Clarke in Hooker (ed.), Fl. Brit. India 4: 538. 1885; Nasir & Ali in Malik & Ghafoor (eds.), Fl. Pakistan 188: 37. 1988; Hu, Deng & Daniel in Wu, Raven & Hong (eds.), Fl. China 19: 452. 2011. ≡ *Rostellaria diffusa* (Willd.) Nees in Wallich (ed.), Pl. Asiat. Rar. 3: 100. 1832. ≡ *Rostellularia diffusa* (Willd.) Nees in de Candolle (ed.), Prodr. 11: 371. 1847; Hu, Fl. Reipubl. Popularis Sin. 70: 303. 2002. ≡ *Ecbolium diffusum* Kuntze, Revis. Gen. Pl. 2: 487. 1891. Type: India, *Klein 724* (lectotype designated by Malik & Ghafoor in Nasir & Ali (eds.), Fl. Pakistan 188: 37. 1988: B).

Rostellaria hedyotidifolia Nees in Wallich (ed.), Pl. Asiat. Rar. 3: 100. 1832. ≡ *Rostellularia hedyotidifolia* (Nees) Nees, Prodr. 11: 370. 1847. ≡ *Justicia diffusa* var. *hedyotidifolia* (Nees) C. B. Clarke in Hooker (ed.), Fl. Brit. India 4: 538. 1885. ≡ *Rostellularia diffusa* var. *hedyotidifolia* (Nees) Kumari, Fl. Tamil Nadu, Ser. 1, Analysis 2: 158. 1987; Hu, Fl. Reipubl. Popularis Sin. 70: 304. 2002. ≡ *Justicia hedyotidifolia* (Nees) Remadevi, Contrib. Fl. Kerala Acanth.: 97. 2009. Type: *Herbs Heyn in Wallich Cat. 2450* [lectotype designated by Soumya & Sunojkumar, Phytotaxa 415: 131. 2019: K-W (K001116056)].

Herbs, 10–30 cm high; stems basally procumbent and diffuse, upwards sacending, quadrangular, sulcate, sparsely pubescent. Petiole 1–2 mm long, pubescent; leaf blade lanceolate-elliptic, subcircular, or ovate, (0.7–)2–5 cm long, 0.7–1.5 cm broad, glabrous or sparsely pubescent, lateral veins 2–4 pairs, base cuneate, margin entire, apex acute. Spikes (1.5–)2.5–5 cm long, sparsely hirsute. Bracts oblong-lanceolate, 2–3 mm long, 0.8–1 mm broad, sparsely pubescent; bracteoles linear, 2–2.5 mm long, ca. 1 mm broad, sparsely pubescent. Calyx 3.5–5 mm long, adaxially sparsely pubescent, 4- or 5-lobed; lobes lanceolate, margin ciliate, apex acute. Corolla purplish red, 5–6 mm long, upper lip emarginate, lower lip sparsely puberulent, 3-lobed. Stamens 2; filaments 3–4 mm long, hirsute; anther-thecae superposed, lower one spurred at base. Ovary pubescent; style 4–5 mm long, pubescent. Capsules oblong, 4–5 mm long, subglabrous to pubescent. Seeds ovate in outline, compressed, 1.5–2 mm long and broad, subsmooth. **Chromosomes** 2n = 18.

Phenology: flowering and fruiting from August to December.

Habitats: grasslands, stream banks, sandy places; 100–1600 m.

Distribution: N Pakistan, Jammu & Kashmir, U Ganga & Indus, W Nepal, C Nepal, E Nepal (Map 46-183); also in Bangladesh, Sri Lanka, Thailand, Vietnam, and other parts of India, Myanmar and China (S Yunnan, Hainan, Guangxi, Guangdong, Fujian, Taiwan).

N PAKISTAN: Hazara, Garhi Habibullah at Chitta Batta, *S. A. Faruqi & M. Qaiser 3413* (KUH, fide Malik & Ghafoor, 1988). **JAMMU & KASHMIR: Jammu**, Nagrota, 8 May 1969, *B.*

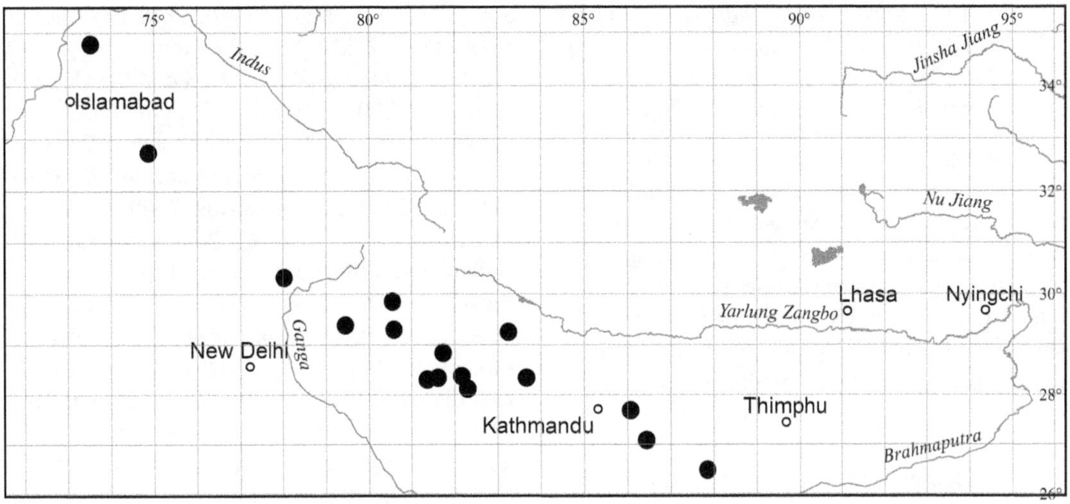

Map 46-183: **Justicia diffusa** Willd.

M. Sharma 211 (RRLH). **U GANGA & INDUS**: **Dehradun**, Matak Majri, 450 m, Nov. 1891, *J. S. Gamble 23184* (K); **Kumaon**, above Ramersar Bridge, May 1845, *T. Thomson 196* (K); Barmdeo, 210 m, *R. Strachey & J. E. Witerbottom 2* (K); **Nainital**, Near Bhim Tal, 1400–1600 m, *K. H. Rechinger 60620* (MO). **W NEPAL**: **Bheri**, Banke, Goddhoi, 210 m, 4 Nov. 1980, *R. N. Shukla 3304* (KATH); Banke, Kharguar, 180 m, Oct. 1981, *R. N. Shunkla 4062* (KATH); 1982, *Rauniyar 1887* (KATH); Banke, Kharaguar, 12 Oct. 1977, *Rauniyar 2313* (KATH); Banke, Santalia, Sep. 1981, *T. K. Bhattacharya 3942* (KATH); Bardiya, Kohalpur, 25 Apr. 1980, *T. K. Bhattacharya 387* (KATH); Bardiya, Kohalpur, 180 m, 19 Nov. 1978, *T. K. Bhattacharya 3658* (KATH); Dailekh, Dandamadi-Panipokhari, 3 Aug. 1991, *M. Suzuki 9160194* (TI); **Karnali**, Dolpa, Bheri River, 760 m, 23 Aug. 1952, *O. V. Polunin, W. R. Sykes & L. H. J. Williams 549* (BM, E); **Mahakali**, Dadeldhura, Kalina to Gagan, 590–1420 m, 14 Sep. 1981, *I. Sharma et al. 352* (KATH); Darchula, Mau, Oct. 1981, *T. K. Bhattachaya 3772* (KATH); **Rapti**, Dang, Chorkholi, 350 m, 26 Aug. 1982, *N. P. Manandhar & N. K. Bhattarai 8427* (KATH); Dang, Koilabas, 21 Mar. 1973, *N. P. Manandhar 9789* (KATH); Dang, Kohalpur, 30 Nov. 1973, *N. P. Manandhar & K. J. Malla 17148* (KATH); Dang, Babai Nadi, 330–960 m, 9 Dec. 1983, *T. K. Bhattacharya et al. 2543* (KATH); Dang, Chorkholi, 300 m, 13 Sep. 1979, *T. K. Bhattacharya & K. J. Malla 2919* (KATH); Salyan, Gapta Khola, 960 m, 20 Aug. 1983, *P. P. Kurma 5608* (KATH). **C NEPAL**: **Bagmati**, Rasuwa, 2490 m, 15 Jul. 1977, *N. P. Manandhar 77* (KATH); **Dhawalagiri**, Parbat, Kusma, 29 May 1985, *T. K. Bhattacharya 5243* (KATH); **Janakpur**, Dolka, Kandra, 600 m, 21 Nov. 1975, *L. R. Sharma 826* (KATH). **E NEPAL: Mechi**, Jhapa, Dolaghat, 900 m, 30 Jul. 1953, *J. B. Tyson 143* (E).

2. Justicia procumbens L., Sp. Pl. 1: 15. 1753; Clarke in Hooker (ed.), Fl. Brit. India 4: 539. 1885; Nasir & Ali in Malik & Ghafoor (eds.), Fl. Pakistan 188: 36. 1988; Hu, Deng & Daniel in Wu, Raven & Hong (eds.), Fl. China 19: 452. 2011. ≡ *Rostellaria procumbens* (L.) Nees in Wallich (ed.), Pl. Asiat. Rar. 3: 101. 1832. ≡ *Ecbolium procumbens* (L.) Kuntze, Revis. Gen. Pl. 2: 488. 1891. ≡ *Rostellularia procumbens* (L.) Nees in de Candolle (ed.), Prodr. 11: 371. 1847; Hu, Fl. Reipubl. Popularis Sin. 70: 305. 2002. Type: *LINN-28.14* (lectotype designated by Malik & Ghafoor in Nasir & Ali (eds.), Fl. Pakistan 188: 37. 1988: LINN).

Justicia japonica Thunb. in Murray, Syst. Veg., ed. 14: 63. 1784; Thunberg, Fl. Jap. 20.

1784. ≡ *Rostellularia japonica* (Thunb.) J. L. Ellis, Bull. Bot. Surv. India 22: 196. 1980. Type: Japan, *C. P. Thunberg s. n.* [holotype: UPS (UPS-THUNB-000386)].

Herbs 20–50 cm high, procumbent; stems quadrangular, sulcate, pubescent. Petiole 3–8 mm long, pubescent; leaf blade elliptic, ovate-elliptic, or elliptic-oblong, 1.5–4 cm long, 0.8–1.5 cm broad, subglabrous to sparsely hispid, lateral veins 3–6 pairs, base broadly cuneate to subrounded and slightly decurrent onto petiole, margin entire or slightly undulate, apex acute to obtuse. Spikes terminal or axillary, cylindric, 1–6 cm long, dense-flowered; peduncle 0.5–7 cm long, densely pilose. Bracts ovate to elliptic-lanceolate, 2.5–8 mm long, 0.6–1.3 mm broad, abaxially pilose, margin ciliate; bracteoles lanceolate, 2.5–5 mm long, abaxially pilose, margin ciliate. Calyx 4–6 mm long, 4-lobed to base; lobes linear, abaxially pilose along veins, 1-veined, margin yellowish white and ciliate. Corolla pink or white and red-spotted on lower lip, 5–8.5 mm long; upper lip 2.5–3 mm long, apex emarginated, lower lip 2.5–3 mm long, 3–3.5 mm broad, 3-lobed, lobes ovate, the middle lobe slightly larger, 1–1.3 mm long, 1–1.5 mm broad. Stamens 2, exserted; filaments 3–4 mm long, glabrous; anther-thecae superposed, lower one spurred at base, upper one muticous. Ovary pubescent; style 4–5 mm long, pubescent. Capsules 4–6 mm long, 4-seeded. Seeds ovate in outline, compressed, 1–1.2 mm long and broad, rugose. **Chromosomes** 2n = 28. Figure 46-107.

Phenology: flowering from June to November; fruiting from September to December.

Habitats: wastelands, roadsides, lawns, open fields, littoral forests; below 3100 m.

Distribution: U Ganga & Indus, U Yarlung Zangbo, C Nepal, E Nepal, N Hengduan, S Hengduan, U Irrawaddy (Map 46-184); also in Bangladesh, Cambodia, Indonesia, Japan, Laos, Malaysia, the Philippines, Sri Lanka, Thailand, Vietnam, and other parts of Pakistan, India, Myanmar and China.

U GANGA & INDUS: **Kumaon**, Ramari, 1350 m, *R. Strachey & J. E. Winterbottom 1* (K); **Simla**, 1800 m, 7 Sep. 1886, *H. Collett 177* (K). **U YARLUNG ZANGBO**: **Gyirong**, Mangmu, 12 Jul. 1975, *Qinghai-Xizang Exped. 6307* (KUM); Mangmu, 2200 m, 16 Jul. 1975, *Qinghai-Xizang Exped. 6925* (KUM, PE); Zhirechang, 2433 m, 6 Aug. 2010, *PE Xizang Expedition 280* (PE); Jiangcun, 2700–2922 m, 8 Aug. 2000, *PE Xizang Expedtion 510* (PE). **C NEPAL**: **Bagmati**, Kathmandu, Taudaha, 1330 m, 14 Jul. 2002, *N. R. Acharya 147* (TUCH); Lilatpur, Godavari, 1500 m, 8 Sep. 1970, *C. Chama 64786* (TI); Nuwakot, Kakani, 2073 m, 9 Aug. 2002, *T. Watanabe et al. s. n.* (TI); Rasuwa, Thare to Sim Chautara, 970–2200 m, 6 Jun. 1969, *H. Hara et al. 69112* (TI); Rasuwa, Dhunche to Deolari, 21 Jul. 1995, *T. Hoshino et al. 9535025* (TI); Rasuwa, Dhunche to Deolari, 2000 m, 21 Jul. 1995, *T. Hoshino et al. 9537006* (TI); Rasuwa, Sing Gomba to Dhunche, 31 Jul. 1995, *T. Hoshiro et al. 9539236* (TI); **Dhawalagiri**, Myagdi, Shika to Narchang (Nracheng), 1730 m, 26 Aug. 1988, *M. Suzuki et al. 8860745* (A, TI); Myagdi, Beni to Babichor, 2 Sep. 1996, *M. Mikage et al. 9685049* (A, E); Myagdi, Shika to Narchang (Nracheng), 1700 m, 26 Aug. 1988, *M. Suzuki et al. 8881402* (A, E, L, MO, TI); **Gandaki**, Kaski, Tikhedhunga to Kande, 1240 m, 7 Sep. 1988, *M. Suzuki et al. 88000845* (TI); Kaski, Nayapur-Tirkhedhunga, 15 Sep. 1995, *M. Mikage et al. 9550052* (E, TI); **Janakpur**, Ramechhap, Bhandar to Kinja, *H. Ohba et al. 8530869* (MO). **E NEPAL**: **Koshi**, Dhankuta, Hile to Pakhribas, 1830 m, 12 Aug. 1997, *S. Noshiro et al. 9760002* (TI); Dhankuta, Diyale to Hile, *S. Noshiro et al. 9830238* (MO); Sankhuwasabha, Khandbari to Botebus, 1150–1800 m, 27 Jul. 1977, *H. Ohashi et al. 770102* (TI); Sankhuwasabha, Num to Shemma, 850–1600 m, 31 Jul. 1977, *H. Ohashi et al. 771811* (TI); Sankhuwasabha, higho Wollung to Apsuwa Doban, 860 m, 30 Aug. 1997, *S. Noshiro et al. 9760481* (MBK, TI); **Mechi**, Ilam, Mayum, 2100 m, 5 Sep. 1962, *K. Nishioka 857* (KYO); Taplejung, Yamphodim to Telok, 1800 m, 27 Jun. 1969, *L. H. J. Williams 988* (TI); Taplejung, Papung to Sangrati Pati, 1900 m, 26 Aug. 1977, *H. Ohba et al. 772733* (TI); **Sagarmatha**, Bhojpur, Deurali Danda to Deurali, 2080 m, 31 Oct. 1995, *M. Mikage et al. 9558240* (E, TI). **N HENGDUAN**: **Wenxian**, Bikou, 650

Figure 46-107: **Justicia procumbens** L.: **a**, habit; **b**, stem; **c**, bract, bracteole and calyx; **d**, corolla; **e**, stamen; **f**, capsule. Drawn by Quan LIN; reproduced from *Icon. Cormophyt. Sin.* 4: 176, fig. 5765. 1975.

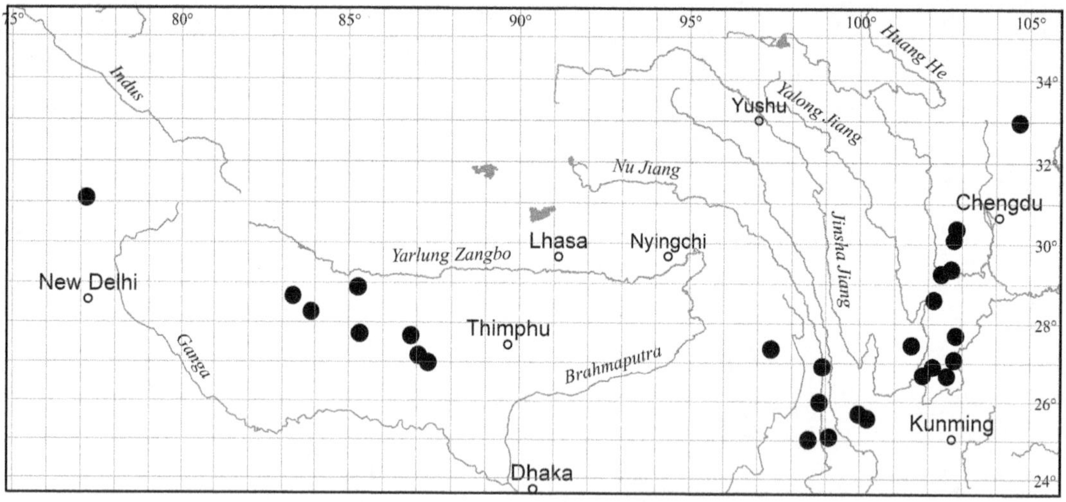

Map 46-184: **Justicia procumbens** L.

m, 18 Oct. 1958, *Z. P. Wei 3154* (SZ, WUK). **S HENGDUAN: Baoxing**, Malushan to Lingguan, 1000 m, 7 Oct. 1936, *K. L. Chü 3976* (IBSC, PE, SZ, WUK); from the city to Linguan, 1300 m, 19 Sep. 1938, *C. Pei 8240* (PE); **Butuo**, Niujiaowan, Xixihe, 680 m, 11 Oct. 2014, *Y. F. Deng 25873* (IBSC); **Dali**, Hedso prope Gwangdung, 1830 m, 6 Sep. 1914, *H. Handel-Mazzetti 4883* (WU); 2300–2500 m, *C. Schneider 3786* (A); nearby the city, 5 Sep. 2013, *Y. F. Deng et al. 24136* (IBSC, KUN); Xizhou, 31 Jul. 2019, *Y. F. Deng 28633* (IBSC); Cangshan, near Wuweisi, 2300 m, 9 Sep. 2009, *Z. J. Yin et al. 1796* (KUN); **Fugong**, Wawaqiao, 1150 m, 12 Sep. 1996, *Exped. Gaoligong Shan 7071* (KUN); **Hanyuan**, Chengguan, 800m, 15 Sep. 1993, *Z. C. Zhao 0383* (PE); Dashu Qu, 1100 m, 17 Sep. 1993, *Z. C. Zhao 0423* (PE); **Huidong**, Dacong, 11 Oct. 2014, *Y. F. Deng 25879* (IBSC); **Lanping**, Xiawenshuimiao, 15 Sep. 2013, *Y. F. Deng et al. 24967* (IBSC); **Longyang**, Baihualing, 1530 m, 23 Aug. 1995, *X. C. Shi & S. X. Yang 514* (KUN); Mangkuan, Baihualing, Hanlong, 1550 m, 23 Nov. 20003, *H. Li 13358* (CAS, KUN); Lujiang, Bawan, 1560 m, 30 Aug. 2003, *Gaoligong Shan Biodiversity Survey 18131* (CAS, KUN); Lujiang, Bawan, Mopanshi, 1800 m, 1 Sep. 2003, *Gaoligong Shan Biodiversity Survey 18386* (CAS, KUN); Mangkuan, Baihualing, Manghuang, 1170 m, 9 Sep. 2003, *Gaoligong Shan Biodiversity Survey 19021* (CAS, KUN); Mangkuan, Baihualing, Dajie Shan, 1525 m, 10 Sep. 2003, *Gaoligong Shan Biodiversity Survey 19065* (CAS, KUN); Mangkuan, 702 m, 1 Jan. 2005, *Gaoligong Shan Biodiversity Survey* 26177 (KUN); Wayao, 1160 m, 23 Sep. 1977, *Y. H. Li & P. H. Yu 20412* (HITBC); Baihualing, 12 Sep. 2013, *Y. F. Deng 24777* (IBSC); **Lushui**, Luzhang, 1800 m, 24 Sep. 1996, *Exped. Gaoligong Shan 8003* (KUN); on road N of Liuku along the W bank of the Nujiang, 950 m, 16 Oct. 1998, *H. Li 10448* (KUN); Liuku, Laimao, 850 m, 31 Oct. 1990, *Exped. Dulongjiang 057* (MBK); Liuku, Yuejinqiao, 13 Sep. 2013, *Y. F. Deng et al. 24907* (IBSC); **Mianning**, Dapo, 270 m, 6 Dec. 1963, *Chuan Mian 152* (CDBI); **Miyi**, Xiaohe, Longceshan, 1200 m, Oct. 1963, *Miyi Gongzuozu 247* (CDBI); Puwei, 1700 m, 13 Jul. 1958, *S. Y. Chen et al. 10727* (NAS, SZ); **Ningnan**, Hulukou, 750 m, 1 Sep. 2012, *Y. F. Deng 23361* (IBSC); Luoge, Lishu, 21 Oct. 2014, *Y. Q. Hao CSH12871* (CSH); **Shimian**, Chengguan, 900 m, 27 Sep. 1993, *Z. C. Zhao 1183* (PE); Kaluogou, 1400 m, 29 Sep. 1993, *Z. C. Zhao 1442* (PE); **Tianquan**, Wenbifeng Park, 16 Oct. 2008, *Y. F. Deng & H. J. Xin 20904* (IBSC); s. loc., 15 Oct. 1935, *Y. Y. Ho 6892* (NAS); 850 m, 11 Sep. 1963, *Exped. W. Sichuan (K. C. Kuan et W. T. Wang) 3317* (HIB, PE); Shiyan, Zhoucunganng, Xiaolonggang, 980 m, 18 Oct. 1982, *D. Y. Peng 46816* (CDBI, IBSC); Chengxiang, Xiaolonggang, 970 m, 27 Oct. 1982, *D. Y. Peng 46869* (CDBI, IBSC);

Chengxiang, Xiaolonggang, 965 m, 27 Oct. 1982, *D. Y. Peng 46871* (CDBI); Chengxiang, Xiaolonggang, 965 m, 27 Oct. 1982, *D. Y. Peng 46875* (CDBI, IBSC); s. loc., Oct. 13, 1939, *F. C. Tai & C. M. Teng 5299* (IBSC); **Xichang**, Lushan, 1600 m, 2 Aug. 2006, *Y. F. Deng 19066* (IBSC); **Yanbian**, s. loc., 1600 m, 10 Oct. 1932. *T. T. Yü 1678* (A); **Yangbi**, Pingpo, 5 Sep. 2013, *Y. F. Deng 24167* (IBSC, KUN); Cangshanxi, 5 Sep. 2013, *Y. F. Deng 24189* (IBSC, KUN); **Yanyuan**, Mopanshan, Yakou, 3050 m, 11 Oct. 1982, *Y. B. Yang et al. 58* (CDBI, NAS); Yongsheng, 18 Sep. 2013, *Y. F. Deng 25184* (IBSC). **U IRRAWADDY**: Myitkyina, camp Landis area, 8 Oct. 1945, *R. O. Belcher 740* (K); **Putao**, Mularshidi, 460 m, 11 Oct. 2014, *Putao Expedition Team (X. H. Jin et al.) PT-ET 3* (PE); **Tengchong**, Gudong, 1700 m, 17 Oct. 1977, *Y. H. Li 20282* (HITBC); Qingshui, Rehai hot spring, 1500 m, 24 Oct. 1998, *H. Li 10876* (KUN); Mazhang, Xiaokong Shan, 1930 m, 25 Oct. 1998, *H. Li 10936* (KUN); Mazhang, Xiaokong Shan, 1930 m, 25 Oct. 1998, *H. Li 10950* (KUN); Jietou, between Jietou and Datang, 27 Oct. 1998, *H. Li 11072* (KUN); Qushui, Longkou, between Qushui and Jietou, 1510 m, 29 Oct. 1998, *H. Li 11208* (KUN); Qushi, Longkou, between Qushui and Jietou, 1510 m, 29 Oct. 1998, *H. Li 11229* (KUN); Wuhe, Huangcaoling area, near Zhengding village, 1900 m, 26 Aug. 2003, *Gaoligong Shan Biodiversity Survey 17962* (KUN); Ruidian, Dashuiqing, 2000 m, 18 Oct. 1977, *P. H. Yu & Y. H. Li 20498* (HITBC); Qushi, Daba, Songguanzhayingchu, 1900 m, 6 Sep. 1995, *X. C. Shi & S. X. Yang 741* (KUN); hills around Tengyueh, *G. Forrest 27191* (AU, IBSC, K); Laifengshan, 10 Sep. 2013, *Y. F. Deng 24589* (IBSC).

3. Justicia simplex D. Don, Prodr. Fl. Nepal.: 118. 1825; Clarke in Hooker (ed.), Fl. Brit. India 4: 539. 1885; Wood in Grierson & Long (eds.), Fl. Bhutan 2(3): 1288. 2001. ≡ *Ecbolium simplex* (D. Don) Kuntze, Revis. Gen. Pl. 2: 488. 1891. ≡ *Justicia procumbens* var. *simplex* (D. Don) Yamazaki, Fl. E. Himalaya: 302. 1966. Type: "Hab. In Nepalia, *Wallich*" (not located).

Rostellaria mollissima Nees, Pl. Asiat. Rar. 3: 101. 1832. ≡ *Rostellularia mollissima* (Nees) Nees in de Candolle (ed.), Prodr. 11: 373. 1847. ≡ *Justicia mollissima* (Nees) Y. F. Deng & T. F. Daniel, Fl. China 19: 451. 2001. Type: Nepal, *Wallich Cat. 2445* [lectotype **here designated**: K-W (K001B116046); isolectotypes: E, K-W (K001116047), LD, NY, P (P00719782), S].

Justicia procumbens var. *latispica* C. B. Clarke in Hooker (ed.), Fl. Brit. India 4: 539. 1885. ≡ *Rostellularia khasiana* var. *latispica* (C. B. Clarke) C. Y. Wu [Index Fl. Yunnan 2: 1684. 1984, nom. invald.] ex C. C. Hu, Fl. Reipubl. Popularis Sin. 70: 305. 2002. ≡ *Rostellularia latispica* (C. B. Clarke) Bremek., Verh. Kon. Ned. Akad. Wetensch., Afd. Natuurk., Sect. 2. 45(2): 71. 1948. Type: [*icon*] Wight, Icon. Pl. Ind. Orient 4: t. 1539. 1850.

Rostellaria rotundifolia Nees in Wallich (ed.), Pl. Asiat. Rar. 3: 100. 1832, non *Justicia rotundifolia* Nees in Wallich (ed.), Pl. Asiat. Rar. (Wallich). 3: 108. 1832. ≡ *Rostellularia rotundifolia* (Nees) Nees in de Candolle (ed.), Prodr. 11: 370. 1847. ≡ *Justicia orbiculata* Wall. ex T. Anderson, J. Linn. Soc., Bot. 9: 512. 1867. ≡ *Justicia diffusa* var. *orbiculata* (Wall. ex T. Anderson) C. B. Clarke in Hooker (ed.), Fl. Brit. India 4: 538. 1885. ≡ *Justicia nepalensis* V. A. W. Graham, Kew Bull. 43: 597. 1988, nom. illeg. (Art. 52.1 & 52.2). ≡ *Rostellularia diffusa* var. *rotundifolia* (Nees) Ellis, Bull. Bot. Surv. India 30: 133. 1988. Type: Nepal, 1821, *Herbs Wight in Wallich Cat. 2489* [lectotype **here designated**: K-W (K001116176)].

Justicia khasiana C. B. Clarke in Hooker (ed.), Fl. Brit. India 4: 537. 1885. ≡ *Ecbolium khasianum* (C. B. Clarke) Kuntze, Revis. Gen. Pl. 2: 980. 1891. ≡ *Rostellularia khasiana* (C. B. Clarke) J. L. Ellis, Bull. Bot. Surv. India 20: 172. 1979. Type: India, Khasia, 400–600 ft., *J. D. Hooker s. n.* [lectotype **here designated**: K (K000884094); isolectotype: K (K000884095)].

Herbs, 60–100 cm high; stems erect, somewhat zigzag, quadrangular, hispid. Leaves sessile or shortly petiolate; petiole 0–2 mm long, hispida; blade elliptic-lanceolate, oblong to broadly ovate, 2.5–4.5 cm long, 0.5–1.2 cm broad, both surfaces hispid, lateral veins 4–6 pairs, base broadly cuneate, margin entire, apex acute to obtuse. Spikes terminal or axillary, cylindric, 2.5–6 cm long, sessile or stalked. Bracts elliptic-ovate, 3–5 mm long, 1.5–2 mm broad, hispid; bracteoles linear, 5–6 mm long. Calyx ca. 6 mm long, abaxially hispid, 4-lobed almost to base; lobes linear, margin ciliate. Corolla purplish red, 0.8–1.2 cm long, glabrous outside; tube 2.5–3 mm long; upper lip 2.5–3 mm long, slightly 2-lobed, hairy outside, lower lip ovate-suborbicular, 2.5–3 mm long, hairy outside, shortly 3–lobed. Stamens 2; filaments 2.5–3 mm long, basally dilated, sparsely hirsute; anther-thecae superposed, lower one white spurred at base. Ovary oblong, 1.2–1.5 mm long, hirsute towards apex; style 3–4 mm long, basal portion hirsute. Capsules ellipsoid, 3–5 mm long, pubescent. Seeds ovate in outline, compressed, 1–1.2 mm long and broad, minutely reticulate. **Chromosomes** 2n = 36.

Phenology: flowering and fruiting from June to December.

Habitats: trailsides; 800–2500 m.

Distribution: N Pakistan, Jammu & Lashmir, U Ganga & Indus, U Yarlung Zangbo, W Nepal, C Nepal, E Nepal, Bhutan, S Hengduan, U Irrawaddy (Map 46-185); also in Sri Lanka, Thailand, and other parts of Pakistan, India, Myanmar and China (SW Yunnan).

Notes: The species was treated as *Justicia japonica* Thunb. in some references (viz. Malik & Ghafoor, 1988). However, examination of the type material, it reveals that *Justicia japonica* has petiolate leaves and might be conspecific with *J. procumbens* L.

N PAKISTAN: Abbottabad, Sherwar Road Abbottabad, *s. coll. 731* (KUH, fide Malik & Ghafoor, 1988); from Garhi Habibullah to Balakot, *S. A. Faruqi & M. Qaiser 3208* (KUH, fide Malik & Ghafoor, 1988). **JAMMU & KASHMIR**: **Jammu**, Katra, 830 m, 16 Sep. 1979, *S. K. Kapoor 15902* (RRLH); **Rajouri**, Ujhan, 1000 m, 27 Sep. 1988, *G. H. Dar 9783* (KASH, fide Dar et al., 2014). **U GANGA & INDUS**: Dharmsala, 1500 m, 21 Sep. 1896, *G. A. Gammie 18657* (K); **Dalhousie**, 1500 m, 20 Aeptember 1874, *C. B. Clarke 22830* (K); **Dehradun**, *K. M. Balapure 541* (A); **Kumao**, Ramari, 1350 m, *R. Strachey & J. E. Winterbottom 1* (K); **Musoorie**, *M. B. Saxena 288* (DD, fide Raizada & Saxena, 1978); **Simla**, Jakko, 2 Oct. 1915, *H. H. Rich 75* (K); Mahasu, 1885, *J. R. Drummond 26686* (K); 2100 m, 12 Sep. 1877, *N. Gamble 4988A* (K); Elyscum Hill, 2250 m, 14 Sep. 1876, *N. Gamble 1375A* (K). **U YARLUNG ZANGBO**: **Gyirong**, Jiangcun to Resuoqiao, 1840 m, *X. F. Gao et al. 14733* (CDBI). **W**

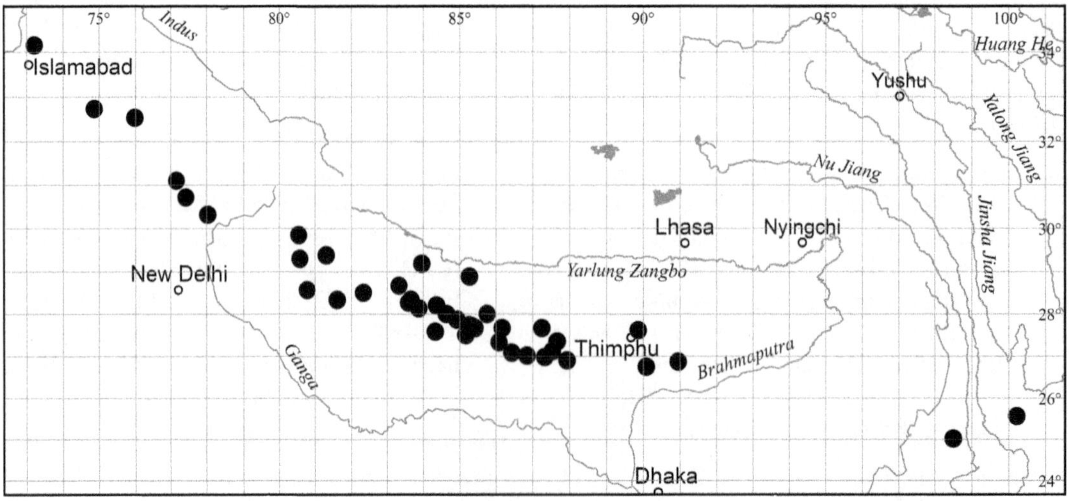

Map 46-185: **Justicia simplex** D. Don

NEPAL: **Bheri**, Banke, Shamshergang, Napelganj, 2 Oct. 1977, *R. Rana 1887* (KATH); Bardiya, Chisapani to Gainikanda, 6 Oct. 1996, *M. Mikage et al. 9689098* (TI); **Mahakali**, Dadeldhura, Sahukharka, 1995 m, 11 Nov. 2012, *B. Adhikari & G. D. Bhatt 46* (E); Darchula, Khakanga, 840 m, 26 Aug. 1982, *M. M. Amatya & P. M. Regmi W484/82* (KATH); Darchula, Darchula to Dalilekh, 840–1810 m, 28 Aug. 1982, *s. coll. W507/82* (KATH); Darchula, Maiekholi to Sipti, 1810–1660 m, 30 Aug. 1982, *M. M. Amatya & P. M. Regmi W575/82* (KATH); Darchula, Darchula to Huti, 1000 m, 24 Aug. 1980, *K. R. Rajbhandari & K. J. Malla 5252* (KATH); Darchula, Darchula to Huti, 1000 m, 24 Aug. 1980, *K. R. Rajbhandari & K. J. Malla 5521* (KATH); **Rapti**, Rukum, Sezigaon, 1390 m, 9 Sep. 1982, *N. P. Manandhar & N. K. Bhattarai 8855* (KATH); **Seti**, Bajura, Serigaon to Bashala, 14 Aug. 1991, *M. Suzuki et al. 9193713* (KATH, TI); Bajura, Kaudegaon to Babali, 29 Aug. 1991, *M. Suzuki 9193834* (KATH, TI); Bajura, Baudi Khola, S of Kuru, 1921 m, 21 Aug. 2017, *s. coll. B173* (KATH); Kailali, Budhetola, 1137 m, 10 Nov. 2012, *B. Adhikari & D. D. Bhatt 38* (E). **C NEPAL**: **Bagmati**, Bhaktapur, Nagarkot, 1950 m, 28 Aug. 1966, *Prabha & Ramola 6317* (KATH); Dhading, Terhethum to Mulpani, 1650 m, 5 Aug. 1976, *N. P. Manandhar 186* (KATH); Kabhve, Drapcha, 1740 m, 16 Nov. 1988, *N. P. Manandhar 12051* (KATH); Kathmandu, Kathmandu Airport, 13 Oct. 1963, *H. Hara et al. 6303619* (TI); Kathmandu, Badigaon, 1400 m, 10 Aug. 1971, *J. F. Dobremez 753* (E, KATH); Kathmandu, Chobar Gorge, south of Kathmandu, 21 Sep. 1963, *H. Kanai 6303622* (K); Kathmandu, Chandragiri, 2300 m, 23 Oct. 1994, *C. B. Baniya 21* (TUCH); Kathmandu, Gokarna, 1320 m, 6 Sep. 1971, *R. Ranan & Bajracharya 12272* (KATH); Kathmandu, Sundarijar to Mulkhanla, 20 Aug. 1969, *Malla & Kanai 673461* (KATH); Kathmandu, Sundarijal, 20 Aug. 1969, *S. B. Malla & H. Kanai 673472* (KATH); Kathmandu, Banepa, 1500 m, 5 Sep. 1954, *A. Zimmermann 1029* (K); Kathmandu, Swyambhu, 1300 m, 26 Sep. 1970, *N. P. Manandhar 9426* (KATH); Kathmandu, Gokarna, 6 Aug. 1966, *P. Pradhan & R. Ihapa 6419* (KATH); Kathmandu, Gokarna, 1320 m, 21 Apr. 1967, *P. Pradhan & R. Ihapa 6497* (KATH); Kathmandu, Nagarjun, 1500 m, 11 Sep. 1960, *S. B. Malla & S. B. Rajbhandari 85* (KATH); Kavrepalanchok, Khardarpati, 1550 m, 19 Nov. 1988, *N. P. Manandhar 12175* (KATH); Kavrepalanchok, Bhumlndanda, Dolalghat, 1500 m, 10 Sep. 1964, *Herbarium Banerji 1310* (A); Kavrepalanchok, Dalalghat-Nayopati, 1500 m, 10 Sep. 1964, *Bennerji et al. 2602* (KATH); Lalitpur, Godawari, N Mt. Phulchoki, 1650 m, *M. Mikage et al. 9552633* (A, E); Lalitpur, Lele, 1540 m, 8 Sep. 1990, *I. Shrestha & N. Pradhan 196* (KATH); Lalitpur, Godawari, 12 Oct. 1992, *N. Joshi & I. Shrestha 243* (KATH); Lalitpur, Godawari, Royal Botanical Garden, 1515 m, 9 Jul. 2001, *N. Joshi 469* (KATH); Lalitpur, Godawari, 1650 m, 29 Jul. 1967, *N. P. Manandhar et al. 7141* (KATH); Lalitpur, Godawari, 1515 m, 24 Nov. 1974, *Indina et al. 74/16* (KATH); Lilatpur, Godavari, 1500 m, 8 Sep. 1970, *C. Chuma 674784* (TI); Nuwakot, Kakani, 2073 m, 9 Aug. 2002, *T. Watanabe et al. NPL-SP020809(004)* (MBK); Nuwakot, village de Salme, 2150 m, 24 Aug. 1981, *J. Wiart 179* (E); Nuwakot, village de Salme, 1600 m, 26 Jun. 1981, *Programme "VERSANT" du GRECO Himalaya-Karakorum 697* (E); Nuwakot, Lampata to Shange, 1220–1270 m, 8 Aug. 1973, *D. P. Joshi & M. M. Amatya 73/331* (KATH); Rasuwa, Dhunche to Syabru, 10 Jul. 1992, *H. Takayama et al. 9239013* (E); Rasuwa, Dunche to Thulo Bhargu to Syabru Bensi, *F. Miyamoto et al. 9410002* (KATH, MO); Rasuwa, Dunche–Thulo Bhargu–Syabru Bensi, 1800 m, 23 Jul. 1994, *F. Miyamoto et al. 9400004* (A, E, KATH); Rasuwa, Lingju to Tibling, *F. Miyamoto et al. 9420267* (A, E, KATH); Rasuwa, Gram to Dunche, 21 Aug. 1972, *H. Kanai et al. 4911* (KATH, TI); Rasuwa, Sim Chotala to Gram, 1700 m, 20 Aug. 1972, *H. Kanai et al. 721672* (TI); Rasuwa, Bhorlang to Sundarijar, 1400–2400 m, *H. Kanai et al. s. n.* (TI); Rasuwa, Nangjul, 2460 m, 15 Jul. 1977, *N. P. Manandhar 77* (KATH); Sindhupalchok, Sim Chotala to Ramche, 1900 m, 20 Aug. 1972, *H. Kanai et al. 4910* (A, TI); Sindhupalchok, SE Thagaon, above Bhotekoshi Nadi, 1340 m, 9 Sep. 2011, *M. F. Watson et al. EKSIN5* (E, KATH, MBK); **Dhawalagiri**, Baglung, Bhuji Khola, S Burtiwang, 1060 m, 20 Aug. 2014, *s. coll. A007*

(E); Baglung, Ratnechour, 840 m, 1 Sep. 1996, *M. Mikage et al. 9682111* (TI); Myagdi, Gorepani Deorali to Shikha, 14 Jul. 1983, *H. Ohba et al. 8330428* (TI); Myagdi, Gorepani Deorali to Shikha, 14 Jul. 1983, *H. Ohba et al. 8330435* (TI); Myagdi, Tatopani to Ghasa, 18 Sep. 1995, *M. Mitage et al. 9550181* (TI); Myagdi, Beni to Babichor, 2 Sep. 1996, *M. Mitage et al. 9681052* (TI); Myagdi, Muri to Boghara, 5 Sep. 1996, *M. Mikate et al. 9681161* (TI); Myagda, below Baglung, 870 m, 24 Sep. 1996, *M. Mitage et al. 9683057* (TI); Myagdi, Babichor to Dharapani, 3 Sep. 1996, *M. Mikage et al. 9685122* (TI); Myagdi, Dharapani to Takum, 1560 m, 4 Sep. 1996, *M. Mikage et al. 9687101* (TI); Myagdi, Mayagdi Khola, 900 m, *J. D. A. Stainton, W. R. Sykes & L. H. J. Williams 4178* (L); Myagdi, near Lumsum, 1950 m, *J. D. A. Stainton, W. R. Sykes & L. H. J. Williams 4017* (A, E); Myagdi, Chinkhola Village, 7 Aug. 1986, *J. J. Metz 412* (KATH); Myagdi, Chhamarke, 1000 m, 10 Sep. 1991, *N. P. Manandhar 1081-91* (KATH); Mustang, Tatopani, Dana to Ghasa, 1440 m, 5 Aug. 1999, *M. Mikage et al. 9962908* (KATH, TI); Parbat, Ulleri, 2000 m, 20 Jul. 1974, *D. P. Joshi & T. K. Bhattachaya 74/1742* (KATH); Parbat, Kusma, 13 Sep. 1991, *N. P. Manandhar 1092/91* (KATH); **Gandaki**, Gorkha, Macha Khola to Syaule Bhatti, 910 m, *M. Suzuki et al. 9470136* (A, E); Gorkha, Machha Khola, 960 m, 26 Jul. 2008, *s. coll. 20811036* (E); Gorkha, Machha Khola, before Tatopani, 880 m, 26 Jul. 2008, *s. coll. 20815023* (E); Gorkha, Machha Khola, 920 m, 26 Jul. 2008, *s. coll. 20816032* (E); Gorkha, Below Bih, 1970 m, 7 Jun. 1983, *P. R. Shakya et al. 7775* (KATH); Kaski, Tistan Deorali, Indo-Nepal Highway, 26 Jul. 1969, *H. Kanai 673307* (KATH); Kaski, Bakhri Kharka, Madi Khola, 1800 m, 17 *Jul.* 1954, *J. D. A. Stainton, W. R. Sykes & L. H. J. Williams 6429* (E); Kaski, Pokhara, 910 m, 12 Aug. 1954, *J. D. A. Stainton, W. R. Sykes & L. H. J. Williams 6764* (E); Kaski, Pokhara to Suike, 950 m, 6 Jul. 1983, *H. Ohba et al. 8310041* (TI); Kaski, Pokhara to Suike, 6 Jul. 1983, *H. Ohba et al. 8350073* (TI); Kaski, Nayapur to Tirkhedhunga, 15 Sep. 1995, *M. Mikage et al. 9550052* (TI); Kaski, Nayapur to Tirkhedhunga, 15 Sep. 1995, *M. Mikage et al. 9550053* (TI); Kaski, Nayapur to Tirkhedhunga, 960–1460 m, *M. Mikage et al. 9552099* (A, BM, E, MO); Kaski, Pokhara, Bagar, 20 Nov. 1977, *N. P. Manandhar 603* (KATH); Kaski, Pardi, Pokhara, 24 Jun. 1986, *s. coll. 10811* (KATH); Kaski, Naudanda Phedi to Dhampus Jhaure, 1100 m, 19 Aug. 1988, *M. Suzuki et al. 8810765* (KATH); Kaski, on the way to Naudanda, 1170–1470 m, 27 Oct. 1971, *Malla et al. 13811* (KATH); Kaski, Kaski Deawrali, 1650 m, 28 Jun. 1986, *N. P. Manandhar & L. P. Ratel 10965* (KATH); Lamjung, Syange, near Ghermu Phant, 1180 m, 17 Aug. 2008, *s. coll. 20818162* (E); Lamjung, Jagat to Bahundanda, 1100 m, 9 Aug. 1983, *H. Ohba et al. 8311207* (TI); Manang, Marsyandi Khola, Jagat *to* Bahundanda, *H. Ohbaet al. 82311207* (A); **Janakpur**, Dolakha, Tamba Koshi, 28 Aug. 1983, *H. Ohba et al. 8351060* (KYO); Ramechhap, Nhandar to Shivalaya, 6 Aug. 1985, *H. Ohba et al. 8571368* (A, KATH); Ramechhap, Patkare to Bhandar, 5 Aug. 1985, *H. Ohba et al. 8530764* (A, E); Ramechhap, Shivalaya to Deorali, 5 Jul. 1985, *H. Ohba et al. 8580105* (KATH, TI); Ramechhap, Kyama to Patkare, 4 Aug. 1985, *H. Ohba et al. 8571262* (KATH); Ramechhap, Shivalaya to Bhandar, 1800 m, 28 Jul. 1997, *M. Wakabayashi et al. 9710022* (KATH); **Narayani**, Chitwan, Jandola, 1650 m, 16 Nov. 1989, *N. P. Manandhar 14018* (KATH); Makwanpur, Chuniya to Purano Gaon, Hetauda to Daman, 1220 m, 28 Nov. 2004, *T. Miller et al. (DNEP2) A179* (E, KATH); Makwanpur, Tistung, 1 Oct. 1963, *P. R. Shakya 1524* (KATH). **E NEPAL**: **Koshi**, Dhankuta, 1200 m, 19 Oct. 1963, *H. Hara et al. 6303615* (TI); Dhankuta, 1200 m, 19 Oct. 1963, *H. Hara et al. 6303616* (TI); Dhankuta, Hile to Shidua, 12 *Jul.* 1991, *H. Ohba et al. 9120011* (E); Dhankuta, Jile (Hille) to Gholi Kharka (Dholikharka), *M. Suzuki et al. 8820078* (A); Dhankuta, Hile to Gholi Kharka, 1800 m, 2 Jul. 1988, *M. Suzuki et al. 8880040* (KATH); Dhankuta, Hile, near Hille, 1760 m, 16 Sep. 1981, *P. R. Shakya 6506* (KATH); Sankhuwasabha, Arun Valley, Sabhaya Khola, N of Chainpur, 6 Sep. 1956, *J. D. A. Stainton 1550* (A, E); Sankhuwasabha, Khandbari to Bhotebas, 1320 m, 23 Sep. 1981, *P. R. Shakya 6603* (KATH); Sankhuwasabha, Num, below Num, 1360 m, 9 Oct. 1981, *P. R. Shakya 6955* (KATH); Sankhuwasabha, Sekaha to Chipe Gaun, 1470 m,

7 Jul. 1988, *M. Suzuki et al. 8860073* (KATH); Sankhuwasabha, Semma, 1500 m, 14 Aug. 1998, *S. Noshiro et al. 9830011* (A, BM, E); Sankhuwasabha, Num to Shemma, 850–1600 m, 31 Jul. 1977, *H. Ohashi et al. 773369* (E, TI); Sankhuwasabha, Fururu to Num, 1600–2100 m, 29 Jul. 1977, *H. Ohashi et al. 774567* (TI); Terhathum, between Jiri Khimty and Terhathum, 1650 m, 7 Oct. 1981, *C. Grey-Wilson et al. 4693* (BM, K); Terhathum, Mulpani, 1670 m, 5 Aug. 1976, *N. P. Manandhar et al. 186* (KATH); **Mechi**, Ilam, Bheri River, 1800 m, *O. V. Polunin, W. R. Sykes & L. H. J. Williams 3222* (L); Panchitar, Chyangthapu to Dabale Deurali, 1700–2000 m, *S. Noshiro et al. 9241014* (A, E); Panchitar, Memeng, 2000 m, 30 Sep. 2007, *s. coll. 293* (E); Taplejung, 1700 m, 10 Aug. 1976, *N. P. Manandhar 258* (KATH); Taplejung, Helok to Ilandanda, 8 Nov. 1963, *H. Hara et al. 6303612* (TI); Taplejung, Tuwa to Taplethok, 5 Nov. 1963, *H. Hara et al. 6303614* (KATH, TI); Taplejing, Papung to Sangrati Pati, 1050–2000 m, 26 Aug. 1977, *H. Ohashi et al. 775344* (TI); Taplejung, Iladanda to Helok, 13 Nov. 1963, *H. Kanai et al. 6303618* (TI); Taplejung, Yamphodim to Telok, 1800 m, 27 Jun. 1969, *s. coll. 15948* (KATH); **Sagarmatha**, Khotang, 1250 m, 30 Oct. 1995, *M. Mikage 9554223* (E). **BHUTAN**: **Punakha**, near Punakha Dzong, E Pho Chu River, 1100 m, 3 Oct. 1987, *J. R. I. Wood 5921* (E); **Sankoshi**, between Daga Dzong and around Bazaar, 1600 m, 6 Aug. 1989, *J. R. I. Wood 7041* (E); **Tongsa**, Mangde Chu valley, near Langthel, 900 m, Nov. 1989, *J. R. I. Wood 7219* (E). **S HENGDUAN: Dali**, *K. K. Tsoong 2275* (SYS); Xiaguan Pool, *P. Y. Chiu 60817* (CDBI, HIB); Fodingfeng, 2400 m, *K. J. Luo et al. 614238* (YIM). **U IRRAWADDY**: **Putao**, Kachi Hills, 9 Dec. 1930, *F. Kingdon-Ward 9032* (E); **Tengchong**, Haikou, 1000 m, *Z. Y. Wang & R. W. Zhang 6006030* (YIM); Jiangcao, Yanyao, 1700 m, 11 Sep. 1960, *Z. Y. Wang & R. W. Zhang 6006121* (YIM); Qushi, Daba, 24 Sep. 1965, *Tengchong Working Group 653121* (YIM); Heshun, 9 Sep. 2013, *Y. F. Deng 24576* (IBSC); Laifengshan, 10 Sep. 2013, *Y. F. Deng 24583* (IBSC); Qushi, 10 Sep. 2013, *Y. F. Deng 24652* (IBSC); Qushi, 10 Sep. 2013, *Y. F. Deng 24686* (IBSC).

4. Justicia peploides (Nees) T. Anderson, J. Linn. Soc., Bot. 9: 511. 1867; Nasir & Ali in Malik & Ghafoor (eds.), Fl. Pakistan 188: 33. 1988. ≡ *Rostellaria peploides* Nees in Wallich (ed.), Pl. Asiat. Rar. 3: 101. 1832. ≡ *Rostellularia peploides* (Nees) Nees in de Candolle (ed.), Prodr. 11: 375. 1847. ≡ *Rostellaria quinqueangularis* var. *peploides* (Nees) C. B. Clarke in Hooker (ed.), Fl. Brit. India 4: 537. 1885. Type: India, Ripa Gangis ad Bhagulpur, 1820, *Wallich Cat. 2441b* [lectotype **here designated**: K-W (K001115997)].

Annual herbs, erect, up to 45 cm high; stems 4–5-angled geniculate, glabrous. Petiole 8–10 mm long, glabrous; blade oblong-elliptic to ovate or obovate, 2–4 cm long, 1–2 cm broad, glabrous on both surfaces, base attenuate to petiole, margin entire, apex obtuse. Inflorescences a terminal spike, 2–5 cm long. Bracts linear-lanceolate, 6–8 mm long, scabrid, margin scarious, apex obtuse; bracteoles similar to bracts, smaller in size. Calyx 4-lobed to base; lobes linear-lanceolate, 3–4 mm long, margin whitish and setulose, apex obtuse, mucronulate. Corolla pinkish-purple, 6–7 mm long; tube cylindric, 2–2.5 mm long, upper lip 2.5–3 mm long, apex emarginate, lower lip patent, 3–3.5 mm long, glabrous, apex 3-lobed. Stamens 2; filaments 3–3.5 mm long. Ovary oblong, hairy towards apex; style 3.5–4 mm long. Capsules oblong, 4–5 mm long, glabrous, mucronate, stipitate. Seeds orbicular, 1.5–1.8 mm in diameter, rugose. **Chromosomes** 2n = 18.

Phenology: flowering from August to November.

Habitats: open places; 700–1000 m.

Distribution: N Pakistan, Jammu & Kashmir, U Ganga & Indus (Map 46-186); also in other parts of Pakistan and India.

N PAKISTAN: **Azad Kashmir**, Muzaffarabad, *M. N. Chaudhri et al. 324* (A); 30 Sep. 1944, *R. L. Badhwar 5761* (RRLH). **JAMMU & KASHMIR**: Gachi, Thelum Valley Road, 750 m, *R. R. Stewart 8481* (MO). **U GANGA & INDUS:** Tanri Valley, 900 m, 25 May 1891, *G. G. Gammie s. n.* (K).

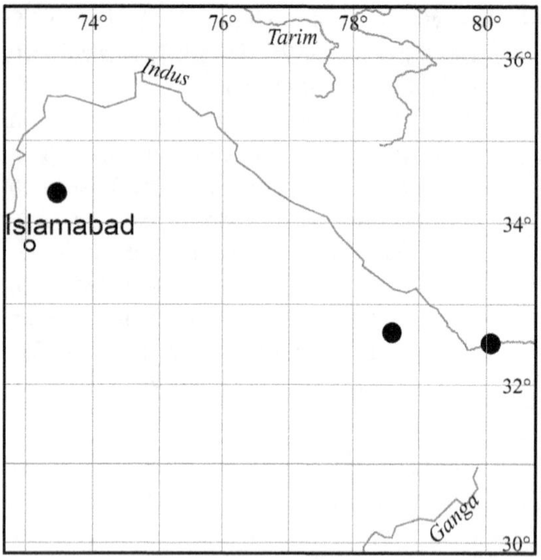

Map 46-186: **Justicia peploides** (Nees) T. Anderson

5. Justicia quinqueangularis J. Koenig ex Roxb., Fl. India (ed. Carey) 1: 134. 1820; Clarke in Hooler (ed.), Fl. Brit. India 4: 536. 1885. ≡ *Rostellaria quinqueangularis* (J. Koenig ex Roxb.) Nees in Wallich (ed.), Pl. Asiat. Rar. 3: 101. 1832. ≡ *Rostellularia quinqueangularis* (J. Koenig ex Roxb.) Nees in de Candolle (ed.), Prodr. 11: 375. 1847. ≡ *Ecbolium quinquangulare* (J. Koenig ex Roxb.) Kuntze, Revis. Gen. Pl. 2: 981. 1891. Type: India, s. d., *Banks s. n.* [holotype: C (C10004992)].

Justicia vahlii Roth, Nov. Pl. Sp. 14. 1821; Nasir & Ali in Malik & Ghafoor (eds.), Fl. Pakistan 188: 35. 1988. ≡ *Rostellularia vahlii* (Roth) Nees, Pl. Asiat. Rar. 3: 102. 1832; Nees in de Candolle (ed.), Prodr. 11: 376. 1847. ≡ *Justicia diffusa* var. *Vahlii* (Roth) C. B. Clarke in Hooker (ed.), Fl. Brit. India 4: 538. 1885. Type: India, *Rotteler 585* [lectotype **here designated**: C (C10005023)].

Herbs, erect, to 45 cm high, much branched; stems erect or ascending, woody basally, quadrangular, glabrous or pubescent. Petiole 1–4 (–6) mm long; blade linear or lanceolate to elliptic-lanceolate, 2.5–8 cm long, 0.5–1.5 cm broad, glabrous, base narrowly cuneate, margin entire, margin entire or subentire, apex subacute. Spikes axillary or terminal, 2.5–15 cm long, 0.5–1 cm broad. Bracts narrowly ovate, 5–6 mm long, 1.5–2 mm broad, membranous, margin ciliate, apex acute; bracteoles linear-lanceolate, 3–4 mm long, smaller than calyx, hairy. Calyx 4-lobed nearly to base; lobes lanceolate, 5–6 mm long, subequal, margin scarious and ciliate, apex acute. Corolla purple, 0.8–1 cm long; tube funnel-shaped, 3–4 mm long, outside pubescent; upper lip nearly as long as tube, apex emarginate, lower lip slightly longer than upper lip, patent, 3-lobed. Stamens 2; filaments 3–4 mm long, pilose towards base. Ovary sparsely hairy towards apex; style filiform, hairy on lower portion. Capsules oblong, 4–5 mm long, somewhat quadrangular, pilose towards apex, mucronate. Seeds concentric, papillate-rugose. **Chromosomes** 2n = 18.

Phenology: flowering from August to November.

Habitats: open places; below 300 m.

Distribution: U Ganga & Indus, W Nepal, C Nepal, E Nepal (Map 46-187); also in other parts of Pakistan and India.

U GANGA & INDUS: **Dehradun**, Sep. 1882, *J. Rithie 2329* (E). **W NEPAL: Bheri**, Banke, 21 Nov. 1972, *N. P. Manandhar 15016* (KATH); **Bardiya**, Bandar, 180 m, 18 Dec. 1982, *R. N. Shukla 5163* (KATH); **Bardiya**, Gularia, Zabadi, 1981, *T. K. Bhattacharya 3274* (KATH); **Rapti**, Dang, Gadhawa, 280 m, 11 Mar. 1976, *N. P. Manandhar & P. M. Regmi 309* (KATH); **Seti**, Kailali, Balmi, Beladevipur, 182 m, 10 Nov. 2012, *B. Adhikari & G. D. Bhatt 31* (E). **C NEPAL**: **Dhawalagiri**, Dhaugari, 3 Apr. 1074, *N. P. Manandhar et al. 55* (KATH); **Narayani**, Bara, Lower Mijgard forests near Tawa, 140 m, 14 Feb. 1993, *L. P. Kattel 137/93* (KATH); Chitwan, Sauraha- Harda Khola, 19 Jan. 1996, *M. Mikage 9614210* (E). **E NEPAL**: **Mechi**, Jhapa, Sibganja- Gauriganja, 12 Dec. 1963, *H. Hara et al. 6303677* (K, KYO); Morang, Mahra Bahara to Kathgare, 13 Dec. 1963, *H. Kanai et al. 6303676* (KATH, KYO).

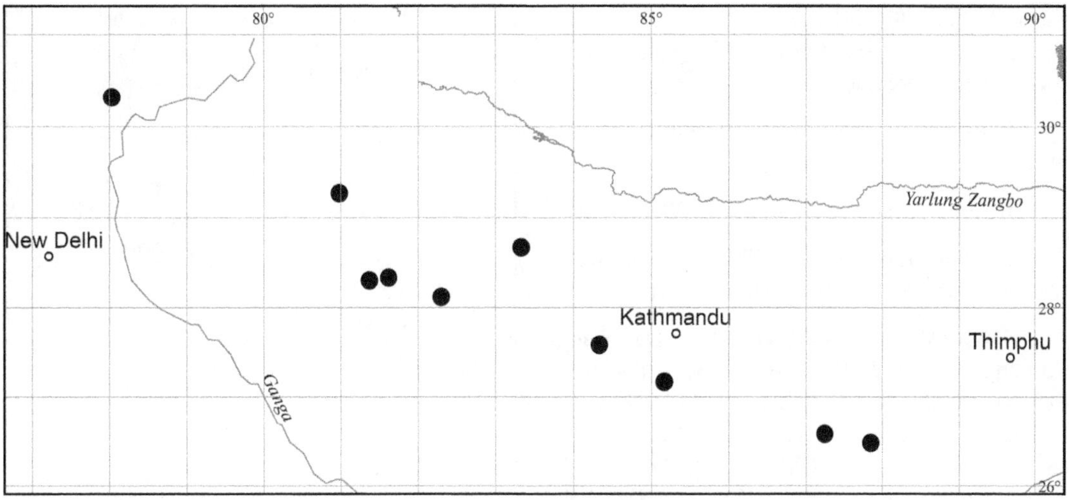

Map 46-187: **Justicia quinqueangularis** J. Koenig ex Roxb.

II. Justicia sect. **Betonica** (Nees) T. Anderson, J. Proc. Linn. Soc., Bot. 7: 38. 1863; Clarke in Hooker (ed.), Fl. Brit. India 4: 525. 1885. ≡ *Adhatoda* sect. *Betonica* Nees in Mart., Fl. Bras. 9: 147. 1847; et in de Candolle (ed.), Prodr. 11: 384. 1847. Type: *Justicia betonica* L.

Nicoteba Lindau, Bot. Jahrb. Syst. 18: 56. 1893; Lindau in Engler & Prantl (eds.), Nat. Pflanzenfam. 4(3B): 329. 1895. Type (designated by Graham, Kew Bull. 43: 585. 1988): *Nicoteba betonica* (L.) Lindau (≡ *Justica betonica* L.).

Herbs, or rarely shrubs, perennials. Inflorescences a simple spike with 1 flower per node, axis eglandular. Bracts ovate and exceeding calyx. Calyx 5-lobed; lobes equal. Corolla white or pale pink or blue. Anther-thecae equal or lower one larger, slightly to clearly superposed, parallel, slightly oblique to oblique, appendages large. Ovary pubescent. Capsules densely pubescent or pubescent to glabrous. Seeds not compressed and rounded-rugulose-tuberculate.

About 30 species: Old World, mainly in Africa and Madagascar; one species in the Pan-Himalaya.

6. Justicia atkinsonii T. Anderson, J. Linn. Soc., Bot. 9: 509. 1867; Clarke in Hooker (ed.), Fl. Brit. India 4: 528. 1885; Wood in Grierson & Long (eds.), Fl. Bhutan 2(3): 1287. 2001. ≡ *Ecbolium atkinsonii* (T. Anderson) Kuntze, Revis. Gen. Pl. 2: 980. 1891. Type: [Sikkim & Darjiling] India, Sikkim, "inter vicos Rinchingpong et Yoksun", *T. Anderson herb. Sikkimensis 1066* (K).

Shrubs, 0.5–1 m high; stems bifariously pubescent, then glabrescent. Petiole 1–5 cm; leaf blade ovate, 8–22 cm long, 4–12.5 cm broad, abaxially pilose along veins, base attenuate, margin subentire, apex shortly acuminate. Spikes terminal, simple or in panicles, 4–12 cm long; peduncles 1–3 cm long. Bracts densely imbricate, ovate or ovate-lanceolate, 7–14 mm long, 5–10 mm broad, glandular-pubescent, purple-margined, apex rounded; bracteoles subulate, 1–1.3 cm long. Calyx 7–9 mm long, puberulent, 5-lobed; lobes linear. Corolla yellowish-white with rose markings, 1.7–2 cm long, pilose; tube base cylindric; upper lip emarginate, lower lip 3-lobed; lobes ovate-lanceolate, apex rounded. Stamens 2; filaments glabrous; anthers thecae spurred at base. Ovary glabrous; style glabrous.

Capsules clavate, 1.4–1.7 cm long, 4-seeded. Seeds ovate in outline, 1.6–2 mm long and broad, rugose, subtubercular.

Phenology: flowering from September to November.

Habitats: subtropical forests; 600–1550 m.

Distribution: Bhutan, Sikkim & Darjiling (Map 46-188); also in other parts of India.

SIKKIM & DARJILING: **Darjiling**, Pornong, 31 Oct. 1870, *C. B. Clarke 13634* (K); Pornong, 600m, *C. B. Clarke 13634* (K, LE); Pornong, 900 m, 23 Aug. 1869, *C. B. Clarke 8787* (K, LE); **Sikkim**, near Chakung, 29 Oct. 1959, *H. Kihara & S. Nakao s. n.* (KYO). **BHUTAN**: Ralimpoong, 21 Sep. 1869, *C. B. Clarke 9168* (K).

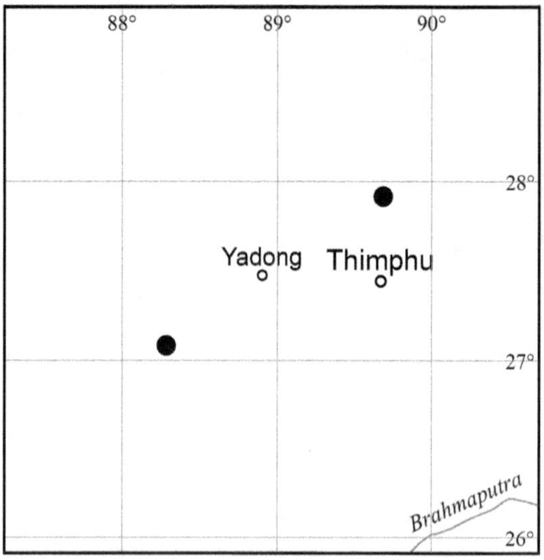

Map 46-188: **Justicia atkinsonii** T. Anderson

III. Justicia sect. **Vasica** Lindau in Engler & Prantl (eds.), Nat. Pflanzenfam. 4(3b): 349. 1895. Type (designated by Graham, Kew Bull. 43: 584. 1988): *Justicia adhatoda* L..

Adhatoda P. Mill., Gard. Dict. Abr. ed. 4: 28. 1754; Hu, Fl. Reipubl. Popularis Sin. 70: 275. 2002. Type (designated by Heine, Fl. Gabon 13: 173. 1966): *Adhatoda vasica* Nees (≡ *Justicia adhatoda* L.).

Shrubby perennials, usually large. Inflorescences a simple spike with 2 flowers per node. Bracts ovate, exceeding calyx. Calyx 5-lobed; lobes equal. Corolla white or cream; anther-thecae equal or lower one smaller or larger, slightly superposed, oblique, appendages small. Ovary densely pubescent or rarely glabrous. Capsules stalked at base, sterile portion 2/5–3/5 of its length, pubescent or glabrous. Seeds scarcely compressed and rounded-rugulose-tuberculate, or compressed and rugulose-alveolate, or smooth.

About ten species: Old World; one species in the Pan-Himalaya.

7. Justicia adhatoda L., Sp. Pl. 1: 15. 1753; Stearn in J. Arnold Arbor. 52: 640. 1971; Nasir & Ali in Malik & Ghafoor (eds.), Fl. Pakistan 188: 39. 1988; Wood in Grierson & Long (eds.), Fl. Bhutan 2(3): 1287. 2001; Hu, Deng & Daniel in Wu, Raven & Hong (eds.), Fl. China 19: 456. 2011. ≡ *Adhatoda vasica* Nees in Wall., Pl. Asiat. Rar. 3: 103. 1832; Clarke in Hooker (ed.), Fl. Brit. India 4: 540. 1885; Hu, Fl. Reipubl. Popularis Sin. 70: 277. 2002. ≡ *Ecbolium adhatoda* (L.) Kuntze, Revis. Gen. Pl. 2: 487. 1891. ≡ *Adhatoda adhatoda* (L.) Huth, Helios 11(9): 132. 1893, nom. invalid. (Art. 23.4). Type: 'Habitats in Zeylona', *Herb. Hermann 2: 43, No. 16* [lectotype designated by Manning & Getliffe Norris, S. Afr. J. Bot. 51: 483. 1985: BM (BM000621656)]; [*icon*] Herb. Hermann. 5: Icon. 226 [epitype (as "lectotype") designated by Graham, Kew Bull. 43: 584. 1988: BM].

Adhatoda zeylanica Medik., Hist. Commantat. Acad. Elect. Sci. Theo.-Palat. 6: 369. 1790. Type: not designated.

Shrubs, 1.5–4 m high. Branches quadrangular, lenticellate, pubescent when young, soon glabrescent. Petiole 0.8–3 cm long, puberulent; leaf blade ovate to elliptic-ovate, 7–18 cm long, 2–7 cm broad, abaxially puberulent, adaxially densely tomentose when young, soon glabrescent except tomentose along veins, lateral veins 9–12 pairs, base broadly cuneate, margin entire, apex acuminate and sometimes slightly falcate. Spikes terminal or axillary, ovoid to broadly ovoid, 3–7 cm long; peduncle 3–7 cm long. Bracts imbricate, ovate-oblong, 1–2 cm long, 0.5–1.5 cm broad, puberulent, 3–7-veined, margin ciliate, apex acute; bracteoles linear-lanceolate, 1–1.4 cm long, 0.4–0.5 cm broad, puberulent, 3–5-veined, margin ciliate, apex acute. Calyx 9–11 mm long, 5-lobed; lobes linear-oblong, 2.5–3 mm broad, margin narrowly scarious and ciliate. Corolla white or pink with purplish or pinkish stripes outside, broadly tubular, 2.5–3 cm long, outside pilose; tube base cylindric, 4–5 mm long, 4–5 mm broad, upwards slightly inflated and bent; upper lip ovate-oblong, 1.5–1.8 cm long, erect, shallowly 2-lobed, lower lip oblong-circular, spreading, 3-lobed, the middle lobe subcircular, 7–9 mm long, 4–5 mm broad, lateral lobes ovate, 6–8 mm long, 4–4.5 mm broad. Stamens 2, exserted; filaments 1.2–1.5 cm long, declinate, glabrous except basally sericeous; anther-thecae ellipsoid, 3–3.5 mm long, equal, superposed, lower one spurred at base. Ovary pubescent especially towards apex; style 2.2–2.5 cm long, recurved, basal part pubescent; stigma simple. Capsules obovoid, 2.5–4 cm long, 4–5 mm broad. Seeds orbicular in outline, 3–4 mm long and broad. **Chromosomes** 2n = 34. Figure 46-108.

Phenology: flowering from January to March; fruiting from April to June.

Habitats: roadsides, thickets; below 2200 m.

Distribution: N Pakistan, Jammu & Kashmir, U Ganga & Indus, W Nepal, C Nepal, E Nepal, Sikkim & Darjiling, Bhutan, S Hengduan (Map 46-189); also in Sri Lanka, Bangladesh, Thailand, Laos, Vietnam, Cambodia, Malaysia, Indonesia, and other parts of Pakistan, India, Myanmar and China (S Yunnan, Guizhou, Guangxi, Hainan, Guangdong); widely cultivated and naturalized in tropics, obscuring native distribution.

Uses: medicine.

N PAKISTAN: Azad Kashmir, Mirpur, Jhelum Dist., 1 May 1937, *K. L. Malhotra 15849* (RAW, fide Malik & Ghafoor, 1988); **Chitral**, Dir, Dir-Bajur near Nanargari Bazar, 20 Feb. 1973, *S. M. A. Kazmi 4797* (KUH, fide Malik & Ghafoor, 1988); **Hazara**, Havalian, 20 Aug. 1972, *M. Qaiser & A. Ghafoor 4843* (KUH, fide Malik & Ghafoor, 1988); Thal, 19 May 1978, *S. Nazimuddin, S. Abedin & Hamidullah 988* (KUH, fide Malik & Ghafoor, 1988); **Swat**, Ugarh, ca. 8 miles from Mingora, 11 Apr. 1970, *S. I. Ali & S. A. Faruqi 1658* (KUH, fide Malik & Ghafoor, 1988). **JAMMU & KASHMIR**: **Jammu**, Ram Nagar, 21 Apr. 1970, *B. M. Sharma 526* (RRLH); **Rajouri**, Sunderbani, 600 m, 6 Jan. 2010, *Akhtar 1103* (KASH, fide Hassan Dar et al., 2014). **U GANGA & INDUS**: **Chamba**, Bakairi Nala, 1500 m, 17 May 1897, *J. H. Lace 1538* (E); **Dehradun**, 690 m, *J. S. Gamble 24442* (K); *D. W. D. Siha s. n.* (FHO); *U Singh 191* (A); 28 Mar. 1923, *H. N. Puri 94* (LE); Jun. 1930, *E. K. Kingham 40* (A); 16 Mar. 1921, *Bahoan & Singh 16* (A); Dehradun to Mussoorie Rd., 960 m, 21 Mar. 1975, *D. Chambulaim 1* (E); **Kumaon**, Jalikoti, 1200 m, 30 Jan. 1910, *G. Nill 6* (E); **Mussorie**, *M. B. Saxena 186* (DD, fide Raizada & Saxena, 1978); **Simla**, Bashahr, 24 Apr. 1891, *J. H. Lace 839* (E). **W NEPAL**: **Bheri**, Banke, Napelganj, Gharbari Tol, 1 Jan. 1973, *N. P. Manandhar 10383* (KATH); Bardiya, Balapur, 180 m, 22 Feb. 1993, *P. P. Kurmi KB639* (KATH); Dailekh, Lohari, 790 m, 28 Feb. 1991, *N. P. Manandhar 514/91* (KATH); Surkhet, 670 m, 27 Oct. 1975, *T. B. Shrestha & N. P. Manandhar 499* (KATH); Surkhet, Surkhet to Katuka, 720–1380 m, 23 Dec. 1983, *T. K. Chatthacharya & P. P. Kurmi 2588* (KATH); Surkhet, Kholta Pani, 550 m, 31 Dec. 1982, *P. P. Kurmi & D. K. Shrestha 4962* (KATH); Surkhet, Jun Khola, 700 m, 13 Apr. 1974, *J. F. Dobremez 2479 (N. P. Manandhar 74-108)* (BM, E, KATH); Surkhet, Solighopte, 700 m, 21 Feb. 1991, *N. P. Manandhar & N. Acharya 1-91* (KATH); Surkhet, Kalyan, 870 m, 23 Feb. 1991, *N. P. Manandhar & N. Acharya 148-91* (KATH); **Karnali**, Dolpa, Jarap, 1200 m, 23 Apr. 1965, *T. B. Shrastha 4012* (KATH); **Mahakali**, Dadeldhura, Jogbudha, 350 m, 18 Jan. 1983, *K. R. Rajbhandari et al. 7066* (KATH); Dadeldhura, Okhadhunga, 1500 m, 15 Apr. 1982, *N. Rawal 741* (KATH); Dadeldhura, Pokhara, 1500 m, 27 Dec. 1980, *L. P. Kattel 147* (KATH);

Figure 46-108: **Justicia adhatoda** L.: **a**, portion of plant; **b**, opened corolla showing stamen; **c**, capsule; **d**, seed. Drawn by Quan LIN; reproduced from *Icon. Cormophyt. Sin.* 4: 174, fig. 5761. 1975.

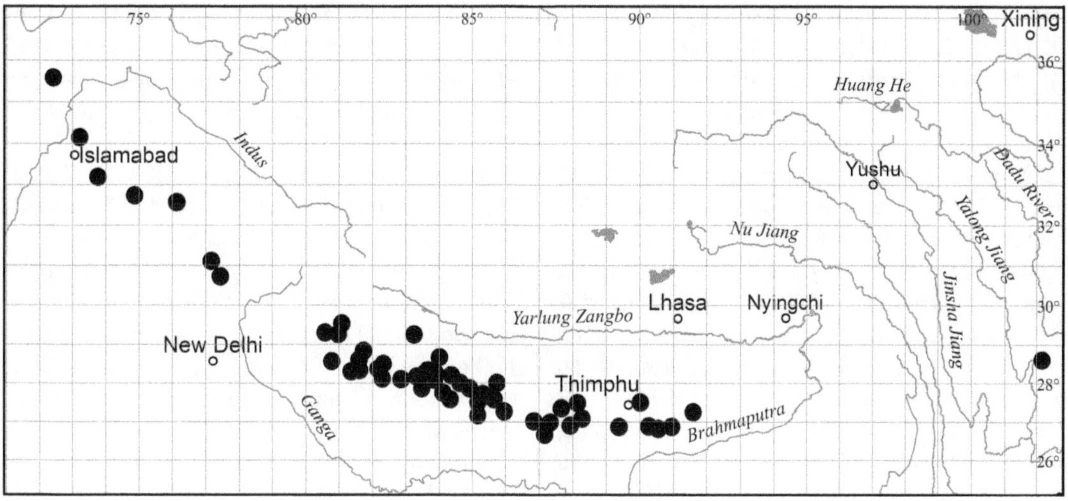

Map 46-189: **Justicia adhatoda** L.

Darchula, Mallaj, 29 Jan. 1964, *J. Kamakita s. n.* (KYO); **Rapti**, Dang, Patu Khola, 730 m, 26 Mar. 1952, *O. V. Polunin, W. R. Sykes & L. H. J. Williams 605* (BM, E); Dang, Ghorahi, 630 m, 15 Apr. 1985, *P. M. Regmi & P. P. Kurmi 4302* (KATH); Dang, Gadhawa, 250 m, 28 Aug. 1982, *N. P. Manandhar & N. K. Bhattarai 8511* (KATH); Dang, Koilabas, 21 Mar. 1973, *N. P. Manandhar et al. 9789* (KATH); Dang, Ghorahi to Lepe, 25 Dec. 1984, *T. K. Bhattacharya & Shrivastava 84.55/10* (KATH); Dang, Dang Valley, 300 m, 7 Dec. 1978, *T. K. Bhattacharya 514* (KATH); Dang, 26 Feb. 1974, *K. J. Malla 14579* (KATH); Pyuthan, Bang Sal, 460 m, 6 Mar. 1991, *P. P. Kurmi 0055* (KATH); Pyuthan, Lampapokhari, 700 m, 12 Feb. 1983, *Poudyal et al. 5304* (KATH); Rolpa, Sallebot, 1800 m, 24 Sep. 1988, *N. K. Bhattarai & S. R. Baral 88/479* (KATH); Salyan, Phalwang to Salyan, 1429–1841 m, 18 Jan. 1982, *s. coll. W225/82* (KATH); Salyan, Ram Dada, 1400 m, 29 Nov. 1991, *P. P. Kurmi KB108* (KATH); **Seti**, Bajhang, 1100 m, 4 Sep. 1990, *N. K. Bhattarai 90/1281* (KATH); Doti, Kaher Khola, 1000 m, 16 Apr. 1973, *J. F. Dobremez 1841* (BM, E, KATH); Kailali, Thuligad, 25 Mar., 1929, *B. Ram 84* (LE). **C NEPAL: Bagmati**, Bhaktapur, Nagarkot, 1950 m, 11 Mar. 1975, *D. P. Joshi & Rajbhandari 75/473* (KATH); Bhaktapur, Cipadol, 1700 m, 14 Jul. 1986, *N. K. Bhattarai 86/649* (KATH); Dhading, on the way from Naubise to Palung, 959 m, 12 Feb. 2003, *T. Watanabe et al. s. n.* (MBK, TI); Dhading, Gajuri, 600 m, 19 Jun. 1986, *N. P. Manandhar & L. P. Katel 10552* (KATH); Dhading, Tarke, 1170 m, 21 Jun. 1986, *N. P. Manandhar & L. P. Katel 10660* (KATH); Dhading, Majuwa, 900 m, 8 Nov. 1989, *N. P. Manandhar 13128* (KATH); Dhading, Chyangadi, 1100 m, 10 Nov. 1989, *N. P. Manandhar 13260* (KATH); Kathmandu, Sankhu, 1400 m, 7 Jul. 2002, *T. Watanabe et al. 70* (MBK, TI); Kathmandu, Naubise to Mungling, 5 Nov. 2000, *T. Watanabe et al. s. n.* (MBK); Kathmandu, near Vishnumati River, Dalu Thole, 1280 m, 19 Apr. 1992, *D. G. Long & S.E. McDermott 21899* (E); Kathmandu, Parping, 1600 m, 31 Dec. 1966, *D. H. Nicolson 2902* (BM, KATH, TI); Kathmandu, Nagarjun Forests, 3 Jul. 1961, *A. B. Shrestha et al. 115* (KATH); Kathmandu, Nagarjun, 1500 m, Feb. 1964, *Shrestha & Upadhyay 628* (KATH); Kathmandu, Balaju to Swayambhu, 1300 m, 7 Feb. 1969, *H. Kanai 15046* (KATH); Kathmandu, Balaju, 1300 m, 7 Feb. 1969, *H. Kanai 15096* (KATH); Kathmandu, 18 Feb. 1969, *H. Kanai et al. s. n.* (TI); Kathmandu, Tika Bhairab to Champi, 1450–1560 m, 16 Mar. 1973, *M. Amatya & T. K. Bhatthacharya 73-52* (KATH); Kathmandu, Gokarna, 1260 m, Apr. 1966, *P. Prabhe & R. Ihapa 4398* (KATH); Kathmandu, Swayanbhu Nath, 1500 m, 15 Jun. 1967, *H. Hara et al. 9127* (TI); Kavre Palanchok,

Phulchoki, 2100 m, 5 Jul. 1968, *s. coll. 10802* (KATH); Kavre Palanchok, Khardar Pati, 1550 m, 19 Nov. 1988, *N. P. Manandhar 12192* (KATH); Kavre Palanchok, Palanchok, 1450 m, 6 Oct. 1984, *N. K. Bhattarai 84679* (KATH); Lalitpur, Bajrabasahi, Chapgaon, 1400 m, 16 May 1976, *D. P. Joshi & R. Joshi 1016* (KATH); Lalitpur, Godavari, 1650 m, Mar. 1966, *P. Pradhan 4315* (KATH); Lalitpur, Godawari, Royal Botanical Garden, 20 Apr. 2002, *B. Silwal & R. K. Bista 442* (KATH); Nuwakot, Sikre, 1100 m, 24 Nov. 2012, *Forrest Himal Brandt 60* (CMUB); Palanchok, 1450 m, 6 Oct. 1984, *N. K. Bhattarai 84/679* (KATH); Sindupalchok, Syaule, 1650 m, 30 Sep. 1984, *N. K. Bhattarai 84/560* (KATH); Sindupalchok, Nurbu Gaon, 1600 m, 5 Nov. 1985, *N. K. Bhattarai 85/847* (KATH); Sindupalchok, Lamosangu to Thulopakhra, 720–1750 m, 18 Apr. 1978, *I. Sharma et al. 78/116* (KATH); **Dhawalagiri**, Baglung, Jimire, 1600 m, 7 Sep. 1991, *N. P. Manandhar 920-91* (KATH); Baglung, Burtibang, 1100 m, 30 May 1992, *N. P. Manandhar & S. K. Acharya 534-92* (KATH); Baglung, Nwar, 820 m, 31 May 1992, *N. P. Manandhar & S. K. Acharya 566-92* (KATH); Parbat, Kusma, 13 Sep. 1991, *N. P. Manandhar 1096/91* (KATH); **Gandaki**, Gorkha, 940 m, 23 Sep. 1995, *M. E. Olsen 372* (KATH); Gorkha, Jaunbari, 930 m, 11 May 1987, *N. P. Manandhar & L. P. Katel 11589* (KATH); Gorkha, Pandoung Deanrali, 1100 m, 12 May 1987, *N. P. Manandhar & L. P. Katel 11739* (KATH); Kaski, 880 m, 29 Jun. 1986, *N. P. Manandhar & L. P. Katel 11133* (KATH); Lamjung, Phalan Sangu, 760 m, 22 Dec. 1979, *N. P. Manandhar 3307* (KATH); Manang, Dharpani Sangra, 900 m, 31 Jan. 1984, *T. K. Bhatthacharya 5792* (KATH); Syangja, Dhale, 1300 m, 16 Mar. 1974, *D. P. Joshi & M. M. Amatya 74/1592* (KATH); **Janakpur**, Sindhuli, Mulkot, 1450 m, 21 Nov. 1988, *N. P. Manandhar 12312* (KATH); Sindhuli, Khurkot, 1200 m, 22 Nov. 1988, *N. P. Manandhar 12399* (KATH); **Lumbini**, Gulmi, Sardhi Kheska to Tarnghas, 984–1587 m, 3 Feb. 1982, *s. coll. W451/82* (KATH); Nawalparasi, Tribeni, 500 m, 10 Apr. 1973, *J. F. Dobremez 1785* (BM, E, KATH); Palpa, Masure, 1200 m, 24 Nov. 1973, *D. P. Joshi & M. M. Amartya 73-961* (KATH); Palpa, Hathikot, 1000 m, 27 Feb. 1974, *D. P. Joshi & M. M. Amatya 74/1278* (KATH); Palpa, Sarakot, Kaseni Panchayet, 1400 m, 5 Mar. 1974, *D. P. Joshi & M. M. Amatya 74/1425* (KATH); Palpa, Shreepur Danda, 1400 m, 21 Dec. 1975, *D. P. Joshi et al. 75/3651* (KATH); Palpa, 1250 m, 1 Mar. 1976, *N. P. Manandhar & P. M. Regmi 6* (KATH); Palpa, Dumure, 1100 m, *K. Yoda s. n.* (KYO); **Narayani**, Bara, Gumi, 130 m, 18 Fenruary 1993, *L. P. Kattel 212/93* (KATH); Chitwan, Sauraha, 160–200 m, 15 Jan. 1996, *K. Yoda & T. Kurosawa 9614067* (E, TI); Makwanpur, Suparitar, 650 m, 12 Feb. 1987, *N. K. Bhattarai 87/126* (KATH); Makwanpur, Chainpur, 13 Nov. 1989, *N. P. Manandhar 13905* (KATH); Makwanpur, Bhimphedi, between Bimphedi & Katmandu, 31 Mar. to 3 Apr. 1953, *P. C. Gardner 1623* (BM). **E NEPAL: Koshi**, Dhankuta, Phedi, 1200 m, 12 Nov. 1978, *P. Pradhan et al. 738* (KATH); Sunsari, Baraha Chetra, 170 m, 22 Jan. 1982, *K. R. Rajbhandari et al. 6041* (KATH); Sunsari, Phusri to Dhara Pani, 16 Oct. 1963, *H. Hara et al. 6303613* (TI); **Mechi**, Ilam, near Ilam, 800 m, 21 Apr. 1962, *K. Nishioka 1226* (KYO); Ilam, Mahabarat, 1100 m, 16 Apr. 1973, *AMA 733* (E); Ilam, Ilam to Ranga Pani, 8 Dec. 1963, *H. Hara et al. 6303624* (KATH, TI); Ilam, Dhara Pani to Illam, 5 Dec. 1963, *H. Hara et al. 6303625* (A, BM, K, KYO, TI); Ilam, Lanu to Chisa Pani, 8 May 1984, *T. K. Bhatthacharya & S. B. Malla 4210* (KATH); Taplejung, Barundin, 25 Nov. 1963, *H. Hara et al. 6303623* (KATH, KYO, TI); **Sagarmatha,** Khotang, Regmitar to Khani Danda, 28 Oct. 1995, *M. Mikage et al. 9551203* (TI); Khotang, Regmitar to Khani Danda, 390 m, 28 Oct. 1995, *M. Mikage et al. 9555060* (A, E, KATH, TI). **SIKKIM & DARJILING**: **Darjiling**, *J. M. Cowan s. n.* (E); **Sikkim**, *Griffith 2441* (K). **BHUTAN**: **Chukha**, on the footpath in Tectona plantation, 200 m, 1 May 1979, *A. J. C. Grierson & D. G. Long 710* (E); **Gaylephug**, Taklai Khola E of Gaylephug, 320 m, 23 Mar. 1982, *A. J. C. Grieson & D. G. Long 3955* (E); **Mongar**, Mongar to Lhuntshi, 1370 m, 21 Feb. 1987, *Bigger 2808* (E); **Phuntsoling**, Torsa River, Phuntsoling, 200 m, 11 Feb. 1982, *A. J. C. Grieson & D. G. Long 2866* (E); **Punakha**, near Wache, Tang Chu, 1610 m, 20 Apr. 1982, *A. J. C. Grieson & D. G. Long 4533* (E); **Sarbhang**, near Sarbhang Bazaar, 500 m, 23 Feb. 1988, *J. R. I.*

Wood 6158 (E); **Shamgong**, 27 km S of Shamgong between Shamgong and the Mangde Chu river, 950 m, *B. Bartholomew & D. E. Boufford 3807* (A, CAS); ca. 20 km north of Shemgang towards Kharila, 1500 m, 26 Feb. 1988, *J. R. I. Wood 6177* (E); **Tongsa**, Dakpai, 1200 m, 15 Mar. 1992, *J. R. I. Wood 7545* (E). **S HENGDUAN**: **Longyang**, Baihualing, 1600 m, 24 Jan. 2008, *Y. F. Deng 20484* (IBSC); **Mianning**, Taniushan, 1450 m, *T. T. Yü 18203* (IBSC, KUN).

IV. Justicia sect. **Rhaphidospora** (Nees) T. Anderson, J. Proc. Linn. Soc., Bot. 7: 43. 1863; Clarke in Hooker (ed.), Fl. Brit. India 4: 535. 1885. ≡ *Rhaphidospora* Nees in Wallich (ed.), Pl. Asiat. Rar. 3: 77, 115. 1832; Nees in de Candolle (ed.), Prodr. 11: 49. 1847; Hu, Fl. Reipubl. Popularis Sin. 70: 253. 2002. ≡ *Rhaphidosperma* G. Don in Sweet, Hort. Brit. ed. 3: 559. 1839, nom. illeg. (Art. 52.1). Type: *Justicia glabra* Koenig ex Roxb.

Hemichoriste Nees in Wallich (ed.), Pl. Asiat. Rar. 3: 76, 102. 1832. ≡ *Justicia* sect. *Hemichoriste* (Nees) Benth. & Hook. f., Gen. Pl. 2:1109. 1876. Type: *Hemichoriste montana* Nees [≡ *Justicia montana* (Nees) Wall. ex T. Anderson].

Gendarussa Nees in Wallich (ed.), Pl. Asiat. Rar. 3: 76, 103. 1832; Nees in de Candolle (ed.), Prodr. 11: 410. 1847; Hu, Fl. Reipubl. Popularis Sin. 70: 299. 2002. ≡ *Justicia* sect. *Gendarussa* (Nees) T. Anderson, J. Proc. Linn. Soc., Bot. 7: 41. 1863; Clarke in Hooker (ed.), Fl. Brit. India 4: 532. 1885. ≡ *Justicia* subg. *Gendarussa* (Nees) Lindau in Engler & Prantl (eds.), Nat. Pflanzen-fam. 4(3B): 347. 1895. Type: *Gendarussa vulgaris* Nees (≡ *Justicia gendarussa* Burm. f.).

Shrubby, or more rarely herbaceous, perennials. Inflorescences dichasial, axis glandular. Bracts absent in simple dichasia, or subulate to ovate and exceeding to equalling or shorter than calyx in compound dichasia. Calyx 5-lobed; lobes equal. Corolla white or cream to more rarely pale pink. Anther-thecae superposed, slightly oblique or oblique, equal or lower one larger, appendages large. Ovary densely pubescent, or more rarely pubescent to glabrous. Capsules stalked at base, sterile portion about 1/2 of its length, densely pubescent or pubescent. Seeds not compressed to scarcely compressed or compressed, without margin, rounded-rugulose-tuberculate.

About 50 species: Old World, centred in West Tropical Africa, extending through tropical Africa to South Africa with a few species in India, Malaysia, New Caledonia and Australia; four species in the Pan-Himalaya.

8. Justicia pubigera (Nees) C. B. Clarke, Fl. Brit. India 4: 536. 1885; Clarke in Hooker (ed.), Fl. Brit. India 4: 536. 1885; Nasir & Ali in Malik & Ghafoor (eds.), Fl. Pakistan 188: 42. 1988. ≡ *Peristrophe pubigera* Nees in Wallich (ed.), Pl. Asiat. Rar. 3: 114. 1832; Nees in de Candolle (ed.), Prodr. 11: 498. 1847. ≡ *Ecbolium pubigerum* Kuntze, Revis. Gen. Pl. 2: 981. 1891. ≡ *Justicia pubigera* Wall., Numer. List no. 2490. 1830, nom. invalid. Type: India, Budhrinath Kamroop, *Wallich Cat. 2490* [lectotype designated by Malik & Ghafoor (eds.), Fl. Pakistan 188: 42. 1988: K].

Rhaphidospora ciliata Nees, Prodr. 11: 500. 1847, non *Justicia ciliata* Jacq., Hort. Bot. Vindob. 2: 47. 1772. ≡ *Justicia edgeworthii* T. Anderson, J. Linn. Soc., Bot. 9: 516. 1867. Type: India, Oct. 1838, *E. Edgeworth 255* [lectotype **here designated**: GZU (GZU000250810); isolectotype: GZU].

Herbs, 30–60 cm high; stems terete, striate, pubescent when young. Petiole 0.5–1.5 cm long, pubescent; leaf blade variable in size, lower ones elliptic to ovate-lanceolate, 4.5–11 cm long, 2–4.5 cm broad, pubescent on both surfaces especially along veins, lateral veins 6–8 pairs, base cuneate, margin entire, apex acute or bluntly acuminate. Panicle of cymes axillary, 3–8 cm long; peduncles 1.2–6 cm long. Bracts linear, 2–3 mm long, pubescent; bracteoles rudimentary or absent. Calyx 5-lobed; lobes linear, 4–6 mm long, glandular-hairy outside. Corolla 1.5–1.8 cm long, pinkish-red, lower lip with a conspicuous white branched veins, pubescent; tube base 6–8 mm long; upper lip 7–9 mm long, emarginate, lower lip 6–8 mm long, deeply 3-lobed. Capsules oblong, clavate, 1.2–1.3 cm long, pubescent. Seeds ovate in outline, compressed, 2–2.3 mm long, 1.8–2 mm broad. **Chromosomes** 2n = 28.

Phenology: flowering from August to October.

Habitats: roadside, shady places; 800–2100 m.

Distribution: N Pakistan, Jammu & Kashmir, W Nepal, C Nepal (Map 46-190); also in other parts of India and Pakistan.

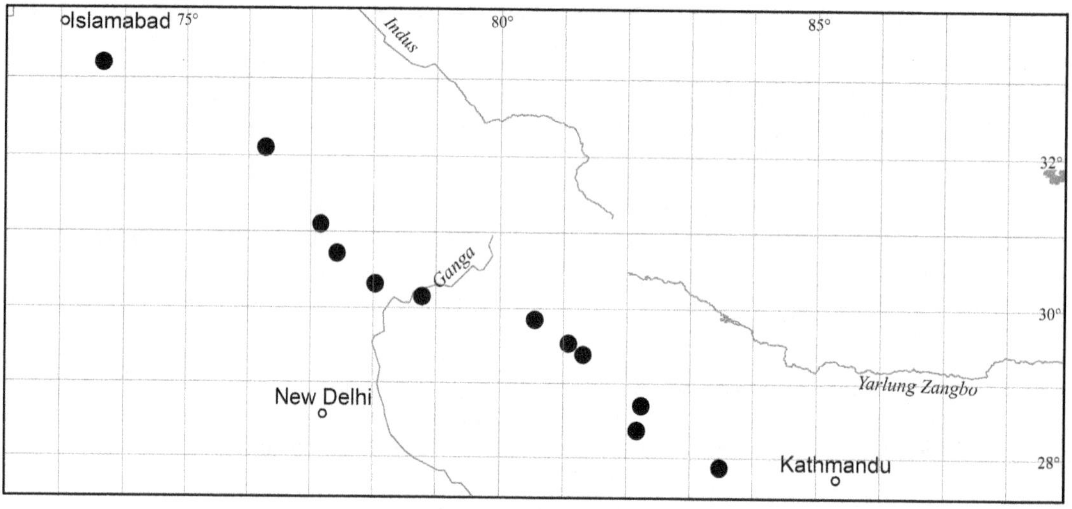

Map 46-190: **Justicia pubigera** (Nees) C. B. Clarke

N PAKISTAN: **Azad Kashmir**, **Poonch**, Pail, under Pinus excelsa, *A. Rashid Khan 109* (RAW, fide Malik & Ghafoor, 1988). **JAMMU & KASHMIR**: **Dehradun**, Jashimath, North Garhwal, 1650 m, *Office Slip 1149* (K); **Kangra**, *M. P. Edgeworth 253* (K); **Kumaon**, Banks of Sarju, 1350 m, 10 Aug. 1848, *R. Strachey & J. E. Winterbottom 1* (K); **Mussoorie**, 2100 m, Oct. 1873, *W. S. Atkinson 19491* (K); **Garhwal**, Govindgath, Bhyundar Valley, between Josimath and Badrinath, 1900 m, 13 Aug. 1975, *F. Billiet & J. Leonard 6497* (K); *Royle s. n.* (LIV); **Simla**, *Madden s. n.* (K). **W NEPAL**: **Bheri**, Jajarkot, Gad Pangchi, 1680 m, 16 Oct. 1952, *O. V. Polunin, W. R. Sykes & L. H. J. Williams 5749* (BM, E); **Mahakali**, Darchula, Makarigad, 1400 m, 2 Aug. 2015, *S. K. Ghimire et al. ANCA 1596* (TUCH); Darchula, Sipti–Serigaon, 1660–850 m, 31 Aug. 1982, *M. M. Amatya & P. M. Regmi W607/82* (KATH); **Rapti**, Salyan, Kapurkot, 1442 m, 8 Sep. 2014, *B. Adhikari 53* (E); **Seti**, Bajura, Kala Gad, 1984 m, 23 Aug. 2017, *B168* (E); Dajhang, Trail from Patiudi to Surma, 2157 m, 7 *Jul.* 2009, *H. Ikeda et al. 20913069* (E). **C NEPAL**: **Lumbini**, Palpa, Batase, 1478 m, 13 Jul. 2016, *B. Adhikari & G. Parmar BG53* (E).

9. Justicia gendarussa Burm. f., Fl. Ind.: 10. 1768; Clarke in Hooker (ed.), Fl. Brit. India 4: 532. 1885; Nasir & Ali in Malik & Ghafoor (eds.), Fl. Pakistan 188: 43. 1988; Wood in Grierson & Long (eds.), Fl. Bhutan 2(3): 1287. 2001; Hu, Deng & Daniel in Wu, Raven & Hong (eds.), Fl. China 19: 457. 2011. ≡ *Gendarussa vulgaris* Nees in Wall. Pl. Asiat. Rar. 3: 104. 1832; Nees in de Candolle (ed.), Prodr. 11: 410. 1847; Hu, Fl. Reipubl. Popularis Sin. 70: 300. 2002. ≡ *Ecbolium gendarussa* (Burm. f.) Kuntze, Revis. Gen. Pl. 2: 487. 1891. Type: Indonesia, Java, *s. coll. s. n.* [lectotype **here designated**: LINN (LINN-HS46-7)].

Subshrubs, 0.7–1.5 cm high; stems subterete, glabrous. Petiole 3–10 mm long; leaf blade narrowly lanceolate, 6–10 cm long, 1–1.5 cm broad, glabrous, lateral veins 5–8 pairs, base cuneate to attenuate, margin subsinuate, apex acute to shortly acuminate. Spikes terminal or axillary, 3–12 cm long, interrupted, usually in a leafy panicle; peduncle 0.5–1.5 cm long. Bracts triangular, 2–6 mm long, 1–2.5 mm broad, basal ones longer than calyx upwards gradually smaller with apicalmost ones shorter than calyx, margin ciliate, apex acute; bracteoles elliptic to linear-lanceolate, 2.5–3 mm long, 0.8–1 mm broad, margin ciliate, apex acute. Calyx 4–5 mm long, 5-lobed; lobes linear-lanceolate, 3–4 mm long, 0.5–0.7 mm broad, subequal, apex acuminate. Corolla creamy white, 1.2–1.5 cm long; tube base cylindric, 1.7–2 mm broad, 8–9 mm long; upper lip violet blotched, triangular, 5–7 mm long, 3–3.5 mm broad, 2-cleft, lower lip violet dotted basally, cuneate-obovate, 6–10 mm broad, 3-lobed, lobes oblanceolate and 3–5 mm long, 3–3.5 mm broad. Stamens 2, exserted; filaments 3–6 mm long, glabrous; anther-thecae oblong,1–1.2 mm long, superposed, lower one spurred at base, upper one muticous. Ovary glabrous; style 1–1.2 cm long, glabrous; stigma capitate, shortly 2-lobed. Capsules clavate, 1–1.2 cm long. **Chromosomes** 2n = 30. Figure 46-109.

Phenology: flowering from January to April.

Habitats: roadsides, thickets.

Distribution: U Ganga & Indus, E Nepal, Bhutan (Map 46-191); native to S and SE Asia but is widely cultivated and naturalized.

U GANGA & INDUS: **Mussorie**, *M. B. Saxena 236* (DD, fide Raizada & Saxena, 1978). **E NEPAL**: **Mechi**, Jhapa, Dhulabari to Bhadrapur, 200 m, 20 Mar. 2003, *P. P. Kurmi et al. s. n.* (MBK). **BHUTAN**: **Phuntsholing**, Puntsholing, 200 m, 11 Feb. 1982, *A. J. C. Grierson & D. G. Long 2858* (E).

10. Justicia virgata (Nees) T. Anderson, J. Linn. Soc., Bot. 9: 516. 1867. ≡ *Leptostachya virgata* Nees in Wallich (ed.), Pl. Asiat. Rar. 3: 105. 1832; Nees in de Candolle (ed.), Prodr. 11: 379. 1847. ≡ *Dianthera virgata* (Nees) Benth., Gen. Pl. 2: 1114. 1876; Clarke in Hooker (ed.), Fl. Brit. India 4: 542. 1885. ≡ *Ecbolium virgatum* (Nees) Kuntze, Revis. Gen. Pl. 2: 981. 1891. Type: Myanmar, Taong Dong, 26 Nov. 1826, *Wallich Cat. 7177* [lectotype (as "holotype") designated by Graham, Kew Bull. 43: 590. 1988: K-W (K001126828); isotype: BM].

Herbs or small shrubs, 60–90 cm high; stems slender. Petiole 2–3 cm long, pubescent; leaf blade elliptic, 15–18 cm long, 8–10 cm broad, pubescent along veins abaxially, base attenuate or rhomboidal, margin entire, apex acuminate. Spikes terminal, 1–3 congested or sometimes in panicles, 15–25 cm long; flowers distinct on rachis. Bracts ovate or elliptic, 1.8–2 mm long, pubescent, apex acute; bracteoles subulate, 1.7–2 mm long, pubescent. Calyx 5–6 mm long, pubescent; lobes linear-lanceolate, apex acuminate. Corolla 8–11 mm long, white or creamy white, sometimes yellow tinged, pubescent, 2-lipped, upper lip bilobed, lower lip trilobed. Stamens 2; anther-thecae superposed, spherical, lower one spurred at base. Capsules clavate, pubescent, 4-seeded. Seeds ovate in outline, compressed, glabrous; tubercular-verrucose.

Phenology: flowering from November to January of next year.

Habitats: moist shady places; 300–900 m.

Distribution: Sikkim & Darjiling, Yarlung Zangbo-Brahmaputra (Map 46-192); also in other parts of India and Myanmar.

SIKKIM & DARJILING: **Sikkim**, *Griffith s. n.* (K).

11. Justicia vasculosa (Nees) T. Anderson, J. Linn. Soc., Bot. 9: 515. 1867; Clarke in Hooker (ed.), Fl. Brit. India 4: 533. 1885; Wood in Grierson & Long (eds.), Fl. Bhutan 2(3): 1288. 2001. ≡ *Gendarussa vasculosa* Nees in Wallich (ed.), Pl. Asiat. Rar. 3. 104. ≡ *Adhatoda vasculosa* Nees in de Candolle (ed.), Prodr. 11: 407. 1847. ≡ *Ecbolium vasculosum* (Nees) Kuntze, Revis. Gen. Pl. 2: 981. 1891. ≡ *Mananthes vasculosa* (Nees) Bremek., Verh. Kon. Ned. Akad. Wetensch., Afd. Natuurk., Sect. 2 45(2): 59. 1948; Hu, Fl. Reipubl. Popularis Sin, 70: 298. 2002. ≡ *Justicia vasculosa* Wall., Numer.

Figure 46-109: **Justicia gendarussa** Burm. f.: **a**, portion of plant; **b**, corolla with bract, bracteoles and calyx; **c**, stamen. Drawn by Quan LIN; reproduced from *Icon. Cormophyt. Sin.* 4: 174, fig. 5762. 1975.

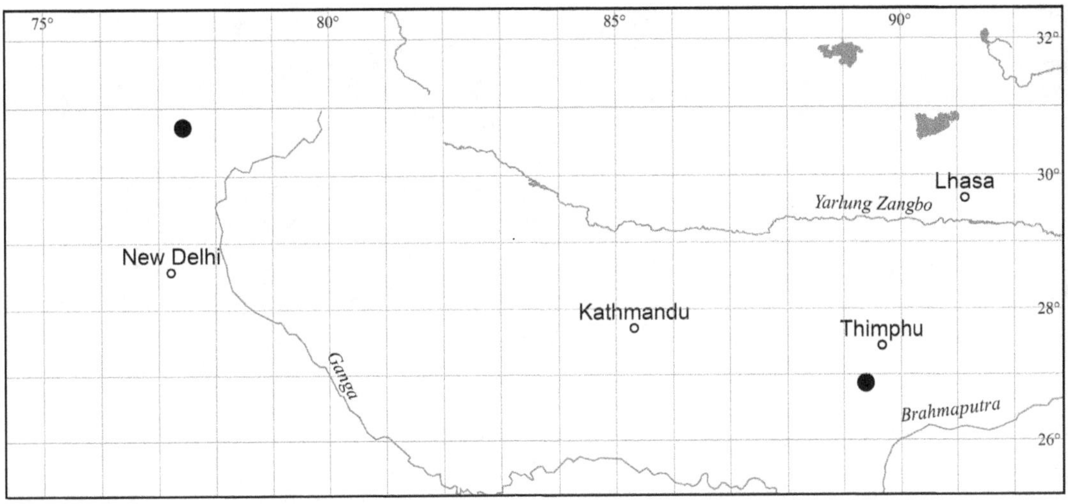

Map 46-191: **Justicia gendarussa** Burm. f.

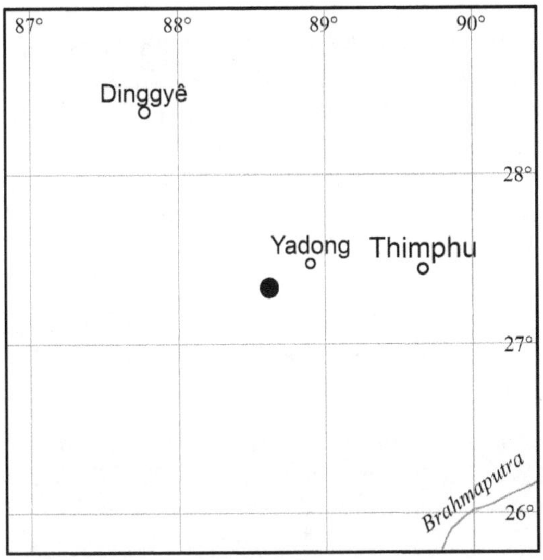

Map 46-192: **Justicia virgata** (Nees) T. Anderson

List no. 2469. 1830, nom. invalid. Type: Bangladesh, Silhet, *F. de Silva in Wallich Cat. 2469a* [lectotype **here designated**: K-W (K001116145)].

Perennial subshrubs, 0.6–1 m high; stems glabrous. Petiole 1–5 cm long, glabrous; leaf blade ovate-elliptic, 5–20 cm long, 2–8 cm broad, glabrous, lateral veins 7–9 pairs, base attenuate, margin entire, apex shortly acuminate. Spikes axillary and terminal, simple or in panicles, 2–12 cm long; flowers arranged in pairs along rachis, usually interrupted below; peduncle 0–3 cm long. Bracts linear-lanceolate, 2.5–3 mm long, pubescent; bracteoles linear-lanceolate, 3–4 mm long, 1.2–1.5 mm broad. Calyx 4–5 mm long, pubescent, becoming scarious in fruit, 5-lobed; lobes lanceolate, apex purple-tipped. Corolla yellow-brown with purple markings, 1–1.3 cm long, pubescent; tube cylindric 1–1.2 cm long, 5–6 mm broad; upper lip 1–1.2 cm long, 1.7–2 cm broad, lower lip bent downwards, broadly elliptic, 1–1.2 cm long, 2–3 mm broad. Stamens 2; filaments 5–6 mm long; anther-thecae superposed, lower one indistinctly spurred. Ovary glabrous; style 1–1.2 cm long, pubescent on lower portion. Capsules clavate, 1.2–1.8 cm long, glabrous.

Phenology: flowering from January to April.

Habitats: stream banks in moist forests; 100–900 m.

Distribution: Sikkim & Darjiling, Bhutan, Yarlung Zangbo-Brahmaputra (Map 46-193); also in Thailand, and other parts of India, Myanmar and China (S Yunnan).

SIKKIM & DARJILING: **Sikkim**, Baumanpolin, 450 m, *J. S. Gamble 10255* (CAL, K). **BHUTAN**: **Gaylegphug**, N Gaylegphug, 800–900 m, 14 Feb. 1988, *J. R. I. Wood 6145* (E); Gaylegphug to Surey, above the Shershong

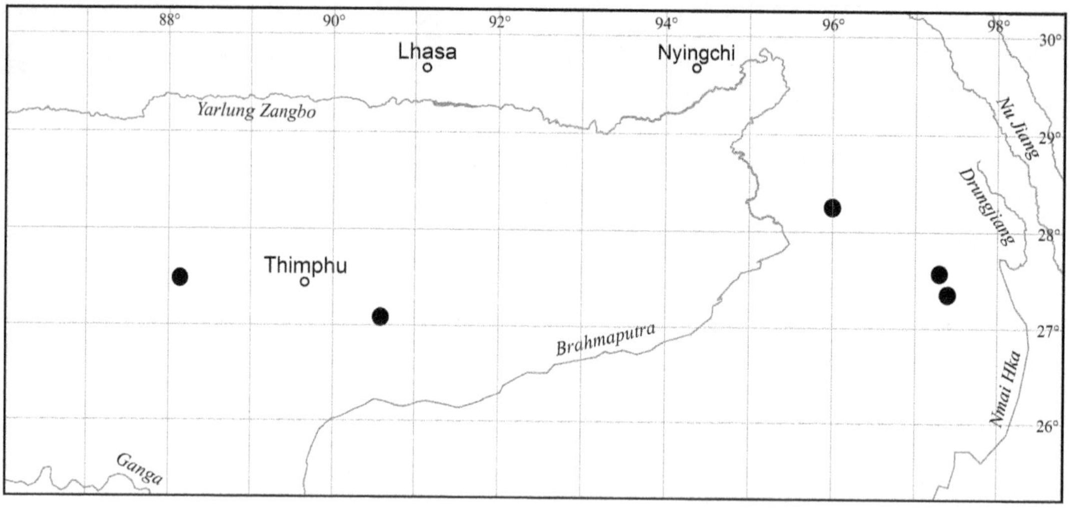

Map 46-193: **Justicia vasculosa** (Nees) T. Anderson

Bridge, 700 m, 19 Feb. 1989, *J. R. I. Wood 6869* (E). **YARLUNG ZANGBO-BRAHMAPUTRA**: **Aprunyi**, Pasighat, Dihang Valley, 150–300 m, 9 Feb. 1928, *F. Kingdon-Ward 7818* (K); near Pasighat, *I. H. Burkill 37115* (CAL); between Rammi Dam Bang & Janakmuth, 270 m, *I. H. Burkill 36392* (CAL); Takarpami near Pasighat, *I. H. Burkill 37438* (CAL); **Duffa Hills**, Jan. 1890, *King 38* (CAL); **Mishimee**, *Herbs Griffith 6163* (K). **U IRRAWADDY**: **Putao**, Wasadam to Ziyadam, 15 Oct. 2014, *Putao Expedtion Team (X. H. Jin et al.) PT-ET 465* (PE); Ziyadam, 21 Oct. 2014, *Putao Expedtion Team (X. H. Jin et al.) PT-ET 2014* (PE).

V. Justicia Sect. **Harnieria** (Solms-Laub.) Benth. in Bentham & Hooker (eds.), Gen. Pl. 2: 1109. 1876. ≡ *Harnieria* Solms-Laub., Sitzungsber. Ges. Naturf. Freunde Berlin 1864: 1. 1864. Type: *Justicia dimorphocarpa* Solms-Laub. (= *Justicia heterocarpa* T. Anderson).

Justicia sect. *Calophanoides* C. B. Clarke in Hooker (ed.), Fl. Brit. India 4: 530. 1885. ≡ *Calophanoides* (C. B. Clarke) Ridl., Fl. Malay Penins. 2: 592. 1923; Hu, Fl. Reip. Popularis Sin. 70: 277. 2002. Type (designated by Bremekamp, 1955): *Calophanoides quadrifaria* (Nees) Ridl. (≡ *Gendarussa quadrifaria* Nees).

Shrubby or herbaceous perennials or sometimes annuals. Inflorescences a sessile, axillary cluster of more or less sessile flowers. Bracts normally absent. Calyx 5-lobed; lobes equal. Corolla white or yellow to lilac or purple. Anther-thecae unequal, superposed, parallel or oblique, lower one larger, appendages large. Ovary glabrous. Capsules clavate, pubescent or glabrous. Seeds scarcely compressed, pointed-rugulose-tuberculate.

About 40 species Old World; nine species in the Pan-Himalaya.

12. Justicia tukuchensis V. W. Graham, J. Jap. Bot. 56: 117. 1981. Type: [W] Nepal, Dhaulagiri, Mustang, Ghasa, South of Tukucha, Kali Gandaki, 2270 m, 28 Aug. 1954, *J. D. A. Stainton, W. R. Sykes & L. H. J. Williams 7515* [holotype: BM (BM000521514); isotypes: KATH, KUN].

Herbs, 20–35 cm high; stems quadrangular, pubescent, base procumbent, upwards ascending, rooting at lower nodes. Leaves sessile or shortly petiolate; petiole 0–3 mm long; blade ovate, 3–7 cm long, 1.5–3 cm broad, adaxially pubescent, abaxially glabrous, base subcordate, margin entire, apex acute. Cymes axillary, 1–3-flowered. Inflorescence bracts spatulate, 3.5–4.5 mm long, 2.3–2.6 mm broad, margin pubescent; bracts and bracteoles subulate, 1–1.2 mm long. Calyx 6–8 mm long, 5-lobed; lobes equal, narrowly triangular-lanceolate, glabrous, margin ciliate. Corolla 1.3–1.6 cm long, white with purple markings, sparsely pubescent outside; tube base cylindric, 6–8 mm long; upper lip 5–6 mm long, triangular, apex emarginate, lower lip 5–6 mm long, 3-lobed, lobes 0.8–1 mm long and broad. Stamens 2; filaments 4–5 mm long; anther-thecae superposed, equal, 1–1.1 mm long, lower one spurred at base. Ovary glabrous; style 1.2–1.4 cm long; stigma 2-lobed. Capsules 8–9 mm long, glabrous, 4-seeded. Seeds ovate in outline, 1.6–1.9 mm in diameter, lateral compressed, rugulose-tuberculate.

Phenology: flowering from August to September.

Habitats: thickets, slopes; 1400–2300 m.

Distribution: endemic; C Nepal (Map 46-194).

C NEPAL: **Dhawalagiri**, Mustang, Ghasa, 2120 m, 26 Aug. 1994, *M. Mikage et al. 9485550* (BM, E); Mustang, Ghasa, S Tukucha, Kali Gandaki, 2270 m, 28 Aug. 1954, *J. D. A. Stainton, W. R. Sykes & L. H. J. Williams 7515* (BM, KATH, KUN); Mustang & Myagdi, Tatopani to Dana, Kopchepani to Ghasa, 1820 m, 15 Aug. 1999, *M. Mikage et al. 9967082* (KATH); Myagdi, Shika to Narcheng (Nracheng), 1400 m, *M. Suzuki et al. 8881396* (MO); Myagdi, Tatopani to Dana, 1910 m, 26 Jul. 1996, *T. Hoshino et al. 9666055* (A,

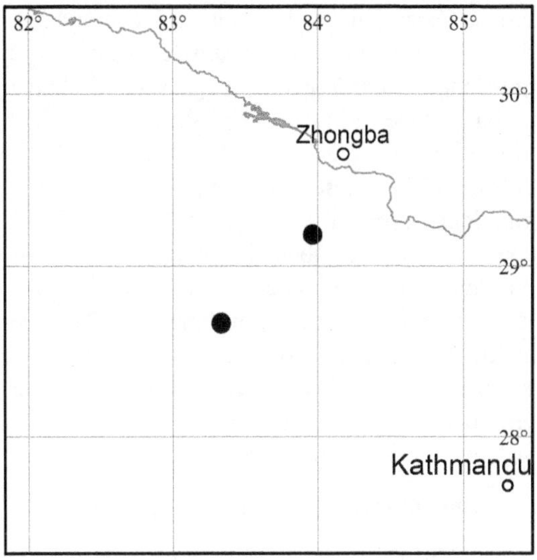

Map 46-194: **Justicia tukuchensis** V. W. Graham

CBM, E); Myagdi, Kopchepani, *T. Hoshino et al. 9662064* (A, E); Myagdi, Shika to Narchang (Nracheng), 1700 m, 26 Aug. 1988, *M. Suzuki et al. 8811059* (A, BM, E, KUN, L); Myagdi, Dana, 1470 m, 30 Aug. 1983, *N. P. Manandhar 10279* (KATH).

13. Justicia wardii W. W. Sm., Not. Bot. Gard. Edinb. 10: 184. 1918; Hu, Deng & Daniel in Wu, Raven & Hong (eds.), Fl. China 19: 453. 2011. ≡ *Calophanoides wardii* (W. W. Sm.) C. Y. Wu [Index Fl. Yunnan. 2: 1671. 1984, nom. invalid. (Art. 41.5)] ex C.C. Hu, Fl. Reipubl. Popularis Sin. 70: 268. 2002. Type: [S Hengduan] China, Yunnan, Yolong, by streams in arid region of Yangtze, 10,000 ft., May 1913, *F. Kingdon-Ward 301* [holotype: E (E265765)].

Shrubs, 30–60 cm high, diffuse; stems straw-colored, flexuous, densely grayish tomentose, soon glabrescent. Petiole 3–5 mm long, gray pilose; leaf blade elliptic, elliptic-lanceolate, oblanceolate, or sometimes obovate, 3–4.5 cm long, 1.5–2 cm broad, both surfaces white tomentose when young, then becoming sparsely pilose, base cuneate, margin slightly undulate, apex shortly acuminate, acute, or rarely retuse. Spikes axillary, 1–1.5 cm long, (1–)3–5-flowered; peduncles 0–3 mm long. Inflorescence bracts linear-lanceolate, 8–9 mm long, densely white villous. Bracts linear-lanceolate, 5–6 mm long, densely white villous; bracteoles linear, 1.5–2 mm long. Calyx 7–8 mm long, 5-lobed almost to base; lobes oblanceolate, abaxially densely villous, apex acute. Corolla 2–2.2 cm long, outside white villous; tube base cylindric, 5–6 mm long; upper lip triangular, 7–8 mm long, apex shallowly 2-lobed, lower lip oblong, 5–7 mm long, 4–8 mm broad, 3-lobed to the middle, lobes oblong, 1.5–2.5 mm long, 1.5–2.2 mm broad. Stamens 2; filaments whitish pubescent towards base; anther-thecae superposed, lower one spurred at base. Ovary glabrous; style white villous. Capsules unknown.

Phenology: flowering from April to May.

Habitats: rocks, thickets; ca. 3000 m.

Distribution: endemic; S Hengduan (Map 46-195).

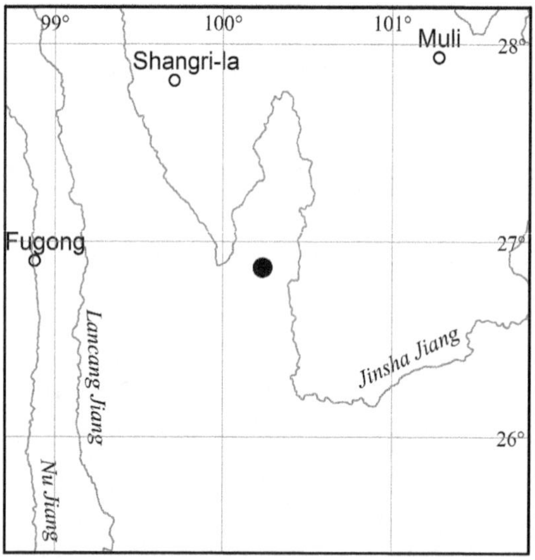

Map 46-195: **Justicia wardii** W. W. Sm.

S HENGDUAN: **Yulong** (Lijiang), region of Yangtze, 3000 m, May 1913, *F. Kingdon-Ward 301* (E).

14. Justicia siccanea W. W. Sm., Not. Bot. Gard. Edinb. 10: 43. 1918; Graham, Kew Bull. 43: 594. 1988; Hu et al. in Wu et al., Fl. China 19: 456. 2011. ≡ *Calophanoides siccanea* (W. W. Sm.) C. Y. Wu [Index Fl. Yunnan. 2: 1671. 1984, nom. invalid. (Art. 41.5)] ex C. C. Hu, Fl. Reipubl. Popularis Sin. 70: 285. 2002. Type: [S Hengduan] China, Yunnan, Weixi, Mekong Valley, Lat. 27°40′N, 7000 ft., Aug. 1914, *G. Forrest 13118* [lectotype **here designated**: E (E00047306); isotypes: E, K, IBSC].

Shrubs, 30–60 cm high, prostrate; stems grayish tomentose, soon glabrescent. Petiole 3–8 cm long, villous; leaf blade ovate to lanceolate, 1.5–4 cm long, 1–2 cm broad, both surfaces sparsely pilose, lateral veins 4 or 5 pairs, base cuneate, margin entire, apex rounded to acute. Spikes axillary, 0.5–1 cm long, 1–3-flowered; peduncle 1–3 mm long, grayish pubescent. Inflorescence bracts elliptic, 3–5 mm long, grayish pubescent g. Bracts linear-lanceolate, 2–3 mm long, abaxially grayish pubescent; bracteoles similar to bracts. Calyx 7–8 mm long, abaxially pubescent, 5-lobed to base; lobes linear. Corolla white tinged with red, 1.3–1.5 cm long, outside pubescent; upper lip shorter, apex emarginated, lower lip very short, 3-lobed, lobes broadly ovate. Stamens 2; filaments 3–4 mm long, whitish villous towards base; anther-thecae superposed, lower one spurred at base. Ovary glabrous; style 6–8 mm long, white villous. Capsules 6–7 mm long, glabrous. Seeds ovate in outline, compressed, 1–1.2 mm long and broad, verrucose, glabrous.

Phenology: flowering from August to October.

Habitats: on rocks by trailsides, open slopes; 2100–2500 m.

Distribution: endemic; S Hengduan (Map 46-196).

S HENGDUAN: **Muli**, 2500 m, 27 Oct. 1959, *S. K. Wu 3561* (KUN); **Weixi**, Mekong valley, 2100 m, *G. Forrest 13118* (E, IBSC, K); **Yulong** (Lijiang), mountains on the N. E. Yangtze bend, Sep. 1913, *G. Forrest 11273* (A, BM, E, K).

Notes: This species is very close to *J. wardii* W. W. Sm. and differs from the latter mainly in the size of bract and corolla. Here, I treat them as independent species due to limited material seen.

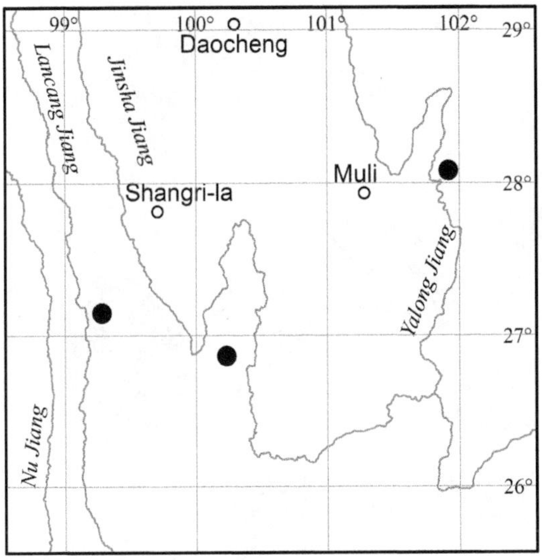

Map 46-196: **Justicia siccanea** W. W. Sm.

15. Justicia longipetiolata Y. Tong & Y. F. Deng, Phytotaxa 413: 282. 2019. Type: China, Yunnan, Lincang Shi, Cangyuan Xian, Mengjiao Xiang, alt. 1200 m, 6 Oct. 2013, *Y. Tong, Z. L. Lin, Y. L. Li & X. D. Ma 13100614* (holotype: IBSC; isotypes: IBSC, CSH).

Perennial herbs or subshrub, 20–30 cm tall; stems terete, densely pubescent. Petiole 2–5.5 cm long, sparsely pubescent; leaf blade ovate to oblong, 4–18 cm long, 1–5 cm broad, sparsely papery, adaxially usually glabrous, abaxially pubescent along veins, lateral veins 6–8 pairs, base cuneate, decurrent onto petiole, margin slightly undulate, apex caudate to shortly caudate. Spikes axillary, with several sessile flowers clustered. Inflorescence bracts leaf-like, sometimes absent, shortly petiolate, obcordate, 4–6.5 mm long, 3–5 mm broad, sparsely pubescent, lateral veins 1 or 2 pairs, base decurrent onto petiole, apex emarginate or sometimes obtuse. Bract triangular, densely pubescent, 0.7–1 mm long; bracteoles linear, 1–1.2 mm long, pubescent. Calyx 4–5.5 mm long, 5-lobed almost to base; lobes linear-lanceolate, subglabrous, with conspicuous white membranous margin. Corolla 7–8 mm long, white on tube and upper lip, outside pubescent, especially on throat; tube slightly shorter than limb; upper lip triangular, erect, minutely 2-lobed; lower lip white with purple patches, 3-lobed, lobes imbricate, suborbicular. Stamens 2, attached to corolla tube, exserted; filaments 0.8–1 cm long, hirsute toward base; anthers bithecous, thecae superposed, lower one white spurred at base; connective pubescent. Ovary fusiform, glabrous; style 5–7 mm long, sparsely hirsute, stigma un-lobed. Capsules fusiform, 1–1.2 cm long, glabrous, base stipitate, apex acute. Seeds 4, compressed, heart-shaped, squircle or obovate in outline, brown, 1.4–1.9 mm long, 1.0–1.6 mm broad; tuberculate. Figure 46-110.

Phenology: flowering from August to October; fruiting from October to November.

Habitats: in thickets, forests valley or on slope; 1100–1600 m.

Distribution: S Hengduan (Map 46-197); also in other parts of China (S & SW Yunnan).

S HENGDUAN: **Longyang**, Mangkuan, Baihualing, Zaotang, 1590 m, 9 Sep. 2003, *Gaoligong Shan Biodiversity Survey 18980* (CAS, KUN); Mangkuan, Baihualing, Manghuang, 1170 m, 9 Sep. 2003, *Gaoligong Shan Biodiversity Survey 19007* (CAS, KUN); **Yunlong**, Baofeng, 14 Sep. 2013, *Y. F. Deng 24949* (IBSC).

16. Justicia quadrifaria (Nees) T. Anderson, J. Linn. Soc. Bot. 9: 514. 1867; Clarke in Hooker (ed.), Fl. Brit. India 4: 530. 1885; Hu, Deng & Daniel in Wu, Raven & Hong (eds.), Fl. China 19: 455. 2011. ≡ *Gendarussa quadrifaria* Nees in Wallich (ed.), Pl. Asiat. Rar. 3: 105. 1832. ≡ *Adhatoda quadrifaria* (Nees) Nees in de Candolle (ed.), Prodr. 11: 396. 1847 ≡ *Ecbolium quadrifarium* (Nees) Kuntze, Revis. Gen. Pl. 2: 981 (1891). ≡ *Calophanoides quadrifaria* (Nees) Ridl., Fl. Mal. Pen. 2: 593. 1923. ≡ *Justicia quadrifaria* Wall., Numer. List n. 2479, 1830, nom. invalid. Type: Bangladesh, Sylhet, Mont Silhet, *de Silva in Wallich Cat. 2479a* [lectotype (as "holotype") designated by Graham, Kew Bull. 43: 594. 1988: K-W (K001116159); isolectotypes: BM, GZU (GZU000251567)].

Adhatoda chinensis Benth., Kew Journ. Bot. 5: 134. 1853. ≡ *Justicia championii* T.

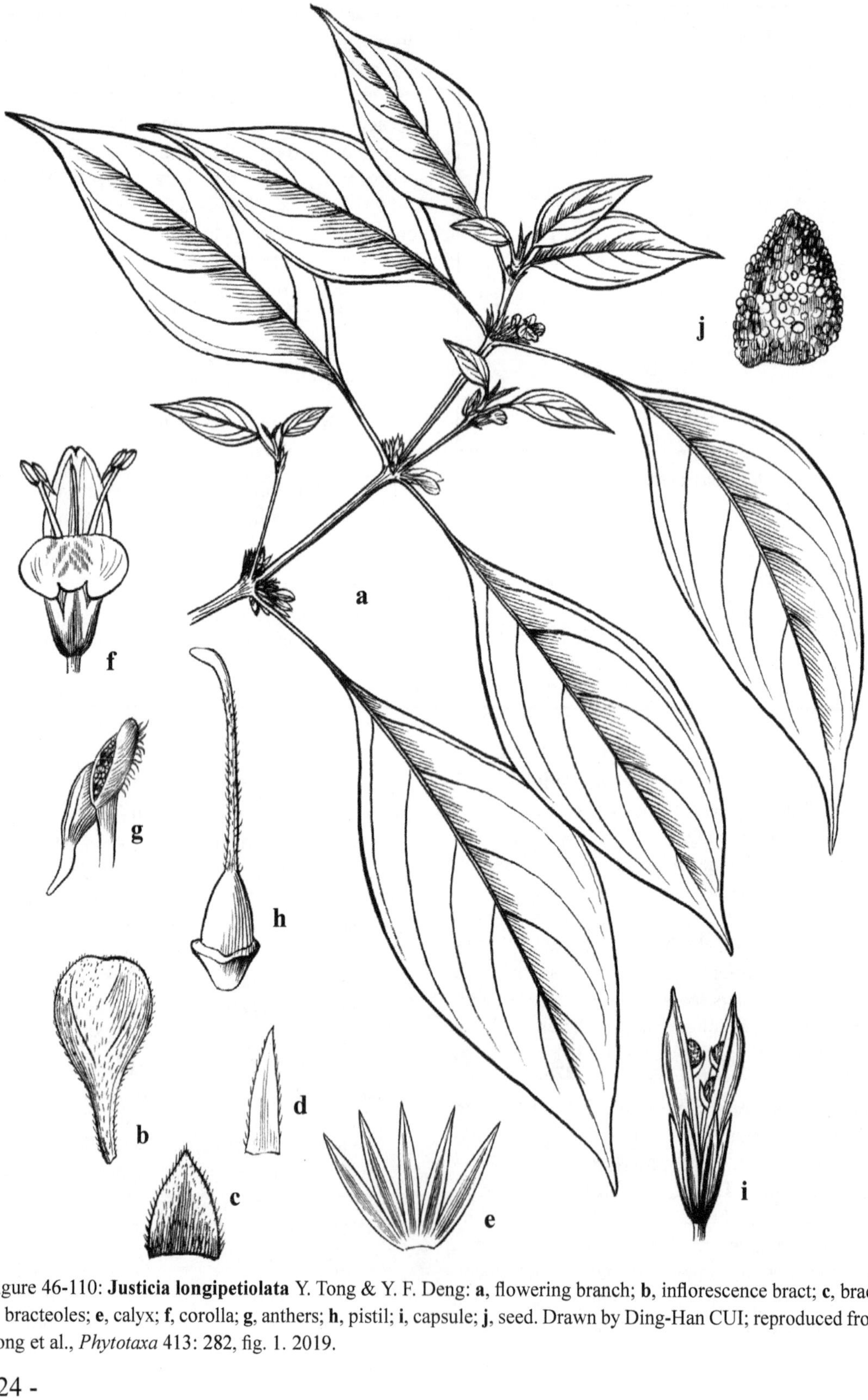

Figure 46-110: **Justicia longipetiolata** Y. Tong & Y. F. Deng: **a**, flowering branch; **b**, inflorescence bract; **c**, bract; **d**, bracteoles; **e**, calyx; **f**, corolla; **g**, anthers; **h**, pistil; **i**, capsule; **j**, seed. Drawn by Ding-Han CUI; reproduced from Tong et al., *Phytotaxa* 413: 282, fig. 1. 2019.

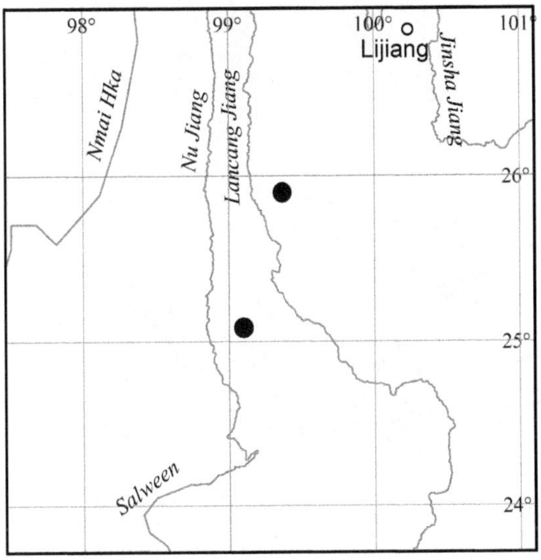

Map 46-197: **Justicia longipetiolata** Y. Tong & Y. F. Deng

Anderson in Bentham (ed.), Fl. Hongk.: 264. 1861; Anderson, J. Linn. Soc., Bot. 9: 514. 1867; Hu, Deng & Daniel in Wu, Raven & Hong (eds.), Fl. China 19: 453. 2011. ≡ *Ecbolium championii* (T. Anderson) Kuntze, Revis. Gen. Pl. 2: 980. 1891. ≡ *Justicia chinensis* (Benth.) Druce, Bot. Exch. Club Brit. Sci. Rep. 4: 629. 1914, non *Justicia chinensis* L., 1753. ≡ *Calophanoides chinensis* (Benth.) C. Y. Wu & H. S. Lo, Fl. Hainan. 3: 596. 1974; Hu, Fl. Reipubl. Popularis Sin. 70: 281. 2002. Type: China, Hong Kong, *Champion 210* [lectotype (as "holotype") designated by Graham, Kew Bull. 43: 592. 1988: K].

Justicia salicifolia T. Anderson, J. Linn. Soc., Bot. 9: 514. 1867, non *Justicia salicifolia* Blume ex Steud., 1840; Clarke in Hooker (ed.), Fl. Brit. India 4: 531. 1885. ≡ *Ecbolium salicifolium* Kuntze, Revis. Gen. Pl. 2: 981. 1891. ≡ *Calophanoides salicifolia* (Kuntze) Bremek., Dansk Bot. Ark. 20: 84 1961. ≡ *Justicia barapaniensis* P. Soumya & Sunojk., Phytotaxa 332(1): 98. 2017. Type: India, Assam, Khasia, Jul. 1850, *Adhatoa no. 20, Herb India of J. D. Hooker & T. Thompson* [lectotype designated by Soumya & Sunojkumar, Phytotaxa 332: 98. 2017: K (K000884122); isolectotypes: C (C10004991), CAL, E (E00273547), K (K000884123), L].

Herbs; stems quadrangular, sulcate, bifariously pubescent, soon glabrescent, basally decumbent and usually rooting at nodes, upwards ascending. Petiole 0.6–2.5 cm long, sparsely pubescent; leaf blade oblong to lanceolate, or rarely orbicular, (2–)5.5–11.5 cm long, (1–)2–3.5 cm broad, both surfaces glabrous or only pubescent along veins, lateral veins 5–7 pairs, base narrowly cuneate and decurrent onto petiole, margin entire or slightly serrate, apex acute. Spikes axillary, 1–1.4 cm long, usually 3-flowered, solitary or several in a cluster. Inflorescence bracts leaf-like, ovate to obovate, 3–8 mm long, 3–5 mm broad, with a petiole for 3–4 mm long, sparsely pubescent, pinnately veined, apex obtuse. Bracts sublate, 1–1.3 mm long, pilose; bracteoles linear, 1–1.4 mm long, glabrous. Calyx 5–6 mm long, abaxially puberulent, 5-lobed almost to base; lobes linear-lanceolate. Corolla white with purple spots on lower lip, 6–8 mm long, outside pubescent; upper lip oblong, erect, apex 2-lobed, lower lip 3–4 mm long, 2.5–3 mm broad, patent, apex 3-lobed. Anthers with thecae superposed, lower one spurred at base. Ovary glabrous; style 5–6 mm long, sparsely hirsute. Capsules 6–8 mm long, glabrous. Seeds ovate in outline, compressed, 1.2–1.5 mm long, 1–1.2 mm broad, verruculose, glabrous. Figure 46-111.

Phenology: flowering and fruiting from August to December.

Habitats: forests valley or on slope; 800–1600 m.

Distribution: E Nepal (Map 46-198); also in Laos, Vietnam, Thailand, Bangladesh, and other parts of India, Myanmar and China.

E NEPAL: Mechi, Jhapa, Hokse to Dolaghat, 1000 m, 7 Oct. 1960, *S. B. Malla & S. B. Rajbhandari 328* (KATH).

Figure 46-111: **Justicia quadrifaria** (Nees) T. Anderson: **a**, portion of plant; **b**, corolla; **c**, stamen; **d**, capsule. Drawn by Han-Ping YU; reproduced from *Fl. Reipubl. Popularis Sin.* 70: 281, pl. 44: 1–4. 2006.

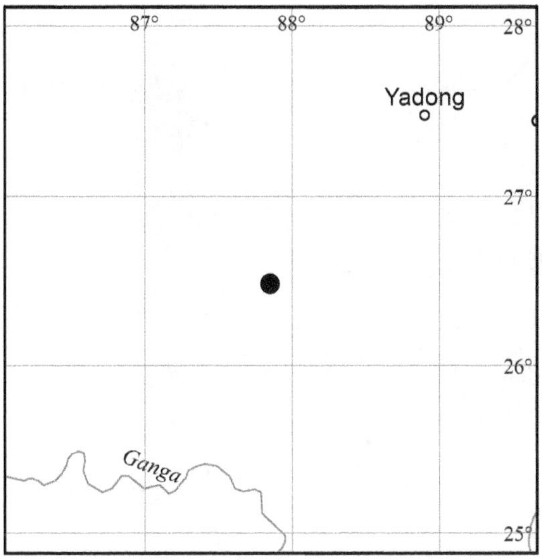

Map 46-198: **Justicia quadrifaria** (Nees) T. Anderson

17. Justicia cyclostegia (Hand.-Mazz.) Y. Tong & Y. F. Deng, Phytotaxa 415: 140. 2019. ≡ *Dicliptera cyclostegia* Hand.-Mazz., Anz. Akad. Wiss. Wien, Math.-Naturwiss. Kl. 62: 235. 1925. Type: [S Hengduan] China, Sichuan, "in regione subtropica convallis fluminis Yalung ad affluentem versus Yenyüen", 27°12'–27°20', 1250–1700 m, 28 Sep. 1914, *H. Handel-Mazzetti 5351* [lectotype designated Tong & Deng, Phytotaxa 415: 140. 2019: W (W1926-0002034); isolectotypes: E, LE, WU (WU0037699)].

Subshrubs; stems teret, sulcate, pilose. Leaves petiolate; petiole 1–1.5 cm long; blade chartaceous to thick chartaceous, elliptic, 3–6.5 cm long, 1–2.5 cm broad, pilose on both surfaces, lateral veins 3 or 4 pairs, base cuneate, margin undulate, apex acute. Spikes axillary, few-flowered. Inflorescence bracts orbicular or broadly ovate, 8–10 mm long, 4–5 mm broad, pilose, pinnately 1–2-veined, apex rounded or obtuse. Bracts and bracteoles narrowly triangular, 1–1.3 mm long, pilose. Calyx 5–6 mm long, glabrescent, 5-lobed; lobes linear-lanceolate, margin membranous, white. Corolla white, 6–8 mm long; tube 3–4 mm long, pilose outside and inside; upper lip erect, apex 2-lobed, lower lip 3-lobed, lobes purple striped, ovate, 1.5–2 mm long, 1.2–1.6 mm broad. Stamens 2; filaments 4–6 mm long; anther-thecae superposed, lower one spurred at base. Ovary cylindric, glabrous; style 4–5 mm long, pilose towards base. Capsules clavate, 6–8 mm long, glabrous. Seeds suborbicular, 1.5–1.8 mm long and broad, lateral compressed, rugose-tuberculate. Figure 46-112.

Phenology: flowering from September to November; fruiting from September to December.

Habitats: thicket on rocky slope; 1200–1900 m.

Distribution: S Hengduan (Map 46-199); also in other parts of China (S Sichuan, N Yunnan).

S HENGDUAN: **Muli**, Eya, 1900 m, 11 Aug. 1978, *Q. S. Zhao & F. Q. Lü 7945* (CDBI); **Yanyuan**, Yenyüen, 1250–1700 m, 28 Sep. 1914, *H. Handel-Mazzetti 5351* (E, LE, WU).

18. Justicia albovelata W. W. Sm., Notes Roy. Bot. Gard. Edinburgh 10: 182. 1918; Hu, Deng & Daniel in Wu, Raven & Hong (eds.), Fl. China 19: 454. 2011. ≡ *Calophanoides albovelata* (W. W. Sm.) C. Y. Wu ex H. P. Tsui in Wang (ed.), Vasc. Pl. Hengduan Mount. 2: 1876. 1994; Hu, Fl. Reipubl. Popularis Sin. 70: 279. 2002. Type: China, Yunnan, Yong Shan in the Yangtze bend, lat. 27°20'N, alt. 9000 ft., Sep. 1913, *G. Forrest 10996* [holotype: E (E00765315); isotypes: BM (BM001191510, BM001191511), K (K000884046), PE (PE01545299)].

Herbs, perennial, 20–50 cm high; stems basally decumbent then erect and zigzag, densely white villous for 3 mm long. Petiole 3–5 mm long, densely white villous; leaf blade lanceolate, 3.5–4.5 cm long, 0.8–1.5 cm broad, abaxially villous, adaxially densely white villous, lateral veins 4 or 5 pairs, base cuneate, margin entire, apex obtuse. Spikes axillary, 1–1.3 cm long, usually 3(or 4)-flowered. Inflorescence bracts spatulate-circular to obovate, 5–10 mm long, 3–7 mm broad. Bracts and bracteoles oblanceolate to linear. Pedicels 1–2 mm long. Calyx 5–6 mm long, 5-lobed almost to base; lobes linear-lanceolate, densely white pilose along veins, apex acuminate.

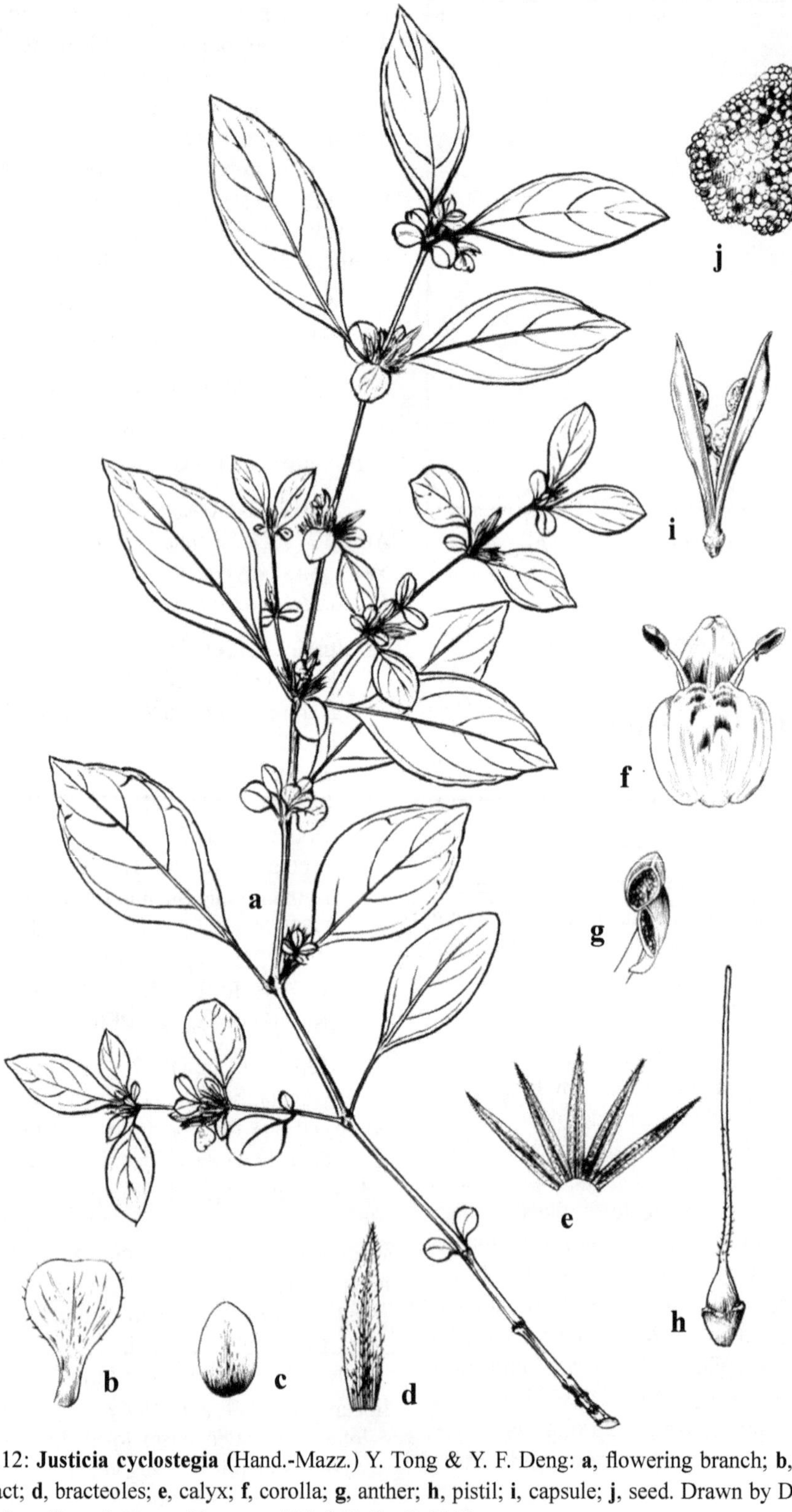

Figure 46-112: **Justicia cyclostegia** (Hand.-Mazz.) Y. Tong & Y. F. Deng: **a**, flowering branch; **b**, inflorescence bract; **c**, bract; **d**, bracteoles; **e**, calyx; **f**, corolla; **g**, anther; **h**, pistil; **i**, capsule; **j**, seed. Drawn by Ding-Han CUI; reproduced from Tong & Deng, *Phytotaxa* 415: 141, fig. 3. 2019.

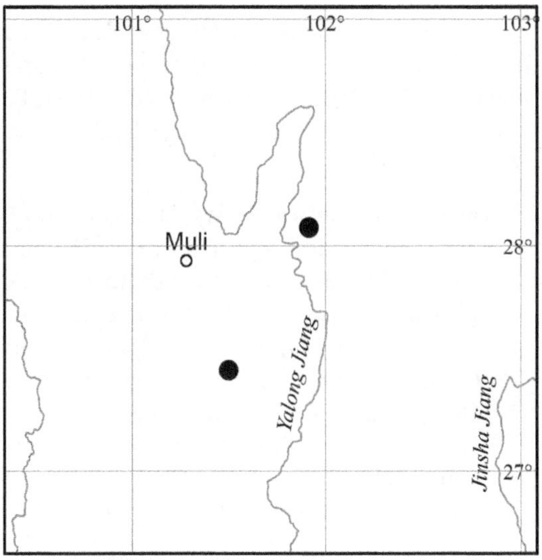

Map 46-199: **Justicia cyclostegia** (Hand.-Mazz.) Y. Tong & Y. F. Deng

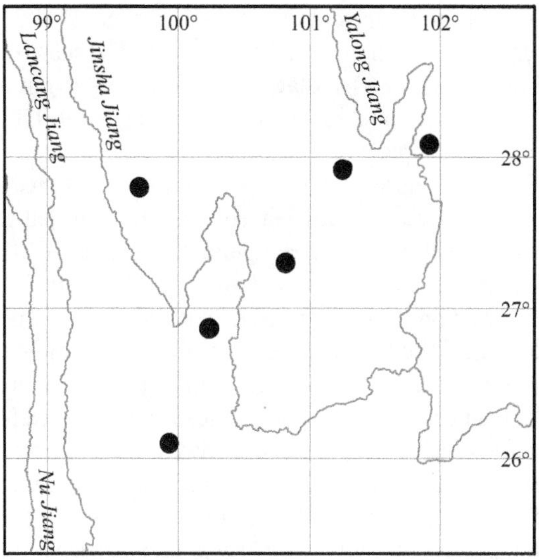

Map 46-200. **Justicia albovelata** W. W. Sm.

Corolla whitish-yellow to yellow, 7–9 mm long; tube slightly longer than limb, outside pilose; upper lip ovate 4–6 mm long, 3–4.5 mm broad, 2-lobed, lower lip oblong, 4–5 mm long, 3.5–4.5 mm broad, 3-lobed, lobes ovate, 1.5–2 mm long, 1.2–1.6 mm broad. Stamens 2; filaments basally hirsute; anther-thecae superposed, lower one white spurred at base. Ovary villous. Capsules clavate, 6–8 mm long, villous, 4-seeded. Seeds ovate in outline, compressed, 1.2–1.5 mm long and broad, verrucose.

Phenology: flowering from August to October; fruiting from September to November.

Habitats: thickets, stream banks; 1400–2000 m.

Distribution: S Hengduan (Map 46-200); also in other parts of China (NE Yunnan).

S HENGDUAN: **Bo'ao**, 1830 m, 11 Sep. 1959, *S. K. Wu 2559* (KUN); **Eryuan**, *R. C. Ching 24657* (KUN, PE); **Muli**, Chongtianhe, 1900 m, 25 Sep. 1983, *Qinghai-Xizang Exped. 14388* (KUN, PE); **Ninglang**, Yongning, Tuodian, bank of Jinshajiang, 1600 m, 21 Nov. 1959, *S. K. Wu 2941* (KUN); **Shangri-la** (Zhongdian), N flank of Haba Snow Range, on open hill side, Sep. 2, 1939, *K. M. Feng 2247* (A, KUN); **Yulong** (Lijiang), Daju, 1400 m, 16 Sep. 2013, *Y. F. Deng 25100* (IBSC); Daju, 1400 m, 16 Sep. 2013, *Y. F. Deng 25103* (IBSC); Daju, 1500 m, 16 Sep. 2013, 16 Sep. 2013, *Y. F. Deng 25104* (IBSC); Daju, 1500 m, 16 Sep. 2013, *Y. F. Deng 25105* (IBSC); Daju, 1500 m, 16 Sep. 2013, *Y. F. Deng 25106* (IBSC); Daju, Hutiaoxia, 1800 m, 15 Aug. 1956, *Yunnan University 731* (IBSC, KUN, YUKU), Daju, Hutiaoxia, 1930 m, 18 Aug. 1962, *Zhongdian Exped. 1258* (KUN, PE).

19. Justicia yunnanensis W. W. Sm., Notes Roy. Bot. Gard. Edinburgh 11: 215. 1919; Hu, Deng & Daniel in Wu, Raven & Hong (eds.), Fl. China 19: 455. 2011. ≡ *Calophanoides yunnanensis* (W. W. Sm.) C. Y. Wu [Ind. Fl. Yunnan. 2: 1671. 1984, nom. invalid., Art. 45.1] ex H. P. Tsui in Wang (ed.), Vasc. Pl. Hengduan Mount. 2: 1877. 1994. Type: China, Yunnan, in the Yung-pe mountams, lat. 26°45′N, alt. 10,000 ft., Sep. 1913, *G. Forrest 11051* [holotype: E (E00045273); isotypes: BM (BM001191508), K (K000884028), PE (PE01545404)].

Shrubs, 20–50 cm high; stems basally decumbent; upwards ascending, whitish sericeous. Petiole 2–5(–7) mm long, pilose; leaf blade lanceolate to oblanceolate, 1.5–2.5 cm long, 0.7–1 cm broad, papery, both surfaces pilose, base narrowly cuneate and decurrent onto petiole, margin entire, apex obtuse. Spikes axillary, 1–1.3 cm long, usually 3–5-flowered. Inflorescence bracts orbicular-spatulate to obovate, 5–10 mm long. Bracts and bracteoles lanceolate, 0.9–1.1 mm long, sparsely pubescent, apex acuminate. Calyx light grayish green, 4–5 mm long, 5-lobed to base; lobes linear-lanceolate, midvein pale whitish-yellow, apex acuminate. Corolla light whitish-yellow, 6–8 mm long, outside sparsely pilose; upper lip triangular, 2-lobed, lower lip broad, 3-lobed, center one larger than lateral ones, ovate, 1.5–2 mm long, 1–1.8 mm broad. Stamens 2; filaments 4–5 mm long, pilose towards base; anther-thecae superposed, lower one white spurred at base. Capsules glabrous, 4-seeded. Seeds ovate in outline, compressed, 1.2–1.5 mm long, 1–1.2 mm broad, glabrous, verrucose.

Phenology: flowering and fruiting from August to November.

Habitats: limestone slopes; ca. 1600 m.

Distribution: S Hengduan (Map 46-201); also in other parts of China (C Yunnan).

Notes: *Justicia yunnanensis* appears remarkably similar to *J. albovelata*, if it is indeed distinct. We have refrained from making taxonomic alterations because few specimens of *J. yunnanensis* were studied.

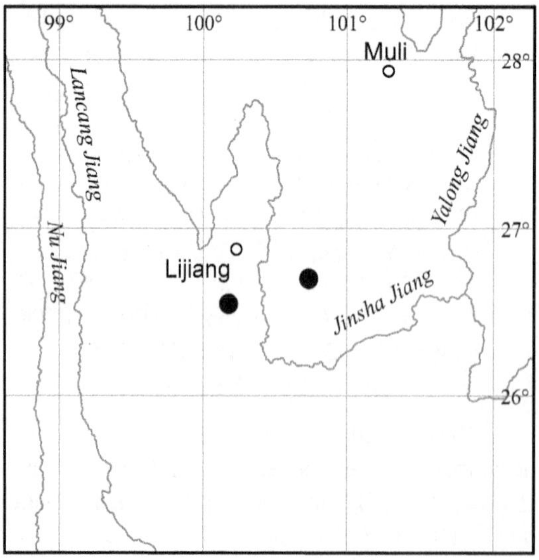

Map 46-201: **Justicia yunnanensis** W. W. Sm.

S HENGDUAN: Heqing, Ta-pin-tze, *M. Delavay s. n.* (K, P); **Yongsheng**, Yungpe mountains Sep. 1913, *G. Forrest 11051* (BM, E, K, PE); Sanchuan to Yongbei, 31 Oct. 2014, *Y. P. Chen EM148* (IBSC).

20. Justicia xerobatica W. W. Sm., Notes Roy. Bot. Gard. Edinburgh 11: 213. 1919; Hu, Deng & Daniel in Wu, Raven & Hong (eds.), Fl. China 19: 454. 2011. ≡ *Calophanoides xerobatica* (W. W. Sm.) H. P. Tsui in Wang (ed.), Vasc. Pl. Hengduan Mount. 2: 1876. 1994; Hu, Fl. Reipubl. Popularis Sin. 70: 287. 2002. Type: China, Yunnan, La-kou, 2400 m, Oct. 1914, *E. E. Maire 382* [holotype: E (E00045275); isotype: BM]

Justicia xerophila W. W. Sm., Notes Roy. Bot. Gard. Edinburgh 11: 214. 1919, **syn. nov.**; Hu, Deng & Daniel in Wu, Raven & Hong (eds.), Fl. China 19: 454. 2011. ≡ *Calophanoides xerophila* (W. W. Sm.) C. Y. Wu ex C. C. Hu, Fl. Reipubl. Popularis Sin. 70: 287. 2002. Type: China, Yunnan, La-kou, 2400 m, Jul. 1913, *E. E. Maire 196* [holotype: E (E00045274)].

Justicia xylopoda W. W. Sm., Notes Roy. Bot. Gard. Edinburgh 11: 214. 1919, **syn. nov.**; Hu, Deng & Daniel in Wu, Raven & Hong (eds.), Fl. China 19: 454. 2011. ≡ *Calophanoides xylopoda* (W. W. Sm.) C. Y. Wu [Ind. Fl. Yunnan. 2: 1671. 1984] ex H. P. Tsui in Wang (ed.), Vasc. Pl. Hengduan Mount. 2: 1876. 1994; Hu, Fl. Reipubl. Popularis Sin. 70: 288. 2002. Type: [S Hengduan] China, Yunnan, Yangtze valley at Ta-Ku, lat. 27°30'N, 8000 ft., Aug. 1914, *G. Forrest 13069* [lectotype **here designated**: E (E00265764); isolectotype: K (K000884027)].

Subshrubs, 10–30 cm high, prostrate; stems basally woody, slender, flexuous and often rooting at nodes, apically ascending, grayish to white sericeous. Petiole 1–3 cm long, white pilose; leaf blade subcircular, broadly obovate, or narrowly lanceolate to oblanceolate, 0.8–2.5 cm long, 0.5–1.2 cm broad, softly membranous, both surfaces pilose along veins, lateral veins 3 or 4 pairs, base cuneate, margin entire and ciliate, apex rounded to obtuse. Spikes axillary, 1–1.2 cm long, 1–3-flowered. Inflorescence bracts broadly obovate to subcircular, 3–5 mm long, pubescent. Bracts elliptic, 1–1.5 mm long, pilose, margin ciliate, apex acute; bracteoles ovate to subulate-lanceolate, 1.5–2 mm long, pilose. Calyx green, 5–6 mm long, abaxially slightly pilose to subglabrous, 5-lobed almost to base; lobes linear-lanceolate, pilose along midveins, margin scarious, apex acuminate. Corolla white, 0.7–1 cm long, outside sparsely pilose; upper lip triangular, apex emarginate, lower lip 3-lobed with center one larger. Stamens 2; filaments 4–5 mm long, pilose; anther-thecae superposed, lower one spurred at base. Ovary glabrous; style 5–7 mm long, subglabrous. Capsules 5–6 mm long, glabrous. Seeds ovate in outline, compressed, 1–1.2 mm long and broad; tuberculate.

Phenology: flowering from September to October.

Habitats: thickets on the rock on dry slopes; 600–3300 m.

Distribution: S Hengduan (Map 46-202); also in other parts of China (NE Yunnan).

S HENGDUAN: **Butuo**, Niujiaowan, Xixihe, 680 m, 11 Oct. 2014, *Y. F. Deng 25859* (IBSC); **Dali**, E. flank of Tali Range, 2100 m, Aug. 1929, *G. Forrest 27903* (BM, E); **Hanyuan**, Xigou, 940 m, 15 Sep. 1993, *Z. C. Zhao 0372* (PE); Dashu Qu, 900 m, 17 Sep. 1993, *Z. C. Zhao 0454* (PE); **Muli**, Ninglang, Chongtianhe, 1900 m, 25 Sep. 1983, *Qinghai-Xizang Exped. 14388* (PE); Eya, 2300 m, 29 Sep. 1983, *Qinghai-Xizang Exped. 14862* (PE); Erya, 3300 m, 19 Aug. 1978, *Q. S. Zhao et al. 7933* (SZ); **Panzhihua**, Panzhihua

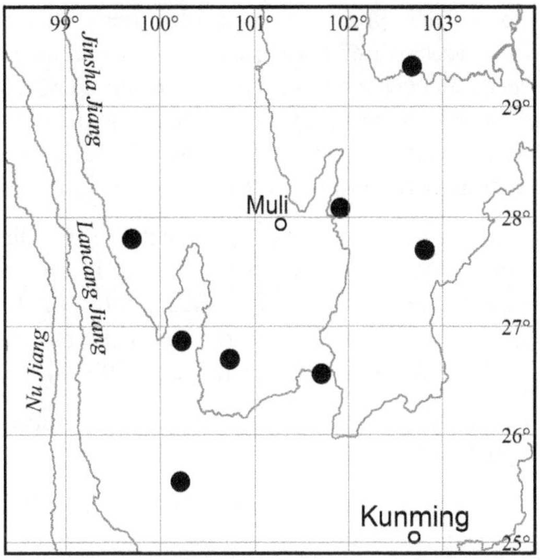

Map 46-202: **Justicia xerobatica** W. W. Sm.

Cycas Natural Reserve, Biaoshuiya, 1604 m, 15 Sep. 2010, *Expedition to Cycas Panzhihuaensis 510* (IBSC, PE); **Shangri-la** (Zhongdian), *Zhongdian Exped. 2221* (KUN, PE); *Zhongdian Exped. 1210* (KUN, PE); moutains in the NE of the Yangtze heud, Sep. 1913, *G. Forrest 11004* (E, K); Yongk, road from Haba to Hutiaoxia on Jinsha Jiang, 2010 m, 20 Jul., 2006, *D. E. Boufford et al. 35064* (A); **Yongsheng**, Yungpe, *S. Tén 10* (E); **Yulong** (Lijiang), *G. Forrest 13069* (E, K); Ndaku, 1750–2250 m, 1 Aug. 1914, *H. Handel-Mazzett 4387* (LE, WU), Duinaoko, 1500–2000 m, 25 Oct. 1916, *H. Handel-Mazzetti 12995* (LE, WU); Daju, 16 Sep. 2013, *Y. F. Deng 25055* (IBSC); Daju, *Y. F. Deng 25107* (IBSC); La-kou, 2400 m, Jul. 1914, *E. E. Maire s. n.* (E); La-kou, 2400 m, Oct., 1914, *E. E. Maire 382* (E); Ta-Ku, 2400 m, Aug. 1914, *G. Forrest 13069* (E, K).

30. RHINACANTHUS Nees

Nees in Wallich (ed.), Pl. Asiat. Rar. 3: 76, 108. 1832; de Candolle, Prodr. 11: 442. 1847; Clarke in Hooker (ed.), Fl. Brit. India 4: 541. 1885; Wood in Grierson & Long (eds.), Fl. Bhutan 2(3): 1290. 2001; Hu, Fl. Reipubl. Popularis Sin. 70: 267. 2002; Hu, Deng & Daniel in Wu, Raven & Hong (eds.), Fl. China 19: 461. 2011. Type: *Rhinacanthus communis* Nees, nom. illeg. [≡ *Justicia nasuta* L. ≡ *Rhinacanthus nasutus* (L.) Kurz].

Shrubs, subshrubs or herbs, with cystoliths. Leaves opposite, petiolate or subsessile; leaf blade entire or slightly undulate. Inflorescences axillary or terminal, spikes or racemes, sometimes branched into panicles. Bracts and bracteoles shorter than calyx. Calyx deeply 5-lobed; lobes subequal. Corolla white, greenish or purple; tube cylindric, elongate, narrow, limb 2-lipped; lower lip 3-lobed; lobes ascending cochlear in bud, upper lip entire or 2-cleft, inside rugulose. Stamens 2, exserted from mouth of corolla; anthers bithecous; thecae unequally inserted to superposed, lower theca with or without a basal appendage; staminodes absent. Capsules with a solid stalk, clavate, 4-seeded; retinacula present. Seeds discoid, variously ornamented, lacking trichomes.

Recent references: Kantachot, 2015. The genus *Rhinacanthus* (Acanthaceae) in Thailand. *Thai J. Bot.* 7(1): 65–70; Hai et al., 2017. Morpholohical chracters and taxonomy of *Rhinacanthus* (Acanthaceae) in the flora of Vietnam. In: Va et al. (eds.), *Proseedings of the 7th Mational Scientific Conference on Ecology and Biological Resources, Hanoi, 20 October 2017*. Nha Xuat Ban Khoa Hoc Tu Nhien Va Cong Nghe, Hanoi. pp. 140–144.

About 25 species: tropical and subtropical regions of Africa and Asia; three species (one endemic) in the Pan-Himalaya.

1a. Petiole 1.5–3 cm long; leaf blade ovate; inflorescences 3–5 cm long .... 1. *R. grandiflorus*
1b. Petiole to 5 mm long; leaf blade oblanceolate; inflorescences longer than 10 cm.
  2a. Leaf blade glabrous; bracts sublate, 1–2 mm long .... 2. *R. calcaratus*
  2b. Leaf blade pubescent abaxially; bracts leaf-like, lanceolate, 0.8–1.2 cm long .... 3. *R. beesianus*

1. Rhinacanthus grandiflorus Dunn, Kew Bull. 1920: 135. 1920. Type: Yarlung Zangbo-Brahmaputra, Outer Arbor Hills, common in shade of Shingkeng and ither woods at Janakmukh and Ramidambang, *I. H. Burkill 37117* [lectotype **here designated**: K (K000884359); isolectotype: K (K000884361)]

Shrubs, 0.6–1 m high; stems glabrous. Petiole 1.5–3 cm long, glabrous; blade ovate, 13–17 cm long, 5–8 cm broad, glabrous, lateral veins 6–8 pairs, base cuneate, apex acuminate. Inflorescences terminal, 3–5 cm long; peduncle very short; bracteoles triangular, 1–1.2 mm long. Pedicel 1.5–2 mm long. Calyx campanulate, 2–3 mm long; lobes sublate. Corolla 5–7 cm long; tube cylindric, 1.5–2 mm in diameter; upper lip 1.2–1.5 cm long, apex acute, lower lip 3-lobed; lobes 1.7–2 cm long. Stamens 2; anther-thecae oblong. Ovary glabrous. Capsules oblong, 2.5–3 cm long, solid stalked at base, 4-seeded. Seeds ovate in outline, compressed, 4–5 mm in diameter, rugose.

Phenology: flowering from December to January of next year.

Habitats: shade places.

Distribution: endemic; Yarlung Zangbo-Brahmaputra (Map 46-203).

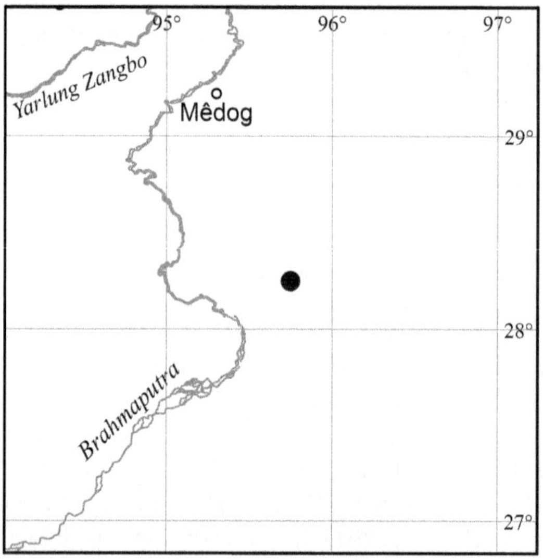

Map 46-203: **Rhinacanthus grandiflorus** Dunn

YARLUNG ZANGBO-BRAHMAPUTRA: Outer **Arbor Hills**, Shingkeng, *H. I. Burkill 37117* (K); Myain Ghat, 300–600 m, 30 Nov. 1928, *F. Kingdon-Ward 8739* (K).

2. Rhinacanthus calcaratus (Wall.) Nees in Wallich (ed.), Pl. Asiat. Rar. 3: 109. 1832; de Candolle, Prodr. 11: 444. 1847; Clarke in Hooker (ed.), Fl. Brit. India 4: 541. 1884; Lindau in Engl. & Prantl, Nat. Pflanzenfam. 4(3b): 339. 1895; Wood in Grierson & Long (eds.), Fl. Bhutan 2(3): 1290. 2001. ≡ *Justicia calcarata* Wall., Pl. Asiat. Rar. 2: 9, t. 113. 1830; Hochst., Flora 26: 73. 1843. Type: Bangladesh, Sylhet, *F. de Silva in Wallich Cat. 2426a* [lectotype **here designated**: K-W (K001115912); isolectotypes: CGE, E, K, P].

Shrubs, 1–3 m high; stems quadrangular, glabrous. Petiole 1–3 mm long; leaf blade narrowly elliptic to lanceolate, 10–23(–40) cm long, 3–8(–14) cm broad, glabrous, lateral veins 10–15 pairs, base gradually tapering to petiole, apex shortly acuminate, margin entire. Panicles terminal or axillary, 12–18 cm long; branches 2–12 cm long; rachis glandular-pubescent; peduncles pubescent. Bracts sublate, 1–2 mm long; bracteoles sublate, 0.8–1.2 mm long. Calyx 5–6 mm long, glandular-pubescent; lobes lanceolate. Corolla greenish-white, 4–5 cm long, pubescent; tube cylindric, 3–4 cm long, 1–2 mm broad; upper lip greenish-white, 9–10 mm long, lower lip white, 1–1.5 cm long, 3-lobed, lobes elliptic, 8–10 mm long, 6–8 mm broad. Stamens 2, included; filaments 3–5 mm long, glabrous; anthers oblong, greenish-white. Ovary pubescent; style 2.5–3.5 cm long; stigma capitate. Capsules oblong, 4–4.5 cm long, glandular-pubescent. Seeds discoid, 4–5 mm in diameter, rugulate.

Phenology: flowering from March to April.

Habitats: thickets; 500–700 m.

Distribution: Bhutan, U Irrawaddy (Map 46-204); also in Thailand, and other parts of Myanmar, India and China (S Yunnan).

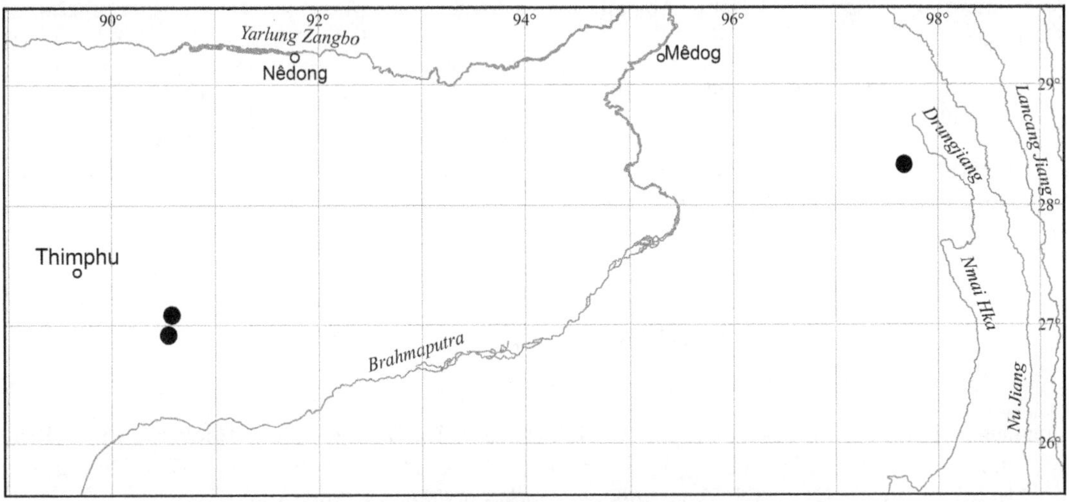

Map 46-204: **Rhinacanthus calcaratus** (Wall.) Nees

BHUTAN: **Gaylegphug**, above the Shershong Bridge, 700 m, 3 Apr. 1989, *J. R. I. Wood 6891* (E); Karai Khola above Aie Bridge, Gaylegphug, 510 m, 22 Mar. 1982, *A. J. C. Grierson & D. G. Long 3933* (E). **U IRRAWADDY**: **Chaungmaw Kyaun**, North Trangle (Hkinlum), 500 m, 21 Sep. 1953, *F. Kingdon-Ward 21348* (A).

3. Rhinacanthus beesianus Diels, Notes Roy. Bot. Gard. Edinburgh 5: 164. 1912; Hu, Fl. Reipubl. Popularis Sin. 70: 267. 2002; Hu, Deng & Daniel in Wu, Raven & Hong (eds.), Fl. China 19: 462. 2011. Type: [S Hengduan] China, Yunnan, "western slopes of the Yung-ping-Hsicn Yalley, on the Teng-YuehTalifu Road", lat. 25°25'N, alt. 7000–8000 ft., Sep. 1903, *G. Forrest 1053* (holotype: E).

Rhinacanthus calcaratus var. *maxima* Prain, J. Asiat. Soc. Bengal. 67(2): 300. 1898. Type: Myanmar, Kachin, *King s. n.* (holotype: CAL).

Shrubs to 1.5 m high; stems quadrangular, subglabrous. Petiole 0–4 mm long; leaf blade narrowly obovate to oblanceolate, 20–24 cm long, 6–8 cm broad, abaxially glabrous, adaxially slightly pubescent, lateral veins 8–10 pairs, base attenuate and decurrent onto petiole, margin slightly undulate to subentire, apex acute. Panicles terminal, 12–15 cm long, 8–9 cm broad. Bracts leaf-like, lanceolate, 0.8–1.2 cm long, 3–4 mm broad. Flowers pedicellate. Calyx lobes linear, 8–10 mm long, 1.2–1.5 mm broad, pubescent, 1-veined, apex acute. Corolla white, fragrant, outside glandular-pubescent; tube 3.5–4 cm long, lower lip lobes elliptic, 2–2.5 cm long; upper lip 1.6–2 cm long, apex 2-cleft. Stamens 2; filaments 1–1.2 cm long. Ovary glabrous. Capsules oblong, 2.4–2.8 cm long, glabrous, solid stalked at base, 4-seeded. Seeds ovate in outline, compressed, 3.5–4.5 mm in diameter, rugose. Figure 46-113.

Phenology: flowering from August to September.

Habitats: open dry situations on mountain slopes; 1100–2400 m.

Distribution: S Hengduan, U Irrawaddy (Map 46-205); also in other parts of Myanmar (Kachin) and China (SW Yunnan).

S HENGDUAN: Longyang, Mangkuan, Baihualing, Wangjiawa, 1600 m, 8 Dec. 2000, *H. Li 13516* (KUN); Mangkuan, Baihualing, Manghuang Area, 1170 m, 9 Sep. 2003, *Gaoligong Shan Survey 19009* (CAS, KUN); Mangkuan, Baihualing, Hanlong, Mucheng Shan, 1625 m, 10 Sep. 2003, *Gaoligong Shan Survey 19024* (CAS, KUN); Mangkuan, Baihualing, 12 Sep. 2013, *Y. F. Deng 24748* (IBSC); **Yongping**, Yungping Hsien valley, on the Tengyueh-talifu road, 2100–2400 m, Sep. 1905, *G. Forrest 1053* (E, K). **U IRRAWADDY**: **Drungjiang**, Kiukiang Valley, S. of Kongpong, 1200 m, 26 Sep. 1938, *T. T. Yü 20459* (A, E, KUN); Dudu, Dulonghe, Laimenmianlamu, 1300 m, 15 Sep. 1956, *P. Y. Mao 457* (KUN, SZ). Dadieshui, 1250 m, 18 Dec. 1990, *Exped. Dulongjiang 1175* (CAS, KUN, MBK); Qinglangdang, 1280 m, 10 Mar. 1991, *Exped. Dulongjiang 4494* (KUN); **Tengchong**, Puchuan, Xiajiajie, Yibasan, 1350–1650 m, 15 Oct. 1960, *W. Q. Yin 60-1454* (KUN); Machang, on the way to Mangbang, 11 Sep. 2013, *Y. F. Deng 24694* (IBSC); Tiantian, Qingliangshan, 2067 m, 24 Oct. 2015, *Tengchong Expedition 53052215102482LY* (IMDY).

31. SPHINCTACANTHUS Benth.

Bentham in Bentham & Hooker (eds.), Gen. Pl. 2: 1070, 1118. 1876; Clarke in Hooker (ed.), Fl. Brit. India 4: 544. 1885; Lindau in Engler & Prantl (eds.), Nat. Pflanzenfam. 4(3B): 337. 1895; Hansen, Nordic J. Bot. 5: 226. 1985. Type: *Sphinctacanthus griffithii* (T. Anderson) Benth.

Shrub. Leaves large; blade toothed or subentire. Flowering arranged in intterupted spikes forming a terminal panicles; flowers opposite. Bracts and bracteoles small. Calyx 5-lobed; lobes linear. Corolla tube ovoid, inflated, constricted at throat, limb 2-lipped, upper lip emarginate, lower lip patent, 3-lobbed; lobes imbricated in bud. Stamens 2; anthers bithecous, thecae oblong, muticous. Staminodes absent. Disc caplular. Ovary with 2 ovules per locule; style linear, minutely bifid. Capsules unknown.

Recent references: Hansen, 1985. Notes on the genus *Sphinctacanthus* (Acanthaceae). *Nord. J. Bot.* 5: 225–228.

Monospecific genus: Himalaya and neighboring regions.

Figure 46-113: **Rhinacanthus beesianus** Diels: **a**, portion of plant; **b**, bracteole, calyx and pistil; **c**, opened corolla showing stamens; **d**, anther; **e**, capsule. Drawn by Yun-Xiao LIU.

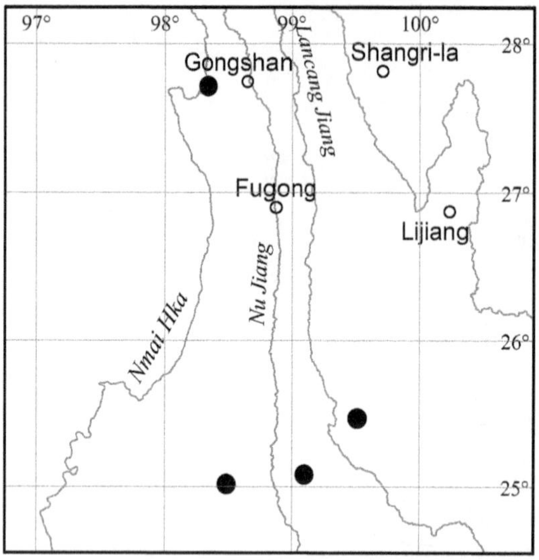

Map 46-205: **Rhinacanthus beesianus** Diels

1. **Sphinctacanthus griffithii** (T. Anderson) Benth. in Bentham & Hooker (eds.), Gen. Pl. 2: 1118. 1876; Bentham, Hooker's Icon. Pl. 13: 5. pl. 1205. 1877; Hansen, Nordic J. Bot. 5: 226. 1985. ≡ *Justicia griffithii* T. Anderson, J. Linn. Soc., Bot. 9: 516. 1867. ≡ *Dianthera griffithii* (T. Anderson) C. B. Clarke in Hooker (ed.), Fl. Br. India 4: 542. 1885. Type: Yarlung Zangbo-Brahmaputra, Mishime, *Griffith (E.I.C. 6161)* (lectotype designated by Hansen, Nordic J. Bot. 5: 226. 1985: CAL; isolectotype: K)

Shrub, isophyllous; stems glabrous. Leaves equal in each pair; petiole 1–2.5 cm long, puberulent; blade elliptic-lanceolate, 8.5–20 cm long, 3.5–8.5 cm broad, adaxially glabrous, abaxially puberulent along veins, lateral veins 8–11 pairs, base cuneate, margin subentire, apex acuminate. Panicles of 2–4-branched spikes terminal, 7–13 cm long, 4–6 cm broad; peduncle 1–1.5 cm long; flowers interrupted on rachis. Bracts triangular, 1.5–1.8 mm long, 0.6–0.8 mm broad, glabrous; bracteoles triangular, 1–1.2 mm long, 0.4–0.5 mm broad, glabrous. Calyx 2–2.5 mm long; lobes 5, equal. Corolla bilabiate; tube slightly inflated, 3–4 mm long, with constriction at throat, inside pubescent; limb 4.5–5 mm long; upper lip shortly bilobed, lower lip deeply trilobed with an inward bulge just proximal to each incision; lobes imbricate, 7–9 mm long, reflexed. Stamens 2, inserted in throat; filament 5–6 mm long, united to the middle by a narrow membrane, pubescent; anthers 2–3 mm long, thecae almost equal or one slightly longer than another, parallel, at same level, glabrous, base muticous. Ovary glandular-pubescent; style pubescent. Capsules subclavate, 9–11 mm long, apex acute. Figure 46-114.

Phenology: flowering from November to February of next year; fruiting from January to March.

Habitats: thickets; 400–900 m.

Distribution: Yarlung Zangbo-Brahmaputra, U Irrawaddy (Map 46-206); also in other parts of India.

YARLUNG ZANGBO-BRAHMAPUTRA: **Mishime**, *Griffith (E.I.C. 6161)* (CAL, K). **U IRRAWADDY**: Myitkyina, Namma to Nammun, 1 Mar. 1910, *J. H. Lace 5148* (E, K); **Chaungmaw Kyaun**, North Triangle (Hkrang Kha Valley), 900 m, 30 Nov. 1953, *F. Kingdon-Ward 21648* (BM); **Putao**, Tangkhul to Ningma Daw, valley of the Mali Hka, 450–600 m, 22 Mar. 1953, *F. Kingdon-Ward 22050* (BM).

32. ISOGLOSSA Oerst.

Oersted, Vidensk. Meddel. Dansk Naturhist. Foren. Kjøbenhavn 1854: 155. 1854, nom. cons.; Lindau in Engler & Prantl (eds.), Nat. Pflanzenfam. 4(3B): 344. 1895; Wood in Grierson & Long (eds.), Fl. Bhutan 2(3): 1289. 2001; Hu, Fl. Reipubl. Popularis Sin. 70: 231 2002; Hu, Deng & Daniel in Wu, Raven & Hong (eds.), Fl. China 19: 441. 2011. Type: *Isoglossa ciliata* Oerst.

Strophacanthus Lindau in Engler & Prantl (eds.), Nat. Pflanzenfam. 4(3b): 344. 1895; Backer & Bakhuizen f., Fi. Java 2: 585. 1965. Type: *Strophacanthus collinus* (T. Anderson) Lindau [≡ *Isoglossa collina* (T. Anderson) B. Hansen].

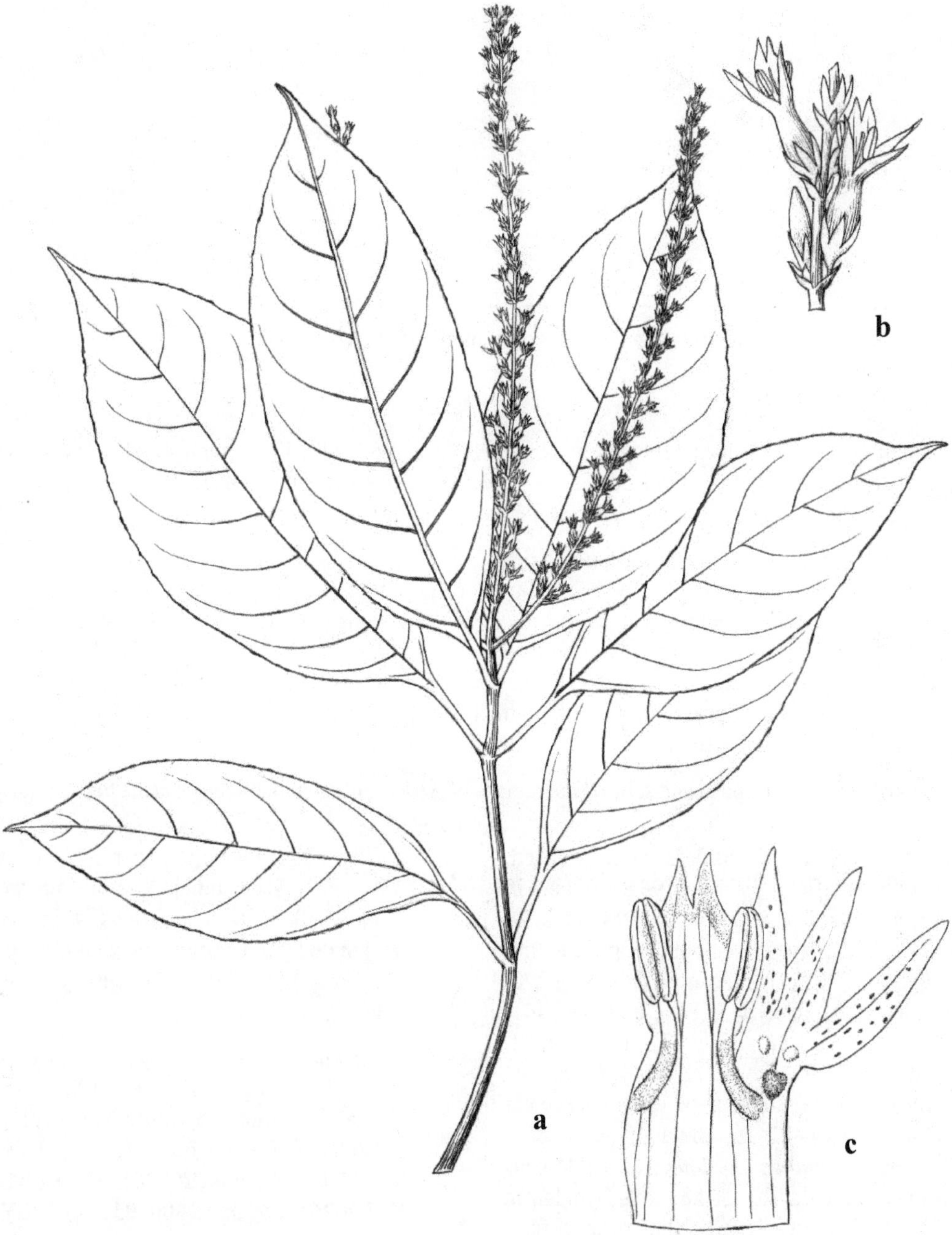

Figure 46-114: **Sphinctacanthus griffithii** (T. Anderson) Benth.: **a**, portion of plant; **b**, inflorescence; **c**, opened corolla showing stamens. Drawn by Yun-Xiao LIU.

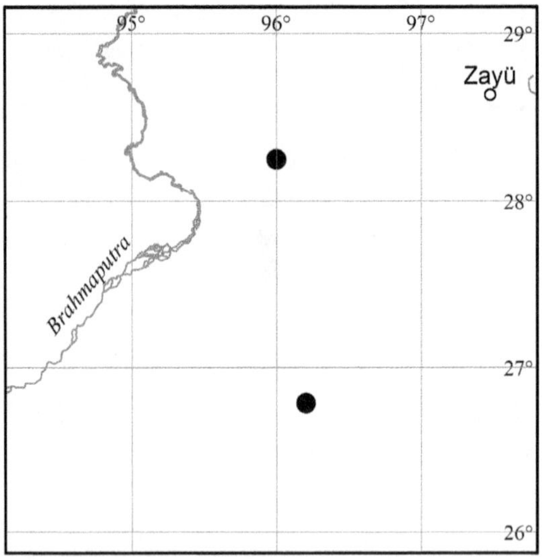

Map 46-206: **Sphinctacanthus griffithii** (T. Anderson) Benth.

Chingiacanthus Hand.-Mazz., Sinensia 5: 11. 1934; Li in Wu (ed.), Fl. Xizang 4: 417. 1985. Type: *Chingiacanthus patulus* Hand.-Mazz. [= *Isoglossa collina* (T. Anderson) B. Hansen].

Shrubs or herbs, isophyllous or slightly anisophyllous; with cystoliths. Leaves petiolate. Inflorescences terminal and/or sometimes axillary, open panicles or thyrses. Bracts small, shorter than calyx. Calyx deeply 5-lobed; lobes similar. Corolla funnel-shaped, 2-lipped, lower lip 3-lobed, upper lip shortly 2-lobed, lacking a stylar furrow; lobes ascending cochlear in bud. Stamens 2; anthers bithecous; thecae equal or unequal, parallel to perpendicular, subequally to unequally inserted, base muticous; staminodes absent. Ovary with 2 ovules per locule. Capsules clavate with a solid sterile basal stalk, 4-seeded; retinacula present; septa with attached retinacula remaining attached to inner wall of mature capsule. Seeds subdiscoid.

About 25 species: tropical and subtropical regions of Africa and Asia; one species in the Pan-Himalaya.

1. Isoglossa collina (T. Anderson) B. Hansen, Nordic J. Bot. 5: 12. 1985; Wood in Grierson & Long (eds.), Fl. Bhutan 2(3): 1290. 2001; Hu, Fl. Reipubl. Popularis Sin. 70: 232. 2002; Hu, Deng & Daniel in Wu, Raven & Hong (eds.), Fl. China 19: 441. 2011. ≡ *Justicia collina* T. Anderson, J. Linn. Soc., Bot. 9: 515. 1867. ≡ *Dianthera collina* (T. Anderson) C. B. Clarke in Hooker (ed.), Fl. Br. India 4: 543. 1885. ≡ *Strophacanthus collinus* (T. Anderson) Lindau, Bot. Jahrb. 18: 58. 1893; Lindau in Engler & Prantl (eds.), Nat. Pflanzenfam. 4(3b): 344. 1895; Bremekamp, Verh. Kon. Ned. Akad. Wetensch., Afd. Natuurk., Sect. 2 45(2): 44. 1948. ≡ *Ptyssiglottis collina* (T. Anderson) Moore, J. Bot. 60: 356. 1922. Type: India, Khasia, Surureem, 1 Nov. 1835, *W. Griffith 6164* [neotype designated by Hansen, Nordic J. Bot. 5: 12. 1985: K (K000195161); isoneotype: P (P00719884)].

Dianthera sinensis W. W. Sm., Notes Roy. Bot. Gard. Edinburgh 12: 204. 1920. Type: China. Yunnan, Shweli-Salween divide, 25°20′N, 9000 ft., Jul. 1918, *G. Forrest 17574* [holotype: E (E00265766); isotypes: E (E00265766), K (K000195161), WSY].

Chingiacanthus patulus Hand.-Mazz., Sinensia 5: 11. 1934; Li in Wu (ed.), Fl. Xizang 4: 417. 1985. Type: China, Guangdong,

Conghua, Wan-Tong-Shan, 15 Sep. 1931, *H. Y. Liang 61074* [lectotype **here designated**: IBSC (IBSC0004955); isotypes: IBK (IBK0055721), IBSC (IBSC0004956), PE (PE01432622, PE01432623, PE01432622), W (W1940-0010794)].

Herbs, 40–100 cm high; stems prostrate and rooting at base, upwards ascending, terete, sparsely puberulent when young, then glabrescent. Petiole 1–3 cm long, brownish pubescent; leaf blade ovate to ovate-elliptic, 3.5–11 cm long, 2–4.8 cm broad, abaxially sparsely puberulent especially along veins, adaxially glabrous, lateral veins 5–7 pairs, base cuneate or uppermost ones rounded, margin entire or subentire, apex acuminate. Panicles terminal or axillary, 5–18 cm long, 3–10 cm broad; rachis glandular-puberulent or subglabrous with few sessile glands; branches slightly flattened. Bracts lanceolate, 1.5–2 mm long, 0.4–0.6 mm broad, glabrous or slightly glandular-pubescent. Calyx 5–8 mm long; lobes subulate, glabrous or glandular-puberulent. Corolla white with purplish dots, 1.7–4.5 cm long, outside glabrous; tube base cylindric for about 1/3 of its length, upwards gradually widened into a funnel-shaped throat; lips 8–10 mm long; lower lip 3-lobed, lobes oblong-elliptic, 3–6 mm long, unequal, central one broader then lateral ones. Stamens 2; filaments 4–10 mm long, glabrous; anther-thecae 3.5–4.5 mm long, equal, parallel, overlapping for ca. ½ their length or more. Ovary glabrous; style glabrous. Capsules 1–1.4 cm long, glabrous. Seeds subdiscoid, 2–2.5 mm long, 1.8–2.2 mm broad, coarsely corrugated with irregular ridges and often with minute acute protuberances. Figure 46-115.

Phenology: flowering from August to November; fruiting October to December.

Habitats: evergreen broad-leaved forests, wetlands by streams; 1800–2200 m.

Distribution: Sikkim & Darjiling, Bhutan, Yarlung Zangbo- Brahmaputra, S Hengduan (Map 46-207); also in Thailand, Vietnam, and other parts of India and China (S Yunnan, Guangxi, Guangdong, Hunan, Jiangxi, Fujian).

SIKKIM & DARJILING: **Darjiling**, *J. M. Cowan s. n.* (E); **Sikkim**, Yoksun, 1800 m, *C. B. Clarke 25209* (K, LE). **BHUTAN**: **Tashigang**, Forests Huest House, ca. 10 km below Kori La, 2140 m, 7 Sep. 1998, *H. J. Noltie et al. 109* (E). **YARLUNG ZANGBO-BRAHMAPUTRA**: **Gutang**, 2200 m, 27 Aug. 1974, *Qinghai-Xizang Exped. 74-4917* (KUN). **S HENGDUAN**: **Longyang**, Heishuitang, 2146 m, 8 Dec. 2012, *Y. J. Guo et al. 12CS5067* (CSH).

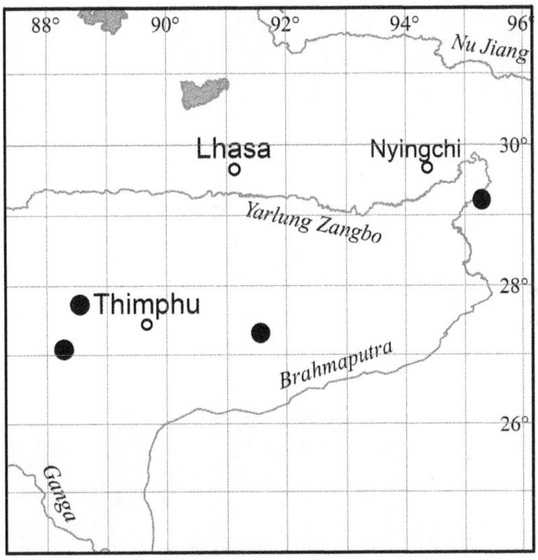

Map 46-207: **Isoglossa collina** (T. Anderson) B. Hansen

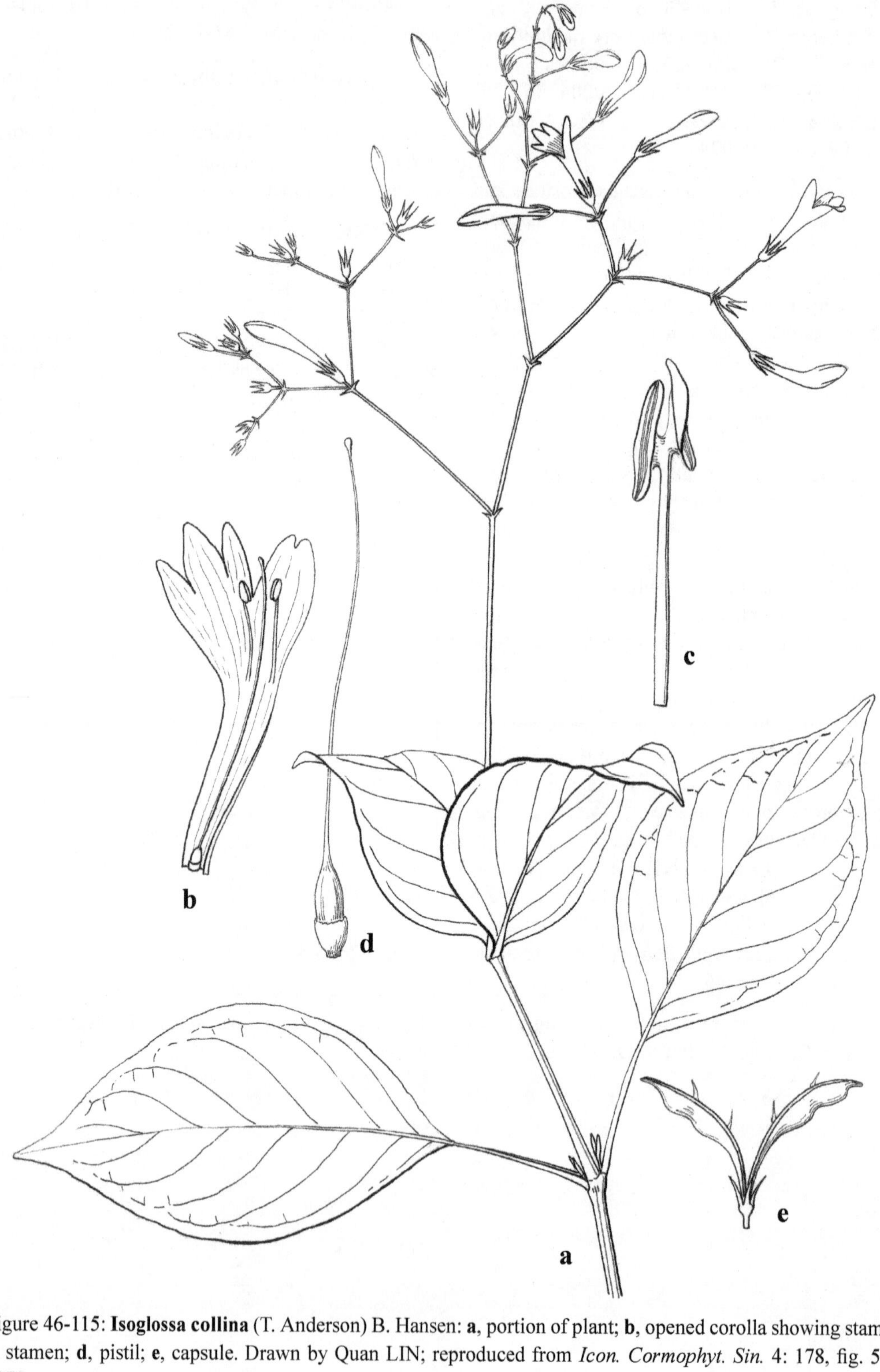

Figure 46-115: **Isoglossa collina** (T. Anderson) B. Hansen: **a**, portion of plant; **b**, opened corolla showing stamens; **c**, stamen; **d**, pistil; **e**, capsule. Drawn by Quan LIN; reproduced from *Icon. Cormophyt. Sin.* 4: 178, fig. 5770. 1975.

33. LEPTOSTACHYA Nees

Nees in Wallich (ed.), Pl. Asiat. Rar. 3: 76, 105. 1832, nom. cons., non *Leptostachia* Adanson, Fam. 2: 201. 1763; de Candolle, Prodr. 11: 376. 1847; Wood in Grierson & Long (eds.), Fl. Bhutan 2(3): 1289. 2001; Hu, Fl. Reipubl. Populasris Sin. 70: 228. 2002; Hu, Deng & Daniel in Wu, Raven & Hong (eds.), Fl. China 19: 447. 2011. Type: *Leptostachya wallichii* Nees, type cons.

Herbs, perennial, with cystoliths. Leaves petiolate. Inflorescences terminal spikes, simple or several in panicle. Bracts mostly shorter than calyx; bracteoles linear, minute. Calyx deeply 5-lobed; lobes similar or posterior lobe shorter. Corolla tube cylindric, short; limb 2-lipped; lower lip 3-lobed, upper lip emarginate and galeate, lacking a stylar furrow; lobes ascending cochlear in bud. Stamens 2; anthers bithecous; thecae equally to subequally inserted, parallel, muticous at base; staminodes absent. Ovary with 2 ovules per locule. Capsules clavate with a solid sterile stalk at base, 4-seeded; retinacula present. Seeds lenticular.

Recent references: Hansen, 1985. A taxonomic revision of the Asian genus *Leptostachya* (Acanthaceae). *Nord. J. Bot.* 5: 469–473; Deng & Xia, 2005. (1668) Proposal to conserve the name *Leptostachya* Nees (*Acanthaceae*) against *Leptostachia* Adans. (*Phrymaceae*) with a conserved type. *Taxon* 54: 192–193.

One species: tropical Asia, also in the Pan-Himalaya.

1. **Leptostachya wallichii** Nees in Wallich (ed.), Pl. Asiat. Rar. 3: 105. 1832; Wood in Grierson & Long (eds.), Fl. Bhutan 2(3): 1289. 2001; Hu, Fl. Reipubl. Populasris Sin. 70: 229. 2002; Hu, Deng & Daniel in Wu, Raven & Hong (eds.), Fl. China 19: 447. 2011. ≡ *Justicia wallichii* (Nees) T. Anderson, J. Linn. Soc., Bot. 9: 516. 1867. ≡ *Dianthera wallichii* (Nees) Vidal, Phan. Cuming. Philipp. 8: 1885. ≡ *Dianthera leptostachya* C. B. Clarke in Hooker (ed.), Fl. Brit. India 4: 542. 1885, nom. illeg. (Art. 52.1). ≡ *Odontonemella leptostachya* Lindau in Engler & Prantl (eds.), Nat. Pflanzenfam. 4(3b): 336. 1895, nom. illeg. (Art. 52.1). ≡ *Ptyssiglottis leptostachya* S. Moore, J. Bot. 60: 358. 1922, nom. illeg. (Art. 52.1). ≡ *Ruellia leptostachya* Wall., Numer. List n. 2411. 1830, nom. invalid. Type: *Wight in Wallich Cat. 2411* [lectotype designated by Hansen, Nordic J. Bot. 5: 472. 1985: K-W (K001115880); isolectotypes: GZU (GZU000250124), K-W (K001115879)].

Dianthera debilis C. B. Clarke in Hooker (ed.), Fl. Brit. India 4: 542. 1885, non *Justicia debilis* (Forssk.) Lam., Encycl. 1(2): 633. 1785. ≡ *Leptostachya debilis* (C. B. Clarke) C. B. Clarke & Stapf ex Hoss., Bot. Jahrb. Syst. 41: 73. 1907. ≡ *Ptyssiglottis debilis* (C. B. Clarke) Moore, J. Bot. 60: 356. 1922. ≡ *Justicia muticitheca* J. B. Imlay, Kew Bull. 1939: 138. 1939. Type: India, Shaila, Khasia Terai, 500 ft., 10 Dec. 1871, *C. B. Clarke 14973* [lectotype **here designated**: K (K000884373); isolectotype: BM (BM000950089)].

Herbs, 30–100 cm high; stems prostrate at base, upwards ascending. Branches subterete, striate, bifariously pubescent, then glabrescent. Petiole 1.5–4 cm long, puberulent; leaf blade ovate, lanceolate, or elliptic, 2–11(–17) cm long, 1–5 cm broad, abaxially sparsely pubescent along veins, adaxially glabrous or with few scattered trichomes, lateral veins 4–7 pairs, base rounded to broadly cuneate, margin entire, apex acute. Spikes terminal, simple, or 2 or 3 in panicles, 4–12 cm long; rachis glandular-pubescent. Bracts oblong-lanceolate, oblanceolate, or obovate, 1.2–2.5 mm long; bracteoles lanceolate, 1–3 mm long, apex acute. Calyx 2.5–3 mm long; lobes subulate, abaxially glandular-pubescent. Corolla white, 5–8 mm long, outside pilose; tube base narrowly cylindric, 4–5 mm long, 0.4–0.6 mm broad; lower lip 3-lobed, lobes triangular-ovate, subequal. Stamens inserted in corolla throat; filaments 1.5–4 mm long, glabrous or puberulent; anther-thecae 1–1.5 mm long. Ovary glabrous or hirsute towards apex; style glabrous or hirsute at basal portion; stigma 2-cleft. Capsules 1–1.5 cm long, pubescent. Seeds 1.8–2 mm long and broad, with concentric ridges of short linear warts. Figure 46-116.

Phenology: flowering from June to September.

Habitats: moist forests; 600–800 m.

Distribution: Sikkim & Darjiling, Bhutan (Map 46-208); also in Indonesia, Laos, Malaysia, Thailand, Vietnam, and other parts of India, Myanmar and China.

SIKKIM & DARJILING: **Darjiling**, Pornong, 600 m, 31 Oct. 1870, *C. B. Clarke 13642* (K, LE); Mangnang, from Samsing, 11 Aug. 1923, *J. M. Cowan s. n.* (E). **BHUTAN**: **Chuka**, on the escarpment above Kharbaudi, 12 Nov. 1987, *J. R. I. Wood 6081* (E); **Phuntsholing**, between Kharbandi and Sorchen, 800 m, 26 Nov. 1991, *J. R. I. Wood 7525* (E).

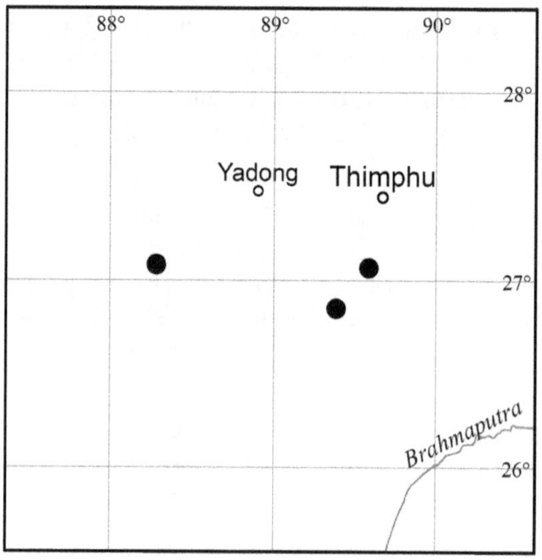

Map 46-208: **Leptostachya wallichii** Nees

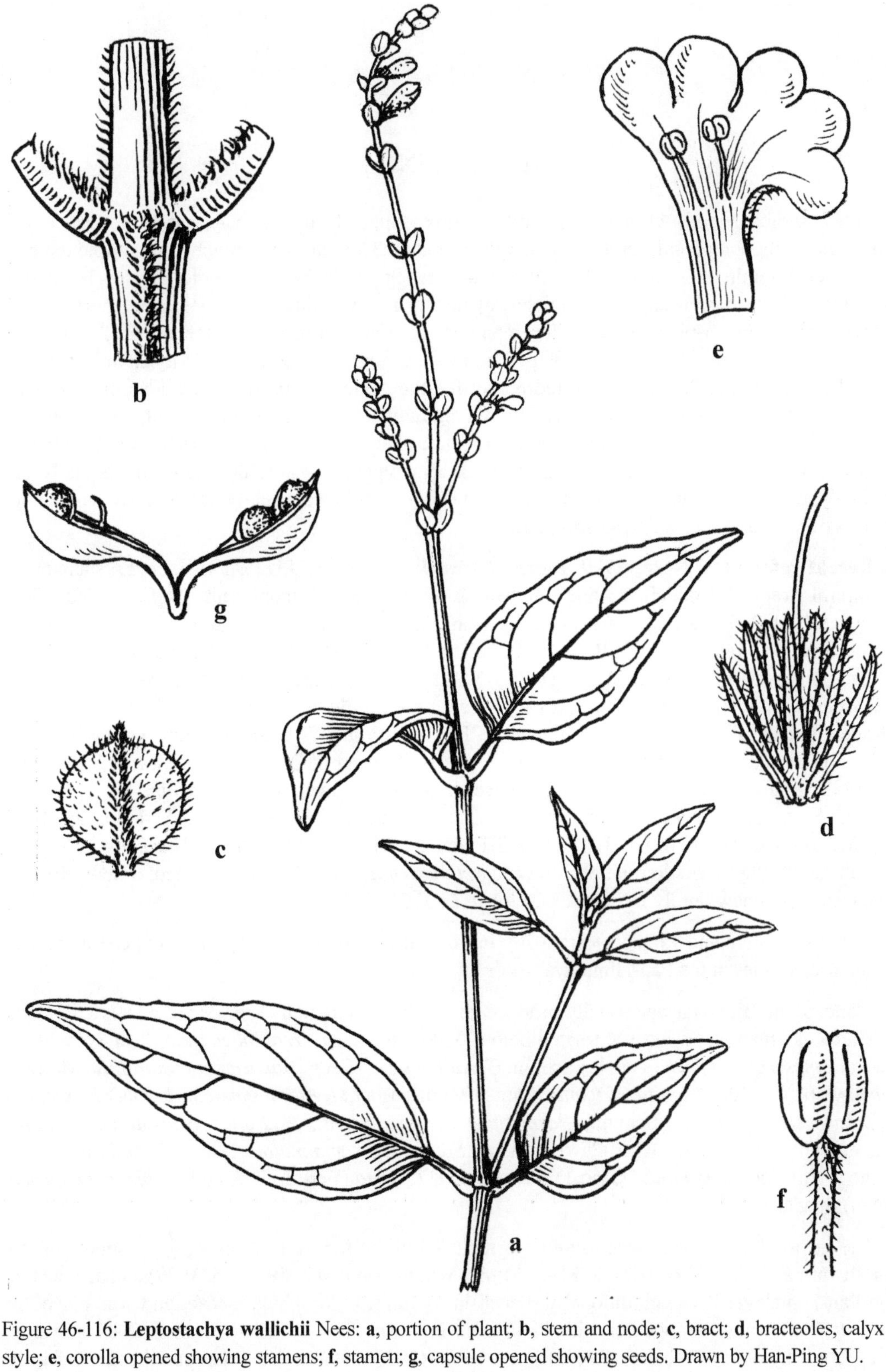

Figure 46-116: **Leptostachya wallichii** Nees: **a**, portion of plant; **b**, stem and node; **c**, bract; **d**, bracteoles, calyx and style; **e**, corolla opened showing stamens; **f**, stamen; **g**, capsule opened showing seeds. Drawn by Han-Ping YU.

BIGNONIACEAE

Yun-Fei DENG

Trees, shrubs; woody climbers, or herbs. Leaves opposite or alternate, rarely in whorls or rosettes, pinnately compound, rarely palmately compound, 3-foliate with branched or simple tendril often replacing terminal leaflet or undivided; glands often present at base of petiole. Flowers bisexual, zygomorphic, 5-merous, arranged in panicles, cymes, or rarely solitary, terminal or on short lateral branches. Bracts inconspicuous. Calyx campanulate, gamosepalous, spathaceous or 2–5-lobed. Corolla campanulate or tubular, with 2-lipped mouth or 5 slightly unequal lobes; lobes imbricate or valvate. Fertile stamens 2, 4 or 5, inserted on corolla-tube, alternate to lobes, included or exserted; anthers bithecous, thecae divergent, dehiscing by longitudinal slits; staminodes absent, 1 or 3. Ovary superior, subsessile, with nectariferous disc, 2-locular; placentation axile or parietal; ovules many; style filiform; stigma 2-lobed; lobes elliptic. Fruit an elongate capsule, dehiscing loculicidally or septicidally, rarely fleshy and indehiscent. Seeds many, flat or ridged, usually prominently winged or with tufts of hairs at both ends; endosperm absent.

Recent references: Nasir, 1979. *Flora of West Pakistan: no. 131. Bignoniaceae*. Islamabad: Agricultural Research Council. 22 pp.; Santisuk & Vidal, 1985. Bignoniaceés. In: Leroy (ed.), *Fl. Cambodge, Laos & Vietnam* 22: 1–63, Paris: Muséum National D'Histoire Naturelle; Tao & Yin, 1990. Bignoniaceae. *In*: Wang (ed.), Fl. Reipubl. Popularis Sin. 69: 1-62, Beijing: Science Press; Aitkin, 2001. Bignoniaceae. In: Grierson & Long (eds.), *Fl. Bhutan* 2(3): 1236–1244, Edinburgh & Thimphu: Royal Botanic Garden, Edinburgh & Royal Government of Bhutan; Fischer, 2004. Bignoniaceae. In: Kadereit (ed.), *The Families and Genera of Vascular Plants* 7: 9–38, Springer-Verlag, Berlin; Olmstead et al., 2009. A molecular phylogeny and classification of Bignoniaceae. *Amer. J. Bot.* 96: 1731–1743; Zhang & Santisuk, 1998. Bignoniaceae. In: Wu & Raven (eds.), *Fl. China* 18: 213–225, Beijing & St. Louis: Science Press & Missouri Botanic Garden Press; Lohmann & Taylor, 2014. A new generic classification of tribe Bignonieae (Bignoniaceae). *Ann. Missouri Bot. Gard.* 99: 348–489; Liu et al., 2020. Phylogenetic relationships of *Cyrtandromoea* and *Wightia* revisited: a new tribe in Phrymaceae and a new family in Lamiales, *J. Syst. Evol.* 58: 1–17.

About 80 genera and 750 species; mostly in tropical and subtropical regions; six genera and 23 (11 endemic) species in the Pan-Himalaya.

Plants of the Bignoniaceae usually produce large flowers and many species are widely cultivated in this area as ornamentals but not treated below. These includes: *Arrabidaea magnifica* Sprague ex Steenis, *Campsis grandiflora* (Thunb.) Schum, *C. radicans* (L.) Seem, *Jacaranda cuspidifolia* Martius, *J. mimosifolia* D. Don, *Kigelia africana* (Lam.) Benth., *Macfadyena unguis-cati* (Linn.) A. Gentry, *Pandorea jasminoides* (L.) Schum., *Parmetiera cerifera* Seem., *Podranea ricasoliana* (Tanfini) Sprague, *Pyrostegia venusta* (Ker Gawl.) Miers., *Spathodea campanulata* Bauv., *Stenolobium stans* (L.) Seem. *Tabebuia chrysantha* (Jacq.) G. Nicholson, *T. rosea* (Bertol.) DC. and *Tecomaria capensis* (Thunb.) Spach.

The genus *Wightia* Wall. was sometimes included in the family Bignoniaceae by some authors (Maheshwari, *Bull. Bot. Surv. India* 3: 31–35. 1961; Wu, *Fl. Yunnan* 2: 691–725. 1979) and also occurs in the Pan-Himalaya. It was traditionally placed in the family Scrophulariaceae, and was placed in the family Paulowniaceae together with *Paulownia* in recent APG systems (Stevens, 2001 onwards. Angiosperm Phylogeny Website). Recently, Liu et al. (2020) established a new monogeneric family

Wightiaceae for *Wightia*. It will be treated in volume 45.

1a. Stamens 5; capsules dehiscing septicidally (tribe 1. **Oroxyleae**) .... 1. *Oroxylum*
1b. Stamens 2 or 4; capsules dehiscing loculicidally.
  2a. Trees; leaves simple; stamens 2; seeds with tufts of hairs at both ends (tribe 2. **Catalpeae**).......
  ....2. *Catalpa*
  2b. Herbs, trees, or climbers; leaves pinnately or palmately compounded, rarely reduced to simple leaves; stamens 4; seeds membranous-winged or wingless.
    3a. Herbs; leaves alternate (tribe 3. **Tecomeae**) .... 3. *Incarvillea*
    3b. Trees or shrubs; leaves opposite (tribe 4. **Coleeae**).
      4a. Leaf axis winged; capsules septum membranous; seeds wingless, compressed globose
      .... 4. *Pauldopia*
      4b. Leaf axis usually wingless; capsules septum thick, usually woody; seeds winged.
        5a. Deciduous trees; leaves 2–3-pinnately compound; ovules in 2 rows per locule; capsules 2-anglar, septum compressed terete .... 5. *Radermachera*
        5b. Evergreen trees; leaves usually 1-pinnately compound; ovules in several rows per locule; capsules quadrangular, septum sub terete .... 6. *Stereospermum*

Tribe 1. Oroxyleae A. H. Gentry, Fl. Neotrop. 25: 48. 1980. Type: *Oroxylum* Vent.

Trees, shrubs or lianas, interpetiolar glands and ridges absent, pseudostipules absent. Leaves pinnate. Calyx broadly campanulate or cupular, apex truncate. Corolla campanulate, limb slightly bilabiate; lobes 5. Stamens 4 or 5, thecae parallel to divaricate, with one fertile theca, sterile theca spur-like, or both thecae fertile. Ovary bilocular, placentation axile. Fruit dehiscing parallel to septum (septicidal).

Four genera: tropical Asia; one genus and one species in the Pan-Himalaya.

1. OROXYLUM Vent.

Ventenat, Dec. Gen. Nov. 8. 1808; Bentham & Hooker, Gen. Pl. 2: 1040. 1874; Clarke in Hooker (ed.), Fl. Brit. India 4: 377. 1884; Schumann in Engler & Prantl (eds.), Nat. Pfalnzenfam. 4(3b): 225. 1894; Tao & Yin in Wang (ed.), Fl. Reipubl. Popularis Sin. 69: 10. 1990; Zhang & Santisuk in Wu & Raven (eds.), Fl. China 18: 215. 1998. ≡ *Calosanthes* Blume, Bijdr. Fl. Ned. Ind.: 760. 1826, nom. illeg. (Art. 52.1); de Candolle, Prodr. 9: 177. 1845. ≡ *Hippoxylon* Rafin. Sylv. Tellur. 78. 1838, nom. illeg. (Art. 52.1). Type: *Oroxylum indicum* (L.) Kurz. [≡ *Calosanthes indica* (L.) Blume ≡ *Hippoxylon indica* (L.) Raf.].

Trees, glabrous, few-branched. Leaves opposite, 2- or 3-pinnately compound; leaflet blade ovate, margin entire. Inflorescences racemose, terminal, erect. Calyx broadly campanulate, leathery, fleshy, apex truncate or obscurely toothed. Corolla campanulate-ventricose, creamy white or purplish, limb slightly bilabiate; lobes spreading, apex rounded, margin sinuous. Stamens 5; filaments slender, unequal; anthers bithecous, thecae ellipsoid. Ovary sessile, 2-locular; style filiform, stigma lingulate, compressed. Capsules linear, wooded, compressed, dehiscing septicidally, 2-valved; septum woody, flat. Seeds several in several rows, orbicular, compressed, surrounded by a transparent broad wing.

One species: Tropical Asia; also in the Pan-Himalaya.

1. **Oroxylum indicum** (L.) Benth. ex Kurz, Forests Fl. Burma 2: 237. 1877; Clarke in Hooker (ed.), Fl. Brit. India 4: 378. 1884; Tao & Yin in Wang (ed.), Fl. Reipubl. Popularis 69: 11. 1990; Zhang & Santisuk in Wu & Raven (eds.), Fl. China 18: 215. 1998. ≡ *Bignonia indica* L., Sp. Pl. 2: 625. 1753. ≡ *Calosanthes indica* (L.) Blume, Bijdr. Fl. Ned. Ind. 14: 760. 1826. ≡ *Hippoxylon indica* (L.) Raf., Sylva Tellur.: 78. 1838. ≡ *Spathodea indica* (L.) Pers., Syn. Pl. 2: 173. 1807. Type: [*icon*] Rheede, Hort. Malab., 1: 77, t. 43. 1678 (lectotype designated by Santisuk & Vidal, Fl. Canbodge, Laos & Vietnam 22: 10. 1985).

Trees 6–10 m high. Trunk 15–20 cm in diameter; bark gray-brown. Leaves 2- or 3-pinnately compound, crowded nearly at stem apex, 60–130 cm long; petiolules 5–20 mm long; leaflets triangular-ovate, 5–13 cm long, 3–10 cm broad, glabrous, lateral veins 5 or 6 pairs, base subrounded or cordate, oblique, margin entire, apex acute or short acuminate. Inflorescences 0.4–1.5 m long. Flowers usually open at night, with foul smell. Pedicel 3–7 cm long. Calyx purple, campanulate, 2.2–4.5 cm long, 2–3 cm broad, glabrous, membranous, apex truncate. Corolla tube reddish-purple outside, yellow inside, fleshy, 3–9 cm long, 1–1.5 cm broad; upper lip 2-lobed, lower lip 3-lobed; lobes creamy, suborbicular, 2–4.5 cm long, 2–4.5 cm broad, slightly reflexed, base attenuate, margin crenate and irregularly lacinulate. Stamens inserted at the middle of corolla tube; filaments 3–4.5 cm long, slightly exserted, tomentose towards base; anthers ellipsoid, 8–10 mm long, slightly divergent. Disc large, fleshy, 5-lobed, 4–5 mm thick, 1.2–1.5 cm in diameter. Ovary with several rows of ovules; style 5–7 cm long; stigma 2-parted, 5–7 mm long, 4–5 mm broad. Capsules woody, 40–120 cm long, 4–9 cm broad, 8–10 mm thick; valves with midrib, margin convex. Seeds orbicular, including papery wing 6–9 cm long, 3.5–4 cm broad. **Chromosomes** 2n = 28, 30, 38. Figure 46-117.

Phenology: flowering from June to September; fruiting from September to December.

Habitats: in moist forests; 0–1400 m.

Distribution: U Ganga & Indus, W Nepal, C Nepal, E Nepal, Bhutan, Sikkim & Darjiling, S Hengduan (Map 46-209); also in Thailand, Laos, Vietnam, Cambodia, Malaysia, Indonesia, the Philippines, and other parts of India, Myanmar and China (SE Yunnan, S Sichuan, Guizhou, Guangxi, Guangdong, Fujian, Taiwan).

Uses: The seeds and bark are used medicinally for alleviating pain and as an antiphlogistic medicine.

U GANGA & INDUS: **Chur Pass**, 600 m, *s. coll. 15150* (K); **Dehradun**, 1920–21, *S. N. Nantiyal 6* (E); Aug.–Oct. 1923, *S. P. Sharma 14* (TI); **Kangra**, 600 m, 10 Jul. 1906, *Forester Chowoori Ram G. S. H. 204* (E); s. loc., 1888, *J. R. Drummond 21483* (K). **W NEPAL**: **Bheri**, Dailekh, Matela to Chupra, 670 m, 1985, *P. R. Shakya et al. 8383* (KATH); Surkhet, 450 m, 1982, *S. R. Adhikeri & D. P. Manandhar 9173* (KATH); **Karnali**, Kalikot, Dhangadi, Khailad, 200 m, 1995, *N. Pandey 15* (KATH); **Mahakali**, Dadeldhura, Kataunje, 1000 m, 1980, *K. R. Rajbhandari et al. 5363* (KATH); Darchula, Rikheswer, 1976, *Raucola et al. 3655* (KATH). **C NEPAL**: **Bagmati**, **Dhading**, Charaugi Pauwa, 960 m, 16 Jul. 1994, *M. Suzuki et al. 9455024* (E, KATH); Dhading, Mayuad, 1989, *N. P. Manandhar 13166* (KATH); Dhading, Mathillo Dbrang, 870 m, 1989, *N. P. Manandhar 14228* (KATH); Shidhupalchok, below Helumbu, 1400 m, 1966, *D. H. Nicolson 2672* (BM, KATH); **Dhauwalagiri**, Baglung, Ratnechour, 820 m, 1 Sep. 1996, *M. Mikage et al. 9686006* (TI); Myagdi, Beni, 840 m, 1956, *J. D. A. Stainton, W. R. Sykes & L. H. J. Williams 4131* (BM); between Darban and Sasdhara, 25 Sep. 1976, *H. Tsbata et al. 3952* (KYO); Parbat, Pipaltari, 1050 m, 1991, *N. P. Manandhar 866-01* (KATH); Parbat, Karkineta, 1340 m, 1985, *N. P. Manandhar 10390* (KATH); **Gandaki**, Lidi Khola, 450 m, 11 Aug. 1954, *J. D. A. Stainton, W. R. Sykes & L. H. J. Williams 6809* (BM, E); Syangja, Dhandhunge, 1180 m, 1991, *N.P. Manadhar 756-91* (KATH); **Janakpur**, Dolakha, Malephu, Tama Kosi Valley, 970 m, 1977, *K.R. Rajbhamdari & B. Roy 1188* (KATH); **Sindhuli**, Sitalpati, Sarda Khola, 1000 m, 1952, *O. V. Polunin, W. R. Sykes & L. H. J. Williams 5815* (BM, E); Sindhaniaghat,

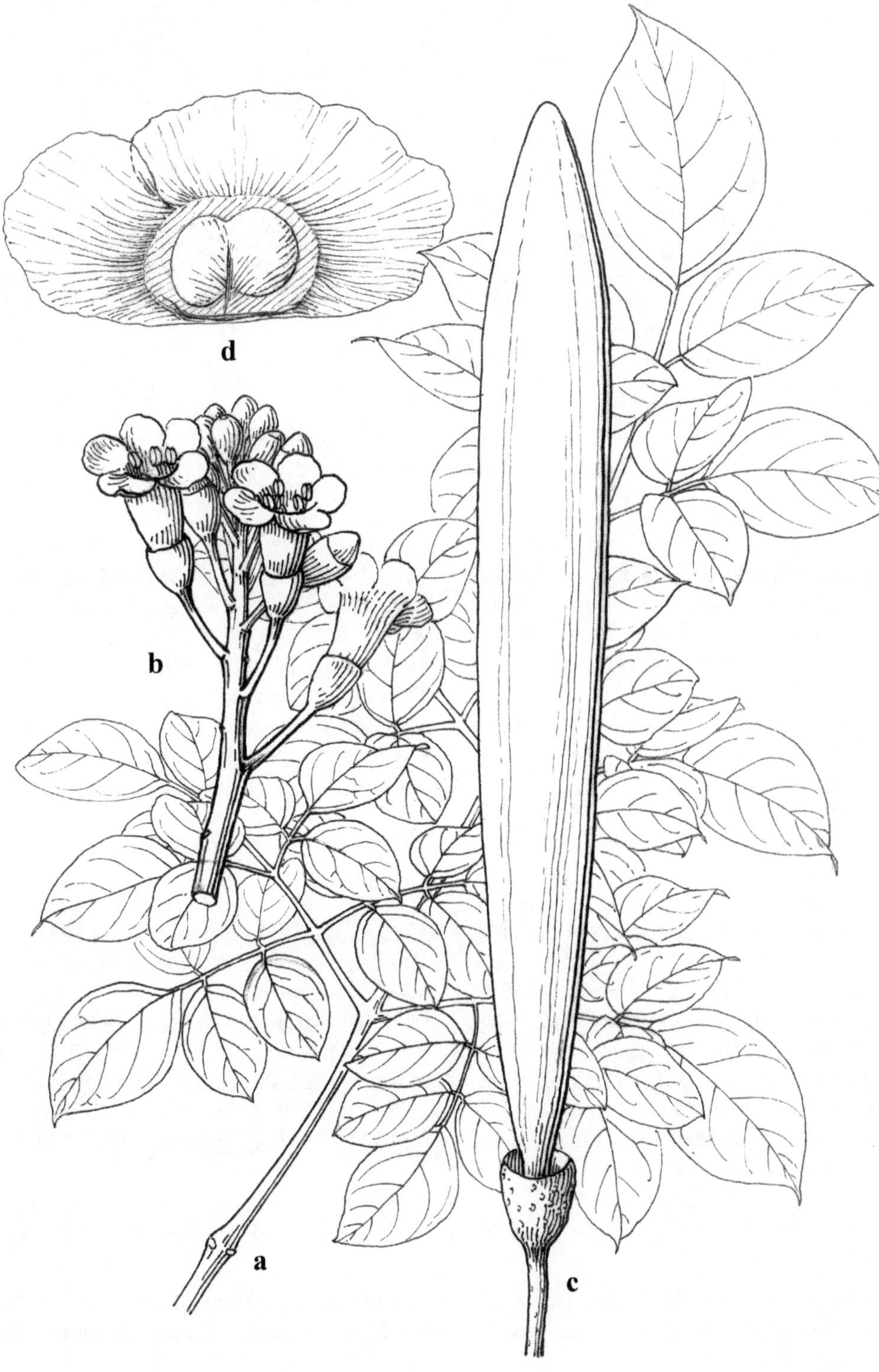

Figure 46-117: **Oroxylum indicum** (L.) Benth. ex Kurz: **a**, leaf; **b**, inflorescence; **c**, capsule; **d**, seed. Drawn by Jin-Yong FENG; reproduced from *Icon. Cormophyt. Sin.* 4: 107, fig. 5627. 1975.

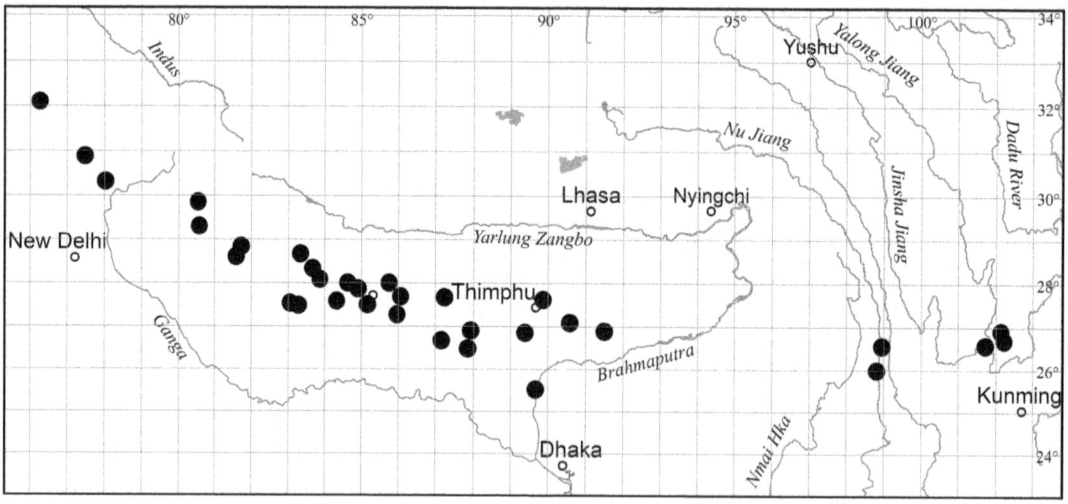

Map 46-209: **Oroxylum indicum** (L.) Benth. ex Kurz

700 m, 1981, *Bhattachaya 3752* (KATH); **Lumbini**, Kapilvastu, Tilaurakot, 165 m, 2005, *V. Manandhar & P. P. Kurmi 36* (KATH); Rupandehi, Butwal-Nawakot, 1973, *D. P. Joshi & M. M. Amatya 73/891* (KATH); **Narayani**, Chitwan, south of Rapti River, 250 m, 13 May 1976, *R. G. Troth 799* (BM, KATH, TI); Chitwan, Kasara, 1975, *D. P. Joshi et al. 75/3442* (KATH); Chitwan, Sauraha to Harda Khola, 1996, *M. Mikage et al. 9615136* (KATH); Makwanpur, Hatiya, 500 m, 1987, *N. K. Bhattarai 87/73* (KATH); Chitwan, Sauraha to Devi Tal, 21 Jan. 1996, *M. Mikage et al. 9612092* (TI). **E Nepal**: **Koshi**, Sankhuwasabha, 340 m, 14 Aug. 1997, *S. Noshiro et al. 9755049* (E, TI); Sunsari, Koshitappu, 70 m, 2005, *K. R. Bhettarai & N. Acharya 200584* (KATH); **Mechi**, Ilam, Ranga Pani to Ghorwa, 9 Dec. 1963, *H. Hara et al. 6303470* (BM, KYO, TI); Jhapa, Biring Khola to Caroowa, 240–290 m, 1974, *P. Pradhan et al. 198/74* (KATH); Jhapa, between Koshi & Kankai Mai, Majuwa, near Mandre, 750 m, 1972, *T. B. Shrestha & T. K. Bhattachaya 71/151* (KATH). **SIKKIM & DARJILING**: **Darjiling**, Tista, 350 m, 20 Jun. 1960, *H. Kanai et al. 6011* (BM, E, KATH, KYO, TI); Tista, 350 m, 26 Apr. 1960, *H. Kanai et al. 6012* (KATH, TI). **BHUTAN**: **Deothang**, Samdrup Jongjhar, 21 Jun. 1979, *A. J. C. Grierson & D. G. Long 2102a* (E); **Gaylegphug**, Near Tori Bari, 400 m, 30 May 1979, *A. J. C. Grierson & D. G. Long 1476* (E, K); **Phuntsholing**, Torsa River, 200 m, 2 May 1979, *A. J. C. Grierson & D. G. Long 783* (E); **Punakha**, 1500 m, 29 Aug. 1914, *R. E. Cooper et al. 3399* (E, BM). **S HENGDUAN**: **Bijiang**, 1200 m, Jul. 1978, *Exped. Nujiang 1944* (KUN); **Huili**, 1300 m, 21 Sep. 1914, *H. Handel-Mazzetti 850* (E); **Lushui**, Liuku, 8 Jun. 1978, *S. coll. s. n.* (HITBC); **Miyi**, Puwei, 8 Jul. 1980, *Q. S. Zhao 6002* (SZ); Anninghe, 5 Jun. 1958, *S. Y. Chen et al. 10054* (SZ); **Panzhihua**, near Jinjiang Station, 29 Jun. 1980, *S. Y. Bao 353* (KUN); Ertan, 1200 m, 16 Jan. 1981, *Q. Q. Wang 23867* (CDBI).

Tribe 2. Catalpeae DC. ex Meisn., Pl. Vasc. Gen.: Tab. Diagn. 300, Comm.: 208. 1840. Type: *Catalpa* Scop.

Trees, interpetiolar glands and ridges absent, pseudostipules absent. Leaves simple. Calyx membraceous, bilabiately split to near base. Stamens 2; staminodes 3, thecae divaricate. Ovary bilocular, placentation axile.

Two genera: North America, the Greater Antilles, East Asia; one genus and three species in the Pan-Himalaya.

2. CATALPA Scop.

Scopoli, Intr. Hist. Natur.: 170. 1777; de Candolle, Prodr. 9: 226. 1845; Tao & Yin in Wang (ed.), Fl. Reipubl. Popularis 69: 13. 1990; Zhang & Santisuk in Wu & Raven (eds.), Fl. China 18: 215. 1998. Type: *Catalpa bignonioides* Walter (= *Bignonia catalpa* L.).

Trees or rarely shrubs; deciduous. Leaves simple in whorls of 3 or opposite, petiolate; blade ovate, elliptic to elliptic lanceolate, abaxially with purple glandular-punctate in axes of lateral veins, margin entire or bi- or trilobed. Inflorescences paniculate, corymbose, or racemose, terminal. Flowers hermaphroditic. Calyx calyptriform, bilabiate or irregularly divided, globose in buds. Corolla campanulate, bilabiate, upper lip 2-lobed, lower lip 3-lobed. Fertile stamens 2, included, inserted at base of corolla tube; staminodes 3. Disc conspicuous. Ovary 2-locular; ovules several. Capsules dehiscing loculicidally, long terete; valves thin; septum slender, terete. Seeds in several rows, rounded, thin membranous, with hair tufts at both ends. **Chromosomes** 2n = 40.

Recent references: Paclt, 1952. Synopsis of the genus *Catalpa*. *Candollea* 13: 241–285; Yao & Huang, 1980. Synopsis of the *Catalpa* and *Gymnocladus* in eastern Asia and northern America. *J. Nanjing Techn. Coll. Forests Prod.* 1980(1): 122–125; Li, 2008. Phylogeny of *Catalpa* (Bignoniaceae) inferred from sequences of chloroplast *nghF* and nuclear ribosomal DNA. *J. Syst. Evol.* 46: 341–348; Kirkbride & Olsen, 2011. Identity of *Catalpa tibetica* (Bignoniaceae). *J. Bot. Res. Inst. Texas* 5: 625–631; Olsen & Kirkbride, 2017. Taxonomic revision of the genus *Catalpa* (Bignoniaceae). *Brittonia* 69: 387–421.

Eight species: N America and E Asia; three species in the Pan-Himalaya.

Notes: Paclt (1952) divided the genus into two sections, sect. *Catalpa* and sect. *Macrocatalpa* Paclt. Yao & Huang (1980) further separated sect. *Catalpa* into two sections, sect. *Sinocatalpa* C. W. Yao & P. C. Huang and sect. *Catalpa*. Yao & Huang's system is not supported by recent molecular studies (Li, 2008). According to Paclt's system, all Pan-Himalayan species belong to sect. *Catalpa*.

1a. Inflorescences corymbose or racemose; flowers pale red .... 1. *C. bungei*
1b. Inflorescences cymose-paniculate or paniculate; flowers pale yellow or white.
   2a. Corolla yellow 2-striate at throat and purple spotted adaxially; leaves broadly ovate, 3-dentate above the middle, scabrous on both sides... 2. *C. ovata*
   2b. Corolla yellow-white, pale purple spotted at throat; leaves broadly ovate, densely pubescent abaxially .... 3. *C. bignonioides*

1. Catalpa bungei C. A. Mey., Bull. Acad. Imp. Sci. Saint-Pétersbourg 2: 51. 1837; Tao & Yin in Wang (ed.), Fl. Reipubl. Popularis 69: 16. 1990; Zhang & Santisuk in Wu & Raven (eds.), Fl. China 18: 215. 1998; Olsen & Kirkbride, Brittonia 69: 401. 2017. Type: China, "China Borealis", 1831, *A. A. von Bunge s. n.* [lectotype designated by Olsen & Kirkbride, Brittonia 69: 400. 2017: P (P00608163); isolectotypes: GH, P (P00608165)].

Catalpa bungei var. *heterophylla* C. A. Mey., Bull. Sci. Acad. Imp. Sci. Saint-Pétersbourg 2: 51. 1837. ≡ *Catalpa heterophylla* (C. A. Mey.) Dode, Bull. Soc. Dendrol. France 1907: 203. 1907. Type: China, "China borealis", 1831, *A. A. von Bunge s. n.* [lectotype designated by Olsen & Kirkbride, Brittonia 69: 400. 2017: P (P00608164)].

Catalpa duclouxii Dode, Bull. Soc. Dendrol. France 1907: 201. 1907. ≡ *Catalpa fargesii* f. *duclouxii* (Dode) Gilmour, Bot. Mag. 159: t. 9458. 1936. Type: [S

Hengduan] China: Yunnan, Mo-so-yn, Apr. 1888, *M. Delavay 3352* [lectotype first-step designated by Paclt, Candollea 13: 257. 1952, second-step designated by Olsen & Kirkbride, Brittonia 69: 401. 2017: P (P00753637); isolectotypes: P (P00753638, P00753639)].

Trees, deciduous, 8–15 m high, stems glabrous to stellate, with elliptic lenticles. Leaves in whorls of 3; petiole 2–10 cm long; blade triangular-ovate or ovate-oblong, 6–20 cm long, 6–13 cm broad, adaxially glabrous to densely stellate, abaxially glabrous to hispidulous and stellate, lateral veins 4–6 pairs, with dark brown, triangular area in axils of base lateral veins, base truncate or cordate, margin entire, apex long acuminate. Inflorescences corymbose or racemose, terminal, 2–12-flowered; peduncle 1.5–5 cm long, glabrous to densely stellate, with a few scattered lenticles; rachis 3–7 cm long, terete, glabrous to densely stellate, with a few lenticles. Bracts narrowly elliptic, 1.1–2.1 cm long, 1.5–4 mm broad, glabrous to densely stellate, apex narrowly acute. Pedicel 2–4 cm long, glabrous to densely stellate, with spherical, glandular trichomes; bracteoles narrowly elliptic, 5–9 mm long, 0.5–1 mm broad, glabrous to densely stellate, apex acute. Calyx broadly turbinate, 9–12 mm long, glabrous to stellate, splitting into 2 or 3 lobes; lobes broadly elliptic to circular, apex obtuse, mucronate. Corolla pale red, yellow 2-striate and dark purple spotted on throat, 3–3.5 cm long; tube 1.7–2.5 cm long, 1.8–2.2 mm broad, upwards gradually widened to 1.5–1.8 cm broad at mouth; lobes circular to transversely broadly elliptic, 1.3–2.3 cm long, 1.3–2.2 cm broad, apex obtuse. Stamens 2; filaments 1.5–2.5 cm long, white, glabrous; staminodes 1–5 mm long, white, glabrous. Ovary 3–5 mm long, green; style 1.7–2.2 cm long, glabrous, white; stigma 1–3 mm long, glabrous. Capsules linear, 25–45 cm long, 5–6 mm broad. Seeds narrowly ellipsoid, 9–10 mm long, 2–2.2 mm broad, villous at both ends. **Chromosomes** 2n = 40. Figure 46-118.

Phenology: flowering from March to June; fruiting from June to November.

Habitats: roadside; 400–2800 m.

Distribution: N Hengduan, S Hengduan, U Irrawaddy (Map 46-210); also in other parts of China, and cultivated in temperate regions of N America and Europe.

N HENGDUAN: **Jiuzhaigou** (Nanping), Junfengshan, 450 m, 6 May 1958, *M. X. Nie & S. S. Lai 2498* (KUN); **Lixian**, 1952, *C. Ho & Z. L. Zhou 13912* (IBSC, NAS, PE); Longxi, Madun, 9 May 1952, *C. Ho 12298* (IBSC); **Wenchuan**, Keku, 8 May 1952, *C. Ho & Z. L. Zhou 12289* (IBSC, NAS, PE, SZ); **Wenxian**, Liujiaping, Dalingliang, 1550 m, 12 May 2007, *Baishuijiang Expedition 4678* (PE); **Wudu**, Luotang, 1250 m, 4 May 1959, *Z. Y. Zhang 1269* (WUK). **S HENGDUAN**: **Baoxing**, 530 m, Jul. 1939, *S. Y. Hu 860* (SZ); **Bijiang**, Nongzhong, 2600 m, 26 May 1978, *Exped. Bijiang 0083* (KUN); **Dali**, 1946, *T. N. Liou 13700* (IBSC, KUN, PE); 5 May 1946, *T. N. Liou 13887* (IBSC, KUN, PE); 17 Mar. 1946, *T. N. Liou 15645* (IBSC, KUN, PE); *T. N. Liou 16715* (IBSC, KUN, PE); 26 Sep. 1946, *T. N. Liou 20045* (IBSC, KUN, PE); Fengyi, Huaying, 11 Jan. 1942, *H. C. Wang 1620* (IBSC); Lanfeng, 10 May 1942, *H. C. Wang 2082* (PE); Wuliqiao, 9 Jun. 1942, *H. C. Wang 2300* (IBSC); near Zhonghe Temple, 28 May 1929, *R. C. Ching 22796* (KUN, PE); Tali Valley, 2000–2500 m, May–Jul. 1906, *G. Forrest 4679* (BM, E); **Dêqên**, Yunling, 2500 m, 11 Jun. 1961, *S. Jiang et al. 9179* (KUN, PE); **Gongshan**, Village of Hse Zo, Shupa Valley, 2600 m, Yangtzse-Mekong Divide, Feb. 1902, *G. Forrest 7612* (E); **Heqing**, Songgui to Lianping, 17 Apr. 1939, *K. M. Feng 688* (PE); **Shangri-la** (Zhongdian), Hutiaoxia, Chongjianghe, 2430 m, 19 Apr. 2007, *L. M. Gao GLM-07401* (KUN); s. loc., 1825 m, *X. K. Ou Ouxk-YS-229* (IBSC); **Tianquan**, Xinhua, 860 m, Oct. 1982, *s. coll. 46723* (CDBI); **Weixi**, Hechunhe, 1700 m, 4 May 1960, *Nan Shui Bei Diao Dui 8664* (KUN); Xiaoweixi, 1800 m, 7 May 1940, *K. M. Feng 3694* (IBSC, KUN, PE); **Yulong** (Lijiang), Lichiang Valley, 2450 m, May 1906, *G. Forrest 2121* (E); 25 Apr. 1939, *R. C. Ching 20093* (KUN, PE); Bahe, 3 Oct. 1944, *R. C. Ching 20954* (IBSC, KUN); 1 May 1940, *R. C. Ching 30755* (KUN, PE); near city, 2500 m, 18 Apr. 1938, *T. T. Yü 11154* (KUN, PE); Nguluko, 2800 m, 17 Jun. 1937, *T. T. Yü 15190* (KUN, PE); Yulongshan, 2500 m, Oct.–Nov. 1935, *C.*

Figure 46-118: **Catalpa bungei** C. A. Mey.: **a**, flowering branch; **b**, capsule; **c**, seed. Drawn by Jin-Yong FENG; reproduced from *Icon. Cormophyt. Sin.* 4: 102, fig. 5618. 1975.

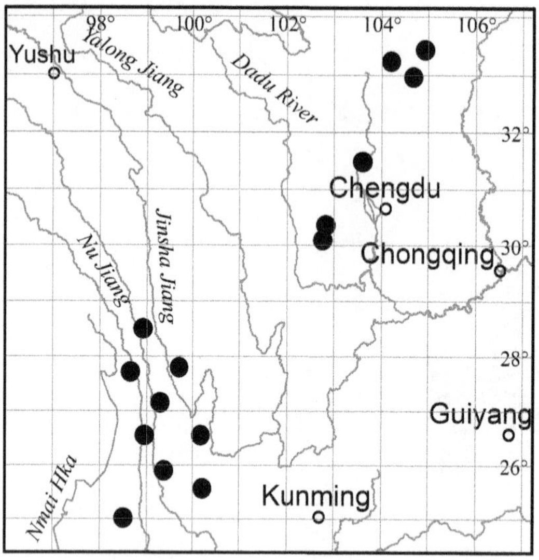

Map 46-210: **Catalpa bungei** C. A. Mey.

W. Wang 71759 (NAS); Yulongshanm Yufengsi, 2750 m, 28 May 1962, *Lijiang Botanical Garden 100456* (KUN); Yulongshan, Xiaoyulonbg, 2650 m, 19 May 1961, *S. W. Yu 24795* (KUN). **U IRRAWADDY**: **Tengchong**, 15 Jan. 1977, *Y. H. Li 20182* (HITBC), Qushi, Gongping, cultivated, 26 Oct. 1998, *H. Li 10965* (E, KUN); Qushi, 12 Sep. 1960, *W. Q. Yin 60-1239* (HITBC, KUN); Tengyueh Valley, 1680 m, May 1912, *G. Forrest 7490* (E, K).

2. Catalpa ovata G. Don, Gen. Hist. 4: 230. 1837; Tao & Yin in Wang (ed.), Fl. Reipubl. Popularis Sin. 69: 14. 1990; Zhang & Santisuk in Wu & Raven (eds.), Fl. China 18: 216. 1998; Olsen & Kirkbride, Brittonia 69: 404. 2017. Type: Japan, s. d., *C. P. Thunberg s. n.* [lectotype first-step designated by Paclt, Candoleea 13: 275. 1952, second-step designated by Olsen & Kirkbride, Brittonia 69: 404. 2017: UPS (UPS-THUNB14230); isolectotype: UPS (UPS-THUNB14231)].

Trees ca. 15 m high; young branches sparsely pubescent. Leaves opposite or nearly so, sometimes whorled; petiole 6–18 cm long; blade broadly ovate, 18–25 cm long, 17–25 cm broad, scabrous, sparsely pubescent or glabrescent, lateral veins 4–6 pairs, palmately 5–7-veined basally, base cordate, margin entire or sinuolate, usually 3-lobed, apex acuminate. Panicles terminal; peduncle sparsely pubescent, 12–28 cm long; rachis 17–25 cm long, glabrous, lenticellate. Bracts narrowly elliptic, 2–11 mm long, 0.5–2 mm braod, glabrous or shortly ciliate, apex acute. Pedicel 2–6 mm long, glabrous. Calyx bilabiate, 6–8 mm long, glabrous, splitting into 2 lobes; lobes elliptic, apex obtuse. Corolla campanulate, pale yellow, yellow 2-striate and purple spotted at throat, 2–2.5 cm long, 1.8–2 cm broad; tube 1–1.2 cm long, 0.7–1 mm broad, upwards gradually widened to 5–8 mm at mouth; lobes broadly elliptic or broadly elliptic to circular, 0.7–1.2 cm long, 0.6–1.2 cm broad, apex obtuse. Stamens 2; filaments 1–1.2 cm long, white, glabrous; anthers divergent; staminodes 1.5–2 mm long, white, glabrous. Ovary 2–2.3 mm long, white, glabrous; style 5–7 mm long, glabrous; stigma 0.8–1.1 mm long, white, 2-lobed. Capsules linear, 20–30 cm long, 5–7 mm broad. Seeds linear-ellipsoid, 6–8 mm long, 2.5–3 mm broad, villous at both ends. **Chromosomes** 2n = 40. Figure 46-119.

Phenology: flowering from May to June; fruiting from August to October.

Habitats: slopes; (500–)800–2700 m.

Distribution: S Hengduan (Map 46-211); also in other parts of Chian, and widely cultivated in temperate areas.

S HENGDUAN: Baoxing, Zhonggang, Qiangeryan, 2125 m, 26 Jul. 1954, *T. P. Soong 39236* (SZ); **Bijiang**, Pihe, 1100 m, 30 May 1978, *Exped. Bijiang 0256* (KUN); **Dêqên**, Chawalongla, 2600 m, 17 Sep. 1938, *T. T. Yü 10291* (IBSC); **Ningnang**, Yongning, 2600 m, 23 Jun. 1960, *S. Jiang et al. 6059* (KUN, PE); **Tianquan**, Yongxing, 800 m, 10 Oct. 1982, *D. Y. Peng 46723* (IBSC); **Yulong** (Lijiang), Lichiang Range, 1933, *H. D. McLaren C110* (BM, E); Basha, Yulong, 2650 m, 30 Jun. 1965, *S. W. Yu 65070* (KUN, PE). **U IRRAWADDY**: **Tengchong**, Dongshan, Tengchong to Rehai Hotspring, 1680 m, 24 Oct. 1998, *H. Li 10873* (E, KUN).

Figure 46-119: **Catalpa ovata** G. Don: **a**, flowering branch; **b**, capsule; **c**, seed. Reproduced from *Icon. Cormophyt. Sin.* 4: 102, fig. 5617. 1975.

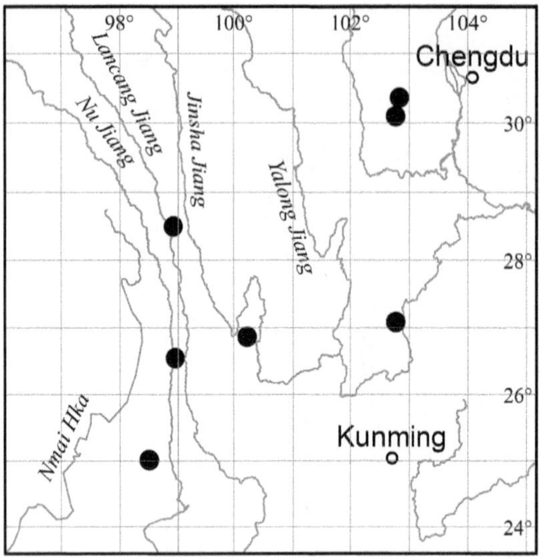

Map 46-211: **Catalpa ovata** G. Don

3. Catalpa bignonioides Walter, Fl. Carol.: 64. 1788; Olsen & Kirkbride, Brittonia 69: 406. 2017. Type: United States, South Carolina, *Fraser 28-D* (neotype as "holotype", designated by Reveal et al., Bartonia 56: 17. 1990: BM-SL); Lexington County, just behind (N of) Truck Stop and "44 Restaurant", on E Side of SC Hwy 34 at I-20, about 9 mi SE of Leesville, 20 May 1997, *J. B. Nelson 18315* (epitype as "neotype", designated by Ward, J. Bot. Res. Inst. Texas 1: 1098. 2007: GH; isoepitype: USCH).

Catalpa tibetica Forrest, Notes Roy. Bot. Gard. Edinburgh 13: 155. 1921; Tao & Yin in Wang (ed.), Fl. Reipubl. Popularis Sin. 69: 14. 1990; Zhang & Santisuk in Wu & Raven (eds.), Fl. China 18: 216. 1998. Type: [S Hengduan] China, Yunnan, Salween-Kiu-chiang Divide, Tsarong, in mixed forests, Jun. 1919, *G. Forrest 18950* [lectotype designated by Kirkbride & Olsen, J. Bot. Res. Inst. Texas 5: 628. 2011: E (E00287957); isolectotypes: E (E00287958), K (K000779196), P].

Shrubs or small trees, to 5 m high. Leaves in whorls of 3, distinctly unequal in size in each whorl; petiole 10–18 cm long, glabrous or villous upwards; blade broadly ovate, 18–23 cm long, 16–22 cm broad, thin leathery, adaxially dark green and glabrous, abaxially pale green and densely pubescent, basally 5-veined, base slightly cordate or truncate, margin entire, apex acuminate or obtuse. Inflorescences corymbose-paniculate, 16–25 cm long, glabrous; peduncle 2–8 cm long, lenticellate; rachis 12–26 cm long, terete, lenticellate; pedicel 3–16 mm long, glabrous. Bracts sublate, narrowly oblong, or narrowly elliptic, 1.3–7 mm long, 0.2–1 mm broad, glabrous, apex acute. Calyx obpyriform, 8–12 mm long, 6–9 mm broad, purple to greenish purple, splitting into 2 or rarely 3 lobes; lobes broadly elliptic to elliptic, or circular, 8–12 mm long, 7–9 mm broad, apex obtuse, mucronate. Corolla yellow-white, pale purple spotted at throat, 4–4.5 cm long; tube base 1.5–2.5 cm long, 1–2 mm broad, upwards gradually widened to 2.5–4 cm broad at mouth; lobes broadly elliptic to orbicular, 1–1.7 cm long, 0.8–1.6 cm broad, glabrous, apex obtuse. Stamens inserted at base of corolla tube; filaments 1.6–2.2 cm long, glabrous; staminodes 3–6 mm long, glabrous. Ovary glabrous; style 1.2–2 cm long, glabrous; stigma 1–2 mm long, 2-lobed. Capsules terete, striate, 25–30 cm long, 0.8–1 cm broad. Seeds ellipsoid, thin membranous, filiform pubescent at both ends. **Chromosomes** 2n = 40.

Phenology: flowering from May to June; fruiting from August to September.

Habitats: slopes, forests; 1600–2700 m.

Distribution: Jammu & Kashmir, S Hengduan (Map 46-212); native to N America, widely cultivated in temperate areas.

JAMMU & KASHMIR: Kashmir Valley, 1600 m, 12 Jun. 1937, *F. Ludlow 86* (BM). **S HENGDUAN**: **Gongshan**, Salwin-Kiu-Chiang Divide, Aug. 1919, *G. Forrest 18926* (E); Tsarong, Jun. 1919, *G. Forrest 18950* (E); **Yulong** (Lijiang), 15 Jun. 1939, *R. C. Ching 20717* (KUN, PE); 4 Jun. 1940, *R. C. Ching 31019* (KUN).

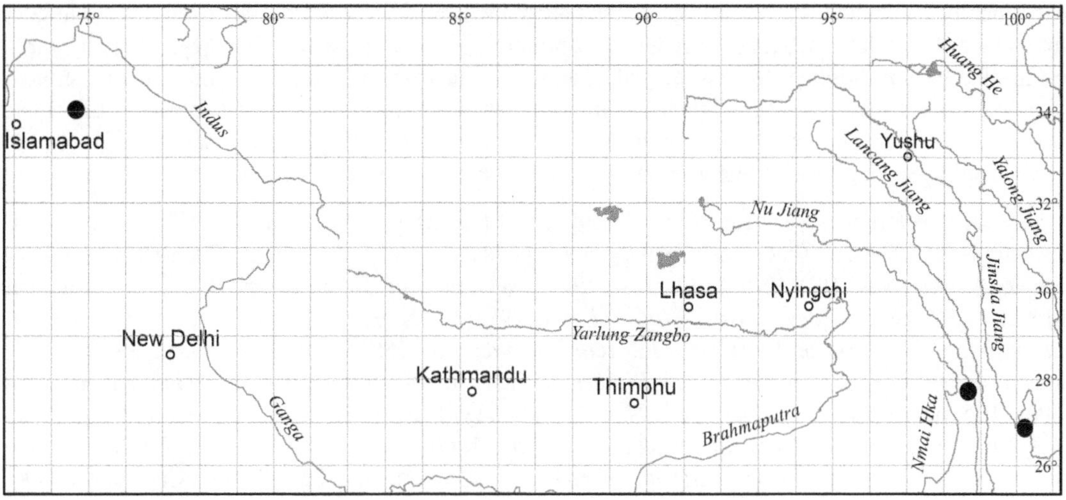

Map 46-212: **Catalpa bignonioides** Walter

Tribe 3. Tecomeae Endl., Gen. Pl.: 711. 1839. Type: *Tecoma* Juss.

Trees, shrubs or lianas, interpetiolar glands and ridges usually absent, pseudostipules usually absent, rarely subulate or foliaceous. Leaves pinnate, palmate or simple. Calyx 5-dentate to irregularly lobed or spathaceous. Stamens 4–5, anthers bithecous, thecae straight, divaricate. Ovary bilocular, placentation axile. Fruit dehiscing perpendicular to septum (loculicidal).

Twelve genera: tropical region of the world; one genus and 14 species (11 endemic) in the Pan-Himalaya.

3. INCARVILLEA Juss.

Jussieu, Gen. Pl.: 138. 1789; de Candolle, Prodr. 9: 237. 1845; Schumann in Engler & Prantl (eds.), Nat. Pflanzenfam. 4(3b): 232. 1897; Tao & Yin in Wang (ed.) Fl. Reipubl. Popularis Sin. 69: 34. 1990; Zhang & Santisuk in Wu & Raven (eds.), Fl. China 17: 220. 1998. Type: *Incarvillea sinensis* Lam.

Incarvillea [unranked] *Amphicome* R. Br. ex Royle, Ill. Bot. Himal. Mts. [9]: 296. 1836. ≡ *Amphicome* (R. Br. ex Royle) G. Don, Gen. Syst. 4: 645, 665. 1838; Schumann in Engler & Prantl (eds.), Nat. Pflanzenfam. 4(3b): 232. 1897. ≡ *Incarvillea* subg. *Amphicome* (R. Br. ex Royle) Grierson, Notes Roy. Bot. Gard. Edinburgh 23: 313. 1961. Type: *Amphicome arguta* Royle [≡ *Incarvillea arguta* (Royle) Royle].

Niedzwedzkia B. A. Fedtschenko, Izv. Imp. Bot. Sada Petra Velikago 15: 399. 1915. ≡ *Incarvillea* subg. *Niedzwedzkia* (B. A. Fedtschenko) Grierson, Notes Roy. Bot. Gard. Edinburgh 23: 350. 1961. Type: *Niedzwedzkia semiretschenskia* B. A. Fedtsch. [≡ *Incarvillea semiretschenskia* (B. A. Fedtsch.) Grierson].

Annual or perennial herbs. Roots simple or branched, woody or tuberous; stems sometimes absent. Basal leaves in rosette, or cauline leaves alternate, rarely opposite, without stipules, pinnate or pinnatisect, rarely undivided, puberulous. Flowers 5-merous, terminal, solitary or racemose, occasionally paniculate, zygomorphic. Pedicel subtended by a short bract. Calyx tube campanulate, lobe ovate, lanceolate, subulate or reduced to minute points. Corolla tube base cylindric, upwards campanulate; limb 5-lobed; lobes subequal, rounded or emarginate. Stamens 4, didynamous, anthers bithecous, thecae with a single bristle; staminode absent or variable. Ovary subsessile, bilocular; stigma horizontally elliptic. Capsules cylindric, quadrangular or 6-winged. Seeds subdiscoid or ovate, surrounded by a wing or with fine hairs at each end. **Chromosomes** 2n = 22.

Recent references: Grieson, 1961. A revision of the genus *Incarvillea.* Notes. Roy. Bot. Gard. Esinburgh 23: 303–354. Zhao, 1985. A study on the genus *Incarvillea* Juss. of the Transverse Mountain Range. *Explor. Nat.* 4(2): 170–174; Zhao, 1983. New taxa of genus *Incarvillea* from Hengduan Mountain Range. *J. Sichuan Univ. Nat. Sci. Ed.* 1983(4): 92–95; Zhao, 1988. A new species of *Incarvillea* (Bignoniaceae) from Sichuan. *Acta Phytotax. Sin.* 26: 78–79; Grey-Wilson, 1998. A new look at the subgenus *Pteroscleris*. *New Plantsman* 5(2): 76–98; Illarionova, 2002. Genus *Incarvillea* Juss. (Bignoniaceae) in Asia Centrali. *Novosti Sist. Vyssh. Rast.* 34: 174–178; Xiao et al., 2002. Karyomorphology of six *Incarvillea* species. *Acta Bot. Yunnan.* 24: 87–93; Chen et al., 2003. Pollen morphology of the genus *Incarvillea* (Bignoniaceae). *Acta Bot. Yunnan.* 25: 457–467; Chen et al., 2004. Karyomorphological study of *Incarvillea* Juss. (Bignoniaceae) and its implications in distribution and taxonomy. *Bot. J. Linn. Soc.* 144: 113–121; Chen et al., 2005. Molecular phylogeny of *Incarvillea* (Bignoniaceae) based on ITS and *trnL-F* sequences. *Amer. J. Bot.* 92: 625–633; Chen et al., 2006. A new subgenus of *Incarvillea* (Bignoniaceae). *Ann. Bot. Fennici* 43: 288–290.

Eighteen species: E and C Asia; 14 species (11 endemic) in the Pan-Himalaya.

Notes: Grierson (1961) divided the genus into four subgenera, subg. *Niedzwedzkia* (B. Fedtsch.) Grierson, subg. *Incarvillea*, subg. *Amphicome* (Rorle) R. Br. and subg. *Pteroscleris* Baill. Chen et al. (2006) further established a new subgenus, subg. *Olgaea* S. T. Chen, K. Y. Guan & Z. K. Zhou, based on a single species *I. olgae* Regel. Three subgenera occur in our area.

1a. Anthers pilose; seeds with a coma of hairs at either end (subg. I. ***Amphicome***).
  2a. Plants up to 1.5 m high; stems branched, leaves cauline; calyx teeth subulate .... 1. *I. arguta*
  2b. Plants up to 50 cm high; stems unbranched, leaves aggregated at base; calyx teeth rounded, apiculate .... 2. *I. emodi*
1b. Anthers glabrous; seeds surrounded by a wing.
  3a. Calyx teeth subulate with thickened base or reduced to minute point, plicae bifid or simple; capsules cylindric; seed wing hyaline (subg. II. ***Incarvillea***) .... 3. *I. sinensis*
  3b. Calyx teeth lanceolate, semiorbicular or triangular, usually broad, base not thickened; capsules more or less quadrangular or ridged; seed wing thick, non-hyaline (subg. III. ***Pterosleris***).
    4a. Plants distinctly stemmed; leaves cauline.
      5a. Leaves simple, blade margin crenate .... 4. *I. forrestii*
      5b. Leaves pinnately divided or compound, blade margin entire or serrate.
        6a. Plants pale brown pubescent; flowers yellow; leaves pinnately divided .... 5. *I. lutea*
        6b. Plants glabrous; flowers red or rose; leaves pinnately compound.
          7a. Calyx teeth semiorbicular, apex rounded to truncate, mucronate; lateral leaflets 5–7 cm broad .... 6. *I. altissima*
          7b. Calyx teeth broadly triangular, apex acute; lateral leaflets ca. 2 cm broad .... 7. *I. beresowskii*
    4b. Plants stemless or shortly stemmed; leaves in a basal rosette.
      8a. Leaves 2-pinnately divided .... 8. *I. dissectifoliola*
      8b. Leaves 1-pinnately compound or divided.

9a. Plants shortly stemmed; leaflets entire .... 9. *I. compacta*
9b. Plants stemless; leaflets serrate.
10a. Leaflets 0–4 pairs; flowers solitary or few in short raceme .... 10. *I. himalayensis*
10b. Leaflets (2–)3–11 pairs; flowers several in panicles.
11a. Leaflets scabrous; plants 10–20 cm high .... 11. *I. younghusbandii*
11b. Leaflets glabrous; plants 20–40 cm high.
12a. Lateral leaflets elliptic-lanceolate, base obtuse to subrounded; corolla 5.5–6.5 cm long; style 2.8–3 cm long .... 12. *I. delavayi*
12b. Lateral leaflets ovate to ovate-lanceolate, base cordate or broadly cuneate; corolla 7–10 cm long; style 5–7 cm long.
13a. Lateral leaflets 2–5 pairs, shiny and bullate adaxially; flowers purple red or pale red .... 13. *I. mairei*
13b. Lateral leaflets 4–9 pairs, matt green and rather smooth above adaxially; flowers deep magenta or crimson-magenta, rarely rose-pink .... 14. *I. zhongdianensis*

I. Incarvillea subg. **Amphicome** (R. Br. ex Royle) Grierson, Notes Roy. Bot. Gard. Edinburgh 23: 313. 1961; Tao & Yin in Wang (ed.), Fl. Reipubl. Popularis Sin. 69: 38. 1990. ≡ *Incarvillea* [unranked] *Amphicome* R. Br. ex Royle, Ill. Bot. Himal. Mts.: 296. 1836; de Candolle, Prodr. 9: 237. 1845; Bentham & Hooker, Gen. Pl. 2: 1048. 1876; Clarke in Hooker (ed.), Fl. Brit. India 4: 385. 1885; Schumman in Engler & Prantl (eds.), Nat. Pflanzenfam. 4(3b): 232. 1894. ≡ *Amphicome* (R. Br. ex Royle) G. Don, Gen. Syst. 4: 645, 665. 1838. Type: *Amphicome arguta* Royle ≡ *Incarvillea arguta* (Royle) Royle.

Erect or ascendent perennial herbs, somewhat woody at base. Leaves subopposite basally, alternate upwards, pinnately compounded. Corolla tube pubescent with long stalked glands at base within; lobes eglandular. Anther-thecae pilose. Capsules elongate, linear, cylindrical, bivalvular- dehiscent, valves thin fibrous. Septum bifid at apex. Seeds ellipsoid or narrowly ovoid with tufts of hairs for 5 mm long at both ends.

Two species in Himalaya and its neighboring regions; both in the Pan-Himalaya.

1. Incarvillea arguta Royle, Ill. Bot. Himal. Mts. 1: 296. 1836; Grierson, Notes Roy. Bot. Gard. Edinburgh 31: 314. 1963; Tao & Yin in Wang (ed.), Fl. Reipubl. Popularis Sin. 69: 38 1990; Zhang & Santisuk in Wu & Raven (eds.), Fl. China 18: 221. 1998. ≡ *Amphicome arguta* (Royle) G. Don, Gen. Syst. 4: 665. 1837; de Candolle, Prodr. 9: 237. 1845; Clarke in Hooker (ed.), Fl. Brit. India 4: 385. 1884. ≡ *Incarvillea diffusa* Royle, Ill. Bot. Himal. Mts. 2: t. 72. fig. 1. 1835. ≡ *Amphicome diffusa* (Royle) Sprague, Bull. Misc. Inform. Kew 1933: 386. 1933. Type: [U Ganga & Indus] India, Turanda in lower Kunawar, s. d., *Royle s. n.* [holotype: BM (BM000649685)].

Incarvillea arguta var. *daochengensis* Q. S. Zhao, J. Sichuan Univ. Nat. Sci. Ed. 1983(4): 94. 1983. Type: [S Hengduan] China, Sichuan, Daocheng Xian, Dongyi Qu, Mofanggou, 2000 m, 26 May 1979, *Dao Zi 88* (holotype: CDBI).

Incarvillea arguta var. *longipedicellata* Q. S. Zhao, J. Sichuan Univ. (Nat. Sci.) 1983(4): 94. 1983, **syn. nov.**; Tao & Yin in Wang (ed.), Fl. Reipubl. Popularis Sin. 69: 39. 1990; Zhang & Santisuk in Wu & Raven (eds.), Fl. China 18: 222. 1998. Type: [S Hengduan] China, Yunnan, Dêqên, bank of Lancangjiang, Yongzi, 2100 m, 25 Jul. 1940, *K. M. Feng 5767* [lectotype

here designated: PE (PE01154708); isolectotypes: IBSC (IBSC0562879), PE (PE01154662)].

Herbs, 0.2–1.5 m high, glabrous or puberulent. Leaves alternate, cauline 1-pinnately compound, 5–20 cm long; leaflets 5–13, opposite or subopposite, blade lanceolate or elliptic, 1–5 cm long, 0.5–2 cm broad, adaxially pubescent, abaxially glabrous, punctate, base oblique, rounded to cuneate, margin acutely serrate, apex acute or acuminate. Racemes terminal, 5–20-flowered; pedicel 8–20 mm long. Bracts subulate, 2–3 mm long; bracteoles 2, 1–1.5 mm long. Calyx campanulate; tube 5–8 mm long, teeth subulate, 1–4 mm long, base subtriangular, not enlarged into glands, margin ciliate. Corolla pink or white, campanulate-funnelform, 3.5–4 cm long, 1.8–2 cm broad; tube 2.5–3.8 cm long, inside glandular pubescent at base; lobes 0.5–1 cm long, 0.9–1.4 cm broad, apex rounded; anther-thecae ovate, 1.6–2 mm long, 0.8–1 mm broad, pilose. Staminode 1–1.2 cm long, filiform tapering to flattened apex. Capsules linear, cylindric, curved, membranous, 8–20 cm long, 2–3 mm broad. Seeds very narrowly ellipsoid, 1.8–2 mm long, 0.8–1 mm broad, pale brown; whitish hairs at both ends for 4–5 mm long. **Chromosomes** 2n = 22. Figure 46-120.

Phenology: flowering from March to July; fruiting from September to December.

Habitats: slopes, thickets, dry stony situations and on limestone ledges; 600–3500 m.

Distribution: U Ganga & Indus, U Yarlung Zangbo, W Nepal, C Nepal, E Nepal, L Yarlung Zangbo, Yarlung Zangbo-Brahmaputra, Tangut, N Hengduan, S Hengduan (Map 46-213); also in other parts of India and China (C Xizang, E & NE Yunnan, SE Sichuan, C & E Qinghai, C & SE Gansu, Guizhou).

Uses: The whole plant is used medicinally for the treatment of injuries.

Notes: In his *Illustrations of the Botany of the Himalayan Mountains*, Royle (Ill. Bot. Himal. Mts. 1: 296. 1836; 2: t. 72, fig. 1. 1835) published the names *Incarvillea diffusa* Royle and *I. arguta* Royle for the same species. The name *Incarvillea diffusa* Royle was published on plate 72 in volume 2 in 1835 and validated by the illustration with analysis according to Art. 38.8 of ICN (Turland et al., International Code of Nomenclature for Algae, Fungi and Plants (Shenzhen Code). Regnum Vegetabile 159. Koenigstein: Koeltz Scientific Books, 2018). *I. arguta* was described in volume 1 which was published in 1836 and referenced to plate 72. Therefore, the name *I. arguta* is a superfluous illegitimate name under Art. 52. The name *Amphicome arguta* G. Don was based on the name *Incarvillea diffusa* Royle and thus is also an illegitimate superfluous name under Art. 52. Therefore, the name *Incarvillea diffusa* Royle might be the correct name for the species. However, the name *I. arguta* was widely used in the references and a proposal to conserve this name will be proposed for the purpose of the nomenclature stability.

Zhao (1983) distinguished *Incarvillea arguta* var. *longipedicellata* from *I. arguta* var. *arguta* by inflorescences 4–10-flowered, pedicel 2–4 cm long, bracts borne at the middle of pedicel and calyx teeth without ciliae. However, flower number, pedicel length and trichomes of calyx are variable in the species. The bract on the pedicel is actually borne at base of pedicel but not at the middle as described in the protologue. However, this variety may represent a form with smaller leaflets. I do not separate it as an independent taxon because further observations on variation of leaflets is necessary.

U GANGA & INDUS: Sarabaw, Sep. 1886, *J. R. Drummond 22132* (K); **Badrinath**, Bhyundar Valley, Josimath to Badrinath, between Gangharea and Govindgath, 2300 m, 18 Aug. 1975, *F. Billiet & J. Leonard 6644* (K); **Chamoli**, Dhauli River, 2100 m, *R. Strachey & J. E. Winterbottom 2* (BM, K); Dhauli River, 2130 m, *R. Strachey & J. E. Winterbottom* 2 (K, BM); **Kinnaur**, Wangta, Sep. 1864, *D. Brandis 3676* (K); s. loc., 1886, *J. R. Drummond 22129* (K); s. loc., 1885, *J. R. Drummond 22341* (E, K); Turanda in lower Kunawar, *Royle s. n.* (BM); Lower Kunawar, *Role 296* (K); Kunawar, 2100 m, 15 May 1847, *s. coll. 1855* (K); **Kulu**, 1800 m, 27 Jun. 1916, *R. E. Cooper 5066* (E); Parbati Nollah, 1800 m, 8 Jun. 1976, *J. D. A. Stainton*

Figure 46-120: **Incarvillea arguta** (Royle) Royle: **a**, proximal portion of plant; **b**, opened corolla showing stamens; **c**, stamens; **d**, capsule; **e**, seed. Drawn by Jin-Yong FENG; reproduced from *Icon. Cormophyt. Sin.* 4: 99. fig. 5611. 1975.

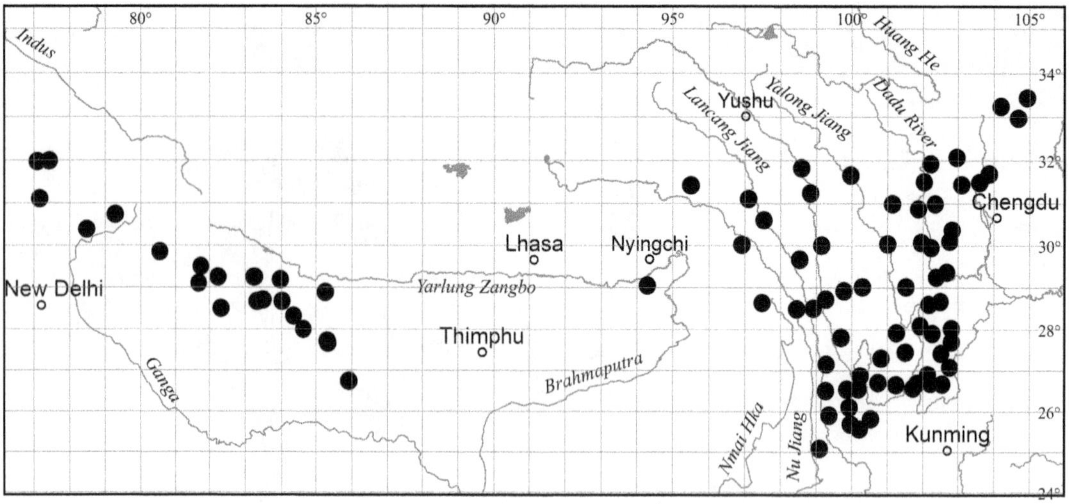

Map 46-213: **Incarvillea arguta** (Royle) Royle

7598 (BM); **Lahul & Spiti**, Leoray, 1888, *J. R. Drummond 23094* (E, K); Manihami to Pulga, 30 May 1888, *J. R. Drummond 23095* (E, K); **Simla**, Bashahr, Ralli to Barary, 1980 m, *Lace 210* (E); Bashahr, Wangtu to Chagaon, 1675 m, *J. H. Lace 701* (E); Wangtu, Sutlej Valley, Simla Hill State, 2100 m, 22 Jun. 1939, *Ludlow & G. Sherriff 7328* (BM, TI); Wangtu, 1560 m, 26 May 1889, *G. Watt 3253* (E); Wangtu, Bushahr, 10 Jul. 1934, *N. Parmanand 423* (E); Theog rocks towards Mahan, 2440 m, Jul. 1885, *Collett 208* (K); Theog to Matiana, 15 Sep. 1917, *H. H. Rich 698* (K); Simla hills, 1886, *J. R. Drummond 22130* (E, K, G); Mattiana, 10 Aug. 1885, *J. R. Drumond 26284* (K); Simla hills, 9 Oct. 1886, *J. R. Drummond 22131* (K); Theog rocks, 2133 m, 8 Sep. 1878, *J. S. Gamble 6495* (K); **Tehri Garhwal**, rivers Bhagirathi & Jalandri, S Harsil, 2400 m, 27 Aug. 1952, *P. P. Huggins 425* (BM); Ganga valley near Suki, 2400 m, 17 Jun. 1883, *J. F. Duthie 83* (BM); Gangoltari, 2440 m, Sep. 1897, *J. S. Gamble 13* (K). **U YARLUNG ZANGBO**: **Gyirong**, Jipu Gorge, 2781 m, 25 Jul. 2013, *PE Xizang Expedition 4208* (PE); Jiangcun, 2022–2700 m, 8 Aug. 2010, *PE Xizang Exped. 478* (PE); Gyirong to Mangmu, 2450 m, 12 Jul. 1975, *Qinghai-Xizang Exped. 6309* (KUN, PE); Mangmu, 2500 m, 16 Jul. 1975, *Exped. Qinghai-Xizang 6902* (HNWP, KUN, PE); Mangmu, 2380 m, 7 Jun. 1972, *Xizang Chinese Herd Expedition 454* (HNWP, PE); Mangda to Resuoqiao, 7 Jun. 1972, *Xizang Chinese Herd Expedition 454a* (HNWP); Bangxing, 3000 m, 4 Jul. 1975, *C. Y. Wu et al. 75-579* (KUN); Lendi Khola, 2438–2743 m, 11 Jul. 1949, *O. V. Polunin 853* (BM). **W NEPAL**: **Karnali**, Dolpa, Tarakot, 29 May 1973, *S. Einarsson et al. 230* (BM); Dolpa, between Tibrikot and Dunai, 2450 m, 10 Sep. 1976, *H. Tabata et al. 3418* (KYO); Dolpa, Timikot, 2285 m, 31 May 1952, *O. V. Polunin, W. R. Sykes & L. H. J. Williams 2132* (BM); Dolpa, Kumlik near Timikot, 2440 m, 5 Jun. 1952, *O. V. Polunin, W. R. Sykes & L. H. J. Williams 4276* (BM, E); Dolpa, Ankhe to Durai, 1940 m, 10 Oct. 1991, *M. Minaki et al. 9104457* (E, KATH); Dolpa, 2070 m, 15 Oct. 1991, *M. Minaki et al. 9104514* (E); Dolpa, 2180 m, 14 Oct. 1991, *M. Minaki et al. 9105145* (E, TI); Dolpa, Chepka to Ankhe, 2780 m, 9 Oct. 1991, *M. minaki et al. 9106147* (KATH); Dolpa, Rahagaon, 3020 m, 10 Oct. 1991, *M. Minaki 9106167* (KATH, TI); Dolpa, Dunai to Juphal Airport, 2070 m, 15 Oct. 1991, *M. Minaki et al. 9108229* (E, TI); Dolpa, Dunai to Lukhor, 11 Oct. 1991, *M. Minaki et al. 9109310* (TI); Dolpa, Thuli Bheri River, 2335 m, 2 Sep. 2014, *A204* (KATH); Dolpa, Namala, 4200 m, Jun. 1961, *JEST 45* (E); Jumla, 2285 m, 5 Oct. 1952, *O. V. Polunin, W. R. Sykes & L. H. J. Williams 5507* (BM, E); Jumla, 2700 m, 30 Apr. 1963, *Jtoh & K. R. Rajbhandari 1161* (KATH); Jumla, Kudari Ti Ranku, 2100 m, 15 Jul. 1979, *K. R. Rajbhandari & B. Roy 3282* (KATH); Jumla, Sija

Khola, 2500 m, 5 Aug. 1979, *K. R. Rajbhandari & B. Roy 4350* (KATH); Jumla, Tatopani, 2110 m, 24 Sep. 1982, *N. P. Manandhar & N. R. Bhattarai 9046* (KATH); Jumla, 14 Jun. 1968, *S. B. Malla 10931* (KATH); Kalikot, Nagma, 2000 m, 12 Aug. 1981, *N. P. Manandhar & D. P. Joshi 8215* (KATH); Kalikot, Nagma, 2000 m, 12 Aug. 1981, *N. P. Manandhar & D. P. Joshi 8228* (KATH); **Mahakali**, Darchula, 2450 m, 20 Jul. 1984, *P. R. Shakya et al. 7973* (KATH); **Rapti**, Rukum, Shimi, 1300 m, 23 Apr. 1974, *J. F. Dobremez & N. P. Mananthar 2694* (E, KATH); Rukum, Jyamise, 1380 m, 5 Jun. 1982, *R. N Shukla, 4509* (KATH); Rukum, Jyamire Gaon, 1350 m, 20 Sep. 1982, *K. R. Rajbhandari & K. J. Malla 6642* (KATH). **C NEPAL**: **Bagmati**, Kathmandu, way to Tanjet, 2130 m, Jul. 1949, *O. V. Polunin 1446* (BM, E); Lalitpur, Godawari, 1500 m, 19 Aug. 1975, *R. K. Manandhar et al. 37* (KATH); Rasuwa, Thinigaon, Muktinath Himal, 3048 m, 24 Jun. 1954, *J. D. A. Stainton, W. R. Sykes & L. H. J. Williams1382* (BM, E); **Dhawalagiri**, Myagdi, on trail to Dana from Tatopani, 1800 m, 2 Jun. 1971, *C. Barclay & P. M. Synge 2436* (K); Myagdi, Midway between Tatopani & Garsam 1600 m, 22 Jul. 1973, *C. Grey-Wilson & B. Phillips 338* (K); Mustang, between Ghasa and Lete, 2100 m, 9 Oct. 1976, *H. Tabata et al. 6157* (KYO); Mustang, between Jomson and Marpha, 2550 m, 14 Oct. 1976, *H. Tabata et al. 6389* (KYO); Mustang, between Tukuche and Khobang, 2450 m, 15 Oct. 1976, *H. Tabata et al. 6468* (KYO); Mustang, Kagbeni to Chele (Tsele), 2860–3074 m, 26 Aug. 2004, *T. Watanabe et al. LOM-SP040826(009)* (TI); Mustang, Alibari to Tukuche, 2620 m, 24 Aug. 1994, *S. Noshiro et al. 9470547* (E, KATH); Mustang, 2100 m, 28 Aug. 1988, *M. Suzuki et al. 8881509* (E, TI); Mustang, Kabhre to Kolopani, 2300 m, 28 Aug. 1988, *M. Suzuki et al. 8811121* (TI); Mustang, Kabhre to Kolopani, 2100 m, 28 Aug. 1988, *M. Suzuki et al. 8811125* (TI); Mustang, 2380 m, 27 Sep. 1995, *M. Mikage et al. 9550462* (E); Mustang, Lete, 2450 m, 29 Jul. 1996, *T. Hoshino et al. 9662069* (E, KATH, TI); Mustang, 2540 m, 29 Jul. 1996, *T. Hoshino et al. 9666078* (E, TI); Mustang, Thinigaon, N of Tukucha, Kali Gandaki, 3505 m, 26 Sep. 1954, *J. D. A. Stainton, W. R. Sykes & L. H. J. Williams 8020* (BM, E); Mustang, Kali Kandaki Tal, westlich Tukuche, 3000 m, 1 Sep. 1967, *H. J. Lange 33* (K); Mustang, Kali Gandaki, Tatopani to Ghasa, 1360–2100 m, 16 Jul. 1983, *H. Ohba et al. 8330508* (E, KATH, TI); Mustang, Kali Gandaki, 1900 m, 16 Jul. 1983, *H. Ohba et al. 8310435* (E, KATH, TI); Mustang, Jomoson to Marpha, 2600–2700 m, 9 Sep. 1999, *S. Ishizawa et al. 99909003* (KATH, TI); Mustang, Kali Gandaki, 17 Jul. 1983, *H. Ohba et al. 8350491* (E, TI); Mustang, Dhumpa to Nilgiri South, 3000–3350 m, 1 Sep. 1996, *T. Shimizu 9680353* (TI); Mustang, Kabra, Kuli Gardaki Valley, 1980m, 31 May 1954, *J. D. A. Stainton, W. R. Sykes & L. H. J. Williams 5489* (BM, E); Mustang, Chele, 3050 m, 16 Aug. 2002, *F. Miyamoto 20220150* (KATH, TI); Mustang, Chele, 3000 m, 26 Aug. 2002, *F. Miyamoto et al. 20220277* (KATH, TI); Mustang, Chele to Taklam La Pass, 16 Aug. 2002, *F. Miyamoto et al. 20230064* (KATH, TI); Kagbeni to Chuksang, 2810–3050 m, 9 Jul. 2000, *Y. Iokawa et al. 20020012* (KATH, TI); Mustang, Kagbeni to Chele, 27 Jun. 2003, *Y. Iokawa et al. 15024* (KATH, TI); Mustang, Jomsom to Kagbeni, 2740 m, 2 Aug. 2001, *S. Noshiro et al. 20104084* (KATH, TI); Mustang, Between Syang and Marpha, 2710 m, *K. R. Rajbhandri et al. 201411021* (KATH); Mustang, Tallo Ghnsa (Ghasa) to Kaiku, 2100 m, 28 Aug. 1988, *M. Suzuki et al. 8860769* (TI); Mustang, Marpha to Jomsom, 2690 m, 1 Aug. 2001, *M. N. Subedi 80067* (KATH); Around Ghansa, 2050 m, 28 Aug. 1988, *M. N. Aubedi 8890915* (KATH); Mustang, Jomsom, 2750 m, 12 Oct. 2007, *V. Manandhar 029* (KATH); Mustang, Ghasa, 2400 m, 24 Jul. 1974, *D. P. Joshi & T. K. Bhattachaya 74-1968* (KATH); Mustang, Chhatang to Kobang, 2550 m, 25 Jul. 1974, *D. P. Joshi & T. K. Bhattachaya 74-2017* (KATH); Mustang, Ghasa to Tukuchey, 2100–2700 m, 5 Apr. 1967, *T. B. Shrestha & M. S. Bista 1871* (KATH); Mustang, Tuchuche, 2590 m, 29 Aug. 1983, *N. P. Manandhar 10224* (KATH); Mustang, Tui, 2600 m, 21 May 1979, *J. F. Dobremez 3096* (E); Mustang, Chele (Tsele) to Shyammochen (Syangboche), 3074–3773 m, 29 Jul. 2002, *T. Watanabe et al. LOM-SP020729(051)* (TI); Myagdi, near Gurjakhana, 2285 m, 26 Jul. 1954, *J. D. A. Stainton, W. R. Sykes & L. H. J. Williams 3640* (BM); Mustang, near Larjung, SE

Jomsom, Upper Kali Gandaki, on Pokhara-Jomsom Route, 2550 m, Sep. 1990, *Chadwell & McKelvie 117* (E); Mustang, Ghasa, N. of Dhana, 2133 m, 25 May 1954, *J. D. A. Stainton, W. R. Sykes & L. H. J. Williams 663* (BM, E); **Gandaki**, Gorkha, Gup, 1990 m, 3 Aug. 1994, *M. Suzuki et al. 9485255* (E, KATH); Gorkha, Upper Buri Gandaki, Nayak, 1800 m, 4 Jul. 1962, *J. D. A. Stainton 3933* (BM, E); Phillim to Dyang, 1820 m, 29 Jul. 2008, *H. Ikeda et al. 20816055* (E); Gorkha, 1820 m, *s. coll. 20816055* (KATH); Gorkha, Shiar, Khola Below Chhokang, 2900 m, 15 Jun. 1953, *P. S. G. 789* (BM); Gorkha, Sardu Khola to Saeda Khola, 1810 m, 26 Jul. 1994, *M. Suzuki et al. 9470187* (E, KATH, TI); Manang, Thangjet to Chilime, 5600 m, 28 Jun. 1976, *H. Kanai & P. R. Shakya 61* (KATH); Manang, Jarakot to Dunai, 1800 m, 1 Jun. 1966, *T. B. Shrestha 5063* (KATH); Manang, Jarakot, 2500 m, 29 May 1973, *Einaonon & Wetterhall 230* (KATH); **Janakpur**, Jarakat to Dunai, 1800 m, 1 Jun. 1966, *T. B. Shrestha 5603* (BM). **L YARLUNG ZANGBO**: **Nyingchi**, Palong, Palong to dabo, 1900 m, 28 Nov. 1982, *B. S. Li & S. Z. Cheng 01904* (PE). **YARLUNG ZANGBO-BRAHMAPUTRA**: **Zayü**, Jitang, 14 Aug. 1976, C. Y. Wu *4629* (KUN). **TANGUT**: **Baxoi**, Nujiang Valley, 3100 m, 15 Aug. 1973, *Exped. Xizang 2177* (HNWP); Lashan, 2900 m, 5 Jul. 1976, *Qinghai-Xizang Exped. 12223* (KUN, PE); Nujiang Gorge, 2786 m, 8 Sep. 2008, *L. M. Gao et al. GLM-081416* (KUN); Wa, Pasho, Kham, 2900 m, 21 Aug. 1936, *J. Hanbury-Tracy 54* (BM); Bangda, Genyao, 3010 m, 19 May 2010, *L. M. Gao et al. GLM-102689* (KUN); on the way from Bangda to Zuogong, 4700 m, 5 Sep. 1977, *P. C. Kuo & W. Y. Wang 23417* (HNWP); way to Bangda, 3080 m, 20 Jul. 2000, *D. E. Boufford et al. 29560* (E); way to Bangda, 3300 m, 20 Jul. 1977, *P. C. Kuo & W. Y. Wang 21645* (HNWP); **Dêgê**, Baiya, 2500 m, 20 Jule 1979, *Dêgê Exped. 0461* (CDBI); **Dêngqên**, 24 Sep. 1992, *C. J. Chen 92332* (PE); **Garzê**, Chengranwu, Heida, 2680 m, 16 May 2010, *L. M. Gao et al. GLM-102661* (KUN); **Qamdo** to **Baxoi**, 6 Jun. 1973, *Qinghai-Xizang Exped. 73-77* (KUN, PE); **Qamdo**, 3400 m, 9 Aug. 1982, *Z. J. Zhao 82193* (SZ); **Zhag'yab**, Jitang, 3100–3500 m, 20 Sep. 1976, *Qinghai-Xizang Exped. 13018* (KUN, PE); Qiongniduoma to Yanduo, 18 Sep. 1951, *Y. W. Cui 5727* (CDBI, PE); Yanduo, 25 Sep. 1951, *Y. W. Cui 5803* (PE). **N HENGDUAN**: **Baiyü**, bank of Jinsha Jiang, 3050 m, 26 Aug. 1983, *Y. F. Han et al. 830326* (PE); **Barkam**, 2700 m, 4 Jun. 1957, *X. Li 71242* (PE, SZ); near Dengjiaqiao, 2600 m, 8 Jun. 1957, *X. Li 71375* (IBSC, IFP, KUN, PE, SZ); near Lieshi Mu (Martyrs), 2600 m, 16 Jun. 1957, *X. Li 71480* (IBSC, PE, SZ); Dangba, 2340 m, 18 Sep. 1957, *X. Li 71888* (IBSC, KUN, PE, SZ, WUK); Suomo, 2870 m, 5 Jul. 1957, *Z. R. Zhang et al. 22527* (IBSC, PE, SZ); Kafeiqiao, 2700 m, 11 Jul. 1957, *Z. R. Zhang & H. F. Zhou 22966* (HNWP, IBSC, KUN, PE, WUK); Songgang, 2550 m, 26 Jul. 1957, *Z. R. Zhang & H. F. Zhou 23245* (IBSC, KUN, PE, WU); E Zhegushan, border between Maerkang to Lixian, 2475 m, 8 Jul. 2004, *C. S. Chang et al. SI0139* (PE); **Batang**, 9 Sep. 1977, *P. C. Kuo & W. Y. Wang 23552* (HNWP); 2700 m; 22 Jun. 1983, *Vegetation Group to Hengdaun Mountain 4261* (PE); 29 Jul. 2014, *B. Chen et al. CSH06638* (CSH); Xiaqiong, Lahe, 2650 m, 17 May 2010, *L. M. Gao et al. G. L. Gao GLM-102681* (KUN); Suwalong, 2430 m, 1 Oct. 2008, *L. M. Gao ett al. GLM-082203* (KUN); Cuosongnong, 2600 m, 5 May 1986, *T. Naito et al. 675* (PE); Jinshajiang River, 2500 m, 6 May 1986, *T. Naito et al. 686* (PE); Zhubalong, 2700 m, 20 Aug. 1976, *s. coll. 361* (KUN); Zhubalong, 2500 m, 8 Sep. 1973, *Expedition to Xizang 2831* (HNWP); 16 Aug. 1960, *Yao Yuan Dui 15552* (CDBI); **Danba**, Donggu, Tushuiyan, 2250 m, 9 Jun. 2010, *L. M. Gao & Z. M. Cai GLM-103050* (KUN); Maoniu, 2855 m, 20 Jul. 1959, *S. Jiang et al. 02189* (PE); **Dawu**, 5 Jul. 1977, *P. C. Kuo & W. Y. Wang 20840* (HNWP, KUN, WUK); 16 Aug. 1960, *Yao Yuan Dui 15552* (CDBI); 20 Sep. 1951, *W. K. Hu 13233* (PE, SZ); Longwangmiao, 3100 m, 4 Jun. 1958, *s. coll. 72* (SZ); near the town, 3400 m, 29 Jun. 1974, *Y. T. Wu & B. C. Gao 111661* (CDBI, SZ, WUK); Xianshui, Zilong, 2970 m, 8 Jun. 2010, *L. M. Gao & Z. M. Cai GLM-103028* (KUN); 3300 m, 19 Jul. 1961, *S. Jiang et al. 9479* (PE); **Heishui**, Weigu, 1600 m, 23 May 1959, *Heishui Group 1143* (PE); Shidonglou, Pangmahe, 1650 m, 31 May 1959, *Heishui Group (Chuan Jing A) 1264*

(KUN, PE); Senkanqiaogou, 2800 m, 26 Jul. 1957, *X. Li 73289* (IFP, KUN, PE, SZ); Luhua, 2400 m, 12 Jul. 1957, *X. Li 73545* (IBSC, IFP, PE, SZ); Luhua, 2550 m, 27 Jul. 1957, *X. Li 73645* (IBSC, KUN, PE, SZ, WUK)' Luhua, 2420 m, 29 Jul. 1957, *X. Li 73847* (IBSC, KUN, PE, SZ, WUK); Maqioshan, 3100 m, 12 Aug. 1957, *X. Li 73954* (IBSC, KUN, PE, SZ, WUK); **Jinchuan**, Bianyanshan, 2160 m, 26 Jun. 1957, *X. Li 75008* (IBSC, PE, SZ); Longjiashan, 2280 m, 8 Jul. 1957, *X. Li 75194* (IBSC, PE, SZ); Liewu, 2180 m, 10 Jul. 1957, *X. Li 75230* (IBSC, PE, SZ); Moer, 2180 m, 12 Jul. 1957, *X. Li 75259* (IBSC, PE, SZ); Liewu, Hangou, 2120 m, 16 Jul. 1957, *X. Li 75318* (IBSC, IFP, PE, SZ); Erlang, 2350 m, 26 Jul. 1957, *H. Li 75438* (IBSC, PE, SZ); Chengxiang, Maocaomu, 2200 m, 10 Aug. 1957, *X. Li 75553* (IBSC, PE, SZ); Liewu, 2150 m, 15 Aug. 1957, *X. Li 75603* (IBSC, PE, SZ); Liewu, Jinjiagou, 2200 m, 17 Aug. 1957, *X. Li 75638* (IBSC, PE SZ); Qingning, Xiawuzhai, 2180 m, 7 Aug. 1957, *X. Li 75868* (IBSC, PE, SZ); Xingfu, 2600 m, 22 Aug. 1957, *X. Li 76255* (IBSC, IFP, KUN, PE, SZ, WUK); Mabang, 2450 m, 16 May 1958, *X. Li 77422* (PE); Kasa, Jiaomutang, 2200 m, 6 Jun. 1958, *X. Li 77610* (IBK, IBSC, PE, WUK); Anning, Baojiaodiaogou, 2300 m, 8 Jul. 1958, *X. Li 78078* (IBK, IBSC, PE, SZ, WUK); Chengxiang, Longjiashan, 3200 m, 4 Aug. 1958, *X. Li 78328* (IBSC, PE, WUK); Anning, Baguadiaogou, 2400 m, 17 Aug. 1958, *X. Li 78378* (IBK, IBSC, PE); **Jiuzhaigou** (Nanping), Shuanghe, Weiziba, 1300 m, 26 Jun. 1963, *Sichuan Economic Plant Expedition 0742* (CDBI); 3 Aug. 1961, *X. N. Tang 00340* (CDBI); **Lixian**, in 1952, *C. Ho & C. L. Chow 12967* (IBK, IBSC, PE, SZ); 27 Aug. 1958, *Z. L. Wu 33704* (PE); 2600 m, 17 Jun. 1956, *S. B. Shao 53-409* (SZ); Dabanzhao, Huangtuliang, 3500 m, 29 Jun. 1957, *D. Z. Deng 1737* (SZ); Putou, 2000 m, 30 Jul. 1989, *Q. S. Zhao 0124* (PE, SZ); Shaba, 2500 m, 5 Aug. 1989, *Q. S. Zhao 1485* (PE); Shaba, 1900 m, 12 Jun. 1982, *W. H. Li et al. H82-0214* (PE); Miyaluo, Jiabi, Luganqiao, 2620 m, 7 Aug. 1958, *S. Y. Chen et al. 5783* (SZ); Dabanzhao, 2550, 17 Jun. 1956, *D. P. He 44219* (SZ); Dabanzhao, Luganqiao, 2600 m, 21 Jun. 1956, *D. P. He 44278* (SZ); Laisugou, Jiujiapeng, 2500 m, 10 Jul. 1956, *D. P. He 44852* (SZ); Miyaluo, 2700 m, 28 Jul. 1956, *D. P. He 45340* (SZ); Miyaluo, Xiabajiaoque, 2850 m, 5 Aug. 1956, *D. P. He 45553* (SZ); Miyaluo, 2850 m, 7 Aug. 1956, *D. P. He 45616* (HNWP, IBSC, SZ, WUK); Dabanzhao, Lamasi, 3000 m, 17 Jun. 1956, *W. K. Hu 53-20* (SZ); Putou, 2000 m, 30 Jul. 1989, *Q. S. Zhao 124* (IBSC); Shaba, 2500 m, 5 Aug. 1989, *Q. S. Zhao 1485* (IBSC); **Maoxian**, Heishuigou, 19 Jul. 1951, *C. Q. Li 2810* (SZ); 1952, *C. Ho & C. L. Chow 13863* (IBK, IBSC, PE); Fengyi, 2040 m, 22 Jun. 1959, *Group Maowen 2891* (CDBI); Tumeng, Fushun, Gangou, 1300 m, 12 Aug. 1975, *s. coll. 8737* (CDBI); **Markam**, Rumei, Duomei, 3050 m, 9 Jun. 2008, *L. M. Gao et al. GLM-081341* (KUN); Rumei, 3020 m, 18 May 2010, *L. M. Gao et al. GLM-102686* (KUN); Rumei, Duiba, 2860 m, 18 May 2010, *L. M. Gao et al. GLM-102687* (KUN); Zhubalong, bank of Jinshajiang, 2600 m, 1 Jun. 1976, *Exped. Qinghai-Xizang 11663* (HNWP, KUN, PE); Yanjing, bank of Lancang Jiang, 2300 m, 3 Jun. 1976, *Qinghai-Xizang Exped. 11703* (KUN, PE); Zhuka, 2500 m, 20 May 1961, *s. coll. 909* (KUN); Yanjing, 3100 m, 28 Sep. 1992, *C. J. Chen 92367A* (PE); **Wenchuan**, 28 Jul. 1943, *Y. R. Kuo 23* (SZ); 13 Jun. 1958, *s. coll. 31803* (PE); 25 Jul. 1928, *W. P. Fang 1501* (E, K, IBSC, PE); W slope of Mingjiang, 1400 m, 22 Aug. 1984, *Y. L. Cao 027* (CDBI); 1400 m, 29 Jul. 1975, *s. coll. 8429* (CDBI); 1400 m, 29 Jul. 1975, *s. coll. 8492* (CDBI, PE); Yanmen, 1330 m, 28 May 1958, *S. Y. Chen et al. 5056* (SZ); Yanmen, 1350 m, 24 Aug. 1958, *S. Y. Chen et al. 5884* (SZ); **Wenxian**, Liujiaping, Dalingliang, 16 Aug. 2006, *Baishuijiang Expedition 2209* (PE); Fanba, Tianchiping, 1200 m, 30 Aug. 2006, *Baishuijiang Expedition 2820* (PE); Fanba, Tianchiping, 1200 m, 30 Aug. 2006, *Baishuijiang Expedition 2823* (PE); Liujiaping, 1200 m, 14 May 2007, *Baishuijiang Expedition 4887* (PE); 18 May 1971, *Zang Yao Dui 31* (HNWP); 650 m, 3 Jun. 1930, *S. K. Hao 269* (WUK); 650 m, 3 Jun. 1930, *S. K. Hao 369* (PE); 650 m, 3 Jun. 1930, *Hopkingson 371* (WUK); 960 m, 17 Aug. 1967, *C. L. Tang 1716* (HNWP, WUK); Yuzishan, 3 Jul. 1983, *Y. Shi et al. 023* (WUK); Shifang, 1100 m, 30 Oct. 1937, *K. T. Fu 2278* (PE, WUK); 1000 m,

24 Apr. 1964, *T. P. Wang 19017* (WUK); 1000 m, 27 Apr. 1964, *T. P. Wang 19054* (HNWP, WUK); Shifang, 920 m, 9 Jul. 1959, *Z. Y. Zhang 6010* (HNWP, WUK); Zhongzhai, Songping, 1000 m, 12 Jul. 1959, *Z. Y. Zhang 6122* (WUK); Shifang, 1200 m, 21 Jul. 1959, *Z. Y. Zhang 7011* (WUK); Shifang, 1800 m, 6 Aug. 1959, *Z. Y. Zhang 7972* (HNWP, WUK); Shifang, 1600 m, 8 Aug. 1959, *Z. Y. Zhang 8067* (WUK); 950 m, 14 Aug. 1959, *Z. Y. Zhang 8179* (HNWP, WUK); Chengguan to Xiyuan, 950 m, 17 Aug. 1959, *Z. Y. Zhang 8205* (HNWP, WUK); Danbao, Liujianping, 1745 m, 27 Jul. 1959, *Z. Y. Zhang 9560* (WUK); Baishuijiang, 800 m, 1 Oct. 1968, *Y. Q. He 878* (PE, WUK); Baishuijiang, 850 m, 2 Oct. 1958, *Y. Q. He 897* (PE, WUK); Linjiang, Jianshan, 900 m, 24 Sep. 1958, *Z. P. Wei 2799* (WUK); **Wudu**, 885 m, 25 Jun. 2000, *Sino-British Qinghai Alpine Garden Society Expedition (SQAE) 424* (E); **Xiaojin**, near city, 2150 m, 11 Aug. 1975, *Sichuan Vegetation Group 9608* (CDBI, IBSC); 3100 m, 27 Aug. 1975, *Sichuan Vegetation Group 9680* (CDBI); Hanping, Wodi, Fan'an, 2400 m, 6 Jun. 1959, *Chuan Jing A 225* (CDBI, KUN, PE, SZ); Fan'an, 2300 m, 19 Aug. 1958, *X. S. Zhang & Y. X. Ren 6710* (PE, SZ); Shuangbai, 2990 m, 8 Jul. 2004, *H. N. Qin et al. 17094* (PE); **Yajiang**, along first tributary of the Yalong Jiang E of the city, 2875 m, 6 Aug. 2006, *D. E. Boufford et al. 35954* (PE); 2560 m, 25 Mar. 1967, *Z. T. Guan 57-0249* (IBSC); 2680 m, 1969, *Z. T. Guan C-156* (IBSC); 4 Jul. 1930, *Z. P. Huang et al. 542* (PE); 4 Jul. 1930, *Z. P. Huang et al. 560* (PE); 1930, *K. Z. Huang 560* (SZ). **S HENGDUAN**: **Baoxing**, 28 Aug. 1938, *C. Pei 8067* (NAS, PE); 2 Jul. 1936, *K. L. Chü 3068* (BM, E, IBSC, PE, SZ); Ganyang, Diaofanggou, 1800 m, 25 Aug. 1936, *K. L. Chü 3678* (BM, E, IBSC, K, PE, SZ, WUK); 10 Oct. 2014, *Y. Q. Hao BX-42* (CSH); Yanjing, 1500 m, 3 Sep. 1963, *K. C. Kuan & W. T. Wang 2801* (PE); near Yanbi, 1300 m, 25 May 1958, *X. S. Zhang & Y. X. Ren 4952* (CDBI, PE, SZ); Qiaoqi, Yutong, 2500 m, 26 Jun. 1958, *X. S. Zhang & Y. X. Ren 5617* (CDBI, PE, SZ); 10 Oct. 2014, *Y. Q. Hao BX-42* (CSH); 1112 m, 16 Jun. 1982, *Z.B. Cheng et al. 116525* (SZ); Ximenwai, 1250 m, 19 Jun. 1933, *T. H. Tu 4201* (IBSC, PE); s. loc., 1933, *T. H. Tu 4378* (IBSC); Qiaoqi, 2290 m, 9 Jul. 1933, *T. H. Tu 4379* (PE); Donghe, Dachigou, 1500 m, 13 Aug. 1961, *S. Jiang 10159* (PE); **Bingchuan**, Xiayang to Bangying, 14 Oct. 1946, *T. N. Liou 21566* (IBSC, PE); **Bo'ao**, 3000 m, 21 Jul. 1978, *Y. B. Yang & K. H. Mou 6689* (CDBI); **Butuo**, Jiaojihe, 1900 m, 19 Jul. 1976, *Sichuan Vegetation Group 14130* (CDBI, PE); Jiaojihe, Dagenba, 1300 m, 20 Aug. 1959, *Chuan Jing Liang 5611* (CDBI, KUN, PE); Jiaojihe, 12 Jun. 1984, *Z. E. Zhao 1884* (HIB); **Cawalong** (Tsawarong), Salween gorge, 2700–3300 m, 3 Aug. 1933, *F. Kingdon-Ward 10681* (BM); Songta, 2000 m, 1 Jun. 1960, *s. coll. 8862* (CDBI, KUN, PE); **Dali**, Diancangshan, 8 Sep. 1919, *s. coll. 2259* (PE); Yinqiao, Heyang, Cangshan, 16 May 1929, *R. C. Ching 22661* (IBSC, KUN, PE); 2000 m, 17 Oct. 1999, *N. Fijii et al. 109* (KUN); 14 May 1999, *W. X. Wu 10241* (NAS); Drainage Basin of Erhai, Tsangshan Range, 2000 m, 15 Apr. 1922, *J. F. Rock 3116* (E); Huadianba, 2900–3100 m, 30 Aug. 1984, *S. K. Wu et al. 1541* (KUN); Tianshenqiao, 1900 m, 22 Jun. 1984, *Xiangliao Zhiwu Kaochadui 89* (KUN); road to Huadianba, Sep. 1987, *B & L 12496* (E); Tali Valley, 2032 m, Jun., *G. Forrest 4730* (BM, E); E. flank of Tali range, 1980–2285 m, *G. Forrest* 6795 (BM, E, K); Taiyi, Dajiao, 1200 m, 4 May 1988, *Y. H. Li et al. 00516* (KUN); Xiaguang, 2150 m, 17 Oct. 1987, *Y. H. Li et al. 00329* (KUN); Cang, Qingbixi, 2100 m, 12 Sep. 1983, *Sino-Amer. Bot. Exped. 110* (KUN); Diancang Shan mountain range, Huadianba to Xizhou, 2350 m, 20 Jul. 1984, *1984 Sino-Amer. Bot. Exped. 1218* (E, KUN); the old city wall of Dali, 2060 m, 20 Jul. 1984, *1984 Sino-Amer. Bot. Exped. 1222* (E, KUN); road to Huadianba, N end of Cangshan, 2200 m, 17 May 1981, *1981 Sono-British Expedition to Cangshan 707* (E, K, KUN); Xiaguan, 2100 m, 26 Aug. 1976, *N-W Institute of Botany & Sichuan Institute of Biology 81* (CDBI); Cangshan, 2100 m, 26 Aug. 1976, *Y. Q. He et al. 081* (WUK); Xiaguan, 26 Nov. 1940, *T. N. Liou 17402* (PE); Oct. 1940, *T. N. Liou 19560* (IBSC, KUN); Fengyi, Xiaoshao, 27 Sep. 1946, *T. N. Liou 20890* (IBSC, PE); on the way to Pingpo, 5 Nov. 1946, *T. N. Liou 22421* (IBSC, KUN, PE); Cangshan, Fengyandong, 25 Nov. 1946, *T. N. Liou 23020* (PE); Cangshan, Fengyandong, 25 Nov. 1946, *T. N. Liou 23021* (IBSC, PE); 1500 m, May 1935, *C. W. Wang 63378* (KUN, PE); E

Chongshan, Tsoongwutze, 2500 m, 17 Aug. 1933, *Y. Tsiang 11358* (IBSC); **Daocheng**, Gongling, 3300 m, 4 Jul. 1998, *D. E. Boufford et al. 28211* (E); Dongyi Qu, Zhishui to Qika, 2700 m, 22 Jun. 1971, *Z. G. Liu 0054* (CDBI); Julong, Bulang, 3200 m, 7 Jul. 1973, *Sichuan Vegetation Exped. 1983* (CDBI, KUN, PE); Riwa, 3250 m, 19 Jun. 1973, *Sichuan Vegetation Exped. 2205* (CDBI, KUN, PE); Riwa, 3200 m, 2 Sep. 1971, *X. H. Hu 506* (CDBI); Dongyi, Mofanggou, 2 May 1979, *Dao 0088* (CDBI); **Dêqên**, Benzilan, 2100–2700 m, 16 Aug. 1981, *Exped. Hengduanshan 3099* (PE); Benzilan, 2080 m, *Qinghai-Xizang Exped. 1633* (KUN, PE); 15 Sep. 1970, *Z. H. Yang 7337* (HITBC); Shuichanpan, 2000 m, 21 May 1964, *X. Zhou 1021* (KUN); Chongjianghe, Sanjia, 2400 m, 26 Jun. 1965, *9502* (KUN, PE); Atuntze, Wuyen, 2400 m, 2 Jun. 1937, *T. T. Yü 8385* (KUN, PE); Atuntze, 2650 m, 14 Sep. 1937, *T. T. Yü 10210* (BM, E, KUN, PE); Shuaquchang, 2100 m, 28 Aug. 2013, *Y. H. Yan & X. P. Qi YYH12709* (CSH); s. loc., 2000 m, 30 Jul. 1981, *W. H. Li & Y. Hu 435* (PE); Benzilan, 23 Aug. 1981, *W. H. Li & Y. Hu 81-1786* (PE); Benzilan, 2630 m, 22 Jul. 2009, *K. X. Song 090658* (PE); 3300 m, 22 Sep. 1986, *H. Sun & Z. G. Qian 697* (KUN); Benzilan, Dongzhulin, 2800 m, 8 Jul. 1981, *Kunming Institute of Botany 1961* (KUN); Benzilan, 2200 m, 12 Aug. 1981, *Z. X. Tang 856* (PE); Foshan, 3140 m, 5 Sep. 2008, *L. M. Gao et al. GLM-081314* (KUN); way to Weixi, 3000 m, 23 Jul. 2007, *K. X. Song 090702* (PE); Benzilan, 3 Oct. 1989, *T. M. Ho 1307* (HNWP); 2100–3000 m, 1 Jul. 1983, *Vegetation Group to Hengduan Mountain 4322* (PE); Diling, 18 Jul. 1981, *s. coll. s. n.* (HITBC068781); Yangzha to Jiabi, 2200 m, 8 Jun. 1960, *S. Jiang et al. 10083* (KUN, PE); bank of Lancangjiang, Yongzi, 2100 m, 25 Jul. 1940, *K. M. Feng 5767* (IBSC, PE); Atuntze, 2700 m, Sep. 1935, *C. W. Wang 70257* (IBSC, KUN, LBG, PE, WUK); **Dêrong**, Dongwanhe, 2145 m, 11 Jun. 1993, *B. Alden et al. 1128* (E); 2400 m, 1 Oct. 2008, *L. M. Gao et al. GLM-082207* (KUN); Zhigen, 20 Jun. 1981, *s. coll. s. n.* (HITBC); bank of Dingquhe, 2400 m, 2 Aug. 1981, *Qinghai-Xizang Exped. 3148* (KUN, PE); 20 Aug. 1984, *X. W. Tian 100* (PE); **Eryuan**, way to Liantie, 2080 m, 19 Jul. 2009, *K. X. Song 090424* (PE); Niujie, 28 Jul. 1960, *Exped NW Yunnan 10548* (KUN); Liantie, 16 Jun. 1942, *H. C. Wang 2377* (IBSC, KUN, PE); Fengxiang, Yingjiaqing, 9 Jul. 1929, *R. C. Ching 23199* (IBSC, KUN, PE); Zhujiaying, Dawan, 20 Jun. 1929, *R. C. Ching 23313* (IBSC, KUN, PE); **Hanyuan**, Sianglingshan, 1700 m, 2 Sep. 1938, *T. P. Wang 9481* (PE, WUK); Daxiangling, 9 Aug. 1939, *X. J. Yang 3532* (PE); Shunhe, 780 m, 14 Sep. 1993, *Z. C. Zhao 027* (PE); Dashu Qu, bank of Daduhe, 17 Sep. 1993, *Z. C. Zhao 0434* (PE); 5 km from Han Yuen, 1740 m, 9 Aug. 1939, *C.Y. Chiao 1715* (KYO); **Heqing**, Lankong-Hoching Divide, 2440–2740 m, May 1913, *G. Forrest 9958* (BM, E, K, PE); Tapingtze, 9 Jun. 1888, *M. Delavay s. n.* (KUN); **Huaping**, Xianrendong, 28 May 2012, *X. F. Jin 0189* (PE); **Huidong**, Haiba, 2470 m, 1 Aug. 2009, *K. X. Song 090818* (PE); Chuanyu, 1700 m, 23 Jun. 1965, *Woody Oil Plant Expedition 65-0169* (HITBC, KUN); Baiyan, 1800 m, 2 Jul. 1978, *Q. S. Zhao & G. Hu 5784* (CDBI, SZ); Rizhi, Gongye, 2550 m, 20 Jul. 1960 m, *L. Z. Hu et al. 60-1388* (SZ); Qianjin, 1873, 19 Jul. 1960, *C. F. Xia & W. X. Jiang 60-1433* (SZ); **Huili**, on the way to Huidong, 1800 m, 24 May 1984, *W. L. Chen et al. 5476* (PE); Taiping, 2150 m, 18 Jul. 1976, *Sichuan Vegetation Investigation Group 11896* (CDBI, PE); Daqiao to Xiaochang, 1920 m, 15 Jun. 1959, *S. K. Wu 269* (CDBI, KUN, SZ); 1700 m, 21 Aug. 2001, *C. M. Hu 9077* (IBSC); Baitashan, 1900 m, 8 Aug. 2006, *C. M. Hu 20165* (IBSC); Xiacun, 1930 m, 26 Aug. 1960, *C. F. Xia 60-1076* (SZ); in the city, 21 Sep. 1958, *C. Ho et al. 11475* (SZ); Fengying, 1920 m, 31 Jul. 2009, *S. D. Zhang & K. X. Song 090804* (IBSC); **Jianchuan**, Shizhongshan, 27 Sep. 1975, *B. Y. Qiu 61073* (CDBI, KUN); Madeng Qu, Maliqing, 25 Sep. 1995, *Z. S. Yue et al. 94* (KUN); **Jiulong**, Sanyan, Longjinhe, 2460 m, 17 Jun. 1960, *J. S. Ying et al. 4009* (PE); between the city and the Yalong Jiang, 2250–2300 m, 16 Jul. 2005, *D. E. Boufford et al. 33003* (PE); Wulaxi to Fachanggou, 2000 m, 11 Jun. 1979, *X. H. Hu 20174* (CDBI); Naiqu, 2400 m, 11 Aug. 1979, *Q. Q. Wang 21511* (CDBI); Wulaxi, 1900 m, 17 Jun. 1974, *Z. G. Liu 4554* (CDBI, PE); 30 May 1933, *K. Huang et al. 270* (KUN); **Kangding**, 7 Jul. 1930, *Z. P. Huang et al. 1307* (PE); 2640 m, 14 Aug. 1934, *C. S. Liu 1053* (IBSC, PE, SZ); Noqiao River, N Jichou Pass, 2980 m, 1 Oct.

1991, *Exped. Sichuan (SICH) 730* (K); Songyu, 3100 m, 12 May 1960, *J. S. Ying 3395* (PE); Shade, 2880 m, 28 May 1961, *S. Jiang 5014* (KUN, PE); Jichou Pass, 2980 m, 1 Oct. 1991, *Simmons et al. SICH730* (K); Tatsienlu, 14 Jul. 1893, *Potanin s. n.* (K); Tatsienlu, 2740–3960 m, *A. E. Pratt 270* (BM, K); Tatsienlu, 2740–3960 m, *A. E. Pratt 725* (K, BM); Tatsienlu, in 1923, *R. Ningham 161* (E); Tatsienlu, 1938, *H. D. McLaren AC187* (E); Tatsienlu, 2700–3950 m, *A. E. Pratt* 804 (BM); Tatsienlu, in 1893, *J. A. Soulie 416* (K, G); Tatsienlu, 2600–2700 m, 22 Sep. 1928, *W. P. Fang 3503* (E, IBSC, K, PE); Shade to Baiwanqiao, 3100 m, 6 Sep. 1980, *Z. A. Liu 23037* (CDBI, IBSC, WUK); Wasi, 1900 m, 31 May 1974, *D. W. Wang 6151* (CDBI, PE); Zhonggu to Sandaoqiao, 2800 m, 4 Aug. 1963, *K. C. Kuan & W. T. Wang 1462* (HNWP, K, PE); Guza, Jiangju, 1620 m, 29 May 1974, *Sichan Vegetation Group 5254* (CDBI); Guza, Jiangju, 1620 m, 29 May 1974, *s. coll. 5354* (CDBI, PE); Ridi to Wasigou, 1400 m, 15 May 1974, *Y. T. Wu & Q. S. Zhao 111004* (CDBI, PE, SZ); Jintang, Bank of Daduhe, 29 Apr. 1981, *Z. J. Zhao et al. 113425* (CDBI, SZ); Guza, Daduhe, 1350 m, 18 May 1981, *Z. J. Zhao et al. 114090* (CDBI, SZ); Erdaoqiao, 2600 m, 25 Jun. 1981, *Z. J. Zhao et al. 114693* (CDBI, SZ); Daduhe, 1570 m, 26 Jun. 1981, *Z. J. Zhao et al. 114710* (CDBI, SZ); Erdaoqiao, 2550 m, 12 Sep. 1981, *Z. J. Zhao et al. 115300* (CDBI, SZ); Yala, 18 Sep. 1951, *W. K. Hu & C. Ho 11216* (PE, SZ, WUK); Wasi, 9 Oct. 1951, *W. K. Hu & C. Ho 11396* (PE, SZ, WUK); 23 Sep. 1951, *W. K. Hu 13243* (PE); Dapaoshan, Wangmu to Sandaoqiao, 2900 m, 1 Oct. 1953, *W. K. Hu 37008* (PE, SZ); Dapaoshan, 11 Aug. 1940, *K. L. Chü 7652* (IBSC, PE); **Lanping**, 2700 m, 9 Sep. 1933, *H. T. Tsai 56067* (IBSC, KUN, PE, SZ, WUK); Yingpan, Lajijing, 2000 m, 19 Sep. 1964, *S. K. Wu 8979* (KUN); Yingpan, Enluo, 2000 m, 20 Jun. 1981, *Exped. Hengduanshan 630* (PE); **Longyang**, Mangkuan, 13 Sep. 2013, *D. K. Tian & C. Li TDK01502* (CSH); **Luding**, 1320 m, 18 May 1980, *Z. G. Liu & Y. B. Yang 21891* (CDBI, IBSC, WUK); E bank of Daduhe, Wuligou, 1550 m, 17 Aug. 1963, *K. C. Kuan & W. T. Wang 1537* (HIB, PE); Lengji, 1300 m, 25 May 1983, *Q. S. Zhao & Z. M. Tan 119101* (SZ); Dewei, 2100 m, 25 Sep. 1983, *Vegetation Group 31932* (CDBI); Detuo, Wandong, roadside, 2 Aug. 1982, *Vegetation Group 42013* (CDBI); Moxi, Gonggashan, 1790 m, 23 Aug. 1997, *D. E. Boufford et al. 27448* (E); W Erlang Shan, 1800 m, 19 Sep. 1991, *D. Chamberlain et al. CEE260* (E); Yuling, 1443 m, 9 Aug. 2010, *J. I. Jeon et al. SI1237* (PE); bank of Daduhe, 1330 m, 15 May 1980, *Z. Y. Chen et al. 112077* (CDBI, SZ); 1800 m, 10 Aug. 1938, *K. L. Chü 6605* (IBSC, PE, SZ); s. loc., 23 Jun. 1977, *P. C. Kuo & W. Y. Wang 20291* (HNWP, KUN, WUK); bank of Daduhe, 1350 m, 27 Jun. 1965, *Y. T. Zhang & K. Y. Lang 19* (PE); Erlangshan, Qingganpo, 2600 m, 19 Jun. 1953, *H. L. Tsiung et al. 34503* (IBK, IBSC, PE, SZ); 1242 m, 8 Sep. 1953, *H. L. Tsiung 35483* (IBK, IBSC); Dabashan, 8 Jul. 1951, *W. K. Hu & C. Ho 10302* (PE, SZ); 1240 m, 8 Sep. 1953, *W. K. Hu 35483* (PE, SZ); bank of Dadu He, 26 Jun. 1983, *Y. H. Liu s. n.* (PE); **Mianning**, Lugu Qu, Sunshuiguang, 1850 m, 30 Jun. 1976, *s. coll. 12284* (CDBI); **Miyi**, Puwei, 2100 m, 17 Jul. 2001, *Z. M. Tan 84*(SZ); Mianhuashan, 1300 m, 13 Jun. 1978, *Q. S. Zhao & J. S. Liu 5378* (CDBI, SZ); **Muli**, Sihe, 2300 m, 20 Jul. 1978, *Q. S. Zhao 6132* (CDBI, SZ); Taoba Qu, 2600 m, 19 Jul. 1978, *J. S. Liu 7604* (CDBI, SZ); 247 Gong Zhuang, 2100 m, 7 Jun. 1978, *Q. S. Zhao & Y. B. Yang 4737* (CDBI, SZ); Zhali, 2750 m, 20 Oct. 1982, *M. Z. Wen & G. Yao 322* (CDBI); Litang River, 220 m, 3 Jun. 1937, *T. T. Yü 5911* (KUN, PE); Ketse, 2200 m, 13 Jul. 1937, *T. T. Yü 7200* (IBSC, KUN, PE); Gutu, 2300 m, 8 Sep. 1937, *T. T. Yü 14210* (BM, E, KUN, PE); near Lama-sery, 2500 m, 9 Dec. 1937, *T. T. Yü 14855* (BM, E, KUN, PE); Shanghetaowan, 3050 m, 29 Jul. 2009, *K. X. Song 090774* (PE); Kala, Bank of Yalongjiang, 24 Aug. 1983, *s. coll. 40* (HITBC); Zhang'ao, 3000 m, 21 Jul. 1978, *K. H. Mou & Y. B. Yang 4489* (SZ); from the city to Dasi, 2500 m, 7 Oct. 1959, *S. K. Wu 2818* (KUN); Kala, bank of Yalong Jiang, 24 Aug. 1983, *s. coll. 40* (HITBC); **Ninglang**, *s. coll. 240504* (KUN); Cuiyi, Yangjiancao, 3400 m, 12 Jul. 1981, *Y. F. Han et al. 81-1182* (PE); Yongning, Awar, 2400 m, 1 May 1937, *T. T. Yü 5217* (KUN, PE); Luguhu, 2900 m, 27 Jul. 2009, *S. D. Zhang & K. X. Song 090754* (IBSC); **Ningnan**, bank of Jinshajiang, 750 m, 4 Jul. 1964, *Sichuan Economic Plant Expedition 604* (CDBI); Xiyao, 1200 m, 24 Apr. 2007, *Y. Y. Geng et al. 20070204* (HX); Huadan,

Longtanwan, 720 m, 19 Oct. 2015, *Y. Q. Hao CSH12811* (CSH); Jinshajiang, 640 m, 12 Jun. 1982, *H. Li et al. 1157* (KUN); Jul. 1939, *S. Y. Hu 972* (SZ); **Panzhihua**, Renhe, Wuben, 2300 m, 5 Aug. 2001, *Z. M. Tan 633* (SZ); **Puge**, Tuomujing, 2300 m, 26 Aug. 1959, *Chuan Jing Liang 5541* (KUN, PE); on the way from the city to Ningnan, 1420 m, 5 May 1959, *Z. T. Guan 8091* (IBSC, PE); **Shangri-la** (Zhongdian), Nixi, 3400 m, 13 May 2009, *S. S. Zhou 5566* (PE); State Road 214, 2840 m, 12 Aug. 2009, *Southeast Tibet Expedition Team SET-ET 40* (PE); s. loc., 20 Junly 1981, *W. H. Li & Y. Hu 81-320* (PE); 1900 m, 20 Jul. 1981, *W. H. Li & Y. Hu 81-320* (PE); Haba, Yingshan, 2600 m, 28 Oct. 1962, *A. L. Zhang 101042* (KUN); Dongwang, 2400 m, 24 Aug. 1988, *J. S. Yang 88-200* (IBSC); Xiahutiao, 2000 m, 20 Oct. 1987, *S. J. Pei et al. 834* (KUN); Nixi, 2385 m, 10 Jul. 2008, *Z. H. Sun et al. Ouxk-BC-0002* (IBSC); Baidi, 2200 m, 18 Aug. 1962, *Exped. Zhongdian 1065* (KUN); Baoshan, Tuguan, 2700 m, 14 Jun. 1963, *Exped. Zhongdian 63-2336* (KUN); Haba, 3100 m, 23 Jun. 1963, *Exped. Zhongdian 63-2649* (KUN); Qiaotou, Lengdongzhai, 2300 m, 1 Jul. 1963, *Exped. Zhongdian 63-3001* (KUN); Chungtien plateau to Yangtze, 2740–3050 m, *G. Forrest 7110* (E, K); Road to Haba Shan, Sanba to Haba, 2650 m, 15 Jun. 1994, *Alpine Garden Society Expedition to China (1994) ACE 288* (E, K); between Tang Dun and Yangtze River, 2860 m, 1 Jun. 1993, *B. Alden et al. 639* (E); s. loc., *Zhongdian Exped. 63-3001* (KUN); Yuejin, 2650 m, 29 Jun. 1981, *Qinghai-Xizang Exped. 1647* (PE); Nixi, 2100 m, 2 Jul. 1981, *Qinghai-Xizang Exped. 2093* (KUN, PE); on the way to Langdatuci, 2500 m, 21 Aug. 1939, *K. M. Feng 2732* (PE); Paiti, 2200 m, 27 May 1937, *T. T. Yü 11445* (KUN, PE); Hapa, Chungtien, 3400 m, 25 Sep. 1937, *T. T. Yü 13586* (KUN, PE); Xiaqiaotou, Hutiaoxia, 1900 m, 9 Aug. 1981, *Exped. Hengduanshan 2755* (PE); Xiaoqiaotou, Hutiaoxia, 1900 m, 9 Aug. 1981, *Exped. Hengduanshan 2786* (PE); Below Haba Xueshan, 2725–2750 m, 19 Jul. 2010, *D. E. Boufford et al. 42052* (TI); **Shimian**, Mamadi, 2300 m, 6 May 1955, *C. C. Hsieh 40340* (PE, WUK); s. loc., 1955, *C. C. Hsieh 41586* (IBSC, PE, WUK); Ximagu, 950 m, 6 Jul. 1955, *T. P. Soong 41686* (SZ); Xiyoufang, Hegangou, 1800 m, 7 Oct. 1993, *Z. C. Zhao 2741* (PE); **Tianquan**, Erlangshan, Ganhaizi, 2730 m, 19 Aug. 1982, *D. Y. Peng 46281* (CDBI, IBSC); Erlangshan, Ganhaizi, 2710 m, 20 Aug. 1982, *D. Y. Peng 46318* (CDBI, IBSC); Erlangshan, 22 Jun. 1977, *P. C. Kuo et al. 20291* (HNWP, KUN, WUK); Erlangshan, Fodouyan, 29 Apr. 1956, *D. P. He 43203* (SZ); Erlangshan, Lengleng, 1000 m, 27 May 1956, *D. P. He 43953* (SZ); Erlangshan, Lengleng, 1000 m, 27 May 1956, *D. P. He 43960* (SZ); Erlangshan, Ludingqiao, 1200 m, 1 Jun. 1956, *D. P. He 43112* (SZ); **Weixi**, Yezhi, 1850 m, 20 May 1940, *K. M. Feng 4307* (PE); Tse-jao, 3300 m, 11 May 1937, *T. T. Yü 8218* (KUN, PE); Yezhi, 23 Jul. 2009, *S. D. Zhang & K. X. Song 090709* (IBSC); **Xichang**, no exact locality, 2500 m, 1 Aug. 1932, *T. T. Yü 1214* (PE); **Xiangcheng**, Zhongxinrong, 2800 m, 2 Aug. 1972, *Z. G. Liu 1563* (CDBI, PE); Qianjin, Pinggu, 3200 m, 14 Jul. 1972, *Z. G. Liu 1082* (CDBI, SZ); Yuejin, 2700 m, 23 Apr. 1974, *B. C. Gao et al. 111277* (CDBI, SZ, WUK); Sangong, 3040–3100 m, 11 Jul. 1998, *D. E. Boufford et al. 28567* (E); Xiarewu, 2840 m, 22 Jul. 1998, *D. E. Boufford 29006* (E); Xiarewu, 2920 m, 22 Jul. 1998, *D. E. Boufford 29035* (E); Shangreling, 3200 m, 6 Jul. 1973, *Sichuan Vegetation Group 2884* (CDBI, KUN, PE); **Yanbian**, Guosheng, Dongbawan, 2000 m, 25 Jul. 2001, *Z. M. Tan 229* (SZ); **Yangbi**, Michang, 1550 m, 25 Aug. 1984, *K. Y. Guan et al. 299* (KUN); **Yanyuan**, Boda to Meiyu, 2315 m, 28 Jul. 2009, *S. D. Zhang & K. X. Song 090765* (KUN); Pingchuan, 1900 m, 23 May 1978, *Q. S. Zhao & J. S. Liu 4585* (CDBI, SZ); Ganhaizi, 2350 m, 7 Jun. 1960, *S. Jiang et al. 5861* (KUN, PE); **Yongsheng**, on the Lijiang-Yongsheng Road, Jinshajiang Valley, 1300 m, 30 Jun. 1981, *Y. F. Han & K. M. Deng 81-877* (PE); Mountains between Yungpeh and Yungning, 3050–3350 m, *G. Forrest 22057* (E, K); 2330 m, 27 Jul. 2007, *K. X. Song 090747* (PE); Jinshajiang Valley, 2000 m, 10 Jun. 1981, *W. H. Li & Y. Hu 81-0127* (PE); Apr. 1940, *R. C. Ching 31164* (KUN); **Yuexi**, S. of the town, 1660 m, 26 Jun. 1959, *Chuan Jing Liang 3410* (CDBI, KUN, PE); 2000 m, 13 Jun. 1932, *T. T. Yü 1017* (PE); bank of Jinshajiang, 1280 m, 11 May 1965, *Exped. to Sichuan, Yunnan and Guizhou, S-W China Compl. Exped. CAS415* (PE); **Yulong** (Lijiang), Daju, Lipei, 16 May 1939, *Z. G. Zhao 30095*

(KUN, PE); Judian, 1850 m, 9 Jun. 1981, *Exped. Qinghai-Xizang 419* (KUN); mountain of Lichiang Glacier, 16 Jun. 1922, *J. W. & C. J. Gregory s. n.* (BM); W Likiang Snow Range, Apr. 1923, *J. F. Rock 8545* (E); Shudi, 2000 m, 2 Jun. 1986, *H. Li 838* (KUN); Tiger's Leap, 1900 m, 1 Jun. 1987, *Sino-British Lijiang Exped. (D. Chamberain et al.) SBLE327* (E, K); between Bai Shui and Hei Shui, N of Lijiang, 2820 m, 6 Oct. 1990, *Chungtien-Lijanng-Dali Expedition (CLD-90) 878* (E, K); Lichiang range, 3050 m, *G. Forrest 10872* (BM, E, K, PE); Lichiang range, Jul.–Sep., *G. Forrest 12793* (E, IBSC, PE); Lichiang Range, Sep. 1929, *G. Forrest 27900* (E, KUN, PE); Lichiang range, 3050 m, *G. Forrest 27930* (E); Gucheng, Jin'an, 2261 m, 30 Jun. 2015, *L. Wei & J. S. Hao 15019* (BNU); Jinan, Sangu, 2545 m, 26 Jul. 2009, *K. X. Song 090731* (PE); s. loc., 2300 m, 24 Oct. 2009, *Z. H. Sun et al. Ouxk-YS-221* (IBSC); Daju, 1650 m, 3 Sep. 1988, *H. J. Guo 2039* (KUN); Judian, 1850 m, 9 Jun. 1981, *Exped. Qinghai-Xizang 419* (PE); Judian, 1900 m, 9 Jun. 1981, *Exped. Qinghai-Xizang 766* (KUN, PE); Rehe, 2500 m, 12 Jul. 1964, *S. W. Yu 10* (KUN); Dajum 1750 m, 18 Aug. 1962, *Exped. Zhongdian 1251* (KUN, PE); Daju, Yingpanjie, 26 Mar. 1939, *K. M. Feng 597* (KUN, PE); Yulongshan, Sandadaowan, 15 Jul. 1960, *K. M. Feng 23142* (KUN); Judian, 1900 m, 12 Sep. 1958, *W. T. Wang 212* (KUN); Jiazi, Wenshan, 2400 m, 18 Sep. 1955, *K. M. Feng 21578* (KUN, PE, SZ); Dongmu, Yucha, 14 May 1939, *W. L. He & Z. Y. Zhao 20341* (KUN, PE); Tak-Serh, 14 May 1939, *R. C. Ching 20341* (PE); Puchitze, 31 Aug. 1939, *R. C. Ching 21386* (KUN); Takoo (Ta-chi); 16 May 1939, *R. C. Ching 30095* (PE); Pujisi, 31 Aug. 1939, *H. C. He 21386* (PE); Daju, 16 May 1939, *Z. G. Zhao 30095* (KUN, PE); Yulongxueshan, 2900 m, 4 Aug. 2019, *Y. F. Deng 28639* (IBSC); **Yunlong**, Biaocun, Xiaodian, 1380–1570 m, 31 May 2009, *L. M. Gao & Z. R. Zhang GLM-092564* (KUN); Lishupo, 2400 m, 16 Jul. 2009, *K. X. Song 090515* (KUN, PE); Lishupo, 2400 m, 16 Jul. 2009, *S. D. Zhang & K. X. Song 090515* (KUN); **Zhaojue**, 2134 m, 6 Aug. 1964, *Sichuan Economic Plant Expedition 120* (CDBI); Wupo, 2100 m, 29 Jun. 1976, *s. coll. 12722* (CDBI, PE); Danglu to Changpo, 13 Jul. 1960, *Zhi 57 Jia 0394* (SZ).

2. Incarvillea emodi (Lindl.) Chatterjee, Kew Bull. 3: 185. 1948; Griersonin Notes Roy. Bot. Gard. Edin. 23: 318. 1963. ≡ *Amphicome emodi* Lindl., Bot. Reg. 24: tab. 19. 1838; de Candolle, Prodr. 9: 237. 1845; Clarke in Hooker (ed.), Fl. Brit. India 4: 385. 1884. Type: [U Ganga & Indus] India. Suen Range, *Wallich Cat.* 487 p. p. (holotype: CGE; isotypes: BM, K).

Ascendent herbs, 15–50 cm high, glabrous or pubescent; stems 1–5 cm long, bearing prominent leaf scars. Leaves somewhat fleshy, pinnately compounded, 15–25 cm long, usually crowded basally; lateral leaflets 4–5 pairs, opposite, blade ovate, 1–4 cm long, 0.6–1.8 cm broad, punctate, glabrous or slightly to densely pubescent, base obliquely rounded, margin dentate or rarely pinnately divided, apex acute; terminal leaflet similar or somewhat smaller, distinct or confluent with uppermost pair of lateral leaflets. Racemes terminal, 6–18-flowered, secund; pedicel 1–3 cm long. Bracts 5–8 mm long; bracteoles 2, 3–6 mm long. Calyx tube 4–7 mm long, campanulate, glabrous or pubescent; teeth rounded, apiculate, 1–2 mm long, 2–3 mm broad. Corolla cyclamen purple with maize yellow throat; tube 3.5–5.8 cm long, inside glandular pubescent near base; lobes orbicular, 1–1.5 cm long, 1.5–2.5 cm broad. Anther-thecae ovate, 0.2–0.3 mm long, 0.1–0.2 mm broad, pilose. Staminode short, crooked terminally. Capsules 6–15 cm long, 2–3 mm broad, elongate, subulate, curved. Seeds very narrowly ellipsoid, 4–5 mm long, 1–1.3 mm broad, testa minutely rugose; coma of hairs 4–5 mm long at both ends. **Chromosomes** 2n = 22.

Phenology: flowering from April to May; fruiting from May to August.

Habitats: rocky grounds and rocky crevices; 400–2800 m.

Distribution: N Pakistan, Jammu & Kashmir, U Ganga & Indus, W Nepal (Map 46-214); also in other parts of Afghanistan, Pakistan and India.

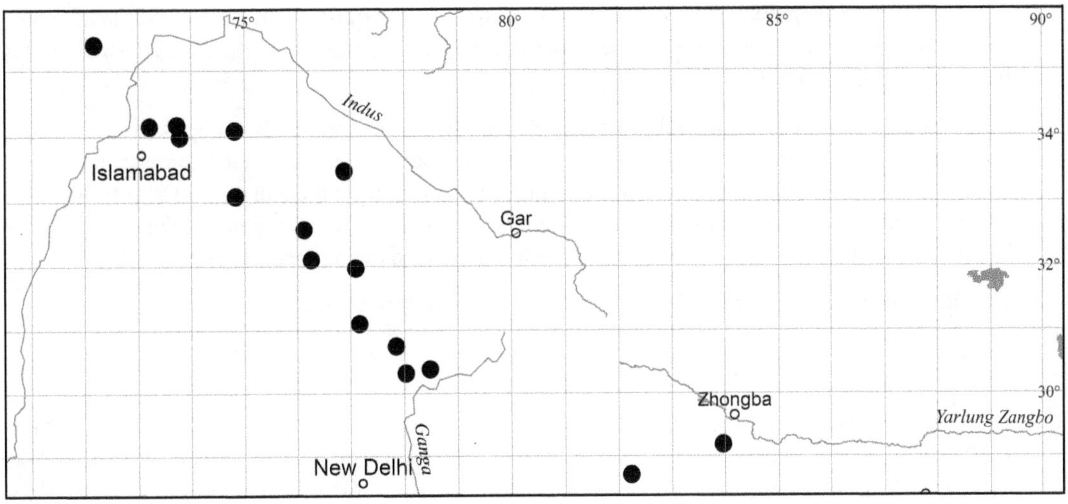

Map 46-214: **Incarvillea emodi** (Lindl.) Chatterjee

N PAKISTAN: **Azad Kashmir**, Bagh, Domel, *Timmins 52* (BM); Bagh, Jhelum valley, Domel, 760 m, 27 Mar. 1939, *G. Sherriff s. n.* (BM); **Hazara**, Abbotabad to Havelian, May 1957, *M. Masami s. n.* (TI); Abbotabad, 1400 m, 24 Mar. 1958, *Burtt B481* (E); Abbotabad, between Abbottabad amd Havelian, above Khota bridge, 6 Apr. 1973, *C. C. Townsend 73/365* (K); Abbotabad, Halokyle Hills, Halibay, 900 m, 25 Feb. 1921, *A. Fleming 37* (K); Abbotabad, *J. F. Duthie 22075a* (K); Abbotabad, Mt. Tilla, 25 Mar. 1893, *Aitchison 15* (K); Abbotabad, Salt Range, Mt. Tilla, Mar. 1874, *Aitchison 1095* (K); Abbotabad, Abbottabad-Havelian, *C. C. Townsend 365* (K); Abbottabad, Hill E. of town, *B. L. Burtt 481* (KUH, E); **Swat**, above Barikot, road to Saidu Sharif, 800 m, 1 Jun. 1965, *J. Lamond 1669* (E); Barikot, 900 m, *S. Ali 26086* (RAW, BM); Barikot to Mingora, 700–900 m, *K. H. Rechinger 30503* (W); Swat, 900 m, *R. R. Stewart 27507* (BM). **JAMMU & KASHMIR**: **Padum**, Dakochara, 610 m, 18 Jan. 1889, *G. Watt 2829* (E); **Reasi**, Thargdi & Batal gla, 17 Mar. 1962, *P. S. Jamwel 8302* (RRLH); **Srinagar**, Suen Range, *Wallich Cat. 487 p. p.* (BM, CGE, K). **U GANGA & INDUS**: **Chamba**, Jikri, 1500 m, *G. Watt 637* (E); Kalel, 1200 m, 22 Jun. 1917, *R. R. Stewart 2350* (K); **Dehradun**, Chakrata, 13 Apr. 1922, *S. D. Parshad s. n.* (K); Banks of Sutlej, near Belaspoor, 609 m, *Madden s. n.* (K); Jaunsar, Chakrata, *Saklani 104* (G); Jaunsar, Konain, 2135–2440 m, 18 Apr. 1898, *Duthie 21004* (BM, E, G, K); Jaunsar, Konain, 2250 m, Apr. 1898, *J. S. Gamble 26526* (K); Jaunsar, Boryar, 6000 ft., May 1891, *J. S. Gamble 23123* (K); **Deoban**, 2440–2740 m, 11 Apr. 1897, *Duthie 19861* (BM); **Kangra**, Mat. Kangra, *J. S. Gamble* 22837 (K); **Kulu**, Larji, 25 Jun. 1888, *J. R. Drummond 23096* (K); **Shimla**, Kalka below Simla, 1609 m, Feb. 1889, *G. Watt s. n.* (E); **Tehri Garhwal**, Deoprayag, 460 m, *R. Strachey & J. E. Winterbottom s. n.* (K). **W NEPAL**: **Karnali**, Jumla, Bheri River, below Gajarkot, 760 m, 1 Apr. 1952, *O. V. Polunin, W. R. Sykes&L. H. J. Williams* 1829 (BM, E).

II. Incarvillea subg. **Incarvillea** ≡ *Incarvillea* subg. *Euincarvillea* Baill., Hist. Pl. 10: 252. 1888, nom. inval. (Art. 24.3); Schumann in Engler & Prantl (eds.), Nat. Pflanzenfam. 4(3b): 232. 1897. Type: *Incarvillea sinensis* Lam.

Herbs, annual or per perennial herbs, often becoming woody at base. Leaves opposite or alternate, pinnate, pinnatisect or bipinnatisect. Inflorescences terminal, racemose. Calyx tube campanulate, teeth swollen at base, subulate or reduced to minute points, sinuses membranous drawn up into bifid teeth, thus calyces appear to be ten toothed. Capsules coriaceous, cylindric, acuminate at apex, dehiscing longitudinally on posterior side only. Septum entire at apex. Seeds wing hyaline, entire.

Two species, in Russia, Mongolia and China; one species in Pan-Himalaya.

3. Incarvillea sinensis Lam., Encycl. 3: 243. 1789; Tao & Yin in Wang (ed), Fl. Reipubl. Popularis Sin. 69: 36. 1990; Zhang & Santisuk in Wu & Raven (eds.), Fl. China 18: 221. 1998. Type: China, Beijing, "Peking and environs, s. d., *d'Incarville s. n.* (holotype: P).

Incarvillea variabilis Batalin, Trudy Imp. S.-Peterburgsk. Bot. Sada 12: 177. 1892. ≡ *Incarvillea sinensis* subsp. *variabilis* (Batalin) Grierson, Notes Roy. Bot. Gard. Edinburgh 23: 324. 1961. ≡ *Incarvillea sinensis* f. *variabilis* (Batalin) Grierson, Notes Roy. Bot. Gard. Edinburgh 23: 324. 1961. Type: China, Sichuan, "prov. Szetschuan septentrionali, Tibet orientale, ad fl. Poi-ho prope Terga", 12 Jul. 1885, *G. N. Potanin s. n.* (lectotype designated by Grierson, Notes Roy. Bot. Gard. Edinburgh 23: 325. 1961: LE; isolectotypes: K, P, PE).

Incarvillea variabilis var. *latifolia* Batalin, Trudy Imp. S.-Peterburgsk. Bot. Sada 12(1): 177. 1892. Type: China, Gansu, "Kansu occidentale, prope pagum Shun-dan-sien", 7 Sep. 1885, *G. N. Potanin s. n.* [holotype: LE; isotypes: K (K000779188), P].

Incarvillea variabilis var. *fumaeriaefolia* Batalin, Trudy Imp. S.-Peterburgsk. Bot. Sada 12(1): 178. 1892. Type: China, Sichuan, "Szechuan septentrionale, prope oppidum Kwantin ad fluv. Pei-shui, prope, oppid, Sigu", 23 Jun. 1885, *G. N. Potanin s. n.* [holotype: LE; isotypes: K (K000779187), PE].

Incarvillea variabilis var. *przewalskii* Batalin, Trudy Imp. S.-Peterburgsk. Bot. Sada 14: 180. 1895. ≡ *Incarvillea przewalskii* (Batalin) Iljin, Mater. Istorii Fl. Rastitel'n. S. S. S. R. 3: 214. 1958. ≡ *Incarvillea sinensis* f. *przewalskii* (Batalin) Grierson, Notes Roy. Bot. Gard. Edinburgh 23(3): 325. 1961. ≡ *Incarvillea sinensis* var. *przewalskii* (Batalin) C. Y. Wu & W. C. Yin, Fl. Reipubl. Popularis Sin. 69: 36. 1990; Zhang & Santisuk in Wu & Raven (eds.), Fl. China 18: 221. 1998. Type: China, Gansu, "regio Tangut (Prov. Kansu), jugum N. Tetung", 1 Aug. 1880, *N. M. Przewalsky s. n.* (holotype designated by Illarionova, Novosti Sist. Vyssh. Rast. 34: 177. 2002: LE; isolectotype: PE).

Herbs annual or perennial, with branched stems, 50–80 cm high. Roots subwoody, branched. Leaves alternate, 2- or 3-pinnately divided, variable in shape and size, 4–6 cm long; leaflets irregularly lobed; lobes linear-lanceolate, margin serrate or entire. Inflorescences sparsely racemose, 15–20 cm long. Bracts elliptic-subulate, 0.5–1 cm long; bracteoles linear, 3–5 mm long. Pedicel 1–5 mm long. Pedicels 0.5–1 cm long. Calyx green with purple-red, campanulate, 4–5 mm long, 4–5 mm broad, teeth subulate, base enlarged into glands. Corolla campanulate-funnelform, 3.5–4 cm long, 2.2–2.5 cm broad, constricted into tubular base; lobes rounded. Stamens inserted at base of corolla tube; anthers adherent. Style pale yellow. Capsules pale green, narrowly terete, 3–6(–10) cm long, 4–5 mm broad, apex caudate-acuminate. Seeds compressed globose, 1.5–2 mm in diameter, surrounded by transparent and membranous wings, apex incised. **Chromosomes** 2n = 22. Figure 46-121.

Figure 46-121: **Incarvillea sinensis** Lam.: **a**, habit; **b**, calyx; **c**, opened corolla showing stamens; **d**, stamen; **e**, capsule. a, c, d drawn by Jin-Yong FENG; reproduced from *Icon. Cormophyt. Sin.* 4: 99, fig. 5612. 1975; b & e drawn by Bao-Fu ZHANG; reproduced from *Fl. Reipubl. Popularis Sin.* 69: 37, pl. 9: 5–6. 1990.

Phenology: flowering from May to September; fruiting from October to November.

Habitats: slopes, thickets, fields; 800–3900 m.

Distribution: Tangut, N Hengduan, S Hengduan (Map 46-215); also in other parts of China.

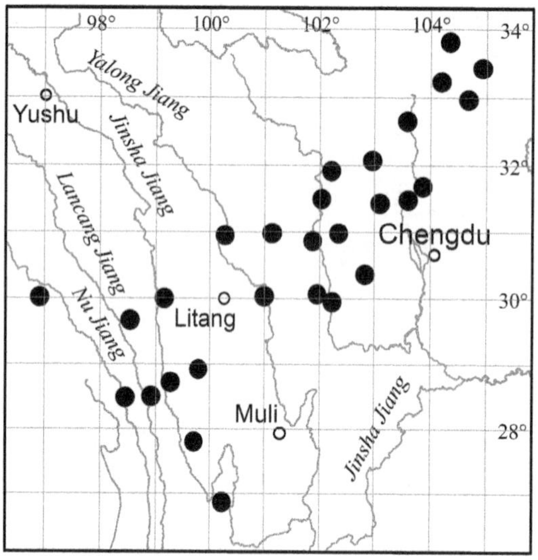

Map 46-215: **Incarvillea sinensis** Lam.

Notes: Grierson (1961) divided the species into two subspecies (subsp. *sinensis* and subsp. *variabilis*) and further separated subsp. *variabilis* into two forms (f. *variabilis* and f. *przewalskii*) on the basis of the flower colour. Tao & Yin (1990) divided the species into two varieties (var. *sinensis* and var. *przewalskii*) and merged *I. variabilis* (*I. sinensis* f. *variabilis*) with var. *sinensis*. The form *przewalskii* differs from f. *sinensis* by its yellow flowers while the the flower colour in the latter is pale rose or pale red, sometimes purple. In present treatment, I do not separate the plant with yellow flowers as the independent infraspecific taxon.

Lin et al. (Acta Bot. Boreal.-Occid. Sin. 37: 1045. 2017) designated the specimens at PE as the lectotype for for the names *Incarvillea variabilis*, *I. variabilis* var. *fumariaefolia* and *I. variabilis* var. *przewalskii*, respectively. Unfortunately, they overlooked that the names have been lectotypified by Grierson (1961) whose choice must be followed under Art. 9.19.

TANGUT: **Baxoi**, Bangda, 3300 m, *P. C. Kuo & W. Y. Wang 21645* (HNWP). **N HENGDUAN**: **Barkam**, 2800 m, 13 Jul. 1984, *W. L. Chen et al. 7615* (PE); Keeryin, Taiyanghe, 2480 m, 15 Jun. 1959, *S. Jiang et al. 1048* (KUN, PE, SZ); Puya, 2800 m, 11 Jul. 1957, *Z. R. Zhang & H. F. Zhou 22788* (PE); s. loc., 2650 m, 10 Jul. 1957, *Z. R. Zhang & H. F. Zhou 22935* (IBSC, SZ); s. loc., 2600 m, 19 Jul. 1957, *Z. R. Zhang & H. F. Zhou 23116* (IBSC, KUN, PE, SZ); s. loc., 7 Jul. 1957, *X. Li 75199* (WUK); **Batang**, s. loc., 10 Sptember 1973, *Xizang Exped. 2841* (HNWP); 17 Jun. 1984, *J. S. Yang 84-145* (KUN); Zhubalong, 2650 m, 29 Jun. 1980, *s. coll. 29165* (CDBI); near the town, 16 Sep. 1975, *Sichuan Veget. Exped. 3912* (CDBI, KUN, PE); Zhubalong, 2400–2430 m, 30 Jul. 2006, *D. E. Boufford et al. 35490* (PE); S of the city, 2900 m, 28 Jul. 1982, *X. Z. Tang 0535* (PE); Bingzhan, 2700 m, 20 Aug. 1976, *s. coll. 76362* (KUN, NAS); 2750 m, 4 Aug. 1959, *S. Jiang et al. 2711* (KUN, PE); bank of Jinshajiang, 2620 m, 2 Jul. 1961, *S. Jiang 5291* (KUN, PE); s. loc., *Xizang Exped. 2841* (HNWP); Zhubalong, E bank of Jinshajiang, 2450 m, 29 Aug. 1981, *Qinghai-Xizang Exped. 5184* (KUN); Zhubalong, 2650 m, 29 Jun. 1980, *s. coll. 28165* (CDBI); **Danba**, Shuiziba, 2080 m, 20 Jul. 1959, *S. Jiang et al. 2219* (PE); Dajinchuan, 30 Aug. 1940, *K. L. Chü 7872* (PE); Donggu, 2250 m, 9 Jun. 2010, *L. M. Gao & Z. M. Cai GLM-103052* (KUN); **Dawu**, Qianning, 2800 m, 7 Aug. 1974, *Sichuan Veget. Exped. 5690* (CDBI, PE); **Heishui**, Shidiaolou, 1600 m, 31 May 1959, *Heishui Group (Chuan Jing A) 1270* (KUN, PE); **Jinchuan**, Chuosijia, 2400 m, 19 Sep. 1957, *X. Li 71911* (IBSC, IFP, PE, SZ); Mengjiaping, 2210 m, 29 Jun. 1957, *X. Li 75068* (IBSC, PE, SZ); Longjiashan, 2750 m, 8 Jul. 1957, *X. Li 75192* (IBSC, KUN, PE, SZ); Anshun, 2250 m, 2 Aug. 1957, *X. Li 75510* (IBSC, PE, SZ); Chonghua, Guandimiao, 1940 m, 1 Oct. 1957, *X. Li 76557* (IBSC, PE, SZ); Zengda, 2300 m, 14 May 1958, *X. Li 77401* (PE); Anning, Congling, 2300 m, 3 Jul. 1958, *X. Li 78019* (IBSC, PE); **Jiuzhaigou** (Nanping), Shuanghe, Weiziba, 14 Jul. 1962, *Sichuan Econ. Plant Exped. 913* (CDBI); s. loc., 1340 m, 10 Aug. 1961, *X. N. Tang et al. 433* (CDBI); Yuwa, Yaojiashan, 2600 m, 29 Jul. 1959, *Nanping Group (Chuan Jing A) 4185* (CDBI); **Lixian**, 1886 m, 18 Jun. 1983, *Zhi Bei Zu 30231*

(CDBI); Li-pan-fu, 17 Jul. 1893, *Potanin s. n.* (E, K, LE, P); Tonghua, 1230 m, 26 Jul. 1979, *s. coll. 79525* (SZ); 2000 m, 17 Aug. 1933, *T. T. Yü 2487* (IBSC, PE); Shaba, 1600 m, 21 May 1987, *s. coll. 1085* (CDBI); **Maoxian**, near the Minjiang, 1700 m, 22 Aug. 1984, *Y. L. Cao 30* (CDBI); Gucheng (old city), 1500 m, 9 Jun. 1959, *S. Jiang 650* (KUN, SZ); 1953, *C. Ho13119* (IBSC, PE, SZ); s. loc., 1952, *C. Ho & C. L. Chow 13917* (IBSC, PE, SZ); s. loc., 1900 m, 2 Sep. 1934, *T. T. Yü 2505* (IBSC, PE); on the way to Songjipu, 1600–2200 m, 12 Jun. 1986, *C. Y. Wu 496* (CDBI, KUN); **Markam**, Xiquhe Valley, 2800 m, 2 Jul. 1976, *Qinghai-Xizang Exped. 11673* (KUN, PE); Zhubalong, near Jinshajiang Bridge, 2500 m, 19 Aug. 2010, *Kham Expedition 10-2287* (PE); Zhubalong, near Xiquhe, 2520–2560 m, 19 Aug. 2010, *Kham Expedition 10-2330* (PE); **Songpan**, 15 Sep. 1928, *W. P. Fang 4455* (E); Zhangla, 3200 m, 30 Jun. 1961, *S. Jiang et al. A7343* (PE); Heihchiao, 2000 m, 22 Oct. 1957, *T. P. Wang 7930* (WUK); Zhongtiantang, 1600 m, 23 Oct. 1937, *K. T. Fu 2210* (PE); **Wenchuan**, 1400 m, 28 Jul. 1975, *Sichuan Veget. Exped. 8400* (CDBI, PE); Muxing, 1600 m, 23 Aug. 1981, *Q. Q. Wang 28688* (CDBI); Yanmen, 1340 m, 28 May 1958, *S. Y. Chen et al. 5058* (SZ); **Wenxian**, way to Shifang, 920 m, 9 Jul. 1959, *Z. Y. Zhang 6061* (HNWP, WUK); Zhongzhai, Songping, 1000 m, 12 Jul. 1959, *Z. Y. Zhang 6116* (HNWP, WUK); Shifang, 1200 m, 21 Jul. 1959, *Z. Y. Zhang 7061* (HNWP, WUK); 950 m, 14 Aug. 1959, *Z. Y. Zhang 8144* (WUK); Chengguan to Xiyuan, 950 m, 17 Aug. 1959, *Z. Y. Zhang 8187* (HNWP, WUK); Danbao, Yangjiagou, 1325 m, 6 Aug. 1959, *Z. Y. Zhang 10255* (WUK); Danbao, 1250 m, 6 Aug. 1959, *Z. Y. Zhang 10368* (WUK); Yuxushan, 1000–1050 m, 6 Aug. 2010, *Z. M. Jiang et al. 1626* (WUK); 1150 m, 1 Aug. 1984, *W. L. Chen et al. 8719* (PE); Chengguan, 870 m, 15 Oct. 1973, *J. X. Yang & Z. X. Hu 3468* (PE, WUK); the the city, bank of Baishuijiang, 970 m, 16 Aug. 1967, *C. L. Tang 1708* (HNWP, WUK); Linjiang Qu, Jianshan, 1000 m, 10 Oct. 1958, *Z. P. Wei 2806* (HIMC, WUK); on the way from Nangping to Wenxian, 1000 m, 29 Oct. 1957, *T. P. Wang 7983* (PE, WUK); Mao-chou, ca. 1800 m. 29 Jun. 1922, *H. Smith 2288* (PE); **Wudu**, along road near Daigusi (85 km), 1800 m, 19 Jun. 2000, *Sino-British Qinghai Garden Society Expedition (SQAE) 313* (E); road from Wudu to Wenxian, 3885 m, 25 Jun. 2000, *Sino-British Qinghai Garden Society Expedition (SQAE) 425* (E); Liangshui, 1200 m, 23 Jun. 1964, *P. C. Kuo 5026* (WUK); Toufang to Huapingli, 1420 m, 15 Jun. 1959, *Z. Y. Zhang 1420* (WUK); near Yaojiazhai, 1590 m, 26 May 1959, *Z. Y. Zhang 2707* (WUK); Toufang to Huaping, 1420 m, 15 Jun. 1959, *Z. Y. Zhang 4391* (WUK); Shimen, 1200 m, 4 Sep. 1939, *W. Y. Hsia 6394* (WUK); S of the city, 1200 m, 14 Sep. 1951, *T. P. Wang 14929* (IBSC, WUK); Liangshui, 1050, 12 Sep. 1958, *Y. Q. He 511* (PE, WUK); s. loc., 20 May 1959, *S. Jiang et al. 00336* (PE, SZ); **Xiaojin**, near Xinjie, 2400 m, 15 Jul. 1958, *X. S. Zhang & Y. X. Ren 5889* (CDBI, PE); near the town, 10 Aug. 1975, 10 Aug. 1975, *Sichuan Veget. Exped. 9606* (CDBI, IBSC, PE); Heping, 2510 m, 30 May 1957, *J. Zhou 46* (PE); **Xinlong**, Zhenge, 3200 m, 26 Jun. 1974, *S. A. Lu 6351* (CDBI, PE); **Yajiang**, Zihe, 2450 m, 5 May 1960, *S. Jiang et al. 3323* (KUN, PE, SZ); **Zhugqu**, Chengguan, 1500 m, 21 Jul. 1964, *P. C. Kuo 5481* (HIB, WUK); Hanbanpu, 1380 m, 11 Jul. 1951, *T. P. Wang 14305* (WUK). **S HENGDUAN**: **Baoxing**, 26 Apr. 1936, *K. L. Chü 2958* (WUK); **Cawalong** (Tsawarong), Salwin Kui-chiang Divide, *G. Forrest 19038* (BM, E, K); **Dêqên**, Chipeh, 3850 m, 18 Aug. 1937, *T. T. Yü 9621* (KUN, PE); Bei ma shan, 3350 m, *G. Forrest 14552* (E, K); *J. R. Rock 9935* (GH, W), near Yunling, 2300 m, 18 Jul. 1981, *Qinghai-Xizang Exped. 3551* (PE); Benzilan, 1960 m, 9 Jun. 1976, *Qinghai-Xizang Exped. 11806* (HNWP, KUN, PE); Yunling, Chilidi, 26 Oct. 1963, *C. C. Lü 63223* (KUN); Cizhong to Yongzi, Guta, 2100 m, 25 Jul. 1940, *K. M. Feng 5771* (PE); Atuntze, Chiapeh, 3850 m, 18 Aug. 1937, *T. T. Yü 9621* (KUN, PE); Tonghua, 6 May 1937, *H. Q. Yang 0014* (SZ); **Dêrong**, Zigen, 2080 m, 4 Jul. 1981, *Qinghai-Xizang Exped. 1769* (KUN, PE); Diquhe, 2400 m, 2 Aug. 1981, *Qinghai-Xizang Exped. 3145* (PE); **Kangding**, Wa Sze Ka, E of Tatsienlu, 1983 m, 10 May 1934, *A. C. Young 44* (IBSC); Ridi to Wasigou, 1400 m, 15 May 1974, *Y. T. Wu & Q. S. Zhao 111002* (CDBI, PE, SZ, WUK); Songyu to Mati, 3100 m, 12 May 1960, *S. Jiang et al. 3393* (KUN, PE, SZ); Shade, Mati, 2700 m,

29 May 1961, *S. Jiang et al. 5033* (KUN, PE); Guza, bank of Daduhe, 1400 m, 28 May 1980, *Q. S. Zhao 116548* (SZ); Tatsienlu, 2400–2550 m, 8 Oct. 1928, *W. P. Fang 3712* (E, K); Ta Tsien Lu, *A.E. Pratt 726* (BM, K); s. loc., 24 Sep. 1934, *C. S. Li 1556* (PE); s. loc., 1934, *T. H. Tu 1556* (IBSC); s. loc., 2830 m, 8 Oct. 1928, *W. P. Fang 3712* (PE); **Luding**, Guoqing, 1500 m, 18 Oct. 1960, *Maowen Group 16530* (CDBI); **Shangri-la** (Zhongdian), mountains of the Chungtien plateau, Jul. 1914, *G. Forrest 12800* (E, IBSC); Dongwang, Mailuo, 24 Jun. 1975, *J. S. Yang 7445* (IBSC); Dongwang, 3280 m, 5 Aug. 1984, *J. S. Yang 84-874* (KUN); **Xiangcheng**, 13 Jul. 1973, *Z. G. Liu 1065* (CDBI); Dongsong, Shage, 2700 m, 20 Jul. 1973, *Sichuan Veget. Exped. 2970* (CDBI, KUN, PE); s. loc., 2800 m, 24 Jun. 1960, *S. Jiang et al. 9609* (KUN, PE); Ranwu, Xiarewu on road betweencheng and Daxue Shan, 2920 m, 22 Jul. 1998, *D. E. Boufford et al. 29024* (E); Shangrewu, Shuoquhe, 2700 m, 31 Jul. 1981, *Qinghai-Xizang Exped. 3097* (KUN, PE); **Yulong** (Lijiang), Judian, 1900 m, 12 Sep. 1958, *W. T. Wang 213* (KUN); Judian, 2000 m, 11 Aug. 1944, *K. M. Feng 9448* (PE).

III. Incarvillea subg. **Pteroscleris** Baill., Hist. Pl. 10: 52. 1888; Schumann in Engler and Prantl, Nat. Pflanzenfam. 4(3b): 232. 1897; Grierson, Notes Roy. Bot. Gard. Edinburgh 23: 328. 1961; Tao & Yin in Wang (ed.), Fl. Reipubl. Popularis Sin. 69: 39. 1990. Type: *Incarvillea delavayi* Bureau & Franch.

Perennial herbs, acaulous or with erect stems. Rootstock often fleshy; stems absent or cylindric, striate. Leaves alternate, undivided, pinnate or pinnatisect. Racemes terminal or sometimes axillary, 1–many-flowered; pedicels short subtended by a bract and bearing two smaller bracteoles. Calyx tube campanulate usually strongly ribbed; teeth lanceolate or triangular often folded, apex acute or obtuse. Corolla lobes covered with shortly stalked glands. Anthers glabrous. Capsules ligno-coriace-elongate, straight or curved, more or less quadrangular, dehiscing longitudinally on both surfaces but only in upper one third on anterior surface. Septum entire at apex. Seeds ovate in outline, compressed, plano-convex; wing opaque, narrow, entire or notched or reduced to a rim.

Eleven species: Nepal, Bhutan and W China; all in the Pan-Himalaya.

4. Incarvillea forrestii H. R. Fletcher, Notes Roy. Bot. Gard. Edinburgh 18: 310. 1935; Tao & Yin in Wang (ed), Fl. Reipubl. Popularis Sin. 69: 40. 1990; Zhang & Santisuk in Wu & Raven (eds.), Fl. China 18: 222. 1998. Type: [S Hengduan] China, Yunnan, Shangri-la, "Chungtien plateau", 27°30′N, 3048 m, Jul. 1914, *G. Forrest 12676* [lectotype designated by Grierson, Notes Roy. Bot. Gard. Edinburgh 23: 338. 1963: E (E00135989)].

Herbs perennial, 15–30(–60) cm high, glabrescent. Leaves alternate, simple; petiole 2–5(–15) cm long; leaf blade ovate to ovate-elliptic, 6–20 cm long, 3–15 cm broad, puberulent along veins on both surfaces, lateral veins 7–9 pairs, base rounded to cuneate, margin crenate, apex obtuse. Racemose terminal, 6–12-flowered; peduncle 2–4 cm long. Bracts linear-lanceolate, 5–12 mm long; bracteoles small. Pedicel 5–10 mm long. Calyx campanulate, 1.4–2.5 cm long, teeth 2–4 mm long, 7–10 mm broad, apex acute or obtuse. Corolla pale rose-pink, 4.5–5.5 cm long, 2–3 cm broad; tube 4.5–5 cm long, purple-red striate and spotted adaxially; lobes rounded, 1.4–1.8 cm long, 1.8–2.2 cm broad. Capsules lanceolate, compressed, quadrangular, 4–6(–9) cm long, 5–7 mm broad, apex acuminate. Seeds ovate in outline, compressed, 5–6 mm long, 3–4 mm broad; wing 0.7–1.2 mm broad. **Chromosomes** 2n = 22. Figure 46-122.

Phenology: flowering from May to July; fruiting from August to November.

Habitats: slopes, grasslands, thickets; 400–4300 m.

Distribution: endemic; N Hengduan, S Hengduan (Map 46-216).

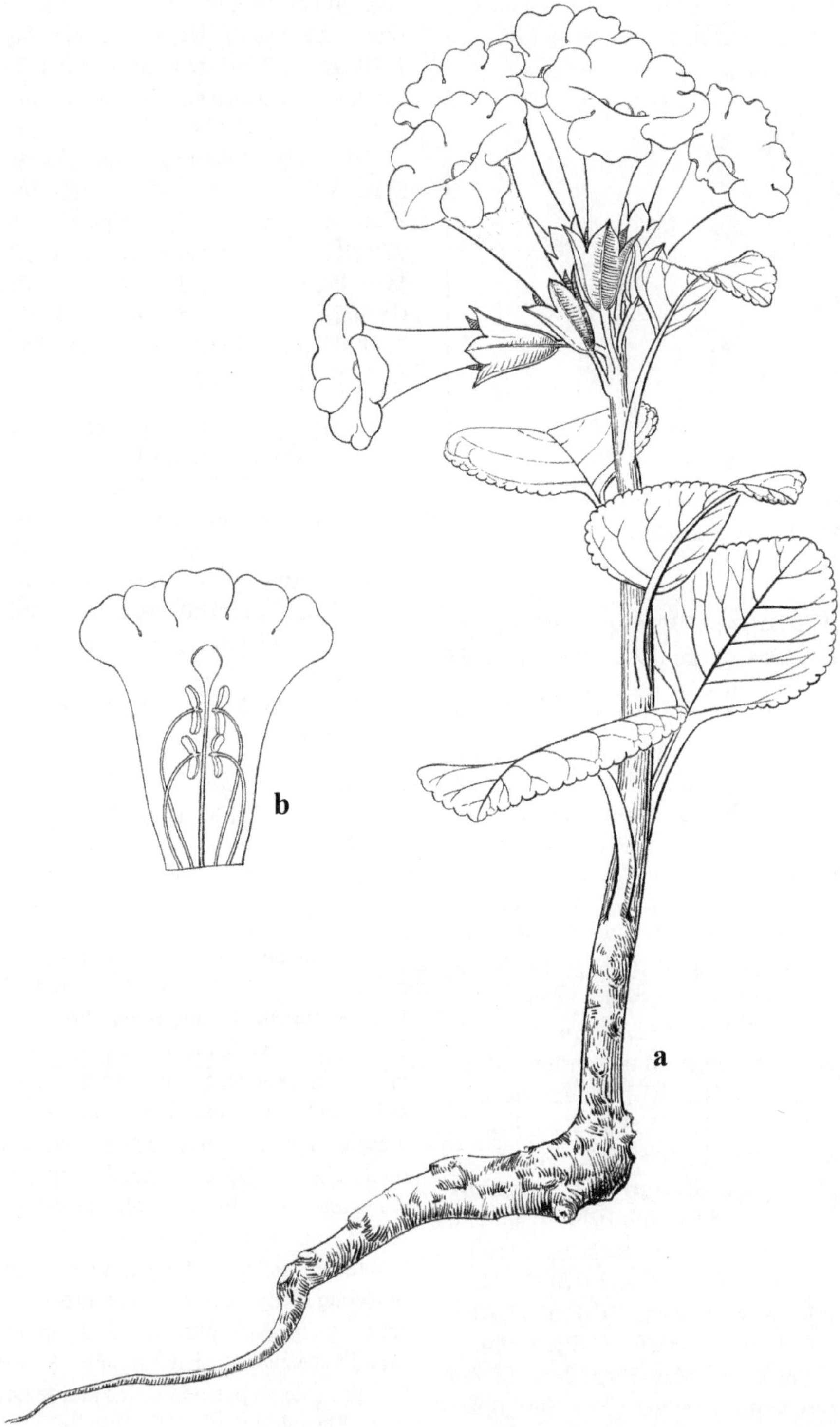

Figure 46-122: **Incarvillea forrestii** H. R. Fletcher: **a**, flowering plant; **b**, corolla opened showing stamens and distal portion of style. Drawn by Bao-Fu ZHANG; reproduced from *Fl. Reipubl. Popularis Sin.* 69: 41, pl. 10: 4–5. 1990.

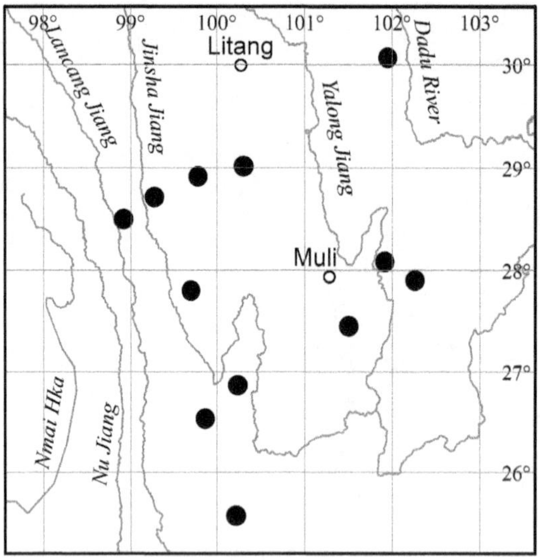

Map 46-216: **Incarvillea forrestii** H. R. Fletcher

N HENGDUAN: **Dêrong**, to Riyuzhege, 4000 m, 4 Aug. 1981, *Qinghai-Xizang Exped. 3213* (KUN, PE); Zhegeshan, 4300 m, 7 Aug. 1981, *Qinghai-Xizang 3465* (KUN). **S HENGDUAN**: **Dali**, Huadian, 3150 m, 4 May 1988, *J. Y. Cui et al. 1446* (KUN); Cangshan, 3100 m, 6 Apr. 1974, *Z. W. Lü 1079* (KUN); **Daocheng**, Dongni Qu, Geka, Kasi, 3800 m, 27 Jun. 1971, *Z. G. Liu 0111* (CDBI); **Dêqên**, Baimashan, 9 Jul. 1984, *J. S. Yang 84-513* (KUN); **Jianchuan**, 3300–3600 m, Jul. 1922, *G. Forrest 21526* (E, K); **Kangding**, on the way to Gaoershi Pass, 4210 m, 20 Jun. 1984, *W. L. Chen et al. 6579* (PE); **Muli**, Muli Range, N flank, 3600 m, 22 Jun. 1913, *F. Kingdon-Ward 5214* (E); W Mt. Mitzuga, Muli Territory, May–Jun. 1932, *J. F. Rock 24051* (E); Muli Area, s. d., *G. Forrest 30655* (E, PE); 3400 m, s. d., *T. T. Yü 5612* (KUN); Mi-tau-ga, 400 m, 23 May 1937, *T. T. Yü 5635* (PE); **Shangri-la** (Zhongdian), Chungtien Plateau, 3050 m, Jul. 1914, *G. Forrest 12676* (E); Dongwang, boder tocheng, 4450 m, 22 Aug. 1988, *J. S. Yang 88-162* (IBSC, KUN, PE); **Xiangcheng**, Maanshan, 4050 m, 11 Aug. 1983, *Y. F. Han et al. 83-206* (PE); **Xichang**, Luojishan, 2600 m, 16 May 1985, *Y. J. Li 729* (CDBI); **Yanyuan**, Yuanbao Qu, Huolushan, 3800 m, 22 Jul. 1983, *Qinghai-Xizang Exped. 12262* (HITBC, KUN); Heshang, Maoniushan, 3400 m, 9 Jun. 1960, *S. Jiang et al. 5896* (KUN, PE); around summit of Huolu Shan, E Guabie, above the Xiaojin He, 4300–4320 m, 13 Aug. 2010, *D. E. Boufford et al. 42858* (TI); **Yulong** (Lijiang), Yulongshan, 3000 m, 30 May 1986, *H. Li 685* (KUN); Ganheba, 3200 m, 3 Jun. 1986, *H. Li 889* (KUN); Shihuiyao, Ganheba, 3060 m, 27 May 1988, *S. J. Pei 1580* (KUN); Yulongshan, 3160 m, 2 Jun. 1981, *Qinghai-Xizang Exped. 279* (KUN); Yulongshan, Ganheba, 3200 m, 26 May 1964, *X. Zhou 1037* (KUN); Yulongshan, Ganhaizi, 3100 m, 25 Sep. 1959, *K. M. Feng 22741* (HNWP, KUN, PE); Yulongshan, 3250 m, *Q. E. Yang 320* (KUN).

5. Incarvillea lutea Bureau & Franch., J. Bot. (Morot) 5: 137. 1891; Tao & Yin in Wang (ed.), Fl. Reipubl. Popularis Sin. 69: 42. 1990; Zhang & Santisuk in Wu & Raven (eds.), Fl. China 18: 222. 1998. Type: [S Hengduan] China, Yunnan, Eryuan, "Coteaux rochailleux au-dessus de sources du Lankien-ho, près du col de Hee-chan-men", 31 May 1889, *M. Delavay 4349* [lectotype **here designated**: P (P00608146); isolectotype: P (P00219424)].

Incarvillea principis Bureau & Franch., J. Bot. (Morot) 5: 136. 1891. Type: [N Hengduan] China, Sichuan, "apres Batang, ravin boise", 15 Jun. 1890, *Prince Henri d'Orleans s. n.* [holotype: P (P00219423)].

Herbs perennial, 0.5–1 m high, pale brown pubescent. Roots fleshy, 1–2 cm in diameter. Leaves pinnate below, pinnatifid above, 12–27 cm long; lateral leaflets 6–9 pairs, sessile, blade elliptic-lanceolate, 5–9 cm long, 1.5–3 cm broad, pubescent along veins, base narrowly cuneate, margin serrate, apex acute; terminal leaflet blade elliptic, 2–5 cm long, 1–2 cm broad, base cuneate or confluent with uppermost lateral leaflet. Inflorescences racemose, 5–12-flowered; bracteoles linear, 1–1.5 cm long. Pedicel 5–10 mm long. Calyx campanulate, green with purple spots, veins dark purple, 1.5–3 cm long; teeth broadly triangular, 5–10 mm long, apex acute. Corolla yellow, purple spotted and striate at base, 5–6(–8) cm long; tube 2.5–4 cm long, mouth 2.8–3.2 cm in diameter; lobes oblong, 1.5–2 cm long, 2–2.5 cm broad, glandular. Staminode 1–1.5 mm long; filaments and anthers pale yellow. Capsules lanceolate, 8–10 cm long, 1.3–1.5 cm broad, 6-angled, woody, apex acuminate. Seeds pale yellow-brown, ovoid, or globose, 4–4.5 mm long, 3–4.5 mm broad, apex rounded or emarginate, usually gray pubescent; wings 0.7–1.2 mm broad. **Chromosomes** 2n = 22. Figure 46-123.

Phenology: flowering from May to August; fruiting from September to November.

Habitats: forests, slopes, grasslands; 2100–4500 m.

Distribution: endemic; Tangut, N Hengduan, S Hengduan (Map 46-217).

TANGUT: **Dêgê**, Gongya, 3075 m, 7 Jun. 2010, *L. M. Gao & Z. M. Cai GLM-103008* (KUN); s. loc., 10 Jul. 1990, *J. S. Yang 134* (KUN); **Zadoi**, Zhaqing, Diqing, 4460 m, 24 Jul. 2005, *Y. H. Wu 35060* (HNWP). **N HENGDUAN**: **Barkam**, Singganghe, 2600 m, 20 May 1957, *H. Li 70756* (IBSC, KUN, SZ, WUK); **Batang**, s. loc., 15 Jun. 1890, *Prince Henri d'Orleans s. n.* (P); **Jinchuan**, Bianyanshan, 2160 m, 26 Jun. 1957, *H. Li 75016* (IBSC); **Litang**, Labo, 4200 m, 4 Aug. 1973, *Sichuan Veget. Exped. 3694* (CDBI, KUN, PE); Litang Rivier Divide, 3300 m, 21 Jun. 1905, *F. Kingdon-Ward 4197* (E); **Luhuo**, Keluodong, 31 Jul. 1951, *Y. W. Cui 5053* (PE); **Yajiang**, Baimenda to Xiadecha, 4000 m, 16 Jun. 1961, *S. Jiang 5222* (KUN, PE). **S HENGDUAN: Daocheng**, Kasi, 3800 m, 28 Jun. 1973, *Sichuan Veget. Exped. 2281* (CDBI, KUN, PE); Geka, Kasi, 27 Jun. 1971, *Z. G. Liu 0111* (CDBI); Gongling, 3300 m, 4 Jul. 1998, *D. E. Boufford et al. 28198* (E); **Heqing**, Machang, 2 Aug. 1940, *R. C. Ching 23517* (KUN, PE); **Jiulong**, Chengguan, 2600 m, 24 Jun. 1974, *R. S. Lü 4690* (CDBI, PE); 11 Jun. 1930, *Z. Y. Luo et al. 355* (IBSC); **Kangding**, Shade, 3200 m, 15 Jun. 1984, *W. L. Chen et al. 6308* (PE); **Muli**, Mahuanggou, 3350 m, 25 Jul. 1978, *Q. S. Zhao et al. 6781* (CDBI, SZ); Baishawan, 2600 m, 21 Aug. 1978, *Q. S. Zhao 8595* (CDBI, SZ); near Yazui, 3300 m, 1 Jul. 1978, *Y. B. Yang 7142* (CDBI); mountains of Kulu, E Muli Gomba, 3400 m, Jun. 1928, *J. F. Rock 16498* (E); mountain of Kulu, 4000 m, Jun. 1929, *J. F. Rock 17991* (E); mountains of Yetsi, Sep.–Oct. 1932, *J. F. Rock 24443* (BM, E, K); Muli Area, *G. Forrest 30642* (BM, E); mountains around Muli, Jul. 1930, *G. Forrest 28415* (BM, E, PE); Kulu, 3300 m, 14 Sep. 1937, *T. T. Yü 14271* (BM, E, KUN); Yazhui Linchang, 3100 m, 11 Jun. 1984, *Q. S. Zhao & J. B. Shi 122398* (SZ); Wachin, 3200 m, 9 Jun. 1937, *T. T. Yü 5959* (KUN, PE); Wachin, 3250 m, 12 Jun. 1937, *T. T. Yü 6176* (KUN, PE); Kopatan, 2500 m, 5 Jul. 1937, *T. T. Yü 7243* (KUN, PE); **Shangri-la** (Zhongdian), Chungtien Area, *G. Forrest 30842* (E); Habaxueshan, Xiabazhi to Habanuo, 3100 m, 30 Jun. 1963, *Exped. Zhongdian 63-2946* (KUN); Yanyuan, Mianyashan, 3000 m, 9 Jun. 1984, *Q. S. Zhao & Z. F. Wang 122382* (SZ); **Yanyuan**, Heshang, Maoniushan, 3450 m, 9 Jun. 1960, *S. Jiang et al. 5905* (KUN, PE, SZ); **Yulong** (Lijiang), Yuhu, Shihuiyao, 2910 m, 27 May 1988, *S. J. Pei 1573* (KUN); 27 Mar. 1934, *H. D. McLaren 36* (BM, E, K); Lichiang Range, *H. D. McLaren s. n.* (K); E slope of Likiang Snown Range, 2895–3050 m, May–Oct. 1922, *J. F. Rock 3482* (BM, E, K, US), Peshmee Po, E slope of Likiang Snown Range, May–Oct. 1922, *J. F. Rock 3560* (E, P, US), Sa Ba, E slope of Likiang Snown Range, 3350 m, May–Oct. 1922, *J. F. Rock 3587* (E, US), Sa Ba road, *J. F. Rock 11412* (US), E slopes of the Likiang snow range (Yulung Shan), Mt. Satseto, 3350 m, Jun.–Aug. 1932, *J. F. Rock 24961* (E, K); Lao-chun Shan, south of Shi-ku and the Yangtze, 3200 m, May 1932, *J. F. Rock 25109* (E); supra vic. Nguluko prope urbem Likiang, 2900 m, *H. Handel-Mazzetti 6637* (W); E flank of the Lichiang Range, 3050–3350 m, Jun. 1906, *G. Forrest 2338* (BM, E, K); E flank of Likiang Range, Jun. 1910, *G. Forrest 5774* (BM, E, IBSC, K, P, PE, TI); Yulongshan, below Gang Ho Ba, 2800 m, flower pale yellow, 23 May 1985, *D. Chamberlain et al. KEY47* (E); Yulongshan, Haligou to Wenghai, 3300 m, 5 Jun. 1985, *D. Chamberlain et al. KEY532* (E); Yulongshan, below Mahuangba, 3250 m, 2 Jun. 1987, *D. Chamberlainet al. SBL354* (E, K); 3500 m, *C. Schneider 2816* (K); Likiang Snow Range, 2600 m, 15 Jun. 1937, *T. T. Yü 15175* (BM, E, GH, KUN); Lichiang Snow Range, 3000 m, 18 Juy 1937, *T. T. Yü 15328* (BM, E, GH, KUN, PE); Likiang hsien, 2300 m, *C. W. Wang70733* (GH, KUN, PE); s. loc., 2300 m, *C. W. Wang 70897* (GH, IBSC, KUN, PE, WUK);Yulongshan, Muzhugou to Ganheba, 2800 m, 23 May 1985, *X. Zhou 1029*

Figure 46-123: **Incarvillea lutea** Bureau & Franch.: **a**, habit; **b**, inflorescence; **c**, calyx; **d**, corolla opened showing stamens and distal portion of style; **e**, frutescence. Drawn by Bao-Fu ZHANG; reproduced from *Fl. Reipubl. Popularis Sin.* 69: 37, pl. 9: 1–3. 1990.

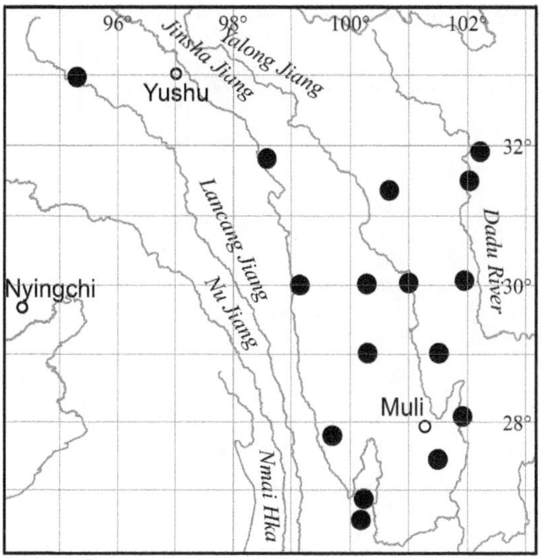

Map 46-217: **Incarvillea lutea** Bureau & Franch.

(KUN); Yulongshan, Anzhage, 3200 m, 26 May 1962, *A. L. Zhang 100292* (KUN); Jiaguxueshan, Siluo, 3500–3800 m, 29 Aug. 1942, *K. M. Feng 9183* (IBSC, KUN, PE); Yulongxueshan, 3100 m, 17 Jul. 1959, *K. M. Feng 22427* (KUN, PE); Xuesong, 2900 m, 2 Sep. 1958, *W. T. Wang 116* (KUN); Likiang range, 29 May 1939, *R. C. Ching 30199* (GH, KUN, PE); Likiang range, 8 Oct. 1940, *R. C. Ching 30693* (GH, KUN); Xueshan, Oct. 1940, *R. C. Ching 31124* (KUN, PE).

6. Incarvillea altissima Forrest, Notes Bot. Gard. Edinburgh 13: 164. 1921; Tao & Yin in Wang (ed.), Fl. Reipubl. Popularis Sin. 69: 40. 1990; Zhang & Santisuk in Wu & Raven (eds.), Fl. China 18: 222. 1998. Type: [S Hengduan] China, Yunnan, Lijiang, "on the Tong-shan", 27°20′N, 2800 m, Aug. 1914, *G. Forrest 13135* [lectotype (as "holotype") designated by Grierson, Notes Roy. Bot. Gard. Edinburgh 23: 338. 1963: E (E00135986); isolectotypes: E (E00135987, E00135988)].

Herbs perennial, 1–2 m high, glabrous. Leaves pinnate, 7–8 cm long; lateral leaflets 2–3 pairs, subequal, blade ovate, 8–12 cm long, 5–7 cm broad, base rounded to truncate, margin serrate, apex acute; terminal leaflet blade elliptic, 5–7 cm long, 3–4 cm broad, base obliquely cordate or cuneate. Racemes terminal, 40–65 cm long, 10–30-flowered. Bracts linear, 1.2–2.5 cm long; bracteoles linear. Pedicel 5–10 mm long. Calyx campanulate, 1.5–1.7 cm long; teeth semiorbicular, 2–2.5 mm long, 5–6 mm broad, apex rounded to truncate, mucronate. Corolla pale red, 5–7 cm long; lobes orbicular, 1.6–2.5 cm long, 2.5–2.8 cm broad, glandular pubescent, apex truncate, mucronate. Staminode 1.2–1.5 mm long. Ovary glabrous. Capsules long horned, 5–6 cm long, 4-angular, apex acuminate. Seeds broadly ovoid, 4.5–5 mm long, 3.5–4 mm broad, brown, surrounding by a wing; wing 1–2 mm broad. **Chromosomes** 2n = 22.

Phenology: flowering from June to August; fruiting September to November.

Habitats: slopes, thickets, forests; 2800–4000 m.

Distribution: endemic; L Yarlung Zangbo, N Hengduan, S Hengduan (Map 46-218).

L YARLUNG ZANGBO: **Gyaca**, Budanglashan, 4000 m, 19 Sep. 1999, *J. S. Yang 583* (KUN). **N HENGDUAN**: **Dawu**, Bamei, Qinda, 4 Jul. 1940, *K. L. Chü 7300* (PE); **Xinlong**, Zituoxi, 3950 m, roadside, 26 Jun. 1974, *J. F. Wang 6393* (CDBI, PE). **S HENGDUAN**: **Ninglang**, mountains east of Yung-ning, Jul. 1926, *G. Forrest 20584* (E, K); **Yongsheng**, Yungpeh, Jul. 1918, *G. Forrest 16651* (E, K); **Yulong** (Lijiang), Region of Tungshan, *J. F. Rock 10455* (E); Tongshan, 2800 m, Aug. 1914, *G. Forrest 13135* (E); Tongshan, Sep. 1918, *G. Forrest 17062* (E); Tungshan, E of Likiang, Aug. 1923, *J. F. Rock 10455* (E, US), Latu, region of Tungshan, *J. F. Rock 10484* (US).

7. Incarvillea beresowskii Batalin, Trudy Imp. S.-Peter-burgsk. Bot. Sada 14: 181. 1895; Tao & Yin in Wang (ed.), Fl. Reipubl. Popularis Sin. 69: 42. 1990; Zhang & Santisuk in Wu & Raven (eds.), Fl. China 18: 222. 1998. Type: [N Hengduan] China, Sichuan, Sungpan ting, 5 May 1894, *M. Beresowsky s. n.* [holotype: LE; isotypes: K (K000779214), PE].

Incarvillea longiracemosa Sprague, Bull. Misc. Inform. Kew 1907: 320. 1907. ≡ *Incarvillea*

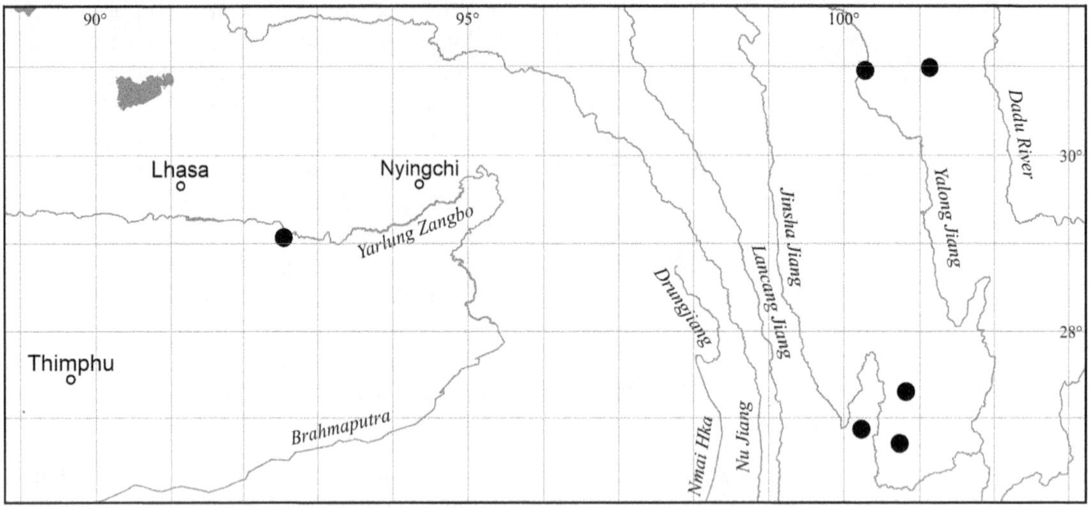

Map 46-218: **Incarvillea altissima** Forrest

lutea subsp. *longiracemosa* (Sprague) Grierson, Notes Roy. Bot. Gard. Edinburgh 23: 335. 1961. Type: [M Yarlung Zangbo] China, Xizang, between Phari & Shigatza, 1882, *King's collector s. n.* [holotype: K (K000779205)].

Incarvillea wilsonii Sprague, Bull. Misc. Inform. Kew 1912(1): 43. Type: [N Hengduan] China, Sichuan, Hsao Chin Ho valley, near Monkong Ting, 2100–2400 m, *E. H. Wilson 3058* [holotype: K (K000779215, K000779216)].

Herbs perennial, 0.5–1 m high, glabrous. Leaves pinnate; lateral leaflets 3–6 pairs, blade lanceolate-elliptic, 6–7 cm long, 1.5–2 cm broad, base obliquely cuneate, margin crenate-serrulate, apex obtuse or acute; terminal leaflet blade elliptic, 2–7 cm long, 1–3.2 cm broad, margin entire or serrate, apex acute. Racemes terminal, more than 17-flowered, 40–55 cm long. Bracts lanceolate, 0.5–1.5 cm long; bracteoles linear. Pedicel 5–10 mm long. Calyx campanulate, yellow-green, 1.5–3.0 cm long, 0.8–1.5 cm broad; teeth broadly triangular, 3–4 mm long, 4–7 mm broad, apex acute. Corolla rose or red, patent; tube 3.5–5 cm long, base constricted; lobes ovate, 1.6–2 cm long, 1.4–2.5 cm broad. Filaments yellow-green; anthers yellow. Ovary 1.6–2 mm long; style pale yellow; stigma compressed, 2-lobed. Capsules quadrangular, 8–10 cm long, 1–1.3 cm broad, dehiscing by 2 leathery valves, apex acute. Seeds pale brown, 4–4.5 mm long, 2.6–3 mm broad, puberulent, surrounded by a wing; wing 0.8–1.2 mm long, crenulate. **Chromosomes** 2n = 22. Figure 46-124.

Phenology: flowering from May to July; fruiting from July to October.

Habitats: slopes; 300–4500 m.

Distribution: **endemic**; U Yarlung Zangbo, M Yarlung Zangbo, L Yarlung Zangbo, Yarlung Zangbo-Brahmaputra, N Hengduan, S Hengduan (Map 46-219).

U YARLUNG ZANGBO: **Tingri**, Karta, 3600 m, 17 Jun. 1935, *F. Kingdon-Ward 11729* (BM). **M YARLUNG ZANGBO**: **Lhasa**, Dechen Dzong, 2 miles E of hasa, 3900 m, 23 Jun. 1943, *F. Ludlow & G. Sherriff 9626* (BM, E); vicinity of Lhasa, 3900 m, Jun. 1939, *H. E. Richardson 209* (BM, E); Katuosi, 4500 m, 19 Aug. 1965, *Y. T. Zhang & K. Y. Lang 1959* (PE); **Lhozhag**, Lakang, Kajisi, 18 Aug. 2013, *Y. S. Chen et al. 13-1542* (PE); **Nagarzê**, Yam Dok Cho, 4500 m, Aug. 1904, *H. J. Walton s. n.* (K); YamdrokTso, 4500 m, 12 Jul. 1975, *C. Y. Wu et al. 75-751* (HITBC, HNWP, KUN, PE); between **Phari** & **Shigatza**, 1882, *King's collector s.n.* (K). **L YARLUNG ZANGBO**: **Gongbo'gyamd**, 24 Jul. 1976, *Exped. Qinghai-Xizang 12467*

Figure 46-124: **Incarvillea beresowskii** Batalin: **a**, proximal portion of plant; **b**, flowering branch; **c**, stamen. Drawn by Zhang-Hua WU; reproduced from *Fl. Reipubl. Popularis Sin.* 69: 43, pl. 11. 1990.

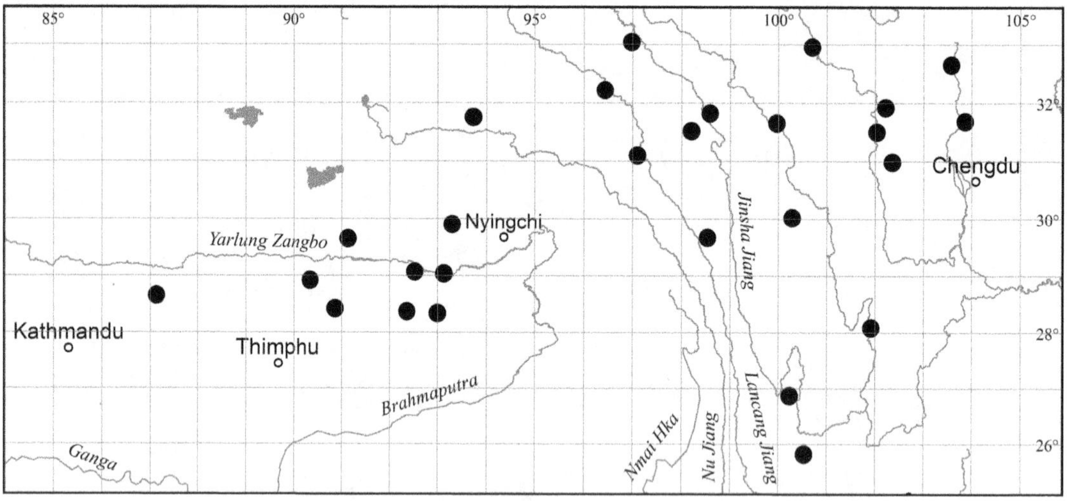

Map 46-219: **Incarvillea beresowskii** Batalin

(HNWP, KUN, PE); **Gyaca**, Poshishan, 4210 m, 25 Jul. 1975, *Suppl. Exped. Qinghai-Xizang 750709* (HNWP, KUN, PE); **Nangxian**, Zhang, 24 Jul. 1997, *B. Lyon 11066* (E); Laduo Qu, 3300 m, 24 Aug. 1972, *Xizang Medic. Herd Exped. 4513* (HNWP, PE). **YARLUNG ZANGBO-BRAHMAPUTRA**: **Chayul Chu**, Karutra, 3900 m, 12 Jul. 1936, *F. Ludlow & G. Sherriff 2342* (BM); **Lhünzê**, Sauga Chöling, 3750 m, 29 Jun. 1936, *F. Ludlow & G. Sherriff 2252* (BM); Kyimdong Dzong, 3600 m, 18 Jun. 1936, *F. Lidlow & G. Sherriff 1827* (BM); Kyimdong Dzong, 3750 m, 26 Sep. 1936, *F. Ludlow & G. Sherriff 2628* (BM); Kyimdong Dzong, Tsangpo Valley, 360 m, 18 Jun. 1936, *F. Ludlow & G. Sherriff 1827* (E). **TANGUT**: **Baima**, valley south of Baima along Makehe River, 3305 m, 9 Jul. 2000, *Sino-British Qinghai Alpine Garden Society Expedition (SQAE) 850* (E); **Dêgê**, Gongya, 3075 m, 7 Jun. 2010, *L. M. Gao & Z. M. Cai GLM-103008* (KUN); **Garzê**, Dongggu, 3600 m, 1974, *Q. S. Zhao et al. 111084* (CDBI, PE, SZ, WUK); **Jomda**, Gangtuo, 3450 m, 19 Jun. 1974, *s. coll. 6958* (CDBI, PE); Gangtuo to Tuoba, 3400 m, 16 Jun. 1974 *Exped. Qinghai-Xizang 74-3538* (KUN, PE); **Nangqên**, Wuguan, 3800 m, *Y. C. Yang 00964* (HNWP, PE); **Qamdo**, Karub, Mianda, 3545 m, 1 Sep. 2008, *B. Q. Xu et al. Xianh0179* (IBSC); **Riwoqê**, W side of Zhuoga-La Pass, 3920 m, 13 Aug. 2004, *D. E. Boufford et al. 32340* (TI); **Soxian**, 4000 m, 15 Aug. 1976, *D. D. Tao 10899* (PE); **Yushu**, E side of Zi Qu, NE of Jiangxi Forests Station and SE of Mozhong, 3570 m, 30 Aug. 1996, *T. N. Ho et al. 2709* (BM, E, HNWP, PE, TI); Xiewu, 3750 m, 5 Jul. 2005, *Y. H. Wu 32559* (HNWP); near Tongtianhe, 8 Jun. 1985, *W. Y. Wang et al. 00062* (HNWP); Gyêgu Lake, 3700 m, 8 Jul. 1977, *S. W. Liu 2242* (HNWP); Xiao Sumang, Jiangxi Linchang, Lalonggou, 3660 m, 1 Aug. 2010, *Kham Expedition 10-1085* (PE); N of the city, 3600 m, 30 Jun. 1995, *D. E. Boufford et al. 26809* (TI). **N HENGDUAN**: **Barkam**, Maomaochu, 3400 m, 16 May 1957, *H. Li 71049* (PE, SZ); **Gonjo**, Luosng Dengzgu, 3750–3900 m, 1 Aug. 2009, *D. E. Boufford et al. 41674* (TI); **Jinchuan**, Zuosijia, 2700 m, 24 May 1961, *S. Jiang et al. 8885* (PE); **Litang**, road to Mount Genyen, 4077 m, 4 Sep. 2010, *Kunming & Edinburgh Expedition to Sichuan 178* (E); Litang Qu Valley between Jiawa and Mula, 4022 m, 6 Sep. 2010, *Kunming & Edinburgh Expedtion to Sichuan 187* (E); **Maoxian**, Min Valley near Mao chon, 2438 m, *E. H. Wilson 4292* (K); **Markam**, Haitong, 3670 m, 17 Aug. 2010, *Kham Expedition 10-2109* (PE); Rumei to Zogang, 4193 m, 7 Jun. 2013, *X. X. Huang & Z. B. Zhong 15789* (IBSC); between Mangkang and the Jinsha Jiang, along the Xi Qu, 3600–3650 m, 16 Jul. 2009, *D. E. Boufford et al. 40563* (TI); **Songpan**, Sungpan ting, 5 May 1894, *M. Beresowsky s. n.* (K, LE, PE); Minjiang, Guihuagou, 2390 m, 5 Jun. 1962, *s. coll. 725* (CDBI); **Xiaojin**, valley of Hsiao Chin Hoo, 2100–2400 m, Jun. 1908, *E. H. Wilson 3058* (E). **S HENGDUAN**: **Bingchuan**, 2400 m, 19 Jul. 1933, *H. T. Tsai 52984* (IBSC, KUN, PE, SZ); **Muli**, Qianggang, 3520 m, 6 Jun. 1982, *Z. X. Tang 0075* (PE); **Yulong** (Lijiang), Shitougou, 3000 m, 31 May 1986, *H. Li 758* (KUN).

8. Incarvillea dissectifoliola Q. S. Zhao, Acta Phytotax. Sin. 26: 78. 1988; Zhang & Santisuk in Wu & Raven (eds.), Fl. China 18: 221. 1998. Type: [S Hengduan] China, Sichuan, Yanyuan, on the slope of Xiaogao, 3000 m, 8 Jun. 1984, *Q. S. Zhao & Z.F. Wang 122377* (holotype: SZ).

Herbs perennial, stemless. Rhizome fleshy, robust. Leaves basal, 2-pinnately divided; rachis 20–30 cm long, 1-sulcate, puberulent; lateral leaflets 7 or 8 pairs, opposite or subopposite, sessile, blade lanceolate or ovate lanceolate, 3–6 cm long, 2–3 cm broad, puberulent, margin pinnatifid, crenate. Racemes 30–50(–80) cm long, 10-sulcate, 5–8-flowered. Bracts lanceolate, 1–1.2 cm long; bracteoles linear to lanceolate, 3–5 mm long. Pedicel 0.5–1.5 cm long. Calyx broadly campanulate, 1–1.5 cm long, 5-veined; lobes triangular, 5–7 mm long, 6–8 mm broad, puberulent, apex reflexed mucronate. Corolla narrowly funnel-form; tube purple outside, 3.5–4 cm long, 4–5 mm broad, 1.2–1.5 cm broad; lobes suborbicular, purplish, 1–1.2 cm in diameter. Filaments inserted at base of corolla tube. Ovary glabrous. Stigma flabellate. Capsules unknown. **Chromosomes** 2n = 22. Figure 46-125.

Phenology: flowering from June to July.

Habitats: slopes; ca. 3000 m.

Distribution: endemic; S Hengduan (Map 46-220).

S HENGDUAN: **Yanyuan**, Xiaogao, 3000 m, 8 Jun. 1984, *Q. S. Zhao & Z. F. Wang 122377* (SZ).

9. Incarvillea compacta Maxim., Bull. Acad. Imp. Sci. Saint-Petersbourg 27: 521. 1881; Tao & Yin in Wang (ed.), Fl. Reipubl. Popularis Sin. 69: 49. 1990; Zhang & Santisuk in Wu & Raven (eds.), Fl. China 18: 223. 1998. Type: [Tangut] China, Tangut, *N. M. Przewalski s. n.* [holotype: LE; isotypes: K (K000779209), PE].

Incarvillea bonvalotii Bureau & Franch., J. Bot. (Morot) 5: 141–142. 1891. Type: China, “Thibat, Sur la route de Lhassa à Batang”, 22 May 1890, *P. G. E. Bonvalot s. n.* [holotype: P (P00608144)].

Incarvillea compacta var. *qinghaiensis* Grey-Wilson, New Plantsman 5: 84. 1998, **syn. nov**. Type: China. Qinghai, Zekog, Tangzhen Area, Doufudong, 35°7′8″N, 101°42′34″E, 3680 m, 17 Jul. 1997, *Sino-British Qinghai Expedition 1997 No. 790* [holotype: E (E00136349)].

Herbs perennial, 20–30 cm high, shortly stemed. Roots fleshy, conical, 15–23 cm. Leaves pinnate, clustered at stem base, 8–15 cm long; lateral leaflets 2–6 pairs, blade ovate, 2–3.5 cm long, 1–2 cm broad, base rounded, margin entire, apex acuminate; terminal leaflets ovate, 0.7–4 cm long, 0.3–2 cm broad, margin entire, apex rounded. Racemes terminal, densely flowered; flowers congested at apex of rachis. Bracts 1.8–3 cm long. Pedicel 1–4 cm long, linear. Calyx 1.2–1.8 cm long, green or purple-red, with dark purple spots, campanulate; teeth triangular, 6–12 mm long, 5–7 mm broad. Corolla red or purplish-red, 3.5–4 cm long, 1.8–2 cm broad; lobes orbicular or oblate, 1.7–2.8 cm long, 2–3.9 cm broad, apex emarginate, with glands. Stamens inserted at base of corolla tube; anthers adherent; staminode small, curved. Style 3.5–4 cm long; stigma flabellate. Capsules narrowly lanceolate, distinctly quadrangular, 7–11 cm long, 0.8–1.3 cm broad, acute at both ends. Seeds dark greyish-brown, 3.5–5 mm long, 2–3.5 mm broad; wings 0.4–0.6 mm broad. Figure 46-126.

Phenology: flowering from May to July; fruiting from August to December.

Habitats: slopes, thickets, grasslands; 2700–4900 m.

Distribution: endemic; L Yarlung Zangbo, Tangut, N Hengduan, S Hengduan (Map 46-221).

L YARLUNG ZANGBO: **Gyaca**: Gyaca Shan, 3800 m, 17 Aug. 1988, *S. W. Liu 3470* (HNWP). **TANGUT**: **Baima**, Zhiqin Valley, 3575 m, 7 Jul. 2000, *Sino-British QInghai Alpine*

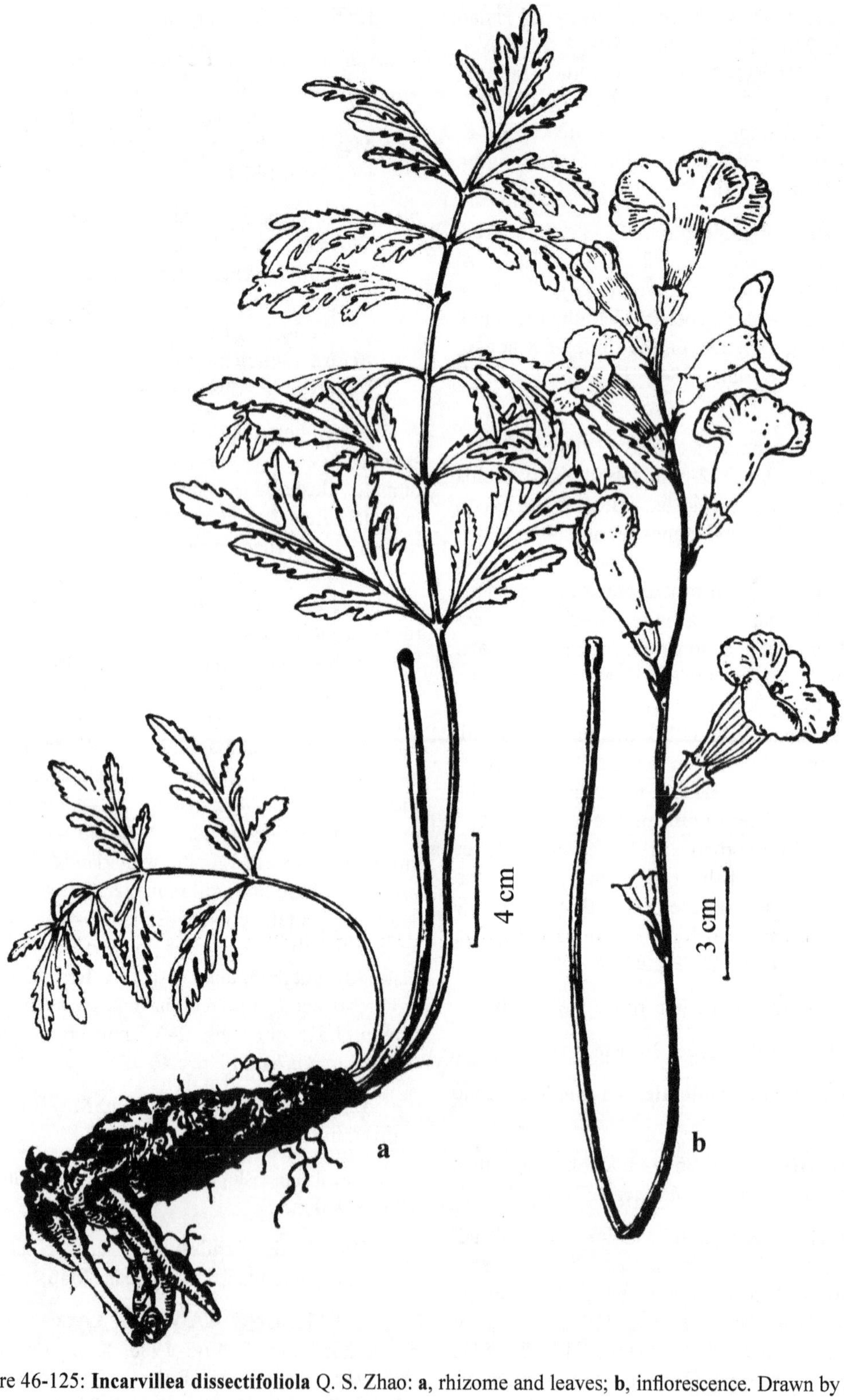

Figure 46-125: **Incarvillea dissectifoliola** Q. S. Zhao: **a**, rhizome and leaves; **b**, inflorescence. Drawn by Jin-Bin SHI; reproduced from zhao, *Acta Phytotax. Sin.* 26: 79, fig. 1. 1983.

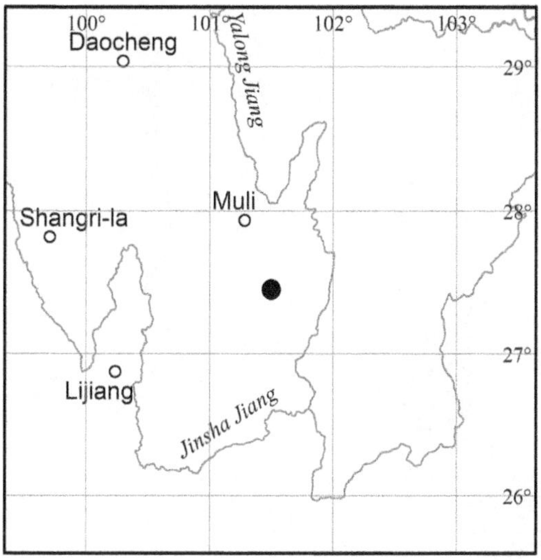

Map 46-220: **Incarvillea dissectifoliola** Q. S. Zhao

Garden Society Expedition (SQAE) 691 (E); **Chindu**, Zhujiesi, 4300 m, 4 Jul. 2005, *Y. H. Wu 32534* (HNWP); Zhujiesi, 4300 m, 4 Jul. 2005, *Y. H. Wu 32548-A* (HNWP); s. loc., 4460 m, 10 Aug. 1978, *P. C. Kuo 423* (HNWP); Oba, 3500 m, 4 Jul. 1995, *D. E. Boufford et al. 26981* (TI); Chewu, 4350 m, 2 Jul. 1995, *D. E. Boufford et al. 26915* (TI); **Garzê**, Laima, 3775 m, 8 Jun. 2010, *L. M. Gao & Z. M. Cai GLM-103019* (KUN); **Jomda**, Jomda, Sulaicong, 3490 m, 7 Jun. 2010, *L. M. Gao & Z. M. Cai GLM-10307* (KUN); Qu'nyido, Yulong, 4310 m, 7 Jun. 2010, *L. M. Gao & Z. M. Cai GLM-103004* (KUN); Tuoba, 4000 m, 30 Jun. 1965, *Y. T. Zhang & K. Y. Lang 206* (PE); Gangtuo, 3300 m, 31 May 1973, *Expeed. Qinghai-Xizang 73-28* (PE); **Nagqu**, W of the city, 4800 m, 26 Aug. 1990, *J. S. Yang 90-423* (IBSC); N Amdo, 5000 m, 2 Aug. 1989, *R. F. Huang 89-041* (HNWP); Amdo, Xigetangcuo, 4700 m, 25 Aug. 1989, *Y. P. Shen 89-848* (HNWP); **Nangqên**, Kandaxia, 3800 m, 20 May 1999, *Y. H. Wu 17189* (HNWP); Maozhuang, 3900 m, 15 Jun. 1965, *Y. C. Yang 927* (KUN, PE); Banwu, 3800 m, 16 Jun. 1965, *Y. C. Yang 964* (KUN); **Qamdo**, Toba, 4200 m, 10 Aug. 1976, *Exped. Qinghai-Xizang 12750* (HNWP, KUN, PE); **Qumarlêb**, Qiuzhi, 4500 m, 18 Jul. 1978, *H. T. Cui s. n.* (KUN); **Sêrxü**, Hongqi, 4200 m, 22 Jun. 2004, *Y. H. Wu 59462* (HNWP); Derongma, Longren, 4240 m, 27 Jul. 2005, *D. E. Boufford et al. 33590* (TI); **Sogxian**, 4000 m, 15 Aug. 1976, *Exped. Qinghai-Xizang 10897* (HNWP, KUN, PE); **Yushu**, Baizha Linchang, 4100 m, 14 Jun. 1985, *W. Y. Wang et al. 0176* (HNWP); Baizha Linchang, 4300 m, 16 Jul. 1985, *W. Y. Wang et al. 224* (HNWP); Baizha, 3900 m, 18 Jun. 1985, *W. Y. Wang et al. 361* (HNWP); Baizha, 4100 m, 20 Jun. 1985, *W. Y. Wang et al. 470* (HNWP); Batang, 4200 m, 23 Jun. 1971, *P. C. Kuo & W. Y. Wang 8362* (HNWP); Batang, 4100 m, 23 Jun. 1971, *P. C. Kuo & W. Y. Wang 8364* (HNWP); Shangbatang, Paerkala, 4100 m, 14 Jun. 2964, *s. coll. 37* (HNWP); Shangbatang, 4300 m, 15 Jun. 1964, *Y. C. Yang 73* (HNWP, PE); Jiangxigou, 4000 m, 1 Jul. 1971, *P. C. Kuo & W. Y. Wang 8424* (HNWP); **Zadoi**, 4150 m, 19 Jun. 1965, *S. W. Liu 0030* (HNWP, PE, WUK); Angsai, 4600 m, 10 Jul. 2005, *Y. H. Wu 33345* (HNWP); Angsai, 4600 m, 10 Jul. 2005, *Y. H. Wu 33350* (HNWP); Angsai, 4100 m, 14 Jul. 2005, *Y. H. Wu 33924* (HNWP); Angsai, Guodaogou, 4100 m, 14 Jul. 2005, *Y. H. Wu 33934* (HNWP); Angsai, Longdengshan, 4585 m, 15 Jul. 2005, *Y. H. Wu 34188* (HNWP); Angsai, Longdengshan, 4585 m, 15 Jul. 2005, *Y. H. Wu 34207* (HNWP); Angsai, Longdengshan, 4585 m, 15 Jul. 2005, *Y. H. Wu 342229* (HNWP); Sulu, 4260 m, 19 Jul. 2005, *Y. H Wu 34287* (HNWP); Sulu, 4260 m, 19 Jul. 2005, *Y. H. Wu 34457* (HNWP); Jieduo, 4445 m, 21 Jul. 2005, *Y. H. Wu 35121* (HNWP); Zhaqing, Diqing, 4460 m, 24 Jul. 2005, *Y. H. Wu 35060* (HNWP); Zhaqing, Ranzhegala Shenshan, 25 Jul. 2005, *Y. H. Wu 35186* (HNWP); Aduo, 4600 m, 6 Aug. 2005, *Y. H. Wu 35374* (HNWP); Aduo, 4600 m, 6 Aug. 2005, *Y. H. Wu 35396* (HNWP); Aduo, Aquqiao, 4660 m, 7 Aug. 2005, *Y. H. Wu 36075* (HNWP); Aduo, Aquqiao, 4660 m, 7 Aug. 2005, *Y. H. Wu 36090* (HNWP); **Zhidoi**, s. loc., 4400 m, 10 Jun. 1966, *L. H. Zhou 54* (HNWP). **N HENGDUAN**: **Barkam**, Shuajingsi, 3800 m, 12 Jun. 1958, *P. X. Li 10189* (PE); **Batang**, 8 Jun. 1893, *Kashkarow s. n.* (K); **Dawu**, Yuke Qu, 4300 m, 18 Jun. 1974, *Y. T. Wu & B. C. Gao 111585* (CDBI, PE, SZ, WUK); Qianning, 1 Jul. 1977, *P. C. Kuo & W. Y. Wang 20713* (HNWP, KUN); Qianning, 1 Jul. 1977, *P. C. Kuo & W. Y. Wang 20771* (HNWP); **Dêgê**, Dama, Yandagou, *S. X. Jia 216* (PE); Dama, Yandagou, *S. X. Jia 221* (PE); Dumuling, 3800 m, 12 Jun. 1977, *Sichuan Econ. Pl. Exped. 15074* (CDBI); Manigange, 3900 m, 23 Jun. 1968, *Sichuan Econ. Pl. Exped.*

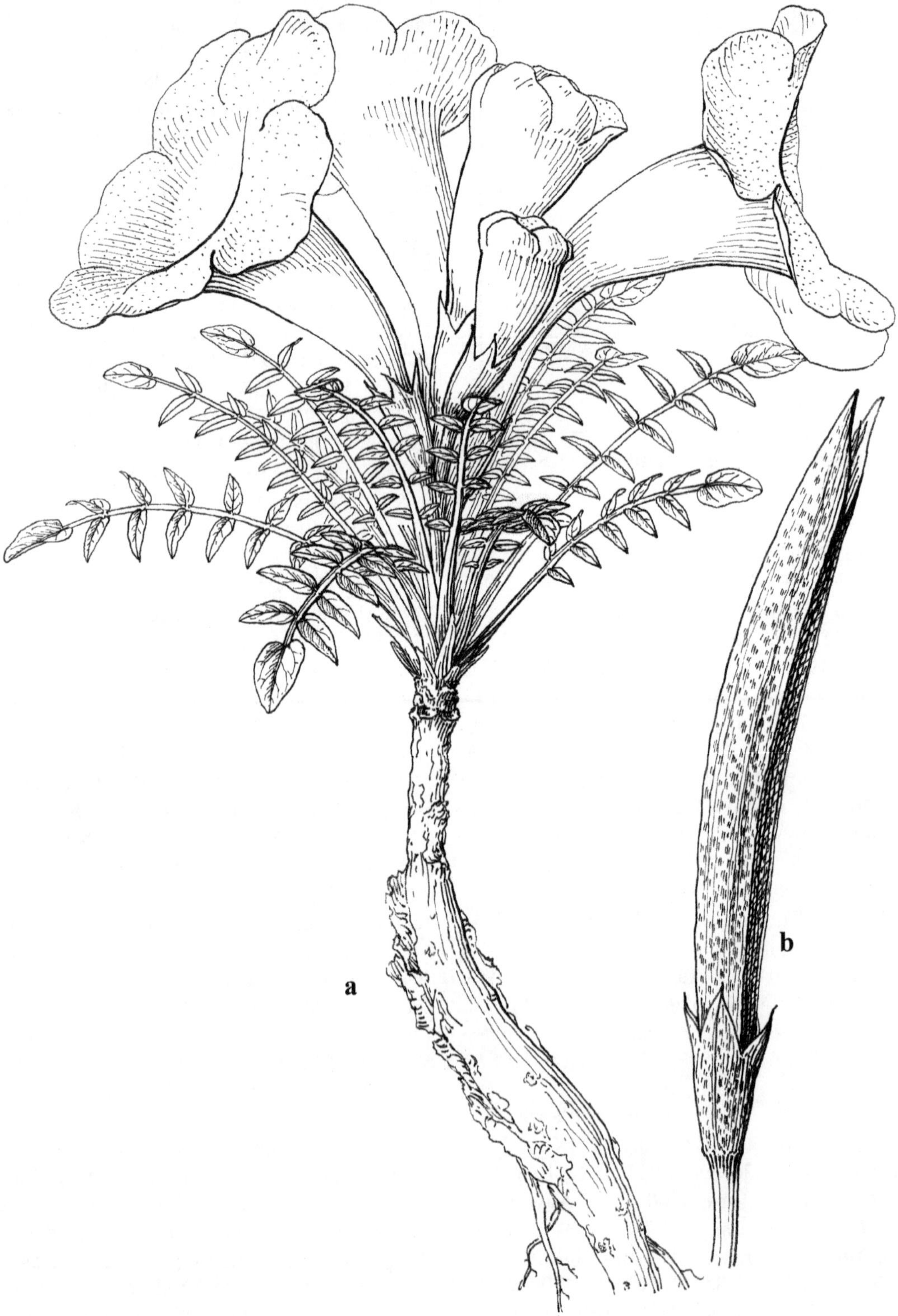

Figure 46-126: **Incarvillea compacta** Maxim.: **a**, habit; **b**, capsule. Reproduced from *Icon. Cormophyt. Sin.* 4: 100, fig. 5614. 1975.

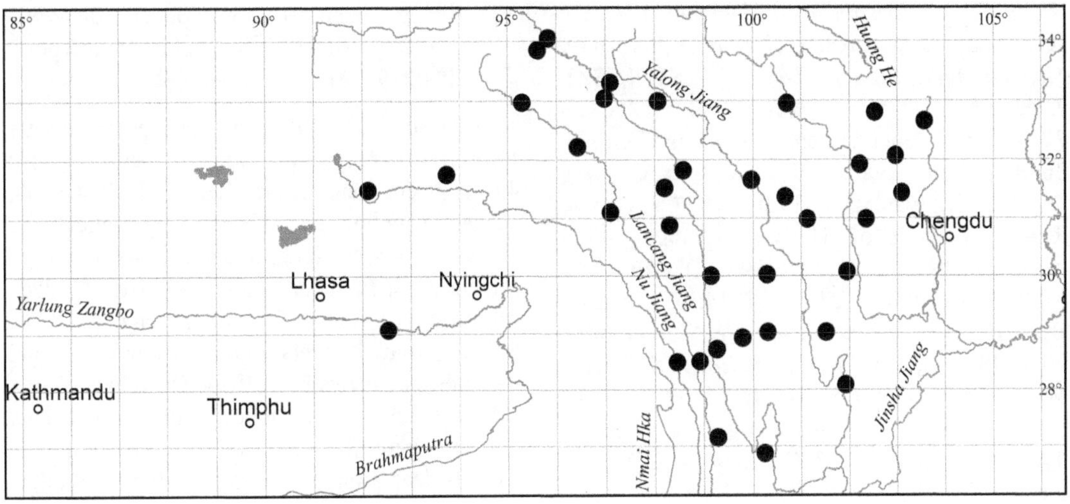

Map 46-221: **Incarvillea compacta** Maxim.

2 (CDBI); Babang, Shangbawu, 5 Jul. 1974, *Sichuan Veget. Exped. 7302* (CDBI); 4000 m, 8 Jul. 1961, *S. Jiang et al. 9247* (PE); Songlinkou, 15 Jul. 1951, *Y. W. Cui 4456* (PE); **Garzê**, 3800 m, 8 Jun. 1974, *Q. S. Zhao et al. 111114* (CDBI, SZ, WUK); Zuodashan, 4600 m, 9 Sep. 1965, *K. T. Fu 16550* (WUK); **Gonjo**, 4000 m, 6 Jul. 1991, *J. S. Yang 553* (KUN); 3800 m, 13 Aug. 2010, *Kham Expedition 10-1746* (PE); **Heishui**, Shidiaolou, Kuguazhai, 3000 m, 27 May 1959, *Heishui Group 1225* (CDBI, KUN, PE); **Hongyuan**, Amukehe, 3700 m, 3 Jun. 1964, *B. J. Ni et al. 42* (CDBI); road from Hongyuan to Jighzi, 3675 m, 5 Jul. 2000, *Sino-British Qinghai Alpine Garden Society Expedition (SQAE) 674* (E); **Litang**, Litang Qu valley, 4021 m, 6 Sep. 2010, *Kunming & Edinburgh Expedition to Sichuan 194* (E); N slope of pass on Jiawa-Dewu road, 4400 m, 7 Sep. 2010, *Kunming & Edinburgh Expedition to Sichuan 214* (E); Qudeng, Huncha, 4225–4775 m, 26 Jul. 2010, *D. E. Boufford et al. 42310* (TI); **Lixian**, Miyaluo, Jiabi Linchang, 3600–4200 m, *S. Jiang & C. L. Jin 850* (KUN, PE, SZ); **Luhuo**, Xiongjiling, 3500 m, 13 Jun. 1974, *Qinghai-Xizang Exped. 3517* (KUN, PE); **Songpan**, Top of Pass "Huangshen Guan", 3700 m, 1 Jul. 2000, *Sino-British Alpine Garden Society Expedition (SQAE) 577* (E); Top of Pass "Huangshen Guan", 3700 m, 1 Jul. 2000, *Sino-British Alpine Garden Society Expedition (SQAE) 579* (E); Galete, 3700 m, 16 Jun. 1983, *Q. S. Zhao & Z. J. Zhao 120515* (CDBI, SZ); Sep. 1937, *K. T. Fu 1351* (WUK); 25 Sep. 1937, *K. T. Fu 1937* (WUK); **Xiaojin**, Tianlong, Yangtianwan, 3800 m, 15 May 1959, *Z. G. Liu et al. (Chuan Jing A) 35* (CDBI, KUN, PE, SZ). **S HENGDUAN**: **Cawalong** (Tsawalung), Mountains of Londjre, Mekong-Salween watershed, 1923, *J. F. Rock 10368* (E); **Daocheng**, Sangdui Qu, 3900 m, 21 Jun. 1980, *s. coll. 28087* (CDBI); **Dêqên**, Baimaxueshan, 4500 m, 22 Aug. 1999, *N. Fujii et al. 615* (KUN); Bei-Ma Shan, 4200–4500 m, Sep. 1921, *G. Forrest 20180* (E); Baima Shan, 4322 m, 28 Jun. 1994, *Alpine Garden Society Expedition to China (1994) ACE 730* (E); Baima Shan, 4450 m, 4 Jun. 1993, *B. Alden et al. 812* (E); Baimashan, 6 Jul. 1938, *T. T. Yü 8781* (KUN, PE); Baimashan, Munela, 4060 m, 27 Jul. 1937, *T. T. Yü 9065* (KUN, PE); Auntze, Bai-Ma Shan, 4100 m, 15 Nov. 1937, *T. T. Yü 10777* (BM, E, KUN, PE); Peima Shan, Mekong-Yangtze divide, southeast of Atuntze, May–Jun. 1932, *J. F. Rock 22785* (BM, E, K); Baima Shan, 4050 m, Sep. 1932, *J. F. Rock 23383* (BM, E, K); Baimashan, 3400 m, Sep. 1935, *C. W. Wang 69443* (NAS, PE); Baimashan, 3500 m, Sep. 1935, *C. W. Wu 69544* (IBSC, PE); **Dêrong**, Mekong-Yangtse Divide, Quyagong, 4200 m, Oct. 1921, *G. Forrest 20611* (E); **Jiulong**, Karlong, 3400 m, 2 Aug. 1922, *Harry Smith 3981* (PE); Tanggu, Jizhuishan, 3800 m, 30 May 1979, *Q. Q. Wang 20070* (CDBI); Pass north of Jiulong on road to Take, 4600 m, 16 Sep. 2010, *Kunming*

& Edinburgh Expedition to Sichuan 327 (E); **Kangding**, Huangcaozhai, 2700 m, 19 Jun. 1959, *Sichuan Agriculture College 132* (KUN, SZ); Zheduoshan, 3700 m, 28 May 1980, *Z. Y. Chen 122551* (SZ); Gonggashan, Old Gongga Temple, 3630 m, 9 Jun. 1980, *Z. J. Zhao & Z. X. Xiong 112804* (SZ); Xinduqiao, 30 Jun. 1977, *P. C. Kuo & W. Y. Wang 20662* (HNWP); Yeniugou, 4000 m, 25 Jul. 1933, *T. H. Tu 4459* (PE); **Muli**, SW Muli, 4200–4500 m, 21 Jun. 1914, *F. Kingdon-Ward 4072* (E); Mount Mitzuga, W Muli Gomba, 3050–4875 m, Jun. 1928, *J. F. Rock 16077* (E); **Weixi**, Tsejao, 3600 m, 11 May 1937, *T. T. Yü 8230* (KUN, PE); **Xiangcheng**, Wumingshan, 4900 m, 26 May 1971, *Z. G. Liu 0004* (CDBI); s. loc., 4900 m, 26 May 1971, *Z. G. Liu 4* (CDBI); Xiangcheng to Rede, 4163 m, 9 Sep. 2010, *Kunming & Edinburgh Expedition to Sichuan 223* (E); Xiangcheng to Derong, 3930 m, 26 Jun. 1994, *Alpine Garden Society Expedition to China (1994) ACE 682* (E,K); Maanshan, 3 Aug. 1983, *Hengduan Veget. Group 4781* (PE); **Yulong** (Lijiang), N Lichiang valley, 2700 m, *Forrest 2131* (K); E flank of the Lichiang Range, 3300 m, *G. Forrest 2464* (K); s. loc., *J. F. Rock 2462* (IBSC).

10. Incarvillea himalayensis Grey-Wilson, New Plantsman 5(2): 88. 1998; Aitken in Grierson & Long (eds.), Fl. Bhutan 2(3): 1239. 2001. Type: Bhutan, Chojo Dzong, Upper Pho Chu, Eastern Branch, 13,500 ft., 3 Jul. 1949, *F. Ludlow, G. Sherriff & J. H. Hicks 16722* [holotype: BM; isolectotype: E (E00273398)].

Incarvillea grandiflora Bureau & Franch., J. Bot. (Morot) 5: 138. 1891, non *Incarvillea grandiflora* Poir., Dict. Sci. Nat., ed. 2. 23: 53 (1822). 1821, non *Incarvillea grandiflora* Spreng., Syst. Veg., ed. 16 2: 836. 1827. ≡ *Incarvillea compacta* Maxim. var. *grandiflora* Wehrh., Die Garten-Stauden 2: 947. 1931. ≡ *Incarvillea mairei* var. *grandiflora* (Wehrh.) Grierson, Notes Roy. Bot. Gard. Edinburgh 23: 344. 1961; Tao & Yin in Wang (ed.), Fl. Reipubl. Popularis Sin. 69: 46. 1990; Zhang & Santisuk in Wu & Raven (eds.), Fl. China 18: 223. 1998. Type: [N Hengduan] China, Entre Batang et Litang, gazon secs apres Batang, *Bonranot et Princfe Henri d'Orleans 21* [holotype: P (P00219425)].

Perennial, acaulescent. Leaves in rosette from fleshy root-stock, pinnate; petiole (1.5–) 4.0–10.0 cm long; lateral leaflets in 0–4 pairs, subsessile, petiolules 1–1.5 mm long, leaflet blade ovate, 0.7–4.0 cm long, 6–2.0 cm broad, margin serrulate, apex acute; terminal leaflet blade orbicular, (1.5–)2.5–4.0(–7.0) cm long, 3.0–4.5(–6.5) cm broad. Flowers solitary or few in terminal racemes; peduncles 2.2–2.8(–4.5) cm long. Calyx tube 1.0–1.5 cm long; lobes ovate to lanceolate, 3–6 mm long, 2–6 mm broad, apex acuminate. Corolla deep pink, base of tube yellow or orange, throat with white streaks; tube 4.0–6.5 cm long; lobes orbicular, 1.0–1.5(–2.0) mm long and broad. Capsules woody, quadrangular in section, curved, 4.5–7.0 cm long, 0.5–l.0 cm broad. Seeds subdiscoid or ovate, densely covered with thick glandular hairs, 5–7 mm long, 5–6 mm broad; wing narrow.

Phenology: flowering from June to July; fruiting from July to September.

Habitats: slopes, grasslands, mountain pastures; 2400–4700 m.

Distribution: endemic; W Nepal, Bhutan, L Yarlung Zangbo, Tangut, N Hengduan (Map 46-222).

W NEPAL: Karnali, Dolpa, Barbung Khola, Kaya Khola, 4050 m, 6 Jun. 1952, *O. V. Polunin, W. R. Sykes & L. H. J. Williams 1089* (E); Dolpa, near Rohagain, Suli Gad, 3690 m, 5 Jun. 1952, *O. V. Polunin, W. R. Sykes & L. H. J. Williams 2160* (E); Dolpa, Rohagain, Suli Gad, 3450 m, 14 Sep. 1952, *O. V. Polunin, W. R. Sykes & L. H. J. Williams 3399* (E); Dolpa, Jumla Karnali, 3300 m, 15 Jun. 1952, *O. V. Polunin, W. R. Sykes & L. H. J. Williams 4306* (E); Dolpa, Phijor, 4650 m, 22 Jun. 1963, *J. D. A. Stainton 4346* (E); Dolpa, Tarap, 4350 m, 3 Jul. 1963, *J. D. A. Satinton 4398* (E); s. loc., 3300 m, 11 Jun. 1980, *P. R. Shakya & B. Roy 5623* (KATH); Dolpa, Rahagaon to Rhyenje, 3200 m, 29 Jun. 1980, *P. R. Shakya & B. Roy 5979* (KATH); Dolpa, Phoksundo, 3690 m, 11 Jul. 1966, *T. B. Shrestha 5327* (KATH); **Rapti**, Rukum, near Dogadi Khola, 3900 m,

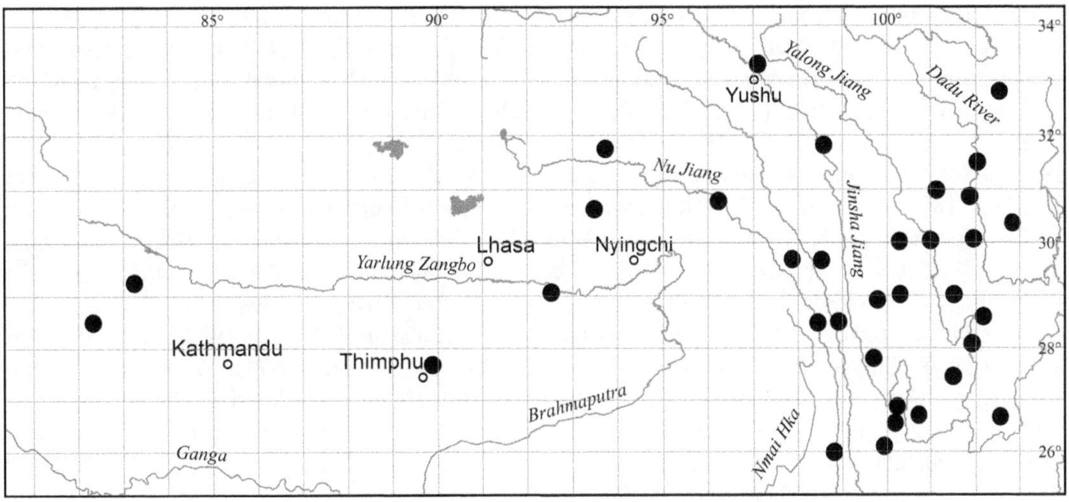

Map 46-222: **Incarvillea himalayensis** Grey-Wilson

20 Jun. 1954, *J. D. A. Stainton, W. R. Sykes & L. H. J. Williams 3201* (E). **BHUTAN**: **Upper Pho Chu**, Tranga, Gaffoo Na, 3900 m, 16 Sep. 1949, *F. Ludlow & G. Sherriff 17250* (BM). **L YARLUNG ZANGBO**: **Gyaca**, Dalasha, 4210 m, 25 Jul. 1975, *Suppl. Exped. Qinghai-Xizang 750758* (KUN, PE); Hongqi Qu, 4000 m, 19 Aug. 1972, *s. coll. 4271* (HNWP, PE); **Lhari**, Sangba, 4600 m, 6 Aug. 1976, *Exped Qinghai-Xizang 10480* (KUN). **TANGUT**: **Dêgê**, 14 Jul. 1977, *P. C. Kuo 21404* (HNWP, WUK); **Chindu**, Xiwu, S of Chumda, along E side of the Tongtian He (upper Changjiang), 3400 m, 16 Aug. 1996, *T. N. Ho et al. 1912* (BM, E); Maowaying, 4000 m, 19 Jul. 1977, *S. W. Liu 2390* (HNWP); **Sogxian**, 3950 m, 24 Aug. 2015, *J. W. Zhang et al. ZJW-3405* (KUN); Jilutong, 3800 m, 21 May 1961, *J. T. Wang 3211* (PE). **N HENGDUAN**: **Danba**, Donggu, 3900 m, 11 Jun. 1974, *Z. M. Tan 111217* (CDBI, SZ, WUK, ZM); **Dawu**, Qianning, 3600 m, 2 Jun. 1980, *Q. S. Zhao 116563* (SZ); Gexi, Bakaka, 3100 m, 12 May 1958, *s. coll. 20* (SZ); Wangjia, 3360 m, 4 May 1957, *H. Li 70430* (SZ); **Hongyuan**, Shuajingsi, 3400 m, 16 Jun. 1957, *W. P. Fang et al. 22176* (IBSC, SZ); Shuajingsi Primary School, 3350 m, 22 Jun. 1957, *W. P. Fang et al. 22306* (IBSC, NAS, PE, SZ); **Jinchuan**, Chuosijia, 3400 m, 24 May 1961, *S. Jiang et al. 8886* (KUN, PE); Chuosijia, Ribule, 3000 m, 28 May 1958, *s. coll. 5423* (IBSC); **Litang**, 3300 m, 21 Jun. 1905, *F. Kingdon-Ward 4196* (E); between Litang and Hochuka, 15 Jun. 1893, *Kashikarow s. n.* (K); the the city, 4000 m, 30 May 1991, *J. S. Yang 210* (KUN); Qudeng, Huncha, 4200–4500 m, 9 Aug. 2006, *D. E. Boufford et al. 36148* (TI); **Markam**, Qiaoyaohu, 4200 m, 14 Jun. 1976, *Exped. Qinghai-Xizang 11851* (HNEP, KUN, PE); Rongxi, 4000 m, 26 Jun. 1976, *Exped. Qinghai-Xizang 12007* (HNWP); **Yajiang**, Shangmayi, 4030 m, 11 Jun. 1961, *S. Jiang et al. 5138* (PE); Gongcun, 4000 m, 12 Jun. 1961, *S. Jiang 5170* (KUN, PE); **Zogang**, to Jomda, 3950 m, 5 Sep. 1973, *Exped. Xizang 2762* (HNWP); Soldier station, 4000 m, 3 Jul. 1976, *Exped. Qinghai-Xizang 12107* (KUN, PE). **S HENGDUAN**: **Baidiao**, Zhang'ao to Baidong, 3500 m, 9 Sep. 1959, *S. K. Wu 2510* (CDBI, KUN); Teao to Baidiao, 3500–3800 m, 9 Sep. 1959, *S. K. Wu 2520* (CDBI); **Baoxing**, 3250 m, 25 Jun. 1959, *S. K. Wu 1569* (CDBI, KUN); **Cawalong** (Tsawalong), Lajue, 3400 m, Aug. 1935, *C. W. Wang 65857* (NAS, PE, WUK); **Daocheng**, 3800 m, 29 May 1973, *Sichuan Veget. Exped. 1721* (CDBI, KUN, PE); **Dêqên**, Beima Shan, 4320 m, 28 Jun. 1994, *Alpine Garden Society Expedition to China ACE 732* (E); Baimaxueshan, 4350 m, 17 Oct. 1982, *Z. W. Lü 118* (KUN); Baimashan, 4600 m, 14 Jul. 1981, *Qinghai-Xizang Exped. 3054* (PE); **Dêrong**, Ciwu, Dimu Caodi, 4070 m, 20 Jul. 2004, *D. E. Boufford et al. 30970* (A, KUN, TI); **Eryuan**, Luoping, 13 Jul. 1942, *H. C. Wang 2484* (KUN); **Heqing**, *s. coll. 21300*

(KUN); **Huidong**, Pantang, Neidaniu, 24 Jun. 1959, *S. K. Wu 1569* (CDBI, KUN); **Jiulong**, Chahualingzi to FangMaping, 28 May 1960, *S. Jiang & J. S. Ying 3659* (KUN, PE); 25 Jun. 1930, *Z. Y. Luo 449* (IBSC, PE); **Kangding**, 1934, *Z. S. Liu 846* (PE); Ta-tsien-lu, Proncipaute de Kiala, 1893, *J. A. Soulie 315* (K); Ta-tsien-lu, Proncipaute de Kiala, 1893, *J. A. Soulie 806* (K, P); Tachienlu, *A. E. Pratt 669* (BM, K); Tachienlu, *A. E. Pratt 855* (BM, K); Tachienlu, *A. E. Pratt 24* (K); Tatsienlu, 1938, *H. D. McLaren Ae154* (E); Xinduqiao, Pengbuxi, Nati, 3 Jun. 1974, *Q. Q. Wang 4441* (CDBI, PE); Zhongge, 4000 m, 27 Jul. 1963, *K. C. Kuan & W. T. Wang 480* (K); Tachienlu, *A. F. Partt 300* (BM, K); Jedo Pass, NW Tachienlu, 4500 m, 2 Jun. 1934, *A. Chanfoung 124* (IBSC); Zheduoshan, 3800 m, 30 May 1983, *Z. M. Tan & Q. S. Zhao 119237* (CDBI, SZ); Liuma, Xiacheng, 3640 m, 8 Jun. 1981, *Z. J. Zhao et al. 114408* (SZ); Zhonggu, Dagaigou, 3900 m, 25 Jul. 1963, *K. C. Kuan & W. T. Wang 347* (PE); Zhonggu, Dagai, 4200 m, 27 Jul. 1963, *K. C. Kuan & W. T. Wang 480* (PE); Djesi-La and Djesi-Longba, S Tatsienlu, 4450 m, Jun. 1929, *J. F. Rock 17728* (PE); Jichou Shan, 4410–4650 m, 5 Aug. 2010, *D. E. Boufford et al. 42639* (TI); **Kangwu**, 3580 m, 8 Jun. 1982, *Z. X. Tang 0084* (PE); **Lushui**, inta Kalapa et Liuku, 3700–4000 m, *C. Schneider 1245* (K); **Mianning**, Luoguo, 3900 m, 16 Jul. 1959, *S. K. Wu 2236* (KUN, PE); **Muli**, Labu, 3600 m, *T. T. Yü 5469* (IBSC, KUN, PE); Gumadian to Dipobianni, 3500 m, *T. T. Yü 5724* (IBSC, KUN, PE); Kulu, 3400 m, 13 Sep. 1937, *T. T. Yü 14254* (BM, E, KUN, PE); Wachin, Deongomba, 3500 m, 11 Nov. 1937, *T. T. Yü 14762* (BM, E, KUN, PE); Mountai Mitzuga, W of Muli Gomba, 3050–4875 m, *J. F. Rock 16072* (E); Mountains of Yetsi, north of Kulu, May 1932, *J. F. Jock 23909* (BM, E, K); mountain of Kulu, 3600 m, Jun. 1932, *J. F. Rock 23930* (BM, E, K); Mountains of Yetsi, 3600 m, Sep.–Oct. 1932, *J. F. Rock 24424* (BM, E, K); **Shangri-la** (Zhongdian), Dongwang, Yuejin, 31 Jul. 1976, *J. S. Yang 8243* (IBSC); on the way to Bitahai, 3300 m, 17 Jun. 2001, *H. Sun 8143* (KUN); Habaxueshan, 2600 m, 5 Jun. 1938, *T. T. Yü 11591* (KUN, PE); Pica, 3400 m, 6 Jul. 1938, *T. T. Yü 11947* (KUN, PE); Haba, 3500 m, 8 Sep. 1937, *T. T. Yü 13512* (BM, E, KUN, PE); Haba, 3600 m, 16 Sep. 1937, *T. T. Yü 13554* (BM, E, KUN, PE); Kungsinshuh, 3000 m, 7 Nov. 1937, *T. T. Yü 13974* (BM, E, KUN, PE); Napa Hai, 3970 m, 12 Jun. 1994, *Alpine Garden Society Expedition to China ACE 177* (E); Wengsui, 4150 m, 23 Jun. 1994, *Alpine Garden Society Expedition to China ACE 630* (E); Napa Hai, 4000 m, 12 Jun. 1993, *B. Alden 1192* (E); Bi Ta Hai, 3490 m, 14 Jun. 1993, *B. Alden 1327* (E); Napa Hai, 3910 m, 15 Jun. 1993, *B. Alden 1381* (E); Haba Shan, 3600 m, in 1932, *J. F. Rock 24766* (BM, E, K); Bitahai, 3600 m, 9 Jun. 1994, *G. Yao 94-1094* (NAS); Habaxueshan, 28 May 1939, *K. M. Feng 1106* (PE); Tuguan to Haba Xuemenkan, 4000 m, 16 Jun. 1962, *Exped. Zhongdian 2410* (KUN); Tuguan to Haba Xuemenkan, 4000 m, 16 Jun. 1962, *Exped. Zhongdian 2414* (KUN); Habaxueshan, Nikezhua, 3800 m, 28 Jun. 1963, *Exped. Zhongdian 2803* (KUN); Qiaotou, Lengdongzhai, 2600 m, 1 Jul. 1963, *Exped. Zhongdian 63-2967* (KUN); Haba, Zhazhamu to Nanbuzi, 3700 m, 6 Jul. 1963, *Exped. Zhongdian 63-3268* (KUN); Annan to Shangjiunuo, 3450 m, 12 Jul. 1963, *Exped. Zhongdian 63-3356* (KUN); **Xiangcheng**, Dumuling, 4500 m, 27 Jun. 1973, *Sichuan Veget. Exped. 2780* (CDBI, KUN, PE); Rewu, Balang to Qingda, 3900 m, 9 Aug. 1981, *Qinghai-Xizang Exped. 4002* (KUN); s. loc., 3500 m, 11 Aug. 1981, *Qinghai-Xizang Exped. 4585* (KUN); Gujiazhawo to Rizhaoshan, 3700 m, 12 Aug. 1981, *Qinghai-Xizang Exped. 4746* (KUN); Daxueshan, 19 Jul. 2010, *KHam Expedition 10-0212* (PE); Maan Shan, 3950–3980 m, 8 Jul. 2004, *D. E. Boufford et al. 30437* (TI); **Yanyuan**, in mountis Liuku-liangdse, 3600 m, 17 May 1914, *H. Handel-Mazzetti 458* (E); **Yongsheng**, Yongbei, 4 Oct. 1940, *s. coll. 33170* (KUN, PE); **Yulong** (Lijiang), 3000 m, Jul. 1935, *C. W. Wang 70967* (KUN, NAS, PE); 2500 m, Jul. 1935, *C. W. Wu 71171* (HIB, IBSC, LBG, NAS, NKU, WUK); Yulongxueshan, 2400–4500 m, 4 Jun. 2013, *H. Sun et al. KC-0149* (KUN); Feijichang, 7 May 1962, *A. L. Zhang 100425* (KUN, PE); Baishuihe, 19 May 1939, *R. C. Ching 30135* (KUN, PE); W flank of the Lichiang Range, Jun. 1910, *G. Forrest 5917* (BM, E, IBSC, K, PE); Likiang Range, Jun. 1913, *G. Forrest 10271* (BM, E, IBSC, K, KUN, PE, TI).

11. Incarvillea younghusbandii Sprague, Bull. Misc. Inform. Kew 1907: 320. 1907; Zhang & Santisuk in Wu & Raven (eds.), Fl. China 18: 223. 1998; Aitken in Grierson & Long, Fl. Bhutan 2(3): 1239. 2001. Type: China. Xizang, Rhamba Jong, 5 Jul. 1903, *Major F. E. Younghusband 21* [lectotype first-step designated by Grierson, Notes Roy. Bot. Gard. Edinburgh 23: 347. 1961, second-step **here designated**: K; isolectotype: E (E00135983)].

Herbs, 10–20 cm high, stemless. Roots fleshy, 6–11 mm in diameter. Leaves basal, pinnate; lateral leaflets 2–5 pairs, sessile, blade ovate-elliptic, 1–2 cm long, 0.7–1 cm broad, scabrous, margin serrate; terminal leaflet ovate to orbicular, 3–5(–7) cm long, 3–5(–7) cm broad, base cordate, apex obtuse or rounded. Racemose 3–6-flowered or reduced to solitary. Pedicel 6–9 mm long. Calyx campanulate, 0.8–1.2 cm long, glabrous; teeth 5, unequal, smooth, 5–7 mm long. Corolla funnelform, 4–7 cm long; tube orange-yellow, 3–5 cm long, 2–3 mm broad, upwards widened to 8 mm at mouth; lobes patent, rounded. Anthers versatile, connective with a basal spur 0.8–1.2 mm long. Ovules 1- or 2-rowed on each placenta. Stigma flabellate, 2-lobed. Capsules 3–4.5 cm long, strongly curved, quadrangular. Seeds nearly black, ellipsoid, 4–5 mm long, 2–2.5 mm broad, concave on one side, convex on another, surrounded by wings, scaly. **Chromosomes** 2n = 22. Figure 46-127.

Phenology: flowering from May to August; fruiting from August to October.

Habitats: gravel, slopes, thickets; 1300–9000 m.

Distribution: **endemic**; U Yarlung Zangbo, W Nepal, M Yarlung Zangbo, L Yarlung Zangbo, Tangut, N Hengduan (Map 46-223).

Notes: There are several gatherings were cited in the protologue and none of them was designated as the type. Grierson (1961) designated *Younghusband 21* as the lectotype, but he specified two herbaria. Hence, it is further narrowed here to one herbarium by second-step lectotypification.

U YARLUNG ZANGBO: **Dinggyê**, Sar, 12 Jun. 1975, *Exped. Qinghai-Xizang 5710* (HNWP, KUN, PE); Kuma, Shaligunba, 4770 m, 20 Jul. 1961, *Exped. Xizang 1637* (PE); **Gyirong**, Rongha, 4200 m, 8 Jul. 1972, *s. coll. 716* (HNWP, PE); **Lhazê**, Mange Chu, 1380–1500 m, 29 Apr. 1984, *G. Miehe 1284* (BM); **Nyalam**, 3950 m, 12 Jun. 1966, *Y. T. Zhang & K. Y. Lang 3890* (KUN, PE); 4200 m, 2 Jun. 1974, *C. Z. Zhang 266* (PE); near Selong, 4800 m, 12 Jun. 1966, *Y. T. Zhang & K. Y. Lang 4001* (KUN, PE); near Selong, 4850 m, 14 Jun. 1966, *Y. T. Zhang & K. Y. Lang 4050* (KUN, PE); Niexiongla Pass, 5000 m, 20 Jun. 1966, *Y. T. Zhang & K. Y. Lang 4328* (PE); Nyalam (Shekar), 3900 m, 12 Jun. 1980, *J. D. A. Satinton 8211* (E); Sela, 4400 m, 6 Jul. 1975, *Exped. Qinghai-Xizang 6187* (HNWP, KUN, PE); Yaruxiongla, 4500 m, 2 Jun. 1990, *Sino-Jap. Exped. T214* (PE); 3800 m, 30 May 1974, *Xizang Medic. Herd Exped. 1721* (HNWP, PE); Nyalam, Zhaxigang, 4696 m, 26 Jul. 2013, *PE Xizang Exped. 4235* (PE); Upper Manga Chu, 4600–5000 m, 29 Aug. 1984, *G. Miehe 1284* (BM, HNWP); Upper Sun Kosi (Bo Chu), 4420–4700 m, 2 Sep. 1984, *G. Miehe 1364* (HNWP); Paikiuco (Paiku Tso), 5020 m, 8 Sep. 1984, *G. Miehe 1658* (BM, HNWP); **Shigatse-Shekar Road**, 4350 m, 10 Jun. 1980, *J. D. A. Stainton 8204* (E); **Tingri**, Khata Valley, 3150 m, 6 May 1938, *P. Lloyd 6* (K); between 1 and Base Camp, East Rongbuk Glacier, s. d., *N. Humphreys 5059* (BM, E); Fushila, 4800 m, 14 Jun. 1990, *Sino-Jap. Exped. T558* (PE); Camp place, 4600 m, 17 Jun. 1990, *Sino-Jap. Exped. T595* (PE); Jiabula, 5050 m, 21 Jun. 1990, *Sino-Jap. Exped. T772* (PE); Jiabula, 5150 m, 21 Jun. 1990, *Sino-Jap. Exped. T783* (PE); Camp place, 4600 m, 23 Jun. 1990, *Sino-Jap. Exped. T825* (PE); Jiabula, 5000 m, 30 May 1974, *Climbing Mountain Sci. Invest. 227* (HNWP); Qomolangma, Rongbude Temple, 5400 m, 1 Jun. 1960, *Y. T. Zhang & K. Y. Lang 3741* (PE); Quzong, 4800 m, 3 Jun. 1966, *Y. T. Zhang & K. Y. Lang 3801* (KUN, PE); Rongbuk Valey, 4800 m, 6 Jun. 1922, *s. coll. 23* (K); Rongbuk valley, Juset, 16 Jun. 1924, *R. W. G. Hingston 394* (K); Rongbak, 4800 m, 18 Jun. 1933, *L. R. Wager 11* (K); **Zhongba**, near Guisang, 4800 m, 24 Jul. 1975, *Exped. Qinghai-Xizang 6444* (HNWP, KUN, PE); 4600 m, 8 Aug. 1975, *Exped. Qinghai-Xizang 7240* (KUN, PE).

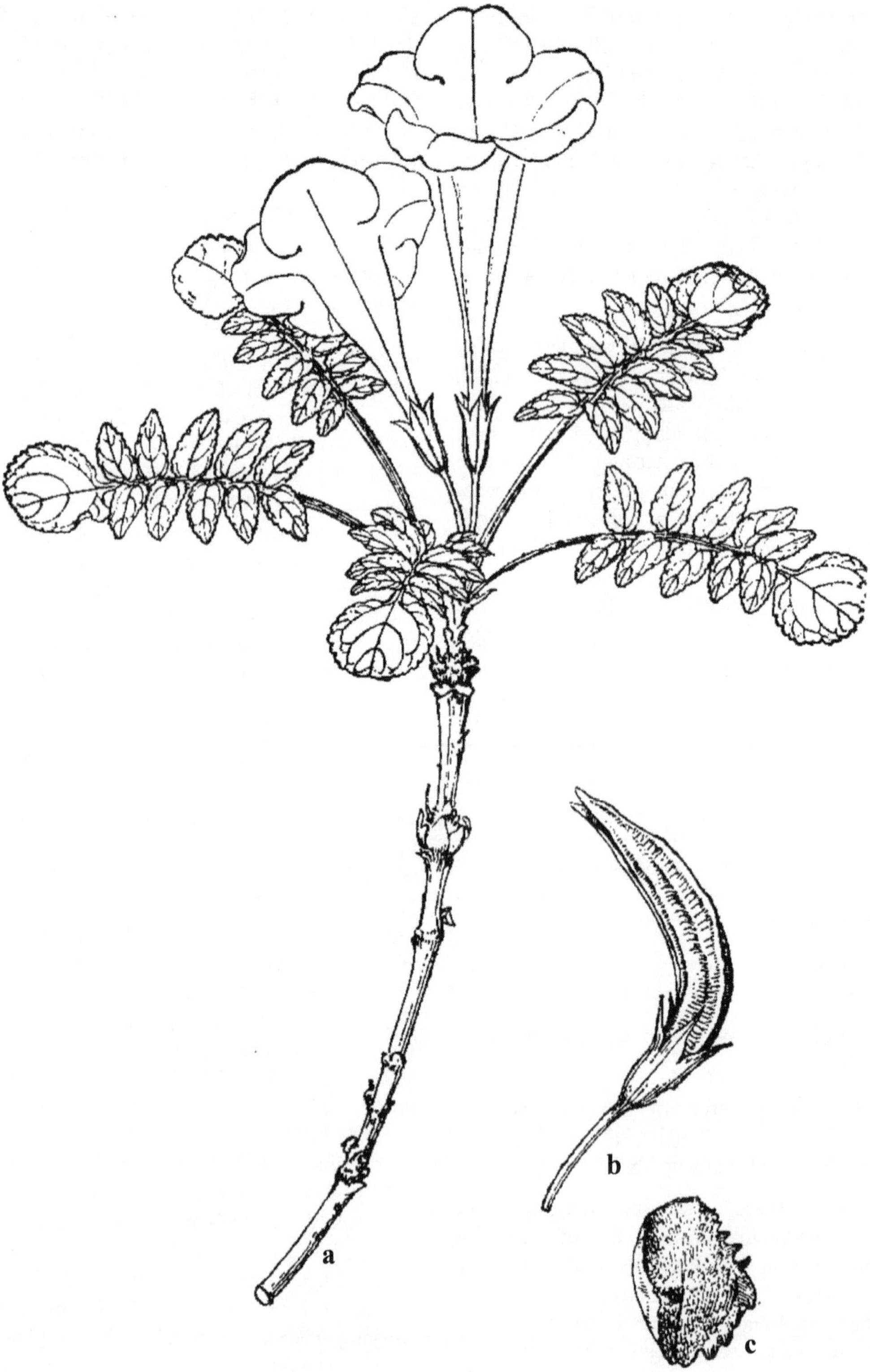

Figure 46-127: **Incarvillea younghusbandii** Sprague: **a**, flowering plant, **b**, capsule with calyx, **c**, seed. Drawn by Xi-Chou LI; reproduced from *Fl. Reipubl. Popularis Sin.* 69: 48, pl. 13: 1–3. 1992.

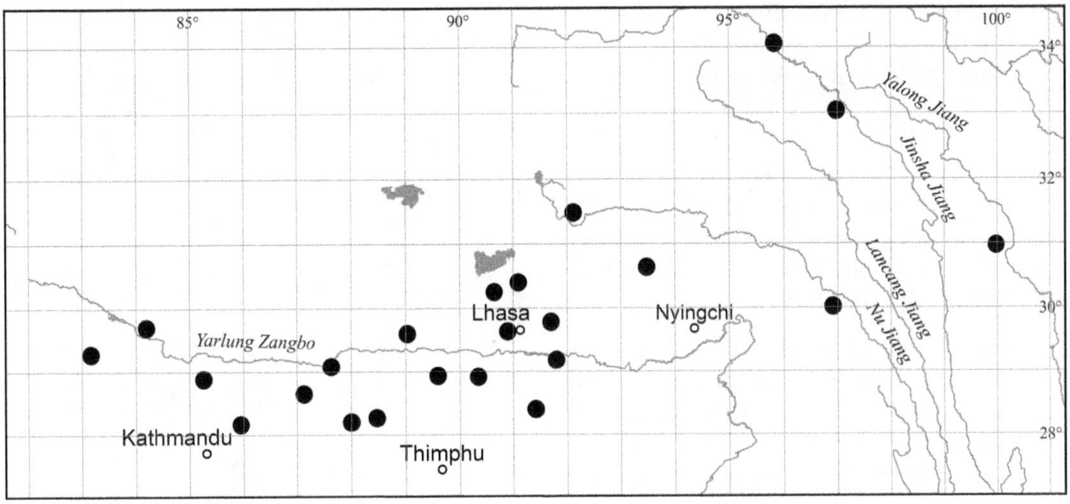

Map 46-223: **Incarvillea younghusbandii** Sprague

W NEPAL: **Karnali**, Dolpo, Koma, 4350 m, 26 Jun. 1963, *J. D. A. Stainton 4365* (E, TI). **M YARLUNG ZANGBO**: **Comai**, Trigu Dzong, 4800 m, 17 May 1947, *F. Ludlow & G. Shrriff 12452* (BM, E); **Damxung**, Nyainqentanglha Shan, Yangbajain to Damxung, MW of Lhasa, Piedmont SE of Nyainqentanglha Feng, 8 Aug. 1989, *B. Dickore 3460* (K); Shengli, 7 Aug. 1985, *D. X. Su 1-068* (PE); **Doilungdêqên**, 3800 m, 26 May 1875, *s. coll. 75-63* (PE); **Gamba**, Khamba Jong, 4500 m, 5 Jul. 1903, *F. E. Younghusband 21* (K); Khamba Jong, 4500 m, 7 Sep. 1903, *F. E. Younghusband 316* (K); **Gyangzê**, between Phari and Gyangtze, 4500–9000 m, Jun. 1904, *H. J. Walton s. n.* (K); **Lhasa**, Hills N of Lhasa, 4350 m, 10 Jun. 1942, *F. Ludlow & G. Sherriff 8689* (BM, E); Reting, 60 miles N of Lhasa, 4200 m, 17 Jul. 1944, *F. Ludlow & G. Sherriff 11015* (BM, E); Lhasa and hills South of Lhasa, 3600–4500 m, 31 May 1943, *F. Ludlow & G. Sherriff 9530* (BM, E); vicinity of Lhasa, s. d., *P. Aufschaniter s. n.* (BM); Hills near Lhasa, Minjen thang La Plain, 4350 m, May 1939, *H. E. Richardson 57* (BM); Longjiaocaika, 5100 m, 1 Sep. 1965, *s. coll. 2581* (KUN); **Gyangzê**, near Gyantse, 4350 m, 10 Jun. 1924, *F. Ludlow 10* (BM); Gyantse, Jul.–Sep. 1904, *H. J. Watson 103* (BM, K); 30 miles from Gyangtse, 3750 m, 9 Jun. 1980, *J. D. A. Stainton 8183* (E); **Maizhokunggar**, 4100 m, 28 May 1960, *K. X. Fu 121* (PE); **Namling**, Weilong, 4600 m, 4 Aug. 1972, *Xizang Med. Herd Exped. 1015* (HNWP, PE); **Nagarzê**, Yamdrok Tso, 4800 m, 20 Jun. 1975, *C. Y. Wu et al. 75-372* (HNWP, KUN, PE); **Nêdong**, Zedang to Lüntzê, 4800 m, 13 Aug. 1977, *P. C. Kuo & W. Y. Wang 22380* (HNWP); Yadui, 4750 m, 3 Aug. 2013, *Y. S. Chen et al. 13-0189* (PE); **Yadong**, Phari to Tuna, ca. 4500 m, 21 Jun. 1939, *B. J. Gould 2177* (K); Kangbu to Gangba, 4552 m, 11 Jul. 2013, *PE Xizang Exped. 2986* (PE). **L YARLUNG ZANGBO**: **Lhari**, on the way to Nagqu, 4500 m, 10 Jun. 1976, *Exped. Qinghai-Xizang 10499* (KUN, PE). **TANGUT**: **Baxoi**, Demulashan, 4200 m, 23 Jun. 1998, *Z. C. Ni et al. 0167* (PE); Kangsha, 4000 m, 9 Jun. 1982, *Z. J. Zhao et al. 82007* (SZ); **Nagqu**, Nira, 4650 m, 24 Jun. 1928, *F. M. Beiley s. n.* (K); Nyima, 4630 m, 15 Sep. 1974, *Exped. Xizang 4375* (HNWP); Kormang, Gyablung, 4496 m, 23 Jul. 2002, *G. Berg 2002-01* (KUN); Kormang, Gyablung, 4496 m, 24 Jul. 2002, *G. Berg 2002-29* (KUN); Liangdaohe, 4640 m, 14 Jun. 1975, *C. Y. Wu et al. 75-283* (HNWP, KUN, PE); **Qamdo**, Gyri, 4650 m, 20 Jul. 1903, *F. E. Younghusband s.n.* (K); **Qumarlêb**, 4300 m, 12 Jun. 1966, *S. W. Liu & R. F. Huang 620* (HNWP); **Yushu**, 4670 m, 14 Jun. 1975, *C. Y. Wu et al. 75-0268* (HNWP, KUN, PE). **N HENGDUAN**: **Baiyü**, Milashan Yakou, 4500 m, 24 Jul. 2010, *Kham Expedition 10-0614* (PE).

12. Incarvillea delavayi Bureau & Franch., J. Bot. (Morot) 5: 138. 1891; Tao & Yin in Wang (ed.), Fl. Reipubl. Popularis Sin. 69: 44. 1990; Zhang & Santisuk in Wu & Raven (eds.), Fl. China 18: 223. 1998. Type: China, Yunnan, "pâturages de montagnes, Mt. Hee-chan-men (Langkong)", 2 Jan. 1984, *M. Delavay 1027* [lectotype designated by Grey-Wilson, New Plantsman 5(2): 85. 1998: P (P00608145)].

Herbs perennial, acaulescent. Leaves basal, 8–25 cm long, pinnate below, pinnatifid above; leaf rachis 12–20 cm long; lateral leaflets 4–11 pairs, blade elliptic-lanceolate, 4–7 cm long, 1–3 cm broad, glabrous, base obtuse to subrounded, margin serrate, apex acuminate; terminal leaflet blade 1.5–3.5 cm long, 1–2.5 cm broad, subsessile, base cuneate, distinct of confluent with uppermost lateral leaflets, margin crenate, apex obtuse or acute. Racemes terminal, 2–6-flowered; peduncle 20–30 cm long. Bracts linear, 0.8–1 cm long; bracteoles 3.5–5 mm long. Pedicel 0.5–1.5 cm long. Calyx campanulate, 1.3–2 cm long; teeth 5–7 mm long, 5–7 mm broad, apex caudate-acuminate. Corolla red, campanulate, 5.5–6.5 cm long, 3.2–3.5 cm broad; tube 4–5 cm long; lobes semiorbicular, spreading. Filaments 2–2.5 cm long; anthers ovoid. Style 2.8–3 cm long; stigma compressed, flabellate. Capsules gray-brown, quadrangular, 5–7.5 cm long, woody. Seeds broadly ovoid, glabrous on one side, pubescent on another side, 4–5 mm long, 3–4 mm broad; wings ca. 1 mm broad, margin entire. **Chromosomes** 2n = 22.

Phenology: flowering from May to August.

Habitats: grasslands, slopes; 2000–3500 m.

Distribution: endemic; N Hengduan, S Hengduan (Map 46-224).

N HENGDUAN: **Dawu**, 12 Jul. 1930, *K. Huang et al. 954* (KUN). **S HENGDUAN**: **Bingchuan**, Jizushan, Zhushengsi, 2900 m, 17 May 1964, *X. Zhou 1019* (KUN); Kouty region de Pitchoaen, Jun. 1911, *F. Ducloux 7015* (KUN); **Dali**, Xiaohuadianba, 18 May 1981, *Sino-British Expedition to Cangshan 774* (E, K); W flank of the Tali Range, 3300 m, Aug. 1913, *G. Forrest 11658* (E, K); Cangshan, Luoshage, 13 Jun. 1940,

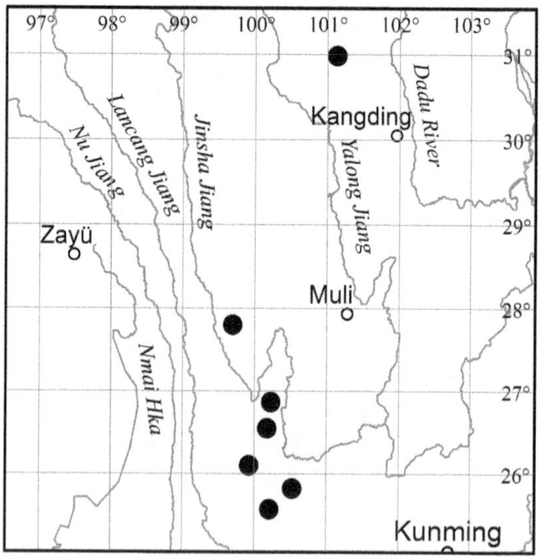

Map 46-224: **Incarvillea delavayi** Bureau & Franch.

R. C. Ching 22929 (IBSC, KUN, PE); **Eryuan**, Junmachang, 2600 m, 20 Jun. 2001, *H. Sun 8148* (KUN); Junmachang, 2600 m, 22 Aug. 2000, *H. Sun et al. 9015* (KUN); **Heqing**, Che tcho tze, Ta pin tze, 2000 m, 27 Aug. 1885, *M. Delavay s. n.* (KUN1230253); **Shangri-la** (Zhongdian), Bi Ta Hai Forests, 3450 m, 24 Sep. 1990, *Lijiang & Dali Expedition CLD233* (E); N of the town, 2 Jul. 1994, *Alpine Garden Society Expedition to China ACE894* (K); **Yulong** (Lijiang), s. d., *Kunming-Edinburgh Yulong-Shan Expedition 47* (KUN); Xianjingyan, 2950 m, 3 Jun. 1985, *Kunming-Edinburgh Yulong-Shan Expedition 459* (KUN); Yulongshan, 11 May 1971, *s. coll. 7* (KUN).

13. Incarvillea mairei (H. Lév.) Grierson, Notes Roy. Bot. Gard. Edinburgh 23: 341. 1961; Tao & Yin in Wang (ed.), Fl. Reipubl. Popularis Sin. 69: 45. 1990; Zhang & Santisuk in Wu & Raven (eds.), Fl. China 18: 223. 1998. ≡ *Tecoma mairei* H. Lév., Cat. Pl. Yun Nan: 20. 1915. Type: China. Yunnan, "rochers sous brousse mont de Pe-long-tsin", 3200 m, 1914, *E. E. Maire 824* [holotype: E (E00135984); isotypes: BM, E (E00135985)].

Incarvillea grandiflora var. *brevispes* Sprague, Bull. Misc. Inform. Kew. 1909: 263. 1909. ≡ *Incarvillea compacta* var. *brevipes* (Sprague) Wehrh., Die Garten-stauden 2:

947. 1931. Type: [S Hengduan] China, Yunnan, on barren, dry, stony ground on plain at the north end of the Lichiang valley, 2700 m, s. d. *G. Forrest 2131* (lectotype **here designated**: K; isolectotype E).

Incarvillea racemosa Q. S. Zhao, J. Sichuan Univ., Nat. Sci. Ed. 4: 92, pl. 1. 1983. Type: [S Hengduan] China, Sichuan, Huidong Xian, 3250 m, 20 Jun. 1959, *S. K. Wu 1560* (holotype: CDBI).

Herbs perennial, 30–40 cm high. Leaves basal, 1-pinnately compound; lateral leaflets 2–5 pairs, blade ovate to ovate-lanceolate, glabrous, shiny and bullate adaxially, base slightly cordate to broadly cuneate, margin serrate, apex obtuse to acuminate; terminal leaflet broadly ovate or oblong, base cordate or rounded, margin serrulate, apex obtuse. Racemose subterminal, 2–4-flowered; peduncle subequaling or much longer than pedicels. Bracteoles linear, 0.8–1.2 cm long. Calyx campanulate, 2–2.5 cm long; teeth triangular, apex acuminate. Corolla purple-red or pale red, 7–10 cm long, 5–7 cm broad; tube 5–6 cm long; lobes rounded. Anthers adherent and surrounding style, strongly divergent. Style 5–7 cm long; stigma flabellate, thin membranous, 2-lobed. Capsules conical, 6–8 cm long, 0.8–1 cm broad, indistinctly striate. Seeds broadly ovoid, 4–6 mm long, 2–3.7 mm broad; wings 0.8–1 mm broad, margin entire. **Chromosomes** 2n = 22. Figure 46-128.

Phenology: flowering from May to August; fruiting from July to November.

Habitats: slopes, roadsides, grasslands, forests; 2400–4500 m.

Distribution: endemic; M Yarlung Zangbo, L Yarlung Zangbo, Tangut, N Hengduan, S Hengduan (Map 46-225).

M YARLUNG ZANGBO: **Lhasa**, 5 Sep. 1970, *P. C. Kuo s. n.* (HNWP). **L YARLUNG ZANGBO**: **Rawu**, on the way to Demula Pass, 4300 m, 15 Jun. 1973, *Exped. Qinghai-Xizang 73-155* (KUN, PE); Quchi to Laigu, 4200 m, 14 Aug. 1973, *Exped. Qinghai-Xizang 1148* (KUN, PE); **Zayü**, Sangjiu, 3400 m, *X. H. Jin et al. ST2613* (PE); Zhuwageng, Zhaku, 4000 m, 20 Jul. 2010, *South Tibet Expedition Team (X. H. Jin et al.) STET0922* (PE). **TANGUT**: **Baxoi**, Siwu Qu, 4000 m, 13 Aug. 1973, *Exped. Xizang 2113* (HNWP); **Biru**, Nujiang, 3800–4000 m, 12 Sep. 1976, *Exped. Qinghai-Xizang 11328* (KUN, PE); **Chindu**, E Tongtian He, 3400 m, 16 Aug. 1996, *T. N. Ho et al. 1912* (HNWP, PE); **Nagqu,** Chanlong Power Station, 4480 m, 3 Oct. 2010, *B. Q. Xu & Y. H. Tong Xianh-07ZX-0591A* (IBSC); **Riwoqê**, 4600 m, 4 Aug. 2010, *Kham Expedition 10-1273* (PE); **Sogxian**, Rongbu, 3850 m, 17 Aug. 1976, *Exped. Qinghai-Xizang 10933* (HNWP, KUN, PE); Rongbu to Yaduo, 4500 m, 20 Aug. 1976, *Exped. Qinghai 11021* (KUN, PE). **N HENGDUAN**: **Barkam**, Shuajingsi, 3360 m, 12 Jun. 1957, *W. P. Fang et al. 22041* (IBSC, SZ); Matang, 17 May 1988, *Q.M. Hu 8852* (HX, IBSC); **Dawu**, Qianning, 17 Jun. 1974, *Sichuan Veget. Exped. 5402* (CDBI, PE); on the way to Garzê, 3200 m, 22 Jun. 1984, *W. L. Chen et al. 6658* (PE); 12 Jul. 1930, *Z. P. Huang et al. 951* (PE); **Heishui**, Shuajinsi, 16 Jun. 1957, *C. Y. Chang & H. F. Chow 22176* (KUN, PE); **Hongyuan**, Waqie, Ningqinggou, 3550 m, 20 Aug. 1975, *s. coll. 10444* (IBSC); Ougushan, 18 Jun. 1983, *Q. S. Zhao 120559* (CDBI, SZ); **Jinchuan**, Xiaoaoluo, Gouweizi, 3850 m, 12 Jun. 1983, *Z. X. Tang 1021* (PE); **Lixian**, Dabanzhao, Zhaishougou, 3750 m, 10 Aug. 1994, *D. P. He 44425* (SZ); Dabanzhao, 3320 m, 19 Jun. 1956, *W. K. Hu 53-499* (SZ); 10 Jul. 1956, *W. K. Hu 53-750* (SZ); 9 Jul. 1956, *W. K. Hu 739* (SZ); Dabanzhao, Huangtuliang, 3700 m, 22 Jun. 1956, *H. J. Hu 53-564* (SZ); Daniuchang, 3600 m, 3 Jun. 1956, *S. F. Yan 53-339* (SZ); **Litang**, road to Mount Genyen, 4320 m, 3 Sep. 2010, *Kunming & Edinburgh Expedition to Sichuan 154* (E); 30 May 1991, *J. S. Yang 209* (KUN); 3900 m, 28 May 1991, *J. S. Yang 91-173* (IBSC); **Markam**, Lengmo, 2900 m, 3 Jun. 1984, *T. A. Zhu Mu 195* (CDBI); Dongshan, 3800 m, 22 Jun. 1976, *Exped. Qinghai-Xizang 11936* (HNWP, KUN, PE); **Xinlong**, Jialaxugou, Lamasi, 3980 m, 8 Jun. 1961, *Q. L. Zhang 30361* (IBSC); Yixigou, 3400 m, 10 Jun. 1974, *Q. H. Li 6277* (CDBI); **Yajiang**, Jianziwan, 4180 m, 26 Jun. 1983, *Hengduan Veget. Group 4207* (PE); Jianziwan, 4300 m, 2 Aug. 1959, *S. Jiang & C. L. Jin 2699* (KUN, PE, SZ); **Zogang**, Soldiers Station, 3900

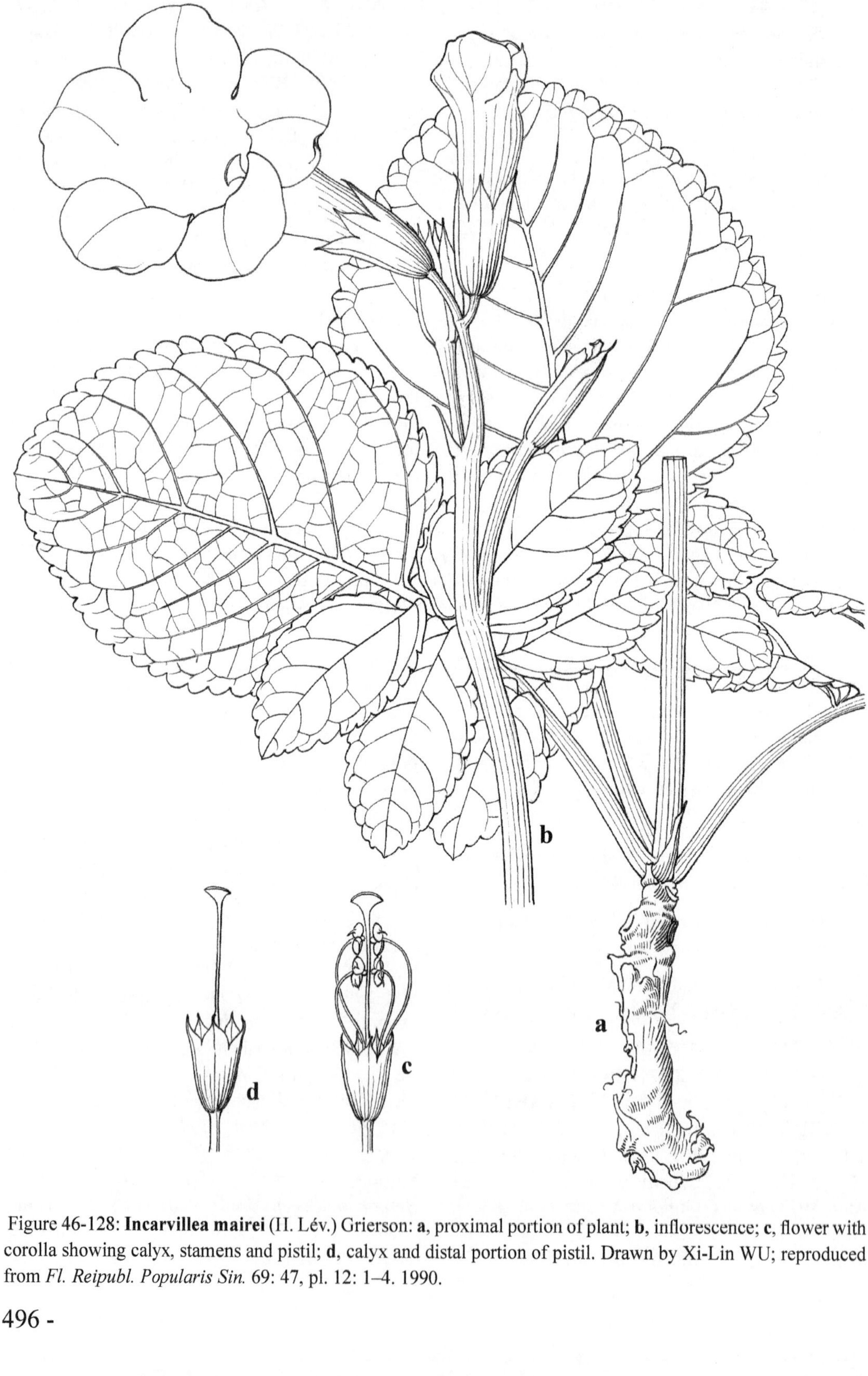

Figure 46-128: **Incarvillea mairei** (H. Lév.) Grierson: **a**, proximal portion of plant; **b**, inflorescence; **c**, flower with corolla showing calyx, stamens and pistil; **d**, calyx and distal portion of pistil. Drawn by Xi-Lin WU; reproduced from *Fl. Reipubl. Popularis Sin.* 69: 47, pl. 12: 1–4. 1990.

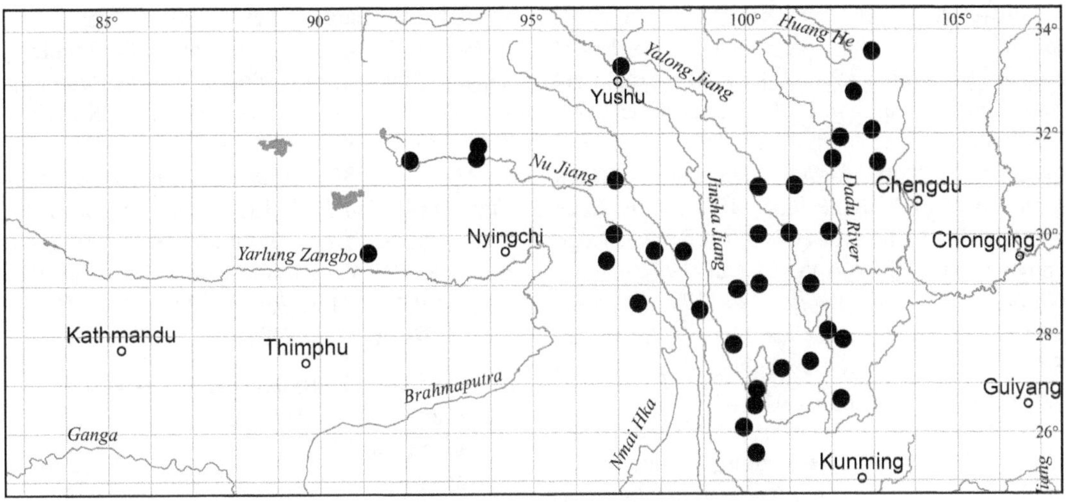

Map 46-225: **Incarvillea mairei** (H. Lév.) Grierson

m, 1 Jul. 1976, *Exped. Qinghai-Xizang 12088* (HNWP, KUN, PE); **Zoigê**, 3300 m, 3 Aug. 1993, *Z. X. Tang 0392* (PE). **S HENGDUAN**: **Dali**, Cangshan, Huadianba to Xiaohuadian, 3050–3150 m, 18 May 1981, *Kunming-Edinburgh Yulong-Shan Expedition 774* (KUN); Diancang Shan mountain range, Xiaohuadianba, 2900–3200 m, 19 Jul. 1984, *1984 Sino-Amer. Bot. Exped. 1165* (E, KUN, KYO, TI); Huadianba, 3280 m, 21 Oct. 1987, *Y. H. Li et al. 00368* (KUN); **Daocheng**, Minzhu, 3850 m, 16 Jul. 1982, *T. C. Wei 28778* (CDBI); Sangdui, 3900 m, 21 Jun. 1980, *s. coll. 28087* (CDBI); Riwa, Lamuge, 3800 m, 18 Sep. 1971, *X. H. Hu 675* (CDBI); between Sandui and Litang, SW of Haizi Shan, 1 Jul. 1998, *D. E. Boufford et al. 28101* (E); E Daocheng, MunaE, 3800 m, 10 Sep. 2010, *Kunming & Edinburgh Expedition to Sichuan 240* (E); Near Honggang, 4158 m, 4 Aug. 2010, *T. Zhang et al. 10CS2241* (K); **Dêqên**, Bamaxueshan, 27 May 1974, *J. S. Yang 6845* (KUN); Sunyongka, 4000 m, 16 Nov. 1937, *T. T. Yü 10722* (IBSC, KUN, PE); Beima Shan, 4500 m, 29 Sep. 1994, *Alpine Garden Society Expedition to China ACE 1546* (E); **Eryuan**, Luoping, 30 Jun. 1929, *R. C. Ching 23050* (KUN, PE); **Heqing**, Baiya, Shaxi, 3100 m, 6 Sep. 1940, *R. C. Ching 24368* (IBSC, KUN, PE); **Huidong**, 3250 m, 20 Jun. 1959, *S. K. Wu 1560* (CDBI); **Huili**, Hongqi, Waibai, 2500 m, 20 Jul. 1976, *Xi Shi Dui 12339* (CDBI, PE, SZ); **Jiulong**, Tanggu, 4400 m, 16 Jun. 1960, *Z. T. Guan 25-0238* (IBSC); Tanggu, Jichoushan, 3800 m, 30 May 1979, *Q. Q. Wang 20070* (CDBI); Jichouhe, 3880, 8 Jul. 1983, *Z. X. Tang 0229* (PE); Metikonga, Lamashai, 3600 m, 6 Jul. 1937, *T. T. Yü 6867* (KUN, PE); **Kangding**, Tachienlu, 3300 m, 15 Jul. 1934, *H. Smith 10508* (TI); Jedo Pass, NW of Tachienlu, 4575 m, 2 Jun. 1934, *A. C. Yong 124* (IBSC); Gonggashan, Moxigou, 3300 m, 6 Jun. 1980, *Z. Y. Chen et al. 112702* (CDBI, SZ); Zheduoshan, 3800 m, 30 May 1983, *Z. M. Tan & Q. S. Zhao 119237* (CDBI); Yuling, Yuyagou, 3500 m, 2 Jun. 1953, *H. L. Tsiang & T. H. Hsiung 35838* (IBSC); Yulin, 3600 m, 6 Jun. 1953, *H. L. Tsiang & T. H. Hsiung 35872* (IBSC); Zheduoshan, Zheduotang, 3600 m, 9 Jul. 1953, *H. L. Tsiang 36250* (IBSC); **Muli**, Wachin, Deonbomba, 3900 m, 13 Jun. 1937, *T. T. Yü 6210* (KUN, PE); Mairi, 4600 m, 30 Jul. 1978, *Q. S. Zhao 6301* (CDBI, SZ); Baer, 5000 m, 13 Aug. 1978, *Q. S. Zhao 6592* (CDBI, SZ); Dingdonggou, 3400 m, 23 Jul. 1978, *K. H. Mou & Y. B. Yang 6699* (CDBI, SZ); Yazui, Weipeng, 3450 m, 2 Jul. 1978, *Y. B. Yang 7174* (CDBI, SZ); Yazui Yuchang, 3500 m, 10 Jul. 1978, *Y. B. Yang 7297* (CDBI, SZ); No. 912 Linchang, 3400 m, 5 Jul. 1978, *Y. B. Yang 8772* (CDBI, SZ); Yazui, Paifang, 1 Jun. 1976, *s. coll. 114* (KUN); mount Mitzuga, west of Muli Gomba, 3050–4875 m, Jun. 1928, *J. F. Rock 16165* (E); SW of Muli, 3300–3600 m, 21 Jun. 1913, *F. Kingdon-Ward 4060* (E); Tangyang, Buchang, 4150 m, 10 Sep. 1983,

Qinghai-Xizang Exped. 13939 (KUN); Guzeng, to Shuiluo Yakou, 4070 m, 31 Aug. 2015, *FLPH Sichuan Expedition 151854* (PE); **Ninglang**, East Mountain in Yungning, *H. D. McLaren 33* (BM, E, K); **Shangri-la** (Zhongdian), Napazhang, 3350 m, 31 Sep. 2000, *H. Sun 9019* (KUN); Haba shan, north of Yangtze loop, 1932, *J. F. Rock 24808* (BM, E, K, KUN); North of town, ridge above Napa Hai, 3966 m, 12 Jun. 1994, *Alpine Garden Society Expedition to China (1994) ACE177* (K); Degen, Beima Shan, 4400 m, 26 Sep. 1994, *Alpine Garden Society Expedtion to China ACE1337* (E); S Zhongdian, 3125 m, 2 Oct. 1994, *Alpine Garden Society Expedition to China ACE1600* (E); **Xiangcheng**, Hongqi, 3900 m, 30 Jun. 1973, *Z. G. Liu 994* (CDBI); between Reda and Xiangcheng on the Reda side of pass, 3450–3800 m, 19 Jul. 1998, *D. E. Boufford et al. 28956* (E); Wensui to Xangchen, 3805 m, 25 Jun. 1994, *Alpine Garden Society Expedtion to China ACE664* (E, K); Xiangbala, Lenglong, 3930 m, 26 Jul. 1994, *Alpine Garden Society Expedtion to China ACE 681* (E, K); near Chaoyang Daoban, 4200 m, 11 Aug. 1973, *Sichuan Veget. Exped. 3089* (CDBI, KUN, PE); Rewu, Balang to Qingda, 3900 m, 9 Aug. 1981, *Qinghai-Xizang Exped. 3926* (KUN); **Xichang**, Luojishan, 3600 m, 2600 m, 16 May 185, *Y. H. Li 729* (CDBI); **Yanyuan**, 3600 m, 17 May 1914, *H. Handel-Mazzetti 2264* (K); **Yulong** (Lijiang), s. d., *C. Schneider 3872* (K); Yulongshan, 2700–3000 m, 22 Jun. 1975, *K. L. Chü 928* (SZ); Plain ca. 20 km N of the city, 2500 m, 30 Apr. 1987, *P. J. Cribb C.42* (K); E slopes of Likiang Snow Range, May–Oct. 1922, *J. F. Rock 3324* (E, K); E slopes of Likiang Snow Range, May–Oct. 1922, *J. F. Rock 3380* (E); E slopes of Likiang Snow Range, 12–29 May 1922, *J. F. Rock 4127* (BM, E); E slopes of Likiang Snow Range, 12–29 May 1922, *J. F. Rock 4128* (E); E slopes of Likiang Snow Range, May–Oct. 1922, *J. F. Rock 6122* (E); E slope of Likiang Snow Range, Yangtze watershed, 1923–1924, *J. F. Rock 8683* (E); E slope of Mount Dyinaloko, 1923, *J. F. Rock 8991* (E); Yu-lung shan, Mt. Satseto, Jun.–Aug. 1932, *J. F. Rock 24852* (BM, E, K, KUN); Yu-lung shan, Sep.–Oct. 1932, *J. F.Rock 25338* (E); Likiang Snow Range, 2400 m, 12 Jun. 1937, *T. T. Yü 15165* (BM, E, KUN); s. loc., 2800 m, Jul. 1935, *C. W. Wang 70904* (IBSC, KUN, PE); N of the Lichiang Valley, 2700 m, May 1906, *G. Forrest 2131* (E, K); E flank of the Lichiang Range, 3300 m, Jun. 1906, *G. Forrest 2464* (E); Kan-hai-tsi, 3000 m, May 1913, *F. Kingdon-Ward 239* (E); Yulong Shan, Gang Ho Ba, 3000 m, 23 May 1985, *Kunming-Edinburgh Yulong-Shan Expedition KEY 3* (E, KUN); Yulongshan, Xianjiyan, 3500 m, 15 Jun. 1959, *K. M. Feng 22257* (KUN); Yulong Snow Rang, Yulongzhai to Mahuangba, 3414 m, 31 May 2013, *X. X. Huang & Z. B. Zhong 15686* (IBSC).

14. Incarvillea zhongdianensis Grey-Wilson, New Plantsman 5(2): 96. 1998. Type: [S Hengduan] China, Yunnan, Zhongdian, plain south of Zhongdian, 27°32′N, 99°52′E, 3312 m, 10 Jun. 1994, *Alpine Garden Society Expedtion to China ACE 64* [holotype: E (E00040609); isotype: K].

Incarvillea mairei f. *multifoliolata* C. Y. Wu & W. C. Yin, Fl. Yunnan. 2: 721. 1979. ≡ *Incarvillea mairei* var. *multifoliolata* (C. Y. Wu & W. C. Yin) C. Y. Wu & W. C. Yin, Fl. Reipubl. Popularis Sin. 69: 45. 1990; Zhang & Santisuk in Wu & Raven (eds.), Fl. China 18: 223. 1998. Type: China, Yunnan, Zhongdian, Tapose, 3200 m, 1 Aug. 1937, *T. T. Yü 12556* [holotype: KUN (KUN0210121); isotypes: (IBSC0563092), PE (PE01154871, PE01155024)].

Herbs, acaulescent. Leaves 10–25 cm long, pinnate; petiole 7.4–14 cm long, 1.5–2 mm broad; lateral leaflets 4–9 pairs, the uppermost pair sometimes incompletely separated from the rachis and the terminal leaflet, upper leaflet pairs with a narrow wing running onto the rachis, blade lanceolate to elliptic, 1.2–3.8 cm long, 0.6–2 cm broad, matt green and rather sooth adaxially; terminal leaflet blade broadly ovate to elliptic, 2.2–5.0 cm long, 1.7–4.0 cm broad, margin serrulate to subentire, apex acuminate. Racemes 1–3 (rarely 4)-flowered. Peduncle erect, 7–17.5 cm long; pedicel 1–1.6 cm long. Calyx narrow-campanulate, 2.2–3.1 cm long, teeth narrowly triangular, straight, 8–10 mm long. Corolla deep magenta or crimson-magenta, rarely rose pink, with two white oblong flares at base of each corolla-lobe and a yellow throat; tube 4.5–5.3 cm long, slightly curved, outside pale magenta or whitish; lobes suborbicular, margin undulate or entire. Capsules asymmetrically fusiform, slightly curved or straight, 6–11 cm long, 0.8–1.3 cm broad, puberulent and with some scattered black dots, apex acute, rectangular in cross-section. Seeds entire with slightly undulate wings. Figure 46-129.

Phenology: flowering from June to August; fruiting from August to October.

Habitats: forests, slopes, grasslands; 300–4400 m.

Distribution: endemic; Tangut, N Hengduan, S Hengduan (Map 46-226).

TANGUT: **Dêgê**, 4400 m, 20 Jun. 1991, *J. S. Yang 405* (KUN); **Riwoqê**, 4645 m, 7 Aug. 2010, *Kham Expedition 10-1454* (PE); **Sêrxü**, Hongqi, 4200 m, 22 Jul. 2004, *Y. H. Wu 29420* (HNWP). **N HENGDUAN**: **Barkam**, Lengmo, 2900 m, 3 Jun. 1984, *T. C. Wei 195* (CDBI); Shua-Ma Road, 3050 m, 8 May 1957, *H. Li 70459* (IBSC, SZ); Dalangjiaogou, 2850 m, 2 Jun. 1957, *H. Li 70977* (IBSC, KUN, PE, SZ); 3200 m, 7 Jun. 1957, *H. Li 71351* (IBSC, KUN, PE, SZ, WUK); near Matang, 3180 m, 17 Jun. 1957, *H. Li 71508* (IBSC, PE, SZ); near Matang, 3200 m, 21 Jun. 1957, *H. Li 71644* (IBSC, PE, SZ); 2800 m, 1 Jul. 1957, *H. Li 76508* (YFP); **Chindu**, Xiewu, Anbala Shankou, 4600 m, 31 Jul. 2010, *Kham Expedition 10-0993* (PE); **Danba**, Haiziping, 3800 m, 24 Aug. 1958, *Sichuan 8th Forests Exped. 2602* (IBSC); **Dawu**, 5 Jul. 1977, *P. C. Kuo & W. Y. Wang 20986* (HNWP); Geka, 3900 m, 23 Jul. 1961, *S. Jiang et al. 9593* (PE); Qianning, 3450, 20 Jul. 1959, *S. Jiang & C. L. Jin 2115* (KUN, PE, SZ); **Hongyuan**, Waqie Muchang, Ningqinggou, 3550 m, 20 Aug. 1975, *s. coll. 10444* (CDBI, IBSC, PE); Shuajingsi, 3400 m, 16 Jun. 1957, *W. P. Fang et al. 22174* (SZ); **Jinchuan**, Zuosijia, 3600 m, 10 Jun. 1958, *Sichuan 8th Forests Exped. 2185* (IBSC); **Lixian**, Dabanzhao, Huangtuliang, 3600 m, 16 Jun. 1957, *D. Z. Deng 1212* (SZ); Maoxian or Lixian, 1952, *C. Ho & C. L. Chow 13727* (NAS, PE, SZ); **Markam**, Honglashan, 3600–3800 m, 1 Jun. 1976, *Exped. Qinghai-Xizang 11634* (HNWP, KUN); E of the city, 3900 m, 8 Sep. 1973, *Exped. Xizang 2817* (HNWP); Bangda, Honglashan, 4060 m, 20 Aug. 2010, *Kham Expedition 10-2433* (PE). **S HENGDUAN**: **Dêqên**, Bai-ma Shan, Jun. 1917, *G. Forrest 13825* (BM, E, K); 3600 m, Lashakou, 3600 m, 10 Aug. 1976, *C. Y. Wu 4444* (KUN); Baimashan, 3500 m, Sep. 1935, *C. W. Wang 69701* (KUN, PE); **Heqing**, 3400 m, 26 Jun. 1960, *S. Jiangsu et al. 9515* (KUN); 1960, *Nanshui Beidiao Exped. (S. Jiangsu et al.) 10122* (KUN); **Kangding**, Yala, Yusi, 3911 m, 20 Jul. 2008, *Y. L. Peng et al. Gaoxf-0406* (IBSC); Xinduqiao, 3600 m, 19 Jul. 1963, *K. C. Kuan & W. T. Wang 778* (PE); Xindianzi to Zhongggu, 3150 m, 27 May 1981, *Z. J. Zhao et al. 114202* (CDBI, SZ); Jintang, Heikaliangzi, 4000 m, 1 Jul. 2007, ca. *Zhang 20071093* (HX), Zheduoshan, Zheduotang, Guangou, 9 Jul. 1953, *H. L. Tsiung 36250* (IBK, IBSC, PE, SZ); Yulin, Zhangyuyagoum 3500 m, 2 Jun. 1953, *H. L. Tsiung & C.H. Hsiung 35838* (IBSC, PE, SZ); Zheduoshan, Zheduotang, Guangou, 9 Jul. 1953, *H. L. Tsiung 36250* (IBK, SZ); Zheduoshan, 30 Jun. 1951, *Y. W. Cui 4195* (PE); **Shangri-la** (Zhongdian), Lanapahai, 24 May 2012, *X. F. Jin 0002* (PE); S Zhongdian, 3312 m, 10 Jun. 1994, *Alpine Garden Society Expedtion to China ACE 57* (E, K); S edge of plateau, 7 Jun. 1992, *Sino Scottish Exped. to NW Yunnan SSY 92* (E); Nixi, 3279 m, 2 Jul. 1994, *Alpine Garden Society Expedtion to China ACE894* (E); W Hsiao Chungtien, 380–390 m, Apr.–May 1932, *J. F. Rock 24642* (BM, E, IBSC,K); W Hsiao Chungtien, 380–390 m, 1932, *J. F. Rock 25262* (E); near Napa Hai, 3350 m, 1 Jun. 1993, *B. Alden et al. 627* (E); N Xiaozhongdian, 3300 m, 17 Jun. 1993, *B. Alden et al. 1557* (E); Bitahai, 3650 m, 15 Jun. 1994, *G. Yao 94-1051* (NAS); 3300 m, 25 Jun. 1981, *Rhododendron Illustration Group 81-0038* (KUN); Shikashan, 3730 m, 3 Jul. 1981, *X. Z. Tang 528* (PE); Xiaozhongdian, Jishan, 3400 m, 14 Jun. 1981, *Qinghai-Xizang Expedition 988* (KUN); Wufengshan, Daoshipo,

Figure 46-129: **Incarvillea zhongdianensis** Grey-Wilson: **a**, proximal portion of plant; **b**, inflorescence; **c**, capsule. Drawn by Jin-Yong FENG; reproduced from *Icon. Cormophyt. Sin.* 4: 101, fig. 5615. 1975.

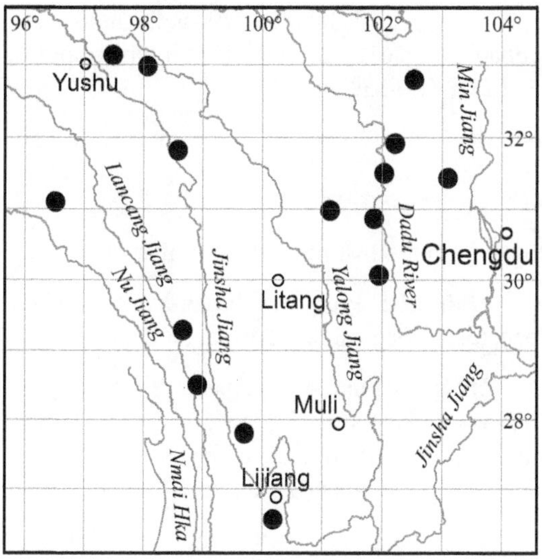

Map 46-226: **Incarvillea zhongdianensis** Grey-Wilson

3300–3400 m, 14 Aug. 1962, *Exped. Zhongdian 783* (KUN); Shangjiunuo to Tianshengqiao, 3400 m, 14 Jul. 963, *Exped. Zhongdian 63-3450* (KUN); Zhongdianba, 3300 m, 20 Jun. 1960, *S. Jiang et al. 10125* (KUN, PE); Biluo, near Hongtupo, 3400 m, 24 Jun. 1960, *S. Jiang et al. 10225* (KUN, PE); Jinagu, 3500 m, 25 Aug. 1959, *K. M. Feng 23210* (KUN, PE); Habaxueshan, 8 Jun. 1939, *K. M. Feng 1223* (PE); Haba, 3200 m, 1 Jun. 1938, *T. T. Yü 11481* (KUN, PE); Pica, 3200 m, 5 Jul. 1937, *T. T. Yü 11914* (KUN, PE); Tapose, 3200 m, 1 Aug. 1937, *T. T. Yü 12556* (IBSC, KUN, PE); Pica, 3200 m, 4 Nov. 1937, *T. T. Yü 13955* (BM, E, KUN, PE); 3300 m, 21 Jul. 2001, *H. G. Ye 5897* (IBSC); Qianhushan, 3400 m, 5 Jun. 2006, *J. S. Zhou & F. W. Xing 467* (IBSC); Nixi, 3300 m, 8 Jun. 2006, *J. S Zhou & F. W. Xing 639* (IBSC).

Tribe 4. Coleeae Bojer ex Meisn., Pl. Vasc. Gen. 1: 301. 1840. Type: *Colea* Bojer ex Meisn.

Trees or shrubs; interpetiolar glands and ridges absent, pseudostipules present or absent. Leaves opposite, pinnate or simple. Calyx cupular or campanulate, 5-dentate. Stamens 4, anthers monothecous or dithecous, thecae divaricate or not. Ovary unilocular with 2 parietal placentae or bilocular, placentation axile. Fruit indehiscent, fleshy or rarely fibrous with hard cortex.

About 35 genera; three genera and five species in the Pan-Himalaya.

The tribe Coleeae was previously included five genera (*Phylloctenium* Baill., *Phyllarthron* DC. ex Meisn., *Colea*, *Ophiocolea* H. Perrier and *Rhodocolea* Baill.) restrict to Madagascar and its neighboring areas. However, recent molecular evidence (Olmstead et al., 2009) indicated that several taxa placed in Tecomeae might be moved to the tribe Coleeae which is expended to include at least thirty genera.

4. PAULDOPIA Steenis

van Steenis, Acta Bot. Neerl. 18: 425. 1969; Tao & Yin in Wang (ed.), Fl. Reipubl. Popularis Sin. 69: 18. 1990; Zhang & Santisuk in Wu & Raven (eds.), Fl. China 18: 216. 1998. Type: *Pauldopia ghorta* (Buch.-Ham. ex G. Don) Steenis.

Radermachera sect. *Exalatae* Steenis, Act. Bot. Neerl. 2: 307. 1953. Type: *Radermachera bipinnata* (Coll. & Hemsl.) Steenis ex Chatterjee [= *Pauldopia ghorta* (Buch.-Ham. ex G. Don) Steenis].

Shrubs or small trees. Leaves opposite, 2-pinnately compound; rachis narrowly winged; leaflets subsessile, white scaly pubescent adaxially, sparsely glandular abaxially. Inflorescences paniculate, many flowcrcd. Calyx campanulate, 5-lobed. Corolla tube tubular; lobes 5, semiorbicular. Stamens 4, didynamous. Disc cupular. Ovary 2-locular; ovules borne on septum. Capsules dehiscing loculicidally, terete; septum membranous. Seeds compressed globose, thick, nearly wingless.

Recent references: van Steenis, 1969. *Pauldopia*, a new genus of Southeast Asian Bignoniaceae. *Acta Bot. Neerl.* 18(3): 425–428; Madhukar & Bandyopadhyay, 2017. Correction of a typographical error in *Bignonia 'ghorta'* (Bignoniaceae). *Phytotaxa* 331 (1): 147–150.

One species: China, India, Laos, Myanmar, Nepal, Sri Lanka, Thailand, Vietnam; also in Pan-Himalaya.

1. Pauldopia ghorta (Buch.-Ham. ex G. Don) Steenis, Acta Bot. Neerl. 18: 427. 1969; Tao & Yin in Wang (ed.), Fl. Reipubl. Popularis Sin. 69: 18. 1990; Zhang & Santisuk in Wu & Raven (eds.), Fl. China 18: 216. 1998. ≡ *Bignonia ghorta* Buch.-Ham. ex G. Don, Gen. Hist. 4: 222. 1838; de Candolle, Prodr. 9: 169. 1845. ≡ *Stereospermum ghorta* (Buch.-Ham. ex G. Don) C. B. Clarke in Hooker (ed.), Fl. Brit. India 4: 384. 1884. ≡ *Bignonia ghonta* Buch.-Ham. in Wallich, Numer. List n. 6510. 1832, nom. invalid. Type: India, "Deviduba", *Buchmann.-Hamilton in Wallich Cat. 6510* [lectotype designated by Steenis, Acta Bot. Neerl. 18: 427. 1969: K (K001124075)].

Shrubs, 1.5–2.5(–6) m high. Branchlets copiously lenticellate. Leaves 20–40 cm long; leaflet blade ovate-lanceolate, 3–7.5 cm long, 1.5–2.5 cm broad, sparsely pubescent, base cuneate, margin entire, ciliate, apex long acuminate. Panicles terminal, 8–12 cm long, nodding, sometimes flowers densely clustered at peduncle apex; peduncle 15–20 cm long. Pedicel 1–2 cm long, sparsely pubescent. Calyx 1.2–1.5 cm long, less than 1 cm in diameter, apex subtruncate. Corolla lobes semiorbicular, spreading, red-brown, 1.2–1.5 cm long; tube dark yellow, 3–6 cm long, slightly curved. Filaments filiform, 2–2.5 cm long, glabrous; anthers divergent, bithecous, connective subulate. Style 2.5–3 cm long, smooth; stigma lingulate. Capsules long terete, 18–25 cm long, 0.8–1.2 cm broad, long acuminate at both ends; valves thin leathery; septum membranous. Figure 46-130.

Phenology: flowering from May to June; fruiting from November to December.

Habitats: evergreen broad-leaved forests, roadsides, slopes; 800–1500 m.

Distribution: C Nepal, E Nepal, U Irrawaddy (Map 46-227); also in Laos, Sri Lanka, Thailand, Vietnam, and other parts of India, Myanmar and China (S Yunnan, SE Guangxi).

C NEPAL: no exact locality, *Wallich Cat. 6510* (K). **E NEPAL**: Dlam, 1200 m, 6 Jun. 1969, *L. H. J. Williams 432* (BM). **U IRRAWADDY**: **Chaungmaw Kyaun**, North Triangle, Khinlum, 1200 m, 2 May 1953, *F. Kingdon-Ward 20769* (BM).

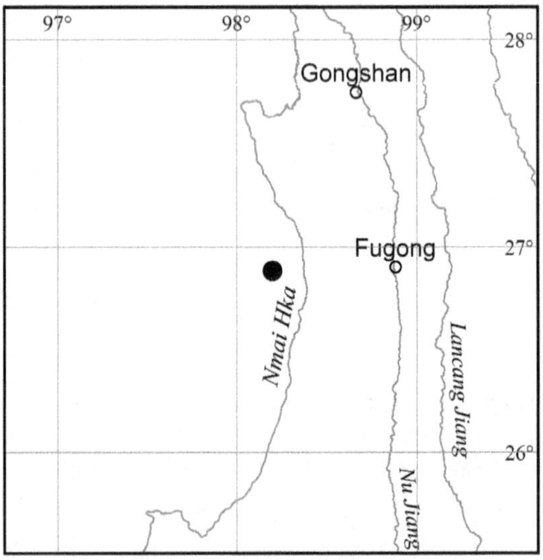

Map 46-227: **Pauldopia ghorta** (Buch.-Ham. ex G. Don) Steenis

Figure 46-130: **Pauldopia ghorta** (Buch.-Ham. ex G. Don) Steenis: **a**, habit; **b**, young calyx; **c**, mature flower; **d**, anthers; **e**, gynoecium surrounded by disk; **f**, ovules on septum; **g**, mature pod; **h**, seeds. Drawn by Yun-Xiao LIU.

5. RADERMACHERA Zoll. & Moritz

Zollinger & Moritz, Syst. Verz. 3: 53. 1855; Tao & Yin in Wang (ed.), Fl. Reipubl. Popularis Sin. 69: 26. 1990; Zhang & Santisuk in Wu & Raven (eds.), Fl. China 18: 218. 1998. ≡ *Stereospermum* sect. *Radermachera* (Zoll. & Moritz) Benth. in Bentam & Hooker (eds.), Gen. Pl. 2: 1047. 1876. Type: *Radermachera stricta* Zoll. & Moritzi.

Mayodendron Kurz, Prelim. Rep. Forest Veg. Pegu App. D: [1]. 1875; Forest Fl. Burma 2: 232. 1875. Type: *Mayodendron igneum* Kurz.

Trees. Leaves opposite, 1–3-pinnately compound; leaflets entire, petiolate. Inflorescences paniculate or reduced almost to fascicles, terminal or lateral. Bracts and bracteoles linear or leaflike. Calyx campanulate, apex 5-lobed or truncate. Corolla funnelform-campanulate or salverform; tube short or long, limb slightly bilabiate; lobes rounded, spreading. Stamens 4, didynamous or subequal; staminode usually present. Disc annular, slightly fleshy. Ovary terete; ovules numerous, 2-rowed per locule. Style included; stigma ligulate, 2-lobed. Capsules dehiscing loculicidally, long, terete, 2-angular; septum compressed, woody. Seeds compressed, transparent membranous winged at both ends.

About 16 species: tropical Asia; two species in Pan-Himalaya.

1a. Capsules thick leathery, without lenticle..1. *R. sinica*
1a. Capsules subwoody, densely white lenticellate .... 2. *R. yunnanensis*

1. Radermachera sinica (Hance) Hemsl., Hooker's Icon. Pl. 28: t. 2728. 1902; Tao & Yin in Wang (ed.), Fl. Reipubl. Popularis Sin. 69: 30. 1990; Zhang & Santisuk in Wu & Raven (eds.), Fl. China 18: 219. 1998. ≡ *Stereospermum sinicum* Hance, J. Bot. 20: 16. 1882. Type: China, Guangdong, "Lienchau", Jan. 1879, *C. Gerlach 20797* (holotype: K).

Radermachera borii C. E. C. Fisch., Bull. Misc. Inform. Kew 1940: 197. 1940. Type: India, Assam, Naga Hills, Khonoma, 1850 m, Nov.–Dec. 1939, *N. L. Bor 2699* [holotype: K (K000779277, K000779277); isotype: L (L0003405)].

Trees, to 10 m high. Leaves 2 (or 3)-pinnately compound; leaf rachis 20–30 cm long; lateral petiolules less than 5 mm long, terminal one 1–2 cm long; leaflets ovate to ovate-lanceolate, 2.5–8 cm long, 1–3.5 cm broad, glabrous, lateral veins 5 or 6 pairs on each side of midrib, base broadly cuneate, margin entire, apex caudate-acuminate. Panicles terminal, 25–35 cm long, 2.2–2.5 cm long, erect. Calyx 5-lobed; lobes ovate-lanceolate, 5–12 mm long, apex obtuse. Corolla white to pale yellow, campanulate-funnel-form, 5–8 cm long; lobes rounded, 1.5–2 cm long, 2–2.5 cm broad, margin crenate. Stamens 4, didynamous; staminode present, filiform. Ovules 2-rowed per locule; style exserted; stigma 2-lobed. Capsules terete, nodding, angular, 45–85 cm long, 0.5–1 cm broad, not ridged; pericarp thin leathery, indistinctly lenticellate; septum 0.3–0.4 mm in diameter, terete, slightly compressed. Seeds ellipsoid, including wing 2–3 cm long, 5–10 mm broad. Figure 46-131.

Phenology: flowering from May to September; fruiting from October to December.

Habitats: slopes and forests; 700–1700 m.

Figure 46-131: **Radermachera sinica** (Hance) Hemsl.: **a**, flowering branch; **b**, corolla with calyx; **c**, capsule; **d**, seeds. Drawn by Jin-Yong FENG; reproduced from *Icon. Cormophyt. Sin.* 4: 106, fig. 5625. 1975.

Distribution: Bhutan, M Yarlung Zangbo, Yarlung Zangbo-Brahmaputra, S Hengduan, U Irrawaddy (Map 46-228); also in Thailand, Vietnam, and other parts of India, Myanmar and China (S Yunnan, Guizhou, Guangxi, Guangdong, Hainan, Taiwan).

BHUTAN: **Deothang**, *Griffith 87* (K); **Punakha**, Rinchu, 1500 m, 2 Jun. 1915, *R. E. Cooper et al. 3947* (BM, E); **Shamgong**, Wangde Khola, 980 m, 4 Apr. 1982, *A. J. C. Grierson & D. G. Long 4302* (E); **Tashigang**, E bank of Kulong Chu near Shali, S of Tashi Yangtsi, 1675 m, 28 Jun. 1979, *A. J. C. Grierson & D. G. Long 2346* (E); **Thimphu**, Marichema, 900 m, 3 Jul. 1914, *R. E. Cooper, A. K. Bulley & E. Cheshire 1129* (BM, E); **Tongsa**, Mangde Chu Valley, 1000 m, 9 May 1937, *F. Ludlow & G. Sherriff 3003* (BM, TI). **M YARLUNG ZANGBO**: **Nyam Jang Chu**, Karteng, 1200 m, 28 May 1947, *F. Ludlow, G. Sherriff & H. H. Elliot 12574* (BM, TI). **YARLUNG ZANGBO-BRAHMAPUTRA**: **Baibung**, Xirang, 1500 m, 7 Dec. 1992, *H. Sun et al. 1962* (KUN); **Bangxing**, Pangxing, 1300 m, 28 Feb. 1993, *H. Sun et al. 4186* (KUN); **Gutang**, Gudeng to Gandai, 14 Dec. 1982, *B. S. Li & S. Z. Cheng 2194* (PE); **Luhit Valley**, Kahao, 1200 m, 23 May 1950, *F. Kingdon-Ward 19499* (BM); **Mêgog**, Mani, 850 m, 12 Aug. 1974, *Qinghai-Xizang Exped. 4183* (KUN, PE); **Mishmi Hills**, Glo, 1050 m, 18 Apr. 1949, *F. Kingdon-Ward 18521* (BM). **S HENGDUAN**: **Dali**, Taiyi, 1500–1700 m, 4 May 1988, *Y. H. Li 504* (KUN); **Fugong**, Emadi, 22 Jun. *1986, s. coll. s. n.* (HITBC068869); **Lushui**, Binggong, 1500 m, 15 Apr. 1960, *S. Jiang et al., 8159* (KUN, PE); Yangbi, 8 Nov. 1946, *T. N. Liou 22666* (PE). **U IRRAWADDY**: **Chaungmaw Kyaun**, North Triangle, Hkinlum, 1200 m, 25 Apr. 1953, *F. Kingdon-Ward 20715* (BM); **Tengchong**, Jietou, Jun. 1924, *G. Forrest 24491* (E, PE).

2. Radermachera yunnanensis C. Y. Wu & W. C. Yin, Fl. Yunnan. 2: 712. 1979; Tao & Yin in Wang (ed.), Fl. Reipubl. Popularis Sin. 69: 30. 1990; Zhang & Santisuk in Wu & Raven (eds.), Fl. China 18: 219. 1998. Type: China, Yunnan, Jingdong, *Nan Shui Bei Diao Dui 8159* [holotype: KUN; isotypes: PE (PE01155676, PE01155693)].

Trees, 10–16 m high. Barks gray-black. Leaves 2- (or 3-) pinnately compound, 55–70 cm long; lateral petiolule 1–5 mm long, terminal one 1–2.5 cm long; leaflet blade ovate, 4–9 cm long, 2–5 cm broad, adaxially densely white glandular, lateral veins 5–7 pairs, base cuneate, oblique, margin entire, apex caudate-acuminate. Panicles terminal, cymose. Calyx 2.6–3 cm long, 6–8 mm broad; teeth 5, narrowly triangular, 1–1.2 cm long. Corolla white to pale yellow, 7–9 cm long. Stamens 4, inserted nearly at base of corolla tube. Capsules linear-terete, gray-black, subwoody,

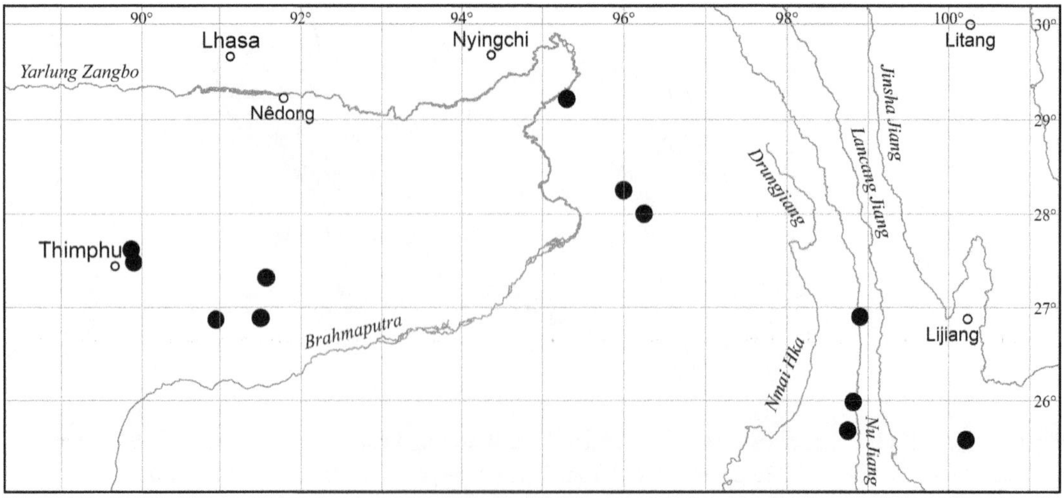

Map 46-228: **Radermachera sinica** (Hance) Hemsl.

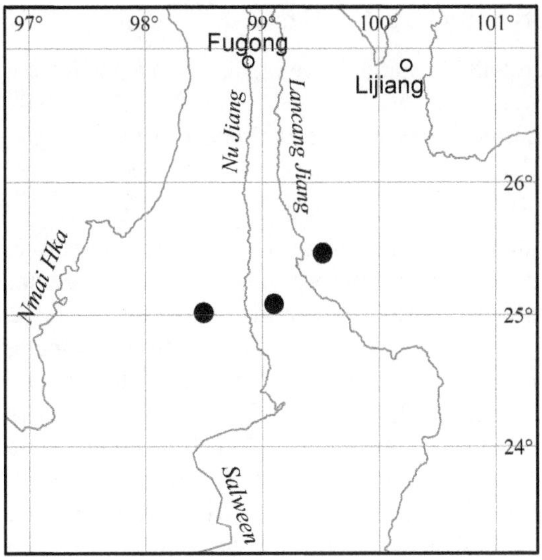

Map 46-229: **Radermachera yunnanensis** C. Y. Wu & W. C. Yin

40–50 cm long, 1–1.2 cm broad, densely white lenticellate, scabrous. Seeds ellipsoid, 1.4–1.7 cm long, 3–5 mm broad, wings narrow.

Phenology: flowering from April to May; fruiting from August to November.

Habitats: slopes, forests; 1400–2400 m.

Distribution: S Hengduan and U Irrawaddy (Map 46-229); also in other parts of China (C & SW Yunnan).

Notes: *Radermachera yunnanensis* C. Y. Wu & W. C. Yin differs from *R. sinica* (Hance) Hemsl. mainly in its lenticellate capsules. It is probable to be merged with the latter and the further study in future is necessary.

S HENGDUAN: Longyang, Gandang, Baihualing, Hanlongzhai, 1820 m, 23 Jul. 2010, *T. Zhang et al. 10CS 2050* (K); **Yongping**, Shan Yang, 2380 m, 24 May 2015, *Z. L. Dao DZL3764* (KUN). **U IRRAWADDY: Tengchong**, Nantien valley west of Tengyueh, May 1921, *G. Forrest 7552* (E); Sanyun, Chuanlong, 1700 m, 3 Nov. 1998, *H. Li 11456* (E, KUN); Wutai, Longtanghe to Jintang village, 1440 m, 24 May 2005, *Gaoligong Shan Biodiversity Survey 24778* (E, KUN); Wutai, 1750–1870 m, 27 May 2005, *Gaoligong Shan Biodiversity Survey 24964* (E, KUN).

6. STEREOSPERMUM Champ.

Champion, Linnaea.: 720. 1832; de Candolle, Prodr. 9: 210. 1845; Benthan & Hooker, Gen. Pl. 2: 104. 1875; Clarke in Hooker (ed.), Fl. Brit. India 4: 382. 1884; Schumann in Engler & Prantl (eds.), Nat. Pflanzenfam. 4(3b): 242. 1895; Tao & Yin in Wang (ed.), Fl. Reipubl. Popularis Sin. 69: 25. 1990; Zhang & Santisuk in Wu & Raven (eds.), Fl. China 18: 217. 1998. Type: *Stereospermum kunthianum* Champ.

Dipterosperma Hassk., Flora 25: 28. 1842. Type: *Dipterosperma personatum* Hassk.

Trees, deciduous. Leaves opposite, 1-pinnately compound; leaflets entire or dentate. Flowers in large lax terminal or axillary panicles. Calyx ovoid in bud, campanulate with short, rounded unequal lobes when open. Corolla tubular-ventricose or subcampanulate, narrowing abruptly in lower part, white with purple streaks, or crimson; lobes ovate or lacerate, spreading, nearly equal, toothed or laciniate. Stamens 4, didynamous, included; anthers divergent; staminode usually rudimentary. Disc cushion-like. Ovary sessile; ovules numerous in several rows. Capsules dehiscing loculicidally, elongate, terete, more or less quadrangular, slightly compressed or quadrangular, with small tubercules on surface; septum thickened, subterete, corky, transversely notched. Seeds with flat, oblong, membranous wing on each side; embryo with ridge embedded in notch in septum; wings of adjacent seeds not overlapping.

Recent references: Santisuk, 1973. Notes on Asiatic Bignoniaceae IV. Nomenclatural and taxonomic notes concerning Indian species of *Stereospermum*. *Kew Bull.* 28: 176–182; Mabberley, 1978. The Latin name of the 'Padri' Tree (Bignoniaceae), *Taxon* 26: 553.

About 20 species: tropical Asia, tropical Africa; two species in Pan-Himalaya.

1a. Young leaves and panicles glabrous or rarely puberulous abaxially; leaves apex caudate-acuminate; capsules twisted, 0.8–1.2 cm diameter .... 1. *S. colais*
1b. Young leaves and panicles glandular or pubescent; leaves apex shorty acuminate; capsules straight, 1.5–2 cm diameter ....2. *S. chelonoides*

1. Stereospermum colais (Buch.-Ham. ex Dillwyn) Mabb., Taxon 27: 553. 1978; Tao & Yin in Wang (ed.), Fl. Reipubl. Popularis Sin. 69: 24. 1990; Zhang & Santisuk in Wu & Raven (eds.), Fl. China 18: 217. 1998; Aitken in Grierson & Long, Fl. Bhutan 2(3): 1240. 2001. ≡ *Bignonia colais* Buch.-Ham. ex Dillwyn, Rev. Hortus Malab. 6: 28. 1839. Type: [*icon*] Rheede, Hort. Malab. 6: t. 26. 1684 (holotype).

Dipterosperma personatum Hassk., Flora 25: 2. 1842. ≡ *Stereospermum personatum* (Hassk.) Chatterjee, Bull. Bot. Soc. Bengal 2: 70. 1948. Type: not designated.

Bignonia caudata DC., Prodr. 9: 166. 1845. ≡ *Stereospermum caudatum* (DC.) Miq., Ann. Mus. Bot. Lugduno-Batavi 1: 200 (1864). Type: India, Mts. de Cottalan, sud de la peninsulae de l'Indiae, *Leschenault s. n.* [holotype: P; isotypes: G-DC (G00133420), L].

Stereospermum tetragonum A. DC., Prodr. 9: 210. 1845; Yin & Tao in Wu (ed.), Fl. Yunnan 2: 708. 1979. ≡ *Bignonia tetragona* DC., Bibl. Univ. Geneve II. 17 (Rev. Bignon. 8): 124. 1838, nom. invalid. Type: Myanmar, Taong Dong, *Wallich Cat. 6501D* [lectotype designated by Santisuk, Kew Bull. 28: 179. 1973: K (K001124056)].

Trees, 15–20(–35) m high; trunk 15–25(–80) cm in diameter; bark thick, rough, grayish-yellow. Leaves 25–50 cm long, broadly obovate to obovate-oblong in outline; petiole and rachis cylindric, leaflets 3–9 pairs; petiolule conspicuous, 0.3–1.2(–2)cm long; blade elliptic-oblong, 5–14 cm long, 2.5–6 cm broad, abaxially glabrous or rarely finely puberulous, base attenuate, truncate, or oblique, margin entire, apex narrowly caudate. Panicles terminal, 20–40 cm long; peduncle pubescent or glabrous. Bract and bracteoles caducous. Pedicel 3–4 mm long, pubescent or glabrous. Calyx campanulate, yellowish-purple, pubescent or glabrous, 4–7 mm long, 4–5 mm broad, 3–5-lobed. Corolla pale yellow, 1.5–2 cm long, slightly curved, tomentose at mouth, upper lip 2-lobed, lower lip 3-lobed; tube terete. Filaments 0.8–1 cm long, glabrous; anthers 1.2–1.5 cm long. Disk 1.2–1.5 mm long. Style 1–1.2 cm long; stigma 0.8–1 mm long, 2-lobed. Capsules terete, 4-angular, twisted when mature, 30–70 cm long, 0.6–1 cm in diameter; pericarp thick, subwoody; septum 4–7 mm in diameter. Seeds ovate in outline, compressed, white membranous winged at apices, including wing 2.5–2.8 cm long, 4–5 cm broad. **Chromosomes** 2n = 40. Figure 46-132.

Phenology: flowering from May to July; fruiting from September to November.

Habitats: forests, slopes; 100–1900 m.

Distribution: C & E Nepal, Sikkim & Darjiling, Bhutan, S Hengduan (Map 46-230); also in Bangladesh, Cambodia, Indonesia (Sumatra), Laos, Malaysia, Sri Lanka, Thailand, and Vietnam, and other parts of India, Myanmar and China (S Yunnan, Guizhou, Guangxi).

C NEPAL: Lumbini, Rupandehi, Khasyauli Forests, 100 m, May 1971, *J. Makin 154* (BM); **Narayani**, Chitwan, Royal Chitwan National Park, W Bhawanipur Choki, 250 m, 14 May 1976, *R. G. Troth 818* (KATH). **E NEPAL**: Maewa Khola, 1050 m, 1 Jul. 1969, *L. H. J. Williams 1058* (BM). **SIKKIM & DARJILING**: **Darjiling**, Chumbati, 600, Apr. 1876, *J. S. Gamble 652A, B* (BM, K); Garidura, Jul. 1973, *J. S. Gamble 3579* (K); **Sikkim**, 340–1000 m, s. d., *J. D. Hooker & T. Thomson s. n.* (L, P). **BHUTAN**: **Monggar**, Above Zimgaon, Shongar Chu, 1500 m, 14 Jul. 1979, *A. J. C. Grierson & D. G. Long 2469* (E); **Phuntsholing**, 350 m, 2 May 1979, *A. J. C. Grierson & D. G. Long 754* (E); **Sankosh**, Balu Khola 7 km W of Phipsoo, 200 m, 19 Mar. 1982, *A. J. C. Grierson & D. G. Long*

Figure 46-132: **Stereospermum colais** (Buch.-Ham. ex Dillwyn) Mabb.: **a**, fruiting branch; **b**, capsule septum; **c**, seeds. Drawn by Shun-Qing HE; reproduced from *Fl. Guangxi*. 4: 764, pl. 211. 2017; with the permission from Committee of *Flora of Guangxi*.

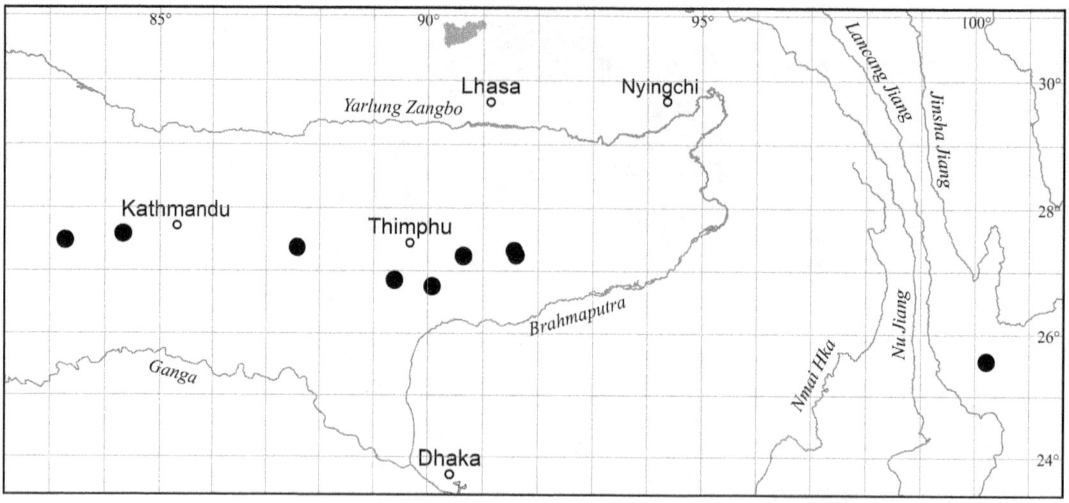

Map 46-230: **Stereospermum colais** (Buch.-Ham. ex Dillwyn) Mabb.

3848 (E, K); **Shamgong**, near Wangdegong, 1200 m, 4 Jun. 1979, *A. J. C. Grierson & D. G. Long 1615* (E, K); **Tashigang**, E bank of Dangme Chu between Cha Zam & Duksum, 1050 m, 29 Jun. 1979, *A. J. C. Grierson & D. G. Long 2356* (E, K). **S HENGDUAN**: **Dali**, Gonglang, Shale to Lingbaoshan National Forests Park, 1850 m, 1 Aug. 2014, *X. K. Cang NJ034* (KUN).

2. Stereospermum chelonoides (L. f.) DC., Biblioth. Universelle Geneve 17: 124. 1838; Aitken in Grierson & Long (eds.), Fl. Bhutan 2(3): 1240. 2001. ≡ *Bignonia chelonoides* L. f., Suppl.: 282. 1781, pro parte, excl. Rheede, *Hort. Malab.* 6: 47, t. 26. 1684; Willd., Sp. Pl. 3: 304 (1801), pro parte; Roxb., Fl. Ind., ed. Carey, 3: 106 (1832), pro parte. Type: India, *Koenig s. n.* (lectotype designated by Santisuk, Kew Bull. 28: 176. 1973: LINN; isolectotype: LD).

Bignonia suaveolens Roxb., Fl. Ind. (ed. 1832) 3: 104. 1832. ≡ *Tecoma suaveolens* (Roxb.) G. Don, Gen. Hist. 4: 224. 1838. ≡ *Stereospermum suaveolens* (Roxb.) A. DC., Prodr. 9: 211. 1845. ≡ *Heterophragma suaveolens* (Roxb.) Dalzell & Gibs., Bombay Fl.: 161. 1861. ≡ *Spathodea suaveolens* (Roxb.) Benth. & Hook. f., Gen. Pl. 2: 1047. 1876. Type: [*icon*] Roxburgh Icon. 1147 (lectotype designated by Santisuk, Kew Bull. 28: 177. 1973: K).

Trees. Petiole and rachis tomentose; leaflets 2–3 pairs; petiolule stout or subsessile, (0–)2–5 mm long, pubescent; blade broadly elliptic or broadly ovate, 10–15 cm long, 6–9.5 cm broad, densely glandular-tomentose, base truncate to subcordate, margin entire, apex shortly acuminate, glabrous adaxially, pubescent abaxially. Panicles terminal, densely glandular-tomentose. Calyx 0.6–0.8 cm long, tomentose. Corolla red; tube 1.8–2.2 cm long; lobes 0.5–1.2 cm long, 4–9 mm in diameter, pubescent outside, densely tomentose within. Capsules 25–35 cm long, 1.5–2 cm broad, not twisted, not prominently ridged; septum 1.2–1.5 cm in diameter. Seeds 7–9 mm long, 2.0–2.5 cm broading, wing narrow.

Phenology: flowering from May to July; fruiting from September to November.

Habitats: hillside and forests; 100–1300 m.

Distribution: U Ganga & Indus, W Nepal, C Nepal, E Nepal, Sikkim & Darjiling, U Irrawaddy (Map 46-231); also in Sri Lanka, and other parts of India and Myanmar.

Uses: The bark and roots are used for coughs and renal complaints.

U GANGA & INDUS: **Dehradunn**, Lachiwala, 540 m, Nov. 1897, *H. H. Rich 71* (K); New forests, Jun. 1935, *M.B. Razada 5027* (BM);

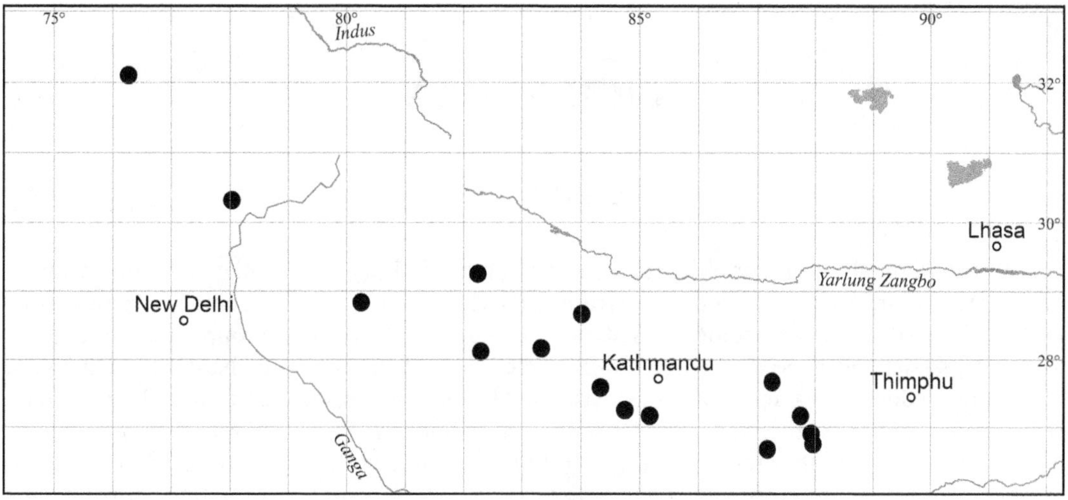

Map 46-231: **Stereospermum chelonoides** (L. f.) DC.

Kangra, 600 m, 5 May 1900, *C. Ram 123* (E); **Kumaon**, *A. G. Hobart-Hampaden 75* (E). **W NEPAL**: **Bheri**, Bardia, Royal Bardia National Park, Chisapani to Gainikanda, 980 m, 6 Oct. 1996, *M. Mikage et al. 9684432* (TI); **Mahakali,** Kanchapur, Mahakali, 25 miles NW of Dhangarhi, 250 m, 7 Dec. 1966, *D. H. Nicolson 2853* (BM, KATH); **Rapti**, Dang, Bijuli, Piuthan, 1300 m, 8 Mar. 1991, *P. P. Kurmi KB0077* (KATH). **C NEPAL**: **Gandaki**, Manang, Marsyandi Khola, 12 Aug. 1983, *H. Ohba et al. 8331462* (E, TI); **Janakpur**, Sindhuli, Hariharpur, Garhi, 350 m, 12 Febuary 1967, *J. D. A. Stainton 5629* (BM); **Lumbini**, Gulmi, Ridhi, 450 m, 2 Mar. 1976, *N. P. Manandhar & P. M. Regmi 68* (KATH); Nawalparasi, Tomaspur forests, 150 m, Mar. 1971, *J. Makin 90* (BM); Nawalparasi, Nawal Farasi, Tomaspur forests, 150 m, Mar. 1971, *J. Makin 90* (TI); **Narayani**, Bara, Patraia to Sakti Mohar Khola, 200 m, 18 Apr. 1969, *H. Kanai & M. S. Bista 11442* (KATH, KYO, TI); Bara, Bakeya, 240 m, 6 May 1978, *H. Tabata et al. 9629* (KYO); Chitwan, Chitwan National Park, 120 m, 20 Nov. 2004, *C. A. Pendry et al. DNEP2 A33* (E, KATH); Chitwan, Royal Chitwan National Park, 250 m, 24 Mar. 1976, *R. G. Troth 701* (KATH); Chitwan, Sauraha to Kasara, 13 Jan. 1996, *M. Suzuki et al. 9615006* (KATH); Chitwan, Kasara, Royal Chitwan National Park, 200 m, 10 Dec. 1975, *D. P. Joshi et al. 75/3467* (KATH); Parsa, Parsa Wildlife Reserve, along Mahadev Khola, 190–200 m, 9 Oct. 1995, *M. Mikage et al. 9550574* (BM, KATH); Chitwan, Tiger Tops, 130–150 m, 21 Jan. 1996, *M. Mikage et al. 961208* (BM, KATH); Chitwan, 500 m, Jun. 1994, *K. Wesche 47* (BM); Chitwam, SW of Patch 4, 250 m, 24 Mar. 1976, *R. G. Troth 701* (BM). **E NEPAL**: **Koshi**, Sankhuwasabha, 340 m, 14 Aug. 1997, *S. Noshiroet al. 9755050* (E, TI); Sunsari, above Chyuribas, 400 m, 12 Nov. 1978, *N. P. Manandhar et al. 736* (KATH); **Mechi**, Ilam, Chulachuli, 370 m, 21 Sep. 1972, *T. B. Shrestha & T. K. Bhattachaya 72-40* (KATH); Ilam, Tarhara to Ghapa, 300 m, 30 Sep. 1972, *T. B. Shrestha & T. K. Bhattachaya 72-283* (KATH); Jhapa, Gamjbari, 300 m, 2 Apr. 1967, *J. D. A. Stainton 5760* (BM); Jhapa, Sibganja to Mahara Bahara, 12 Dec. 1963, *G. Murata & M. Togashi 6303489* (KATH, KYO, TI); Panchthar, Hangthuwa, Phidim, 700 m, 17 Nov. 1978, *P. Pradhau et al. 983* (KATH). **SIKKIM & DARJILING**: **Darjiling**, 26 Nov. 1926, *J. M. Cowan s. n.* (E); Darjiling Terni, 750 m, Apr. 1876, *J. S. Gamble 652* (BM); s. loc., *J. S. Gamble 3521A* (K). **U IRRAWADDY**: **Putao**, valley of the Mali Kha, Suphka Ga, 450–600 m, 3 Jul. 1937, *F. Kingdon-Ward 12780* (BM).

VERBENACEAE

Qiang WANG

Herbs, annual or perennial, shrubs, sometimes climbing, or small trees. Stems terete or quadrangular, erect, suberect, prostrate, or twining, sometimes aromatic, glabrous or with various indumentum, sometimes with prickles or spines. Leaves opposite and usually decussate, sometimes alternate, whorled or fascicled; leaf blade simple, pinnatifid to pinnatisect, sometimes reduced to a scale or spine; stipules absent. Inflorescences terminal or axillary, racemose or spicate, sometimes capitate; simple or compound. Bracts showy or inconspicuous. Flowers bisexual, rarely dioecious. Calyx tubular or campanulate, persistent. Corolla 4- or 5-lobed, zygomorphic or subactinomorphic; corolla lobes usually spreading; corolla tube sometimes enlarged in the middle or at throat. Fertile stamens reduced to 4 or 2, with one or more staminodes, inserted on corolla tube, alternate with petals, included or slightly exserted; filaments free; anthers 2-locular, basifixed or dorsifixed, dehiscing longitudinally. Ovary superior, 2 carpellate, carpel 2-locular; ovules 1 or 2 per locule, erect or pendulous. Style terminal, simple, entire or 2-lobed, rarely weakly 4-lobed. Fruit drupaceous with one, two or four 1-seeded pyrenes, or with two or four 2-seeded pyrenes, or a dry or fleshy schizocarp separating at maturity into two 1- or 2-seeded mericarps, or four 1-seeded mericarps.

Recent references: O'Leary et al., 2009. Reassignment of species of paraphyletic *Junellia s. l.* to the new genus *Mulguraea* (Verbenaceae) and new circumscription of genus *Junellia*: Molecular and morphological congruence. *Syst. Bot.* 34: 777–786; Marx et al., 2010. A molecular phylogeny and classification of Verbenaceae. *Amer. J. Bot.* 97(10): 1647–1663; Angiosperm Phylogeny Group, 2016. An update of the Angiosperm Phylogeny Group classification for the orders and families of flowering plants: APG IV. *Bot. J. Linn. Soc.* 141: 399–436.

The family consists of 36 genera and about 1000 species; distributed in tropical, subtropical to temperate regions, mainly in the New World; eight genera and 11 species in the Pan-Himalaya.

Note: In the present account, tribes are not used, because there are only a few taxa of Verbenaceae distributed in the Pan-Himalaya region, most of which are introduced.

1a. Calyx corolla-like, large and showy; calyx lobes longer than corolla; calyx scales present at inner base of calyx lobes and alternate with them .... 1. *Petrea*
1b. Calyx not corolla-like; calyx lobes shorter than corolla; calyx scales absent.
  2a. Inflorescence elongated and lax; some stamens reduced to sterile staminodes, inserted nearly the middle of corolla tube.
    3a. Indumentum not uncinulate; anther thecae without glandular connectives.
      4a. Flowers with short but distinct pedicel; bracts inconspicuous; fertile stamens 4, staminode one or absent; thecae parallel; fruit a drupe with four 2-seeded pyrenes .... 2. *Duranta*
      4b. Flowers sessile; bracts conspicuous, woody and reflexed after anthesis; fertile stamens two, staminodes two; thecae divaricate; fruit a schizocarp, separating into two 1-seeded mericarps..3. *Stachytarpheta*

3b. Indumentum uncinulate; anther thecae with enlarged glandular connectives ....4. *Priva*

2b. Inflorescence short and dense, often capitate; all stamens fertile, staminodes absent.

5a. Calyx not enclosing fruit after anthesis; fruit a schizocarp, separating into four 1-seeded mericarps ....5. *Verbena*

5b. Calyx enveloping fruit after anthesis; fruit a drupe with two 1-seeded pyrenes, or a schizocarp separating into two 1-seeded or 2-seeded mericarps.

6a. Calyx conspicuously 2- or 4-lobed; fruit a schizocarp, separating into two 1-seeded mericarps.

7a. Calyx hair uncinulate; corolla hypocrateriform, 5-lobed............................6. *Phyla*

7b. Calyx hair not uncinulate; corolla infundibular, 4-lobed...........................7. *Lippia*

6b. Calyx truncate or shallowly lobed; fruit a drupe containing two 1-seeded pyrenes ....8. *Lantana*

1. PETREA L.

Linnaeus, Sp. Pl. 2: 626. 1753; Yao in P'ei & Chen, Fl. Reipubl. Popularis Sin. 65 (1): 21. 1982; Grierson & Long, Fl. Bhutan 2 (2): 918. 1999. Type: *Petrea volubilis* L.

Scandent shrubs, lianas, or small trees. Leaves decussately opposite or verticillate; leaf blade coriaceous, margin entire. Inflorescences terminal or axillary, racemose and many flowered. Bracts inconspicuous or absent. Flowers subactinomorphic, pedicellate. Calyx corolla-like, tubular or campanulate, ribbed, blue to purple, sometimes white, 5-lobed with 5 scales alternately situated at inner base of each lobe, lobes reticulately veined. Corolla purple or white, hypocrateriform, unequally 5-lobed, corolla tube sometimes urceolate or infundibular. Fertile stamens 4, didynamous, inserted above the middle of corolla tube, included, single staminode usually present. Ovary 1-carpellate, 2-locular with 1 ovule in each locule; style persistent or deciduous, obscurely 2-lobed, stigma oblique. Fruit drupaceous with two 1-seeded pyrenes (or one 1-seeded pyrene by abortion), enclosed within persistent calyx. **Chromosomes** 2n = 34.

Recent references: Rueda, 1994. Systematics and evolution of the genus *Petrea* (Verbenaceae). *Ann. Missouri Bot. Gard.* 81: 610–652.

About 11 species, native in tropical America; one species cultivated and naturalized in the Pan-Himalaya.

1. **Petrea volubilis** L., Sp. Pl. 2: 626. 1753; Moldenke, Phytologia 7: 446. 1961; Yao in P'ei & Chen (eds.), Fl. Reipubl. Popularis Sin. 65 (1): 22. 1982; Grierson & Long, Fl. Bhutan 2(2): 918. 1999. Type: *Herb. Clifford. 319* [lectotype designated by Moldenke, Repert. Spec. Nov. Regni Veg. 43: 35. 1938: BM (BM000646189)].

Lianas, up to 6 m long. Branches grey and hairy. Leaves decussately opposite; petiole 0.4–1.5 cm long; leaf blade elliptic, ovate-elliptic to oblong, 5–20 cm long, 2.5–10 cm broad, asperous and coriaceous, base truncate to rounded, margin entire and usually sinuate, apex obtuse or acute. Inflorescences terminal, racemose, 10–30 cm long, usually nutant. Bracts linear, small and inconspicuous. Flowers pedicellate, pedicel 7–8 mm long. Calyx corolla-like, deeply 5-lobed, densely tomentose, calyx tube 2–3 mm long, lobes 1.2–2 cm long, 4–6 mm broad, narrowly oblong; scales 5, alternately situated at inner base of calyx lobe, triangular, apex acute. Corolla hypocrateriform, blue or pale violet, deeply 5-lobed; corolla tube infundibular. Stamens 4, inserted above the middle of corolla tube. Fruit drupaceous, obovoid-oblong, enclosed within persistent calyx. Figure 46-133.

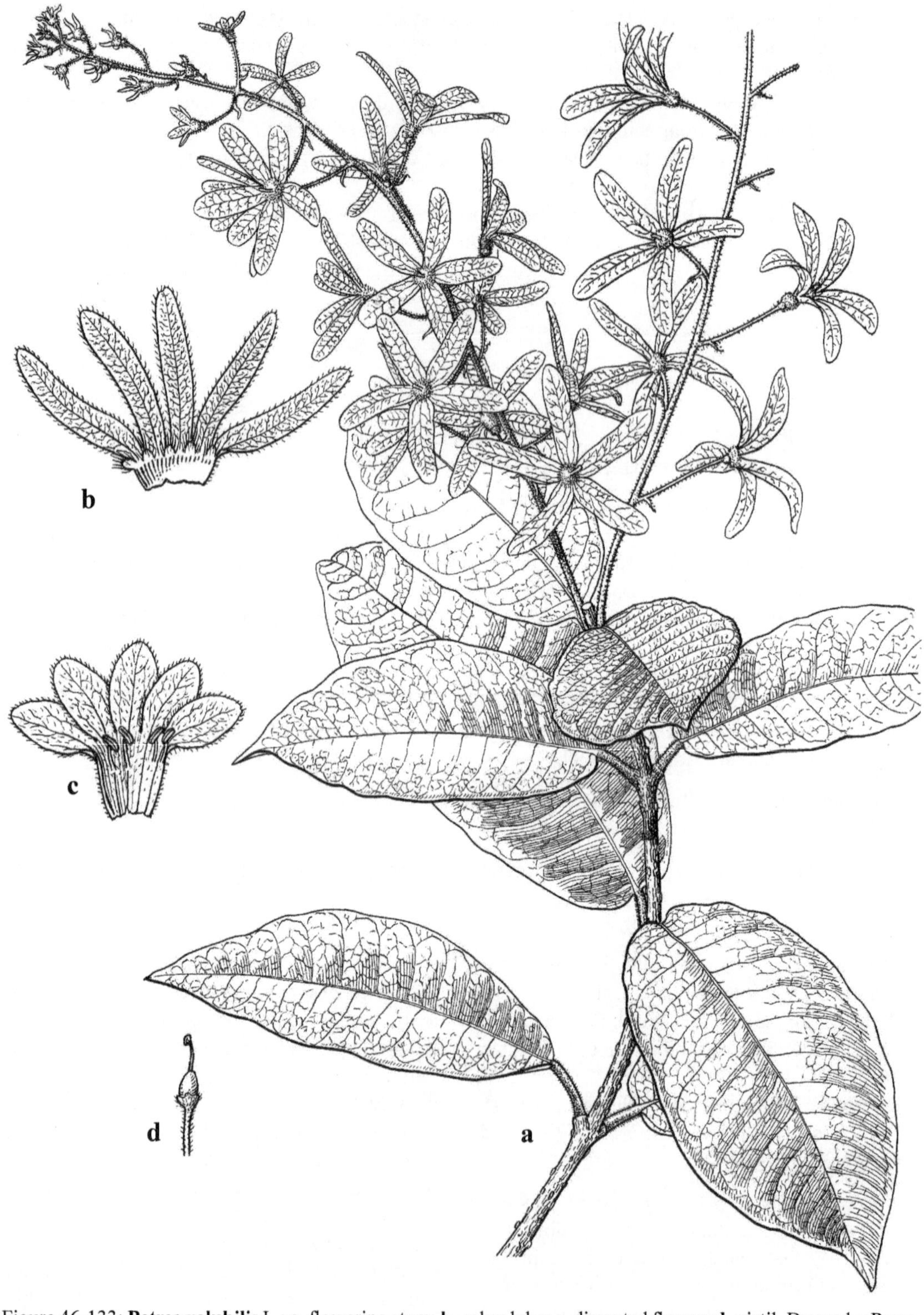

Figure 46-133: **Petrea volubilis** L.: **a**, flowering stem; **b**, calyx lobe; **c**, dissected flower; **d**, pistil. Drawn by Rong-Dao CHEN and redrawn by Chao-Zhen JI; reproduced from *Fl. Reipubl. Popularis Sin.* 65 (1): 23, pl. 4. 1982.

Phenology: flowering from March to May.

Habitats: woods or thickets.

Distribution: cultivated and sometimes naturalized in Sikkim & Darjiling of the Pan-Himalaya (Map 46-232); native to tropical America.

SIKKIM & DARJILING: **Sikkim,** Mar. 1913, *D. C. Prier. XI. 618* (BM).

Map 46-232: **Petrea volubilis** L.

2. DURANTA L.

Linnaeus, Sp. Pl. 2: 637. 1753; Chen & Lan in P'ei & Chen (eds.), Fl. Reipubl. Popularis Sin. 65(1): 22. 1982; Chen & Gilbert in Wu & Raven (eds.), Fl. China 17: 2. 1994; Grierson & Long, Fl. Bhutan 2 (2): 918. 1999. Type: *Duranta erecta* L. (lectotype designated by M. L. Green in Hitchcock & Green, Prop. Brit. Bot.: 169. 1929).

Shrubs or small trees, erect or subscandent, usually spinose. Leaves opposite or whorled, leaf blade margin entire, serrate or dentate. Inflorescences axillary or terminal, racemose. Bracts small, sometimes inconspicuous. Calyx tubular or campanulate, 5-lobed, persistent and enveloping fruit. Corolla blue to purple, sometimes white, subactinomorphic and hypocrateriform, 5-lobed, lobes slightly unequal. Stamens 4, short, inserted above the middle of corolla tube, thecae parallel, single staminode sometimes present. Ovary 8-locular; ovules 8, 1 for each locule. Style weakly 4-lobed; stigma capitate. Fruit drupaceous with four 2-seeded pyrenes. **Chromosomes** 2n = 16.

Recent references: Sanders, 1984. Provisional synopsis of the species and natural hybrids in *Duranta* (Verbenaceae). *Sida* 10: 308–318.

About 20 species, distributed in tropical and subtropical America; one species cultivated and naturalized in the Pan-Himalaya.

1. Duranta erecta L., Sp. Pl. 2: 637. 1753; P'ei, Mem. Sci. Soc. China 1(3): 13. 1932; Chen & Gilbert in Wu & Raven (eds.), Fl. China 17: 2. 1994; Grierson & Long, Fl. Bhutan 2 (2): 918. 1999. Type: [*icon*] *"Duranta inermis"* in Plumier in Burman, Pl. Amer.: t. 79. 1756 (lectotype designated by Caro in Revista Argentiana Agron. 23: 5. 1956).

Duranta repens L., Sp. Pl. 2: 637. 1753. [*icon*] "*Duranta inermis*" in Plumier in Burman, Pl. Amer.: t. 79. 1756 (lectotype designated by Caro in Revista Argentiana Agron. 23: 5. 1956). Type: *Herb. Linn.* No. 806.2 (neotype designated by Sanders in Howard, Fl. Lesser Antilles 6: 225. 1989: LINN).

Shrubs, 1.5–3 m tall, erect or subscandent. Branches spiny, pubescent when young. Leaves opposite, rarely verticillate, petiolate; petiole short, ca. 1 cm long; leaf blade ovate to lanceolate, 1.5–6.5 cm long, 1–3.5 cm broad, papery and pubescent, base cuneate, margin entire to distally crenate, apex acute or obtuse. Inflorescences terminal and axillary, racemose, lax and many flowered. Bracts tiny and inconspicuous. Calyx tubular, 5-lobed, both surfaces pubescent. Corolla blue to purple, 5-lobed, lobes slightly unequal. Stamens 4, included. Ovary glabrous; style included. Fruit drupaceous with four 2-seeded pyrenes, 5–8 mm in diameter, enclosed in persistent calyx. Figure 46-134.

Phenology: flowering and fruiting throughout the year.

Habitats: woods or thickets; 200–400m.

Distribution: cultivated and naturalized in U Ganga & Indus, C Nepal, E Nepal, Sikkim & Darjiling and S Hengduan of the Pan-Himalaya (Map 46-233); native to tropical America, widely cultivated and naturalized in tropical Africa, Asia and Australia.

Note: This species has long and complex synonymy. Here only the involved names from Pan-Himalaya region are treated. *Duranta plumieri* is an illegitimate name because its protolog cited the names of *Duranta* published in Species Plantarum 1753.

U GANGA & INDUS: **Darcha**, Dacca station, 5 Dec. 1868, *C. B. Clarke 8121* (BM). **C NEPAL**: **Bagmati**, Kathmandu, Rabi Bawan, SW of Kathmandu, 1300 m, 30 Oct. 1966, *D. H. Nicolson 2711* (BM); Zowar Hill, Himal, Bee-Concern, 1290–1380 m, 15 Sep. 1993, *s. col. 93-114* (K); **Dhawalagiri**, Baglung, Kali Gandaki Valley, 760 m, 19 Apr. 1954, *J. D. A. Stainton et al. 33* (BM); Baglung, Tatopani, S of Dana, Kali Gandaki, at the edge of fields, 1520 m, 29 Aug. 1954, *J. D. A. Stainton et al. 7563* (BM); **Gandaki**, Kaski, Pokhara, 1000 m, 15 Aug. 1959, *J. D. A. Stainton et al. 6894* (BM). **E NEPAL**: **Mechi**, Ilam, 1200 m, 6 Dec. 1963, *H. Hara et al. 6303504* (BM). **SIKKIM & DARJILING**: **Sikkim**, *s. d., D. Junslu, s. n.* (K). **S HENGDUAN**: **Panzhihua**, Cycas Natural Reserve, Minzheng, 11 Jul. 2010, *Panzhihua Cycas Exped. 1* (PE).

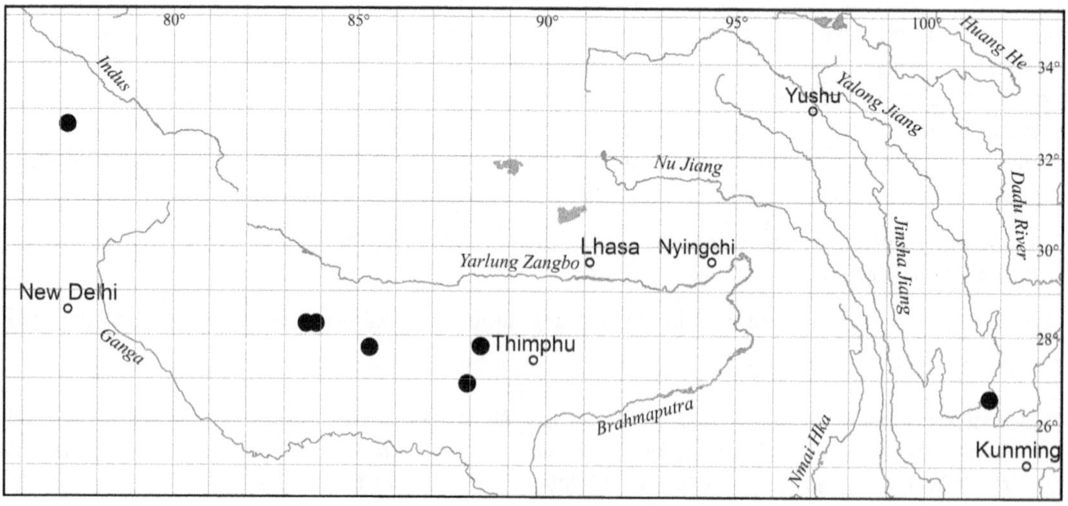

Map 46-233: **Duranta erecta** L.

Figure 46-134: **Duranta erecta** L.: **a**, flowering stem; **b**, branch with thorns; **c**, flower; **d**, dissected flower; **e**, persistent calyx; **f**, fruit. Drawn by Wei-Qing SHI; reproduced from *Fl. Reipubl. Popularis Sin.* 65 (1): 24, fig. 7. 1982.

3. STACHYTARPHETA Vahl

Vahl, Enum. Pl. 1: 205. 1804; Clarke in Hooker (ed.), Fl. Brit. India 4: 564. 1885; Chen & Lan in P'ei & Chen (eds.), Fl. Reipubl. Popularis Sin. 65(1): 20. 1982; Chen & Gilbert in Wu & Raven Hong (eds.), Fl. China 17: 3. 1994; Grierson & Long, Fl. Bhutan 2 (2): 917. 1999. Type: *Stachytarpheta jamaicensis* (L.) Vahl.

Herbs or shrubs. Leaves petiolate, opposite; leaf blade usually rugose; leaf margin dentate or crenate. Inflorescence terminal, spicate. Bracts conspicuous, usually woody and reflexed after anthesis. Flowers sessile or embedded in the inflorescence axis. Calyx tubular, 2–5-lobed, membranous, persistent and enveloping fruit. Corolla white, pink, red, blue, violet to nearly black, hypocrateriform or infundibular, 5-lobed; corolla tube straight or curved. Fertile stamens 2, posterior, thecae divaricate, staminodes 2, anterior, all included. Style persistent, stigma capitate. Fruit a schizocarp, separating at maturity into two 1-seeded mericarps.

About 90 species, widely distributed in tropical and subtropical America, with a few weedy species introduced and naturalized in tropical and subtropical Australia, Asia, and Africa; one species introduced and naturalized in the Pan-Himalaya.

1. Stachytarpheta jamaicensis (L.) Vahl, Enum. Pl. 1: 206. 1804; P'ei, Mem. Sci. Soc. China 1(3): 11. 1932; Chen & Lan in P'ei & Chen (eds.), Fl. Reipubl. Popularis Sin. 65(1): 20. 1982; Chen & Gilbert in Wu & Raven (eds.), Fl. China 17: 3. 1994. ≡ *Verbena jamaicensis* L., Sp. Pl. 1: 19. 1753. Type: *Herb. Linn. 7.13* (lectotype designated by Fernandes, Bol. Soc. Brot., sér. 2 57: 100. 1984: LINN).

Herbs, perennial and robust, or subshrubs, 0.6–2 m tall. Branches quadrangular, sparsely pubescent to subglabrous. Leaves petiolate, petioles 1–3 cm long, winged; leaf blade elliptic to oblong-ovate, 2.4–8 cm long, 3.5–5 cm broad, sparsely pubescent, base cuneate, margin coarsely serrate, apex shortly acuminate. Inflorescences terminal, spicate, 10–30 cm long. Bracts conspicuous, ca. 5 mm long, membranous. Calyx tubular, ca. 6 mm long, membranous, glabrous. Corolla dark blue, 7–12 mm long, 5-lobed. Stamens 2, short, thecae divaricate, staminodes 2, all included. Ovary glabrous; style exserted; stigma capitate. Fruit a schizocarp, enclosed in persistent calyx, separating at maturity into two 1-seeded mericarps. Figure 46-135.

Phenology: flowering in August; fruiting from September to December.

Habitats: shaded grassy slopes; 300–600m.

Distribution: cultivated and sometimes naturalized in Sikkim & Darjiling of the Pan-Himalaya (Map 46-234); native to tropical and subtropical America, widely cultivated and naturalized in tropical Africa, Asia and Australia.

SIKKIM & DARJILING: Sikkim, Fulhaigusi, 300 m, Sep. 1898, *J. J. Wood s. n.* (K).

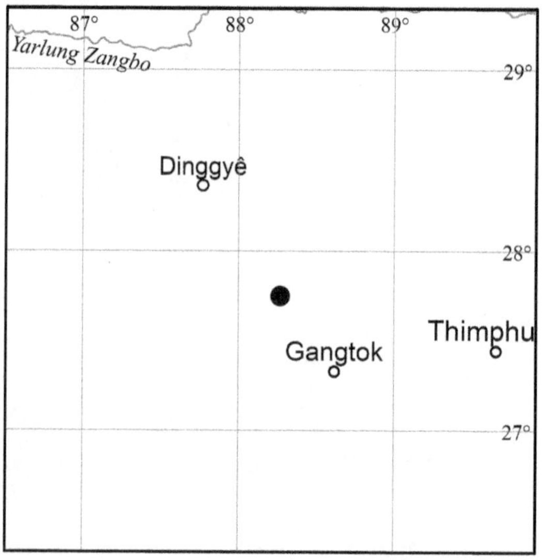

Map 46-234: **Stachytarpheta jamaicensis** (L.) Vahl

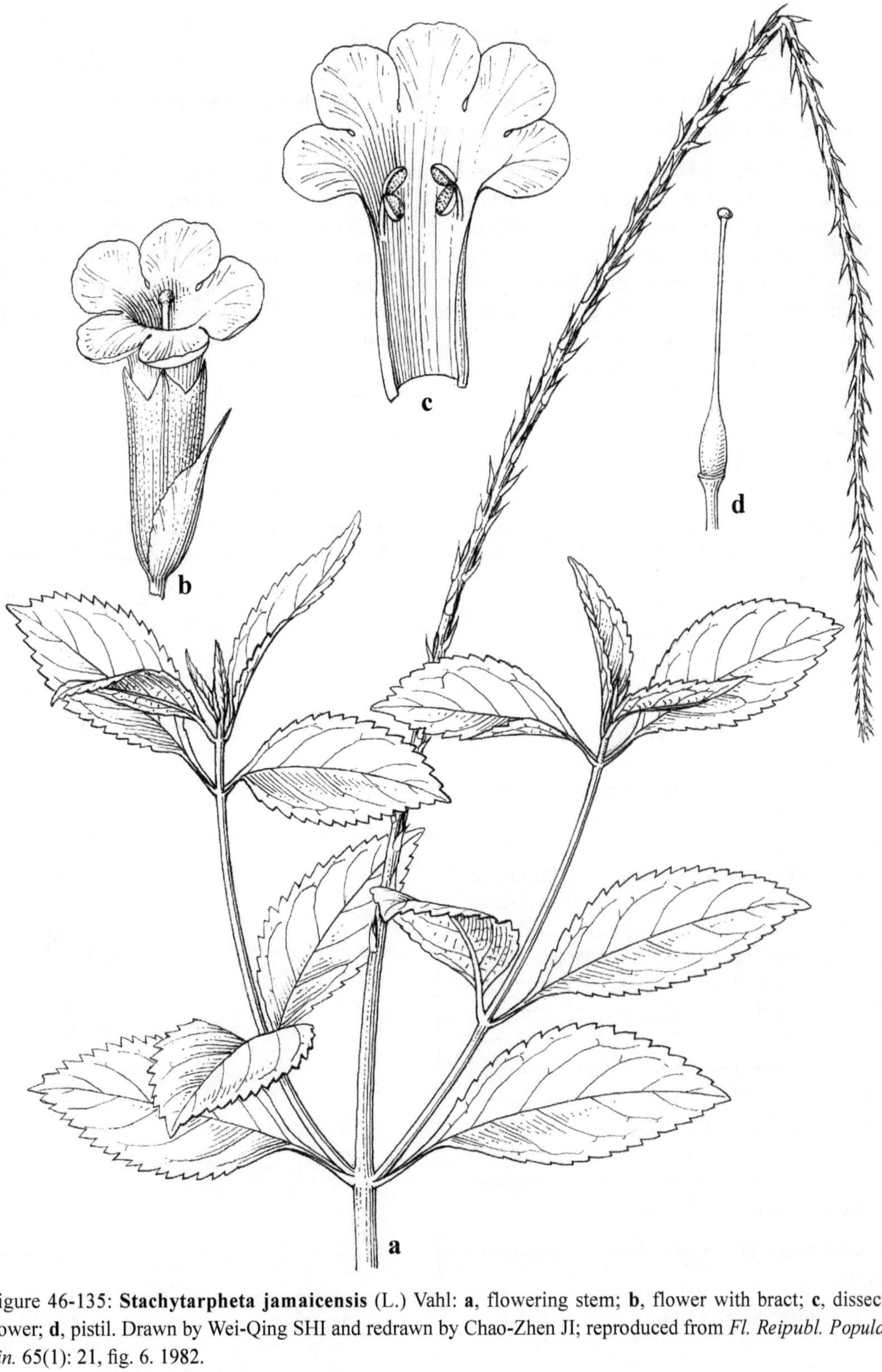

Figure 46-135: **Stachytarpheta jamaicensis** (L.) Vahl: **a**, flowering stem; **b**, flower with bract; **c**, dissected flower; **d**, pistil. Drawn by Wei-Qing SHI and redrawn by Chao-Zhen JI; reproduced from *Fl. Reipubl. Popularis Sin.* 65(1): 21, fig. 6. 1982.

4. **PRIVA** Adans.

Adanson, Fam. Pl. 2: 505. 1763; Clarke in Hooker (ed.), Fl. Brit. India 4: 565. 1885; Moldenke, Ann. Missouri Bot. Gard. 60(1): 77. 1973. Type: *Priva lappulacea* (L.) Pers.

Perennial herbs, sometimes woody at base. Stem erect or decumbent, indumentum of uncinulate hairs. Leaves opposite or subopposite; leaf margin dentate. Inflorescence terminal or axillary, spicate, lax and elongate. Bracts small and inconspicuous. Calyx tubular or tubular-campanulate, sometimes urceolate, 5-lobed, persistent and enveloping fruit, indumentum sometimes uncinulate. Corolla white or mauve, hypocrateriform or infundibular, somewhat zygomorphic, 5-lobed, lobes unequal. Stamens 4, didynamous, inserted slightly above the middle of corolla-tube, posterior pair of stamens with glandular connectives, thecae divaricate; staminode sometimes present. Style persistent or deciduous, anterior lobe larger and stigmatiferous, posterior lobe reduced. Fruit a schizocarp separating at maturity into two 2-locular, dorsally reticulate, verrucose or spiny, 2-seeded mericarps. **Chromosomes** 2n = 12.

About 20 species, mostly distributed in tropical and subtropical Asia, Africa and America; one species in the Pan-Himalaya.

1. Priva cordifolia (L. f.) Druce in Rep. Bot. Exch. Club Brit. Isles. 1916: 641. 1917. ≡ Buchnera cordifolia L. f., Suppl. Pl. 287. 1781[1782]. Type: India, Coromandel, Tanchaur, 1776, *I. G. Koenig s. n.* (**lectotype designated here**: BM000950133; isolectotypes: BM000950134, BM000950135)

Herbs, perennial, erect, up to 1 m tall, sometimes woody at base. Stem erect, quadrangular, uncinulate. Leaves opposite, petiolate; petiole 1–3 cm long; leaf blade ovate, ovate-triangular to elliptic, indumentum uncinulate, 2–8 cm long, 1–5 cm broad, base subcordate or subtruncate, margin serrate, apex obtuse. Inflorescences terminal, spicate, 10–20 cm long, lax, elongated after anthesis. Bracts tiny and inconspicuous. Calyx tubular, shallowly 5-lobed, 6–8 mm long, 2–2.5 mm broad, persistent, enlarged, urceolate and enveloping fruit. Corolla white, rarely mauve, 8–12 mm long; 5-lobed, lobes unequal; corolla tube slightly exceeding the calyx. Stamens 4, didynamous, posterior pair stamens with glandular connective; thecae divaricate. Fruit a schizocarp, separating into two 2-seeded mericarps. Figure 46-136.

Phenology: flowering and fruiting throughout the year.

Habitats: margins of thickets and woods.

Distribution: U Irrawaddy (Map 46-235); also distributed in other parts of India and Pakistan.

Note: In the *Flora of Pakistan*, this species is said to be distributed in Nepal. However, I have not found any specimen to support its distribution in Nepal.

U IRRAWADDY: **Mahtum**, Jul. 1889, *H. Collett 851* (K).

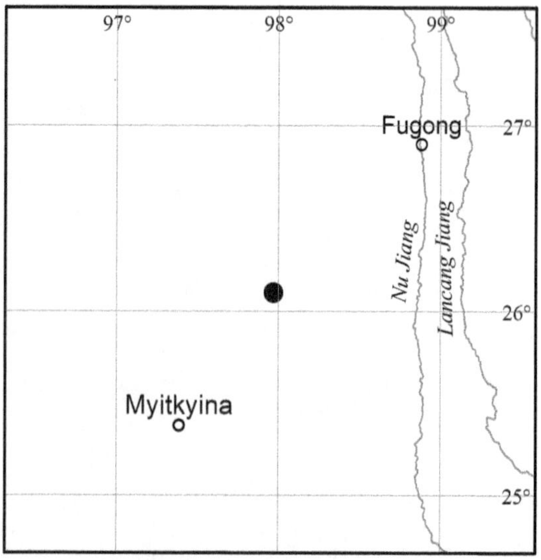

Map 46-235: **Priva cordifolia** (L. f.) Druce

Figure 46-136: **Priva cordifolia** (L. f.) Druce: **a**, flowering stem; **b**, flower; **c**, fruit. Drawn by Ai-Li LI.

5. VERBENA L.

Linnaeus, Sp. Pl. 1: 18. 1753; Clarke in Hooker (ed.), Fl. Brit. India 4: 565. 1885; P'ei & Chen, Fl. Reipubl. Popularis Sin. 65(1): 15. 1982; Chen & Gilbert in Wu & Raven (eds.), Fl. China 17: 3. 1994; Grierson & Long, Fl. Bhutan 2(2): 913. 1999. Type: *Verbena officinalis* L. (designated by Britton & Brown, Ill. Fl. N.U.S. ed. 2. 3: 94. 1913; affirmed by Hitchcock, Prop. Brit. Bot. 116. 1929) .

Herbs or subshrubs, annual or perennial. Stems quadrangular. Leaves opposite, shortly petiolate or sessile, margin dentate, lobed, pinnatifid to pinnatisect. Inflorescences terminal, rarely axillary, spicate, unbranched or in a panicle. Bracts small and inconspicuous. Calyx membranous, tubular, 5-lobed. Corolla white, pink, blue to purple, hypocrateriform or infundibular, 5-lobed, slightly zygomorphic. Stamens 4, fertile, didynamous, inserted in the middle of corolla tube and included; anthers ovate, connectives usually glandular, thecae parallel or slightly divaricate. Ovary 4-locular; ovules erect, 1 per locule. Fruit a schizocarp, included in persistent calyx, separating into four 1-seeded mericarps. **Chromosomes** 2n = 14.

About 250 species, mostly distributed in tropical and temperate America, a few species in Europe, Asia and N Africa; three species in the Pan-Himalaya.

1a. Leaf blade unlobed, subentire to serrate; leaf base slightly auriculate; inflorescences paniculate, spikes ca. 1 cm long; anther connectives not glandular .... 3. *V. bonariensis*

1b. Leaf blade deeply lobed, pinnatifid to pinnatisect, or tripinnatisect; leaf base attenuate; inflorescences usually unbranched, spikes more than 5 cm long; anther connectives glandular.

2a. Stem quadrangular; leaves deeply lobed, pinnatifid to pinnatisect; bracts oblong-ovate; calyx shallowly lobed, lobes broadly triangular .... 1. *V. officinalis*

2b. Stem cylindrical; leaves tripinnatisect, divided into linear segments; bracts linear; calyx deeply lobed, lobes subulate ....2. *V. tenuisecta*

1. Verbena officinalis L., Sp. Pl. 1: 20. 1753; Clarke in Hooker (ed.), Fl. Brit. India 4: 565. 1885; P'ei, Mem. Sci. Soc. China 1(3): 7. 1932; Liu in P'ei & Chen (eds.), Fl. Reipubl. Popularis Sin. 65(1): 15. 1982; Chen & Gilbert in Wu & Raven (eds.), Fl. China 17: 3. 1994; Grierson & Long, Fl. Bhutan 2(2): 913. 1999. Type: *Herb. Clifford: 11, Verbena 6* (lectotype designated by Verdcourt, Regnum Veg. 127: 98. 1993: BM).

Herbs, annual or weakly perennial, erect, 20–140 cm tall. Stem quadrangular, pubescent to subglabrous, sometimes woody at base. Leaves opposite, mostly sessile, lower ones shortly petiolate; leaf blade ovate, obovate, or oblong in outline, pinnatifid to pinnatisect, 2–8 cm long, 1–5 cm broad, hirsute especially on abaxial veins, base attenuate, margin deeply lobed. Inflorescences spicate, lax and many flowered, 10–20 cm long, sometimes in a panicle. Bracts oblong-ovate, about 2 mm long. Calyx 1–4 mm long, shallowly 5-lobed, pubescent and glandular. Corolla pale pink, mauve to blue, 4–8 mm long, pubescent, unequally 5-lobed. Stamens 4, short, inserted in the middle of corolla tube, included; anthers ovate, connectives usually glandular. Ovary glabrous, 4-lobed. Fruit a schizocarp, separating into four 1-seeded mericarps. Figure 46-137.

Phenology: flowering and fruiting July to October.

Habitats: widespread weeds; 100–2000 m.

Distribution: N Pakistan, Jammu & Kashmir, U Ganga & Indus, W Nepal, Sikkim & Darjiling, Bhutan, Yarlung Zangbo-Brahmaputra, N Hengduan, S Hengduan and U Irrawaddy (Map 46-236); worldwide weeds in temperate,

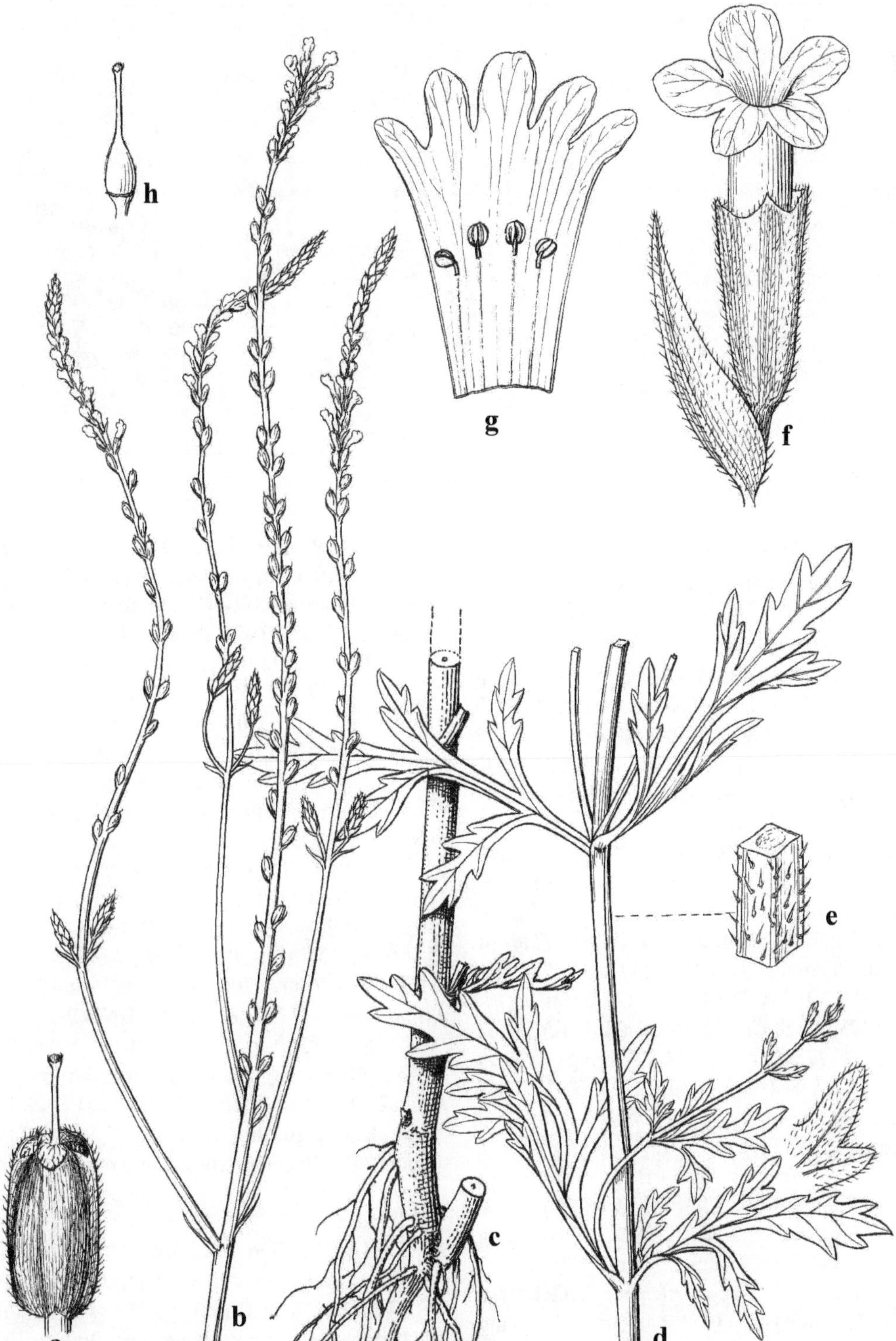

Figure 46-137: **Verbena officinalis** L.: **a**, ovary; **b**, inflorescence; **c**, plant base; **d**, stem leaves; **e**, stem indumentum; **f**, flower with bract; **g**, dissected flower; **h**, pistil. Drawn by Wei-Qing SHI; reproduced from *Fl. Reipubl. Popularis Sin.* 65(1): 16, fig. 3. 1982.

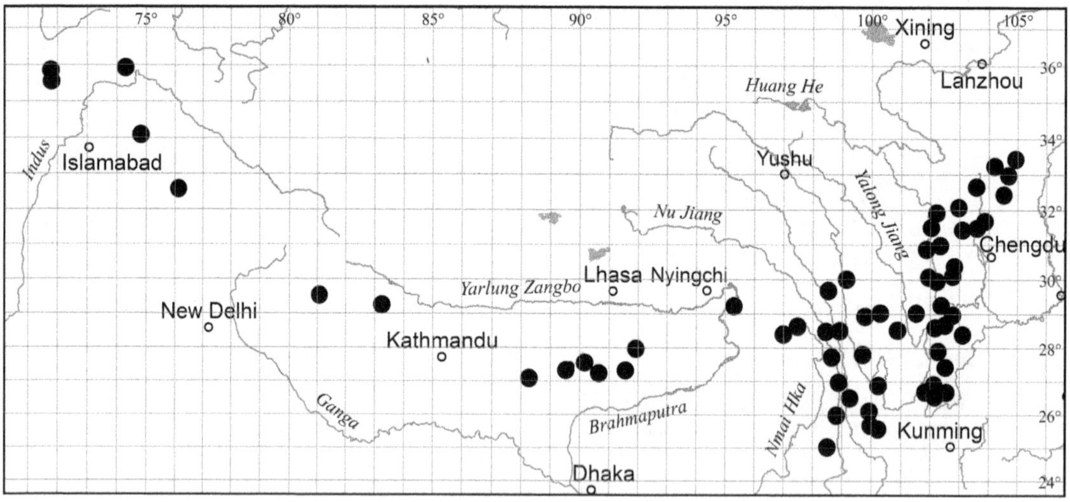

Map 46-236: **Verbena officinalis** L.

subtropical and tropical regions.

N PAKISTAN: **Chitral**, Drosh, 1372 m, 3 Jun. 1958, *J. D. A. Stainton 2578* (BM); Drosh, Khyber Pakhtunkhwa, 1250 m, 22 May 1958, *S. A. B. Lyon 705* (BM); Chitral Village, 1680 m, 30 May 1958, *J. D. A. Stainton 2539* (BM); **Gilgit**, Gilgit Valley, 1370 m, 7 Aug. 1892, *J. F. Duthie s. n.* (BM). **JAMMU & KASHMIR**: **Srinagar**, Drained lake basin of Kashmir, 10 Aug.–30 Sep. 1856, *A. Varks 4436* (BM). **U GANGA & INDUS**: **Chamba**, 1219 m, 4 Sep. 1896, *G. A. Gammie 18482* (K). **W NEPAL**: **Karnali**, Dolpa, 2130–2440 m, 1930, *L. Dauvoj 053* (BM); **Seti,** Bajhang, Seti Zone, Talkot, 1660 m, 18 Aug. 1991, *M. Suzuki, et al. 9160700* (BM). **SIKKIM & DARJILING**: **Darjiling**, 11 Aug. 1875, *C. B. Clarke 26953* (BM, K); *s. loc.* 1830 m, *F. D. He s. n.* (BM); Lebong, 1830 m, Jun. 1875, *J. S. Gamble 535A* (K). **BHUTAN**: **Paro**, 2290 m, 18 Aug. 1971, *R. Bedi 687* (K); **Ritang**, Nabzi, Jirgang Chu, 6 May 1937, *F. ludlow & G. Sherriff 2993* (BM); **Shamgong**, Dzona, Mangoe Chu, 1220 m, 31 Mar. 1949, *F. ludlow & G. Sherriff 18577* (BM); **Tashigang**, Khaling, roadside, 2250 m, 27 Jun. 1979, *A. J. C. Grierson & D. G. Long 2307* (K). **YARLUNG ZANGBO-BRAHMAPUTRA**: **Cona**, Leixiang, to 2500 m, 21 Sep. 2011, *J. Luo et al. 194* (HNWP); **Mêdog**, Renqingshan, 1100 m, 22 Apr. 1993, *H. Sun et al. 5236* (KUN); **Zayü**, pastures, waste ground, hedges and ditches, 1310 m, 11 Jul. 1950, *F. Kingdon-Ward 19547* (BM); Xiazayü, 1650 m, 28 Aug. 1983, *B. S. Li et al. 6953* (PE). **N HENGDUAN**: **Barkam**, Dangba, 2500 m, 17 Sep. 1957, *X. Li 71874* (IBSC, NAS, PE); **Batang**, Baisong, 2900 m, 1 Sep. 1972, *s. coll. 1299* (CDBI, PE); **Danba**, Donggu, 23 Aug. 1940, *K. L. Chü 7818* (NAS, PE); **Heishui**, Shidiaolou, 1800 m, 31 May 1959, *s. coll. 1269* (PE); **Jinchuan**, Anning, Zengda, 2080 m, 20 Oct. 1960, *Sichuan Med. Pl. Exped. 23147* (NAS); **Jiuzhaigou** (Nanping), Bazigou, 1970 m, 1 Aug. 1959, *Sichuan Econ. Pl. Exped.-Nanping team 4229* (CDBI); **Lixian**, Puqing, 1900 m, 3 Aug. 1959, *Q. S. Zhao 1016* (PE); **Maoxian**, 1800 m, 22 Jul. 2002, *X. M. Gao et al. M046* (HX); **Markam**, 2200 m, 6 Jun. 1976, *Qinghai-Tibet Exped. 11753* (KUN, PE, HNWP); **Pingwu**, Laohegou, 1300 m, 26 Oct. 2014, *W. G. Wang et al. 2388* (PE); **Songpan**, Shuikouba, 1700 m, 22 Oct. 1937, *K. T. Fu 2188* (PE, WUK); **Wenchuan**, Yulong, 1500 m, 15 Sep. 1983, *s. coll. 1967* (PE); **Wenxian**, Liujiaping, 1320 m, 4 Sep. 2011, *L. Q. Li et al. 105* (PE); **Wudu**, 1000 m, 17 May 1971, *Tibet Med. Exped. 10* (HNWP); **Xiaojin**, 2150 m, 1 Aug. 1975, *Sichuan Vegetation Exped. 9598* (CDBI, IBSC, PE). **S HENGDUAN**: **Baoxing**, 1954, *T. P. Soong 16168* (KUN, NAS); **Cawarong**, 3000 m, Aug. 1935, *C. W. Wang 65192* (IBSC, NAS, KUN, PE); **Dali**, 2300 m, 16 Oct. 1991, *X. F. Gao et al. 93* (KUN); **Daocheng**, Riwa, 3200 m, 1 Sep.

1971, *s. coll. 595* (CDBI); **Dêqên**, 2100 m, 17 Jul. 1981, *Qinghai-Tibet Exped. 2853* (HITBC, KUN); **Eryuan**, Zhongjie, 2000 m, 22 Oct. 1958, *C. W. Wang 486* (KUN); **Fugong**, Guquan, 1500 m, 25 Jun. 1986, *s. coll. s. n.* (HITBC); **Ganluo**, Haitang, 2000 m, 1 Jul. 1959, *s. coll. 3999* (CDBI, PE); **Gongshan**, Bingzhongluo, 1650 m, 4 Jul. 1982, *Qinghai-Tibet Exped. 7902* (KUN); **Huidong**, Xincun, 2100 m, 13 Sep. 1961, *s. coll. 11929* (NAS, SM); **Huili**, 27 Nov. 1930, *W. P. Fang 9474* (K); **Jiulong**, Naiqu, 2300 m, 1 Aug. 1979, *Q. Q. Wang 21532* (CDBI, IBSC); **Kangding**, Dajianlu, by a stream, 2590–2740 m, 28 Sep. 1928, *W. P. Fang 3661* (K); Shagu, 2850 m, 28 May. 1961, *S. Jiang 5016* (KUN); **Lanping**, 0-2800 m, 13 Sep. 1933, *H. T. Tsai 56092* (IBSC, KUN, LBG, PE); **Lijiang**, between Baishui and Heishui, roadside, 2900 m, 6 Oct. 1990, *Chungtien-Lijiang-Dali Expedition 904* (K); Yuhu, 2650 m, 29 May. 1988, *S. J. Pei 1637* (KUN); **Luding**, 2000 m, 2 Sep. 1983, *Vegetation Exped. 31938* (CDBI); **Lushui**, Laowo, 1600 m, 27 May. 1981, *Z. X. Tang 00065* (PE); **Meigu**, 2 Aug. 1959, *Sichuan Eco. Pl. Exped. 1881* (CDBI, KUN, SM, PE); **Mianning**, 21 Sep. 1930, *W. P. Fang 5050* (K); Sunshuiguan, 1850 m, 2 Jun. 1976, *Xichang Exped. 12271* (CDBI, PE); **Miyi**, Anning River, 4 Jun. 1958, *S. Y. Chen et al. 10035* (NAS, SM); **Muli,** Waerzhai, 3300 m, 26 Oct. 1981, *L. R. Xu 224* (WUK); **Puge**, Juemugou, 2000 m, 19 Aug. 1960, *Sichuan Med. Pl. Exped. 025447* (NAS, SM); **Shangri-la** (Zhongdian), 20 Jul. 1981, *W. H. Li & Y. Hu 306* (PE); **Shimian**, 1955, *C. C. Hsieh 42320* (IBSC, PE); **Tianquan**, Dahe, 850 m, 19 Aug. 2004, *P. Zhuang T010* (HX); **Xiangcheng**, Sangbiling, 2600 m, 16 Sep. 1937, *T. T. Yü 13328* (KUN, PE); **Xichang**, 27 Sep. 1930, *W. P. Fang 5201* (K); Qionghai, 1550 m, 29 May 1964, *L. N. Zhao 2184* (KUN, PE); **Yanbian**, Zemulong, 2000 m, 16 Sep. 2002, *X. F. Gao et al. 3604* (PE); **Yangbi**, Meixi, 2500 m, 24 Jun. 1964, *S. Chow 1268* (KUN); **Yuexi**, 11 Sep. 1930, *W. P. Fang 4928* (K); Zhongheba, 1700 m, 29 Jun.1959, *Sichuan Econ. Pl. Exped. 3574* (KUN, PE). **U IRRAWADDY**: **Tengchong**, Mingguang, 2000 m, 2 Aug. 1980, *S. T. Li 404* (KUN).

2. Verbena tenuisecta Briq. in Ann. Conser. et Jard. Bot. Geneve. 7–8: 294. 1904. Type: not seen.

Herbs perennial, 20–35 cm tall, erect or suberect. Stem cylindrical, glabrescent. Leaves subsessile, 1.5–3.5 cm long; leaf blade tripinnatisectly divided into narrowly linear segments, abaxially sericeous on veins. Inflorescences terminal, spicate, unbranched, 6–12 cm long. Bracts linear, ca. 3 mm long. Calyx tubular, sericeous, about 8 mm long, 5-lobed, lobes subulate. Corolla purple, 1–1.5 cm long, subequally 5-lobed, lobes obovate, apex emarginate. Stamens 4, connectives of anthers glandular. Fruit a schizocarp, ca. 4 mm long, enclosed by persistent calyx, separating into four 1-seeded mericarps.

Phenology: flowering and fruiting from July to October.

Habitats: grassy slopes, thickets margin.

Distribution: cultivated and sometimes naturalized in U Ganga & Indus of the Pan-Himalaya (Map 46-237); native to South America.

U GANGA & INDUS: **Dehra Dun**, Chandbagh, 20 Apr. 1927, *R. N. Parker s. n.* (K).

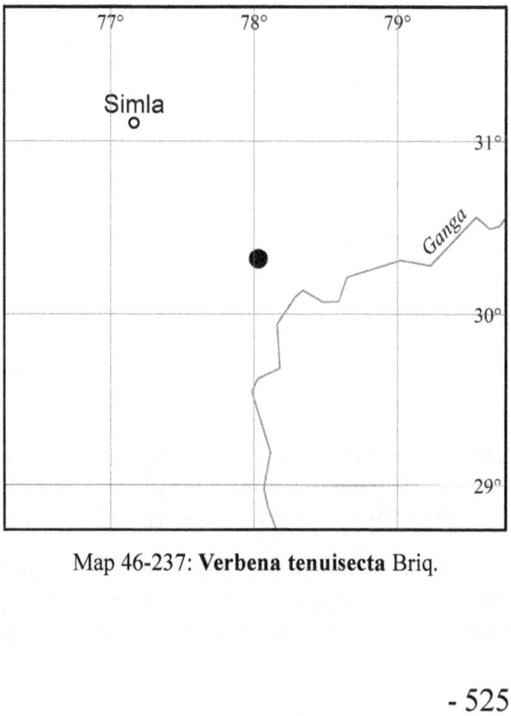

Map 46-237: **Verbena tenuisecta** Briq.

3. Verbena bonariensis L., Sp. Pl. 1: 20. 1753. Type: *LINN 35.11* (lectotype designated by Yeo, Kew Bull. 45: 105. 1990: LINN).

Herbs perennial, 0.3–1 m tall, erect. Stem quadrangular, hispid. Leaves opposite and sessile; leaf blade oblong-lanceolate, unlobed, 4–10 cm long, 0.5–1.5 cm broad, base slightly auriculate, margin subentire to serrate, apex acute. Inflorescences a terminal panicle, flowers in densely spikes. Bracts lanceolate, acuminate, ca. 4 mm long. Calyx tubular and pubescent, 3–4 mm long, 5-lobed, lobes subulate. Corolla violet, 6–8 mm long, subequally 5-lobed, lobes emarginate. Stamens 4, connectives of anthers without appendage. Fruit a schizocarp, ca. 4 mm long, enclosed by persistent calyx, separating into four 1-seeded mericarps.

Phenology: flowering and fruiting from June to October.

Habitats: streamside or thicket margin.

Distribution: cultivated and sometimes naturalized in N Pakistan, Jammu & Kashmir, U Ganga & Indus and Sikkim & Darjiling of the Pan-Himalaya (Map 46-238); native to South America.

N PAKISTAN: **Hazara**, Abbottabad, Jun. 1907, *H. Deane s. n.* (K). **JAMMU & KASHMIR**: **Chumar**, 18 Jun. 1850, *C. B. Clarke 415* (K). **U GANGA & INDUS**: **Dehra Dun**, Mothronwala river side, 610 m, Dec. 1899, *U. Kanjilal 808* (K); Mothronwala, 2 miles above Chauky, 610 m, Jul. 1900, *U. Kanjilal 809* (K); **Dharmsala**, 1524 m, 15 Oct. 1874, *C. B. Clarke 24476* (BM); 1220 m, 20 Oct. 1886, *H. Collett s. n.* (K); 1524 m, 15 Oct. 1874, *C. B. Clarke 24476* (K). **SIKKIM & DARJILING**: **Sikkim**, *s. loc.* 1830 m, 13 Aug. 1874, *Treutler 575* (K).

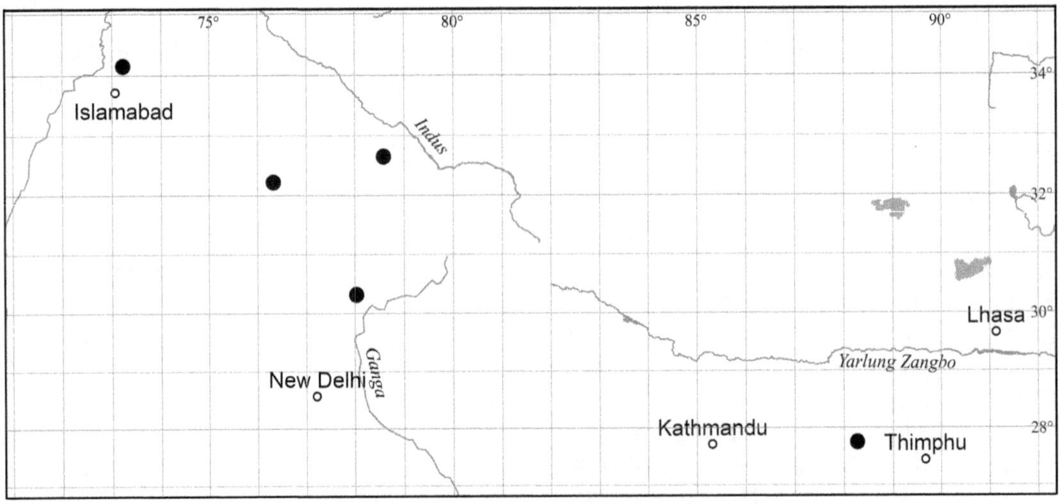

Map 46-238: **Verbena bonariensis** L.

6. PHYLA Lour.

Loureiro, Fl. Cochinch. 1: 66. 1790; P'ei & Chen, Fl. Reipubl. Popularis Sin. 65(1): 18. 1982; Chen & Gilbert in Wu & Raven (eds.), Fl. China 17: 3. 1994; Grierson & Long, Fl. Bhutan 2(2): 916. 1999. Type: *Phyla nodiflora* (L.) E. L. Greene.

Herbs perennial. Stems quadrangular, decumbent and trailing, rooting at nodes, usually in dense mats. Leaves opposite, sometimes verticillate. Inflorescences axillary, long-pedunculate, densely spicate, sometimes capitate, elongated after anthesis. Bracts obovate. Calyx short and tubular, membranous, indumentum uncinulate, 2- or 4-lobed. Corolla white to purple, hypocrateriform, somewhat zygomorphic, 5-lobed, lobes unequal. Stamens 4, didynamous, inserted in distal part of corolla tube, thecae parallel, without glandular connective. Ovary 2-locular; one ovule per locule, style short, stigma capitate. Fruit a schizocarp, separating into two 1-seeded mericarps.

Recent references: O'Leary & Múlgura, 2012. A taxonomic revision of the genus *Phyla* (Verbenaceae). *Ann. Missouri Bot. Gard.* 98: 578–596.

About 15 species, mostly distributed in the Americas, one weedy species widespread in tropical, subtropical and temperate regions; one species in the Pan-Himalaya.

1. Phyla nodiflora (L.) E. L. Greene, Pittonia 4: 46. 1899; P'ei & Chen, Fl. Reipubl. Popularis Sin. 65(1): 18. 1982; Chen & Gilbert in Wu & Raven (eds.), Fl. China 17: 3. 1994; Grierson & Long, Fl. Bhutan 2(2): 916. 1999. ≡ *Verbena nodiflora* L., Sp. Pl. 1: 20. 1753 ≡ *Lippia nodiflora* (L.) Michx. Fl. Bor.-Amer 2: 15. 1803. Type: *Herb. Clifford: 11, Verbena 3* (lectotype designated by Townsend in Davis, Fl. Turkey 7: 32. 1982: BM000557561).

Phyla chinensis Lour., Fl. Cochinch: 66. 1790. Type: China, Guangdong, "Canton", *s. d. J. de Loureiro 66-1* (**lectotype designated here:** P00150899).

Herbs perennial, sometimes with woody rootstock. Stems quadrangular, minutely strigose to glabrescent, decumbent and trailing, rooting at nodes, usually in dense mats. Leaves subsessile, opposite, sometimes verticillate; leaf blade oblanceolate, obovate, or spatulate, minutely strigose to glabrous, 0.5–4 cm long, 0.4–2 cm broad, base cuneate, margin distally serrate, apex acute. Inflorescences axillary, densely spicate to capitate; long pedunculate, peduncle 1–7 cm long. Bracts obovate, apex mucronate or acuminate, about 2 mm long. Calyx short and tubular, membranous, indumentum uncinulate, usually deeply 2-lobed. Corolla pinkish purple or white, hypocrateriform, zygomorphic, 5-lobed, lobes unequal. Fruit a schizocarp, ca. 1.5 mm in diameter, separating into two 1-seeded mericarps. Figure 46-138.

Phenology: flowering and fruiting throughout the year.

Habitats: weedy plants, widespread, usually in wet places.

Distribution: N Pakistan, U Ganga & Indus, W Nepal, C Nepal, E Nepal, S Hengduan (Map 46-239); also widely distributed in tropical, subtropical and temperate regions of the world.

N PAKISTAN: **Azad Kashmir,** Muzaffarabad, Cachemire, 700 m, 16–23 Jun. 1953, *F. Schmid 184* (BM). **U GANGA & INDUS**: **Dehra Dun**, Aug. 1890, *J. S. Gamble 24527* (K); **Chamba**, 920 m, 5 Sep. 1896, *G. A. Gammie s. n.* (K). **W NEPAL**: **Bheri**, Surkhet, Bheri Zone, 1600 m, 30 Jul. 1991, *M. Suzuki et al. 9170092* (BM); **Karnali**, Kalikot, 1220 m, 25 Jul. 1953, *J. B. Freyn 122* (BM); **Rapti**, Salyan, Bheri-Sangi Khola, 550 m, *J. F. Dobremez 2556* (BM); **Seti**, Bajura, Seti Zone, Tolebhir, by path in open grassy place, 1700 m, 8 Aug. 1991, *M. Suzuki et al. 9160402* (BM). **C NEPAL**: **Dhawalagiri**, Baglung, Kali Gandaki Valley, 760 m, 19 Apr. 1954, *J. D. A. Stainton et al. 39* (BM). **E NEPAL**: **Kosi**, Dhankuta, Tamur Valley, 310 m, 17 Apr. 1956, *J. D. A. Stainton 63* (BM); Koshi Zone, 370 m, 3 Jul. 1988, *M. Suzuki et al. 8880089* (BM); Sankhuwa Sabha, on banks by path side, 300 m, 25 Jul. 1990, *M. Minaki et al. 9020117* (BM). **S HENGDUAN**: **Heqing**, Jiangying, 12 Sep. 1929, *R. C. Ching 24556* (KUN, PE); **Huidong**, 0–1700 m, 6 Aug. 1961, *T. L. Dai & S. P. Tang 11821* (NAS, SM); **Miyi**, 1100 m, 1 Apr. 1963, *Xiangmao Group, 0039* (CDBI); **Panzhihua**, Cycas Natural Reserve, Jinjiacun, 1810 m, 12 Jul. 2010, *Panzhihua Cycas Exped. 300* (PE); **Xichang**, Qionghai, 1550 m, 29 May 1964, *L. N. Zhao 2192* (KUN, PE); **Yanbian**, 1200 m, 30 Jun. 1983, *Qinghai-Tibet Exped. 11724* (KUN).

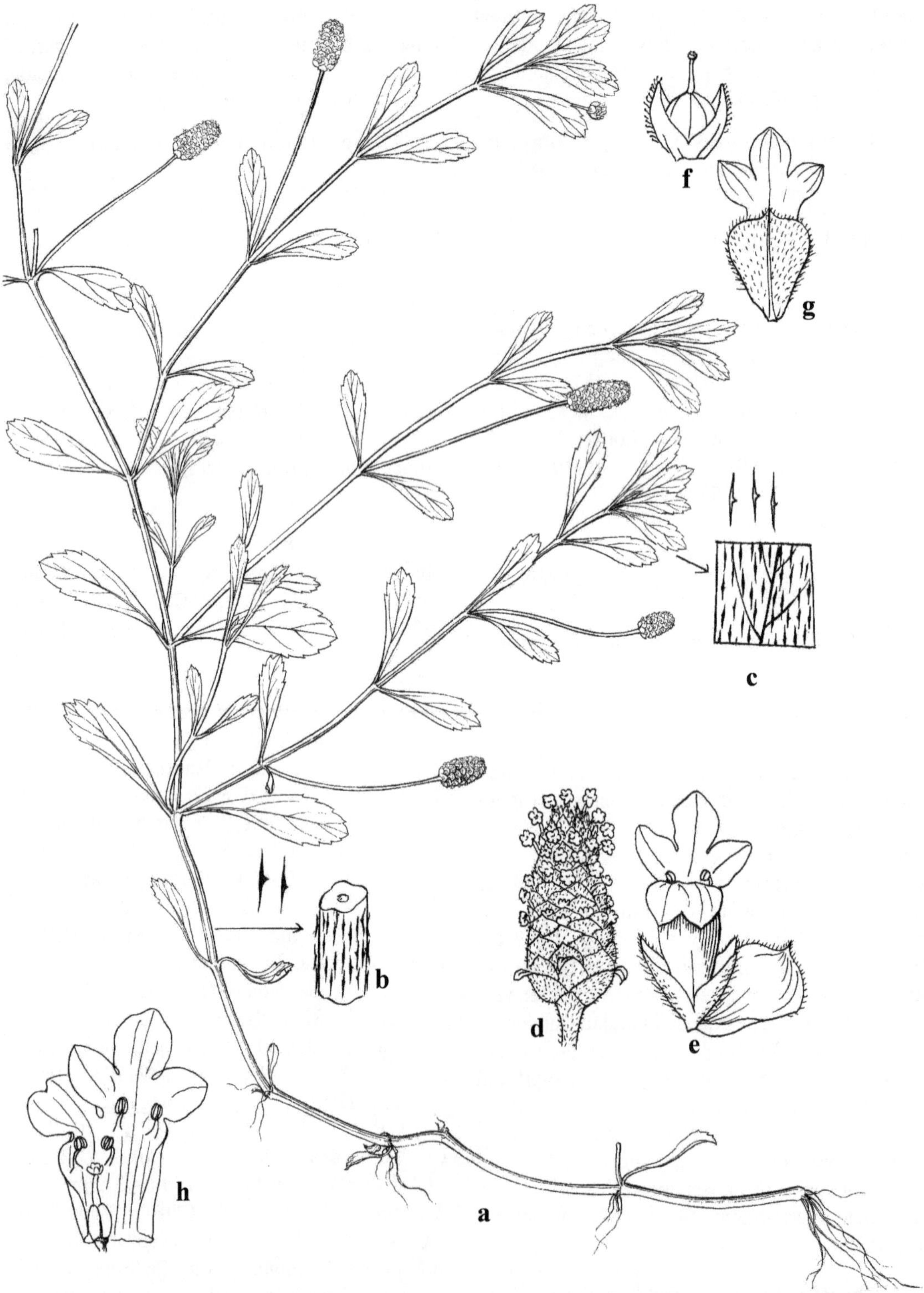

Figure 46-138: **Phyla nodiflora** (L.) Greene: **a**, flowering plant; **b**, stem indumentum; **c**, leaf surface with hairs; **d**, inflorescence; **e**, flower with bract; **f**, fruit embraced by calyx; **g**, dorsal view of flower; **h**, open corolla showing stamens. Drawn by Xing-Qiang JIANG; reproduced from *Fl. Reipubl. Popularis Sin.* 65(1): 19, fig. 5. 1982.

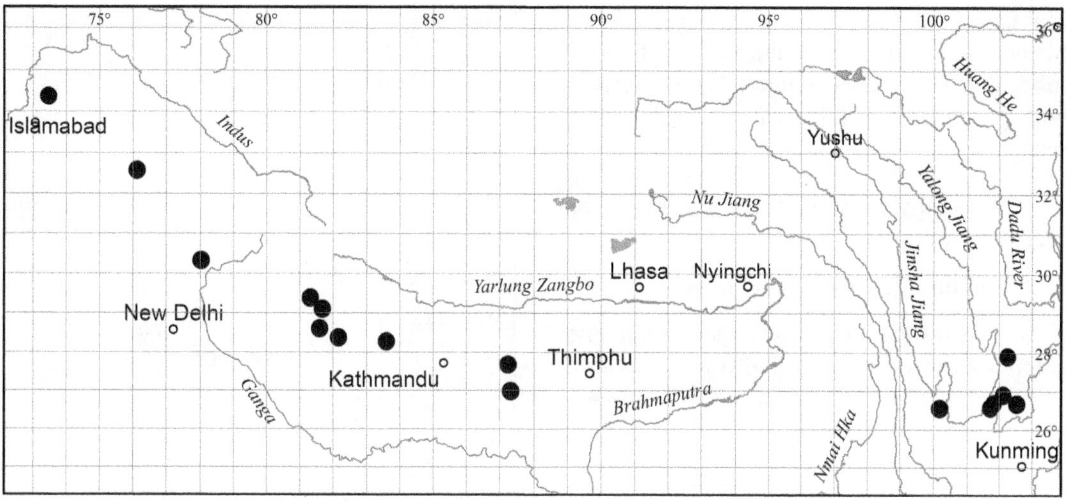

Map 46-239: **Phyla nodiflora** (L.) E. L. Greene

7. LIPPIA L.

Linnaeus, Sp. Pl. 2: 633. 1753; Clarke in Hooker (ed.), Fl. Brit. India 4: 563. 1885; Grierson & Long, Fl. Bhutan 2(2): 916. 1999. Type: *Lippia americana* L.

Herbs, subshrubs, or shrubs, perennial, usually aromatic. Stems quadrangular. Leaves opposite, margin entire or variously dentate. Inflorescences axillary or terminal, densely spicate or capitate. Bracts conspicuous, sometimes enlarged and showy, usually imbricate. Calyx tubular-campanulate, 2- or 4-lobed, persistent. Corolla white, pale yellow, pink to red, infundibular, somewhat zygomorphic, 4-lobed. Stamens 4, inserted above the middle of corolla tube, thecae parallel, connective sometimes with glandular appendage. Stigma oblique or lateral. Fruit a schizocarp separating into two 1-seeded mericarps.

Recent references: Troncoso, 1980. Taxonomic novelties in the genera *Lantana* and *Lippia* (Verbenaceae). *Hickenia* 1: 227–231; Múlgura et al., 1998. Morfología de las inflorescencias en *Lippia* (Verbenaceae). *Darwiniana* 36: 1–12.

About 200 species, distributed in tropical and subtropical America and Africa; one species cultivated and naturalized in the Pan-Himalaya.

1. Lippia alba (Mill.) N. E. Br. ex Britton & P. Wilson, Bot. Porto Rico 6(1): 141. 1925; Grierson & Long, Fl. Bhutan 2(2): 916. 1999. ≡ *Lantana alba* Mill. Gard. Dict. (ed. 8): 1768. Type: *W. Houston s. n.* (**lectotype designated here:** BM).

Shrubs or subshrubs. Stems quadrangular, pubescent. Leaves opposite, shortly petiolate; petioles ca. 1 cm long; leaf blade lanceolate, 2.5–6 cm long, 1–3 cm broad, base cuneate, margin serrate, sometimes crenate, apex acute. Inflorescences densely spicate to capitate, axillary, pedunculate; peduncles 1–3.5 cm long. Bracts ovate, imbricate. Calyx tubular-campanulate, persistent. Corolla white, 4-lobed. Stamens 4, inserted above the middle of corolla tube, thecae parallel, connective sometimes with glandular appendage. Fruit a schizocarp, separating into two 1-seeded mericarps.

Phenology: flowering and fruiting from March to July.

Habitats: valley or grassy slope.

Distribution: cultivated and sometimes naturalized in U Ganga & Indus and Sikkim & Darjiling of the Pan-Himalaya (Map 46-240); native to tropical America.

U GANGA & INDUS: **Darcha**, Dacca, 4 May 1873, *C. B. Clarke 20076* (K). **SIKKIM & DARJILING**: **Sikkim**, *s. loc.*, 22 Mar. 1875, *Treutler 1259* (K).

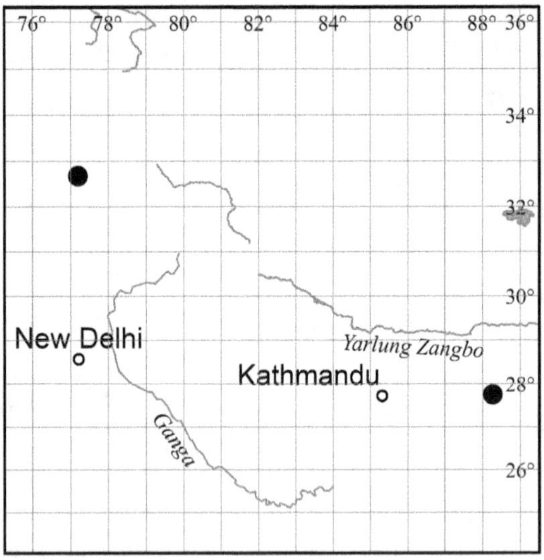

Map 46-240: **Lippia alba** (Mill.) N. E. Br. ex Britton & P. Wilson

8. LANTANA L.

Linnaeus, Sp. Pl. 2: 626. 1753; Clarke in Hooker, Fl. Brit. India 4: 562. 1885; P'ei & Chen, Fl. Reipubl. Popularis Sin. 65(1): 17. 1982; Chen & Gilbert in Wu & Raven (eds.), Fl. China 17: 2. 1994; Grierson & Long, Fl. Bhutan 2(2): 914. 1999. Type: *Lantana camara* L.

Shrubs or subshrubs, perennial, usually aromatic. Stems quadrangular, pubescent or glabrous, sometimes armed with prickles. Leaves opposite and petiolate; leaf blade entire or variously dentate, usually rugose. Inflorescences axillary, capitate or densely spicate, pedunculate, or fascicled and sessile. Bracts conspicuous, linear to triangular. Calyx tubular, membranous, persistent, truncate or shallowly lobed. Corolla white, yellow, pink, red, with colours mixed within one inflorescence, hypocrateriform; corolla tube slender, straight or curved, slightly enlarged above the middle; subzygomorphic, 4- or 5-lobed; corolla lobes unequal, apex obtuse or emarginate. Stamens 4, didynamous, inserted in the middle of corolla tube, included; thecae parallel, connective without appendages. Ovary 2-locular. Style shorter than corolla tube; stigma capitate, lateral or oblique. Fruit drupaceous with two 1-seeded pyrenes.

Recent references: Troncoso, 1980. Taxonomic novelties in the genera *Lantana* and *Lippia* (Verbenaceae). *Hickenia* 1: 227–231; Forman, 1997. Notes concerning the typification of names of William Roxburgh's species of phanerogams. *Kew Bull.* 52: 513–534; Múlgura et al., 2002. Morfología de las inflorescencias en Verbenaceae–Verbenoideae II: Tribu Lantaneae p. p. *Darwiniana* 40: 1–15.

About 150 species, mainly distributed in tropical and subtropical America, a few in Old World tropics; one species native and one species cultivated and naturalized in the Pan-Himalaya.

1a. Base of leaf blade rounded to subcordate; inflorescences capitate; bracts linear or lanceolate .... 1. *L. camara*

1b. Base of leaf blade cuneate to subcuneate; inflorescences densely spicate; bracts ovate .... 2. *L. indica*

1. Lantana camara L., Sp. Pl. 2: 627. 1753; Clarke in Hooker (ed.), Fl. Brit. India 4: 562. 1885; P'ei & Chen, Fl. Reipubl. Popularis Sin. 65(1): 17. 1982; Chen & Gilbert in Wu & Raven (eds.), Fl. China 17: 2. 1994; Grierson & Long, Fl. Bhutan 2 (2): 914. 1999. Type: *LINN 783.4* (lectotype designated by Moldenke & Moldenke, *Rev.* Handb. Fl. Ceylon 4: 220.1983: LINN).

Shrubs, 1–3 m tall, evergreen. Branches pubescent, usually armed with recurved prickles. Leaves opposite, petiolate, petioles 1–3.5 cm; leaf blade pubescent and sometimes rugose, ovate to oblong-ovate, 3–10 cm long, 1.5–5 cm broad; leaf base rounded to subcordate; leaf apex acute, sometimes shortly acuminate; leaf margin crenate to serrate. Inflorescences capitate, terminal and axillary, long pedunculate, peduncles 3–6 cm long. Bracts linear or lanceolate. Calyx short, 2–3 mm long, tubular, membranous, shallowly lobed. Corolla yellow, orange to dark red, often with colours mixed within one inflorescence; hypocrateriform; corolla tube 5–15 mm long, pubescent, slightly enlarged and curved above the middle; usually 4-lobed, subzygomorphic, corolla lobes unequal, upper lobe usually with emarginate apex, the two lateral lobes with rounded apex and lower lobe with acute apex. Stamens 4, didynamous, included. Ovary glabrous. Fruit a drupe, dark purple, globose, about 4 mm in diameter, with two 1-seeded pyrenes. Figure 46-139.

Phenology: flowering and fruiting throughout the year.

Habitats: margins of thickets or woods, usually along stream side; 100–1500 m.

Distribution: widely cultivated and naturalized in N Pakistan, U Ganga & Indus, W Nepal, C Nepal, E Nepal, Bhutan and S Hengduan of the Pan-Himalaya (Map 46-241); distributed in tropical America, widely cultivated and naturalized in tropical and subtropical regions of the world.

Note: This species is widely cultivated as ornamental plants, because of its beautiful flowers which are highly variable in colour. In Hengduan Mountain region and south side of Himalayas, naturalized individuals are common in the field.

N PAKISTAN: **Azad Kashmir**, Muzaffarabad, Rishenganga, 610 m, 12 Apr. 1956, *D. Mohd 146* (BM). **U GANGA & INDUS**: **Chamba**, 1070 m, 8 Sep. 1896, *G. A. Gammie18545* (K). **W NEPAL**: **Rapti**, Dang, 610 m, 26 Oct. 1952, *O. V. Polunin, et al. 5880* (BM); **Seti**, Bajhang, Seti Zone, 25 Aug. 1991, *M. Suzuki et al. 9161014* (BM). **C NEPAL**: **Bagmati**, Kathmandu, 1370 m, 17 May 1969, *L. H. J. Williams 19* (BM); 1350 m, 19 Aug. 1970, *J. F. Dobremez 615* (BM); Kathmandu valley, 1400 m, 20 Aug. 1966, *Lange 130* (K); **Narayani**, Bara, Narayani Zone, on sunny riverside forest margin, 330 m, 10 Oct. 1995, *M. Mikage et al. 9552808* (BM). **E NEPAL**: **Kosi**, Sankhuwa Sabha, Koshi Zone, 1200 m, 7 Jul. 1988, *M. Suzuki et al. 40072* (BM). **BHUTAN**: **Samchi**, Kalapani, hillside, 360 m, 2 Mar. 1982, *A. J. C. Grierson & D. G. Long 3376* (K); **Sarbhang**, roadside, 400 m, 31 May 1979, *A. J. C. Grierson & D. G. Long 1486* (K). **S HENGDUAN**: **Dali**, Weishan, 1952, *R. C. Ching 50419* (KUN, PE); **Hanyuan**, Xigou, 800 m, 15 Sep. 1993, *Z. C. Zhao 0406* (PE); Chengguan, 780 m, 15 Sep. 1993, *Z. C. Zhao 0420* (PE); **Panzhihua**, Cycas Natural Reserve, Minzheng, 1610 m, 11 Jul. 2010, *Panzhihua Cycas Exped. 5* (PE); Suburb, 1000 m, 28 Apr. 2007, *Y. Y. Geng et al. 20070251* (PE); **Xichang**, Qionghai, 1600 m, 7 Apr. 1985, *Y. J. Li 615* (CDBI); **Yongsheng**, Chenghai, 27 May 2012, *X. F. Jin 0169* (PE).

2. Lantana indica Roxb., Fl. Ind. 3: 89. 1832; Clarke in Hooker, Fl. Brit. India 4: 562. 1885. Type: *Wallich 1823 B* [**lectotype designated here**: K (K001114295)].

Shrubs, 1–2.5 m tall. Stems erect and branched, pubescent, sometimes armed with prickles. Leaves opposite, shortly petiolate; petioles ca. 1 cm long, sometimes subsessile; leaf blade ovate to sublanceolate, 3–6 cm long, 1–4 cm broad; leaf base cuneate to subcuneate; leaf apex acute; leaf margin densely crenate to serrate. Inflorescences axillary, densely spicate, ovoid, 0.5–3 cm long; long pedunculate, peduncles 1.5–5 cm long. Bracts ovate, apex acuminate, densely pubescent. Calyx about 0.5 cm long, pubescent. Corolla white to pale pink, 1–1.5 cm

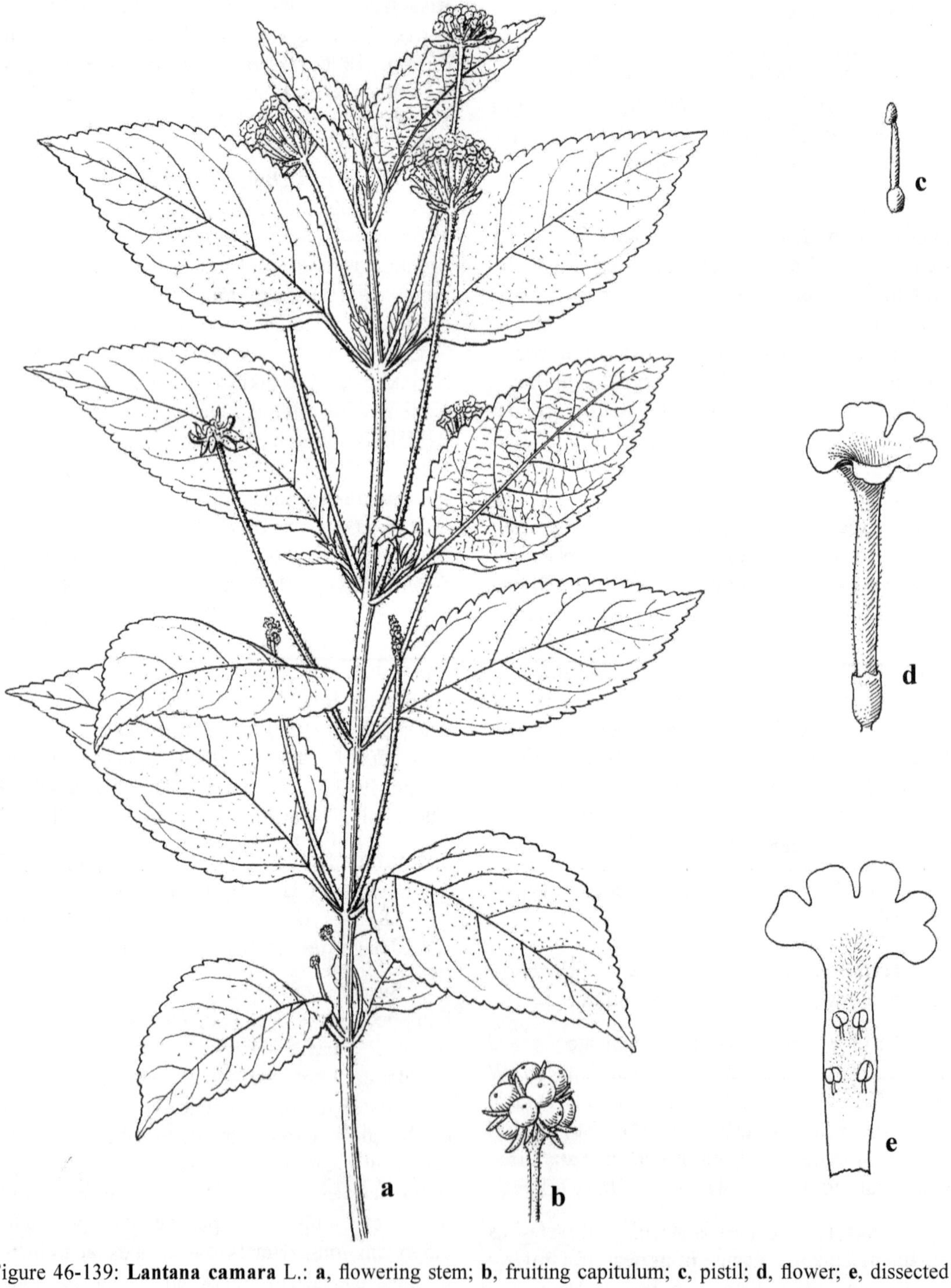

Figure 46-139: **Lantana camara** L.: **a**, flowering stem; **b**, fruiting capitulum; **c**, pistil; **d**, flower; **e**, dissected flower with stamens. Drawn by Xing-Qiang JIANG; reproduced from *Fl. Reipubl. Popularis Sin.* 65(1): 18, fig. 4. 1982.

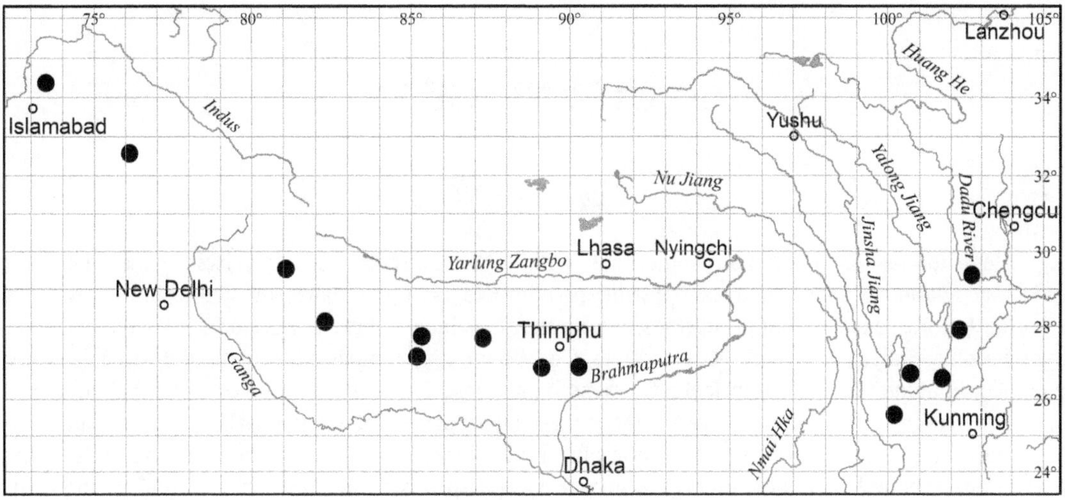

Map 46-241: **Lantana camara** L.

long, pubescent outside; 4-lobed. Fruit a drupe, 3–5 mm in diameter, globose, with two 1-seeded pyrenes.

Phenology: flowering and fruiting July to October.

Habitats: margins of thickets or woods.

Distribution: distributed in N Pakistan and U Ganga & Indus (Map 46-242); also distributed in other parts of Pakistan and India.

Note: According to Forman, Wallich Catalogue 1823B belongs to the original material of *Lantana indica*. I have checked 1823B, and here designate K001114295, which well represents this species, as the lectotype.

N PAKISTAN: **Azad Kashmir**, Mirpur, near water, 15 Sep. 1987, *Tehistli et al. 141* (BM). **U GANGA & INDUS**: **Darcha**, Dacca, 29 Mar. 1868, *C. B. Clarke 6700* (K).

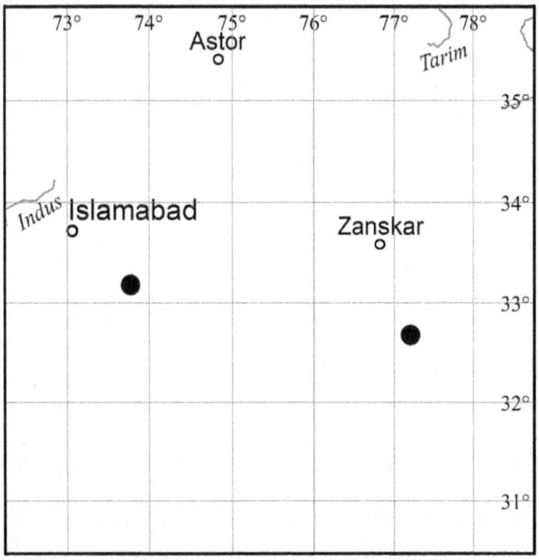

Map 46-242: **Lantana indica** Roxb.

MARTYNIACEAE

Yun-Fei DENG

Herbs annual or perennial, sometimes with tuberous roots, usually viscid pubescent with uniseriate glandular hairs. Leaves opposite or alternate on upper parts of stem, simple, without stipules. Inflorescences terminal racemes. Bracts small, caducous; bracteoles 1 or 2. Flowers bisexual, zygomorphic. Calyx 5, free or partly connate. Corolla tubular, campanulate, or funnel-form, laterally enlarged, bilabiate; lobes 5, imbricate. Stamens 2 or 4, inserted at the base of corolla tube, adherent; anthers spreading. Disc annular. Ovary superior, 1-locular; parietal placentas T-shaped in section, false septa usually present; ovules few to numerous. Fruit oblong, usually, especially in the upper part, curved and incompletely loculicidally deshiscent; exocarp fleshy, peeling off at maturity, endocarp woody. Seeds 4–many, black, rugose, embryo straight, endosperm very thin.

Recent references: Humbert, 1971. Pedaliacees. In: *Fl. Madagasc.* 179: 5–46. Tananarive: Imprimerie Officielle; Gibson, 1974. Martyniaceae. In: Standley, Williams & Gibson (eds.), *Flora of Guatemala* 10(3). *Fieldiana, Bot.* 24(10/3): 233–238; Nasir, 1972. Martyniaceae. In: Nasir & Ali (eds.), *Fl. W Pakistan* 13: 1–3. Rawalpindi: Stewart Herbarium, Gordon College; Thieret, 1977. The Martyniaceae in the southeastern United States. *J. Arnold Arbor.* 58: 25–39; Theobald & Grupe, 1981. Martyniaceae. pp. 202–205. In: Dassanayake (ed.), *A Revised Handbook to the Flora of Ceylon*. vol. 3. New Delhi: Amerind for Smithsonian Inst; Tao, 1985. On the occurrence of two Martyniaceae in China. *Acta Bot. Yunnanica* 7(3): 292; Zhang & Hartmann, 1998. Martyniaceae. In: Wu & Raven (eds.), *Fl. China* 18: 228. Beijing & St. Louis: Science Press & Missouri Botanic Garden Press; Burger, 2000. Family 196. Martyniaceae. In: Burger (ed.), *Flora Costaricensis. Fieldiana, Bot., n. s.* 41: 164–165.

Five genera and 16 species distributed mainly in warm and arid regions from the southern USA to Argentina; one genus and one species naturalized in the Pan-Himalaya.

Notes: The family sometimes was included in the family Pedaliaceae.

1. MARTYNIA L.

Linnaeus, Sp. Pl. 2: 618. 1753; de Candolle, Prodr. 9: 253. 1854; Benthan & Hooker, Gen. Pl. 2: 1056. 1876; Stapf in Engler & Prantl (eds.), Nat. Pflanzenfam. 4(3b): 269. 1895; Zhang & Hartmann in Wu & Raven (eds.), Fl. China 18: 228. 1998. Type: *Martynia annua* L.

Herbs, erect, viscid pubescent throughout. Leaves opposite, broadly ovate, palmately veined. Inflorescences racemose, terminal and subterminal; bracteoles 2, membranous, attached to calyx base. Calyx tube ventrally deeply incised; lobes unequal. Corolla campanulate, constricted at base; lobes unequal, rounded. Stamens 2; staminodes 2; anthers versatile. Ovary superior; style slender; stigma bilobed. Capsules with thin, easily deciduous exocarp and woody, striate endocarp; horns short. Seeds 4, one in each of the 4 peripheral compartments. **Chromosomes** 2n = 30, 36.

Recent references: Hevly, 1969. Nomenclatural history and typification of *Martynia* and *Proboscidea* (Martyniaceae). *Taxon* 18: 527–534.

One species: native in Central America, naturalized everywhere.

1. **Martynia annua** L., Sp. Pl. 2: 618. 1753; Backer, Fl. Males. Ser. 1. 4: 221. 1951; Tao, Fl. Reipubl. Popularis Sin. 69: 67. 1990; Zhang & Hartmann in Wu & Raven (eds.), Fl. China 18: 228. 1998. ≡ *Martynia diandra* Gloxin, Obs. 14. t. 1. 1785, nom. illeg. (Art. 52); de Candolle, Prodr. 9: 253. 1854; Hooker, Fl. Brit. India 4: 386. 1884. ≡ *Martynia angulosa* Lam., Encycl. 2(1): 112. 1786, nom. illeg. (Art. 52). ≡ *Disteira angulosa* Raf., Fl. Tellur. 4: 68. 1836 [1838], nom. illeg. (Art. 52.1). ≡ *Vatkea diandra* O. Hoffm, Verh. Bot. Brandenb. 73: 45. 1881, nom. illeg. (Art. 52). Type: [*icon*] "*Martynia annua villosa et viscosa, folio subrotundo, flore magno rubro*" in Martyn, Hist. Pl. Rar. 42: t. 42. 1731 (lectotype designated by Nafday, Bull. Bot. Soc. Coll. Sci. Nagpur 4: 65. 1964.).

Herbs annual, erect, 30–100 cm tall; stems terete, usually woody at base, 1.5–2 cm in diameter, densely tomentose-glandular. Leaves opposite; petiole stout, 6–45 cm long, glandular-hairy; leaf blade broadly ovate to triangular-ovate, 9–22 cm long, 9–20 cm broad, densely glandular-hairy, base cordate, margin sinuolate-dentate, apex acute. Racemes axillary, 5–15 cm long, 10–20-flowered. Bracts pale red, broadly ovate, 1.2–2.5 cm long, 0.7–1.3 cm broad, membranous, glandular-hairy, caducous; bracteoles ovate-oblong, 0.6–1.5 cm long, 0.4–1.0 cm broad, petaloid or lilac, glandular-hairy. Pedicels 1–3 cm long, nodding, glandular-hairy. Calyx lobes pale yellow-green, 1.5–2 cm long, glandular-pubescent, 5-lobed almost to base. Corolla 3–5 cm long, outside glandular-pubescent and pale pink, inside yellow with stalked glands and purple dots on anterior side; lobes broadly rounded, pink with a dark purple blotch; lower lip with a yellow blotch at base. Filaments white, 1–1.5 cm long, glabrous. Staminodes 2 or 3, clavate. Ovary globose, 2 mm broad, glabrous; style ca. 2 cm long, slender, persistent; stigma 3 sensitive, flat. Capsules ovoid, 3.5–4 cm long, 2–2.5 cm broad, 0.5–1.5 cm thick, with hook-like apical beak for 4–6 mm long, densely glandular pubescent, spiny along suture. Seeds black, hard, with serrate-dentate ribs, 2.5–3 cm long. **Chromosomes** 2n = 36. Figure 46-140.

Phenology: flowering throughout the year.

Habitats: forests or roadsides, naturalized; 100–2400 m.

Distribution: N Pakistan, Jammu & Kashmir, U Ganga & Indus, W Nepal, C Nepal, E Nepal (Map 46-243); native in Central America, introduced and naturalized elsewhere.

N PAKISTAN: **Azad Kashmir**, Kotli, *R. R. Stewart s. n.* (E); Mirpur, Watala, 700 m, 21 Nov. 1966, *J. Mohd s. n.* (RAW, fide Narir, 1972); Muzaffarabad, 6 Nov. 1955, *A. H. Khan s. n.* (RAW, fide Narir, 1972). **JAMMU & KASHMIR**: **Jammu**, Canal Road, Aug. 2003, *Baleshwar & Meena 17287* (RRLH); 900 m, Sep. 1928, *B. O. Coveuty B* (K). **U GANGA & INDUS**: **Chamba**, Daneria to Bawikher, 900–1200 m, 14 Aug. 1898, *J. H. Lace 1819* (E); **Dehradun**, 670 m, 8 Aug. 1898, *J. S. Gamble 27444* (K); road to Devaprayag along the Ganga, 500 m, 12 Aug. 1975, *F. Billiet & J. Leonard 6488* (K); **Simla**, 600 m, 20 Oct. 1886, *H. Collett s. n.* (K). **W NEPAL**: **Bheri**, Banke, Chatar, 10 Feb. 1980, *R. N. Shulka 4026* (KATH); Banke, Nepalgangj, Shamsherganj, 2 Oct. 1977, *R. Rana 2105* (KATH); Banke, Nepalgangj, 30 Oct. 1972, *N. P. Manandhar 9204* (KATH); Bardiya, Mau, 210 m, 17 Jun. 1980, *T. K. Bhattachaya 3769* (KATH); Bardiya, Gularia, 180 m, 26 Jun. 1980, *T. K. Bhattachaya 3842* (KATH); Jajarkot, 700 m, 15 Aug. 1979, *K. R. Rajbhandari & B. Roy 4729* (KATH); Jajarkot, Base of Jajarkot, 900 m, 6 Aug. 1968, *S. B. Malla 14103* (KATH); Jajarkot, 1060 m, 16 Oct. 1952, *O. Polunin, W. R. Sykes & L. H. J. Williams 5696* (BM, E); Jajarkot, Bheri River below Jajalot, 750 m, 23 Aug. 1952, *O. Polunin, W. R. Sykes & L. H. J. Williams 548* (BM); Surkhet, Imlibas, 300 m, 25 Apr. 1981, *T. K. Bhattachaya 362* (KATH); **Mahakali**, Dadeldhura, Kalena, 700 m, 30 Aug. 1982, *L. P. Kattel 950* (KATH); Darchula,

Figure 46-140: **Martynia annua** L.: **a**, flowering branch; **b**, opened corolla showing stamens; **c**, pistil; **d**, capsule. Drawn by Xi-Chou LI; reproduced from *Fl. Reipubl. Popularis Sin.* 69: 65, pl. 18:1–4. 1990.

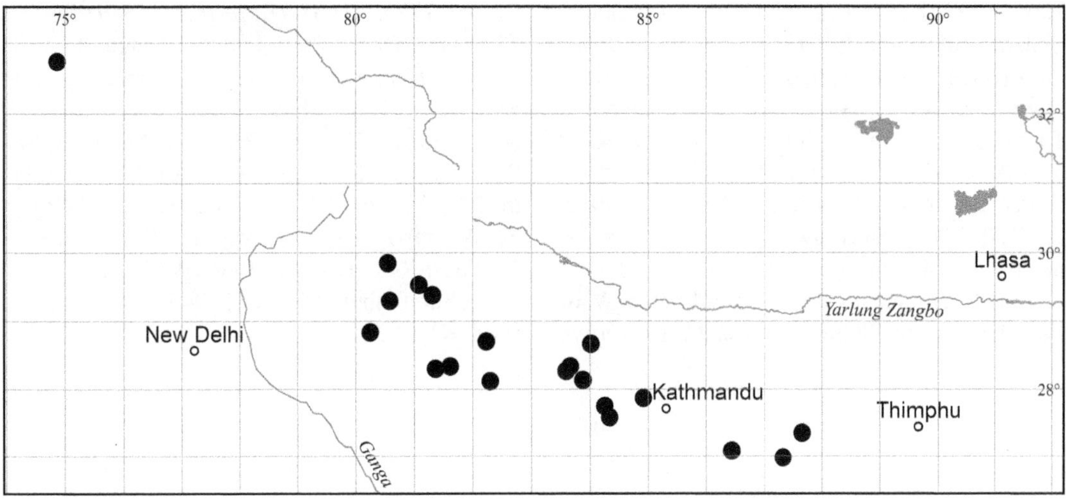

Map 46-243: **Martynia annua** L.

Pathenagar, Baukey, 180 m, 21 Nov. 1979, *L. R. Shama 88* (KATH); Kanchanpur, Mahendra Nagar, 150 m, 4 Aug. 1980, *K. R. Rajbhandari et al. 5051* (KATH); Kanchanpur, Gauja, 220 m, 18 Oct. 1981, *L. P. Kattel 701* (KATH); **Rapti**, Dang, Koilabas, 180 m, 1974, *M. K. Ghemen 108* (KATH); Dang, Koilabas, 19 Apr. 1981, *T. K. Bhattachaya 156* (KATH); Dang, Koilabas, 230 m, 26 Aug. 1982, *N. P. Manandhar & N. R. Bhattarai 8414* (KATH); **Seti,** Bajhang, Simad Bagar, 12 Sep. 1982, *L. P. Kattel 1209* (KATH); Bajura, Baba Kuti, 180 m, 10 May 1980, *T. K. Bhattachaya & T. K. Bhattaya 3934* (KATH); Bajura, Rila to Martari, 1200 m, 11 Aug. 1972, *M. S. Bista & D. P. Joshi 322* (KATH); Doti, Mahapani, 2400 m, 3 Nov. 1974, *K. M. Sharma E. 189* (BM). **C NEPAL**: **Bagmati**, Dhading, Todake, 1150 m, 10 Nov. 1989, *N. P. Manandhar 13236* (KATH); Nuwakot, Trisuli Bajar, 600 m, 2 Sep. 1971, *J. F. Dobremez 903* (E); Muwakot, Trisuli Bazar, 630 m, 13 Jul. 1994, *M. Suzuki et al. 9480002* (E); Rasuwa, Bainse to Sim Ghotala, 19 Aug. 1972, *H. Janai et al. 721615* (KATH); **Dhawalagiri**, Baglung Baglung to Retnechour, 820–960 m, 1 Sep. 1996, *M. Mikage et al. 9687011* (TI); Baglung, Galeshwor to Baglung, 900 m, 9 Aug. 1996, *K. R. Rajbhandari 9672454* (KATH); Myagdi, Rahughat, 890 m, 8 Aug. 1996, *T. Hoshino et al. 9662133* (E, TI); Myagdi, Mayangdi Khola, 900 m, 30 Aug. 1954, *J. D. A. Stainton, W. R. Sykes & L. H. J. Williams 4092* (BM); Myagdi, Yuranti, 600 m, 9 Aug. 1954, *J. D. A. Stainton, W. R. Sykes & L. H. J. Williams 6854* (BM); Myagdi, Kuma, Kali Gandaki River, 900 m, 230 Aug. 1954, *J. D. A. Stainton, W. R. Sykes & L. H. J. Williams 6988* (BM); Myagadi, Tatopani, S of Dana, Kali Gandaki Valley, 1200 m, 30 Aug. 1954, *J. D. A. Stainton, W. R. Sykes & L. H. J. Williams 7594* (BM, E); Parbat, Kushma, 950 m, 13 Sep. 1991, *N.P. Mamandhar 1102-91* (KATH); **Gandaki**, Manang, Marsyandi Khola, Baiszager–Turture Dhara–Chambas–Barabise, 14 Aug. 1983, *H. Ohba et al. 8311306* (KATH); Tanahu, Bais Jaangar–Barabise, 600 m, 13 Aug. 1983, *K. R. Rajbhandari 9263* (KATH); **Janakpur**, Dolpa, Dunge, 1580 m, 12 Apr. 1963, *Itoh & Rajnhandari 1109* (KATH); **Lumbini**, 7 Oct. 1982, *N. P. Manandhar & N. K. Bhattarai 9151* (KATH); **Narayani**, Chitwan, Rampur, 4 Aug. 1992, *N. Joshi & I. Shrestha 176* (KATH); Chitwan, Mugling, 430 m, 2 Jul. 1973, *D. P. Joshi & M. M. Amatya 73/141* (KATH). **E NEPAL**: **Koshi**, Dhankuta, Lagowa to Leguwa, 13 Aug. 1997, *S. Noshiro et al. 9760048* (E, TI); Dhankuta,

East bank of Arun Kosi, near Rithabus, 23 Sep. 2001, *D. G. Long et al. 23* (E, TI); Dhankuta, Belhara, 1090 m, 6 Jul. 1969, *L. H. J. Williams 1174* (BM); Dhankuta, 500 m,14 Aug.1972, *J. F. Dobremez 1461* (E); Sankhuwasabha, Khahare to Tumlingtar, 420 m, 26 Jul. 1990, *M. Minaki et al. 9080039* (TI); Sankhuwasabha, Tumlingtar to Baireni Bagar, 370 m, 31 Aug. 1998, *S. Noshiro et al. 9830214* (E); Sankhuwasabha, Tumlingtar, W of Chainpur, 460 m, 2 Sep. 1956, *J. D. A. Stanton 1501* (BM); Sankhuwasabha, Tumlingtar, W of Chainpur, 310 m, 28 Sep. 1956, *J. D. A. Stainton 1785* (BM); **Mechi**, Ilam, Dhara Panito Tambur Bridge, 17 Oct. 1963, *H. Hara et al. 6303609* (K, TI); Ilam, Dhara Pani–Tambur Bridge, 17 Oct. 1963, *H. Hara et al. 6303906* (TI); Taplejung, near Mul Ghat, 210 m, 28 Aug. 1971, *P. R. Shakya & M. Ohsawa 792* (KATH, TI); **Sagarmatha**, Solukhumbu, Bhote Kosi, Khari Khola, 1200 m, 22 Jun. 1964, *J. D. A. Stainton 4658* (BM); Solukhumbu, Thulaghat, 900 m, 30 Jul. 1953, *J. B. Tyron 148* (BM). **U IRRAWADDY**: **Putao**, along Sumprabun Rd, 3 Sep. 1945, *B. E. Juan 147* (BM, K).

STEMONURACEAE

Yun-Fei DENG

Evergreen trees or shrubs. Buttresses absent, stilt-roots. Leaves simple, alternate, sometimes appearing distichous, entire, penninerved, rarely triplinerved, exstipulate. Inflorescences axillary, terminal or leaf-opposed, occasionally ramiflorous, cymose, rarely reduced to 1 or few flowers. Flowers bisexual or functionally unisexual, actinomorphic, usually 5-merous. Calyx persistent, articulated with pedicel, small, generally cupular, sepals 4–5, connate, calyx rim entire, apiculate or shortly lobed. Petals 4–5(–7), minute to conspicuous, free or rarely coherent, sometimes united, often inflexed at apex; lobes valvate, midrib usually thickened ventrally. Male flowers: stamens isomerous, alternipetalous, filaments filiformor broad and flattened, glabrous or with papillose to clavate hairs ven-trally, dorsally and/or at apex of connective; anthers bisporangiate, basifixed, introrse or extrorse, longitudinally dehiscent. Staminodes in pistillate flowers present or absent, resembling stamens or much smaller but pollen sacs sterile; disk absent or present, cupular, hippocrepiform, or squami-form and unilateral, lobed or entire, persistent in fruit. Female flowers: ovary superior, carpel apparently 1 but likely pseudomonomerous, unilocular, often with exterior lateral appendage above disk, ovules 2, apical, pendulous, anatropous, uniteg-mic; stigma on short style or more often sessile, peltate, subcapitate, or punctiform only in bisexual flowers, entire to lobed, crenate, or ridged, often eccentric, persistent in fruit. Fruit a drupe of various shape. Seeds 1, testa vascularised with post-chalazal or pachychalazal bundles. Embryo small.

Recent references: Sleumer, 1969. Materials Icacinaceae of Asia, Malesia and adjacent areas. *Blumea* 17: 181–264; Chuang, 1981. *Gomphandra*. In: Fang (ed.), *Fl. Reipubl. Popularis Sin.* 46: 40–42. Beijing: Science Press; Meijer & Verdcourt, 1996. Icacinaceae. In: Dassanayake (ed.), *A Revised Handbook to the Flora of Ceylon*, 10: 249–263. A. A. Balkema, Potterdam; Kårehed, 2001. Multiple origin of the tropical forests treefamily Icacinaceae. *Am. J. Bot.* 88: 2259–2274; Peng & Howard, 2008. *Gomphandra*. In: Wu, Raven & Hong (eds.), *Fl. China* 11: 506–507, Beijing & St. Louis: Science Press & Missouri Botanic Garden Press; Schori & Furness, 2014. Pollen diversity in Aquifoliales. *Bot. J. Linn. Soc.* 175: 169–190; Schori, 2016. Stemonuraceae. In: Kadereit & Bittrich (eds.), *The Families and Genera of Vascular Plants* 14: 131–136. Berlin: Springer.

Twelve genera with about 90 species, distributed mainly in the tropics of the Old World, rapidly decreasing in number towards the subtropics; one genus and one species in the Pan-Himalaya.

1. **GOMPHANDRA** Wall. ex Lindl.

Lindley, Nat. Syst. Bot., ed. 2: 439. 1836; Bentham & Hooker, Gen. Pl. 1: 350. 1846; Masters in Hooker (ed.), Fl. Brit. India 1: 585. 1875; Chuang in Fang Fang (ed.), Fl. Reipubl. Popularis Sin. 46: 40. 1981; Peng & Howard in Wu, Raven & Hong (eds.), Fl. China 11: 506. 2008. Type (designated by Sleumer, in Engler et Prantl, Nat. Pflanzenfam. ed. 2. 20b: 364. 1942): *Gomphandra tetrandra* (Wall. ex Roxb.) Sleumer.

Trees or shrubs, deciduous; branchlets smooth. Leaves alternate, margin entire, often tubercled beneath, penninerved. Plants dioecious. Flowers small, unisexual, in di- or trichotomous cymes, axillary, terminal, or opposite to leaves; staminate cymes many flowered; pistillate cymes few flowered. Bracts small. Sepals united into a 4- or 5-lobed cup. Petals 4 or 5, free or cohering in a short tube, valvate. Male flowers: stamens 4 or 5, usually exserted; filaments fleshy, clavate, 2 or 3 times as long as anthers, pilose, rarely glabrous, free from corolla tube; disk pulvinate, coherent with ovary or reduced ovary. Female flowers: staminodes present or absent; disk often absent; ovary terete or obovoid; stigma capitate or discoid, sometimes 2- or 3-lobed, sessile. Drupe ellipsoid or oblong, sometimes shortly beaked, base sometimes swollen; endocarp crustaceous or hard, with 8–10(–14) longitudinal ridges especially prominet in dry fruit.

About 33 species: tropical Asia to NE Australia; one species in the Pan-Himalaya.

1. **Gomphandra tetrandra** (Wall.) Sleumer, Notizbl. Bot. Gart. Berlin-Dahlem 15: 238. 1940; Chuang in Fang (ed.), Fl. Reipubl. Popularis Sin. 46: 42. 1981; Peng & Howard in Wu, Raven & Hong (eds.), Fl. China 11: 506. 2008. ≡ *Lasianthera tetrandra* Wall. in Roxburgh, Fl. Ind. 2: 328. 1824; Masters in Hooker (ed.), Fl. Brit. India 1: 585. 1875. 6: 48. 1931. ≡ *Stemonurus tetrandrus* (Wall.) Alston, Handb. Fl. Ceylon 6: 48. 1931. ≡ *Gomphandra axillaris* Wall., Numer. List n. 3718. 1831, nom. illeg. (Art. 52.1); Voigt, Hort. Suburb. Calcut.: 31. 1845; Masters in Hooker (ed.), Fl. Brit. India 1: 586. 1875. ≡ *Stemonurus axillaris* Miers, Ann. Mag. Nat. Hist. II, 10: 40. 1852, nom. illeg. (Art. 52.1). ≡ *Platea axillaris* Thwait., Enum. Pl. Ceyl. 44. 1852, nom. illeg. (Art. 52). Type: Bangladesh, Sylhet, *M. R. Smith ex Wallich Cat. 3718* (lectotype designated by Meijer & Verdcourt, Revised Handbook to the Flora of Ceylon, 10: 251. 1996: K-W; isolectotypes: BM, CAL, CGE).

Gomphandra polymorpha Wight, Illustr. Ind. Bot. 1: 103. 1840; Icon. Pl. Ind. Orient. 3(3): 3, t. 953 & 954. 1845; Masters in Hooker (ed.), Fl. Brit. India 1: 586. 1875. ≡ *Stemonurus polymorphus* (Wight) Miers, Ann. Mag. Nat. Hist., Ser. 2, 10: 37. 1852. Type: India, Western Peninsula from the Concans southwards, *Wight K. D. 428* [lectotype designated by Sleumer, 1969: K; isolectotypes: C, L (L0014847), M (M0212148)].

Shrubs or small trees, 2–10 m tall. Bark gray; young branches green, densely or sparsely yellow pubescent. Petiole 0.5–1.5 cm long, sparsely pubescent; leaf blade shiny, glabrous, narrowly lanceolate to elliptic, 6–15 cm long, 2–6 cm broad, adaxially glabrous, abaxially yellowish pubescent, lateral veins 6–8 pairs, base cuneate, margin entire, apex acuminate or caudate. Cymes opposite leaves, sometimes axillary, 2–4 cm long, densely yellow-white pubescent; peduncle 2–5 mm long. Staminate flowers: calyx 0.4–0.6 mm long, shallowly 5-lobed; corolla yellow-white or white-green, campanulate, 3–4 mm long, lobes subtriangular, apex abruptly acuminate, incurved; stamens slightly longer than petals, 3.5–4.5 mm long; filaments flat, 0.8–1 mm broad, fleshy, apex white clavate pilose; anthers yellow-white, ovoid, 0.4–0.6 mm long; ovary undeveloped, small, 0.5–1 mm long. Pistillate flowers: calyx as in staminate flowers; corolla yellow-white, campanulate, 4–5 mm; lobes narrowly triangular, margin involute, apex incurved; staminodes shorter than corolla; filaments as in staminate flowers; ovary glabrous or sometimes pubescent; stigma small, 5-lobed to base. Drupe berry-like, yellow to white, ellipsoid, (1.2–)2–2.5 cm long, (5–)7–12 mm broad, conspicuously longitudinally ribbed when dry; fruiting peduncle slightly pubescent. Figure 46-141.

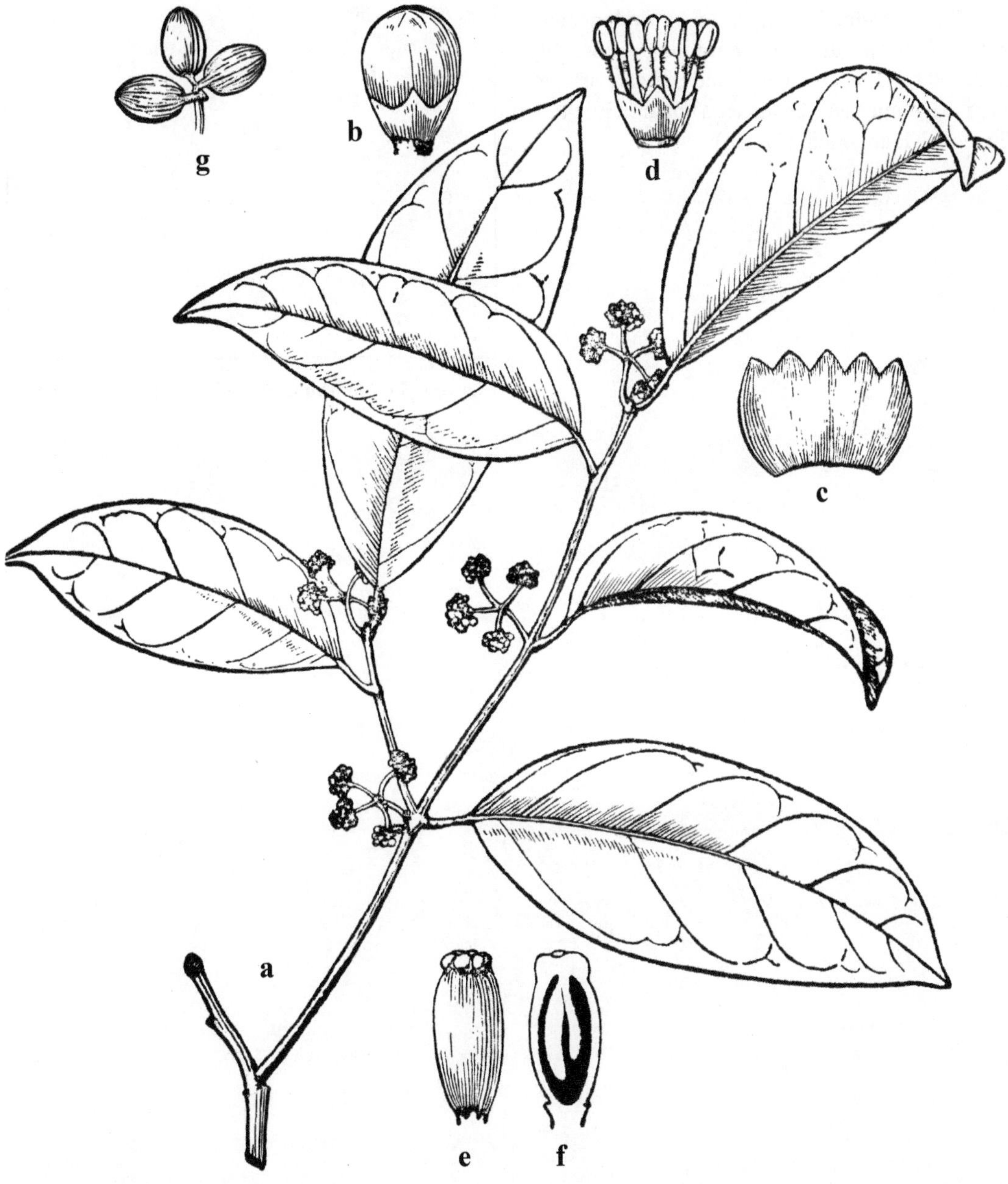

Figure 46-141: **Gomphandra tetrandra** (Wall.) Sleumer: **a**, flowering branch; **b**, female flower; **c**, opened female corolla; **d**, stamens; **e**, pistil; **f**, opened pistil showing ovules; **g**, infrutescence. From *Icon. Cormophyt. Sin.* 2: 694, fig. 3118. 1975.

Phenology: flowering and fruiting all seasons.

Habitats: sparse woods, dense forests, roadside scrub, valleys; 500–2200 m.

Distribution: Yarlung Zangbo-Brahmaputra (Map 46-244); also in Cambodia, Laos, Sri Lanka, Thailand, Vietnam, and other parts of India, Myanmar and China (S Yunnan, Guizhou, Guangxi, Guangdong, Hainan).

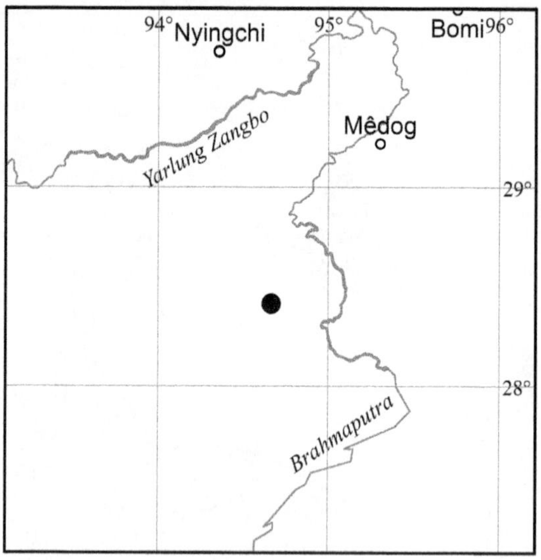

Map 46-244: **Gomphandra tetrandra** (Wall.) Sleumer

CARDIOPTERIDACEAE

Yun-Fei DENG & Hua PENG

Lianes. Leaves simple; petiole bundles annular; leaf blade toothed or spiny, lateral veins palmate. Plant dioecious andromonoecious. Inflorescences branched, ultimate units clearly cymose or not. Bracts absent; calyx (4 or) 5, free, or basally connate, cupular, quincuncial; lobes imbricate. Petals (4 or) 5, caducous, lower half forming a widely funnel-shaped corolla, imbricate in bud. Stamens (4 or) 5, epipetalous, inserted on the upper part of corolla; filaments very short, glabrous; anthers medifixed, introrse, bithecous, longitudinally dehiscent. Disk absent. Ovary superior, ovoid-oblong, slightly 4-ribbed, rudimentary in staminate flowers, 1-loculed, apically with 2 pendulous ovules; style usually slender, about as long as ovary, or ± laterally positioned, branched to base; stigma truncate or capitate; ovule 1, integument not vascularized. Fruit a 2-winged samara, asymmetric, wings horizontally striate, stout style accrescent, endocarp C-shaped in transverse section. **Chromosomes** n = 14.

Recent references: Sleumer, 1942. Peripterygiaceae. In: Engler & Prantl (eds), *Die Natürlichen Pflanzenfamilien*, 20b: 397–400. Berlin: Duncker & Humblot; Scholz, 1964. Celastrales. In: Melchior (ed), A Engler's Syllabus der Pflanzenfamilien. 2: 289–300. Berlin: Gebrüder Borntraeger, Berlin; Sleumer, 1971. Cardiopteridaceae. In: van Steenis (ed), *Flora Malesiana* 7: 93–96. Leyden: Noordhoff International Publishing; Sleumer, 1972. Cardiopteridaceae. *Fl. Thailand* 2(2): 93–94; Hutchinson, 1973. *The Families of flowering plants*, 3rd ed. Oxford: Oxford University Press; Subramanyam, 1973. Materials for the flora of Bhutan. *Rec. Bot. Surv. India* 20(2): 1–278; Cronquist, 1981. *An integrated system of classification of flowering plants*. New York: Columbia University Press; Thorne, 1992. Classification and geography of the flowering plants. *Bot Rev* (*Lancaster*) 58: 225–348; Takhtajan, 1997. *Diversity and classification of flowering plants*. New York: Columbia University Press; Long, 1991. Cardiopteridaceae. In: Grierson & Long (eds.), *Fl. Bhutan* 2(1): 135–136, Edinburgh: Royal Botanical Garden Edinburgh & Royal Government of Bhutan; Takhtajan, 1997. Diversity and classification of flowering plants; Mathur, 2000. Cardiopteridaceae. In: Singh et al. (eds.), *Fl. India* 5: 47–49. Culcutta: Government of India; Kärehed, 2001. Multiple origin of the tropical forest tree family Icacinaceae. *Amer. J. Bot*. 88: 2259–2274; Stevens, 2001 onwards. Angiosperm phylogeny Website. Version 12, July 2012. http://www.mobot.org/MOBOT/research/APweb/. Angiosperm Phylogeny Group, 2003. An update of the Angiosperm Phylogeny Group classification for the orders and families of flowering plants: APG II. *Bot. J. Linn. Soc.* 141: 399–436; Peng & Howard, 2008. Cardiopteridaceae. In: Wu, Raven & Hong (eds.), *Fl. China* 11: 358. Beijing & St. Louis: Science Press & Missouri Botanic Garden Press; Takhtajan, 2009. Flowering plants, 2nd ed. Springer, New York. Angiosperm Phylogeny Group, 2009. An update of the Angiosperm Phylogeny Group classification for the orders and families of flowering plants: APG III. Bot. J. Linnean Soc. 161: 105-121; Tank & Donoghue, 2010. Phylogeny and phylogenetic nomenclature of the Campanulidae based on an expanded sample of genes and taxa. *Syst. Bot.* 35:425 -441; Soltis et al., 2011. Angiosperm phylogeny: 17 genes, 640 taxa. *Amer. J. Bot.* 98:704 -730.

Five genera and about 45 species: tropical region of the world; one genus and one species in the Pan-Himalaya.

1. CARDIOPTERIS Wall. ex Royle

Royle, Ill. Bot. Himal. Mts.: 136. 1834; Long in Grierson & Long (eds.), Fl. Bhutan 2(1): 135. 1991; Peng & Howard in Wu, Raven & Hong (eds.), Fl. China 11: 358. 2008. Type: *Cardiopteris lobata* Wall. ex Benn. & R. Br. [≡ *Cardiopteris quinqueloba* (Hassk.) Hassk.].

Peripterygium Hassk., Tijdschr. Natuurl. Gesch. Physiol. 10: 142. 1843; Chuang in Fang, Fl. Reipubl. Popularis Sin. 46: 64. 1981. Type: *Peripterygium quinquelobum* Hassk. [≡ *Cardiopteris quinqueloba* (Hassk.) Hassk.].

Sioja Buch.-Ham. ex Lindl., Intr. Nat. Syst. Bot. ed. 2: 82. 1836. Type: *Sioja sanguinaria* Buch-Ham. [= *Cardiopteris quinqueloba* (Hassk.) Hassk.].

Sinistrorsely twining herbs, with abundant milky juice, glabrous. Leaves spirally arranged, estipulate, long petiolate; blade cordate or cordate-hastate, simple or lobed to varying degrees, glabrous, palmately 3–7-veined. Panicles axillary, lax, loose, scorpioid at apex. Bracts early caducous. Flowers bisexual or polygamous, small, subsessile. Sepals 4 or 5, coherent at base, imbricate in bud whether or not slightly accrescent in fruit, persistent. Petals 4 or 5, caducous, lower half forming a widely funnel-shaped corolla, imbricate in bud. Stamens 4 or 5, epipetalous, inserted on the upper part of of corolla; filaments very short; anthers medifixed, introrse, bithecous, longitudinally dehiscent. Disk absent. Ovary superior, ovoid-oblong, slightly 4-ribbed, rudimentary in staminate flowers, 1-loculed, apically with 2 pendulous ovules; style 2, one with a deep longitudinal groove and two unequal ovate rather obtuse distal divisions, accrescent in fruit caducous; stigma 2-lobed, one lobe capitate and early caducous, one lobe elongate in fruit, apex 2-lobed, caducous. Fruit 2-winged samara, broad, orbicular or obcordate, indehiscent, compressed, 2-seeded; wings membranous, horizontally veined. Seeds linear, longitudinally grooved; testa thin; endosperm densely granular, fleshy.

Recent references: Bullock, 1957. Nomenclatural Notes: II. *Kew Bull.* 12(2): 356; Bakhuizen van den Brink & van Steenis, 1960. Nomenclature note on *Cardiopteris*. *Fl. Males. Bull.* 15: 725–726; Maheshwari, 1961. Proposal for the conservation of the generic name *Peripterygium* Hasskarl (1843) against *Sioja* Buchanan-Hamilton ex Lindley (1836) and Rejection of the later synonym *Cardiopteris* Wallich (1847). *Taxon* 10(8): 249; Bakhuizen van den Brink & van Steenis, 1962. *Cardiopteris* or *Peripterygium*? (Icacinaceae). *Taxon* 11(1): 28–29; Chuang, 1981. *Peripterygium*. In: Fang (ed.), *Fl. Reipubl. Popularis Sin.* 46: 63–65. Beijing: Science Press; Chen et al., 1994. Discovery and significance of the genus *Cardiopteris*. *Acta Bot. Sin.* 36(7): 552–556; Kong et al., 2002. A new type of embryo sac in *Cardiopteris* and its systematic implication. *Acta Bot. Sin.* 44(4): 496–498; Hiroshi, 2012. Floral structure of *Cardiopteris* (Cardiopteridaceae) with special emphasis on the gynoecium: systematic and evolutionary implications. *J. Pl. Res.* 125(3): 361–369; Kong et al., 2014. Floral development of *Cardiopteris*, with emphasis on gynoecial structure and ovular morphology. *J. Syst. Evol.* 52(5): 629–642; Hiroshi, 2016. Embryology of *Cardiopteris* (Cardiopteridaceae, Aquifoliales), with emphasis on unusual ovule and seeds development. *J. Pl. Res.* 129(5): 883–897; Peng & Howard, 2008. Cardiopteridaceae. pp. 358. In: Wu, Raven & Hong (eds.), *Fl. China* 11. Beijing & St. Louis: Science Press & Missouri Botanic Garden Press.

Previously Angiosperm Phylogeny Website the genus was assigned to Olacaceae (Bentham & Hooker, Gen. P1. 1: 355. 1862), Icacinaceae (Engler, Nat. Pflanzenfam 3(5): 233–257. 1897), or its own family Cardiopteridaceae (e.g. Blume, Rumphia 3:205–207. 1847; Cronquist, 1981; Hutchinson, 1973; Mabberley, 2008; Scholz, 1964; Sleumer, 1942, 1971; Takhtajan, 1997, 2009; Thorne, 1992). However, based on molecular evidence (Kårehed, 2001; Soltis et al., 2011; Tank & Donoghue, 2010), the Angiosperm Phylogeny Group (2003, 2009) has accepted the placement of *Cardiopteris*, along with *Citronella* D. Don, *Gonocaryum* Miq. and *Leptaulus* Benth., in a broadly circumscribed family Cardiopteridaceae that is sister to Stemonuraceae in Aquifoliales (see also Reveal, 2012; Stevens, 2001 onwards).

1. Cardiopteris quinqueloba (Hassk.) Hassk., Natuurk. Tijdschr. Ned.-Indie. 10: 64. 1855; Long in Grierson & Long (eds.), Fl. Bhutan 2(1): 135. 1991; Peng & Howard in Wu, Raven & Hong (eds.), Fl. China 11: 358. 2008. ≡ *Peripterygium quinquelobum* Hassk., Tijdschr. Natuurl. Gesch. Physiol. 10: 142. 1843; Masters in Hooker (ed.), Fl. Brit. India 1: 597. 1875; Chuang in Fang, Fl. Reipubl. Popularis Sin. 46: 64. 1981. ≡ *Cardiopteris lobata* Wall. ex Benn. & R. Br., Pl. Jav. Rar. 246. t. 49. 1852. ≡ *Cardiopteris rumphii* var. *lobata* Baill., Adansonia 10: 281. 1872. ≡ *Cardiopteris rhumphii* Baill., Adandonia 10: 280. 1871, nom. illeg. (Art. 52.1). Type: Myanmar, Hilsa/Ava, 1826, *Wallich Cat. No. 8033a* [lectotype **here designated**: K-W (K001128801); isolectotypes: BM, K].

Deciduous twining herbs; stems terete, twining and striate. Leaves polymorphic; petiole 3–12 cm long; blade broadly ovate, (4–)7–12(–17) cm long, (3–)5–8(–16) cm broad, glabrous, base usually cordate, sometimes truncate, margin usually deeply or irregularly 3–5(–9)-lobed, apex commonly acute, basal lobes auriculate-orbicular or hastate, sometimes sinuous, apical lobe larger, palmately 5- or 7-veined from base, widely divergent, midvein with 1 or 2 pairs of lateral veins, veins slightly prominent, reticulate veins inconspicuous. Panicle solitary, pedunculate, (3–)5–16(–27) cm long, 1–3-forked, cincinni rather few-flowered. Flowers small, bisexual; flower buds yellow-green, subglobose. Pedicel 0–1 mm long, pubescent. Calyx lobes ovate, 1.5–2 mm long, glabrous or outside sparsely puberulent, coherent at base, margin sometimes ciliate. Petals white, ovate to narrowly ovate, 1.8–2 mm long, united at base, apex obtuse or rounded. Filaments filiform, slightly widened at base; anthers oblong-ovoid, longer than filaments. Ovary ellipsoid or narrowly conical, 1.5–3 mm long, slightly 4-ribbed, glabrous; style exserted, terete. Samara golden, shiny, suborbicular or obcordate, 2–3.5 cm long, 1.5–3 cm broad, with 1–8 mm elongate accrescent style at top; wings yellowish. Figure 46-142.

Phenology: flowering from May to November; fruiting from October to March of next year.

Habitats: sparse woods, scrubs, forests, dense valley forests; 100–1300 m.

Distribution: Bhutan, Yarlung Zangbo-Brahmaputra (Map 46-245); also in Thailand, Vietnam, Indonesia, Malaysia, and other parts of India, Myanmar and China (S Yunnan, Guangxi, Hainan).

Uses: the leaves are used as vegetable.

Notes: The species was reported from Bhutan (Gaylegphug) by Subramanyam (1973) and Yarlung Zangbo-Brahmaputra (Dirang, Mishmi Hills) by Mathur (2009). But we have not seen any specimens collected from the Pan-Himalaya.

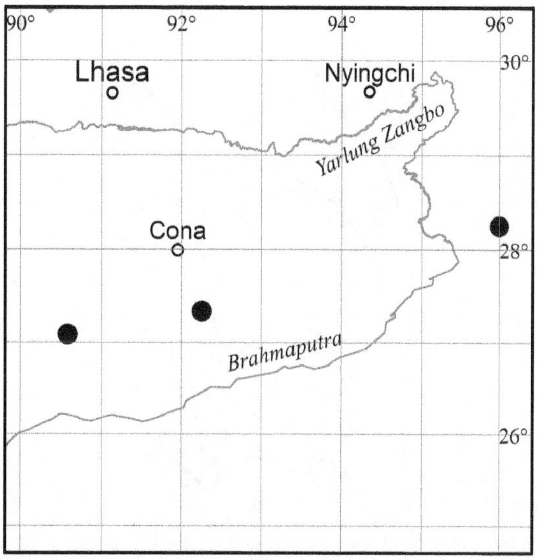

Map 46-245: **Cardiopteris quinqueloba** (Hassk.) Hassk.

Figure 46-142: **Cardiopteris quinqueloba** (Hassk.) Hassk.: **a**, flowering branch; **b**, leaf; **c**, flower; **d**, petals spreading out; **e**, sepals spreading out; **f**, fruit. Drawn by Xi-Chou LI; reproduced from *Fl. Reipubl. Popularis Sin.* 46: 65, pl. 14. 1981.

Nomenclatural novelty appearing in this volume

Phlogacanthus sect. **Cystacanthus** (T. Anderson) Y. F. Deng, st. nov. ≡ *Cystacanthus* T. Anderson

Phlogacanthus sect. **Pubinervia** Y. F. Deng, sect. nov.

Peristrophe sect. **Yunnanenses** Y. F. Deng & Z. P. Hao, sect. nov.

Peristrophe sect. **Paniculati** Y. F. Deng & Z. P. Hao, sect. nov.

Typification appearing in this volume

Acanthus carduaceus Griff., lectotype designation

Acanthus leucostachyus Wall. ex Nees, lectotype designation

Adenosma affinis Griff. ≡ **Strobilanthes affinis** (Griff.) T. Terao ex J. R. I. Wood & J. Benn., lectotype designation

Adenosma Nees, generitype designation = **Hygrophila** R. Br.

Asystasia calycina Nees, lectotype designation ≡ **Strobilanthes helicta** T. Anderson

Asystasia macrocarpa Nees, lectotype designation

Barleria caerulea Roxb., lectotype designation = **Barleria strigosa** Willd.

Barleria ciliata Roxb., lectotype designation = **Barleria cristata** L.

Barleria dichotoma Roxb., lectotype designation = **Barleria cristata** L.

Barleria hexacantha Moris, lectotype designation = **Hygrophila schulli** (Buch-Ham.) M. R. Almeida & S. M. Almeida

Barleria laciniata Nees, lectotype designation = **Barleria cristata** L.

Barleria napalensis Nees, lectotype designation = **Barleria cristata** L.

Barleria trinervis Wall., lectotype designation ≡ **Lepidagathis trinervis** Nees

Buchneria cordifolia L. f., lectotype designation ≡ **Priva cordifolia** (L. f.) Druce

Chingiacanthus patulus Hand.-Mazz., lectotype designation = **Isoglossa collina** (T. Anderson) B. Hansen

Cystacanthus T. Anderson, generitype designation≡ **Phlogacanthus** sect. **Cystacanthus** (T. Anderson) Y. F. Deng

Daedalacanthus ciliatus Craib, lectotype designation ≡ **Eranthemum ciliatum** (Craib) Benoist

Dianthera bicalyculata Retz., lectotype designation = **Peristrophe paniculata** (Forssk.) Brummitt

Dianthera debilis C. B. Clarke, lectotype designation = **Leptostachya wallichii** Nees

Dicliptera buergeriana Miq., lectotype designation = **Peristrophe japonica** (Thunb.) Bremek.

Dicliptera bupleuroides var. *sessilis* Nees, lectotype designation = **Dicliptera bupleuroides** Nees

Dicliptera mairei Benoist, lectotype designation = **Dicliptera elegans** W. W. Sm.

Dipteracanthus urophyllus Nees, lectotype designation ≡ **Strobilanthes urophylla** (Nees) Nees

Duranta plumieri Jacq., neotype designation = **Duranta erecta** L.

Ebermaiera humilis Wall. ex Nees, lectotype designation ≡ **Staurogyne humilis** (Wall. ex Nees) Kuntze

Ebermaiera staurogyne Nees, lectotype designation = **Staurogyne argentea** Wall.

Echinacanthus andersonii C. B. Clarke, lectotype designation = **Strobilanthes violifolia** T. Anderson

Echinacanthus andersonii var. *viscosa* C. B. Clarke, lectotype designation =

Strobilanthes violifolia T. Anderson

Echinacanthus longistylus C. B. Clarke, lectotype designation = **Strobilanthes violifolia** T. Anderson

Echinacanthus pumilio C. B. Clarke, lectotype designation = **Strobilanthes violifolia** T. Anderson

Endopogon macrostegius Nees, lectotype designation = **Strobilanthes fimbriata** Nees

Eranthemum edgeworthianum Nees, lectotype designation = **Eranthemum pulchellum** Andrews

Eranthemum leptanthus C. B. Clarke, lectotype designation ≡ **Pseuderanthemum leptanthus** (C. B. Clarke) Lindau

Eranthemum polyanthum C. B. Clarke ex Oliv., lectotype designation ≡ **Pseuderanthemum polyanthum** (C. B. Clarke ex Oliv.) Merr.

Flemingia grandiflora Roxb. ex Rottl., lectotype designation = **Thunbergia fragrans** Roxb.

Gendarussa vasculosa Nees, lectotype designation ≡ **Justicia vasculosa** (Nees) T. Anderson

Goldfussia biceps Nees, lectotype designation = **Strobilanthes capitata** (Nees) T. Anderson

Goldfussia capitata Nees, lectotype designation ≡ **Strobilanthes capitata** (Nees) T. Anderson

Goldfussia crinita Nees, lectotype designation = **Strobilanthes hamiltoniana** (Steud.) Bosser & Heine

Goldfussia cusia Nees, lectotype designation ≡ **Strobilanthes cusia** (Nees) Kuntze

Goldfussia divaricata Nees, lectotype designation ≡ **Strobilanthes divaricata** (Nees) T. Anderson

Goldfussia edgeworthiana Nees, lectotype designation = **Strobilanthes angustifrons** C. B. Clarke

Goldfussia extensa Nees, lectotype designation ≡ **Strobilanthes extensa** (Nees) Nees

Goldfussia flexuosa Nees, lectotype designation = **Strobilanthes pentstemonoides** (Nees) T. Anderson

Goldfussia glomerata Nees var. *speciosa* Hook., lectotype designation = **Strobilanthes glomerata** (Nees) T. Anderson

Goldfussia pentstemonoides Nees, lectotype designation = **Strobilanthes pentstemonoides** (Nees) T. Anderson

Haplanthus tener Nees, lectotype designation = **Haplanthus laxiflorus** (Blume) Gnanasek., G. V. S. Murthy & Y. F. Deng

Hygrophila phlomoides Nees, lectotype designation

Hypoestes wallichii Nees, lectotype designation = **Hypoestes triflora** (Forssk.) Roem & Schult.

Incarvillea arguta var. *longipedicellata* Q. S. Zhao, lectotype designation = **Incarvillea arguta** Royle

Incarvillea grandiflora var. *brevispes* Sprague, lectotype designation = **Incarvillea mairei** (H. Lév.) Grierson

Incarvillea lutea Bureau & Franch., lectotype designation

Incarvillea younghusbandii Sprague, lectotype designation

Justicia asperula Lodd., lectotype designation = **Phlogacanthus vitellinus** (Roxb.) T. Anderson

Justicia calcarata Wall., lectotype designation ≡ **Rhinacanthus calcaratus** (Wall.) Nees

Justicia curviflora Wall., lectotype designation ≡ **Phlogacanthus curviflorus** (Wall.) Nees

Justicia gendarussa Burm. f., lectotype designation

Justicia guttata Nees, lectotype designation ≡ **Phlogacanthus guttatus** (Nees) Nees

Justicia khasiana C. B. Clarke, lectotype designation = **Justicia simplex** D. Don

Justicia lanceolaria Roxb., lectotype designation ≡ **Peristrophe lanceolaria** (Roxb.) Nees

Justicia nervosa Vahl, lectotype designation =

Eranthemum pulchellum Andrews

Justicia parviflora Retz., lectotype designation = **Rungia pectinata** (L.) Nees

Justicia quadrangularis Hook., lectotype designation = **Phlogacanthus vitellinus** (Roxb.) T. Anderson

Justicia siccanea W. W. Sm., lectotype designation

Justicia speciosa Roxb., lectotype designation ≡ **Peristrophe speciosa** (Roxb.) Nees

Justicia thyrsiflora Roxb., lectotype designation = **Phlogacanthus thyrsiformis** (Roxb. ex Hardw.) Mabb.

Justicia triflora Forssk., lectotype designation ≡ **Hypoestes triflora** (Forssk.) Roem & Schult.

Justicia tubiflora Buch.-Ham. ex Wall., lectotype designation ≡ **Phlogacanthus tubiflorus** Nees

Justicia vahlii Roth, lectotype designation = **Justicia quinqueangularis** J. Koenig ex Roxb.

Justicia vitellina Roxb., lectotype designation ≡ **Phlogacanthus vitellinus** (Roxb.) T. Anderson

Justicia xylopoda W. W. Sm., lectotype designation = **Justicia xerobatica** W. W. Sm.

Lantana alba Mill., lectotype designation ≡ **Lippia alba** (Mill.) N.E. Br. ex Britton & P. Wilson

Lepidagathis hyalina Nees, lectotype designation = **Lepidagathis incurva** Buch.-Ham. ex D. Don

Lepidagathis purpuricaulis Nees, lectotype designation

Lepidagathis semiherbacea Nees, lectotype designation = **Lepidagathis incurva** Buch.-Ham. ex D. Don

Lepidagathis ustulata Nees, lectotype designation = **Lepidagathis incurva** Buch.-Ham. ex D. Don

Peripterygium quinquelobum Hassk., lectotype designation ≡ **Cardiopteris quinqueloba** (Hassk.) Hassk.

Peristrophe bicalyculata var. *rivinoides* C. B. Clarke, lectotype designation = **Peristrophe paniculata** (Forssk.) Brummitt

Peristrophe fera C. B. Clarke, lectotype designation

Phlogacanthus curviflorus var. *brevicalyx* C. B. Clarke, lectotype designation = **Phlogacanthus curviflorus** (Wall.) Nees

Phlogacanthus Nees, generitype designation

Phlogacanthus parviflorus T. Anderson, lectotype designation

Phlogacanthus pubinervius T. Anderson, lectotype designation

Phlogacanthus pulcherrimus T. Anderson, lectotype designation = **Phlogacanthus vitellinus** (Roxb.) T. Anderson

Phyla chinensis Lour., lectotype designation = **Phyla nodiflora** (L.) E. L. Greene

Rhaphidospora ciliata Nees, lectotype designation = **Justicia pubigera** (Nees) C. B. Clarke

Rhinacanthus grandiflorus Dunn, lectotype designation

Rostellaria mollissima Nees, lectotype designation = **Justicia simplex** D. Don

Rostellaria peploides Nees, lectotype designation ≡ **Justicia peploides** (Nees) T. Anderson

Rostellaria rotundifolia Nees, lectotype designation = **Justicia simplex** D. Don

Ruellia bracteata Roxb., lectotype designation = **Petalidium barlerioides** (Roth) Nees

Ruellia crenata Benth. ex Hohen, lectotype designation = **Strobilanthes pavala** (Roxb.) J. R. I. Wood

Ruellia cuspidata Wall., lectotype designation ≡ **Lepidagathis cuspidata** Nees

Ruellia diffusa Wall. ex Nees, lectotype designation = **Strobilanthes pavala**

(Roxb.) J. R. I. Wood

Ruellia hirta Vahl, lectotype designation ≡ **Strobilanthes hirta** (Vahl) Blume

Ruellia lyi H. Lév., lectotype designation = **Strobilanthes henryi** Hemsl.

Ruellia neesiana Wall., lectotype designation ≡ **Asystasia neesiana** (Wall.) Nees

Ruellia sarmentosa Nees, lectotype designation = **Strobilanthes hirta** (Vahl) Blume

Staurogyne argentea Wall., lectotype designation

Strobilanthes aenobarba W. W. Sm., lectotype designation ≡ **Strobilanthes inflata** var. **aenobarba** (W. W. Sm.) J. R. Wood & Y. F. Deng

Strobilanthes amplectens Nees, lectotype designation = **Strobilanthes auriculata** Nees

Strobilanthes anisandrus Benoist, lectotype designation = **Strobilanthes henryi** Hemsl.

Strobilanthes arenicola W. W. Sm., lectotype designation

Strobilanthes auriculata Nees, lectotype designation

Strobilanthes boerhaavioides T. Anderson, lectotype designation = **Strobilanthes rubescens** T. Anderson

Strobilanthes discolor var. *nudicalyx* C. B. Clarke, lectotype designation = **Strobilanthes discolor** (Nees) T. Anderson

Strobilanthes duclouxii C. B. Clarke ex Benoist, lectotype designation = **Strobilanthes extensa** (Nees) Nees

Strobilanthes fimbriata Nees, lectotype designation

Strobilanthes fimbriata var. *majuscula* W. W. Sm., lectotype designation = **Strobilanthes fimbriata** Nees

Strobilanthes flexa Benoist, lectotype designation = **Strobilanthes versicolor** Diels

Strobilanthes henryi Hemsl., lectotype designation

Strobilanthes hygrophiloides var. *subnudus* Benoist, lectotype designation = **Strobilanthes yunnanensis** Diels

Strobilanthes hygrophiloides C. B. Clarke ex W. W. Sm., lectotype designation = **Strobilanthes yunnanensis** Diels

Strobilanthes inflata T. Anderson, lectotype designation

Strobilanthes laevigata C. B. Clarke, lectotype designation

Strobilanthes maclellandii var. *latipes* C. B. Clarke, lectotype designation = **Strobilanthes auriculata** Nees

Strobilanthes mastersii T. Anderson, lectotype designation

Strobilanthes nobilis C. B. Clarke, lectotype designation

Strobilanthes paupera C. B. Clarke, lectotype designation = **Strobilanthes oligocephala** T. Anderson ex C. B. Clarke

Strobilanthes petiolaris var. *tubiflos* C. B. Clarke, lectotype designation ≡ **Strobilanthes tubiflos** (C. B. Clarke) J. R. I. Wood

Strobilanthes plumulosa Nees, lectotype designation = **Strobilanthes auriculata** Nees

Strobilanthes pterygorrhachis C. B. Clarke, lectotype designation

Strobilanthes rubescens var. *microsperma* C. B. Clarke, lectotype designation = **Strobilanthes rubescens** T. Anderson

Strobilanthes sabiniana var. *argentea* Nees, lectotype designation = **Strobilanthes saccata** J. R. I. Wood

Strobilanthes schweliensis W. W. Sm., lectotype designation = **Strobilanthes mastersii** T. Anderson

Strobilanthes secunda T. Anderson, lectotype designation

Strobilanthes stolonifera Benoist, lectotype designation

Strobilanthes urophylla var. *sikkimensis* C. B. Clarke, lectotype designation = **Strobilanthes urophylla** (Nees) Nees

Strobilanthes wallichii var. *microphylla* Nees, lectotype designation = **Strobilanthes atropurpurea** Nees

Strobilanthes wardii W. W. Sm., lectotype designation = **Strobilanthes inflata** T. Anderson

Strobilanthes xanthantha Diels, lectotype designation = **Strobilanthes lachenensis** C. B. Clarke

Strobilanthes yunnanensis Diels, lectotype designation

Thunbergia grandiflora var. *racemosa* Nees, lectotype designation = **Thunbergia grandiflora** Roxb.

Thunbergia lacei Gamble, lectotype designation = **Thunbergia grandiflora** Roxb.

Thunbergia salwenensis W. W. Sm., lectotype designation = **Thunbergia lutea** T. Anderson

Utricularia wallichiana var. *firmula* Oliv., lectotype designation ≡ **Utricularia scandens** subsp. **firmula** (Oliv.) Z. Y. Li

New synonyms appearing in this volume

Clarkeasia parvifora (Anderson) J. R. I. Wood var. *albescens* J. R. I. Wood, **syn. nov.** = **Strobilanthes violifolia** T. Anderson

Dicliptera bupleuroides var. *sessilis* Nees, **syn. nov.** = **Dicliptera bupleuroides** Nees

Dicliptera bupleuroides var. *ciliata* Malik & Ghafoor, **syn. nov.** = **Dicliptera bupleuroides** Nees

Dicliptera bupleuroides var. *nazimii* Malik & Ghafoor, **syn. nov.**= **Dicliptera bupleuroides** Nees

Dicliptera bupleuroides var. *qaiseri* Malik & Ghafoor, **syn. nov.** = **Dicliptera bupleuroides** Nees

Eranthemum tapingense W. W. Sm., **syn. nov.** = **Mackaya indica** (Nees) Ensermu Kelbessa

Hygrophila polysperma f. *albiflora* Kuntze, **syn. nov.** = **Hygrophila polysperma** (Roxb.) T. Anderson

Incarvillea arguta var. *longipedicellata* Q. S. Zhao, **syn. nov.** = **Incarvillea arguta** Royle

Incarvillea compacta var. *qinghaiensis* Grey-Wilson, **syn. nov.**= **Incarvillea compacta** Maxim.

Justicia xerophila W. W. Sm., **syn. nov.** = **Justicia xerobatica** W. W. Sm.

Justicia xylopoda W. W. Sm., **syn. nov.** = **Justicia xerobatica** W. W. Sm.

Lepidagathis prostrata Dalzell, **syn. nov.** = **Lepidagathis trinervis** Nees

Phlogacanthus tubiflorus var. *angustifolius* Nees, **syn. nov.** = **Phlogacanthus tubiflorus** Nees

Strobilanthes atropurpurea var. *berevifolia* Kuntze, **syn. nov.** = **Strobilanthes atropurpurea** Nees

Strobilanthes sabiniana var. *argentea* Nees, **syn. nov.** = **Strobilanthes saccata** J. R. I. Wood

Strobilanthes flexa Benoist, **syn. nov.** = **Strobilanthes versicolor** Diels

Ruellia crenata Benth. ex Hohen., **syn. nov.** = **Strobilanthes pavala** (Roxb.) J. R. I. Wood

Thunbergia maculata Lace, **syn. nov.** = **Thunbergia lutea** T. Anderson

Index of scientific names

www.sciencep.com

ISBN 978-7-03-062355-3

9 787030 623553 >

Price: 528.00 Yuan

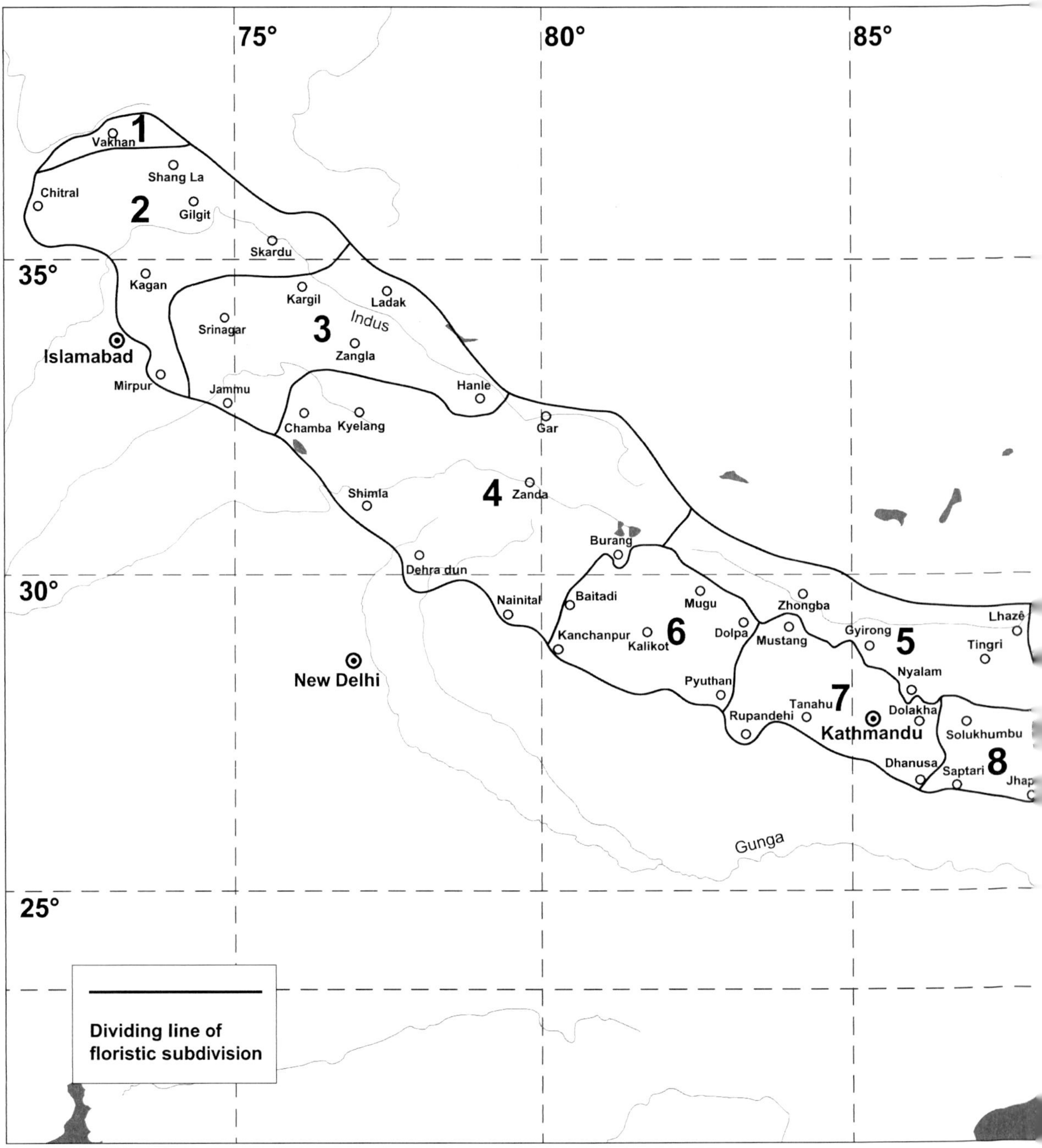

Subdivisions o

1.Vakhan; 2.N Pakistan; 3. Jammu & Kashmir; 4. U Ganga & Indus; 5.
10. Bhutan; 11. M Yarlung Zangbo; 12. L Yarlung Zangbo; 13. Yarlung Zan

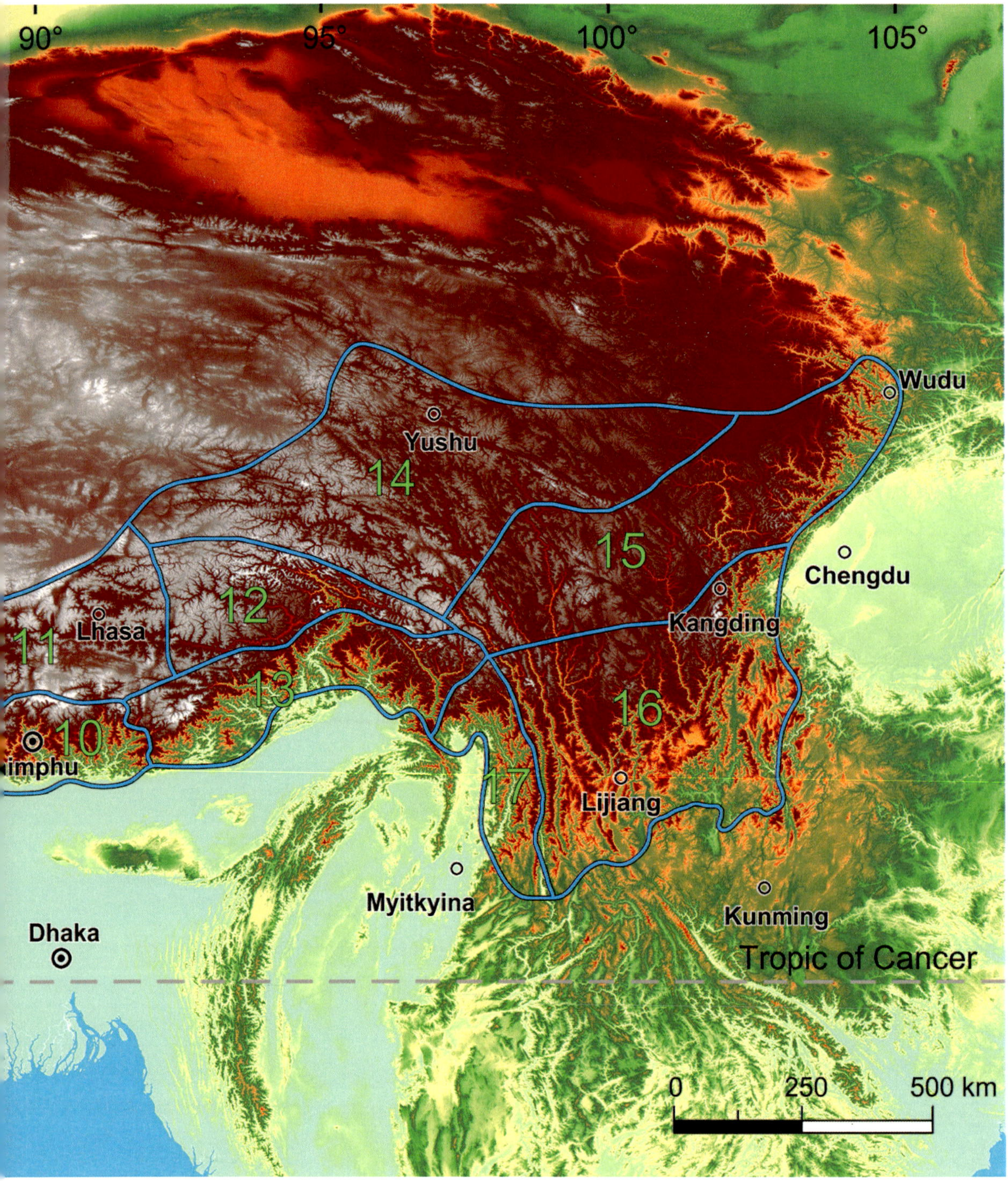

Pan-Himalaya

ung Zangbo; 6. W Nepal; 7. C Nepal; 8. E Nepal; 9. Sikkim & Darjiling;
rahmaputra; 14. Tangut; 15. N Hengduan; 16. S Hengduan; 17. U Irrawaddy.